LABORATORY MANUAL

ROBERTA M. MEEHAN, PH.D.

MARTINI

FUNDAMENTALS OF Anatomy & Physiology

FIFTH EDITION

Upper Saddle River, NJ 07458

A Note to the Student:
Exercise 1 contains a list of safety precautions. Read these carefully before you attempt to perform any activities in this manual. Ask your instructor to explain any precautions you do not understand.

Editor-in-Chief for Biology: ***Sheri Snavely***
Senior Acquisitions Editor: ***Halee Dinsey***
Project Manager: ***Don O'Neal***
Asst.Vice President, ESM Production & Manufacturing: ***David W. Riccardi***
Special Projects Manager: ***Barbara A. Murray***
Production Editor: ***Meaghan Forbes***
Manufacturing Manager: ***Trudy Pisciotti***
Composition: ***The Davidson Group***
Supplement Cover Manager: ***Paul Gourhan***
Illustrations: ***William C. Ober, M.D.; Claire W. Garrison, R.N.; Kandis Elliot***
Cover photograph: ***Jamey Hampton, Daniel Ezralow, Ashley Roland, and Sheila Lehner***
ISO Dance Company Photo© Lois Greenfield, 1991

Acknowledgments appear on pages 767–768 which constitute a continuation of the copyright page.

Printed in the United States of America

10 9 8 7 6 5 4 3 2 1

ISBN 0-13-019693-2

Prentice-Hall International (UK) Limited, *London*
Prentice-Hall of Australia Pty. Limited, *Sydney*
Prentice-Hall Canada, Inc., *Toronto*
Prentice-Hall Hispanoamericana, S.A., *Mexico*
Prentice-Hall of India Private Limited, *New Delhi*
Prentice-Hall of Japan, Inc., *Tokyo*
Pearson Education Asia Pte. Ltd., *Singapore*
Editora Prentice-Hall do Brasil, Ltda., *Rio de Janeiro*

This book is dedicated to my very dear and special friends,
Hoan Ribera and Ruth Mills.
In the Spirit of Love and Peace, I thank you.

Table of Contents

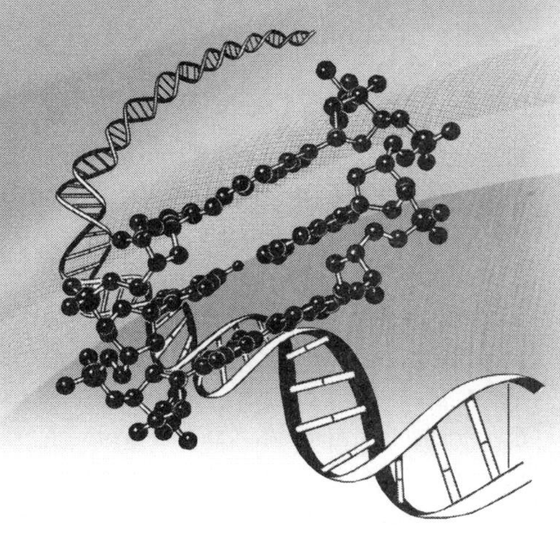

Preface *ix*

UNIT I ANATOMICAL BACKGROUND

Exercise 1: Introduction to the Laboratory 1
Exercise 2: Metric Analysis 7
Exercise 3: Microscopy 13
Exercise 4: Anatomical Orientation and Terminology 23
Exercise 5: Nucleic Acids 35
Exercise 6: Protein Synthesis 43

UNIT II BASIC ORGANIZATIONAL COMPONENTS

Exercise 7: Gross Anatomical Structure 49
Exercise 8: Cell Anatomy 63
Exercise 9: Membrane Transport 73
Exercise 10: Cell Cycle: Mitosis and Cytokinesis 83
Exercise 11: Histology 91
Exercise 12: Membranes 101
Exercise 13: Integument 109

UNIT III SKELETAL SYSTEM

Exercise 14: The Skeleton: An Overview 117
Exercise 15: Skeletal Terminology 129
Exercise 16: Axial Skeleton 133
Exercise 17: Appendicular Skeleton 147
Exercise 18: Articulations 159
Exercise 19: Fetal Skeleton 171

UNIT IV MUSCLE SYSTEM

Exercise 20: Muscle Tissue: An Overview 177
Exercise 21: Human Musculature 185
Exercise 22: Muscle Physiology 219
Exercise 23: Exercise and Stress Physiology 227
Exercise 24a: Introduction to the Cat 233
Exercise 24b: Introduction to the Fetal Pig 241
Exercise 25a: Dissection of the Muscles: Cat 249
Exercise 25b: Dissection of the Muscles: Fetal Pig 263

UNIT V NERVOUS SYSTEM

Exercise 26: Neural Tissue: An Overview 273
Exercise 27: Anatomy of the Spinal Cord 287
Exercise 28: Spinal Nerves 295
Exercise 29: Reflex Physiology 303
Exercise 30: Anatomy of the Human Brain 311
Exercise 31: Electroencephalography 321
Exercise 32: Sheep Brain Dissection 327
Exercise 33: Cranial Nerves 335
Exercise 34: Autonomic Nervous System 343
Exercise 35: Dissection of the Mammalian Nervous System 351

UNIT VI SENSORY SYSTEM

Exercise 36: Anatomy of the Eye 361
Exercise 37: Aspects of Vision 373
Exercise 38: Anatomy of the Ear 385
Exercise 39: Physiological Aspects of Hearing 395
Exercise 40: Physiological Aspects of Equilibrium 405
Exercise 41: Anatomy and Physiology of Taste 413
Exercise 42: Anatomy and Physiology of Smell 421
Exercise 43: Anatomy and Physiology of Touch 427

UNIT VII ENDOCRINE SYSTEM

Exercise 44: Endocrine System: An Overview 437
Exercise 45a: Dissection of the Endocrine System: Cat 451
Exercise 45b: Dissection of the Endocrine System: Fetal Pig 457

UNIT VIII CARDIOVASCULAR SYSTEM

Exercise 46: Anatomy of the Blood 463
Exercise 47: Blood Testing Procedures 473
Exercise 48: Anatomy of the Human Heart 481
Exercise 49: Dissection of the Sheep Heart 491
Exercise 50: Human Vascular System 499
Exercise 51: Cardiovascular Physiology 515
Exercise 52: Electrocardiography 527
Exercise 53a: Dissection of the Cardiovascular System: Cat 535
Exercise 53b: Dissection of the Cardiovascular System: Fetal Pig 547
Exercise 54: Fetal Circulation 559

UNIT IX RESPIRATORY SYSTEM

Exercise 55: Anatomy of the Respiratory System 565
Exercise 56: Respiratory Physiology 575
Exercise 57a: Dissection of the Respiratory System: Cat 587
Exercise 57b: Dissection of the Respiratory System: Fetal Pig 593

UNIT X DIGESTIVE SYSTEM

Exercise 58: Anatomy of the Digestive System 599
Exercise 59: Digestive Physiology 613
Exercise 60: Enzymatic Action in Digestion 625
Exercise 61a: Dissection of the Digestive System: Cat 633
Exercise 61b: Dissection of the Digestive System: Fetal Pig 641

UNIT XI URINARY SYSTEM

Exercise 62: Anatomy of the Urinary System 649
Exercise 63: Dissection of the Mammalian Kidney 659
Exercise 64: Urinalysis 665
Exercise 65a: Dissection of the Urinary System: Cat 673
Exercise 65b: Dissection of the Urinary System: Fetal Pig 681

UNIT XII REPRODUCTIVE SYSTEM

Exercise 66: Anatomy of the Male Reproductive System 689
Exercise 67: Anatomy of the Female Reproductive System 699
Exercise 68: Reproductive Physiology 709
Exercise 69a: Dissection of the Reproductive System: Cat 719
Exercise 69b: Dissection of the Reproductive System: Fetal Pig 731

UNIT XIII CONTINUATION OF THE SPECIES

Exercise 70: Fertilization and Early Development 741
Exercise 71: Heredity: Principles of Inheritance 753

Photo Credits 767

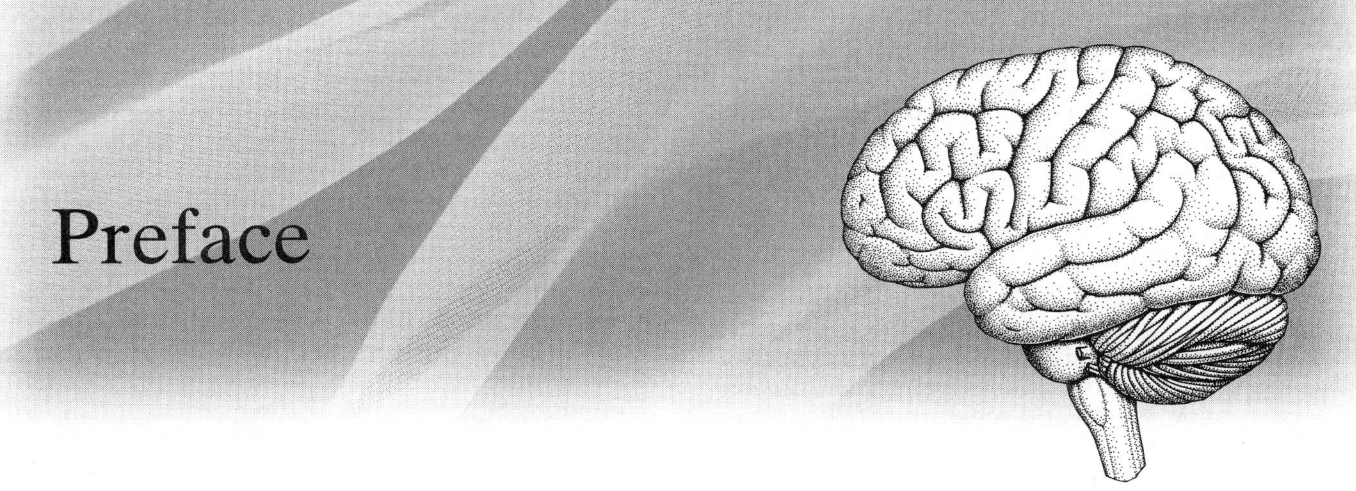

Preface

This laboratory textbook is the embodiment of the thoughts and ideas of my students and colleagues alike, all of whom have spent countless hours discussing how best to conceptualize anatomy and physiology—not only for the benefit of health care students, but for biological science and general education students as well.

Although this conceptual laboratory textbook is written primarily to accompany *Fundamentals of Anatomy and Physiology, Fifth Edition,* by Frederic Martini, it can be used as a stand alone text. Both cat and fetal pig dissection exercises are included, in back-to-back format.

❑ Organization

The organization of the exercises in this text parallels the order of topics in the Martini text. Concept Links to specific pages or topics within the Martini text are included where appropriate throughout this manual.

Anatomy and physiology are presented as separate exercises to allow for greater flexibility in adapting to the constraints of various classroom and laboratory settings. Dissection exercises are separate from the human-oriented exercises for the same reason. We assume that certain exercises will be combined and in the *Instructor's Manual* we offer suggestions for logical exercise combinations.

In addition to the basic anatomy and physiology labs, certain exercises—such as those in the introductory lessons—can be used in class or assigned as outside review material. Other exercises—such as those on anatomical terminology and skeletal terminology—can be referred to throughout the course. Clinically focused exercises—such as the EEG, blood testing, and urinalysis labs—are included where appropriate.

Unit I provides background information. It is designed specifically for those students who may need to review or conceptualize some basic scientific ideas. **Unit II** supplies the foundation for anatomy and physiology today by examining both the microscopic and the macroscopic structures and functions of the human body. **Units III** through **XII** examine the major systems of the body. Each unit begins with basic systemic anatomy and proceeds through the developmental and physiological aspects of the system. Dissection is included where appropriate. **Unit XIII** deals with the future of the human species. This unit is recommended particularly for those students who have not had a course in introductory biology prior to studying anatomy and physiology.

❑ Pedagogical Features

The pedagogical features of this laboratory manual have been designed not only to highlight and integrate the essential concepts and terminology of anatomy and physiology, but to give the student an understanding of the corresponding scientific processes.

Most anatomy and physiology labs are hands-on learning centers where the student has the opportunity to use a variety of learning methods unique to the laboratory setting, such as examining models or slides, dissecting real animals, and performing experiments. Laboratory time is also used to practice vocabulary (often putting terms and concepts together for the first time), to create models, or to perform demonstrations which help master essential concepts. We have attempted to emphasize to the student that there are different ways of approaching new material, that there are different kinds of questions that can be answered using each method.

Exercise Format

We have created an **Advance Organizer** for each exercise which integrates traditional exercise objectives into a framework of different learning activities. Objectives have been formulated as questions, or **Procedural Inquiries**. These **Socratic Objectives**, based on the time-tested Socratic method, help the student see that science is about asking as well as answering questions. This Socratic framework serves as an overview of what is to follow in the laboratory period, and as a vehicle to show that the question asked often determines the procedures used. The student discovers the answers to the inquiries as s/he completes the lab exercise.

The primary learning categories used throughout this lab manual to help the student focus on a particular aspect of an exercise are these:

- **Preparation** Includes physical preparation and safety instructions, as well as certain conceptual overviews and vocabulary background information.
- **Examination** Includes observation of models or slides, in addition to gross or histological specimens.
- **Dissection** Includes all animal and organ dissections.
- **Model-building** Includes certain demonstrations and models made by students.
- **Practice** Includes skill mastery drills and procedures.
- **Experimentation** Includes all tests and actual experiments.
- **Additional Inquiries** Includes important objectives that are not achieved directly by examination or experimentation, but rather by synthesizing information the student has learned or read.

Each exercise is organized in a modified **Outline Format**. The Roman numerals denote the major topics, and the capital letters introduce the subtopics. The procedures themselves are written in a short, step-by-step manner within the outline. All procedures (and only the procedures) are indicated with Arabic numerals. This design helps the student distinguish between background material and the actual tasks at hand, and ensures that all steps in each procedure are accomplished in the correct order.

Where appropriate, **Drawing Boxes** are provided for the student to record gross or microscopic observations. These boxes are labelled for easy reference when the student reviews the exercise.

Clinical Comments are included throughout the laboratory manual to provide interesting information on diseases and disorders as they apply to concepts under consideration in the lab.

Frequent **Concept Check Questions** are found in each exercise. These questions quiz the student about the concept under consideration, ask the student to consider the implications of the laboratory procedures, and urge the student to recall relevant personal experience or personal insight. Some of these questions are checked with a distinctive icon. At the end of each exercise, these checked questions are answered in the **Answers to Selected Concept Check Questions**. In some cases, additional paper may be required.

An **Additional Activities** section is included for enrichment at the end of most laboratory exercises. Some of these additional activities require outside research; others require internal investigation or an extension of what has been covered in the laboratory exercise. Often the needs of the class will dictate the manner in which the additional activities are utilized.

The **Lab Report** begins with a **Box Summary**, in which the student is asked to organize the factual material presented in the exercise. Numbers given with each box correspond to the numbers found on the Advance Organizer inquiries. After completing the laboratory exercise, the student should be able to fill in these boxes. Should difficulty arise, however, the student can refer to the opening inquiries as "hints." All questions are answered in the main body of the exercise.

The second part of the lab report includes a series of questions which the student should be able to answer after working through the procedures. Some of these questions are similar to given inquiries posed in the advance organizer. Some reports may require extra paper.

❑ Supplements

Instructor's Manual

In writing the *Instructor's Manual*, I have focused on flexibility and conceptualization. Each laboratory is bound by a unique set of constraints, and it is important that each exercise exhibit a great deal of flexibility without sacrificing the central theme of the lesson. Some schools teach anatomy and physiology as two separate courses, while other schools integrate anatomical and physiological concepts throughout the span of a one-, two-, or even three-term sequence. Some schools have as little as one hour per week for laboratory work while other schools have as many as four; and some schools have extensive equipment while at other schools, equipment is quite limited.

Because of this need for flexibility, the *Instructor's Manual* offers numerous suggestions for combining or modifying different laboratory exercises to meet specific academic agendas.

Recommendations are made to help the instructor with his or her laboratory ideas according to the defined needs of a particular program. Instructors wishing to use equipment other than that prescribed in the text will find suggestions for alternatives or substitutions throughout.

Adaptation is further enhanced by the modified outline format used in presenting the exercises. Parts of the outline can be enhanced or omitted according to need.

This outline format is also a part of our theme of conceptualization. The philosophy of the manual itself is that an understanding of anatomy and physiology is essential for today's student. In the *Instructor's Manual* we discuss conceptualization and explain what we believe are the primary concepts or ideas of the individual exercises. I have tried to offer suggestions for the implementation of these concepts by pointing out the essence of each exercise and demonstrating the logical and sequential framework around which each exercise is constructed.

Finally, we stress conceptualization by exploring ways in which the instructor can coordinate the inquiry-based objectives in the advance organizer with the concluding Lab Report which is specifically designed to bring together the various aspects of the laboratory exercise.

❑ Acknowledgments

I would like to express my special thanks to Jane Marks, Scottsdale Community College, who reviewed the entire manuscript and who served as a consultant and contributor throughout the production of the third edition.

I would also like to thank the following people who reviewed all or part of previous editions of this manual:

Maxine A'Hearn
Prince George Community College

Steven Basset
Southeast Community College

Linden Haynes
Hinds Community College

Yvette Huet-Hudson
University of North Carolina—Charlotte

Angie Huxley
Pima Community College

Jane Marks
Scottsdale Community College

Lois Peck
Philadelphia College of Pharmacy & Science

Mark Shoop
Macon College

Jeffrey Smith
Delgado Community College

Eric Sun
Macon College

Patricia Turner
Howard Community College

I would also like to express my gratitude to the members of the publication team for their consistent and informed support, especially my very patient immediate editor, Don O'Neal. Other indispensable members of the immediate team include Barbara Murray and Meaghan Forbes of Prentice Hall, and Dave Munger and Heather Hulett of The Davidson Group.

Of course, this project would not be complete without the exquisite art work of Bill Ober and Claire Garrison. Thank you.

The contributions of the reviewers and producers alike have been invaluable. Any remaining errors are solely my responsibility.

In addition, I would like to thank

- My students—past and present—whose ideas permeate these pages.
- Dr. Vickie Cook, for her numerous ideas and insights in neurophysiology.
- Dr. Bruce Rengers, Nutritionist and Dietician, for his assistance with the digestive physiology and gustatory physiology exercises.
- Sam Heen, Associate Professor, Physical Education, Aims Community College, for his thoughts and ideas on muscles and his help with the exercise and stress physiology section.
- Dr. John T. Weigandt, DVM, and Dr. Gerald O'Keeffe, DVM, at the River Bend Animal Clinic, Moline, Illinois, for their help and guidance in putting together the exercises on animal anatomy and physiology.
- Moline (Illinois) Police Department, for background on the practical applications of equilibrium testing.

I would especially like to thank:

Dr. Dorothy R. Martin, whose inspiration prompted the first edition of this book,

Thomas W. Redling, RN, MA, CEN, who offered moral support - as well as the answers to all the clinical questions,

Roberta Newberg, Hans Pieren, Mike Campbell, Donna Lee, as well as all of my other friends who have encouraged me along the way.

Without MJM and JAH, this book would not have been possible. I am extremely grateful to both of you.

And finally, I would like to thank my family: Theodore, Julie, and Daniel Meehan; Annette Meehan; Deborah, Chris, Samantha, Casey, and Heidi Wagner; and Philip Meehan. All of you, your patience has been remarkable.

Your comments and suggestions are welcome.

Roberta M. Meehan

Roberta M. Meehan
P.O. Box 1674
Greeley, CO 80632
Biology@ctos.com

EXERCISE 1

Introduction to the Laboratory

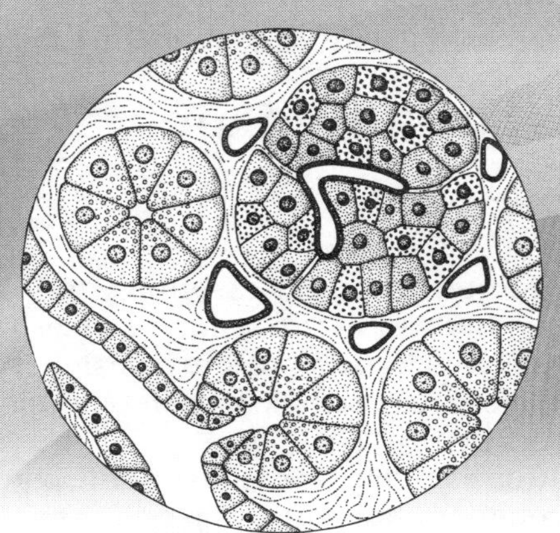

PROCEDURAL INQUIRIES

Laboratory Safety

1. What is the essence of laboratory safety?
2. What are some prominent safety features in your laboratory?
3. What types of clothing should be worn in the laboratory?
4. What should you do about handling blood in the laboratory?
5. How do you read the diamond-shaped safety code?
6. What precautions should pregnant students take?

Laboratory Equipment and Instruments

7. What are some of the basic pieces of equipment found in your laboratory?
8. What are some of the basic instruments found in your laboratory?

Additional Inquiries

9. What are the unique features in your laboratory?
10. What additional regulations should you follow in your laboratory?

Key Terms

Beaker
Cover Slip
Dissecting Kit
Dropping Bottles
Eye Droppers
Eye Wash
Fire Blanket
Fire Extinguisher
Flask
Graduated Cylinder
Indicators
Kymograph
Limb Models
Microscope
Organ Models
Physiograph
Pipet
Plastic or Latex (Surgical) Gloves
Safety Code
Safety Glasses
Shower
Slide
Test Tube
Test Tube Holder
Thermometer
Torsos
Water Bath

Materials Needed

Laboratory Equipment

Every laboratory is unique, from the layout and design down to the types of equipment and particular points of interest found within the room. The specifics of your anatomy/physiology laboratory will depend a great deal on the instructors who use the lab, the various courses that may be taught in your school, and the needs of the individuals who are served by your academic programs.

In this exercise you will be introduced to various aspects of laboratory safety, laboratory equipment, and laboratory terminology. All these points will be covered in greater detail as the course progresses.

Some points in this section may seem quite simplistic to you, particularly if you have had some good laboratory experiences in the past. Keep in mind, however, that this exercise is designed to give everyone an overview of the laboratory. Particulars, supplemental equipment, and additional details for many of the points included here will be covered by your instructor as the need arises.

❑ Laboratory Safety

I. Overview

Laboratory safety is essential, and the essence of laboratory safety is common sense. The laboratory exercises in this book are designed to be safe and efficient. Nevertheless, safety must always be your concern.

In this exercise we discuss some points of safety. Throughout the remaining exercises in this book we stress additional points of safety regarding particular laboratory investigations. We must point out, however, that many schools have adopted uniform safety guidelines. Those guidelines supersede anything we say in this book. In addition to school regulations, many jurisdictions—whether local, regional, or state—have set up safety standards. Again, those regulations take precedence over whatever regulations we explain in this book.

When in doubt about any point of safety, ask for advice. It is far better to ask a question about something that might seem obvious to everyone else than to have a problem caused by doing something in an unsafe manner.

Always let common sense be your guide.

A. Safety Features

1➡ Look around your lab. You should be able to see some rather prominent safety features. Locate the **fire extinguisher** and the **fire blanket**. You should never have to use either of these items, but you should still know where they are located. Should there ever be a need, however, don't hesitate, just use them.

2➡ Notice the **eye wash** and the **shower**. Your instructor will explain these safety features to you. Again, you will probably never have to use them, but you should still be prepared.

B. Precautions

1➡ Always tie your hair back during dissecting or Bunsen burner experiments. Although you should not need to wear a lab coat or a lab apron, common sense also dictates that you not wear good clothes to the laboratory.

2➡ Do not wear wool when dissecting preserved specimens because the preservative is picked up by the wool, and it is often very difficult to eliminate the odors.

3➡ Plan on wearing **plastic or latex gloves** or **surgical gloves** for all exercises in which you are asked to work with live or preserved specimens. The reason is more than aesthetic. Some preservatives are mildly carcinogenic; others cause allergic reactions; still others, with repeated exposure, can cause loss of nerve sensation. With live specimens there is always the danger of parasitic contamination.

4➡ Unless the laws of your state or the specific regulations of your school dictate otherwise, you will not need **safety glasses** for any of the exercises in this book. If you have allergies or sensitive eyes, you may wish to use safety glasses for the more odiferous exercises.

5➡ Be aware of regulations regarding blood. The handling of blood is a major concern in this age of serum-transmitted diseases. If you use human blood for any of the experiments in this book, you must use *extreme care* in following your instructor's directions, both for drawing the blood and for disposing of any blood-tainted materials.

For almost all experiments in this book, animal blood—either freshly drawn or purchased from a supply house—can be safely substituted since animal blood and human blood exhibit the same basic anatomical or physiological principles.

Specifics for safe handling of human blood are covered in Exercises 46 and 47 of this book.

6➡ If you are pregnant or if you become pregnant while taking this course, be assured that these exercises are safe for both you and your baby. You may wish to check with your doctor to confirm this. You may also wish to check with your doctor before performing certain of the physiological exercises.

If the dissecting experiments make you nauseous, be certain that your work area is well ventilated. You may wish to ask your instructor if your dissecting work can be kept to a minimum. The wearing of surgical masks or dissecting under the hood may help.

C. The Safety Code

As you examine various points in your laboratory, you may notice the diamond-shaped **safety codes**. One may be posted on the wall. In addition, you will find these codes on all the newer chemical bottles and jars. Some chemical containers have the safety codes in a box-shaped chart. Although you will probably not encounter any dangers in your laboratory exercises, it is still important that you be familiar with these codes. Examine the safety code in Figure 1–1●.

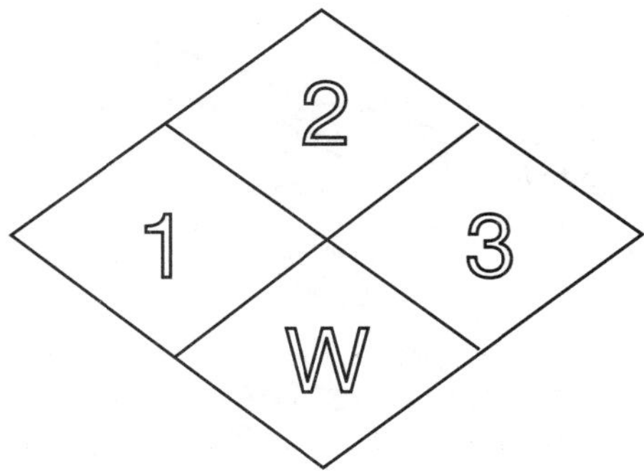

● **FIGURE 1–1**
Safety Code.

The left-hand number is often on a blue background and represents the health risk. The upper number is on red and represents the fire hazard. The right-hand number is on yellow and represents the reactivity (stability) of the substance. The bottom part of the diamond indicates either storage or radioactivity information.

The number in each box represents the severity of the risk and thus gives an indication of the caution necessary for handling the item safely.

0 = Minimal danger
1 = Slight danger
2 = Moderate danger
3 = Serious danger
4 = Extreme danger

Accompanying this code you may also see a safety code for clothing, eye guards (or goggles or shields), proper gloves, or vent hoods. Keep in mind that although you will not be doing caustic or dangerous experiments, some of the chemicals used in your labs are often used in more dangerous experiments, experiments beyond the scope of your course of study here.

❑ Laboratory Equipment and Instruments

To complete many anatomy/physiology laboratory experiments, you will need access to some basic laboratory equipment as well as certain laboratory instruments. This exercise is designed to help you in identifying those items with which you may not be familiar. Diagrams of some pieces of equipment can be found in Figure 1–2●. Your instructor will have out assorted equipment for you to examine.

I. Basic Equipment

Below is a list of some of the items you may have in your lab.

Beaker Open-mouthed container with straight sides. Many beakers have volumetric markings that include "±5%." This means that any given volume in the beaker may be 5% off from what the reading indicates. Beakers should *not* be used for any type of exact measurement.

Cover Slip May be glass or plastic. The cover slip stabilizes specimens for viewing under the microscope.

Dissecting Kit
Pins
Scalpel
Scissors
Blunt nose probe
Sharp nose probe
Forceps

Dropping Bottles May contain any type of stain, dye, or liquid you may need in small quantities.

Eye Droppers May be glass or plastic

Flask Open-mouthed container with a narrowed neck.

Graduated Cylinder Cylinder used specifically for measuring liquids. Graduated cylinders are scientifically calibrated to give an exact measurement at a specific temperature. Experiments should not be performed in graduated cylinders.

Indicators Substance that changes color under given conditions.
Dyes
Papers

Limb Models

Organ Models

Pipet Narrow measured tube into which fluid is suctioned for experimental dispensing.

Slide Assorted types for different puposes. Most of yours will probably be plain glass.

Test Tube Glass (or occasionally plastic) tube for holding and examining liquids and for observational experiments.

Test Tube Holder Wire prong for holding a test tube.

Torsos

II. Basic Instruments

These are some of the instruments you may encounter in your lab. Common sense would dictate that because of their delicate nature, these instruments should be handled with care.

Kymograph Instrument designed for measuring the results of breathing experiments.

Microscope Any of several types of instrument used for magnifying small objects. Exercise 3 deals with the principles of microscopy.

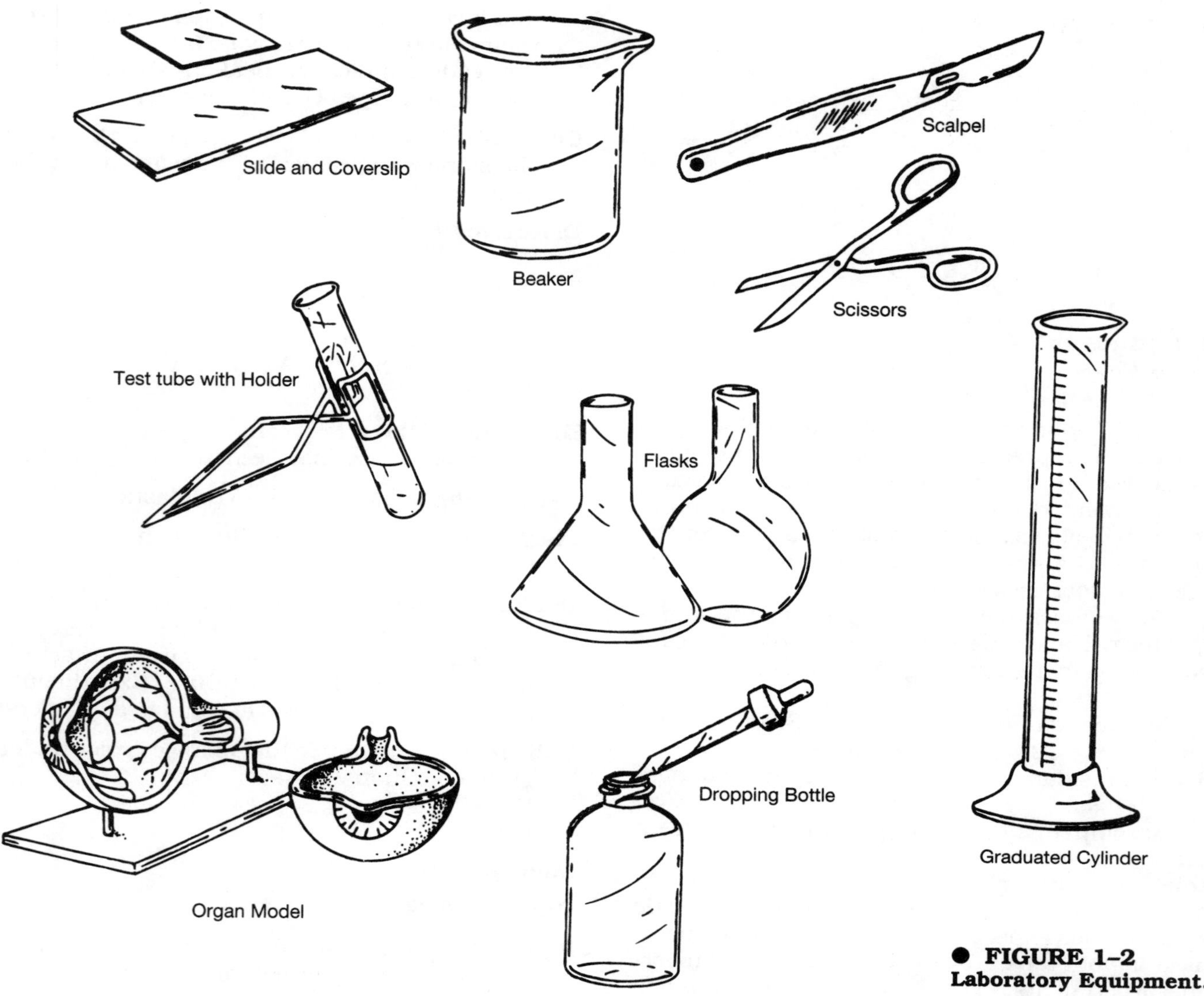

● **FIGURE 1–2**
Laboratory Equipment.

Physiograph Any of several instruments designed for measuring the results of certain physiological experiments.

Thermometer Instrument for measuring temperature. In anatomy and physiology celsius measurements are used.

Water Bath Heated water tub used for holding substances at a constant temperature.

III. Additional Equipment

If your instructor has set out additional items, use the space at the end of this exercise to identify the particular equipment and instruments that may be in your lab.

❑ Additional Activities

NOTES

1. Investigate other equipment you may have in your laboratory.
2. Research the use of computers in the introductory anatomy/physiology laboratory.

❑ Lab Report

1. Referring to the numbered inquiries at the beginning of this exercise, complete the following Box Summary:

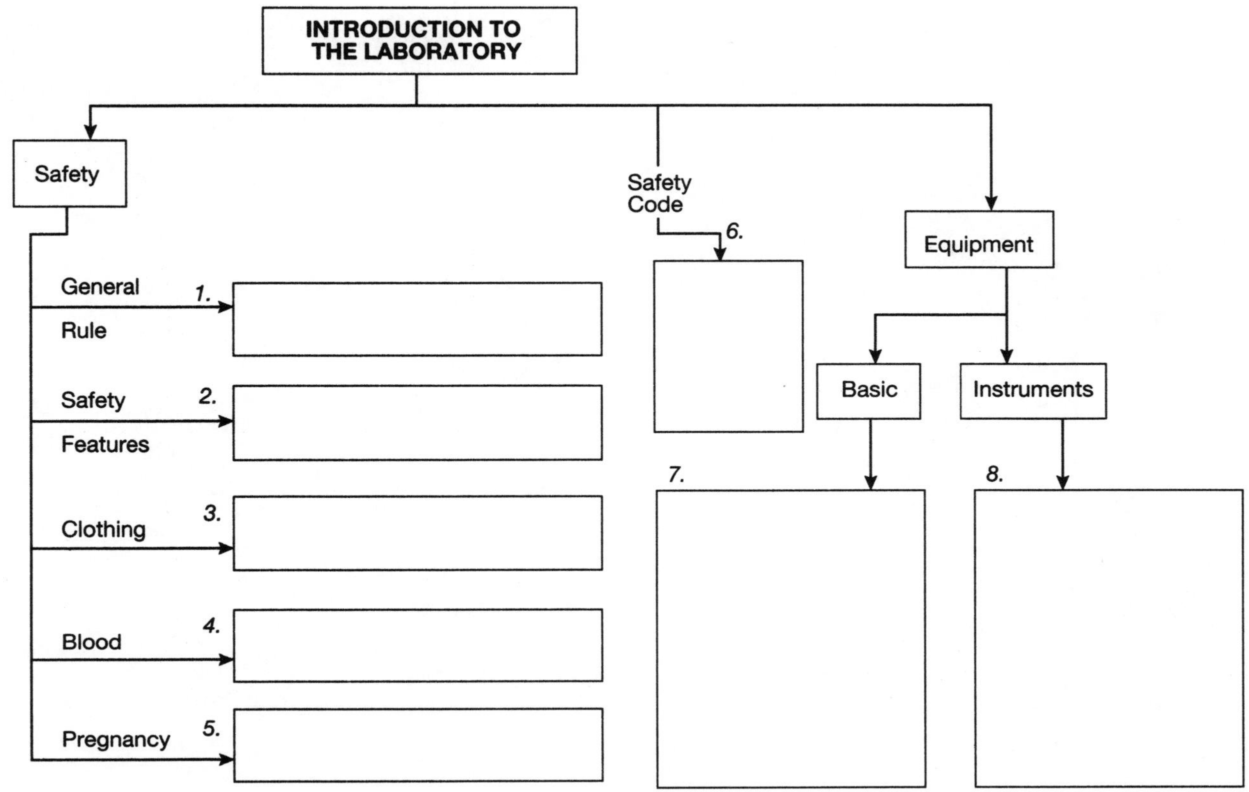

General Questions

NOTES

1. What is unique about your laboratory?

2. What additional regulations should you follow in your laboratory?

EXERCISE 2

Metric Analysis

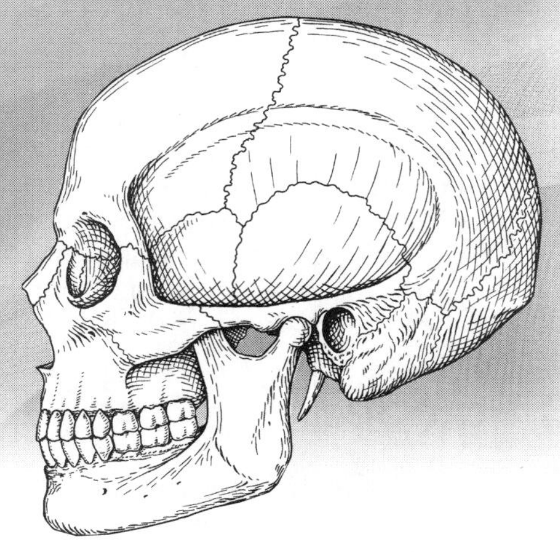

PROCEDURAL INQUIRIES

Preparation

1. Why do we need the metric system in anatomy and physiology?
2. What is the International System?
3. What are the basic units of the International System?
4. Which prefixes are used with the International System?
5. What are the basic units of the US System?

Practice

6. How do we convert quantities to and from scientific notation?
7. How do we use a chart to convert US weights and measures to metric weights and measures?
8. How do we use a chart to convert metric weights and measures to US weights and measures?
9. How do we convert quantities within the metric system?

Key Terms

International System (SI)
Metric Measurement
Scientific Notation
US System of Weights and Measures

Materials Needed

Rulers ***Metric Measurement***
Rulers ***US Measurement***

Metric measurement is a part of the scientific terminology we encounter in studying anatomy and physiology. If you have taken other science courses, you already know that metric measurement is based on the number 10. The major problem many people have with metrics is visualizing the quantities involved. For instance, we all have a good idea of what a quart is; but many of us are never quite sure whether a liter is a bit more or a bit less than a quart, especially since soft drinks come in 2-liter containers. And we are so used to putting gallons into our cars that we have trouble thinking in terms of modified quarts (meaning liters). Even though the United States is the only major country that has not officially adopted the **International System (SI)** of weights and measures, in science we need metric measurements. In health care alone virtually every quantity is expressed metrically. This includes the ion concentrations in the blood, the medications in an IV bag, the fluid intake and output of a hospital patient, and even the vital capacity of the average lung.

In this laboratory exercise we will review a few basic metric concepts.

❑ Preparation

I. Background

A. Metric Awareness

1➡ Keep in mind that scientific notation and metric conversion are methods of simplifying the mathematical aspects of our scientific inquiry.

2➡ Try to visualize what you are doing. If this review is not sufficient for you, consult a good math book. Mathematical concepts play an integral role in your study of anatomy and physiology. It is best to clarify your difficulties now before you begin using these ideas in your daily work.

❑ Practice

II. Scientific Notation

A. Powers of 10

Dilutions and concentrations in anatomy and physiology are often expressed as whole numbers between 1 and 10 raised to powers of 10. This is **scientific notation**, a convenient way of expressing long and involved numbers. Scientific notation decreases the possibility of mathematical error by making complex numbers less burdensome.

1➡ Study Figure 2–1●. This chart shows exponential values. Notice that numbers greater than 1 have positive exponents while numbers less than 1 have negative exponents. (1 itself is 10 raised to the zero power. By mathematical definition, any number raised to the zero power is 1. Other numbers in an equation are not affected—only the number actually raised to the zero power. For instance, $3 \times 10^0 = 3 \times 1 = 3$.)

a. 8.73×10^4 means that 8.73 is multiplied by $10 \times 10 \times 10 \times 10$ (or 10,000).

Number	Exponential Form	Exponent
10,000,000	1×10^7	7
1,000,000	1×10^6	6
100,000	1×10^5	5
10,000	1×10^4	4
1,000	1×10^3	3
100	1×10^2	2
10	1×10^1	1
1	1×10^0	0
0.1	1×10^{-1}	–1
0.01	1×10^{-2}	–2
0.001	1×10^{-3}	–3
0.000 1	1×10^{-4}	–4
0.000 01	1×10^{-5}	–5
0.000 001	1×10^{-6}	–6
0.000 000 1	1×10^{-7}	–7

● **FIGURE 2–1**
Scientific Notation.

The number of zeros equals the exponential value. Since $8.73 \times 10^4 = 87{,}300$, we see that positive exponents tell us how many places to the *right* we must move the decimal point.

b. 7.62×10^{-2} means that 7.62 is multiplied by $1/10 \times 1/10$ (or 1/100). Again, the number of zeros equals the exponential value. Since $7.62 \times 10^{-2} = 0.0762$, we see that negative exponents tell us how many places to the *left* we must move the decimal point.

2➡ Convert these scientific notation numbers to standard form numbers.
3.52×10^3
5.776×10^{-5}
2.01×10^0
3.9×10^{-8}
4.444×10^4

3➡ Convert these standard form numbers to scientific notation numbers.
56822
387.8
0.232
0.00000000034
1.24

III. US ⟷ SI Conversions

A. Examples

We live in a pluralistic society—sometimes we use the **US system of weights and measures** and sometimes we use the International System (SI). Ideally you should practice thinking metric and you should not need to shift back and forth between US and SI units. In anatomy and physiology, metric conversions are often necessary, and it is nice to be able to make ballpark conversions from one system to the other.

1➡ Check Figure 2–2● if you need to review the basic US system of weights and measures.

2➡ Recall the basic metric measurements: meter (length), liter (volume), gram (mass). Recall, too, the basic prefixes that are used to describe how many or how much of the basic measurement you have.

kilo-	(k)	one thousand
hecto-	(h)	one hundred
deca-	(da)	ten
deci-	(d)	one tenth
centi-	(c)	one hundredth
milli-	(m)	one thousandth
micro-	(μ)	one millionth

Thus a *kilo*gram is 1000 grams, a *centi*liter is one one hundredth of a liter, and a *milli*meter is one one thousandth of a meter.

Additional prefixes and symbols as well as additional conversion information can be found in Figures 2–2● and 2–3●.

3➡ When you are comfortable with the information in Figure 2–2●, work through the following problems.

a. The human body averages about 5 ℓ blood. Approximately how many quarts of blood is this?

Since 1 ℓ = 1.06 qt, our equation is 1.06 qt/1 ℓ × 5 ℓ

The liters cancel, so the answer is 5.3 qt.

b. Many health experts claim we should drink 8 cups of water per day. How many liters is this?

Physical Property	*Unit*	*Relationship to Standard Metric Units*	*Conversion to U.S. Units*	
Length	nanometer (nm)	1 nm = 0.000000001 m (10^{-9})	= 4 × 10^{-8} in.	25,000,000 nm = 1 in.
	micrometer	1 µm = 0.000001 m (10^{-6})	= 4 × 10^{-5} in.	25,000µm = 1 in.
	millimeter (mm)	1 mm = 0.001 m (10^{-3})	= 0.0394 in.	25.4 mm = 1 in.
	centimeter (cm)	1 cm = 0.01 m (10^{-2})	= 0.394 in.	2.54 cm = 1 in.
	decimeter (dm)	1 dm = 0.1 m (10^{-1})	= 3.94 in.	0.25 dm = 1 in.
	meter (m)	standard unit of length	= 39.4 in.	0.0254m = 1 in.
			= 3.28 ft	0.3048m = 1 ft
			= 1.09 yd	0.914m = 1 yd
	dekameter (dam)	1 dam = 10 m		
	hectometer (hm)	1 hm = 100 m		
	kilometer (km)	1 km = 1000 m	= 3280 ft	
			= 1093 yd	
			= 0.62 mi	1.609km = 1 mi
Volume	microliter (µℓ)	1 µℓ = 0.000001 ℓ (10^{-6})		
		= 1 cubic millimeter (mm^3)		
	milliliter (mℓ)	1 mℓ = 0.001 ℓ (10^{-3})	= 0.03 fl oz	5 mℓ = 1 tsp
		= 1 cubic centimeter (cm^3 or cc)		15 mℓ = 1 tbsp
				30 mℓ = 1 fl oz
	centiliter (cℓ)	1 cℓ = 0.01 ℓ (10^{-2})	= 0.34 fl oz	3 cℓ = 1 fl oz
	deciliter (dℓ)	1 dℓ = 0.1 ℓ (10^{-1})	= 3.38 fl oz	0.29 dℓ = 1 fl oz
	liter (ℓ)	standard unit of volume	= 33.8 fl oz	0.0295ℓ = 1 fl oz
			= 2.11 pt	0.473ℓ = 1 pt
			= 1.06 qt	0.946ℓ = 1 qt
Mass	picogram (pg)	1 pg = 0.000000000001 g (10^{-12})		
	nanogram (ng)	1 ng = 0.000000001 g (10^{-9})		
	microgram (µg)	1 µg = 0.000001 g (10^{-6})	= 0.000015 gr	66,666 µg = 1 gr
	milligram (mg)	1 mg = 0.001 g (10^{-3})	= 0.015 gr	66.7 mg = 1 gr
	centrigram (cg)	1 cg = 0.01 g (10^{-2})	= 0.15 gr	6.7 cg = 1 gr
	decigram (dg)	1 dg = 0.1 g (10^{-1})	= 1.5 gr	0.67 dg = 1 gr
	gram (g)	standard unit of mass	= 0.035 oz	28.35 g = 1 oz
			= 0.0022 lb	453.6 g = 1 lb
	dekagram (dag)	1 dag = 10 g		
	hectogram (hg)	1 hg = 100 g		
	kilogram (kg)	1 kg = 1000 g	= 2.2 lb	0.453kg = 1 lb
	metric ton (kt)	1 mt = 1000 kg	= 1.1 t	
			= 2205 lb	0.907 kt = 1 t

Temperature	*Centigrade*	*Fahrenheit*
Freezing point of pure water	0°	32°
Normal body temperature	36.8°	98.6°
Boiling point of pure water	100°	212°
Conversion	°C→°F: °F = (1.8 × °C) + 32	°F→°C: °C = (°F – 32) × 0.56

● **FIGURE 2–2**
Metric Conversion.

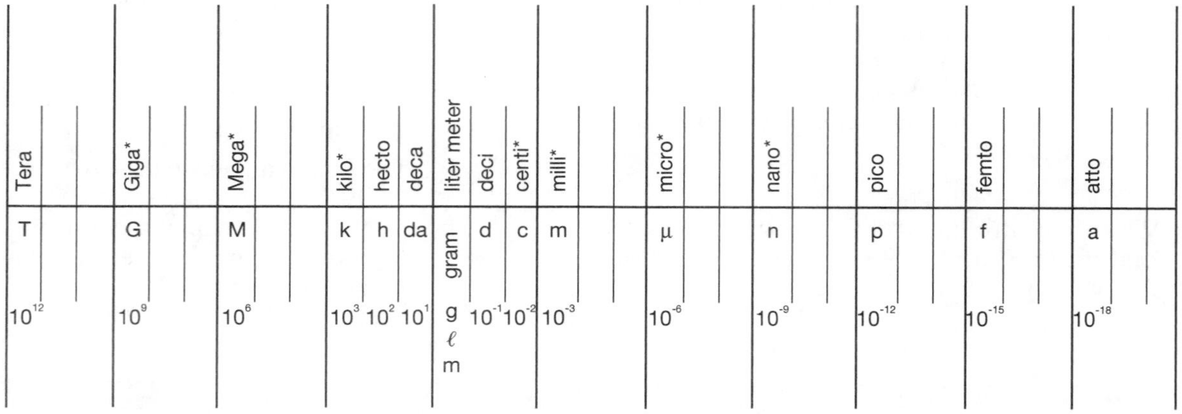

Tera	Giga*	Mega*	kilo*	hecto	deca	liter meter / gram	deci	centi*	milli*	micro*	nano*	pico	femto	atto
T	G	M	k	h	da		d	c	m	µ	n	p	f	a
10^{12}	10^{9}	10^{6}	10^{3}	10^{2}	10^{1}	g ℓ m	10^{-1}	10^{-2}	10^{-3}	10^{-6}	10^{-9}	10^{-12}	10^{-15}	10^{-18}

● **FIGURE 2–3**
Quick and Easy Conversion Chart.

8 cups × 1 pt/2 cups × 0.473 ℓ/1 pt = 1.892 ℓ

Note that cups and pints both cancel.

2➡ Do the following:

a. Figure your weight in kilograms. (Low estimates are acceptable! We are striving for understanding the concept!)

____

____

b. Measure your work space in US units. Convert these units to SI units. Check your results by measuring in metric units.

____

____

c. How many cups of liquid will a 500 ml beaker hold?

____

____

d. Look up the concentration of sodium in the human body. Express this concentration in quarts.

____

____

IV. Metric (SI) Unit Conversions

A. Examples

Conversions within different metric units are amazingly simple. These conversions can be done by using the conversion factors. Study Figure 2–2● again and then work through this example: Convert 50 ℓ to ml. First, note that 1 ℓ = 1,000 ml. Therefore, 50 ℓ × 1,000 ml/ℓ = 50,000 ml. (Again, the number of zeros equals the number of decimal places.)

1➡ Use the conversion factors to do the following problems:
5.4 µm to mm
6782 mg to kg
3.44 mm to cm
66.97 ℓ to µl
66.97 ℓ to kl

2➡ Convert your answers to scientific notation.

B. Alternate Method

We can also use the "quick and easy" method of conversion.

1➡ Use the chart in Figure 2–3●. All units listed on this chart are not in common usage. Commonly used prefixes are marked with an asterisk.

Note the spacing represents the number of decimal places. Suppose you want to convert 35.2 decameters to micrometers. Put the 35.2 in the column under deca. Count the spaces to micro. Seven spaces. That means you will need to move your decimal point 7 spaces. Since you are moving right, your decimal point will move right. Your answer will be 352,000,000. So, 35.2 dam = 352,000,000 µm.

2➡ Use this method to perform the following conversions:
5.687 hm to mm
345.432 µg to ng
53,552 ml to dl
2442.6 cg to kg
0.009786 Gm to am

3➡ Convert your answers to scientific notation.

❑ Additional Activities

NOTES

1. Make up and solve some problems dealing with scientific notation.
2. Make up and solve some problems dealing with US to SI conversions and SI to US conversions.
3. Make up and solve some problems dealing with unit conversion within the metric system.
4. Research the history of different systems of weights and measures.

❑ Lab Report

1. Referring to the numbered inquiries at the beginning of this exercise, complete the following box summary:

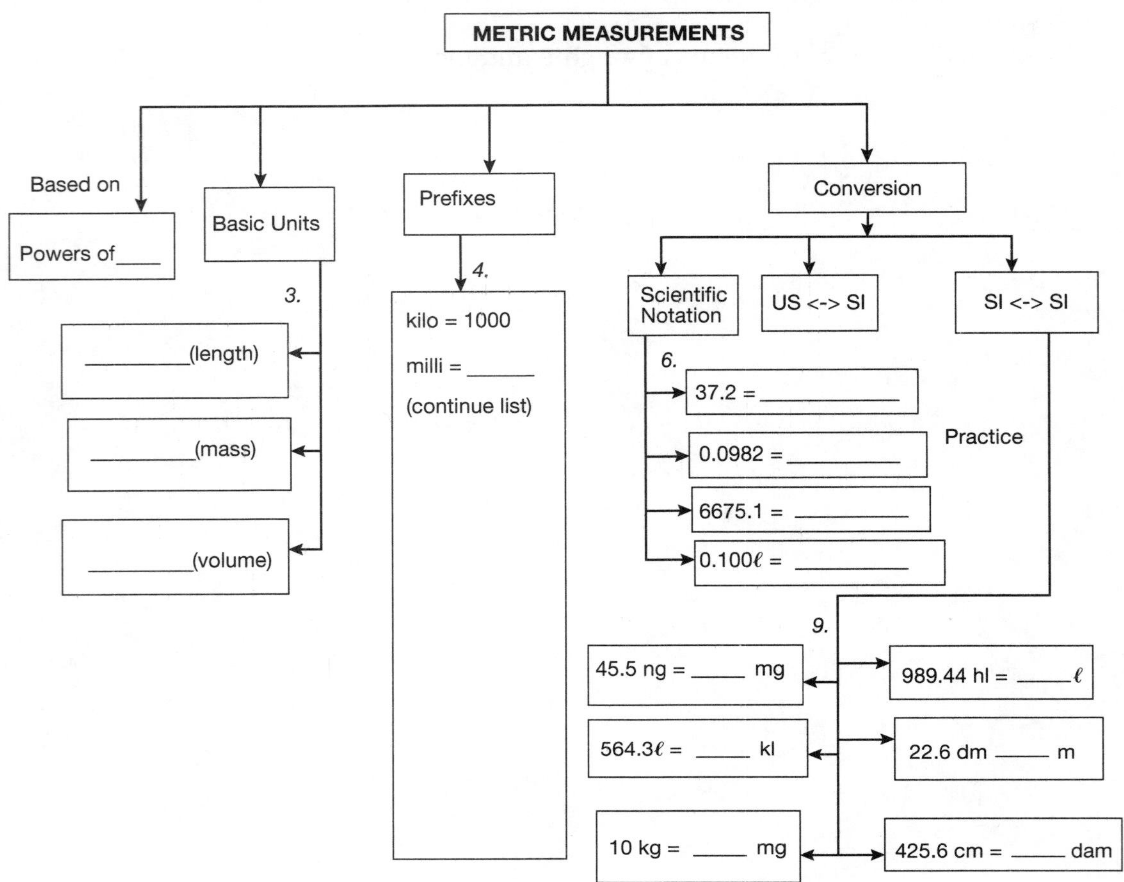

General Questions

NOTES

1. Why do we need the metric system?
2. Explain the value of using scientific notation.
3. List the principles involved in US ⟷ SI conversions.
4. What are the principles involved in metric unit conversions?
5. Why does the "quick and easy" method of metric conversion work as well as the conventional method of metric conversion?

EXERCISE 3

Microscopy

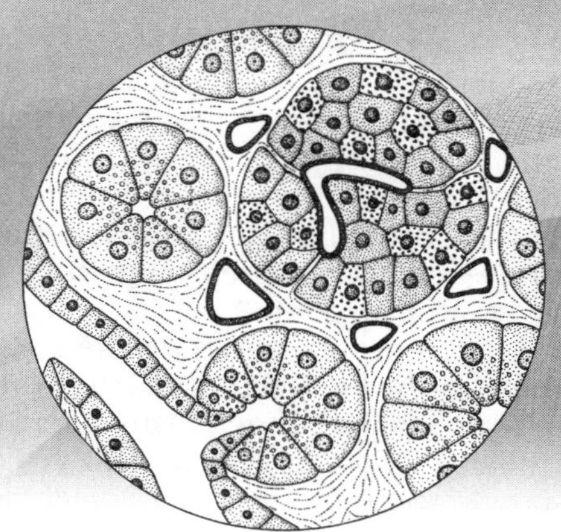

PROCEDURAL INQUIRIES

Examination

1. How do you prepare a wet mount?
2. How do you focus the microscope?
3. How do you use the oil immersion lens?

Experimentation

4. Which direction does the newspaper letter move when you move the slide to the right?
5. What principle does the crossed hairs on the slide demonstrate?
6. What can be seen when viewing the pond culture?
7. Why do you use the oil immersion lens to view the bacteria?

Additional Inquiries

8. Who are some of the people responsible for the development of the microscope?
9. What are the major types of microscopes?
10. What are the principal parts of the compound light microscope?
11. What are the principal rules for the care and handling of the microscope?
12. What are some additional terms connected with microscopy?
13. What are the types of electron microscopes?

Key Terms

[This Key Terms section does not include the parts of the microscope identified in Section IV.A.]

Binocular (Stereoscopic) Dissecting Microscope
Compound System
Compound Light Microscope
Dark Field Microscopy
Depth of Field
Field
Light (Bright) Field Microscopy
Magnification
Nonoptical
Optical
Parcenter
Parfocal
Phase Contrast Microscopy
Photomicrograph
Prepared (Permanent Mount) Slide
Resolution (Resolving Power)
Scanning Electron Microscope (SEM)
Simple Microscope
Total Magnification
Transmission Electron Microscope (TEM)
Wet Mount

Materials Needed

Hand Lens (Magnifying Glass)
Compound Light Microscope
Binocular Microscope (Stereoscopic or Dissecting Microscope)
Microscope Slides and Cover Slips
Newspaper
Scissors
Hair
Water
Pond Culture or Mixed Protozoan Culture
Prepared Mixed Bacterial Slide
Immersion Oil

In 1590 Hans Janssen (also spelled Jan Janszoon) and his son Zacharias, spectacle-makers in Middelburg, Holland, discovered that using combinations of concave and convex lenses could produce both a clarity and a magnification not possible by using either type of lens independently.

The Janssens, whose lives have been variously augmented throughout recent history, probably did not invent the microscope (as they were apparently more interested in astronomy than in biology). Nevertheless, their contributions to lens crafting cannot be denied.

After the Janssens, such people as Galileo and van Leeuwenhoek perfected the microscope. Antony van Leeuwenhoek, the linen merchant from Holland who carried on a 50-year scientific correspondence with the Royal Society in London, probably did more to revolutionize the world of microscopy than did anyone else in history.

The microscope is without doubt the single most important tool used in the study of anatomy and physiology. Not only is the microscope the instrument for the minute observations of tissues and structures, the microscope today is also vital in surgery, diagnostic evaluations, and the analysis of medical and chemical reactions. And, with the advent of electron microscopy, no longer is the world of microscopy limited by the confines of the optical lens.

❑ Examination

Microscopes, which are either optical or nonoptical, are instruments designed for magnifying the miniature world, the world too small for us to see with the naked eye. Before embarking on actual microscopic procedures, familiarize yourself with this preparation section of the exercise.

I. Optical Microscopes

Optical microscopes use a glass lens system to magnify an object, which is then viewed either directly through the lenses or indirectly via a series of mirrors. The object may be living or nonliving, preserved or fresh.

A. Simple Microscope

The **simple** microscope is a magnifying system using only one lens. If you stretch the definition, any magnifying glass is a simple microscope.

1➡ Obtain a hand lens. If the magnification of that hand lens is known, record the number. __________

2➡ Observe objects on your work bench. What limitations do you find in working with the magnifying glass?

____

____

3➡ Take another hand lens. Now try to put the two lenses together. Examine an object such as a pencil point. What problems do you face in increasing your magnification?

____

If you know the magnification of both lenses, multiply these numbers to figure out the total magnification of your object.

Mag 1 × Mag 2 = Total Mag __________

The problem you just experienced in increasing the magnification in your lens system was the major optical problem facing those optical scientists of the 16th century who would have invented both the microscope and the telescope had they known a few more optical principles.

The hand lens (or magnifying glass) is a convex lens and, as you have just demonstrated, when magnification is increased, acuity is decreased. Enter Hans and Zacharias Janssen, who figured out how to use concave and convex lenses in combination, thus giving us the **compound** system. (For a more complete explanation of lenses, consult a physics book or see Exercise 36 in this book.)

B. Compound Microscope

The **compound light microscope** is the technical name given the standard microscope.

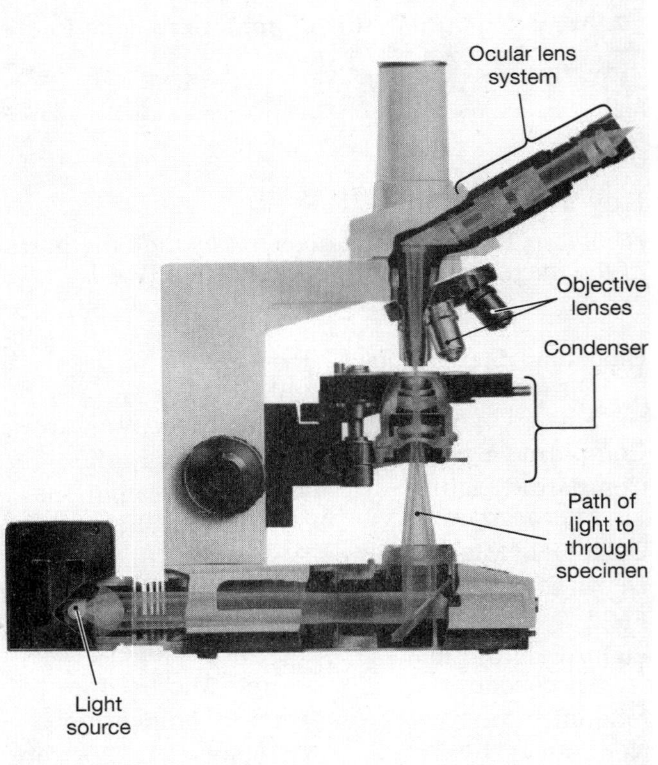

● **FIGURE 3–1**
Optical System of the Microscope.

1➜ Study Figure 3–1● and trace the path of the image as it travels from the object to your eye.

2➜ Note the different parts of the lens system. The compound microscope uses a series of concave and convex lenses in combination to produce optical magnification and clarity. It is beyond the purpose of this lesson to delve into the optical principles involved in fine microscopy. Nevertheless, you should be aware of the complexity involved in producing that clear, magnified image you see when you look through a microscope.

In introductory anatomy and physiology, you will be working with **light** (or **bright**) **field microscopy**. In light field microscopy, the field (the area you are working with) is illuminated because the light waves are passing directly through the field. The objects in the field appear darker or colored.

The compound light microscope can be easily modified for **dark field microscopy**. In dark field microscopy, the field is dark and the objects appear bright because the light source is not direct and the light waves are reflected off the illuminated object at an angle. Dark field microscopy is useful in viewing certain types of live microorganisms (such as the spirochaete bacteria). Your microscope may have a dark field adapter.

The compound light microscope can also be modified for **phase contrast microscopy**, a useful tool for studying the three-dimensional aspects of certain tissues. In phase contrast microscopy, the light waves are deliberately put "out of synch" with each other.

C. Binocular Stereoscopic Microscope

A **binocular stereoscopic** microscope (sometimes called a stereoscope, sometimes called a dissecting microscope), like the type you have in your laboratory, is used for viewing larger objects. Magnification is generally low and depth of field is generally high.

1➜ Study Figure 3–2● and compare this diagram with the stereoscopic microscopes in your lab. You may find certain structural variations between the diagram and your microscope, but you should have no difficulty in identifying the analogous components.

II. Nonoptical Microscopes

Nonoptical microscopes use either nonvisible waves (sound waves, radio waves, microwaves, etc.) or subatomic particles (usually electrons) to define a microscopic object. Electron microscopes are nonoptical microscopes.

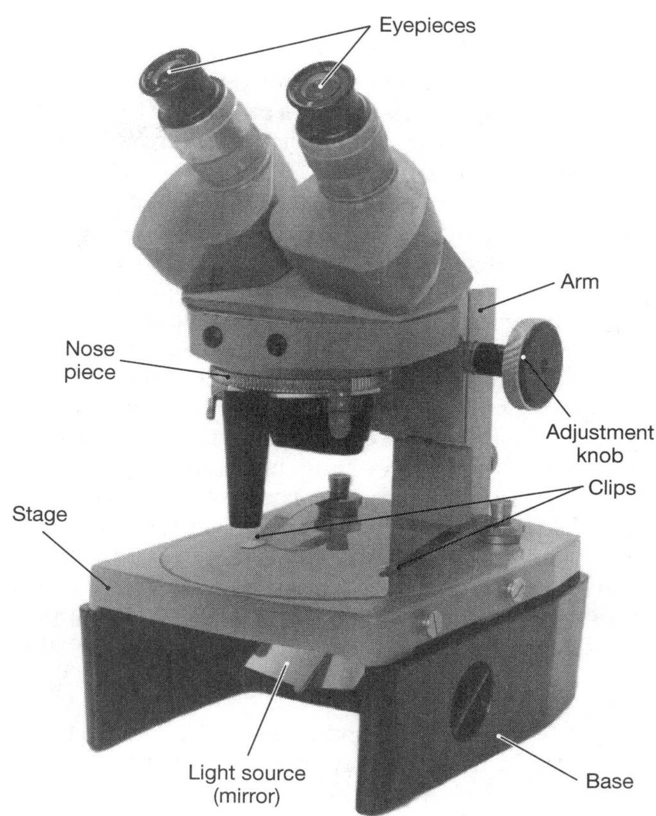

● **FIGURE 3–2**
Dissecting (Stereoscopic) Microscope.

A. Transmission Electron Microscope (TEM)

In a **transmission electron microscope (TEM)** a beam of electrons, directed by appropriate magnets, passes through an ultrathin section of a specimen. If the electrons reach a fluorescent screen, an image is produced. The deviations in the beam caused by the internal structures of the specimen are recorded, and a flat image of these internal structures is depicted.

B. Scanning Electron Microscope (SEM)

In a **scanning electron microscope (SEM)** a beam of electrons strikes the surface of an object causing the emission of secondary electrons which, in turn, activates a screen. A three-dimensional image of these external portions of the object is thus recorded.

III. Care and Handling

Many of the rules for handling the microscope are based on common sense. A few rules, however, deserve some explanation. Except for rule B-3 below, these rules apply to the use of all microscopes, regardless of type. [Certain modifications may be in order for using the particular microscope models in your lab. Your instructor's directions should supersede any rules given here.]

A. Handling and Storage

1➡ Always carry the microscope with two hands, one on the arm and the other under the base. Do not carry two microscopes at the same time.

2➡ After your microscope is set up, make sure the electrical cord is on your work bench and not dangling over the edge. It is very easy for you to catch your knee on the cord as you get up.

3➡ Before putting your microscope away, remove all slides, make certain that the low power objective is in place, and, unless instructed otherwise, be sure that the coarse adjustment knob is at its lowest point. This provides the least possible tension on the gear threads.

4➡ Store your microscope with the cord wrapped around the body according to your instructor's directions.

5➡ If your microscope has a dust jacket, be sure the jacket is in place before you put the microscope away.

6➡ Clean the lens only with approved no-lint optical paper. Never use paper towels or facial tissues as these can scratch the lenses.

B. Usage

1➡ Always begin your viewing with the objective in a position closest to the stage. (If your microscope does not have a safety stop, you must determine this lowest point by observing the action from the side.)

2➡ Never turn the coarse adjustment down (so as to decrease the distance between the objective and the stage) while looking through the ocular.

3➡ Always use a cover slip with a wet mount. The cover slip not only decreases distortion of the image, but also protects the objective from becoming wet. Water on or in the objective can cause serious damage to the delicate lens system.

4➡ If your microscope is binocular, adjust the oculars so that you can view the field comfortably with both eyes. Some binocular microscopes have a secondary focus on one of the oculars. If this is the case, focus the microscope according to your eye *without* the secondary focus. Then use the secondary focus to clarify the image in the other eye.

5➡ Many experts believe you will decrease muscle and eye strain if you keep both eyes open while looking through a monocular microscope. To do this, choose one eye, look through the microscope, and concentrate your energy on that visual field while ignoring the images from your other eye. Try this with each eye to determine which is most comfortable. Although this may seem difficult at first, with a little practice you should be able to do it quite easily.

Some people with visual or muscular weaknesses are better microscopists with one eye closed. We recommend you try to use the monocular microscope with both eyes open, but if you must close one eye, don't worry about it.

6➡ Do not attempt to fix any microscope malfunction. Report all mechanical errors immediately to your instructor.

IV. Optical Microscopy

A. Parts of the Microscope

1➡ Obtain a compound light microscope and identify these parts by comparing your microscope with Figure 3–3●.

Base Lower part of the microscope.

Substage Light Microscope light, found behind or beneath the stage. In some microscopes this light is built in. In other microscopes the light is portable. Some microscopes have no light, using instead natural or room light and substage mirrors.

Stage Platform where the specimen is placed.

Condenser Apparatus for concentrating the light on a specimen. The condenser may be equipped with a height adjustment knob in order to vary the delivery of light. For our work, the condenser should be closest to the lower surface of the stage. (Your microscope may not have a condenser.)

Iris Diaphragm Apparatus for adjusting the amount of light the specimen receives. Instead of an iris diaphragm, some microscopes may have a series of apertures (openings) on a revolving substage disk.

Coarse Adjustment Knob Knob for gross movements of the tube or stage. Used for gross focusing.

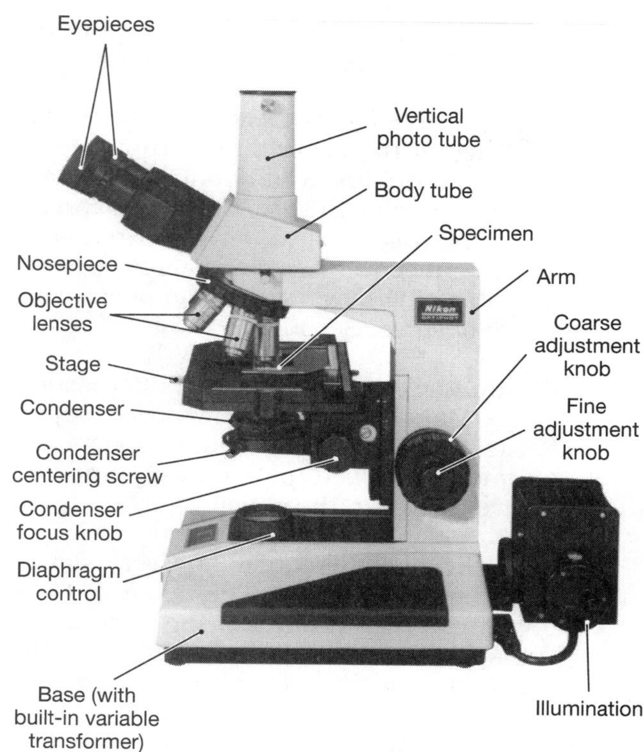

● **FIGURE 3–3**
Parts of the Compound Microscope.

Fine Adjustment Knob Knob for fine movements of the tube or stage. Used for fine-tuning the focus.

Head (Body Tube) Front of microscope housing the lenses and mirrors.

Arm Back part of microscope connecting the head to the base.

Ocular (Eyepiece) Lens closest to the eye. Usual magnification is 10×.

Monocular Microscope having only one ocular.

Binocular Microscope having two oculars.

Dual (Teaching) Microscope with two heads or two sets of oculars. Can be used so more than one person can view the same object.

Pointer Needle-like structure within the ocular. Can be used to point to specific structures by rotating the ocular.

Ocular Micrometer Ruler within the ocular.

Nosepiece Revolving structure housing the objective lenses.

Objective Lens Lens closest to the object being viewed. The tube of this lens is usually etched with a number indicating its magnification. The objective lens in use will click into place as you revolve the nosepiece.

Low Objective with a usual magnification of 10×.

High Objective with a usual magnification of 40×, 43×, 44×, or 45×. Often called the high-dry objective.

Oil Immersion Objective with a usual magnification of 97×, 98×, 99×, or 100×. When in use, this objective is immersed into a drop of oil. Oil concentrates the light rays by decreasing refraction, thus increasing clarity. See Figure 3–4●.

Scanning Objective with the usual magnification of 4× or 5×. Many microscopes do not have this very low power objective, which is generally used to gain an overview of the field.

Light Switch Switch for turning on the light.

Clips or **Mechanical Retainer (Mechanical Stage)** Manual or gear-driven apparatus for anchoring the slides on the stage.

Your microscope may also have a number of other knobs for various types of adjustments. Your instructor will explain any of these microscope parts as you need them to work in your particular class.

B. Microscope Terminology

Although most of the terms connected with basic microscopy have been covered in the above sections, some terms bear special explanations.

1➡ Study the following list of terms. Refer back to these concepts as needed.

Field Area seen under a microscope.

Depth of Field Three-dimensional depth perceived or observed looking through a microscope.

Resolution (Resolving Power) Ability to discern two objects as two distinct points. The

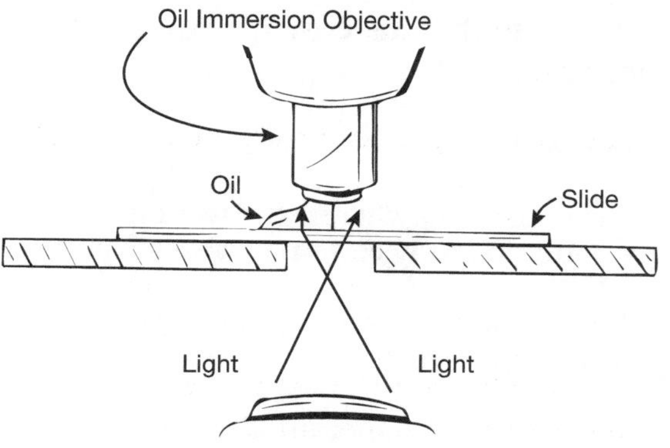

● **FIGURE 3–4**
Comparison of Oil and Air as Refractive Media.

human eye can distinguish two objects approximately 100 μm apart (μm = micrometer). With the aid of the compound microscope, two objects can be identified as close as 0.2 μm.

Photomicrograph (Micrograph) Photograph taken through a microscope.

Wet Mount Nonpermanent slide. Specimen is usually in water and covered with a cover slip.

Prepared (Permanent Mount) Slide Slide purchased or prepared in advance in such a way that specimens are not destroyed after use.

Magnification Number of times larger than itself an object appears to be. A magnification of 10× means the object is seen as 10 times larger than it actually is.

Total Magnification Term taking into account the magnifying power of the ocular and the magnifying power of the objective. Total magnification is the product of the ocular magnification and the objective magnification. For example: If the ocular magnification is 10× and the objective magnification is 43×, the total magnification is 430×.

Parfocal Design characteristic whereby if an object or structure is in focus under one power, then that object or structure will be in focus under all other powers on that microscope.

Parcenter Design characteristic whereby if an object or structure is in the center of a field under one power, that object or structure will be in the center of the field under all other powers on that microscope.

❑ Experimentation

V. Practice with the Compound Light Microscope

A. Newspaper Letters

In this exercise you will use the wet mount of a newspaper letter to study certain optical properties.

1➡ Obtain a slide, cover slip, newspaper, and scissors.

2➡ Take your newspaper and from the regular small print cut either a single letter (preferably an "a" or an "e") or a small strip of letters. Place the letter in the middle of a clean slide. Put a very small drop of water on top of the letter. Cover with a cover slip.

3➡ Cover the letter with a cover slip by placing the edge of the cover slip in the drop of water and holding it at about a 45-degree angle to the slide. The water should spread along the edge of the cover slip. Now, gently drop the cover slip down across the remainder of the water and the object you are to observe. See Figure 3–5●.

4➡ Clamp the slide onto your microscope stage. Make certain the low power objective is clicked into place and the coarse adjustment is turned to its lowest point (objective closest to the stage).

5➡ Look through the ocular and very slowly turn the coarse adjustment up so that the objective is gradually moving away from the stage. Your newspaper letter should gradually come into focus.

6➡ Remember not to turn the coarse adjustment down while looking through the ocular. When you have your letter in focus, fine-tune the focus with the fine adjustment knob.

7➡ Experiment with the iris diaphragm so that you are using the optimum amount of light for your specimen. If you use too much light, your object will fade. With too little light, you cannot see detail.

8➡ Move the slide around. What do you notice about the image?

____

____

Is it in the expected position? ________
What happens as you move the slide? Does the image seem to go up when you move the slide down? Why is this?

____

____

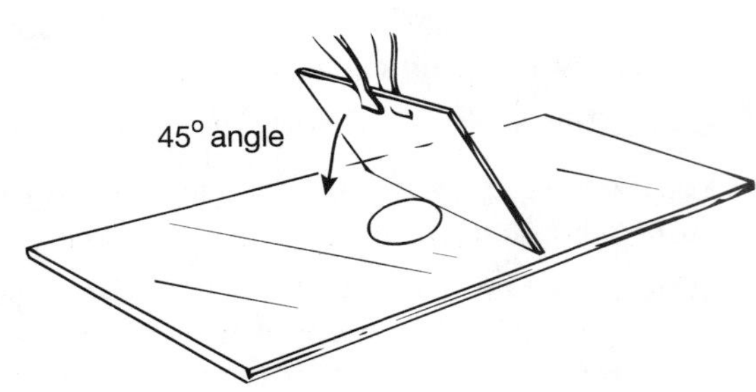

● **FIGURE 3–5**
Coverslip Placement.

9➡ Once you are comfortable with your newspaper letter under low power, switch to high power. DO NOT MOVE THE FOCUS KNOBS! Simply move the nosepiece. If your letter is not in focus, adjust the fine adjustment knob only. Do not adjust the coarse adjustment!

10➡ How does your high power letter compare with your low power letter?______________

____

11➡ Sketch what you see.

Letter

12➡ Clean your slide according to your instructor's directions.

B. Crossed Hairs

In this exercise you will use a wet mount to examine depth of field.

1➡ Obtain a slide and cover slip. Take one hair from your head and one hair from someone else's head. Lay these hairs across your slide so they cross in the middle. Put a small drop of water on the cross and cover with a cover slip.

2➡ Use low power to focus your microscope on the crossed hairs at the center point. Notice the difference between the colors and textures of the hairs. Experiment with the amount of light you are using. What do you observe? ______________________

____

3➡ Move the fine adjustment up and down and gain a perspective on the depth of your field. Did you expect to find such an apparent distance between the two hairs? Whose hair was on top? Repeat under high power.

____

____

4➡ Sketch what you see.

Crossed hairs

5➡ Clean your slide for the next observation.

C. Pond Culture

In this exercise you will view moving objects.

1➡ Take a drop of the pond culture and put it in the center of your slide. You do not need to add water this time. Put the cover slip on the same way you did in the previous exercises. View your slide under low power. Experiment with the amount of light you need.

2➡ Switch to high power when you find some interesting organisms and study the small creatures more closely. You may need more light to study the details under high power.

3➡ Depending on what is in your pond culture, your instructor may ask you to give at least *form identifications* of 5 or more organisms. (Form identification is a nonspecific grouping of organisms based on certain characteristics such as size, shape, movement, coloration, etc.) Because heat from the microscope may damage certain organisms, keep the light off as much as possible.

4➡ Sketch what you see.

Pond culture

5➡ Clean your slide and cover slip and put them away according to directions.

D. Mixed Bacteria

In this exercise you will be using the oil immersion lens.

1➡ Take the prepared bacterial slide. DO NOT USE WATER; DO NOT USE A COVER SLIP.

2➡ Focus the slide under low power.

3➡ Take your eyes away from the ocular and move the nosepiece so it is halfway between the low power objective and oil immersion objective.

4➡ Now put one generous drop of immersion oil in the middle of the slide. DO NOT TOUCH THE COARSE OR FINE ADJUSTMENT KNOBS.

5➡ Click the oil immersion lens into place. (Yes, the oil immersion lens will be right in the oil. Look at Figure 3–4● again.)

6➡ Now look through the ocular. If necessary, use the fine adjustment knob. You may also need more light. Your bacteria should now be in focus.

7➡ Sketch what you see.

Mixed bacteria

8➡ To clean the oil off your lens and to clean your slide, follow your instructor's directions.

❑ Additional Activities

NOTES

1. Check into the lives of these scientists:
   a. Giambattista della Porta (1535–1615), Italian optical scientist whose observations led the way to the development of the microscope.
   b. Hans and Zacharias Janssen.
   c. Galileo Galilei (1564–1642), Italian astronomer and physicist who did much to perfect the microscope.
   d. Antony van Leeuwenhoek (1632–1723), Dutch linen merchant and amateur naturalist who perfected the microscope and whose observations and writings in biological microscopy laid the foundation for much of what we do in biology today.
2. With the help of a physics book or a book on simple optics, construct a simple compound microscope using concave and convex lenses and appropriate cardboard tubes. Find out why brass tubes produce greater optical accuracy than do cardboard tubes.
3. If your school has an electron microscope, discuss the workings of this microscope with the people in charge of it.

❑ Lab Report

1. Referring to the numbered inquiries at the beginning of this exercise, complete the following box summary:

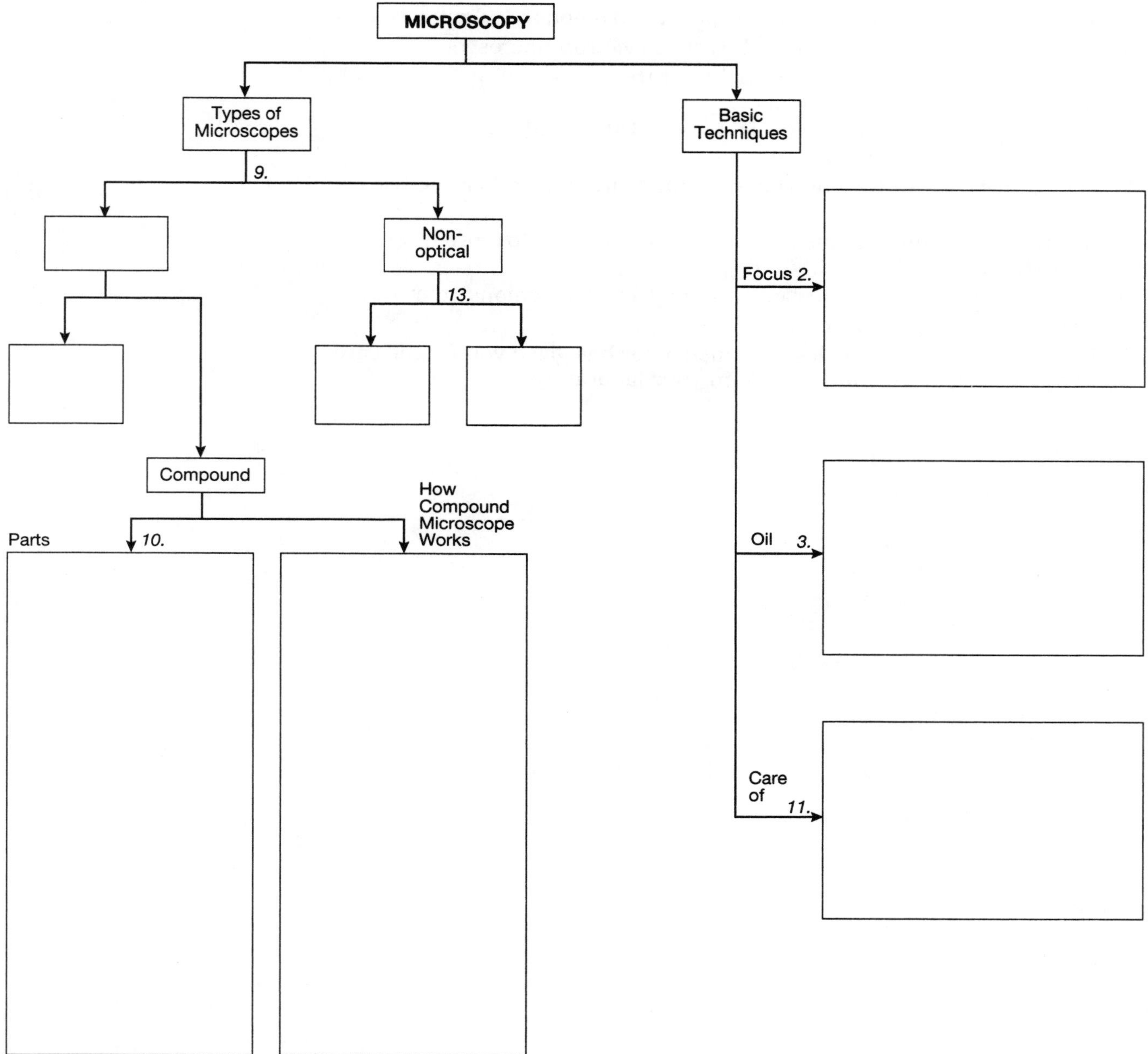

General Questions

NOTES

1. List some of the people responsible for the development of the microscope and give the contributions of each.
2. Explain the difference between the simple and the compound microscope.
3. What is the difference between an optical and a nonoptical microscope?
4. What advantage is there for each type of electron microscope?
5. Why should you never go "down" with the coarse adjustment knob while looking through the ocular?
6. If the objective magnification is 45× and the total magnification is 405×, what is the ocular magnification?
7. Give two ways the oil immersion lens differs from the other microscope objectives.
8. In which direction does the newspaper letter appear to move when you move the slide to the right?
9. What principle did the crossed hairs on the slide demonstrate?
10. How do you prepare a wet mount?
11. Record any additional rules your instructor has given you for the care and handling of the microscopes in your laboratory.

EXERCISE 4

Anatomical Orientation and Terminology

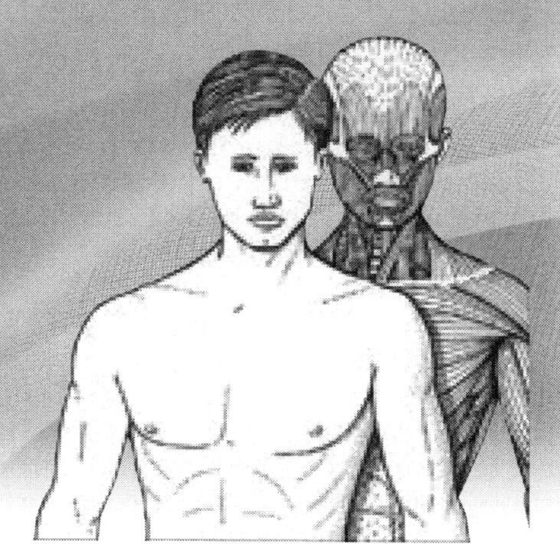

PROCEDURAL INQUIRIES

Examination

1. What is anatomical position?
2. Why is anatomical position important?
3. What are the abdominopelvic quadrants?
4. What are the abdominopelvic regions?
5. What are seven pairs of directional terms used to describe anatomical relationships?
6. What are the major planes of the body?
7. What are the major cavities of the body?

Practice

8. What are some common anatomical and physiological prefixes and suffixes?
9. How can a knowledge of the prefixes and suffixes help our study of anatomy and physiology?

Additional Inquiries

10. What is the value of a common anatomical language?
11. What are the *Nomina Anatomica* and the *Nomina Anatomica Veterinaria*?

Key Terms

This list does not include the terms found in Figures 4–1● and 4–2●.

Abdominal Cavity
Abdominopelvic Cavity
Anterior
Biped
Caudad
Caudal
Cephalad
Cranial
Cranial Cavity
Deep
Diaphragm
Distal
Dorsal
Dorsal Body Cavity
Frontal
Inferior
Lateral
Medial
Median
Mediastinum
Midsagittal
Parasagittal
Pelvic Cavity
Pericardial Cavity
Peritoneal Cavity
Plane
Pleural Cavities
Posterior
Proximal
Quadrants
Quadruped
Sagittal
Section
Spinal Cavity
Superficial
Superior
Thoracic Cavity
Transverse
Ventral
Ventral Body Cavity
Vertebral Cavity

Materials Needed

Models or Diagrams
Laboratory Animals
Dictionaries (Standard and Medical)

A common anatomical language is essential if we are to communicate relative locations and conditions of structure and function. For this reason reference points and conventional definitions have been established. It is important that you become comfortable with the terminology in this lesson.

Anatomical terminology today is the culmination of over 2,000 years of anatomical history. Most modern anatomical terms, many dating to ancient times, come from Latin and have remained rather uniform throughout history and throughout the world. Nevertheless, differences, particularly as related to specific specialties, have arisen.

In 1895 German anatomists proposed a standardized list of terms that became known as the *Basle Nomina Anatomica* (BNA). The BNA formed the basis for the *Nomina Anatomica* (NA). The NA was adopted by the International Congress of Anatomists in Paris in 1955. The NA, which has been revised several times, is essentially what we use today.

The companion list to the NA is the *Nomina Anatomica Veterinaria*, which contains term variations applicable to the study of domestic animals.

❑ Examination

I. Study Hints

A. We offer the following suggestions to help you learn the vocabulary.

1➡ Say the words out loud.

2➡ Use the words out loud in meaningful sentences.

3➡ Write the words in meaningful sentences.

4➡ Listen to the words as you say them and as your classmates say them.

5➡ Quiz your classmates often.

B. In the following descriptions, the term ***biped*** refers to humans and other two-legged creatures. The ***quadrupeds,*** such as dogs, cats, and pigs, are four-legged.

II. Anatomical Position

A. Common Reference

Anatomical position is the standard position that we use as a reference point for all anatomical descriptions, locations, and directions.

1➡ Study Figure 4–1●. The individual in this figure is in anatomical position. Note the position of the hands (palms forward) and the feet (toes pointed straight ahead). Note too that the subject is upright and facing directly forward.

When discussing any relative anatomical consideration our reference is always anatomical position, even if the subject is in an inverted yoga position.

2➡ Relate Figure 4–1● to the models or charts in your laboratory.

3➡ Stand in anatomical position. Describe the position of your hands, arms, feet, and head. If you are working with a partner, describe points about that person's anatomical position.

____

____

____

III. Surface Terminology

A. Identifying the Landmarks

We have specific medical or scientific terms for many of the landmarks of the body.

1➡ Refer again to Figure 4–1●. Quiz a classmate on the anatomical landmark terms. Make certain you have a solid understanding of how the different terms relate to one another.

B. Using "Left" and "Right"

In using the terms "left" and "right," we always refer to the subject's left and right. For example, to locate a cat's left brachial artery, look on the animal's left side, even though the animal's left may be on your right.

1➡ Practice using "left" and "right" with the surface terms. For instance, "left brachium," "right popliteal region," etc.

2➡ How many of these areas can you identify on the laboratory models? If laboratory animals are available, identify as many of the animals' corresponding surface areas as possible.

____

____

3➡ Sketch a line on Figure 4–1● from the right thumb to the shoulder, across the chest and down to the left ankle. Using "right"

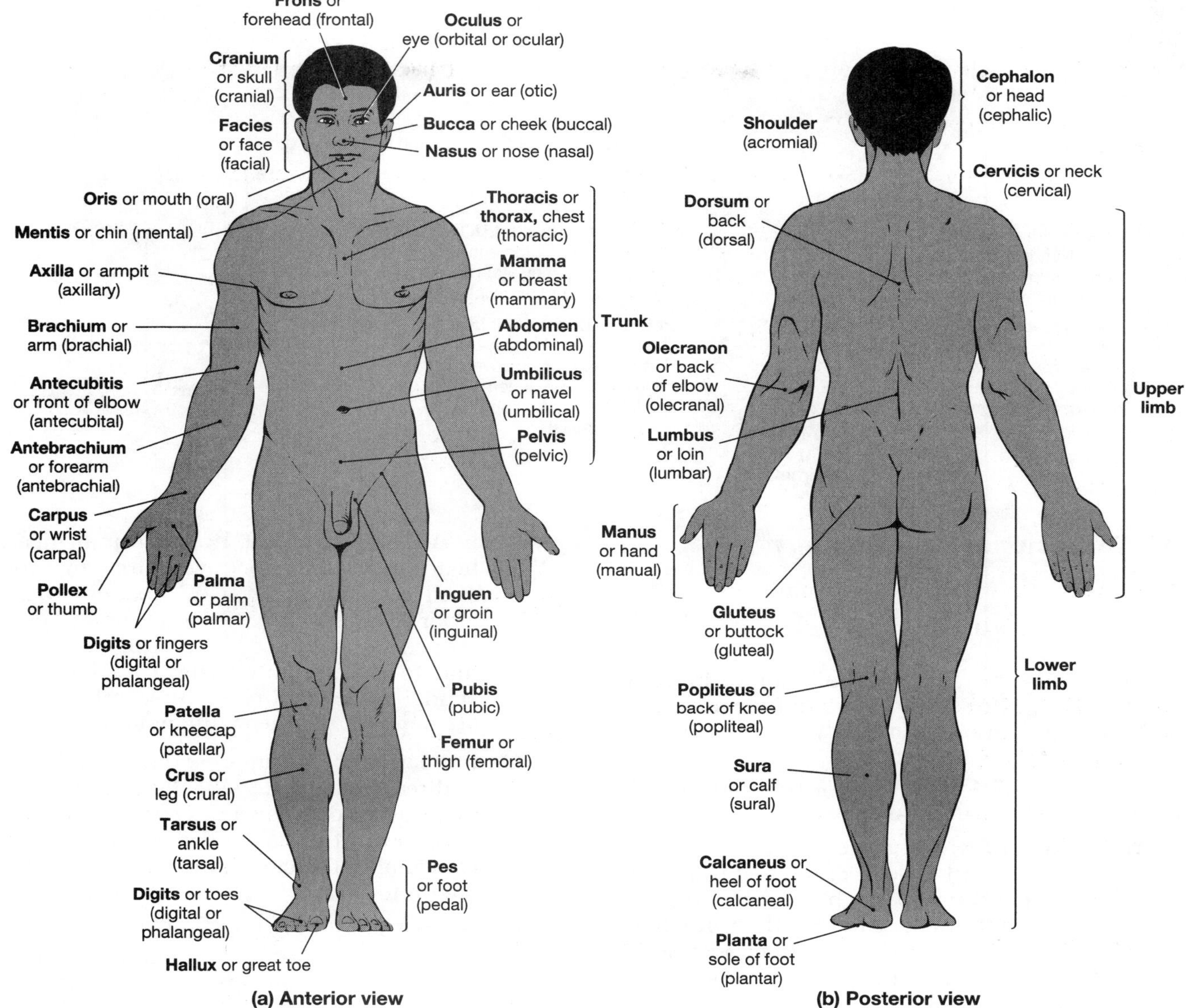

● **FIGURE 4–1**
Anatomical position/anatomical landmarks
(a) Anterior **(b)** Posterior

and "left" appropriately, name the surface regions you have traversed.

C. QUADRANTS AND REGIONS [∞ MARTINI, P. 17]

It is often convenient to divide the abdominal surface into anatomical sections. Abdominopelvic **quadrants** are useful in describing general aches, pains, and injuries. Quadrant identification is used by clinicians as well as by the public at large.

1 ➡ Identify the quadrants shown in Figure 4–2a●.

____

____

2 ➡ For more precise regional descriptions, anatomists commonly divide the abdominopelvic surface into nine **regions**. Use Figure 4–2b● to identify these regions. Do you see the logic in the regional terms?

____

____

3 ➡ Sketch an imaginary line from the nose to the umbilicus and across the abdomen on a diagonal to the area behind the left knee. Name the surface regions.

____

____

● **FIGURE 4–2**
(a) Abdominopelvic quadrants;
(b) Regions.

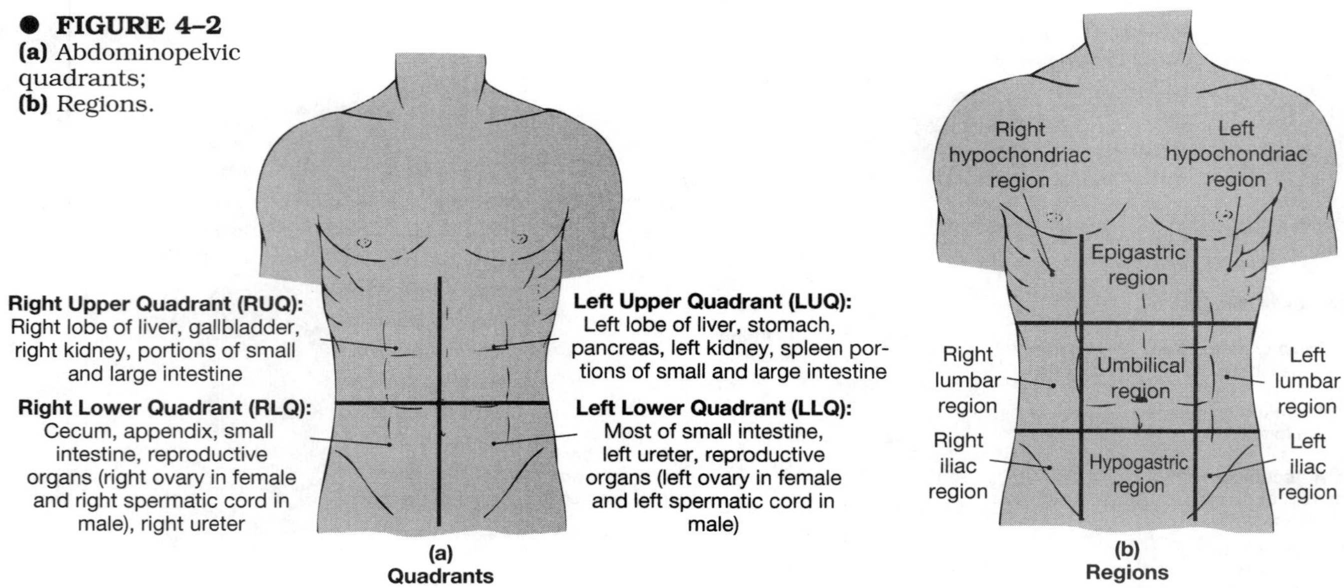

IV. Directional Terminology [∞ *Martini, p. 18*]

A. Directional Pairs

Directional terms are usually found in pairs because they show the directional location of a body part in relation to the body as a whole.

1➡ Use Figure 4–3● to work through the following terminology.

Superior–Inferior: Superior and inferior refer to relative positions along the vertical axis. Thus, in the biped, the eyes are superior to the mouth and the umbilicus is inferior to the sternum. In quadrupeds, the vertebral column is superior to the stomach.

Anterior–Posterior: The anterior is the front and the posterior is the back of the body. In humans, the nose, abdomen, and knees are on the anterior surface, and the buttocks and nape of the neck are on the posterior surface. Also, the abdomen is anterior to the buttocks and the vertebral column is posterior to the heart. In quadrupeds, the anterior is the head region and the posterior is the tail region.

Dorsal–Ventral: The back is dorsal and the front is ventral. Although dorsal and ventral are more commonly used with quadrupeds, in bipeds dorsal and ventral are often synonymous with posterior and anterior. Dorsal and ventral are synonymous with superior and inferior in quadrupeds. The vertebral column is always dorsal to the heart.

Cranial–Caudal: Cranial and caudal mean head and tail, respectively, and may be used interchangeably with superior and inferior in bipeds and with anterior and posterior in quadrupeds. The lungs are cranial to the stomach; the intestines are caudal to the diaphragm. Alternative terms are **cephalad** for head and **caudad** for tail.

Medial–Lateral: Medial describes a location or direction toward the midline of the body and lateral describes a location or direction toward the side of the body. The nose is medial to the cheek and the shoulder is lateral to the clavicle (collar bone).

Proximal–Distal: Proximal and distal are generally used to describe portions of a limb in relation to the limb's point of attachment. The knee is proximal to the ankle, but the toe is distal to the ankle.

Superficial–Deep: Superficial and deep refer to position relative to the surface of the body. The muscle is deep to the skin but the epidermis is superficial to the dermis. The dermis is the skin proper. The epidermis is the outer layer of the skin or, more correctly, the covering on the surface of the skin proper.

2➡ Use directional terminology to describe these relationships:
Throat to esophagus in humans ________
Throat to esophagus in cats ________
Hair to skeletal muscle ________
Thumb to ring finger ________
Heart to lungs ________
Ear to eye ________
Abdomen to thorax ________

3➡ Use the surface terminology explained in section III to describe ten directional relationships.

________________ ________________
________________ ________________
________________ ________________
________________ ________________
________________ ________________

V. Sectional Terminology

A. Planes [∞ Martini, p. 19]

A slice extending through a body or an organ produces an imaginary division called a **plane**. The parts produced by the plane are the **sections**. We can make sections along numerous planes, but normally we use the system where each of the three planes forms a right angle with each of the other two planes.

1➡ Study Figure 4–3a● for the human planes and Figure 4–3b● for the quadruped planes. Identify the following planes in the human and the dog.

The **sagittal** plane cuts the body into right and left sections. If the sagittal plane is centered and the sections are equal or are mirror images of each other, the plane is a **midsagittal** or **median** plane. If the sections are unequal, the plane is **parasagittal**.

The **frontal** plane divides the biped body into anterior and posterior sections and the quadruped body into dorsal and ventral sections. The frontal plane is alternately called the **coronal** plane. (In quadrupeds the frontal plane is also sometimes called the **dorsal** plane.) Terms such as **midfrontal** and **midcoronal** are not generally used because the sections are neither identical nor are they mirror images of each other.

A **transverse** plane (sometimes called a **cross-sectional** plane) divides a body or a body part into superior and inferior, cranial and caudal, or distal and proximal portions. In bipeds a transverse plane is a horizontal plane whereas in quadrupeds a transverse plane is vertical. Keep in mind that the horizontal and vertical planes are always in relation to anatomical position!

2➡ How do the planes relate to the term pairs anterior/posterior and dorsal/ventral?

____
____

B. Cavities [∞ Martini, p. 20]

1➡ Observe Figure 4–4● and identify the body's two major sets of cavities, the **dorsal body cavity** and the **ventral body cavity**. These cavities surround most of the body's vital organs.

The fluid-filled dorsal body cavity holds the central nervous system and can be divided into the **cranial cavity** for the brain and the **spinal** or **vertebral cavity** for the spinal cord. The spinal cavity is continuous with the cranial cavity.

The ventral body cavity is generally subdivided into the **thoracic cavity** of the chest and the **abdominopelvic cavity** of the abdomen. The thoracic cavity can be further divided into a pair of **pleural cavities**, each housing a lung, and the **pericardial cavity** surrounding the heart. The

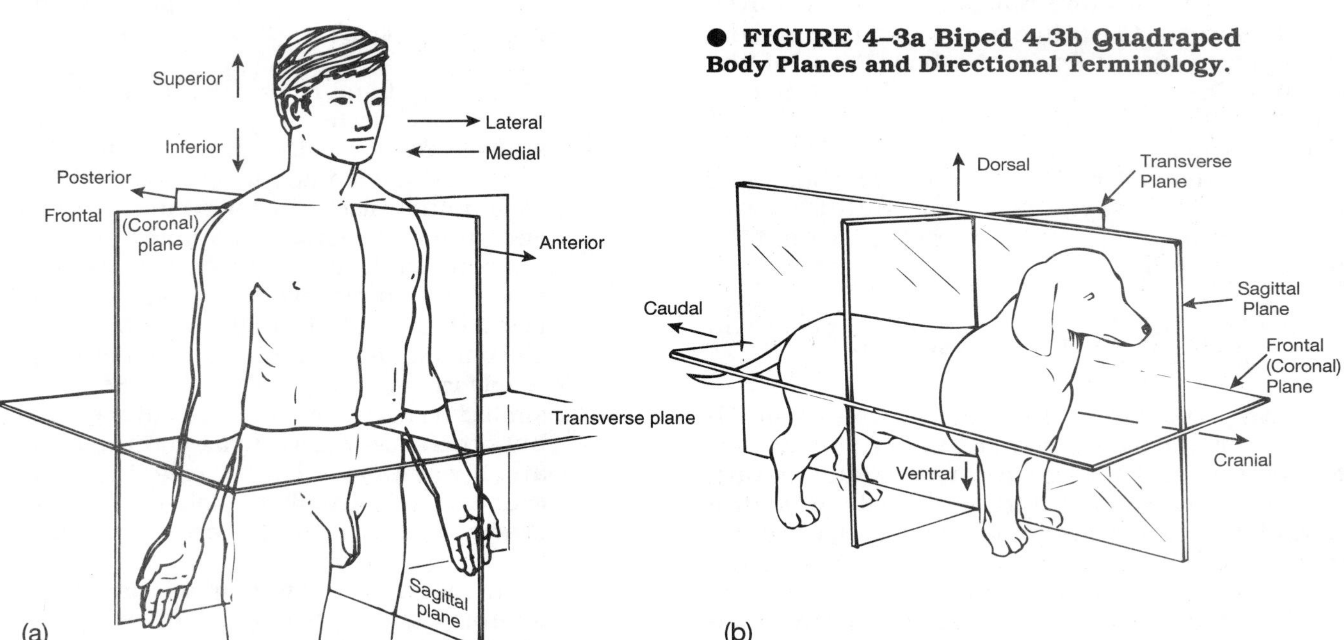

● **FIGURE 4–3a Biped 4-3b Quadraped Body Planes and Directional Terminology.**

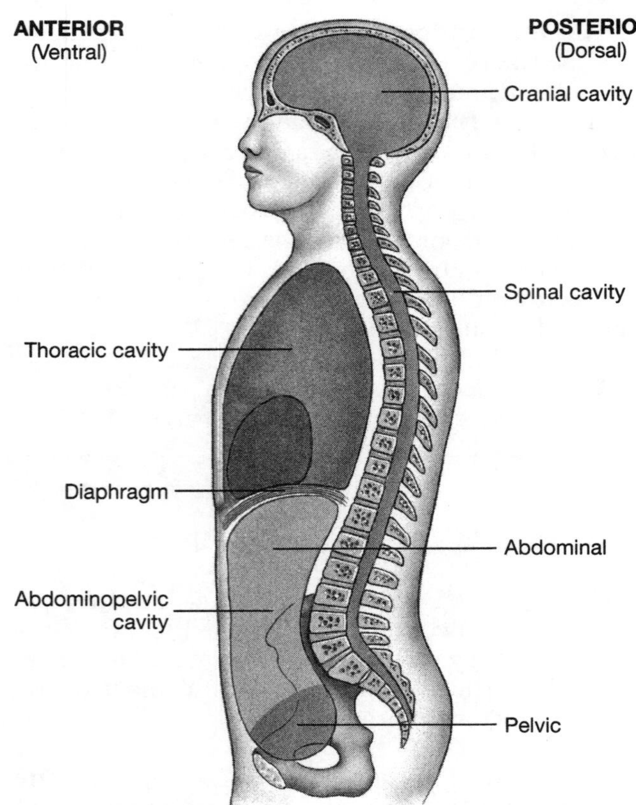

● **FIGURE 4–4**
Body Cavities.

pericardial cavity is part of the **mediastinum**. (The mediastinum, which is the space in the thorax between the pleural sacs, also contains the trachea, esophagus, and various surrounding structures.)

Separating the thoracic cavity from the abdominopelvic cavity (also known as the **peritoneal cavity**) is the **diaphragm**. The superior portion of this cavity (the **abdominal cavity**) extends from the diaphragm to an imaginary line drawn from the symphysis pubis to the base (top) of the sacrum. The area below this line is the **pelvic cavity**. The stomach, liver, spleen, pancreas, gallbladder, small intestine, and parts of the large intestine are considered to be within the abdominal portion of the cavity. The urinary bladder, lower colon, rectum, uterus, and ovaries are within the pelvic portion of the cavity.

2➡ Quiz your lab partner on which organs can be found in which cavity.

Also present in the body are numerous smaller functional cavities, such as the orbital cavities for the eyeballs, the nasal cavity extending outward from the nose, and the oral cavity or mouth. (The allergy-prone sinus cavities are spaces within specific facial bones.)

3➡ If a torso model is available, remove the internal organs and determine which cavity houses which structures. If no torso is available, sketch the dorsal and ventral cavities without looking at Figure 4–4●.

❑ Practice

VI. Terminology

A. Background

Sometimes the beginning anatomy and physiology student feels a bit overwhelmed by the amount of medical vocabulary. New words of medical significance seem to constitute an entirely new language that must be mastered along with a whole series of equally new scientific concepts.

The purpose of this part of the lesson is to expand on some basic anatomical vocabulary by offering practice in working with contemporary medical terminology.

If you are going into a health care profession, this lesson should serve as a springboard for your future exposure to medical terminology. If you are not planning on a health care profession, this exercise should provide basic background material for some of the terminology you will encounter as you explore the fields of anatomy and physiology.

1➡ Study the following list of prefixes and suffixes common to anatomical and physiological terms. The forms given here are in their most common prefix or suffix form. Sometimes a prefix or suffix can be used as a different part of a word and still retain the original meaning. Sometimes, too, you will notice that an additional letter or syllable has been added to a known stem to smooth out the pronunciation of the word. Don't let that worry you!

2➡ Please keep in mind, too, that this list is not complete. Unfortunately, some prefixes and suffixes are not included. Be certain to add to this list any additional prefixes or suffixes you find. Also, in your studies you may encounter numerous examples of words that you think you might know but that do not seem to be exactly what you have learned. Carefully examine such terms and be confident in determining whether the altered form is what you think it is. Some very similar terms have very different meanings but others are exactly what you think they are. If you know the basic meanings, however, you will usually be able to determine exactly what the derived meanings are.

3➡ Remember, very few new words exist in anatomy and physiology. Most words are combinations of words you already know.

B. Vocabulary

a-, *an-*. Without
ab-. From
-ac. Pertaining to
ad-, *To*. Toward
aden-, *adeno-*. Gland
adipo-. Fat
af-. Toward
-al. Pertaining to
-algia. Pain
ana-, *Up*. Back
andro-. Male
angio-. Vessel
anti-, *ant-*. Against
apo-. From
arachn-. Spider
arthro-. Joint
-asis, *-asia*. State, Condition
astro-. Star
atel-. Imperfect
auto-. Self
baro-. Pressure
bi-. Two
blast-, *-blast*. Precursor
brachi-. Arm
brachy-. Short
brady-. Slow
bronch-. Windpipe
cardi-, *cardio-*, *-cardia*. Heart
caud-. Tail
-centesis. Puncture
cephal-. Head
cerebro-. Brain
chole-. Bile
chondro-. Cartilage
chrom-, *chromo-*. Color
circum-. Around
-clast. Broken
co-. Together
coel-, *-coel*. Cavity
contra-. Against
cranio-. Skull
cribr-. Sieve
-crine. To separate
cyst-, *-cyst*. Sac
cyto-, *-cyte*. Cell
dactyl-. Finger, Toe
de-. Down from
deca-. Ten
deci-. Tenth
demi-. Half
dens-, *dent-*. Tooth
derma-, *-derm*. Skin
desmo-. Band
dextro-. Right
di-. Twice
dia-. Through
dipla-, *diplo-*. Double
dis-. Negative, Double, Apart
diure-. To urinate
-dynia. Pain
dys-. Painful
-ecstasis. Expansion
ecto-. Outside
ef-. Away from
emmetro-. In proper measure
encephalo-. Brain
end-, *endo-*. Inside
entero-. Intestine
epi-. On
erythema-. Flushed (skin)
erythro-. Red
esophago-. Esophagus
ex-. Out, Away from
ferr-. Iron
-gen, *-genic*. To produce
genicula-. Kneelike structure
genio-. Chin
glosso-, *-glossus*. Tongue
glyco-. Sugar
-gram. Record
-graph, *-graphia*. To write
gyne-, *gyno-*. Woman
helminth-. Worm
hem-, *hemato-*. Blood
hemi-. Half
hepato-. Liver
hept-. Seven
hetero-. Other
hex-. Six
histo-. Tissue
holo-. Entire
homeo-, *homo-*. Same
hyal-, *hyalo-*. Glass
hydro-. Water
hyo-. U-shaped
hyper-. Above
hypo-. Below
hystero-. Uterus
-ia. State
idio-. Self
ili-, *ilio-*. Of the ilium or flanks
infra-. Beneath
inter-. Between
intra-. Within
ipsi-. Itself (same as)
iso-. Equal
-itis. Inflammation
karyo-. Body (kernel)
kerato-. Horn
kino-, *-kinin*. To move
lacri-. Tear (secretion)
lact-, *lacto-*, *-lactin*. Milk
laparo-. Loin
laryngo-. Larynx
latero-. Side
-lemma. Husk (covering)
lepto-. Small, soft
leuko-. White
liga-. Bind together

lip-, lipo-. Fat
-logia, -logy. Study or science of
lyso-, -lysis, -lyze. Dissolution
macro-. Large
mal-. Bad, abnormal
mammilla-. Breast
mast-, masto-. Breast
medi-. Middle
mega-, megal-. Large
melan-, melano-. Black
meningo-. Of the meninges
mero-. Part
meso-. Middle
meta-. After, Beyond
-meter. Measure
micro-. Small
mio-. Less, Smaller
mono-. Single
morpho-. Form
multi-. Many
-mural. Wall
myelo-. Marrow
myo-. Muscle
natri-. Sodium
neo-. New
nephr-. Kidney
neur-, neuro-. Nerve
non-, not-. No, Not
ob-. Against
oct-. Eight
oculo-. Eye
-oid. Form
oligo-. Little, Few
-ology. Study of
-oma. Swelling
onco-. Mass, tumor
o-, oo-. Egg, ovum
-opia. Eye
orchid-. Testicle
os-. Mouth, Bone
-osis. State, Condition
osteo-. Bone
ota-, oto-. Ear
pachy-. Thick
pan-. All, Entire
para-. Alongside, Beyond
patho-, -path, -pathy. Disease
pedia-. Child
-penia. Lack
pent-. Five
per-. Excessive, Through
peri-. Around
phago-. Eat
pharyno-. Pharynx
-phasia. Speech
-phil, -philia. Love
-phobe, -phobia. Fear
-phylaxis. Guard, Protect
physio-. Nature
pino-. Drink, Take in liquid
-plasia. Formation
-plastic, -plasty. Mold, Form
platy-. Flat
-plegia. Blow, Paralysis
-plexy. Strike
pneu-. Lung, Air
pod-, podo-. Foot
-poiesis. Making
poly-. Many
post-. After
pre-. Before
presby-. Old
pro-. Before, On behalf of
proto-. First
pseudo-. False
psych-. Mind, Soul
pterygo-. Wing
pulp-. Flesh
py-, pyo-. Pus
pyro-. Heat, Fire
quad-. Four, Quarters
retro-. Backward
-(r)rhage. Flow, Hemorrhage
-(r)rhaphy. Suture, Stitch
-(r)rhea. Flow, Discharge
rhino-. Nose
sacchar-. Sugar
sacro-. Sacrum
salphin-, salphingo-. Tube (uterine tube)
sarco-. Flesh
scler-, sclero-. Hard
-scopy. View, See
semi-. Half
-septic. Putrid
-sis. State, Condition
som-, some. Body
spino-. Spine, Vertebral column
-stomy. Mouth, Opening
stylo-. Stake, Pole.
sub-. Below
super-, supra-. Above
syn-. Together
tachy-. Swift
telo-. End
tetra-. Four
therm-, thermo-. Heat
-tomy. Cut
trans-. Through
tri-. Three
-trophic, -trophin, -trophy. Nourishing
tropho-. Nutrition
tropo-. Turning
uni-. One
uro-, -uria. Urine
vaso-. Vessel (blood vessel)
ventro-. Abdomen

C. Using the Terminology

1➜ Use the prefix and suffix definitions given above. Take apart each of the following words and define the components. Figure out what the word means. Check the dictionary after you have arrived at a logical definition.

a. Megakaryocyte
b. Rhinoplasty
c. Antiseptic
d. Psychosomatic
e. Muscular dystrophy
f. Aseptic
g. Pyorrhea
h. Osteocyte
i. Mesoderm
j. Polysaccharide
k. Myeloblast
l. Hepatocyte
m. Atelectasis
n. Autolysis
o. Lipoma
p. Neuralgia
q. Cholecystitis
r. Glycosuria
s. Arthroscopy
t. Physiology
u. Otorhinolaryngology
v. Orchidectomy
w. Interdental
x. Hyperglycemia
y. Hydrocephalia

2➜ Use the definitions given and supply the correct word, words, or definitions. Check a dictionary if necessary. Underlined words or phrases should be replaced. It may be necessary to reword some sentences after you have supplied the proper word(s).

a. What is the difference between a hysterectomy and a hysterotomy?
b. The arachnoid layer is found in the brain. What would you expect the arachnoid layer to look like?
c. What is the difference between an electroencephalogram and an electroencephalograph?
d. Blood in the (afferent/efferent) vessels flows into the kidney.
e. Strenuous exercise might cause (bradycardia/tachycardia).
f. The "top" of the tongue would be a (hyperglossal/hypoglossal) structure.
g. The (osteoblast/osteoclast) produces new bone cells.
h. The living system is a series of checks and balances designed to maintain (homeostasis/heterostasis/both).
i. Waxes are (hydrophobic/hydrophilic).
j. A person with <u>presbyopia</u> might have problems with (near/far) vision.
k. If a blood vessel bifurcates, it....
l. A person suffering from agraphia is....
m. Originally the osteopath was concerned with....
n. If you are suffering from a <u>cardiovascular malady</u>, the doctor might order an ______________. After that you might need ______________.
o. The ______________ studied the <u>shape</u> of the <u>nerve cells</u>.
p. The pain is <u>on the same side as</u> the injury.
q. In a precancerous condition, the cells might be <u>poorly formed</u>.
r. Joe developed an <u>inflamed ear</u> from a <u>fever-producing</u> infection.
s. Because of the accident, she had her <u>voice box removed</u>.
t. Having <u>short fingers</u> is a genetic trait.
u. A head injury might cause <u>double vision</u>.
v. Many <u>intestinal</u> infections begin with <u>low white blood cell counts</u>, followed by <u>very high white blood cell production</u>.
w. Both the dog and his owner suffered from <u>multiple fears</u>.
x. Kidney infections might cause <u>excessive urination</u>, <u>painful urination</u>, <u>cessation of urination</u>, or <u>severely reduced urination</u>.
y. Can you come up with logical definitions of anatomy and physiology?

3➜ *Using the vocabulary in Section VI.B, make a list of everyday words that use the same prefixes and suffixes.*

❑ Additional Activities

1. Set aside a section in your notebook and begin a list of terms you see or hear that are not included in the above sections.

2. Section off VI.B above into groups of five prefixes and suffixes. Each day make a point of using each word segment out loud in a sentence.

NOTES

❑ Lab Report

1. Referring to the numbered inquiries at the beginning of this exercise, complete the following Box Summary:

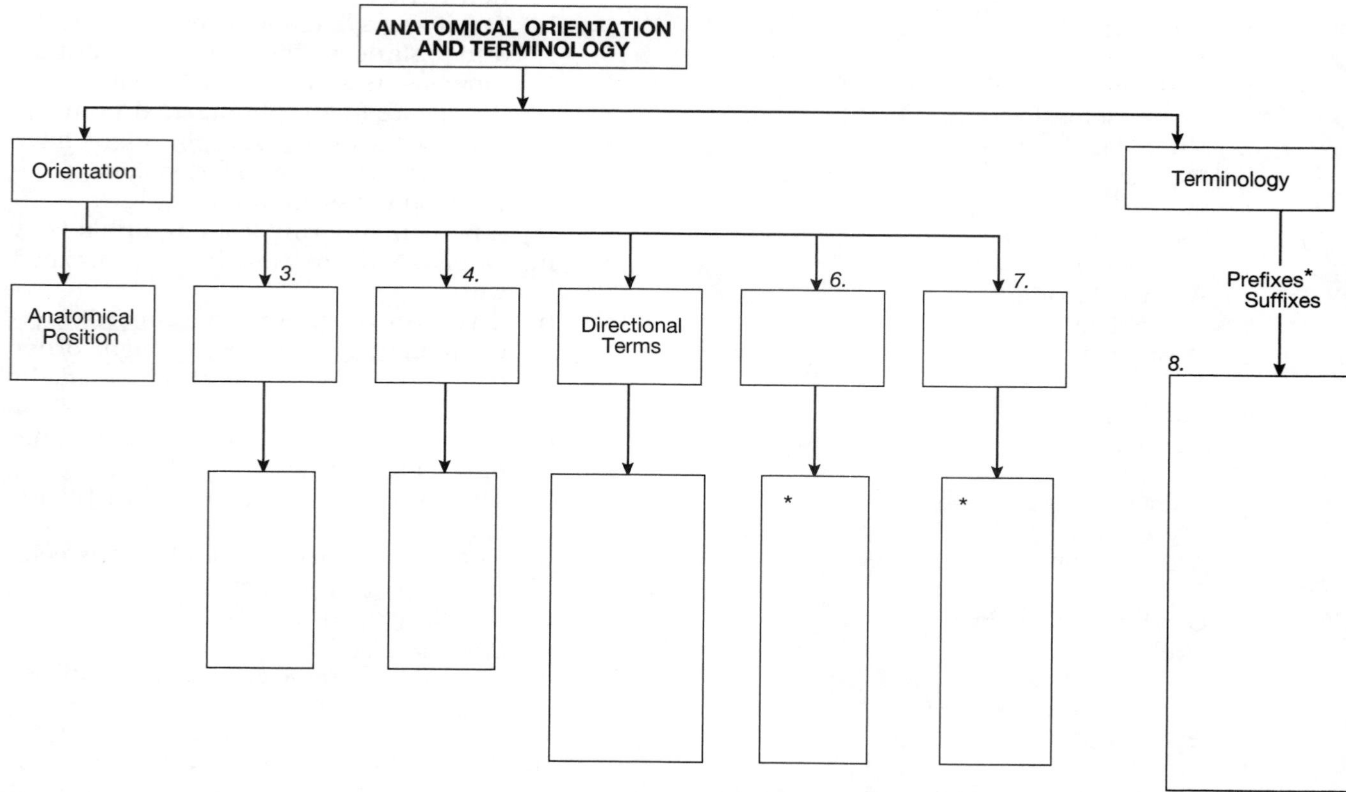

*List according to your instructor's directions.

General Questions

NOTES

1. Which plane would be involved?
   The magician sawing his lovely assistant in half.
   The amputation of the arm at the shoulder.
   The amputation of a finger.
   The tip of the nose being cut off.
   The optometrist checking the distance of each eye from the nose.
2. Which abdominopelvic quadrant is commonly associated with appendicitis?
3. Use directional terminology to describe the relative positions of the nine abdominopelvic regions.
4. Which plane would demonstrate the continuum of the cranial and spinal cavities?
5. Which quadrants are associated with which parts of the abdominopelvic cavity?

6. Using anatomical terminology, describe the relationship between the dorsal and ventral body cavities.
7. You are in a swimming pool floating on your back with your hands stretched out above your head.
   a. Your elbows are (superior/inferior) to your shoulders. Explain.
   b. True or False. You clasp your fingers directly above your head so your fingers become medial to your shoulders. Explain your answer.
8. List the most important points you should be aware of when confronting a new medical or anatomical/physiological term.
9. From what work do we derive our common anatomical language?

NOTES

EXERCISE 5

Nucleic Acids

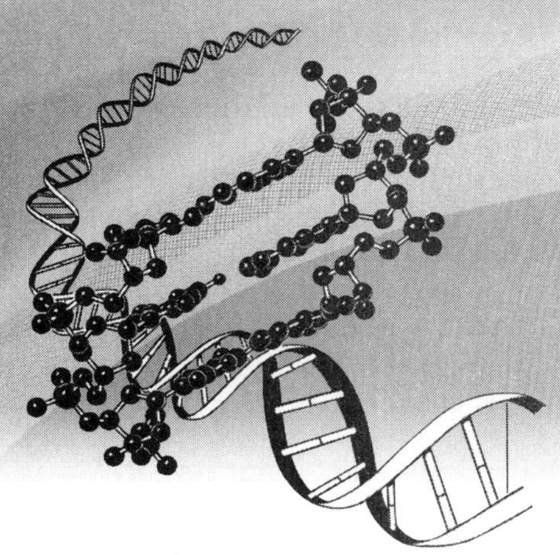

PROCEDURAL INQUIRIES

Examination

1. What is DNA?
2. What is RNA?
3. What are the parts of a nucleotide?
4. What is the difference between a purine and a pyrimidine?
5. What does the phrase 5' ⟶ 3' direction mean?

Model Building

6. What is the pairing pattern between the nitrogen bases?
7. What is semi-conservative replication?
8. What is the difference between the sense strand and the non-sense strand?

Additional Inquiries

9. What are the differences in sugars and the differences in nitrogen bases between RNA and DNA?

Key Terms

Adenine
Antiparallel
Chromosome
Cytosine
Deoxyribose
DNA
Gene
Guanine
Nitrogen base
Non-sense strand
Nucleotide
Purine
Pyrimidine
Replication
Ribose
RNA
Semi-conservative
Sense strand
Thymine
Uracil

Materials Needed

Pipestem Cleaners (8 of one color, 8 of another color, and 24 of a third color)
Toothpicks
Glue

DNA, deoxyribonucleic acid, is the genetic material of which chromosomes are made. DNA is the blueprint of life, is common to all organisms, and dictates the physical properties of the individual organism today and the future generations of that organism's species. **RNA**, ribonucleic acid, is the intermediary between the blueprint and the life structures. Together, DNA and RNA are the nucleic acids that link the past to the future.

The biochemical nature of nucleic acids was identified in 1869. Speculations were made about the relationship between DNA and the observable movements of chromosomes that geneticists studied, but the specific connection between the nucleic acids and heredity was not confirmed until 1944. In 1953 Watson and Crick published their now famous explanation of the structure of the DNA molecule.

The purpose of this laboratory exercise is to work toward understanding the physical structure of the nucleic acids by constructing models of DNA and RNA. Indeed, Watson and Crick used rather simple models such as these to develop their theory of how DNA works.

❑ Examination

I. Genetic Material

A. Nucleic Acids [∞ Martini, p. 56]

A nucleic acid, such as DNA and RNA, is a polynucleotide. That is, DNA and RNA are long chains of connected nucleotides. A **nucleotide** is a complex molecule composed of a phosphate molecule, a sugar molecule, and a nitrogen-base molecule. DNA is a double-stranded polynucleotide; RNA is single-stranded.

1➡ Study the DNA in Figure 5–1●, the diagram of DNA and RNA. As you examine the **chromosomes**, keep in mind that each chromosome is actually a single DNA molecule. Functional units of this DNA molecule are called **genes**. Each gene is a code for a specific structural or regulatory protein. These proteins determine the physical characteristics of the organism. The message of the code is determined by the order of the **nitrogen bases** of the DNA molecule. Human somatic cells have 46 chromosomes (46 DNA molecules); each with many genes along its length.

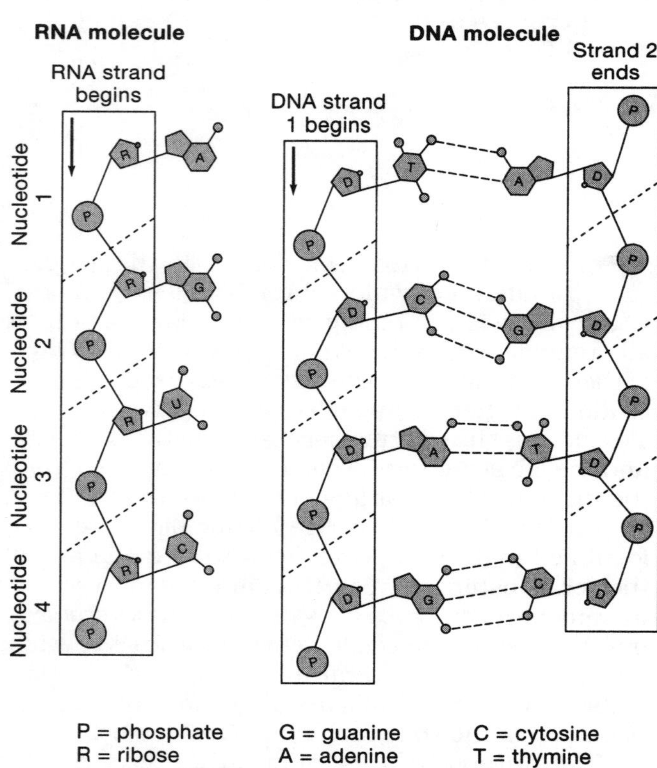

● **FIGURE 5–1**
RNA/DNA.

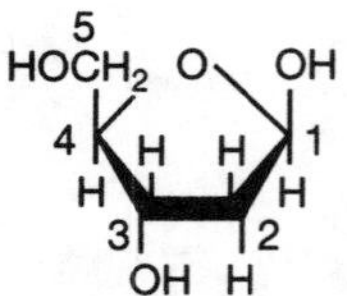

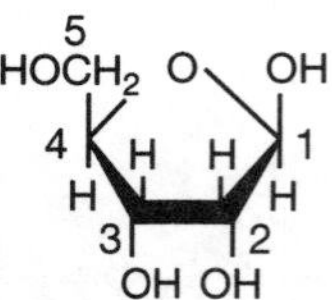

● **FIGURE 5–2**
Deoxyribose/Ribose.

2➡ Note the RNA, also in Figure 5–1●. RNA, which is structurally similar to DNA, enables the genetic messages of the DNA to be actualized into specific proteins in a process known as protein synthesis. (See Exercise 6.)

B. Nucleotides

1➡ Return to Figure 5–1● and identify the nucleotides. Note how the phosphates and sugars alternate to form an interlocking backbone while each nitrogen base is connected to a single sugar molecule. The phosphate molecules are identical in all nucleotides. The sugars, however, are different. The DNA sugar is **deoxyribose** and the RNA sugar is **ribose**.

2➡ Study Figure 5–2●. Note the carbon numbering system. Note that the difference between deoxyribose and ribose can be found at the #2 carbon; deoxyribose has a single hydrogen (–H) at this point and ribose has a hydroxyl (–OH) group.

3➡ Now study Figure 5–3●, which shows the nitrogen bases. **Adenine, guanine, cytosine**, and **thymine** are the four bases found in DNA. In RNA the four nitrogen bases are adenine, guanine, cytosine, and **uracil**. Uracil takes the place of thymine. Guanine and adenine, the double-ringed structures, are **purines**. Cytosine, thymine, and uracil, which have only one ring, are **pyrimidines**. Look at Figure 5–3● in relation to Figure 5–1●. Notice that adenine always pairs with thymine (or uracil) using two hydrogen bonds and that guanine always pairs with cytosine using three hydrogen bonds. We normally abbreviate these nitrogen bases as A, G, C, T, U for adenine, guanine, cytosine, thymine, and uracil, respectively.

❑ Model Building

II. The DNA Molecule

The following steps will guide you through the construction of a model of a DNA molecule. This model

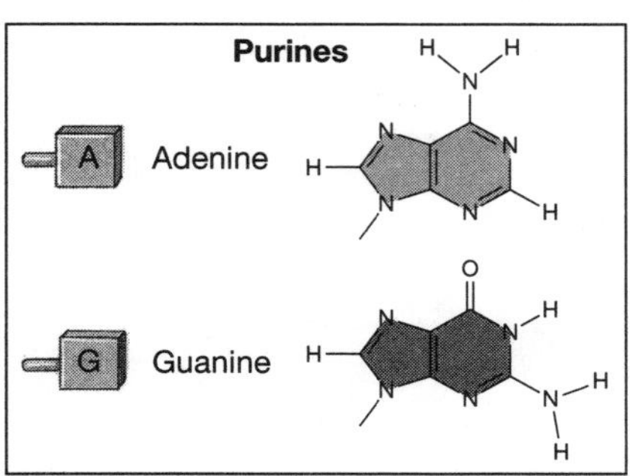

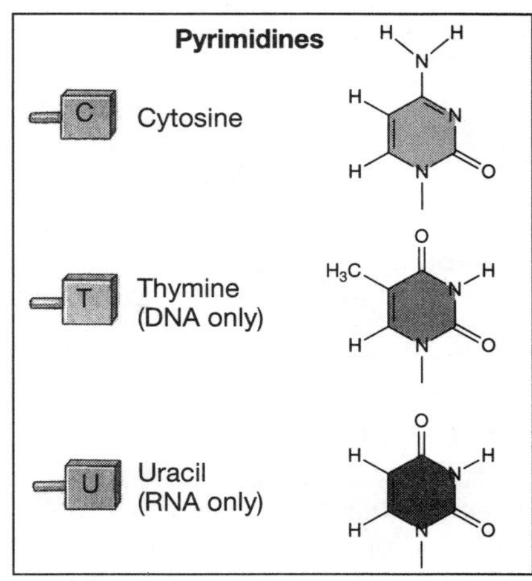

● **FIGURE 5–3**
Nitrogen Bases.

can be used to simulate how DNA makes copies of itself (replication) and how DNA is used as a template for creating RNA (transcription).

A. THE DNA NUCLEOTIDE

1➡ Recall that DNA is a polynucleotide, a group of connected nucleotides, each composed of a phosphate, a sugar, and a nitrogen base.

2➡ Before beginning your construction of a hypothetical DNA molecule, note the hydrogen, oxygen, and hydroxyl groups on the nucleotides in Figures 5–2● and 5–3●. Keep these locations in mind as you put your molecules together because we will not be adding any side groups to those locations.

3➡ Start with the phosphate molecule. Take 8 pipestem cleaners, all of one color, and twist each as shown in Figure 5–4●. The loop represents the phosphate group itself. The ends represent the bonds that will eventually attach the phosphate to the sugar molecules.

4➡ Take 8 more pipestem cleaners (of a different color) and twist each as shown in Figure 5–5●. This is your deoxyribose. Compare the pipestem cleaner figure with the diagram. Identify the corresponding parts. Your little twist should represent the oxygen. Your loose end is the #5 carbon. Keep in mind the carbon numbering system for this sugar molecule. (Refer back to Figure 5–2●.)

5➡ Now take 24 pipestem cleaners (third color) and simulate two copies of each of the four DNA nitrogen bases as indicated in Figure 5–6●. One set of nitrogen bases will be used for each DNA strand. (Recall that DNA is a double-stranded molecule.) Note that guanine and adenine (the purines) require two pipestem cleaners each because they have double rings, whereas cytosine and thymine (the pyrimidines) require only one pipestem cleaner apiece because they have single rings.

[Note the point where you should leave a loose end dangling to represent one nitrogen atom. You will need this dangling end for attaching the nitrogen base to the #1 carbon of the deoxyribose.]

6➡ Break some toothpicks in half and glue them to the nitrogen bases as indicated by the dotted lines in Figure 5–6●. The toothpicks represent potential hydrogen bonds. Notice that adenine and thymine each have two hydrogen bonding sites. Guanine and cytosine each have three sites. You should now see why adenine must always pair with either thymine (in DNA) or uracil (in RNA) and why guanine must always pair with cytosine.

B. CONSTRUCTING THE FIRST DNA STRAND

Now that you have the nucleotides, you can connect them to form the DNA molecule.

1➡ Start your DNA construction by placing a deoxyribose molecule at the base of your work area. Identify the deoxyribose carbon atoms and turn the deoxyribose molecule so that the #5 carbon is on your left. It is important that you build your DNA from the bottom up. Attach a thymine base by twisting the loose end from the nitrogen atom to the #1 carbon of the sugar. (Using

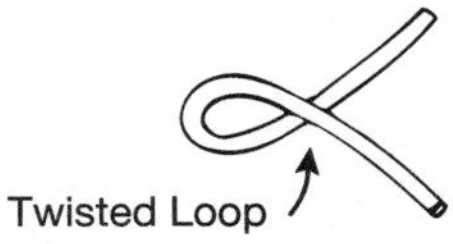

● **FIGURE 5–4**
Twisted Pipestem Cleaner Representing Phosphate Group.

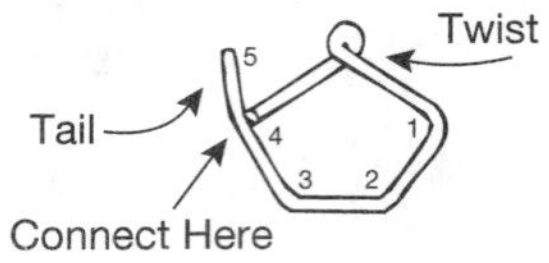

● FIGURE 5–5
Pipestem Cleaner Representing Deoxyribose.

thymine at this point is strictly arbitrary.) Take a phosphate molecule and attach one branch to the #5 carbon of the sugar.

2 ➡ Take another deoxyribose molecule and attach the other free end of the phosphate molecule to the #3 carbon of this second sugar molecule. You have just made a 5' ⟶ 3' connection from the #5 carbon of one sugar to the #3 carbon of the next sugar.

3 ➡ Attach the guanine and the next phosphate in the same manner. Continue with the cytosine nucleotide and finish your strand with the adenine nucleotide at the top as indicated in Figure 5–7●.

C. Constructing the Antiparallel Strand

The **antiparallel** DNA strand is the complement of the strand you just constructed. This strand is antiparallel because it seems to be going in a direction "opposite" to the first strand. A complementary strand is one that contains bases that can pair up with those in the original strand. For example, the complement of ACTTG would be TGAAC.

1 ➡ Begin constructing the antiparallel strand by working with the adenine molecule at the top of your work area. Take the other thymine base and attach a deoxyribose molecule as indicated. Overlap your adenine and thymine toothpick parts but do not glue them together. Your adenine and thymine toothpicks are held together by hydrogen bonds.

2 ➡ Now attach the first phosphate group to the #5 carbon of the deoxyribose molecule. It is important that you work from the top down in constructing this antiparallel strand. Note that your developing nucleotide chain seems to be "upside down" from your first nucleotide strand. It is; the #5 carbon is on the "bottom."

3 ➡ Construct your entire antiparallel strand as indicated in Figure 5–8●. Build downward from the top of your molecule so that as you attach your phosphates to your deoxyribose molecules you are always making 5' ⟶ 3' connections. The 5' ⟶ 3' attachment always occurs as the strand is produced in nature, regardless of the direction of the construction. Be certain to attach the guanine to the cytosine, the cytosine to the guanine, and the adenine to the thymine.

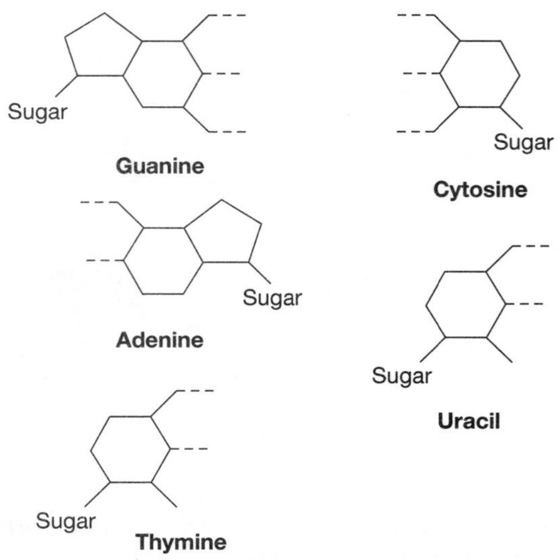

● FIGURE 5–6
Simulated Nitrogen Bases.

You have just produced a DNA fragment. Keep in mind that an actual chromosome is many thousands of nucleotides long. Also, the double-stranded chromosome is twisted to form a double helix. *[∞ Martini, p. 57]*

III. Replicating the DNA

During the S phase of the cell cycle (see Exercise 10), the amount of DNA in a cell doubles by a

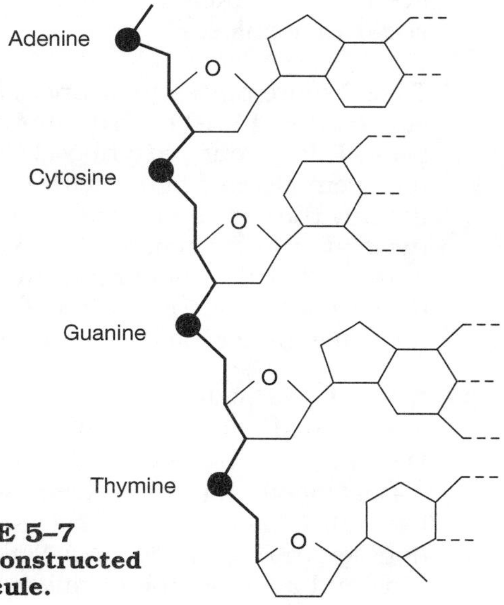

● FIGURE 5–7
Partially Constructed DNA Molecule.

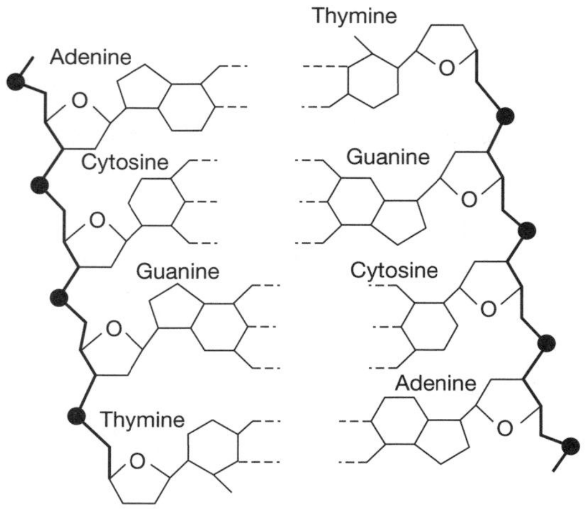

● **FIGURE 5–8**
Double-stranded DNA Model.

process called **replication.** When DNA replicates, a series of enzymes functions sequentially to produce two double-stranded pieces of DNA. Replication is frequently termed **semi-conservative** because half the original DNA is conserved in each new molecule; each daughter chromosome is composed of one parent strand and one newly synthesized strand.

In this section, we will take the DNA and simulate the duplication of an entire chromosome. (Your instructor may direct you to simulate this part of the experiment rather than work it through with additional pipestem cleaner nucleotides.)

A. Assembling the New Nucleotides

1➡ Make eight additional nucleotides — two adenine, two guanine, two cytosine, and two thymine. Be certain that your phosphates extend "up" from the #5 carbon.

B. Constructing the New Polynucleotide

1➡ Begin at the bottom of the DNA fragment you produced in Section II. Gently separate your pieces. As you separate your strands, you should have thymine on your left and adenine on your right. Making certain you properly align the hydrogen bonds, place a new adenine next to your old thymine and a new thymine next to your old adenine. Note that your new adenine is "upside down."

2➡ Continue separating your DNA strand and adding the new nucleotides as indicated. Be aware that you are adding nucleotides in the 5' ⟶ 3' direction and that these nucleotides must be attached in that manner on both sides of the chromosome. As you add the nucleotides and attach the phosphates, you are simulating the ligase enzymes that "sew" the nucleotides together.

IV. Transcribing the DNA

A. The RNA Nucleotide

1➡ Remember that RNA is a single-stranded molecule. RNA is coded from a specific functional section of the DNA molecule, a gene. Several types of RNA are known. The most common of these are transfer RNA (tRNA), messenger RNA (mRNA), and ribosomal RNA (rRNA). Since the components of all RNAs are the same, for this exercise we do not need to specify which type of RNA we are synthesizing.

2➡ Begin your RNA nucleotides with the construction of four sugars. This time, designate your sugar as ribose (instead of deoxyribose). Compare your structure with the ribose pictured in Figure 5–2●. Attach a phosphate to the #5 carbon of each ribose molecule. Complete your RNA nucleotides by attaching an adenine molecule to one, a cytosine to another, and a guanine to a third. For the fourth molecule, construct a uracil nucleotide instead of a thymine nucleotide.

3➡ Take your DNA fragment and designate the left strand as the **sense** strand and the right strand as the **non-sense** strand. Recall that in an actual piece of DNA, usually only one strand of the DNA, the sense strand, can be used to code for RNA. You will be working only with your sense strand. Keep in mind that when an actual piece of DNA is used to code for RNA, only a segment is functional, not the entire chromosome. As previously stated, the functional segment is the gene.

4➡ Separate your strands slowly. As you do, place the RNA adenine nucleotide next to your thymine, the cytosine next to your guanine, the guanine next to your cytosine, and the uracil next to your adenine. (Recall that uracil replaces thymine in RNA.) Again, connect your nucleotides by attaching the phosphates to the sugars in the 5' ⟶ 3' direction.

5➡ Take your RNA segment out and reconnect your DNA strands. Your single-stranded piece of RNA is now ready to actualize the specific protein encoded in the DNA, as will be demonstrated in Exercise 6.

❑ Additional Activities

NOTES

1. Use the library to research the history of our understanding of the structure and function of the nucleic acids.
2. Section III deals with DNA replication. Find out the names and functions of all the various enzymes involved in this process.

❑ Lab Report

1. Referring to the numbered inquiries at the beginning of this exercise, complete the following box summary:

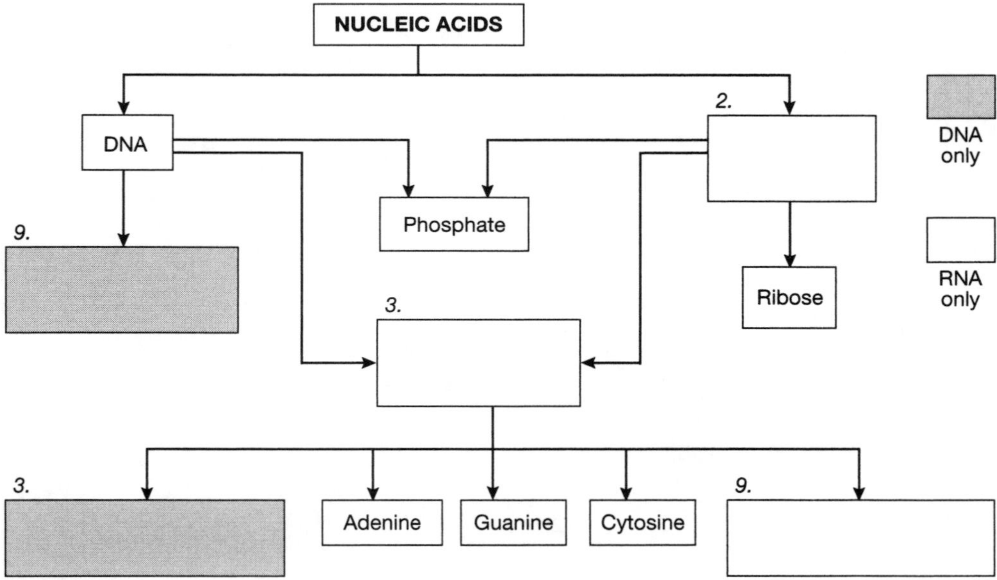

General Questions

NOTES

1. Name the three parts of a nucleotide.
2. Fill in this chart:

Nitrogen Bases	Found in	Purine or Pyrimidine Rings	Pairs With

3. List three ways DNA and RNA are alike. List three ways they are different.
4. Antiparallel refers to the way in which ______________________________

   ____
5. The term 5' ⟶ 3' refers to the way in which ______________________________

   ____
6. Because each new daughter chromosome contains one strand of parent DNA, we call DNA replication ______________________.
7. The DNA template for the coding of RNA is the ______________________ strand.

EXERCISE 6

Protein Synthesis

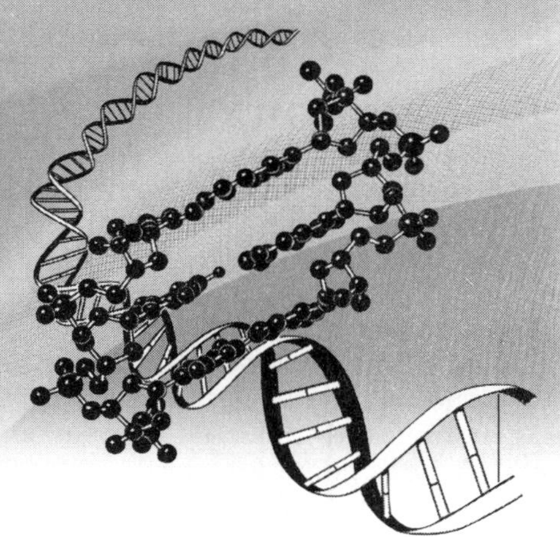

PROCEDURAL INQUIRIES

Preparation

1. What is protein synthesis?
2. What are the two functions of proteins?
3. What is the transcription template for protein synthesis?
4. What is the structure that carries the message in protein synthesis?
5. What is the structure that carries the amino acid in protein synthesis?
6. What is transcription?
7. What is translation?

Model Building

8. What are the basic steps in transcription?
9. What are the basic steps in translation?

Additional Inquiries

10. How are the codon and anticodon related?
11. What is the final site of protein synthesis (the structure in which the polynucleotide is assembled)?
12. What is the Genetic Code?

Key Terms

Adenine
Anticodon
Codon
Cytosine
DNA
Guanine
Non–sense strand
Peptide bond
Protein synthesis
RNA
Sense strand
Template
Thymine
Transcription
Translation
Triplet
Uracil

Materials Needed

Pipestem Cleaners (6 per student)
Colored paper (small pieces, cut into 3 different shapes)
Marking pens (5 colors—suggestion: red, green, yellow, blue, black)
Tape
Plain Paper

Protein synthesis, in its more global sense, is the process whereby genetic information is transcribed from a chromosome (**DNA**) and translated into a chain of amino acids, with several types of **RNA** serving as intermediaries. Proteins, which are composed of one or more chains of amino acids, are classified either according to positional structure or regulatory function. We usually think of the structural proteins as those proteins forming an integral part of the cytoskeleton, the cell membranes, and the cellular organelles. Regulatory proteins, on the other hand, are the facilitators of cellular action. Enzymes constitute the major group of regulatory proteins, and all enzymes are at least part protein. The generalized process of protein synthesis is common to both structural and regulatory proteins.

The purpose of this exercise is to gain an understanding of protein synthesis by reviewing the process at the molecular level and then by modeling the generalized process from the DNA **template** to the hypothetical protein.

❑ Preparation

I. Process of Protein Synthesis

A. Review

1➡ Recall from Exercise 5 that both DNA (deoxyribonucleic acid) and RNA (ribonucleic acid) are polynucleotides. A nucleotide is a compound molecule composed of a phosphate, a sugar (deoxyribose for DNA and ribose for RNA), and a nitrogen base. A polynucleotide is a chain of chemically connected nucleotides. The DNA is double-stranded,with **adenine**, **guanine**, **cytosine**, and **thymine** as its nitrogen bases. The RNA is single-stranded, with adenine, guanine, cytosine, and **uracil** as its nitrogen bases.

A key principle of the genetic code is that each nitrogen base can only pair up with its complementary base. Adenine pairs with thymine (in DNA) or uracil (in RNA). Guanine always pairs with cytosine. (Note that we generally identify nucleotides by their nitrogen bases—i.e., we say "guanine" rather than "guanine-based nucleotide.")

The phosphate and sugar form the backbone of the molecule, and the nitrogen base is the functional "end" of the nucleotide. In eukaryotic cellular organisms DNA is found within the nucleus, where it is a double-stranded molecule twisted into a helical shape; RNA is a helically shaped single-stranded molecule. In protein synthesis one type of DNA is important and three types of RNA are needed: messenger RNA (mRNA), transfer RNA (tRNA), and ribosomal RNA (rRNA).

B. Protein Synthesis: The 10-Step Process

Protein synthesis can be described as a 10-step process, although it is important to remember the overview of the process: the genetic information is transcribed first from DNA to mRNA, which travels to the ribosome (rRNA) and, with the help of tRNA, is translated into protein.

1. Within the nucleus, the double strand of DNA separates, and one strand acts as the genetic blue print, the template. Groups of three nucleotides are called **triplets**.
2. The mRNA is coded off the DNA by a process called **transcription**. Because mRNA nucleotides are assembled opposite the DNA template, the nitrogen bases must be complementary. We often use the term "correspond" to indicate complementary base pairs. The mRNA triplet is called a **codon**.
3. The mRNA leaves the nucleus and becomes associated with the ribosome, the "protein-manufacturing organelle" of the cell. In the eukaryotic cell each ribosome is produced in two parts in the nucleolus of the cell, which is located within the nucleus. The ribosomal parts—as well as the mRNA—leave the nucleus via the nuclear pores.
4. The tRNA, which has been produced by a similar transcription process elsewhere in the genome, goes to the cytoplasm where various amino acids can be found. The identifying triplet at the base of the tRNA is known as the **anticodon**. The anticodon is complementary to a particular mRNA codon.
5. The tRNA picks up an amino acid corresponding to its anticodon. Keep in mind that the codon is the complement of the anticodon.
6. The tRNA carries the amino acid to the ribosome.
7. At the ribosome the codon and its corresponding anticodon join up and the first amino acid of the chain is brought into place. When a subsequent tRNA brings in the next amino acid, the two amino acids are connected by a **peptide bond**.
8. As the incoming amino acids are joined to the growing chain by peptide bonds, they are released from the tRNA. This process, in which the specific sequences of codons in the RNA are used to link up specific sequences of amino acids, is **translation**.
9. The tRNA is released from the mRNA and is then available to pick up another amino acid.
10. When the final amino acid is in place, the two parts of the ribosome separate and release the mRNA and the amino acid chain. Because of its multiple peptide bonds, the resulting amino acid chain is called a polypeptide. A functional polypeptide is a protein.

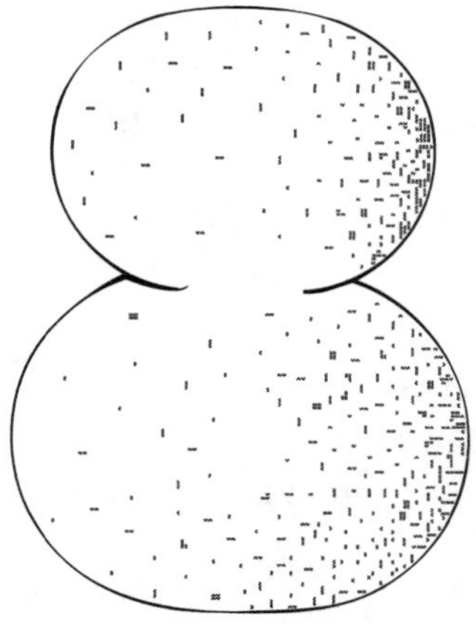

● **FIGURE 6–1**
Simulated Ribosome.

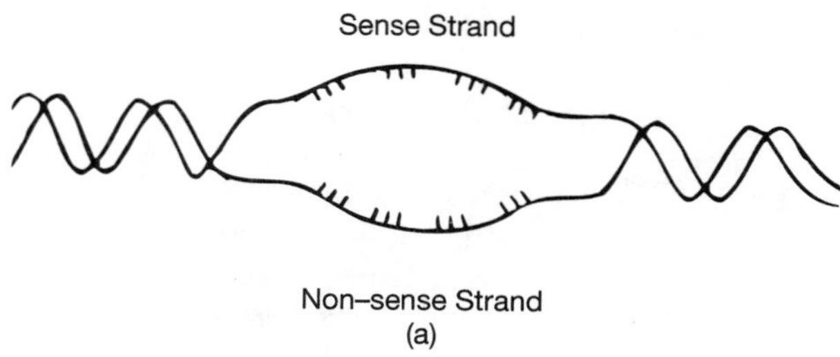

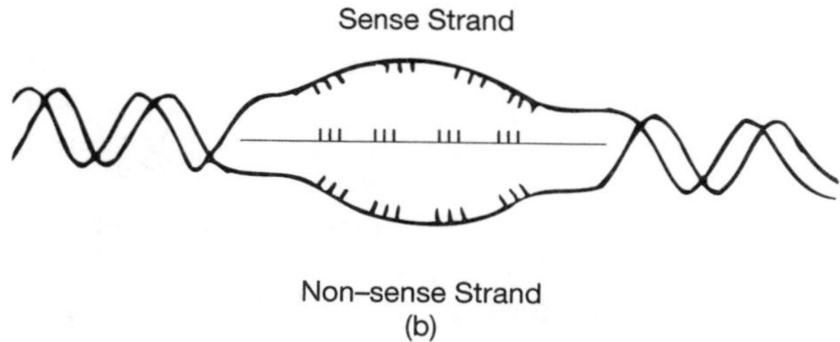

● **FIGURE 6–2**
(a) Simulated Chromosome Separated for Protein Synthesis; **(b)** Simulated Chromosome with mRNA.

❑ Model Building

To begin this demonstration, designate your colored pens as follows: red = adenine, green = guanine, yellow = thymine, blue = cytosine, black = uracil. The nitrogen bases are commonly abbreviated as adenine (A), guanine (G), thymine (T), cytosine (C), uracil (U). Also obtain your six pipestem cleaners, your three pieces of colored paper (one of each shape), and your tape.

II. DNA

A. Simulating a DNA Molecule

1➡ Take two pipestem cleaners and loosely twist them together. You now have a chromosome (a DNA molecule) twisted as a double helix.

2➡ Place your chromosome on the right side of your work space. This will be your nucleus. On the left side of your work space put a piece of white paper on which you have drawn two overlapping 3/4 circles (Figure 6–1●). This is your ribosome.

3➡ Pull your chromosome apart in the middle (Figure 6–2a●). Designate the upper strand the **sense strand**. This is the strand you will use to transcribe your protein. The other strand is often called the **non–sense strand** (or the noncoding strand). We are not concerned with the non–sense strand at this time.

4➡ Mark your sense strand left to right in triplets as follows: TAC (yellow, red, blue), TTG (yellow, yellow, green), CAT (blue, red, yellow), ACT (red, blue, yellow). Separate your triplets as indicated to facilitate your model building. Keep in mind that in an actual DNA strand, no space would exist between the triplets.

III. RNA

A. Simulating an mRNA molecule

1➡ Now, take another pipestem cleaner and place it across the open area of your chromosome (Figure 6–2b•). You are now coding your messenger RNA (mRNA). Working left to right, mark your mRNA to correspond

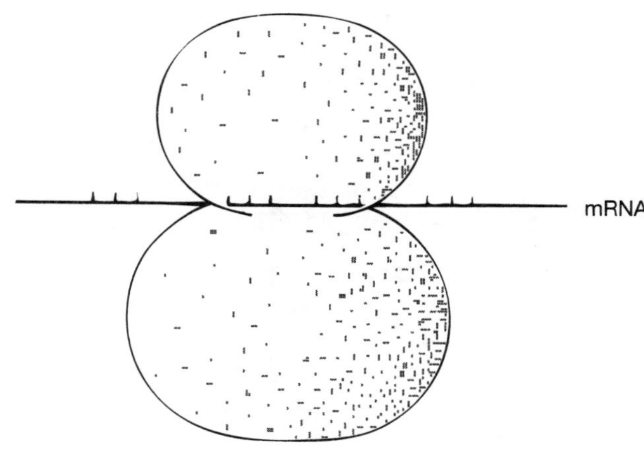

● **FIGURE 6–3**
mRNA across Ribosome.

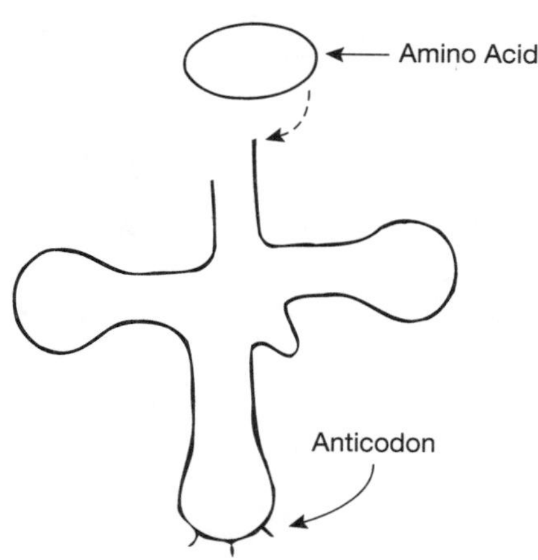

● **FIGURE 6–4**
tRNA.

with the triplets on your chromosome — AUG (red, black, green), AAC (red, red, blue), GUA (green, black, red), UGA (black, green, red). These nitrogen base triplets are the codons. The process described here is transcription—the message has been transcribed from the blue print (the DNA of the chromosome) to the mRNA

2➡ Move your mRNA to the left, away from your chromosome and toward your ribosome. Place your mRNA across the center of the ribosome (Figure 6–3●).

B. SIMULATING A tRNA MOLECULE

1➡ Take three more pipestem cleaners and twist them into modified clovers (Figure 6–4●). This is your transfer RNA (tRNA). On each tRNA mark the bottom three nitrogen bases. The first should be UAC, the second UUG, the third CAU. These nitrogen base triplets are the anticodons. The tRNA is called transfer RNA because it transfers the amino acids from the cytoplasm to the ribosome.

2➡ Check the genetic code (Figure 6–6●)—the chart of amino acids that is written for mRNA codons—and notice that AUG codes for the amino acid methionine. Therefore, you should now label one of your colored papers methionine and attach it to the top of the tRNA with the UAC bases. Remember that the codons and anticodons will join up so the nitrogen bases must be complementary. Your second tRNA should carry asparagine, and the third tRNA should have valine attached to it.

IV. Synthesis

A. CONSTRUCTING THE POLYPEPTIDE

1➡ Move your tRNAs to the ribosome and join the codons with the anticodons. Join your amino acids with tape. Remove your tRNA (Figure 6–5●). The tapes between the amino acids represent peptide bonds, and thus your chain of amino acids becomes a polypeptide. This entire process is translation—the process of translating the message carried by the messenger into a chain of amino acids. If your polypeptide were functional, it would be called a protein.

All proteins begin with methionine. The codon for methionine actually brings the two parts of the ribosome together. Although all protein translation begins with methionine, not all proteins necessarily contain methionine. Proteins can be edited within the cell before assuming their final structural or regulatory function.

2➡ Notice that the last mRNA codon, UGA, does not code for an amino acid. Rather, UGA is one of three codons coding for STOP. As the mRNA passes through the ribosome, the stop codon marks the completion of the making of the protein. The parts of the ribosome separate and synthesis ceases. Notice further on the Genetic Code (Figure 6–6●) that two other codons (UAA and UAG) also signal for a cessation of protein synthesis.

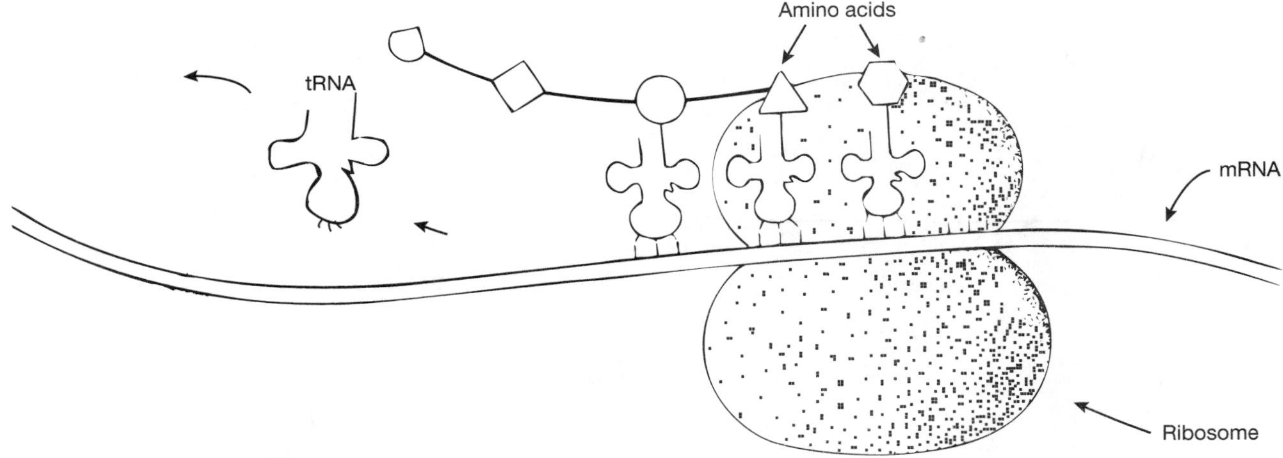

● **FIGURE 6–5**
Growing Polypeptide.

Codon/ Amino Acid	Codon/ Amino Acid	Codon/ Amino Acid	Codon/ Amino Acid
UUU Phenylalanine	UCU Serine	UAU Tyrosine	UGU Cysteine
UUC Phenylalanine	UCC Serine	UAC Tyrosine	UGC Cysteine
UUA Leucine	UCA Serine	UAA STOP	UGA STOP
UUG Leucine	UCG Serine	UAG STOP	UGG Tryptophan
CUU Leucine	CCU Proline	CAU Histine	CGU Arginine
CUC Leucine	CCC Proline	CAC Histine	CGC Arginine
CUA Leucine	CCA Proline	CAA Glutamine	CGA Arginine
CUG Leucine	CCG Proline	CAG Glutamine	CGG Arginine
AUU Isoleucine	ACU Threonine	AAU Asparagine	AGU Serine
AUC Isoleucine	ACC Threonine	AAC Asparagine	AGC Serine
AUA Isoleucine	ACA Threonine	AAA Lysine	AGA Arginine
AUG Methionine	ACG Threonine	AAG Lysine	AGG Arginine
GUU Valine	GCU Alanine	GAU Aspartic acid	GGU Glycine
GUC Valine	GCC Alanine	GAC Aspartic acid	GGC Glycine
GUA Valine	GCA Alanine	GAA Glutamic acid	GGA Glycine
GUG Valine	GCG Alanine	GAG Glutamic acid	GGG Glycine

● **FIGURE 6–6**
The Genetic Code. Codons for mRNA.

❑ Additional Activities

NOTES

1. Check library sources for the current thinking on the evolutionary history of DNA and RNA. Which may have come first? Why?
2. Relate your knowledge of protein synthesis to genetic mutations. Refer to Exercise 5 if necessary. You might also look ahead to Exercise 71.

❑ Lab Report

1. Referring to the numbered inquiries at the beginning of this exercise, complete the following Box Summary:

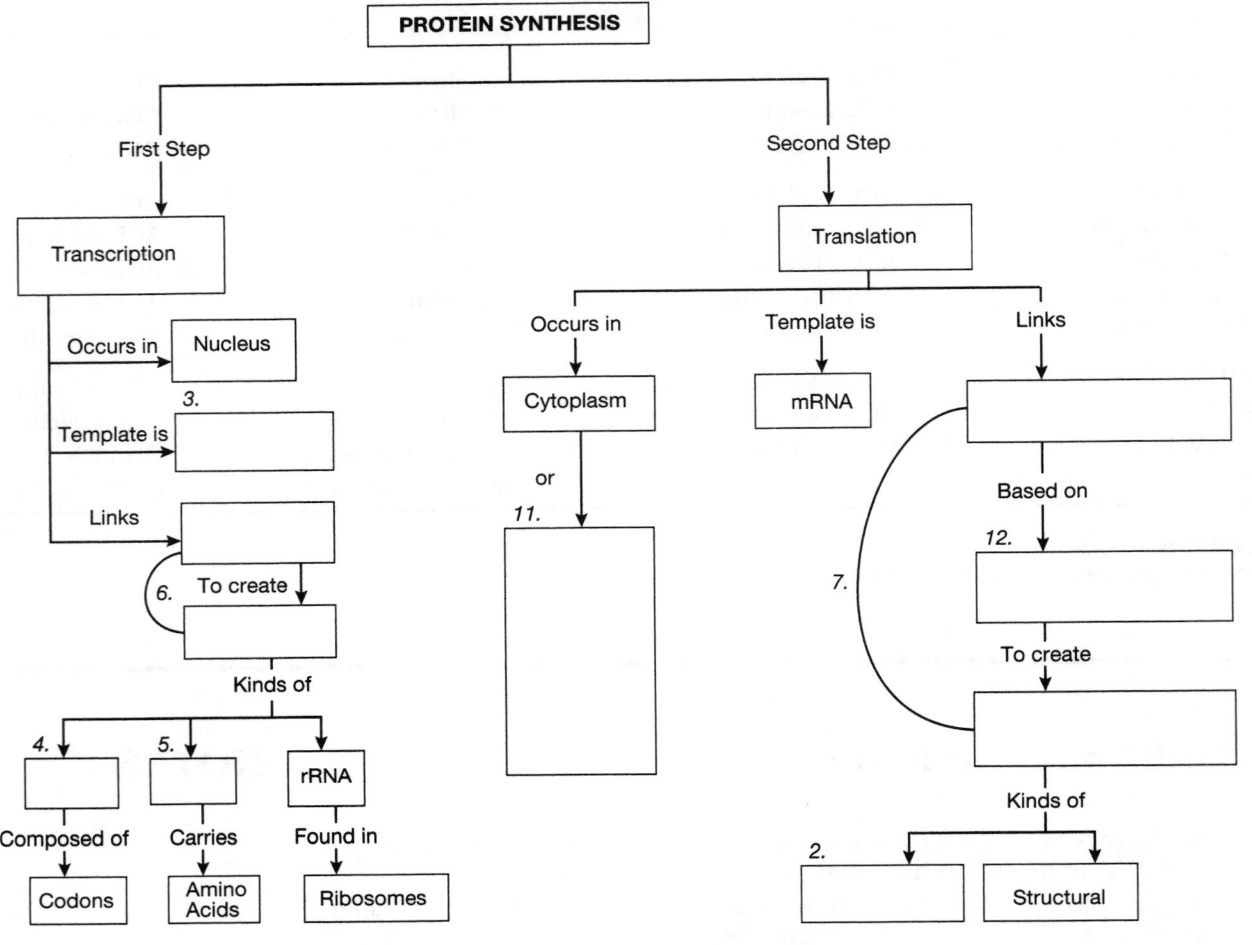

General Questions

NOTES

1. DNA does not code directly for carbohydrates and lipids, and yet the body makes these molecules. Based on what you know about enzymes, how do you suppose protein synthesis is involved in the making of carbohydrates and lipids?
2. Write the genetic code for the following polypeptide: glycine, valine, threonine, lysine, arginine, glutamic acid. Write the anticodons that would correspond to these amino acids. What would the sense strand of the DNA have been? What would the non–sense strand of the DNA have been?
3. What advantage can you see to having more than one codon coding for the same amino acid?
4. Go back to the 10-step process for protein synthesis. What part of your experiment corresponded to each of the steps?
5. What is the difference between transcription and translation?
6. What is the relationship between the codon and the anticodon?

EXERCISE 7

Gross Anatomical Structure

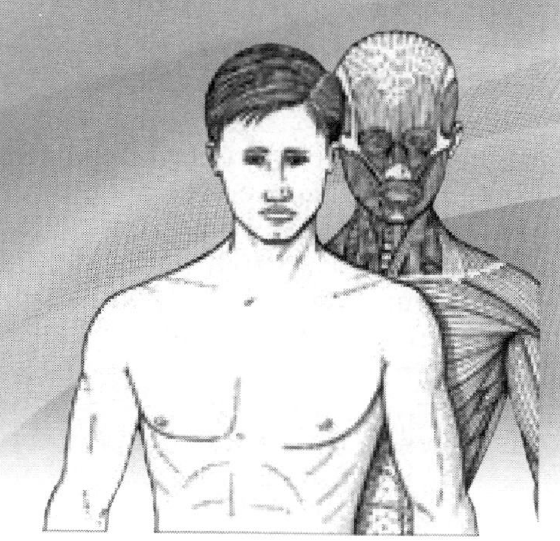

PROCEDURAL INQUIRIES

Preparation

1. What is the hierarchical organization of the human body?
2. What is the difference between macroscopic and microscopic anatomy?

Examination

3. Which organs and organ systems can be identified from gross observation of the external (or surface) anatomy?
4. What are the eleven major organ systems in the human body?
5. What are the principal organs in each of the major organ systems?

Additional Inquiry

6. Why are some organs considered to be a part of more than one organ system?

Key Terms

Cardiovascular System
Cell
Chemical
Digestive System
Endocrine System
Integumentary System
Lymphatic System
Muscular System
Nervous System
Organ
Organelle
Organ System
Reproductive System
Respiratory System
Skeletal System
Tissue
Urinary System

Materials Needed

Hand Lens or Dissecting (Stereoscopic) Microscope
Mirror
Torso
Plastic Models of Organs
Skeleton (Human or Animal—Annotated if available)
Surface Diagrams (as available)
Internal Diagrams (as available)
Cross-Sectional Diagrams (as available)

One of the marvels of the human body is the intricate pattern of organization permeating literally every facet of physical existence. Although the body is composed of numerous readily identifiable parts, those parts are but pieces in an unbelievably complex superstructure. One focus of our study of anatomy and physiology deals with the interactions between the parts of this superstructure. The anatomical components of the body follow a hierarchal plan beginning with the chemical elements and culminating in the human body itself.

From chemistry you know that living structures are composed of the same **chemicals** — atoms, molecules, compounds, the primary units of matter — that are found in the rest of nature. Therefore, we predict that the body will follow the laws of chemistry. In the living organism the chemical workplace is the **cell** — or more specifically, the **organelles** of the cell.

The cell is the basic, or smallest, functional unit of life. Certain organisms, such as amoebas and euglenas, are composed of just a single cell. This single cell carries on all the life functions we normally associate with larger and more complex organisms, functions like metabolism, growth, and reproduction. In a multicellular organism, such as the human body, each cell is a living unit and each cell carries on the same basic life functions. Although these cells work in cooperation with each other, each cell tends to specialize in one function or another. We can identify about 200 different cell types in the human body, each performing the basic life functions, yet each with a unique specialization. We will study cells more thoroughly in Exercises 8, 9, and 10.

A group of cells (and their cell products) specialized toward a common purpose or function is called a **tissue**. Blood, bone, and muscle are common tissues because, as we shall see in Exercise 11, we place a particular conglomerate of cells into one of the four primary tissue types (epithelial, connective, muscle, or neural) according to the structure and function of the cells in question.

Organ is the term used to designate two or more tissues functioning together to perform an essential task. We sometimes mistakenly think an organ is made up of only one type of cell. Not so. For instance, we all know that the liver is an organ. The liver is composed of more than just the hepatocytes (liver cells). The hepatocytes constitute the primary cell type. However, the liver also contains blood, neural tissue, adipose tissue, and several other types of connective tissue.

The same tissue type can be found in different organs, too. For instance, while blood cells (vascular connective tissue) are definitely found in the liver, vascular connective tissue is also found in every other region and organ of the body. Other connective tissues can also be found within and between every other tissue and organ.

The liver, the heart, and the brain are all organs, structures made up of various tissue types. In this particular book we will be dealing with organs only as parts of **organ systems**. (Organ systems are sometimes just called systems.) An organ system is a group of organs performing a common function or series of functions. For instance, the respiratory system is composed of the lungs, the bronchioles, the trachea, the nose, and various accessory structures.

Organ systems function together to make up the organism, in this case the human being. Despite their structural and functional interactions (such as the role of the cardiovascular system in gas exchange, which is the major function of the respiratory system), all organ systems share certain characteristics. These characteristics include: specialization for performing certain functions, functional independence with respect to the environment, dependence on other organ systems, and an integration of activity through neural and hormonal mechanisms. *[∞ Martini, p. 3]*

Many organs, such as the ovary and pancreas, have functions in more than one system. The ovary is part of the endocrine system because it secretes hormones but it is also part of the reproductive system because of its role in the continuance of the species. Another example is the pancreas, which produces hormones (endocrine system) and digestive juices (digestive system).

For our purposes, we recognize eleven major organs systems. Each of these systems will be studied in greater detail throughout the remainder of this book. In this laboratory exercise, however, we will gain a perspective of the body by pulling together background information needed to visualize where parts of the body are located and how these different parts are physically related to one another. To accomplish this end we will examine the gross anatomical structures of the body both as parts of organ systems and as interrelated functional units.

❑ Preparation

I. Background Information

A. Microscopic Anatomy

Microscopic anatomy deals with those structures best studied with the aid of a microscope. Microscopic anatomy is covered specifically in Exercises 8–12 and is touched on with all the major macroscopic topics examined in the rest of the book.

B. Macroscopic Anatomy

Macroscopic anatomy — also called gross anatomy — is concerned with the larger anatomical structures, those structures visible with the unaided eye. This exercise deals with gross anatomy and is designed to lay a foundation for the more detailed studies of the individual systems that you will encounter later in this course.

C. Hierarchical Organization [∞ Martini, p. 4]

1➡ Recall the hierarchical organization from chemical to cell to tissue to organ to system to organism discussed in the introduction to this exercise.

2➡ Study Figure 7–1●, which diagrams this organizational progression. Keep this scheme in mind as you work through the different parts of this exercise.

● FIGURE 7–1
Levels of Organization.

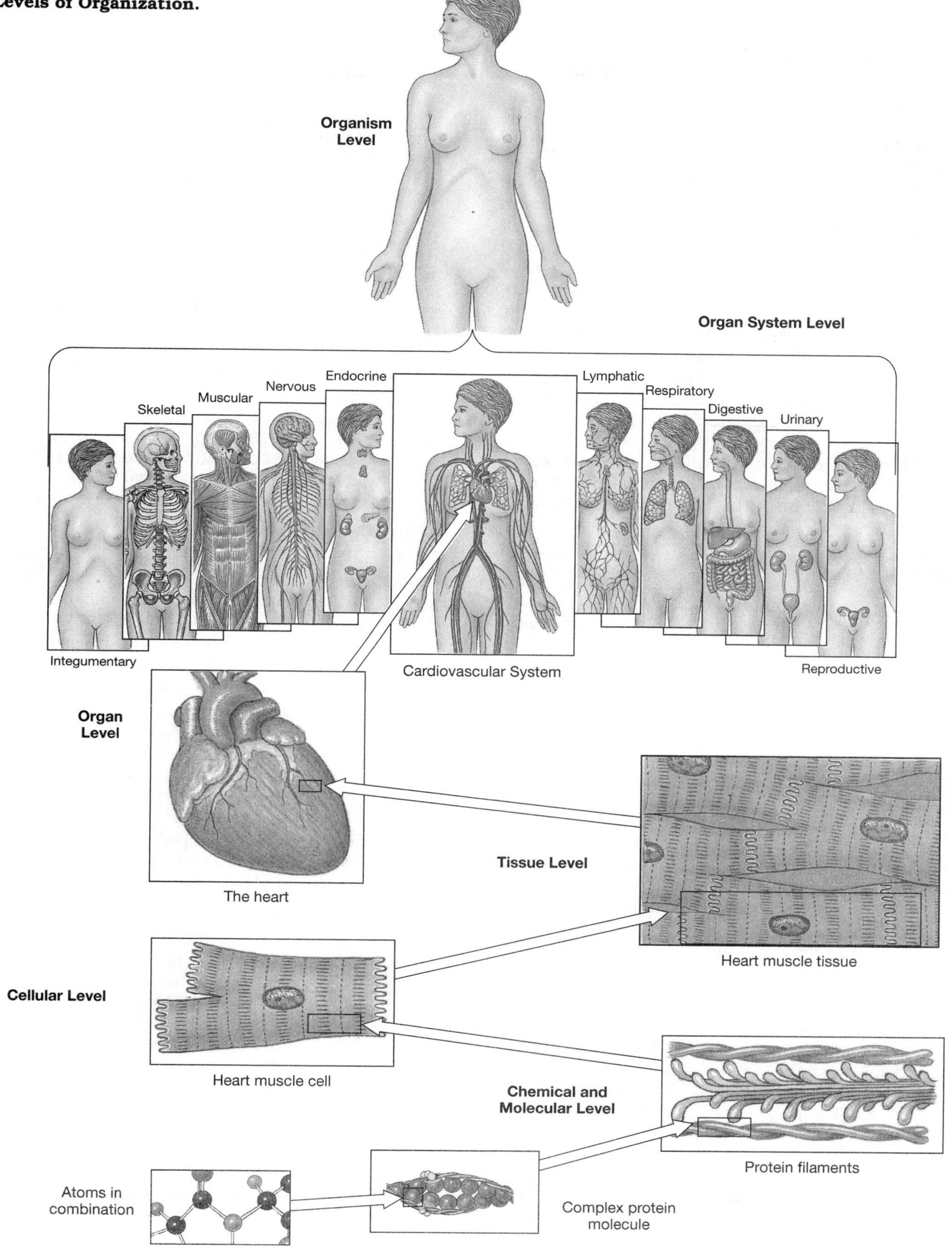
Organism Level
Organ System Level
Integumentary
Skeletal
Muscular
Nervous
Endocrine
Cardiovascular System
Lymphatic
Respiratory
Digestive
Urinary
Reproductive
Organ Level
The heart
Tissue Level
Heart muscle tissue
Cellular Level
Heart muscle cell
Chemical and Molecular Level
Protein filaments
Atoms in combination
Complex protein molecule

❑ Examination

II. Surface Anatomy

A. Anatomical Overview

This section is designed to help you begin thinking about gross anatomical structures. As you go through this brief introductory exercise on surface anatomy, be aware of how many different physical structures you actually encounter.

1➡ Start your examination of the body with a quick scan of your own external anatomy.

2➡ Feel the skin on different parts of your hands, arms, face, and neck. What differences do you perceive?

____

____

Note that despite the differences in texture, the skin is essentially a continuous barrier, a moisture barrier keeping the body fluids within the body and protecting the inside organs from both desiccation and hydration, and an immunological barrier protecting the body from biological, physical, and chemical invaders.

3➡ Take the hand lens (or put your hand under the dissecting microscope). Examine your skin, hair follicles, and fingernails. Describe what you see.

____

____

4➡ Lift the skin on the posterior surface of your hand. Try to do the same with your arm and your neck. You should be able to feel many of the subsurface structures, such as muscles, tendons, and perhaps certain blood vessels. Tense the muscles in these areas and describe what you feel beneath the surface of the skin.

____

____

5➡ Look at the posterior surface of your hand and the anterior surface of your forearm. Notice the blood vessels, a part of the cardio-

Organ System	*Principal Components*	*Major Functions*
Integumentary system	Skin and associated structures (glands, hair, nails)	Protection from environmental hazards, temperature control
Skeletal system	Bones of the skeleton	Support, protection of soft tissues, mineral storage, blood formation
Muscular system	Skeletal muscles	Locomotion, support, heat production
Nervous system	Brain, spinal cord, and nerves	Directing immediate responses to stimuli, usually by coordinating the activities of other organ systems
Endocrine system	Glands such as the thyroid, pituitary, and adrenals, that secrete hormones	Directing slow responses to environmental stimuli by modifying the activities of other organ systems
Cardiovascular system	The heart, blood, and blood vessels	Internal transport of cells and dissolved materials, including nutrients, wastes, and gases
Lymphatic system	Lymph nodes, lymphatic vessels, and organs such as the thymus and spleen	Defense against infection and disease
Respiratory system	Nasal chambers, airways, and lungs	Delivery of air to sites where gas exchange can occur between the air and circulating blood
Digestive system	Teeth, salivary glands, digestive tract, liver, and pancreas	Processing of food and absorption of nutrients, minerals, vitamins, and water
Urinary system	Kidneys, urinary bladder, and conducting passageways	Elimination of excess water, salts, and waste products
Reproductive system	Male: testes, ducts, accessory glands and structures; female: ovaries, ducts, accessory glands and structures	Production of future generations

●FIGURE 7–2
Organ System Overview.

vascular system. Most of the blood vessels you see will be veins. Veins, which appear blue, carry blood toward the heart, and are generally closer to the surface than are arteries. Note the differences between your right hand and arm and your left hand and arm. Compare your posterior hand and arm with the posterior hands and arms of your classmates. Describe any differences you observe.

____

____

6➡ Hyperextend (bend back) your wrist. On the anterior surface, you may be able to notice a pulsating point. This pulse is from your radial artery. Check the location of your temporal (side of the forehead) and carotid (neck) pulse points.

7➡ Look in the mirror and make note of the parts of the different systems you see. Your eyes and ears are parts of the nervous system. Your nose is part of the respiratory system. Your mouth is part of the digestive system. You may also notice blood vessels and even evidence of adipose tissue. All of this is covered with skin, which is part of the integumentary system.

III. Systems [∞ *Martini, p. 6–11*]

A. Orientation

1➡ Begin this section of the exercise by studying Figure 7–2●, the table of the organ systems. Be certain you have a good overview of the systems before continuing.

2➡ Use the figures indicated below along with any classroom models, charts, and diagrams you may have available to identify the structures listed with each system. If your laboratory is equipped with internal or cross-sectional diagrams, use these to gain an added three-dimensional perspective of the structures you are studying.

3➡ Identify each individual organ, reminding yourself that that organ is composed of several tissue types and that tissues are composed of individual cells and cell products.

4➡ Check off the structures on the following lists as you identify them. Since these lists are not all-inclusive, add any other structures you may locate.

5➡ Answer questions 3 and 4 for each system. Use Figure 7–2● and your lecture text, if necessary. (At this time your answers to these questions should be quite general. These answers will become more inclusive as this course progresses.)

B. Integumentary System

1➡ Study Figure 7–3●. Identify the following structures (the asterisks denote structures not identified on the figure):

Hair
Epidermis
Dermis*
Adipose Tissue*
Nerve Ending*
Arteries, Veins, Capillaries*
Fingernail

2➡ Add any additional structures you may locate.

____

____

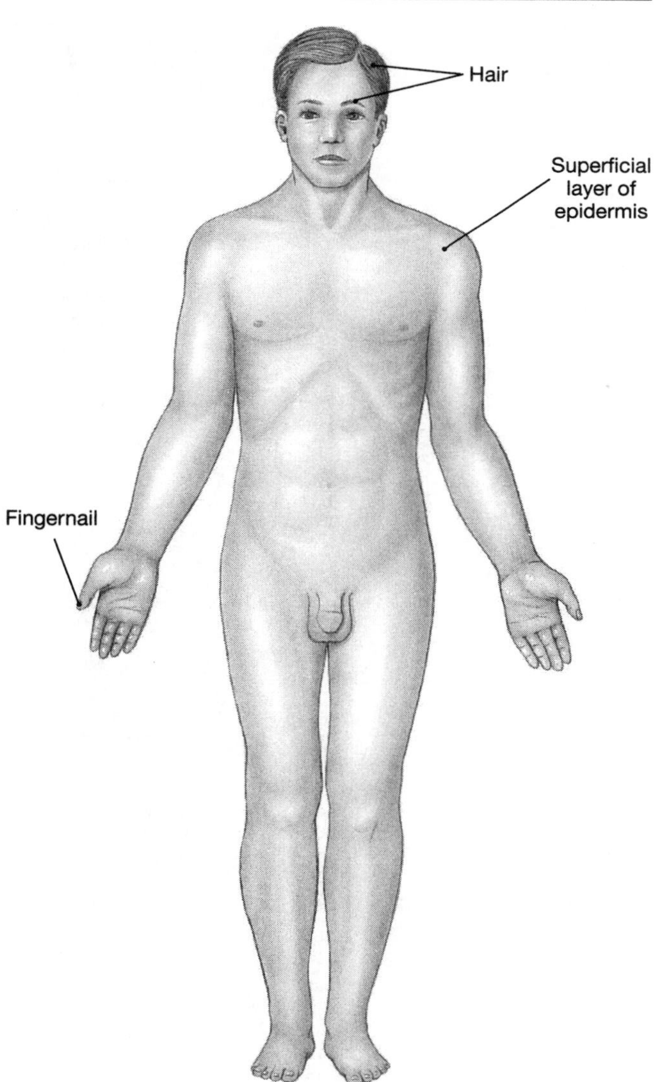

● FIGURE 7–3
The Integumentary System.

3➡ Where is the integumentary system located?

____

____

4➡ What is the general function of the integumentary system?

____

____

C. Skeletal System

1➡ Study Figure 7–4● and the skeletons in your laboratory. Identify the following structures:

Skull
Rib
Pectoral Girdle (Scapula and Collar Bone)
Pelvic Cavity (Hip)
Vertebral Column
Sternum
Sacrum

2➡ Add any additional structures you may locate.

____

____

If you have an annotated skeleton, observe how and where the muscles are attached to the skeleton.

3➡ Where is the skeletal system located?

____

____

4➡ What is the general function of the skeletal system?

____

____

D. Muscular System

1➡ Study Figure 7–5●. Identify the following structures:

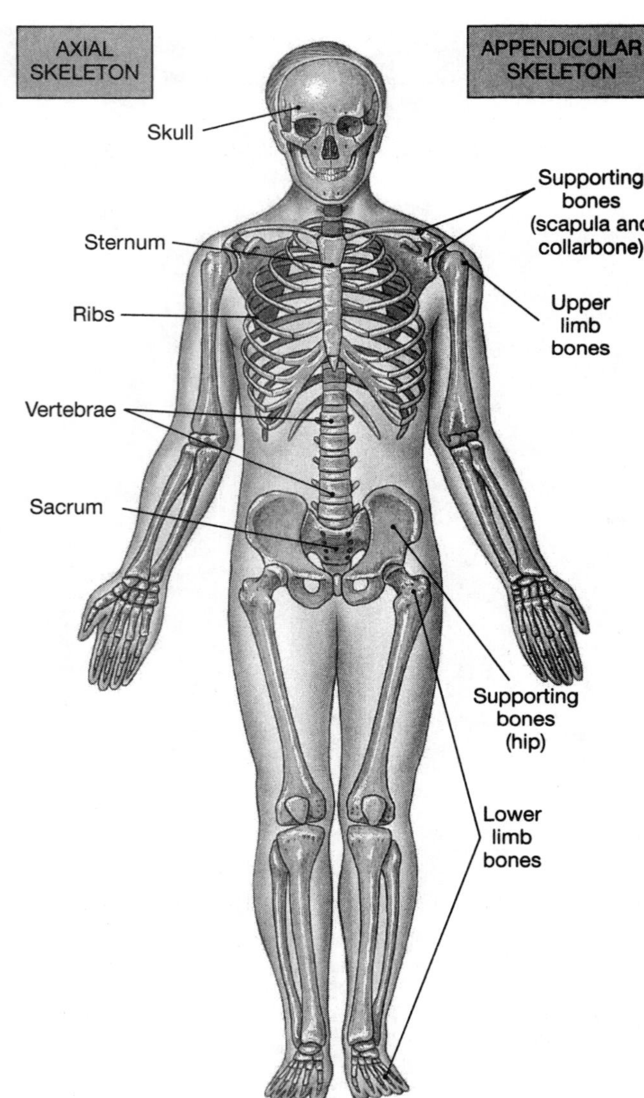

● FIGURE 7–4
The Skeletal System.

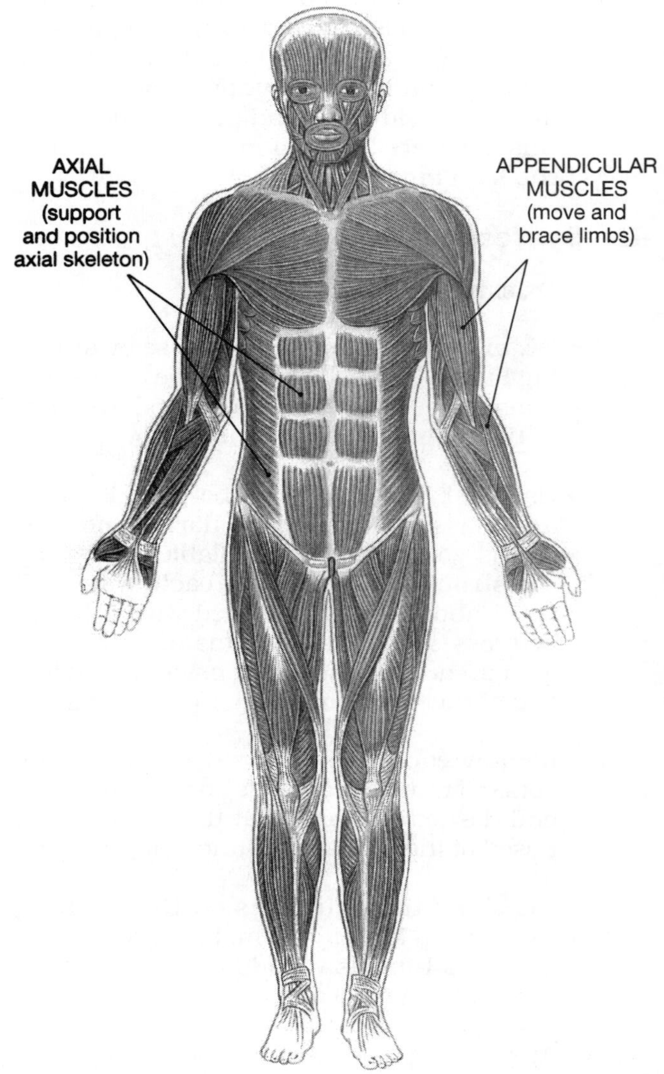

● FIGURE 7–5
The Muscular System.

Skeletal Muscle
Connective Tissue (not identified on drawing).

2➡ Add any additional structures you may locate.

___

___

It should be noted that in most areas of the body, the skeletal muscle is several layers thick. Also, in areas such as the ribs and vertebrae, the individual muscles are not the large and obvious structures you see here.

Usually when we consider the muscular system, we are referring only to the skeletal muscles, those voluntary muscles attached directly or indirectly to the skeleton. We usually think of smooth muscle as tissue components within other systems, such as layers of the digestive system or layers of the vascular system. Cardiac muscle forms the wall of the heart, an organ of the cardiovascular system.

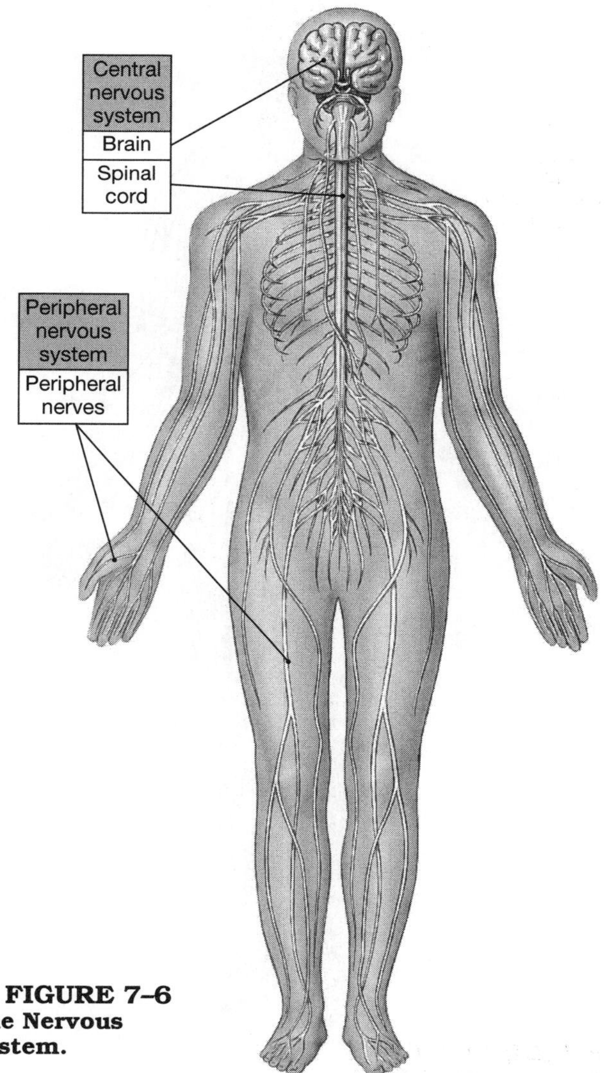

● **FIGURE 7–6**
The Nervous System.

3➡ Where is the (skeletal) muscular system located?

___

___

4➡ What is the general function of the (skeletal) muscular system?

___

___

E. Nervous System

1➡ Study Figure 7–6●. Identify the following structures:

Brain
Spinal Cord
Peripheral Nerves

2➡ Add any additional structures you may locate.

___

___

3➡ Where is the nervous system located?

___

___

4➡ What is the general function of the nervous system?

___

___

F. Endocrine System

1➡ Study Figure 7–7●. Identify the following structures:

Pituitary Gland
Hypothalamus
Thyroid Gland
Parathyroid Glands
Thymus
Pancreas
Adrenal Glands
Kidney
Ovaries (Female Gonads)
Testes (Male Gonads)

2➡ Add any additional structures you may locate.

___

___

3➡ Where is the endocrine system located?

___

___

4➡ What is the general function of the endocrine system?

___

___

G. Cardiovascular System

1➡ Study Figure 7–8●. Identify the following structures:

- Heart
- Arteries
- Veins
- Capillaries

2➡ Add any additional structures you may locate.

____

____

3➡ Where is the cardiovascular system located?

____

____

4➡ What is the general function of the cardiovascular system?

____

____

H. Lymphatic System

1➡ Study Figure 7–9●. Identify the following structures:

- Tonsils (not shown)
- Spleen
- Thymus
- Lymph Nodes
- Lymphatic Vessels

2➡ Add any additional structures you may locate.

____

____

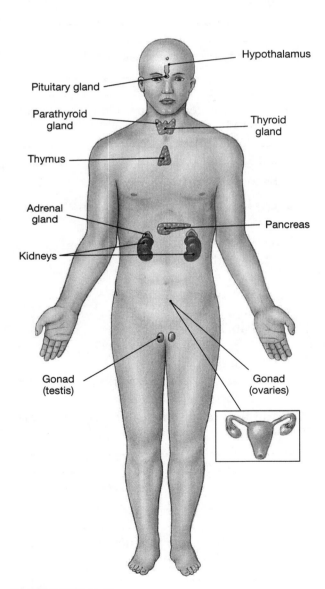

● **FIGURE 7–7**
The Endocrine System.

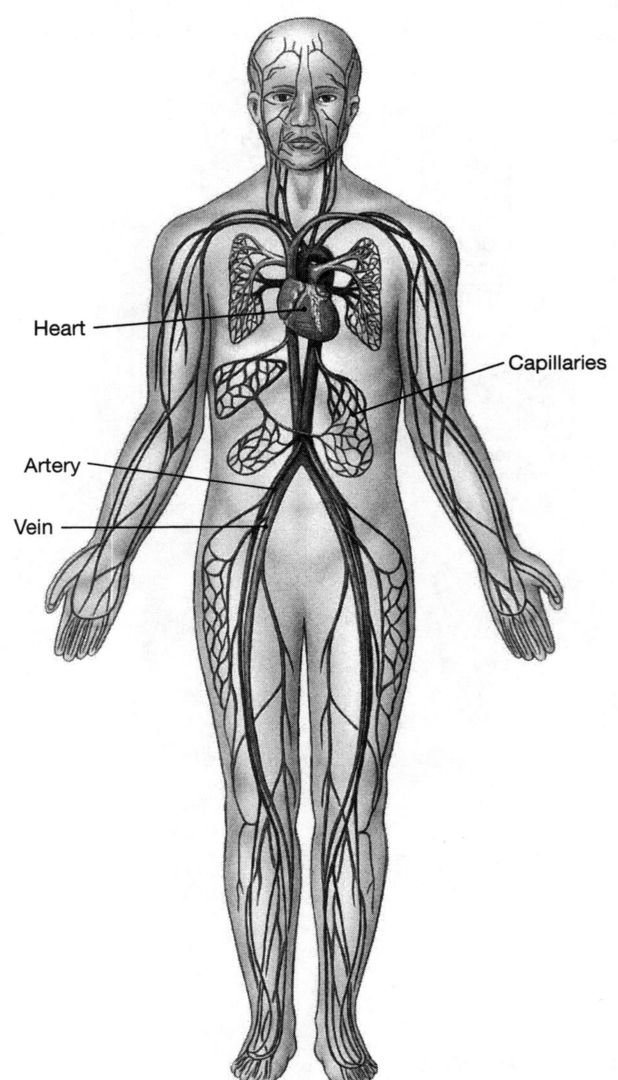

● **FIGURE 7–8**
The Cardiovascular System.

3➡ Where is the lymphatic system located?

____

____

4➡ What is the general function of the lymphatic system?

____

____

I. Respiratory System

1➡ Study Figure 7–10●. Identify the following structures:
- Nasal Cavity
- Trachea
- Lung
- Pharynx
- Larynx
- Bronchi
- Sinuses

2➡ Add any additional structures you may locate.

____

____

3➡ Where is the respiratory system located?

____

____

4➡ What is the general function of the respiratory system?

____

____

J. Digestive System

1➡ Study Figure 7–11●. Identify the following structures:
- Pharynx
- Salivary Gland

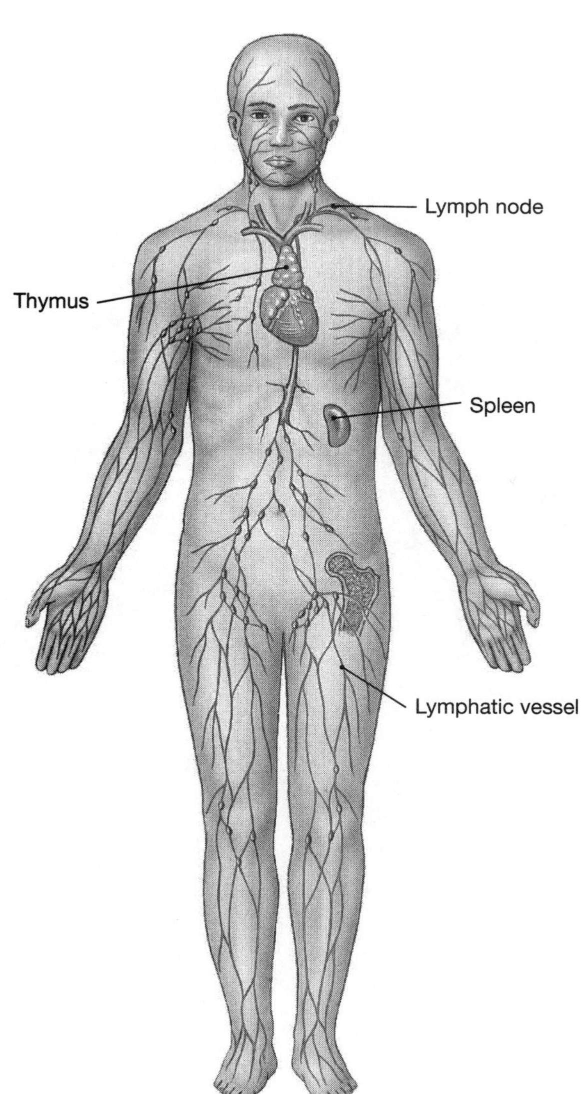

● **FIGURE 7–9**
The Lymphatic System.

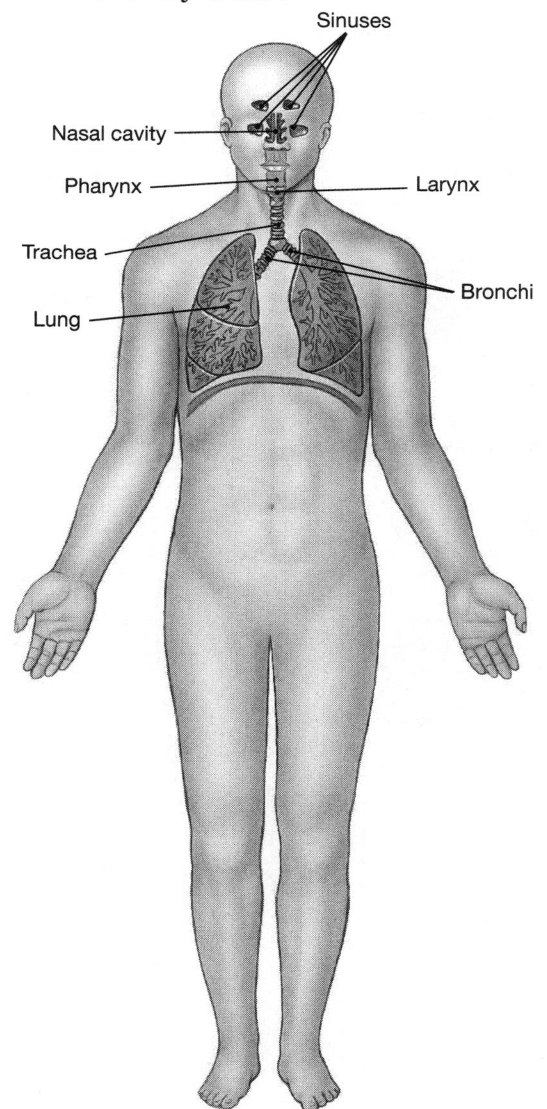

● **FIGURE 7–10**
The Respiratory System.

Esophagus
Stomach
Small Intestine
Large Intestine
Liver
Gallbladder
Pancreas (Figure 7–7●)
Anus

2➡ Add any additional structures you may locate.

______

______

3➡ Where is the digestive system located?

______

______

4➡ What is the general function of the digestive system?

______

______

K. Urinary System

1➡ Study Figure 7–12●. Identify the following structures.
Kidney
Ureter
Urinary Bladder
Urethra

2➡ Add any additional structures you may locate.

______

______

3➡ Where is the urinary system located?

______

______

4➡ What is the general function of the urinary system?

______

______

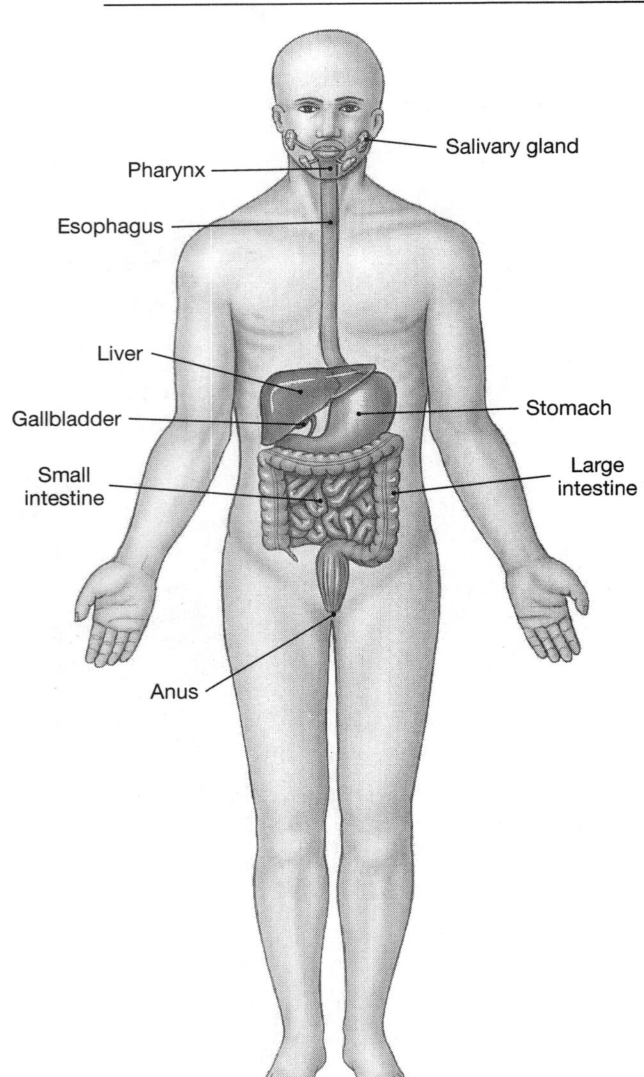

● **FIGURE 7–11**
The Digestive System.

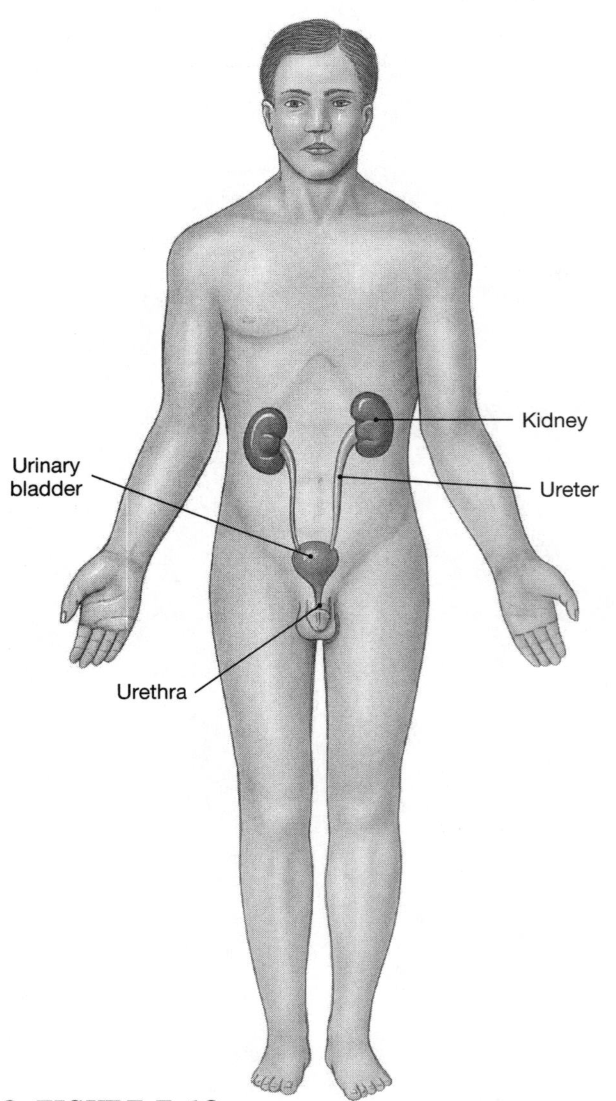

● **FIGURE 7–12**
The Urinary System.

L. Reproductive System

1➡ Study Figure 7–13●. Identify the following structures:

Female: Ovary
Uterine Tube
Uterus
Vagina
External Genitalia
Mammary Gland

Male: Seminal Vesicle
Prostate Gland
Penis
Sperm Duct
Testis
Scrotum
Urethra
Epididymis

2➡ Add any additional structures you may locate.

____

____

3➡ Where is the reproductive system located?

____

____

4➡ What is the general function of the reproductive system?

____

____

Because all systems of the body function together, it is often possible to examine one system and discern a problem in another system. For instance, a blood test could indicate difficulties with the immune system, the lymphatic system, or with the endocrine system. A urine test could yield results indicative of problems with the digestive or respiratory system.

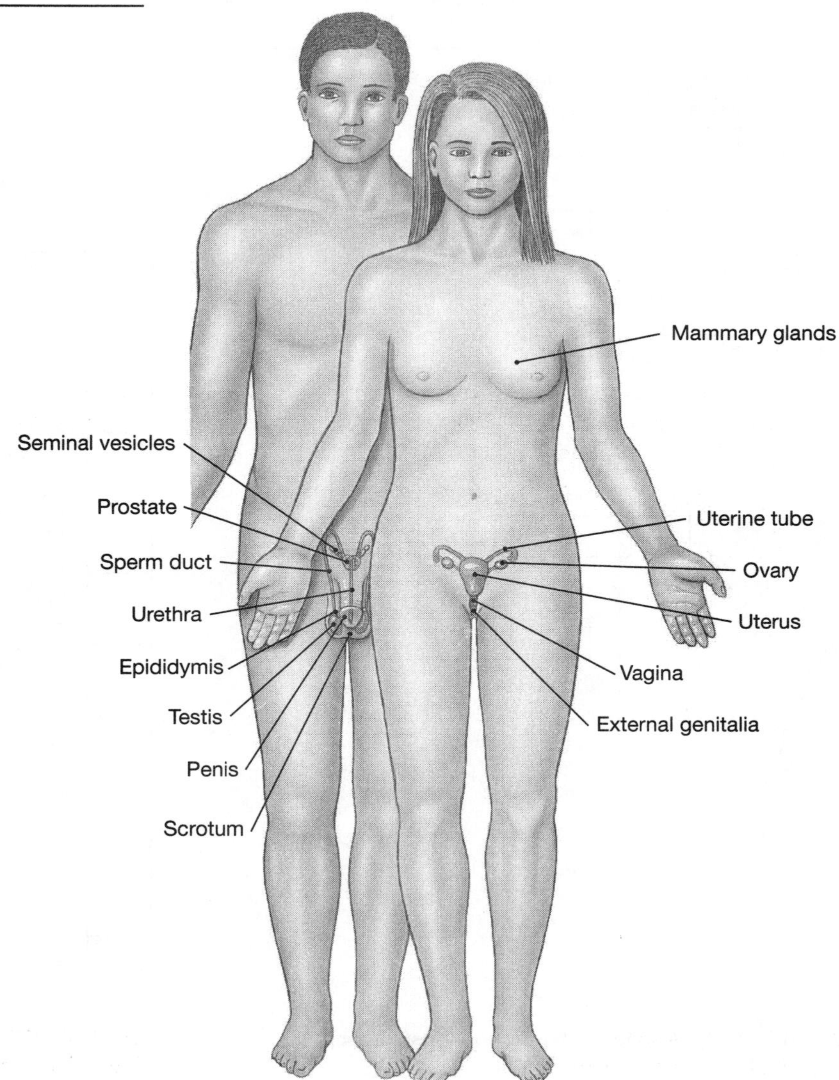

● **FIGURE 7–13**
Reproductive System.

❑ Additional Activities

NOTES

1. Explore the embryology of different body systems. This information is included in your lecture text.
2. Explain what we mean when we say, "All higher animals share a basic body plan."
3. Explain any gross anatomical observations or comparisons you can make after examining various lab animals (or your own pets). How do these animals' structures compare with your own?

❑ Lab Report

1. Referring to the numbered inquiries at the beginning of this exercise, complete the following box summary:

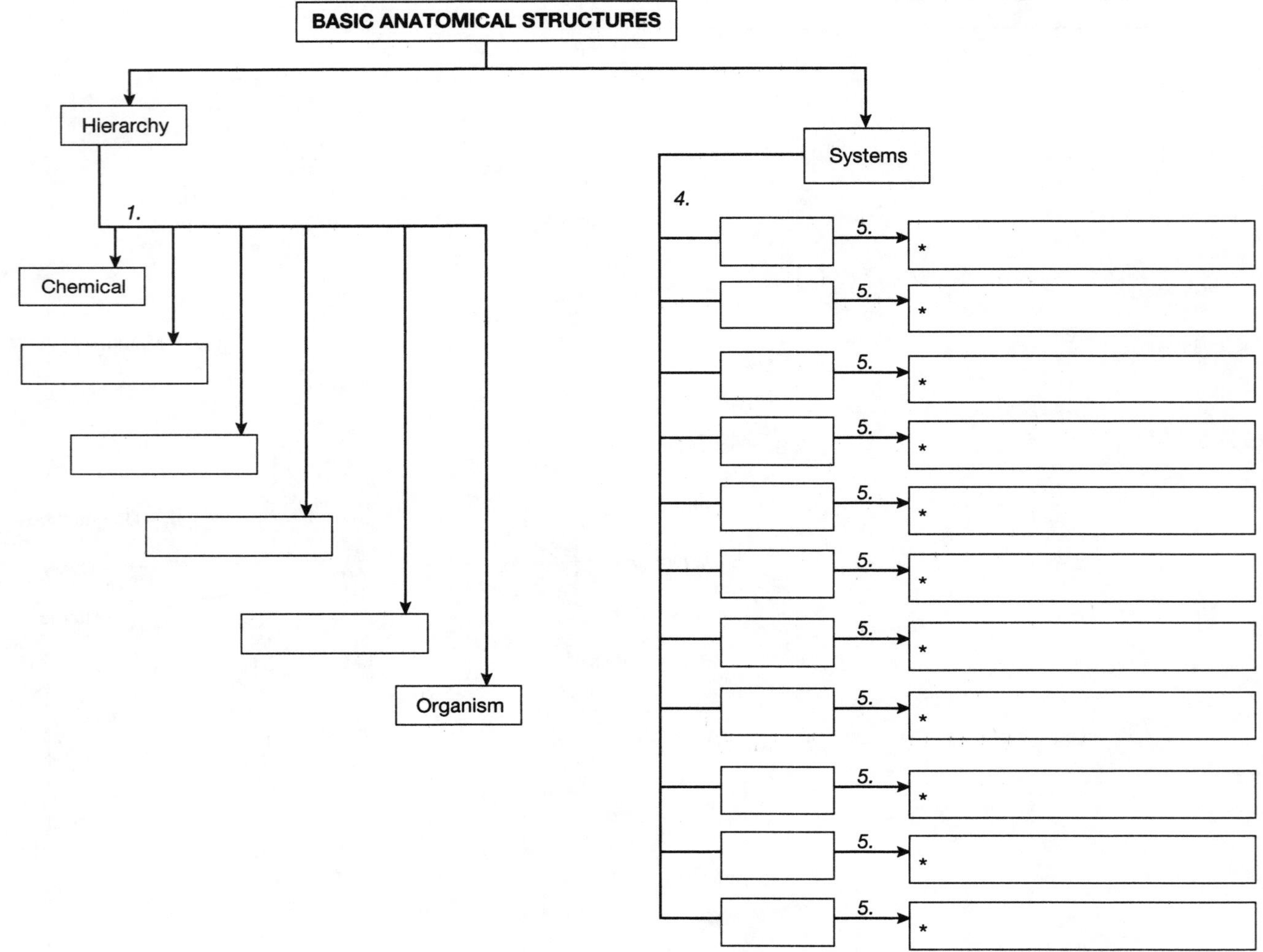

*List Organs as indicated by your Instructor.
Additional space may be required.

General Questions

NOTES

1. What is the difference between microscopic and macroscopic anatomy?
2. Put the following terms in order, beginning with the least inclusive: cell, organism, system, tissue, chemical, organ. Now define each of these terms.

3. In the space below, list the organs you have identified in this exercise. Quiz a classmate on placing these organs into systems.

______________	______________	______________
______________	______________	______________
______________	______________	______________
______________	______________	______________
______________	______________	______________
______________	______________	______________
______________	______________	______________
______________	______________	______________
______________	______________	______________
______________	______________	______________
______________	______________	______________

4. Based on what you already know, explain how the cardiovascular system interacts with the respiratory system, the digestive system, and the urinary system.

   Respiratory System ____

   ____

   Digestive System ____

   ____

   Urinary System ____

   ____

EXERCISE 8

Cell Anatomy

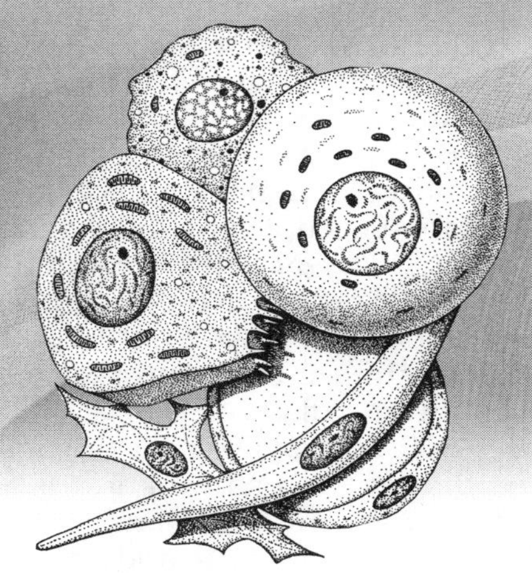

PROCEDURAL INQUIRIES

Examination

1. What are the nuclear structures of the cell?
2. What are the cytoplasmic structures of the cell?
3. What are the membrane structures of the cell?
4. What cellular structures can be readily seen on standard prepared tissue slides?

Experimentation

5. What is a method for making a stained wet mount of living cells?
6. What cellular structures can be seen using a standard wet mount?

Additional Inquiry

7. Who helped shape our present-day cell theory?

Key Terms

Centrioles
Centrosome
Chromatin
Cilia
Cristae
Cytoplasm
Cytoskeleton
Endoplasmic Reticulum
Filaments
Flagella
Golgi Apparatus
Inclusions
Intermediate Filaments
Lysosome
Matrix
Microfilaments
Microtubule
Microvilli
Mitochondrion
Nuclear Membrane
Nuclear Pores
Nucleolus
Nucleoplasm
Nucleus
Organelle
Peroxisome
Plasma Membrane
Ribosome
Rough Endoplasmic Reticulum
Smooth Endoplasmic Reticulum
Thick Filaments
Transport Vesicles
Vacuole

Materials Needed

Cell Models

Cheek or Gingival Smear

- Slides
- Cover Slips
- Methylene Blue
- Toothpicks

Prepared Slides

- Liver
- Respiratory Epithelium
- Intestinal Epithelium
- Nerve Smear

The cell was first conceptualized by Aristotle (384–322 B.C.), who observed that living structures were composed of smaller and smaller functional parts. He concluded that there must be an ultimate fundamental unit of life.

Nevertheless, 2,000 years passed before Robert Hooke saw the first cellular shells in 1665 when he was examining a piece of cork for a demonstration for the Royal Society of London. Hooke was the man who named the cell. The evenly spaced rows of "cork boxes" reminded him of the "cells" (or living quarters) occupied by monks in a monastery. More than 300 years later we still use the term.

Hooke did not see actual cells because the cork itself was dead. It wasn't until 1673 that Antony van Leeuwenhoek saw, and recognized as such, the first living cells.

In 1805 Lorenz Oken, a German naturalist, stated that the cell as we know it was the basic biological unit. The world was not quite ready for such a revolutionary idea, and it was more than 30 years before this early cell theory was formalized in 1839 by two other Germans scientists, Matthias Schleiden, a botanist, and Theodor Schwann, a zoologist. Schleiden and Schwann stated that all living things are composed of cells.

In 1855 Rudolf Virchow stated *OMNIS CELLULA A CELLULA*, all cells arise from preexisting cells. From this statement we have today's cell theory, which includes the following *[∞ Martini, p. 65]*:

1. All parts of the living world are composed of cells or cell products.
2. Cells arise only from existing cells.
3. The cell is the structural and functional unit of life.
4. Each cell maintains independent homeostasis.
5. The organism as a whole depends on the combined and coordinated activities of the individual cells.

Cell size is highly variable. We usually think of the cell as microscopic. Most cells are. But the human ovum can actually be seen by the unaided eye. The human ovum is about the size of the period at the end of this sentence. (Volume-wise, the ostrich egg is the largest single cell.) Some neurons are too tiny to be visible without a high-powered microscope, yet they extend from the spinal cord to the tips of the extremities and can easily reach over a meter in length. (The corresponding giraffe neurons are the longest cells.) At the other end of the size spectrum, we have the red blood cells, so small that literally hundreds of thousands of them can fit in a cubic millimeter.

Cell shapes are just as diverse. As you will see when we study the tissues in Exercise 11, red blood cells are round, some epithelial cells are cuboidal, neurons have long filamentous projections, and many of the muscle cells resemble extended rectangles.

Despite differences in size and shape, however, the cell is the basic anatomical and physiological unit of the human body, and all cells have the same basic structures and processes. Certain anatomical generalizations can be made about the "generic" cell even though most human cells are physiologically highly specialized. There are roughly 200 different cell types in the human body.

In this exercise, we will examine the anatomy of the generalized mammalian cell and briefly discuss that anatomy in relation to basic cellular functions. We will do this first by making some **theoretical observations** of the cell, including identifying the basic **nuclear, cytoplasmic,** and **membrane** structures. We will study some **prepared slides** of typical human cells in order to see commonality among the cellular structures. Then we will examine some fresh cells by preparing a stained wet mount of a cheek or gingival smear.

❑ Examination

To begin this exercise, note Figure 8–1●, the hypothetical cell. As you work through this section, you will be asked to identify prominent structures of the nucleus, the cytoplasm, and the cell membrane.

I. Cellular Structures — Theoretical Observations *[∞ Martini, pp. 66–67, 80–90]*

A. Nuclear Structures

The nuclear structures are the basic components of the nucleus.

1 ➡ Label the nuclear structures in Figure 8–1● with the following boldface terms. Use the descriptions and any available cell models as your guide.

Nucleus The prominent structure often depicted at the center of the cell that directs the activities of the cell.

Nucleolus Organelle within the nucleus responsible for ribosome synthesis. Cells often have more than one nucleolus. (An **organelle** is an intracellular structure with a specific function or group of functions. Some organelles are membrane-bound; some are not.)

Nucleoplasm Gel-like nuclear matrix; the fluid contents of the nucleus.

Chromatin Genetic material (DNA) in nondividing cells. (Chromosome and chromatid are the terms for dividing genetic material. See Exercise 10.)

Nuclear Envelope The double phospholipid bilayer forming the boundary of the nucleus.

Nuclear Pores Openings in the nuclear membrane that allow for the passage of ions and small molecules.

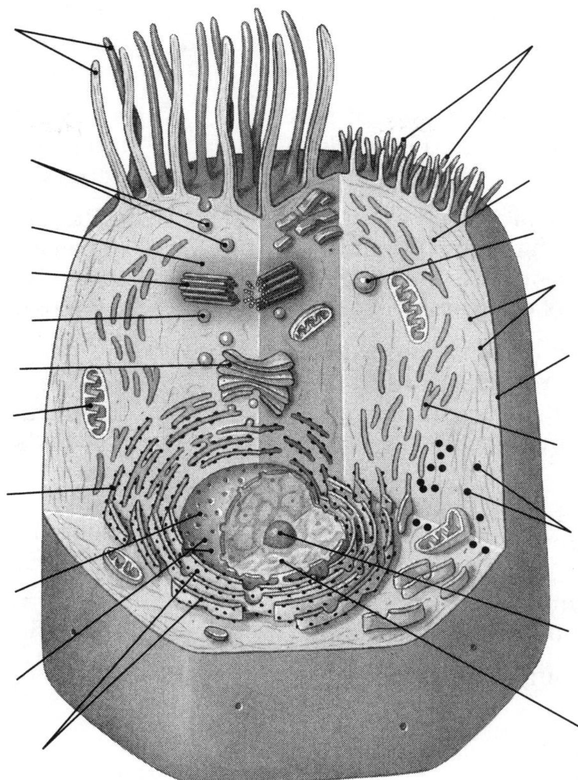

● **FIGURE 8–1**
Typical Cell.

B. CYTOPLASMIC STRUCTURES

The cytoplasmic structures are found outside the nucleus.

1 ➡ Label the cytoplasmic structures in Figure 8–1● with the following boldface terms. Use the descriptions and any available cell models as your guide.

Cytoplasm Gel-like cell matrix. The fluid portion of the cytoplasm is the cytosol. The organelles are also found in the cytoplasm.

Endoplasmic Reticulum (ER) Folded membrane involved in molecular transport and storage as well as in certain biochemical reactions.

Rough Endoplasmic Reticulum Continuous with the outer phospholipid bilayer of the nuclear envelope, the rough ER is dotted with ribosomes and is the site of secretory protein synthesis reactions.

Smooth Endoplasmic Reticulum Often continuous with the rough ER, the smooth ER does not have ribosomes and is the site of lipid and carbohydrate synthesis. In certain cells, parts of the smooth endoplasmic reticulum becomes highly specialized.

Transport Vesicles Pinched-off portions of the ER that fuse with the Golgi Apparatus.

Golgi Apparatus Series of flattened sacs (saccules) and end-point vesicles responsible for packaging and secretion of proteins and lipids, synthesis of glycoproteins, renewal of cell membranes, and packaging of enzymes for use within the cytosol.

Lysosome Vesicle pinched off from Golgi Apparatus filled with digestive enzymes.

Peroxisome Vesicle involved in absorbing and neutralizing toxins.

Ribosome (Free in cytoplasm or fixed to rough ER) two-part structure responsible for protein synthesis.

Mitochondrion (pl. mitochondria) Structure of various shapes (often oblong), containing its own DNA and ribosomes, responsible for energy production in the cell.

Cristae Inner folds of the mitochondrion.

Matrix Gel-like mitochondrial background material.

Vacuole Pinocytic or phagocytic vesicle, storage compartment, or cellular waste repository. (Pinocytic vesicles engulf fluids, while phagocytic vesicles engulf solids.)

Centrosome Non-membrane-bound area (often not seen) containing centrioles.

Centrioles Two microtubular structures at right angles to each other that function in cellular division.

Inclusions Masses of insoluble substances such as crystals, pigments, and lipid globules in cytoplasm.

Cytoskeleton Internal protein framework giving cytoplasm strength and flexibility. (The cytoskeleton as such is not specifically identified on this diagram.)

Microtubule Tubular part of the cytoskeleton, composed of the protein tubulin. Key structural component in centrioles, cilia, and flagella. Microtubules are included on this diagram.

Filaments As parts of the cytoskeleton, three types of interactive protein structures in the cytoplasm. On this diagram you are not asked to distinguish among the three. A nonspecific filament is shown on the diagram.

Microfilaments Composed primarily of actin.

Intermediate filaments Many composed of myosin.

Thick filaments Myosin. Abundant in muscle cells.

Concept Check 1 Why would a cell without a nucleolus be unable to synthesize proteins?

____

____

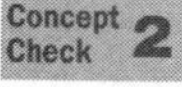
Would you expect a liver cell or an adipose (fat) cell to have more mitochondria?

____

____

TAY-SACHS DISEASE (named for Warren Tay, an English physician, and Bernard Sachs, an American neurologist) is a lysosomal storage disease. The enzyme hexosaminidase A, a lipid-metabolizing enzyme, is not produced and is thus not found in the lysosomes. Because the lysosomes cannot degrade certain lipids, these lipids build up within the cells, causing neurological deterioration and eventual death.

C. Membrane Structures

In this part of the exercise we are concentrating on the cell membrane.

1➜ Label the membrane structures in Figure 8–1● with the following boldface terms. Use the descriptions and any available cell models as your guide.

Plasma Membrane Phospholipid bilayer forming the boundary between the cell and its environment.

Cilia Small hairlike projections anchored to a basal body that beat to move substances across a surface.

Flagella Long, whiplike structures that propel a cell through a medium. (The sperm is the only flagellated cell in the human; the composite cell shown here does not include a flagellum.)

Microvilli Slender projections of the cell membrane that often function to increase the cell's surface area.

Concept Check 3 Would you expect cilia or flagella to be on cells lining the respiratory tract?

____

____

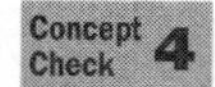
Check your labelled Figure 8–1● with the correct answer on p.70.

II. Prepared Slides

For each of the slides in this exercise, start first with your microscope on low power and, if necessary, change to high power. You should not need the oil immersion lens for these slides.

A. Liver

1➜ Obtain a prepared slide of the liver and identify the following structures:

- Nucleus
- Nucleolus (pl. nucleoli) You may find more than one nucleolus per nucleus.
- Nucleoplasm
- Nuclear Envelope
- Chromatin
- Cytoplasm
- Inclusions (dark-brown-pigment granules)
- Plasma Membrane

2➜ Compare your slide with Figure 8–2a●. Sketch and label what you see.

Liver cell

● **FIGURE 8–2a**
Liver Cell.

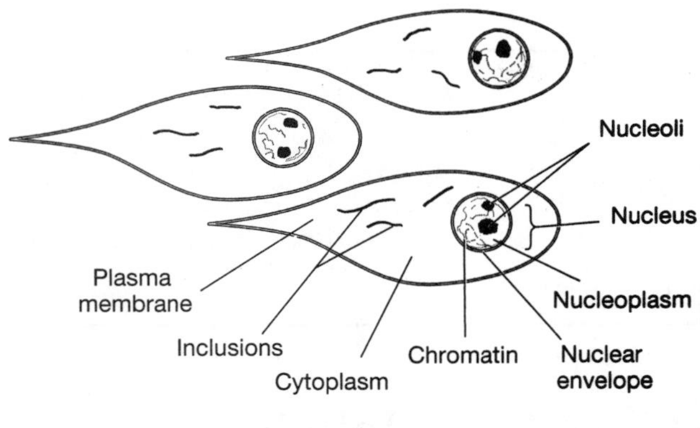

(a)

● **FIGURE 8–2c**
Intestinal Epithelium.

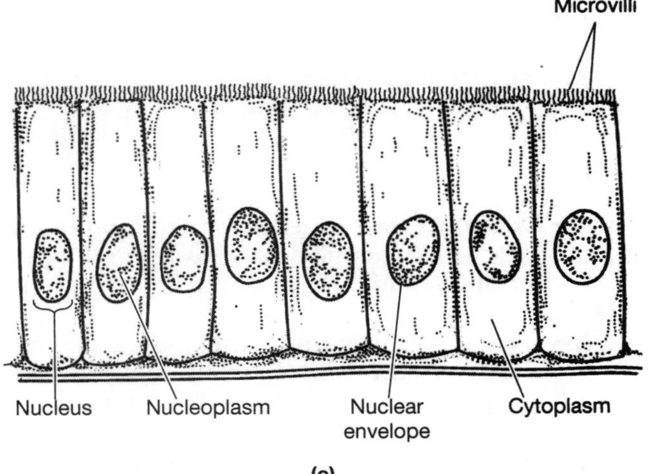

(c)

B. Respiratory Epithelium

1 ➡ Obtain a prepared slide of the respiratory epithelium and identify the following structures:

Nucleus
Nucleoplasm
Nuclear Envelope
Cytoplasm
Vesicles (in goblet cells)
Cilia (on lumen border)

2 ➡ Compare your slide with Figure 8–2b●. Sketch and label what you see.

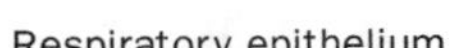

Respiratory epithelium

● **FIGURE 8–2b**
Respiratory Epithelium.

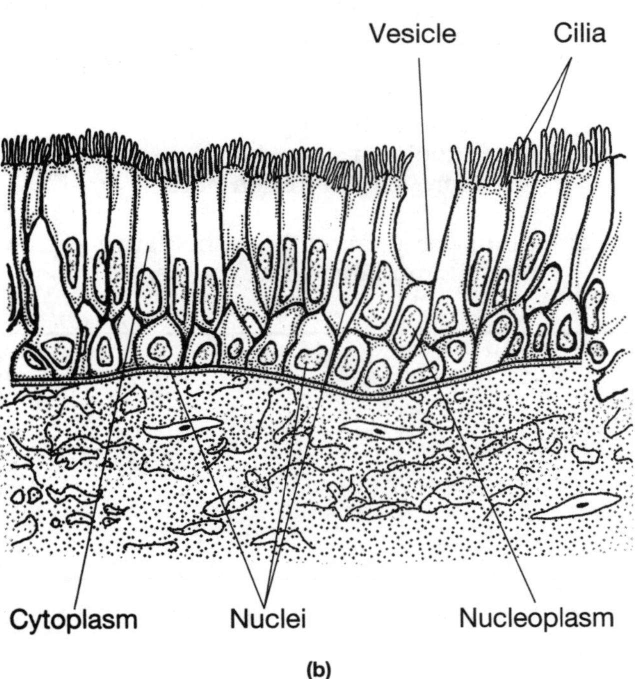

(b)

● **FIGURE 8–2d**
Nerve Smear.

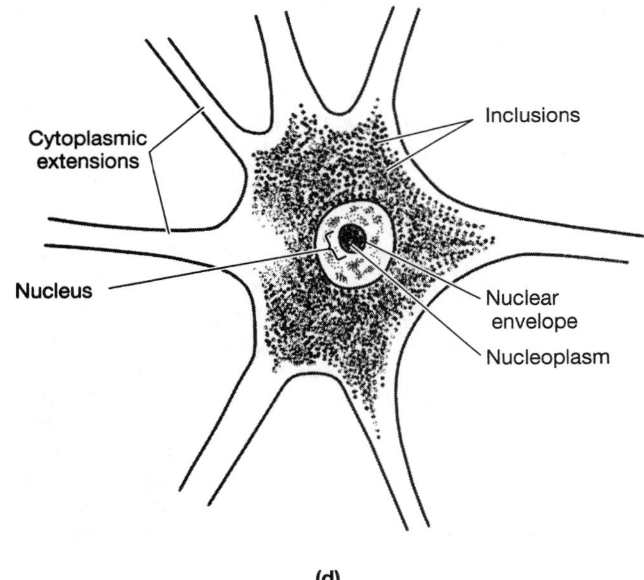

(d)

C. Intestinal Epithelium

1➡ Obtain a prepared slide of the intestinal epithelium and identify the following structures:
- Nucleus
- Nucleoplasm
- Nuclear Envelope
- Cytoplasm
- Microvilli (on lumen border)

2➡ Compare your slide with Figure 8–2c●. Sketch and label what you see.

Intestinal epithelium

D. Nerve Smear

1➡ Obtain a prepared slide of a nerve smear and identify the following structures:
- Nucleus
- Nucleoplasm
- Nuclear Envelope
- Cytoplasm
- Inclusions
- Cytoplasmic Extensions (actually, axons and dendrites)

2➡ Compare your slide with Figure 8–2d●. Sketch and label what you see.

Nerve smear

❑ Experimentation

III. Cheek or Gingival Smear

We will make a stained wet mount from the cheek or gum in order to view a basic living cell.

A. Wet Mount Procedure

If you need a refresher on making a wet mount, refer back to Exercise 3.

1➡ Obtain a clean slide, cover slip, and toothpick. Place a small drop of water in the center of the slide.

2➡ Either gently scrape the inside of your cheek or run the toothpick between your gum (gingival area) and a back molar.

3➡ Dab the scraping into the drop of water.

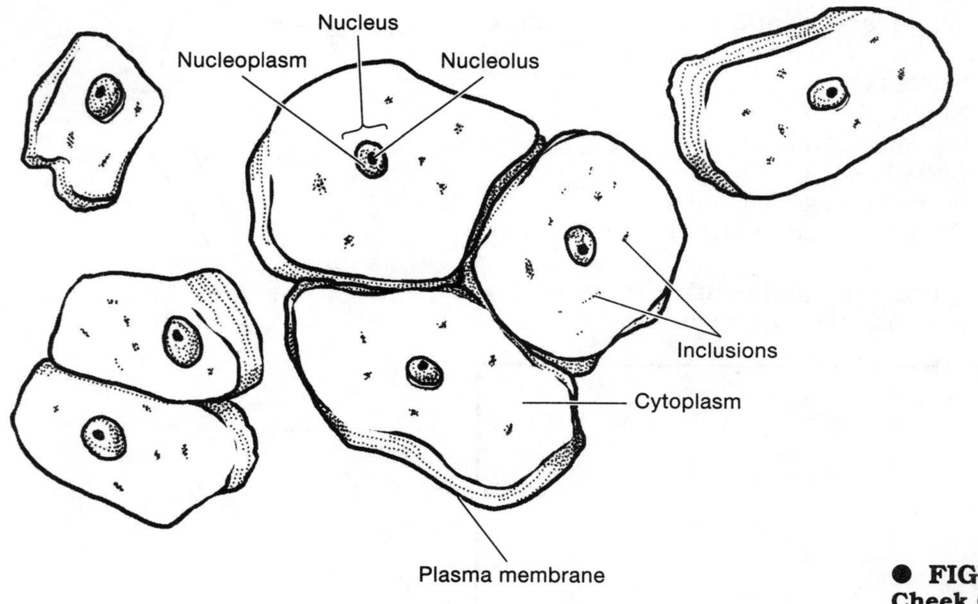

● **FIGURE 8–3**
Cheek Cells.

4➡ Place the cover slip over the drop of water.

5➡ Put a small drop of methylene blue on the side of the cover slip and allow the stain to seep under the cover slip.

[Note: The reason we put the scraping in a drop of water and then put the methylene blue outside the cover slip is to dilute the methylene blue and assure that the dye is more evenly distributed to the sample.]

B. Examining the Wet Mount

1➡ Begin your examination with your slide under low power. When you are comfortable with what you are viewing under low power, switch to high power. Your cheek or gingival cells will look flat or (if you have a good sense of three-dimensional vision) like slightly misshapen cubes. They should be light blue with several dark blue spots in the interior. Use your fine adjustment knob to gain a sense of depth. Keep in mind that these cells are three-dimensional structures.

2➡ Choose a cell on your slide and identify these structures:

Nucleus Darkly stained structure usually centrally located.

Nucleolus Very dark, tiny structure within nucleus (may not be visible in all cells or in heavily stained cells).

Nucleoplasm Nuclear matrix.

Cytoplasm Lightly stained.

Inclusions (may not be visible in all cells).

Plasma Membrane Dark, outer boundary of the cell.

(Technically, with a light microscope, you cannot see either the plasma membrane or the nuclear envelope. Rather, because of the area darkened by staining a three-dimensional structure, you see the extra stain in the place where the membrane would be.)

3➡ Observe the simulated drawing in Figure 8–3●. Sketch and label a cell or group of cells from your own slide. Can you account for any differences between Figure 8–3● and your drawing?

Cheek or gingival smear

❑ Additional Activities

NOTES

1. Compare the generic human cell with typical plant, fungal, protistan, and/or moneran cells.
2. Keep this gelatin cell in mind if you have a relative or friend in junior high or middle school who is studying cell structure.

   Use your ingenuity at home to construct a gelatin cell. Try yellow jello for the matrix and various fruits, vegetables, and pastas for the organelles. A cut-away plum could be the nucleus and the seed the nucleolus. Lasagna could be the endoplasmic reticulum, while spaghetti could be the microfilaments. Grapes make good lysosomes and raisins are good ribosomes. According to what is available, use different foods for the other cellular parts. Put all of this on lettuce (cell membrane) in a glass baking dish.

Answers to Selected Concept Check Questions

NOTES

1. The cell would be unable to synthesize proteins because the cell would not have any ribosomes. Ribosomes are produced in the nucleolus.
2. The liver cell would have more mitochondria.
3. Cilia line the respiratory tract.
4. Labels for cellular components

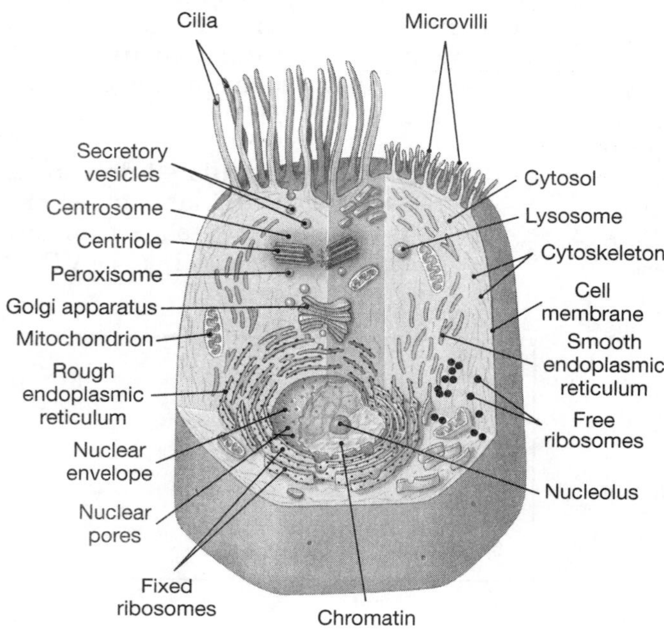

❑ Lab Report

1. Referring to the numbered inquiries at the beginning of this exercise, complete the following Box Summary:

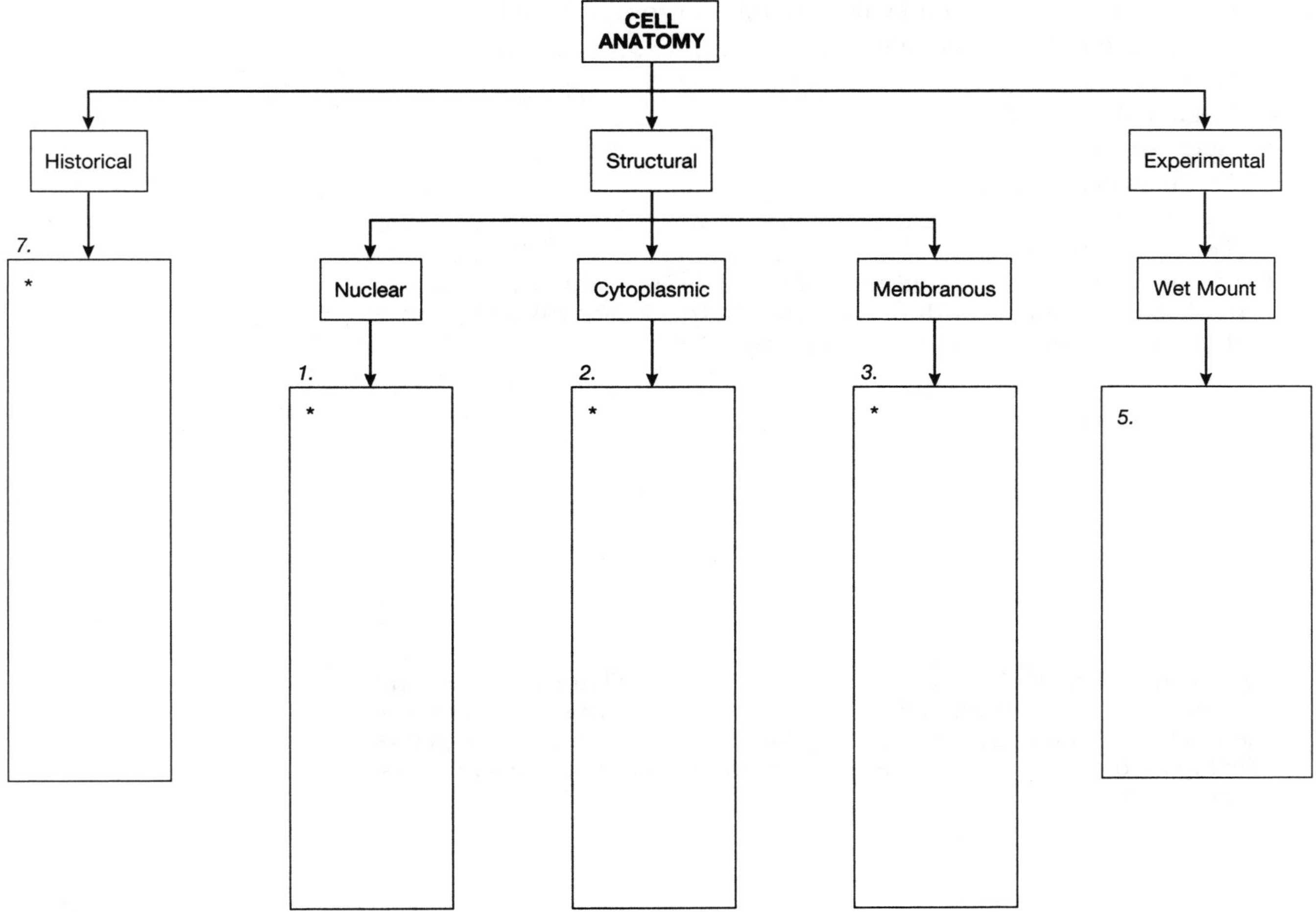

*List as directed by your instructor.
You may need additional paper.

General Questions

NOTES

1. What contributions did each of the following make to our present understanding of the cell?
   Aristotle ______
   Hooke ______
   van Leeuwenhoek ______
   Schleiden and Schwann ______
2. Identify the structures:
   a. Nuclear matrix ______
   b. Structural tube found in centrioles ______
   c. For protein synthesis ______
   d. Structure that engulfs fluids ______

NOTES

e. Carbohydrate synthesis ______
f. Openings in nuclear membrane ______
g. Increases cell surface area ______
h. Contains digestive enzymes ______
i. Nonmetabolic crystal ______
j. Energy production ______

3. Identify the function: (You may need to refer to your lecture text.)
a. Rough Endoplasmic Reticulum ______
b. Chromatin ______
c. Plasma Membrane ______
d. Golgi Apparatus ______
e. Centrosome ______
f. Microfilaments ______
g. Transport Vesicles ______

4. Consider your wet mount and each of the slides you examined. How do your sketches compare with the sketches in this lesson? Were you able to find all requested structures? If not, why not?

5. Again consider each of the five slides you examined. What similarities and differences do you see among the specialized structures and shapes of the nuclei, nucleoli, and cytoplasm? Can you observe any other points of comparison? Can you make a generalization about the size and shape of the "average" cell?

6. Use your lecture text to determine which of the cellular structures you identified were membrane-bound and which were not membrane-bound.

EXERCISE 9

Membrane Transport

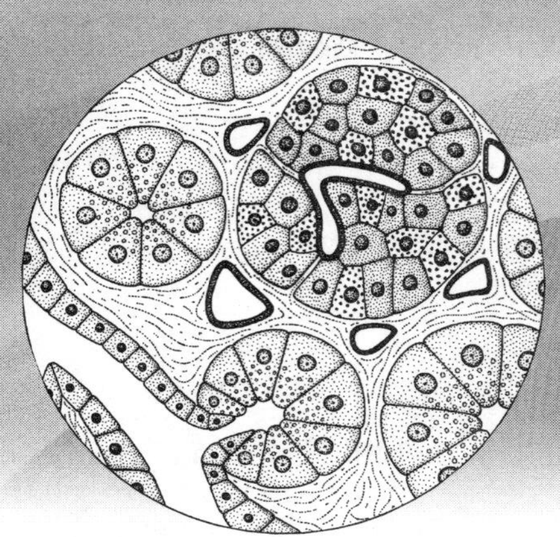

PROCEDURAL INQUIRIES

Preparation

1. Referring to energy, what two ways can substances enter a cell?
2. What is active transport?
3. What is passive transport?
4. How is osmosis related to diffusion?

Experimentation

5. How can we demonstrate active transport?
6. How can we demonstrate Brownian movement?
7. How can we demonstrate diffusion (2 ways)?
8. How can we demonstrate osmosis (3 ways)?

Additional Inquiries

9. In terms of relationships between substances, how can we define "hypertonic," "isotonic," and "hypotonic"?
10. What is the relationship between the size of a molecule and its rate of diffusion?

Key Terms

Active Transport
Brownian Movement
Cell Physiology
Dialysis
Diffusion
Gradient
Hypertonic
Hypotonic
Isotonic
Membrane Transport
Osmosis
Passive Transport

Materials Needed

Active Transport

- Baker's Yeast
- 0.75% Na_2CO_3
- 0.02% Neutral Red
- Erlenmeyer Flasks
- Flame or Hot Plate

Brownian Movement

- India Ink (Carmine Dye or Whole Milk may be substituted)
- Slides and Cover Slips

Diffusion

- Beaker of Distilled Water
- 1.5% Agar-agar (or Gelatin) in petri plate
- Potassium Permanganate ($KMnO_4$) Crystals
- Potassium Dichromate ($K_2Cr_2O_7$) Crystals
- Methylene Blue Crystals
- Metric Ruler (to measure in mm)

Osmosis — Thistle Tube

- Thistle Tube Osmometer
- Salt
- Sugar
- Distilled Water

Artificial Cell

- Dialysis Tubing or Sacs
- Sugar
- String
- Beakers
- Distilled Water
- Scale

Osmosis and Red Blood Cells

- Distilled Water
- Salt Solutions
  - 12–15% Salt
  - 0.85% Salt (This is human, isotonic saline.)

Blood (Human or Animal)
 Lancets and Alcohol if Human Blood is used
Slides and Cover Slips
Microscope

The individual cell is a dynamic microcosm, demonstrating in miniature all the processes and events that occur in the macrocosm, making the apparent function of the whole organism the actual function of many individual cells working in unison.

It is important that we understand how the individual cell, the microcosm, functions so that we can more fully understand how the organism as a whole, the macrocosm, functions. For instance, from Exercise 30 we can say that if an action potential (or nerve impulse) is to be generated in a part of the nervous system, a certain electrical stimulation must be present and certain ions must be moving through appropriate channels. What we sometimes forget is that these events — in this case the electrical stimulation and the ionic movement — occur at the cellular level. What seems to be happening in the organism is happening only because the events are occurring at the level of the individual cell. We could use similar analogies for every system in the body.

In lecture you examined the molecular intricacies of the phospholipid bilayer known as the cell membrane, and you became aware that the cell membrane is selectively permeable, meaning that only certain substances can enter and leave the cell by freely crossing the membrane [∞ *Martini, p. 71*].

You know, for instance, that the membrane is replete with channels, gates, and carrier molecules that either facilitate, inhibit, or repel assorted ions and molecules as they randomly approach the demarcation barrier. This demarcation barrier, the cell membrane, is functional in maintaining cellular integrity.

You also know that since the cell is the microcosm, ions and molecules must cross the barrier, both as nutrients entering the cell and as wastes leaving the cell. In **cell physiology**, we examine how events occur within the cell. Cellular functions follow the basic principles of physiology. Many of these functions do not lend themselves to easy demonstration, particularly this early in an introductory course. However, at this point we can demonstrate a number of functions directly related to **membrane transport**. And membrane transport is one of the main keys to cell physiology. (We will examine other aspects of cell physiology when we study the nervous system and the digestive system.)

❑ Preparation

I. Background and Protocol

A. Explanation of Terms

1➜ Be certain you have a working knowledge of the following terms:

Membrane Transport
Active Transport
Passive Transport
Diffusion
Osmosis
Brownian Movement
Gradient

Membrane transport is any process, active or passive, by which a substance moves from one side of a membrane to the other side of the membrane. We have already established that the membrane acts as a barrier that can facilitate, inhibit, or repel substances, and yet certain substances must cross the barrier.

Active transport requires cellular energy because the substance traversing the barrier either cannot do so without a push or cannot do so in sufficient quantity to maintain cellular integrity. Active transport is absolutely essential in maintaining the functioning organism. Active transport is accomplished via ionic pumps. Many of our carrier systems work because of active transport. Gradients, which enable work to be done, are maintained because of active transport. Active transport can best be demonstrated by visualizing some very common events, both in the laboratory and in daily life.

Passive transport requires no input of cellular energy. Several categories of passive transport can be identified. Nevertheless, the most common readily identifiable form of passive transport is **diffusion**, the net movement of molecules from an area of higher concentration to an area of lower concentration. **Osmosis** is a form of diffusion. In the human system, osmosis is the diffusion of water. Passive transport can be easily demonstrated by both diffusion and osmosis experiments.

Brownian movement, while technically not diffusion, is covered here with the passive transport experiments because Brownian movement demonstrates molecular motion. It is precisely because of this innate movement that molecules are able to move passively from one point to the next point — including from one side of a membrane to the other side.

A **gradient** is any difference in intensity between two sides of a demarcation. For instance, a concentration gradient would be a difference in concentrations of a substance on each side of a demarcation. A physical barrier may or may not be present. We can also have temperature gradients, electrical gradients, flow gradients, and pressure gradients.

B. General Procedural Points

Your instructor may set up parts of this exercise as a demonstration. If so, answer the questions from your observations.

1➡ Work with a partner in Part II.

2➡ Divide the work between you and your partner in Part III so that each of you is doing one of the experiments.

3➡ If you use human blood, READ THE CAUTIONS GIVEN IN EXERCISES 46 AND 47.

❑ Experimentation

II. Active Transport [∞ *Martini, p. 76*]

Active transport is the energy-requiring movement of a substance across a barrier or a gradient. This transport could not take place without the input of energy. In the biological world, energy is supplied by the living cell. We can observe the principles of active transport by examining whether or not the uptake of a substance will take place in living and/or nonliving cells.

Neutral red is a dye that is actively transported into the living yeast cell. Neutral red is red in an acidic solution and yellow in a basic solution. A sodium carbonate (Na_2CO_3) solution is basic.

A. Experiment #1: Active Transport

1➡ Obtain two Erlenmeyer flasks, 2 g yeast, 50 ml sodium carbonate, and 50 ml neutral red.

2➡ Into Flask #1 place half the yeast and half the sodium carbonate. Mix well and heat gently. Boil the solution for 2 or 3 minutes. When the solution is cool, add about half the neutral red. The solution should be yellow.

3➡ Into Flask #2 add the remaining Na_2CO_3 and neutral red. Gently swirl in the remaining yeast and watch for a color change.

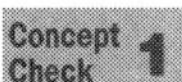

Jot down your observations and discuss with your partner the differences in color between Flask #1 and Flask #2.

____

____

[Hint: In Flask #2 either the dye entered the cell or some substance was given off by the cell to change the color to red. Which answer seems most logical? Why? How might you test the solutions to see exactly what happened?]

B. Additional Observations

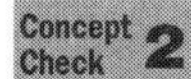

Discuss with your partner the principle involved in windmills and water pumps. What is the role of energy in these apparatuses? How can you relate these functions to the role of active transport in the cell? [∞ *Martini, pp. 76–77*]

____

____

III. Passive Transport (Diffusion)

[∞ *Martini, pp. 72–76*]

Passive transport is the movement of a substance across a membrane or a gradient without the expenditure of cellular energy. We have several experiments that demonstrate passive transport.

A. Experiment #2: Brownian Movement

All molecules with a temperature above absolute zero are in a state of constant motion. Because of this motion, molecules collide regularly and randomly. The manifestation of these collisions is called Brownian movement. Some disagreement exists as to the exact nature of Brownian movement. Nevertheless, for our purposes, the results are the same and we can see Brownian movement whenever we have a liquid mixture composed of differently sized molecules.

Brownian movement is not actually a part of diffusion, but if the molecules were not in motion, they could not diffuse across a membrane or a gradient. Thus, in order to understand diffusion we should gain some insight into molecular movement.

India ink is a suspension of carbon in water, along with ethyl alcohol and acetone. (Carmine dye and whole milk, the alternate substances for this experiment, are also molecular suspensions.)

1➡ Obtain a slide, cover slip, and microscope.

2➡ Place a drop of India ink on a slide with a small drop of water. Cover carefully with a cover slip and allow the solution to rest for 2 or 3 minutes.

3➡ Observe the suspension under both low and high power. If possible, also observe the suspension under oil. Keep the light as low as possible so you maintain maximum contrast. If your suspension is too dark, you may wish to start over using a larger drop of water.

4➡ Sketch what you see.

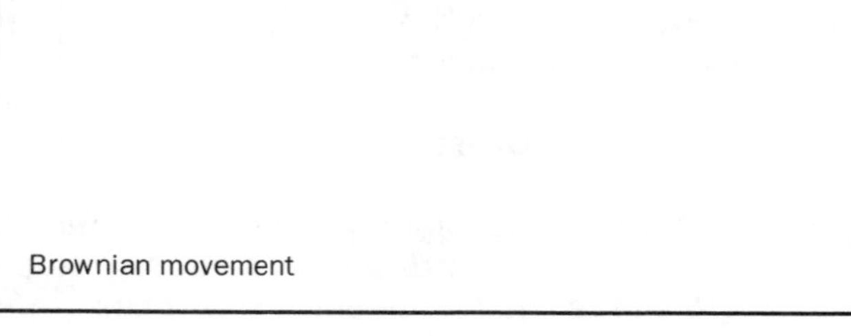

5➡ Describe Brownian movement based on your observations.

______

______

B. Experiment #3: Diffusion in Liquids

Diffusion — the net movement of molecules from an area of higher concentration to an area of lower concentration — can be demonstrated in two different ways.

1➡ Obtain a beaker of water and a crystal of potassium permanganate ($KMnO_4$).

2➡ Drop the crystal into the water and watch the diffusion process. Use Figure 9–1● to sketch the initial diffusion path. Note the time and observe the beaker at 5-minute intervals. Note the time when you think diffusion is complete. How long did it take?

______

C. Experiment #4: Diffusion in Solids

1➡ Obtain an agar-agar or gelatin petri plate and a ruler. Also obtain equal crystals of potassium permanganate ($KMnO_4$), potassium dichromate ($K_2Cr_2O_7$), and methylene blue.

2➡ Construct an imaginary equilateral triangle on the petri plate and put one crystal at each angle, per Figure 9–2●. At 15-minute intervals measure the radius (in millimeters) that each substance has diffused from the original crystal mark. Use Figure 9–3● to record these distances.

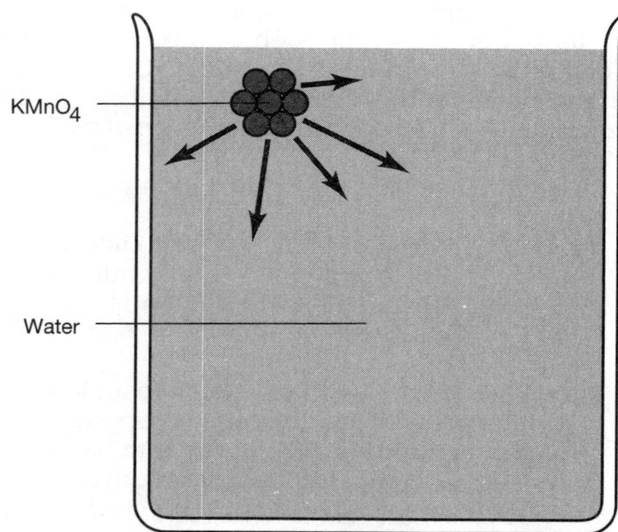

● **FIGURE 9–1**
Diffusion in Water.

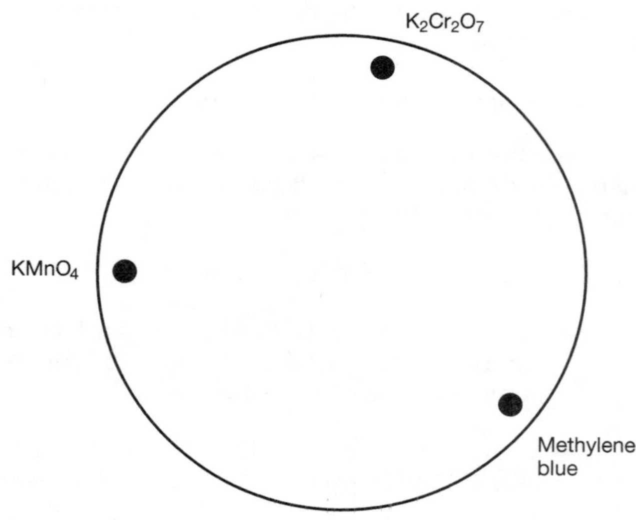

● **FIGURE 9–2**
Petri Plate.

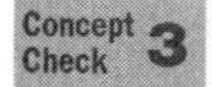

Concept Check 3 The molecular weight for $KMnO_4$ is 158, for $K_2Cr_2O_7$ is 294, and for methylene blue is 320. In general terms, what is the relationship between the distance traveled and the molecular weight of each crystal?

______

______

[Challenge: Although your measurements are probably not exact enough to derive a mathematical formula, do you see that the diffusion rate of any substance might well be inversely proportional to the square root of the molecular weight?]

Time	*$KMnO_4$*	*$K_2Cr_2O_7$*	*Methylene Blue*
15 min			
30 min			
45 min			
60 min			
75 min			
90 min			

● **FIGURE 9–3**
Petri Plate Diffusion Time.

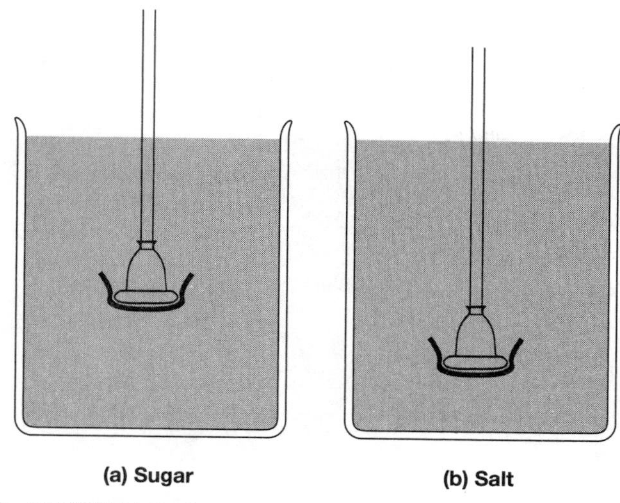

● **FIGURE 9–4**
Thistle Tube Osmometers. (a) Sugar; **(b)** Salt.

D. Osmosis Overview [∞ *Martini, p. 73-74*]

Osmosis is the movement of water from an area of higher water concentration, across a selectively permeable membrane, to an area of lower water concentration. That is, from an area of lower solute concentration to an area of higher solute concentration. Movement is always toward equilibrium. Dilution is toward equilibrium.

Dialysis is the movement of a nonwater particle across a barrier. This movement is based on the size of the particle in relation to the size of the pores within the barrier. In the living system this barrier is the cell membrane.

We can perform three experiments to demonstrate the principles of osmosis and dialysis. Work with a partner.

E. Experiment #5: Osmosis (Thistle Tube) Osmometer

A thistle tube osmometer is a thistle tube with a dialysis sheet across the open end. This osmometer is suspended in a liquid (usually water).

1➡ Set up two thistle tube osmometers. See Figure 9–4●.

2➡ In the first osmometer put a solution of about 40% sugar. Mark the meniscus on the tube. Suspend the osmometer in a beaker of distilled water.

3➡ In the second osmometer put a solution of about 15% sodium chloride. Again mark the meniscus and suspend the osmometer in distilled water.

4➡ Check the tubes at 15-minute intervals. Note the height of the water in each thistle tube.

Depending on the length of your class time and the directions from your instructor, the recording of this data may continue for some time after your laboratory period has ended.

5➡ Use the graph below to record your results.

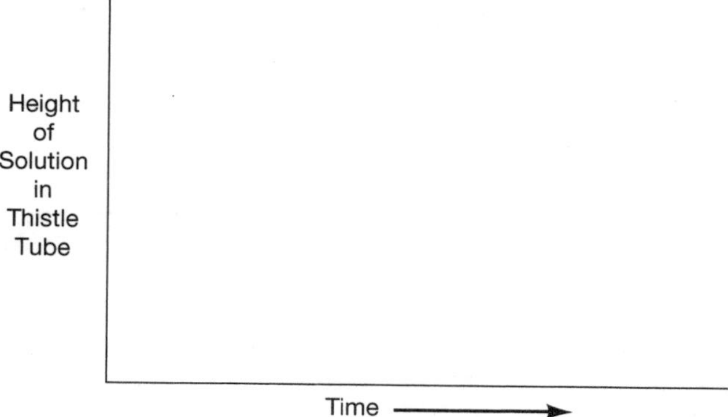

Concept Check 4 Explain your results by answering these questions: At what height did each solution stop rising? Then what happened? What differences did you notice between the sugar and salt osmometers? How do you explain those differences? If you had tasted the water at the end of the experiment, what would you have noticed? How would you explain this?

[Hint: Keep in mind the sizes of various ions, molecules, and membrane pores.]

____
____
____
____

F. Experiment #6: Osmosis (Artificial Cell)

This experiment is outlined in Figures 9–5● and 9–6●.

1➡ Obtain 5 beakers, 5 artificial cells (dialysis tubing or sacs), string, sugar, scale.

2➡ Make 3 sugar solutions, one 40%, one 20%, and one 10%. Set up 5 artificial cells. Into cell #1 put some of the 40% solution, into cell #2 the 20% solution, into cell #3 the 10% solution, and into cell #4 and cell #5 put distilled water.

3➡ Take the 5 beakers. Into beaker #1 and #2 put distilled water, into #3 put 10% sugar, into #4 put 20 % sugar, and into #5 put 40% sugar. Weigh each artificial cell and place it into its corresponding beaker.

4➡ At the end of the laboratory period — or later in the day, depending on the direc-

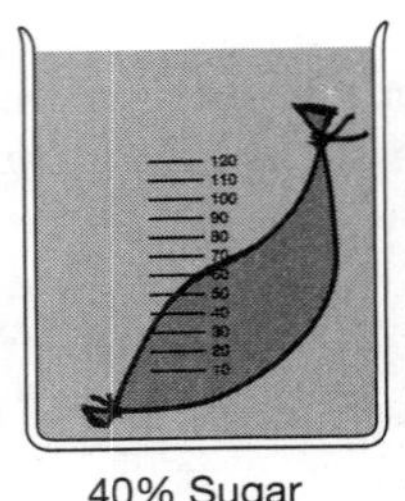
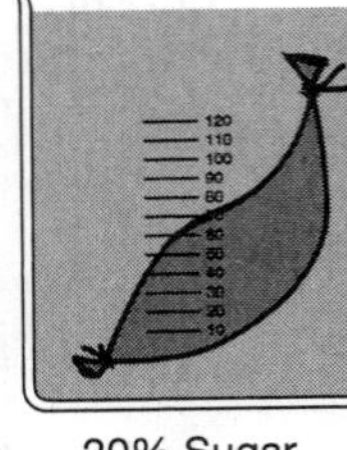
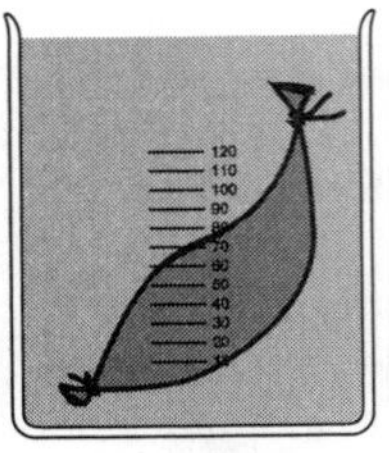
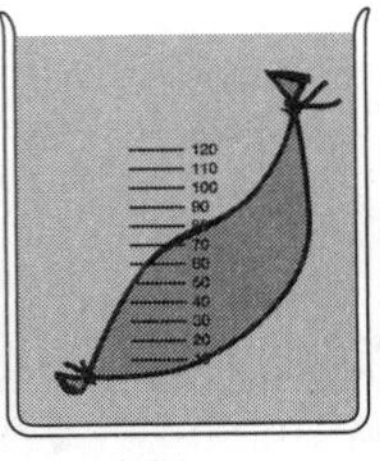
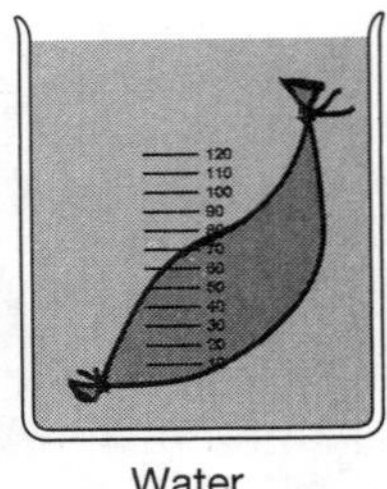

SAC / BEAKER	40% Sugar / Water	20% Sugar / Water	10% Sugar / 10% Sugar	Water / 20% Sugar	Water / 40% Sugar

● **FIGURE 9–5**
Artificial Cells.

	Weight at Start	*Weight at Finish*
#1 Sac—40% Sugar Beaker—Water		
#2 Sac—20% Sugar Beaker—Water		
#3 Sac—10% Sugar Beaker—10% Sugar		
#4 Sac—Water Beaker—20% Sugar		
#5 Sac—Water Beaker—40% Sugar		

● **FIGURE 9–6**
Artificial Cell Results.

tions from your instructor — remove the cells from the beakers, wipe the cells with a paper towel, and reweigh. Calculate the percent difference in weight for each cell. Chart your results on Figure 9–6●. What types of differences do you see in the different cells? How would you explain these differences?

___
___
___

5➡ Below is a partially constructed bar graph. Complete this bar graph to show the relative changes in the weights of the sacs.

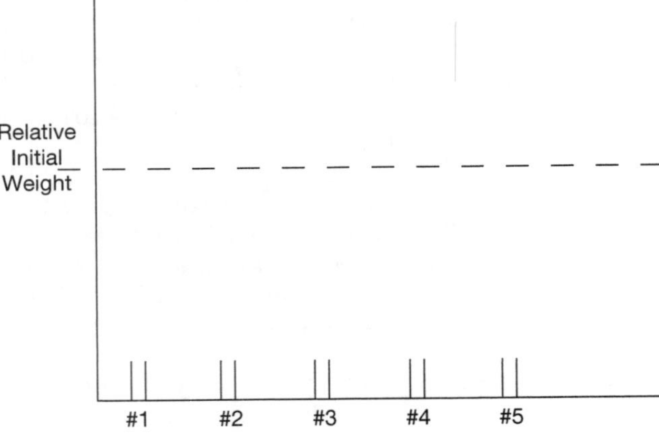

6➡ Repeat #1–4 above using time as a variable. Use the space at the end of this exercise to construct a graph demonstrating your findings.

G. Experiment #7: Osmosis (Red Blood Cells) [∞ *Martini, p. 74*]

A solute is any substance dissolved in any solvent. In biology, the solute is usually ionic, molecular, or particulate. The solvent is generally water or water-based. **Hypertonic**, **isotonic**, and **hypotonic** are terms used to describe the relationship between the solute concentrations on two sides of a membrane or gradient. Keep in mind that these are relationship terms. We shouldn't simply state, "Solution X is hypertonic." To show the relationship, we should say, "Solution X is hypertonic to Y."

Hyper- means greater (more, larger), so a hypertonic solution has a higher particulate concentration than the substance with which it is being compared. Iso- means same, so the two isotonic solutions would be of the same concentration. Hypo- means beneath (less, below), so a hypotonic solution has a lower particulate concentration than the substance with which it is being compared.

When particulate movement occurs between two concentrations, the net movement is always toward equilibrium. Note the words "net movement." If pore size is adequate, some movement against equilibrium does occur because molecules

● **FIGURE 9–7**
Osmosis and Red Blood Cells. White arrows indicate the direction of osmotic water movement. **(a)** Because these red blood cells are immersed in an isotonic saline solution, no osmotic flow occurs and the cells have their normal appearance. **(b)** Immersion in a hypotonic saline solution results in the osmotic flow of water into the cells. The swelling may continue until the cell membrane ruptures. **(c)** Exposure to a hypertonic solution results in the movement of water out of the cells. The red blood cells shrivel and become crenated. (SEM × 833)

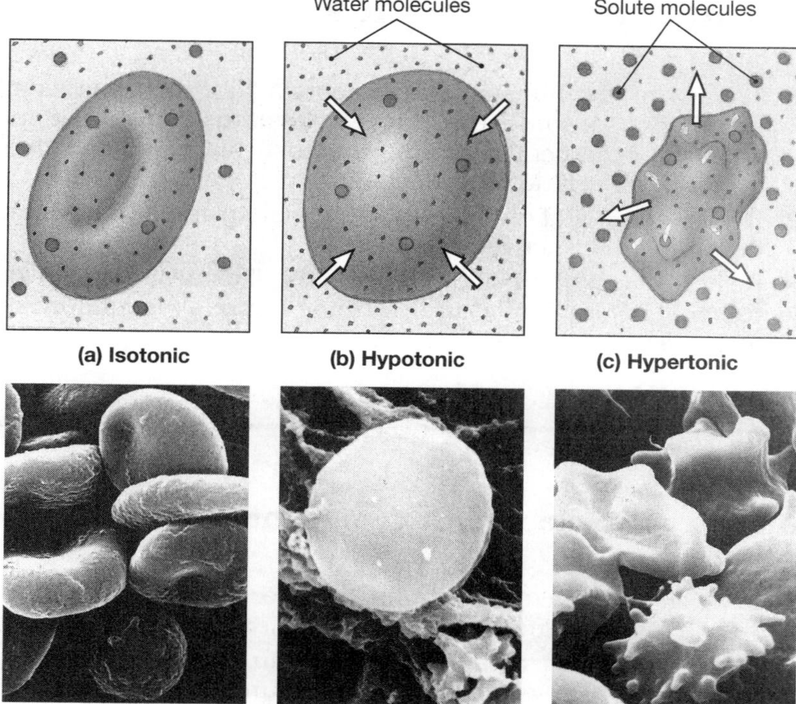

are in constant motion. We are concerned here with net movement. We are also concerned here only with passive movement. (Occasionally, cell membrane integrity or active transport mechanisms may influence net movement. We are not considering those aspects of movement here.)

Diseases such as cholera disrupt osmotic equilibrium. The cholera toxin causes changes in the membranes of the cells of the intestinal epithelium. These cells lose fluid to the lumen. Because the intestinal cells have become more hypertonic to their environment, water from the interstitium enters these cells. This water is also lost to the lumen. Because of the increased hypertonicity of the interstitium, water leaves adjoining cells and enters the interstitium and eventually the epithelial cells. Dehydration progresses rapidly. Diarrhea is a characteristic of cholera because of the massive water loss. The large intestine (which is basically unaffected by the cholera toxin) cannot reabsorb the water as rapidly as it is lost from the small intestine.

This experiment can be done with either animal or human blood. **IF YOU USE HUMAN BLOOD, PLEASE READ THE CAUTIONS GIVEN IN EXERCISES 46 AND 47.** The osmotic principles can be demonstrated equally well with animal blood, though you should be aware that the red blood cells of some animals, such as sheep, are smaller than human red blood cells.

1 ➡ Obtain the water and salt solutions, slides with cover slips, some blood, and a microscope. [If you are using your own blood for this experiment, EXTREME CARE must be taken.

a. Lance your finger quickly. A diagram is included in Exercise 47.
b. Use the lancet only once.
c. Dispose of everything according to your instructor's directions.]

2 ➡ Place a drop of blood on the slide.

3 ➡ Put a cover slip in place and observe under both low power and high power. This is your control so that you can observe blood cells in their natural state.

4 ➡ Place another drop of blood on a slide. Add several drops of the hypertonic 12–15% saline solution. Put the cover slip on the slide and observe under the microscope. Use low power first and then switch to high power. Since in any type of diffusion, the tendency is toward equilibrium, osmosis will take place as water leaves the cell. Crenation (shrinking) will occur. You should be able to observe this quite clearly. In which direction is the dilution occurring?

____

5 ➡ Repeat the above procedure with the isotonic and hypotonic solutions. What happened? Why? In which solution will swelling and subsequent hemolysis (lysis — bursting — of the red blood cell) occur? Why? See Figure 9–7●.

____

❑ Additional Activities

NOTES

1. Try any of the diffusion experiments again, this time using exact measurements. Compute the exact rate of diffusion for each substance. Use a variety of mathematical and/or statistical formulas to discover whether or not other relationships exist.
2. Design additional experiments to demonstrate movement across a membrane — either active or passive.
3. Refer to the experiment on red blood cell osmosis. What would you need to know if you wanted to determine the relative amount of water moving across the cell membrane?

Answers to Selected Concept Check Questions

1. In Flask #2 the neutral red must have been transported into the living yeast cell. In Flask #1 the yeast cells were killed so no transport could occur.
2. These pumps correspond to the sodium/potassium pump in the cell membrane. The sodium/potassium pump actively pumps three sodium ions out of the cell for every two potassium ions pumped into the cell, thus maintaining the membrane potential. You may wish to figure out other macroscopic situations where active energy input is required.
3. The potassium permanganate should have traveled about twice as far as the potassium chromate. The potassium chromate should have traveled just a bit further than the methylene blue.
4. Water will continue to enter the thistle tube containing the sugar. Eventually (perhaps not until the next day) the sugar water may spill out the top of the tube because equilibrium is never reached. The sugar molecule is too large to pass through the membrane so the water in the beaker will remain pure.

   Water will enter the salt thistle tube and you will note an initial rise in the water level, perhaps reaching half way up the tube. Then the water level will start to drop. The sodium and chloride ions of the salt are small enough to cross the membrane and enter the beaker. The water in the beaker will eventuallly be salty to the taste. Equilibrium will eventually be reached.

❑ Lab Report

1. Referring to the numbered inquiries at the beginning of this exercise, complete the following Box Summary:

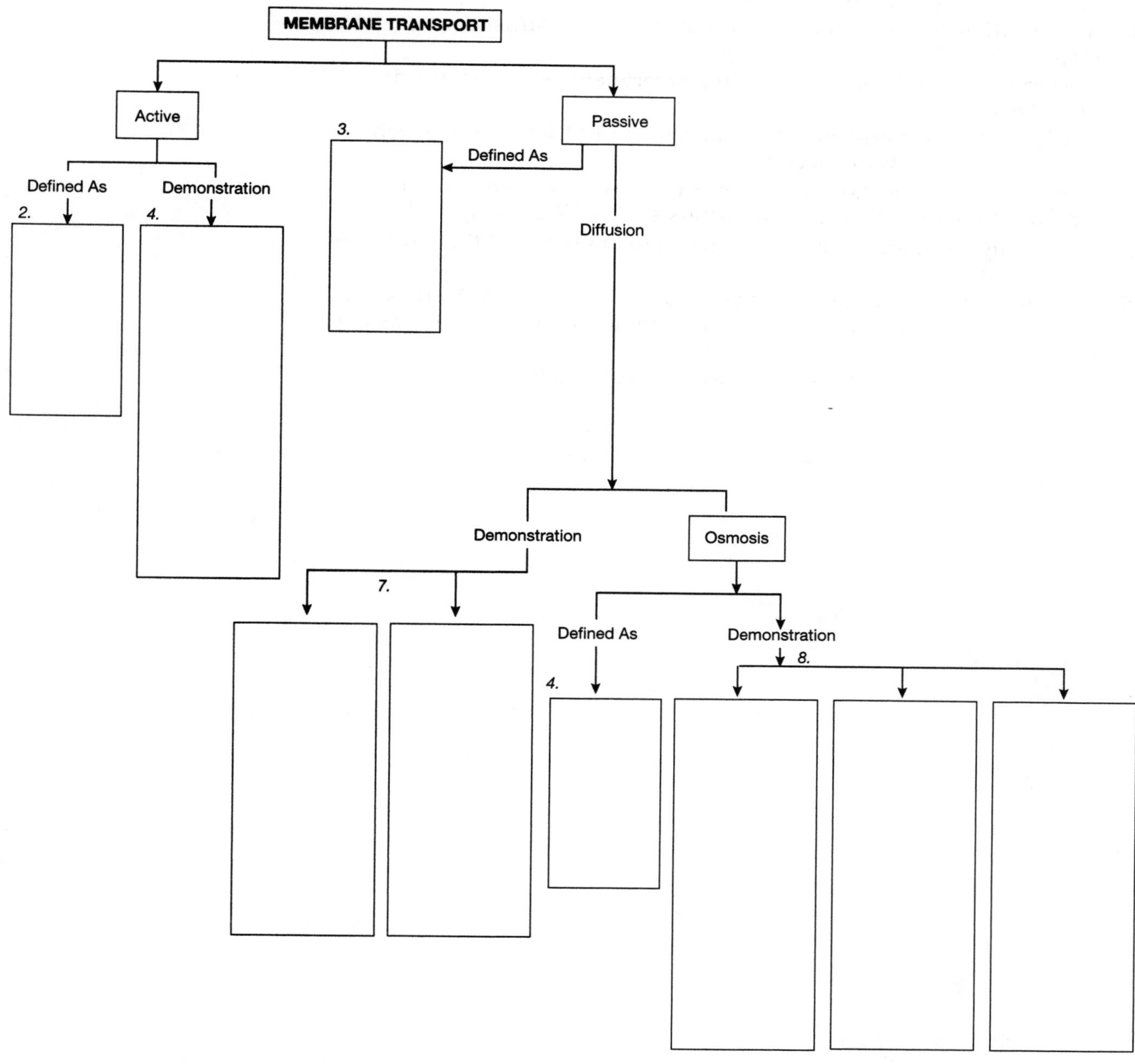

General Questions

NOTES

Use your own paper if necessary.

1. Define active and passive movement across a cell membrane.
2. Explain what happened to the yeast solutions. Why was the solution in Flask #2 a different color than the solution in Flask #1?
3. What caused the Brownian movement you observed?
4. Explain what happened when you dropped the potassium permanganate ($KMnO_4$) crystals into the water.
5. If your diffusion rates do not match the expected diffusion rates, how might you explain the discrepancy?
6. How was the thistle tube osmometer experiment related to our definition of osmosis?
7. In which of the experiments did osmosis occur? Dialysis? In which direction did these events occur?
8. Check your lecture text and find out what effective osmotic pressure is? How does it relate to cellular osmosis? *[∞ Martini, pp. 73]*
9. Explain hypertonic, isotonic, and hypotonic in terms of the red blood cell experiment.
10. List several factors (other than the ones we demonstrated by these experiments) that might influence the rate of either active or passive transport.
11. Why are sugar and salt used for food preservation?

EXERCISE 10

Cell Cycle: Mitosis and Cytokinesis

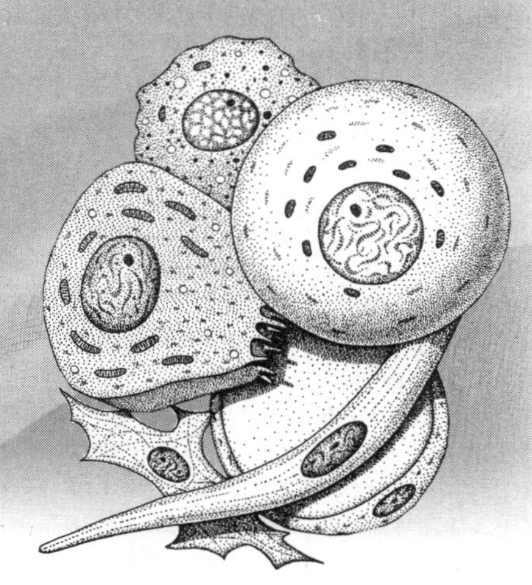

PROCEDURAL INQUIRIES

Preparation

1. What is the cell cycle?
2. What are the phases of the cell cycle?
3. What events occur during each phase of the cell cycle?
4. How does mitosis relate to the cell cycle?
5. How does interphase relate to the cell cycle?

Examination

6. What are the characteristics of prophase?
7. What are the characteristics of metaphase?
8. What are the characteristics of anaphase?
9. What are the characteristics of telophase?
10. What are the characteristics of cytokinesis?

Additional Inquiries

11. Why is the whitefish blastula a good specimen to use to study mitosis?
12. What is the difference between mitosis and cytokinesis?

Key Terms

Anaphase	G_1
Aster	G_2
Cell Cycle	G_m
Cell Plate	G_0
Centriole	Interphase
Centromere	Metaphase
Centrosome	Metaphase Plate
Chromatid	Mitosis
Chromatin	Prophase
Cleavage Furrow	S (Synthesis)
Cytokinesis	Spindle Fiber
Daughter Chromosomes	Telophase

Materials Needed

Compound Microscope
Plastic Models of Mitotic Stages
Prepared Whitefish Blastula Slides

The **cell cycle** is the life cycle of any cell — from the time that cell is formed until and including the time it goes through cell division.

The cell cycle is divided into **interphase** and **mitosis**. Until the development of the electron microscope and the perfecting of certain biochemical tests, all nonmitotic cells were said to be in interphase. We still use the term interphase for cells whose nucleus is not dividing. However, we know today that distinct events taking place during interphase can be identified.

The nonreplicating part of interphase is called $\mathbf{G_0}$. The other four major divisions of the cell cycle are: $\mathbf{G_1}$, where most of the growth, function, replication of extra organelles, and cellular development takes place; **S** (synthesis), where DNA is synthesized (or replicated); $\mathbf{G_2}$, where the products needed for mitosis are synthesized; and mitosis itself, where nuclear division actually occurs. Mitosis is sometimes called $\mathbf{G_m}$. It is in G_1 that "extra" organelles and intracellular material are produced so that the cell has the nongenetic components necessary to divide into two functional daughter cells.

Originally the "G" in G_1 and G_2 stood for "gap" because cytologists were uncertain about what was taking place during these phases. "Gap" has now been replaced with "growth" as the meaning for "G" because the "G's" are two very distinct growth phases.

Cytokinesis, the division of the cytoplasm into two separate cells, is a separate event that usually accompanies mitosis. Cytokinesis normally begins during late anaphase or early telophase and is thus usually studied along with mitosis. Keep in mind, however, that mitosis is a nuclear event and cytokinesis is a cytoplasmic event. Mitosis gives us two nuclei; cytokinesis gives us two daughter cells.

Mitosis is the only phase of the cell cycle that can be identified using the standard light microscope, which is why in former times the events of the G_1, S, and G_2 phases were lumped together under the term interphase. We also use the term interphase today when studying nonmitotic cells under the classroom microscope because we cannot distinguish G_1, S, and G_2 using light microscopy.

Cells that do not seem to be on the replication track are said to be in the G_0 phase. These cells continue to function as active dynamic cells, but they do not prepare for mitosis. This G_0 phase can be found just prior to the G_1 phase. Cells in this phase have the standard diploid number of chromosomes, and they carry on all the normal cellular metabolic activities but they do not produce "extra" organelles and "extra" cellular materials, as do cells preparing for mitosis. Cells in this G_0 phase include the nerve cells, which are normally in permanent arrest, and the bone cells, which replicate only when necessary. If a bone is broken, injured, or subjected to undue stress, the bone cells are stimulated to leave the G_0 phase and reenter the mainstream of the G_1 phase and shortly thereafter to undergo mitosis. Other cells, such as mitotic stem cells, are dividing continuously and thus never have a G_0 phase.

A few cells in the body seem to be in the G_0 phase but are apparently connected with the G_2 phase. These cells, which include the cells of the kidney tubules, have undergone DNA synthesis (and thus already have duplicated chromosomes) and have produced certain essential mitotic products. They are arrested but ready for mitosis. From a time standpoint, this could be vital should an injury to the kidney occur. Think about the need for immediate repair in a kidney as compared to the nonimmediate need for repair in a bone.

❑ Preparation

I. Background

A. Cell Cycle

1➡ Study Figure 10–1●, the cell cycle. Note the hours given on this figure. It is difficult to put the events of the cell cycle into a specific time frame. Some actively reproducing cells complete the cycle in just a few hours, while others take weeks, days, or even years. The times given on Figure 10–1● are to give you a perspective of what some typical cells do and are not meant to be absolutes. *[∞ Martini, p. 95]*

2➡ Notice where mitosis falls in the cell cycle. Mitosis can be divided into four stages or phases: **prophase, metaphase, anaphase,** and **telophase.** Although the primary emphasis of this exercise is mitosis, keep in mind that mitosis is only one part of the cell cycle.

❑ Examination

II. Mitosis and Cytokinesis

Use the plastic models of the stages of mitosis, along with the descriptions given here to work through the phases of mitosis and cytokinesis.

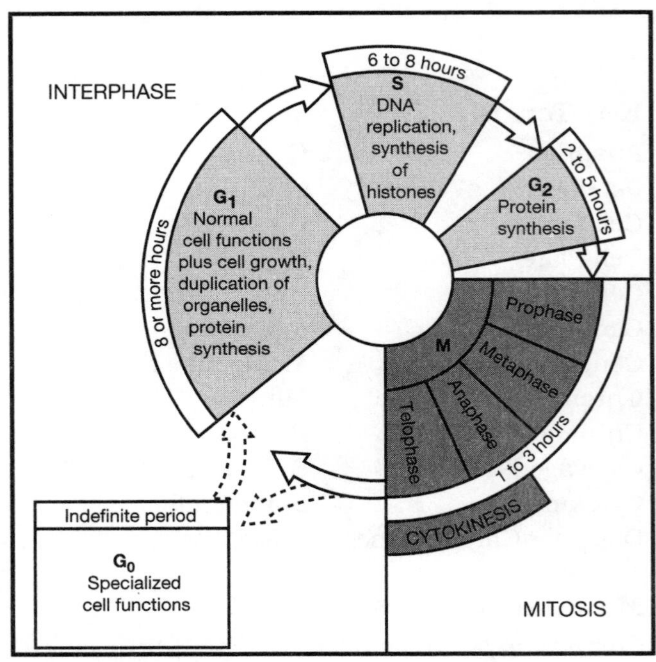

● **FIGURE 10–1**
Cell Cycle.

A. Mitosis [∞ Martini, p. 98]

1 ➡ Identify **prophase**. During prophase the DNA (which has been in a dispersed form called **chromatin**) condenses to become visible, short, thick, rodlike structures. Each chromosome appears to be two strands — **chromatids** — tied together by a knot-like structure, the **centromere**. (This makes sense since the chromatic material duplicated during the S phase of the cell cycle.) As prophase progresses, the nuclear membrane and the nucleolus gradually disappear.

Often prophase is subdivided into early, middle, and late prophase. Prophase is a proportionately long phase, and the events of prophase should be thought of as parts of a continuum and not as individual happenings.

Centrosomes are nonmembrane-bound areas containing **centrioles**. Centrioles are two complex microtubular structures at right angles to each other. In cells possessing **centrosomes** (virtually all animal cells and almost no plant cells), one centriole pair moves toward each pole during prophase. Gradually, additional microtubular structures known as **spindle fibers** and **asters** become visible, forming first at the centriole and extending toward the central plane of the cell. Toward the end of prophase, the duplicated chromosomes migrate toward this cellular equator.

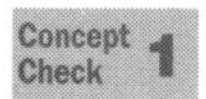

Recall: When did the DNA (the chromosomes) duplicate?

____

____

2 ➡ Identify **metaphase.** Metaphase is a very short and specific phase. During metaphase the duplicated chromosomes are randomly lined up on the **metaphase plate**. The spindle fibers attach to the centromeres and the asters radiate from the centrioles.

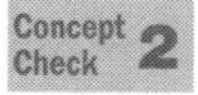

Logically, where did the metaphase plate get its name?

____

____

3 ➡ Identify **anaphase.** The key point in anaphase is the separating of the chromatids. As these chromatids pull apart, they are technically known as **daughter chromosomes**. Although the centromere actually divided during the S phase, it is only during anaphase that this separation is obvious. The centromeres with attached spindle fiber lead the way as the daughter chromosomes begin to migrate toward the poles.

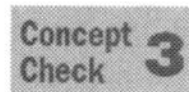

What would happen if a pair of chromatids did not pull apart correctly?

____

____

4 ➡ Identify **telophase**. Telophase begins as the new nuclei begin to form. The chromosomes become less distinct, the new nuclear membranes form, and the nucleoli reappear. The spindles and aster fibers disappear. Telophase is essentially the reverse of prophase.

B. Cytokinesis [∞ Martini, p. 99]

1 ➡ Identify **cytokinesis**. Cytokinesis usually begins during late anaphase or early telophase. In the animal cell cytokinesis is marked by a "pinching in" of the cell membrane. This pinching in is a constriction or **cleavage furrow**. Gradually the constriction is complete and the two daughter cells become separate entities. [In the plant cell cytokinesis begins with the formation of a **cell plate**, which originates in the center of the cell.]

Mitosis is said to be complete when the nuclei have reformed and the chromatin (the dispersed DNA) in these nuclei is nondistinct. In other words, the new daughter cells now appear to be in interphase.

III. The Cell Cycle in the Whitefish Blastula

A. Background Information

The blastula is the ball of cells formed when the zygote (the single cell resulting from the union of the egg and sperm) undergoes continuous rapid nuclear and cellular division. Because this division is nonsynchronous in most animals, examining a cross-sectional slice of a blastula will reveal cells in all stages of the cell cycle. Whitefish blastula are numerous, easy to stain, and easy to use. Whitefish blastula events are also highly representative of animal (including human!) mitosis.

B. Observing Whitefish Blastulae

1 ➡ Obtain a whitefish blastula slide and focus it under low power. Use low power to scan the slide. Use Figure 10–2● to orient yourself.

2 ➡ Identify the individual cells. Also identify the chromosomes. Your slides have been specifically stained so the chromatic (chromosomal) material will stand out, usually as a dark purple. Some cells may appear to be in-

● **FIGURE 10–2**
Whitefish Blastula.

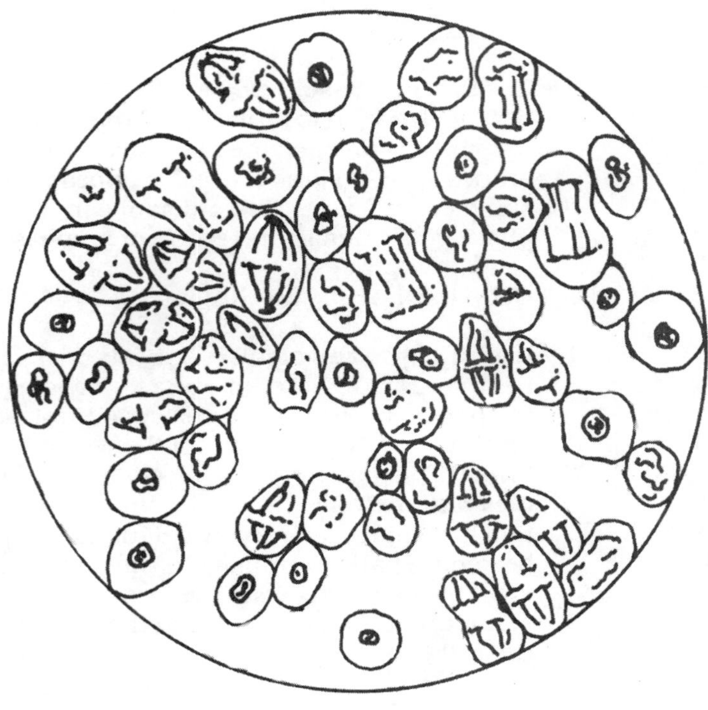

complete. Keep in mind that the blastula is a ball of three-dimensional cells. To make your slide, a slice was taken through the blastula ball. Different cells were sliced at different points in their three-dimensional structure. Note the cell membrane. Note too the nuclear membrane. Be sure you are clear about what is the cell itself and what is the nucleus. Again, use Figure 10–2● as your guide.

3➡ Find examples of each of the four stages of mitosis. Use the following descriptions and Figure 10–3● to help you. Make note of when you are first able to detect cytokinesis. You should be able to find all mitotic phases on the same slide.

4➡ Identify the other cells on the slides as being in interphase. You should be able to distinguish the nucleolus in many of the interphase cells.

5➡ Use the spaces provided on blank Figure 10–4● (p. 88) to sketch the phases of mitosis. Since prophase is a proportionately long phase, you might be able to distinguish more than one part of prophase. If so, use two circles.

6➡ Label each sketch with the appropriate boldface terms found in Section IIA.

● **FIGURE 10–3**
Mitosis.

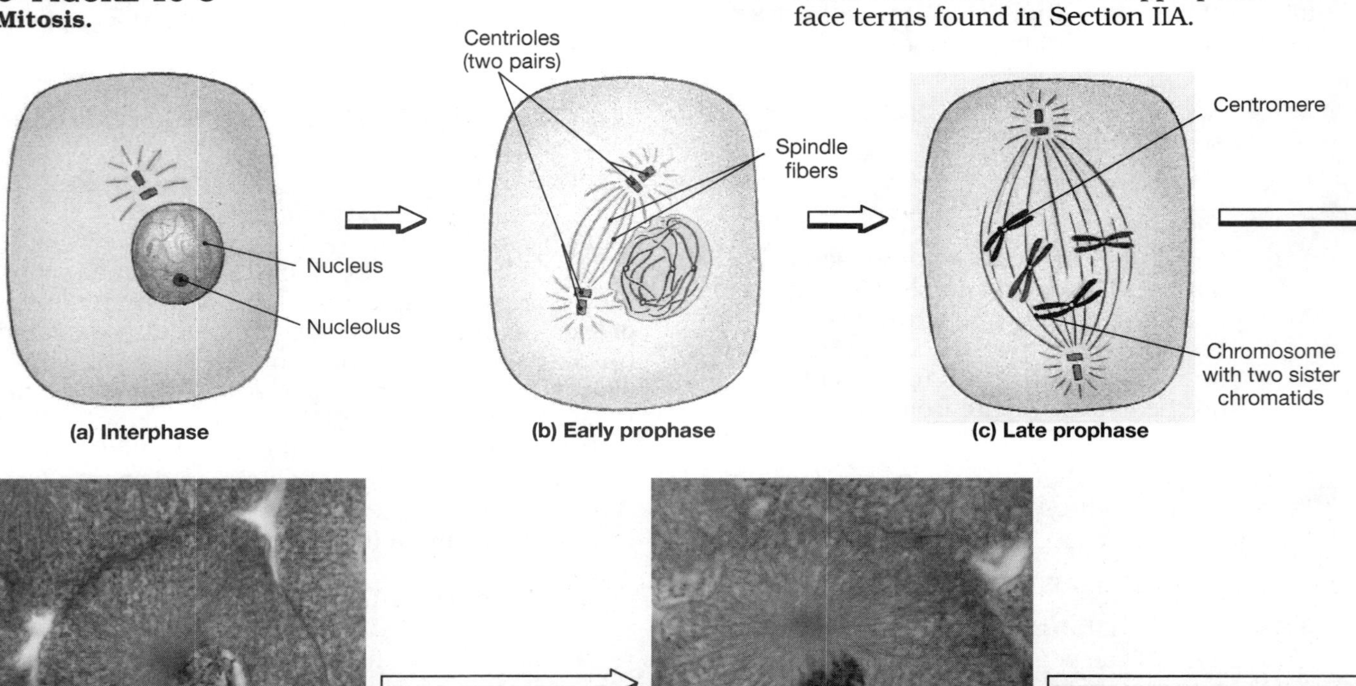

❑ Additional Activities

NOTES

1. Contrast mitosis and cytokinesis in animals, plants, fungi, and protistans. Check out similarities and differences between different species of animals that are closely related to humans.
2. Research the G_0 phase and explain the advantages to the organism that this phase might have.
3. If you are having difficulty conceptualizing mitosis, use toothpicks and glue to construct the phases of mitosis in sequence.

Answers to Selected Concept Check Questions

1. DNA duplicated during the S phase of the cell cycle.
2. The **equatorial plate** seems to form an equator around the middle of the cell.
3. A chromosome mutation — a cell with too many or too few chromosomes — would result. This incomplete separation of the chromosomes is sometimes known as anaphase lag.

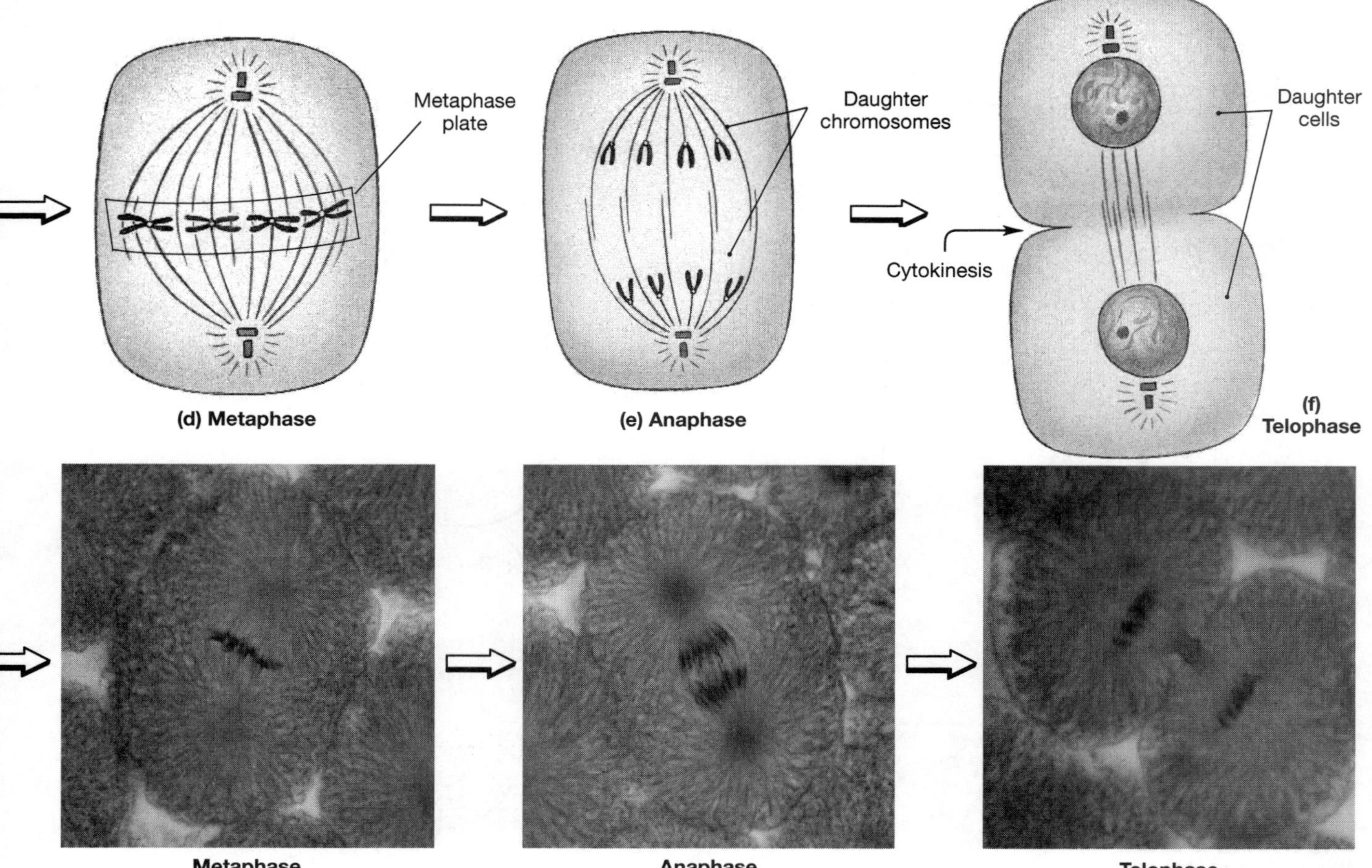

(d) Metaphase (e) Anaphase (f) Telophase

Metaphase Anaphase Telophase

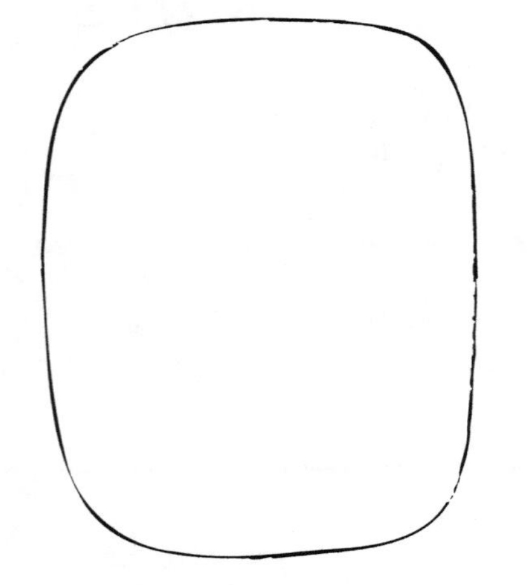

1. ______________________

2. ______________________

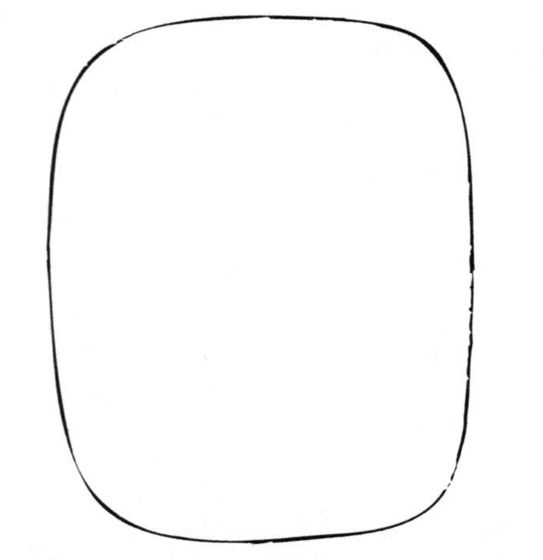

3. ______________________

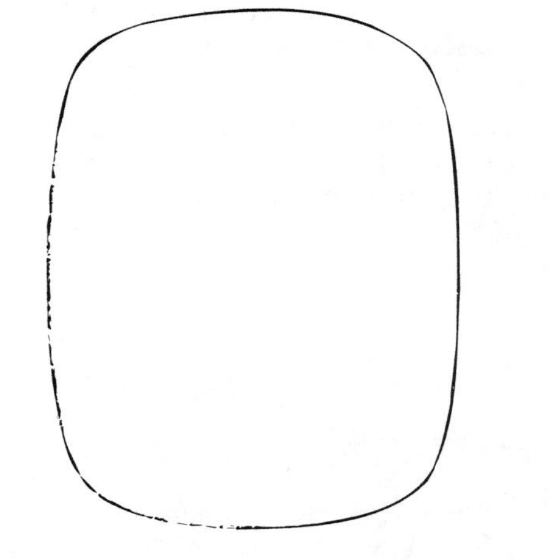

4. ______________________

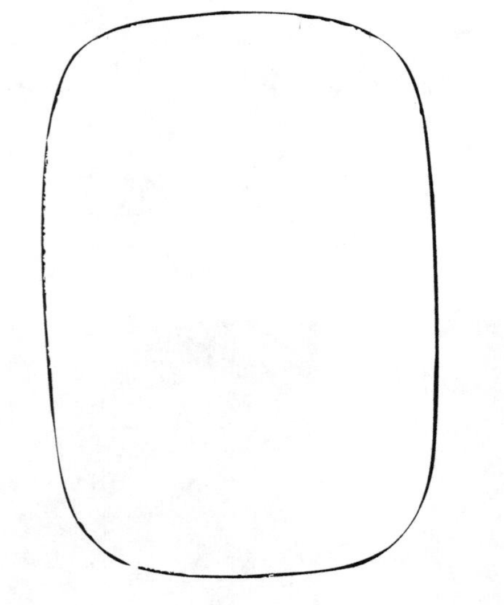

5. ______________________

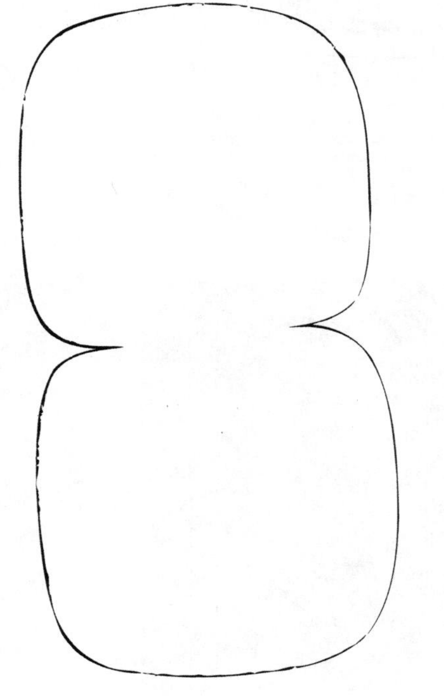

6. ______________________

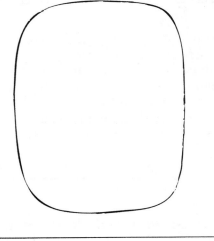

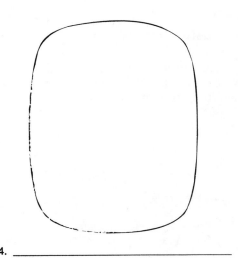

● FIGURE 10–4
Mitosis Sketches.

❑ Lab Report

1. Referring to the numbered inquiries at the beginning of this exercise, complete the following box summary:

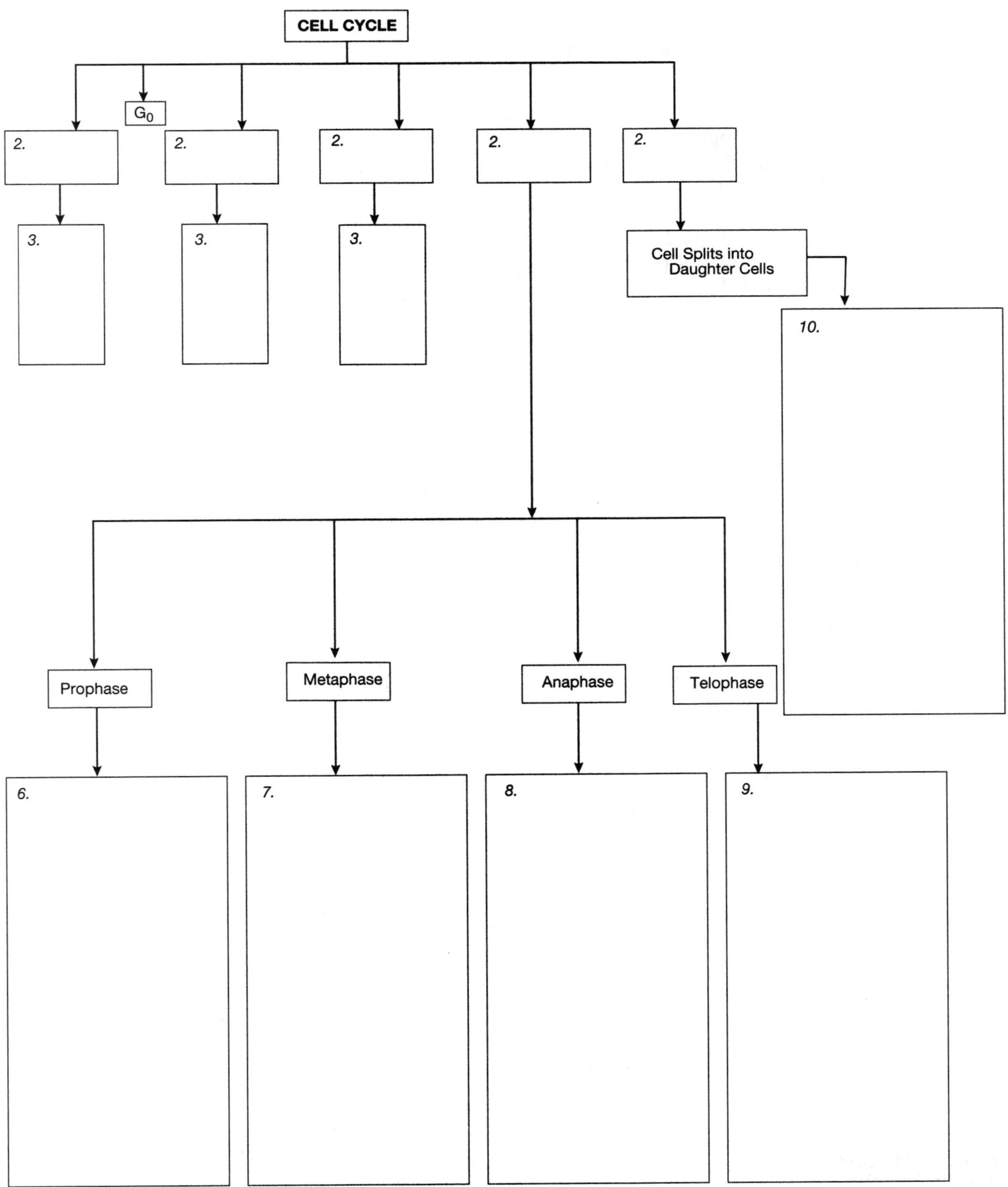

General Questions

NOTES

1. Explain the difference between mitosis and cytokinesis.
2. Why should mitosis and cytokinesis be considered separately?
3. During which phase of mitosis do the following occur?
   a. Chromatids separate
   b. Chromatids lined up on equatorial plane
   c. Nuclear membrane disappears
   d. Nuclear membrane reforms
   e. Nucleoli reappear
   f. Chromatids begin to appear as thick, rod-like structures
4. During which phase or phases of mitosis should the following terms be used? How should the terms be used?
   a. Chromosome
   b. Equatorial plate
   c. Chromatid
   d. Spindle fibers
5. Distinguish between centrosome, centromere, and centriole.
6. What is a cleavage furrow?
7. What is interphase?
8. What is the whitefish blastula and why do we use it to study human mitosis?

EXERCISE 11

Histology

PROCEDURAL INQUIRIES

Preparation

1. What procedure should be followed in working with histological slides?
2. What should histological drawings reflect?
3. What terms are used to describe the distribution of layers of epithelial tissue?
4. What are the four major tissue types?

Examination

5. What are the types of epithelial tissue?
6. What are the major types of connective tissue?
7. What are the types of muscle tissue?
8. What are the types of neural tissue?

Additional Inquiries

9. What is histology?
10. Where would you find a basement membrane?
11. How is the basement membrane formed?
12. What is the difference between keratinized and non-keratinized tissue?

Key Terms

Adipose
Basement Membrane
Blood Cells
Bone
Cardiac Muscle
Cartilage
Collagen
Columnar
Connective
Cuboidal
Dense Connective
Elastic
Embryonic
Epithelial
Histology
Keratinized
Loose Connective
Muscular
Neural
Neuroglia
Neuron
Nonkeratinized
Pseudostratified Columnar Epithelial
Reticular
Simple
Skeletal Muscle
Squamous
Stratified
Tissues
Transitional Epithelial
Visceral Muscle

Materials Needed

Compound Microscope
Prepared Microscope Slides of

Epithelial Tissue

Simple squamous, Simple cuboidal, Simple columnar, Stratified squamous, Stratified cuboidal, Stratified columnar, Pseudostratified columnar, Transitional

Connective Tissue

Loose connective (areolar), Dense connective, Adipose, Elastic (if available), Reticular (if available), Hyaline cartilage, Fibrous cartilage, Elastic cartilage, Bone, Blood

Muscle Tissue

Striated, Visceral, Cardiac

Nerve Tissue

Neural

Histology is the study of **tissues**. A tissue is a group of similar cells and cell products carrying out a common function. The term histology comes from the Greek word "histos" meaning tissue or web and the prefix "histo-" can be used to indicate any fact or condition relating to a tissue.

We generally think in terms of four major tissue types — **epithelial**, **connective**, **muscular**, and **neural**. Epithelial tissue functions as the covering or the lining for most body structures. Thus, epithelial tissue always has a free surface. Connective tissue forms the body's structural framework. Muscle is the contractile tissue, and neural tissue is the primary conduction or informational tissue.

Each of these tissue types can be divided into numerous subtypes, based on the increased specialization of the cells involved. An individual tissue sample is usually made up of more than one cell type. Consider blood, for instance. Blood includes red blood cells, several types of white blood cells, platelets (which are not cells), and such cell products as dissolved fibers, antibodies, and clotting factors. Blood is obviously not a homogeneous structure.

The concept of tissue also crosses organ lines. For instance, most of the major organs of the body are made up of more than one tissue type. Think about a large muscle. The muscle fibers are made up of muscle tissue. This muscle also has a protective boundary composed of epithelial tissue and tough sheaths of connective tissue. The muscle contracts because of an electrical stimulation transmitted by neural tissue.

❑ Preparation

I. Approaching Histology

A. Purpose

The purpose of this laboratory exercise is to gain a gross understanding of histology by examining prepared slides of the major tissue types and subtypes.

1➡ Use this exercise as a foundation for your forthcoming work on the systems of the body. As you work through the succeeding exercises in this book, you will notice continued reference to histological slides. You may often be asked to reexamine slides or tissue types you have studied in this exercise.

2➡ Keep in mind that tissue types do not function as isolated entities. Therefore, you will rarely view a slide showing only one tissue type. To see this idea clearly, examine the borders of the tissue samples. Various cell types work together for the common tissue function and various tissues work together for the common organ or system function. (Recall Exercise 7.)

B. Microscopy

The following procedure should be used as you work through each section in this histology exercise.

1➡ Begin your microscopic examination of the histological slide with the lowest power lens possible. The lower power lenses will give you a perspective of the entire structure. This perspective is especially important with tissue slides; first, because more than one tissue type may be present on the slide and second, because you often need an orientation as to the deep and superficial aspects of the structure.

2➡ Switch to the higher powers for a more detailed analysis of some aspect of the structure after you have gained the proper histological perspective. Except for the blood slide, you should not need to use the oil immersion lens for any of the histological slides listed in this exercise.

3➡ Compare your slide with the indicated illustration.

4➡ Sketch and label a representative sample of what you are observing. Your tissue sketches should reflect both perspective and detail.

❑ Examination

II. Epithelial Tissue *[∞ Martini, p. 107]*

A. Overview

The epithelium is the layer (or layers) of cells forming the secured boundary between a structure and its environment. Because of the varied nature of the organs of the body, different types of epithelial tissue can be found in different locations. Two terms commonly used with epithelial tissue are **simple** and **stratified**. Simple means the tissue occurs in a single layer of cells, whereas stratified means the cells are stacked in layers or strata.

1➡ Study Figure 11–1●. This classification chart gives you an overview of the subdivisions of epithelial tissue. Regardless of type, however, all epithelial tissues rest on a **basement membrane**. This basement membrane is an acellular structure formed by filaments exuded from the epithelial cells and fibers produced by the connective tissue immediately beneath it. The epithelial portion of the basement membrane is known as the basal lamina and the connective tissue portion is the reticular lamina.

B. Squamous Epithelium [∞ Martini, p. 111]

Squamous epithelial cells are flat and somewhat irregularly shaped. The **simple squamous epithelial** cells, such as those found in the mesothelia and endothelia, occur in a single layer. The

Cell Shape	*Simple*	*Stratified*
Squamous	Type: simple squamous Examples: mesothelia, endothelia Functions: reduces friction, permits absorption/secretion in protective environments	Type: stratified squamous Examples: surface of skin, both ends of digestive tract Function: protection
Cuboidal	Type: simple cuboidal Examples: glands, ducts, urinary tract Functions: limited protection, secretion, and/or absorption	Type: stratified cuboidal Example: lining of some ducts (rare) Functions: protection, secretion, and/or absorption Special variation: transitional Examples: urinary bladder, portions of urinary tract Functions: permits expansion and contraction
Columnar	Type: simple columnar Example: lining of digestive tract Functions: protection, secretion, absorption Special variation: pseudostratified columnar Examples: portions of respiratory and male reproductive tracts Functions: protection, secretion	Type: stratified columnar Examples: large ducts and passageways of respiratory tract Function: protection

● **FIGURE 11–1**
Classification of Epithelial Types.

terms mesothelium and endothelium refer to the linings of the ventral body cavity and the vascular systems, respectively.

The stacked layers of **stratified squamous epithelial** cells are found on the surface of the skin and the linings of the mouth, esophagus, and anus. Skin cells are **keratinized**, which makes them tough and water-resistant. Keratin is the sulfur containing fibrous protein that is characteristic of much of the integument.

Other stratified squamous epithelial cells, such as those lining the oral cavity, esophagus, and pharynx, are **non-keratinized**.

1➡ Obtain and observe slides of the all types of squamous epithelium. Compare your slides with Color Plate A (found in the Histology Atlas immediately following Chapter 11).

2➡ Sketch and label what you see.

Squamous epithelium

3➡ Speculate on why the structures of these cells are ideal for their respective functions.

____

____

C. CUBOIDAL EPITHELIUM [∞ MARTINI, P. 112]

Cuboidal epithelial cells are generally cube-shaped cells occurring in a single layer in glands, ducts, and tubes. These are the **simple cuboidal epithelial** cells. In a few ducts, such as sweat gland ducts and the larger mammary gland ducts, cuboidal cells do occur in strata. These are the **stratified cuboidal epithelial** cells.

1➡ Obtain and observe slides of both types of cuboidal epithelium. Compare your slides with Color Plate B.

2➡ Sketch and label what you see.

Cuboidal epithelium

3➡ Speculate on why the structure of these cuboidal cells is ideal for its function.

____

____

D. COLUMNAR EPITHELIUM [∞ MARTINI, P. 112]

Columnar epithelial cells are elongated rectangular cells. **Simple columnar epithelial** cells line the digestive tract while **stratified columnar**

epithelial cells can be found in the large ducts of the respiratory tract.

1➡ Obtain and observe slides of both types of columnar epithelium. Compare your slides with Color Plate C.

2➡ Sketch and label what you see.

Columnar epithelium

3➡ Speculate on why the structure of the columnar cells is ideal for its function.

____

____

E. Pseudostratified Columnar Epithelium

Pseudostratified columnar epithelial cells line the trachea and portions of the male reproductive tract. This tissue type is distinguished because the cells, although they appear to be stratified, all rest on a common boundary, the basement membrane.

1➡ Refer to Color Plate D and identify pseudostratified cells by looking for the common interior (free) boarder and for at least two distinct rows of nuclei. The pseudostratified epithelial cells also tend to be ciliated. (Pollutants can destroy these cilia!)

2➡ Obtain and observe slides of pseudostratified columnar epithelium. On your slide the cilia border will probably appear thicker than the other cell boundaries. Depending on the stain used, the cilia may appear darker or lighter than the rest of the cell. The cilia border may also appear somewhat fuzzy. Compare your slides with Color Plate D.

3➡ Sketch and label what you see.

Pseudostratified columnar epithelium

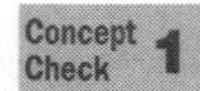

What is the function of the cilia on these epithelial cells?

____

____

F. Transitional Epithelium

Transitional epithelial cells are technically classified as stratified cuboidal cells because epithelial structures are classified according to their most external cell layer. For our purposes, however, we will continue the more conventional "transitional" classification. These cells, which line the urinary bladder, change in appearance from their relaxed to their stretched state.

1➡ Obtain and observe slides of the transitional epithelium. Compare your slides with Color Plate E.

2➡ Sketch and label what you see.

Transitional epithelium

3➡ What is the functional advantage of having transitional epithelial cells in the urinary bladder?

____

____

G. Exocrine Glands

Exocrine glands, the glands that secrete substances into ducts or directly onto the body surface, are epithelial structures because they are specialized invaginations of the surface tissue. We will study these exocrine glands, including the sweat glands, the mammary glands, and the various glands of the digestive system, as we examine the organs and systems with which they are associated.

Endocrine glands, the ductless glands, are either epithelial or neural in origin. We will consider the endocrine glands in Exercises 44 and 45.

III. Connective Tissue [∞ *Martini, p. 117*]

A. Overview

At first glance, connective tissue seems to be a hodgepodge category because it includes so many apparently diverse cell types. However, all connective tissue has a common embryonic origin, and the common functions of the connective tissues are to provide the body with a structural framework; a transport mechanism; a protective, support, and defense system; and a storehouse for energy reserves. In one sense the ubiquitous connective tissues are a unifying factor for the entire body.

1➡ Study Figure 11–2● below. Notice that all connective tissues have three common properties: cells, extracellular fibers, and a fluid (or ground tissue). The fibers and the ground tissues together are often called the **matrix**. The common properties of all connective tissues — the cells, the fibers, and the fluid — can be found in varying proportions in different connective tissue types and in different connective tissue samples.

B. Embryonic Connective Tissue

The **embryonic** connective tissue can be either **mesenchyme** or **mucous**. Mesenchyme is embryonic tissue developing from the mesodermal germ layer and giving rise to all postembryonic connective tissues. (Thus we can make the previous statement that all connective tissue has a common origin.) Mesenchymal cells are star-shaped. The mesenchyme ground fluid is gel-like and contains numerous fine fibers. The adult system no longer includes mesenchymal tissue per se, but mesenchymal cells can be found in certain adult connective tissues and are responsible for connective tissue regeneration.

Mucous connective tissue is a temporary, mesenchymally derived connective tissue appearing in very small quantities in the fetus. Wharton's jelly in the umbilical cord is an example of mucous connective tissue.

1➡ Study embryonic connective tissue as indicated by your instructor. We will be examining certain embryonic connective tissues in more detail in Exercise 19.

C. Loose Connective Tissue [∞ Martini, p. 120]

Loose connective (**areolar**) tissue is the "filler" tissue, taking up the space between the body's organs and tissues.

1➡ Examine a slide of areolar tissue, and identify the thick, white **collagen** fibers; the thin, yellow-to-dark-brown **elastic** fibers; the darkly stained **mast cells** (which release histamine and heparin); and the light-colored, expansive **fibroblasts** (which produce the areolar fibers). You may also see some white blood cells, some adipocytes, and perhaps even some blood vessels replete with red blood cells. Use Color Plate F as your guide.

2➡ Sketch and label what you see.

Areolar tissue

● **FIGURE 11–2**
Major Types of Connective Tissue.

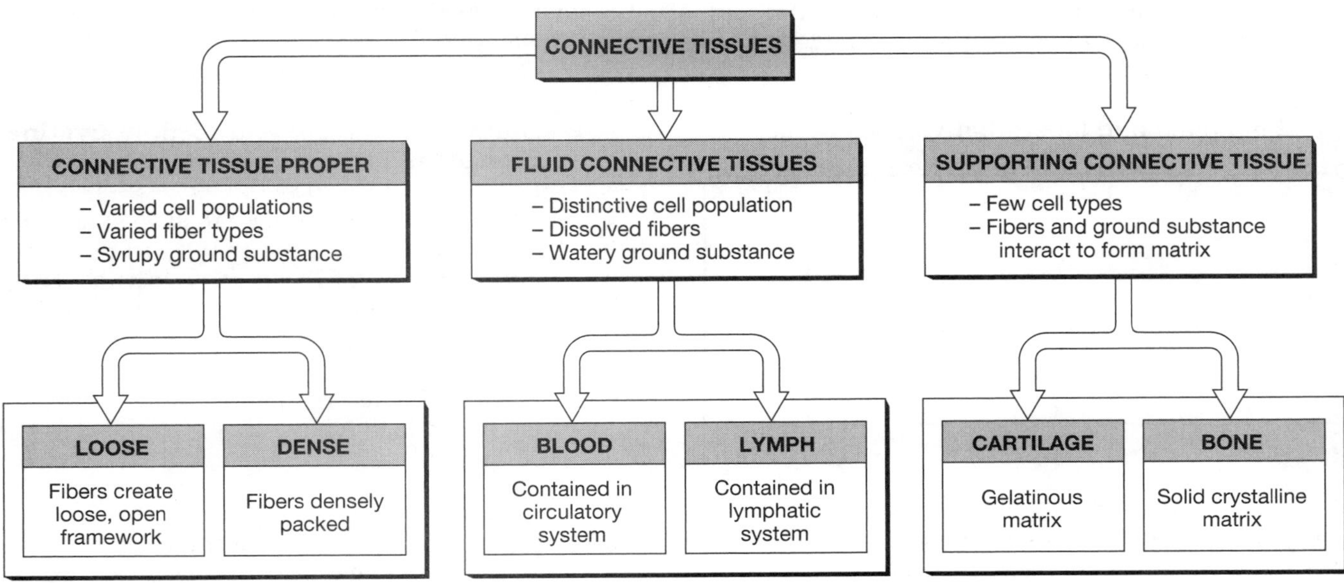

D. Dense Connective Tissue

Dense (white fibrous) connective tissue can be either **regular** (with fibers arranged in an orderly fashion) or **irregular** (with fibers forming an interwoven meshwork). You will find dense connective tissue as a component of ligaments, tendons, aponeuroses, fascia, and organ capsules.

1➡ Study a slide of dense connective tissue. First make note of whether the fibers are regular or irregular. Then identify collagenous fibers and the nuclei of the fibroblasts. Use Color Plate G as your guide.

2➡ Sketch and label what you see.

Dense connective tissue

E. Adipose Tissue [∞ Martini, p. 121]

Adipose (fat) tissue functions to cushion, to insulate, to store energy. You will easily recognize the adipocyte by its large, blandly colored fat vacuole and its minimal quantity of peripheral cytoplasm. An elongated nucleus often seems to be squashed to one side of the cell. Adipocytes are actually modified fibroblasts. Adipocytes often resemble signet rings and are thus sometimes called signet ring cells.

1➡ Obtain a slide of adipose tissue and identify the fat vacuole, cytoplasm, and nucleus of the adipocyte. Use Color Plate H as your guide.

2➡ Sketch and label what you see.

Adipose tissue

Concept Check **2** Do you see any other tissue types on the slide with the adipocytes? ______________

As you continue your histological journey, make note of other tissue slides on which you find adipocytes.

3➡ Speculate on why the structure of the adipocyte is ideal for its function.

____

____

F. Elastic Connective Tissue

Elastic connective tissue lies beneath the transitional epithelium of the urinary system and is a structural component of the respiratory and circulatory systems. Elastic tissue also forms the ligaments interconnecting the vertebrae.

1➡ Follow your instructor's directions for viewing elastic connective tissue.You have already viewed elastic fibers as part of the loose connective tissue smear.

G. Reticular Connective Tissue

Reticular connective tissue forms the **stroma** or framework of the liver, spleen, lymph nodes, and bone marrow. Reticular fibers are essentially collagen fibrils interwoven to form a tough, flexible network.

1➡ Follow your instructor's directions for viewing or identifying specific reticular tissues. If you do not examine specific reticular tissue slides, make note of these tissues when you study those organs just mentioned.

H. Cartilage [∞ Martini, p. 125]

Cartilage is an avascular (without blood vessels) support tissue surrounded by a fibrous, double-layered **perichondrium**. **Chondrocytes** (cartilage cells) dwell in **lacunae** (sing. lacuna), small pockets interspersed in a gel-like matrix.

1➡ Study Color Plate I and identify **hyaline**, **elastic**, and **fibrous** cartilage.

2➡ Obtain the appropriate cartilage slides and locate the matrix, lacunae, and chondrocytes in each of the three cartilage types. The collagen and elastic fibers of the matrix will probably not be visible on your slide, but make note of where they should be. What is the distinguishing characteristic of each cartilage type?

Hyaline ______________________________

Elastic ______________________________

Fibrous ______________________________

3➡ What other tissue types are present on your slide? (The hyaline cartilage slide in particular may also show smooth muscle and adipose cells.)

____

____

4➡ Sketch and label what you see.

Hyaline Elastic Fibrous

Cartilage tissue

I. Bone [Martini, p. 128]

The specifics of **bone** as a highly specialized connective tissue will be considered in the exercises on the skeletal system. At this point we will simply examine the **osteon** or **haversian system**, the structural unit of compact bone tissue.

1➡ Obtain a slide and identify the structures labeled in Color Plate J.

2➡ Sketch and label what you see.

Bone tissue

J. Vascular Tissue

In previous times **vascular tissue** was considered as a separate tissue type. We now think of vascular tissue as one of the fluid connective tissues. **Hemopoietic** cells can be found in the reticular connective tissue of the bone marrow, spleen, and lymph nodes. These are the cells that produce the blood cells.

1➡ Examine a blood smear slide and locate the **red blood cells (erythrocytes)**, which stain as small pink or gray disks, the **white blood cells (leukocytes)**, which stain as larger cells with dark purple nuclei, and **platelets** (thrombocytes), which appear to be fragmented cell pieces. (If you need to review the use of the oil lens, see Exercise 3.) Use Color Plate K as your guide. Mature **blood cells** will be studied in greater detail in a later exercise.

2➡ Sketch and label what you see.

Blood smear

IV. Muscle Tissue *[Martini, p. 132]*

Each of the body's three types of contractile tissue, the muscular tissue, has a unique structure that is directly related to its function. Each of these muscle types will be considered in greater detail in later sections of this book. This section on muscle tissue is an overview.

A. Striated Muscle

Striated or **skeletal muscle**, which is attached either directly or indirectly to the skeleton, has **striations** or **striae**. These distinctive cross bands allow the muscle to function as a unit, thus affording speed, strength, and control over movement. Skeletal muscle cells appear to be long, fibrous, multinucleated strands.

1➡ Obtain a slide of striated muscle and identify the striations and the peripheral nuclei. Use Color Plate L as your guide.

2➡ Sketch and label what you see.

Striated muscle tissue

B. Visceral Muscle

Visceral or **smooth muscle** cells, which basically line most of the involuntary organs of the body, are slender, tapered, and uninucleate. The cells of this tissue may be found either singly or as part of a muscular sheet.

1➡ Obtain a slide of visceral muscle tissue and look for the cell membrane and the centrally located nucleus. No striations are present. Use Color Plate L as your guide.

2➡ Sketch and label what you see.

Visceral muscle tissue

C. Cardiac Muscle

Cardiac muscle cells, which make up the bulk of the heart, are uninucleate, striated, and extensively branched. Between the cardiac cells you will find a darkly stained structure called an **intercalated disk**.

1➡ Obtain a slide of cardiac muscle tissue and identify the striations, the nuclei, the intercalated disks, and any branched cells that may be discernible. Use Color Plate L as your guide. You may also see remnants of the fibrous connective tissue network, which helps to maintain the structural integrity of the heart. On gross examination, cardiac cells appear to be bands of tissue encircling the heart. So, depending on the way your particular heart tissue was sliced, you may see unconnected sections of tissue on your slide. Do not be concerned about this; just look for the single nucleus in the striated cells.

2➡ Sketch and label what you see.

Cardiac muscle tissue

V. Neural Tissue *[∞ Martini, p. 134]*

Neural tissue is the body's primary conducting tissue. The two types of neural tissue are the **neurons**, which transmit electrical information along dendritic and axonic pathways, and the **neuroglia**, which connect, support, and nourish the neurons. We will only consider the neurons in this overview lesson. Neurons, as well as neuroglia, will be studied in greater detail in the nervous system exercises.

A. Generalized Neuron

The generalized neuron is composed of a cell body with a number of short, spindly projections (the dendrites) and a single, longer, more prominent projection (the axon). The dendrite carries electrical information toward the nerve cell body, and the axon carries stimuli away from the nerve cell body.

1➡ Obtain and examine a spinal cord smear slide. You should notice that large neurons are scattered throughout the smear. Identify the cell body, nucleus, axon hillock, axon, and dendrites. The axon hillock is the portion of the neural cell body immediately adjacent to the initial segment of the axon. Under high power you should also be able to identify the neurofibrils — long, thin, darkly stained fibers located in the cytoplasm. Use Color Plate M as your guide.

2➡ Sketch and label what you see.

Spinal cord smear

❑ Additional Activities

NOTES

1. Construct a chart showing *interactions* between tissue structures and tissue functions.
2. Explore the staining techniques used to stain different tissue types. What chemical characteristics make given techniques ideal for given tissue types?

Answers to Selected Concept Check Questions

1. Cilia help move substances along a pathway.
2. Unless your slide is of only adipose tissue, you will see other cell or tissue types. Often in examining slides of other tissues and organs, you will find adipocytes along connective tissue boarders.

❑ Lab Report

1. Referring to the numbered inquiries at the beginning of this exercise, complete the following box summary:

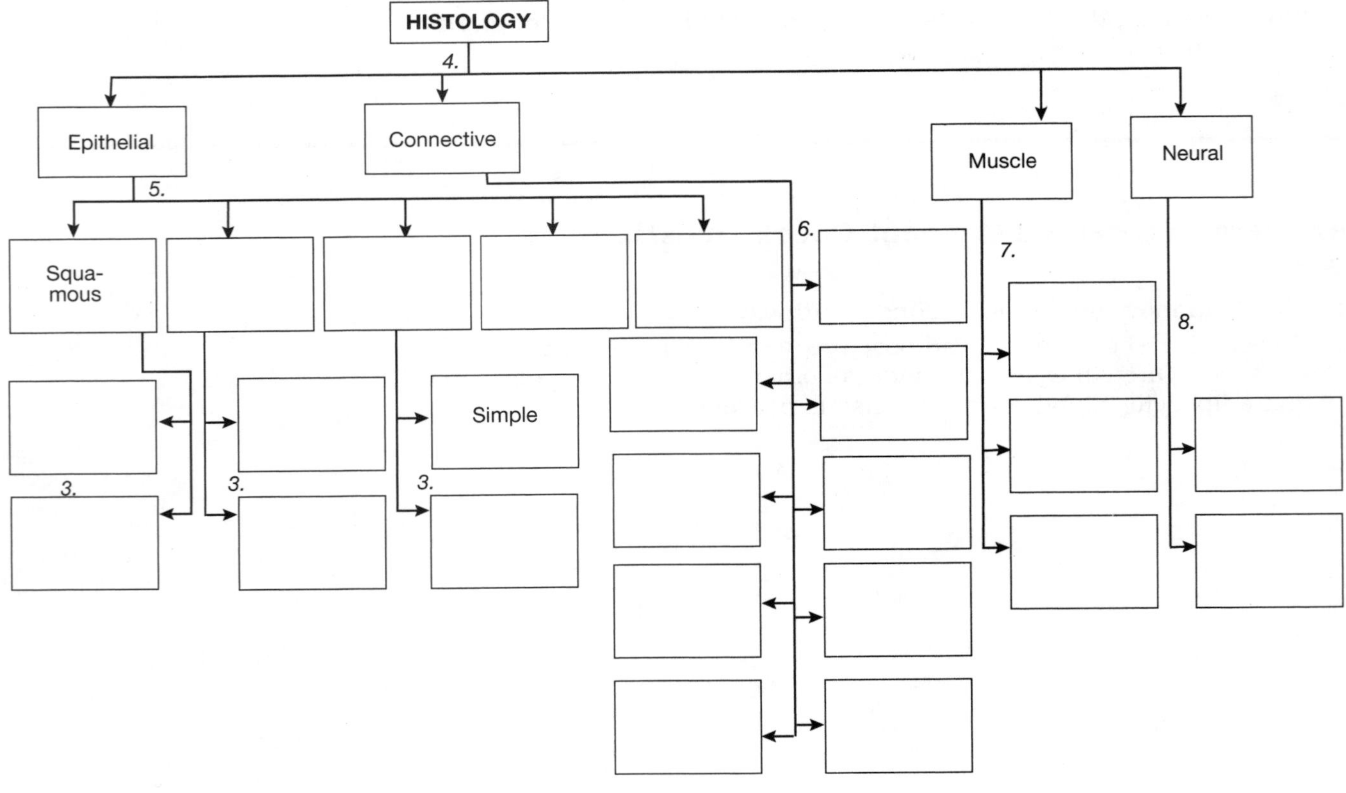

General Questions

1. What is histology? What is the origin of this word?
2. How should you study a histological slide?
3. On a separate sheet of paper, construct a two-column chart. In the first column categorize each of the following structures according to tissue type and subtype. In the second column briefly state the functional advantage of that tissue type for each structure mentioned. You may have to refer to your lecture text to complete this chart.
   a. Cells of the heart proper
   b. White blood cells
   c. Lining of the urinary tract
   d. Tendons
   e. Framework of the liver
   f. Signet ring cells
   g. Lining of the trachea
   h. Tracheal cartilage
   i. Inside of mouth (anterior portion)
   j. Cells nourishing neurons
   k. Skin
   l. Lining of intestine
   m. Lining of body cavity
   n. Exocrine glands
   o. Endocrine glands
4. Where would you find the basement membrane? What two tissue types contribute to the construction of the basement membrane?
5. Where would you find keratinized tissue? Why is this tissue important?

NOTES

ATLAS

Histology

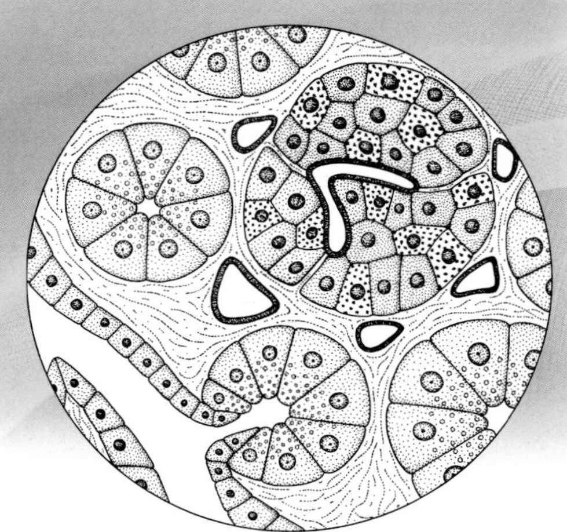

PLATE A Squamous Epithelia

SIMPLE SQUAMOUS EPITHELIUM

LOCATIONS: Mesothelia lining ventral body cavities; endothelia lining heart and blood vessels; portions of kidney tubules (thin sections of loop of Henle), inner lining of cornea, alveoli of lungs

FUNCTIONS: Reduces friction, controls vessel permeability, performs absorption and secretion

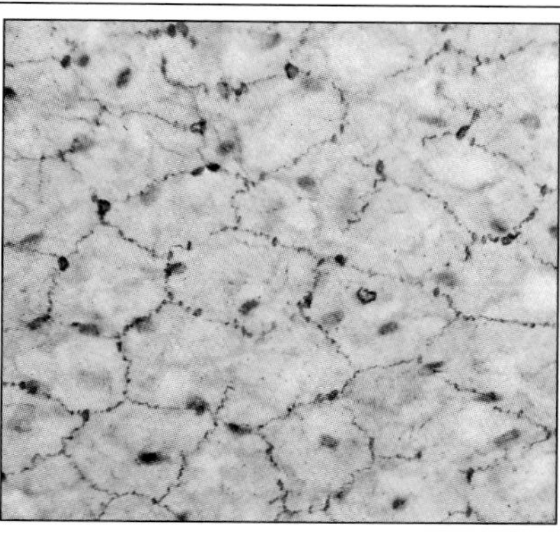

Simple squamous epithelium × 197

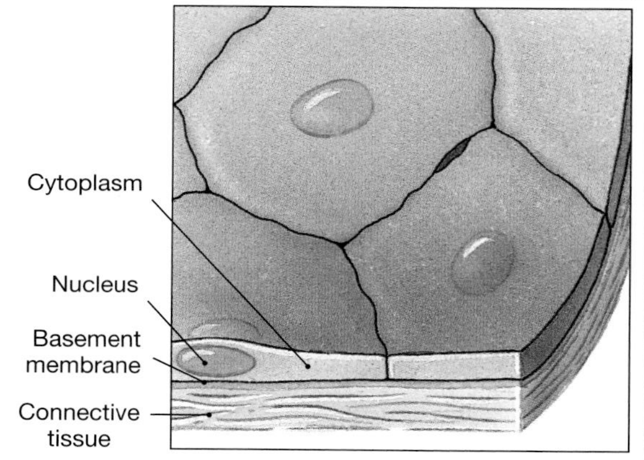

A superficial view of the simple squamous epithelium (mesothelium) that lines the peritoneal cavity. The three-dimensional drawing shows the epithelium in superficial and sectional view.

STRATIFIED SQUAMOUS EPITHELIUM

LOCATIONS: Surface of skin; lining of mouth, throat, esophagus, rectum, anus, and vagina

FUNCTIONS: Provides physical protection against abrasion, pathogens, and chemical attackText--P/U from figs

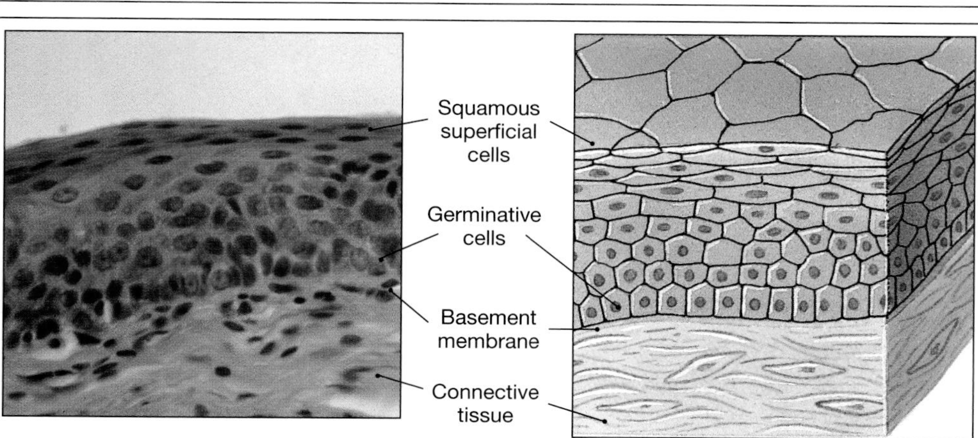

Stratified squamous epithelium × 310

A sectional view of the stratified squamous epithelium that covers the tongue.

PLATE B Cuboidal Epithelia

SIMPLE CUBOIDAL EPITHELIUM

LOCATIONS: Glands, ducts, portions of kidney tubules, thyroid gland

FUNCTIONS: Limited protection, secretion and/or absorption

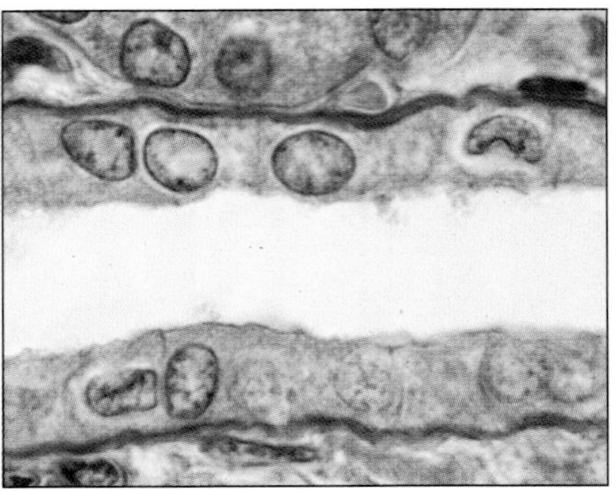

Kidney tubule × 2000

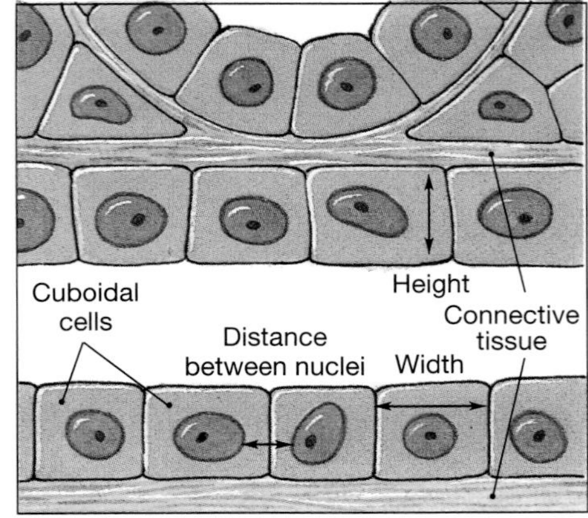

A section through the cuboidal epithelial cells of a kidney tubule. The diagrammatic view emphasizes structural details that permit the classification of an epithelium as cuboidal.

STRATIFIED CUBOIDAL EPITHELIUM

LOCATIONS: Lining of some ducts (rare)

FUNCTIONS: Protection, secretion, absorption

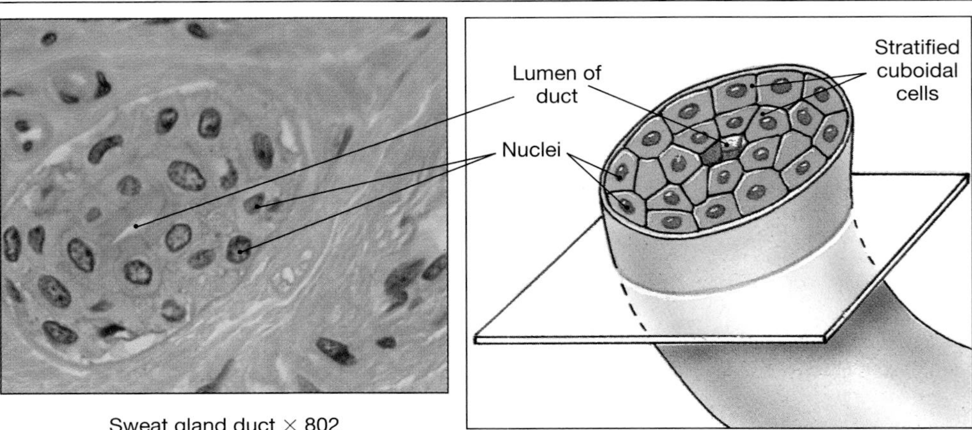

Sweat gland duct × 802

Sectional view of the stratified cuboidal epithelium lining a sweat gland duct in the skin.

PLATE C Columnar Epithelia

SIMPLE COLUMNAR EPITHELIUM

LOCATIONS: Lining of stomach, intestine, gallbladder, uterine tubes, collecting ducts of kidneys

FUNCTIONS: Protection, secretion, absorption

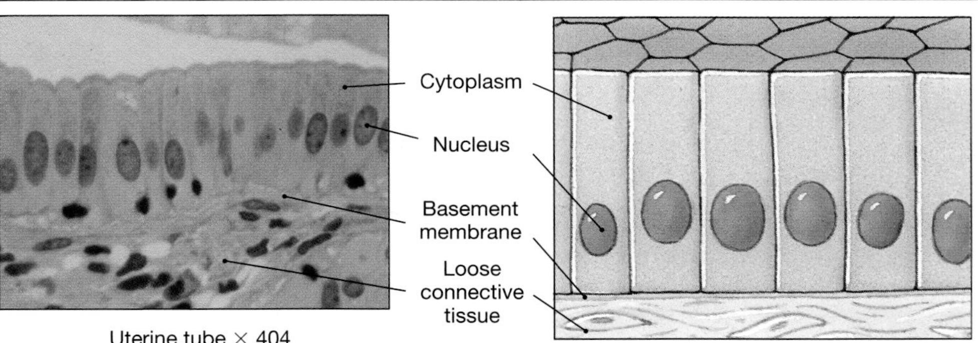

Uterine tube × 404

Micrograph showing the characteristics of simple columnar epithelium. In the diagrammatic sketch, note the relationships between the height and width of each cell; the relative size, shape, and location of nuclei; and the distance between nuclei.

PLATE C Columnar Epithelia *(continued)*

STRATIFIED COLUMNAR EPITHELIUM

LOCATIONS: Small areas of the pharynx, epiglottis, anus, mammary ducts, and urethra

FUNCTION: Protection

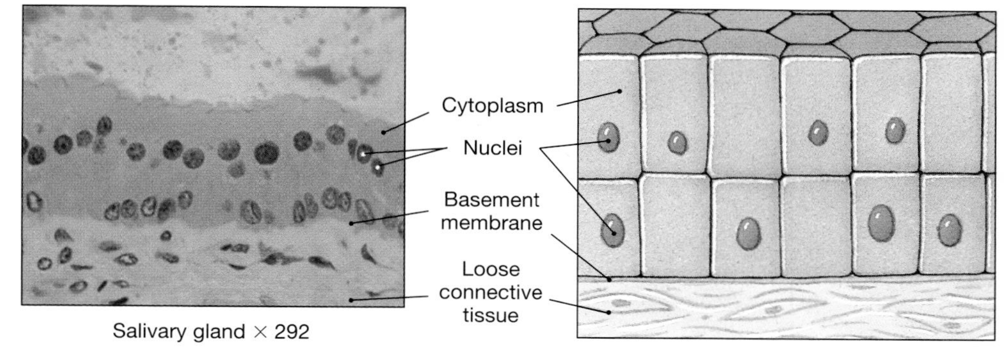

Salivary gland × 292

A stratified columnar epithelium is sometimes found along large ducts, such as this salivary gland duct. Note the thickness of the epithelium and the location and orientation of the nuclei.

PLATE D Pseudostratified Columnar Epithelium

PSEUDOSTRATIFIED COLUMNAR EPITHELIUM

LOCATIONS: Lining of nasal cavity, bronchi

FUNCTIONS: Protection, secretion

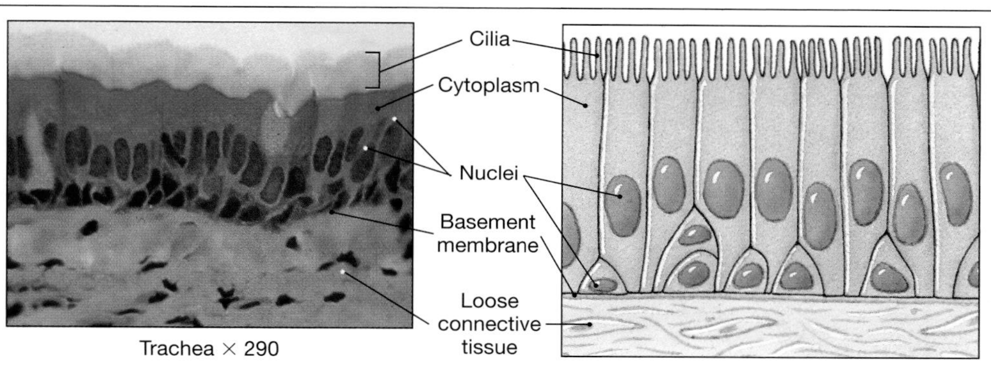

Trachea × 290

A view of the pseudostratified, ciliated, columnar epithelium of the respiratory tract. Note the uneven layering of the nuclei.

PLATE E Transitional Epithelium

TRANSITIONAL EPITHELIUM

LOCATIONS: Urinary bladder, renal pelvis, ureters

FUNCTIONS: Permit expansion and recoil after stretching

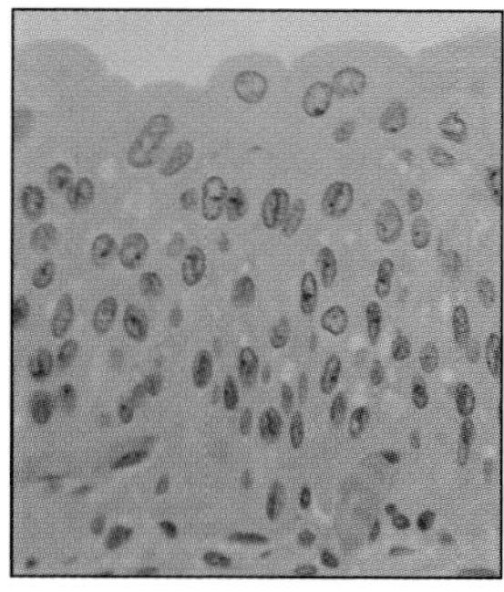

LM × 454

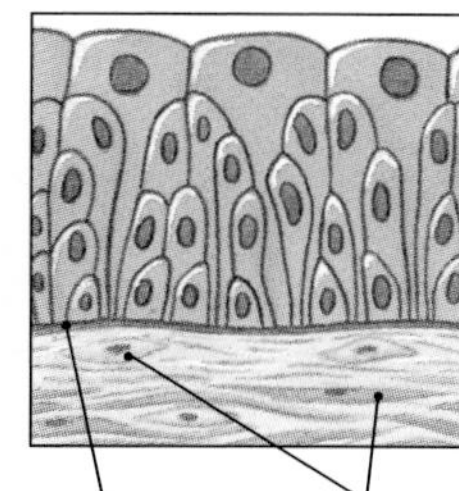

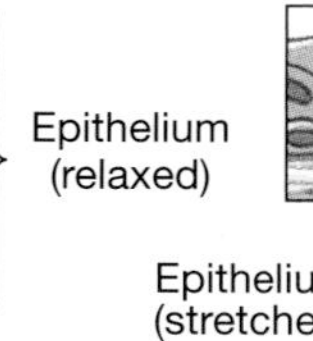

LM × 408

A view of the lining of the *empty* urinary bladder (left), showing relaxed state; a view of the lining of the *full* urinary bladder (right), showing the effects of stretching on the arrangement of cells in the epithelium.

PLATE F Loose Connective Tissue

LOOSE CONNECTIVE TISSUE

LOCATIONS: Throughout the body, between organs

FUNCTION: Protection, support, storage, route for diffusion

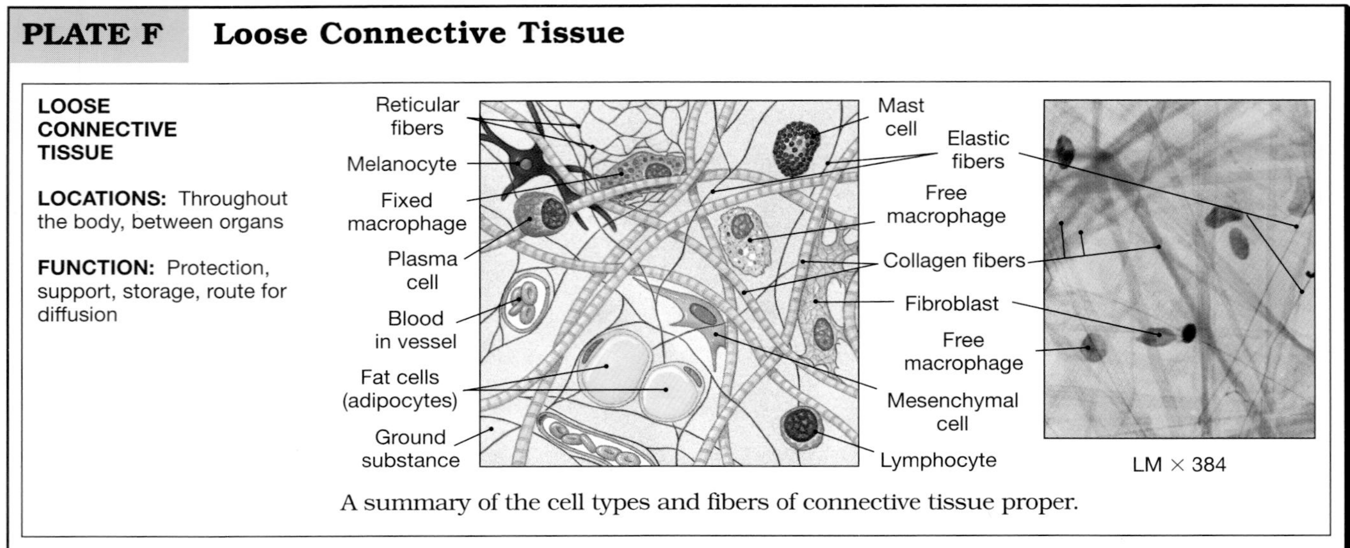

A summary of the cell types and fibers of connective tissue proper.

PLATE G Dense Connective Tissue

DENSE REGULAR CONNECTIVE TISSUE

LOCATIONS: Between skeletal muscles and skeleton (tendons and aponeuroses); between bones (ligaments); covering skeletal muscles (deep fasciae)

FUNCTION: Provides firm attachment; conducts pull of muscles; reduces friction between muscles; stabilizes relative positions of bones

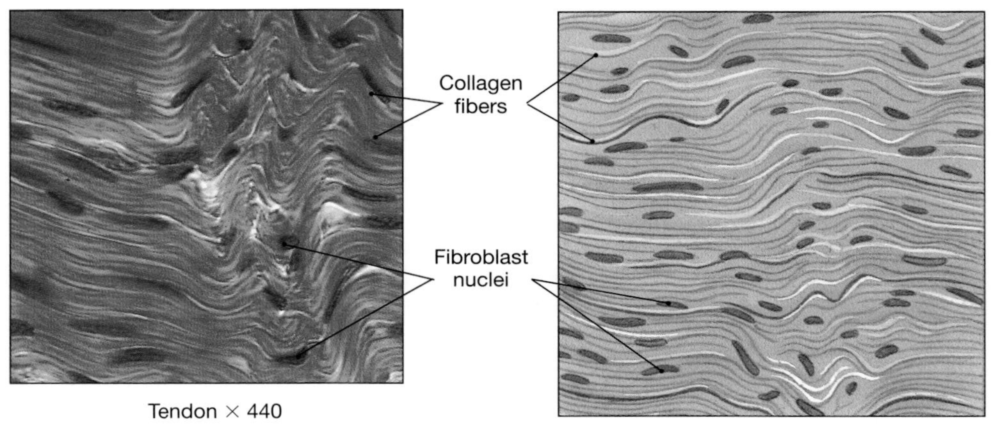

Tendon × 440

The dense regular connective tissue in a tendon. Notice the densely packed, parallel bundles of collagen fibers. The fibroblast nuclei can be seen flattened between the bundles.

DENSE IRREGULAR CONNECTIVE TISSUE

LOCATIONS: Capsules of visceral organs; deep dermis of skin; periostea and perichondria; nerve and muscle sheaths.

FUNCTION: Provides strength to resist forces applied from many directions; helps prevent overexpansion of organs such as the urinary bladder

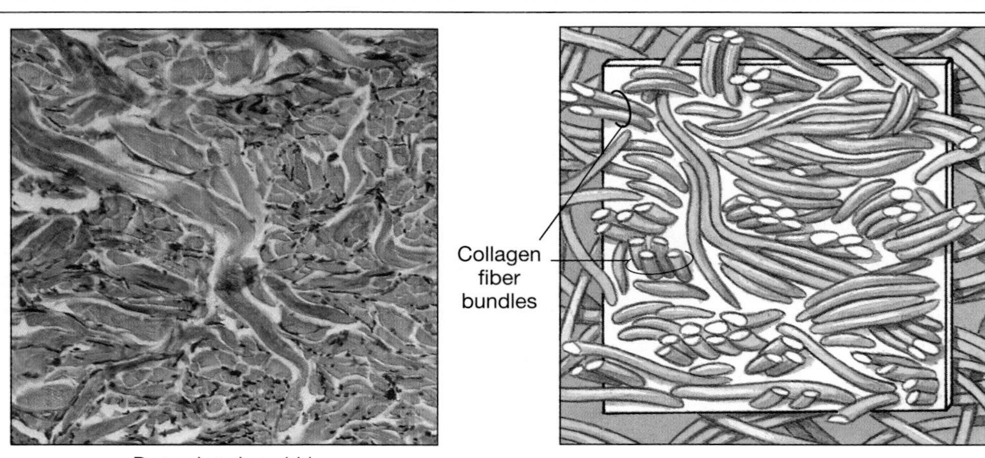

Deep dermis × 111

The deep dermis of the skin contains a thick layer of dense irregular connective tissue.

PLATE H Adipose Tissue

ADIPOSE TISSUE

LOCATIONS: Most parts of the body, especially beneath the skin, surrounding organs, and lining bony sockets

FUNCTION: Padding, cushioning, insulation, filling, storage. Brown fat functions in accelerated metabolic heat generation.

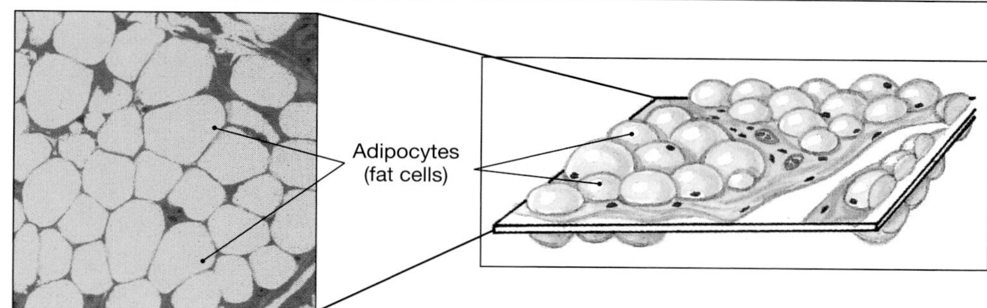

Adipose tissue × 117

Adipose tissue is a loose connective tissue that is dominated by adipocytes. In standard histological preparations, the tissue looks empty because the lipids in the fat cells dissolve in the alcohol used in tissue processing.

PLATE I Types of Cartilage

HYALINE CARTILAGE

LOCATIONS: Between tips of rib and bones of sternum; covering bone surfaces at synovial joints; supporting larynx (voicebox), trachea, and bronchi; forming part of nasal septum

FUNCTION: Provides stiff but somewhat flexible support; reduced friction between bony surfaces

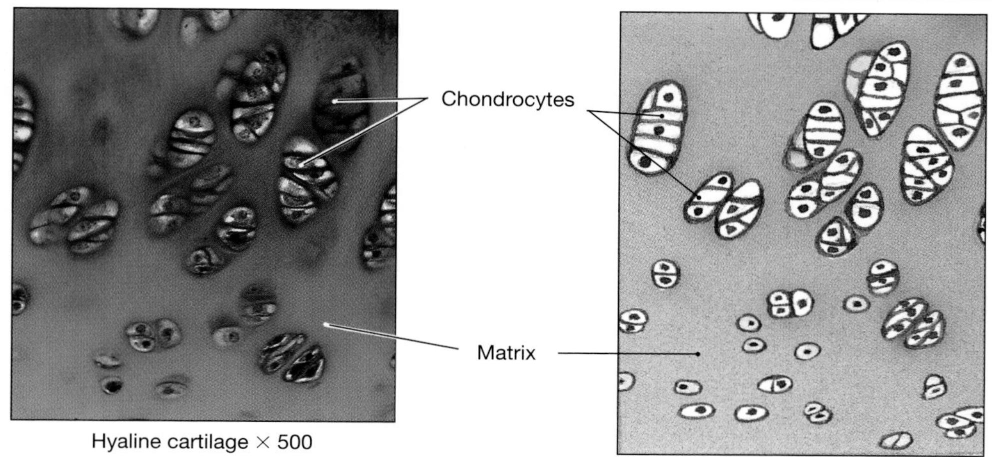

Hyaline cartilage × 500

Hyaline cartilage. Note the translucent matrix and the absence of prominent fibers.

ELASTIC CARTILAGE

LOCATIONS: Pinna of external ear; tip of nose; epiglottis

FUNCTION: Provides support but tolerates distortion without damage and returns to original shape

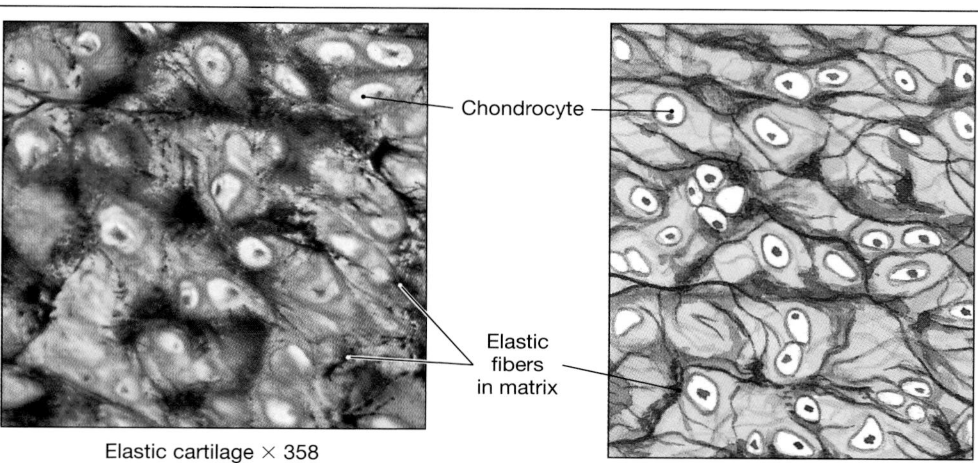

Elastic cartilage × 358

Elastic cartilage. The closely packed elastic fibers are visible between the chondrocytes.

PLATE I Types of Cartilage *(continued)*

FIBROCARTILAGE

LOCATIONS: Intervertebral discs separating vertebrae along spinal column; pads within knee joint; between pubic bones of pelvis

FUNCTION: Resists compression; prevents bone-to-bone contact; limits relative movement

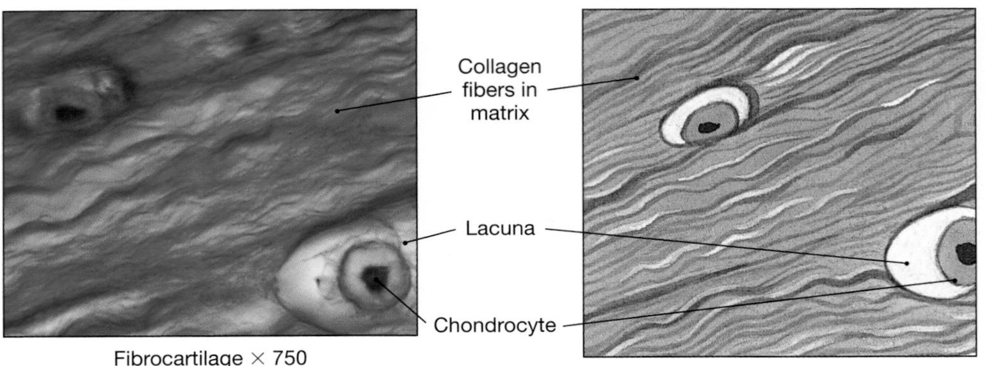

Fibrocartilage. The collagen fibers are extremely dense, and the chondrocytes are relatively far apart.

PLATE J Bone

BONE (OSSEOUS TISSUE)

LOCATION: Skeleton

FUNCTION: Support, protection, storage of minerals and lipids, blood cell production, leverage

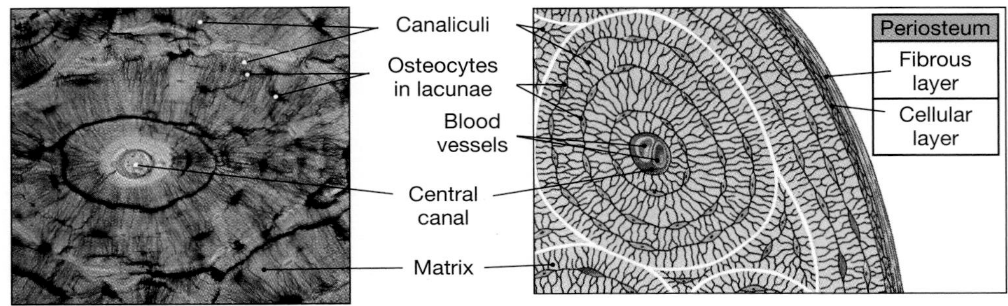

The osteocytes in bone are usually organized in groups around a central space that contains blood vessels. Bone dust produced during grinding fills the lacunae and the central canal, making them appear dark.

PLATE K Blood

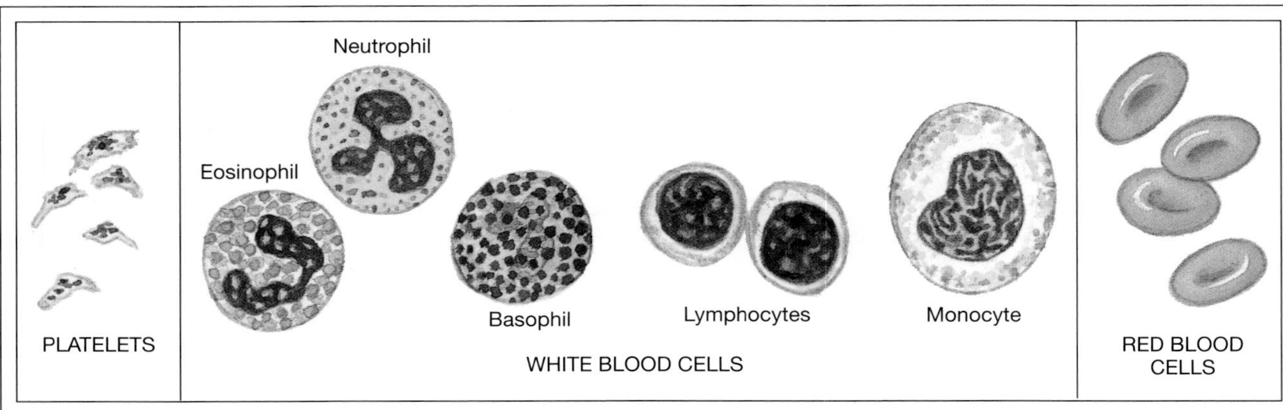

Formed elements of blood.

PLATE K Blood *(continued)*

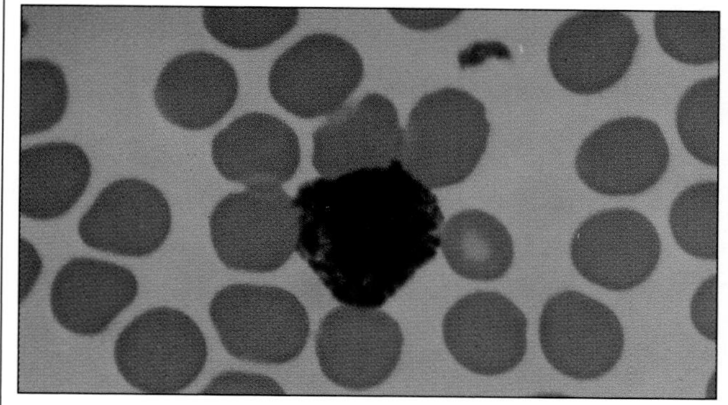
LM × 1500

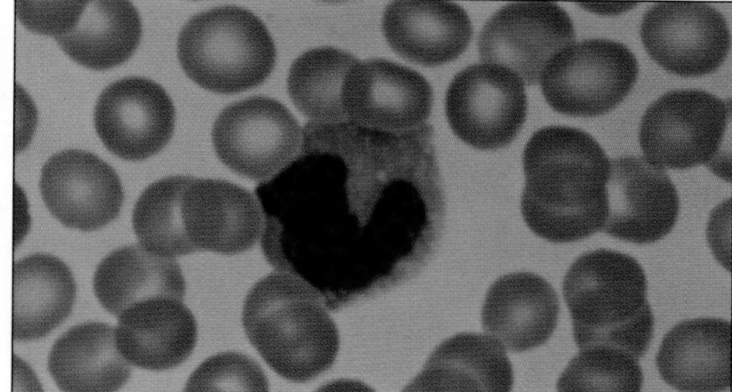
LM × 1500

Blood smears showing numerous RBCs and a basophil (left) and a monocyte (right).

PLATE L Muscle Tissue

SKELETAL MUSCLE TISSUE

LOCATIONS: Combined with connective tissues and nervous tissue in skeletal muscles, organs such as the skeletal muscles of the limbs

FUNCTIONS: Moves or stabilizes the position of the skeleton; guards entrances and exits to the digestive, respiratory, and urinary tracts; generates heat; protects internal organs

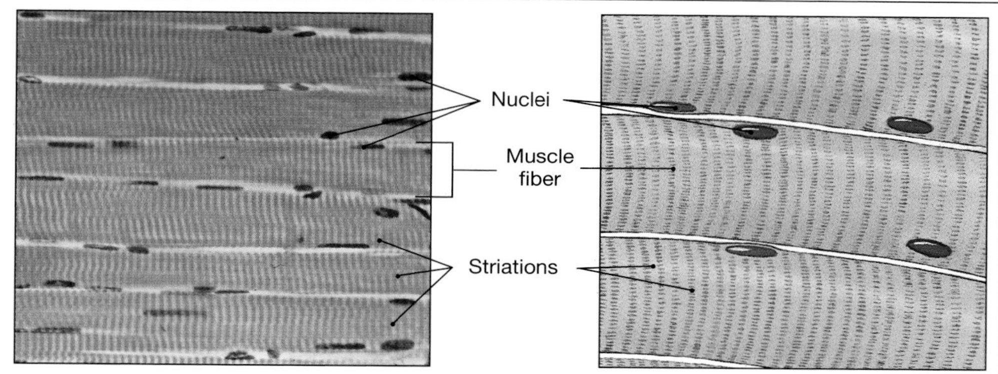

Skeletal muscle × 180

Skeletal muscle fibers. Note the large fiber size, prominent banding pattern, multiple nuclei, and unbranched arrangement.

SMOOTH MUSCLE TISSUE

LOCATIONS: Encircles blood vessels; found in the walls of digestive, respiratory, urinary, and reproductive organs

FUNCTIONS: Moves food, urine, and reproductive tract secretions; controls diameters of respiratory passageways; regulates diameter of blood vessels; and contributes to regulation of tissue blood flow

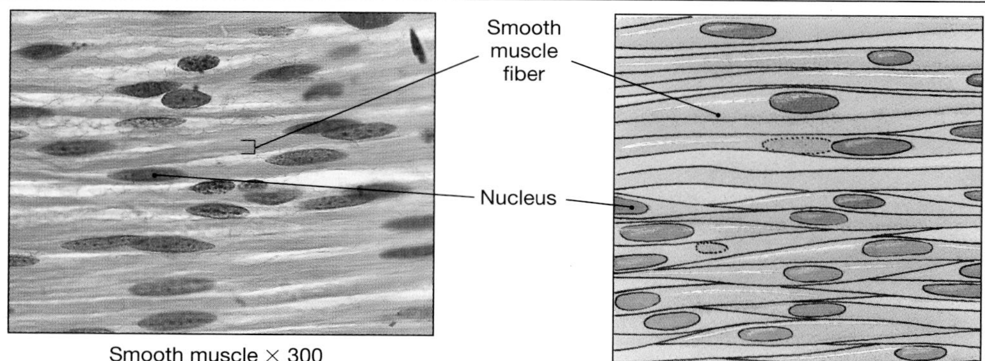

Smooth muscle × 300

Smooth muscle fibers. Smooth muscle fibers are small and spindle-shaped, with a central nucleus. They do not branch, and there are no striations.

PLATE L Muscle Tissue *(continued)*

CARDIAC MUSCLE TISSUE

LOCATION: Heart

FUNCTIONS: Circulates blood; maintains blood (hydrostatic) pressure

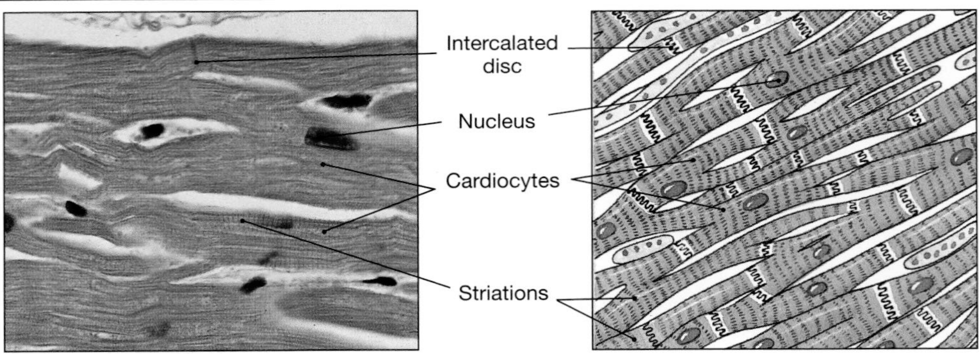

Cardiac muscle × 450

Cardiac muscle cells. Cardiac muscle cells differ from skeletal muscle fibers in three major ways: size (cardiac muscle cells are smaller), organization (cardiac muscle cells branch), and number and location of nuclei (a typical cardiac muscle cell has one centrally placed nucleus). Both contain actin and myosin filaments in an organized array that produces striations.

PLATE M Neural Tissue

NEURONS

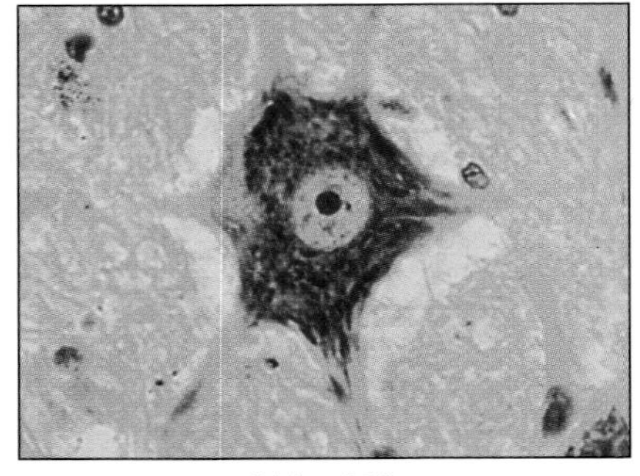

LM × 600

NEUROGLIA (supporting cells)

- Maintain physical structure of tissues
- Repair tissue framework after injury
- Perform phagocytosis
- Nutritional role
- Four major types in CNS, two in PNS

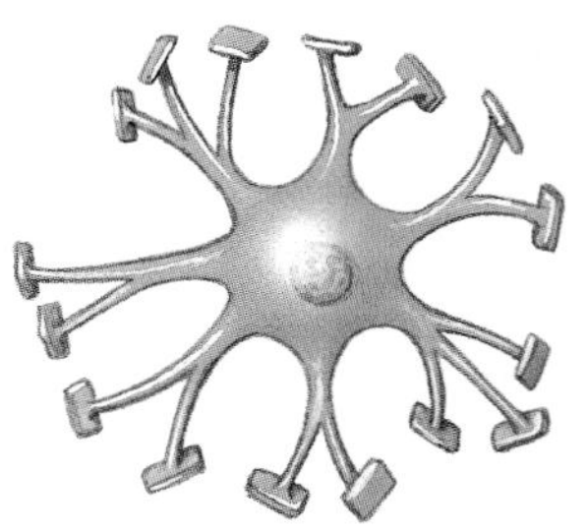

A representative neuroglion (astrocyte)

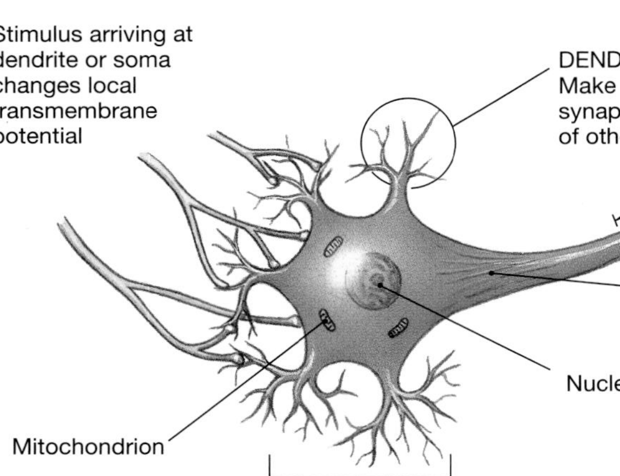

A representative neuron (sizes and shapes vary widely)

EXERCISE 12

Membranes

PROCEDURAL INQUIRIES

Preparation

1. What is the basic function common to all membranes?
2. What are the four major categories of membranes?

Examination

3. What is the composition of the membrane of the cell?
4. What is the composition of the basement membrane? What is the source of each of the component parts of this membrane?
5. What are the characteristics of all epithelial membranes?
6. What are the three major categories of epithelial membranes?
7. What is the identifying characteristic of each category of epithelial membrane?
8. What is the basic structure of the synovial membrane?

Additional Inquiries

9. What is an exudate?
10. How do the membranes in each of the major categories fulfill the basic function of all membranes?

Key Terms

Basement Membrane
Cell Membrane
Cutaneous Membrane
Elastic Lamina
Endothelium
Epithelial Membrane
Exudate
Lamina Propria
Membrane
Mesothelium
Mucous Membrane
Mucus
Parietal
Serous Membrane
Synovial Membrane
Transudate
Visceral

Materials Needed

Compound Microscope

Prepared Slides of

- Stratified Squamous Epithelium (Skin)
  - (to demonstrate basement membrane) or Enhanced Cross Section of Artery
- Mucous Membrane
  - as Cross Section of Trachea
  - as Cross Section of Small Intestine
- Serous Membrane
  - as Mesentery
  - as Pericardium
- Fresh or Preserved Beef Knee
  - (sagittal section if possible — to demonstrate synovial membrane)

2 x 2 Slides (or Overhead Transparencies) of Membranes

A **membrane** is a physical boundary separating a structure from its environment. This definition sounds simple enough, but the term membrane can cause some confusion if we are not clear on the differences between the many types of membranes found in the human body. Some membranes are as fine as the selectively permeable barrier surrounding each cell, while other membranes encompass numerous cell layers and include several different tissue types.

Consider the following annotated outline of the various types of human body membranes.

- Membranes may be components of the cell. These are the **cell membranes** — the phospholipid bilayers — discussed in Exercises 8 and 9. These membranes, common to all cells, function at the molecular level to maintain the integrity of the cell.

- **Basement membranes** are those acellular structures discussed in Exercise 11. Basement membranes, upon which all epithelial borders rest, form the barrier between the organ and its environment.
- **Epithelial membranes** are multicellular structures consisting of both epithelial and connective tissue components. These membranes produce an **exudate** (a fluid expelled by certain membrane cells into the membrane's environment) and offer protection to underlying tissues and organs.
- Epithelial membranes can readily be divided into three categories. The **mucous membranes** line the body cavities that are in direct communication with the outside world. We will encounter these mucous membranes many times during this course. The **serous membranes** line those body cavities not in direct contact with the external environment. These cavities include the pleural, pericardial, and peritoneal cavities. The serous membranes are also characterized by a watery, lubricating exudate. The **cutaneous membrane**, which covers the entire body, is also known as the skin and is discussed in Exercise 13.
- **Synovial membranes** form the synovial sacs, those balloon-like structures that cushion many of the body's highly movable joints. Synovial membranes are sometimes classified as modified epithelial membranes because they are multicellular and because they form boundaries between specific organs (namely, the bones). However, the synovial membranes, while often including some epithelial cells, demonstrate neither an epithelial continuum nor a basement membrane. Thus, the synovial membranes are not true epithelial membranes.

Regardless of structure, however, every membrane functions as a barrier. The permeability of this barrier varies considerably according to its functional components. The barrier formed by the basement membrane of this skin is virtually impermeable to everything from water to bacteria. This is certainly to our advantage when we consider how difficult it is to remain hydrated on a hot summer day and how easy it is for a small cut or hangnail to become infected. One function of the basement membrane is to regulate the passage of materials to and from the epithelium.

On the other hand, the serosal membrane barrier (the squamous epithelial lining of certain body cavities) is highly permeable to a wide range of fluids and ions. This is perfect for reducing friction when the internal organs are jostled together as we pursue our daily activities. Serosal cells are continuously producing and absorbing exudate. Quantities of exudate present at any given time can be dramatically altered with very little notice should conditions warrant more or less of the friction-reducing fluid.

❑ Preparation

I. Overview of the Membranes

A. Exercise Focus

1➡ Keep in mind the outline we established in the introduction:

a. Membranes as cellular components
b. Basement membranes
c. Epithelial membranes
   Mucous membranes
   Serous membranes
   Cutaneous membranes
d. Synovial membranes

As stated, membranes as components of the cell are considered in Exercises 8 and 9. The cutaneous membrane, which is the skin, is discussed in Exercise 13. The other membranes are examined here.

2➡ With each membrane example you study, ask yourself: "How does this membrane fulfill the primary function of all membranes?"

B. Microscopy

1➡ Obtain a compound microscope. With all of the slides in this exercise, begin by examining the structures under low power. Use high power only for comparison. For most of the slides, high power will afford too much magnification for you to gain a proper perspective of the structures you are studying.

❑ Examination

II. Basement Membrane

A. Overview

The basement membrane is the acellular structure upon which every epithelial layer of every organ rests. (Refer back to Exercise 11.) Immediately deep to the basement membrane is a layer of connective tissue. The epithelium provides a meshwork of fine filaments for the membrane, and the connective tissue provides bundles of course fibrils. The meshwork is replete with holes and passageways. Together, the two portions of the basement membrane supply strength, resistance, and structure to the epithelium.

The epithelium usually shows a polarity extending from the basement membrane to the epithelial surface. *[∞ Martini, p. 110]* This polarity, plus the inherent structure of the basement membrane, prevents the passive movement of molecules across the barrier (through the meshwork).

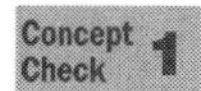

What two structures contribute to the basement membrane?

____

____

How does the basement membrane demonstrate the primary function of any membrane?

____

____

1 ➡ Obtain a slide that demonstrates the basement membrane. Any epithelial membrane slide (whether surface, mucous, or serous) can be used. Your instructor will probably have stratified squamous epithelial cells because when stained, the basement membrane of these cells can be clearly identified.

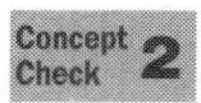

Why can we say that technically any epithelial membrane slide can be used to demonstrate the basement membrane?

____

____

____

2 ➡ If your instructor so indicates, use an enhanced slide of the cross section of a major artery to see the basement membrane. The epithelial tissue layer facing the lumen is the **endothelium**. Immediately deep to the endothelium is the basement membrane. Immediately deep to the basement membrane is a layer of connective tissue known as the **elastic lamina**.

3 ➡ Examine the structures on your slide. Note the basement membrane. The basement membrane usually stains as a sharp dark line. You should be able to distinguish between the stratified cells on the free surface side of the dark line and the fibrous, almost irregular structure (connective tissue) on the interior side of the dark line.

4 ➡ Compare your slide with the diagram, Figure 12–1●. Sketch and label the structures you see.

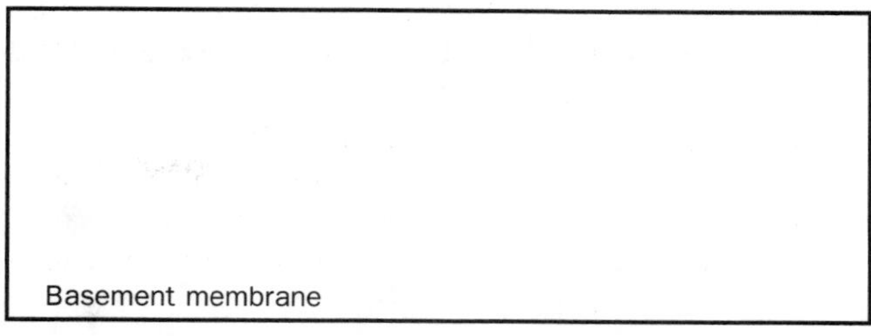

Basement membrane

III. Mucous Membrane *[∞ Martini, p. 129]*

A. Overview

The mucous membrane is the first of the multicellular epithelial membranes, those membranes that line the cavities that communicate directly with the exterior of the body. These cavity surfaces include the lumens of the digestive, respiratory, reproductive, and urinary tracts.

The mucous membrane is composed of an epithelium plus underlying connective tissue (known as the **lamina propria**) that supplies both nourishment and support to the structure. Most mucous membranes are lubricated by a substance called **mucus**, which is secreted by assorted goblet cells and/or assorted multicellular glands along the membrane proper.

1 ➡ Obtain a slide of the trachea and a slide of the small intestine. Theoretically a slide from any of the structures with a mucousal lining can be used because all mucous membranes have an epithelium, a connective tissue layer, and a lubricating exudate. The trachea and small intestine are commonly chosen because of their respective distinct goblet and glandular cells.

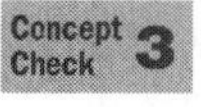

What types of epithelial cells should you be viewing on your trachea and small intestine slides?

____

____

2 ➡ Observe the slides and compare the structures you are viewing with the diagrams, Figures 12–2a● and 12–2b●. Sketch and label what you see.

Trachea vs. small intestine

● **FIGURE 12–1**
Representative Epithelial Tissues and Basement Membranes. **(a)** Columnar epithelial tissue; **(b)** Simple squamous epithelial tissue; **(c)** Stratified squamous epithelial tissue.

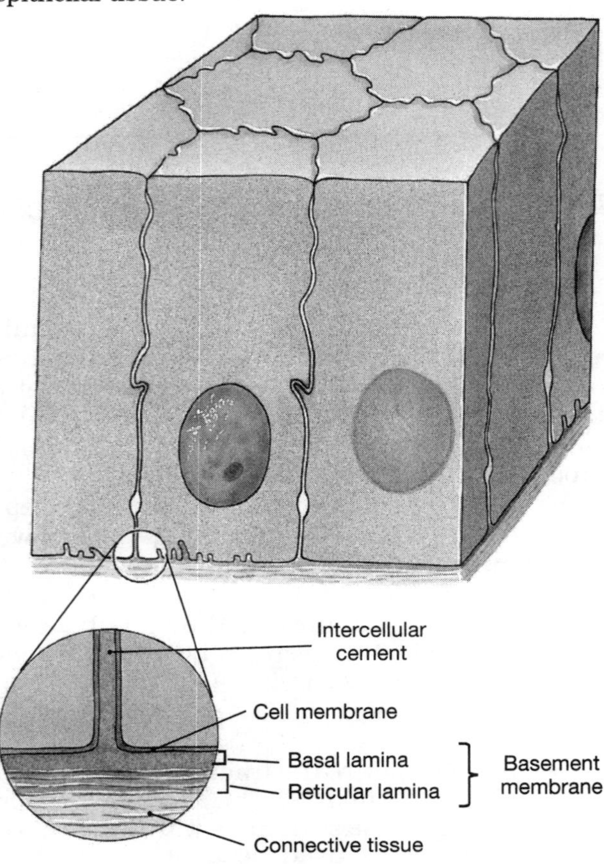

(a) Columnar Epithelial Tissue

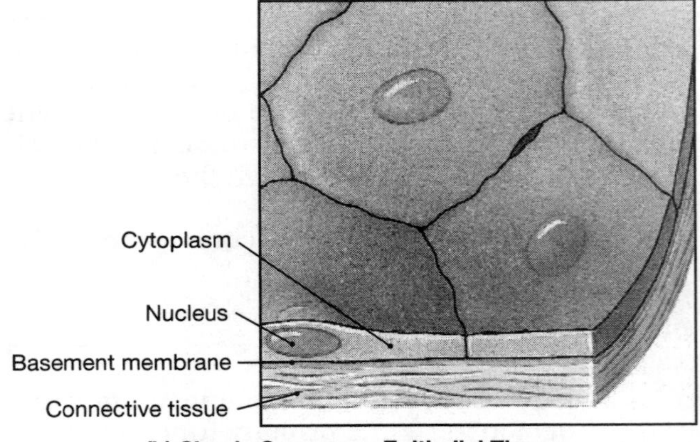

(b) Simple Squamous Epithelial Tissue

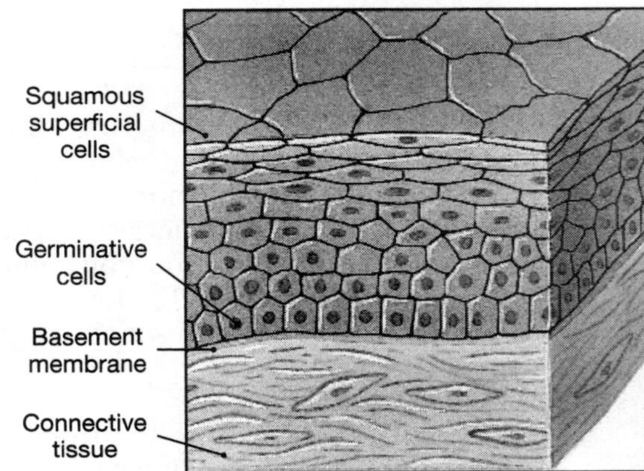

(c) Stratified Squamous Epithelial Tissue

IV. Serous Membrane [∞ *Martini, p. 129*]

A. Overview

Serous membranes, which are also epithelial membranes, line cavities that do not communicate with the exterior of the body. Serous membranes line the internal cavities of the body — the pleural, pericardial, and peritoneal cavities. The epithelium of these internal cavities is called the **mesothelium**. The underlying connective tissue is loose and highly permeable. The serous membrane secretes an exudate, called a **transudate**, which is thin, watery, and minimal in quantity. Since the transudate is named according to its location, the lung transudate is pleural fluid, the abdominal transudate is peritoneal fluid, and the heart transudate is pericardial fluid.

A serous membrane has two parts: a **parietal** portion, which lines the wall of the cavity, and a **visceral** portion, which covers the enclosed organs. The visceral and parietal portions of the serous membrane form a virtual continuum lining the walls of a body sac. The transudate functions to reduce friction between these walls, which would otherwise be in direct contact with each other.

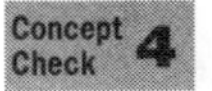
Concept Check 4 In terms of location, what is the difference between a serous membrane and a mucous membrane?

____

____

1➡ Obtain a slide of the mesentery, or whatever serous structure your instructor may have set out for you. The mesentery is often used because the distinct visceral and parietal layers of the serous membrane are easily discernible.

2➡ Identify the visceral and parietal layers and the appropriate body cavity. If you are viewing the mesentery, you should see the peritoneal cavity.

If you cannot discern the two layers and the cavity, at least identify the thin mesothelium and the underlying connective tissue.

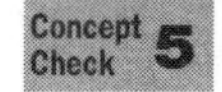
Concept Check 5 What type of cell makes up the mesothelium? ____________________

● **FIGURE 12–2**
Mucous Membrane. (a) Respiratory; **(b)** Digestive.

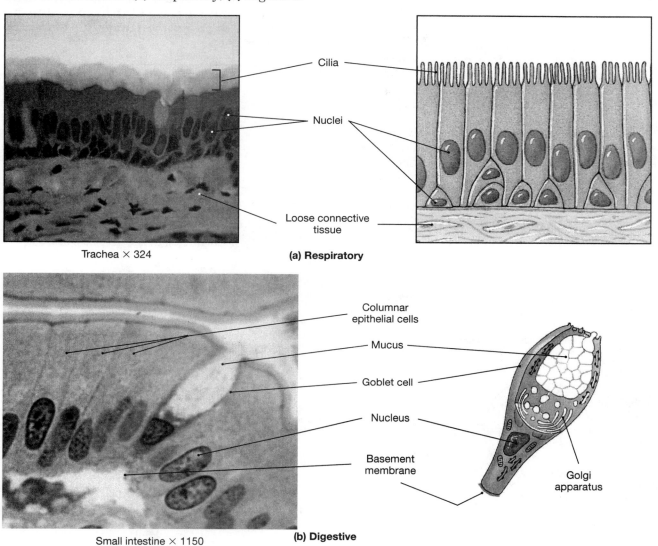

Cilia
Nuclei
Loose connective tissue

Trachea × 324

(a) Respiratory

Columnar epithelial cells
Mucus
Goblet cell
Nucleus
Basement membrane
Golgi apparatus

Small intestine × 1150

(b) Digestive

3➡ Compare your slide with the diagram, Figure 12–3●. Sketch and label what you see.

V. Synovial Membrane *[∞ Martini, p. 131]*

A. Overview

The synovial membrane encases many of the highly movable joints of the body. (See Exercise 18.) The synovial membrane is composed only of connective tissue — no true epithelium, although the structure is sometimes called the synovial epithelium. This synovial membrane consists of a discontinuous layer of squamous or cuboidal cells, no basement membrane and extensive areas of loose connective tissues surrounded by a layer of collagen fibers and fibroblasts. The synovial membrane produces a viscous synovial fluid that serves to protect and separate the cartilagenous ends of the synovial joints.

● **FIGURE 12–3**
Serous Membrane.

Transudate
Epithelium
Loose connective tissue

1 ➡ Observe the beef knee. Follow the membrane around the joint in order to gain a perspective on the dimensions of the synovial sac. Touch the cartilagenous ends of the long bones. Note that these ends are exceptionally smooth. Are you touching the synovial sac or the cartilage?

____

____

Note the relation of the patella to the rest of the joint. Compare Figures 12–4a● and 12–4b●. Sketch and label the parts of the beef knee.

2 ➡ Depending on the condition of your beef knee, your instructor may direct you to try to separate a piece of the synovial membrane for observation under the dissecting microscope.

Beef knee

● **FIGURE 12–4**
Synovial Membrane. (a) Typical synovial membrane; **(b)** Typical sagittal knee joint.

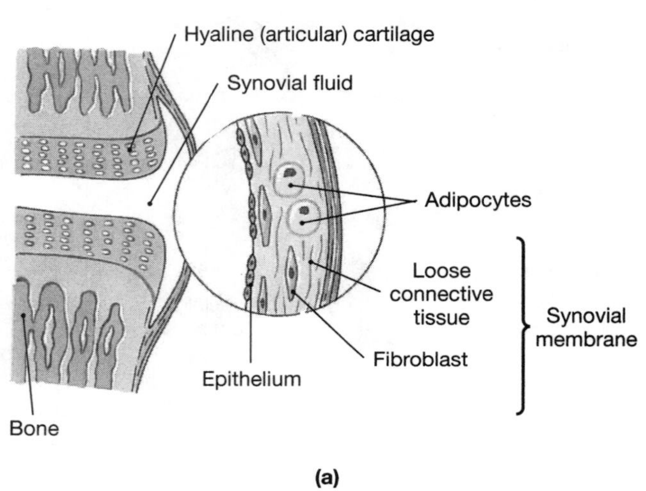

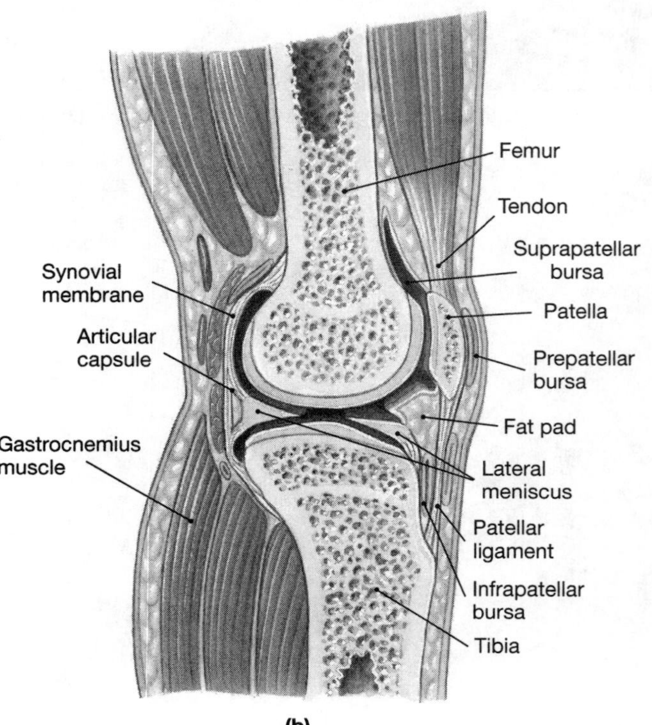

❑ Additional Activities

1. Trace the embryonic origin of the different membranes.
2. Investigate the staining techniques needed to bring out the important characteristics of the membranes listed in this exercise.

NOTES

Answers to Selected Concept Check Questions

NOTES

1. Epithelial cells, connective tissue cells.
2. All epithelia rest on a basement membrane.
3. Trachea ⟶ Pseudostratified columnar epithelium
   Small intestine ⟶ Simple columnar epithelium.
4. Serous membranes do not communicate with the outside of the body; mucous membranes do.
5. Mesothelium ⟶ Simple squamous epithelium.

❑ Lab Report

1. Referring to the numbered inquiries at the beginning of this exercise, complete the following box summary:

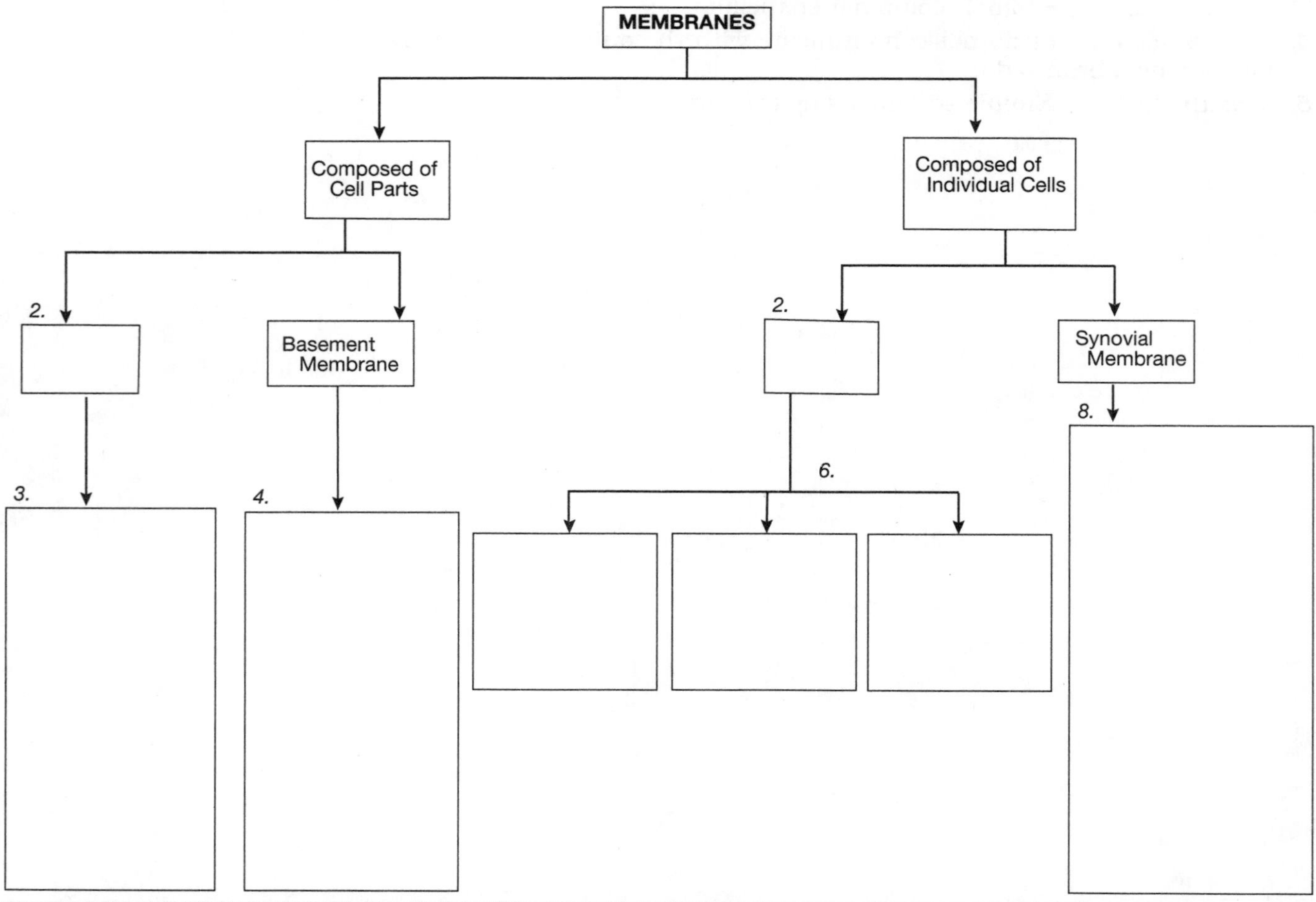

General Questions

1. Complete the following chart:

Membrane Type	Components	Location	Function
Cell (Part of)			
Basement			
Mucous			
Serous			
Synovial			

2. How is each type of membrane well suited for its function?
3. In which type(s) of membrane would you find
   a. Exudate
   b. Fibrous connective tissue
   c. Loose connective tissue
   d. Mesothelium
   e. Epithelial layer
4. What are the differences between the peritoneal (serous) membrane and the mucous membrane?

NOTES

EXERCISE 13

Integument

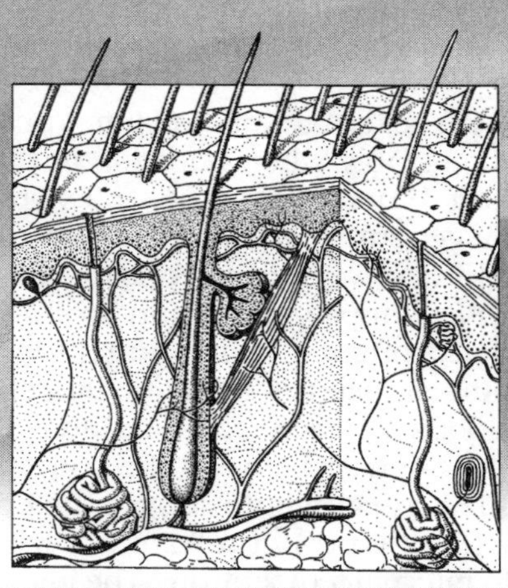

PROCEDURAL INQUIRIES

Examination

1. What are the accessory structures of the epidermis?
2. What are the parts of the nail?
3. What are the parts of the hair and hair follicle?
4. What are the types of glands?
5. What are the layers of the epidermis?
6. What are the structures of the dermis?
7. What are the structures of the subcutaneous layer?

Experimentation

8. What is the distribution of sweat glands on the skin?
9. What is piloerection?

Additional Inquiries

10. What is the overall role of the skin?
11. What is the difference between thick and thin skin?
12. How is the structure of the integument related to its function?

Key Terms

Apocrine
Basement Membrane
Cerumen
Cutaneous Membrane
Dermis
Eccrine
Epidermis
Glands
Keratin
Melanin
Merocrine
Papillae (sing. papilla)
Piloerection
Sebum
Strata
Stratum Corneum
Stratum Germintivum
Stratum Granulosum
Stratum Lucidum
Stratum Spinosum
Subcutaneous Layer
Sweat

Materials Needed

Skin Models (thick and thin, if available)
Hand Lens
Dissecting Microscope (if available — scanning lens or low power lens of compound microscope may be substituted)
Compound Microscope
Prepared Microscope slides
- Dermal Layers — including Melanocytes
- Hair Follicle
- Sweat Glands — Apocrine and Merocrine (Eccrine)
- Sebaceous Gland

Forceps
Emery Board
Microscope Slide
Iodine (preferably IKI or Lugol's; other iodines may be substituted)
Graph Paper (or other Bond Paper) Cut in identical squares about 1 cm/side
Tape
Ice
Plastic Bags
Optional:
Prepared Microscope Slides
- Skin — palmar and plantar
  - Scalp
  - Abdomen
  - Axillary sweat gland

Weighing about 4 kg and encircling the entire body, the integument, consisting of the skin proper plus a host of accessory structures, is the largest and in some ways the most diverse organ in the body — both histologically and functionally. (The terms skin and integument can be used interchangeably when referring to this structure as an *organ*. When referring to the structure as a *system*, we generally only use the word integument.)

Although we tend to consider the skin more for its physical appearance than for its superb physiological attributes, the integument is the body's primary line of defense against chemical, mechanical, and biological invasion. Your skin is that theoretical unbroken barrier between the body and its environment. The integument protects the body from desiccation (drying out) and from experiencing fluctuations in internal temperature. The integument excretes wastes (such as salts, urea, and alcohol), synthesizes vitamin D, stores nutrients, and provides sensitivity sites for the awareness of external stimuli. The biggest threats from second- and third-degree burns are fluid loss and infection, because the skin keeps the water in and the microbial invaders out.

Think of the functional value of the skin in terms of homeostasis. For example, your body temperature is virtually constant whether you are in the middle of a blizzard or in the middle of a heat wave because you have mechanisms for storing and dissipating heat. Some of those mechanisms, such as the ability of your blood vessels to constrict to conserve heat and the ability of your sweat glands to give off water to help you lose excess heat, are functions of the integument.

We say the skin is a "diverse" organ because it contains all the major tissue types — epithelial, connective, muscle, nervous — all functioning together. This complex conglomerate forms not only the barrier protecting the internal body from the hazards of the external world but also the primary interface between ourselves and our sensory experience of the world. (We will examine some of these sensory perceptions in Unit VI.)

The integument has two functional components, the **cutaneous membrane** and the accessory structures. The cutaneous membrane is composed of the superficial epidermis (which can be divided into several layers, or **strata** (singular stratum), and which rests on a basement membrane), the middle dermis, and the underlying subcutaneous layer (or hypodermis).

When we think of skin colloquially, we usually think only of the cutaneous membrane proper; sometimes we think only of the epidermis. We should keep in mind, though, that the integument includes a number of accessory structures. Did you ever stop to think of the hair as a part of the integument? Well, it is. So too are the integumentary **glands**, all three types of them — the sebaceous glands, the sweat glands, and the ceruminous glands. In fact, the mammary glands (discussed in Exercise 67) are modified sweat glands.

The purpose of this laboratory exercise is to develop an understanding of the structural and functional components of the integument by viewing various integumentary models and slides and by examining certain characteristics of the skin.

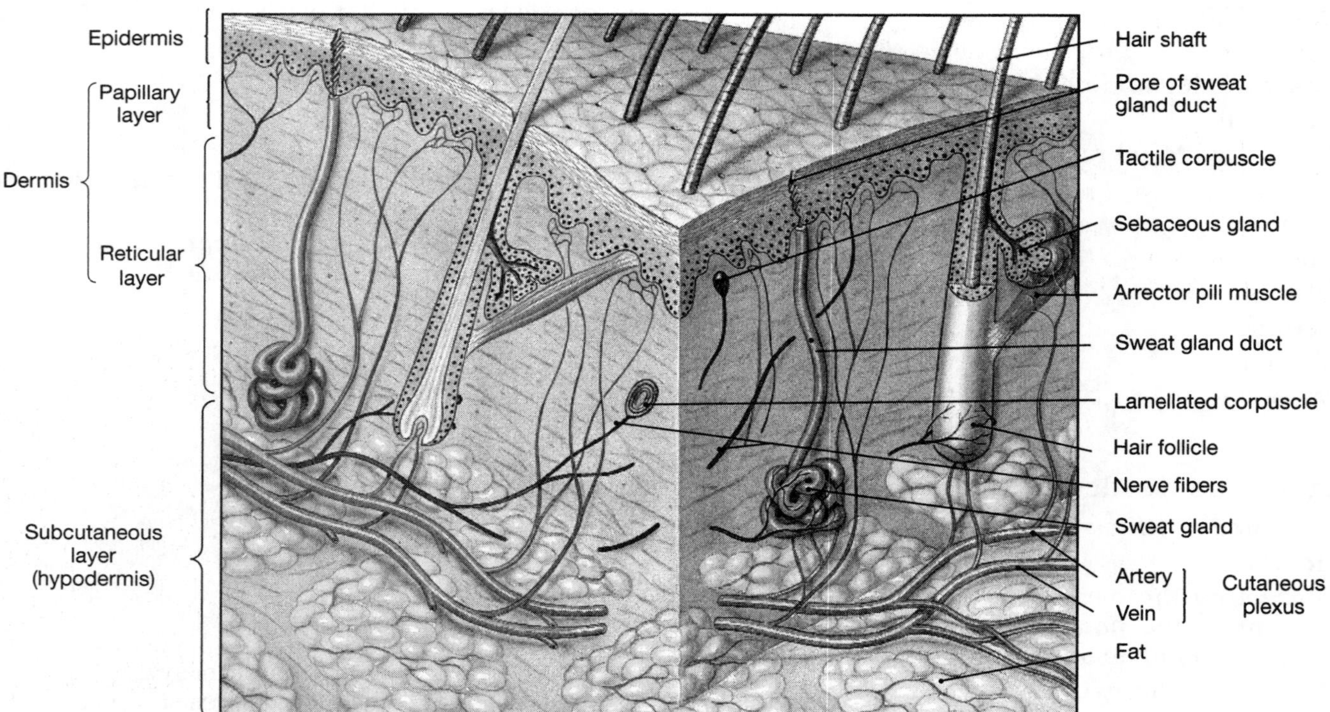

● **FIGURE 13–1**
Components of the Integumentary System.

❑ Examination

[Note: Your instructor may direct you to begin this laboratory period by setting up the iodine experiment in Section III.A. The iodine experiment needs to run for about one hour, during which time you can be working on the rest of the laboratory exercises.]

I. Cutaneous Membrane

A. Epidermis [∞ Martini, p. 145]

1➡ Study both the classroom skin model and Figure 13–1●. When you feel comfortable with the terminology, obtain a compound microscope and a prepared slide showing the layers of the skin. Make note of the identifiable structures discussed as follows:

The **epidermis** consists of stratified squamous epithelial cells resting on a basement membrane. The **basement membrane** is the dark line of demarcation separating the orderly, cellular epidermis from the diversified multitissued dermis. (If necessary, refer back to the section on the basement membrane in Exercise 12.)

The **stratum germinativum** (or stratum basale) rests on the basement membrane. Depending on your slide, you may be able to discern certain mitotic stem cells that replenish the cell population of the entire epidermis and certain irregularly shaped melanocytes that are squeezed into spaces between the epithelial cells. The melanocytes produce the pigment **melanin**, which protects the underlying tissues from the potential dangers of ultraviolet radiation. The amount and location of this melanin — along with **keratin**, pigment molecules, and the blood vessels — give your skin its particular color.

Immediately above the stratum germinativum is the **stratum spinosum,** several tiers of closely bonded cells that lose their mitotic ability as they are displaced upward toward the **stratum granulosum** (grainy layer). Stratum granulosum cells begin manufacturing *keratohyalin,* a precursor of keratin. Recall from Exercise 11 that keratin is that tough, water-resistant protein that gives the skin its protective barrier qualities.

The **stratum lucidum,** normally found only in the palms of the hands and the soles of the feet, is the clear layer (filled with the protein eleidin) immediately above the stratum granulosum and beneath the **stratum corneum.** The stratum corneum is the flattened, dehydrated layer of dead cells that is constantly being shed. If the stratum corneum is keratinized or cornified with large quantities of keratin, the epithelium is highly impermeable to water and electrolytes.

2➡ Observe your classroom skin model and Figure 13–2●. In the following space make note of the differences between thin skin (as found on most of the body) and thick skin (as found on the palms of the hands and soles of the feet), such as the thickness of the epidermal layers, particularly the stratum corneum. Sketch these examples of skin in the following space.

Thick v. thin skin

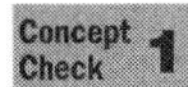

What are the differences between thick and thin skin? What do you think is the advantage of having thick skin on the palms and soles?

____

____

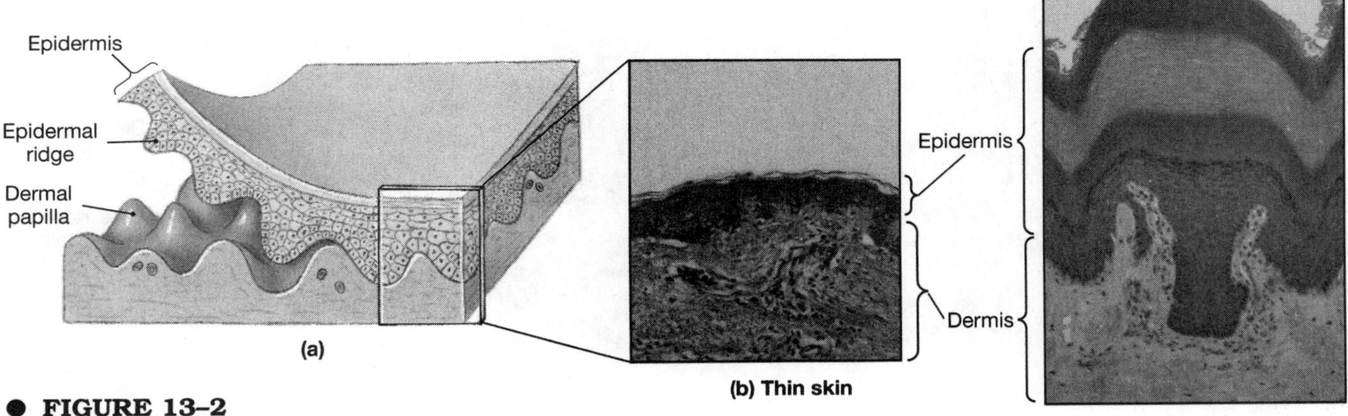

● **FIGURE 13–2**
Thin and Thick Skin.

Corns and callouses are local responses to friction and pressure, involving either thick skin or thin skin. The layers of skin build up to afford additional protection against such onslaughts as ill-fitting shoes, ax handles, and guitar strings.

B. Dermis [∞ Martini, p. 150]

1➡ Look again at Figure 13–1● and observe the **dermis**, a layer of highly varied connective tissue. The upper layer of the dermis seems to form little mounds or hills. Now locate these mounds on Figures 13–2● and 13–3●. These dermal **papillae** (sing. papilla) make up the papillary layer.

2➡ On Figure 13–1● identify the sweat gland, sweat duct, nerve, sebaceous gland, hair follicle, hair shaft, arrector pili muscle, blood vessels, and nerve corpuscles. Try to locate these structures on the skin model and on your integumentary slide.

C. Subcutaneous Layer

1➡ Now examine the **subcutaneous layer** both in Figure 13–1● and on your skin model. Notice that there is no definite line of demarcation between the dermis and the subcutaneous layer.

2➡ Identify the blood vessels and the adipose tissue in Figure 13–1●, on the skin model, and on your slide.

II. Accessory Structures

A. Hair [∞ Martini, p. 152]

1➡ Observe the hair shafts and hair follicles shown in Figure 13–1●. Notice how the epidermal cells line the hair follicle and give rise to the nonliving hair shaft. Identify the arrector pili muscle.

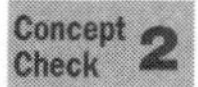

What would happen if this muscle were contracted?

____

____

2➡ Obtain a microscope slide of a sebaceous gland and a hair follicle. Compare your slide with Figure 13–3●. In the space following, sketch and label what you see. Cross-check Figure 13–3● with Figure 13–1●.

Sebaceous gland v. hair follicle

B. Integumentary Glands [∞ Martini, p. 155]

1➡ Return to Figure 13–1● and notice how the ducts of some of these glands empty into the hair follicles. These are the sebaceous glands, which are found on all parts of the body, and which produce a fatty substance called **sebum.** Sebum keeps the keratin flexible and inhibits bacterial growth. The contraction of the arrector pili muscles then forces sebum to the surface of the skin. Other sebaceous glands, called sebaceous follicles, communicate directly with the epidermis. (Sebaceous follicles are not shown in Figure 13–1●.)

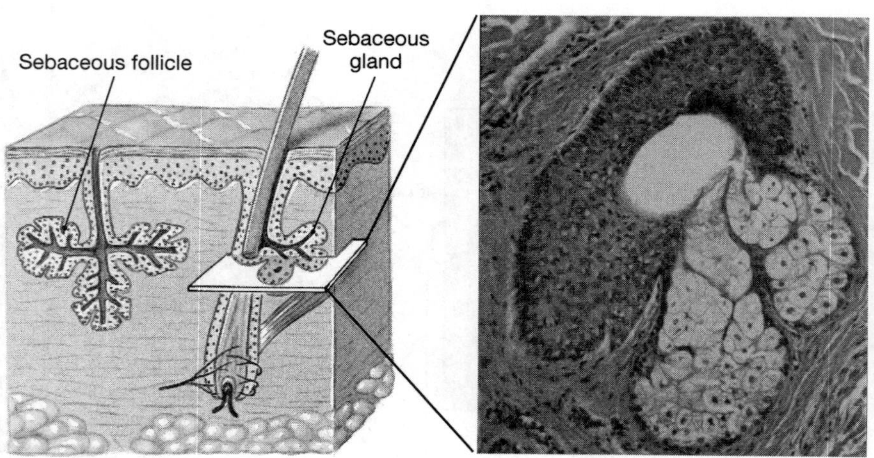

Sebaceous gland (LM × 150)

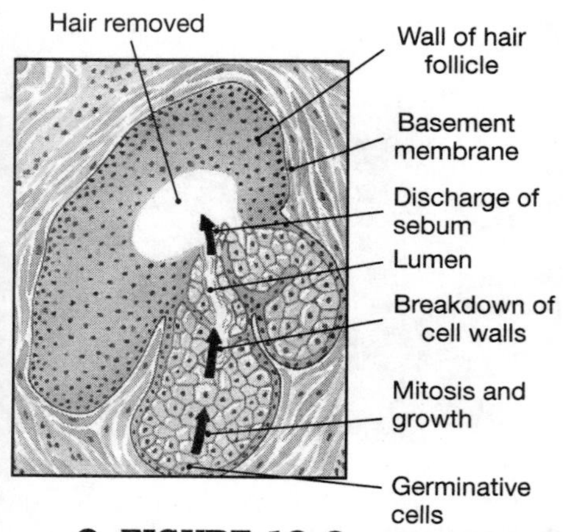

● **FIGURE 13–3**
Sebaceous Glands and Follicles.

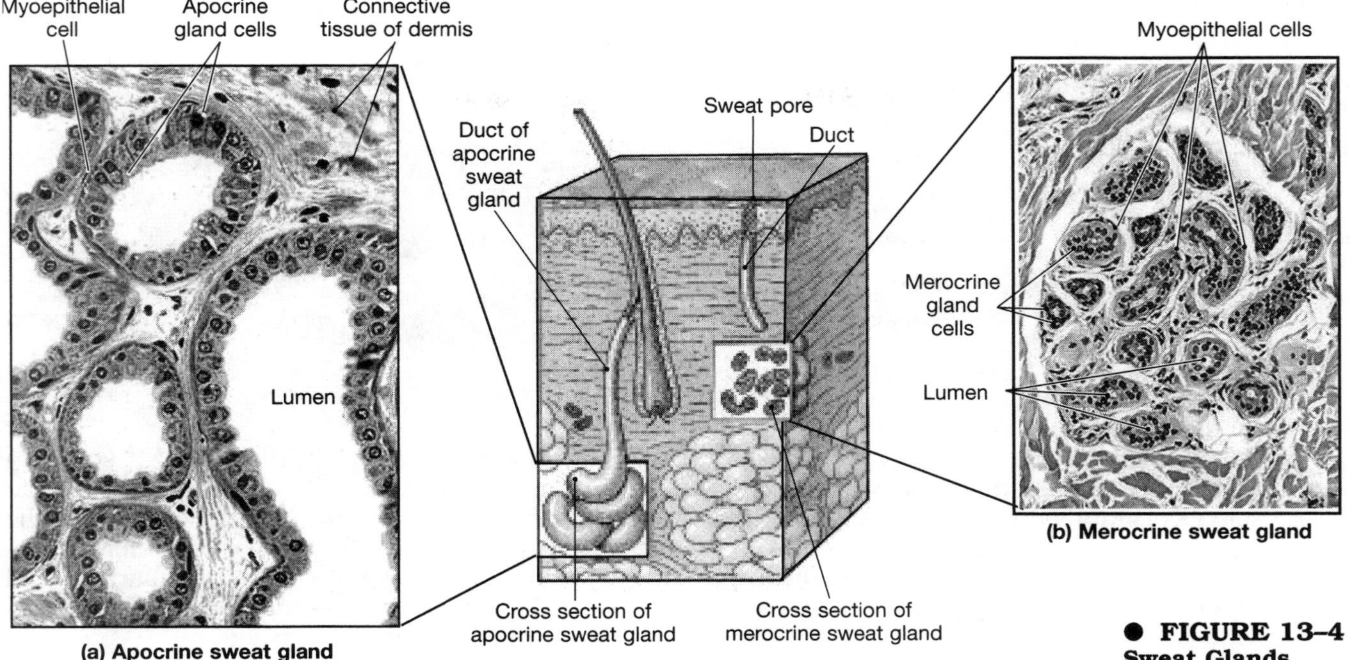

(a) Apocrine sweat gland

(b) Merocrine sweat gland

● FIGURE 13–4 Sweat Glands.

2➡ Identify the sweat glands on Figure 13–1●. **Sweat** glands are usually classified as either **apocrine** or **merocrine** (also known as **eccrine**). The apocrine glands, which begin functioning at puberty, are involved in sexual communication and probably in the "chemistry" between people. Perspiration odor is caused by bacterial action on the substances produced by these apocrine glands.

The three million or so merocrine glands are involved in thermoregulation and antibacterial action. As seen in Figure 13–4●, apocrine glands communicate with hair follicles while merocrine glands communicate directly with the surface. Obtain a slide and observe the glands. In the following space, sketch and label what you see.

Merocrine v. apocrine glands

3➡ Ceruminous glands, modified sweat glands located in the external auditory canal, secrete **cerumen,** ear wax. Refer to your lecture text or to Figures 38–2● and 38–3● in this book. Envision these glands, which are embedded deep within the subcutaneous tissue of the cartilagenous portion of the ear canal.

C. Nail

1➡ Study Figure 13–5●. In the following space, sketch and label the posterior and cutaway views of the nail. Use your own nail as a guide.

Structure of the nail

What is the general function of the nail? *[∞ Martini, p. 158]*

❑ Experimentation

III. Examination of the Integument

A. Iodine Map

The iodine map is a simple experiment demonstrating the distribution of sweat glands. Students with a known sensitivity to iodine should not do this test. (You may have set up this experiment at the beginning of your lab period. If the experiment has run for at least an hour, you should be ready to read your results.)

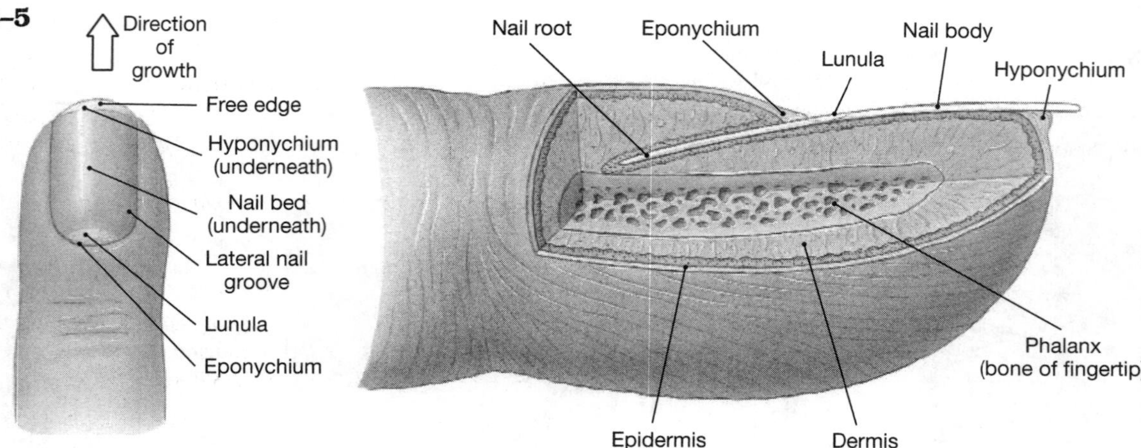

● **FIGURE 13–5 Structures of the Nail.**

1➡ Obtain some iodine, a swab, and two equal squares of bond graph paper. Measure these pieces of graph paper in square centimeters.

2➡ Choose an area on the palm of your hand. Avoid the creases. (If you are right-handed, choose your left hand, and vice versa.) Choose another area on your forearm. Swab both areas with iodine and allow the areas to dry completely. Tape a square of bond graph paper over each area. If you are adventuresome, try taping an identical paper square to some other part of your body.

3➡ After about an hour, remove the paper and count the black dots. The iodine you swabbed on your hand and arm was dissolved in the sweat in the glands. This iodine was carried to the surface where it reacted to the starch in the bond paper. Each dot indicates the location of a sweat gland. Compute the number of sweat glands per square centimeter in each area. Hand __________ Forearm ________ Other areas ______________________________

Concept Check 3 Which area seems to have the most sweat glands? Why do you think this would be the case?

____

____

B. Microscopic Examination of the Skin

1➡ Take the hand lens (or use the dissecting microscope) and examine your hand. Observe the pores, hair follicles, and nails.

2➡ Compare your epidermal ridges (fingerprints) with the epidermal ridges of your classmates. What do you notice?

____

____

3➡ Use the emery board and put a scraping from your finger nail onto a slide. If your skin is at all dry, scrape a bit of the skin onto the same slide. Examine the slide under the dissecting microscope. What similarities and differences do you see? Remember that both structures are keratinized.

____

____

C. Microscopic Examination of the Hair

1➡ With the forceps, carefully grasp a hair on some part of your body. Using a quick sharp motion, pull the hair in the direction it is growing so that you do not injure either the bulb or the shaft. Head and leg hairs are usually best because you will find these coarser hairs easier to work with.

2➡ Examine your hair and a classmate's hair under the dissecting microscope. (If you use the standard microscope instead of the stereoscopic microscope, use the lowest power lens.) Identify the bulb and the shaft. Can you discern any of the internal structures? In the following space, sketch the hair bulb and shaft.

Hair bulb v. hair shaft

Write a brief description comparing your hair with the hair of a classmate.

____

____

D. PILOERECTION

Piloerection occurs when the body hair shifts in position, becoming more perpendicular to the skin. In many animals, piloerection traps a layer of air between the skin and a raised meshwork of hair. This air is then heated by the body and provides an additional layer of insulation against external cold. (You may utilize this same principle in the winter by putting two blankets on your bed. Two thin blankets are warmer than one thick blanket because of the heated layer of air between them.) If your laboratory is cool enough, you should be able to induce piloerection.

1➡ Put a few ice cubes in a plastic bag and rest the plastic bag on your forearm or your thigh. After a minute or two, "goose flesh" — the small raised bumps in the skin — will appear. Goose flesh is caused by the contraction of the arrector pili muscles in the dermis. The contraction of these muscles pulls the hair follicles, which in turn causes the appearance of the bumps and causes the hair to "stand on end."

2➡ Use the hand lens to examine the goose flesh and the piloerected hairs.

Concept Check 4

4. Why is it more difficult to induce piloerection in a warmer laboratory? Why does this experiment usually work better on your thigh than on your forearm?

___

___

___

(Receptor experiments dealing with fine and coarse touch and heat and cold awareness can be found in laboratory Exercise 43.)

❑ Additional Activities

NOTES

1. If you have a dog or cat, bring in a sample of your pet's hair and compare it under the microscope with your own. Observe the claws on your pet and note any similarities or differences between your pet's nails and your own.
2. If your laboratory is exceptionally warm, collect a sample of sensible perspiration for some simple chemical experiments. (Sensible perspiration is released by the merocrine sweat glands to cool the skin.) You might try these experiments: 1) Add a drop of silver nitrate to a drop of perspiration. A precipitate would indicate the presence of the chloride ion. 2) Take the pH. Is the pH higher or lower than the pH of your saliva? 3) Use assorted medical test strips to check for glucose, ketones, urea, blood, or protein that may be present in your perspiration.
3. Examine the palmar, plantar, scalp, and abdominal skin slides as well as the slides of the axillary sweat glands.

Answers to Selected Concept Check Questions

1. In thick skin the stratum corneum is much thicker and can generally be seen as layered. The stratum lucidum is usually only present in the thick skin. The stratum granulosum is generally thinner in thick skin.
2. The arrector pili muscle will cause the hair to stand on end.
3. You will probably have more sweat glands on your hand.
4. It is more difficult to induce piloerection in a warmer lab because the warmer lab has an ambient temperature closer to body temperature. Thus it will be more difficult to "convince the body" of the "need for warmth." The induction of piloerection is easier on the thigh because the thigh has thinner skin.

❑ Lab Report

1. Referring to the numbered inquiries at the beginning of this exercise, complete the following box summary:

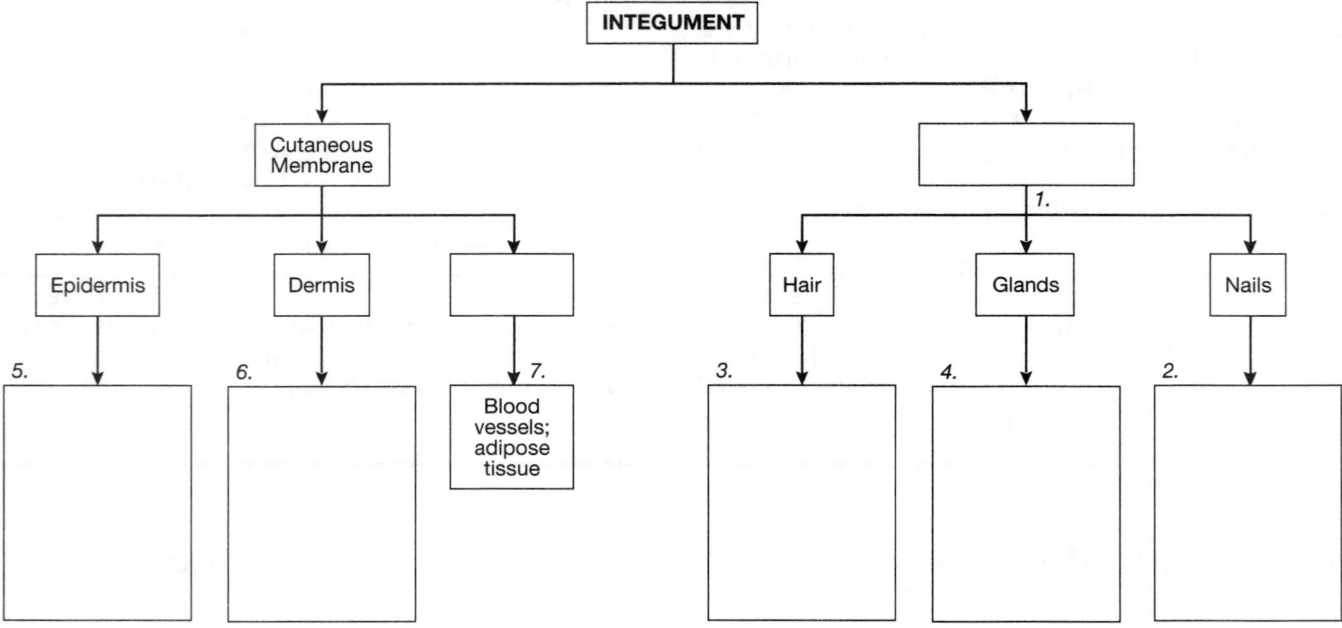

General Questions

NOTES

1. In which layer or layers of the skin would you find each of the following:
   a. Melanocytes
   b. Sebaceous glands
   c. Blood vessels
   d. Arrector pili muscles
   e. Adipose tissue
   f. Nerve endings
2. Relate the arrector pili muscles to piloerection. What function does piloerection serve?
3. When examining the integument, where would you find the basement membrane?
4. What would happen if a person who was deficient in melanocytes spent a great deal of time hiking in the mountains? What precaution should such a person take in everyday life? Why?
5. The hair shaft is produced by ________________ cells which are found ________________________________.
6. Explain why the iodine map is a valid test for determining the location of sweat glands.
7. Where would you find the papillary layer?
8. What product keeps keratin flexible?
9. Where would you find the ceruminous glands?
10. The mammary glands are modified ______________________ glands.
11. The ceruminous glands are modified ______________________ glands.
12. Merocrine glands communicate with the ______________________.

EXERCISE 14

The Skeleton: An Overview

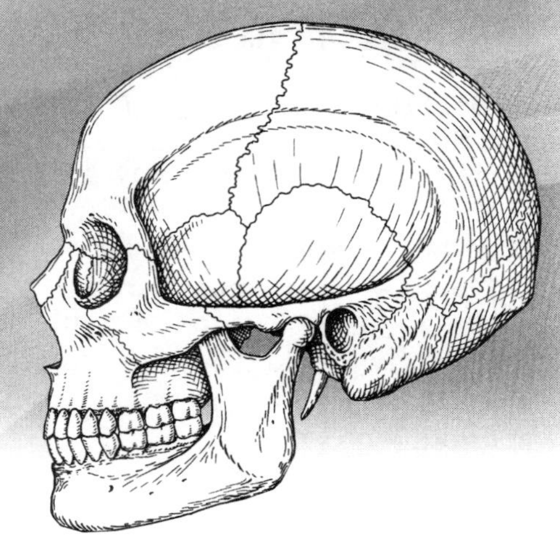

PROCEDURAL INQUIRIES

Examination

1. What are the major connective tissue types found in the skeleton?
2. What are the differences between dense bone and spongy bone?
3. What are the major structural landmarks found in the cross section of a bone?
4. What are the types of bone marrow? What is the function of each?
5. What are the types of cartilage connected with the skeletal system? Where is each located?
6. What are the structural and functional differences between tendons and ligaments?
7. What are the basic bone locations?
8. What are the basic bone shapes?

Practice

9. How do we classify bones according to body location?
10. How do we classify bones according to shape?

Additional Inquiries

11. What are the major chemical components of bone and how is each of these components identified?
12. What is the difference between intramembranous and endochondral ossification?

Key Terms

Bone
Canaliculi
Cancellous
Cartilage
Central Canal
Chondrocytes
Compact
Dense
Diaphysis
Elastic Cartilage
Endosteum
Epiphyseal Line
Epiphyseal Plate
Epiphysis
Fibrocartilage
Flat Bones
Haversian Canal
Haversian System
Heterotopic Bones
Hyaline Cartilage
Irregular Bones
Lacunae
Lamellae
Ligaments
Long Bones
Medullary Cavities
Metaphysis
Osteocytes
Osteons
Perforating Canals
Periosteum
Red Bone Marrow
Sesamoid Bones
Short Bones
Spongy
Sutural Bones
Tendons
Trabeculae
Volkmann's Canals
Wormian Bones
Yellow Bone Marrow

Materials Needed

Blunt Nose Probe
Articulated Skeleton (with muscle delineation, if possible)
Disarticulated Skeleton
Articulated Skulls
Bones previously soaked in Nitric Acid [Prepared the day before]
Bones previously baked dry [Prepared the day before]

Fresh Beef Bone, including a joint — Longitudinal Slice
[This is available from any grocery or meat store.]
Chicken Leg — upper and lower
Round bone — from steak or roast
(preferably cut close to the epiphysis)
[This is available from any grocery or meat store.]
Compound Microscope
Stereoscopic Microscope
Prepared Slides of
Ground Bone
Periosteum (if available)
Hyaline Cartilage
Fibrocartilage
Intervertebral disk (if available)

Although many of us think of the skeleton as just the compilation of bones, we must recognize that the skeleton is more than simply the 206 formally recognized bones of the body. The skeleton (or the skeletal system) is actually composed of four major types of connective tissue: bone, cartilage, ligaments, and tendons. Bone and cartilage are themselves specific connective tissues. Ligaments and tendons are composed of connective tissue fibers. (If necessary, return to Exercise 11 and review the properties of the various types of connective tissues.)

Numerous other organs and tissue types are intimately involved with the skeletal system. We know muscles are essential to skeletal movement. The endocrine system functions in bone growth. Because the skeletal tissues are composed of living material, nutrients are supplied to and waste products are removed from the skeletal system by the cardiovascular system. (The cardiovascular system actually connects the skeletal system with every other system in the body.)

The most obvious functions of the skeletal system are movement, support, and protection. To give a few examples: the bones move as a result of muscle contraction; the vertebral column helps support the thorax; and the skull protects the delicate brain.

Although most of the better-known skeletal functions do indeed center around the bones, it is important to remember that the cartilage, tendons, and ligaments also have specific skeletal functions. The cartilage is involved in the embryonic, connective, and protective activities we will be examining as this unit progresses. The ligaments are the fibers connecting one bone with the next bone. The tendons connect the muscles to the bone. The anatomical synergy between the bones and the cartilage, ligaments, and tendons, and their associated structures, make movement, support, and protection possible.

The bones are also responsible for calcium and phosphate storage. In addition, the yellow bone marrow stores lipids, and blood cell production occurs in the red bone marrow. The bones also operate as levers changing both the magnitude and the direction of forces generated by the skeletal system. *[∞ Martini, p. 168]*

❑ Examination

I. Skeletal Overview

A. Bone

Bone is the connective tissue chemically identified by its crystals, primarily crystals of calcium phosphate, $Ca_3(PO_4)_2$, and, to a lesser extent, crystals of calcium carbonate, $CaCO_3$. These salts make up almost two-thirds of the weight of the bone. Crystalline calcium salts are strong but quite brittle. *[∞ Martini, p. 170]*

The main component of the rest of the bone is collagen. Collagen fibers are long, straight, unbranched, three-ply protein chains that are highly flexible though not nearly as strong as the calcium salts. Since the calcium salts are organized around the collagen fibers, the combined portions of the osseous (bony) tissue exhibit the best of both worlds. The result is a strong, flexible, and relatively shatter-resistant structure.

1 ➡ Examine the chemical properties of bone by comparing the bone soaked in nitric acid with the bone baked in the dry oven. Tap these bones with your probe. Try to bend the bones. Note Figure 14–1●. Recall from chemistry that salts often react with acids and that heat destroys protein.

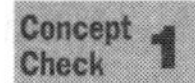

Explain the properties you have observed in these treated bones.

____

____

2 ➡ Study Figure 14–2●. Bone is often described as **compact (dense)** or **spongy (cancellous)**. These terms refer to the architectural arrangements of the osseous components. Dense bone consists of well-aligned, densely packed substructures. Spongy bone is composed of seemingly random plates, **trabeculae**, jutting out at odd angles from one another. Identify these architectural components in Figure 14–2●.

As a generalization, bones have a spongy inner layer and a dense outer layer. The physics of this arrangement allows for the greatest strength to

● FIGURE 14–1
(a) Bone soaked in acid; **(b)** Bone baked in oven.

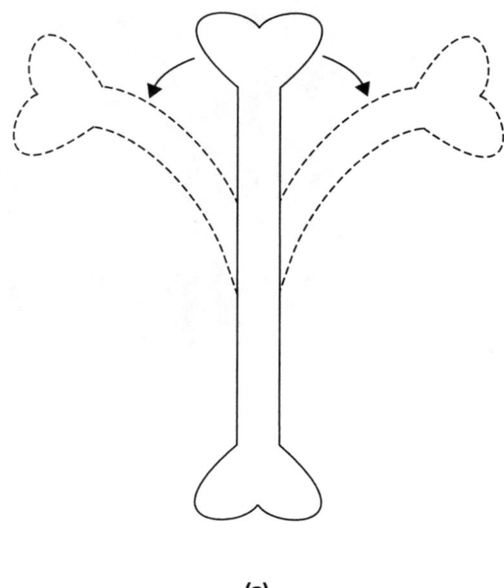

(a)

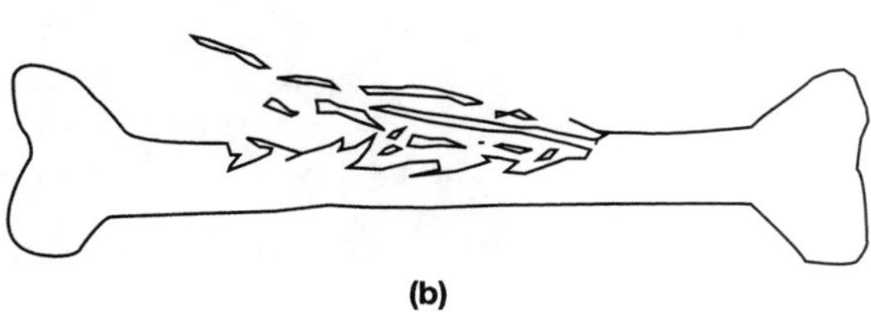

(b)

mass ratio for the specific function of the bone. Why is this logical?

3➡ Obtain a round bone and a stereoscopic microscope. Note the trabeculae. Sketch and label what you see.

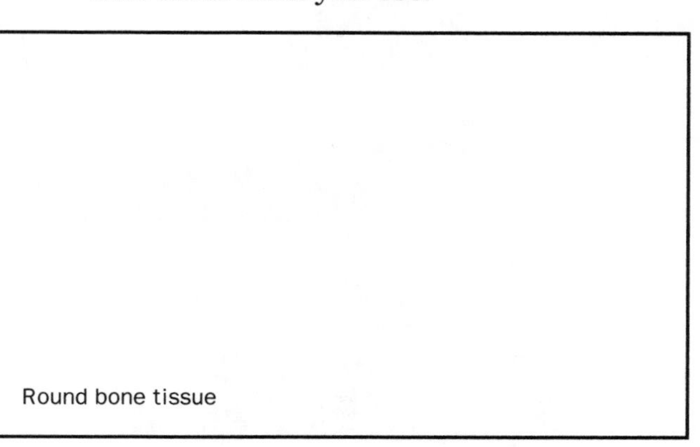

How does your diagram compare with Figure 14–2●?

4➡ Examine the longitudinal cut of the beef bone. In addition to the dense and spongy layers of the bone, the shafts of the long bones are often hollow in the middle. These **medullary cavities** house the lipoidal **yellow bone marrow** (for storing energy reserves in the form of lipids). **Red bone marrow** (which produces red and white blood cells as well as other blood products) is found within the spaces of the spongy-type bone. The marrow may still be in place in your specimen. If not, you can at least observe the medullary cavity. Sketch and label what you see.

Beef bone (longitudinal cut)

5➡ Try to discern the delicate **endosteum** as you examine the medullary cavity. The endosteum is the inner layer separating the bone proper from the marrow. The endosteum also lines the central canals. Check Figure 14–2●. If you cannot identify the endosteum on this longitudinal cut, the next time you have a fresh beef bone at home, gently pull away any marrow and try to find the layer adhering to the marrow cavity. This layer is the endosteum. If you have a magnifying glass at home, try using it to observe your fresh

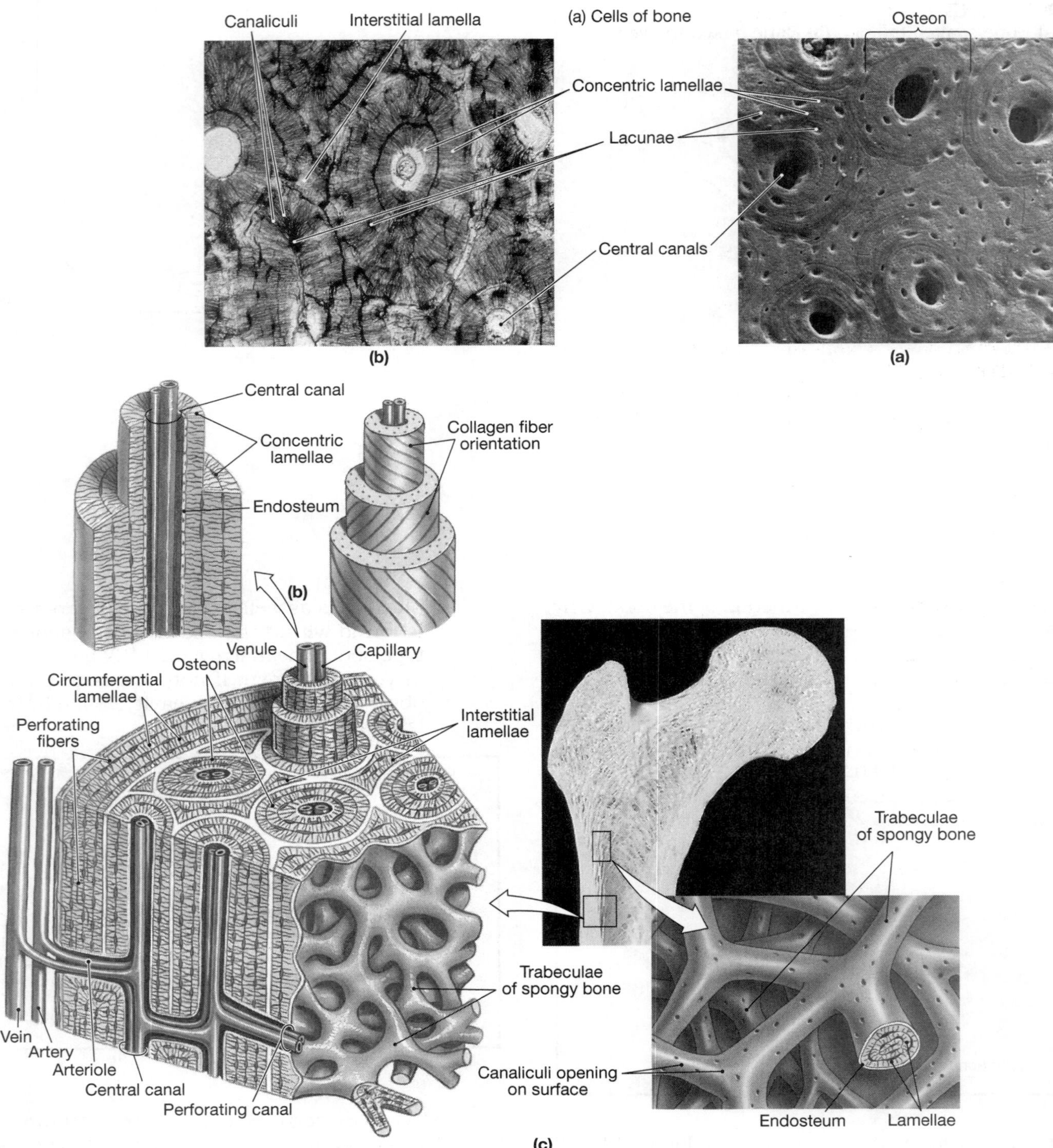

● **FIGURE 14–2**
Structural Organization of a Bone. (a) The cells of osseous tissue and a scanning electron micrograph of several osteons in compact bone. (SEM × 182) **(b)** A thin section through compact bone; in this procedure the intact matrix and central canals appear white, and the lacunae and canaliculi are shown in black. (LM × 272) **(c)** Diagrammatic view of the structure of a representative bone. (Reproduced from R. G. Kessel and R. H. Kardon, *Tissues and Organs: A Text-Atlas of Scanning Electron Microscopy*, W.H. Freeman & Co., 1979.)

beef bone. (Use caution in your identification because as you try to dislodge the marrow; it can "get caught" in the trabeculae and you may not be seeing what you think you are seeing.)

6➡ Refer back to Figure 14–2●. Notice the orderly cylinders making up the bone proper. These are the **osteons** or **Haversian systems**. The concentric circles are the **lamellae** (sing. lamella). The black spaces within the lamellae are the **lacunae** (sing. lacuna), wherein reside the **osteocytes** or bone cells. Lacuna can also be found in the interstitial lamellae. Note the location of the interstitial lamellae.

The **canaliculi** are the canals running between the lacunae. In addition to housing the cytoplasmic streamers of the osteocytes, the canaliculi serve as the passageway for osteocytic nutrients and waste products. Within the **central canal**, the **Haversian canal**, are the blood vessels of the bone. The Haversian vessels and canals are longitudinal. The bone also contains transverse vessels found in transverse **perforating canals** (or **canals of Volkmann**). Check Figure 14–2●.

(We sometimes forget that bone cells are living cells carrying on all normal metabolic functions and, since their osseous matrix prevents the rapid diffusion of the life fluids, a means must exist for substances to get back and forth between the cells. Thus we have this intricate canal system.)

7➡ Obtain a compound microscope and a prepared slide of ground bone tissue. (The term "ground" has two meanings here. The first meaning is that of tissue proper. In other words, in this case, ground bone is a piece of generic bone tissue, complete with matrix and assorted microscopic landmarks. The second meaning of ground comes from the fact that bone must often be "ground up" so that small, light transversing fragments can be produced.)

8➡ Use Figure 14–2● as your guide and identify these structures: lamella, lacuna, osteon, canaliculi, central canal.

Concept Check 2 You will not be able to locate any osteocytes within the lacunae. Why not?

____

____

Sketch and label what you do see.

Ground bone tissue

9➡ Refer back to Figure 14–2●. Observe the fibrous and cellular layers of the **periosteum**, the covering isolating the bone from the surrounding tissue. If you have a slide showing the periosteum, examine that slide. Sketch and label what you see. If you do not have such a slide, rely on the accuracy of Figure 14–2● and make note of the periosteal components.

Periosteum

10➡ Look at Figure 14–3●. Note that near the joints the periosteum is continuous with the joint capsules. The periosteum is also continuous with the tendons attaching the muscles to the bone. The collagen fibers of the tendon actually become cemented into the superficial lamellae, thus weaving the tendon into the general structure of the bone. Some students are amazed at how many muscles are attached to many of the bones. If your classroom skeleton delineates the muscle attachments, examine that now. In Exercises 15 through 18 we discuss some of these muscle attachments.

B. Cartilage

Cartilage is the avascular connective tissue found connected to and in between many bones. We usually identify three major types of cartilage, two of which are directly connected with the bones. (To review cartilage, refer back to Exercise 11.)

Hyaline cartilage

Hyaline cartilage is composed of close collagen fibers, making it tough and slightly flexible. Hyaline cartilage is found on the articular surfaces of bones. Hyaline cartilage also connects the ribs to the sternum and is the chief component of the embryonic skeleton.

1➡ Obtain a slide of hyaline cartilage. Note the **chondrocytes** (cartilage cells) within the lacunae. The collagen fibers are too fine to be seen using ordinary staining procedures and ordinary microscopes. Look around the slide and assure yourself that the cartilage is indeed avascular. Because it is avascular, nutrients and wastes must diffuse through the matrix. Sketch your hyaline cartilage. Label the lacunae and the chondrocytes.

Hyaline cartilage

2➡ Gain a perspective of the hyaline cartilage by taking apart a chicken leg. Study the smooth joint surfaces where the bones meet. Also, return to the beef bone and examine the articulating surfaces of that joint.

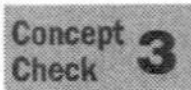

Describe the structure and the apparent function of the cartilage in its present articulating position.

____

____

3➡ Examine the articulated skeleton in your lab. Note the cartilagenous connections between the ribs and the sternum. (This hyaline cartilage is probably plastic on your model, but the principle is the same.)

Fibrocartilage

Fibrocartilage, with little matrix and abundant collagen, is a support tissue able to withstand considerable compression. Fibrocartilage is found at the symphysis pubis (the articulation between the two pelvic bones), the intervertebral discs (between the vertebrae), and the menisci of the knee joint.

4➡ If you have a slide of an intervertebral disc, use this slide to study the wavy collagenous fibers. Use the information in Exercise 11 as your guide. Sketch and label what you see.

Fibrocartilage

5➡ Examine the articulated skeleton. Although the original symphysis pubis and the original intervertebral discs have been replaced by ones more conducive to classroom models, by examining these pads you should gain an understanding of the gross functional anatomy of the fibrocartilage.

Elastic Cartilage

Elastic cartilage, which is a component of the external ear and the epiglottis, does not play a direct role in the skeletal system and is mentioned here only to give you a more complete picture of cartilage.

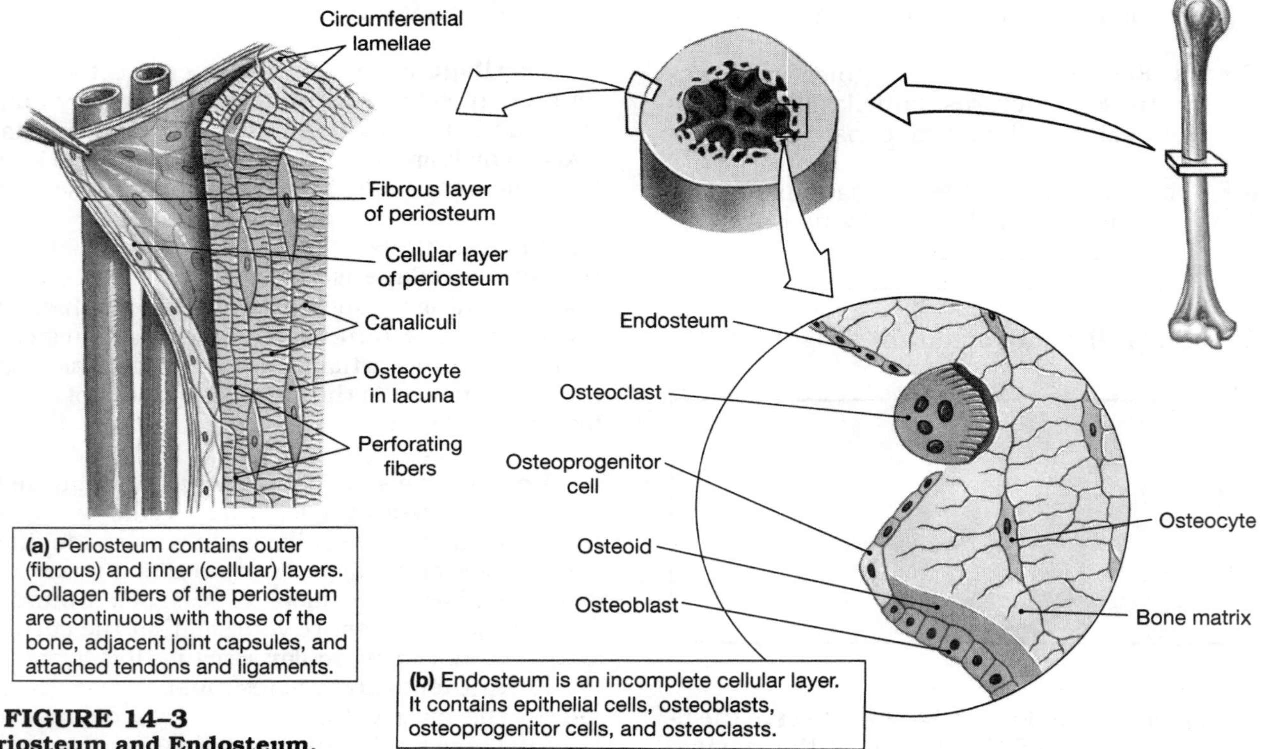

(a) Periosteum contains outer (fibrous) and inner (cellular) layers. Collagen fibers of the periosteum are continuous with those of the bone, adjacent joint capsules, and attached tendons and ligaments.

(b) Endosteum is an incomplete cellular layer. It contains epithelial cells, osteoblasts, osteoprogenitor cells, and osteoclasts.

● **FIGURE 14–3**
Periosteum and Endosteum.

C. Ligaments

Ligaments connect one bone with another. Ligaments are generally composed of elastic and collagen fibers. This allows for some stretch.

1 ➡ Study Figure 14–4●. Your instructor may point out some ligaments on the longitudinal section of the beef bone. If no ligaments can be found there, search through some pictures and diagrams — including those in your lecture text — to gain a perspective on how ligaments hold bones together.

D. Tendons

Tendons are composed of collagenous connective tissue: dense regular connective tissue consisting mostly of collagen fibers. All the collagen fibers are tightly packed and uniformly aligned. Tendons are usually studied with the muscle unit. When examining the skeleton, however, it is important for you to remember that tendons play a vital role in the skeletal system, too. Bones could not move if they were not attached to the muscles via the tendons.

1 ➡ Go back to your fresh beef bone. Do you see any remnants of meat attached directly to the bone? ____________ If possible, examine this phenomenon under the stereoscopic microscope. You may be able to identify the tendons. Sketch what you observe.

Beef bone tendons

The next time you cook steak or beef or pork ribs, examine the muscle-to-bone connections. Do you see why it is sometimes so difficult to get "all the meat off the bone?" ____________

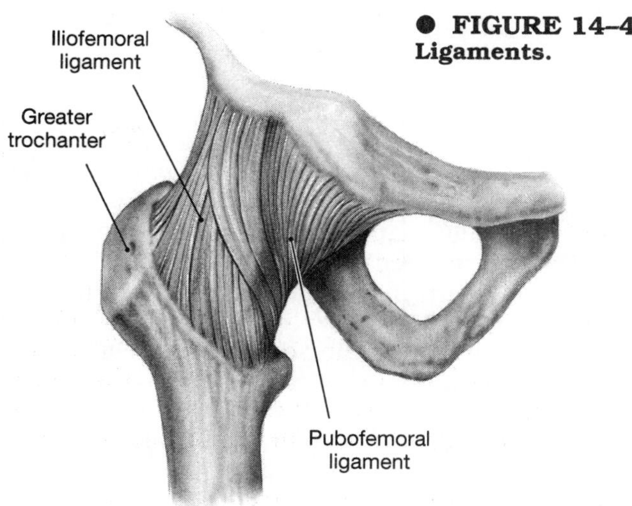

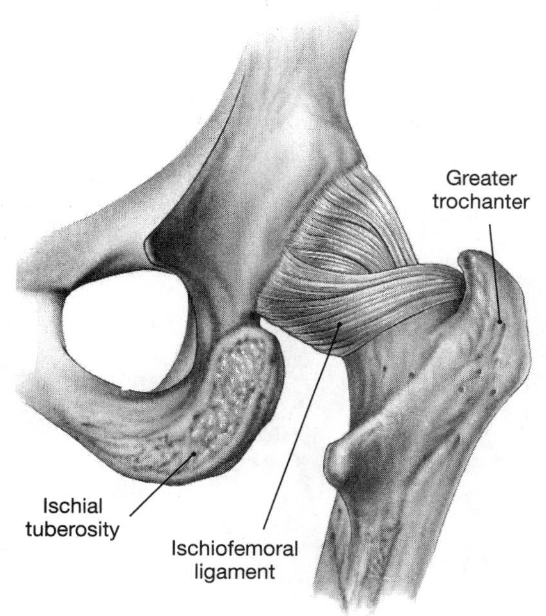

● **FIGURE 14–4 Ligaments.**

II. Classification

The most common ways of classifying bones are according to their formation, location, and shape.

A. Formation

Bones can readily be classified according to their ossification patterns (the way in which they were formed).

Intramembranous Ossification
Intramembranous bones began when connective tissues started producing collagen fibers and bone matrix. The intramembranous bones include the roof of the skull, the lower jaw, the collar bone, and the sesamoid bones. *[∞ Martini, p. 175]*

1 ➡ Locate bones formed by intramembranous ossification on the skull or articulated skeleton.

Endochondral Ossification [∞ Martini, p. 176]
Endochondral bones develop from hyaline cartilage precursors. Chondrocytes hypertrophy and die. The area, an ossification center, is invaded by osteoblasts and blood vessels. Soon bone tissue is being put in place. The ossification process will

continue for another 18 or 20 years. Most of the bones of the body develop this way.

During your second month of prenatal life, your axial and appendicular cartilage began to form. The cartilagenous component of your skeleton will continue forming until you die. This is why the noses and ears in some elderly people seem much larger proportionately than they do in younger people.

Ossification began during your fourth week, and by the third month of your prenatal life, the ossification centers were spreading. The epiphyseal plates began forming during your eighth prenatal month. (Fetal ossification is explained more fully in Exercise 19.)

Many ossification centers are no longer functional after the epiphyseal plates close (at about age 18 or 20 years). Many potential ossification sites, however, continue throughout life to be at least functional on demand.

This is important if we think about how bones readily heal when broken and how bones occasionally fuse, even when we don't want them to. Ossification is also evident when, in certain individuals such as some lumberjacks and some tennis players, stress is applied to certain areas and heterotopic (irregular) bones begin forming.

When you are born you have about 270 separate bones. By your mid-20's, due to fusion (as discussed later in this unit), this number dwindles to the standard number, 206. As you work through this and the succeeding exercises in this unit, make note of how we can justify these numbers and why you yourself may not have exactly 206 bones.

B. Location

We can also classify bones according to body location. The next few exercises in this unit use the **appendicular** and **axial** classification scheme. The appendicular skeleton includes the bones of the pectoral and pelvic girdles, the bones along the horizontal axes of the body. The axial skeleton deals with those bones along the vertical axis of the body.

C. Shape

Another way of classifying bones is according to shape. We generally recognize six basic shape categories.

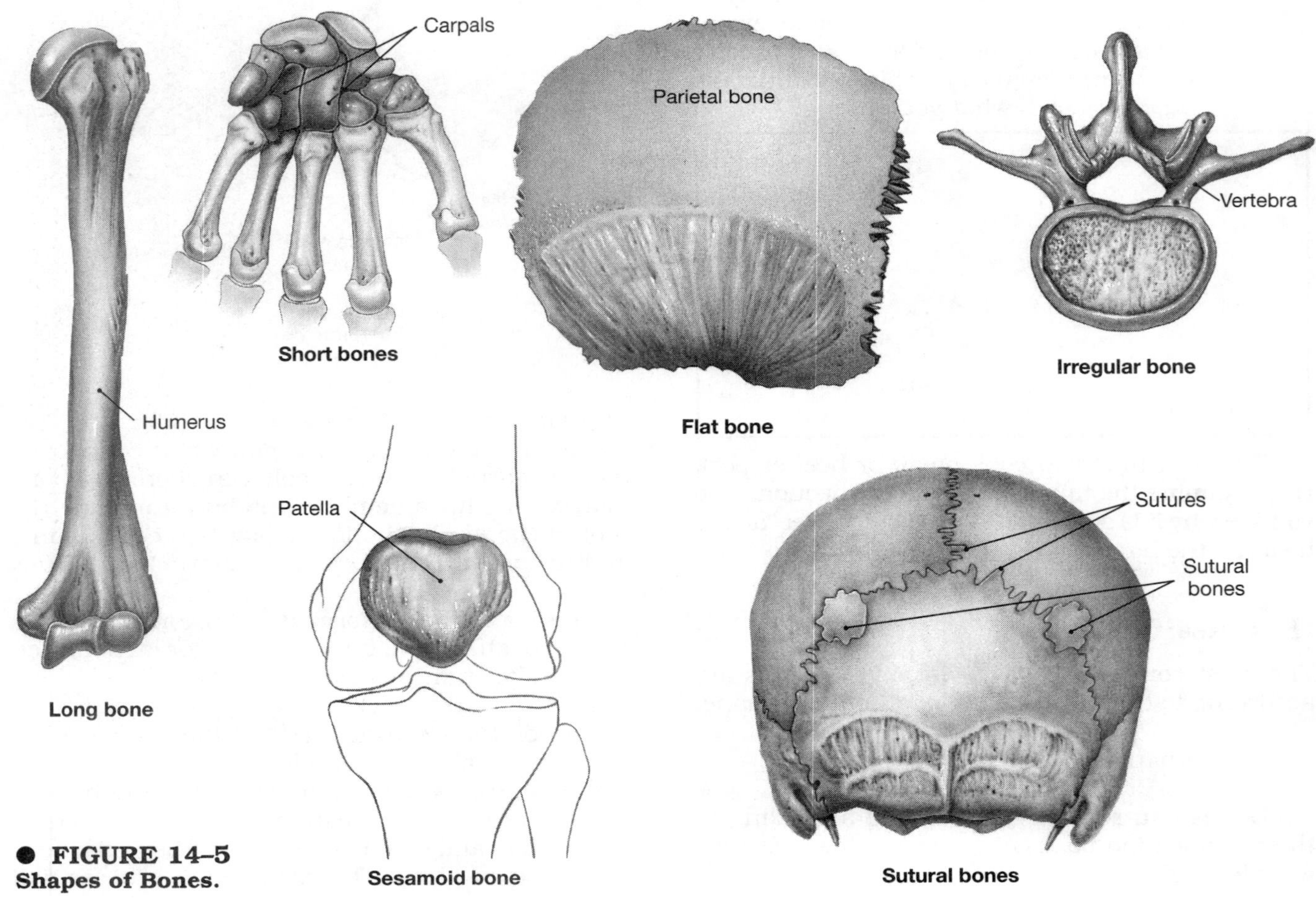

● **FIGURE 14–5**
Shapes of Bones.

1➡ See Figure 14–5● for a delineation of bone shapes.

Long Bones

Long bones, such as the femur, are longer than they are wide.

2➡ Look at the long bone in Figure 14–5●. Identify the two **epiphyses**, the broad ends of the bone. Also identify the **diaphysis**, the shaft of the bone. Which is the proximal and which is the distal epiphysis?

____

____

3➡ Return to the longitudinal beef bone and locate the epiphyses, diaphysis, and the medullary cavity. See Figure 14–6●.

4➡ Observe the cartilagenous streak line between the epiphysis and diaphysis. This line is the **epiphyseal plate**, sometimes called the **metaphysis**. The final ossification of the cartilage on either side of this plate marks the end of long bone growth. The vestigial plate is called the **epiphyseal line**.

Short Bones

Short bones are generally boxy, or cuboidal. The tarsals (ankle bones) and carpals (wrist bones) are short bones.

5➡ Locate some short bones on the articulated skeleton.

Flat Bones

Flat bones are thin and plate-like, such as the flat bones of the skull.

6➡ Locate some flat bones on the skull or articulated skeleton.

Irregular Bones

Irregular bones, such as the vertebrae and the ethmoid bone (an oddly shaped bone forming part of the nose), cannot be categorized as one particular shape. Thus, they are called irregular.

7➡ Locate some of the irregular bones on the articulated skeleton.

Sesamoid Bones

Sesamoid bones are irregular bones that usually develop within a tendon. The patella (knee) is a sesamoid bone. Everyone has patellae, but other sesamoid bones may develop in almost any connective tissue. It is not unusual to find sesamoid bones in the wrists of people engaged in heavy labor. Bones developing in unusual places are sometimes called **heterotopic bones**.

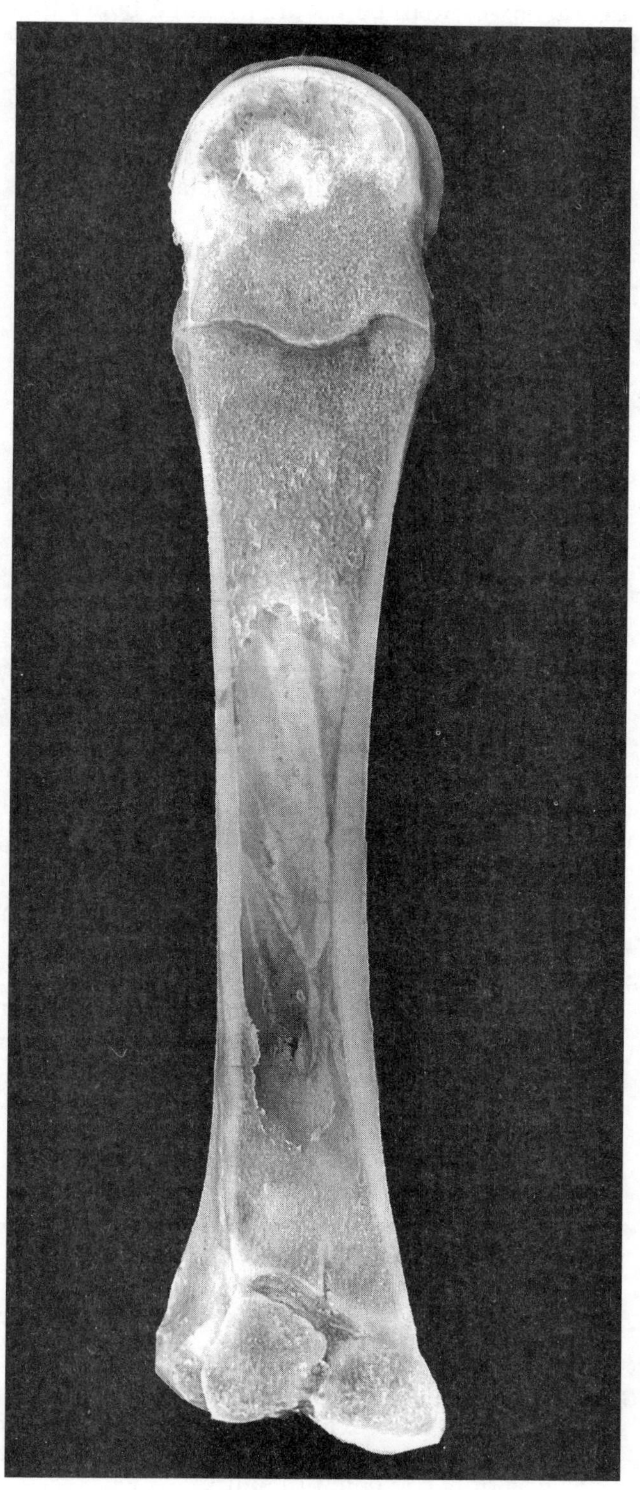

● **FIGURE 14–6**
Mammalian Long Bone.

8➡ Locate some sesamoid bones on the articulated skeleton.

Sutural Bones

Sutural or **wormian bones** are irregular bones that develop along suture lines. The skeleton in your classroom may have sutural bones, particularly along the lambdoidal suture of the skull. Check Figure 14–5●.

9➡ Locate some sutural bones (if possible) on the skull or articulated skeleton.

❑ Practice

III. Application of Classification Principles

A. Classification According to Location

1➡ Work with a partner and classify as many of the bones of the body as you can as a part of either the appendicular or axial skeleton. Work with the articulated skeleton. Use the space below to begin your list. Continue the list according to your instructor's directions.

______________	______________
______________	______________
______________	______________
______________	______________
______________	______________
______________	______________

B. Classification According to Shape

1➡ Work with a partner and classify as many of the bones of the body as you can according to their particular shapes. Work with the disarticulated skeleton. Use the space below to begin your list. Continue the list according to your instructor's directions.

______________	______________
______________	______________
______________	______________
______________	______________
______________	______________
______________	______________

❑ Additional Activities

1. Find out the specific chemical reactions that take place when the bones are placed in nitric acid.
2. What is it about dinosaur bones that leads many paleontologists to believe that dinosaurs were warm-blooded?
3. Research the life and scientific careers of:
   a. Clopton Havers (1650–1702), English anatomist.
   b. Alfred W. Volkmann (1800–1877), German physiologist.
4. Compare the structure and function of human bones with the structure and function of bird bones. Think particularly about the flight patterns of birds.

NOTES

Answers to Selected Concept Check Questions

1. The acid-soaked bone will be rubbery because the crystalline matrix has been destroyed, leaving only the flexible protein. The heated bone will be brittle or flaky because the collagen (protein) fibers have been destroyed, leaving only the calcium salts.
2. You will not see osteocytes in the lacunae because the bone tissue is no longer living. Osteocytes are living cells.
3. The cartilage covers the epiphysis of each bone. Based on our gross observation, the function appears to be protective.

❑ Lab Report

1. Referring to the numbered inquiries at the beginning of this exercise, complete the following box summary:

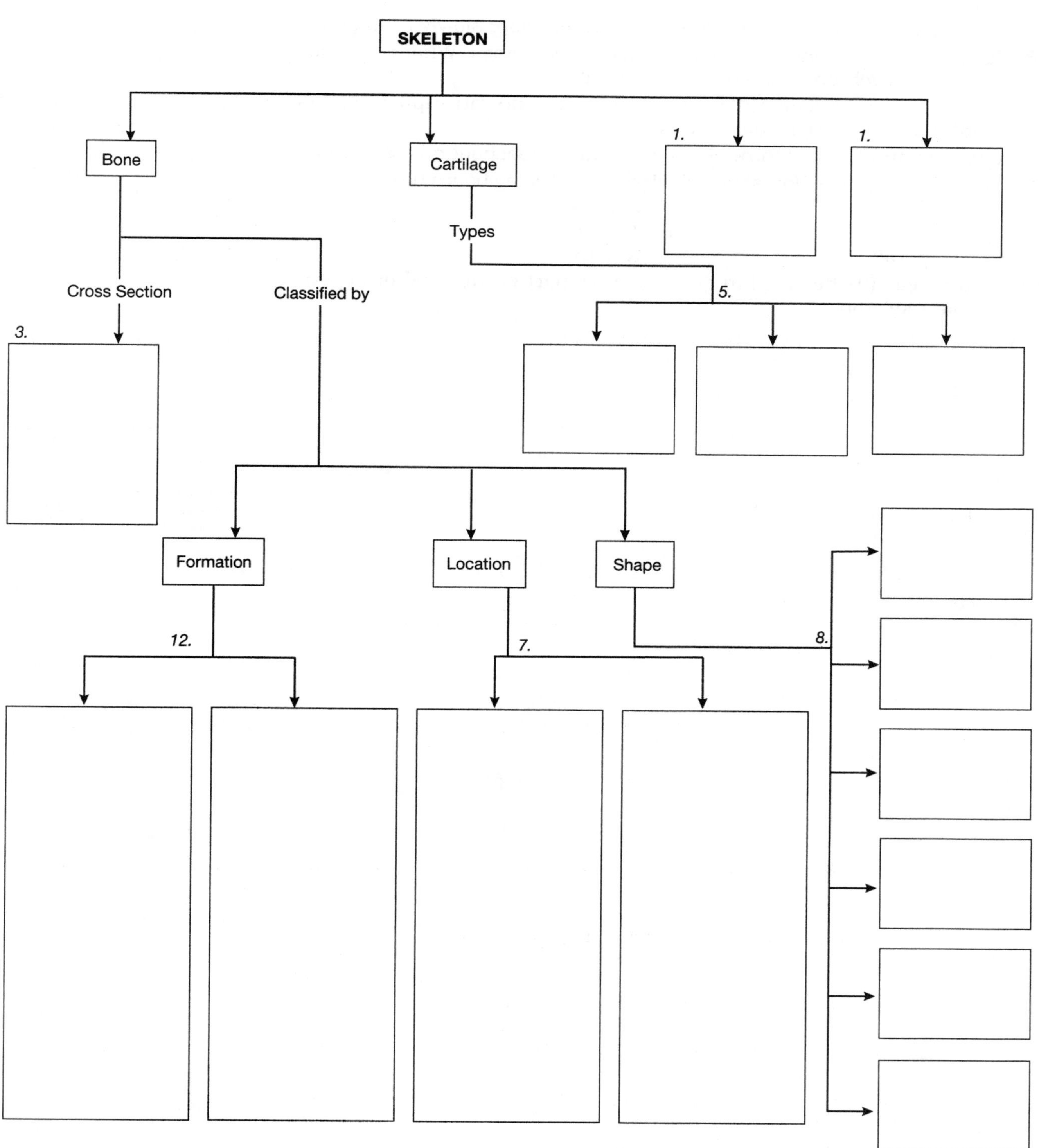

General Questions

NOTES

1. What are the differences between compact bone and spongy bone?
2. What are the types of bone marrow, and what is the function of each?
3. Compare and contrast bone and cartilage.
4. What are the structural and functional differences between the tendon and the ligament?
5. Why do you suppose cartilage heals so much more slowly than bone?
6. Identify the three major bone classification schemes. What is the advantage of each bone classification scheme?
7. Tendinitis is an inflammation of a tendon. Why do you suppose this is painful and potentially dangerous?
8. Compare and contrast intramembranous and endochondral ossification.
9. According to shape, how should each of the following bones be classified?
   a. Tibia
   b. Digit
   c. Temporal
   d. Coccyx
   e. Tarsal
   f. Radius
10. Identify each bone listed in question 9 as part of the axial or appendicular skeleton.

EXERCISE 15

Skeletal Terminology

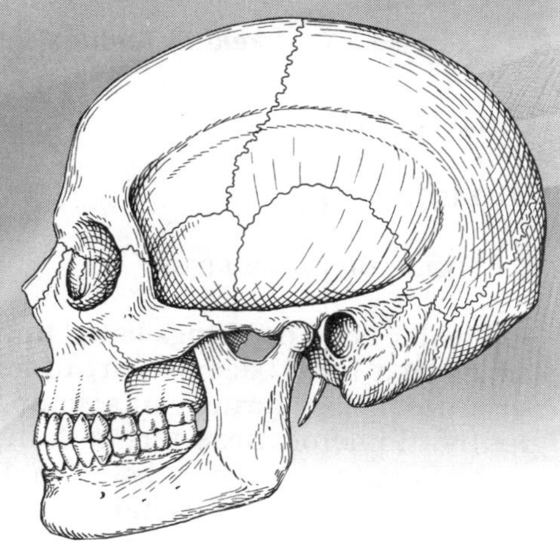

PROCEDURAL INQUIRIES

Preparation

1. What is skeletal terminology?
2. Why is skeletal terminology important?

Practice

3. What are the major bone markings.
4. What are some additional terms used to describe skeletal components.

Materials Needed

None required

Studying the skeletal system can be a formidable task without an understanding of **skeletal terminology**. Skeletal terminology is more than simply the names of the bones. In addition to names, bones have distinctive features, and the anatomical features of the bones are not just whims of nature. Rather, each part of each bone has a functional value related to the entire anatomy and physiology of the skeletal system. (This functional value will become apparent to you as you proceed through this unit and the succeeding units.) It is therefore important that you understand the terminology used to describe both the bone and the parts of the bone.

Since skeletal terminology is the descriptive summation of the bones and their anatomical landmarks, this exercise is included here to introduce you to the "language" of the skeleton and to help you develop a terminology base that can be applied to the rest of your skeletal studies. To accomplish this we have included here a list of **major bone markings**, or major landmarks, plus a list of **additional terms** often associated with the skeletal system.

❑ Preparation

I. Approaching Skeletal Terminology

A. Study Hints

You will probably complete this exercise either as an outside assignment or in conjunction with one or more of the other skeletal exercises in this unit. We urge you not to skimp on your work with this terminology. We suggest the following study procedure:

1➡ Learn these markings first by saying the terms and their definitions out loud and then by quizzing your classmates on the specific meanings of the words. Several of you might sit together and quiz each other by going around the circle.

2➡ Try using the names of the markings in everyday conversation, even in contexts that don't have much to do with the skeleton. The purpose of this is to imprint the terminology in your mind.

3➡ Go through the diagrams in Exercises 14 and 16–19 and highlight the terms you learned in this exercise.

4➡ Add to these lists when you find additional skeletal terminology that may be used in your particular school or place of employment.

5➡ Keep this exercise close at hand while you work on the other skeletal system exercises so you can refer to these lists as needed.

❑ Practice

II. Terminology

A. Major Bone Markings

The major bone markings listed in this section are terms that have become part of formal landmark names. These terms can also be used independently. Some of these terms do have definitions unrelated to the skeletal system. We are only interested in the skeletal terminology here.

Antrum Chamber or pocket within a bone

Alveolus Blind pocket

Condyle Large, knoblike convex process

Crest Prominent ridge on a bone

Epicondyle Process above a condyle

Facet Flattened, smooth articular surface

Fissure Narrow, slit-like opening for passage of nerves and/or blood vessels

Foramen Opening in a bone for passage of nerves and/or blood vessels

Fossa Shallow depression in a bone

Head Rounded articular surface at the end of a bone

Line Low ridge on a bone

Meatus Canal in a bone

Process Extension (of a bone)

Ramus Branch (of a bone)

Sinus Cavity within a bone

Spine Sharp, slender process of a bone

Sulcus Groove in a bone

Trochanter Very large projection, only on the femur

Trochlea Pulley, or structure functioning like a pulley

Tubercle Small roughened process or projection on a bone

Tuberosity Large roughened process or projection on a bone

B. Additional Terms

Some of the following terms are used to identify other skeletal structures or lesser bone markings. All of these terms can be used independently. In some cases, additional definitions for terms may exist. The definitions given here are those pertaining to the human skeletal system. Some of these terms were defined in Exercise 4 but are redefined here because of their relevance to the skeletal system.

Anterior Toward the front

Apex Tip or pointed end

Appendicular Pertaining to the arms or legs

Arch Narrow curved structure

Articular Pertaining to a joint

Auricle Ear-shaped appendage

Axial Pertaining to the longitudinal axis

Base Broad area of a triangular structure

Canal Narrow opening going through a structure

Concha Thin, scroll-like bones

Cornu Shaped like a horn

Eminence Area projecting out from another area

Girdle Structure or band around another structure

Inferior Beneath or below

Inter- Between

Intra- Within

Lateral Toward the side

Medial Toward the middle

Neck Constricted portion of a structure

Notch Indented area

Ovale Oval-shaped structure

Posterior Toward the back

Superior Above or on top of

Suture Immovable joint

Transverse Across

❑ Additional Activities

NOTES

1. Explain why practice with the use of correct skeletal terminology is to your advantage even though it may occasionally seem a bit awkward.
2. Search through your lecture textbook and find some additional uses of the terms included in this exercise.
3. Find additional terms that could be used in a skeletal terminology exercise. Use these terms as appropriate.
4. Refer back to Exercise 4 and correlate the anatomical terminology with the information included here and through Exercises 16–19.

❑ Lab Report

1. Referring to the numbered inquiries at the beginning of this exercise, complete the following box summary:

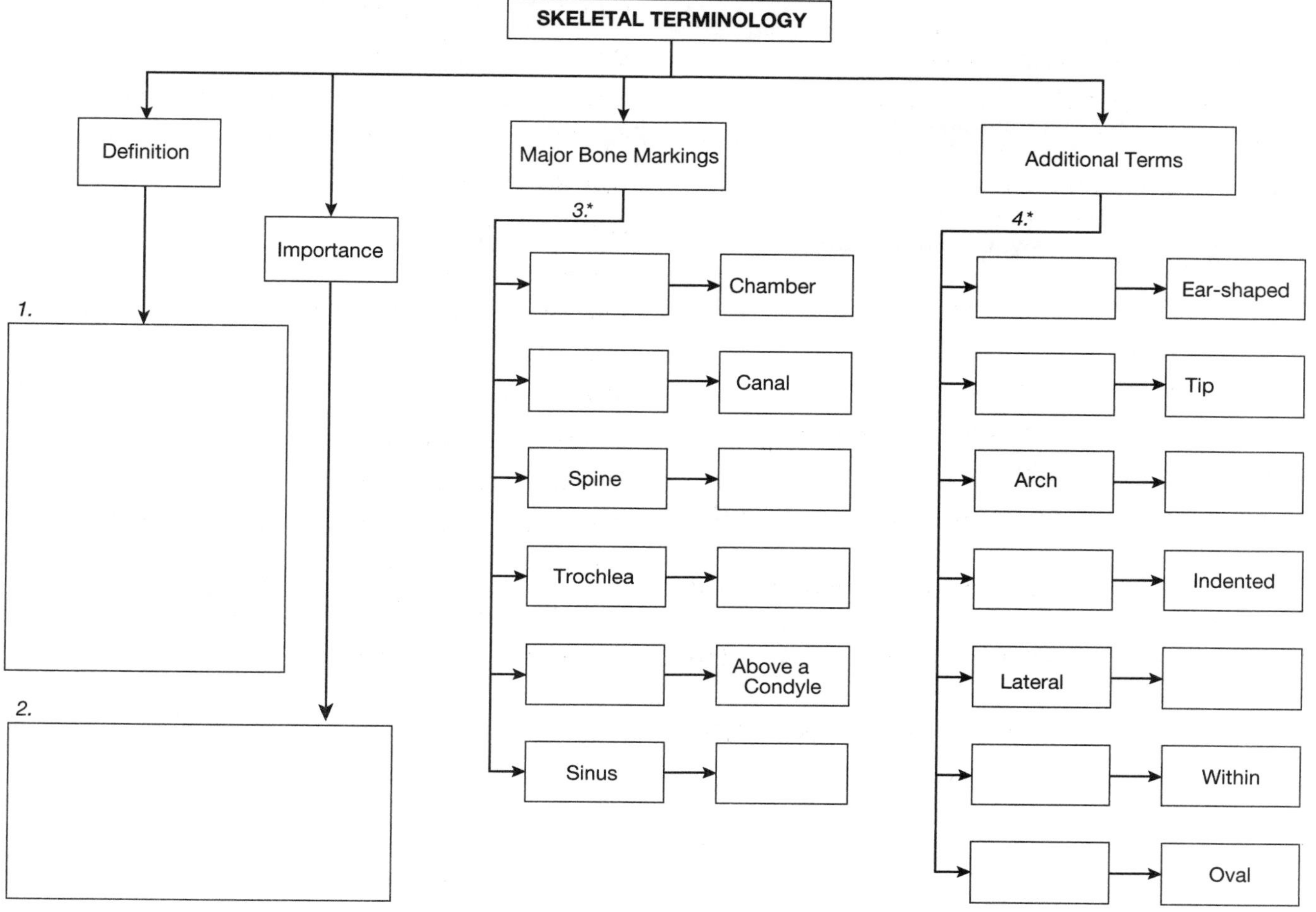

*Continue according to your instructor's directions.

General Questions

NOTES

1. Use the preceding definitions to explain the following locations.
   a. The epicondyle was inferior to the medial notch.
   b. He found the artificial suture on the neck of the bone.
   c. The transverse process is lateral to the ramus of the bone.
   d. The base of the patella is superior to the apex.
   e. The tibial tuberosity is superior to the anterior crest.
   f. The intercondylar eminence articulates with the intercondylar fossa.
   g. The elbow exhibits trochlear motion.
   h. The intertubercular groove is on the anterior humerus.
   i. The external auditory meatus is also known as the ear canal.
   j. The infraspinous fossa is inferior to the spine of the scapula.
2. What is the difference between:
   a. Line and crest
   b. Condyle and epicondyle
   c. Tubercle and tuberosity
   d. Fissure and foramen
   e. Facet and fossa
   f. Meatus and sulcus
   g. Base and apex
   h. Axial and appendicular
   i. Inter- and intra-
   j. Anterior and posterior
   k. Medial and lateral
   l. Superior and inferior
3. Give some real or imaginary locations using the terminology in this exercise.
4. Check through the sections of your lecture textbook that deal with bones and bone markings. Make note of how the preceding terms are used. Are there any terms you do not understand? *[∞ Martini, Chapters 6–9]*

EXERCISE 16

Axial Skeleton

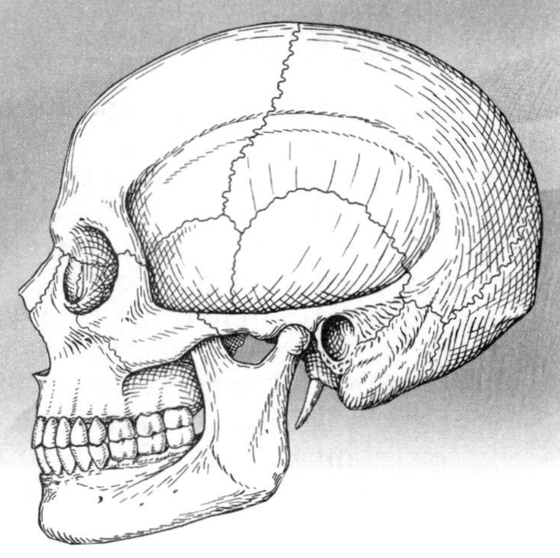

PROCEDURAL INQUIRIES

Preparation

1. What are the divisions of the axial skeleton?
2. What are the divisions of the skull?
3. What are the bones of the cranium?
4. What are the bones of the face?
5. What are the bones of the thoracic cage?
6. What are the distinguishing characteristics of each type of vertebra?
7. What are the principal markings of the bones of the cranium?
8. What are the principal markings of the bones of the face?
9. How are the ossicles related to the rest of the skull?

Examination

10. How can you demonstrate the relationship between the vertebrae?
11. How can you demonstrate the articulations between the ribs and the vertebrae and the ribs and the sternum?

Additional Inquiries

12. What is unique about the hyoid bone?
13. What differences exist between the sacral and the coccygeal vertebrae?

Key Terms

Cranium
Ethmoid
Facial bones
Frontal
Hyoid
Incus
Inferior Nasal Concha
Lacrimal
Malleus
Mandible
Maxilla
Nasal
Occipital
Ossicles
Palatine
Parietal
Rib
Skull
Sphenoid
Stapes
Sternum
Temporal
Thoracic Cage
Vertebra
Vomer
Zygomatic

Materials Needed

Articulated Skeleton
Disarticulated Skeleton
Articulated Spinal Column
Assorted Disarticulated Vertebrae
Skulls (including disarticulated [or Beauchene] skull if available)
Ossicles Set (if available)
Probe
Abnormal Skeleton or Abnormal Axial Skeletal Components (if available)

The axial skeleton, those bones along the perpendicular axis of the human body, normally consists of 80 bones. Some of these bones are paired; some are not. Learning the names and the parts of the bones is a "hands on" task, best accomplished by holding the bones, feeling the markings, and running your probe around the anatomical landmarks. If you are working with a partner, quiz your partner out loud so that you both are bringing as many senses as possible into the learning experience. When you begin to study a particular bone, first locate that bone on the articulated skeleton. Make note of any specific identifying points you may observe.

When you are working with the bones of the skull, use the diagrams along with the articulated skull to help you gain a perspective on the size and location of the individual bones. If your laboratory has a disarticulated skull, pay particular attention to the deep irregular bones of the face.

For the bones other than the skull, after you have located the bone on the articulated skeleton, proceed to the disarticulated skeleton. Take the bone you are studying and examine it thoroughly.

For all bones, read the descriptions given here, then use the diagrams in this lesson to identify the boldface bone markings listed in this exercise. [∞ *Martini, Chapter 7*]

❑ Preparation

As you will learn in your study of anatomy, another name for a joint, the junction between bones, is *articulation*. Thus, an articulated skeleton is one where the bones are in place as they would be in the living organism. In the disarticulated skeleton the bones are separated. Both are valuable in the study of the skeleton.

I. Divisions of the Skeleton as a Whole

A. The Appendicular Skeleton

This division consists primarily of the bones of the limbs and will be the subject of Exercise 17.

B. The Axial Skeleton

The axial skeleton consists of the bones that form the longitudinal axis of the body (from head to foot). Following is an outline of the axial skeleton. The divisions of this skeleton are printed in upper-case letters, and the names of the specific bones are in lowercase letters. The numbers in parentheses indicate how many of that particular bone can be found in the human body.

1➡ Refer to Martini Figure 7–1● to observe the axial skeleton as part of the articulated skeleton.

2➡ Identify the major division of the axial skeleton. Note the specific bones in each section. Compare the articulated skeleton in your laboratory with this outline. We will be studying these bones in greater detail as the exercise progresses.

❑ SKULL

CRANIUM

Frontal	(1)
Parietal	(2)
Temporal	(2)
Occipital	(1)
Sphenoid	(1)

FACIAL BONES

Nasal	(2)
Maxilla	(2)
Lacrimal	(2)
Zygomatic	(2)
Inferior Nasal Concha	(2)
Vomer	(1)
Palatine	(2)
Mandible	(1)
Ethmoid	(1)

Bones Associated with the Skull

Hyoid	(1)

OSSICLES

Incus	(2)
Malleus	(2)
Stapes	(2)

❑ THORACIC CAGE

Sternum	(1)
Ribs	(24)

❑ VERTEBRAL COLUMN

Vertebrae	(26)

❑ Examination

II. Skull [∞ *Martini, p. 195*]

For convenience , the **skull** can be divided into the **cranial bones** and the **facial bones**. To work through the bones of the skull, use Figures 16–1● through 16–6● as your guides. Try to observe the bones from as many angles as possible. Relate your observations to the skulls. If your lab has a disarticulated (Beauchene) skull, you will find this helpful for understanding the relative positions of the bones of the skull.

A. Cranium

The **cranium** consists of the eight large and easily recognizable bones that surround the **cranial cavity**. The cranial cavity, also known as the cranial vault, houses the brain. The superior portion of the cranium forms the skullcap or **calvaria**.

1➡ Use Figure 16–1a● as you view the skull anteriorly. Note the **frontal bone**, the large bone forming the forehead. The inferior edge of this bone can be found deep within the orbit of the eye and at the superior edge of the nose. The raised area located deep to the eyebrow is known as the **superciliary arch**. The inferior edge of this arch or ridge is the **supraorbital margin**, and the small opening on this ridge is the **supraorbital**

foramen. The smooth area above the nose is the **glabella**. The **frontal sinuses** are located within the frontal bone, immediately superior and medial to the orbits.

2➡ Find the **parietal bones**, which are directly posterior to the frontal bone, one on each side of the skull. The articulation between the parietal bones is the **sagittal suture**. By putting your hand on top of your head, you may be able to feel this suture line. The articulation between the frontal bone and the parietal bones is the **coronal suture**, which you may also be able to feel.

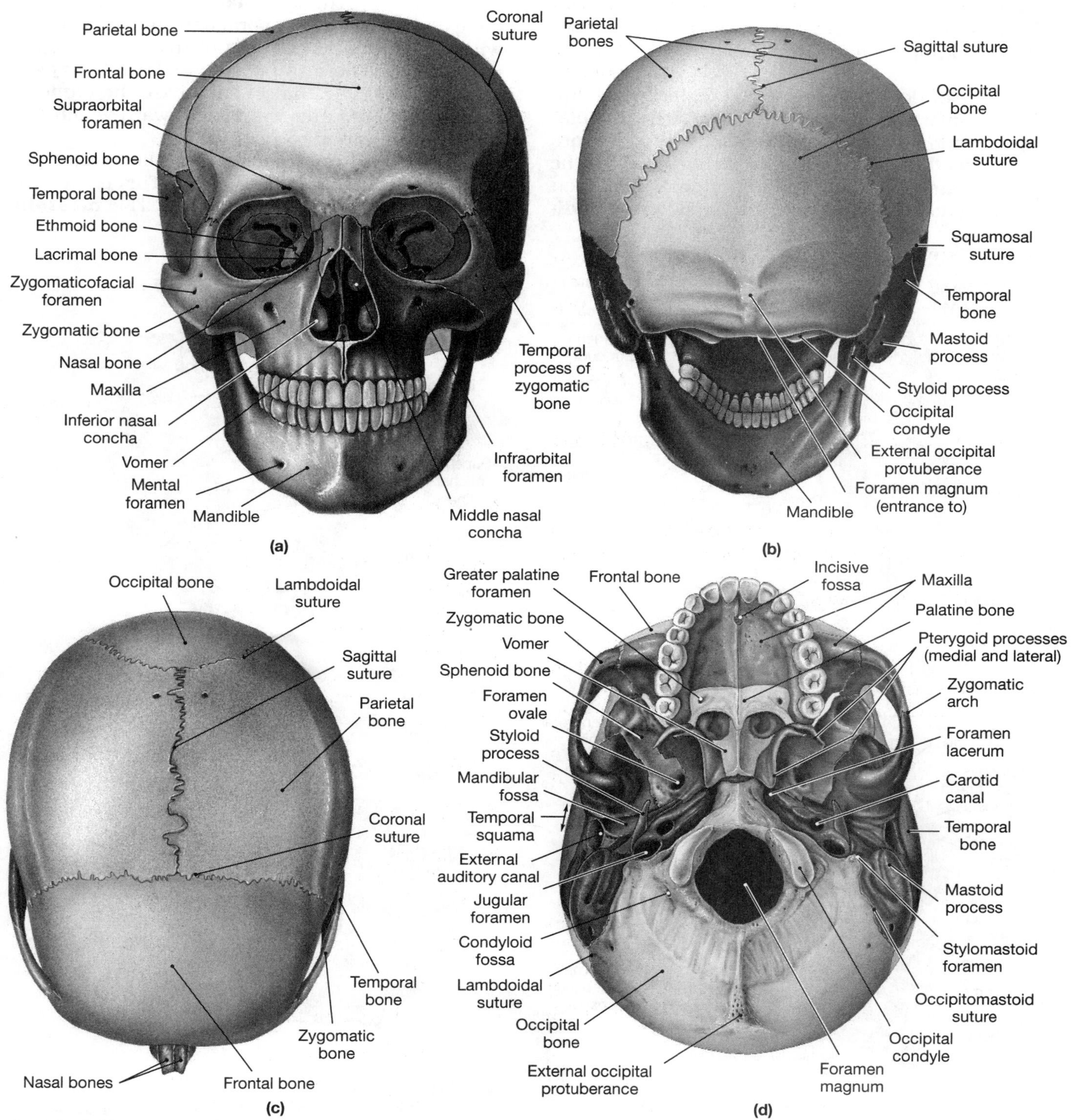

● **FIGURE 16–1**
Anatomy of the Skull. **(a)** Anterior surface; **(b)** Posterior surface; **(c)** Superior surface; **(d)** Inferior surface.

On the lateral surface of the parietal bone, you should notice two semicircular ridges which extend into the frontal bone. These are the **superior** and **inferior temporal lines**. These lines are points of muscle attachment. Superior to these lines is a smooth area known as the **parietal eminence**. Along the posterior third of the sagittal suture you may be able to find the **parietal foramen**. Note that the interior surface of the parietal bone is marked with the impressions of a number of branching veins and arteries.

3➡ Now locate the **temporal bones**, which are inferior to the parietal bones. The articulation between the temporal and parietal bones is the **squamosal suture**. By observing the lateral surface of the temporal bone you should be able to locate the **external auditory canal**, also known as the ear canal. Anterior to this meatus you will notice an oval depression known as the **mandibular fossa**. This fossa articulates with the mandibular condyle, forming the **temporomandibular joint** (the site of the jaw dysfunction TMJ).

Examine the three significant processes of the temporal bone. The **zygomatic process** is a slender extension of the temporal bone that articulates with the **temporal process** of the zygomatic (or cheek) bone. These two processes form the **zygomatic arch**. (Note that in each case the name of the process indicates the bone toward which the process is pointing.)

The **styloid process** is a spine-like projection at the base of the temporal bone. (This process is often lost in classroom models.) The **mastoid**

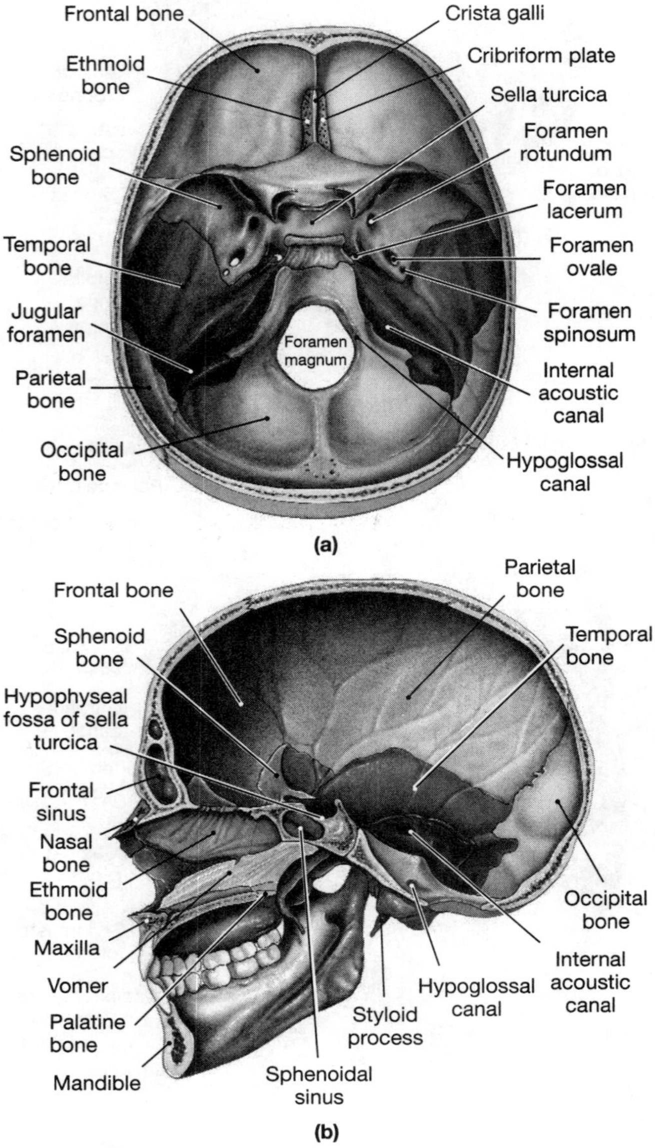

(a)

(b)

● **FIGURE 16–2**
Sectional Anatomy of the Skull.

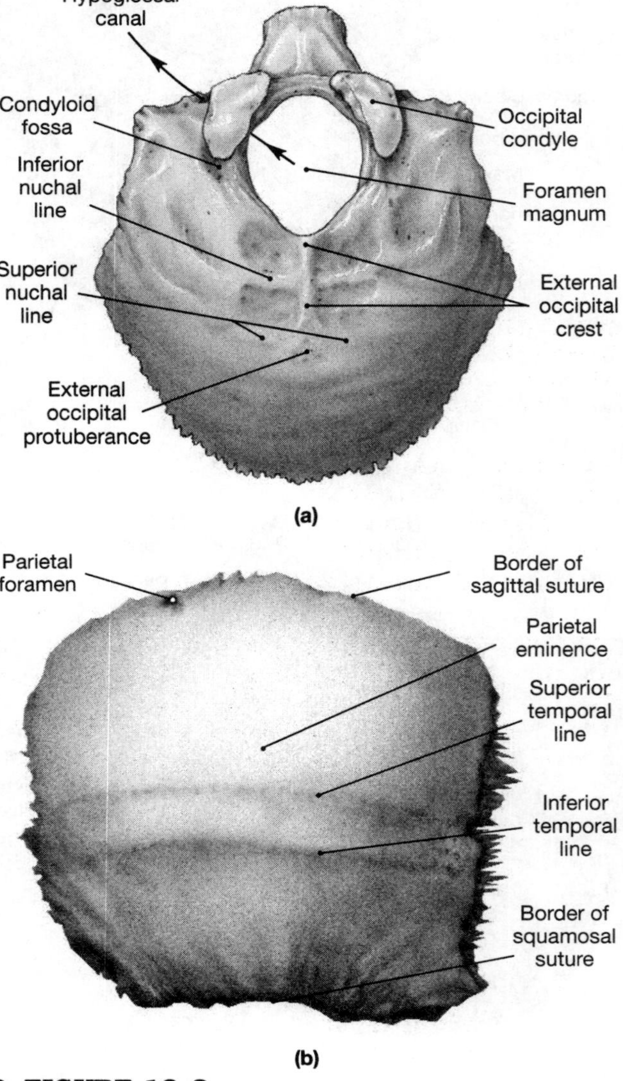

● **FIGURE 16–3**
Occipital and Parietal Bones. (a) Posterior surface of the occipital bone; **(b)** Lateral surface of the parietal bone.

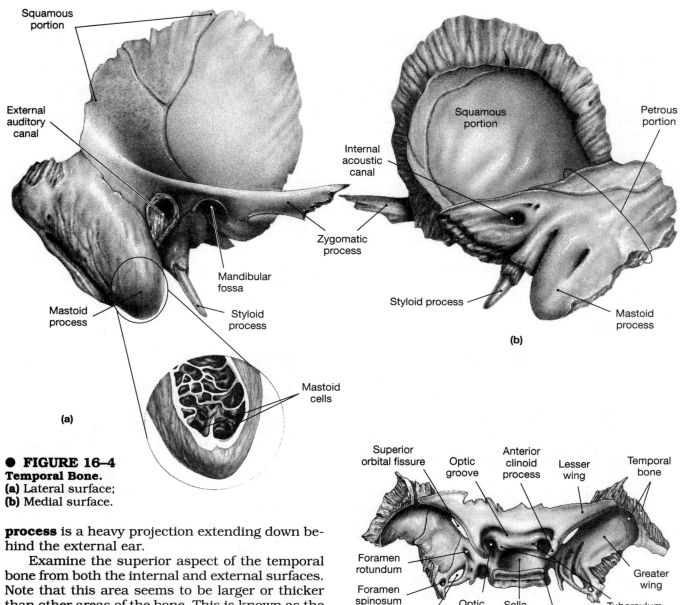

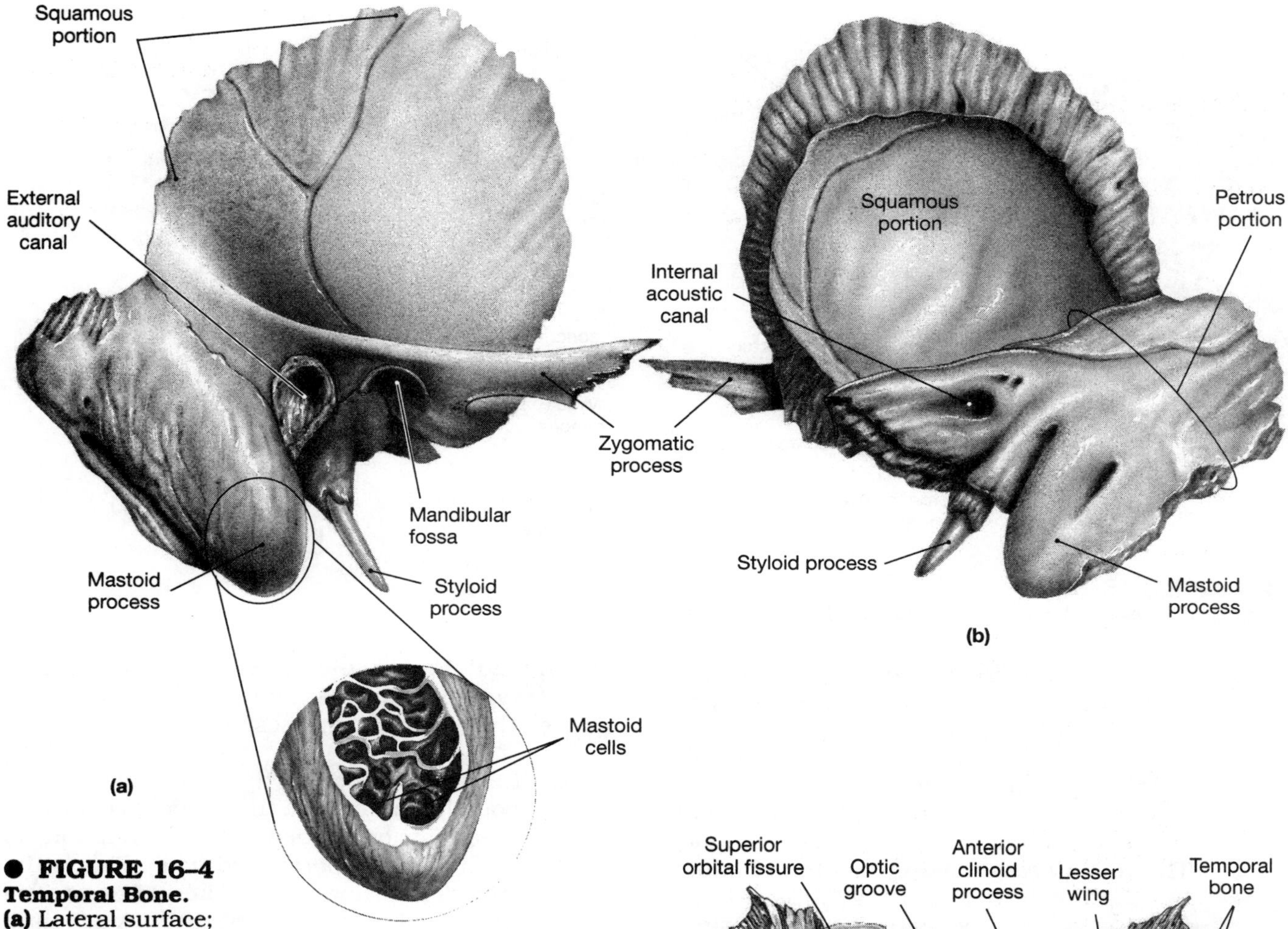

● **FIGURE 16–4**
Temporal Bone.
(a) Lateral surface;
(b) Medial surface.

process is a heavy projection extending down behind the external ear.

Examine the superior aspect of the temporal bone from both the internal and external surfaces. Note that this area seems to be larger or thicker than other areas of the bone. This is known as the **petrous portion**. Petrous means rocky or dense, and this portion of the bone is enlarged to house the ossicles and inner ear structures.

4➡ Move posteriorly and observe the **occipital bone**, which articulates via the **lambdoidal suture** with both of the parietal bones. On the inferior surface of the occipital bone find the **foramen magnum**. Flanking the foramen magnum are the **occipital condyles**, the articulation points with the fossae of the atlas, the most superior bone of the vertebral column. The **occipital crest** extends posteriorly from the foramen magnum and terminates in a centrally located bump, the **external occipital protuberance**. The ridges that intersect the crest are known as the **inferior** and **superior nuchal lines**.

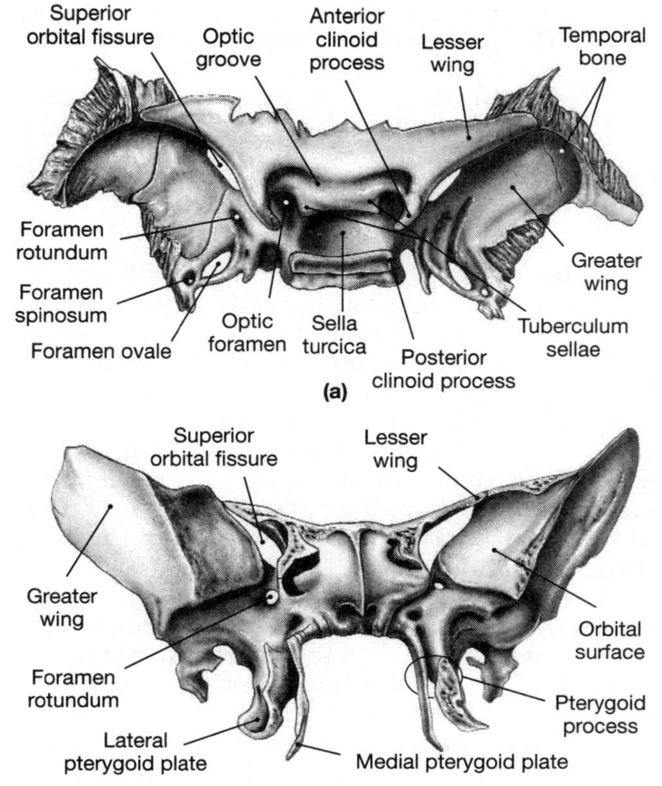

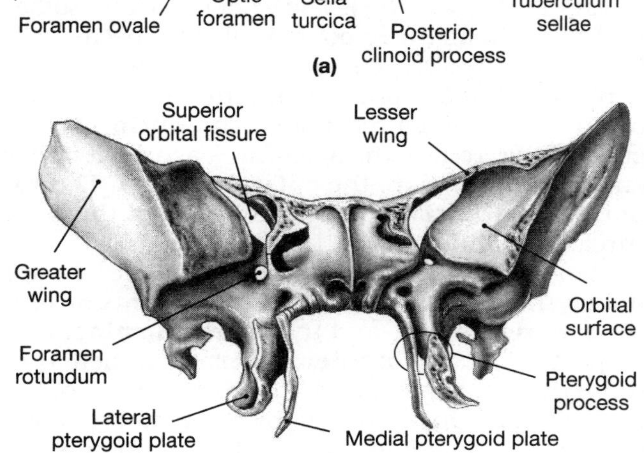

● **FIGURE 16–5**
The Sphenoid. **(a)** Anterior surface; **(b)** Superior surface.

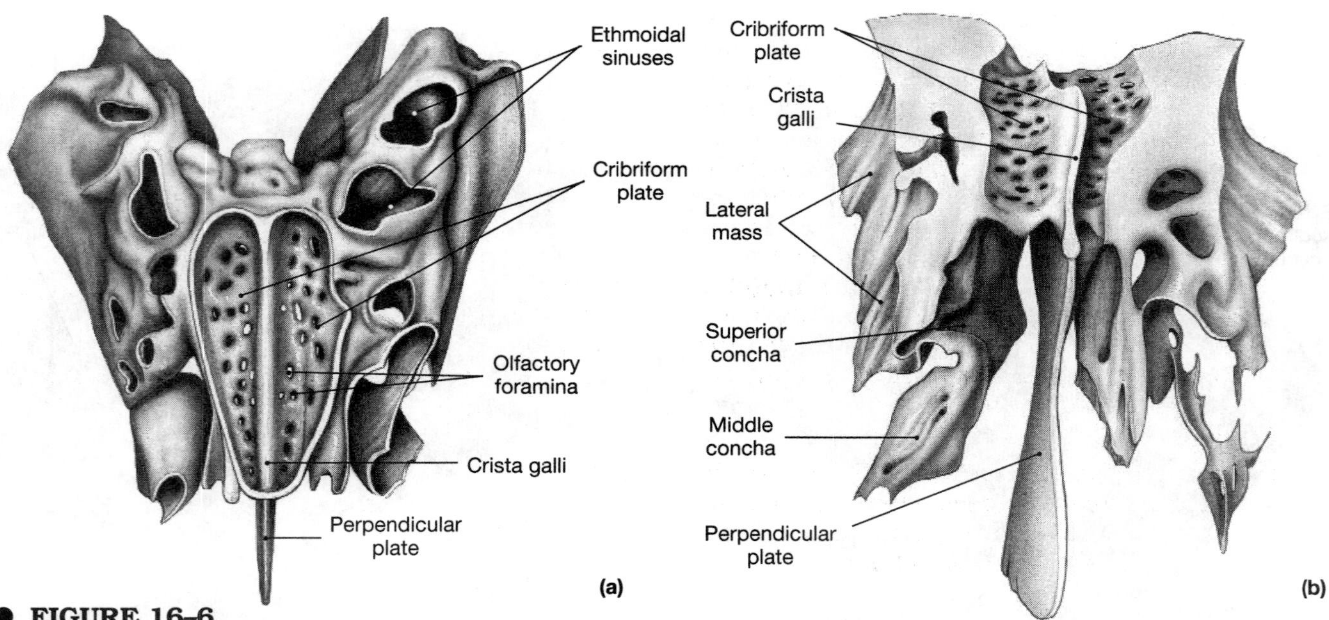

● **FIGURE 16–6**
Ethmoid Bone. **(a)** Superior surface; **(b)** Anterior surface.

5➡ Now find the **sphenoid bone**. Note that only a small portion of the large, bat-shaped sphenoid bone can be seen from either the anterior or lateral aspects of the skull. To view the sphenoid bone, examine the floor of the cranial vault. Note the prominent ridge that articulates with the frontal bone. This is the **lesser wing** of the sphenoid bone. Note how this wing "dips down," forming an area beneath the lesser wing. This larger, inferior, wing-like area is the **greater wing** of the sphenoid bone. Part of this greater wing can be seen from the lateral view of the skull.

In the center of the wings is a smooth saddle-like depression known as the **sella turcica** (which means Turkish saddle). The sella turcica is sometimes known as the hypophyseal fossa because in this indentation rests the pituitary gland (hypophysis). Several prominent foramina can be located in the sphenoid bone. The **optic groove** crosses the front of the sella turcica, and at either end of the groove are the **optic foramina**. The **superior orbital fissure** is to the side of the sella turcica, just "below" the lesser wings.

Posterior to the superior orbital fissure you will notice three foramina, the **foramen rotundum**, the **foramen ovale**, and the **foramen spinosum**, respectively. The **foramen lacerum** is medial to these three foramina, and the posterior wall of the foramen lacerum is formed by the temporal bone and the occipital bone. Thus, this foramen is between the temporal and occipital bone.

By examining the inferior aspect of the skull, you can find three sphenoid processes, the **lateral** and **medial pterygoid plates** and between them the **pterygoid hamulus**. The sphenoid bone also houses the sphenoid sinuses.

6➡ Examine the floor of the cranial cavity, and note that anterior to the sphenoid bone and seemingly embedded in the frontal bone is an apparently perforated bone. This is the superior portion of the **ethmoid bone**. The perforated area is the **cribriform plate**, and the ridge in the center of the plate is the **crista galli**. (The olfactory nerves traverse the cribriform plate.) By viewing the ethmoid bone from the anterior or from a midsagittal section, you can see the **perpendicular plate**, which extends down into the nose forming part of the nasal septum. Lateral to the perpendicular plate are the **superior** and **middle nasal conchae**. These conchae are parts of the ethmoid bone. The lateral portions of the ethmoid bone are known as the **lateral masses**. These masses house the ethmoid sinuses.

B. FACIAL BONES [∞ MARTINI, P. 205]

Of the fourteen facial bones, the large anterior bones are readily visible and easily identified. For an accurate identification of the deep facial bones — those bones forming parts of the oral, nasal, and orbital complexes — you will need to examine the skull from several perspectives.

1➡ Find the **nasal bones**, the small bones located between the eyes just inferior to the frontal bone.

2➡ Locate the **maxilla**, the bones of the upper jaw, which begin lateral to the nasal bones and extend around the nasal openings. Note that the maxilla articulate with the frontal bone at the superior and medial border of the orbit. The **infraorbital foramen** is located inferior to the medial aspect of the orbit. Centrally located on the inferior surface of the maxilla, directly posterior to the upper teeth, is the **incisive foramen**. The posterior portion of the maxilla on the roof of the mouth is called the **palatine process** because it extends toward the palatine bone.

3➡ Now find the **lacrimal bones** on the medial side of the orbit, deep to the maxilla. The lateral margins of the lacrimal bones articulate with the ethmoid bone.

4➡ Note the prominent cheek bones. These are the **zygomatic bones**. Notice how these bones form the "hooks" on the side of the face. The hook, which extends toward the temporal bone, is called the **temporal process**.

5➡ Look inside the nasal cavity. Do you see the three lateral wall protrusions? These are the nasal conchae. The superior and middle nasal conchae you have already examined as part of the ethmoid bone. The **inferior nasal concha** is a separate bone that extends medially from the internal portions of the maxilla.

6➡ Examine the floor of the nasal cavity and notice the small medial bone extending upward from the maxilla. This is the **vomer**. Check the different views of the vomer and note that it also articulates with the sphenoid, ethmoid, and palatine bones.

7➡ Now find the **palatine bones**, which form the posterior third of the hard palate. The portion of the palatine bone extending toward the maxilla is sometimes called the **maxillary process**. On the lateral portions of the palatine part of the hard palate are the **greater palatine foramina**.

8➡ Locate the **mandible** or lower jaw (Figure 16–7●). The angle of the jaw is also known scientifically as the **angle**. The main part of the jaw is the **body**, and the part extending toward the temporal bone is the **ramus**. The large opening just inferior to the premolar is the **mental foramen**.

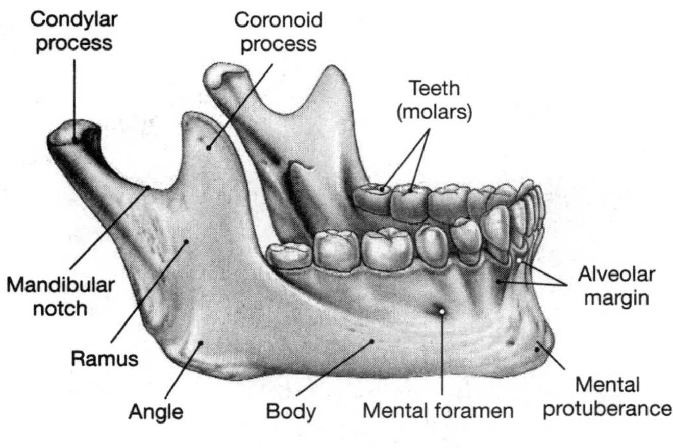

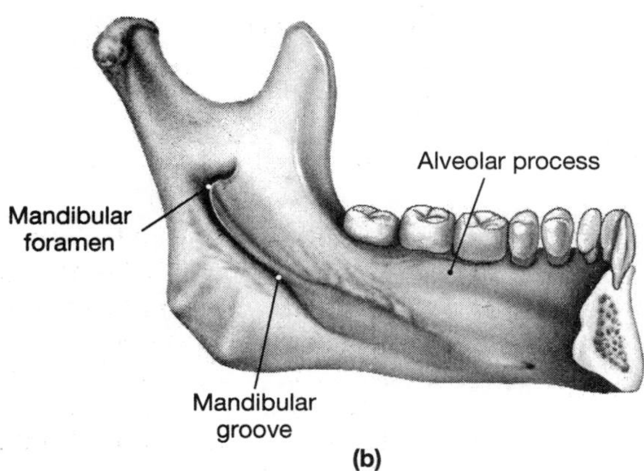

● **FIGURE 16–7**
Mandible. **(a)** Lateral view; **(b)** Medial view — left half.

Notice the two processes at the superior end of the ramus. The more lateral is the **condylar process**, which articulates with the temporal bone. The more medial is the **coronoid process**, which extends up under the zygomatic arch.

On the internal aspect of the mandible, find the **mandibular foramen**, centrally located in the ramus, and the **mandibular groove**, extending medially from the mandibular foramen. The **mylohyoid line** is superior to the groove.

III. Other Bones of the Head Region

A. Hyoid

1➡ Locate the **hyoid bone**, Figure 16–8●. It is the only bone that does not articulate with any other bone in the body and is found inferior to the mandible and superior to the thyroid cartilage. The **greater cornu** consists of the lateral horns of the hyoid, and the **lesser cornu** consists of the superior projections toward the medial part of the hyoid bone.

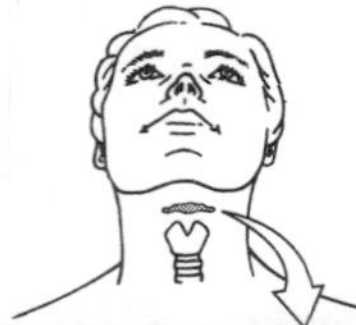

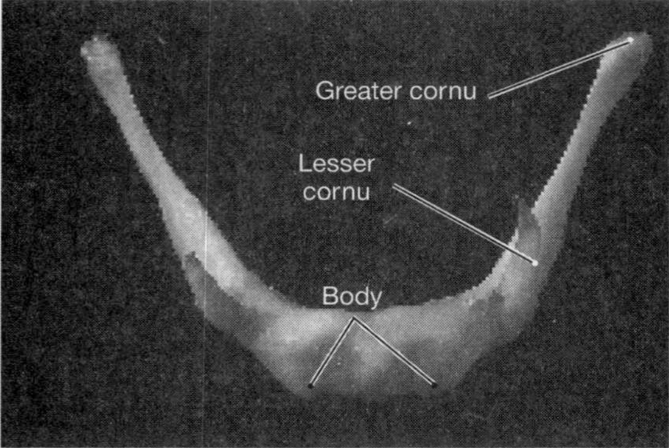

Anterior view

● **FIGURE 16–8**
Hyoid Bone.

B. Ossicles

1➡ Use Figure 16–9● to locate the **ossicles**, the tiny bones of the middle ear housed within the petrous portion of the temporal bone. Note how the **malleus** attaches to the ear drum and the **incus** connects the malleus to the **stapes**. The stapes connects the incus with the inner ear.

2➡ If your laboratory has an ossicles set, note the relative size of these bones. These are the smallest bones in the body. We will study the ossicles in greater detail in Exercises 38 and 39.

IV. Vertebral Column [∞ *Martini, p. 212*]

The **vertebrae** are the osseous elements of the spinal column. In addition to serving as a point of muscle attachment, each vertebra functions to support the body and to protect the spinal cord. The thoracic and sacral vertebrae are also articulation points for the ribs and pelvic bones, respectively. Dissection of the vertebral column is covered in Exercise 35.

A. Vertebral Structures

We have five sets of vertebrae: cervical, thoracic, lumbar, sacral, and coccygeal. We often use the human vertebral pattern as 7-12-5 to identify the cervical, thoracic, and lumbar vertebrae, respectively. Numerically these vertebrae are identified as C_1 to C_7, T_1 to T_{12}, and L_1 to L_5. Numerical identification is not usually given either to the sacral vertebrae (as these five bones are normally fused) or to the coccygeal vertebrae (as these bones are variable in number).

As you work through this section, make careful note of the distinguishing characteristics as well as the functions of each type of vertebrae.

1➡ Take a disarticulated vertebra and locate the principal landmarks. Each vertebra is composed of a **body**, or centrum, and two **pedicles**, each extending dorsolaterally from the centrum toward the dorsally located **spinous process**. Each vertebra also has articulating **facets**, points where the vertebra comes together with other bones.

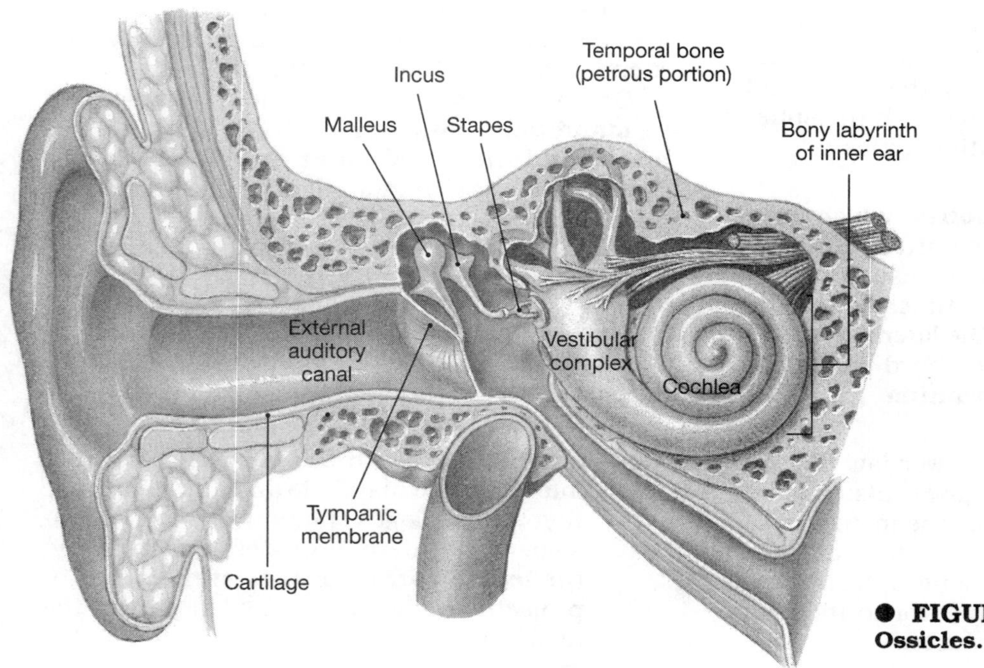

● **FIGURE 16–9**
Ossicles.

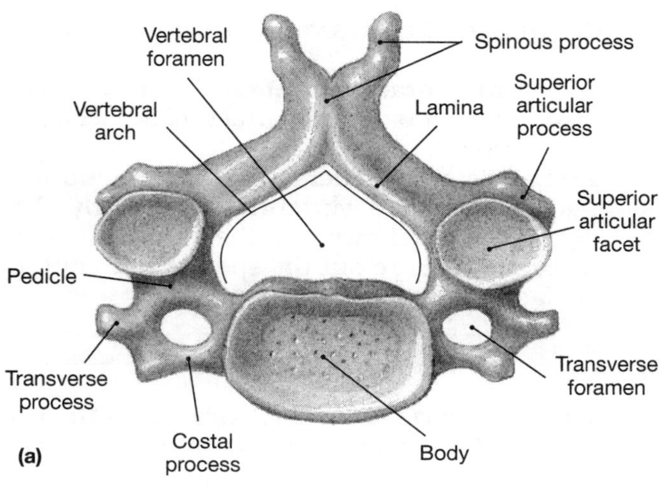

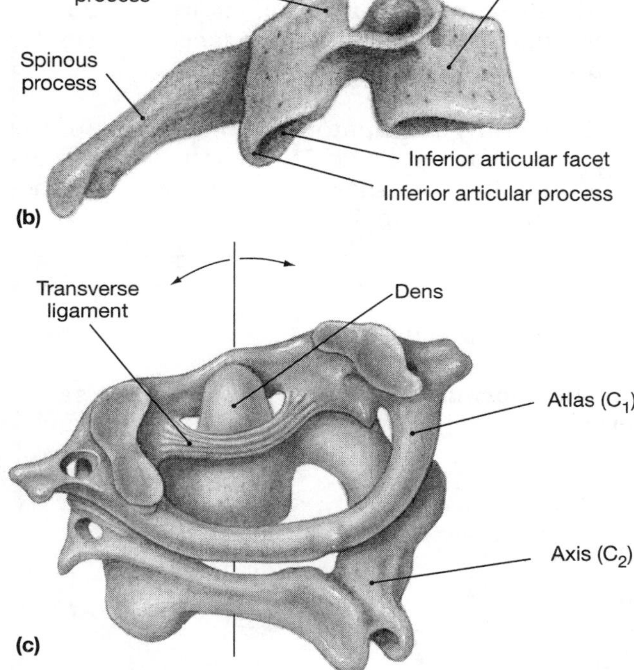

● **FIGURE 16–10**
Cervical Vertebra. **(a)** Superior view; **(b)** Lateral view; **(c)** Atlas/axis complex.

2➜ Take your vertebra to the articulated skeleton (or the articulated spinal column) and align it with the skeletal spinal column. The large "hole" between the centrum and the spinous process is the **vertebral foramen.** Note how these vertebral foramina line up to form the vertebral canal, the passage way for the entire spinal cord.

3➜ As you examine the articulated skeleton or the articulated spinal column, notice that the vertebrae are separated by cartilaginous discs known as **vertebral** or **intervertebral discs.** Notice too that numerous foramina are formed as the vertebrae come together. The spinal nerves pass through these **intervertebral foramina.**

B. Cervical Vertebrae

The seven **cervical vertebrae** are distinguished by their transverse foramina. The first cervical vertebra is the **atlas**. This vertebra has no body but does have **superior articular surfaces,** which articulate with the occipital condyles of the skull. Quite literally, the atlas holds up the weight of the world!

The second cervical vertebra is the axis, which has a superior protrusion known as the **dens** or **odontoid process.** The dens allows for the pivoting of the atlas. (Embryologically, the dens is the body of the atlas.)

1➜ Study Figure 16–10● and identify the cervical vertebral markings. As a point of interest, all mammals have seven cervical vertebrae, mice as well as shrews, giraffes and whales.

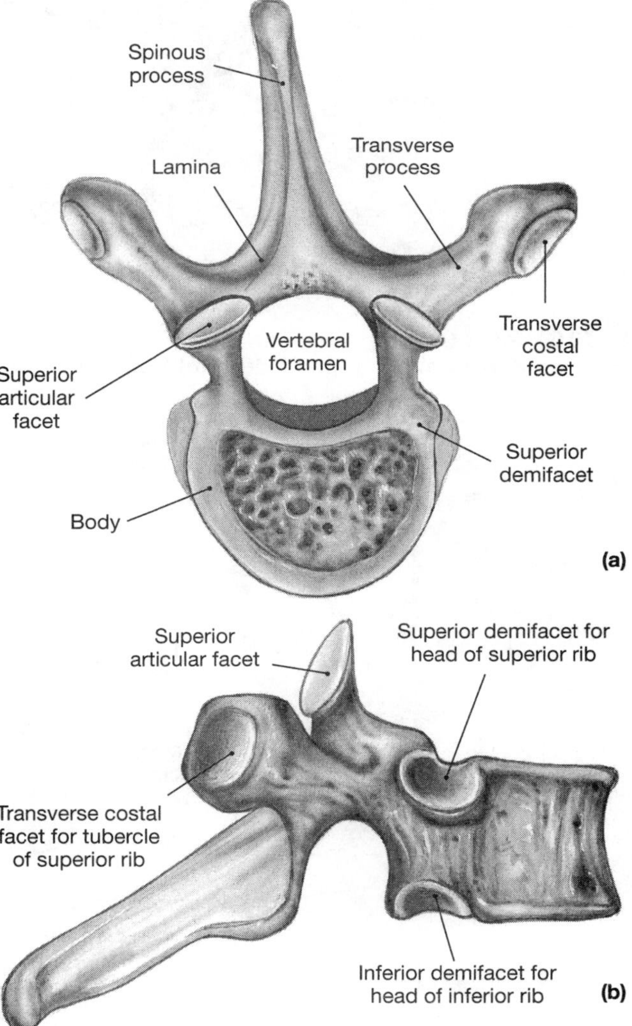

● **FIGURE 16–11**
Thoracic Vertebra. **(a)** Superior view; **(b)** Lateral view.

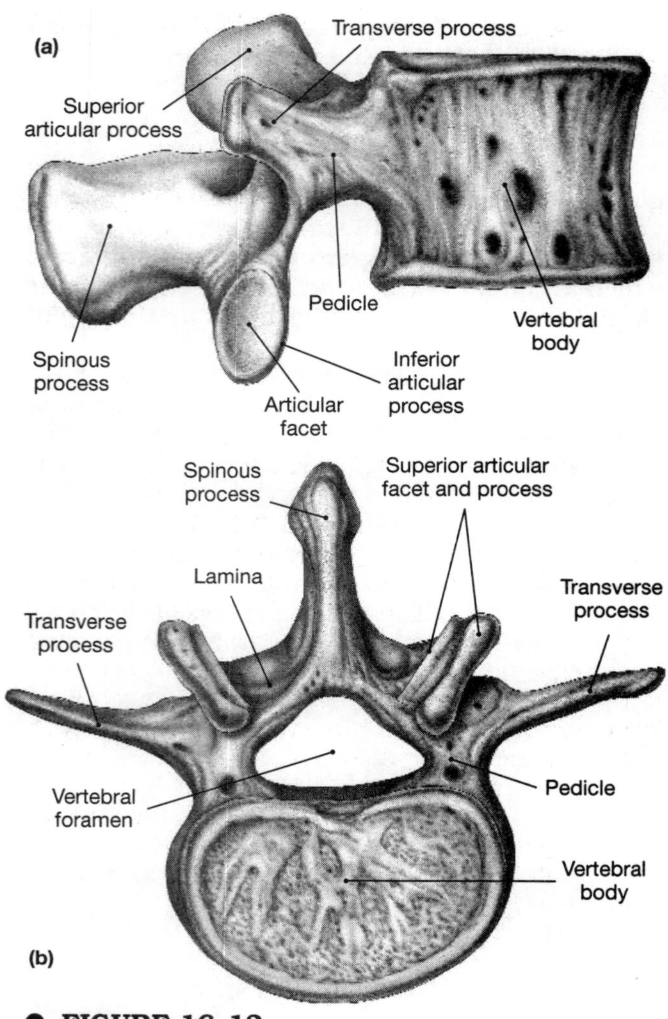

● **FIGURE 16–12**
Lumbar Vertebra. (a) Lateral view; **(b)** Superior view.

C. Thoracic Vertebrae

The twelve **thoracic vertebrae**, Figure 16–11●, are distinguished by the rib attachment facets.

1➡ Use both the articulated skeleton and the disarticulated vertebrae to identify the markings shown on the vertebral diagrams. Figure out the specific roles of the facets and demifacets.

____

____

2➡ Obtain a rib and demonstrate how the rib would articulate with a thoracic vertebra.

D. Lumbar Vertebrae

The five **lumbar vertebrae**, Figure 16–12●, are blocky bones without articular facets (places for rib attachment).

1➡ Identify the anatomical landmarks of the lumbar vertebrae. How do the superior articular processes fit with the inferior articular processes?

____

____

E. Sacrum and Coccyx

The **sacrum** consists of the five fused sacral vertebrae.

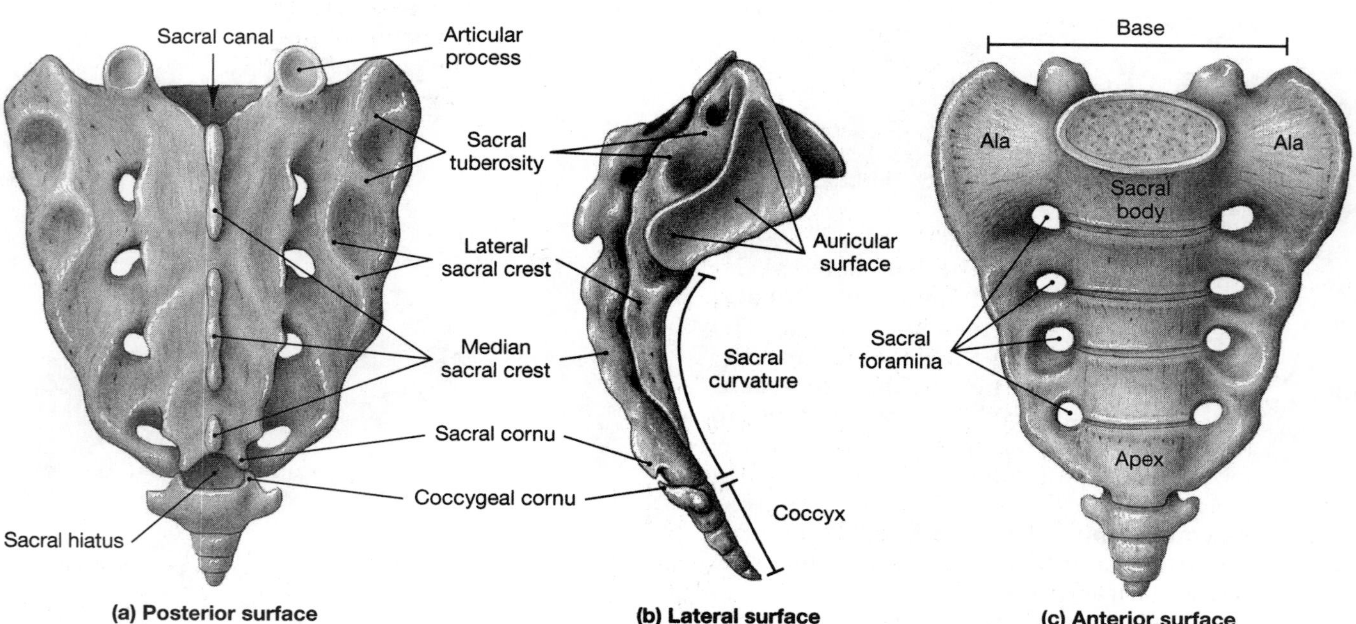

● **FIGURE 16–13**
Sacrum and Coccyx. (a) Posterior view; **(b)** Lateral view; **(c)** Anterior view.

1 ➡ Examine Figure 16–13●. Use a disarticulated sacrum to identify the anatomical markings on both the anterior and posterior surfaces. Where do you find the sacral portion of the sacroiliac joint?

____

____

Find the three to five tiny bones at the caudal end of the sacrum. These bones, which are usually fused, form the **coccyx**. The first two of these little vertebrae do have transverse processes. This is the tail bone, the vestigial remains of the tail so common among other members of the mammalian world. Interesting coccygeal anomalies are occasionally found in humans.

How many coccygeal bones are on your laboratory skeleton? ____________

V. Thoracic Cage *[∞ Martini, p. 219]*

Humans have twelve pairs of **ribs**, all of which have a posterior attachment to the vertebral column. The first seven are the **true ribs**, attached directly to the **sternum**, the bone often called the breast bone. The next three pairs are **false ribs**, attached only indirectly to the sternum. The last two pairs are **floating ribs**, so called because they have no osseous or cartilaginous attachment to the sternum. This rib pattern is often described as 7-3-2.

A. Ribs

1 ➡ Use the articulated skeleton and locate the twelve pairs of ribs. If your laboratory has an abnormal skeleton, you may notice anomalies in the ribs — extra ribs, fused ribs, or unusual rib patterns.

2 ➡ Study the diagram of the rib, Figure 16–14●, and be able to identify the **head**, **angle**, and **neck** of the rib. How can you tell if it is a right rib or a left rib? Where is the **costal groove**?

____

____

B. Sternum

The sternum, centrally located in the anterior chest, has three major components. The **manubrium** is the triangular-shaped superior portion of the sternum. The line of demarcation between the manubrium and the **body** of the sternum is often called the **angle of Louis**. The body of the sternum is composed of several bones that usually are not completely ossified until at least age 25. The slender inferior process of the sternum is the **xyphoid process**. This delicate structure is easily broken, both in living humans and in classroom skeletons.

1 ➡ Obtain a sternum and locate the components mentioned in the prior paragraph. Where would you find the costal (rib) articulation sites?

____

____

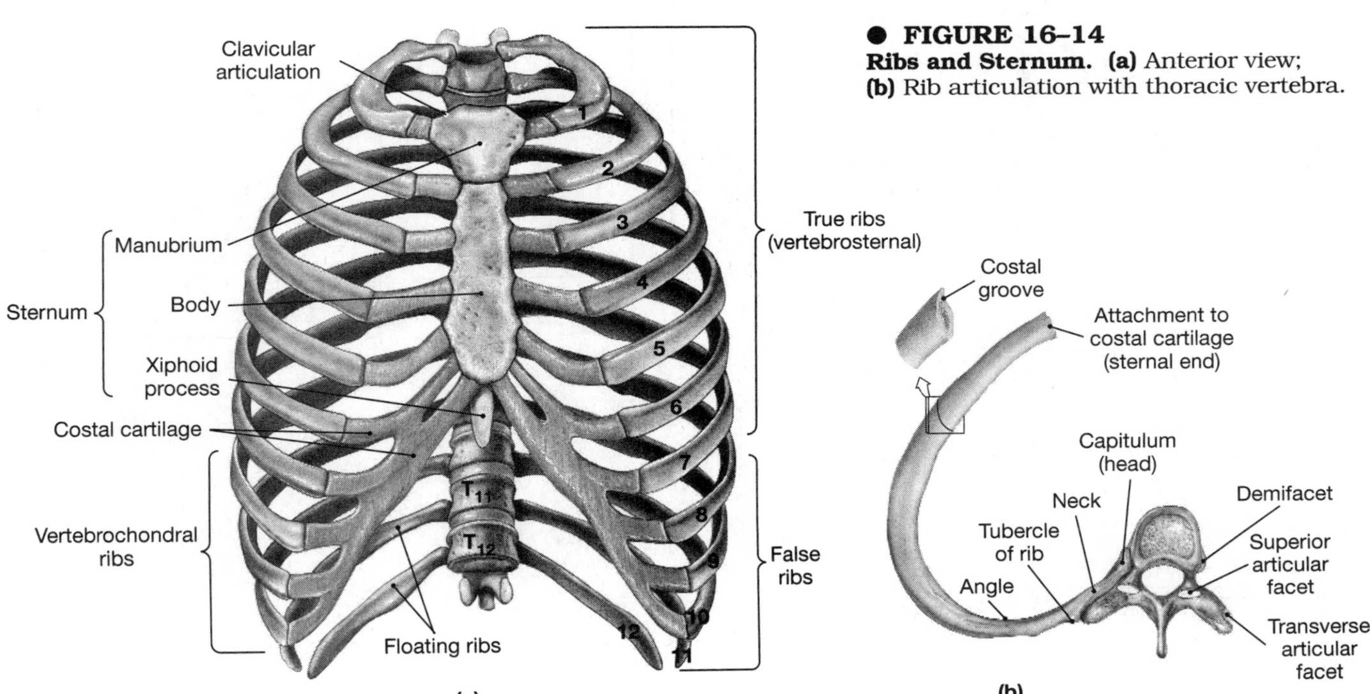

● **FIGURE 16–14**
Ribs and Sternum. **(a)** Anterior view; **(b)** Rib articulation with thoracic vertebra.

❑ Additional Activities

NOTES

1. Compare the bones of the human axial skeleton with the bones of the axial skeletons of other animals, mammals and nonmammals alike. Birds have an especially interesting sternal arrangement. Fish have interesting vertebrae.
2. Explain the difference between a vertebrate and a chordate.
3. Research the embryology of the vertebral column. What is the relationship between the notochord and the intervertebral discs?

❑ Lab Report

1. Referring to the numbered inquiries at the beginning of this exercise, complete the following box summary:

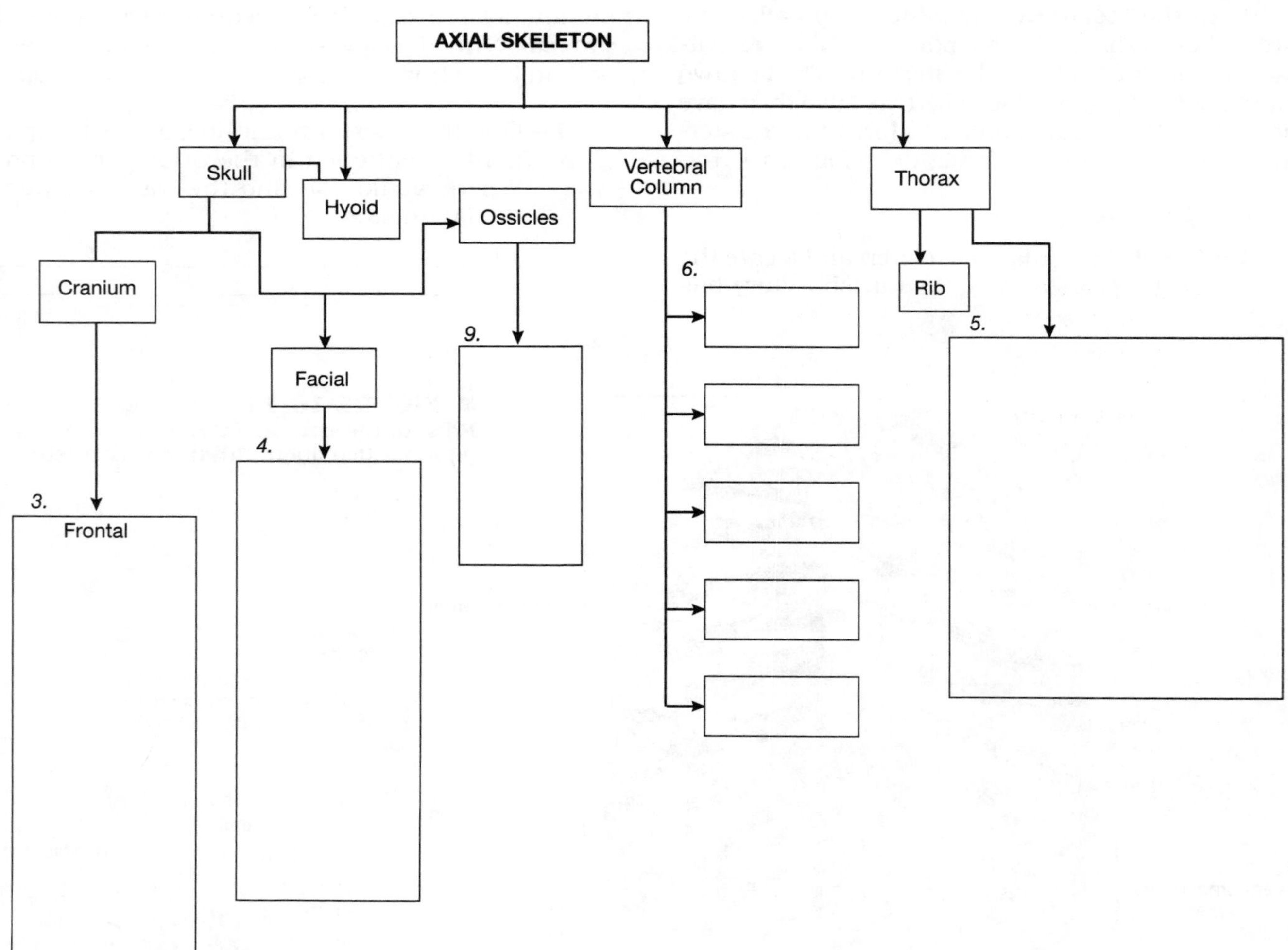

General Questions

NOTES

1. Complete this chart:

Bones of the Vertebral Column

	Number of bones	Processes	Foramina	Other characteristics
Cervical				
Thoracic				
Lumbar				

2. Complete this chart:

Cranial bones	Articulate with
______________________	______________________
______________________	______________________
______________________	______________________
______________________	______________________

3. Complete this chart:

Facial bones	Articulate with
______________________	______________________
______________________	______________________
______________________	______________________
______________________	______________________
______________________	______________________
______________________	______________________
______________________	______________________
______________________	______________________
______________________	______________________

4. Ribs can be classified according to their articulations. How many ribs articulate with the vertebrae? With the sternum?
5. Using the names of the anatomical landmarks found on a rib, explain how you distinguish between a left rib and right rib.
6. Where do you find the hyoid bone?
7. How many bones are in the sacrum? How many bones in the coccyx?

EXERCISE 17

Appendicular Skeleton

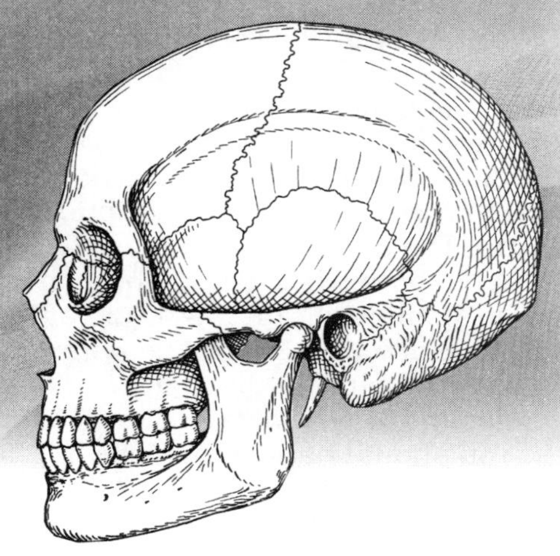

PROCEDURAL INQUIRIES

Preparation

1. What are the divisions of the appendicular skeleton?
2. What are the bones of the pectoral girdle?
3. What are the bones of the pelvic girdle?
4. What are the bones of the upper extremities?
5. What are the bones of the lower extremities?

Examination

6. What are the distinguishing characteristics of each of the appendicular girdles?
7. What are the principal markings of the bones of the pectoral girdle?
8. What are the principal markings of the bones of the pelvic girdle?
9. What are the principal markings of the bones of the upper extremity?
10. What are the principal markings of the bones of the lower extremity?

Additional Inquiries

11. How can we demonstrate the relationship between the bones of the girdles?
12. How can we distinguish between the right-side bones and the left-side bones?
13. What is the sesamoid bone of the lower extremity?
14. What bones fuse to form the pelvic girdle?

Key Terms

Carpals
Clavicle
Femur
Fibula
Humerus
Metacarpals
Metatarsals
Os Coxa
Patella
Pectoral Girdle
Pelvic Girdle
Phalanges
Radius
Scapula
Tarsals
Tibia
Ulna

Materials Needed

Articulated Skeleton
Disarticulated Skeleton
Probes
Male and Female Pelves (for comparison)
Abnormal Skeleton (if available)

The appendicular skeleton, which includes the pectoral and pelvic girdles and the extremities, normally consists of 126 bones. These bones are paired; in addition to learning the names of these bones and their principal anatomical markings, it is important that you learn to distinguish the right-side bones from the left-side bones.

When you begin to study a particular bone, first locate that bone on the articulated skeleton. Make note of the relative size and shape of the bone. Jot down any markings or specific points you observe. Then ask yourself, "How do I know if that bone is the right or left bone?" Explain to yourself and to your lab partner how the bone markings and the physical relationships between the bones help you figure out whether the bone is a right bone or a left bone.

Now proceed to the disarticulated skeleton. Take the bone you wish to study. Make mental note of how this bone in disarticulated form compares with the corresponding bone on the articulated skeleton. Take the bone to the articulated skeleton for comparison. Read the descriptions given here, then use the diagrams in this exercise to identify the major bone markings. These markings are given in boldface. You should be able to find all boldface terms listed. Line the bone up with your own body to help you distinguish right from left.

Learning the bones is indeed a "hands on" task. You should handle the bones, feel the markings, and run your probe around the anatomical landmarks. Say the terms out loud and quiz your partner so you both are bringing as many senses as possible into the learning experience.

❑ Preparation

Keep in mind that another name for a joint, the junction between bones, is *articulation*. Thus, an articulated skeleton is one where the bones are in place as they would be in the living organism. In the disarticulated skeleton the bones are separated. Both are valuable in the study of the skeleton.

I. Divisions of the Skeleton as a Whole

A. The Axial Skeleton

The axial skeleton consists of those bones found along the longitudinal axis of the body. The axial skeleton is covered in Exercise 16.

B. Appendicular Skeleton

The following is an outline of the appendicular skeleton. The divisions of the appendicular skeleton are given in upper case, while the names of the bones are in lower case letters. The numbers in parentheses represent how many of each particular bone can be found in the human body. *[∞ Martini, Figure 8.1]*

❑ PECTORAL GIRDLE

Clavicle	(2)
Scapula	(2)

❑ UPPER EXTREMITIES

Humerus	(2)
Ulna	(2)
Radius	(2)
Carpals	(16)
Metacarpals	(10)
Phalanges	(28)

❑ PELVIC GIRDLE

Os coxa	(2)

❑ LOWER EXTREMITIES

Femur	(2)
Patella	(2)
Fibula	(2)
Tibia	(2)
Tarsals	(14)
Metatarsals	(10)
Phalanges	(28)

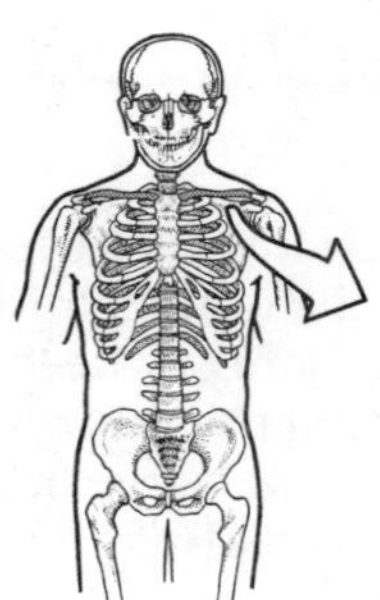

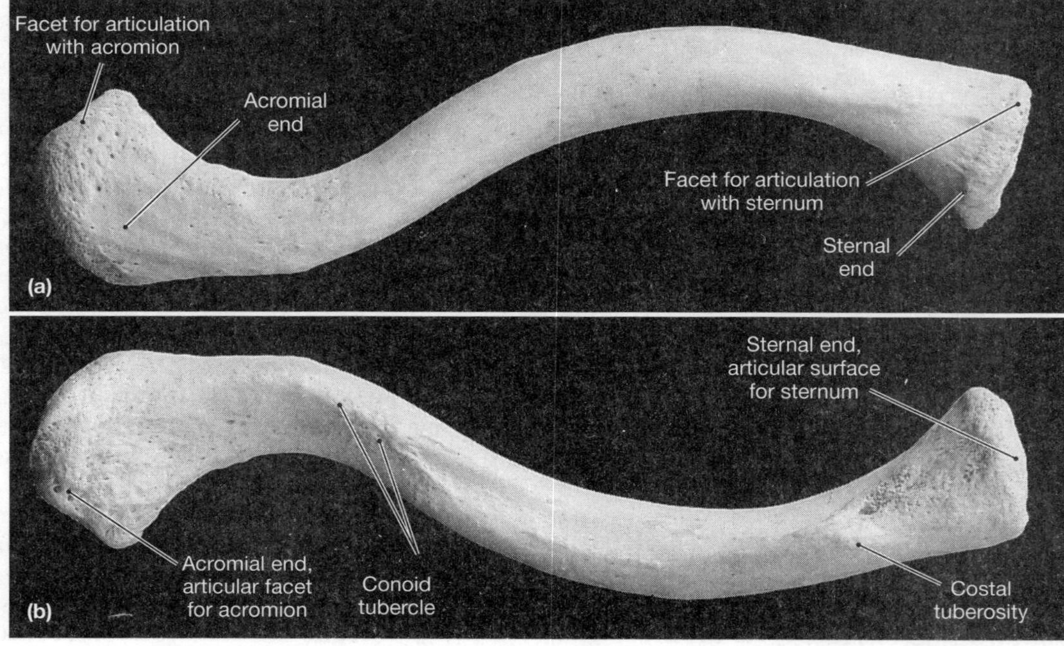

● **FIGURE 17–1 Clavicle. (a)** Superior and **(b)** Inferior view of the right clavicle.

❑ Examination

II. Pectoral Girdle [∞ *Martini, p. 228*]

The **pectoral girdle** consists of the **clavicle** and **scapula**. With each of these bones, consider how the bone articulates with the rest of the skeleton and how the landmarks help determine the anatomical position of the bone.

A. Clavicle

1➡ Find the clavicle (Figure 17–1●). The clavicle has a blunt, triangular **sternal end** (denoting its articulation with the sternum) and a larger, more rounded **acromial end** (denoting its articulation with the acromion of the scapula). The clavicle also articulates with the cartilage of the first rib.

2➡ Feel your own clavicle and, as you move your shoulders forward, note the pronounced S-curve. What is the difference between the left and the right curve?

____

____

3➡ Shift your shoulders forward. Feel the clavicular protrusion near the sternum. That protrusion is the sternal end of the clavicle. It is also the attachment site for the pectoralis major muscle. Can you feel this muscle just inferior to the clavicle? ____________
In addition to the pectoralis major, five other muscles also attach to the clavicle.

B. Scapula

1➡ Now identify the scapula (Figure 17–2●). Most of the markings on the scapula can be observed from the posterior surface. The **spine** protrudes perpendicularly from the **body** and is continuous with the **acromion**, which articulates with the acromial end of the clavicle. The **infraspinous fossa** is immediately inferior to the spine, and the **supraspinous fossa** is superior to the spine.

The **superior border** runs along the top of the scapula. Proceeding laterally along the superior border you will find the **scapular notch**. (The notch serves as the passageway for the suprascapular nerve.) The hooked protrusion lateral to the notch is the **coracoid process**. The **medial**

● **FIGURE 17–2**
Scapula. (a) Anterior view; **(b)** Lateral view; and **(c)** Posterior view of the right scapula.

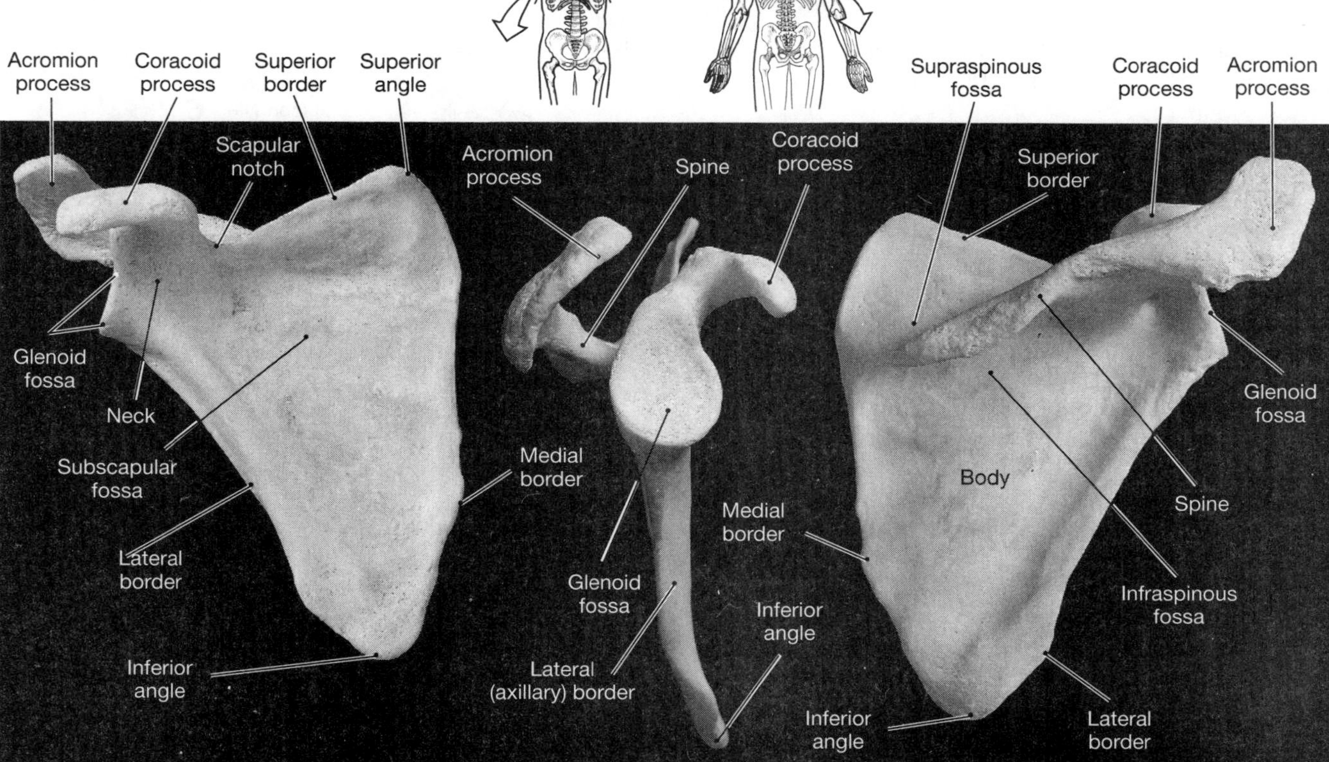

border is on the vertebral side of the body, and the **lateral** border is on the opposite side. The caudalmost point of the scapula is the **inferior angle**.

2➡ Turn the scapula so the lateral side is facing you. The indentation that articulates with the humerus is the **glenoid fossa**. (Note that the spine of the scapula always denotes the posterior surface, and the glenoid fossa always denotes the lateral side.)

3➡ Now turn the scapula so the anterior surface is facing you. On the anterior surface you will notice a large, shallow indentation, the **subscapular fossa**.

The fossas, processes, angles, and borders of the scapula serve as attachment sites for muscles — 17 in all — including the deltoid, trapezius, supraspinatus, biceps brachii, triceps brachii, and pectoralis minor.

C. Development

The clavicle is probably the first bone of the body to begin ossification — perhaps as early as prenatal day 31 — but complete fusion of the ossification centers is usually not complete until about age 25. Although scapular formation is fairly complete by the mid-teens, complete ossification often takes another seven to ten years.

III. Upper Extremities *[∞ Martini, p. 230]*

The upper extremities include the bones of the upper arm, the lower arm, the wrist, and the hand.

A. Humerus

1➡ Locate the **humerus** (Figure 17–3●). The rounded head of the humerus articulates with the scapula. Immediately below the head is the indentation known as the **anatomical neck**. With the anterior surface facing you, note the raised area opposite the head. This is the **greater tubercle**. Facing you is a smaller raised area, the **lesser tubercle**. Note the sulcus or groove between the tubercles. This is the **intertubercular groove**. The area immediately distal to the head and tubercles is the **surgical neck**.

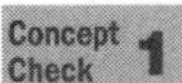

What physical differences do you see between the anatomical and surgical necks?

___

___

The surgical neck is so named because it is the area most prone to breaks and injuries — and thus most in need of repair.

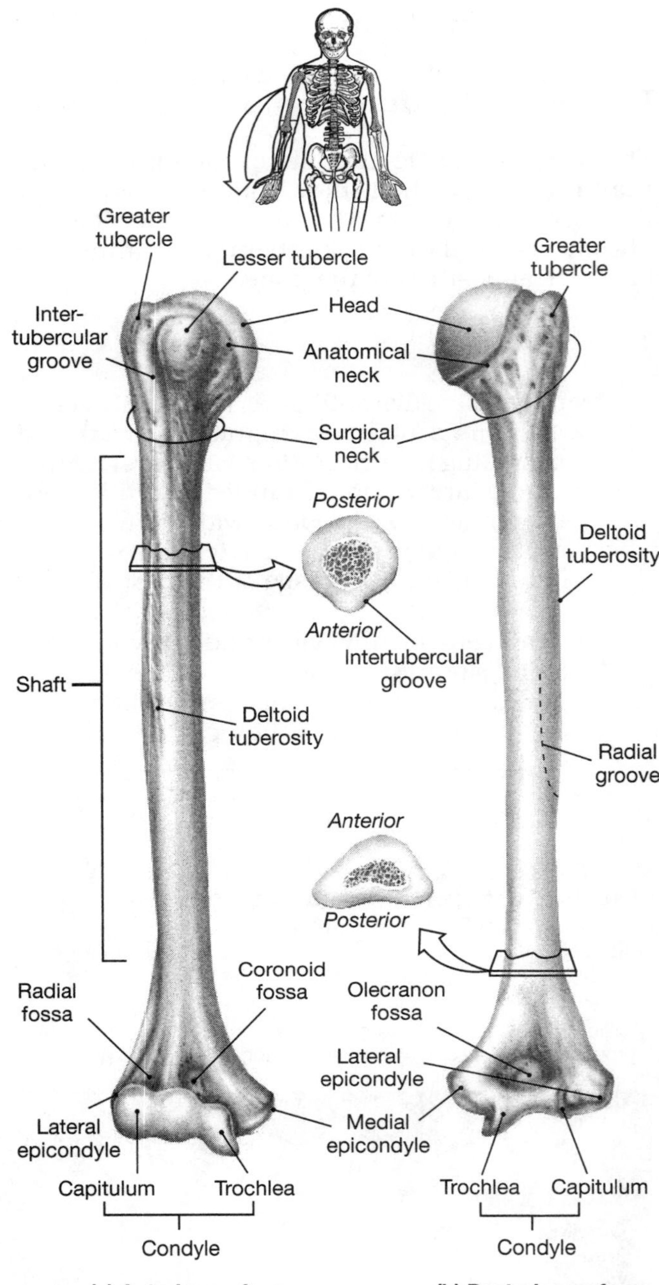

● **FIGURE 17–3**
Humerus. Major landmarks on **(a)** the anterior and **(b)** the posterior surfaces of the right humerus.

2➡ Notice the raised, rough area on the anterior surface of the shaft. This is the **deltoid tuberosity**. The depression on the posterior shaft is the **radial groove.**

3➡ Look at the distal end of the humerus, which is best studied by picturing the bone in anatomical position and by correlating the names of the structures with their locations (both on the body and in relation to the ulna). The **epicondyles** are the protrusions. The **lateral epicondyle** faces toward the outside of the body, and the **medial epicondyle** faces the body proper. The pulley-like **trochlea** extends down toward the ulna, where it articulates with the trochlear notch of the ulna. The extension opposite the trochlea is the **capitulum**, which articulates with the head of the radius.

4➡ Find the two fossas on the anterior surface, the lateral **radial fossa** and the medial **coronoid fossa**. The **olecranon fossa** is on the posterior surface.

Twenty-four different muscles attach to the humerus, including the supraspinatus, the latissimus dorsi, the triceps brachii, and many of the flexors and extensors.

B. Radius and Ulna

1➡ Study the **radius** and **ulna** together (Figure 17–4●). The radius radiates, or makes radiating movements. The radius has a flat **head**, which articulates with the capitulum of the humerus and the ulna. Immediately distal to the head is the **neck**. Note the slight outward bowing of the shaft of the radius. On the anterior surface, note the **radial tuberosity**.

2➡ Locate the distal end of the radius, which articulates with the lunate and scaphoid bones of the wrist. Note the **styloid process** on the lateral side of the radius and the **ulnar notch** marking the distal radioulnar articulation.

Note, too, the medial **styloid process** extending from the **head** of the ulna as it approaches the wrist. The ulna does not articulate directly with the wrist but rather with a triangular interarticular fibrocartilage. This **articular cartilage** rests between the ulna and the lunate and triquetal bones. The styloid process functions in stabilizing the wrist.

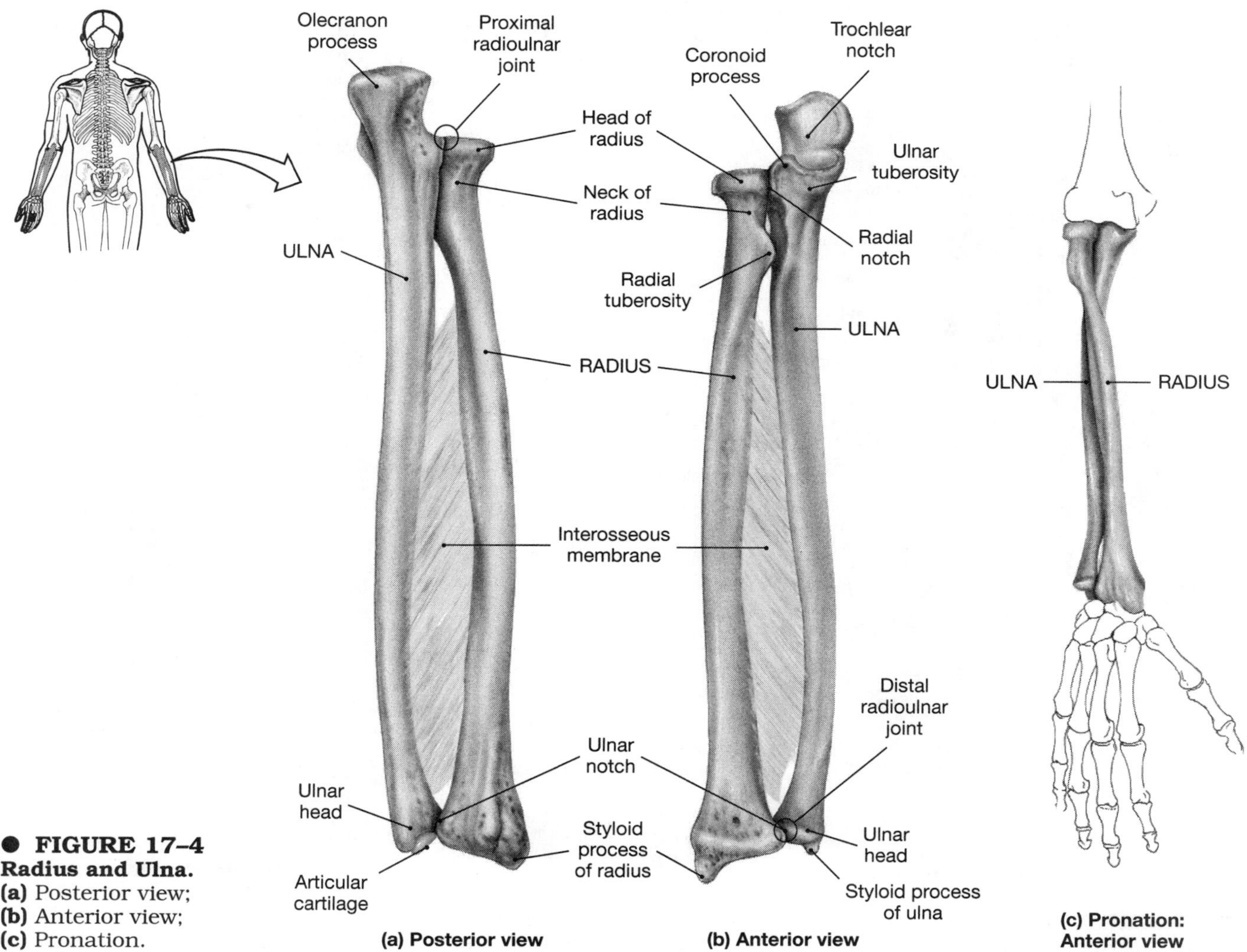

● **FIGURE 17–4**
Radius and Ulna.
(a) Posterior view;
(b) Anterior view;
(c) Pronation.

3➡ Now examine the proximal end of the ulna. The moon-shaped protrusion is the **olecranon**, the end of which moves in and out of the olecranon fossa of the humerus. The notch itself, the **trochlear notch**, articulates with the trochlea of the humerus. (The trochlear notch is sometimes called the **semilunar notch** because it is shaped like a half moon.) The anterior extension of the notch is the **coronoid process**. The coronoid process moves into and out of the coronoid fossa of the humerus. The **ulnar tuberosity** is the roughened area just distal to the coronoid process. Immediately distal to the ulnar tuberosity is the **radial notch**, the point of proximal articulation between the radius and the ulna. This articulation is known as the proximal radioulnar joint.

Nine muscles attach to the radius. Probably the best known of these muscles is the biceps brachii, which inserts on the radial tuberosity and is responsible for decreasing the angle between the lower arm and upper arm. Sixteen muscles attach to the ulna. These include the triceps, the brachialis, and a number of the flexors and extensors.

C. Carpus

1➡ Look at the **carpus**, the wrist, which includes four proximal carpals (**scaphoid**, **lunate**, **triquetal**, and **pisiform**) and four distal carpals (**trapezium**, **trapezoid**, **capitate**, and **hamate**). Study Figure 17–5●. You and your partner might want to make up a mnemonic phrase to help you remember the names of the carpals.

D. Metacarpals and Phalanges

1➡ Note the **metacarpals,** the bones of the hand proper, and the **proximal**, **middle**, and **distal phalanges** (sing. phalanx), the bones of the fingers. Note that the **pollex**, the thumb, has only two phalanges and the other fingers each have three. The proximal, middle, and distal phalanges are often called the first, second, and third phalanges, respectively.

E. Development

Development of all the bones of the upper extremity is usually complete by age 20.

(a) Posterior view

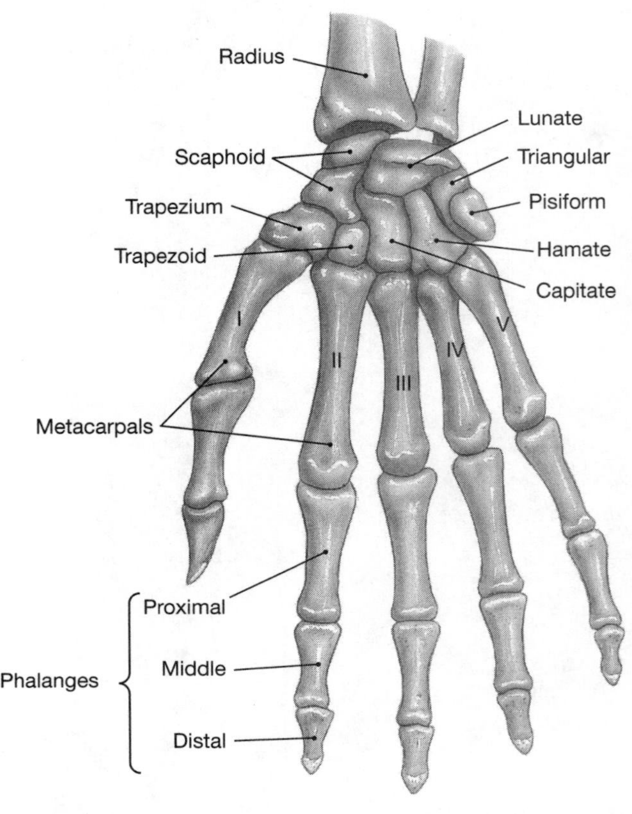

(b) Anterior view

● **FIGURE 17–5**
Wrist and Hand. (a) Posterior view of the right hand. **(b)** Anterior view of the right hand.

IV. Pelvic Girdle [∞ *Martini, p. 234*]

The **pelvic girdle** is composed of two large coxal bones, the **os coxae**. Consider how each os coxa articulates with the rest of the skeleton and how the landmarks help determine the anatomical position of the bone.

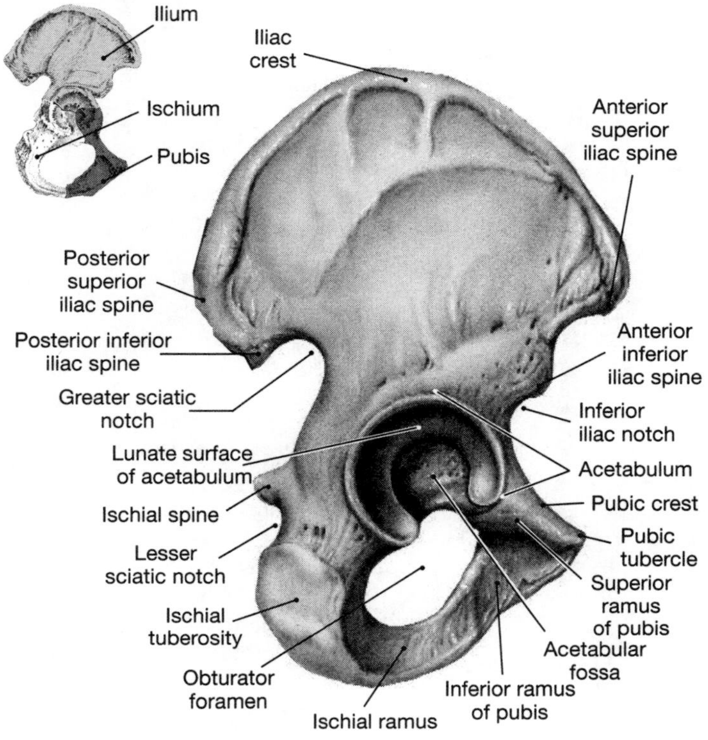

(a) Right coxa, lateral view

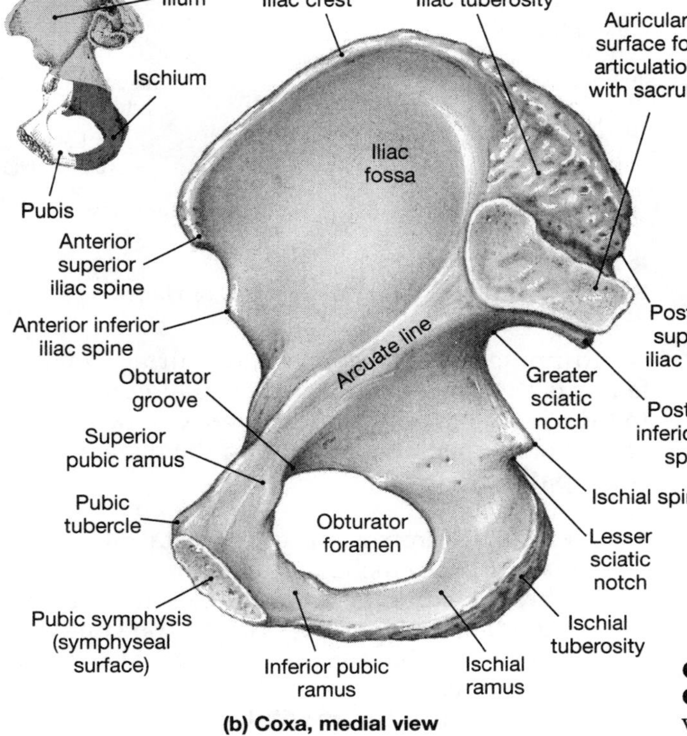

(b) Coxa, medial view

A. Os Coxae

1➡ Identify the pelvic girdle and each of two large coxal bones (Figure 17–6●). Colloquially, these are the hipbones. Each coxa is formed by the fusion of three bones, the **ilium**, the **ischium**, and the **pubis**. These bones begin the ossification/fusion process early in prenatal development; fusion is not complete until about the age of 20.

2➡ Study the anterior view of the coxa first. The shallow indentation wherein the pelvic organs sit is the **iliac fossa**. The ridge along the superior border is the **iliac crest**. The **sacroiliac articulation** is the joint between the coxa and the sacrum. The roughened coxal area of this articulation is simply the

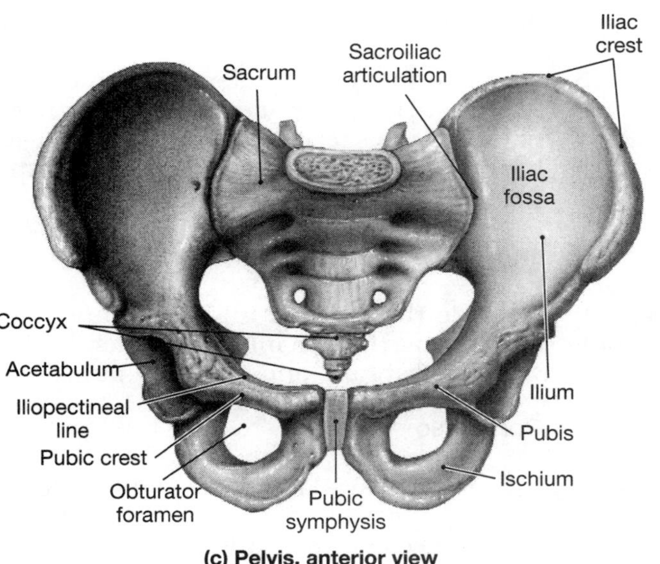

(c) Pelvis, anterior view

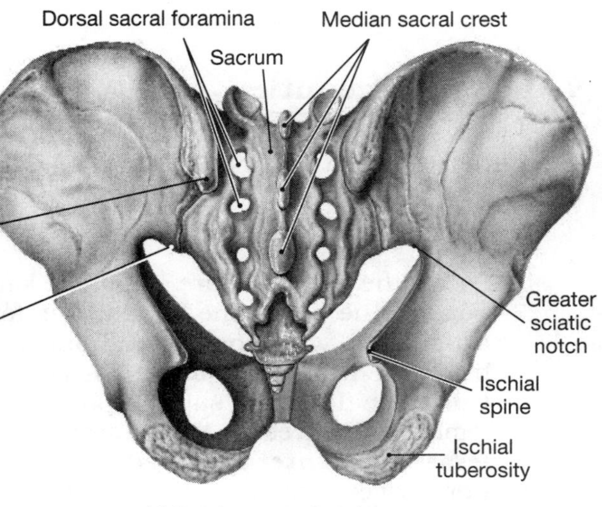

(d) Pelvis, posterior view

● **FIGURE 17–6**
Coxa and Pelvis. (a) Coxa, lateral view; **(b)** Coxa, medial view; **(c)** Pelvis, anterior view; **(d)** Pelvis, posterior view.

articular surface. The large "hole" on the inferior aspect of the coxa is the **obturator foramen**, the opening for the passage of nerves and blood vessels.

3➡ Identify the **symphysis pubis**, the joint between the pubic bones. If you have difficulty differentiating between the left and right coxae, try orienting the symphysis pubis and the iliac crest to your own body.

4➡ Now turn the coxa so that you have a lateral view of the markings. Follow Figure 17–6● and identify the **anterior superior iliac spine**, **anterior inferior iliac spine**, **inferior iliac notch**, **pubic tubercle**, **ischial ramus**, **ischial tuberosity**, **lesser sciatic notch**, **ischial spine**, **greater sciatic notch**, **posterior inferior iliac spine**, **posterior superior iliac spine**.

5➡ Examine the socket where the femur articulates. This is the **acetabulum**. You may be able to discern how the three pelvic bones come together in the acetabulum. Refer again to Figure 17–6●.

B. Muscle Attachment

If we divide the ox coxa into its component bones, we will find that 16 muscles attach to the ilium, 13 to the ischium, and 16 to the pubis.

C. Comparison

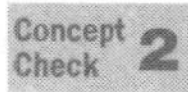

If you have available male and female os coxae, examine those bones now. What differences do you notice?

____

____

V. Lower Extremities [∞ *Martini, p. 237*]

The lower extremities include the bones of the thigh, the leg, the ankle, and the foot.

A. Femur

1➡ Locate the **femur**, Figure 17–7●, the longest bone in the body. The **head** of the femur articulates with the coxa. Immediately below the head is the **neck**. The large protrusion on the lateral superior portion of the femur is the **greater trochanter**, and the **lesser trochanter** is located on the posterior surface, inferior to the head. The raised posterior ridge extending from the greater to the lesser trochanters is the **trochanteric crest**, and its anterior analog is the **trochanteric line**.

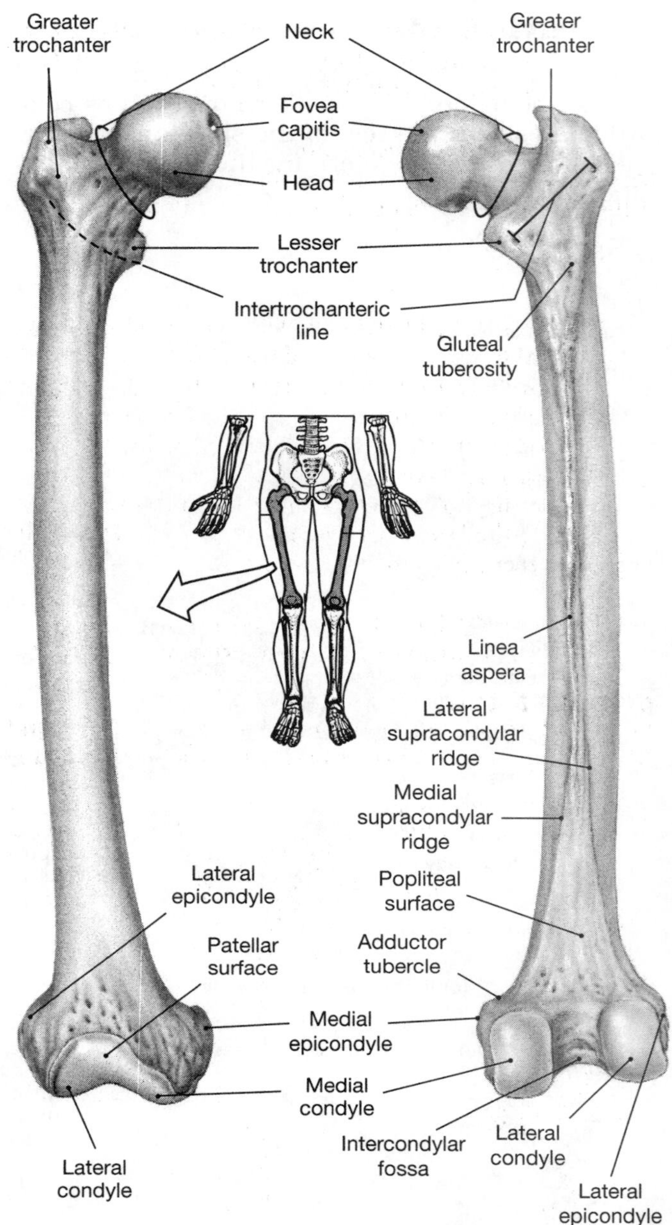

● **FIGURE 17–7**
Femur. (a) Anterior and **(b)** Posterior surfaces.

Running down the posterior surface of the femur is the **linea aspera**, a ridge that culminates in the apron-like **popliteal surface**. On the anterior side of the popliteal surface on the distal femur is the **patellar surface**.

2➡ Note the protrusions at the distal end of the femur. The smooth, articulating protrusions are the **condyles**, and the raised areas extending out just superior to the condyles are the **epicondyles**. The **lateral** condyle and epicondyle are on the lateral side of the femur, and the **medial** condyle and epicondyle are on the medial side of the femur. The posterior indentations between the condyles is the **intercondylar fossa**.

The femur is the attachment site for 23 muscles, including the gluteal muscles, the quadriceps femoris muscles, the gastrocnemius, and the plantaris.

B. PATELLA

1 ➡ Find the **patella**, a sesamoid bone superficial to the distal end of the femur. (Sesamoid bones form intramembranously. Aside from the patella, most sesamoid bones are extra bones occurring irregularly within the population. Wormian bones, which are often studied along with sesamoid bones, are irregularly occurring bones found along the suture lines in the skull.)

2 ➡ Identify the broad proximal part of the patella known as the **base** and the pointed distal part known as the **apex**. Four muscles attach to the patella. Note how the patella articulates with the femur.

C. TIBIA AND FIBULA

1 ➡ Study the large **tibia** and its smaller companion, the **fibula**, as a unit (see Figure 17–8●). Note the **lateral** and **medial** condyles of the tibia. These condyles are just inferior to the respective **articular surfaces** of the same name. The **intercondylar eminence** rises between the articular surfaces.

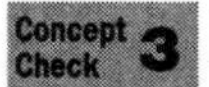

What is the relationship between this eminence and the intercondylar fossa of the femur?

____

____

2 ➡ Examine the anterior surface of the tibia where you can find the **tibial tuberosity**.

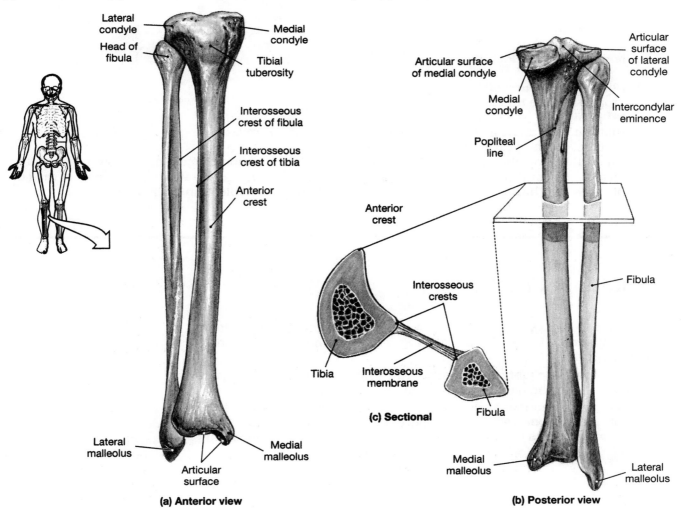

● **FIGURE 17–8**
Tibia and Fibula. (a) Anterior view of the right tibia and fibula. **(b)** Posterior view. **(c)** Sectional view at the level indicated in part (b).

The tibial tuberosity extends down the leg, forming the **anterior crest**. The **interosseous crest** is the raised area on both the tibia and fibula, which serves as the site of membranous attachment between the two bones.

3➔ Now locate the **lateral malleolus**, which is the protrusion on the distal end of the fibula, and the **medial malleolus**, found on the distal tibia. The maleoluses are what we normally think of as the ankle. The **articular surface** of the tibia forms the joint with the talus.

Twelve muscles attach to the tibia and nine attach to the fibula.

D. Tarsus

1➔ Look at the **tarsus**, the ankle proper, which consists of seven bones, the **talus**, **calcaneus**, **navicular**, **cuboid**, and the **medial**, **intermediate**, and **lateral cuneiforms**. Identify these bones on Figure 17–9●. As with the carpus, you and your partner may wish to make up a mnemonic phrase to help you remember the bones of the tarsus.

E. Metatarsals and Phalanges

1➔ Note the **metatarsals**, the bones of the foot proper, and the **proximal**, **middle**, and **distal** phalanges, the bones of the toes. Note that the **hallux**, the great toe, has only two phalanges. The proximal phalanx is often called the first phalanx, the middle phalanx the second phalanx, and the distal phalanx the third phalanx.

● **FIGURE 17–9**
Ankle and Foot.

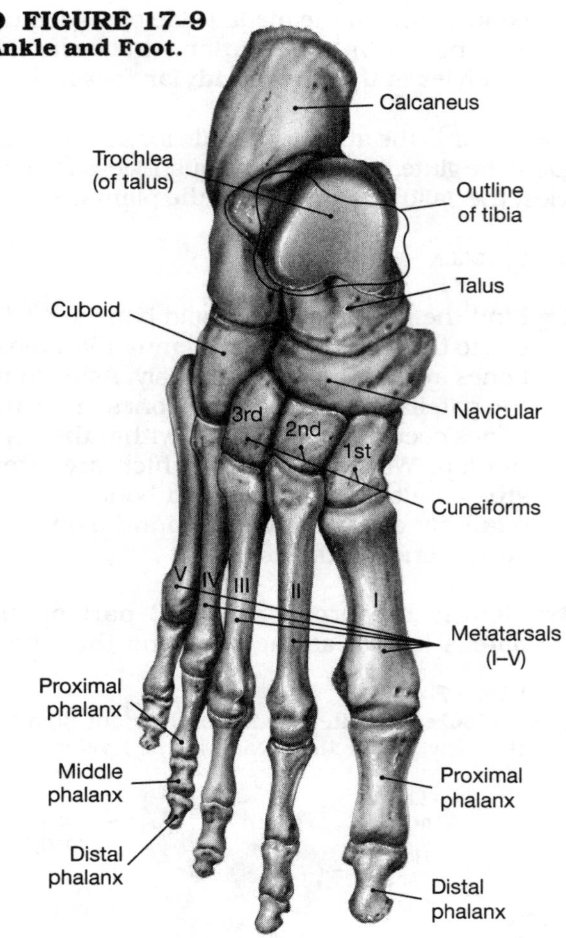

Superior view, right foot

F. Development

Ossification of most of the lower extremity is complete by age 20, although one part of the fibula is not complete until age 25.

❑ Additional Activities

NOTES

1. Compare the bones of the human appendicular skeleton with the bones of the appendicular skeletons of other animals, mammals and non-mammals alike.
2. Explore how bone markings can give pathologists and medical examiners insight into the background of a murder victim.
3. Examine any abnormal skeletons that may be in your laboratory. Make note of unusual bone structures. Speculate on what types of abnormal muscle attachments may coincide with the abnormal skeletal components.

Answers to Selected Concept Check Questions

NOTES

1. The anatomical neck is immediately distal to the rounded head and is at an angle to the axis of the bone. The surgical neck is distal to the tuberosities and is transverse to the bone.
2. [∞ *Martini, p. 236*] In general the female pelvis is smoother, lighter, and the anatomical markings are less prominent. In addition, the female pelvis has a larger pelvic outlet; a wider more circular pelvic inlet; a broader pubic arch (usually with an inferior angle of more than 100°; and less curvature of the sacrum and coccyx. Also, the female ilia usually flare more laterally than do the male ilia.
3. The eminence rises into the fossa between the condyles.

❑ Lab Report

1. Referring to the numbered inquiries at the beginning of this exercise, complete the following box summary:

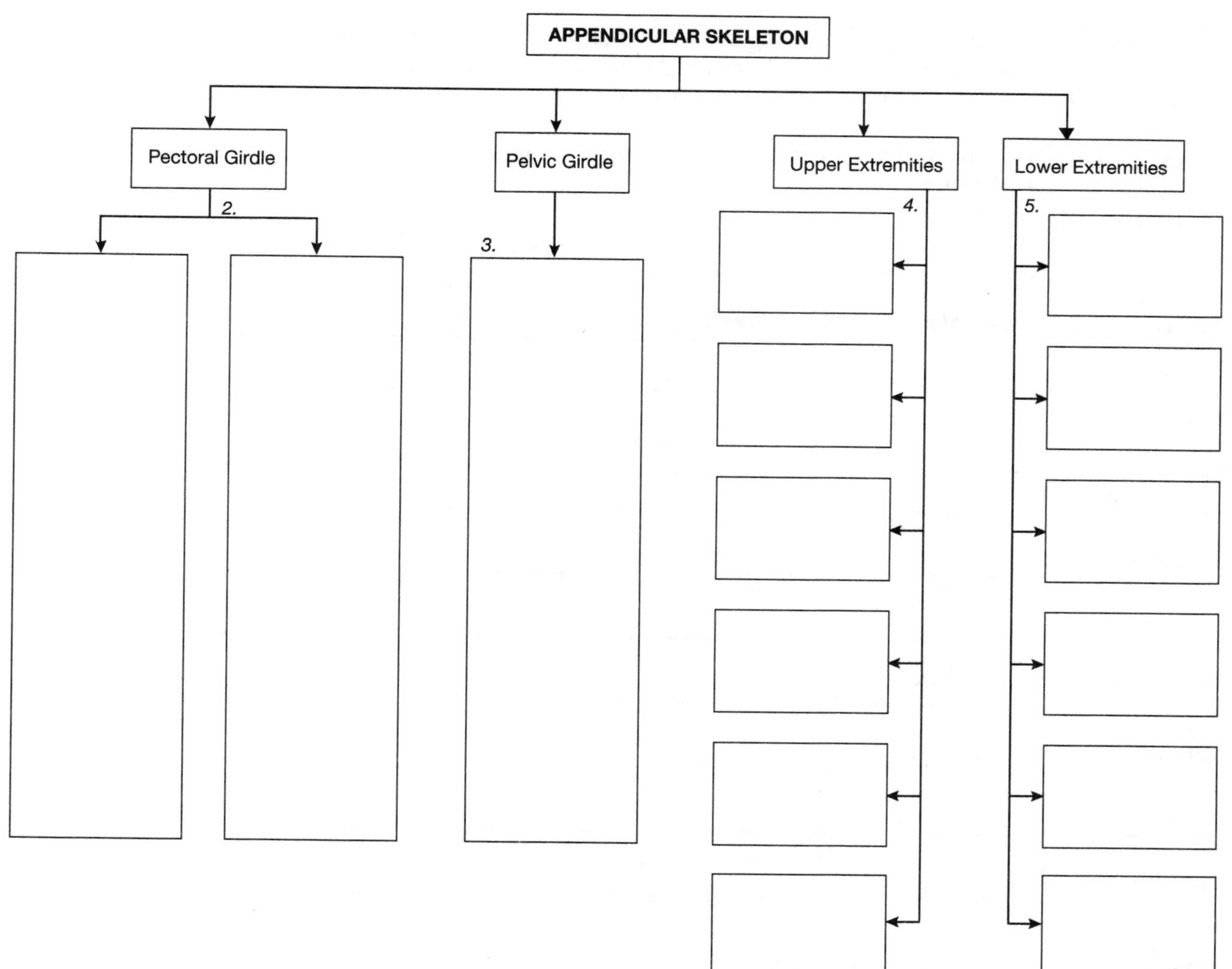

NOTES

General Questions

1. Fill in this chart to show the similarities and differences between the upper and lower extremities.

	Upper Extremity	Lower Extremity
Proximal Long Bones		
Distal Long Bones		
Articulation with Girdle		
Presence/Absence of Sesamoid Bone		
No. of Bones — Carpus/Tarsus		
Numbers of Phalanges		

2. Make a list of structures on articulating bones and explain how the names of the markings can help you remember structural names and locations. Example: the olecranon of the ulna and the olecranon fossa of the humerus.
3. Show how the articulating surfaces of the long bones facilitate the movement of those long bones.
4. Complete this chart:

Bone	Location	Markings	Distinguish Right from Left
Scapula			
Clavicle			
Humerus			
Radius			
Ulna			
Os Coxa			
Femur			
Tibia			
Fibula			

5. List the three bones of the os coxae.

EXERCISE **18**

Articulations

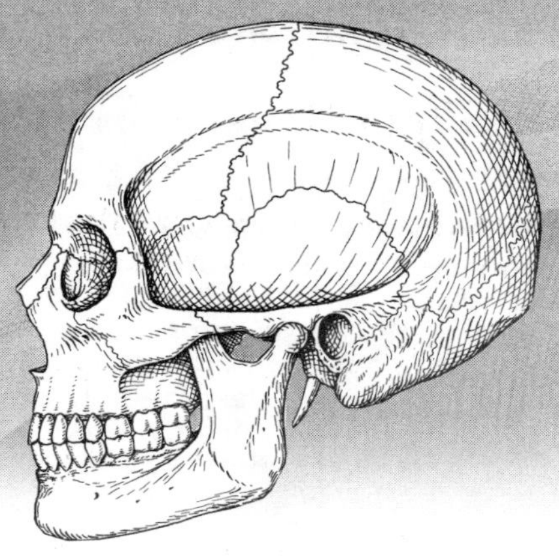

PROCEDURAL INQUIRIES

Preparation

1. What is an articulation?
2. What are the parts of a joint? What is the role of each part?
3. What are the types of joints?
4. What are the types of cartilage associated with the joints?
5. What is the role of each type of joint cartilage?

Examination

6. What are the types of synarthroses? How can we demonstrate each?
7. What are the types of amphiarthroses? How can we demonstrate each?
8. What are the types of diarthroses? How can we demonstrate each?
9. What are the major movements of the movable joints?

Additional Inquiries

10. How can we classify joints according to structure?
11. How can we classify joints according to function?

Key Terms

* Amphiarthrosis
Articulation
Bursa
* Diarthrosis
Function
* Gomphosis
Joint
Suture
* Symphysis
* Synarthrosis
* Synchondrosis
* Syndesmosis
* Synostosis
Synovial

* Note: These terms are listed in the singular. Plural formed by changing "is" to "es".

Materials Needed

Articulated Skeleton

Articulated Skull

Abnormal Skeleton or Abnormal Skeletal Component (if available)

Calf Knee — Sagittal Section

[Available in preserved form from most biological supply houses

Available in fresh form from most grocery and meat markets]

Small Water Balloon Partially Filled

2 Cylindrical Objects About 5 cm Diameter

Tape

The junction of any two or more bones is an **articulation** or **joint**. Articulations are usually classified by either function or structure. *Functional classification* is based on whether the joint moves and, if so, how much and in what manner it moves. *Structural classification* is concerned with the tissues that form the junction of the bones. (Nearly all joints are separated by some type of fibrous or cartilagenous connective tissue, although this may not always be obvious on your laboratory models.) These classifications are outlined in Figure 18–1●.

Synovial joints, such as the knee and shoulder, are what we classically think of as joints. Each of these joints is surrounded by a fluid-filled sac that protects the articular surfaces from injury caused by constant friction. Movement is the hallmark of the freely movable joint. All synovial joints are diarthrotic.

We often think of bone as the first tissue of the joint because we define a joint as the junction of two or more bones. However, cartilage also plays a major role in the anatomy of the articulations. The diarthrotic joints have articular cartilages functioning as protective caps over the ends of the bones. The symphyses have cartilagenous pads between the bones.

Functional Category	*Structural Category*	*Description*	*Example*
SYNARTHROSIS (no movement)	**Fibrous**		
	Suture	Fibrous connections plus interdigitation	Between the bones of the skull
	Gomphosis	Fibrous connections plus insertion in alveolus	Between the teeth and jaws
	Cartilaginous		
	Synchondrosis	Interposition of cartilage plate	Epiphyseal plates
	Bony fusion		
	Synostosis	Conversion of other articular form to solid mass of bone	Portions of the skull
AMPHIARTHROSIS (little movement)	**Fibrous**		
	Syndesmosis	Ligamentous connection	Between the tibia and fibula
	Cartilaginous		
	Symphysis	Connection by a fibrocartilage pad	Between right and left halves of pelvis; between adjacent vertebrae of spinal column
DIARTHROSIS (free movement)	**Synovial**	Complex joint bounded by joint capsule and containing synovial fluid	Numerous; subdivided by range of movement (see Figure 18–6●)
	Monaxial	Permits movement in one plane	Elbow, ankle
	Biaxial	Permits movement in two planes	Ribs, wrist
	Triaxial	Permits movement in all three planes	Shoulder, hip

● **FIGURE 18–1**
Classification of Articulations.

Other connective tissues are also found in and around the joints. The suture joints are composed of thin layers of dense fibrous connective tissue. Tough cords of connective tissue called **ligaments** are found around and between the bones of many synovial joints. (Recall our discussion of ligaments in Exercise 14.) Synovial sacs, as well as the bursae, are also composed of connective tissue. In addition, the **tendons** are sheathes of connective tissue that form the continuum between the bones and the muscles. Although tendons are not specifically articular, this musculo-skeletal connection is the basis of structural motion.

In this exercise, we will explore the different kinds of joints in the body, their parts, and the kinds of movement they accomplish in the skeleton.

❑ Preparation

I. Classification of Joints *[∞ Martini, p. 246]*

A. Function and Structure

1➡ Study Figure 18–1●. This table delineates the joints according to the two different classification schemes, functional and structural. (Note that there is not a direct correlation between the functional and structural components of about half the joints.) Familiarize yourself with the vocabulary of joints as you work through Figure 18–1●.

B. Accessory Structures of the Synovial Joints

1➡ Look at Figures 18–2● and 18–4● and make note of certain accessory structures found primarily in connection with the synovial joints:

Menisci (sing. meniscus) articular disks within the joint.

Fat pads deposits of adipose tissue near the joint.

Synovial capsule joint capsule protective apparatus surrounding the joint.

Bursae (sing. bursa) fluid-filled sacs which reduce friction and function as shock absorbers.

Tendons & Ligaments

2➡ Note too the names and locations of additional structures associated with the synovial joints.

❑ Examination

The following experiments are based on articular motion. Keep the presence or absence of movement in mind and continue to refer to Figure 18–1● as you work through these experiments demonstrating the functional classification of joints.

II. Synarthroses [∞ *Martini, p. 246*]

The **synarthroses** are immovable — either because the bones involved are fused or because the bones are so tightly bound that movement is all but impossible.

A. Suture Joints

1➡ To examine **suture** joints, take the articulated skull. Notice the interdigitation, which refers to the manner in which the jagged edges of each bone border seem to fit neatly together — like your fingers when you fold your hands. Some classroom skulls have pads between the skull bones simulating the fibrous connections found in the living skull. Examine these connections or visualize where such connections would be.

Gently try to move some of these cranial joints. Are you able to move them? Put your finger tips on your own parietal bones near the sagittal suture. See if you can move your parietal bones. Try moving some of the other bones, such as the temporal bones or the occipital bone.

At birth, the suture joints (or more correctly, the presuture joints) are far more movable because bone growth in the infant does not keep pace with brain growth. Large fibrous strips of connective tissue form the sutures. At about age five, brain growth is almost complete but the bones continue to develop. Cranial sutures as we know them begin to appear. (True sutures actually do begin to form at about age 18 months, but structured development is still several years away.) Look at the sutures and speculate on how the interdigitation formed.

____

____

B. Gomphoses

1➡ Identify the **gomphoses**, the joints between each tooth and the mandible or maxilla. These joints are immovable — except in a young child about to lose some baby teeth. But have you ever used dental floss to remove something caught between two teeth, and, after you dislodged it, you felt like your teeth weren't quite in the right place? Those few micrometers felt more like a few miles! You soon forgot about it because the gomphosis moved back to its original position. The next time this happens, remember that you are moving an essentially immovable joint.

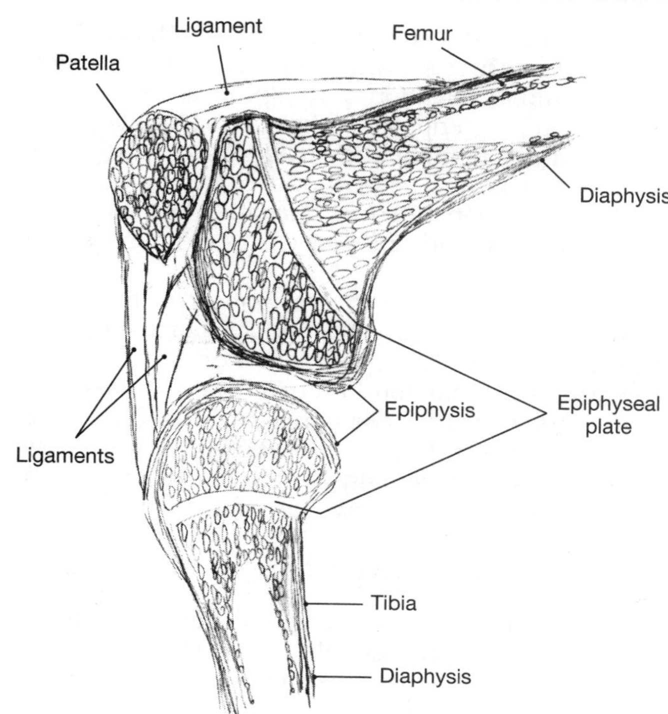

● **FIGURE 18–2**
Sagittal Section—Generalized Mammalian Knee.

C. Synchondroses

1➡ To check for **synchondrosis** movement, observe the tibia of the calf knee joint (see also Figure 18–2●). Do you see the cartilagenous epiphyseal plate? That is a true synchondrotic joint. Try to move the plate. (If the knee joint is in a preservative, the class might wish to appoint a designated mover, who can put on the rubber gloves to demonstrate the immobility of the joint.) The epiphyseal plate seals as the young leg (whether calf or human) reaches maturity. In the mature bone, the synchondrotic joint is all but obliterated, though an epiphyseal line still remains.

Even though the synchondrotic joint is immovable, it is somewhat fragile, and fractures along this joint line in the young can sometimes cause serious bone growth problems in later life. Severe force can actually cause the epiphysis to become separated from the diaphysis — such a separation is termed an epiphyseal fracture.

D. Synostoses

1➡ The ultimate synarthroses are the synostoses. Consider the pelvis. Go back to the skeleton and try to move the coxal components of the pelvic bone. The pelvic bone is a single bone in the adult, but the child is born with three separate bones. Complete fusion and ossification of the ilium, ischium, and pubis do not take place for 20 to 25 years. If you examine the pelvic bone carefully, you may be able to see the lines defining the three coxal bones. These lines may be clearest in the acetabulum. (Refer back to Exercise 17.)

Now consider the frontal bones of the skull. Can you separate the right and left portions of the frontal bone, either on yourself or on the laboratory skulls? At birth you had two frontal bones, but within a few years these bones had fused and the suture between them had disappeared. In a few people the suture remains throughout life; in such cases this suture is known as a metotic suture.

E. Abnormal Synarthroses

1➡ Examine an abnormal skeleton, if one is available in your laboratory. Check to see if any synostoses are present, such as in fused vertebrae or in the fused articular components of certain arthritic joints. These are abnormal manifestations of synostosis; nevertheless, the principle of osseous fusion is the same. (Recall that in Exercise 14 we said the ossification ability of some parts of the skeleton never really ceases.)

III. Amphiarthroses [∞ *Martini, p. 247*]

The **amphiarthroses**, or slightly movable joints, provide an interesting insight into movement. How much movement can you actually get from a "slightly movable" joint? Keep in mind that the slight movement, while difficult to define, still serves an identifiable function.

A. Syndesmoses

1➡ To observe **syndesmosis** in a joint, you will have to return to the articulated skeleton. Look at the way the tibia and fibula come together. Imagine the ligamentous connection between these bones. What types of movement do you think are possible?

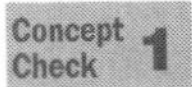

What advantage do you see in the distal tibio-fibular joints being amphiarthritic?

B. Symphyses

1➡ Find the **symphysis** (cartilagenous) joints between the adjacent vertebrae of your laboratory skeleton. The bones of these joints are separated by fibrocartilage pads. These pads have some flexibility.

2➡ Now demonstrate flexibility of the symphyses in your back by moving your own vertebral column in all directions. Concentrate on how much intervertebral motion is actually occurring. Have your lab partner put his/her hand on your back to detect movement in two adjacent vertebrae. You should both feel how little intervertebral motion you actually have. Most of the apparent motion you feel in your back is *cumulative*: each intervertebral joint is moving just a little, but the effect is a wide range of back motion.

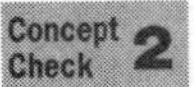

What is the functional advantage of this limited intervertebral motion?

3➡ If the articulated skeleton in your laboratory is well wired, demonstrate intervertebral motion by moving the skeleton's back and vertebral column. Keep your eye on the individual vertebrae.

4➡ Consider the limited movement of the symphysis pubis. Since this articulation on most classroom skeletons is well glued, motion is more constrained than it is in real life. Don't try to move this joint; just imagine what the movement would be like.

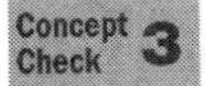

What is the functional advantage of the symphysis pubis being an amphiarthritic joint?

During childbirth it is often necessary for the fibrocartilage of the symphysis pubis to stretch considerably to allow for the passage of the child. Under certain circumstances cartilage is easily torn and not easily mended, so starting about the fourth month of pregnancy, the placenta begins synthesizing the hormone relaxin. Relaxin softens the fibrocartilage of the symphysis pubis, allowing it to stretch in response to fetal pressure.

IV. Diarthroses [∞ *Martini, p. 247*]

The diarthroses, the freely movable joints, are all characterized by a synovial apparatus. Although the specific synovial structure varies to some extent with the different synovial joints, the elements of the structure are common to all diarthrotic joints. To demonstrate diarthrotic articulation, we can construct a simulated synovial joint.

A. Structures of a Typical Synovial Joint

1 ➡ Look at Figure 18–3●. You will notice, deep to the patella, a small sac labeled *bursa*. A **bursa** is a packet of synovial fluid. You will find bursae associated with many synovial joints in the areas of tendons and ligaments, just beneath the skin covering a bone, or within other connective tissues exposed to a great deal of pressure. The bursae of the shoulder are well known for becoming inflamed, particularly after strenuous or unusual physical stress. Inflammation of the bursa is called *bursitis*.

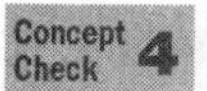

What is the functional advantage of the bursae, especially around the knee and the shoulder? Why is bursitis so painful?

____

____

____

2 ➡ Also in Figure 18–3●, identify the ligaments — the intracapsular ligament connecting the bones from within the synovial sac and the extracapsular ligament affording strength and integrity to the articular capsule.

Note the meniscus in Figure 18.3b. The meniscus is composed of bone, fibrocartilage, and hyaline cartilage. What is the apparent function of the menisci?

____

____

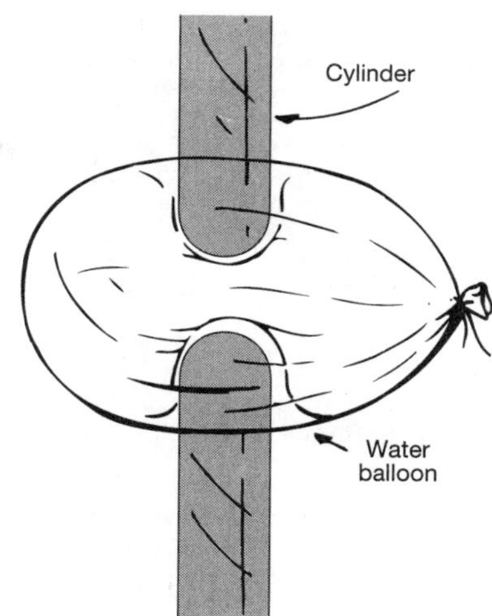

● **FIGURE 18–4**
Simulated Synovial Joint.

❑ Model Building

V. Synovial Joint Model

To demonstrate diarthrotic articulation, we can construct a simulated synovial joint.

A. Making the Model

1 ➡ Tape the end of each of your cylindrical objects. Take your water balloon and push the cylindrical objects into the balloon until they almost touch. See Figure 18–4●. Now

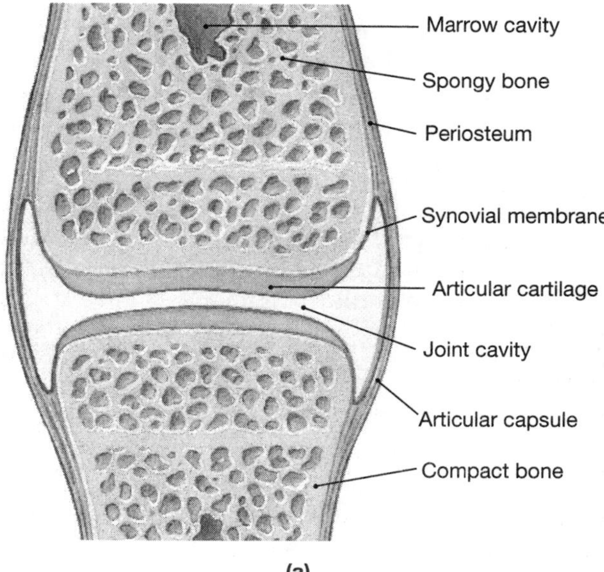

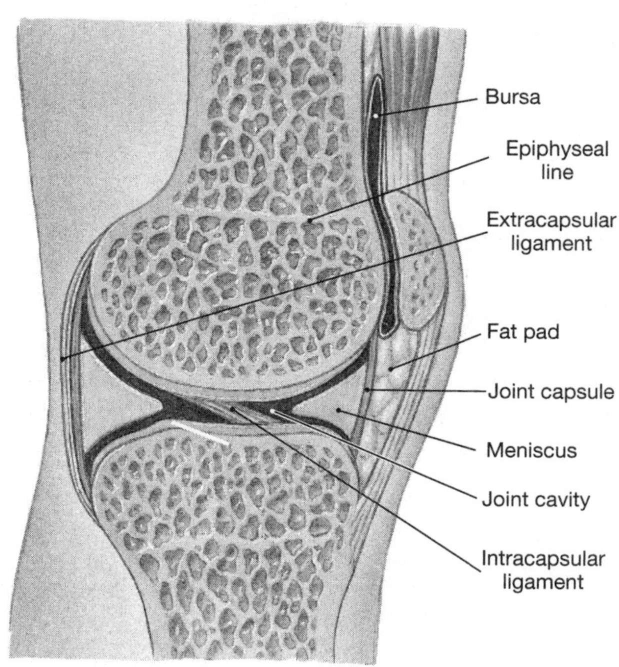

● **FIGURE 18–3**
Structure of a Synovial Joint.
(a) Diagrammatic view of a simple articulation.
(b) A simplified sectional view of the knee joint.

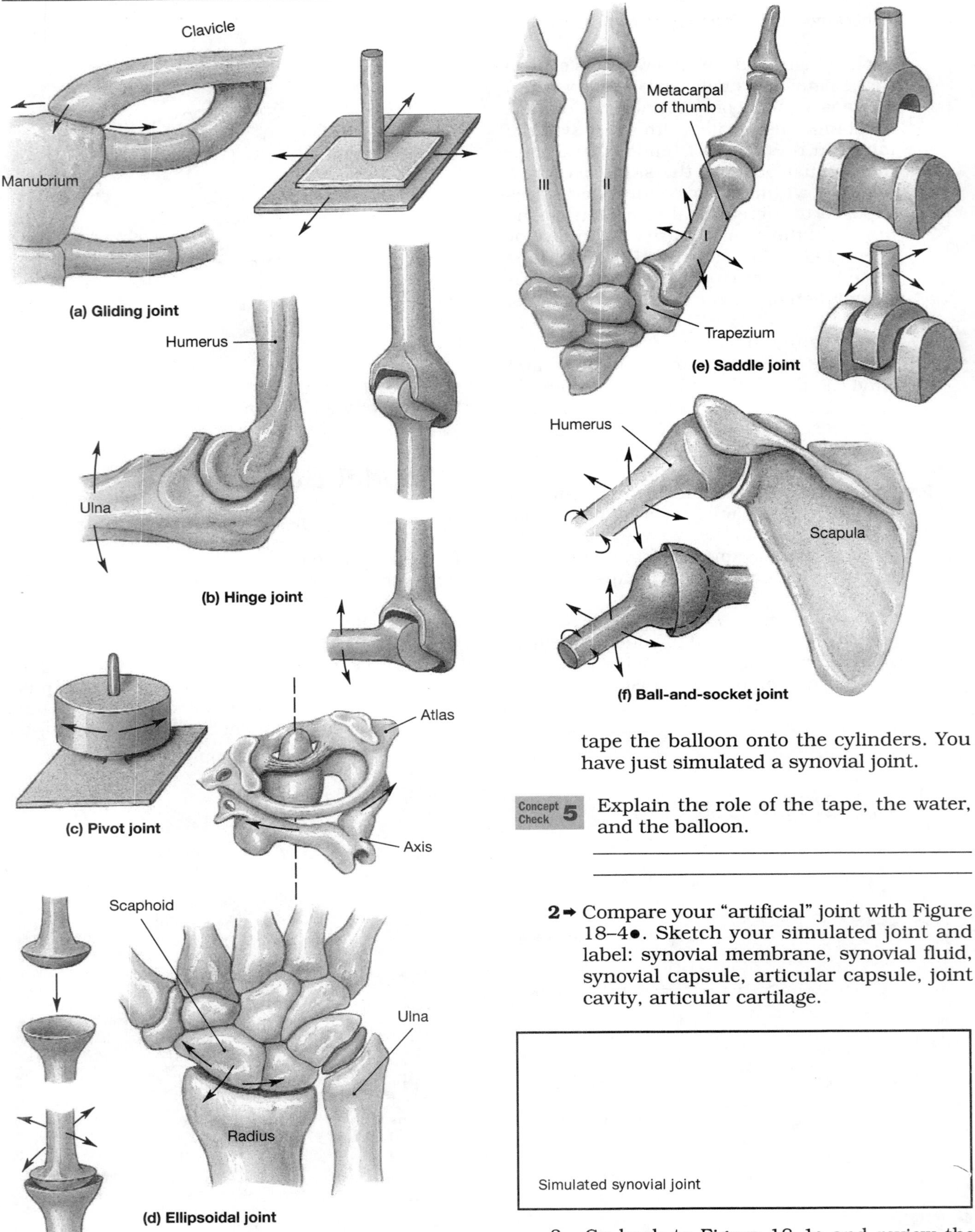

(a) Gliding joint

(b) Hinge joint

(c) Pivot joint

(d) Ellipsoidal joint

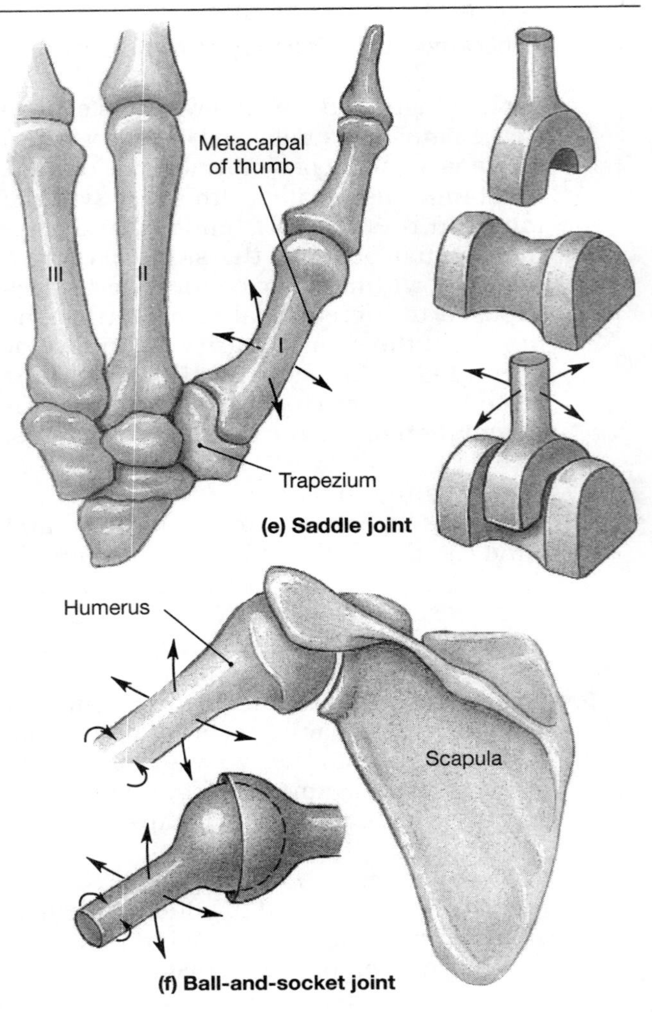

(e) Saddle joint

(f) Ball-and-socket joint

● **FIGURE 18–5**
Functional Classification of Synovial Joints.

tape the balloon onto the cylinders. You have just simulated a synovial joint.

Concept Check 5 Explain the role of the tape, the water, and the balloon.

____

____

2 ➡ Compare your "artificial" joint with Figure 18–4●. Sketch your simulated joint and label: synovial membrane, synovial fluid, synovial capsule, articular capsule, joint cavity, articular cartilage.

Simulated synovial joint

3. Go back to Figure 18–1● and review the synovial joints under both function and structure. Now examine Figure 18–5●, show-

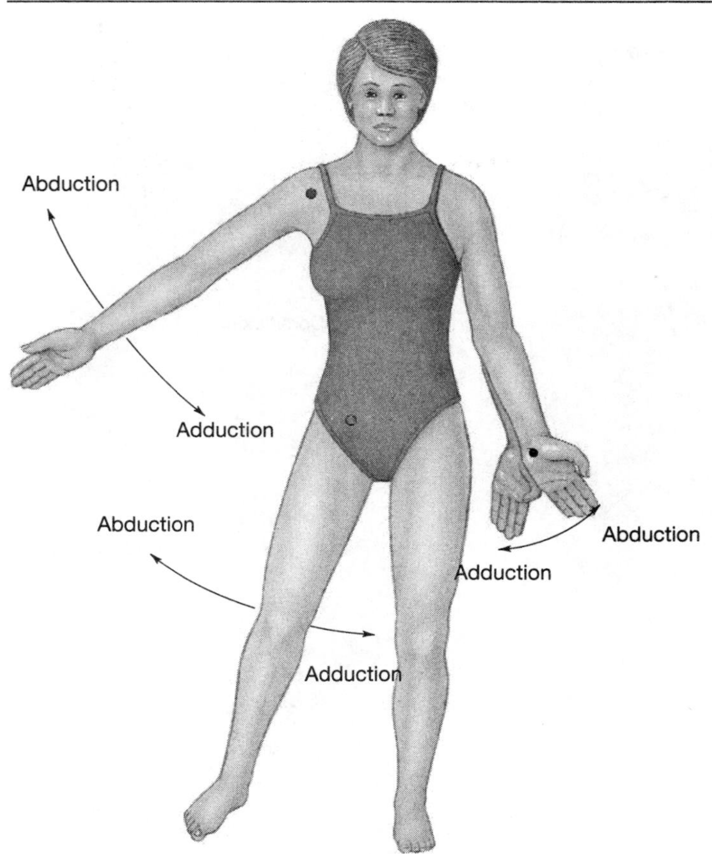

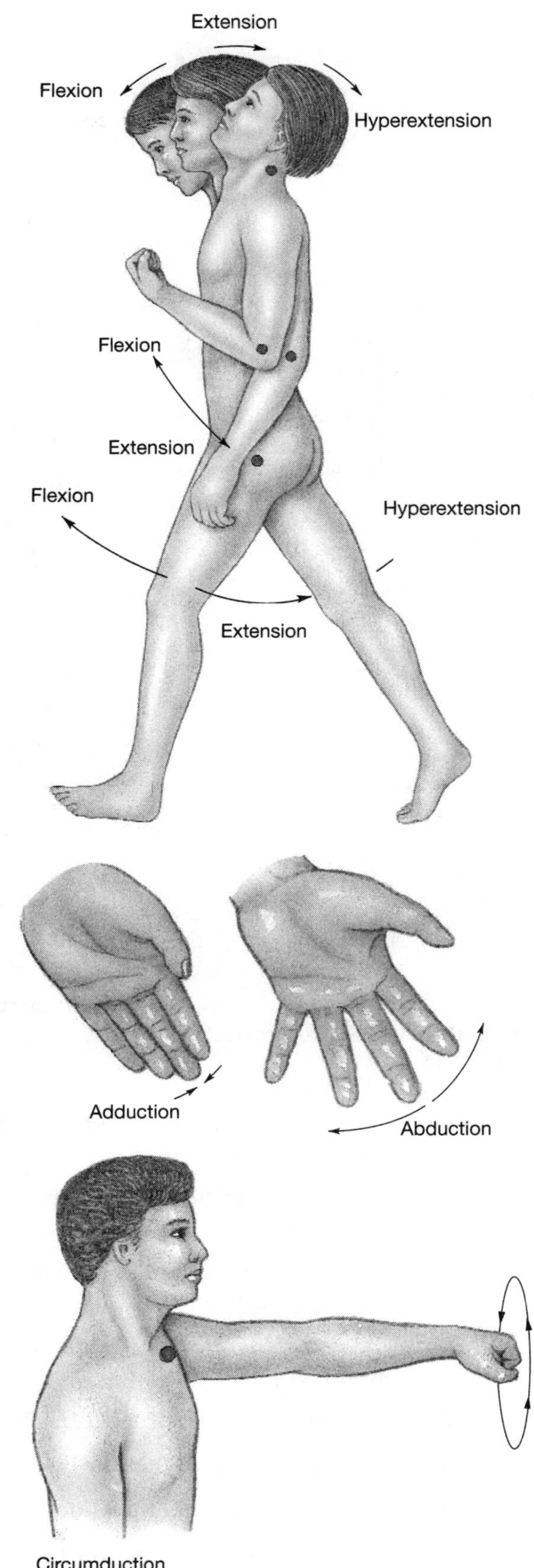

● **FIGURE 18–6**
Movements.

ing the different types of diarthotic joints. Monaxial joints allow movement in one plane. Hinge joints and pivot joints are monaxial. Gliding, ellipsoidal, and saddle joints are classified as biaxial because movement is restricted to two directions. Triaxial joints, the ball-and-socket joints, allow movement in all directions, or all three planes.

4➡ With your partner, demonstrate what is meant by monaxial, biaxial, and triaxial. Use your balloon model to demonstrate these movements. Explain how these three motions correlate with the different types of synovial joints. *[∞ Martini, p. 251]*

❑ Practice

VI. Gross Movements with Movable Joints

The extension of structural classification for the diarthrotic joints — monaxial, biaxial, or triaxial — is the description of gross movements based on these types of structures.

A. Terminology

1➡ Study Figure 18–6●. Since these terms are used throughout anatomy and physiology,

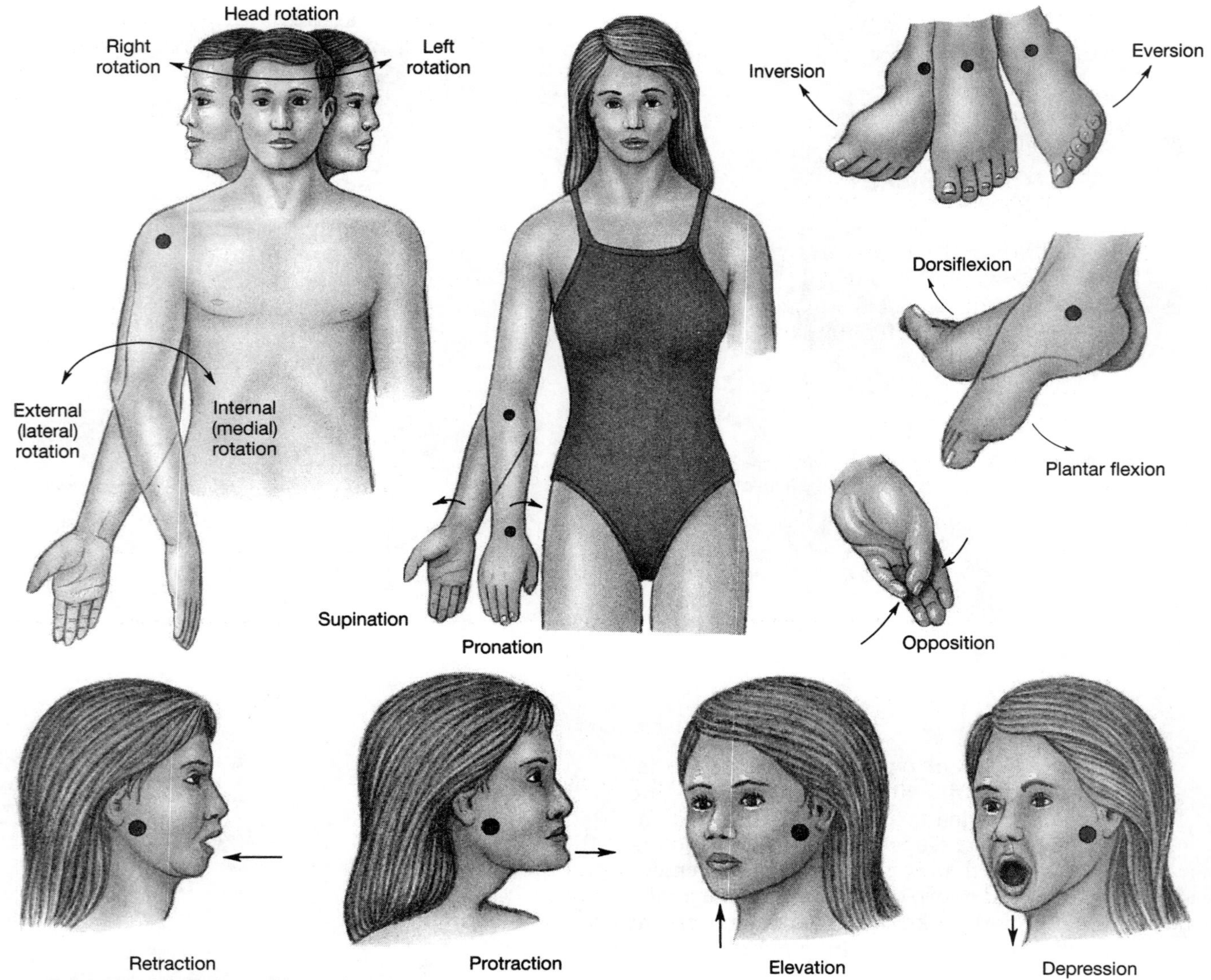

● **FIGURE 18–6 *(continued)***
Movements.

you should strive to understand each meaning: *[∞ Martini, p. 251]*

Flex reduce the angle between articulating surfaces.

Extend increase the angle between articulating surfaces.

Hyperextend increase the angle beyond anatomical position.

Abduct move away from a center line.

Adduct move toward a center line.

Invert turn toward a center line.

Evert turn away from a center line.

Depress move down.

Elevate move up.

Rotate move around a central point (for example, turn head side to side).

Lateral (external) rotation move an anterior surface laterally (for example, turn the leg outward).

Medial (internal) rotation move an anterior surface medially (for example, turn the leg inward).

Circumduction move in a loop or circle.

Opposition (thumb only) — thumb to little finger.

Protract move forward on a horizontal plane.

Retract move backward on a horizontal plane.

Pronate turn palm down.

Supinate turn palm up.

Dorsiflex decrease the upper surface angle of the foot.

Plantarflex decrease the lower surface angle of the foot.

B. IDENTIFYING BODY MOVEMENTS

1➡ Work with your partner to use the preceding terms to describe the following motions: shaking your head, waving goodbye, pitching a baseball, shrugging your shoulders, tapping your foot.

____

____

2➡ Identify the antagonistic movements in the preceding list.

____

____

3➡ Imitate the above motions. Now try some of these movements with parts of your body other than those pictured. For instance, abduct and adduct your toes. Write down what you did.

____

____

4➡ Try performing more than one motion at a time. For instance, can you simultaneously abduct and laterally rotate your leg? Take notes as you think through the action.

____

____

5➡ Discuss this semantic question with your lab partner: If "to flex" means "to decrease the angle," do you flex a muscle or do you flex a bone or do you flex a joint?

____

____

❑ Additional Activities

NOTES

1. Investigate the principle behind using braces to reconstruct the gomphoses.
2. Explore different types of joint dislocations.
3. Double-jointed people don't really have two joints at an articulation point. Instead, double-jointedness is the result of differences in the synovial sac. What types of differences do you suppose these are?

Answers to Selected Concept Check Questions

1. The fact that the distal tibio-fibular joint is amphiarthrotic affords additional protection to the bones.
2. The functional advantages of limited intervertebral motion include increased stability and increased strength.
3. The functional advantages of the symphysis pubis being an amphiarthritic joint is protection during jars and jolts.
4. The bursae act as shock absorbers. Bursitis is painful because as the closed sac becomes inflamed, nerve endings are stimulated.
5. The roles of the components of the simulated synovial joint are as follows:
   Tape = articular cartilage
   Water = synovial fluid
   Balloon = synovial capsule

❑ Lab Report

1. Referring to the numbered inquiries at the beginning of this exercise, complete the following box summary:

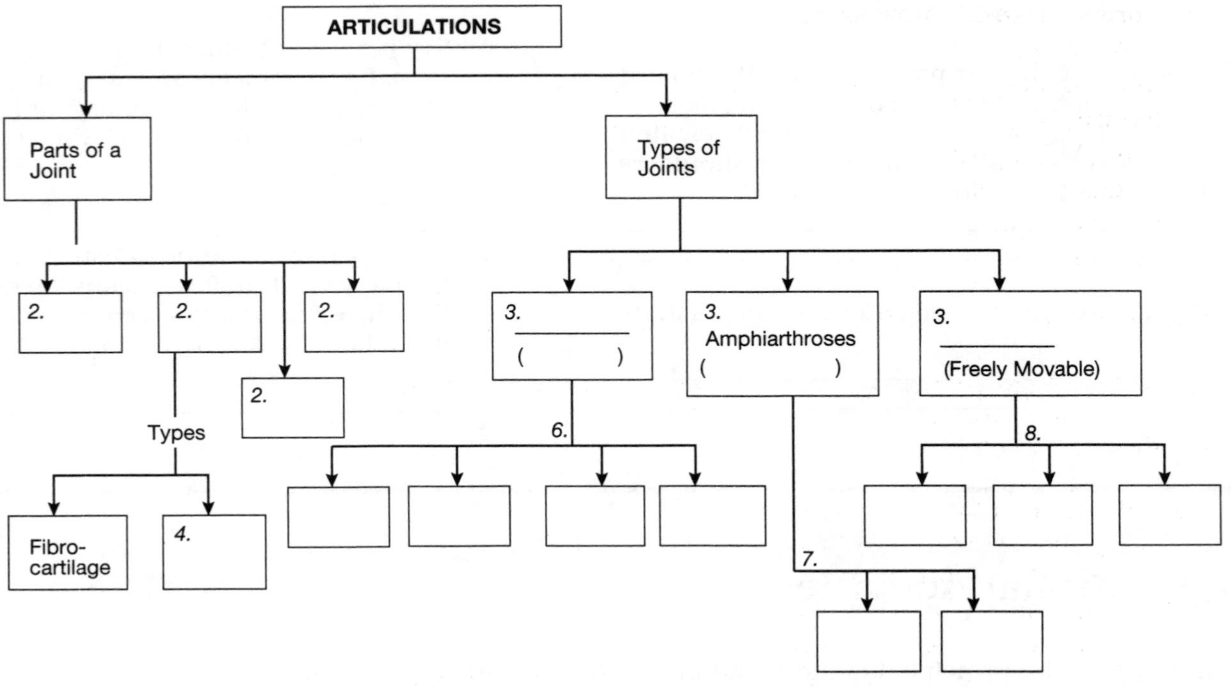

General Questions

1. Describe the fibrous connections and the cartilagenous connections found in the synarthritic and amphiarthritic joints.

2. List the tissues involved in articulations.

3. For each joint on the list below, give the type of joint and the bones involved. Try completing this exercise using just your knowledge of the bones and the given descriptions of the joint characteristics. If necessary, refer to your lecture text.

	Type	Bones
a. Intervertebral	__________	__________
b. Acromio-clavicular	__________	__________
c. Proximal radio-ulnar	__________	__________
d. Proximal tibio-fibular	__________	__________
e. Sterno-clavicular	__________	__________
f. Symphysis pubis	__________	__________
g. Metatarso-phalangeal	__________	__________
h. Atlanto-axial	__________	__________
i. Sacroiliac	__________	__________
j. Knee (patella/femur)	__________	__________
k. Maxillo-lacrimal	__________	__________

l. Interphalangeal ____________ ____________
m. Intraphalangeal ____________ ____________
n. Intercuneiform ____________ ____________
o. Hip ____________ ____________
p. Middle ear ____________ ____________
q. Temporo-mandibular ____________ ____________
r. Coronal suture ____________ ____________
s. Radio-carpal ____________ ____________
t. Intercarpal ____________ ____________
u. Ilio-ischial ____________ ____________
v. Scapulo-humeral ____________ ____________
w. Metaphyseal plate ____________ ____________
x. Sterno-costal ____________ ____________
y. Your choice! ____________ ____________

4. Why won't you get arthritis of the mandibulo-hyoid joint?

5. Go back to Figures 18–5● and 18–1●. In your own words explain the movement as shown in Figure 18–5●. How does this movement relate to the description given in Figure 18–1●?

6. Consider each of the following four actions:
   a. Sitting there reading and turning the pages of this book. First decide how you are holding the book.
   b. Getting a glass from the cupboard and filling it with water.
   c. Reaching down to pick up a coin from the floor.
   d. Walking with your leg in a cast because you have a broken patella.

- For each of the above four actions, answer the following:
  a. Which joints were involved?
  b. What actions were accomplished?
  c. Describe the mechanics involved:

- What action or movement occurred at each joint?
  a. Is there any other way this action could have happened?
  b. Are any compensatory movements involved?

7. What is the difference between:
   a. Extension and hyperextension
   b. Eversion and inversion
   c. Dorsiflexion and plantarflexion
   d. Pronation and supination
   e. Abduction and adduction
   f. Lateral rotation and medial rotation
   g. Protraction and retraction

8. Refer to your synovial joint model. If this were a knee joint, where would the menisci and the patella be located?

EXERCISE 19

Fetal Skeleton

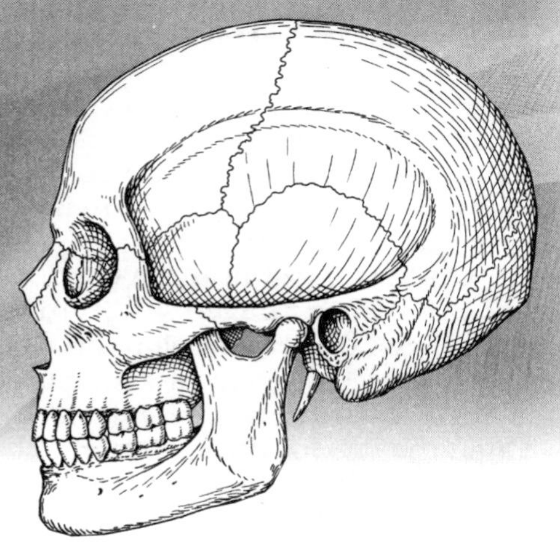

PROCEDURAL INQUIRIES

Preparation

1. What is the primordial tissue of the skeletal system?
2. What is the difference between endochondral and intramembranous ossification?

Examination

3. What are the steps involved in vertebral ossification?
4. What processes are involved in the ossification of the skull?
5. What is the apical ectodermal ridge?
6. What are the steps involved in fetal limb bud development?

Additional Inquiries

7. How are joints formed?
8. How does the fetal skeleton compare with the adult skeleton?

Key Terms

Apical Ectodermal Ridge
Appendicular
Axial
Bone
Cartilage
Condensation Center
Coronal Suture
Cranium
Endochondral Ossification
Fetal Skeleton
Fontanels
Frontal
Frontal (metopic) Suture
Interzonal
Intramembranous Ossification
Joint
Lambdoidal Suture
Mastoid
Mesenchyme
Neurocranium
Occipital
Sagittal Suture
Sphenoid
Spina Bifida
Squamosal Suture
Vertebral Column

Materials Needed

Charts and Diagrams of the Fetal Skeleton
Adult Skull
Fetal Skull
Adult Vertebra
Adult Skeleton
Fetal Skeleton (if available)
2 × 2 Slides of Ossification (if available)

In this laboratory exercise we will examine briefly the early development of the human skeletal system. [∞ AM] Some specific points of later skeletal development can also be found in Exercises 14 & 18.

Skeletal formation is a long and involved process. It is also an ongoing process. The **fetal skeleton**, which is the focus of this exercise, is one part of the developmental continuum.

The precursors of the skeletal system can be identified during the second gestational week, yet the skeleton itself will not be completely formed for another 25 or so years. For some people, the process may take even longer, and certain cartilagenous portions of the skeleton (such as those in the nose) never really stop growing. (In this exercise we are primarily concerned with bone development rather than with the development of the permanent cartilage.)

Bone formation is essentially mesenchymal ossification with or without a chondrification intermediary. **Mesenchyme** is the connective tissue derivative of the mesodermal germ layer. This mesenchyme condenses at specific points or **condensation centers**. Most of these condensation centers give rise to **cartilage** that, in turn, gives rise to **bone**. Some centers develop directly into bone. This developing bone grows in one or more directions until ossification is complete.

Often the cue for the cessation of bone growth is contact inhibition. In other words, two plates or two bones (or their endpoints) come together and stop developing. Sometimes this contact results in the complete fusion of bones (as seen in the frontal bone and in the bones of the os coxae), and sometimes this contact results in the formation of a **joint**, be it the intricate interdigitation found between the bones of the skull or the loose synovial processes found in the knee and elbow. When contact occurs, a cartilagenous line may or may not remain. (Recall the epiphyseal plates discussed in Exercise 18.)

Synovial, cartilagenous, and fibrous joints all develop from **interzonal** mesenchyme, the mesenchyme located between the areas of bone development compaction. The ligaments are also of this interzonal mesenchymal origin.

Recall that in Exercise 14 we stated that the neonate's skeleton has about 270 bones and the adult skeleton has only 206 bones. In the fetus and in the newborn many of the ossified skeletal areas are counted as separate bones (14 for the sphenoid bone alone!). When these bones fuse, they are no longer counted separately and the total bone count for the body goes down.

In Exercise 14 we also mentioned the Wormian and sesamoid bones. These bones can form for many different reasons. If you think about the sutural bones that may occur between the parietal and occipital bones, you can see that one common cause for extra bone formation might be the presence of extra ossification centers and the subsequent incomplete fusion of the newly formed bony tissue.

In studying the fetal bone development, it is appropriate to divide the skeletal system into its **axial** and **appendicular** components because the specifics of ossification are somewhat different for each part. In addition, we often subdivide the axial skeleton and study the **vertebral column** and the **cranium (skull)** separately.

❑ Preparation

I. Overview

A. Mesoderm

The skeleton develops from the mesoderm, the middle layer of the embryonic tissue. The mesenchyme is the embryonic or fetal connective tissue from which such structures as the bones develop. (Recall Exercise 11.)

B. Mesenchyme

Intraembryonic mesoderm gives rise to mesenchymal tissue. This mesenchyme then condenses in specific places in one of two ways.

1➔ In **endochondral ossification** most of these condensations undergo chondrification (cartilage formation). Chondrification begins at certain centers and spreads throughout the mesenchymal condensation. Ossification (bone formation) begins at centers within the cartilage and spreads throughout the tissue. [∞ *Martini, p. 176*]

2➔ In **intramembranous ossification** some of the mesenchymal condensations ossify directly (without first forming cartilage). [∞ *Martini, p. 175*]

❑ Examination

In the following paragraphs keep in mind that the human embryonic period extends from fertilization through the first eight weeks. After two gestational months the embryo is called a fetus.

II. Axial Skeleton

A. Vertebral Column [∞ AM]

Precartilagenous tissue begins to develop during the fourth week as mesenchymal cells surround the notochord to form the vertebral body, the vertebral arch, and the ribs. (A notochord is a dorsal rod, which functions as a support structure in the chordate embryo.) Remnants of the human notochord can be found in the intervertebral discs.

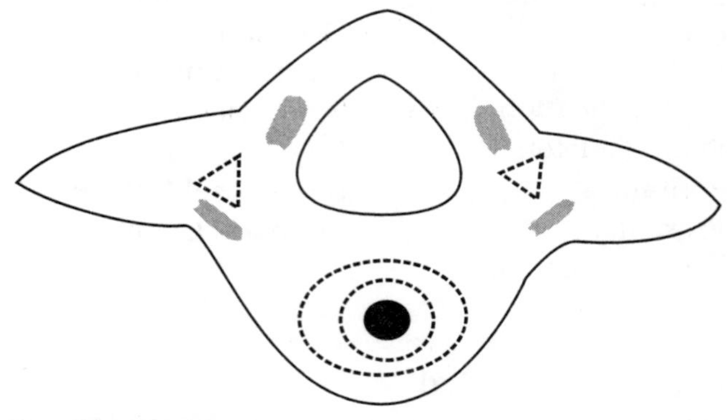

● **FIGURE 19–1**
Developing Vertebra.

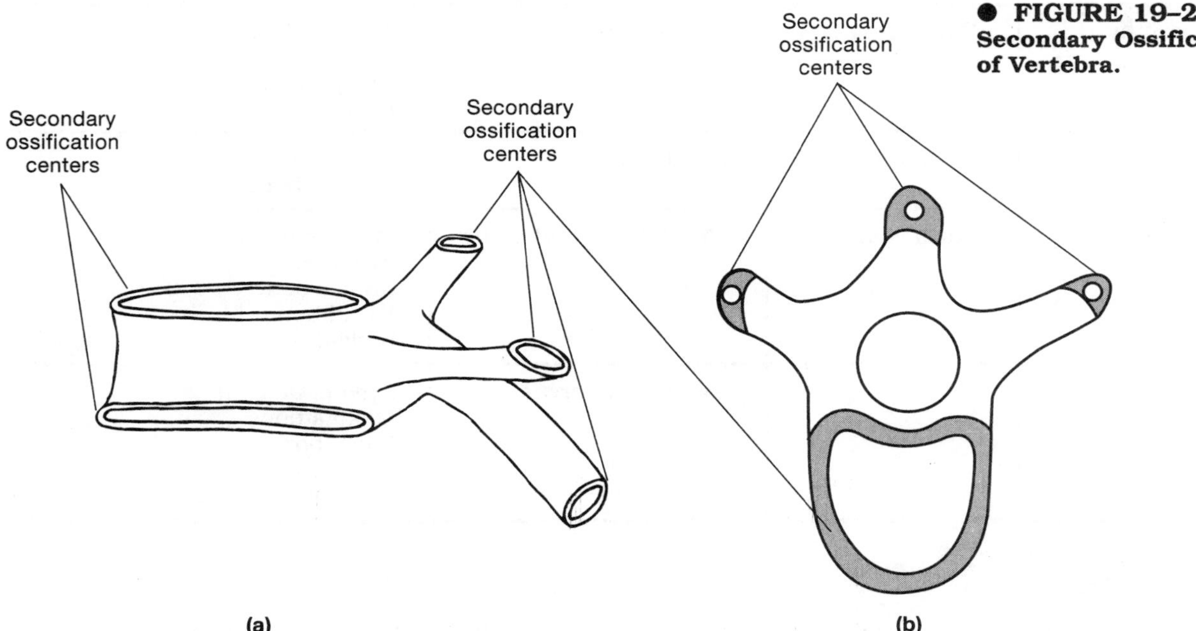

● **FIGURE 19–2**
Secondary Ossification of Vertebra.

1➡ Study Figure 19–1●, a generic developing vertebra. Refer back to Figures 16–10● and 16–12● if necessary as you label the following structures on Figure 19–1●: centrum, vertebral arch, costal process, vertebral foramen, notochord.

2➡ Note the shaded areas on Figure 19–1●. These are the chondrification centers that appear during the sixth gestational week. These centers all expand, and fusion of the chondrification sites takes place at the end of the embryonic period.

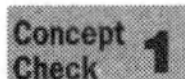

Do the chondrification centers begin during the embryonic or fetal period?

____

____

3➡ Use Figure 19–1● as you darken the areas marked off with the dotted lines — one around the notochord and one on each vertebral arch. You have just identified the primary ossification centers of a generic vertebra. The primary ossification centers are obvious during the seventh week (though ossification itself is still a few days away).

4➡ Study Figure 19–2●. The halves of the vertebral arches will fuse in three to five years and the vertebral arches will fuse with the centrum in three to six years. After puberty, secondary ossification centers appear and ring the centrum and the tips of the spinous and transverse processes of the vertebra. These secondary centers will unite with the rest of the vertebra at about age 25.

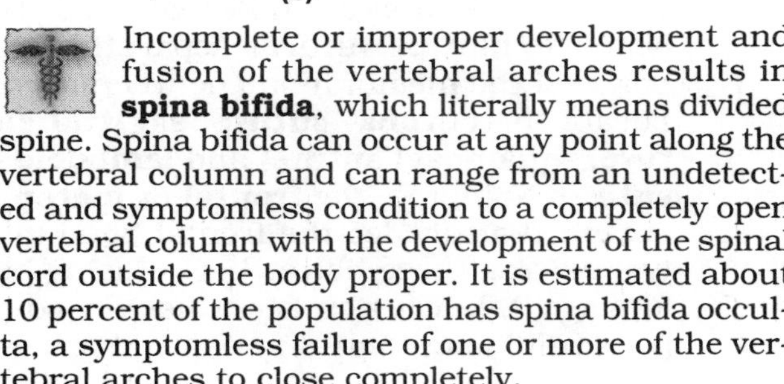
Incomplete or improper development and fusion of the vertebral arches results in **spina bifida**, which literally means divided spine. Spina bifida can occur at any point along the vertebral column and can range from an undetected and symptomless condition to a completely open vertebral column with the development of the spinal cord outside the body proper. It is estimated about 10 percent of the population has spina bifida occulta, a symptomless failure of one or more of the vertebral arches to close completely.

5➡ Study the mesenchymal costal processes. These processes give rise to the ribs, which then undergo standard chondrification and ossification. A synovial joint develops at this point from the interzonal mesenchyme. Interestingly, about five percent of the population has an abnormal number of ribs.

6➡ Examine the laboratory adult vertebrae and identify the locations of the primary and secondary ossification centers. Identify parts as listed.

B. Skull [∞ Martini, p. 211]

In studying fetal skeletal development, we often divide the skull into the **neurocranium**, the protective case for the brain, and the **viscerocranium**, the skeleton of the jaws. Parts of both the neurocranium and the viscerocranium are formed endochondrally, and other parts of both are formed intramembranously.

1➡ Study Figure 19–3●. Locate as many bones as possible on either the fetal skull or the adult skull.

Neurocranium	Cartilaginous	Bones of base of skull	Sphenoid bone Petrous portion of temporal bone Occipital bone Ethmoid bone
	Membranous	Cranial vault	Parietal bones Frontal bones
Viscerocranium	Cartilagenous	First 3 branchial arches	Malleus bone Incus bone Stapes bone Hyoid bone Styloid process of temporal bone
	Membranous	Lower facial areas	Zygomatic bone Maxilla bone Nasal bone Mandible bone Squamous portion of temporal bone

● **FIGURE 19–3**
Cranial Derivatives.

2➡ Study the fetal skull and Figure 19–4●. Note the fibrous sutures separating the bones of the membranous neurocranium. Locate the **coronal suture**, **sagittal suture**, **squamosal suture**, and **lambdoidal suture**. Note, too, the **frontal** or **metopic suture** that divides the frontal bone into two halves. These halves begin fusing before age 2, but fusion is not usually complete until about age 8.

3➡ Use Figure 19–4● (along with a fetal skull, if available) to locate the six fibrous "soft spots," the **fontanels** (also spelled fontanelles). These include: **frontal** (anterior) fontanel, **occipital** (posterior) fontanel, two **sphenoid** (anterolateral) fontanels, and two **mastoid** (posterolateral) fontanels. Find these "former fontanels" on the adult skull. *[∞ AM, p. 211]*

Concept Check 2 Give two functional reasons for the infant's fibrous sutures and fibrous fontanels.

____

____

III. Appendicular Skeleton

A. Development

To understand the fetal development of the appendicular skeleton, we must first understand that arms and legs begin as limb buds, which are discernible about the end of the fourth week. The mesenchyme of this bud is covered by a layer of ectoderm. The end point of the ectoderm

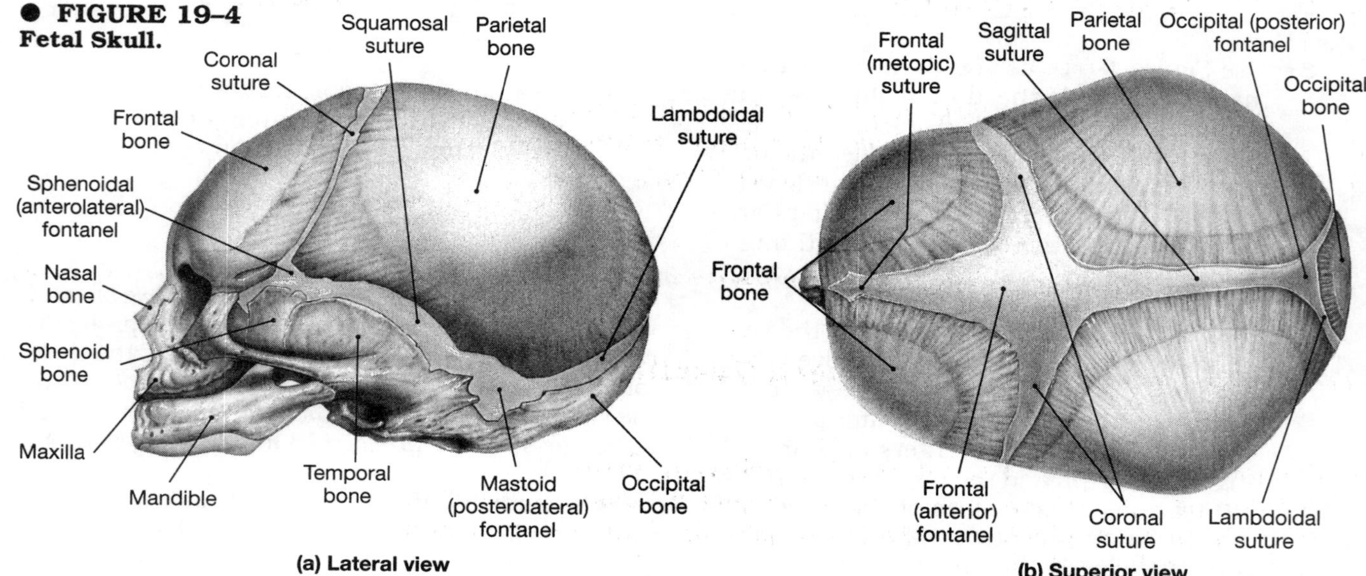

● **FIGURE 19–4**
Fetal Skull.

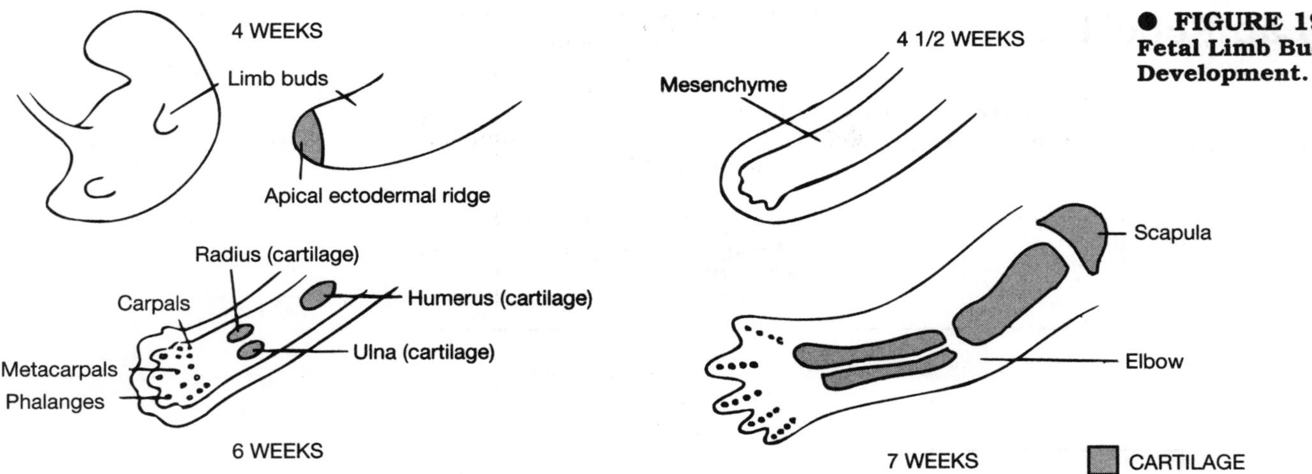

● **FIGURE 19–5 Fetal Limb Bud Development.**

is the **apical ectodermal ridge**, which promotes the growth and development of the limbs, including directing the formation of the cartilage.

The mesenchymal prebone of the limbs undergoes chondrification, forming hyaline cartilage during the fifth week. Ossification begins during the eighth week. By 12 weeks nearly all appendicular bones have primary ossification centers. Secondary ossification centers develop after birth. [∞ AM]

Information about the completion of ossification of the bones of the appendicular skeleton can be found in Exercise 17.

1➡ Follow Figure 19–5●, which shows the fetal development of a limb. Although this figure shows a generic arm, keep in mind that the arm and leg are homologous structures, with the lower limb generally developing a few days after the upper limb. [∞ AM]

IV. Conceptual Extension

A. Fetal Skeleton

1➡ If a fetal skeleton is available, follow the ossification process as outlined in this exercise.

B. Visual Aids

1➡ View any ossification slides that are in your laboratory.

❑ Additional Activities

NOTES

1. Examine any additional charts, models, or diagrams that are in your laboratory.
2. Research the approximate ages when the different sutures and fontanels close.
3. Research some common fetal skeletal malformations.

Answers to Selected Concept Check Questions

1. Chondrification centers begin during the embryonic period.
2. The fibrous sutures and fibrous fontanels offer flexibility of the skull during birth (to avoid injury) and allow for growth of both the brain and the bones of the skull.

❑ Lab Report

1. Referring to the numbered inquiries at the beginning of this exercise, complete the following box summary:

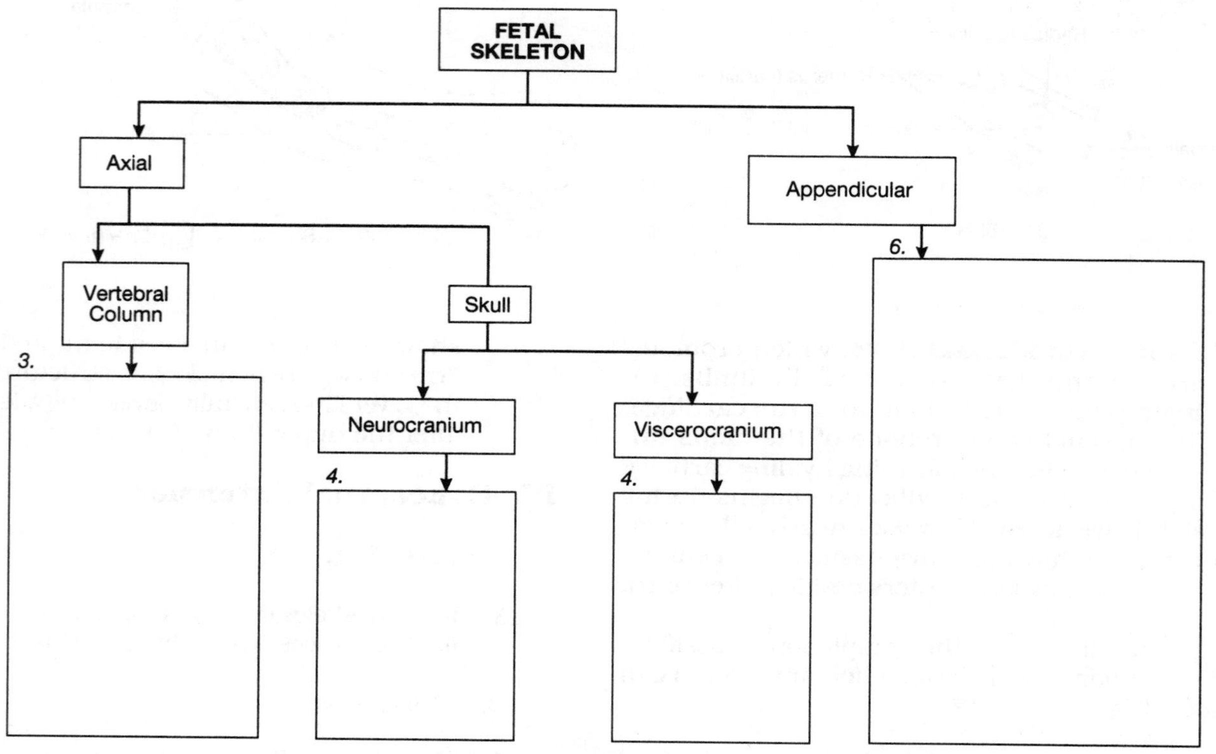

General Questions

NOTES

1. Construct a flowchart using the following terms: secondary ossification, chondrification, mesenchyme, mesoderm, primary ossification. Which of these steps is sometimes not a part of the bone formation process? Consider correlating your information here with the preceeding box summary.
2. Complete the following chart demonstrating the similarities and differences between axial and appendicular skeletal formation. Use your own paper.

	Similarities	Differences
Axial		
Appendicular		

3. List the similarities and differences between the formation of a vertebra and the formation of a long bone, from mesenchyme to old age. Refer to your lecture text if necessary.

EXERCISE 20

Muscle Tissue: An Overview

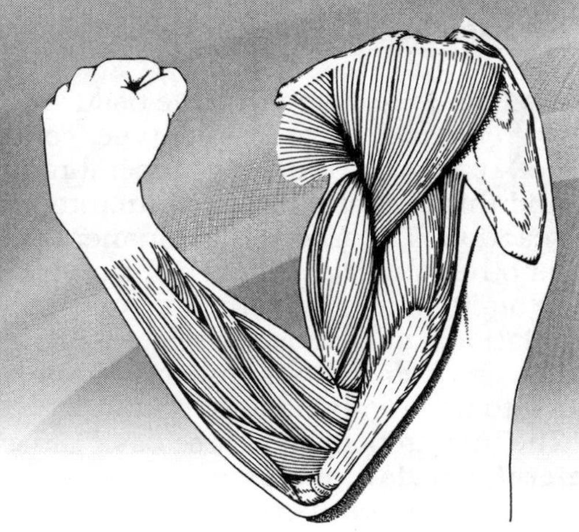

PROCEDURAL INQUIRIES

Examination

1. What are the characteristics of skeletal muscle?
2. What are the components of skeletal muscle?
3. Where is skeletal muscle found?
4. What is the functional importance of skeletal muscle?
5. What are the characteristics of cardiac muscle?
6. What are the components of cardiac muscle?
7. Where is cardiac muscle found?
8. What is the functional importance of cardiac muscle?
9. What are the characteristics of smooth muscle?
10. What are the two types of smooth muscle?
11. Where is each type of smooth muscle found?
12. What is the functional importance of each type of smooth muscle?

Additional Inquiries

13. What are some properties common to all muscle tissue?
14. What is the difference between voluntary and involuntary muscle?
15. What are dense bodies and why are they important?
16. What are striations?

Key Terms

Actin
Atrial Tissue
Cardiac Muscle
Dense Bodies
Endomysium
Epimysium
Fasciculi
Intercalated Disk
Modified Electrical Tissue
Multiunit Smooth Muscle
Muscle Fiber
Myoblasts
Myofibrils
Myofilaments
Myosin
Perimysium
Sarcomere
Smooth Muscle
Striated Muscle
Syncytium
Tendon
Ventricular
Visceral Smooth Muscle
Voluntary Muscle

Materials Needed

Compound Microscope
Dissecting Microscope
Prepared Slides

- Skeletal Muscle — Longitudinal Section and Cross Section
- Cardiac Muscle — Longitudinal Section
- Smooth Muscle
  - Bundle
  - Teased
  - Digestive System — Cross section
  - Blood Vessel — Cross section

Muscle is the contractile tissue of the body. Muscle, regardless of type, contracts because the muscle cell membrane is excited, and this excitation leads to an internal chemical reaction. The internal components slide over one another and the cell shortens or contracts.

In addition to contractibility (due to membrane stimulation), muscle is also extensible and elastic. This means that any muscle can stretch and return to its original length.

The human body has three basic muscle types: **skeletal**, **cardiac**, and **smooth**. Skeletal and cardiac muscle are both **striated**. This means within the muscle cells there are transverse bands called striations that contract simultaneously upon stimulation. Smooth muscle is not striated. The functional units within the smooth muscle cell do not contract in unison. The word **syncytium** is often used to explain this simultanious contraction.

Approximately 50% of the body mass is muscle tissue. (About 40% of the body is skeletal muscle; about 10% is smooth and cardiac muscle.)

In this overview laboratory exercise we will examine the basic anatomy and histology of the different types of muscle tissue.

❑ Examination

I. Striated Muscle — Skeletal

[∞ Martini, p. 271]

A. Anatomy

1➜ Begin by studying Figure 20–1a & b●. Refer to this figure as you work through this section on skeletal muscle tissue.

Skeletal muscles generally develop from embryonic uninucleate **myoblasts**, which fuse to form the multinucleate syncytium we commonly think of as **muscle fiber**. The muscle fiber is the muscle cell, and each muscle fiber is surrounded by a delicate connective tissue called the **endomysium**. Muscle fibers occur in bundles, known as **fasciculi**. Each fasciculus is surrounded by a connective tissue called the **perimysium**. A group of bundles is surrounded by another connective tissue, the **epimysium**. This group of bundles is what we usually refer to as the skeletal muscle. The epimysium is continuous with the **tendon**, and the tendons connect muscles to bones or muscles to other muscles.

Each muscle fiber (muscle cell) is composed of cylindrical **myofibrils**, and each myofibril is composed of the **myofilaments**, **actin** and **myosin**.

2➜ Note the striations in Figure 20–1b●. The functional unit of the striated muscle is the **sarcomere**. The sarcomere runs from striation to striation. The actin and myosin myofilaments interact in unison within the sarcomere to form the strong synchronous muscle contractions we are familiar with. We will examine the specifics of this actin-myosin interaction in Exercise 22.

B. Function

One function of every muscle is to contract. Locomotion, posture, support, balance, guarding orifices, facilitating internal movement, regulating blood flow, and maintaining body temperature are all possible because of muscle contraction.

Skeletal muscles are usually classified as **voluntary**. You have voluntary control over most of these muscles, although you do not have any control over the esophageal or pharyngeal muscles. Certain other muscles, such as the diaphragm, operate without your conscious control, but you can exert some control as you choose. (That is, you can until the body's survival mechanisms take over. For example, if you hold your breath too long, you will pass out and start breathing normally.) You can control other muscles, such as those in your arms and legs, most of the time, except when they are involved in a reflex situation.

1➜ As you are sitting reading this lesson, which general muscle groups are you voluntarily contracting? Which voluntary muscles are contracting without your conscious control?

____

____

C. Distribution

Skeletal muscles, found throughout the body, are so named because most are directly attached to the skeleton. A few, such as the tongue, the external anal sphincter, and some of the esophageal and pharyngeal muscles, are not attached to any bone but, morphologically, these muscles are indistinguishable from other skeletal muscles.

D. Slides and Illustrations

1➜ Study Figure 20–1a & b●. Note the peripheral nuclei. Check off these elements as you locate them:

Endomysium	____	Fasciculus	____
Myofibril	____	Perimysium	____
Muscle fiber	____	Myofilament	____
Epimysium	____	Striation	____
Nuclei	____	Sarcomere	____
Tendon	____		

2➜ Examine the slide of the cross section and longitudinal section of the skeletal muscle. Identify the components described in Section I.A. Sketch and label what you see.

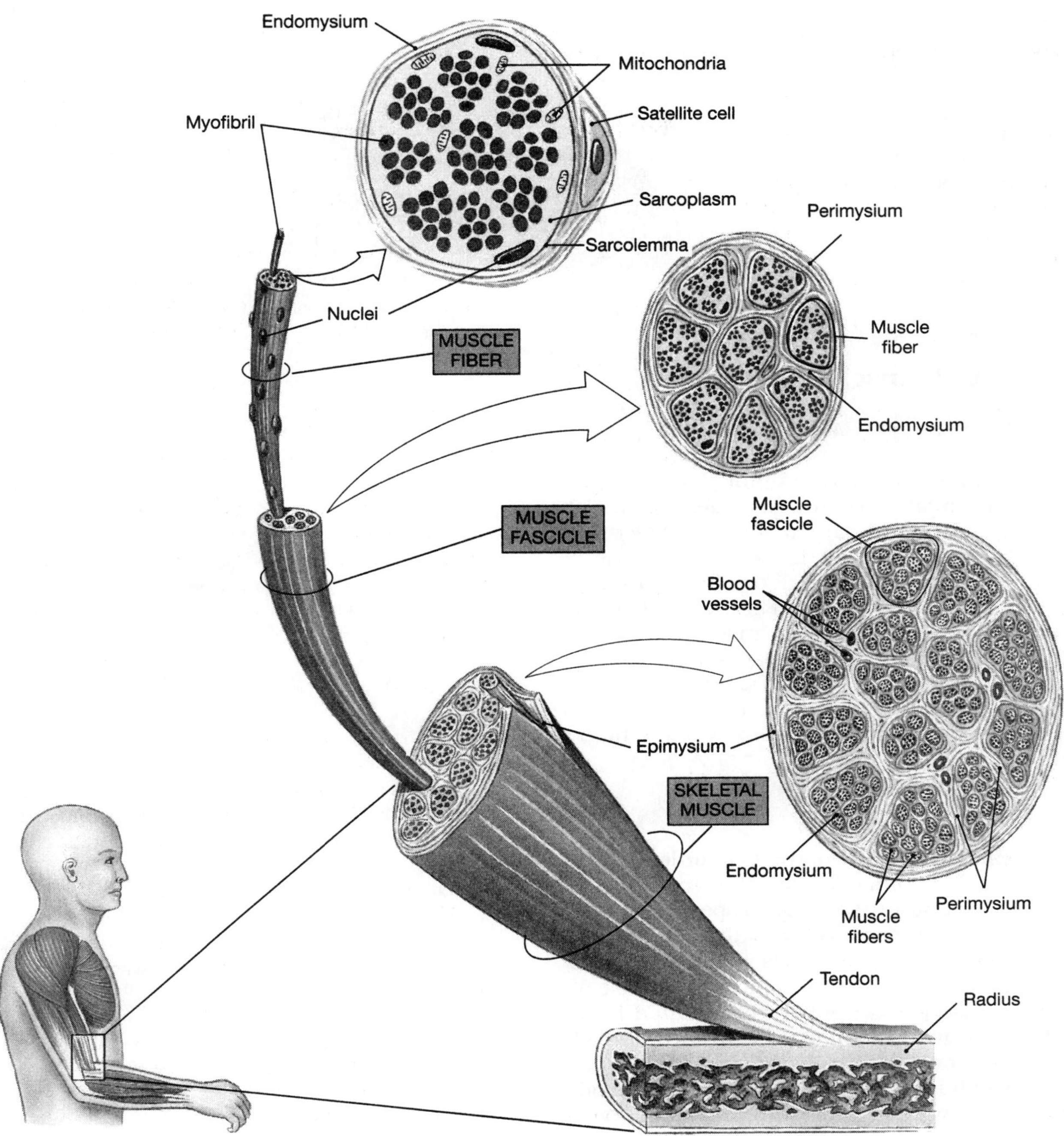

● **FIGURE 20–1a**

Structural Components, Skeletal Muscles. A skeletal muscle consists of fascicles (bundles of muscle fibers) enclosed by the epimysium. The bundles are separated by connective tissue fibers of the perimysium, and within each bundle the muscle fibers are surrounded by the endomysium. Each muscle fiber has many superficial nuclei, as well as mitochondria and other organelles.

Muscle cross section	Muscle longitudinal section

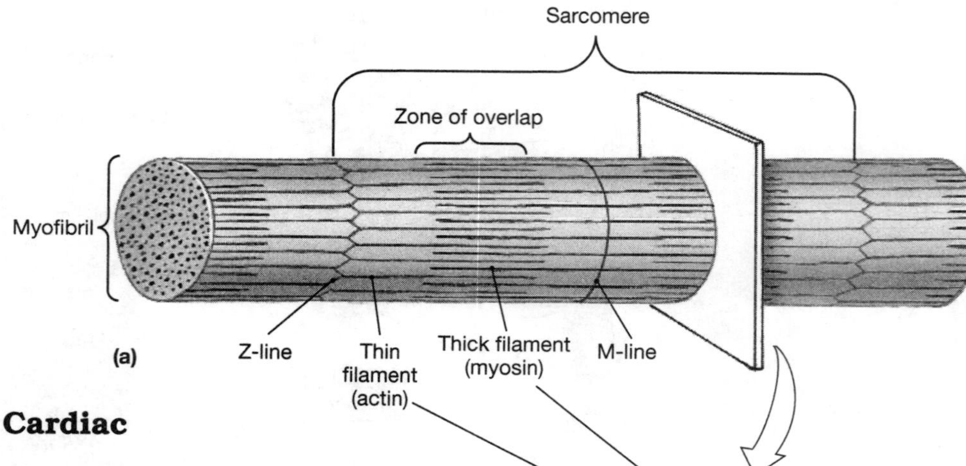

● **FIGURE 20–1b**
Sarcomere Structure. Each myofibril consists of a linear series of sarcomeres. **(a)** Organization of thick and thin filaments. **(b)** A cross-sectional view through the zone of overlap.

II. Striated Muscle — Cardiac

A. Anatomy [∞ Martini, p. 302]

Cardiac muscle cells are uninucleate cells separated by **intercalated disks**. The intercalated disks are actually cell membranes that fuse, giving the appearance of interdigitation. These disks form permeable junctions with only about one four-hundredth the electrical resistance found between other cells. As a result, the cardiac cells form a functional syncytium, carrying electrical messages from one cell to the next. Additional information about cardiac muscle cells can be found in your lecture text and in the laboratory exercises on the anatomy of the human heart and human cardiovascular physiology.

B. Function

The cardiac muscle contracts in order to pump the blood to the lungs for gas exchange with the environment and to the body proper for nutrient and waste exchange with the individual cells.

As we will see in the exercise on cardiovascular physiology, cardiac cells contract because of a nerve impulse (action potential) initiated by the cardiac conduction system. This autorhythmicity can be influenced by the vagus nerve.

Cardiac impulses are not normally controlled voluntarily. By using certain biofeedback techniques, you may be able to slow down or speed up your heart rate. However, this degree of voluntarism is minimal compared with your inability to start and stop your cardiac contractions.

Cardiac muscle cells are striated. As with skeletal muscle, the actin and myosin myofilaments interact synchronously within the functional units, producing syncytial contractions.

C. Distribution

Cardiac muscle, found exclusively in the heart, can be subdivided into three types of cardiac muscle tissue. The upper chambers of the heart are composed of **atrial** tissue, the lower chambers of **ventricular** tissue, and the excitatory and conductive fibers of **modified electrical tissue**. The atrial and ventricular tissues are contractile tissues and the electrical tissue has very little contractibility.

D. Slides and Illustrations

1➡ Study Figure 20–2●. Note the centrally located nuclei and the branching pattern of the cells.

2➡ Examine a slide of cardiac muscle. Sketch and label what you see.

Cardiac tissue

3➡ Now compare the skeletal and cardiac muscle. Look for similarities and differences in the striations, cell membranes, nuclei, size, etc. List the similarities and differences you are able to find.

Similarities ______________________

Differences ______________________

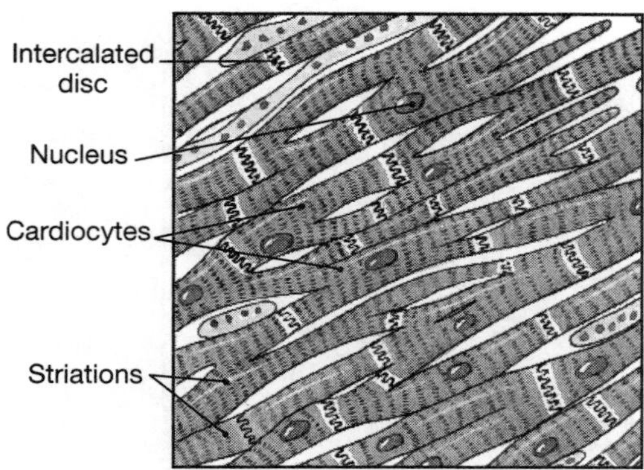

● **FIGURE 20–2**
Cardiac Muscle.

III. Smooth Muscle *[∞ Martini, p. 303]*

A. Anatomy

1➡ Smooth muscle cells, distinctive for each organ in which they are located, vary considerably with regard to morphology, response to stimuli, innervation, and function. Nevertheless, we usually identify two major smooth muscle types.

The **multiunit smooth muscle** is found in discrete fibers that function independently in response to single points of innervation from one or several motor neurons. These include the muscle of the iris of the eye and the muscles responsible for piloerection. These smooth muscle cells include a basement membrane-like glycoprotein for insulation, and they rarely contract spontaneously.

Visceral smooth muscle fibers lack direct contact with any motor neuron. The cells are found in sheets or bundles, with numerous gap junctions between the cells to facilitate ion flow. Action potentials are caused by direct electrical conduction. Some smooth muscle sheets show cyclic rhythmicity. The visceral smooth muscle is a functional syncytium.

2➡ The individual smooth muscle cell is a long, tapered uninucleate structure. As with the other muscle types, smooth muscle cells contain actin and myosin myofilaments. The actin and myosin myofilaments are not arranged in neat packets, as they are in the striated muscles. Smooth muscles are not striated. Since the function of the actin and myosin is identical in all muscle cells, these myofilaments chemically pull across one another (sliding filament model), thus shortening the muscle cell itself. In this case, however, the actin and myosin filaments do not act in unison, so contraction within a cell is not synchronous.

The actin and myosin in the smooth muscle seem to be randomly placed within the cell. Interestingly, though, the actin myofilaments have nodules called **dense bodies**, some of which attach to the cell membranes. Other dense bodies join together, forming a protein scaffold within the cell.

B. Function

Smooth muscle has numerous functions. As with other muscle types, all smooth muscle functions are directly related to contraction. Smooth muscles are responsible for moving substances through canals, guarding openings, and providing support for bladders. Smooth muscles also function in visceral reflexes and certain sensory perceptions, such as pupillary size.

Smooth muscle contractions are far slower than the contractions of either of the striated muscle types. Smooth muscle contractions are also highly energy-efficient. Smooth muscle uses between one twentieth and one four-hundredth the energy required to achieve comparable tension in a skeletal muscle. In the situations of the smooth muscles, slow, sustained, energy-efficient contractions are exactly what is needed.

1➡ Why would a slow, sustained, energy-efficient contraction be appropriate for the muscles of the digestive tract?

____

____

C. Distribution

Smooth muscle fibers are found in virtually all parts of the body. Two examples of extensive smooth muscle presence are the layers of the blood vessels and the layers surrounding the digestive tract.

D. Slide and Illustrations

1➡ Study Figure 20–3a-c●. Note the centrally located nuclei and the cylindrical shape of the cells.

2➡ Examine the slide of the visceral smooth muscle sheet. Sketch and label structures within the individual cells.

Visceral smooth muscle

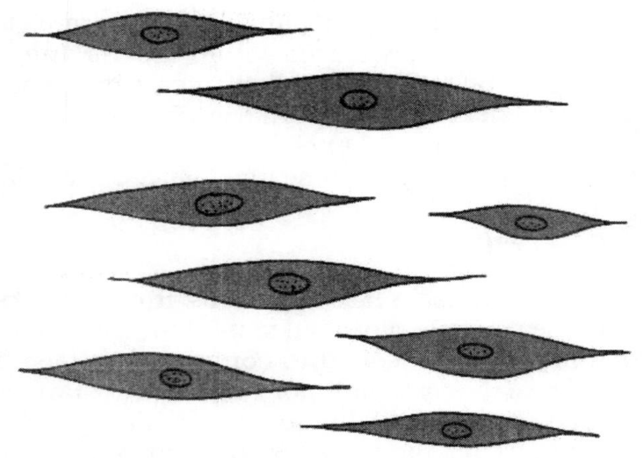

● **FIGURE 20–3a**
Smooth Muscle — Teased.

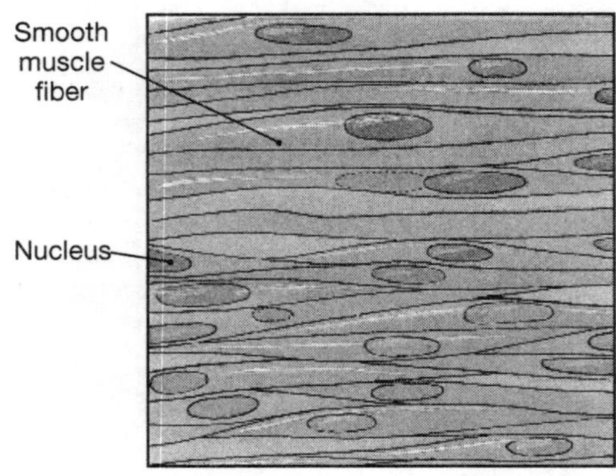

● **FIGURE 20–3b**
Smooth Muscle — Sheet.

3➡ Examine the slide of the teased smooth muscle. Teased means that the muscle cells have been separated out. Note the long thin cells with the elongated nuclei. Your slide probably is not stained in such a way that you can discern the actin and myosin filaments. However, in making your sketch, draw the cell large enough so that you can include the actin and myosin. Check Figure 20–3c●. Draw a dense body on each of the actin filaments. To which cellular parts would these dense bodies connect?

Blood vessel smooth muscle

Teased smooth muscle

4➡ Examine the slide of the cross section of the digestive tract and the cross section of the blood vessel. Identify the smooth muscle layers in each of these tissues as you sketch and label what you see.

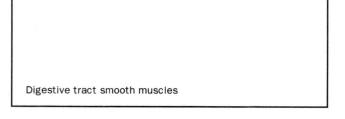

Digestive tract smooth muscles

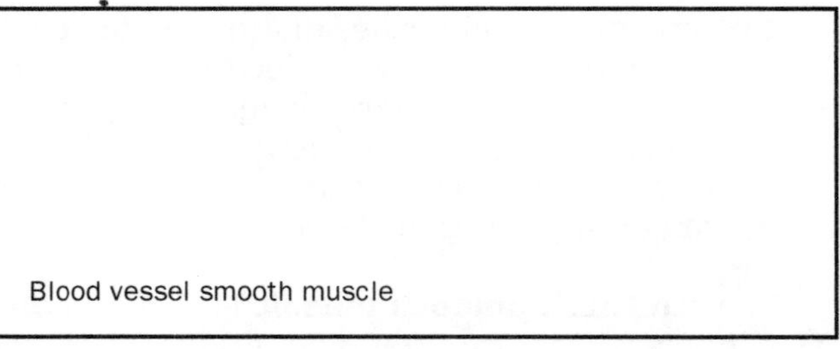

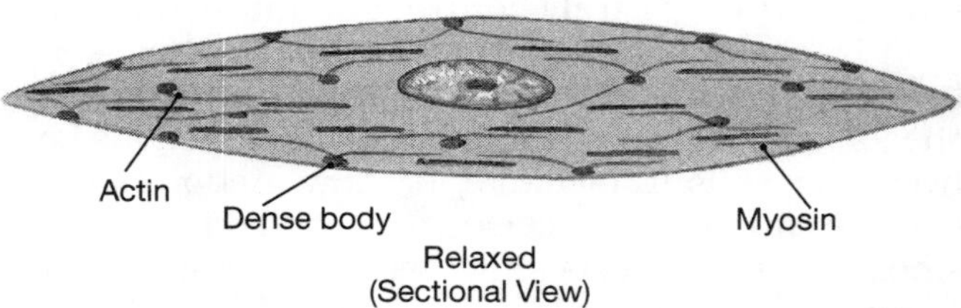

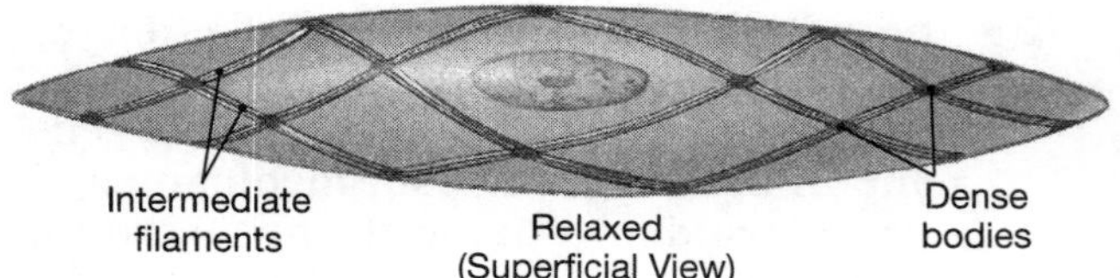

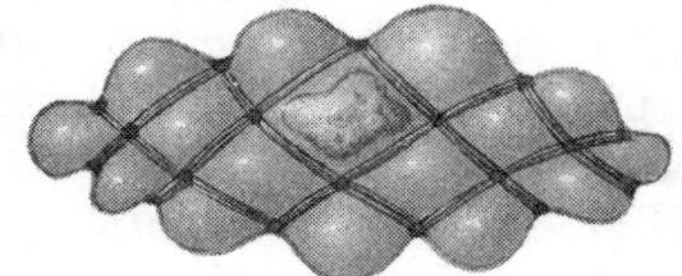

● **FIGURE 20–3c**
Smooth Muscle — Internal Elements.

❑ Additional Activities

NOTES

1. Research methods for studying the muscle actions of the different muscle types.
2. Research the life and times of some of the pioneers in muscle anatomy.

❑ Lab Report

1. Referring to the numbered inquiries at the beginning of this exercise, complete the following box summary:

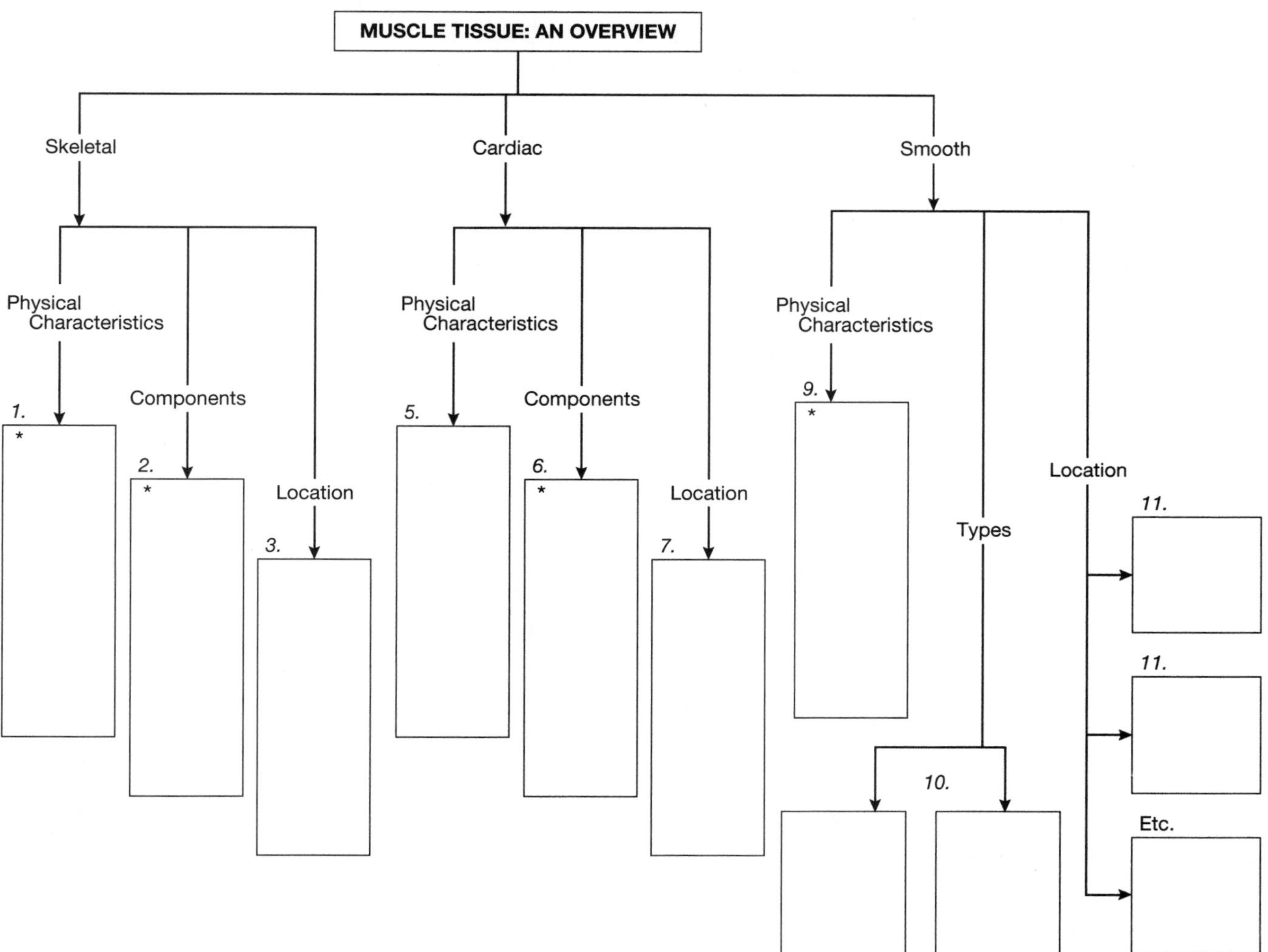

*List according to your instructor's directions

General Questions

NOTES

1. Make a chart delineating the similarities and differences between the three major muscle types.
2. If you have a muscle tissue sample, how can the number and location of the nuclei in the individual cells help you determine the muscle type?
3. On the distant planet Zorch lives a group of unique beings. Zorchian muscle fibers are similar to ours in both structure and function. Interestingly, however, Zorchians have a built-in pouch in which they can transport reserve amounts of their necessary life support fluid, which is similar to our water. The support tissues for this pouch have both striated and smooth muscles. The Zorchians can choose to use either set of muscles, but they cannot use both sets at the same time. A Zorchian is about to set out on a long journey across the desert, so he fills his pouch. Which set of muscles should he use to support this pouch on his journey? Why?

EXERCISE 21

Human Musculature

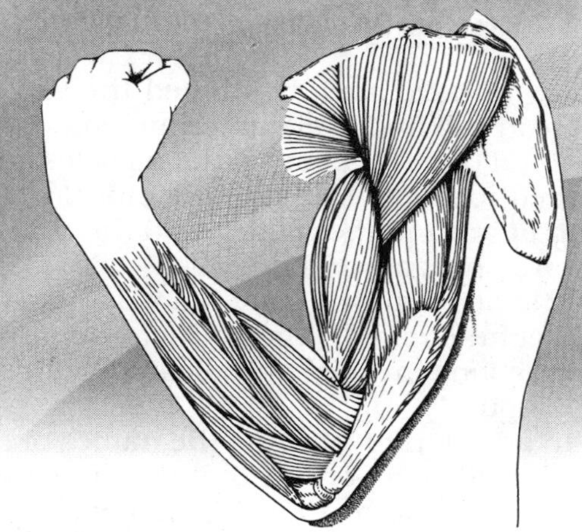

PROCEDURAL INQUIRIES

Preparation

1. How many skeletal muscles are there in the human body?
2. How are the muscles named?
3. How are origin and insertion related?

Examination

4. What are some common terms used to describe muscle morphology?
5. What are some common terms used to describe muscle location?
6. What are some common terms used to describe muscle action?
7. What are the principal muscles of the head?
8. What are the principal muscles of the torso?
9. What are the principal muscles of the upper extremities?
10. What are the principal muscles of the lower extremities?

Additional Inquiries

11. What is the relationship between the agonist and the antagonist?
12. What is the synergist?

Key Terms*

Adductor longus**
Agonist
Antagonist
Aponeuroses
Biceps**
Biceps brachii**
Insertion
Intercostals**
Origin
Prime mover
Radialis**
Rectus femoris**
Synergist
Tibialis**

Materials Needed

Charts and Diagrams Showing Human Musculature
Human Torso Model
Human Limb Models

*Include the names of muscles indicated by your instructor.

**Muscles used in this Exercise as specific examples.

Almost 700 individual skeletal muscles are responsible for the intricacies of human movement. In this exercise about 140 muscles are named. All of these muscles are paired and some of the muscles (such as the **intercostals**) are actually groups of muscles. Therefore, at the end of this lesson, you should be able to identify about half the muscles in the human body. (In Exercises 22 and 23 we will concentrate on the contractile functions of these muscles.)

Learning the names of the muscles can be quite interesting because many of these muscle names come from information you already have about their physical actions, locations, and characteristics.

For instance, in Exercise 18 you learned basic physical action terminology — such as abduction and adduction. Here you can put that terminology to use because the names of many of the muscles indicate exactly what they do. For example, the **adductor longus** muscle adducts the thigh, or moves the thigh toward the midline of the body.

If you have already studied the skeletal system, you will find skeletal terminology particularly helpful because many of the muscles are named for their location. The **radialis** muscles are located on or around the radius, and the **tibialis** muscles are on or around the tibia.

Still other muscles are named for some structural characteristic. The adductor longus muscle is also a long muscle, as indicated by the descriptor "longus."

In addition to learning the names of the muscles, it is also important that you understand the dual concept of the muscle's **origin** and the muscle's **insertion**. The origin is where the muscle begins; it is the fixed, stationary point of the muscle. With some of the muscles, particularly the thoracic muscles, the term "arises from" is often used synonymously with "origin." The most commonly considered muscles originate on bone, although some arise from other muscles or from connective tissues.

The insertion, which may be on a bone, a muscle, or a segment of connective tissue, is the point of action. Every muscle contracts. Contraction shortens the muscle and that shortening moves the point of action closer to the point of origination. In other words, the insertion moves toward the origin.

As an example of origin and insertion, think about flexing the **biceps** (more correctly, the **biceps brachii**). This muscle originates on the coracoid process and the glenoid tuberosity of the scapula and inserts on the tuberosity of the radius. If you flex your biceps, you move your lower arm closer to your upper arm. The point of action does indeed center on the movement of the lower arm. As stated, the insertion is on the radius, and the movement of the insertion is toward the origin.

❑ Preparation

Begin this exercise by first doing this terminology section. Because there are so many muscles to learn, a systematic approach may help you. The following steps outline one way to orient yourself to this body system. This will be the general method used as you go through this exercise.

I. General Method

A. Terminology

1➜ Learn the basic roots, suffixes, and prefixes. These terms, which are often based on action, shape, or location, can be used to correctly identify many of the individual muscles.

2➜ Once you understand the terminology, work on naming the muscles themselves. The approach taken in this book is a regional one. We begin with the head and work through the torso, the upper extremities, and finally the lower extremities.

B. Visualization

1➜ After you are satisfied that you have a good grasp of the names of the muscles, work through the section on muscle action. This will solidify your concept of what muscles do and how muscles work together.

C. Study Suggestions

1➜ Say the names of the muscles out loud. Listen to what you are saying. The more senses involved, the better the learning.

2➜ Concentrate on one section of the body at a time.

3➜ Get a three-dimensional picture of the muscle. Use the torso and/or the classroom limbs. Even if you cannot find a particular muscle on the laboratory body parts, you will be able to gain a perspective on where the muscle is located and how it is related to other muscles and body parts.

4➜ Associate the names of the muscles with what you already know.

5➜ Quiz your partner.

6➜ After you have worked through the muscles, complete the indicated exercises as explained in Section II.

❑ Examination

II. Musculature *[∞ Martini, p. 312]*

A. General Terminology

As stated in the introduction, most muscles are named for their morphology, their action, their location, or some combination of these characteristics. For instance, the **rectus femoris** muscle is a straight muscle on the femur. "Rectus" means straight and "femoris" refers to the femoral region.

Concept Check 1 Use Figure 21–1● to describe a muscle of the following: (Example: Ribs ⟶ Costalis)

1. Skin ______________________

2. Neck ______________________

3. Tongue ______________________

Terms Indicating Direction Relative to Axes of the Body	*Terms Indicating Specific Regions of the Body*	*Terms Indicating Structural Characteristics of the Muscle*	*Terms Indicating Actions*
Anterior (front)	Abdominis (abdomen)	**Origin/Insertion**	**General**
Externus (superficial)	Anconeus (elbow)	Biceps (two heads)	Abductor
Extrinsic (outside)	Auricularis (auricle of ear)	Triceps (three heads)	Adductor
Inferioris (inferior)	Brachialis (brachium)	Quadriceps (four heads)	Depressor
Internus (deep, internal)	Capitis (head)		Extensor
Intrinsic (inside)	Carpi (wrist)	**Shape**	Flexor
Lateralis (lateral)	Cervicis (neck)	Deltoid (triangle)	Levator
Medialis/medius (medial, middle)	Cleido/clavius (clavicle)	Orbicularis (circle)	Pronator
Obliquus (oblique)	Coccygeus (coccyx)	Pectinate (comblike)	Rotator
Posterior (back)	Costalis (ribs)	Piriformis (pear-shaped)	Supinator
Profundus (deep)	Cutaneous (skin)	Platys- (flat)	Tensor
Rectus (straight, parallel)	Femoris (femur)	Pyramidal (pyramid)	
Superficialis (superficial)	Genio- (chin)	Rhomboideus (rhomboid)	**Specific**
Superioris (superior)	Glosso/glossal (tongue)	Serratus (serrated)	Buccinator (trumpeter)
Transversus (transverse)	Hallucis (great toe)	Splenius (bandage)	Risorius (laugher)
	Ilio- (ilium)	Teres (long and round)	Sartorius (like a tailor)
	Inguinal (groin)	Trapezius (trapezoid)	
	Lumborum (lumbar region)		
	Nasalis (nose)	**Other Striking Features**	
	Nuchal (back of neck)	Alba (white)	
	Oculo- (eye)	Brevis (short)	
	Oris (mouth)	Gracilis (slender)	
	Palpebrae (eyelid)	Latissimus (widest)	
	Pollicis (thumb)	Longissimus (longest)	
	Popliteus (behind knee)	Longus (long)	
	Psoas (loin)	Magnus (large)	
	Radialis (radius)	Major (larger)	
	Scapularis (scapula)	Maximus (largest)	
	Temporalis (temples)	Minimus (smallest)	
	Thoracis (thoracic region)	Minor (smaller)	
	Tibialis (tibia)	-tendinosus (tendinous)	
	Ulnaris (ulna)	Vastus (great)	
	Uro- (urinary)		

● **FIGURE 21–1**
Muscle Terminology.

4. Urinary System ____________
5. Wrist ____________
6. Mouth ____________
7. Head ____________
8. Nose ____________
9. Tailbone ____________
10. Clavicle (Collarbone) ____________

Concept Check 2 Use Figure 21–1• to describe muscles with each of these characteristics.

1. Triangular ____________
2. Three-headed ____________
3. Deep ____________
4. Short ____________
5. Outside ____________
6. Above something ____________
7. Toward the side of the body ____________
8. Toward the back of the body ____________
9. Flat ____________
10. Very small ____________

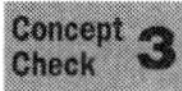
Based on the information in Figure 21–1•, use only the muscle name and give as much information as you can about the

following muscles. Try to do this exercise without looking up the answers on a muscle chart.

1. Rhomboideus ____________
2. Tibialis Anterior ____________
3. Biceps Femoris ____________
4. Extensor Digitorum Longus ____________
5. Adductor Magnus ____________
6. Vastus Medialis ____________
7. Flexor Digitorum Superficialis ____________
8. Abductor Pollicis Brevis ____________
9. Supinator ____________
10. Extensor Carpi Radialis Longus ____________
11. Rectus Abdominis ____________
12. Levator Scapulae ____________
13. Temporalis ____________
14. (—) Capitis ____________ ((—) indicates several muscle names could be used. What does the "capitis" part of the muscle name mean?)
15. Depressor Anguli Oris ____________
16. Orbicularis Oculi ____________
17. Masseter ____________

B. Specific Muscles

As you work through this material, you will come across several **aponeuroses**. Keep in mind that an aponeurosis is a collagenous band that resembles a broad tendon. The aponeuroses cover muscle surfaces and assist in attaching muscles to other structures.

1. Study the charts, torso, and anatomical limbs in your laboratory in relation to Figures 21–2• to 21–16•. Note that each figure covers a specific anatomical region of the body and that each corresponding table offers a description of the muscles in that region.

2➜ Carefully study the descriptions in the tables.

3➜ Label Figures 21–2• to 21–16• with the names of the appropriate muscles. *[∞ Martini, beginning p. 321]*

C. Muscles in Action

Often several muscles work together to move bones. The muscle primarily responsible for the movement is called the **agonist** or the **prime mover**. If more than one muscle is involved in an action, the "assistant muscles" are referred to as the **synergists**. A muscle whose primary action opposes the action of another muscle is an **antagonist**. The biceps brachii and the triceps often function as agonist and antagonist.

1➜ List all muscles involved in each of the following actions. (You will probably want to use a separate piece of paper.)

Concept Check 4 Standing by the sink, reaching in the cupboard, taking a glass, filling the glass with water, drinking the water, setting the glass in the sink for someone else to wash.

Concept Check 5 Kicking a pencil out of the way (without breaking stride) as you walk down the hall.

2➜ Identify the agonist for each of the following actions:

Talking
Doing sit-ups
Throwing a softball

Consider the same actions again. What, if any, antagonist or synergist muscles are involved in these actions?

● FIGURE 21–2 Muscles of the Face and Neck.

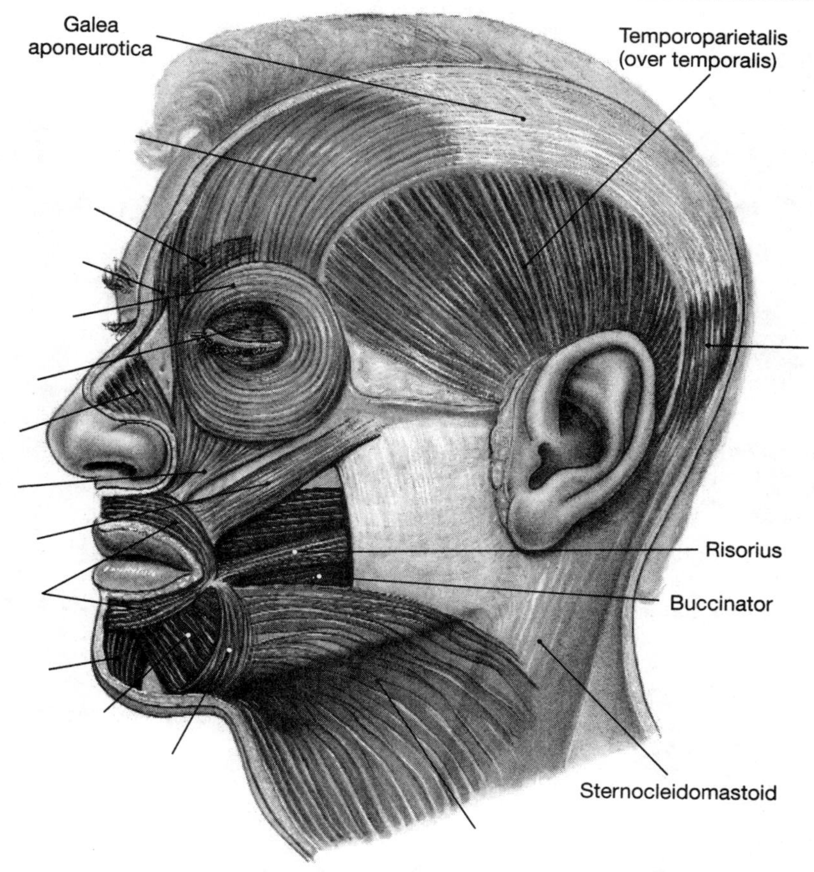

Region/Muscle	Origin	Insertion	Action	Innervation
Mouth				
Buccinator	Alveolar processes of maxillae and mandible	Blends into fibers of orbicularis oris	Compresses cheeks	Facial nerve (N VII)
Depressor labii	Mandible	Lower lip	Depresses lip	As above
Levator labii	Maxillae	Orbicularis oris	Raises upper lip	As above
Mentalis	Mandible	Skin of chin	Elevates and protrudes lower lip	As above
Orbicularis oris	Maxillae and mandible	Lips	Compresses, purses lips	As above
Risorius	Fascia surrounding parotid salivary gland	Angle of mouth	Draws corner of mouth to the side	As above
Depressor anguli oris	Anterolateral surface of mandibular body	Skin at angle of mouth	Depresses corner of mouth	Zygomatic branch of facial nerve (N VII)
Zygomaticus	Zygomatic bone	Angle of mouth	Draws corner of mouth back and up	As above
Eye				
Corrugator supercilii	Frontal bone near nasal suture	Eyebrow	Pulls skin down and forward; wrinkles brow	As above
Levator palpebrae	Tendinous band around optic foramen	Upper eyelid	Raises upper eyelid	Oculomotor nerve (N III)*
Orbicularis oculi	Medial margin of orbit	Skin around eyelids	Closes eye	Facial nerve (N VII)
Nose				
Procerus	Skull	Skin, cartilages of nose	Moves nose, changes position and shape of nostrils	As above
Nasalis	Maxilla	Bridge, inferior corners and tip of nose	Compresses bridge, depresses tip, elevates corners	As above
Ear (extrinsic)				
Temporoparietalis	Fascia around external ear	Galea aponeurotica	Tenses scalp, moves external ear	As above
Scalp				
Frontalis	Galea aponeurotica	Skin of eyebrow and bridge of nose	Raises eyebrows, wrinkles forehead	As above
Occipitalis	Superior nuchal line	Galea aponeurotica	Tenses, retracts scalp	As above
Neck				
Platysma	Upper thorax between cartilage of second rib and acromion of scapula	Mandible and skin of cheek	Tenses skin of neck, depresses mandible	As above

*This muscle originates in association with the extrinsic oculomotor muscles, so its innervation is unusual.

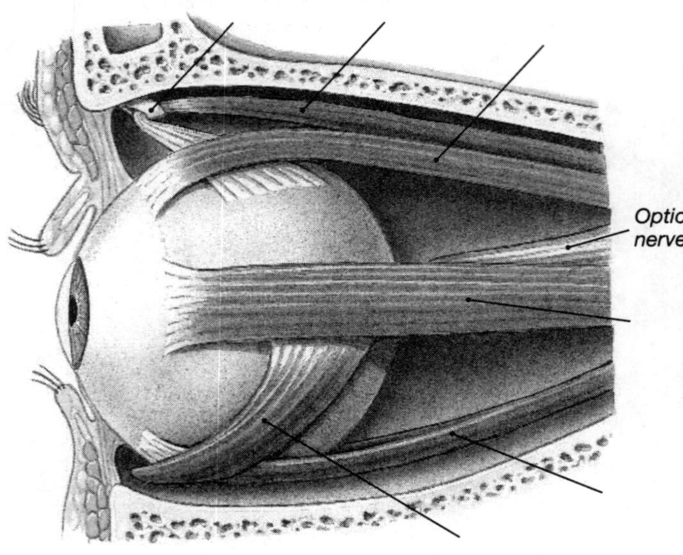

(a) Lateral surface

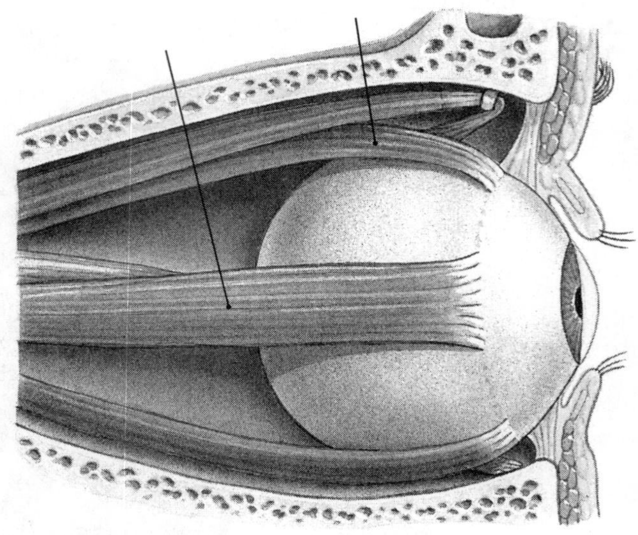
(b) Medial surface

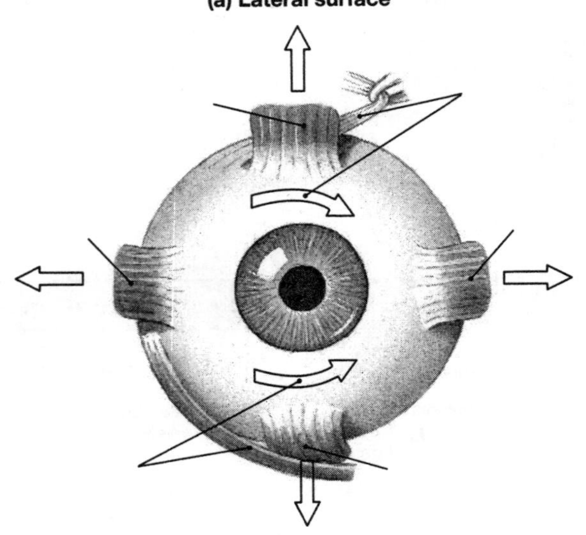
(c) Anterior view

● **FIGURE 21–3**
Extrinsic Oculomotor Muscles.

Muscle	*Origin*	*Insertion*	*Action*	*Innervation*
Inferior rectus	Sphenoid around optic foramen	Inferior, medial surface of eyeball	Eye looks down	Oculomotor nerve (N III)
Medial rectus	As above	Medial surface of eyeball	Eye rotates medially	As above
Superior rectus	As above	Superior, medial surface of eyeball	Eye looks up	As above
Inferior oblique	Maxilla at front of orbit	Inferior, lateral surface of eyeball	Eye rolls, looks up and to the side	As above
Superior oblique	Sphenoid around optic foramen	Superior, medial surface of eyeball	Eye rolls, looks down and to the side	Trochlear nerve (N IV)
Lateral rectus	As above	Lateral surface of eyeball	Eye rotates laterally	Abducens nerve (N VI)

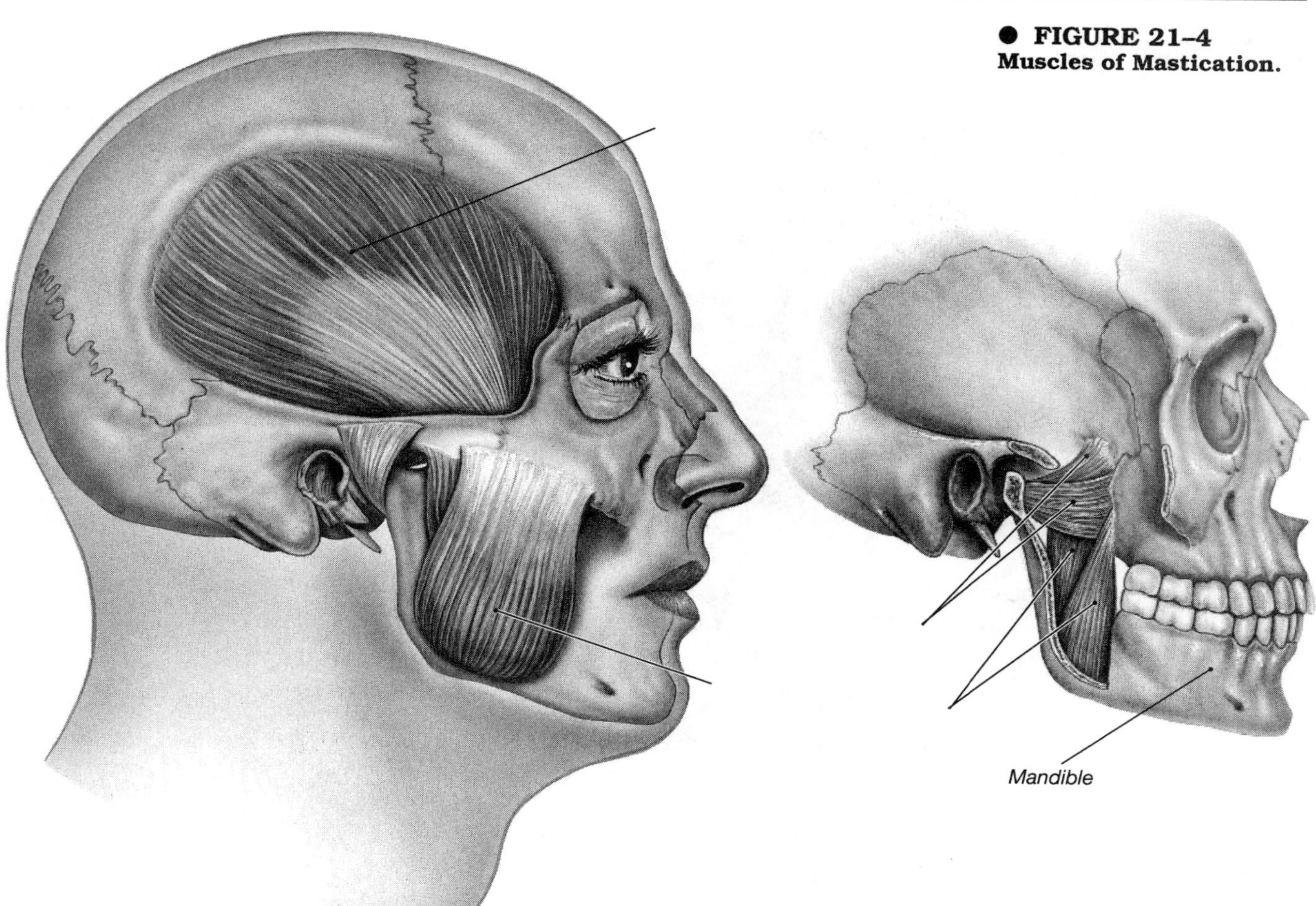

● **FIGURE 21–4**
Muscles of Mastication.

Muscle	*Origin*	*Insertion*	*Action*	*Innervation*
Masseter	Zygomatic arch	Lateral surface of mandibular ramus	Elevates mandible	Trigeminal nerve (N V), mandibular branch
Temporalis	Along temporal lines of skull	Coronoid process of mandible	Elevates mandible	As above
Pterygoideus (medial and lateral)	Medial pterygoid plate and processes	Medial surface of mandibular ramus	Elevates, protracts, and/or moves mandible laterally	As above

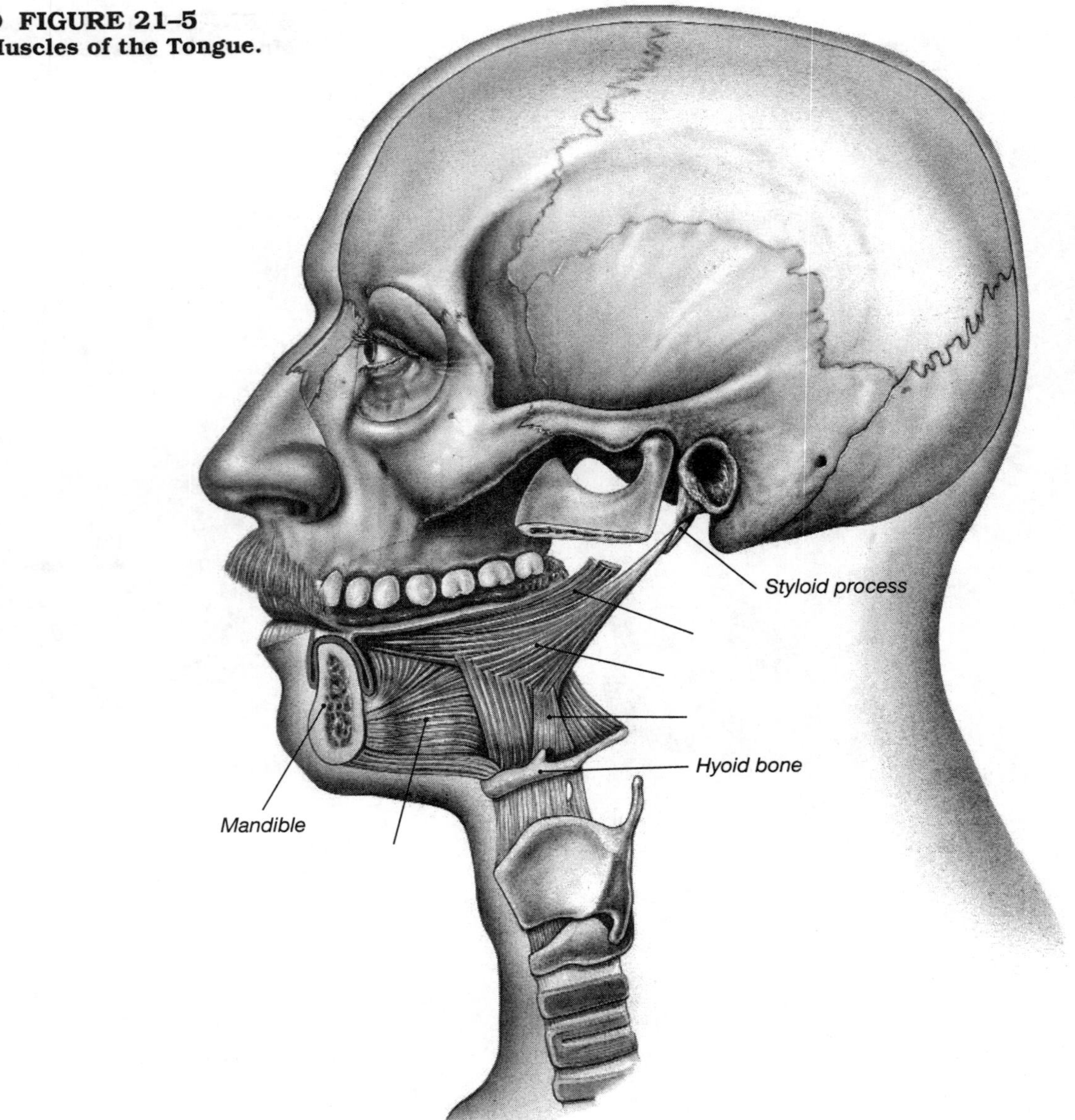

● **FIGURE 21–5**
Muscles of the Tongue.

Muscle	*Origin*	*Insertion*	*Action*	*Innervation*
Genioglossus	Medial surface of mandible around chin	Body of tongue, hyoid bone	Depresses and protracts tongue	Hypoglossal nerve (N XII)
Hyoglossus	Body and greater cornu of hyoid bone	Side of tongue	Depresses and retracts tongue	As above
Palatoglossus	Anterior surface of soft palate	As above	Elevates tongue, depresses soft palate	As above
Styloglossus	Styloid process of temporal bone	Via side to the tip and base of tongue	Retracts tongue, elevates side	As above

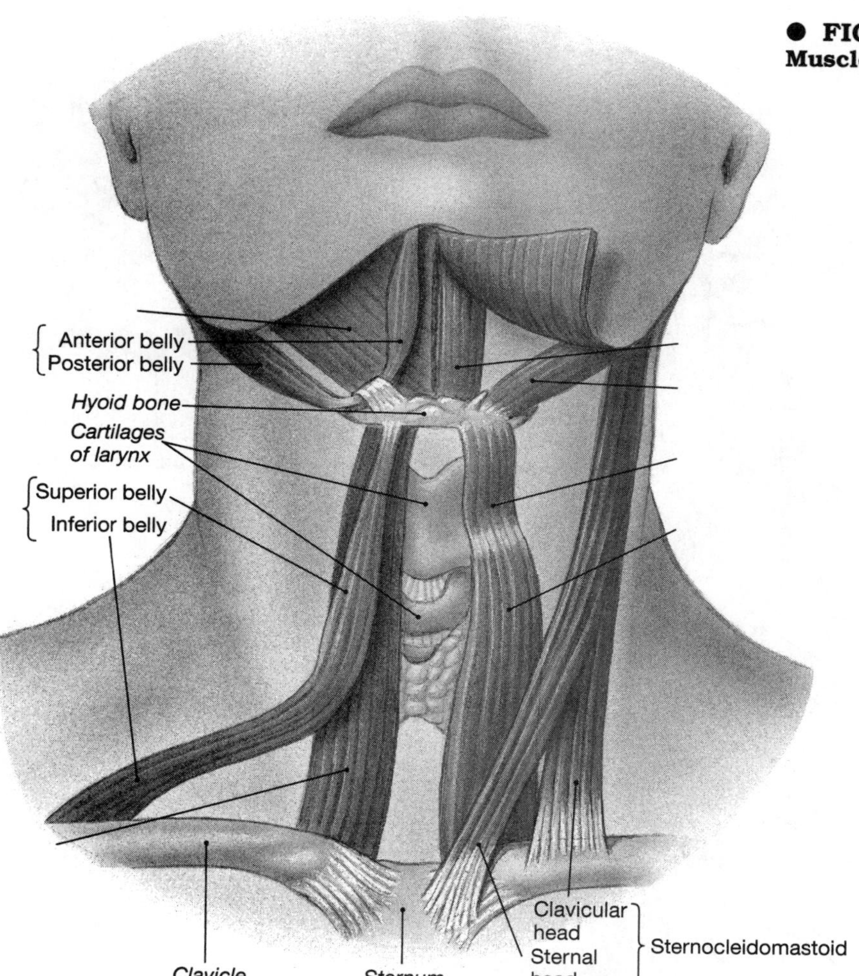

● **FIGURE 21–6**
Muscles of the Anterior Neck.

Muscle	*Origin*	*Insertion*	*Action*	*Innervation*
Digastricus	Two bellies: *posterior* from mastoid region of temporal; *anterior* from inferior surface of mandible at chin	Hyoid bone	Depresses mandible and/or elevates larynx	Hypoglossal nerve (N XII) to posterior belly Trigeminal nerve (N V), mandibular branch, to anterior belly
Geniohyoid	Medial surface of mandible at chin	Hyoid bone	As above and pulls hyoid anteriorly	Cervical nerve C_1 via hypoglossal nerve (N XII)
Mylohyoid	Mylohyoid line of mandible	Medial connective tissue band (raphe) that runs to hyoid bone	Elevates floor of mouth, elevates hyoid, and/or depresses mandible	Trigeminal nerve (N V), mandibular branch
Omohyoid	Central tendon attaches to clavicle and 1st rib	Two bellies: *anterior* attaches to hyoid bone; *posterior* to superior margin of scapula	Depresses hyoid and larynx	Cervical spinal nerves
Sternohyoid	Clavicle and manubrium	Hyoid	As above	As above
Sternothyroid	Dorsal surface of manubrium and 1st rib	Thyroid cartilage of larynx	As above	As above
Stylohyoid	Styloid process of temporal bone	Hyoid	Elevates larynx	Facial nerve (N VII)
Thyrohyoid	Thyroid cartilage of larynx	Hyoid	Elevates thyroid, depresses hyoid	Hypoglossal nerve (N XII)
Sternocleido-mastoid	Superior margins of manubrium and clavicle	Mastoid region of skull	Together they flex the neck; alone one side bends head toward shoulder and turns face to opposite side	Accessory nerve (N XI) and cervical spinal nerves

● **FIGURE 21–7**
Muscles of the Spine.

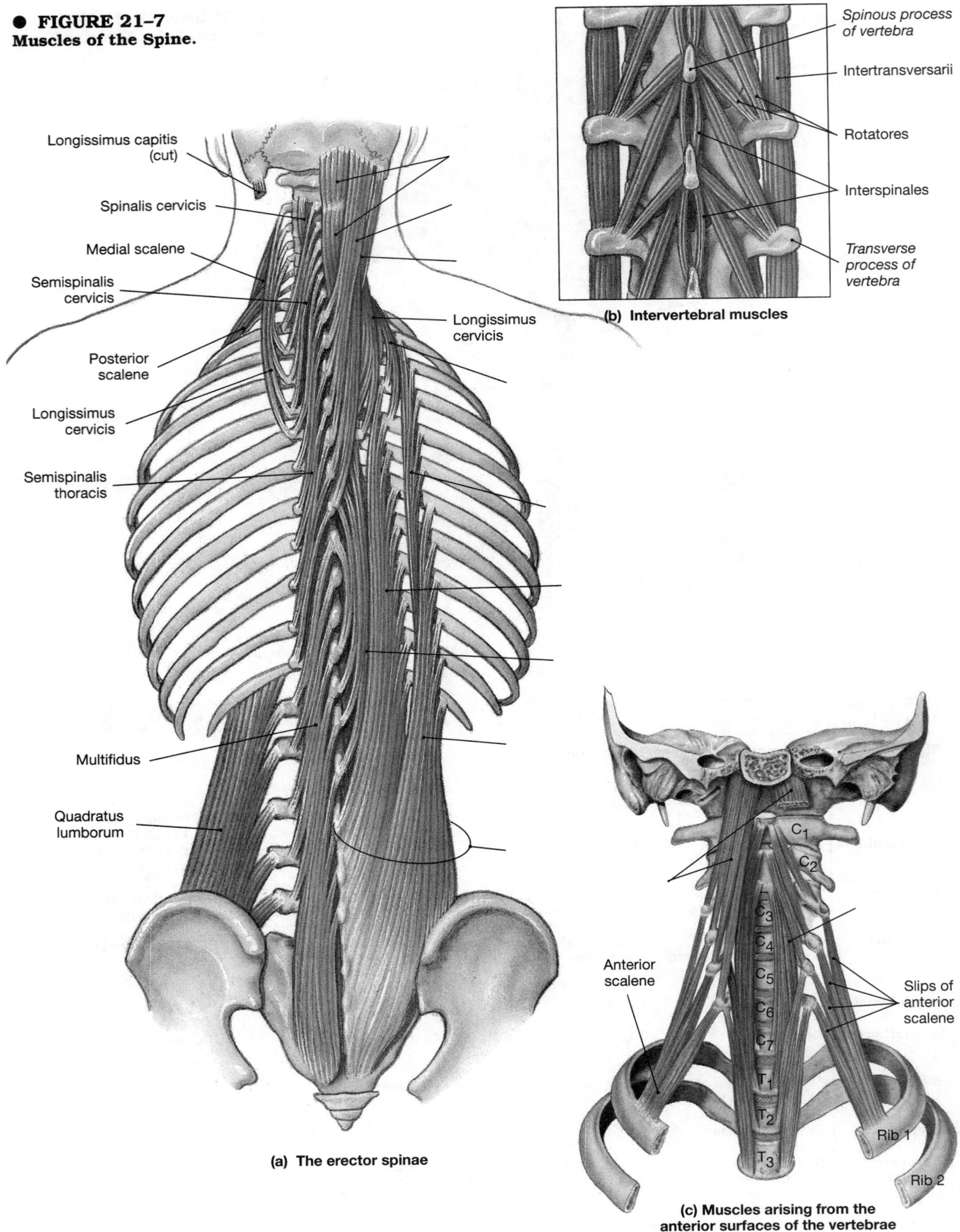

(a) The erector spinae

(b) Intervertebral muscles

(c) Muscles arising from the anterior surfaces of the vertebrae

Group/Muscle	Origin	Insertion	Action	Innervation
SUPERFICIAL SPINAL EXTENSORS				
Spinalis group				
Semispinalis capitis	Processes of lower cervical and upper thoracic vertebrae	Occipital bone, between nuchal lines	The two sides act together to extend head; either alone extends and tilts head to that side	Cervical spinal nerves
Semispinalis cervicis	Transverse processes of T_1–T_5 or T_6	Spinous processes of C_2–C_5	Extends vertebral column and rotates toward opposite side	Cervical spinal nerves
Semispinalis thoracis	Transverse processes of T_5–T_{10}	Spinous processes of C_5–T_4	As above	Thoracic spinal nerves
Splenius (Splenius capitis, splenius cervicis)	Spinous processes and ligaments connecting upper cervical vertebrae	Mastoid process and occipital bone of skull	The two sides act together to extend head; either alone rotates and tilts head to that side	Cervical spinal nerves
Spinalis thoracis	Spinous processes of lower thoracic and upper lumbar vertebrae	Spinous processes of upper thoracic vertebrae	Extends spinal column	Cervical and thoracic spinal nerves
Spinalis cervicis	Inferior portion of ligamentum nuchae and spinous process of C_7	Spinous process of axis	Spinous processes of vertebrae L_1–L_4	Cervical spinal nerves
Spinalis thoracis	Spinous processes of lumbar vertebrae 1, 2, 3, 4, 5	Spinous processes of upper thoracic vertebrae		Thoracic and lumbar spinal nerves
Longissimus group				
Longissimus capitis	Processes of lower cervical and upper thoracic vertebrae	Mastoid process of temporal bone	The two sides act together to extend head; either alone rotates and tilts head to that side	Cervical and thoracic spinal nerves
Longissimus cervicis	Transverse processes of upper thoracic vertebrae	Transverse processes of middle and upper cervical vertebrae	As above	As above
Longissimus thoracis	Broad aponeurosis and at transverse processes of lower thoracic and upper lumbar vertebrae; joins iliocostalis to form "sacrospinalis"	Transverse processes of higher vertebrae and inferior surfaces of ribs	Extends and/or bends spine to the side	As above
Iliocostalis group				
Iliocostalis cervicis	Superior borders of vertebrosternal ribs near the angles	Transverse processes of middle and lower cervical vertebrae	Extends or bends neck, elevates ribs	As above
Iliocostalis thoracis	Superior borders of lower 7 ribs medial to the angles	Upper ribs and transverse process of last cervical vertebra	Stabilizes thoracic vertebrae in extension	Thoracic spinal nerves
Iliocostalis lumborum	Sacrospinal aponeurosis and iliac crest	Inferior surfaces of lower 7 ribs near their angles	Extends spine, depresses ribs	Lumbar spinal nerves
DEEP SPINAL EXTENSORS AND ROTATORS				
Multifidus	Sacrum and transverse processes of each vertebra	Spinous processes of the fourth or fifth more superior vertebrae	Extend vertebral column and rotate toward opposite side	Cervical, thoracic, and lumbar spinal nerves
Rotatores	Transverse processes of each vertebra	Spinous process of adjacent, more superior vertebra	Extend vertebral column and rotate toward opposite side	As above
Interspinales	Spinous processes of each vertebra	Spinous processes of preceding vertebra	Extend vertebral column	As above
Intertransversarii	Transverse processes of each vertebra	Transverse process of preceding vertebra	Bend the vertebral column laterally	As above
SPINAL FLEXORS				
Longus capitis	Transverse processes of cervical vertebrae	Base of the occipital bone	The two sides act together to bend head forward; either alone rotates head to that side	Cervical spinal nerves
Longus cervicis	Anterior surfaces of cervical and upper thoracic vertebrae	Transverse processes of upper cervical vertebrae	Flexes and/or rotates neck; limits hyperextension	As above
Quadratus lumborum	Iliac crest	Last rib and transverse processes of lumbar vertebrae	Together they depress ribs, flex spine; one side alone flexes spine laterally	Thoracic and lumbar spinal nerves

● **FIGURE 21–8**
Oblique and Rectus Muscles.

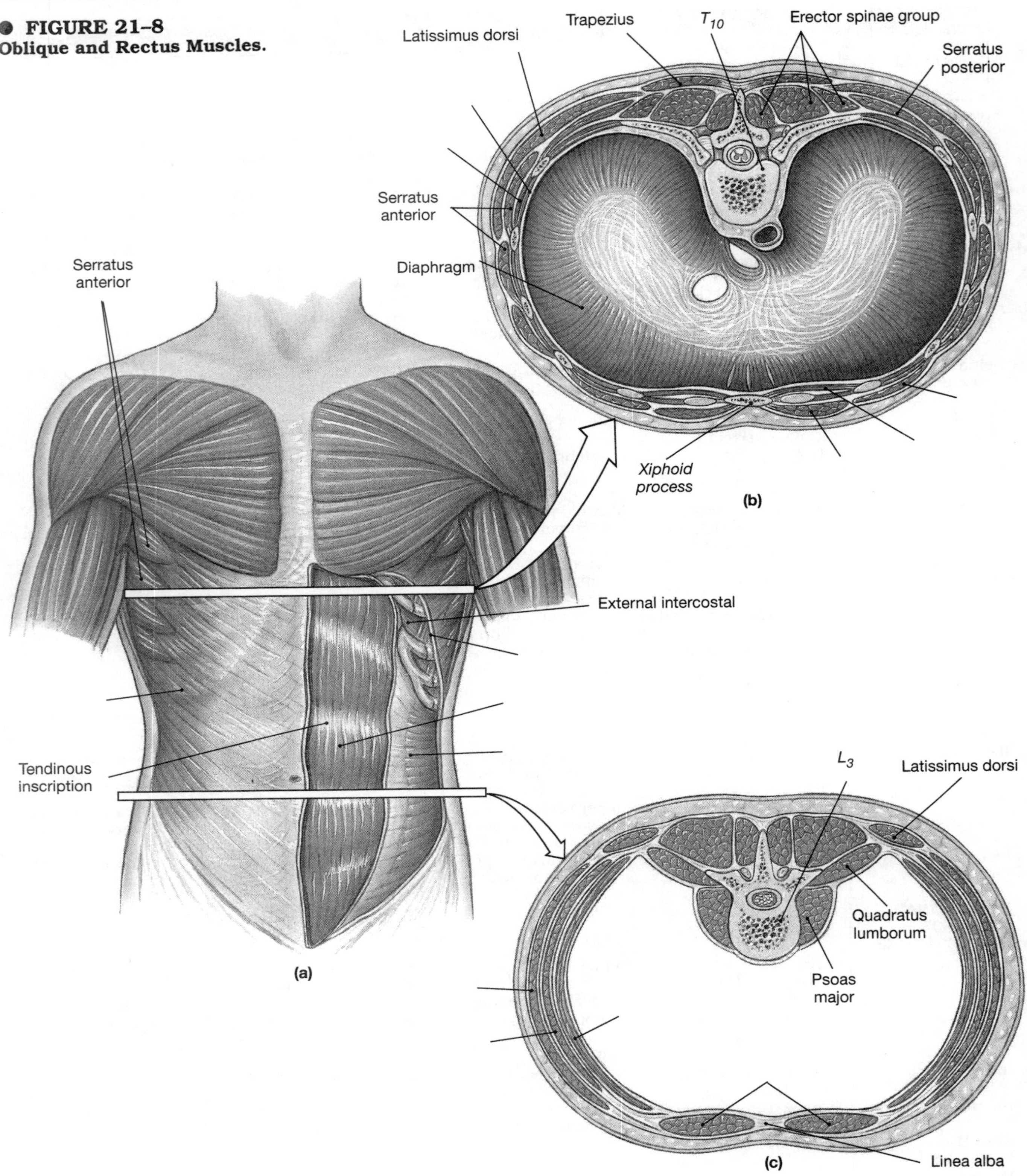

Group/Muscle	Origin	Insertion	Action	Innervation
OBLIQUE GROUP				
Cervical region				
Scalenes	Transverse and costal processes of cervical vertebrae	Superior surfaces of first two ribs	Elevates ribs, and/or flexes neck	Cervical spinal nerves
Thoracic region				
External intercostals	Inferior border of each rib	Superior border of the next rib	Elevate ribs	Intercostal nerves (branches of thoracic spinal nerves)
Internal intercostals	Superior border of each rib	Inferior border of the previous rib	Depress ribs	As above
Transversus thoracis	Medial surface of sternum	Cartilages of ribs	As above	As above
Abdominal region				
External oblique	Lower eight ribs	Linea alba and iliac crest	Compresses abdomen, depresses ribs, flexes or bends spine	Intercostals and iliohypogastric nerves
Internal oblique	Lumbodorsal fascia and iliac crest	Lower ribs, xiphoid of sternum, and linea alba	As above	As above
Transversus abdominis	Cartilages of lower ribs, iliac crest, and lumbodorsal fascia	Linea alba and pubis	Compresses abdomen	Intercostals, iliohypogastric, and ilioinguinal nerves
RECTUS GROUP				
Cervical region	*See muscles on pp. 194–195*			
Thoracic region				
Diaphragm	Xiphoid process, cartilages of ribs 4–10, and anterior surfaces of lumbar vertebrae	Central tendinous sheet	Contraction expands thoracic cavity, compresses abdominopelvic cavity	Phrenic nerves
Abdominal region				
Rectus abdominis	Superior surface of pubis around symphysis	Inferior surfaces of costal cartilages (ribs 5–7) and xiphoid process of sternum	Depresses ribs, flexes vertebral column	Thoracic spinal nerves (T_7–T_{12})

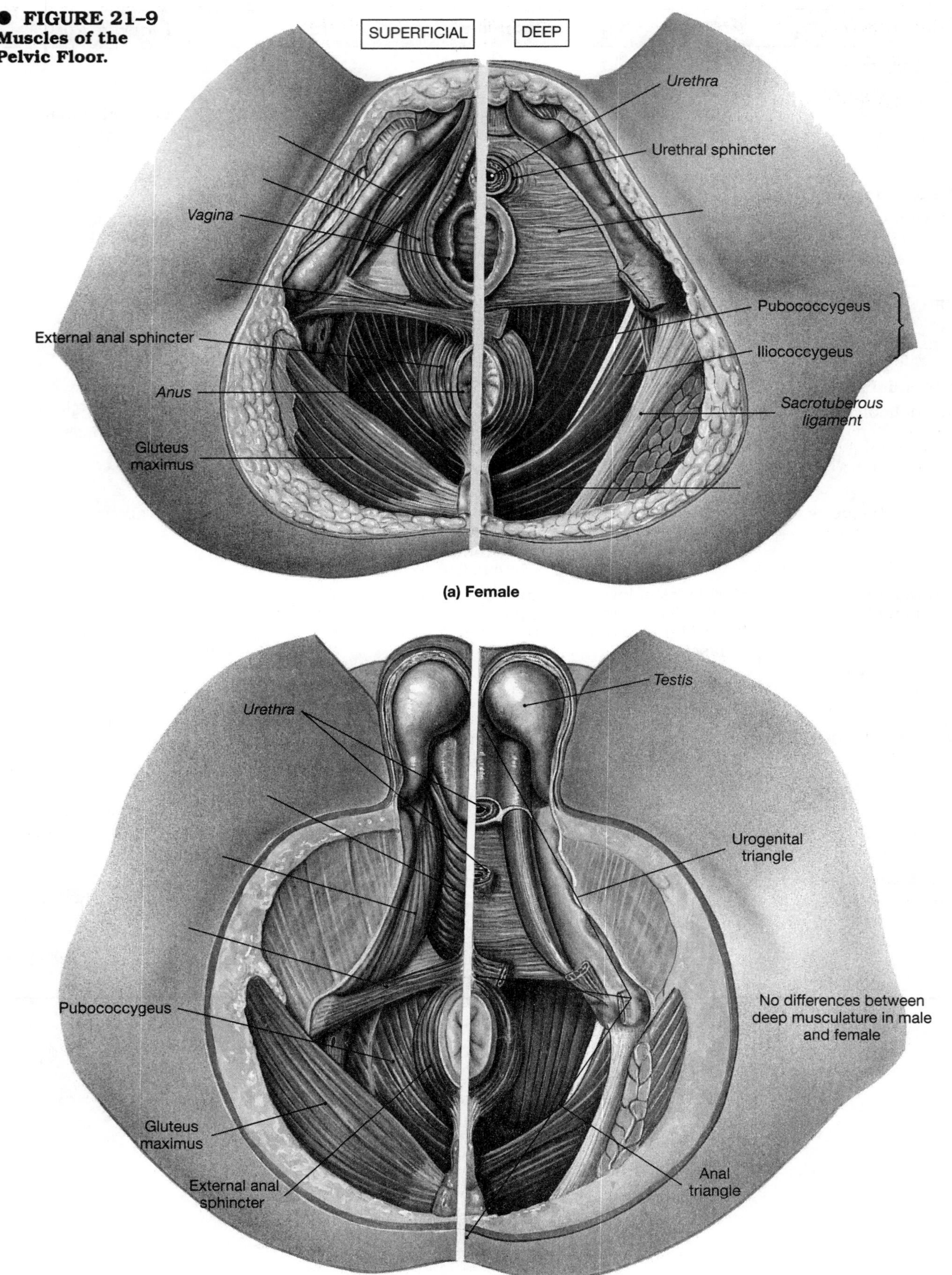

● FIGURE 21–9 Muscles of the Pelvic Floor.

Group/Muscle	Origin	Insertion	Action	Innervation
UROGENITAL TRIANGLE				
Superficial muscles				
Bulbocavernosus:				
male	Collagen sheath at base of penis; fibers cross over urethra	Median raphe and central tendon of perineum	Compresses base and stiffens penis, ejects urine or semen	Pudendal nerve, perineal branch
female	Collagen sheath at base of clitoris; fibers run on either side of urethral and vaginal openings	Central tendon of perineum	Compresses and stiffens clitoris, narrows vaginal opening	As above
Ischiocavernosus	Inferior ramus and tuberosity of ischium	Symphysis pubis anterior to base of penis or clitoris	Compresses and stiffens penis or clitoris	As above
Superficial transverse perineus	Inferior ischial ramus	Central tendon of perineum	Stabilizes central tendon of perineum	As above
Deep muscles: The urogenital diaphragm				
Deep transverse perineus	Inferior ischial ramus	Median raphe of urogenital diaphragm	Stabilizes central tendon of perineum	Pudendal nerve, perineal branch
Urethra sphincter:				
male	Ischial and pubic rami	To median raphe at base of penis; inner fibers encircle urethra	Closes urethra, compresses prostate and bulbourethral glands	As above
female	Ischial and pubic rami	To median raphe; inner fibers encircle urethra	Closes urethra, compresses vagina and greater vestibular glands	As above
ANAL TRIANGLE				
Pelvic diaphragm				
Coccygeus	Ischial spine	Lateral, inferior borders of the sacrum	Flexes coccyx and coccygeal vertebrae	Pudendal nerve
External anal sphincter	Via tendon to coccyx	Encircles anal opening	Closes anal opening	Pudendal nerve, hemorrhoidal branch
Levator ani:				
Iliococcygeus	Ischial spine, pubis	Coccyx	Tenses floor of pelvis, flexes coccyx, elevates and retracts anus	Pudendal nerve
Pubococcygeus	Inner margins of pubis	Coccyx	As above	As above

● **FIGURE 21–10**
Muscles that Move the Shoulder Girdle.

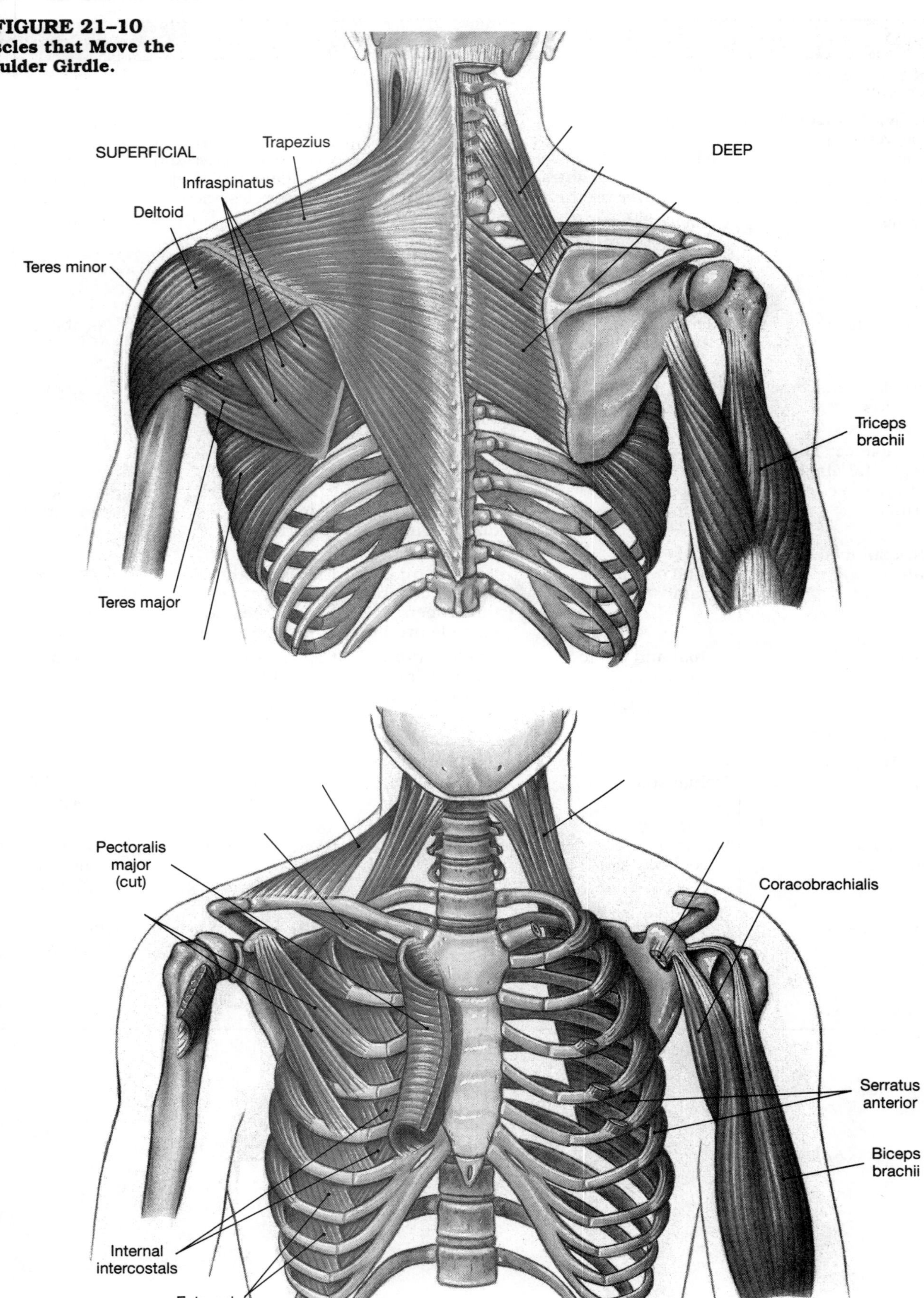

Muscle	*Origin*	*Insertion*	*Action*	*Innervation*
Levator scapulae	Dorsal surfaces of C_1–C_4	Vertebral border of scapula near superior angle	Elevates scapula	Dorsal scapular nerve
Pectoralis minor	Ventral surfaces of ribs 3–5	Coracoid process of scapula	Depresses and protracts shoulder; rotates scapula laterally; elevates ribs if scapula is stationary	Median pectoral nerve
Rhomboideus major	Spinal processes of upper thoracic vertebrae	Vertebral border of scapula from spine to inferior angle	Adducts and rotates scapula laterally	Dorsal scapular nerve
Rhomboideus minor	Spinal processes of vertebrae C_7–T_1	Vertebral border near spine	As above	As above
Serratus anterior	Ventral and superior margins of ribs 1–9	Ventral surface of vertebral border of scapula	Protracts shoulder, abducts and medially rotates scapula	Long thoracic nerve
Subclavius	First rib	Clavicle	Depresses and protracts shoulder	Subclavian nerve
Trapezius	Occipital bone, ligamentum nuchae, and spinal processes of thoracic vertebrae	Clavicle and scapula (acromion and scapular spine)	Depends on active region and state of other muscles; may elevate, adduct, depress, or rotate scapula and/or elevate clavicle; can also extend head and neck	Accessory nerve (N XI) and cervical spinal nerves

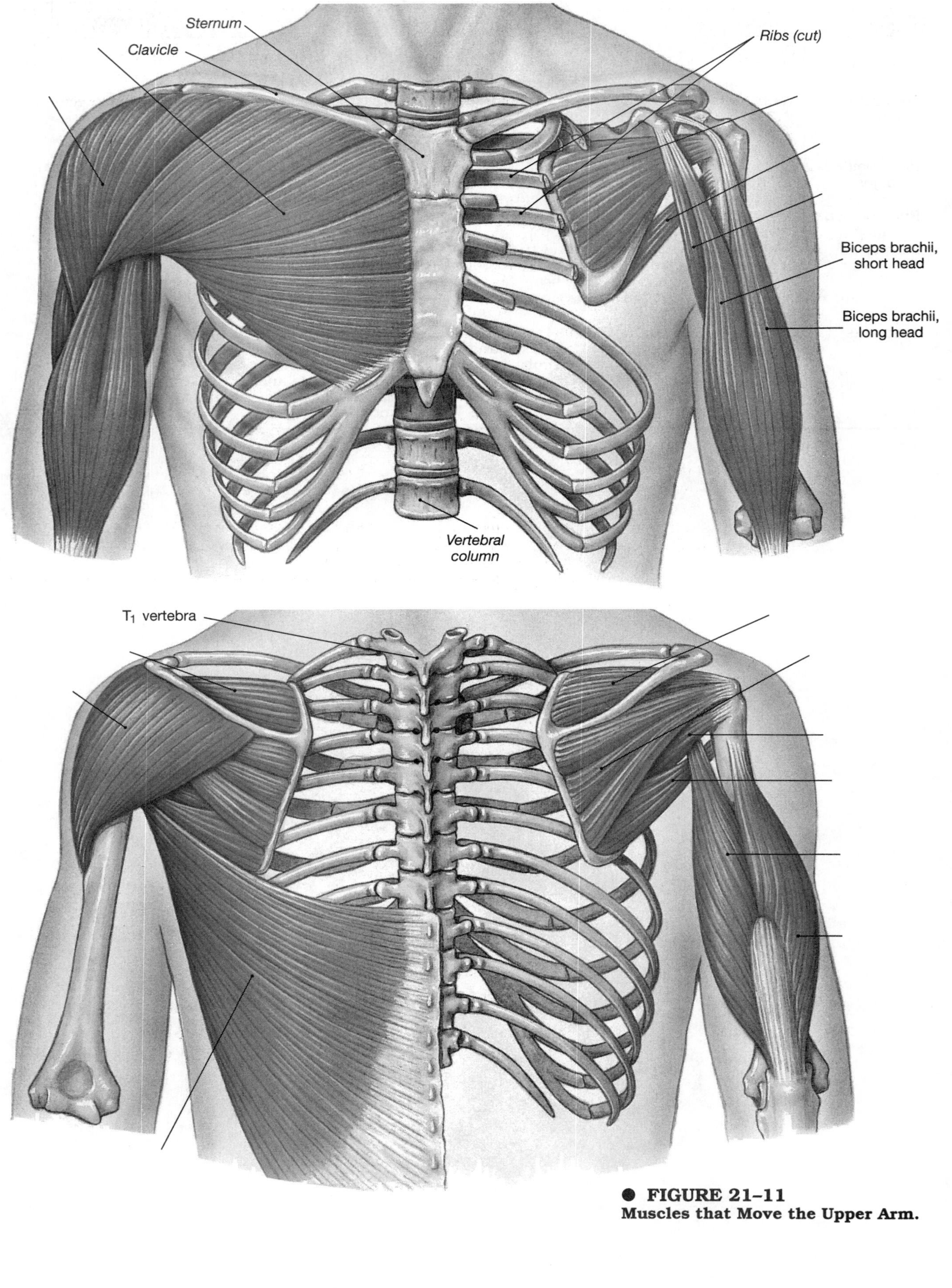

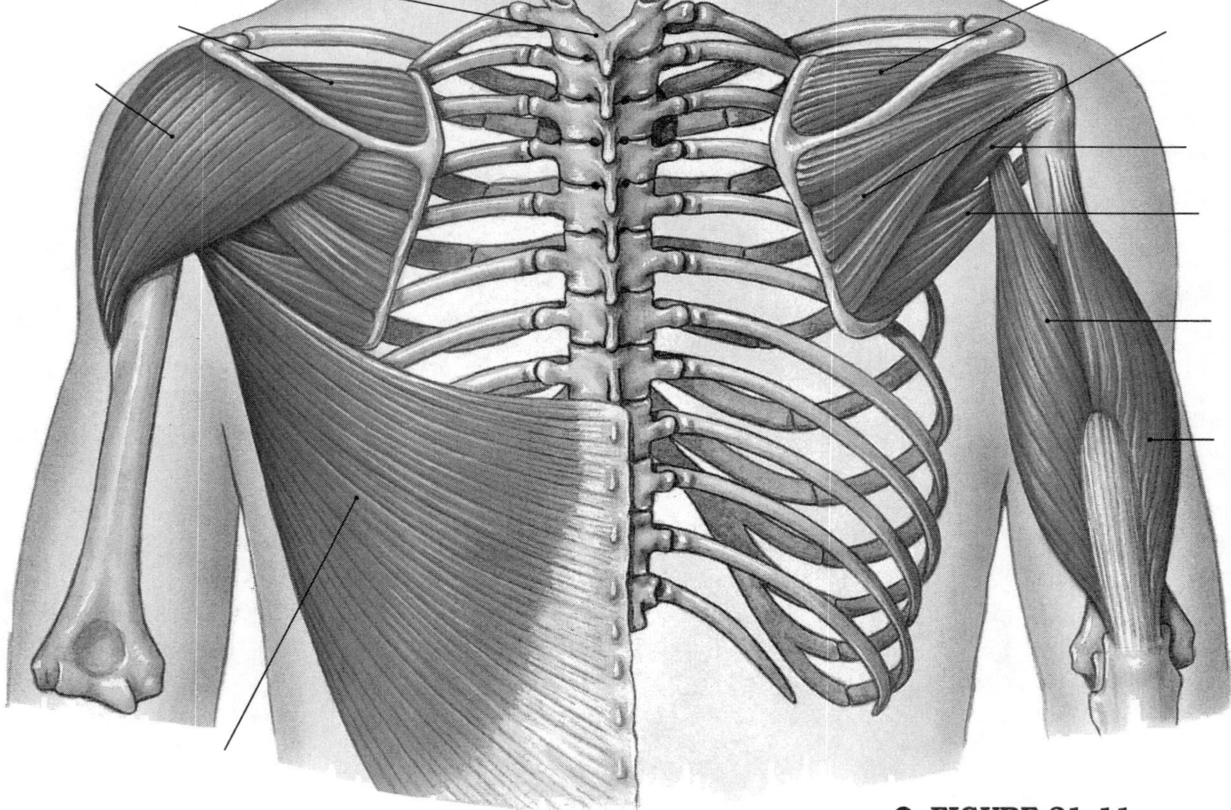

● **FIGURE 21–11**
Muscles that Move the Upper Arm.

Muscle	*Origin*	*Insertion*	*Action*	*Innervation*
Coracobrachialis	Coracoid process	Medial margin of shaft of humerus	Adducts and flexes humerus	Musculocutaneous nerve
Deltoid	Clavicle and scapula (acromion and adjacent scapular spine)	Deltoid tuberosity of humerus	Abducts arm	Axillary nerve
Supraspinatus	Supraspinous fossa of scapula	Greater tubercle of humerus	Abducts arm	Suprascapular nerve
Infraspinatus	Infraspinous fossa of scapula	Greater tubercle of humerus	Lateral rotation of humerus	Suprascapular nerve
Subscapularis	Subscapular fossa of scapula	Lesser tubercle of humerus	Medial rotation of humerus	Subscapular nerve
Teres major	Inferior angle of scapula	Intertubercular groove of humerus	Adducts and medially rotates arm	Lower subscapular nerve
Teres minor	Axillary border of scapula	Greater tubercle of humerus	Lateral rotation of humerus	Axillary nerve
Triceps brachii (long head)	See Figure 21–12●, and p. 205			
Latissimus dorsi	Spinal processes of lower thoracic vertebrae, ribs 8–12, the spines of lumbar vertebrae, and the lumbodorsal fascia	Lesser tubercle, intertubercular groove of humerus	Extends, adducts, and medially rotates humerus	Thoracodorsal nerve
Pectoralis major	Cartilages of ribs 2–6, body of sternum, and inferior, medial portion of clavicle	Greater tubercle of humerus	Flexes, adducts, and medially rotates humerus	Pectoral nerves

● **FIGURE 21–12**
Muscles that Move the Forearm and Wrist.

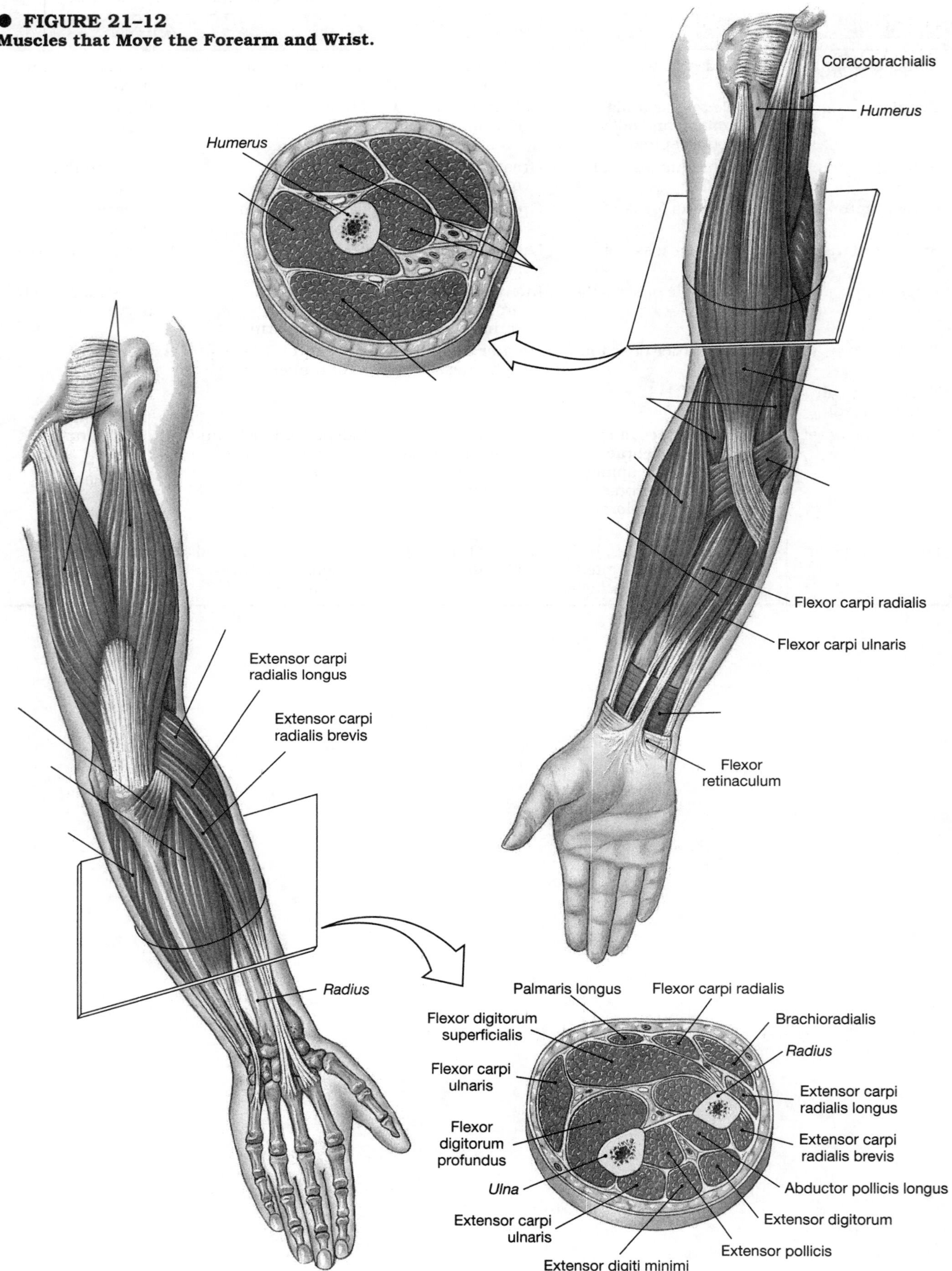
Coracobrachialis
Humerus
Humerus
Flexor carpi radialis
Flexor carpi ulnaris
Flexor retinaculum
Extensor carpi radialis longus
Extensor carpi radialis brevis
Radius
Palmaris longus
Flexor carpi radialis
Flexor digitorum superficialis
Brachioradialis
Radius
Flexor carpi ulnaris
Extensor carpi radialis longus
Flexor digitorum profundus
Extensor carpi radialis brevis
Ulna
Abductor pollicis longus
Extensor carpi ulnaris
Extensor digitorum
Extensor pollicis
Extensor digiti minimi

Muscle	Origin	Insertion	Action	Innervation
PRIMARY ACTION AT THE ELBOW				
Flexors				
Biceps brachii	Short head from the coracoid process; long head from the supraglenoid tuberosity (both on the scapula)	Tuberosity of radius	Flexes and supinates forearm	Musculo-cutaneous nerve
Brachialis	Anterior, distal surface of humerus	Tuberosity of ulna	Flexes forearm	As above
Brachioradialis	Lateral epicondyle of humerus	Lateral aspect of styloid process of radius	As above	Radial nerve
Extensors				
Anconeus	Posterior surface of lateral epicondyle of humerus	Lateral margin of olecranon on ulna	Extends forearm, moves ulna laterally during pronation	As above
Triceps brachii				
lateral head	Superior, lateral margin of humerus	Olecranon process of ulna	Extends forearm	As above
long head	Infraglenoid tuberosity of scapula	As above	As above	As above
medial head	Posterior margin of humerus inferior to radial groove	As above	As above	As above
PRONATORS/ SUPINATORS				
Pronator quadratus	Medial surface of distal portion of ulna	Anterolateral surface of distal portion of radius	Pronates forearm	Median nerve
teres	Medial epicondyle of humerus and coronoid process of ulna	Distal lateral surface of radius	As above	As above
Supinator	Lateral epicondyle of humerus and ulna	Anterolateral surface of radius distal to the radial tuberosity	Supinates forearm	Radial nerve
PRIMARY ACTION AT THE WRIST				
Flexors				
Flexor carpi radialis	Medial epicondyle of humerus	Bases of 2nd and 3rd metacarpals	Flexes and abducts palm	Median nerve
Flexor carpi ulnaris	Medial epicondyle of humerus; adjacent medial surface of olecranon and anteromedial portion of ulna	Bases of 3rd and 4th metacarpals, pisiform, and hamate	Flexes and adducts palm	Ulnar nerve
Palmaris longus	Medial epicondyle of humerus	Palmar aponeurosis	Flexes palm	Median nerve
Extensors				
Extensor carpi radialis longus	Lateral supracondylar ridge of humerus	Base of 2nd metacarpal	Extends and abducts palm	Radial nerve
radialis brevis	Lateral epicondyle of humerus	Base of 3rd metacarpal	As above	As above
ulnaris	Lateral epicondyle of humerus; adjacent dorsal surface of ulna	Base of 5th metacarpal	Extends and adducts palm	Deep radial nerve

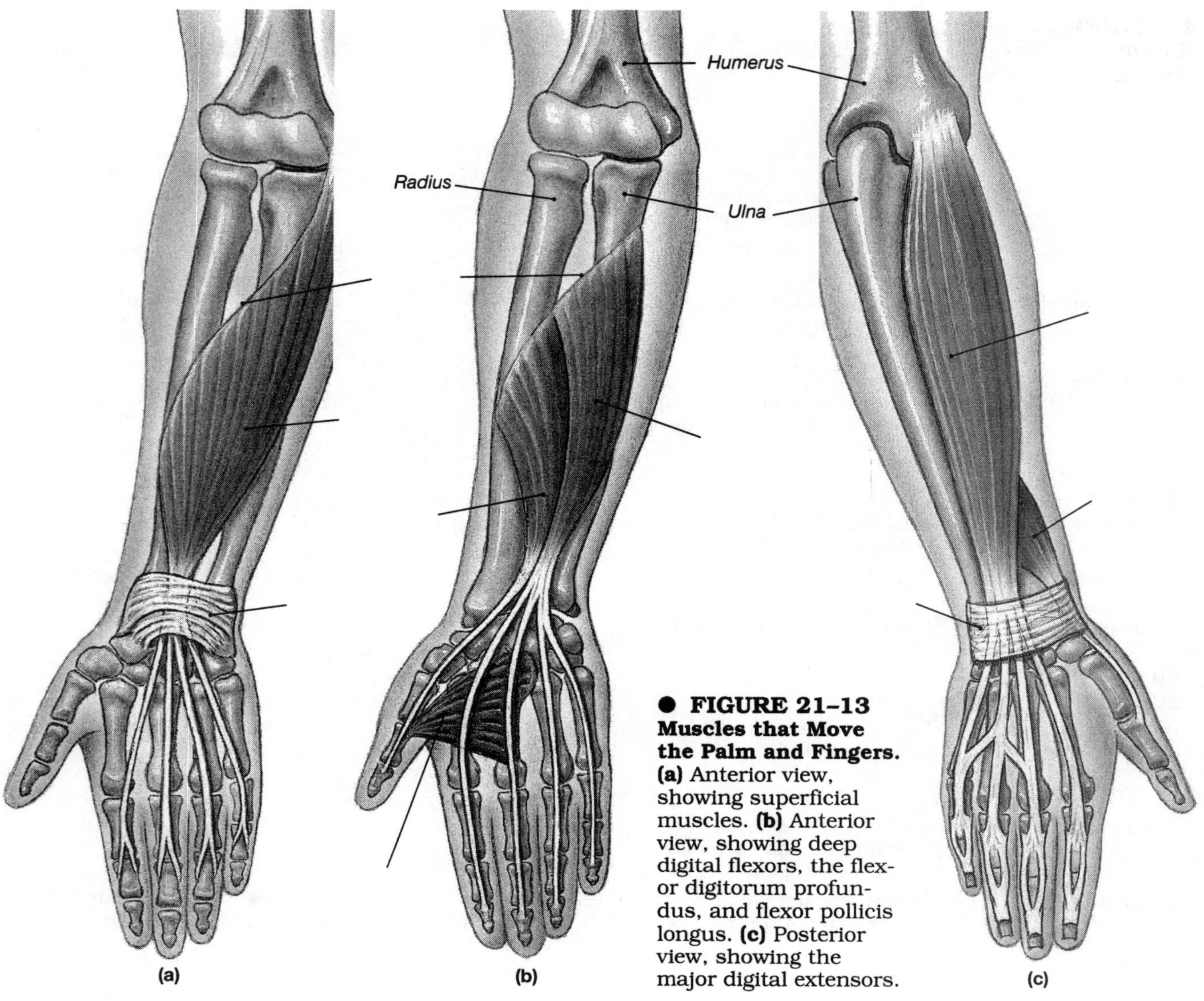

● **FIGURE 21–13 Muscles that Move the Palm and Fingers.** **(a)** Anterior view, showing superficial muscles. **(b)** Anterior view, showing deep digital flexors, the flexor digitorum profundus, and flexor pollicis longus. **(c)** Posterior view, showing the major digital extensors.

Muscle	*Origin*	*Insertion*	*Action*	*Innervation*
Abductor pollicis longus	Proximal dorsal surfaces of ulna and radius	Lateral margin of 1st metacarpal	Abducts thumb	Deep radial nerve
Extensor digitorum	Lateral epicondyle of humerus	Dorsal surfaces of the phalanges	Extends fingers and palms	As above
Flexor digitorum profundus	Proximal anteromedial surface of ulna	Bases of distal phalanges	Flexes fingers, specifically 3rd phalanx on 2nd, and 2nd on 1st; flexes palm	Median nerve
Flexor digitorum superficialis	Medial epicondyle of humerus; adjacent anterior surfaces of ulna and radius	Midlateral surface of 2nd phalanx; connected by ligaments to others	Flexes fingers, specifically 2nd phalanx on 1st; flexes palm	As above
Flexor pollicis longus	Shaft of radius and interosseous membrane	Distal phalanx of thumb	Flexes thumb	As above

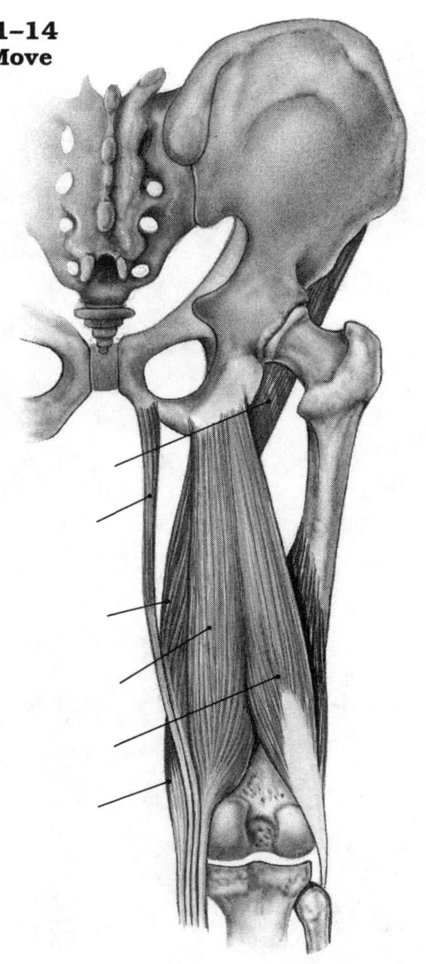

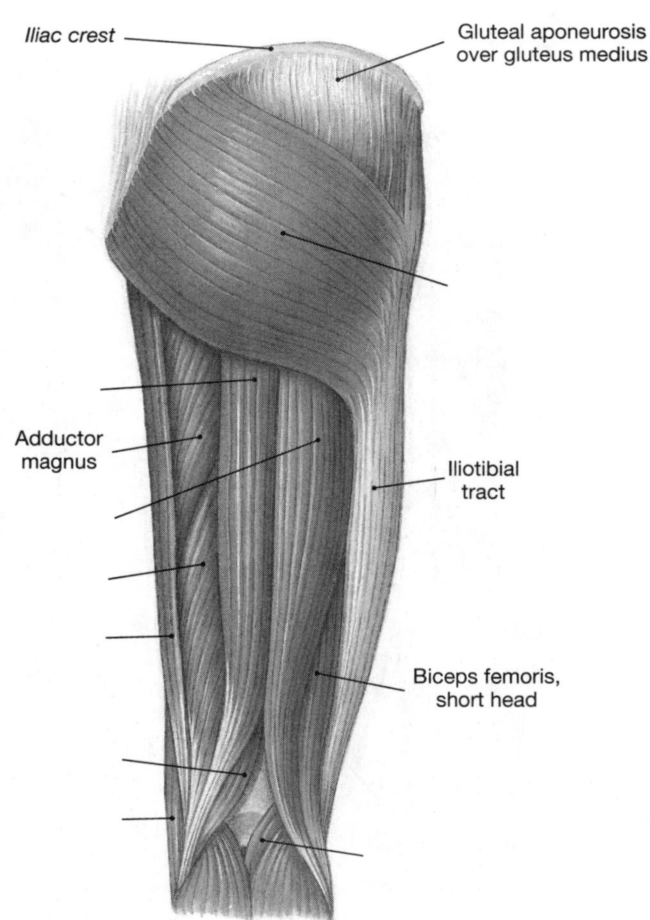

● **FIGURE 21–14 Muscles that Move the Leg.**

Muscle	*Origin*	*Insertion*	*Action*	*Innervation*
Flexors of the leg				
Biceps femoris	Tuberosity of ischium and linea aspera of femur	Head of fibula, lateral condyle of tibia	Flexes leg, extends and adducts thigh	Sciatic nerve, peroneal branch
Semimembranosus	Tuberosity of ischium	Posterior surface of medial condyle of tibia	Flexes leg, extends, adducts, and medially rotates thigh	Sciatic nerve
Semitendinosus	Tuberosity of ischium	Proximal, postero-medial surface of tibia near insertion of gracilis	As above	As above
Sartorius	Anterior superior spine of ilium	Medial surface of tibia near tibial tuberosity	Flexes leg, flexes and laterally rotates thigh	Femoral nerve
Popliteus	Proximal shaft of tibia	Lateral condyle of femur	Medially rotates tibia (or laterally rotates femur)	Sciatic nerve, tibial branch
Extensors of the leg				
Rectus femoris	Anterior inferior spine and superior acetabular rim of ilium	Tibial tuberosity via patellar ligament	Extends leg, flexes thigh	Femoral nerve
Vastus intermedius	Anterolateral surface of femur along linea aspera (distal half)	As above	Extends leg	As above
Vastus lateralis	Anterior and inferior to greater trochanter of femur and along linea aspera (proximal half)	As above	As above	As above
Vastus medialis	Entire length of linea aspera of femur	As above	As above	As above

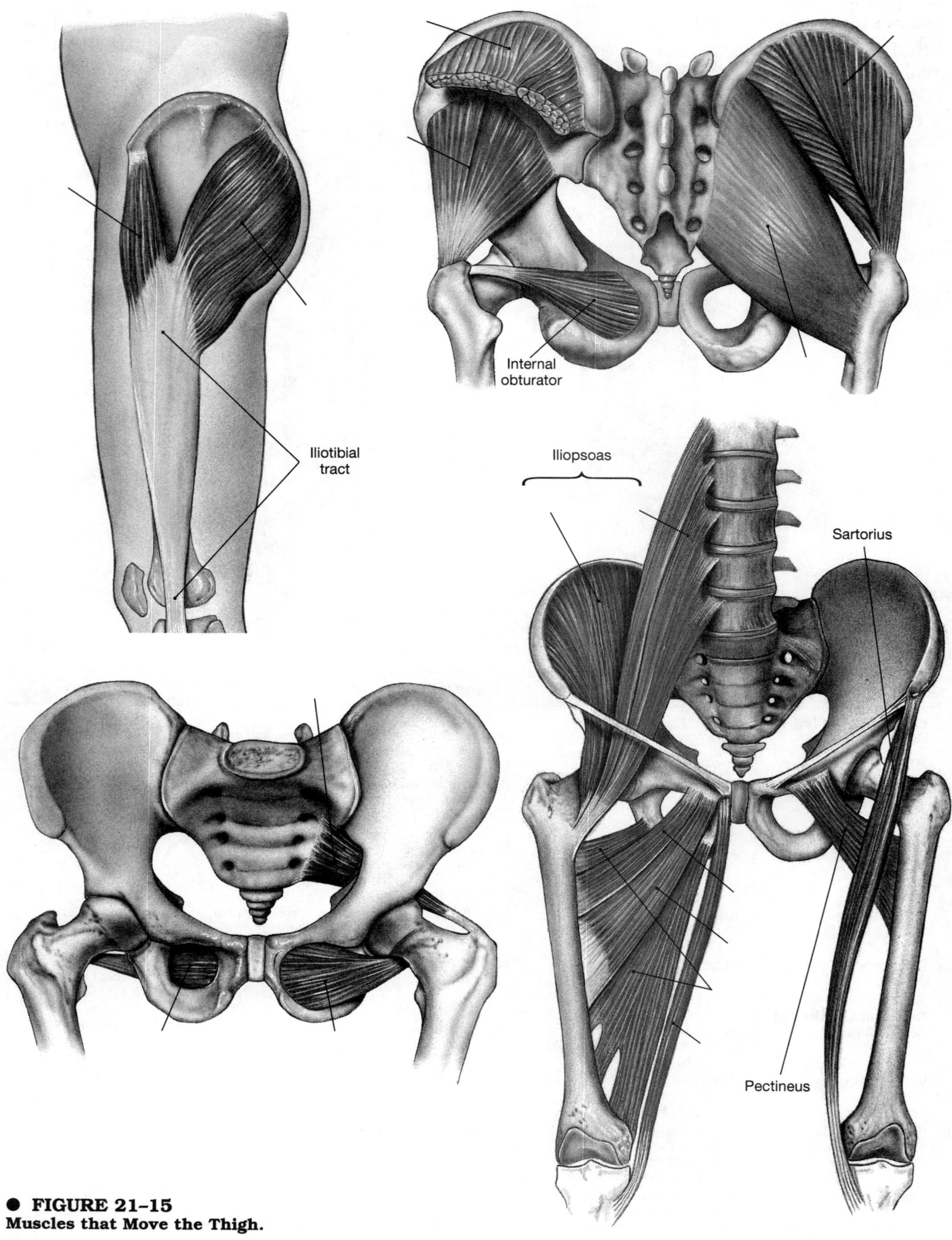

● FIGURE 21–15
Muscles that Move the Thigh.

Group/Muscle	Origin	Insertion	Action	Innervation
Gluteal group				
Gluteus maximus	Iliac crest of ilium, sacrum, coccyx, and lumbodorsal fascia	Iliotibial tract and gluteal tuberosity of femur	Extends and laterally rotates thigh	Inferior gluteal nerve
Gluteus medius	Anterior iliac crest of ilium, lateral surface between superior and inferior gluteal lines	Greater trochanter of femur	Abducts and medially rotates thigh	Superior gluteal nerve
Gluteus minimus	Lateral surface of ilium between inferior and anterior gluteal lines	Greater trochanter of femur	Abducts and medially rotates thigh	As above
Tensor fasciae latae	Iliac crest and surface of ilium between anterior iliac spines	Iliotibial tract	Flexes, abducts, and medially rotates thigh; tenses fasciae latae, which laterally supports the knee	As above
Lateral rotator group				
Obturators (externus and internus)	Lateral and medial margins of obturator foramen	Trochanteric fossa of femur	Laterally rotates thigh	Obturator nerve
Piriformis	Anterolateral surface of sacrum	Greater trochanter of femur	Laterally rotates and adducts thigh	As above
Adductor group				
Adductor brevis	Inferior ramus of pubis	Linea aspera of femur	Adducts thigh	As above
Adductor longus	Inferior ramus of pubis anterior to brevis	As above	Adducts, flexes, and medially rotates thigh	As above
Adductor magnus	Inferior ramus of pubis posterior to brevis	As above	Adducts thigh; anterior portion flexes thigh, posterior portion extends thigh	Obturator and sciatic nerves
Pectineus	Superior surface of pubis	Pectineal line inferior to lesser trochanter of femur	Flexes and adducts thigh	Femoral nerve
Gracilis	Inferior rami of pubis and ischium	Anterior surface of tibia inferior to medial condyle	Flexes leg and adducts thigh	Obturator nerve
Iliopsoas group				
Iliacus	Iliac fossa of ilium	Femur distal to lesser trochanter; tendon fused with that of psoas	Flexes hip and/or lumbar spine	As above
Psoas major	Anterior surfaces and transverse processes of vertebrae T_{12}–L_5	Femur distal to lesser trochanter in company with iliacus	As above	As above

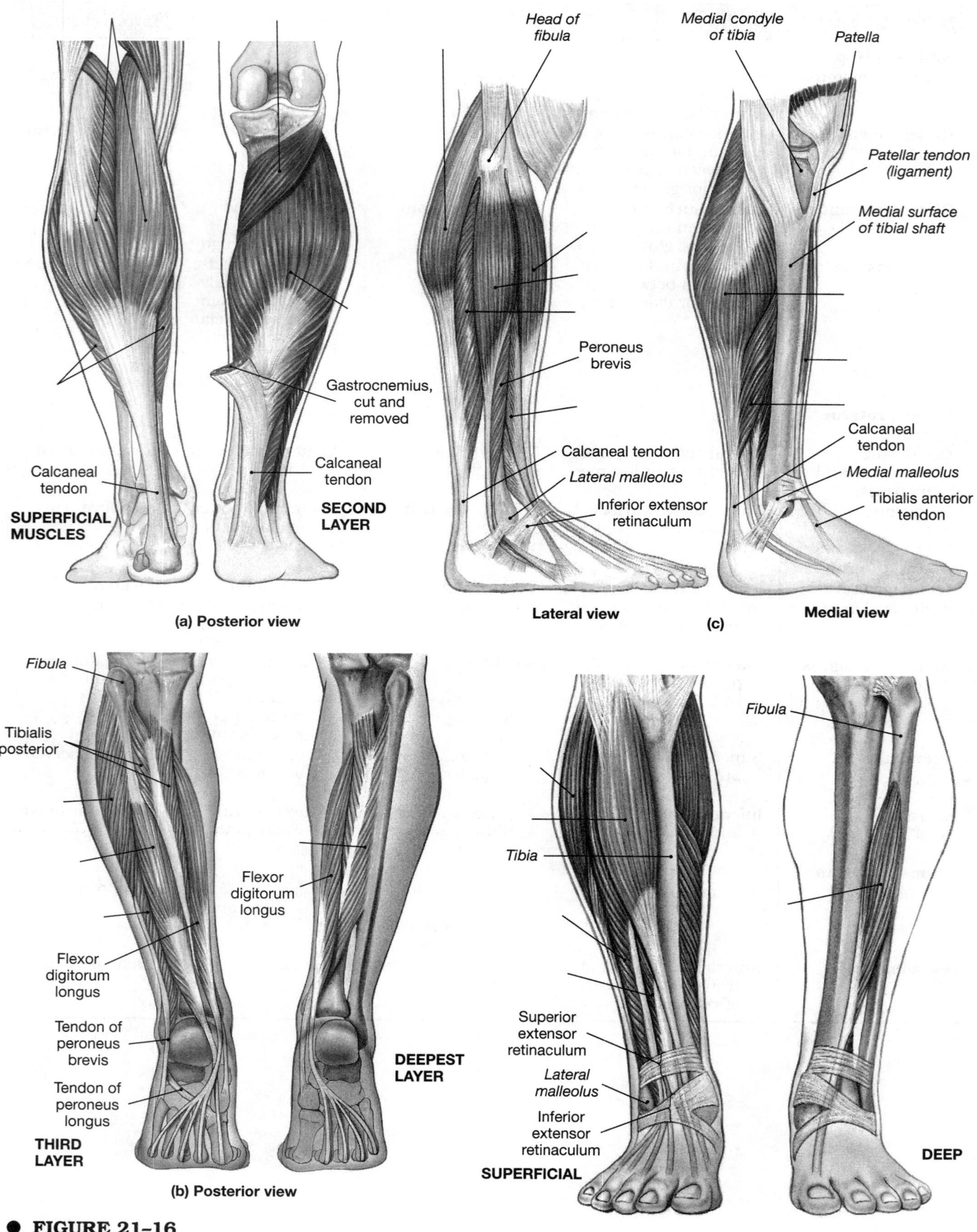
Head of fibula
Medial condyle of tibia
Patella
Patellar tendon (ligament)
Medial surface of tibial shaft
Peroneus brevis
Gastrocnemius, cut and removed
Calcaneal tendon
Calcaneal tendon
Calcaneal tendon
Lateral malleolus
Inferior extensor retinaculum
Calcaneal tendon
Medial malleolus
Tibialis anterior tendon
SUPERFICIAL MUSCLES
SECOND LAYER
(a) Posterior view
Lateral view
(c)
Medial view
Fibula
Tibialis posterior
Flexor digitorum longus
Flexor digitorum longus
Tendon of peroneus brevis
Tendon of peroneus longus
DEEPEST LAYER
THIRD LAYER
(b) Posterior view
Fibula
Tibia
Superior extensor retinaculum
Lateral malleolus
Inferior extensor retinaculum
SUPERFICIAL
DEEP
(d) Anterior view

● FIGURE 21–16
Muscles that Move the Ankle and Foot.

Muscle	Origin	Insertion	Action	Innervation
Dorsiflexor				
Tibialis anterior	Lateral condyle and proximal shaft of tibia	Base of first metatarsal	Dorsiflexes foot	Sciatic nerve, deep peroneal branch
Plantar flexors				
Gastrocnemius	Above femoral condyles	Calcaneus via calcaneal tendon	Plantar flexes, inverts, and adducts foot; flexes leg	Sciatic nerve, tibial branch
Peroneus brevis	Midlateral margin of fibula	Base of 5th metatarsal	Everts foot	Sciatic nerve, superficial peroneal branch
Peroneus longus	Lateral condyle of tibia and head of fibula	Base of 1st metatarsal	Everts and plantar flexes foot; supports longitudinal arch	Sciatic nerve, superficial peroneal branch
Soleus	Head and proximal shaft of fibula, and adjacent posteromedial shaft of tibia	Calcaneus via calcanae tendon (with gastrocnemius)	Plantar flexes, inverts, and adducts foot	Sciatic nerve, tibial branch
Tibialis posterior	Interosseous membrane and adjacent shafts of tibia and fibula	Tarsals and metatarsals	Adducts and inverts foot	As above
Flexors				
Flexor digitorum longus	Posteromedial surface of tibia	Inferior surface of phalanges, toes 2–5	Flexes toes 2–5	Sciatic nerve, tibial branch
Flexor hallucis longus	Posterior surface of fibula	Inferior surface, terminal phalanx of great toe	Flexes great toe	As above
Extensors				
Extensor digitorum longus	Lateral condyle of tibia, anterior surface of fibula	Superior surfaces of phalanges, toes 2–5	Extends toes 2–5	Sciatic nerve, deep peroneal branch
Extensor hallucis longus	Anterior surface of fibula	Superior surface, terminal phalanx of great toe	Extends great toe	As above

❑ Additional Activities

NOTES

1. A sprain is a trauma to a joint with varying degrees of injury to the ligaments involved. A strain is a muscle-related injury. Explore these common problems and explain how each is related to this exercise.
2. Find out how the human musculature compares with musculatures throughout the animal kingdom.
3. Research some common muscle anomalies found in humans. Hint: One common anomaly can be found on the anterior forearm.
4. Name the agonist, antagonist, and/or synergist involved in these actions:
   Moving one's ears back and forth
   Rolling one's eyes toward the upper left
   Playing the piano

Answers to Selected Concept Check Questions

1.
   1. Skin → Cutaneous
   2. Neck → Cervicis or Nuchal (nuchal is back of neck)
   3. Tongue → Glossal
   4. Urinary system → Uro-
   5. Wrist → Carpi
   6. Mouth → Oris
   7. Head → Capitis
   8. Nose → Nasalis
   9. Tailbone → Coccygeus
   10. Clavicle → Cleido or Clavius

2.
   1. Triangular → Deltoid
   2. Three-headed → Triceps
   3. Deep → Profundus
   4. Short → Brevis
   5. Outside → Extrinsic
   6. Above something → Superioris
   7. Toward the side of the body → Lateralis
   8. Toward the back of the body → Posterior
   9. Flat → Platy-
   10. Very small → Minimus

3.
   1. Rhomboid-shaped
   2. On the front of the lower leg
   3. Two-headed muscle on the femur (upper leg)
   4. Long muscle that extends (decreases the angle of) the fingers (or toes)
   5. Large muscle that moves something toward the body's midline
   6. Very large middle muscle
   7. Superficial (outer) muscle that flexes (decreases the angle of) the fingers (or toes)
   8. Short muscle that moves the thumb away from the body's midline

9. Muscle turning hand (or body) face down
10. Long muscle located on the radius that extends (decreases the angle of) the wrist
11. Straight muscle of the abdomen
12. Raises (levitates) the shoulder blade
13. Located on the temporal bone
14. (Something to do with) the head
15. Lowers the angle of the mouth
16. Goes around the eye
17. Chewer muscle (masticate)

NOTES

4. Most of the muscles of the hand, arm, and lower face should be listed.
5. Most of the muscles of the leg, abdomen, and hip should be listed.

❑ Lab Report

1. Referring to the numbered inquiries at the beginning of this exercise, complete the following box summary:

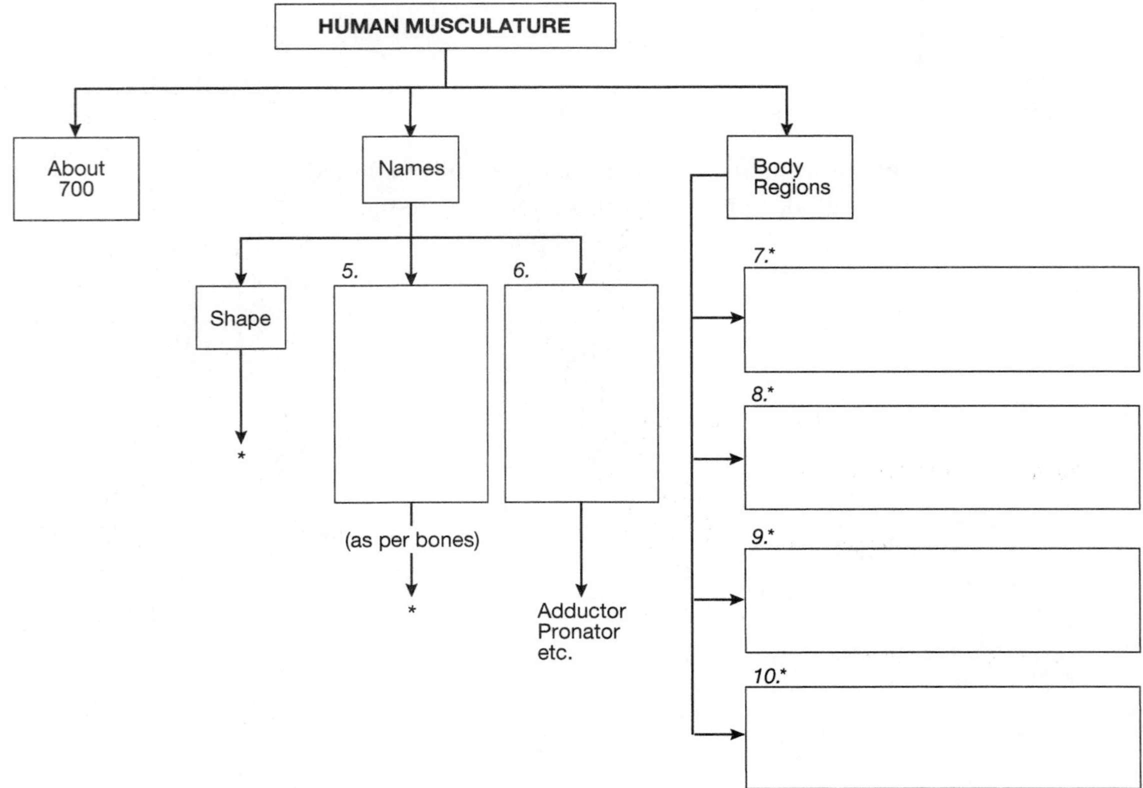

*List according to your instructor's directions
Use your own paper if necessary.

General Questions

I. Muscle Identification — General

1. Identify the anterior muscles as indicated.

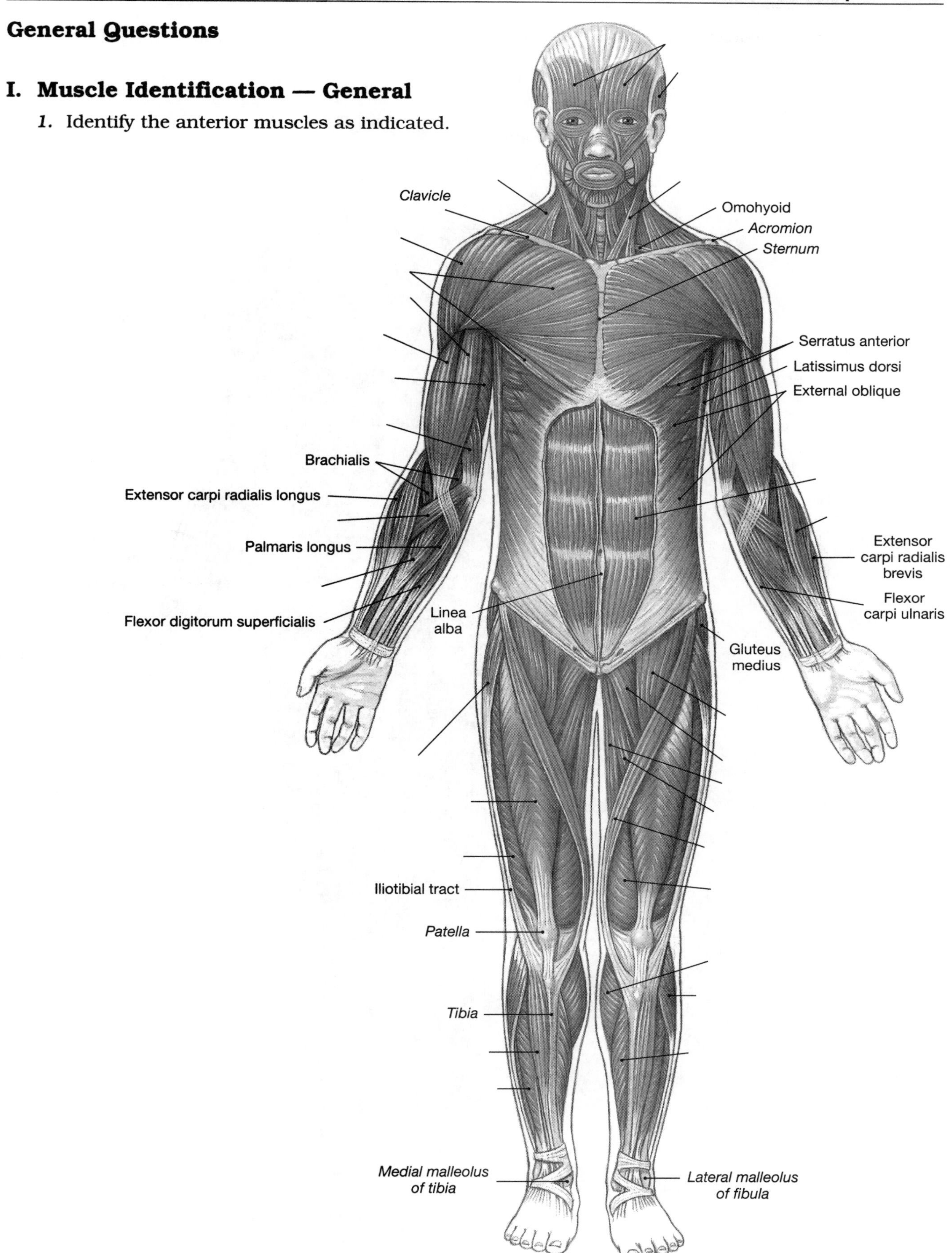

2. Identify the posterior muscles as indicated.

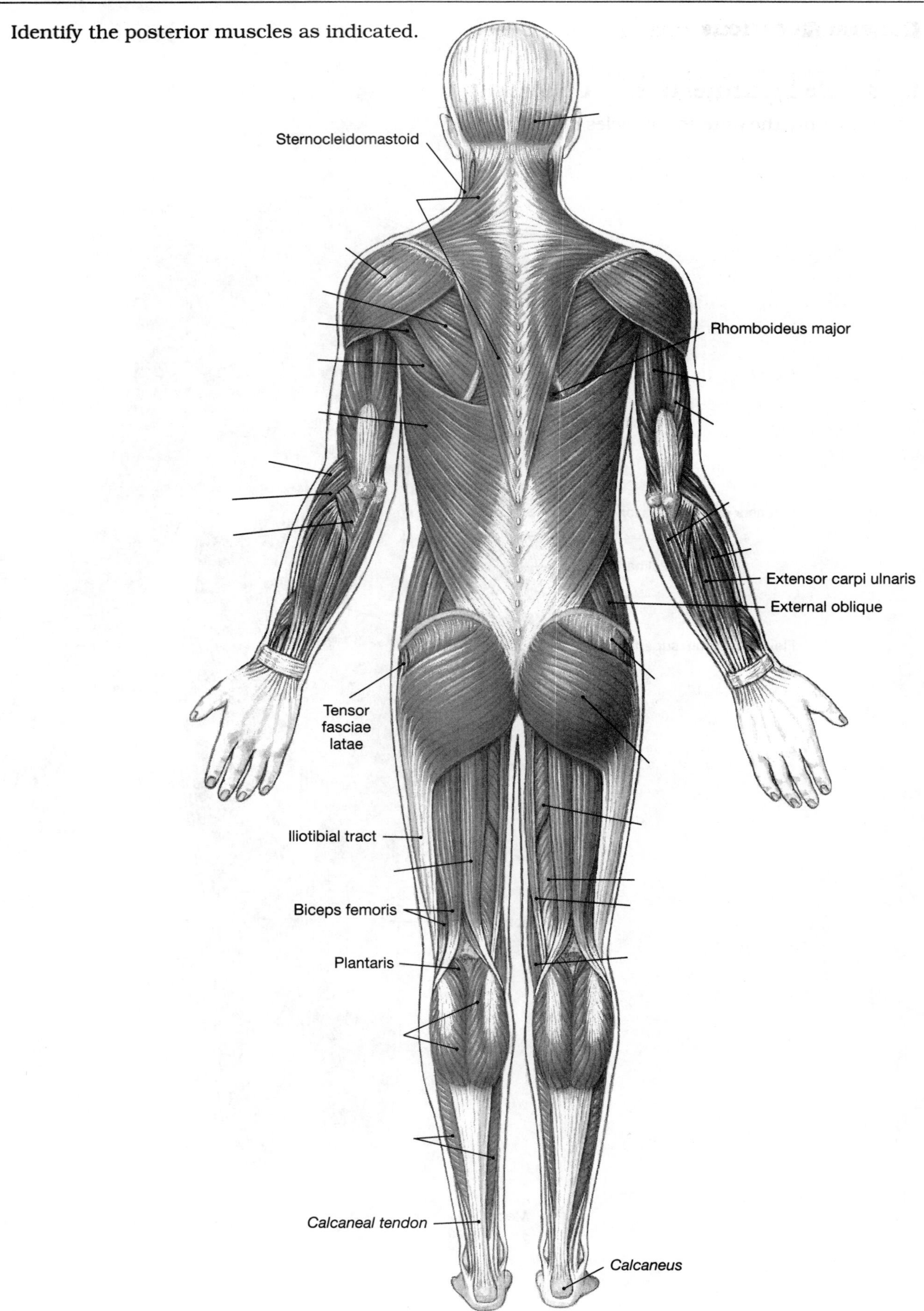

II. Muscle Identification — Specific

NOTES

A. Identify the major muscle(s) involved in each of these actions:

1. Sitting cross-legged ______
2. Closing your eyes ______
3. Kissing ______
4. Turning your head to the right, to the left ______
5. Shrugging your shoulders ______
6. Frowning ______
7. Holding a pencil ______
8. Pulling in your stomach ______
9. Standing on your toes, as a toe dancer might do ______
10. Breathing (in addition to the diaphragm) ______
11. Flex the vertebral column ______
12. Extend the great toe ______
13. Medially rotate the leg ______
14. Supinate the right hand ______
15. Extend the calf ______

B. Go back and identify the origins and insertions of the muscles you listed in the previous section.

C. Identify the muscle(s) that originate on the

1. Linea aspera of the femur ______
2. Cervical vertebrae ______
3. Iliac crest ______
4. Hyoid bone ______
5. Thyroid cartilage ______
6. Scapula ______
7. Occipital bone ______
8. Lateral epicondyle of the humerus ______
9. Proximal shaft of the tibia ______
10. Above femoral condyles ______
11. Maxilla ______
12. Galea aponeurotica ______
13. Zygomatic arch ______
14. Lower thoracic vertebrae ______
15. Sphenoid bone ______

D. Identify the muscle(s) that insert on the

1. Tarsals ______
2. Head of fibula ______
3. Iliotibial tract ______
4. Distal femur ______
5. Distal phalanges (hand) ______
6. Olecranon ______
7. Radial tuberosity ______
8. Greater tubercle of the humerus ______
9. Coccyx ______
10. Clavicle ______
11. Mastoid process ______
12. Central tendinous sheet ______

13. Mandible ____________________
14. Tongue ____________________
15. Lateral surface of eyeball ____________________

NOTES

III. Additional Questions

1. How do you stick your tongue out if muscles only contract? It would seem that the "tongue muscle" can't possibly be contracting if the tongue is extending.
2. Which muscle's origin becomes its insertion, and vice versa, depending on the action taking place?
3. As you sit there trying to answer this question, some of your muscles are acting synergistically. Explain.
4. With your left hand firmly around your right lower arm, first flex, then extend, and then hyperextend your wrist. Keep your hand in place and do the same with your fingers. Repeat the process with your right hand firmly around your left lower arm. Where are your flexor muscles? Where are your extensor muscles?

EXERCISE 22

Muscle Physiology

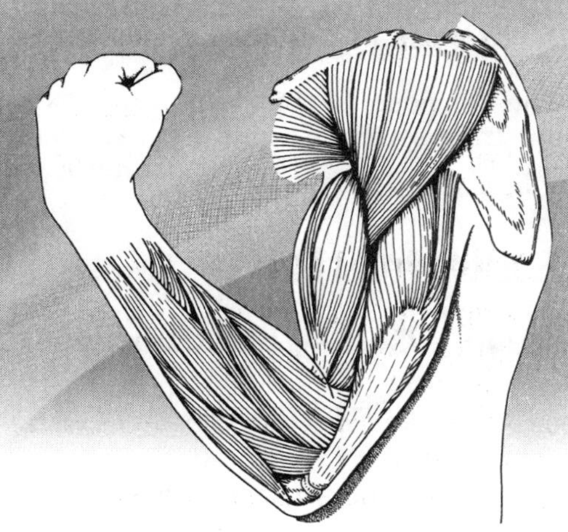

PROCEDURAL INQUIRIES

Examination

1. What are the parts of the muscle cell analogous to the cytoplasm, the cell membrane, and the endoplasmic reticulum?
2. What is a sarcomere?
3. What are actin and myosin and how do they relate to the sliding filament theory?
4. What are the landmarks of the sarcomere?
5. What is the neuromuscular junction?
6. What are the physiological events leading to a muscle contraction?
7. What are some common terms used in describing muscle contraction events?

Experimentation

8. What is the difference between an electromyograph and an electromyogram?
9. How do you set up an electromyograph?
10. How can you demonstrate treppe?
11. How can you demonstrate wave summation?

Additional Inquiries

12. Why can we say that all cells have an electrical potential?
13. What is the difference between a needle electrode and a plate electrode?
14. What are artifacts and interference?

Key Terms

A Band
Action Potential
All or None Principle
Artifacts
Electromyogram
Electromyograph
H Zone
I Band
Interference
Latency Period
M Line
Multiple Motor Unit Summation
Muscle Tone
Neuromuscular Junction
Neurotransmitter
Recruited
Resistance
Sarcomere
Sarcoplasm
Sarcoplasmic Reticulum
Sliding Filament Theory
Summation of Twitches
Tension
Tetanus
Transverse (T) Tubules
Treppe
Twitch
Wave Summation
Z Line

Materials Needed

Diagrams or Models of Sliding Filament
Electromyograph
Rubber Ball
Optional:
- Toothpicks and Glue
- Pipestem Cleaners

All human cells have an electrical potential. That is, an electrical difference exists between the two sides of the cell membrane. When this electrical potential is functional, we call it a membrane potential. The functionality of the membrane potential dictates the excitability of the cells and thus the ability of the cells to contract. This contractibility is the basis of action, the basis of muscle physiology.

The purpose of this laboratory exercise is to demonstrate some of the basic phenomena connected with the physiology of skeletal muscle contraction.

❑ Examination

I. Muscle Cell — Functional Anatomy

A. Muscle Cell Structure [∞ Martini, p. 273]

1. Review the structure of skeletal muscle in Exercise 20. Make note that the cytoplasm of the muscle cell is often called the **sarcoplasm**. The plasma membrane of the muscle cell is the **sarcolemma**, and the specialized endoplasmic reticulum of the muscle cell is the **sarcoplasmic reticulum**. The **transverse (T) tubules** are deep invaginations in the sarcolemma. The functional unit of the skeletal muscle is the **sarcomere**. The sarcomere is composed of **actin** (thin) and **myosin** (thick) filaments.

2. Examine Figure 22–1●. Note the relationships between the actin and myosin myofilaments. Include in your observation the following sectional landmarks:

Z lines (from German *zwischenscheibe*, "between disk"), the darkened lines marking the boundaries between adjacent sarcomeres. The Z lines form the striations readily viewed under the light microscope.

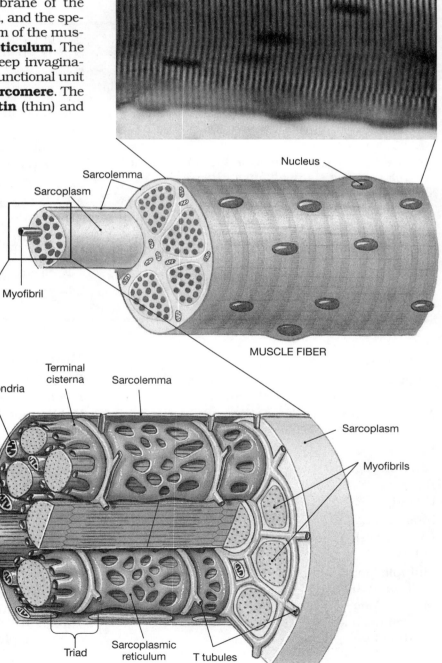

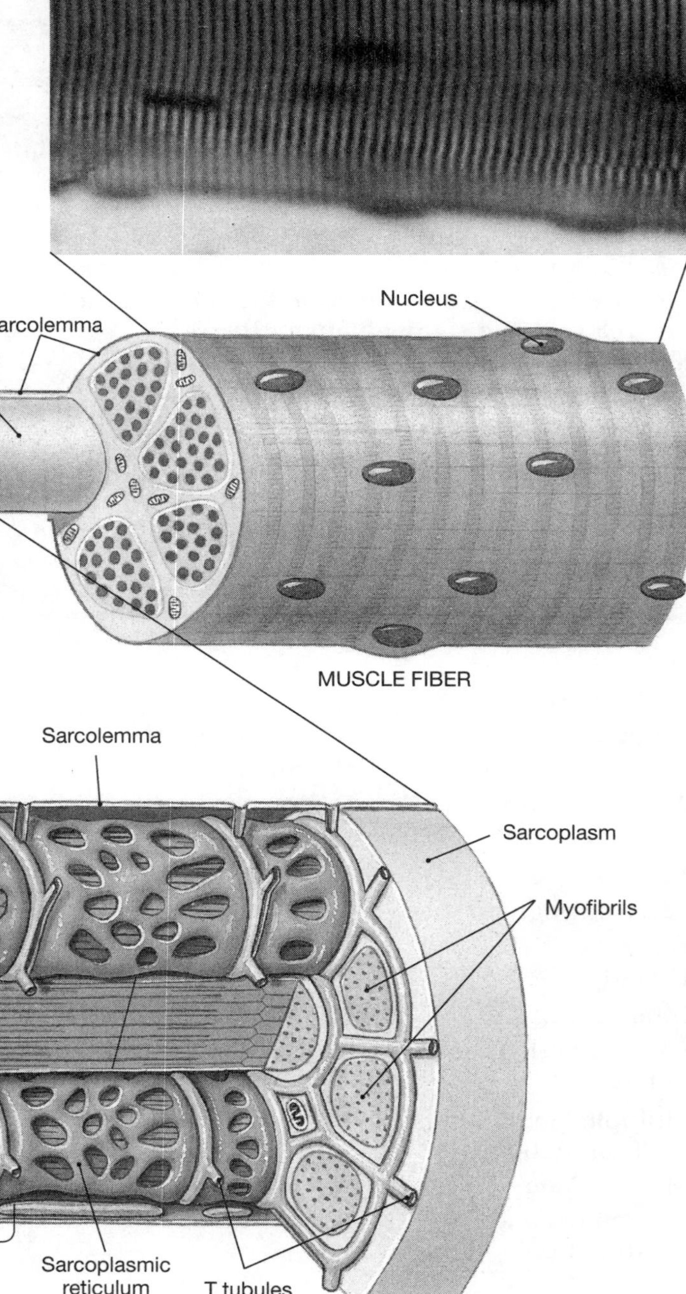

● **FIGURE 22–1a**
Skeletal Muscle Structure. LM × 612

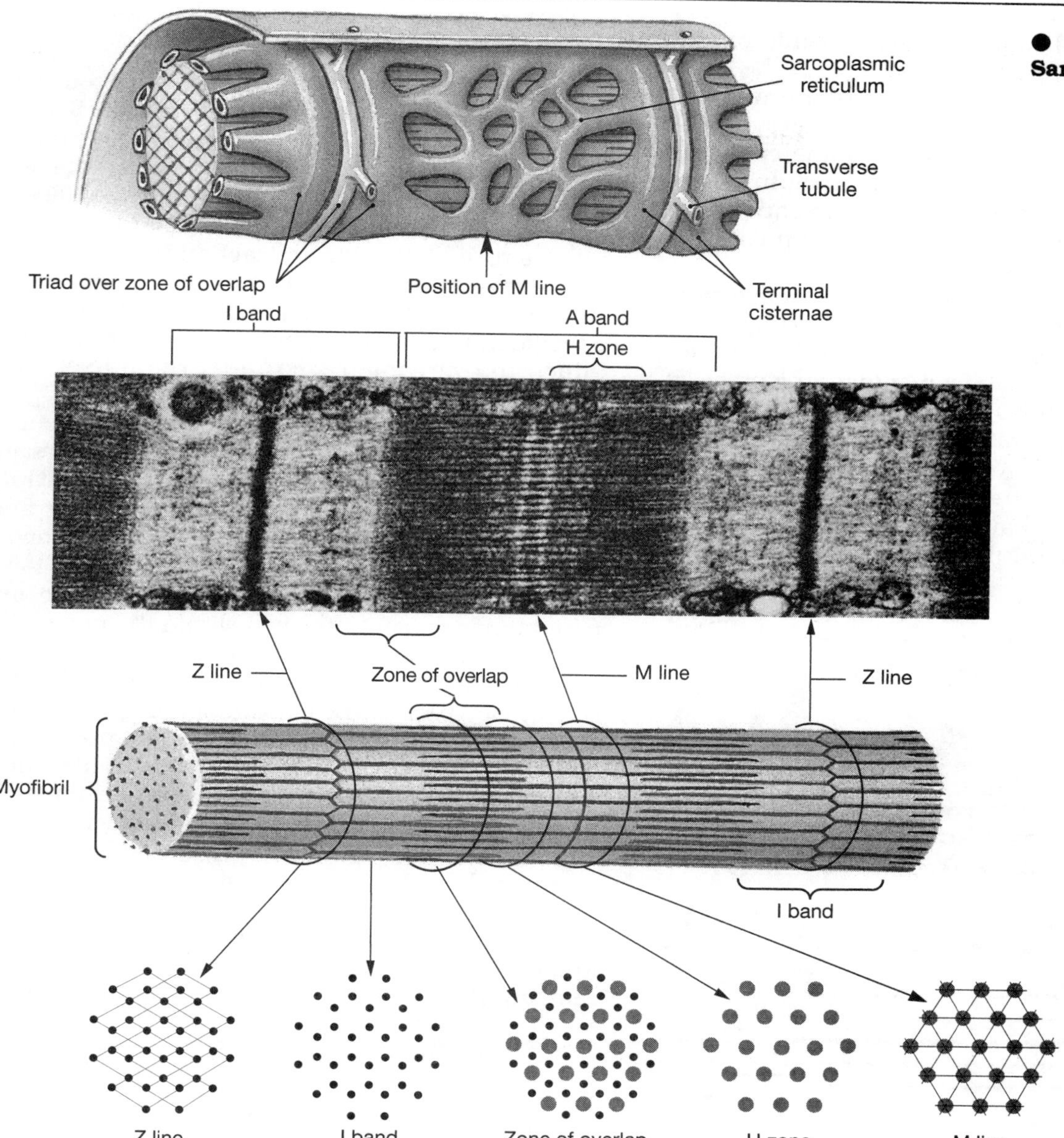

● **FIGURE 22–1b Sarcomere Structure.**

I band (from Isotropic, because these areas do not polarize light), the light area flanking the Z line. The Z line runs through the middle of each I band.

A band (from Anisotropic, because these areas polarize visible light), the darkened area making up most of the central area of the sarcomere.

H zone (from German *helle*, "bright"), the light area in the center of the sarcomere.

M line (from *mitte*), a darkened set of lines in the middle of the H zone.

B. Sliding Filament Theory [∞ Martini, p. 279]

The **sliding filament theory** states that during a contraction, the thin actin filaments slide over the thick myosin filaments, causing a shortening of the sarcomere unit.

To be functional, the sarcomere must be stimulated. This stimulation is the **action potential**. Technically, an action potential is a conducted change in the membrane potential initiated by a change in the permeability of the membrane.

C. Neuromuscular Junction [∞ Martini, p. 280]

A stimulus begins with the moving excitation of a nerve cell. This stimulation at the axon ending of the **neuromuscular junction** (the point where the nerve ending and muscle fibers come together but do not touch) causes the release of a **neurotransmitter** substance (often acetylcholine) from the axon. This substance crosses the cleft (the space between the nerve cell and the muscle cell) and attaches to the sarcolemma.

The action potential travels along the sarcolemma and the T tubules, changing the membrane characteristics of the sarcoplasmic reticulum and initiating a contraction.

Normally the sarcoplasmic reticulum is not particularly permeable to calcium. However, the change in membrane permeability initiated by the action potential causes a rush of calcium out of the sarcoplasmic reticulum and consequently a hundredfold increase in the calcium ion concentration in the sarcoplasm.

The calcium ions interact with the active sites on the actin filaments, and cross bridges are formed between the actin fibers and the paddle-like heads on the myosin filaments. In the process, ATP is converted to ADP, and energy is released.

The contraction continues as long as ATP and calcium are available. When the stimulation stops, the calcium ions are recruited back into the sarcoplasmic reticulum and the fiber rests until excited by another stimulus.

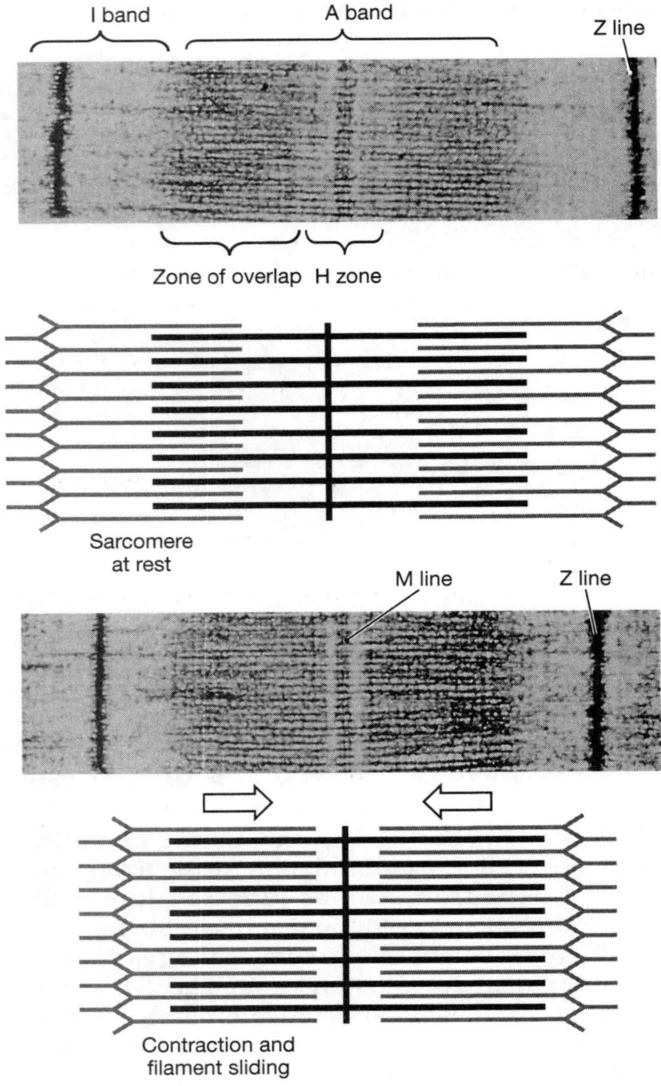

● **FIGURE 22–1c**
Sarcomere During Muscle Fiber Contraction.

D. In-Text Terminology

Terminology is sometimes best understood when used in context. For instance, each muscle fiber either contracts or it does not contract. This is the **all or none principle**. Keep in mind, however, that any individual muscle is composed of numerous fibers. Each of those fibers contracts upon stimulation.

A single stimulus/contraction/relaxation sequence is a **twitch**. Note in Figure 22–2a● that the time required for the twitch varies according to the specific muscle involvement. If enough **tension** is added to overcome **resistance**, the muscle will shorten (see Figure 22–2b●).

Usually a single muscle fiber is not sufficient to perform a task. In this case more and more fibers are **recruited**. This phenomenon is known as **multiple motor unit summation**. In other words, more and more fibers function to perform a certain action or contraction. If you have good **muscle tone**, you have numerous motor units contracting and relaxing and maintaining constant tension within the muscle.

Between the time of the stimulus and the beginning of a contraction is a **latency period**. After the latency period, the contraction begins, peaks, and relaxes. If a second stimulus is applied immediately after relaxation, the second contraction will develop a slightly higher maximum tension. Several such stimuli will give a stair effect, known as **treppe**, with each tension peak being slightly stronger than the one before it.

If the relaxation phase is not complete when the second stimulus is applied, the phenomenon is known as a **summation of twitches**, or **wave summation**. If any relaxation occurs, the condition is

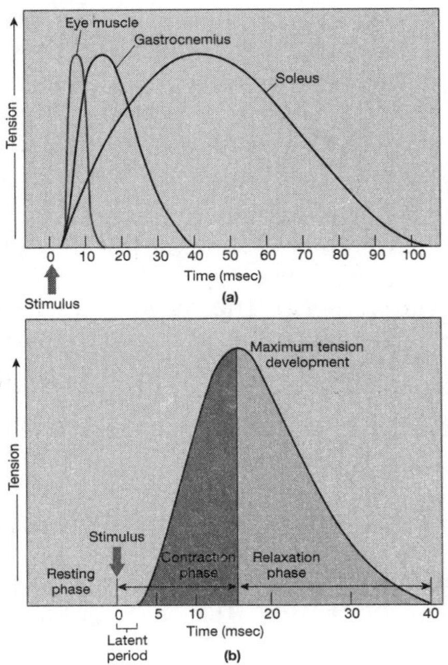

● **FIGURE 22–2**
The Twitch and Development of Tension.

incomplete tetanus. If no relaxation occurs, the situation is **complete tetanus**. See Figure 22-3●.

TETANUS the disease is caused by the toxin produced by the bacterium *Clostridium tetani*. This toxin inhibits the acetylcholine esterase at the neuromuscular junction, and therefore the neurotransmitter acetylcholine remains bound to the sarcolemma. This causes continuous stimulation, and thus continuous contraction, of the muscle involved. This disease state tetanus should not be confused with the normal tetanus resulting from continuous muscle contraction without relaxation.

❑ Experimentation

II. Electromyography

A. Electromyograph

The **electromyograph** is the instrument used to record the **electromyogram (EMG)**. Numerous types of electromyographs are in use around the country. These range from some very old physiographs and rather expensive oscilloscopes to some very modern computer-interfaced EMG modules. Your instructor will give you directions for the particular electromyograph you will be using.

Regardless of the type of myograph you are using, the principle will be the same. The electromyograph records the electrical activities in a particular muscle or group of muscles. In a professional setting, this is normally done by placing *needle electrodes* at strategic points within the specific muscle tissue of interest to the health care professional. Readings are then taken from the electrodes. Changes in specific muscle activity can indicate muscle or nerve dysfunction at a particular point in the system.

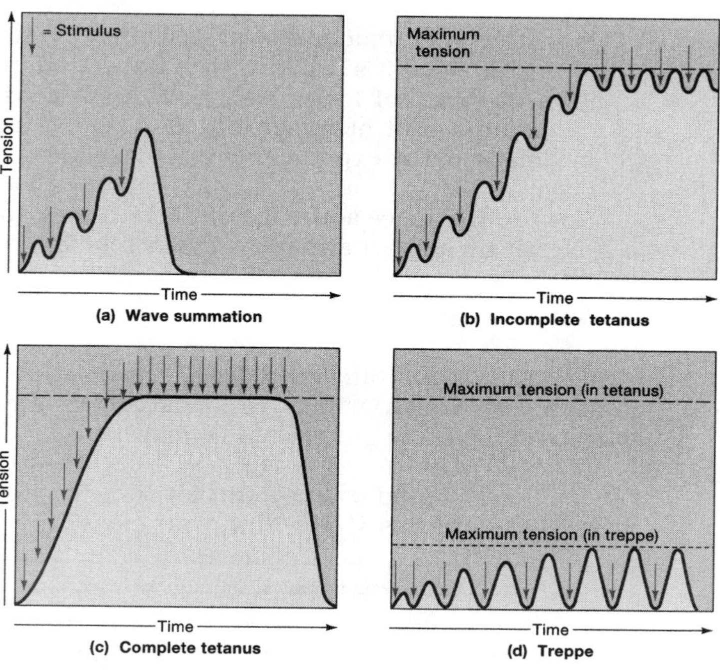

● **FIGURE 22-3**
Effects of Repeated Stimulations.

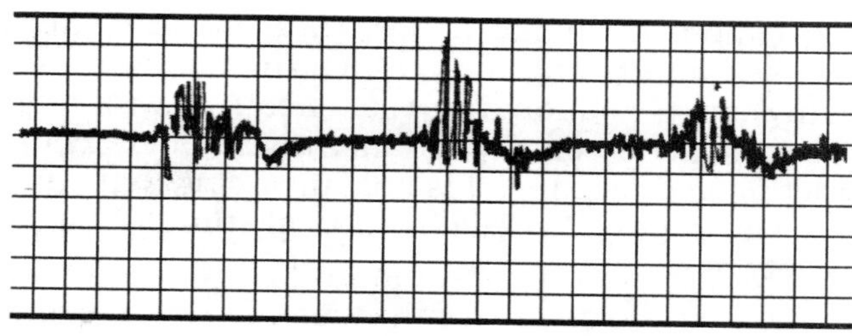

● **FIGURE 22-4**
Electromyography Reading.

In the introductory classroom setting, however, readings are taken across major muscle groups with *plate electrodes*. These electrodes, one positive and one negative, record the activity in particular muscles or muscle groups.

B. Electromyogram

The EMG is traced on quadrille lined paper or on a computer screen. The computer screen may or may not be graphed, depending on the computer program. The darkened blocks on the paper strip are 5 mm square (see Figure 22-4●).

As the readout progresses, the horizontal axis of the strip measures time. The standard run time is 25 mm per second, although the tracing speed can be altered to examine some particular aspect of the EMG.

The vertical axis (amplitude) records electrical activity in microvolts. For most purposes, it is usually adequate to set the electromyograph to record 20 microvolts per millimeter.

The amplitude will spike (increase) with greater tension, as more muscle fibers are recruited to develop the tension. This change in amplitude can easily be seen on the electromyogram.

Since muscle contraction is electrically stimulated, the frequency of the nerve impulses to the muscle fibers also influences how much tension is or can be generated. We cannot demonstrate nerve impulse frequency with plate electrodes. Needle electrodes are necessary to isolate action in fine sections of muscle fiber.

C. Readout Problems

Two problems you may encounter when reading your EMG are **artifacts** and **interference**. Artifacts are bits of "background noise" and interference is any unwanted electrical impulse you may notice. The words can often be used interchangeably. Since the body has so many muscles and since the smooth flow of any action depends on numerous muscle fibers contracting and relaxing, it is not surprising that surface electrodes will detect electrical stimuli from a variety of sources.

D. Experiments

The following experiments should serve to demonstrate changes in electrical activity as different muscle fibers are contracted.

1➡ Prepare your electromyograph according to your instructor's directions. Have your rubber ball ready.

2➡ Set the electromyograph to measure the activity of your *biceps brachii*. Squeeze a rubber ball. The amplitude spikes are due to the summation of action potentials from an increasing number of fibers as they begin conducting.

3➡ Demonstrate the following conditions on the right and left biceps independently. Describe what happened. If you have a paper printout, write on the paper so you have the labeled data all together.

Resting

____

____

Sudden tension (quickly tightening your squeeze on the ball)

____

____

Gradual tension (gradually tightening your squeeze on the ball)

____

____

4➡ Go back to the previous section. Write out the definitions of treppe and wave summation:

____

____

How can you demonstrate treppe?

____

____

Wave summation?

____

____

Is there a difference between treppe and wave summation?

____

____

Are these the same as what you are doing by squeezing the rubber ball? Explain.

____

____

Do you notice a difference between your right- and left-side actions?

____

____

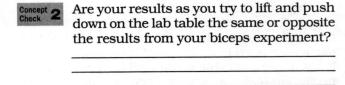

Concept Check 1 With the electromyograph still attached, try lifting the lab table. After noting your results, try pushing down on the table. What differences, if any, do you notice?

____

____

____

____

5➡ Now attach the myograph to your triceps (right and left). Repeat the experiments. What differences do you notice?

____

____

Concept Check 2 Are your results as you try to lift and push down on the lab table the same or opposite the results from your biceps experiment?

____

____

____

____

6➡ Attach the machine successively to your gastrocnemius (calf muscle), tibialis anterior (front of lower leg), and quadriceps group (front of upper leg), and repeat as many of the experiments as is feasible.

7➡ Use the space at the top of the next page to set up a chart and record your results.

E. Analysis

After examining your data, discuss your results with your partner. Your instructor may wish to make some group data comparisons.

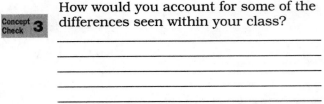

Concept Check 3 How would you account for some of the differences seen within your class?

____

____

____

____

____

1. Using the pipestem cleaners and toothpicks,

Electromyograph results

❑ Additional Activities

NOTES

construct a sarcomere. Use the pipestem cleaners for the actin and myosin. For the myosin heads, break the toothpicks about 1 cm from the end and set the broken part at about a 45-degree angle. Attach the toothpick to the "myosin pipestem." Try to orient your myosin heads according to the figures so that pictorially accurate cross bridges can be formed. Glue as necessary. Start with a two-dimensional model. The actin and myosin filaments should slide over each other and you should be able to identify your active sites. With a little practice you should be able to demonstrate that your sarcomere is a functional sliding filament. You can also demonstrate the anatomy achieved by varying the sarcomere length.

2. If time permits, put your sarcomere together with your classmates' sarcomeres. (When you are comfortable with your two-dimensional model, try to add the third dimension.)
3. The anatomy of the sarcomere has been compared to a test tube brush in a test tube. The bristles would be analogous to the myosin heads. Discuss why this may or may not be a valid comparison.

Answers to Selected Concept Check Questions

1. Your results should be opposite because antagonistic muscle groups are involved.
2. Your results should be opposite because of the attachment to opposite muscle groups.
3. Almost anything can cause the differences: physical condition, genetics, the way the machine was operated.

❑ Lab Report

1. Referring to the numbered inquiries at the beginning of this exercise, complete the following box summary:

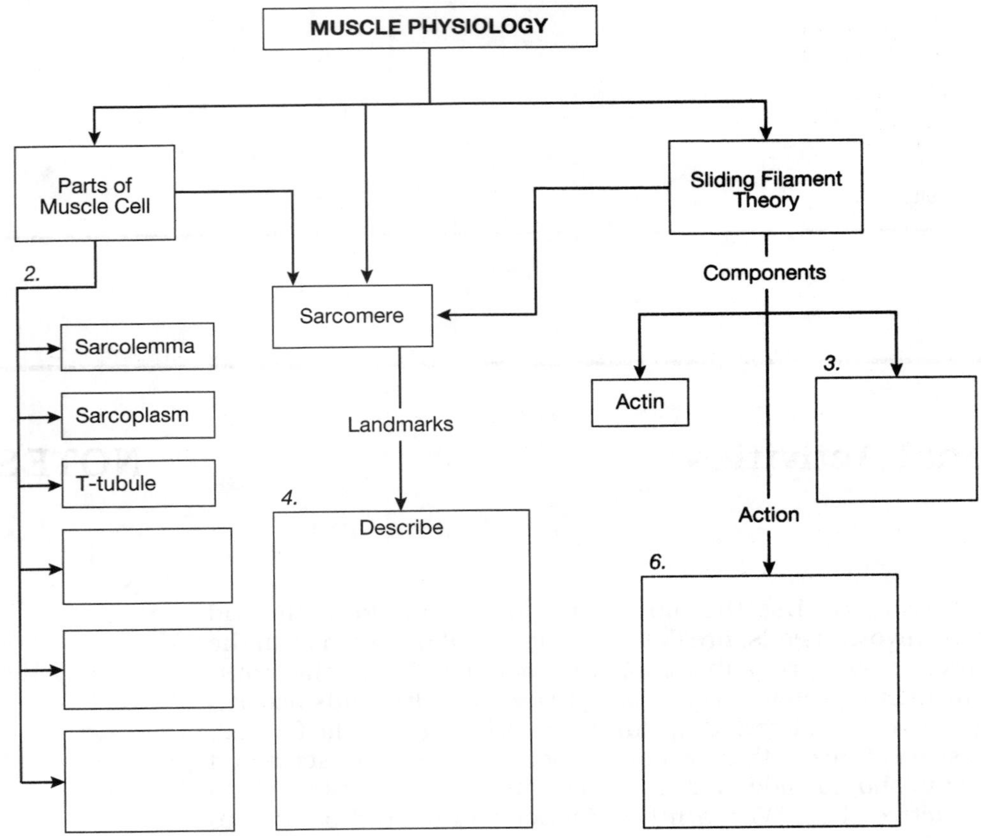

General Questions

NOTES

1. Supply the correct word:
   a. Increasing strength of a muscle contraction ______________________
   b. Sustained contraction without relaxation ______________________
   c. Difference in voltage between sides of a membrane ______________________
   d. Background noise on an electromyograph ______________________

2. What type of electromyogram do you have in your lab and how do you set it up properly?

3. What is the difference between a needle electrode and a plate electrode?

4. In your own words describe the muscle fiber events from the time a transmitter substances reaches the sarcolemma, through the muscle contraction, and culminating in the relaxation of the muscle fiber.

5. How does rigor mortis (temporary rigidity of muscles occurring after death) relate to muscle contraction?

6. Based on what you know about muscle activity, how can you explain the apparent muscle contraction of a person who has been dead for some time?

7. How does muscle fatigue relate to muscle physiology?

EXERCISE 23

Exercise and Stress Physiology

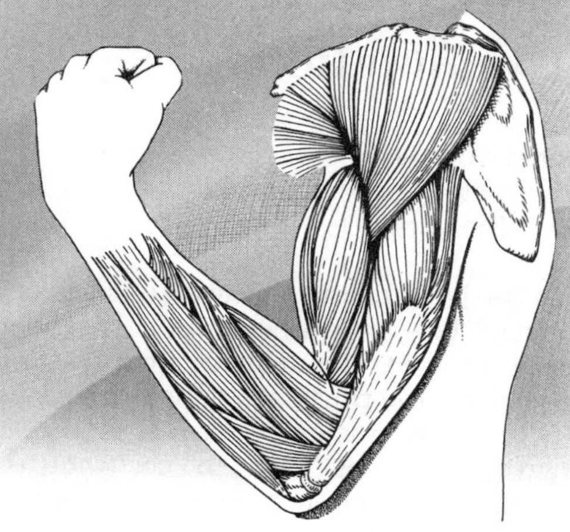

PROCEDURAL INQUIRIES

Examination

1. What is exercise?
2. What is stress?
3. What happens to the muscle when it is "trained"?
4. How do endorphins relate to stress?
5. What does the American College of Sports Medicine recommend?

Experimentation

6. What are graduated exercises?
7. What are progressive resistance exercises?

Additional Inquiries

8. What is the overall purpose of exercise and stress physiology?
9. What is the result exercise has on the cardiovascular system?

Key Terms

American College of Sports Medicine
Endorphins
Exercise
Graduated Exercises
Progressive Resistance Exercises
Stress

Materials Needed

Assorted Weights — 10 g to 5 kg

(Objects, such as books or stones, of varying weight may be substituted)

(This ancillary laboratory exercise is designed to give the non–physical education student and the non-kinesiology student an overview of certain principles of exercise and stress physiology. This lesson is not meant to supersede in any way any of the more detailed exercise and stress physiology principles that may have been learned in connection with formal physical education and kinesiology classes.)

If you are at all familiar with exercise and stress physiology, you know that physical activity has a positive effect on health. Nevertheless, the concept of exercise and stress physiology means different things to different people. The sports enthusiast may think of a well-equipped gym; the health care professional may think in terms of wellness from good nutrition and exercise; and the student may consider exercise and stress physiology to be the compromise between exercise and sleep, food and rest, relaxation and good study habits.

In this exercise we will look at an overview of the principles behind exercise and stress physiology by outlining some exercise and stress generalizations and by performing some basic exercise and stress physiology experiments.

This laboratory exercise can be correlated with Exercises 22, 29, and 51.

No specific Concept Links are given in this Exercise. Throughout the Martini text, particularly in Chapters 10, 11, and 20, references are made, both directly and indirectly, to exercise and stress physiology concepts.

❑ Examination

I. Overview

A. Exercise

In the broad sense, **exercise** is the contraction of a muscle or a muscle fiber. In the narrower sense, exercise is the deliberate contraction of muscle tissue for the purposes of health and well-being.

One function of every muscle is to contract. We know that muscles "have to" contract to stay "intact." We continuously contract our muscles, either consciously or reflexively. For instance, have you ever absolutely had to get up and move around? Have you ever wondered why your dog or cat stretches all its major muscles when it gets up? Think about these examples as you consider why the muscles of comatose persons are exercised for them.

The point is that we are continually using our muscles, exercising our muscles, and trying to improve the health and well-being of our bodies by increasing muscular activity. Even play is a necessary form of muscular activity.

1 ➡ Think about this: If play weren't a necessary physical activity, why would we be so enthused about doing something pleasurable even when we are totally exhausted?

B. Stress

In human biology, **stress** is the resistance placed against a contracting tissue. As we examine stress, we see a continuum of possible scenarios. At one end of the spectrum is normal muscle contraction. Aside from intrinsic stimulation, a skeletal muscle could not contract without the presence of a resistance (such as a bone). At the other end of the spectrum is a resistance beyond the capabilities of the muscle, no matter how many motor units are called into action (such as trying to lift the front end of a car).

The gradual buildup of physical stress increases the muscle's ability to respond to stress. In other words, the more you practice lifting weights, the better you get at it. Several factors are involved in this. First, you are training the muscle itself to respond to your demands. This may mean increasing muscle mass, or it may mean increasing the vascularization of a particular tissue. Whatever the reason, you are increasing the strength and health of the muscle tissue.

Initially it would seem that stress decreases the ability of the muscle to perform. With gradual buildup, the opposite is actually true. Stress increases both muscular ability and muscular efficiency.

Increasing stress on the heart increases cardiovascular efficiency. Cardiovascular exercise, like any exercise, must be built up gradually. Again, as with any muscle, practice increases the ability of the heart muscle to perform well. With the heart, this includes greater efficiency (oxygen-carrying capacity) with less energy output.

Psychological stress can result in physiological change. We generally look at psychological stress as being detrimental. In one sense this is true. However, let us look at this situation from the standpoint of an interrelationship between the psychological and the physical. If psychological stress causes physiological problems, perhaps we can use physiological means to alleviate psychological stress.

C. Endorphins

Endorphins are neurotransmitters released by the nervous system in response to pain, real or anticipated. As we exercise, we adjust to stress. The exercise becomes easier as the muscle becomes more fit. Endorphins are released. The exerciser "feels good" and may even develop a sense of euphoria. If you are a runner, you may have been aware of an endorphin release when you suddenly got that "second wind." Endorphin release is one reason exercise is often recommended for people who are under great emotional or psychological stress.

Stress management becomes a matter of maintaining appropriate levels of physical stress for optimal functioning. This may include reducing psychological stress by physical means.

II. Physiology Principles

A. Goals

Several basic principles are involved in exercise and stress physiology. The overriding factor is that any physical activity has a positive effect on health. Physical activity greatly increases both the length and quality of life. These effects are linear, and therefore anything you do is better than doing nothing.

Physical exercise can be anything that improves or maintains fitness and increases health. The goals of most experts is to convince people to increase their physical activity and thereby increase their personal fitness levels.

Goal-setting in terms of quantity of exercise is not always practical as evidenced by the fact that only about 20 percent of the public ever reaches a physical fitness goal. Far better to set a minor goal like stretching and holding the stretch for a half minute several times a day than become discouraged by grandiose plans. Minor goals are more likely to be met, and the psychological positive feedback can aid in setting additional minor goals.

B. American College of Sports Medicine (ACSM)

The **American College of Sports Medicine** (ACSM) has made several recommendations for total body fitness. Fitness includes both resistance and duration training. The ACSM believes everyone should exercise three to five days per week and that all major skeletal muscle groups should be involved in fitness training.

Many elaborate exercise series programs exist. However, some of the simplest exercises are most effective. For instance, the ACSM recommends walking and hiking, running, cycling, dancing, cross country skiing, skipping rope, rowing, climbing stairs, swimming, and skating.

To specifically increase resistance, the ACSM recommends 8–12 repetitions of 8–10 exercises of every major muscle group at least twice a week. Some experts simply state that we should do a minimum of three sets of exercise for each major muscle group two or three times a week.

For duration training 20 to 60 minutes of continuous aerobic activity is recommended. Since duration is dependent on intensity, lower-intensity exercises require longer periods of activity.

The frequency, intensity, and duration of exercise do increase cardiorespiratory efficiency, muscular strength, and physical endurance.

C. Diet

The general consensus is that a proper diet plays an inseparable role in our physiological well-being. However, discussion of a healthy diet is beyond the scope of this particular exercise.

❑ Experimentation

III. Experiments

The experiments given here do not require any specific physiological equipment. These experiments can be performed either alone or in conjunction with computer-generated experiments. The data gathered here are more qualitative in nature; you will notice that personal judgment accompanies mathematical data. All of these experiments can be done using the computer.

(Although graduated exercises and progressive resistance exercises are discussed separately here, the exercises themselves are variations on a common theme.)

A. Graduated Exercise

To demonstrate some of the principles of exercise physiology, we will first do a few **graduated exercises** (graduated experiments). As you work through these exercises, pay particular attention to which muscles you actually notice and which muscles seem to be involved in the action.

1➡ Take four of the assorted weights. Lift each weight with, successively, your little finger, your whole hand, your toes (with your heel on the floor), your whole foot moving forward, and your ankle kicking backwards. In the space provided or on a separate paper, construct a chart that lists each of your exercises, including a statement about muscle involvement. You may wish to make a graph using one of the blank graphs at the end of this exercise.

Make vertical and horizontal comparative statements. (Statements about increasing weights using the same body part are vertical statements. Statements about using the same weight with different parts of the body are horizontal statements.)

Use your lecture text, or Exercise 22 of this text, to look up the definitions of contraction, summation, and multiple motor units. Referring to these definitions, what type of generalized statements can you make about this weight-lifting demonstration? Consider both the vertical and horizontal results.

____

____

____

2➡ Take one of the heavier weights in your hand. Note the time. Hold this object in your laterally outstretched hand for five minutes (or until muscle fatigue sets in). Don't rest your arm on the desk or shift the object to your other hand, but feel free to move your arm around or change the way you are holding the object. What does your moving the object around tell you about sustained contractions? What does this have to do with exercise physiology?

____

____

____

3➡ Pick up the heaviest object. Note the time. Hold this object in your laterally outstretched hand until fatigue sets in. Rest for two minutes. Repeat the experiment. Were you able to hold the object longer the first or second time? Why? You may wish to make a graph using one of the blank graphs at the end of this exercise.

____
____
____

4➡ Take a lighter object and lift it repeatedly while counting the number of lifts. Continue until your arm is too tired to continue. Record the number of lifts. Rest. Record the length of your rest. Repeat the experiment. Were you able to lift the object more times or fewer times? Repeat the whole experiment, varying the amount of time you rest. You may wish to make a graph using one of the blank graphs at the end of this exercise.

____
____
____

5➡ Based on these experiments, write a definition of graduated exercise.

____
____
____

B. Progressive Resistance

Consider the phenomenon of ascending and descending **progressive resistance**. (In this experiment, you may become somewhat sore if you are presently out of shape.) You may wish to make a graph comparing the next two experiments. Use the blank graphs at the end of this exercise.

1➡ For ascending progressive resistance exercises, start with a light object and do an arm curl with your object in your hand and your elbow firmly on the table. Count the curls as you continue to do curls until you cannot do any more. Record the number. Rest briefly. Now pick up a heavier weight and repeat. Keep your hand and arm in the same position as for the first set of curls. What did you notice? Repeat with a still heavier weight. What did you notice?

____
____
____

2➡ Why is it important to keep your hand and arm in the same position for each of the curling exercises?

____
____
____

3➡ For descending progressive resistance, begin with the heaviest weight and work down. Interestingly, if you can do ten arm curls with the heaviest object, you may well be able to do only seven curls with the lighter object. Why do you suppose that happens? Repeat with an even lighter weight. Speculate on what happened.

____
____
____

4➡ Based on these experiments, write a definition of progressive resistance.

____
____
____

C. Additional Ideas

Changes in exercise routine — even changes as slight as doing an arm exercise laterally instead of medially — change the way the muscle is stressed. This changes the way the muscle responds to the stress. The body is conditioned. This stimulates muscle growth, tissue proliferation, or increased vascularization.

1➡ Brainstorm with your lab partners and come up with a series of exercises that you could do to increase the efficiency of your own muscles.

____
____
____

2➡ Why do these exercise and stress physiology concepts work with cardiac muscle as well as skeletal muscle? Refer back to Section I B. Do you think the same principle applies to smooth muscle?

____
____
____

3➡ If you have already completed Exercise 21, go back through the information in that exercise and related it to the principles discussed here. Which muscle groups are involved in the activities recommended by the ACSM or in the activities you and your partners worked on?

____
____
____

❑ Additional Activities

NOTES

1. Although computerized muscle physiology has been around for several decades, at least in its simulated form, it is only in recent years that hands-on, computer-generated physiology has become readily available throughout the country.

   Because of the variety of computer programs that might be available to your school, we will not attempt to discuss the specifics of any particular program or system. Be assured that many excellent programs exist, programs that allow you to acquire and analyze data specifically gathered from experiments you perform in your classroom laboratory.

   If your school is equipped with any of the computer-generated muscle physiology (or exercise and stress physiology) instruments, your instructor will demonstrate the specifics of your equipment.

   If your exercise and stress physiology experiments are to be computerized, your results will be quantitative. You will obtain specific data that you will be able to compare and analyze mathematically or graphically.
2. If you do not have a regular exercise routine, write out a workable exercise and stress physiology plan for yourself.
3. Use whatever sources you have available to define dynamometer and ergometer. Find out how these instruments relate to this exercise.

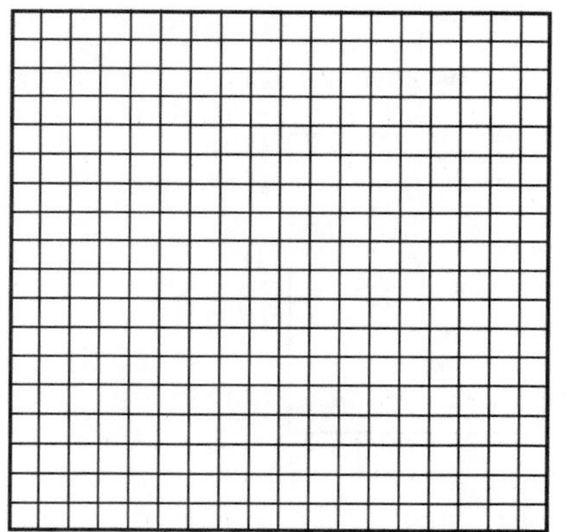
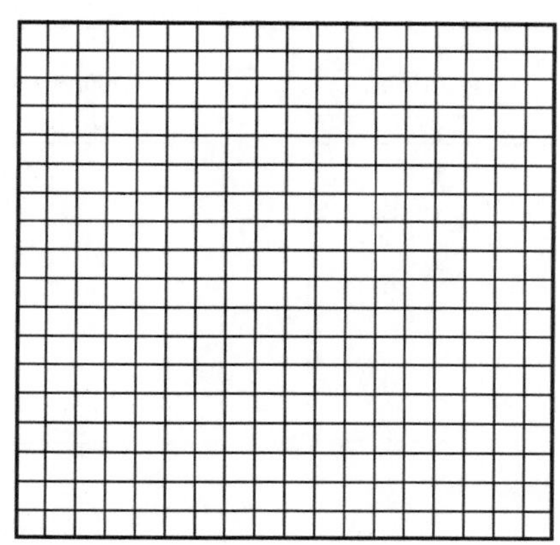
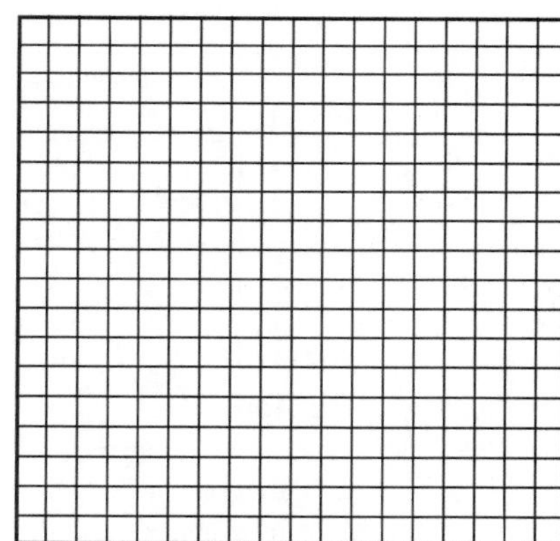
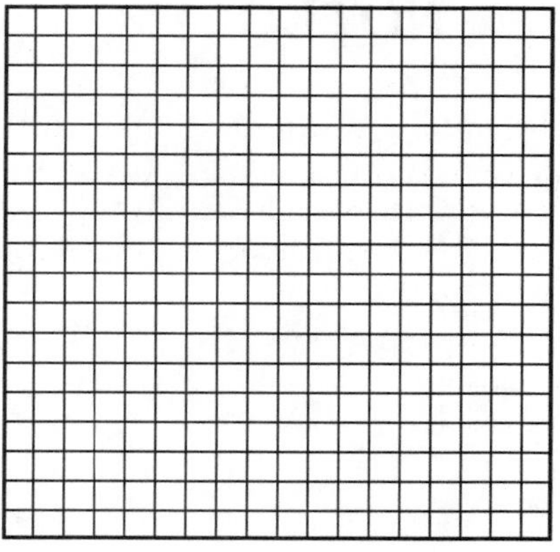
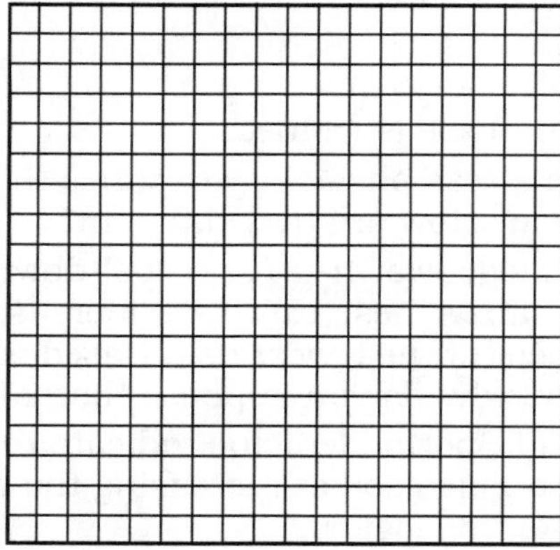
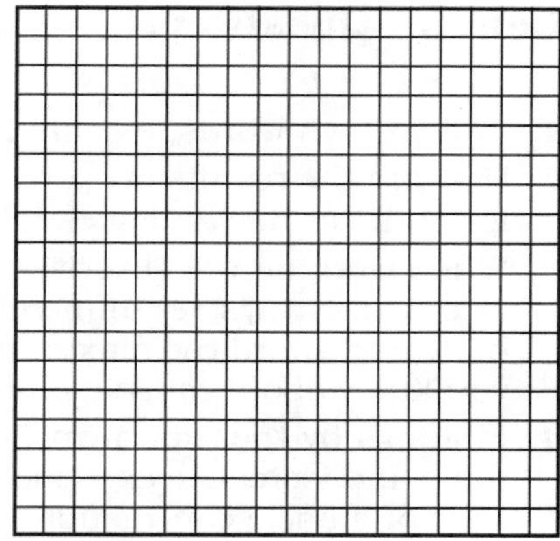

❑ Lab Report

1. Referring to the numbered inquiries at the beginning of this exercise, complete the following box summary:

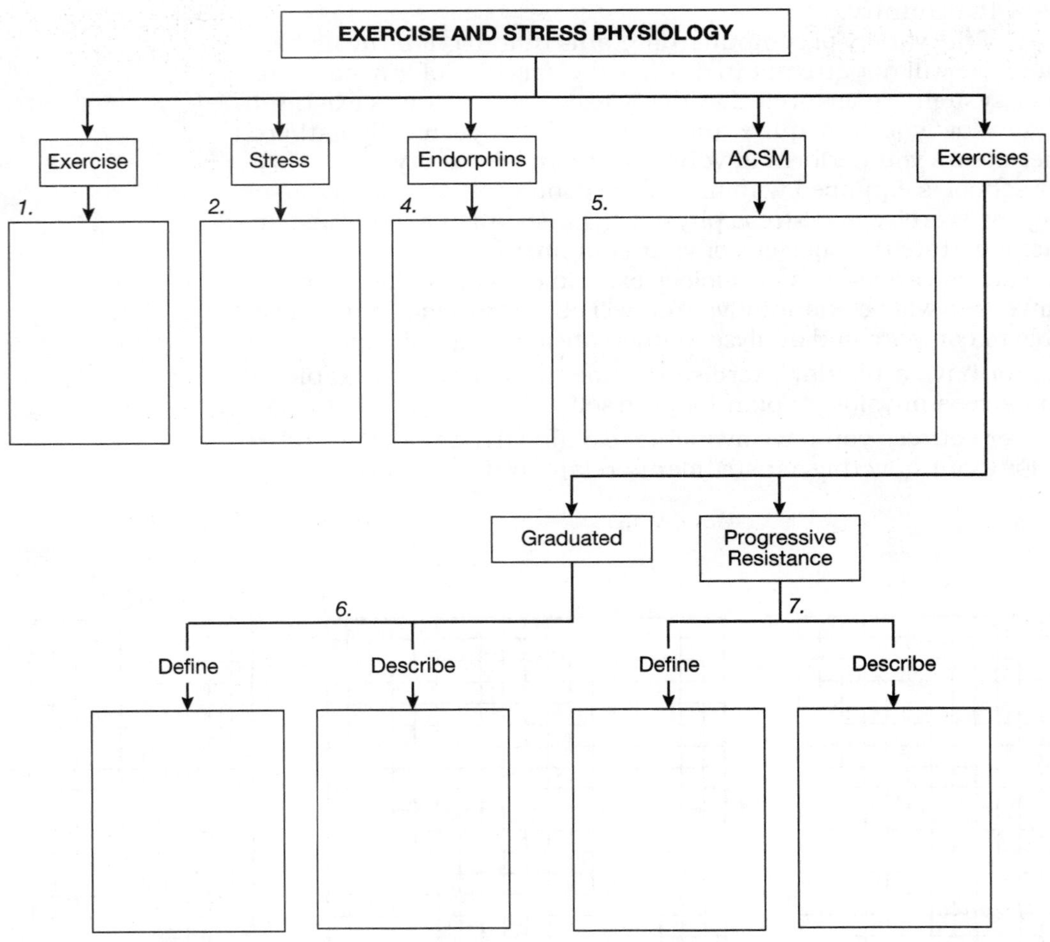

General Questions

NOTES

1. In your own words, explain how a muscle is trained.
2. How are the results of exercise and stress physiology routines the same for both cardiac and skeletal muscle? How are they different?
3. When working with exercise physiology, warm-up and cool-down exercises are always recommended because these exercises increase flexibility and afford the maximum benefit at the lowest risk. Based on what you know about muscle physiology, why do you suppose that is?
4. Findings by the American College of Sports Medicine indicate that isotonic and isometric exercises are equally good at increasing the magnitude of muscle contractions. Why is this logical?
5. Why should we be interested in maintaining an exercise and stress physiology routine?

EXERCISE 24a

Introduction to the Cat

PROCEDURAL INQUIRIES

Preparation

1. How do you determine the sex of a cat?
2. How should you prepare a cat for dissection?
3. What safety precautions should be followed in skinning a cat?

Dissection

4. What is a ventral incision?
5. How do you skin a cat?
6. What are some subcutaneous structures found in skinning a cat?

Additional Inquiries

7. Why is the cat a good animal to study in a human anatomy lab? Which body systems are analogous?
8. What cleanup steps should be taken following dissection?

Key Terms

Cutaneous Maximus
Cutaneous Nerves
Superficial Fascia
Ventral Incision

Materials Needed

Articulated Cat Skeleton (if available)
Gloves
Preserved Cat
Dissecting Pan
Dissecting Kit
Animal Storage Bags
Labels/Tags
Disinfectant

The focal point of the anatomy laboratory is the three-dimensional, hands-on study of the human body. Because it is often not possible to study either an actual human body or a cadaver, substitutes must be found. Plastic and molded models of human torsos and human body parts are valuable additions to the anatomical learning experience. Nevertheless, these artificial bodies do not afford the opportunity of feeling and manipulating living (or formerly living) tissue.

All mammalian bodies are structurally and functionally quite similar. This means we can study the nonhuman mammalian body and gain an anatomical and physiological understanding of the human body.

Each year millions of cats across the country are destroyed because no one can give them a home. Fortunately, some of these animals do end up in biological supply houses, where they are prepared for classroom or laboratory study. Be assured that the cat you are working with was not clandestinely snatched from someone's yard, nor was any pain inflicted on this animal in the preparation process.

The adult cat is an ideal animal for laboratory dissection. Because of its size, the organs and systems of the cat are large enough to see and work with easily. Because of its lifestyle, the body of the cat is normally free of excess fat. (Excess fat can inhibit the study of many structures.)

In later exercises in this book we will be dissecting the cat's muscles as well as its cardiovascular, respiratory, digestive, urinary, endocrine, and reproductive systems.

The muscles of the cat, which correspond closely with the muscles of the human in both name and location, are usually well developed, readily separated, and easily identified.

The feline blood vessels are also well defined and easy to work with. The names and locations of the major cat blood vessels are almost identical to those of the human blood vessels.

The functional anatomy of the other internal systems of the cat is also analogous to the functional anatomy of the human. Although you will notice some differences (e.g., the cat does not have an appendix), you will find that a good understanding of the internal structures of the cat will give you a solid foundation in grasping the three-dimensional nature of human anatomy.

This introductory exercise is designed to orient you to the cat, to help you with the preparation of the cat for future dissection exercises, and to introduce you to some of the basic methodologies of cat dissection. In your particular laboratory setting, this exercise may be combined with your first dissection exercise.

❑ Preparation

I. Safety

A. Preservatives

1➡ Wear protective gloves while dissecting any animal, preserved or fresh. Some animal preservatives are mildly to severely irritating to the skin, other preservatives can be carcinogenic. Live or freshly killed animals may harbor toxins or disease-carrying organisms.

B. Parasites

1➡ Use care to keep your hands away from your face while dissecting. Sometimes cats harbor intestinal parasites. Although the preserving process will kill the mature parasite, the parasite larvae, or eggs, often have a highly resistant protective coat. The round worm, *Ascaris lumbricoides*, for example, sheds eggs that have been known to survive for more than two years in the intestines of cats preserved in formalin.

C. Regulations

1➡ Follow any regulations set down by your state or local government. Certain jurisdictions have enacted safety regulations for the handling of live or preserved animals. Those regulations take precedence over any safety tips written in this book.

2➡ Follow any regulations set down by your school. Some schools have set up a uniform code of safety for the handling of biological specimens. It is imperative, then, that you follow all directions your instructor may give regarding the handling of the animals.

D. Allergies

Some students worry about allergies to cats. If you are allergic to cats, you are probably allergic to the cat dander or to a substance in the cat's body fluids, rather than to the cat itself. The preserved animal has neither dander nor active body fluids and thus should not aggravate your allergies.

1➡ Should the preservative accidentally touch your skin, wash the area thoroughly.

E. Additional Precautions

1➡ Make note here of any additional safety regulations your lab may have regarding dissection and the safe handling of preserved specimens.

____
____
____

II. Preliminary Instructions

A. Identification

1➡ Obtain and tag your cat according to your instructor's directions.

2➡ Orient yourself to the cat by examining the articulated feline skeleton. If no skeleton is available, study the accompanying diagram (Figure 24a–1●). Note particularly the feet, knees, ankles, and elbows of the animal. The cat is a digitigrade animal, meaning that it walks on its digits rather than on its entire foot. Note, too, the angles formed as the bones articulate with one another. How do these angles of articulation correspond with the bones in your body?

____
____

As you work through the rest of this lesson, be aware of the orientation of the different parts of the feline body.

3➡ Determine the sex of your cat. Is your animal male or female? ______________

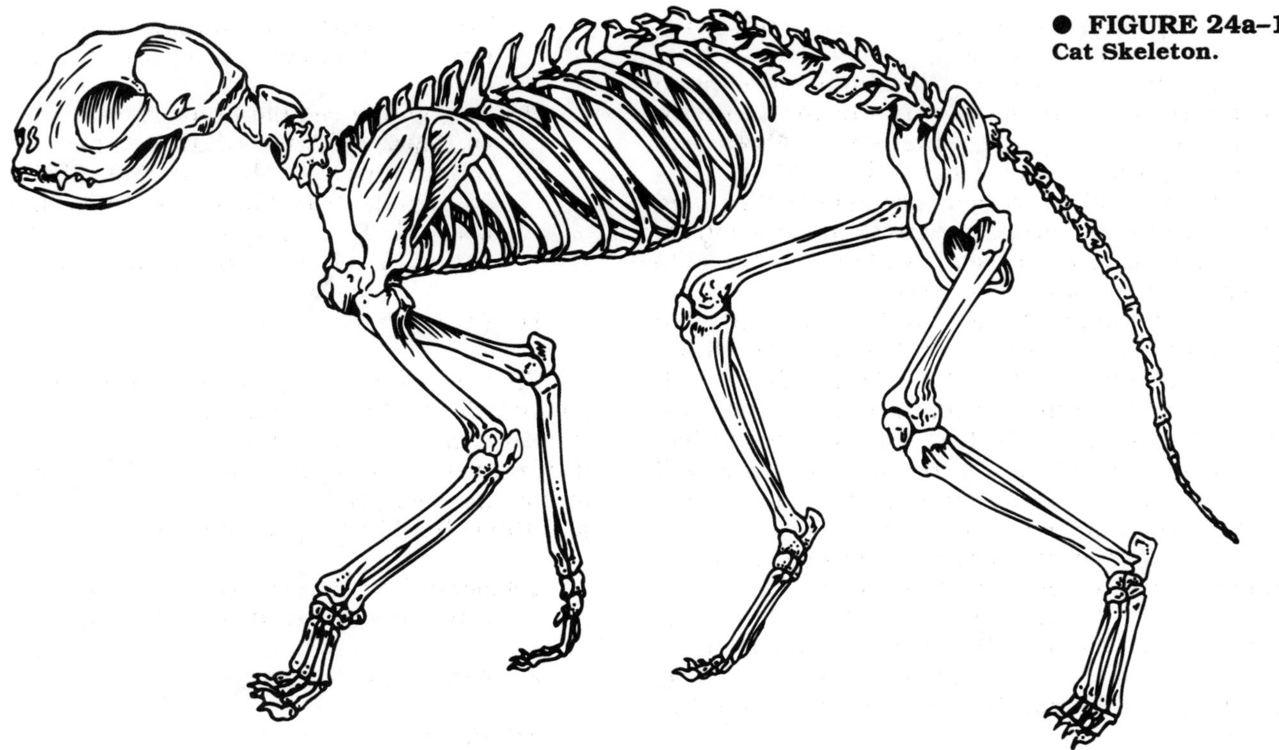

● **FIGURE 24a–1**
Cat Skeleton.

Examine the urogenital area, which, in both the male and female cat, is located at the caudal end directly anterior to the tail. Check for a scrotal sac (male). (If you have a castrated male, no scrotal sac will be present.) The female urogenital opening is much larger than the male penis opening. If you are uncertain about the sex of your cat, check with your instructor because the sex of the animal will be important for parts of the internal dissection.

If your cat is female, check her mammary glands. Is there any evidence of recent lactation? ______________________

B. Preparation for Dissection

Follow these instructions every time you work with your cat.

1➜ Carefully remove the cat from its bag or bags and rinse it thoroughly under running water. This will remove any excess preservatives from the outside of the cat. Unless instructed otherwise, place the cat ventral side up in the dissecting pan. It is not normally necessary to anchor the cat in place.

2➜ If you are dissecting the muscles of the cat, you will want to skin the animal. To prevent drying out, however, your instructor may direct you to skin only one part of the cat at a time. If you are careful, the skin can be wrapped around the animal at the end of each laboratory session and used as an extra protective layer.

(If you are not dissecting the muscles, skinning the cat is not necessary. Your instructor will explain how much of this exercise you should complete.)

3➜ Set aside the dissecting instruments you will need for your dissection. The specifics of which instruments should be used at which stages of your cat dissection will be covered with each dissection exercise. For this exercise you should examine all your instruments.

4➜ Make note of any additional instructions here. ______________________
____
____

Dissection

III. Opening the Cat

A. Ventral Incision

Several methods exist for opening the cat. The **ventral incision** is suggested here because it can be used as a starting point for dissecting all systems, even if you are not dissecting the muscles. (Should your instructor suggest a dorsal incision, follow his or her directions.)

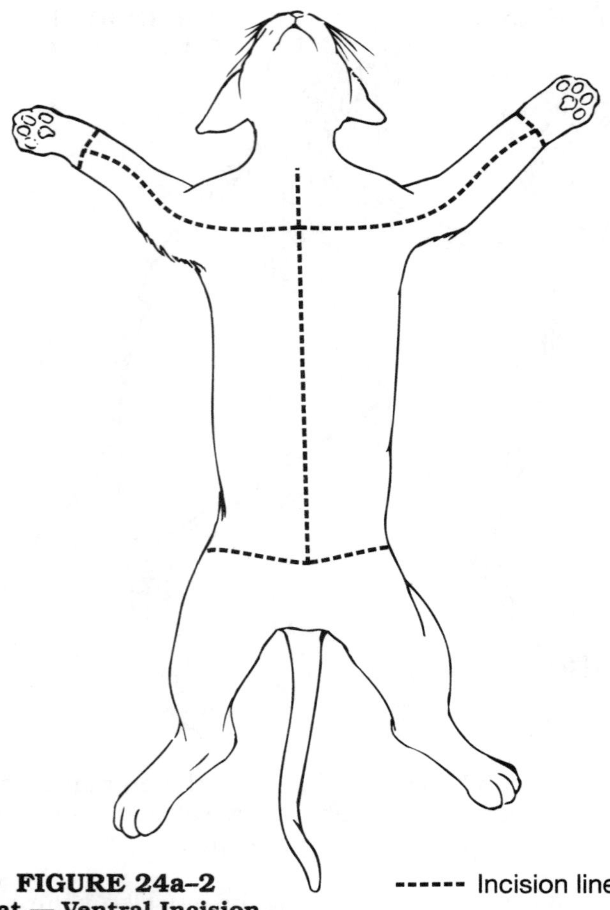

● **FIGURE 24a–2**
Cat — Ventral Incision.

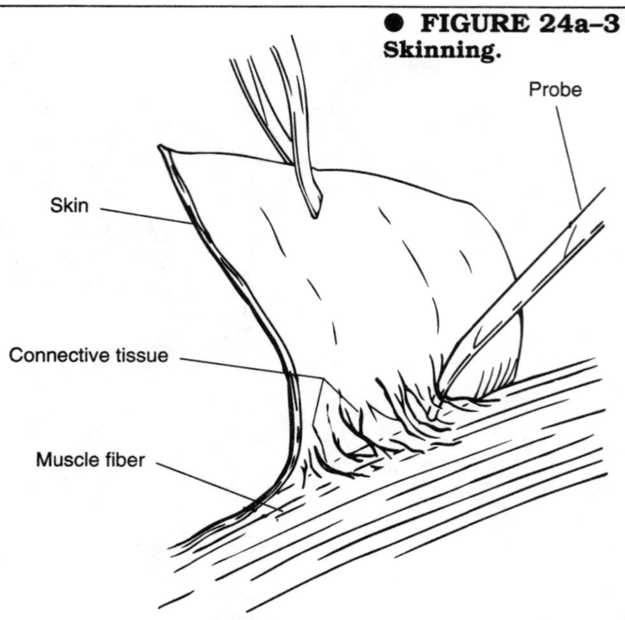

● **FIGURE 24a–3**
Skinning.

1➡ Begin by making an imaginary midsagittal incision on the ventral surface of the animal from the groin up to the neck. Now make an imaginary transverse incision across the shoulders and another across the groin (Figure 24a–2●).

2➡ Lift up the skin in the groin area and make a small (about 0.5 cm) snip with your scissors. Place the blunt end of your scissors into the snip and cut across your imaginary groin line. Keep your scissors pointing at as high an angle as is comfortably possible so that you cut only the skin. Now proceed to your sagittal line and follow this line as you cut from the groin to the neck. Cut the skin on your shoulder line.

3➡ Return to the abdominal area and lift the skin along your incision. Use your blunt nose probe to separate the skin from the subcutaneous connective tissue (Figure 24a–3●). You will notice several different tissue types. The elastic **superficial fascia** will be white and "fluffy." The **cutaneous nerves** will be stocky, cord-like structures extending from the skin to the muscle. Describe the appearance and location of the superficial fascia and the cutaneous nerves.

____

____

Depending on the age and physical condition of the cat, you will find varying degrees of yellowish adipose tissue. As time permits you can remove this fatty material with your hands or forceps.

4➡ Notice a thin sheet of muscle tissue adhering to the undersurface of the skin. This is the **cutaneous maximus** muscle that allows the cat to twitch or move its skin. In which specific areas did you find the cutaneous maximus muscle? ______________

____

The cutaneous maximus is continuous with the sheet-like **platysma** muscle of the neck. Were you able to separate the platysma? ________

5➡ Extend your shoulder and groin incisions down the forelimbs and hind limbs, respectively. Just above the carpals and tarsals, cut the skin completely around the limbs. Separate the skin from the muscle (Figure 24a–3●).

6➡ Extend the shoulder incision around the neck, and extend the groin incision around the lower back just cranial to the tail (Figure 24a–4●). It is not necessary to skin the head, tail, or genital area at this time.

7➡ If your specimen is a recently lactating female, follow your instructor's directions as to whether you should remove the mammary glands now or if you should wait until you dissect the reproductive system.

● **FIGURE 24a–4**
Cat — Dorsal Incision.

B. Other Systems

By the time you are ready to work on the respiratory, digestive, urinary, endocrine, and reproductive systems, you should feel quite comfortable with your dissecting skills and you should have no particular difficulty following the directions for any given lesson.

Specific directions for the dissection of the muscle system, the cardiovascular system, and the systems involving the other internal organs can be found in the exercises on those particular systems.

1➡ Be aware of nervous fiber. According to the wishes of your instructor, you may or may not be dissecting all or parts of the nervous system, but you should avoid destroying the nerves as you dissect other body systems. Most of the peripheral nerves in the cat look like white threads with varying degrees of thickness. (The plexes are like thick cords; the sciatic nerve is like a strong thread; the phrenic nerve is a delicate thread.)

2➡ Keep in mind the skeletal orientation mentioned at the beginning of this lesson. Although direct anatomical observation of the skeletal system is not included in this book, as you dissect the muscles you may have an opportunity to observe the intimate relationship between the muscle and the bone tissue. Refer back to Figure 24a–1● or the articulated feline skeleton in your laboratory as needed.

IV. Clean-Up

A. General Instructions

1➡ Keep your animal moist because as the tissue dries out it becomes more and more difficult to work with. We recommend you prepare the animal for storage as follows:

- ❑ Use moist paper towels to dampen the cat. These may be left with the animal in storage.
- ❑ Wrap the skin around the cat.
- ❑ Double bag if possible.
- ❑ Close the storage bags tightly.

We do not recommend that additional formalin be used to help preserve or help keep the cat moist.

2➡ Wash and dry your instruments and your work area. Follow your instructor's directions concerning which detergents or disinfectants should be used.

3➡ Dispose of animal tissue as directed by your instructor so as not to clog the plumbing or create undue stench in the laboratory.

4➡ Record any additional instructions here.

____

____

❑ Additional Activities

1. Do a comparative study on the ways in which different animals are used in different laboratory situations.
2. Study the visceral and parietal peritoneums. Demonstrate how these structures are related in the body of your specimen.

NOTES

❑ Lab Report

1. Referring to the numbered inquiries at the beginning of this exercise, complete the following box summary:

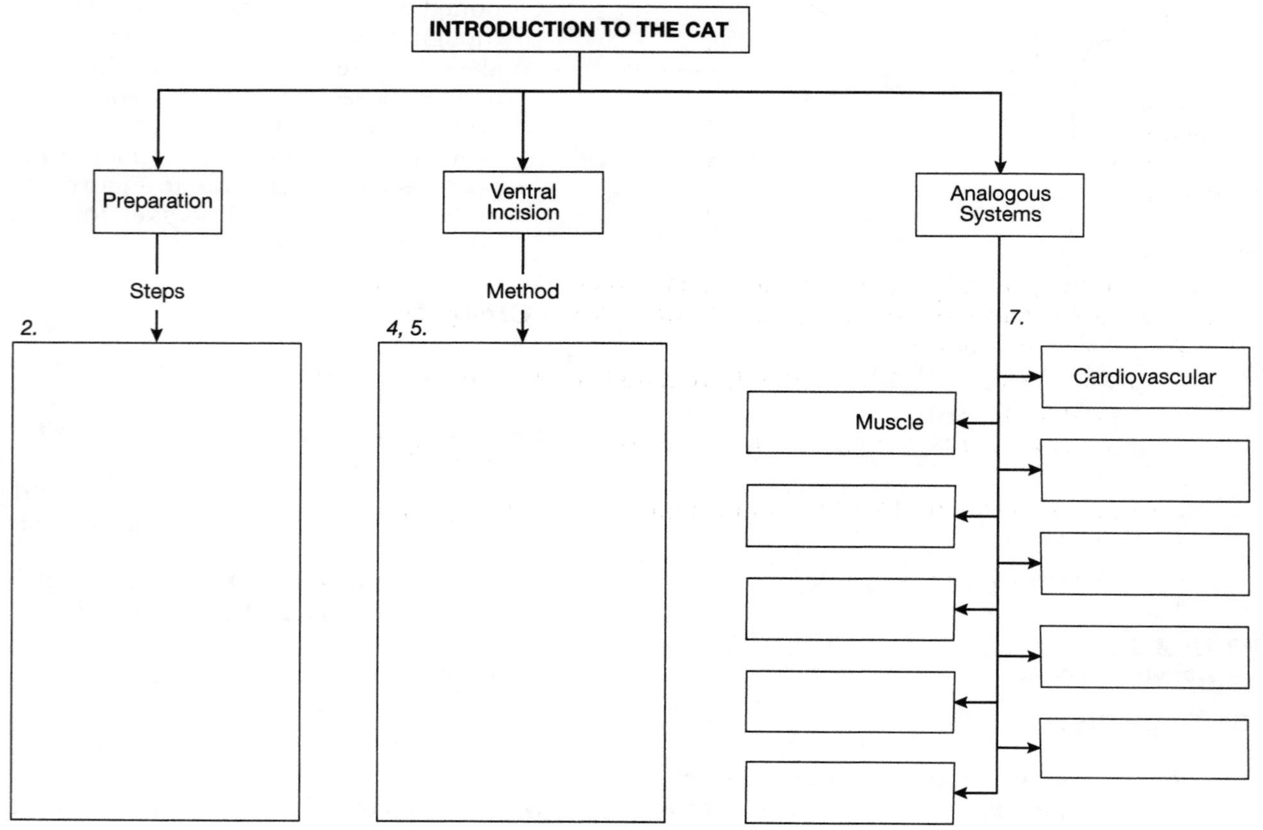

General Questions

NOTES

1. How does the skeletal system of the cat predispose it to its lifestyle?
2. List five safety regulations to be followed when dissecting.
3. Why is it important to know whether your cat is male or female?
4. Regarding skinning the cat:
   a. Which two instruments are suggested in this exercise?
   b. Describe the types and locations of the incisions you made.
   c. Describe the method you used for separating the skin from the body of the cat.

5. Fill in this chart:

	General Appearance	Location
Superficial fascia		
Cutaneous nerves		
Cutaneous maximus		
Platysma		

6. Speculate:
   a. The clavicle is generally absent in the cat. Based on your studies of the human skeleton and human musculature, does this surprise you? Why or why not?
   b. Why does the living cat often seem to have a body that is much smaller than its skin?
   c. What physiological advantage does the cutaneous maximus muscle have for the cat?
   d. Why do you suppose the deep fascia becomes tougher as the animal ages?
7. Explain the clean-up process.

NOTES

EXERCISE **24b**

Introduction to the Fetal Pig

PROCEDURAL INQUIRIES

Preparation

1. How do you determine the sex of a fetal pig?
2. How should you prepare a fetal pig for dissection?
3. What safety precautions should be followed in skinning a fetal pig?

Dissection

4. What is a ventral incision?
5. How do you skin a fetal pig?
6. What are some subcutaneous structures found in skinning a fetal pig?

Additional Inquiries

7. Why is the fetal pig a good animal to study in a human anatomy lab? Which body systems are analogous?
8. What cleanup steps should be taken following dissection?

Key Terms

Cutaneous Maximus
Cutaneous Nerves
Superficial Fascia
Ventral Incision

Materials Needed

Articulated Fetal Pig Skeleton (if available)
Gloves
Preserved Fetal Pig
Dissecting Pan
Dissecting Kit
Animal Storage Bags
Labels/Tags
Disinfectant

The focal point of the human anatomy laboratory is the three-dimensional, hands-on study of the human body. Because it is often not possible to study either an actual human body or a cadaver, substitutes must be found. Plastic and molded models of human torsos and human body parts are valuable additions to the anatomical learning experience. Nevertheless, these artificial bodies do not afford the opportunity of feeling and manipulating real tissue.

All mammalian bodies are quite similar structurally and functionally. This means that we can study the nonhuman mammalian body and gain an anatomical and physiological understanding of the actual human body.

Each year thousands of fetal pigs across the country are acquired by biological supply houses and prepared for classroom (or laboratory) study. Be assured that the fetal pig you are working with was not clandestinely snatched from someone's pet sow, nor was any pain inflicted on the animal.

The fetal pig is an ideal animal for laboratory dissection. Because the animal is still developing, the body of the fetal pig is usually free of excess fat (which can inhibit accurate identification of body parts), and the muscles of the fetal pig are generally well developed, readily separated, and easily identified. Porcine (pig) musculature generally corresponds to human musculature.

The blood vessels of the fetal pig are also well defined and easy to work with. You should have no particular difficulty in identifying the major arteries and veins and in locating the principal branches of these vessels. The names and locations of the major porcine vessels are almost identical to those of human blood vessels.

Internal fetal pig anatomy — meaning the respiratory, digestive, urinary, nervous, endocrine, skeletal, and reproductive systems — is essentially analogous to internal human anatomy. Although you will notice some slight differences (e.g., the fetal pig does not have an appendix), you will find that a good understanding of the internal organs of the fetal pig will give you a solid foundation in grasping the three-dimensional nature of human anatomy.

This introductory exercise is designed to orient you to the fetal pig, to help you with the preparation of the fetal pig for future dissection exercises, and to introduce you to some of the basic methodologies of fetal pig dissection. In your particular laboratory setting, this exercise may be combined with your first dissection exercise.

❑ Preparation

I. Safety

A. Preservatives

1➡ Wear protective gloves while dissecting any animal, preserved or fresh. Some animal preservatives are mildly to severely irritating to the skin; other preservatives can be carcinogenic.

B. Parasites

1➡ Although live or freshly killed animals may harbor toxins or disease-carrying organisms, fetal pigs rarely harbor parasites because in their unborn condition, their environment was virtually sterile and the pigs were preserved at the approximate time they were taken from the sow's uterus. It is remotely possible that some parasite larvae or eggs could be present from some external source. Therefore, use care to keep your hands away from your face while dissecting. Although the preserving process will kill the mature parasite, the parasite larvae or eggs often have a highly resistant protective coat. The round worm, *Ascaris lumbricoides*, for example, sheds eggs which have been known to survive for more than two years in the intestines of animals preserved in formalin.

C. Regulations

1➡ Follow any regulations set down by your state or local government. Certain jurisdictions have enacted safety regulations for the handling of live or preserved animals. Those regulations take precedence over any safety tips written in this book.

2➡ Follow any regulations set down by your school. Some schools have set up a uniform code of safety for the handling of biological specimens. It is imperative that you follow all directions your instructor may give regarding the handling of the animals.

D. Allergies

1➡ Some students worry about allergies to fetal pigs. If you are allergic to pigs in general, you are probably allergic to the pig dander, rather than to the pig itself. The preserved animal has no dander and thus should not aggravate your allergies.

2➡ Should the preservative accidentally touch your skin, wash the area thoroughly.

3➡ Food allergies to pork or pork products should not be manifested in fetal pig dissection.

E. Additional Precautions

1➡ Make note here of any additional safety regulations your lab may have regarding dissection and the safe handling of preserved specimens.

____

____

____

II. Preliminary Instructions

A. Identification

1➡ Obtain and tag your fetal pig according to your instructor's directions.

2➡ Measure your fetal pig and try to determine its approximate stage of gestation. Porcine gestation is generally 114 days. Fetal size is somewhat dependent on breed, but as a general rule, an 8 cm fetus is at about 70 days' gestation. This is a standard size for fetuses found in commercially prepared pregnant pig uteri. Fetal pigs used for dissection are usually about 25 cm. At farrowing (birth) the piglet is usually about 30 to 40 cm. Usually the closer to gestation your pig is, the easier that pig will be to work with. Approximately how close to gestation was your pig? ______________

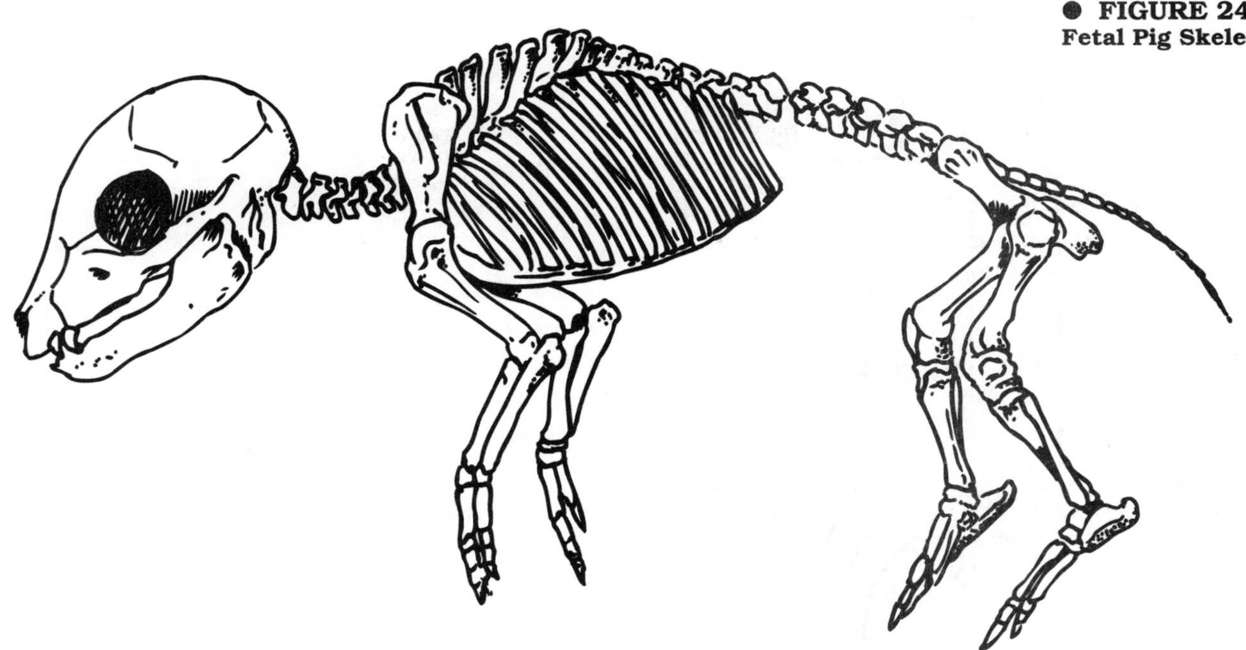

● **FIGURE 24b–1**
Fetal Pig Skeleton.

3➡ Orient yourself to the fetal pig by examining the articulated porcine skeleton. If no skeleton is available, study the accompanying diagram (Figure 24b–1●). Note particularly the feet, knees, ankles, and elbows of the animal. Note too the angles formed as the bones articulate with one another. How do these angles of articulation correspond with the bones of the human body?

____

____

As you work through the rest of this lesson, be aware of the orientation of the different parts of the porcine body.

4➡ Determine the sex of your fetal pig. Is your animal male or female? __________ Examine the urogenital area which, in the female fetal pig, is located directly beneath the tail, and in the male fetal pig is located toward the posterior portion of the ventral surface. If you are uncertain about the sex of your fetal pig, check with your instructor; the sex of the animal will be important for parts of the internal dissection.

B. Preparation for Dissection

Follow these instructions every time you work with your fetal pig.

1➡ Carefully remove the fetal pig from its bag or bags and rinse it thoroughly under running water. This will remove any excess preservatives from the outside of the fetal pig. Unless instructed otherwise, place the fetal pig ventral side up in the dissecting pan. It is usually necessary to anchor the fetal pig in place. Your instructor will demonstrate how this should be done.

2➡ If you are dissecting the muscles of the fetal pig, you will want to skin the animal. To prevent drying out, however, your instructor may direct you to skin only one part of the fetal pig at a time. If you are careful, the skin can be wrapped around the animal at the end of each laboratory session and used as an extra layer to protect against dessication.

(If you are not dissecting the muscles, skinning the fetal pig is not necessary. Your instructor will explain how much of this exercise you should complete.)

3➡ Set aside the dissecting instruments needed for your dissection. The specifics of which instruments should be used at which stages of your fetal pig dissection will be covered with each dissection exercise. For this exercise you should make certain you are familiar with all the instruments you may have available.

4➡ Make note of any additional instructions here. ______________________________

____

____

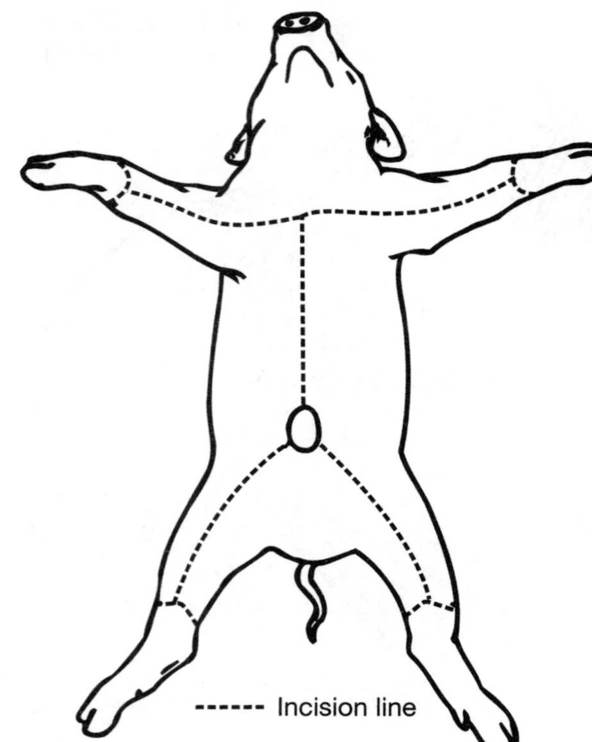

● FIGURE 24b–2
Fetal Pig — Ventral Incision.

Dissection

III. Opening the Fetal Pig

A. Ventral Incision

Several methods exist for opening the fetal pig. The **ventral incision** is suggested here because it can be used as a starting point for dissecting all systems, even if you are not dissecting the muscles. (Should your instructor suggest a dorsal incision, follow those directions.)

1➡ Begin by making an imaginary mid-sagittal incision on the ventral surface of the animal from the umbilicus up to the neck. Now make an imaginary transverse incision across the neck and from the midline down the forelimbs (Figure 24b–2●). Draw another imaginary line from the midline around the umbilicus (and external genitalia, if your pig is male) and down each leg. These imaginary lines represent your incision plan.

2➡ Lift up the skin in the neck area and make a small (about 0.5 cm) snip with your scissors. Place the blunt end of your scissors into the snip and cut across your imaginary lines. Keep your scissors pointing at as high an angle as is comfortably possible so that you cut only the skin. Just above the carpals and tarsals, cut the skin completely around the limbs. (If you have dissected postbirth animals, you will notice that the fetal skin is softer and more pliable than that of the juvenile or adult animal.) If you wish, experiment briefly with your scalpel to see why the scissors are preferred for gross skinning.

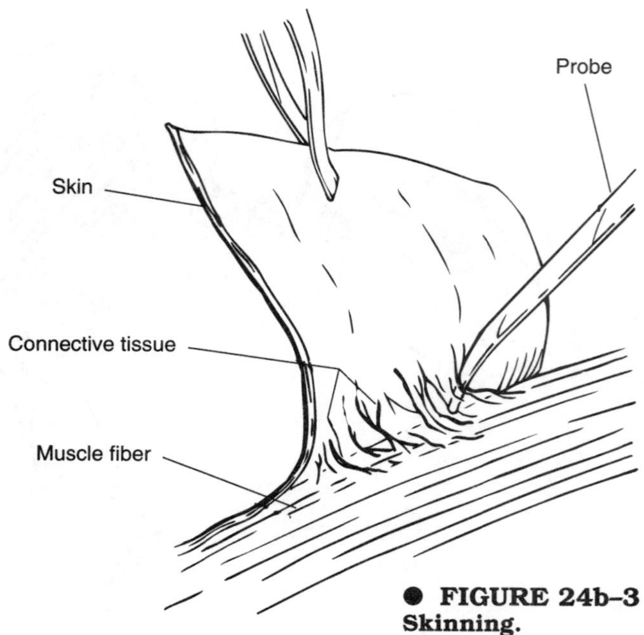

● FIGURE 24b–3
Skinning.

3➡ Return to the abdominal area and lift the skin along your incision. Use your blunt nose probe to separate the skin from the subcutaneous connective tissue (Figure 24b–4●). If you wish, experiment briefly with your sharp nose probe to see why the blunt nose probe is preferred for separating the skin.

4➡ Extend the shoulder incision around the neck and extend the rear leg incision around the lower back just cranial to the tail (Figure 24b–3●). Make another incision from the umbilical incision posteriorly around to the tail incision. It is not necessary to skin the head, tail, or genital area at this time.

You will notice several different tissue types. The elastic **superficial fascia** will be white and "fluffy." The **cutaneous nerves** will be stocky, cordlike structures extending from the skin to the muscle. Describe the appearance and location of the superficial fascia and the cutaneous nerves.

____

____

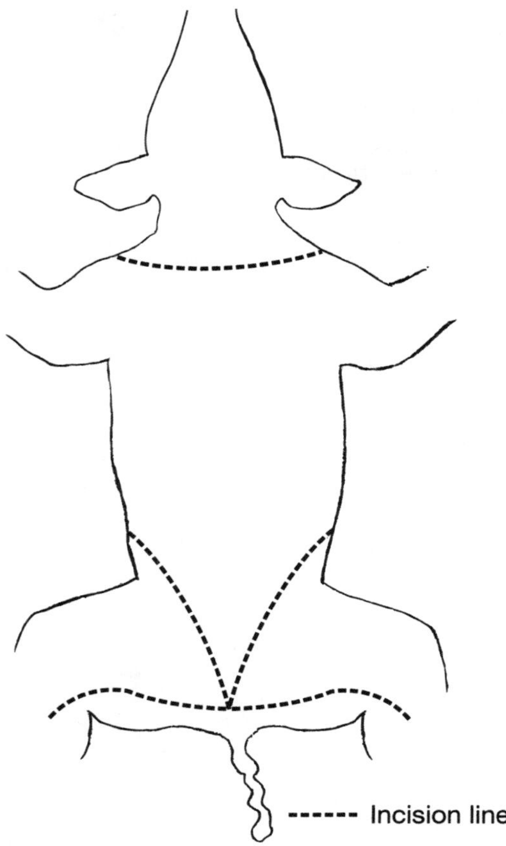

● **FIGURE 24b–4**
Fetal Pig — Dorsal Incision.

Depending on the age and physical condition of the fetal pig, you will find small amounts of yellowish adipose tissue. (Fetal animals have little fat.) As time permits you can remove this fatty material with your hands or forceps.

5➡ Notice a thin sheet of muscle tissue adhering to the under-surface of the skin. These are the cutaneous muscles which allow the pig to twitch or move its skin. The porcine cutaneous muscles are a series of muscles with specific names dependent on physical location. In which specific areas did you find the **cutaneous maximus muscle**? ____________________

B. Other Systems

1➡ Specific directions for the dissection of the muscle system, the cardiovascular system, and the systems involving the other internal organs can be found in the exercises on those particular systems.

2➡ By the time you are ready to work on the respiratory, digestive, urinary, endocrine and reproductive systems, you should feel quite comfortable with your dissecting skills and you should have no particular difficulty following the directions for any given lesson.

3➡ According to the wishes of your instructor, you may or may not be dissecting all or parts of the nervous system. Nevertheless, you should be aware of nervous fiber. Most of the peripheral nerves in the fetal pig look like white threads with varying degrees of thickness. (The plexes are like thick cords; the sciatic nerve is like a strong thread; the phrenic nerve is a delicate thread.)

4➡ Direct anatomical observation of the skeletal system is not included in this book. You should, however, be aware of the skeletal orientation mentioned at the beginning of this lesson. As you dissect the muscles, you may have an opportunity to observe the intimate relationship between the muscle and the bone tissue. Refer to Figure 24b–1● or the articulated porcine skeleton in your laboratory as needed.

IV. Clean-Up

A. General Instructions

1➡ Keep your animal moist. As the tissue dries out it becomes more and more difficult to work with. We recommend that you prepare the animal for storage as follows:

- ❑ Use moist paper towels to dampen the fetal pig. These may be left with the animal in storage.
- ❑ Wrap the skin around the fetal pig.
- ❑ Double bag if possible.
- ❑ Close the storage bags tightly.

We do not recommend that additional formalin be used to help preserve or help keep the fetal pig moist.

2➡ Wash and dry your instruments and your work area. Follow your instructor's directions concerning which detergents or disinfectants should be used.

3➡ Dispose of animal tissue as directed by your instructor so as not to clog the plumbing or create undue stench in the laboratory.

4➡ Record any additional instructions here.
____
____

❑ Additional Activities

NOTES

1. Do a comparative study on the ways in which different animals are used in different laboratory situations.
2. Obtain a cookbook. Identify different cuts of pork according to the location on your fetal pig's body.

❑ Lab Report

1. Referring to the numbered inquiries at the beginning of this exercise, complete the following box summary:

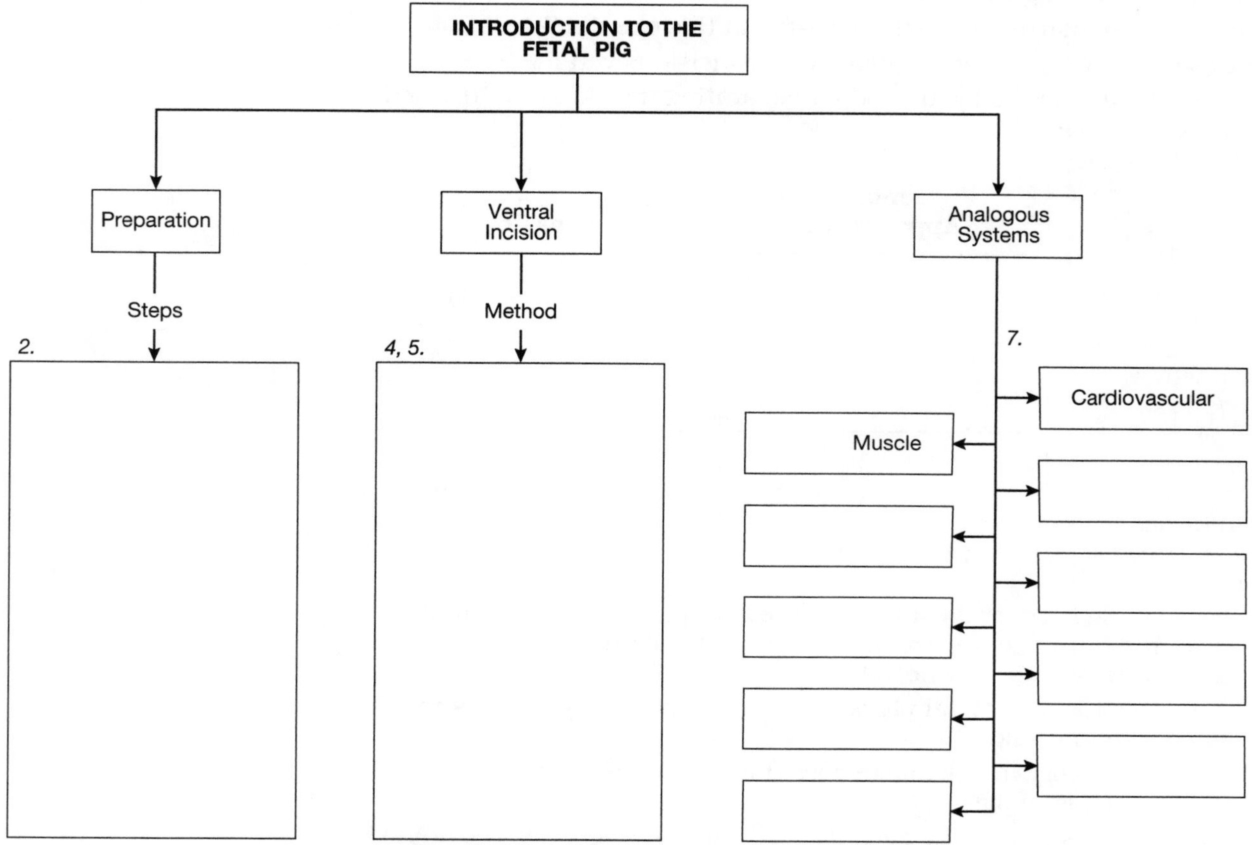

General Questions

NOTES

1. How does the skeletal system of the fetal pig predispose it to its lifestyle?
2. List five safety regulations to be followed when dissecting.
3. Why is it important to know whether your fetal pig is male or female?
4. Regarding skinning the fetal pig:
   a. Which two instruments are suggested in this exercise?
   b. Describe the types and locations of the incisions you made.
   c. Describe the method you used for separating the skin from the body of the fetal pig.
5. Fill in this chart:

	General Appearance	Location
Superficial fascia		
Cutaneous nerves		
Cutaneous maximus muscles		

6. Speculate:
   a. The clavicle is generally absent in the fetal pig. Based on your studies of the human skeleton and human musculature, does this surprise you? Why or why not?
   b. Why does the living fetal pig often seem to have a body that is much smaller than its skin?
   c. What physiological advantage does the cutaneous maximus muscle have for the fetal pig?
   d. Why do you suppose the deep fascia becomes tougher as the animal ages?
7. Explain the clean-up process.

EXERCISE **25a**

Dissection of the Muscles: Cat

PROCEDURAL INQUIRIES

Preparation

1. What terminology should one be familiar with before beginning muscle dissection?
2. Which dissecting instruments should be used in working with cat muscles?
3. What is fascia and how should it be worked with?
4. What guidelines should be used for identification of individual muscles?

Dissection

5. What are the major deep and superficial muscles of the ventral thorax and abdomen?
6. What are the major deep and superficial muscles of the head and neck?
7. What are the major deep and superficial muscles of the dorsal area?
8. What are the major deep and superficial muscles of the forelimb?
9. What are the major deep and superficial muscles of the hind limb?

Additional Inquiries

10. How is the cat musculature analogous to the human musculature?
11. When and how should a muscle be transected?

HUMAN CORRELATION: AM Cadaver Atlas in Applications Manual.

Key Terms

Deep Fascia
Epimysium
Grain
Insertion
Linea Alba
Muscles of the
- Dorsal Area
- Forelimb
- Head and Neck
- Hind Limb
- Ventral Thorax and Abdomen

Origin
Superficial Fascia
Transecting

Materials Needed

Preserved Cat
Dissecting Kit
Dissecting Pan
Gloves
Disinfectant
Optional: Human Torsos or Limb Models for Comparative Purposes
Other Equipment as Indicated by Your Instructor

Muscle systems are basically analogous throughout the mammalian world. This means the muscles of the cat are structurally and functionally similar to our own. Therefore, learning the muscles of one mammal by handling, separating, and identifying them means you have almost learned the muscles for all mammals (including the human).

Muscle variations do exist. Muscle fusion is the variation you should be most aware of. Many of the non-human animals have more individual muscles than we do. For instance, the cat has eight superficial shoulder muscles, and we have only three. The proportionate muscle mass is approximately the same, but developmentally our muscles fuse whereas the cat's do not. Nevertheless, we can readily correlate most feline muscles with human muscles. In working through the muscle dissection, you should concentrate on this correlation. Look for the similarities.

This laboratory exercise is designed to help you develop a conceptual understanding of the human muscle system by dissecting the muscles of the cat. Your goal is to separate and identify the individual cat muscles. You are encouraged to work in pairs, to say the muscle names out loud, to study the diagrams, to quiz your partner extensively on the names and locations of the muscles, and to correlate the cat musculature with your previous knowledge of the human musculature.

In this exercise we are using the regional approach and will thus examine the muscles of the ventral abdomen and thorax, the head and neck, the dorsal area, the forelimbs, and the hind limbs. With each section we will look at both the superficial and the deep muscles. We will dissect all major and readily identifiable muscles, except for those of the head and upper face. In all cases, identifying data such as the **origin** and **insertion** are given to help you locate the muscle.

❑ Preparation

I. Getting Ready

A. Terminology

1➡ Be certain you have a good grasp of skeletal terminology before you begin dissecting the muscles of the cat. (Refer back to Unit III if necessary.) You should also be familiar with the material in Exercise 21. You are encouraged to correlate the feline and human musculatures. Most of the information in Exercise 21 is applicable to the cat as well as to the human. Use the classroom models often as you make your comparisons.

B. Tools

Beginning students are often afraid to attack the deep fascia. Be assured that the deep fascia is tough, and the older your animal, the tougher the fascia. To separate many of the muscles, it will be necessary for you to break through the deep fascia.

1➡ Use your blunt nose probe as your primary muscle-dissecting instrument. You can rip the deep fascia with this probe without worrying about tearing the muscle fibers.

As you separate muscles, always work with the direction or **grain** of the fibers. You can usually distinguish separate muscles by a change in the grain. Individual muscles also separate naturally from one another in all directions. You will be able to run your probe beneath or along a given muscle without any difficulty (after you have penetrated the deep fascia).

2➡ Do not use your scissors or scalpel for muscle dissection unless you are deliberately **transecting** (cutting) a muscle. You may occasionally use your sharp nose probe, but you must be careful that you do not tear a delicate blood vessel or nerve fiber. Your forceps can be used to grip fascia or muscle tissue or to pull out adipose tissue.

3➡ Work with a partner unless you are directed otherwise. Both of you should use your blunt nose probes to feel the individual muscles. Both of you should say the names of the muscles out loud and study the diagrams carefully. Quiz each other often.

4➡ Transect (cut) certain muscles according to directions. This is necessary in order to study the deeper muscles. Before you transect anything, be certain you are transecting in the right place. Reconstructive surgery on preserved cats is not highly successful.

C. Incision

1➡ Refer back to Exercise 24 or follow your instructor's directions if you have not already opened your cat.

2➡ Your instructor will advise you about how much of the skin should be removed at any given time. If you will be working on your muscle dissection for several weeks, we recommend skinning on an "as needed" basis. This prevents excess drying of your specimen. If your muscle dissection will be completed in a relatively short time, skinning the entire cat at once may be appropriate. We also recommend that you wrap the removed skin around your cat before placing the cat back in its storage bag. (Directions for skinning are given in Exercise 24.)

3➡ Recall from Exercise 20 that a bundle of skeletal muscle fiber is surrounded by a layer of coarse connective tissue called the **epimysium**. Covering the entire muscle (exterior to the epimysium) is a tough fibrous membrane, the **deep fascia**, the fibers of which intermesh with the fibers of the **superficial fascia**. (The superficial fascia helps anchor the skin to the body.) The deep fascia is also continuous with the tendons, ligaments, and periosteum.

Dissection

II. Muscles of the Ventral Thorax and Abdomen

Open the cat and lay back (or remove) the skin on the ventral thorax and abdomen. Probe your way through the fascia. Remove any adipose tissue that may be in your way.

A. Separating the Superficial Muscles

1 ➡ Observe the ventral surface of the animal. Follow Figures 25a–1● to 25a–3● and use the right side of your animal to locate the superficial muscles. Separate the muscles with your probe.

Pectoantebrachialis A strip of muscle running from the manubrium to the superficial fascia of the forearm. Humans do not have a pectoantebrachialis.

Pectoralis major From the sternum to the humerus. Note how the pectoralis major forms a sheet deep to the pectoantebrachialis. Verify that despite the pectoantebrachialis, the pectoralis major is only one muscle. The pectoantebrachialis seems to form a superficial strip over the pectoralis major. Note the grain of the two muscles.

Pectoralis minor From the sternum to the humerus. Note how the pectoralis minor seems to glide up under the pectoralis major. The pectoralis minor is also a heftier muscle than the pectoralis major. In humans the pectoralis major is the larger muscle.

Xiphihumeralis From the xiphoid process to the humerus. This muscle, which is not found in humans, fuses with the pectoralis minor.

External oblique Large sheet of muscle with diagonal fibers covering most of the abdomen.

The white abdominal midline is the **linea alba**.

B. Separating the Deep Muscles

1 ➡ Use the left side of your animal. Immediately to the left of the linea alba, carefully transect and reflect the 5 muscles just mentioned. (To reflect means to fold back.) Identify these deep muscles. Follow Figures 25a–2● and 25a–3●.

Rectus abdominis Longitudinal muscle extending from the pubis to the sternum and costal cartilages.

Internal oblique Immediately deep to the external oblique. The fibers of the external and internal obliques form a 90° angle.

Transversus abdominis Transverse fibers immediately deep to the internal oblique. (The external oblique, internal oblique, and transversus abdominis muscles are paper-thin layers and may be difficult to separate.)

Serratus ventralis Large fan-like muscle arising from the ribs and inserting on the vertebral border of the scapula.

Transversus costarum Arising from the sternum and inserting on the first rib. (Not identified on the dissection figures.)

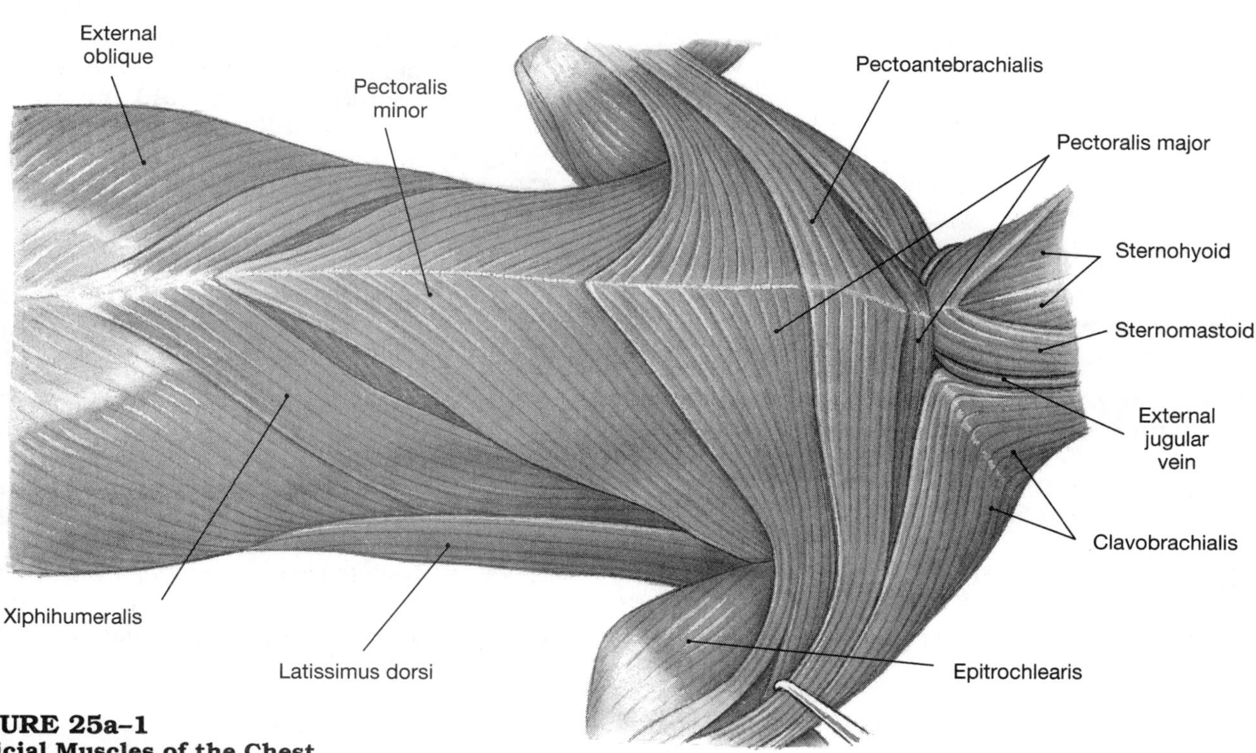

● **FIGURE 25a–1**
Superficial Muscles of the Chest.

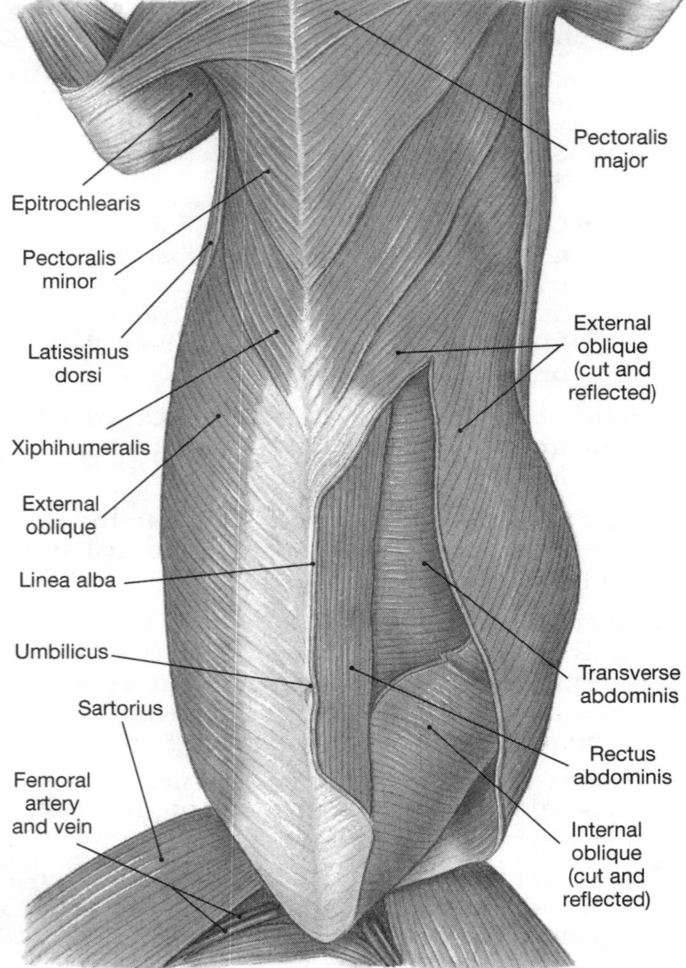

● **FIGURE 25a–2**
Superficial Muscles of Abdomen — Ventral View.

External intercostals Series of muscles between the ribs, immediately deep to the transversus costarum. The external intercostals are superficial to the internal intercostals, another series of muscles also found between the ribs.

Scalenus Two-part muscle lateral to the transversus costarum.

Levator scapulae Lateral muscle used to raise the scapula (not identified on the dissection figures).

Subscapularis On the "back wall" as you look on the inside of the deep muscles. Actually this muscle extends from the subscapular fossa to the lesser tubercle of the humerus. (Not identified on the dissection figures.)

III. Muscles of the Head and Neck

A. Separating the Superficial Muscles

1 ➡ Study Figures 25a–4● and 25a–5●. We will describe a few of these muscles. Because of the injection sites on many cats, it may not be possible for you to find all of the muscles in the diagram. Before beginning this dissection, palpate the hyoid bone and the thyroid cartilage. These landmarks will orient you to the muscles described here. These superficial muscles include:

Clavotrapezius The flat, shoulder-covering muscle originating on the crest of the skull and inserting in the region of the clavicle. (In most cats the clavicle is either absent or present only as a vestigial bone. This muscle is mentioned again as a muscle of the dorsal region.)

Sternomastoid Diagonal muscle medial to the clavotrapezius.

Sternohyoid Longitudinal medial muscle just deep to the sternomastoid.

Stylohyoid Thin transverse muscle strip following the hyoid bone.

Mylohyoid Flat transverse muscle superior to the stylohyoid.

Digastric Diagonal muscle superficial to the mylohyoid.

Masseter Large cheek muscle.

B. Separating the Deep Muscles

1 ➡ Return to Figure 25a–5●. Find as many of the following deep muscles as you can. Not all of these muscles are shown on the diagrams: **geniohyoid** (deep to the mylohyoid), **genioglossus** (deep to the geniohyoid), **styloglossus** (deep to the lateral mylohyoid), **hyoglossus** (lateral to the geniohyoid), **cleidomastoid, thyrohyoid (deep to the sternohyoid), cricothyroid (medial to the sternothyroid), sternothyroid.**

IV. Muscles of the Dorsal Region

A. Separating the Superficial Muscles

1 ➡ Use Figure 25a–6● to help you locate these superficial muscles of the dorsal region. As you work through these muscles, notice that the cat has three separate trapezius muscles, whereas we have only one. (Although three separate trapezius muscles can be identified, most veterinarians today consider these three trapezius muscles as a single trapezius muscle analogous to our own.)

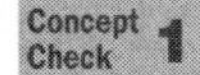

The cat also has three separate deltoid muscles, as compared to our one. Note these three deltoids below. What types of movements does the cat have that make this arrangement advantageous to the animal?

____

____

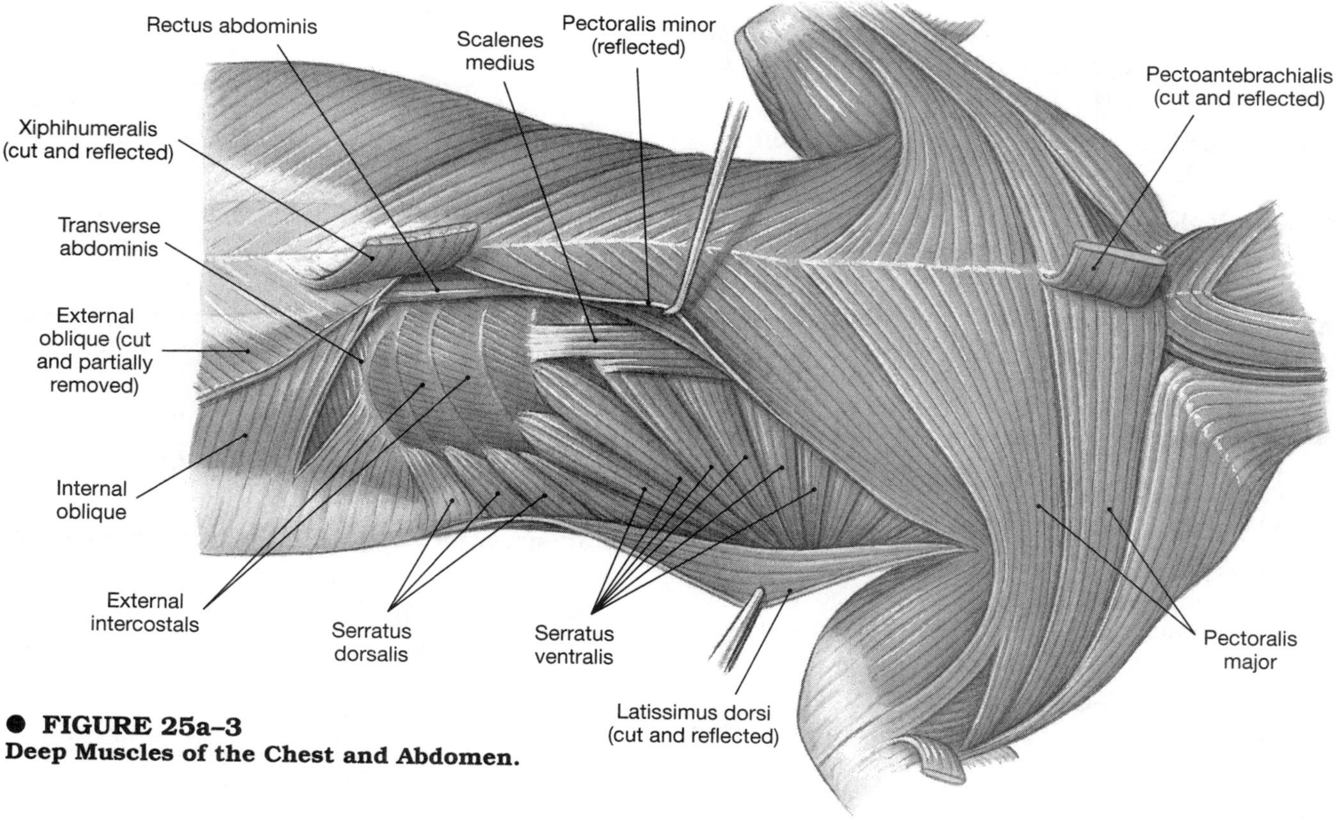

● **FIGURE 25a–3**
Deep Muscles of the Chest and Abdomen.

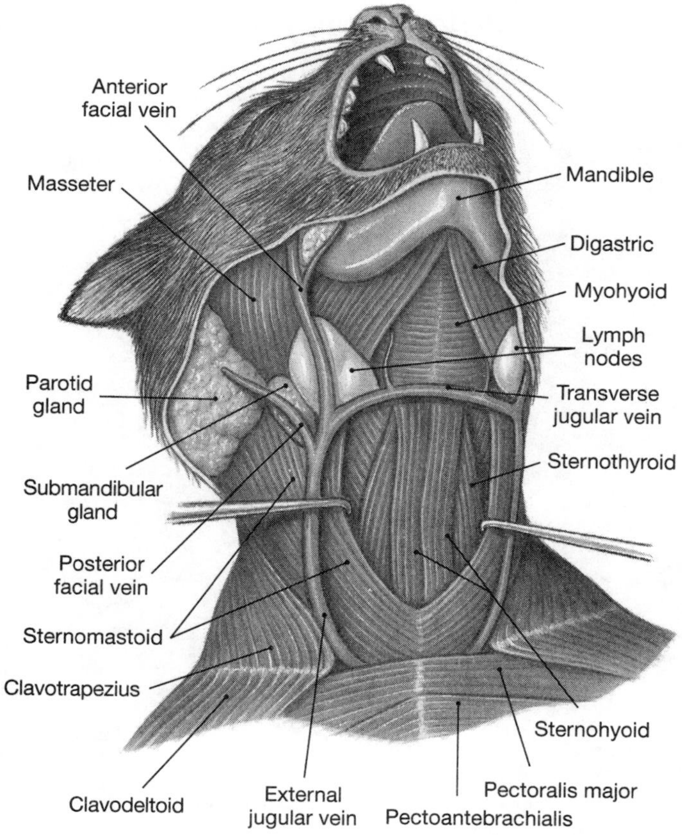

● **FIGURE 25a–4**
Superficial Muscles of the Neck — Ventral View.

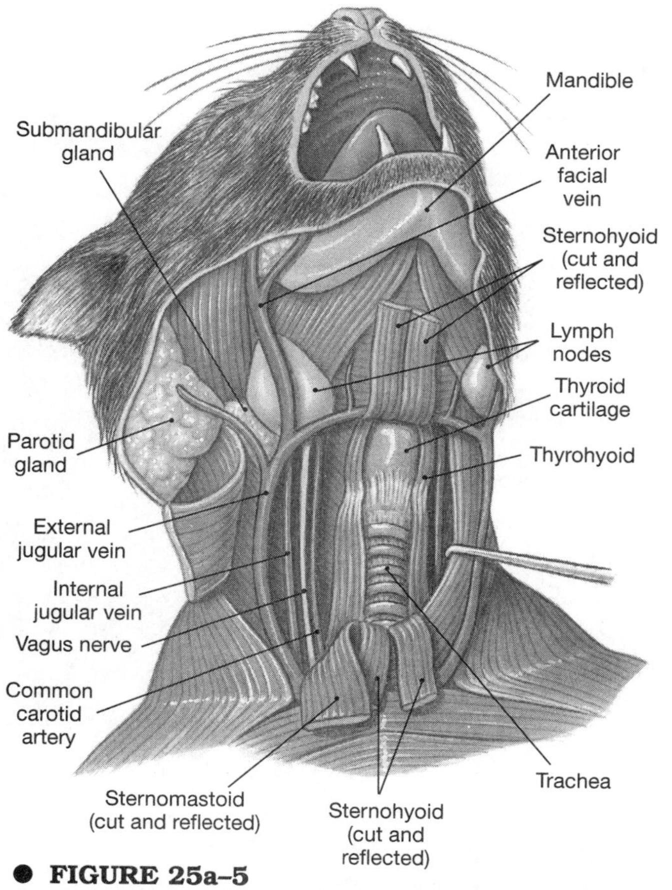

● **FIGURE 25a–5**
Deep Muscles and Vessels of the Neck.

2➡ Identify the superficial muscles on the right side of your cat and the deep muscles on the left side.

Clavotrapezius Extending from the midline of the back to the clavicle region.

Clavodeltoid Extending from the clavicle to the ulna. This muscle almost seems to be an extension of the clavotrapezius.

Acromiodeltoid Just inferior to the clavodeltoid, extending from the scapula to the spinodeltoid muscle.

Acromiotrapezius Flat muscle just caudal to the clavotrapezius, originating from the cervical and thoracic vertebrae and inserting on the scapular spine.

Spinotrapezius Flat muscle just caudal to the acromiotrapezius, originating at the spines of the thoracic vertebrae and inserting on the fascia of the scapular muscles.

Spinodeltoid Muscle seeming to be a continuation of the spinotrapezius, extending up under the acromiodeltoid, from the scapula to the humerus.

Latissimus dorsi Large flat muscle with diagonal fibers extending from the lumbardorsal fascia, around the lateral part of the body, and onto the humerus.

B. Separating the Deep Muscles

1➡ Transect and reflect the left-side muscles. You should have no difficulty identifying the following muscles.

Rhomboideus Muscle extending from midback to the medial border of the scapula. Humans and cats both have two rhomboideus muscles, the major and the minor.

Rhomboideus capitis Longitudinal muscle lateral to the rhomboideus and inserting on the angle of the scapula.

Splenius Flat muscle lateral to the rhomboideus.

Levator scapulae Extending from mid-scapula to the occipital bone.

Supraspinatus Superior of two muscles covering the scapula.

Infraspinatus Inferior of two muscles covering the scapula.

Teres major Running from the medial border of the scapula to the humerus.

Teres minor Deeper muscle extending from the lateral border of the scapula to the greater tubercle of the humerus. (Not Shown).

Serratus dorsalis Two sets of muscles, **superior** and **inferior**, fanning out from the ribs.

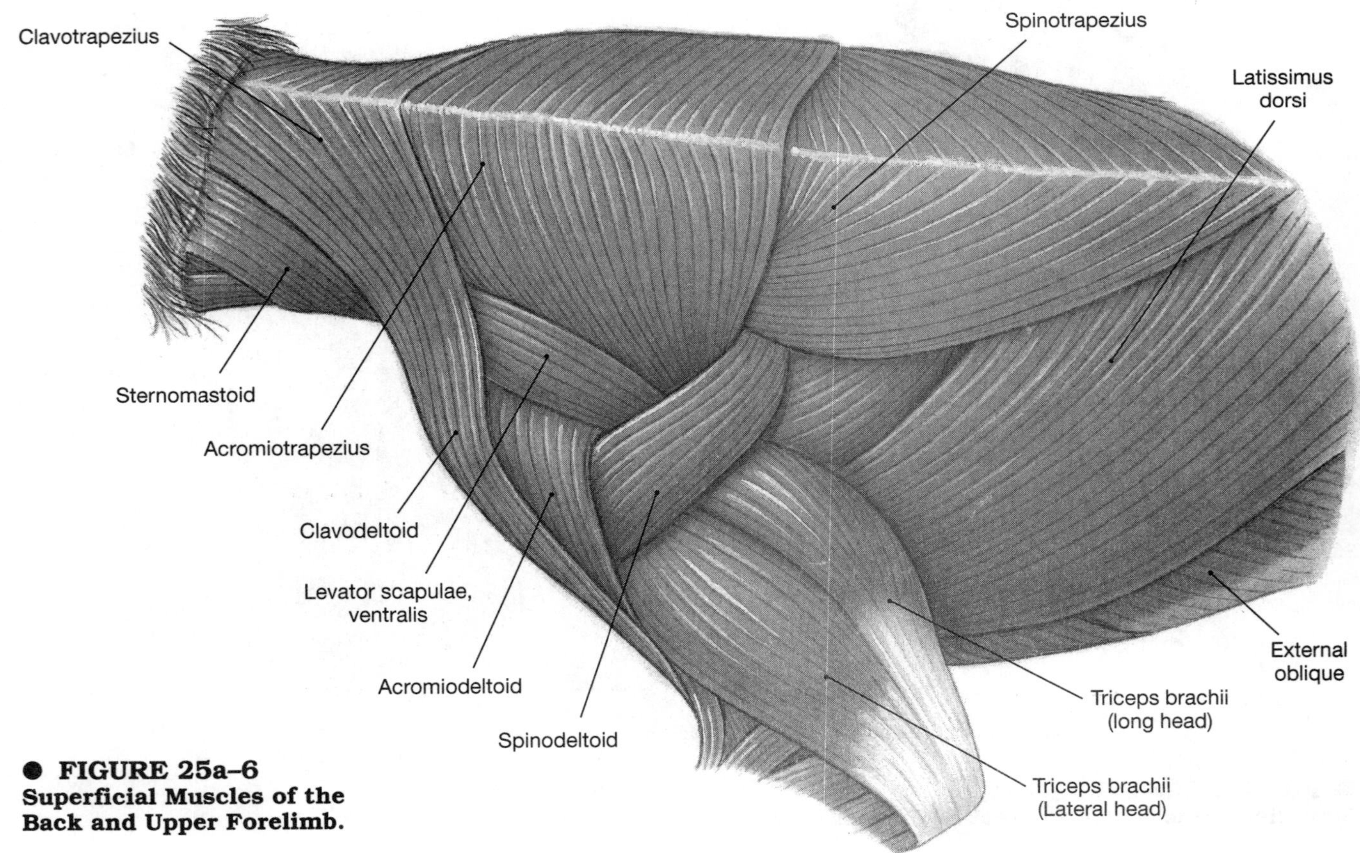

● **FIGURE 25a–6**
Superficial Muscles of the Back and Upper Forelimb.

Sacrospinalis Collective term for several longitudinal muscle masses, **spinalis dorsi**, **longissimus dorsi** and **iliocostalis** (not identified).

V. Muscles of the Forelimb

A. Separating the Major Muscles

1 ➡ Choose either the right or the left forelimb. Use Figures 25a–6● through 25a–9●. Begin with the medial side of the brachium.

Epitrochlearis Sheet-like muscle on the medial side of the brachium (not found in humans).

Biceps brachii Crosses the brachium from the scapula to the radial tuberosity.

Coracobrachialis Small muscle connecting the coracoid process with the proximal humerus.

Triceps brachii Crosses the brachium from the scapula and humerus to the olecranon process of the ulna.

Anconeus Small muscle between the distal humerus and the olecranon process of the ulna (not identified on the dissection pictures).

Brachialis From the lateral surface of the humerus to the proximal end of the ulna.

B. Identifying Additional Muscles

1 ➡ Many of the muscles of the forelimb are often difficult to identify in classroom cat specimens. Based on your instructor's directions, use Figures 25a–9● and 25a–10● to locate the following: **brachioradialis, extensor carpi radialis longus, extensor carpi radialis brevis, extensor digitorum communis, extensor digitorum lateralis, extensor carpi ulnaris, pronator teres, flexor carpi radialis, palmaris longus, flexor carpi ulnaris**.

VI. Muscles of the Hind Limb

A. Separating the Superficial Muscles

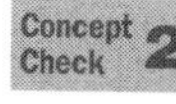

Observe that the cat's hind limb is proportionately far less "round" than the human leg. What functional advantage do you see for the cat having this svelte leg?

____

____

Choose either the right or the left hind limb. Use Figures 25a–11● through 25a–16● to help you isolate and identify these thigh muscles.

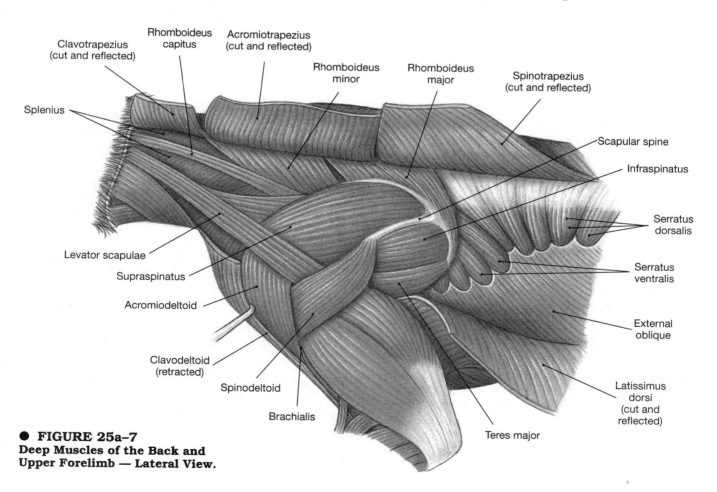

● **FIGURE 25a–7**
Deep Muscles of the Back and Upper Forelimb — Lateral View.

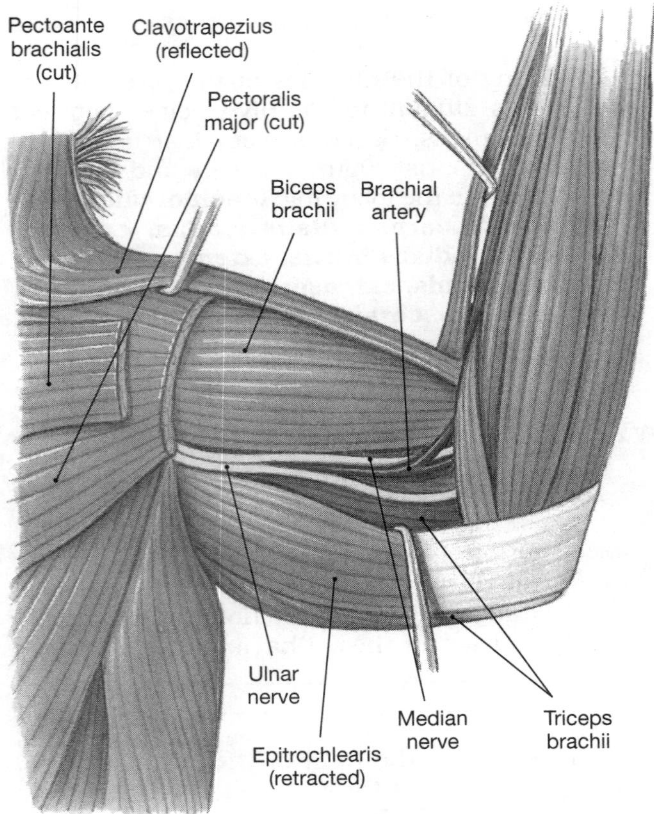

● **FIGURE 25a–8**
Muscles of Upper Forelimb — Medial View.

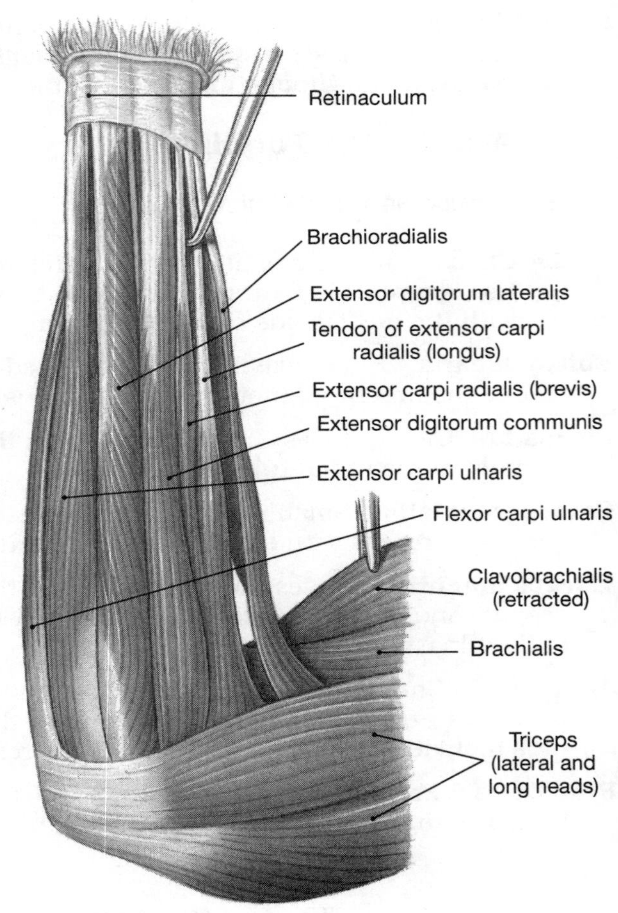

● **FIGURE 25a–9**
Muscles of Left Forelimb — Lateral View.

Sartorius Large plate-like muscle extending from the ilium to the tibia.

Tensor fasciae latae Lateral to the sartorius. The wide muscular portion of the tensor fasciae latae originates on the iliac crest. This muscle narrows to a tough band of fascia that inserts on the proximal tibia.

Biceps femoris Large lateral muscle.

Gracilis Inner thigh muscle extending from the ischium and symphysis pubis to the tibia.

B. Separating the Muscles of the Upper Thigh

1 ➡ Transect and reflect the sartorius, the tensor fasciae latae, the biceps femoris, and the gracilis. Now identify these muscles:

Gluteus medius Lateral to the tensor fasciae latae and extending up into the hip.

Gluteus maximus Lateral to the gluteus medius. As in humans, the gluteus maximus is the buttock muscle. However, in cats the maximus is smaller than the medius. (Cats do have a gluteus minimus and several other small muscles deep to the medius. We will not identify those muscles here.)

Caudofemoralis Long, somewhat triangular muscle lateral to the gluteus maximus and extending down to the patella. Humans do not have a caudofemoralis muscle.

Tenuissimus Long, very thin muscle extending from the second caudal vertebra to the fascia of the biceps femoris. This is one of the hamstrings.

C. Separating the Muscles on the Anterior Surface of the Hind Limb

1 ➡ Locate the following muscles:

Vastus lateralis Large muscle running from the proximal femur to the patella, deep to the tensor fasciae latae.

Rectus femoris Large muscle originating on the ilium and extending to the patella. You can find it on the proximal end of the femur between the vastus lateralis and the vastus medialis.

Vastus medialis Large muscle just medial to the rectus femoris.

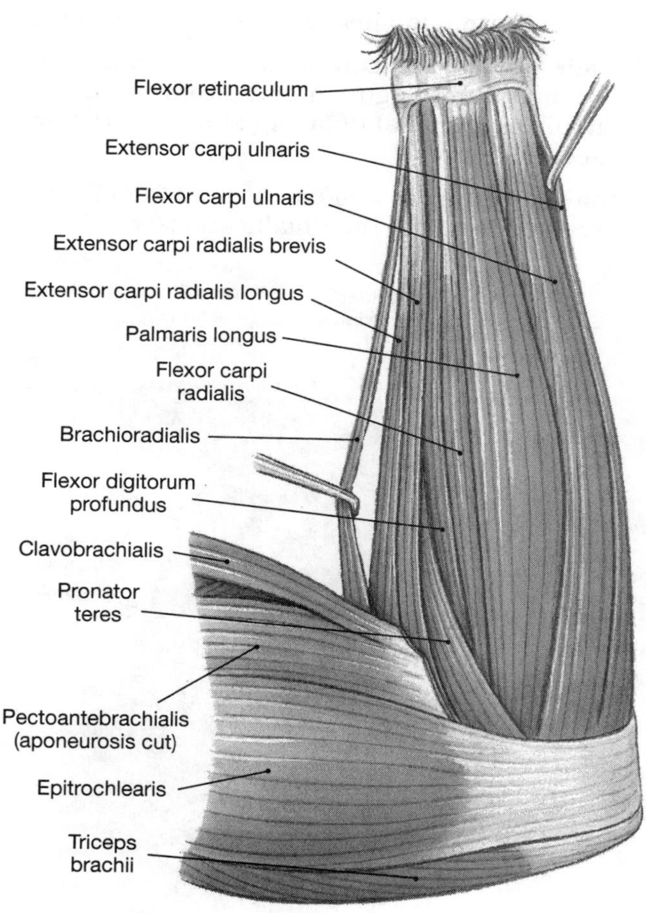

● **FIGURE 25a–10**
Muscles of Forelimb — Medial View.

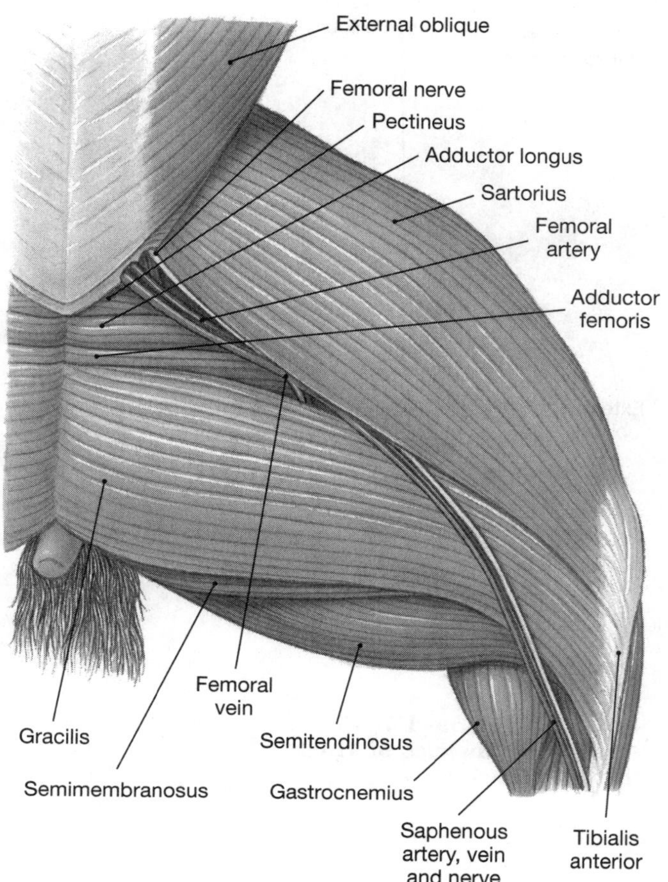

● **FIGURE 25a–11**
Superficial Muscles of Thigh (Ventral/Medial View).

Vastus intermedius Small muscle deep and medial to the vastus lateralis and merging superiorly with the vastus medialis (not identified on the dissection figures).

Iliopsoas Transverse muscle you may not be able to find without dissecting the abdominal wall, located high on the medial leg. This muscle runs from the ilium and vertebrae to the lesser trochanter (not shown on the dissection figures).

Pectineus Small, proximal, centrally located longitudinal muscle on the medial thigh.

Adductor longus Longitudinal muscle immediately medial to the pectineus.

D. Separating the Posterior Medial Muscles

1➡ Now notice a pack of three muscles that seems to spiral around the leg from the posterior toward the center of the medial thigh. Identify these muscles:

Adductor femoris Extending from the ischium and pubis to the femur.

Semimembranosus The next two strips of muscle tissue. One of the hamstrings.

Semitendinosus The most dorsal of the muscle masses. One of the hamstrings.

E. Separating the Muscles of the Lower Leg

1➡ Follow Figures 25a–11● through 25a–16●. You should have no trouble identifying these two muscles of the lower leg.

Gastrocnemius Large calf muscle.

Soleus Deep to the lateral gastrocnemius.

2➡ Separate these muscles of the lower leg according to directions from your instructor. Depending on the time element and the condition of your cat, some of these muscles may not be readily identifiable.

Tibialis anterior Across the anterior tibia from the proximal fibula and tibia to the first metatarsal.

Plantaris Viewed from the anterior, immediately medial to the gastrocnemius (not shown on the dissection figures).

Flexor digitorum longus Anterior to the plantaris.

Tibialis posterior Lateral and deep to the flexor digitorum longus, arising from the medial fibula and ventral tibia (not shown on the dissection figures).

Extensor digitorum longus Viewed from the lateral, posterior to the tibialis anterior.

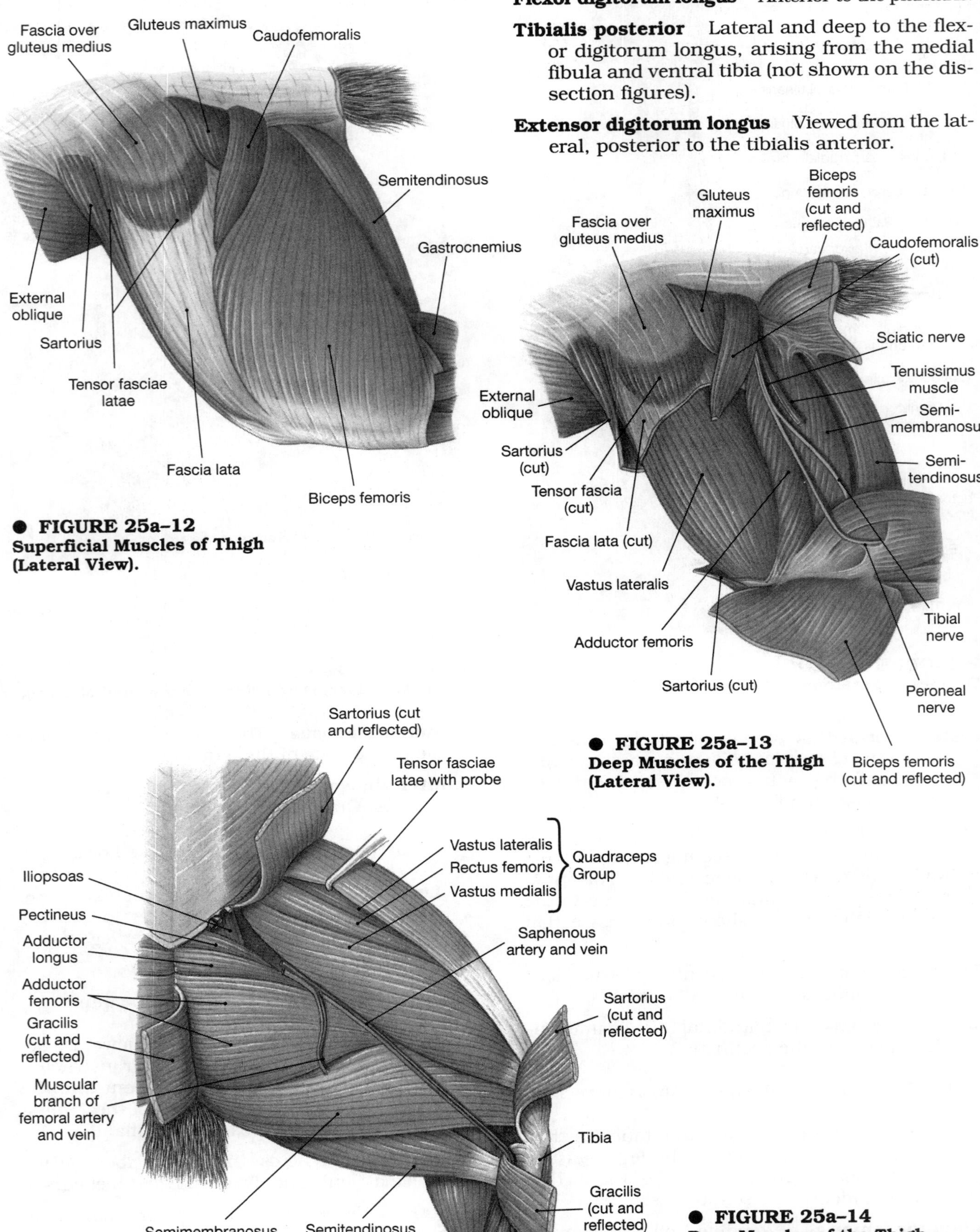

● **FIGURE 25a–12**
Superficial Muscles of Thigh (Lateral View).

● **FIGURE 25a–13**
Deep Muscles of the Thigh (Lateral View).

● **FIGURE 25a–14**
Deep Muscles of the Thigh (Ventromedial View).

Peroneus muscles Viewed from the lateral, a series of muscles found between the soleus and the extensor digitorum longus.

3➡ Refer to the lesson on human muscles for the names of other feline muscles. Most are analogous to human muscles.

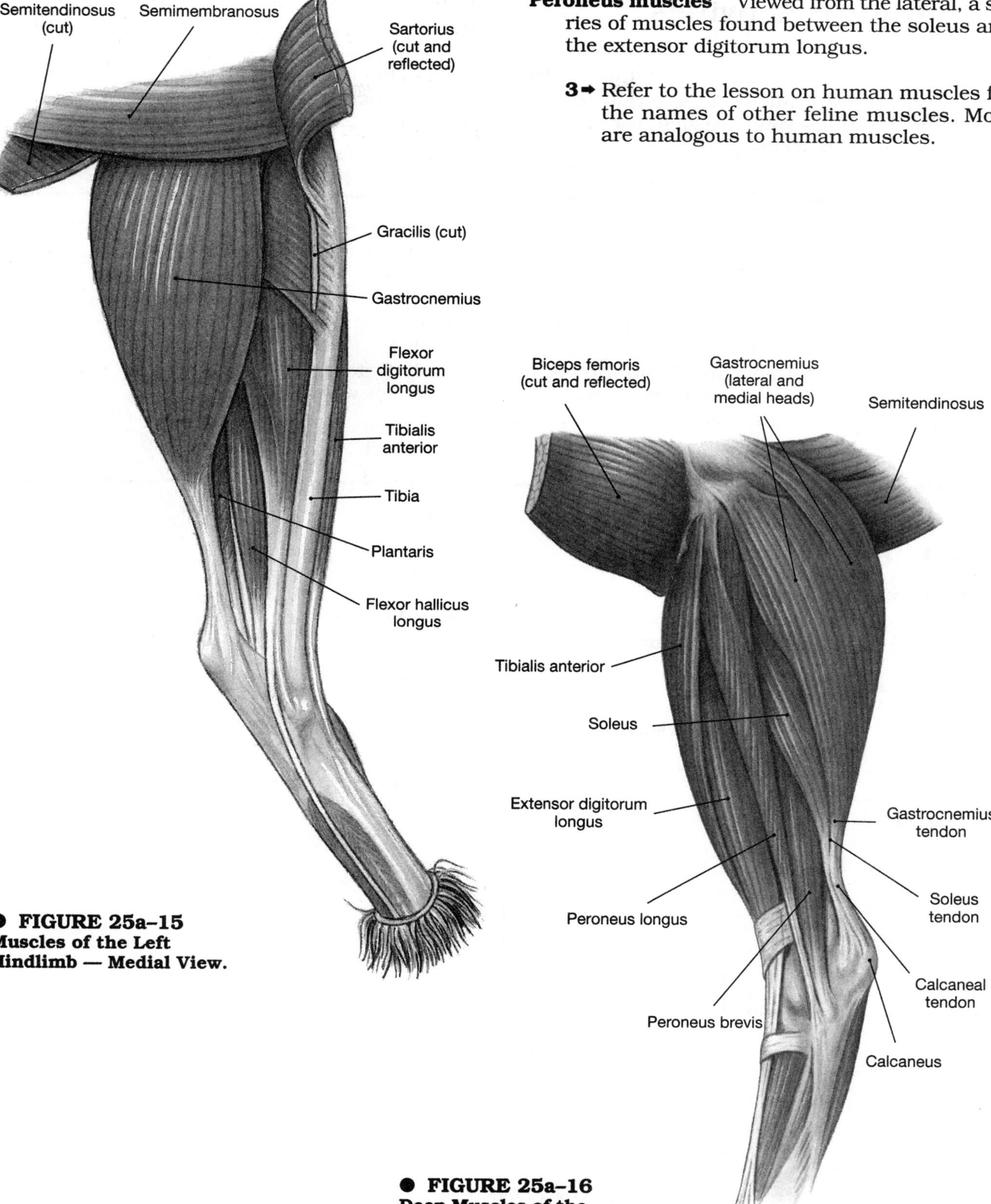

● **FIGURE 25a–15 Muscles of the Left Hindlimb — Medial View.**

● **FIGURE 25a–16 Deep Muscles of the Hindlimb — Lateral View.**

❑ Additional Activities

NOTES

1. Compare the muscles of the cat with the muscles of other animals.
2. By studying the locations of the various muscles, predict the actions of the major muscles you have identified. Use a reference text, or the lesson on human muscles, to verify your answers.

Answers to Selected Concept Check Questions

NOTES

1. Since the cat is a quadruped, the extra muscles may help in fine tuning the shoulder movements for jumping, climbing, and landing.
2. The cat is more "aerodynamic" and thus is comparatively faster than the human. The muscle arrangement is also advantageous for walking on narrow ledges and for power in jumping.

❑ Lab Report

1. Referring to the numbered inquiries at the beginning of this exercise, complete the following box summary:

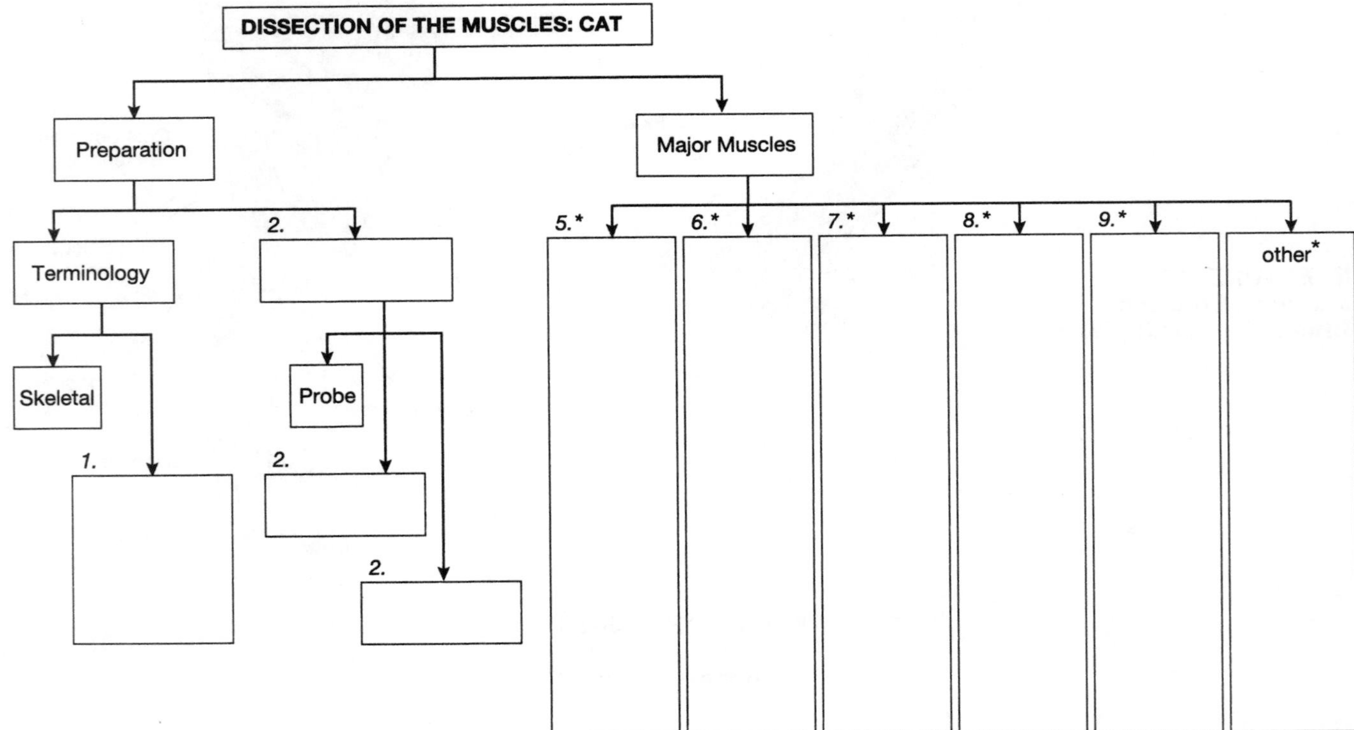

*List according to your instructor's directions

General Questions

NOTES

1. Identify these terms:
   Fascia ______
   Transect ______
   Muscle grain ______
2. What is the functional advantage of the cat having so many muscles performing the same or very similar actions? How does this relate to the human musculature?
3. List the major dissection instruments. How did you use each of these in your cat muscle dissection?
4. Go through the cat muscles and list at least 10 pairs of antagonistic muscles. Why are these muscles antagonists? Use the information in Exercise 21 as your guide.
5. Regarding the differences between feline and human musculature:
   a. List five muscles the cat has that humans don't have.
   b. List five muscles that, while present in both cat and human, are appreciably different between the two.
   c. What is the significance of the differences you find in the cat and human musculatures?
6. How do the origins and insertions of the muscles predict their actions?

EXERCISE 25b

Dissection of the Muscles: Fetal Pig

PROCEDURAL INQUIRIES

Preparation

1. What terminology should one be familiar with before beginning muscle dissection?
2. Which dissecting instruments should be used in working with fetal pig muscles?
3. What is fascia and how should it be worked with?
4. What guidelines should be used for identification of individual muscles?

Dissection

5. What are the major deep and superficial muscles of the ventral thorax and abdomen?
6. What are the major deep and superficial muscles of the head and neck?
7. What are the major deep and superficial muscles of the dorsal area?
8. What are the major deep and superficial muscles of the forelimb?
9. What are the major deep and superficial muscles of the hind limb?

Additional Inquiries

10. How is the fetal pig musculature analogous to the human musculature?
11. When and how should a muscle be transected?

HUMAN CORRELATION: AM Cadaver Atlas in Applications Manual.

Key Terms

Deep Fascia
Epimysium
Grain
Insertion
Linea Alba
Muscles of the
- Dorsal Area
- Forelimb
- Head and Neck
- Hind Limb
- Ventral Thorax and Abdomen

Origin
Superficial Fascia
Transecting

Materials Needed

Preserved Fetal Pig
Dissecting Kit
Dissecting Pan
Gloves
Disinfectant
Optional: Human Torso or Limb Model for Comparative Purposes
Other Equipment as Indicated by Your Instructor

Muscle systems are basically analogous throughout the mammalian world. This means that the muscles of the fetal pig are structurally and functionally similar to our own. Learning the muscles of one mammal by handling, separating, and identifying them means you have almost learned the muscles for all mammals (including the human).

Muscle variations do exist. Muscle fusion is the variation you should be most aware of. Many of the non-human animals have more individual muscles than we do. For instance, the fetal pig has three pectoralis muscles while we have only two. The proportionate muscle mass is approximately the same, but developmentally our muscles fuse whereas the fetal pig's do not. Nevertheless, we can readily correlate most porcine muscles with human muscles. In working through the muscle dissection, you should concentrate on this correlation. Look for the similarities.

This laboratory exercise is designed to help you develop a conceptual understanding of the human muscle system by dissecting the muscles of the fetal pig. Your goal is to separate and identify the individual fetal pig muscles. You are encouraged to work in pairs, to say the muscle names out loud, to study the diagrams, to quiz your partner extensively on the names and locations of the muscles, and to correlate the fetal pig musculature with your previous knowledge of the human musculature.

In this exercise we are using the regional approach and will thus examine the muscles of the ventral abdomen and thorax, the head and neck, the dorsal area, the forelimbs, and the hind limbs. With each section we will look at both the superficial and the deep muscles. We will dissect all major and readily identifiable muscles, except for those of the head and upper face. In all cases identifying data such as the **origin** and **insertion** are given to help you locate the muscle.

❑ Preparation

I. Methodology

A. Terminology

1➡ Be certain you have a good grasp of skeletal terminology before you begin dissecting the muscles of the fetal pig. (Refer to Unit III if necessary.) You should also be familiar with the material in Exercise 21, Human Musculature. You are encouraged to correlate the porcine and human musculatures. Most of the information in Exercise 21 is applicable to the fetal pig as well as to the human. Use the classroom models often, as you make your comparisons.

B. Tools

Beginning students are often afraid to attack the deep fascia. Be assured that the deep fascia is tough, and the older your animal, the tougher the fascia. To separate many of the muscles, it will be necessary for you to break through the deep fascia.

1➡ Use your blunt nose probe as your primary muscle-dissecting instrument. You can rip the deep fascia with this probe without worrying about tearing the muscle fibers.

As you separate muscles, always work with the direction or **grain** of the fibers. You can usually distinguish separate muscles by a change in the grain. Individual muscles also separate naturally from one another in all directions. You will be able to run your probe beneath or alongside a given muscle without any difficulty, after you have penetrated the deep fascia.

2➡ Do not use your scissors or scalpel for muscle dissection unless you are deliberately **transecting** (cutting) a muscle. You may occasionally use your sharp nose probe but you must be careful that you do not tear a delicate blood vessel or nerve fiber. Your forceps can be used to grip fascia or muscle tissue or to pull out adipose tissue.

3➡ Work with a partner unless you are directed otherwise. Both of you should use your blunt nose probes to feel the individual muscles. Both of you should say the names of the muscles out loud and study the diagrams carefully. Quiz each other often.

4➡ Transect (cut) certain muscles according to directions. This is necessary in order to study the deeper muscles. Before you transect anything, be certain you are transecting in the right place. Reconstructive surgery on preserved fetal pigs is not highly successful.

C. Incision

1➡ Refer to Exercise 24 or follow your instructor's directions if you have not already opened your fetal pig.

2➡ Your instructor will advise you about how much of the skin should be removed at any given time. If you will be working on your muscle dissection for several weeks, we recommend skinning on an "as needed" basis. This prevents excess drying of your specimen. If your muscle dissection will be completed in a relatively short period of time, skinning the entire fetal pig at once may be appropriate. We also recommend that you wrap the removed skin around your fetal pig before placing the fetal pig back in its storage bag. (Directions for skinning are given in Exercise 24.)

3➡ Recall from Exercise 20 that a bundle of skeletal muscle fiber is surrounded by a layer of coarse connective tissue called the **epimysium**. Covering the entire muscle (exterior to the epimysium) is a tough fibrous membrane, the **deep fascia**, the fibers of which intermesh with the fibers of the **superficial fascia**. (The superficial fascia helps anchor the skin to the body.) The deep fascia is also continuous with the tendons, ligaments, and periosteum.

Dissection

II. Muscles of the Ventral Thorax and Abdomen

Open the fetal pig and lay back (or remove) the skin on the ventral thorax and abdomen. Probe your way through the fascia. Remove any adipose tissue that may be in your way. (Fetal pigs have very little excess adipose tissue.)

A. Separating the Superficial Muscles

1➡ Observe the ventral surface of the animal. Follow Figure 25b–1● and use the left side of your animal to locate the superficial muscles. Separate the muscles with your probe.

Pectoralis Three muscles, arising from the sternum, somewhat fused, especially in the fetal animal. The **superficial pectoral** forms a band across the upper chest and brachium, inserting on the proximal humerus. Transect and reflect the right superficial pectoral muscle. Locate the **anterior deep pectoral** and its caudal partner, the **posterior deep pectoral**, which arise just beneath the superficial pectoral.

Rectus abdominus Longitudinal muscle extending from the symphysis pubis to the sternum.

External oblique Large sheet of muscle with diagonal fibers covering most of the abdomen.

Linea alba White abdominal midline.

B. Separating the Deep Muscles

1➡ Use the right side of your animal and immediately to the right of the linea alba carefully transect and reflect (fold back) the rectus abdominus and the external oblique. Identify these deep muscles.

Internal oblique Immediately deep to the external oblique. The fibers of the external and internal obliques form a 90° angle.

Transversus abdominis Transverse fibers immediately deep to the internal oblique. (The external oblique, internal oblique, and transversus abdominis muscles are paper-thin layers. Use caution if you have difficulty separating these muscles.)

Serratus ventralis Large, fan-like muscle arising from the ribs and inserting on the vertebral border of the scapula. You will see this muscle again when you dissect the muscles of the back.

External intercostals Series of muscles between the ribs, found in many animals but *not* in the pig. Included here for your information.

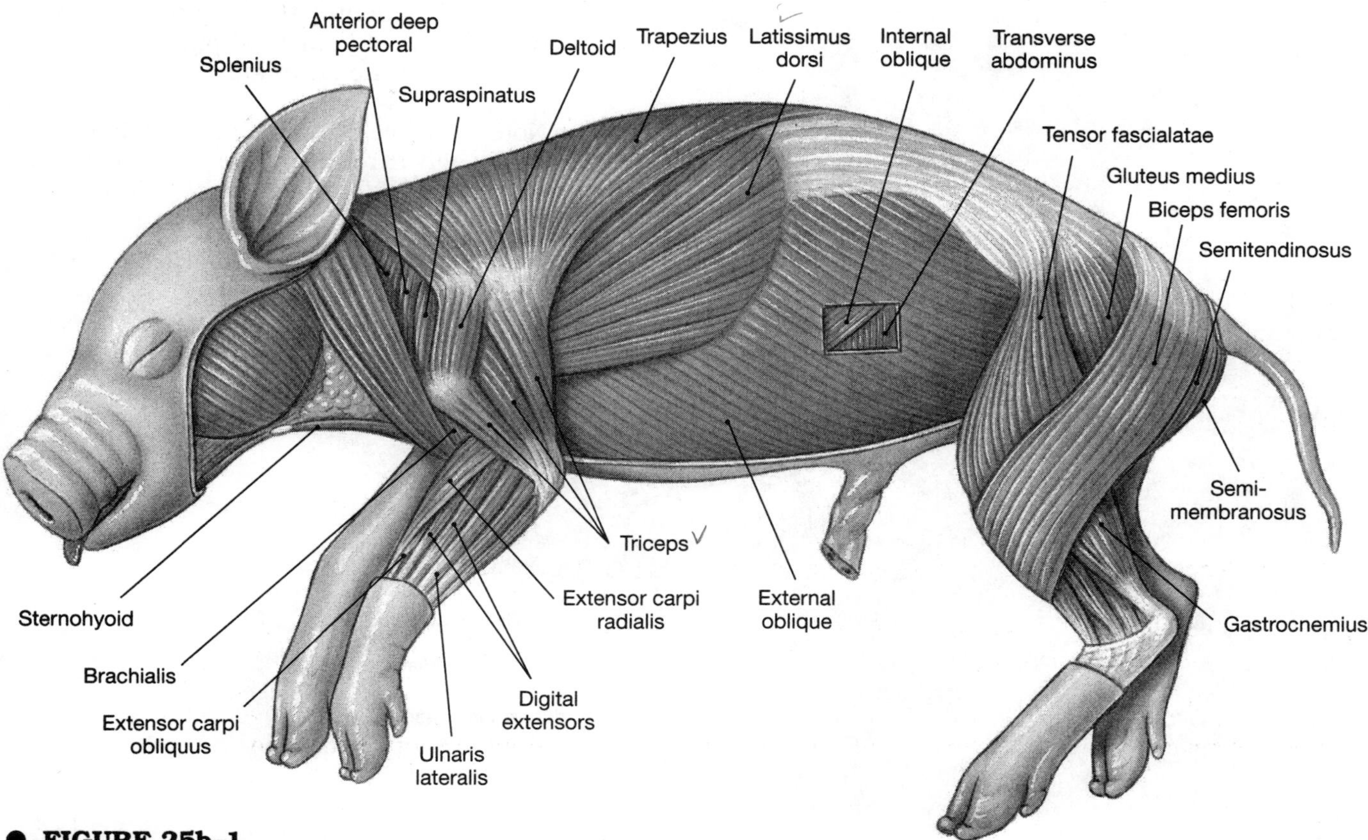

● **FIGURE 25b–1**
Muscles of Fetal Pig — Lateral View.

Other muscles of the abdomen and thorax will not be dissected here.

III. Muscles of the Head and Neck

A. Separating the Superficial Muscles

1 ➡ Study Figure 25b–2●. We will describe a few of these muscles. Because of the injection sites on many fetal pigs, it may not be possible for you to find all of the muscles in the diagram. If an individual muscle has been destroyed, note where it should have been. Keep in mind that you are striving for an understanding of the muscle system. The paired superficial muscles include:

Sternohyoid Longitudinal muscle extending from the sternum to the hyoid bone. The sternal end of this muscle can be found deep to the sternocephalic muscle.

Sternocephalic Also called the **sternomastoid**. Oblique muscle extending from the mastoid process to the sternum.

Mylohyoid Flat transverse muscle superior to the sternohyoid.

Digastric Longitudinal muscle superficial and lateral to the mylohyoid.

Masseter Large cheek muscle, lateral and superior to the digastric muscle. You will have to dissect up into the cheek to find this muscle.

Sternothyroid Medial muscle deep to the sternohyoid originating on the sternum and inserting on the thyroid cartilage of the larynx.

B. Separating the Deep Muscles

1 ➡ Return to Figure 25b–2●. Note that other muscles do exist. We will not locate these muscles at this time. Most other muscles are analogous to the muscles described for other animals.

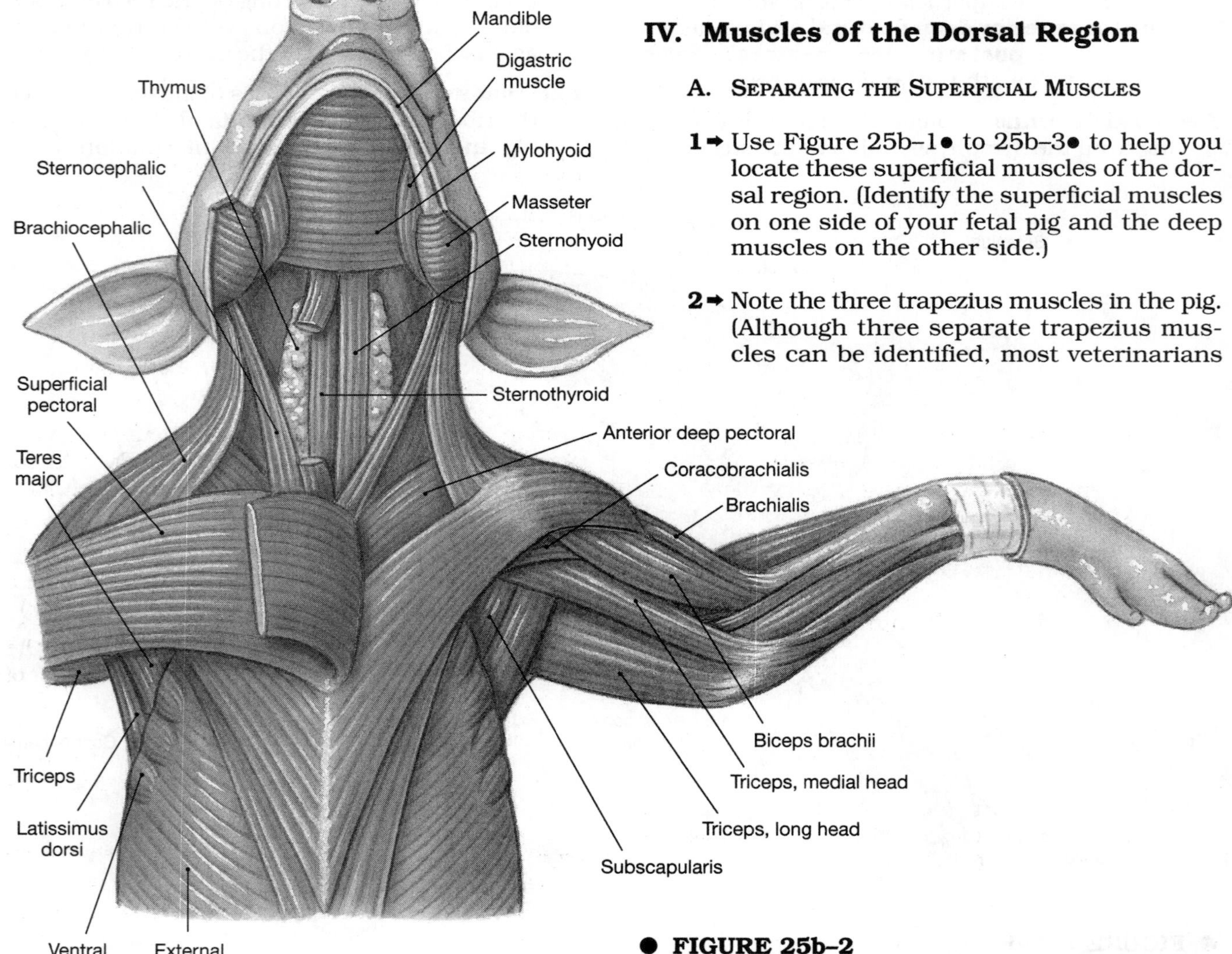

● **FIGURE 25b–2**
Muscles of the Ventral Thoracic Region and Forelimb.

IV. Muscles of the Dorsal Region

A. Separating the Superficial Muscles

1 ➡ Use Figure 25b–1● to 25b–3● to help you locate these superficial muscles of the dorsal region. (Identify the superficial muscles on one side of your fetal pig and the deep muscles on the other side.)

2 ➡ Note the three trapezius muscles in the pig. (Although three separate trapezius muscles can be identified, most veterinarians

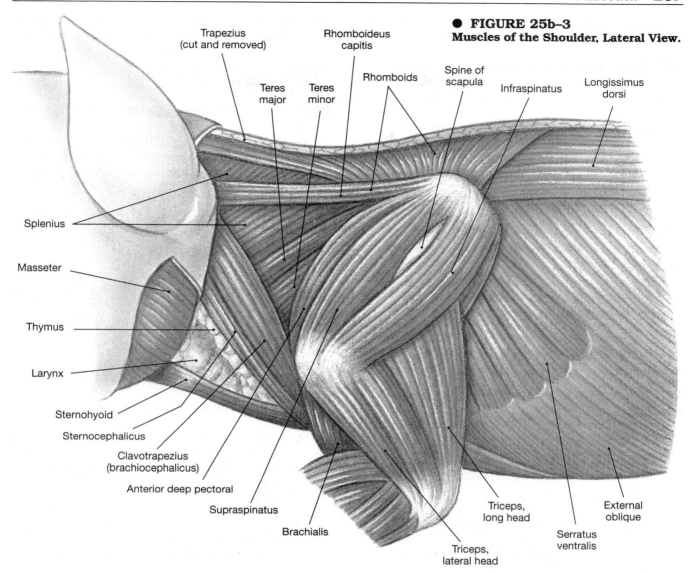

● **FIGURE 25b–3**
Muscles of the Shoulder, Lateral View.

today consider these three trapezius muscles as a single trapezius muscle analogous to our own.) How many trapezius muscles do humans have? ____________

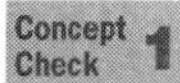

Speculate on the types of movements the pig has that might make this muscle arrangement advantageous to the animal

____

____

Clavotrapezius Also known as the superior portion of the **brachiocephalic**, large shoulder muscle viewed from either the ventral or lateral aspect, extending from the mastoid process and the nuchal crest to the clavicle. (The pig has only a vestigial clavicle.)

If you skin high enough onto the posterior head, it may be possible to find the point where the clavotrapezius divides into the cleidomastoid branch (to the mastoid process) and the cleidooccipitalis branch (to the nuchal crest). The ventral portion of the clavotrapezius is the **clavobrachialis.**

Acromiotrapezius Flat muscle just caudal to the clavotrapezius, originating from the cervical and thoracic vertebrae and inserting on the scapular spine (see trapezius on the figures).

Spinotrapezius Flat muscle just caudal to the acromiotrapezius, originating at the spines of the thoracic vertebrae and inserting on the fascia of the scapular muscles. The line of demarcation between the acromiotrapezius and the spinotrapezius is not distinct (see trapezius on the figures).

Deltoid Thin strap from the aponeurosis covering the infraspinatus muscle of the scapula to the proximal humerus and fascia of the forelimb. In humans, of course, this muscle is thick and fleshy.

Latissimus dorsi Large, flat muscle with diagonal fibers extending from the lumbardorsal fascia, around the lateral part of the body and onto the humerus.

B. Separating the Deep Muscles

1 ➡ Transect and reflect the right side muscles. You should have no particular difficulty identifying the following muscles.

Rhomboids Several small muscles extending from mid-back to the medial border of the scapula. Humans have two rhomboideus muscles, the major and the minor.

Rhomboideus capitis Longitudinal muscle lateral to the rhomboids originating on the occipital bone and inserting on the angle of the scapula. This muscle is not found in humans.

Splenius Flat muscle lateral to the rhomboids.

Supraspinatus Superior (and larger) of two muscles covering the scapula.

Infraspinatus Inferior (and wider) of two muscles covering the scapula.

Teres major Muscle running from the medial border of the scapula to the humerus.

Teres minor Deeper muscle extending from the lateral border of the scapula to the greater tubercle and deltoid tuberosity of the humerus.

Serratus ventralis Fingerlike muscle fanning out from the ribs. The cervical part of this muscle is well developed and easily identified. You previously located this muscle from the ventral surface.

V. Muscles of the Forelimb

A. Separating the Major Muscles

1 ➡ Choose either the right or the left forelimb. Use Figure 25b–1● through 25–3●. Begin with the medial side of the brachium.

Biceps brachii Muscle crossing the brachium from the glenoid fossa of the scapula to the radius (beneath the brachialis tendon) and the proximal ulna. This muscle is deep to the clavobrachialis.

Coracobrachialis Small muscle connecting the coracoid process with the tendon of the subscapularis muscle.

Brachialis From the lateral surface of the humerus to the proximal end of the ulna.

Triceps brachii Muscle crossing the brachium from the scapula and humerus to the olecranon process of the ulna. Observe this muscle from both the lateral and medial aspects.

B. Identifying Additional Muscles

1 ➡ Many of the muscles of the forelimb are often difficult to identify in classroom fetal pig specimens. Based on your instructor's directions, use Figure 25b–2● to locate the

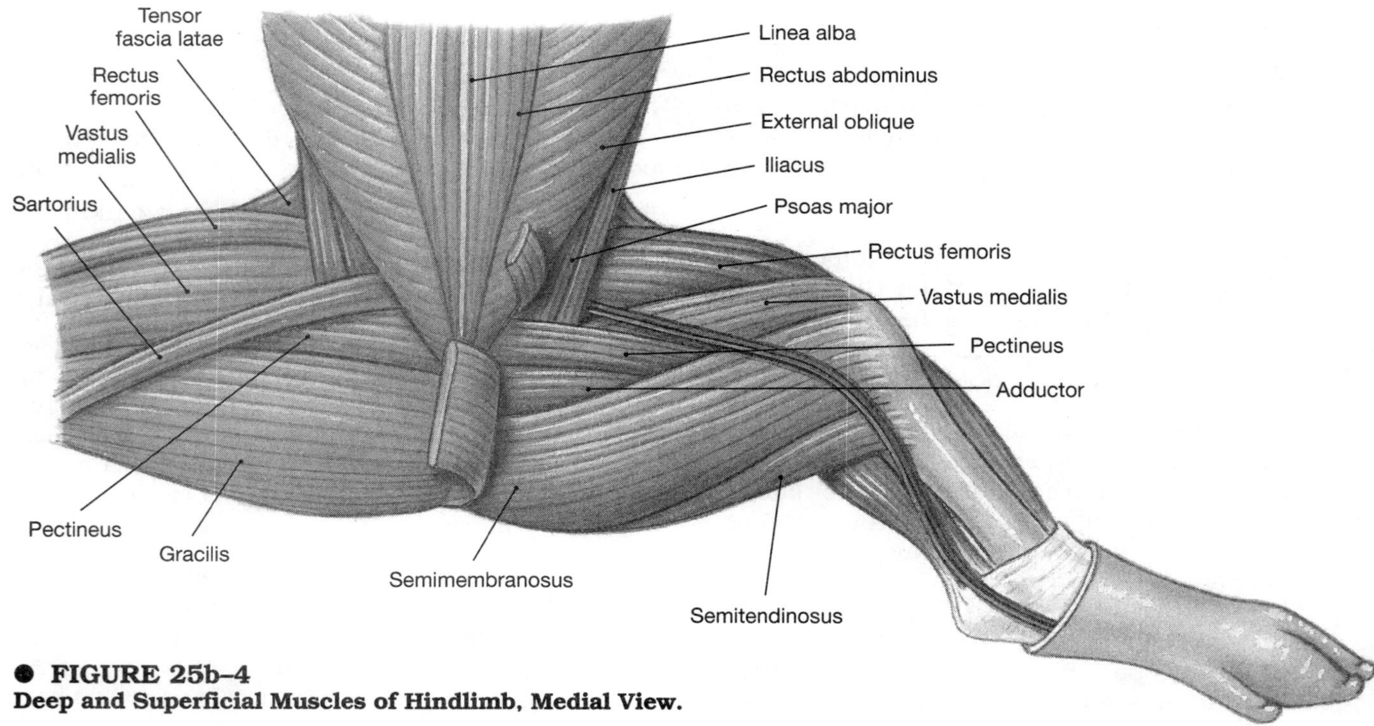

● **FIGURE 25b–4**
Deep and Superficial Muscles of Hindlimb, Medial View.

following: **brachioradialis, extensor carpi radialis, flexor carpi radialis, flexor carpi ulnaris, and flexor digitalis** (several).

The **pronator teres** is a delicate muscle located on the medial surface of the elbow and the proximal portion of the forearm.

From the lateral side, the major muscles from anterior to posterior are: **extensor carpi radialis, extensor digitorum communis, extensor digitorum communis, extensor digitorum longus, extensor carpi ulnaris.**

VI. Muscles of the Hind Limb

A. Separating the Superficial Muscles

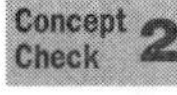

Observe that the fetal pig's hind limb is proportionately far more "round" than the human leg. What functional advantage do you see for the adult pig having this hefty leg?

____

____

● **FIGURE 25b–5**
Deep Muscles of Hindlimb, Lateral View.

Choose either the right or the left hind limb. Use Figures 25b–1●, 25b–4● and 25b–5● to help you isolate and identify these hind limb muscles. To gain a perspective of the hind limb musles, compare these three figures.

Sartorius Large, platelike muscle extending from the iliac fascia and the psoas minor to the tibia.

Tensor fasciae latae Lateral to the sartorius.

Biceps femoris Large lateral muscle.

Gracilis Inner thigh muscle extending from the ishium and symphysis pubis to the tibia.

B. Separating the Muscles of the Upper Thigh

1 ➡ Transect and reflect the sartorius, the tensor fasciae latae, the biceps femoris, and the gracilis. Now identify these muscles:

Gluteus medius Lateral to the tensor fasciae latae and extending up into the hip.

Gluteus maximus Thin muscle overlying the anterior gluteus medius. Unless you are so directed, do not attempt to distinguish the gluteus maximus from the gluteus medius. As in humans, the gluteus maximus is a buttock muscle. However, in fetal pigs the maximus is smaller than the medius. (Fetal pigs do have a gluteus minimus and several other small muscles deep to the medius. We will not identify those muscles here.)

C. Separating the Muscles on the Anterior Surface of the Hind Limb

1 ➡ Locate the following muscles:

Vastus lateralis Large muscle running from the proximal femur to the patella, deep to the tensor fasciae latae.

Rectus femoris Large muscle originating on the ilium and extending to the patella. You can find it on the proximal end of the femur between the vastus lateralis and the vastus medialis.

Vastus medialis Large muscle just medial to the rectus femoris.

Vastus intermedius Small muscle deep to the vastus lateralis and rectus femoris. (The vastus lateralis, rectus femoris, vastus medialis, and vastus intermedius form the **quadriceps femoris** group.)

Iliopsoas Two transverse muscles, **iliacus** and **psoas major**, on the ventral surface just medial to the tensor fascia latae. This muscle runs from the ilium and vertebrae to the lesser trochanter.

Pectineus Small, proximal, centrally located longitudinal muscle on the medial thigh.

D. Separating the Posterior Medial Muscles

1➡ Now notice a pack of three muscles which seem to spiral around the leg from the posterior toward the center of the medial thigh. Identify these muscles.

Adductor magnus Large, undivided muscle extending from the ishium to the femur.

Semimembranosus Caudally, the next two strips of muscle tissue. This muscle (one of the hamstrings) has one origin and two insertions.

Semitendinosus The most dorsal of the muscle masses. This muscle (one of the hamstrings) has two heads.

E. Separating the Muscles of the Lower Leg

1➡ Follow Figures 25b–4● and 25b–5●. You should have no trouble identifying these two muscles of the lower leg.

Gastrocnemius Large calf muscle.

Soleus Deep to the lateral gastrocnemius.

2➡ Separate these muscles of the lower leg according to directions from your instructor. Depending on the time element, the condition of your fetal pig, and the directions from your instructor, you may be able to separate other muscles of the lower leg. From the lateral aspect, you should find the major muscles in this order:

Tibialis anterior Muscle found across the anterior tibia from the proximal fibula and tibia to the first metatarsal.

Peroneus tertius Also known as the **fibularis tertius**, this muscle covers the extensor digitorum longus, with which it is distally united.

Peroneus longus Muscle lateral to the peroneus tertius.

Extensor digitorum (several) Muscles lateral and deep to the peroneus longus.

Extensor digitorum longus Specific digital extensor, viewed from the lateral, posterior to the tibialis anterior.

Flexor digitorum (several) Muscle on the medial surface, anterior to the soleus.

3➡ Refer to the lesson on human muscles for the names of other porcine muscles. Most are analogous to human muscles.

❑ Additional Activities

1. Compare the muscles of the fetal pig with the muscles of other animals.
2. By studying the locations of the various muscles, predict the actions of the major muscles you have identified. Use a reference text, or the lesson on human muscles, to verify your answers.
3. Obtain a book on veterinary anatomy. Dissect additional porcine muscles. The pig has some fascinating muscles associated with the face and ear.

NOTES

Answers to Selected Concept Check Questions

1. Since the pig is a quadruped, the extra muscles may help in strengthening and fine tuning the shoulder movements.
2. The pig is a quadruped with a "wallowing" lifestyle.

❑ Lab Report

1. Referring to the numbered inquiries at the beginning of this exercise, complete the following box summary:

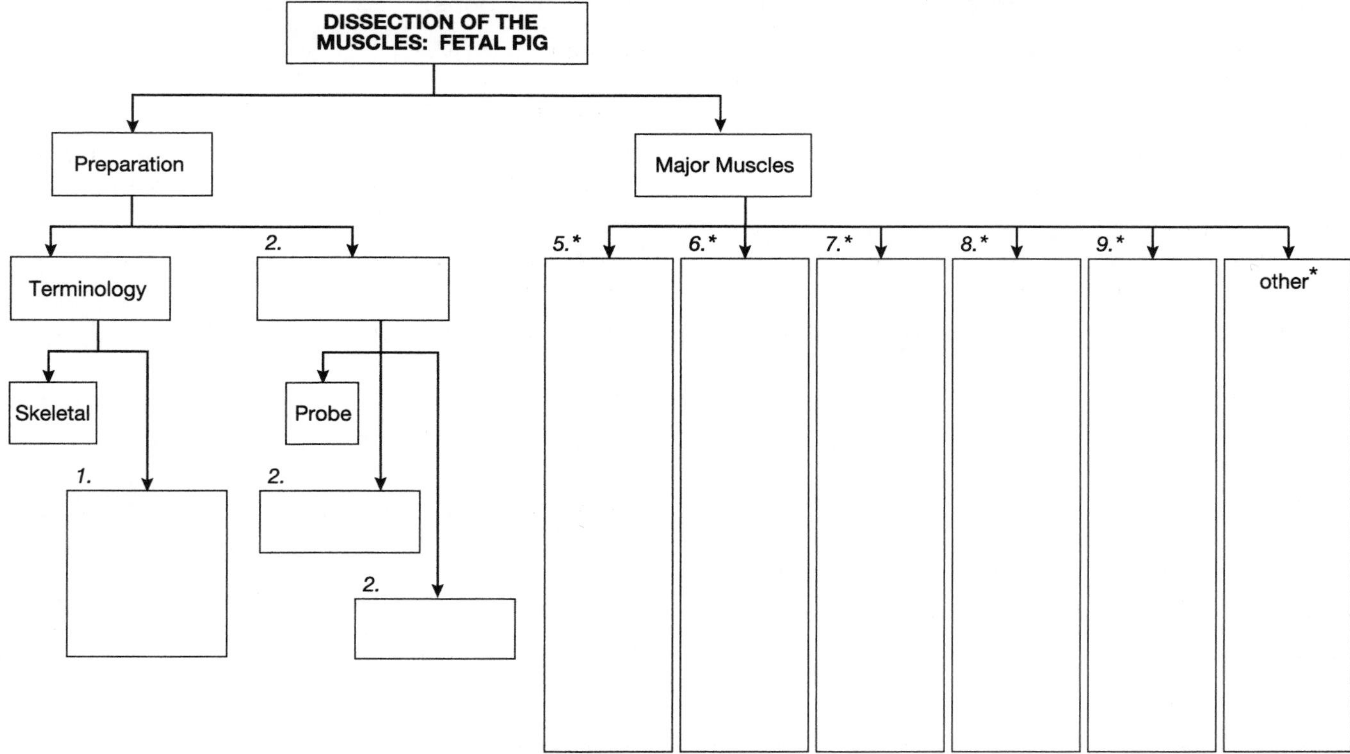

*List according to your instructor's directions
You may need separate paper.

General Questions

NOTES

1. Identify these terms:
   Fascia ____________________
   Transect ____________________
   Muscle grain ____________________
2. What is the functional advantage of the fetal pig having so many muscles performing the same or very similar actions? How does this relate to the human musculature?
3. List the major dissection instruments. How did you use each of these in your fetal pig muscle dissection?
4. Go through the fetal pig muscles and list at least 10 pairs of antagonistic muscles. Why are these muscles antagonists? Use the information in Exercise 21 as your guide.

NOTES

5. Regarding the differences between porcine and human musculature:
   a. List five muscles the fetal pig has that humans don't have.
   b. List five muscles that, while present in both fetal pig and human, are appreciably different.
   c. What is the significance of the differences you find in the fetal pig and human musculatures?
6. How do the origins and insertions of the muscles predict their actions?

EXERCISE 26

Neural Tissue: An Overview

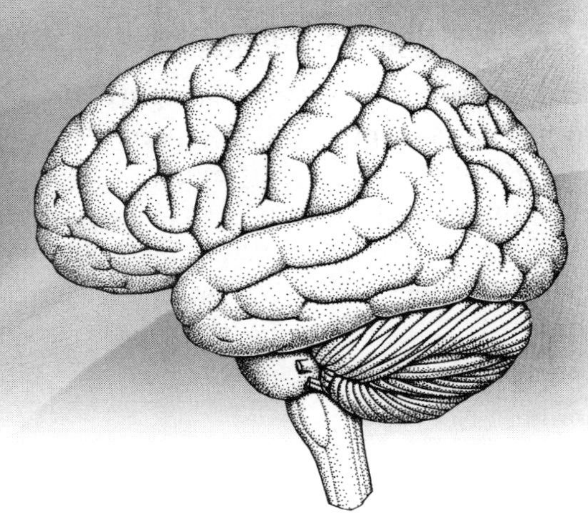

PROCEDURAL INQUIRIES

Preparation

1. What is the embryological origin of the nervous system?
2. What is a common nervous system delineation scheme?
3. What are some of the common terms used to describe nervous system structures and events?

Examination

4. What is the soma? What is the perikaryon? How are the soma and perikaryon related?
5. What structures are found in the perikaryon?
6. What are Nissl bodies and what is their function?
7. What are the cytoplasmic extensions?
8. Where would you find a myelin sheath and what is its function?
9. What is a common axonic and dendritic nerve cell classification scheme?
10. What are the connective tissues of the nervous system?
11. What are the glial cells of the central nervous system? What is the function of each?
12. What are the glial cells of the peripheral nervous system? What is the function of each?
13. What is an action potential?
14. What determines the direction of an electrical impulse through a neuron?
15. What is the role of the ion channels in electrical transmission?
16. How can we explain neuronal excitation?

Experimentation

17. How can we demonstrate the role of time in electrical conduction?

Additional Inquiries

18. Approximately how many neurons and glial cells can be found in the body?
19. What is the difference between IPSP and EPSP?

Key Terms

Action Potential
Afferent
Anaxonic Neurons
Astrocytes
Autonomic Nervous System
Axon
Axon Hillock
Bipolar Neurons
Central Nervous System
Dendrite
Efferent
Endoneurium
Ependymal Cells
Epineurium
EPSP
Excitation Threshold
Ganglion
- Preganglionic
- Postganglionic

Initial Segment
IPSP
Microglia
Multipolar Neurons
Myelin Sheath
Nerve
Neuroglia
Neuron
Nissl Bodies
Nodes of Ranvier
Nucleus
Oligodendrocyte
Parasympathetic Division
Perikaryon

Perineurium
Peripheral Nervous System
Satellite Cells (Amphicytes)
Schwann Cell
Soma
Somatic
Somatic Nervous System
Sympathetic Division
Synapse
  Presynaptic
  Postsynaptic
Unipolar Neurons
Visceral

Materials Needed

Compound Microscope
Clock or Watch with Second Hand
Histological Slides

- Giant Multipolar Motor Neuron (Nissl Body)
- Motor Nerve Endings
- Peripheral Nerve (cross section)
- Neuroglia (Astrocytes)
- Choroid Plexus (4th Ventricle)
- Myelinated (Medullated) Nerve Fiber
- Other Nerve Slides (as available)
  - Spinal Cord Smear (Sections, if available)
  - Pyramidal Cells (Cerebral cortex)
  - Purkinje Cells (Cerebellar cortex)
  - Dorsal Root Ganglion
  - Sympathetic Ganglia
  - Auerbach's or Meissner's Plexus

With 10^{12} neurons and at least as many glial cells, with hundreds (possibly thousands) of different cell types, and with precise patterns and connections unparalleled throughout the rest of the body, the nervous system remains the last great bastion of anatomical and physiological mystery.

Although at present we really know very little about the complexities of the nervous system, we do know that we will not be able to understand the intricacies of the body as a whole until we have successfully grasped the intricacies of the nervous system. Neural tissue is responsible for communication and regulation between all parts of the body. In addition, memory and learning are neural functions, as is the integration of just about every imaginable bodily function.

In this laboratory exercise we will identify the basic anatomical, histological, and physiological components of the neural tissue.

❑ Preparation

I. Basic Definitions

A. Nervous System Divisions [∞ Martini, p. 361]

Neural tissue arises from the neural ectoderm, a part of the ectodermal embryonic germ layer whose fate is determined very early in development. This ectoderm folds over itself, forming a neural tube. The tube becomes the **central nervous system**, the brain and the spinal cord. As the tube seals, certain cells break away, forming the ganglia and supporting cells of the **peripheral nervous system**.

1➡ Be familiar with this common nervous system delineation scheme:

Central Nervous System (CNS) Brain and spinal cord.

Peripheral Nervous System (PNS) All nervous tissue that is not part of the brain and spinal cord.

Somatic Nervous System (SNS) Part of the nervous system that controls the musculoskeletal (voluntary) system.

Autonomic Nervous System (ANS) Part of the nervous system responsible for the visceral (involuntary) system.

Sympathetic Division Part of the nervous system responsible for the visceral "fight or flight" responses.

Parasympathetic Division Part of the nervous system responsible for the visceral "rest and repose" responses.

B. Nervous System Terminology

1➡ Be familiar with this nervous system terminology:

Neuron Message-transmitting cell, nerve cell.

Nerve (When used alone) a cluster of nerve cell fibers (axons) outside the central nervous system. (It is *not* correct to use the word "nerve" when referring to a single nerve cell.)

Sensory Neuron carrying a message from an input (sensory) point toward a central processing station. The sensory neuron is also called the afferent neuron.

Motor Neuron carrying a message to a muscle. The motor neuron is also called the efferent neuron.

Afferent Neuron carrying a message from a sensory point toward the central processing unit.

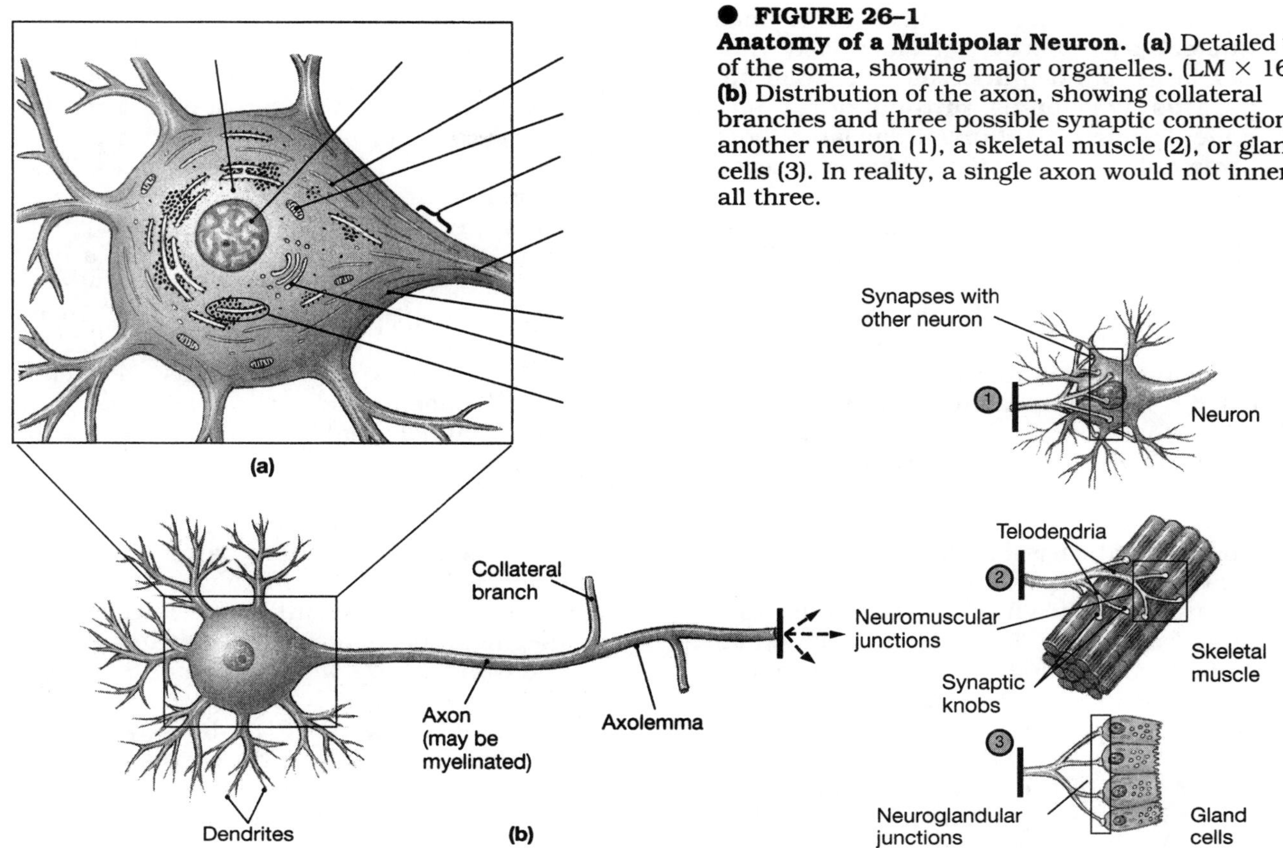

● **FIGURE 26–1**
Anatomy of a Multipolar Neuron. (a) Detailed view of the soma, showing major organelles. (LM × 1600) **(b)** Distribution of the axon, showing collateral branches and three possible synaptic connections: to another neuron (1), a skeletal muscle (2), or gland cells (3). In reality, a single axon would not innervate all three.

Afferent neuron is an alternate term for sensory neuron.

Efferent Neuron carrying a message away from the central processing unit and toward an effector (muscle, gland, etc.). Efferent neuron is an alternate term for motor neuron.

Somatic Body or muscle neurons; neurons associated with the voluntary muscles of the body.

Visceral Pertaining to the viscera or gut neurons; neurons associated with the involuntary muscles of the body.

Synapse Site of communication between a nerve cell and some other cell.

Synaptic Cleft Gap between the axon of one neuron and the point where the neuronal message is heading: a dendrite, a muscle, a nerve cell body, or some other cell.

Presynaptic Cell or portion of a cell (nerve or muscle) occurring before a synapse.

Postsynaptic Cell or portion of a cell (nerve, muscle or gland) occurring after a synapse.

Nucleus Mass of nerve cell bodies occurring within the central nervous system.

Ganglion Mass of nerve cell bodies occurring outside the central nervous system.

Preganglionic Nerve or message traveling toward a ganglion.

Postganglionic Nerve or message traveling away from a ganglion.

Axon That portion of a nerve cell carrying the message away from the nerve cell body.

Dendrite That portion of a nerve cell carrying the message toward the nerve cell body.

Action Potential Nerve impulse.

❑ Examination

II. Functional Unit — Neuron

A. Basic Anatomy of the Nerve Cell Body

1➡ Read the following description and identify the parts of the generic nerve cell (Figure 26–1●) accordingly. Some of the nerve cell parts are not specifically labeled. If a model is available, use it to locate these structures.

2➡ Identify the **soma**, the cell body itself.

3➡ Find the **perikaryon** in the cytoplasmic portion of the cell body proper, the part of the neuron resembling the typical cell. A small **nucleolus** can normally be identified within a large rounded **nucleus**. The **Golgi apparatus**, **lysosomes**, **mitochondria**, and **endoplasmic reticulum** are similar to those commonly found in other eukaryotic cells.

4➡ Identify the tubules, called **neurotubules**, and the finer filaments, the **neurofilaments**. These are located somewhat concentrically around the nucleus. The neurofilaments (as well as Golgi bodies, mitochondria, and endoplasmic reticulum fragments) are also located in the **cytoplasmic extensions** and can even be found toward the **cytoplasmic knobs**, those expanded areas where the neuron synapses with another cell.

5➡ Now locate the **Nissl bodies**, the dark areas also within the perikaryon. The Nissl bodies are now known to be abundant clusters of ribosomes (for protein production) embedded on rough endoplasmic reticulum. The Nissl bodies are what give the gray matter of the spinal cord its grayness.

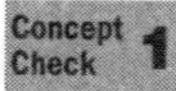

Over 98% of the synaptic proteins are produced in the perikaryon. Why is it logical that the nerve cell body would be packed with ribosomes?

____

____

6➡ Label the dark pigments of **melanin** or **lipofuscin** that can sometimes be found within the perikaryon. Not pictured in Figure 26–1● are **glial cell processes**, those extensions of the glial cells that communicate with the perikaryon or cytoplasmic extensions of the neuron. (Glial cells will be discussed in the section on the neuroglia.)

B. Anatomy of the Cytoplasmic Extensions

1➡ Continue labeling Figure 26–1● as you work through this section.

2➡ Turn your attention to the cytoplasmic extensions. The short, highly branched extensions are the **dendrites**. The long, distinctive branch is the **axon**. The single axon is not highly branched but may have one or more functional **collateral** branches heading specifically toward other cells. The fine extensions at the end of the axonic line are the **telodendria**. The ends of the telodendria form the synaptic knobs. (These synaptic knobs are also sometimes called **terminal knobs, buttons**, or **endfeet**.) The cytoplasm in the axon is called the **axoplasm**, and the cell membrane of the axon is more properly termed the **axolemma**.

3➡ Find the zone of separation between the perikaryon proper and the axon proper. This zone is the **axon hillock**. In a functional neuron, it is important to identify this area because when an action potential (electrical message) arrives at a neuronal body, the new action potential does not originate on the soma (body) membrane. Rather, the new action potential begins at the axon hillock. This area has a high concentration of voltage-gated sodium channels that facilitate the generation of the new impulse.

4➡ Locate the **initial segment**, the first part of the axon immediately after the axon hillock. It is often difficult to draw a line between the axon hillock and the initial segment, particularly when examining the neuron with a standard light microscope. With electron microscopy the usual distinction is that most Nissl clusters are in the axon and that few ribosomes are in the initial segment. In the hillock the microtubules and neurofilaments gradually funnel toward the cytoplasmic extensions. As the tubules come together, the area is officially the initial segment.

5➡ Note that the initial segment is not myelinated. The **myelin sheath** is the series of myelin coatings surrounding most but not all axons. Each "piece of sheath" is formed by an individual **Schwann cell** or an **oligodendrocyte**. The minute spaces between the myelin-forming cells are the **nodes of Ranvier**. These gaps are the enablers of saltatory conduction. In saltatory conduction the jumping of the impulse from node to node increases the speed of electrical transmission. [∞ *Martini, p. 370*]

The myelin sheath could be simulated by a roll of paper towels. Look at the end. The cardboard tube is the neuronal axon. The layers of paper towel are the layers of myelin. Note how that paper towel Schwann cell has wrapped around the central tube axon. Imagine several rolls of paper towels end to end. Note the gaps between the rolls (the myelin sheaths). Those gaps are the nodes of Ranvier.

C. Microscopic Examination of the Neuron

1➡ Obtain a compound microscope and a giant multipolar neuron slide. Identify as many of the preceeding structures as possible.

● **FIGURE 26–2**
An Anatomical Classification of Neurons.

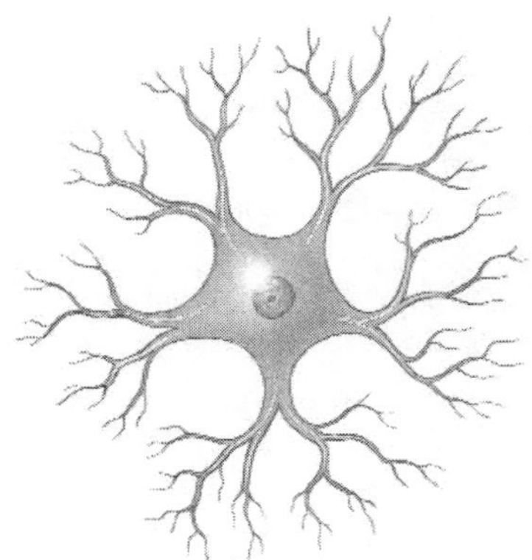

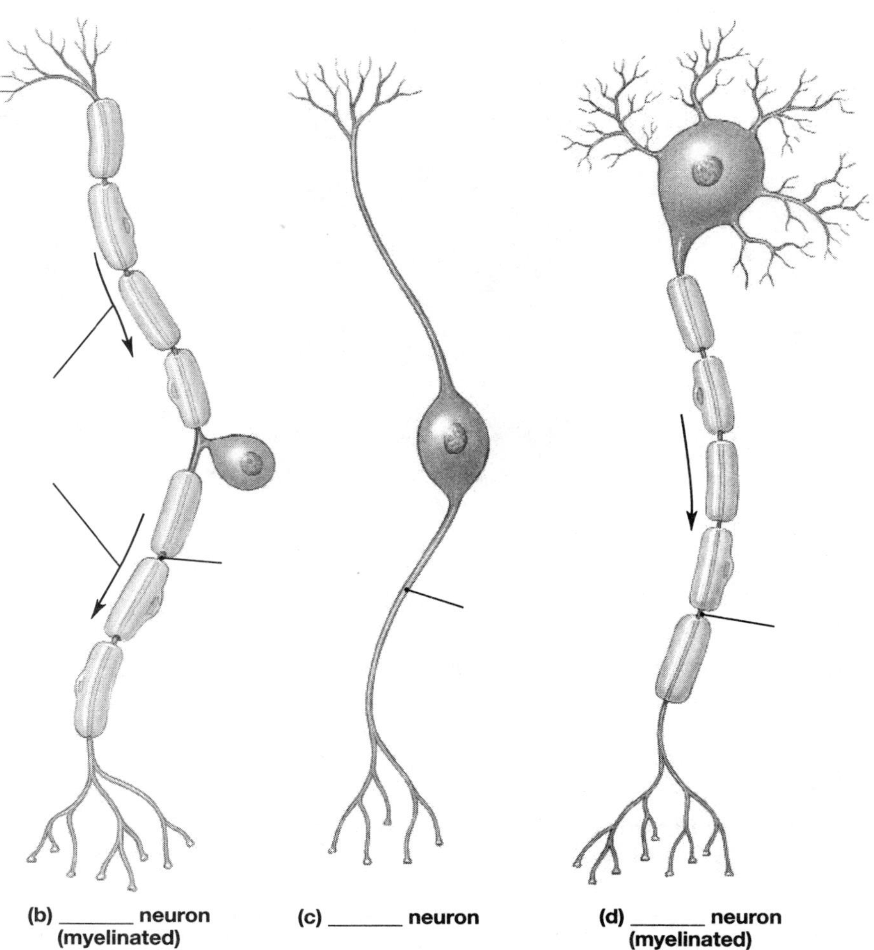

Giant multipolar neurons are often used for this exercise because the structures are easy to see. Depending on how the smear was made, you may or may not be able to identify parts of the axon.

2➡ Find the Nissl bodies. If properly stained, the Nissl bodies should be fairly clear. So, too, should the rather distinct lysosomes (which are occasionally confused with assorted neurosecretory granules). Lysosomes are visible even in unstained preparations. Depending on the type of stain used, you may be able to discern the Golgi apparatus, which in neuronal cells appears to be a mass of tangled strands and small vacuoles.

3➡ Distinguish between the axons and the dendrites. If you are having difficulty differentiating between the axons and the dendrites, identify the dendrites first by looking for extensive branching near the cell body. Sketch and label what you see.

Giant multipolar neuron

4➡ Obtain a slide of motor nerve endings. You should be able to discern the synaptic knobs of the telodendria as they communicate with the muscle fibers. Sketch and label what you see.

Motor nerve ending

D. Neuronal Classification [∞ Martini, p. 365]

Neurons are often classified according to their axonic and dendritic arrangement, as seen Figure 26–2●.

1➡ Read the following descriptions and label these Figure 26–2● neurons appropriately.

Multipolar neurons have several dendrites and one axon. The neuron you looked at in the previous section was a multipolar neuron. All motor neurons controlling skeletal muscle motor units are multipolar.

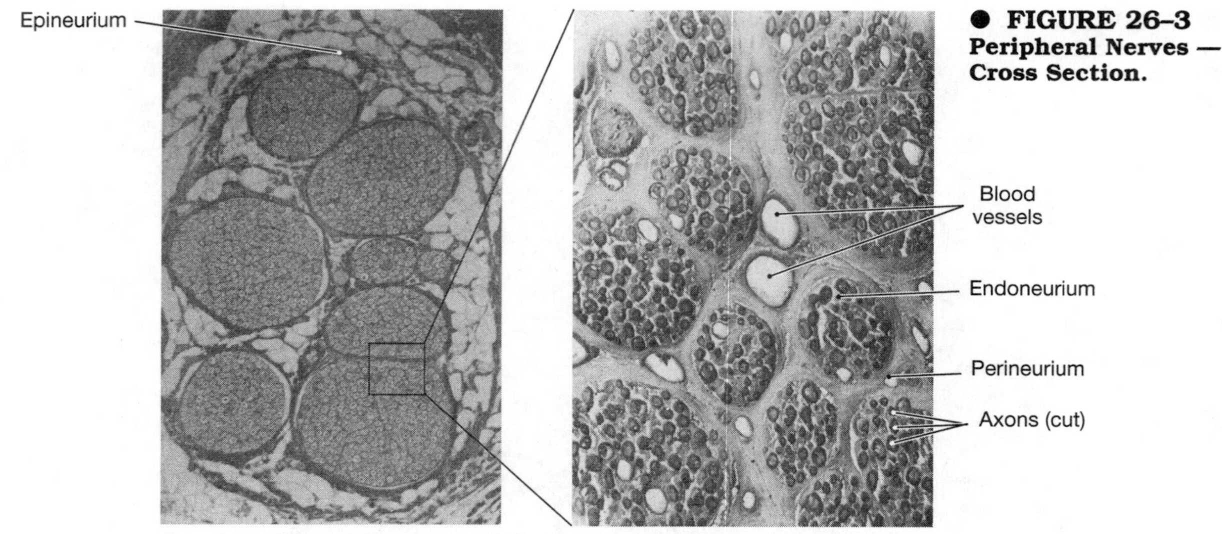

● **FIGURE 26–3 Peripheral Nerves — Cross Section.**

(Reproduced from R. G. Kessel and R. H. Kardon, *Tissues and Organs: A Text-Atlas of Scanning Electron* Microscopy, W.H. Freeman & Co., 1979.)

Bipolar neurons have a soma positioned between one axon and one dendrite. These neurons, found in the CNS and in sensory components of the eye and ear, have no myelination. Refer to Exercise 36, Figure 36–5●.

A **unipolar neuron**, sometimes called a **pseudo-unipolar neuron**, has a fused axon and dendrite. The sensory neurons of the PNS (thus the neurons of touch—Exercise 43) are unipolar. The entire fused shaft of this neuron is myelinated and the impulse racing along this track seems to fly by the PNS ganglion. Refer also to Exercise 29.

Anaxonic neurons have no cytoplasmic clues regarding the identities of the axons or dendrites. These small neurons, whose unmyelinated cytoplasmic extensions are all more or less identical, are found within the CNS and in special sense organs.

E. Connective Tissue

We have used the term neuron to indicate an individual nerve cell. We have also explained that a nerve is a conglomerate of neuronal fibers (axons). These fibers are surrounded by several different layers of connective tissue.

1 ➡ Study Figure 26–3●. Obtain a slide of the peripheral nerve cross section. From the figure, identify the **epineurium** (the collagenous fibers surrounding the peripheral nerve), the **perineurium** (the connective layer separating adjacent bundles of nerve fibers), and the **endoneurium** (the delicate connective layer surrounding the individual nerve fiber). Sketch and label what you see.

Peripheral nerve cross section

What is the overall function of these connective tissue layers?

____

____

How is the neuronal connective tissue arrangement similar to that of the muscle? Refer to Exercise 20, Figure 20–1●.

____

____

III. Supporting Cells — Neuroglia

[∞ Martini, p. 367]

A. Basic Function

The **neuroglia**, or **glial cells**, are the support cells of the nervous system, analogous to the connective tissues surrounding many of the other organs of the body. These cells have the same ectodermal origin as the neurons. Apparently the neuronal and neuroglial lineages split quite early in embryological development.

Although the glial cells have been known for over a century (Virchow recognized these cells in 1846 and subsequently coined the term *neuroglia*, "nerve glue."), new evidence indicates that these

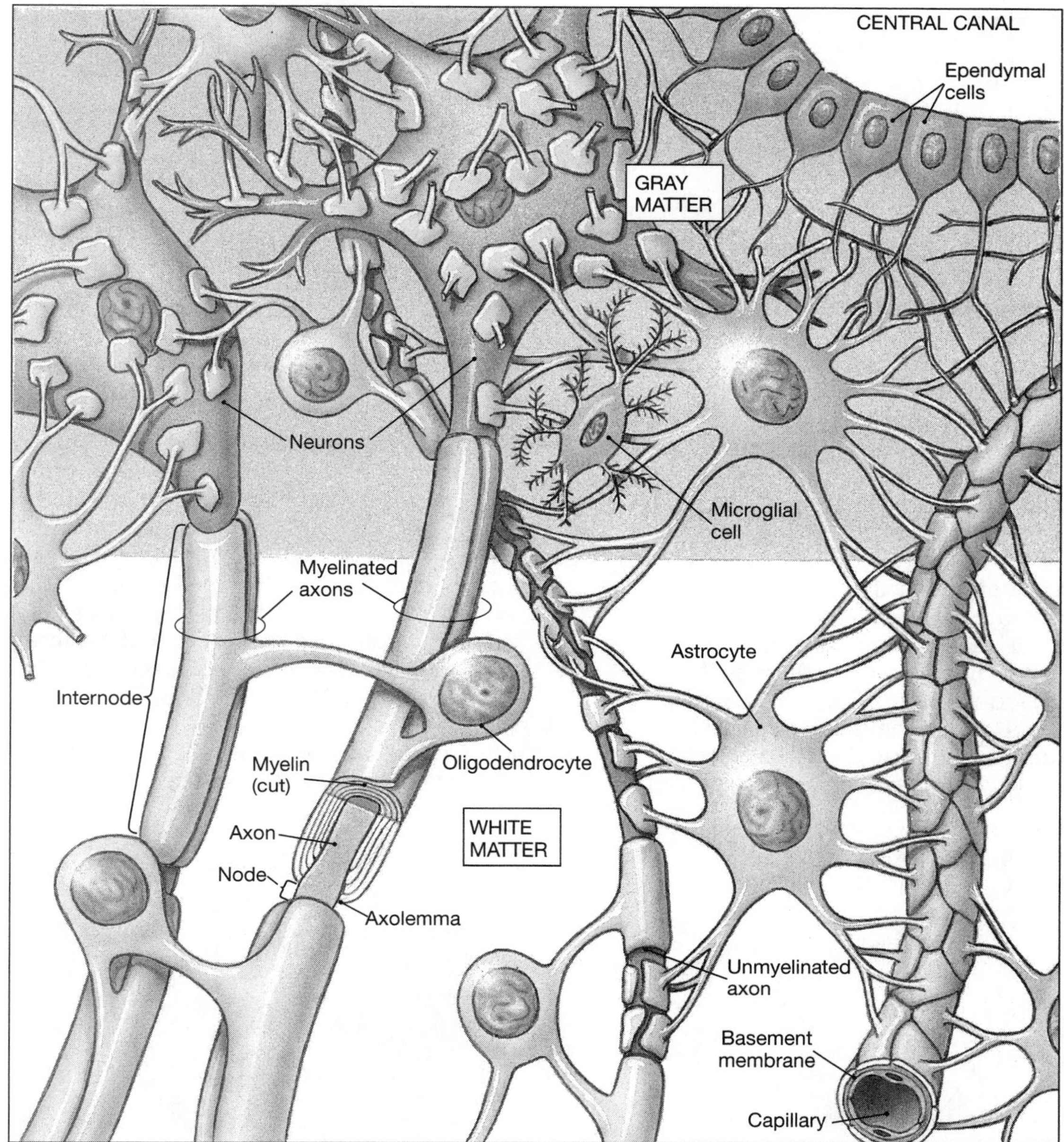

● **FIGURE 26–4**
Neuroglia in the CNS.

cells may play a far greater role in neuronal functioning than had previously been thought. As a generalization we can say that the glial cells serve in a variety of protective capacities ranging from insulation to metabolic transport. They are also apparently involved in the coordinating of information passed between nervous system components.

Of the six major neuroglial cell types, four are in the central nervous system, and the other two are in the peripheral nervous system.

B. Central Nervous System Neuroglia

1➡ Check Figure 26–4● and locate the star-shaped **astrocytes**. Note how these glial cells wrap around the CNS capillaries and how they interact with the neuronal soma. Astrocytes regulate interstitial fluid, which includes the maintenance of the blood-brain barrier, the capillary network separating the brain's blood flow from that of the rest of the body. Astrocytes also function with oligodendrocytes in insulating the neurons from direct contact with one another. *[∞ Martini, p. 368]*

2➡ Obtain the slide that may be labeled neuroglia or astrocytes. Sketch and label what you see.

Astrocytes

3➡ Observe the **oligodendrocytes,** which are also pictured in Figure 26–4●. Notice the flattened, pancake-like somas of these cells. Note, too, the myelin sheathes generated by these cells as they wrap around the CNS axons. As stated previously, myelin sheaths facilitate the speedy transmission of action potential conduction along an axon. *[∞ Martini, p. 368]*

4➡ Find the **microglia**, Figure 26–4●. These microglia are actually phagocytic white blood cells (of mesodermal origin) that crossed the line into the CNS during the neonatal period to act as a wandering clean-up crew. Usually the microglial count is rather constant—about 5% of the neuronal tissue in humans—but in times of necessity massive numbers of reserves are called in. *[∞ Martini, p. 370]*

Actually, under adverse conditions, the microglia multiply by mitosis. Because of the mesodermal origin and the mitosis, some neurobiologists feel that the microglia should not be regarded as true neuroglial elements. In times of injury or infection, other neuronal cells can also function in macrophagic (or clean-up) capacities. No one is quite certain how the mopped-up material from the injury or infection is disposed of from the CNS.

5➡ Examine the **ependymal cells**, Figures 26–4● and 26–5●. These are squamous cuboidal or columnar cells that can be found lining the ventricles and central canal of the CNS. The four ventricles are the expanded chambers within the brain, and the central canal is the narrow passageway within the spinal cord. **Cerebrospinal fluid (CSF)** is produced by specialized ependymal cells. *[∞ Martini, p. 368]*

Ependymal cells throughout the CNS apparently monitor the composition of the CSF. Other

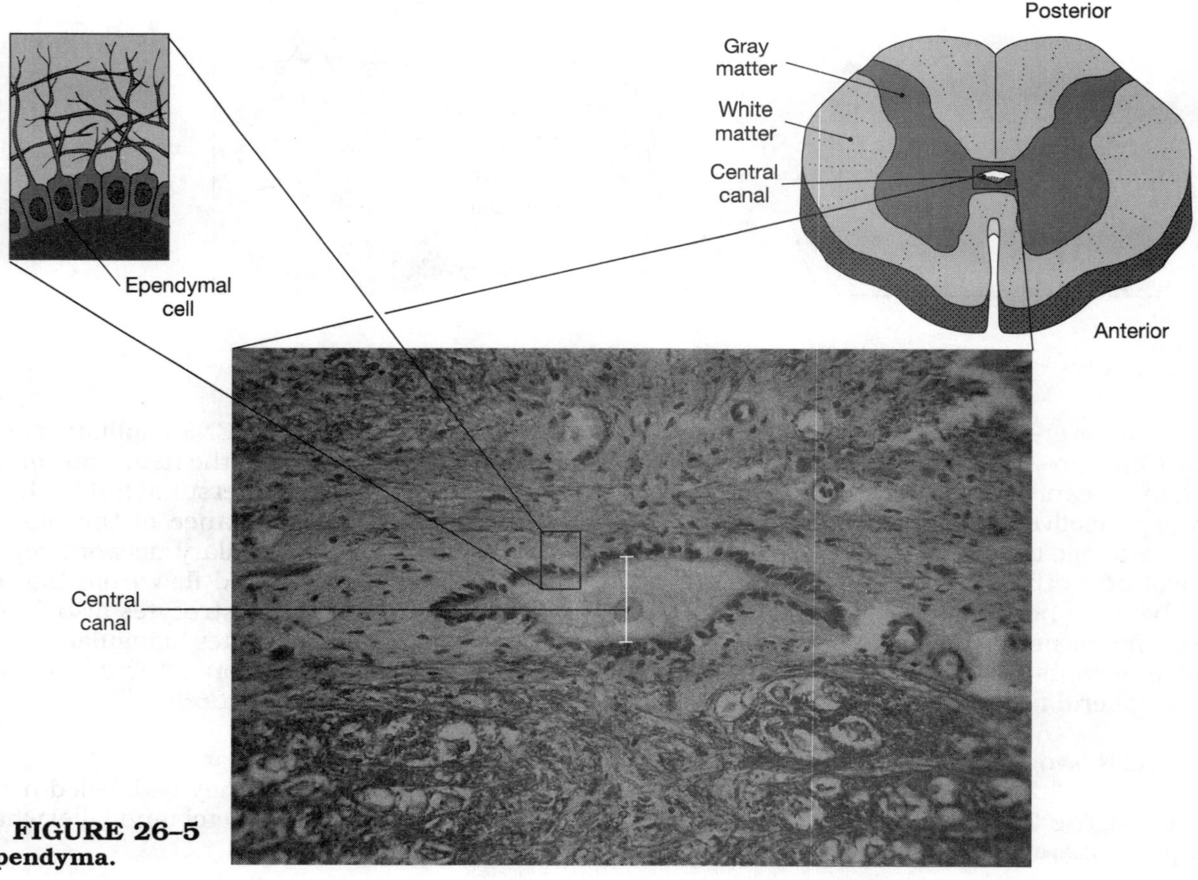

● **FIGURE 26–5**
Ependyma.

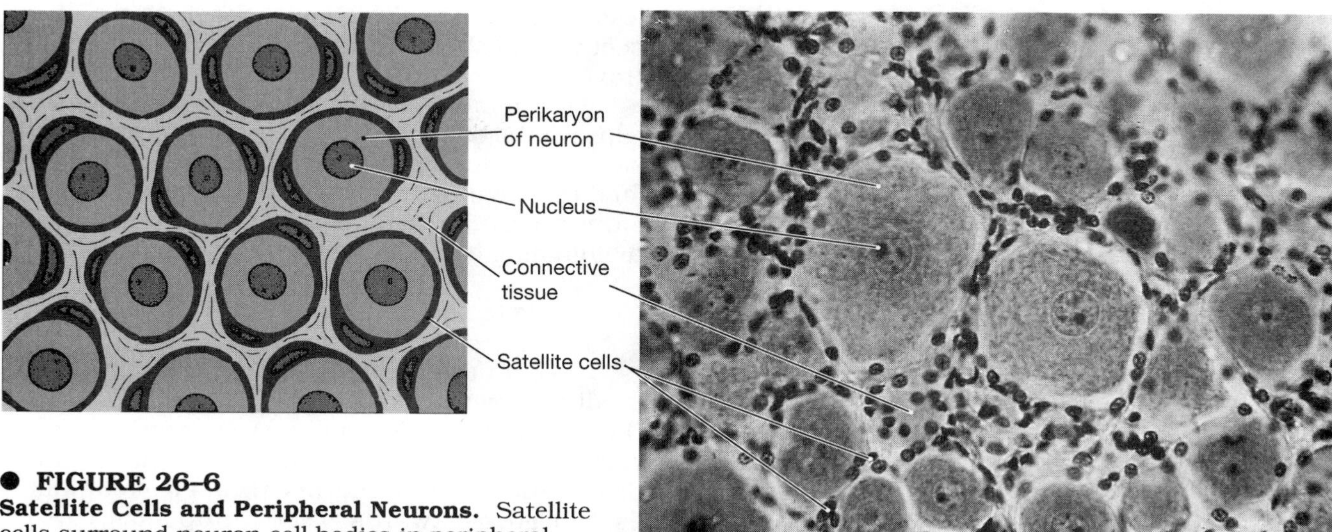

● **FIGURE 26–6**
Satellite Cells and Peripheral Neurons. Satellite cells surround neuron cell bodies in peripheral ganglia. (LM × 128)

functions of the ependyma (support, sensation, and secretion) are still under investigation.

6 ➡ Obtain the slide that may be marked "4th Ventricle" or "ependymal cells" or "choroid plexus." There are four choroid plexuses, one for each ventricle, and each is derived from the ependymal lining. The choroid plexus is actually a vascular fold of the pia mater (a meningeal layer of the CNS). The epithelial surface of this structure is the ependymal plexus. (Technically the vascular fold and the epithelium should be called the **tela choroidea**, but for most purposes calling the entire structure the choroid plexus is quite sufficient.)

From the slide you should be able to identify the elongated (cuboidal to columnar) cells that line the ventricle. You may also be able to identify the microvilli that cover the free surface of the ependymal cells. Sketch and label what you see.

Ependymal cells

C. Peripheral Nervous System Neuroglia

1 ➡ Find the **satellite cells** (or **amphicytes**), Figure 26–6●. These cells surround the nerve cell bodies. Thus, amphicytes are found in the PNS ganglia. *[∞ Martini, p. 370]*

2 ➡ Now locate the Schwann cells, the myelinating cells located outside the CNS. Return to Figure 26–1● and note the diagrammatic structure of the myelin sheath. Obviously, one function of the Schwann cell will be identical to one function of the oligodendrocyte. *[∞ Martini, p. 370]*

Concept Check 2 What is that common function between Schwann cells and oligodendrocytes?

____

____

3 ➡ Obtain the Schwann cell slide. This slide might be labeled "myelinated" or "medullated nerve fiber." Use Figure 26–6● as your guide. Sketch and label as many parts of the cells as you can identify.

Schwann cells

IV. Principles of Electrical Conduction *[∞ Martini, p. 373]*

A. Voltage Potential Across a Membrane

Membranes separate ions. Ions have positive or negative charges; therefore, voltage exists. Since there is a difference in charge across the membrane, ions have the potential to move across. This creates a

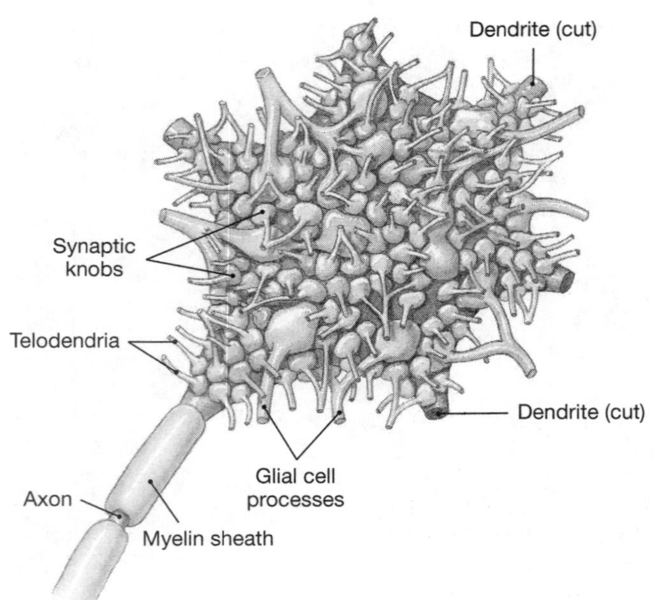

● **FIGURE 26–7**
Synaptic Interactions.

membrane or voltage potential. This voltage potential is requesite for an action potential.

B. Action Potential = Nerve Impulse

The resting membrane potential is the imbalance in charge existing across a membrane. The difference in charge is measured in millivolts. As the difference in charge decreases, the threshold potential is reached. When the threshold potential is reached, the neuron "fires." The "firing" is the quick jump to the action potential.

An action potential can begin anywhere. However, we usually think of an impulse beginning when a dendritic ending is stimulated. We should not forget that neuronal soma also receive information from the synaptic knobs of other nerve cells. See Figure 26–7●. It is the ability to carry a charge — the difference or separation in ionic charges — that allows for the action potential to exist.

C. Ion Channels

Electrical conduction is enhanced by the presence of ion channels. Ion channels are points in a membrane which, upon proper electrical or chemical stimulation, allow the passage of specific ions. Sodium channels allow the passage of sodium ions; potassium channels, potassium; chloride channels, chloride; calcium channels, calcium.

Neuronal axons have a high concentration of sodium channels. When these channels are stimulated, sodium ions rush into the axon. The normal millivolt charge on the inside of the axon is negative; sodium ions are positive. Thus, the net charge on the inside of the membrane becomes less negative. The threshold is reached; excitation occurs and the action potential (nerve impulse) traverses the axon.

Keep in mind that the impulse is propagated because of the difference in ionic charges between the sides of the membrane. This difference is achieved (and restored) by the activity of the sodium/potassium pump, an ionic carrier which functions in maintaining the membrane's resting potential.

D. Direction

Once a new action potential begins, it travels in all directions from the point of initiation (along the axon hillock). This includes spanning the neuronal cell body and even moving back into the dendrites. You might think this means nerve impulses can go backwards. (Nerve impulses can go backwards in jellyfish and sponges and certain other invertebrates.)

Theoretically, impulses could go backwards in humans too. The intrinsic nature of the neuronal tissue makes direction irrelevant. However, impulses do not normally go backwards because the **excitation threshold** is too great.

E. Excitation Threshold

Normal neuronal excitation is about –45 millivolts. (Recall that –45 millivolts means that the "inside" charge — ionic imbalance — is more negative than the "outside" charge.) Normal neuronal resting potential is about –65 millivolts. Thus, for the electrical impulse to travel backwards and jump across that dendritic synapse in the opposite direction, it will have to be +20 millivolts more positive than the membrane. That would be achievable on the axon hillock or along the axon where sodium channels are abundant, where sodium can rush in and decrease the electronegativity of the membrane.

Sodium channels allow for quick shifts in electronegativity and electropositivity. The dendrites and the neuronal soma have very few voltage-gated sodium channels. Remember, initial excitation is achieved from external stimulation, and dendritic excitation from internal stimuli cannot usually be achieved. After a few feeble attempts at backtracking, the action potential becomes a one–way trip straight along the axon and away from the cell body.

F. Identification of Modulating Factors [∞ Martini, p. 392]

If the threshold of excitation is made more positive, or if the internal membrane itself is made more negative, the increase in negativity is called an **inhibitory postsynaptic potential** or **IPSP**.

If the gap between the membrane potential and the excitatory threshold decreases, we call this an **excitatory postsynaptic potential** or **EPSP**. In other words, if a stimulus comes into a neuron, factors can influence whether that stimulus will generate an action potential. Two or more influencing factors can negate each other.

Sodium causes an EPSP because as sodium influxes, the inner membrane becomes less negative; less push is necessary to excite the membrane. Axonic sodium channels are abundant. Action potentials zip along.

G. Transmitters

We mentioned earlier that most synaptic proteins are produced in the perikaryon. These synaptic proteins travel down the axons to the synaptic knobs. Many of these proteins are neurotransmitter substances (of which close to 100 have been identified). Some of these neurotransmitters are ISP's or ESP's.

When the electrical impulse reaches the knob, neurotransmitters are released. The transmitters cross the synaptic cleft and attach to the postsynaptic membrane of a muscle fiber or another neuron. The impulse continues or is modulated accordingly.

❑ Experimentation

V. Nervous System Demonstrations

A. Time and Electrical Conduction

This is a four-part demonstration to show you that electrical conduction is time-related. Your instructor will direct these experiments for the class as a whole and will use the board to record the times involved.

1➡ Everyone stand in a line. If you are right-handed, put your right hand on the shoulder of the person in front of you. If you are left-handed, put your left hand on the shoulder of the person in front of you. If that person is right-handed, put your hand on his or her *right* shoulder. If that person is left-handed, put your hand on his or her *left* shoulder.

When the instructor, while carefully noting the exact time, says, "Go," the last person in line squeezes the shoulder of the person in front of him or her. When that person feels the squeeze, he or she squeezes the next person. The last person to receive the squeeze should immediately signal the instructor, who can then make note of the time. The number of seconds required for the electrical signal to pass through the entire line should be recorded. Do this three times and take an average. (If your class is large enough, your instructor might wish to divide the class in two and see if the time is comparable.)

2➡ Everyone stand in line. If you are right-handed, put your left hand on the shoulder of the person in front of you. If you are left-handed, put your right hand on the shoulder of the person in front of you. Repeat the experiment as above. Do this three times and take an average.

3➡ This time put both hands on the shoulders of the person in front of you. Squeeze together. Do this three times and take an average.

4➡ This time put both hands on the shoulders of the person in front of you. However, the last person starts with a right-hand squeeze, but the next person must pass it along with a left-hand squeeze. That next person does a right hand squeeze, and so forth. Do this three times and take an average.

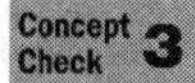

Analyze your results from these experiments.

____

____

❑ Additional Activities

NOTES

1. Certain endocrine glands are believed to be parts of the nervous system. Research the embryology and morphology of this concept.
2. Research the embryology of the nervous system.
3. Research some of the people whose discoveries played a role in our present understanding of the nervous system:
   Alois Alzheimer (1864–1915), German neurologist
   Otto Loewi (1873–1961), German pharmacologist
   Franz Nissl (1860–1919), German neuroanatomist

NOTES

James Parkinson (1755–1824), English physician
Louis A. Ranvier (1835–1922), French pathologist
Bernard Sachs (1858–1944), American neurologist
Theodor Schwann (1810–1882), German histologist
Warren Tay (1843–1927), English physician

4. We generally recognize six types of neuroglia, although by the time all the subdivisions are classified, the numbers of types, subtypes, and sub-sub-types will probably be in the hundreds. Research some of these proposed subdivisions.
5. Perhaps only a fine line of distinction exists between astrocytes and oligodendrocytes. Some neurologists claim that astrocytes and oligodendrocytes exist on a cellular continuum, arising from a common cell line, and that transitional forms of these neuroglia do appear in certain pathological conditions. Research why some scientists believe this.

Answers to Selected Concept Check Questions

1. Ribosomes function in protein production.
2. The Schwann cells and the oligodendrocytes both produce myelin sheathes and thus both protect and facilitate electrical transmission.
3. In Experiment #1 everyone is using his/her dominant brain hemisphere. The message must still cross over.

Experiment #2 should take a longer time than #1 because everyone is using the nondominant hemisphere.

Experiment #3 should be as fast as or faster than #1 because the message is reaching both sides of the brain simultaneously, and no real concentration is required.

Experiment #4 should be the slowest because not only does the message have to cross over in your brain, but also you must make a conscious effort to use the opposite side.

Keep in mind too that conduction time is longer if the information has to pass through the corpus callosum (e.g., when a signal to your right shoulder requires a response from your left hand; in that case, the signal reaches the left hemisphere, but the command to the left hand must come from the right hemisphere). See Exercise 30.

❑ Lab Report

1. Referring to the numbered inquiries at the beginning of this exercise, complete the following box summary:

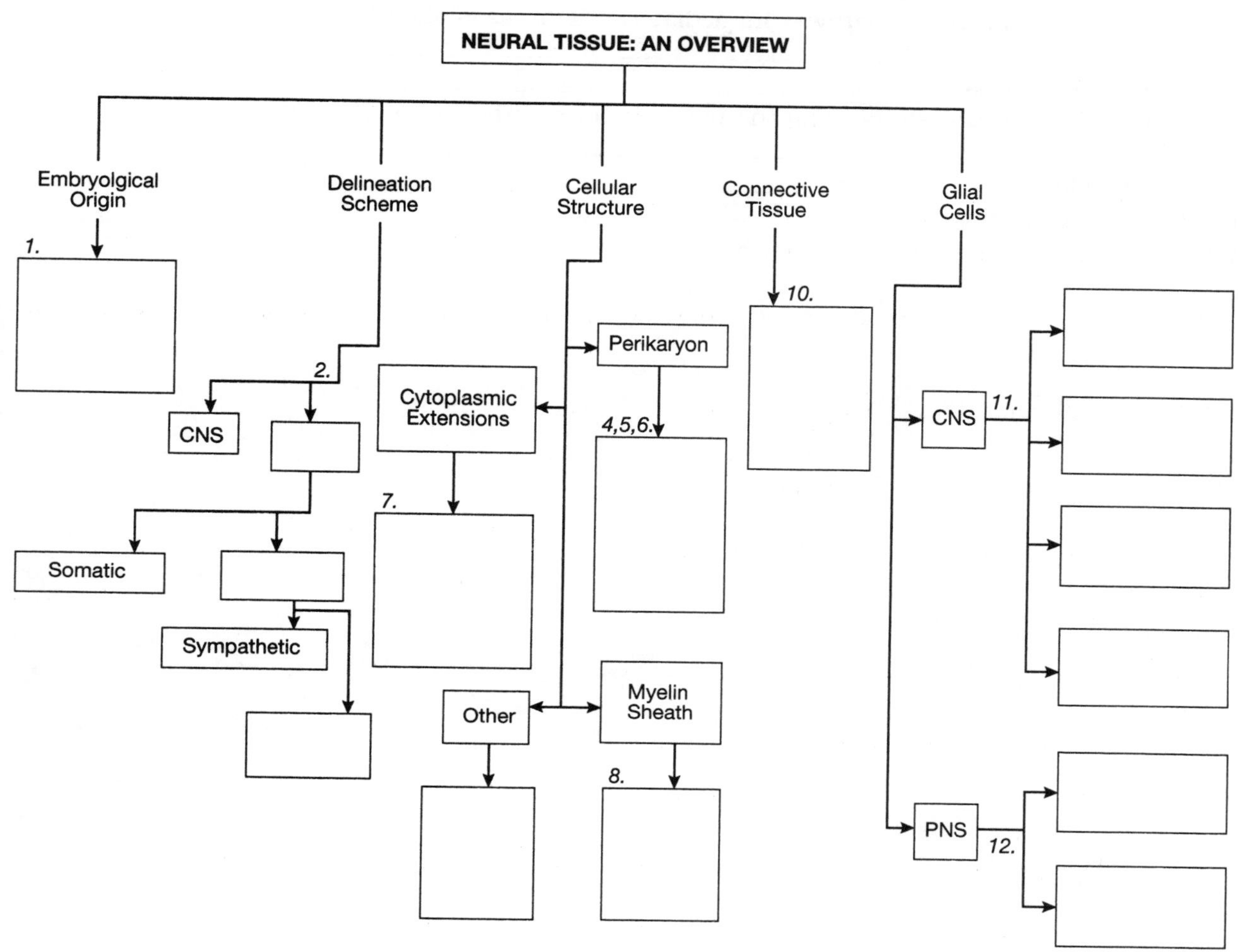

General Questions

NOTES

1. In what ways are neuronal cells similar to any other eukaryotic cells? In what ways are they different?
2. According to axonic and dendritic arrangement, what are the four major types of neurons, and what are the distinguishing characteristics of each?
3. Explain: In theory nervous tissue can conduct in any direction. In practice, it does not.
4. Explain the similarities and differences between the neurons and the glia.
5. Why is gray matter gray?
6. What is the difference between a nerve and a neuron?
7. What is the difference between a nucleus and a ganglion?

8. Use the correct term for each of the following definitions or events:
   a. The space between two conducting cells ____________
   b. Occurring in the nerves at the beginning of a series of electrical events ____________
   c. Neurons associated with the digestive tract ____________
   d. Neuron carrying message to the sartorius muscle ____________
   e. Occurring on a nerve cell pathway immediately after a mass of nerve cell bodies ____________
   f. The electrical message itself ____________
   g. Portion of the nerve cell carrying the message toward the cell body ____________
   h. Synonym for sensory neuron ____________
9. Describe electrical conduction.
10. What is the difference between IPSP and EPSP?

NOTES

EXERCISE 27

Anatomy of the Spinal Cord

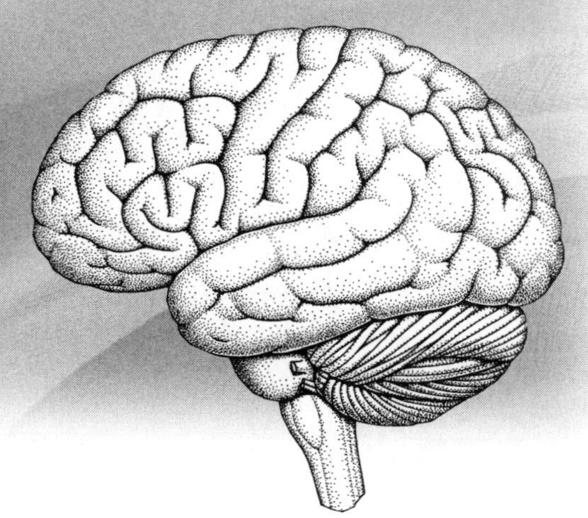

PROCEDURAL INQUIRIES

Examination

1. What is the bony structure encasing the spinal cord?
2. What are the names and locations of the meningeal layers?
3. Where are the spinal cord enlargements found?
4. What terms are used to describe the caudal portions of the spinal cord?
5. What are the major landmarks found on a spinal cord cross section?
6. What are the anatomical and physiological differences between the ascending and descending tracts of the spinal cord?

Additional Inquiries

7. What are two functions of the spinal cord?
8. How can we compare the spinal cord of a child with the spinal cord of an adult?

Key Terms

Arachnoid Layer
Cauda Equina
Cerebrospinal Fluid (CSF)
Conus Medullaris
Dura Mater
Epidural Space
Filum Terminale
Pia Mater
Subarachnoid Space
Subdural Space

Materials Needed

Skeleton (or vertebral column)
Classroom Models of Spinal Cord — cross section
Charts and Diagrams
Compound Microscope
Stereoscopic Microscope
Prepared Slides
- Conus Medullaris (if available)
- Cauda Equina (if available)
- Spinal Cord — Smear
- Spinal Cord — Serial Sections

The spinal cord, an aggregate of neural and support tissues, is a relay system composed of millions of very different cells. Neural tissues, which include tiny neural cell bodies with elongated cytoplasmic extensions called axons and dendrites, begin forming when the neural portion of the ectoderm folds over itself very early in embryonic development.

Recall from Exercise 26 and from Martini, Chapter 12, that an axon carries an impulse away from a nerve cell body and a dendrite carries an impulse toward a nerve cell body. Spinal cord anatomy is found in Chapter 13 of Martini.

The spinal cord is a continuation of the brain. Many reflexes, such as the classic example of moving one's hand upon touching a hot stove, operate through the spinal cord, independent of the brain. However, any action requiring conscious or unconscious coordination cannot occur without a relay of information, via the spinal cord, to or from one or more brain centers. In this laboratory exercise we will examine various anatomical components of the spinal cord.

❑ Preparation

Although no direct reference is made in this lesson to classroom charts and diagrams, you are encouraged to use these aids plus any other aids you may have in your lab to help you understand the anatomy of the spinal cord.

❑ Examination

I. Spinal Cord, Longitudinal View

A. Protective Coverings of the Spinal Cord

[∞ Martini, p. 409]

1➡ Observe the classroom skeleton. The human spinal cord is encased in the vertebral column. Make note of the column of vertebral foramina through which the spinal cord passes. (Refer to Exercise 16 if necessary.) Note, too, the intervertebral foramina that are passageways for the 31 pairs of spinal nerves.

2➡ Note that the spinal cord is also protected from homeostatic disruptions by the same three meningeal layers that can be found surrounding the brain. The outermost layer is the tough, fibrous **dura mater** (literally, "tough mother"). Exterior to the dura mater is the **epidural space**, which cushions the cord from onslaught by the vertebrae. The epidural space contains loose connective tissue, blood vessels, and adipose tissue.

Deep to the dura mater is the **subdural space**, which houses small quantities of lubricating lymphatic fluid. The middle meningeal layer is the **arachnoid** (spider) **layer**, and deep to the arachnoid layer is the **subarachnoid space**, which contains **cerebrospinal fluid** (CSF). The CSF functions both as a shock absorber and as a homeostatic medium.

The innermost layer, the **pia mater** (literally, "gentle mother"), is a meshwork of elastin and collagen tightly bound to the neural tissue.

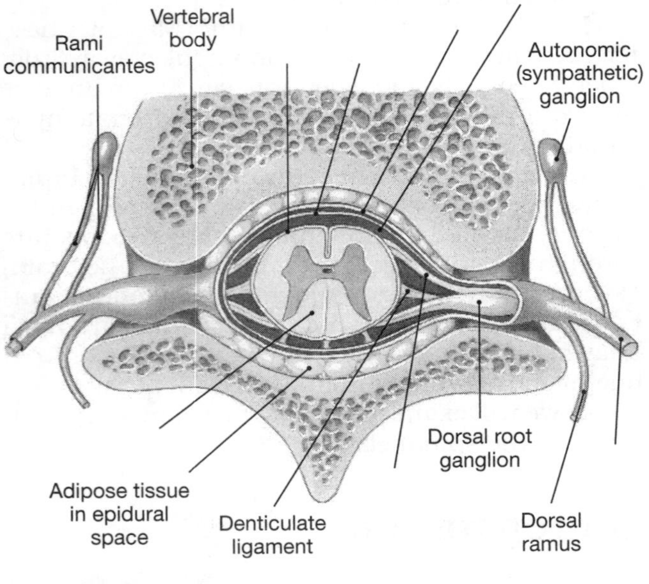

● FIGURE 27–1
Cross Section of the Vertebral Column, Meningeal Layers, and Spinal Cord.

3➡ Write labels on Figure 27–1● for the structures listed in bold print in the preceding section.

B. The Cauda Equina

1➡ In the adult, the spinal cord proper extends only to about the first or second lumbar vertebra. Locate this point on the skeleton. Refer also to Figure 27–2●. Although the cord ceases with a tapering area called the **conus medullaris**, the more caudal nerve roots radiate inferiorly from the lower part of the cord. These fibers are collectively known as the **cauda equina** because they resemble a horse's tail.

2➡ Note on Figure 27–2● how the cauda equina continues giving off fibers on through the sacral foramina. The last remaining piece of the spinal cord is the **filum terminale**, a fibrous strand composed mostly of pia mater.

3➡ Were you able to compare these three areas on the skeleton with Figure 27–2●?

____

C. Slides

1➡ If available, examine the prepared microscope slides of the cauda equina and the conus medullaris. Sketch and label what you see.

Cauda equina	Conus medullaris

2➡ If you have not previously examined a spinal smear slide, or if your instructor feels a review is in order, study this slide to gain a perspective on the size and complexity of the spinal cord. Recall that a spinal cord smear is a "piece" of spinal cord that has been spread out and stained for viewing. Sketch and label what you see.

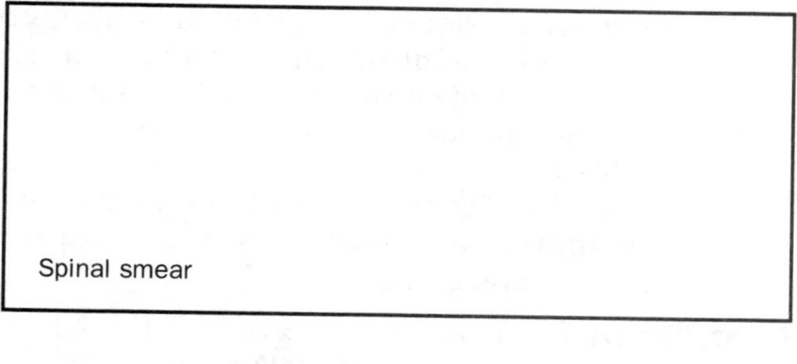

D. Other Features of the Spinal Cord

1 ➡ Reexamine Figure 27–2●. Notice the **cervical enlargement** and the **lumbar enlargement**, and try to determine the logical reason for these enlargements. Write out your ideas here.

____

____

2 ➡ Again referring to Figure 27–2●, list the differences you see between the spinal cord of the adult and the spinal cord of the child.

____

____

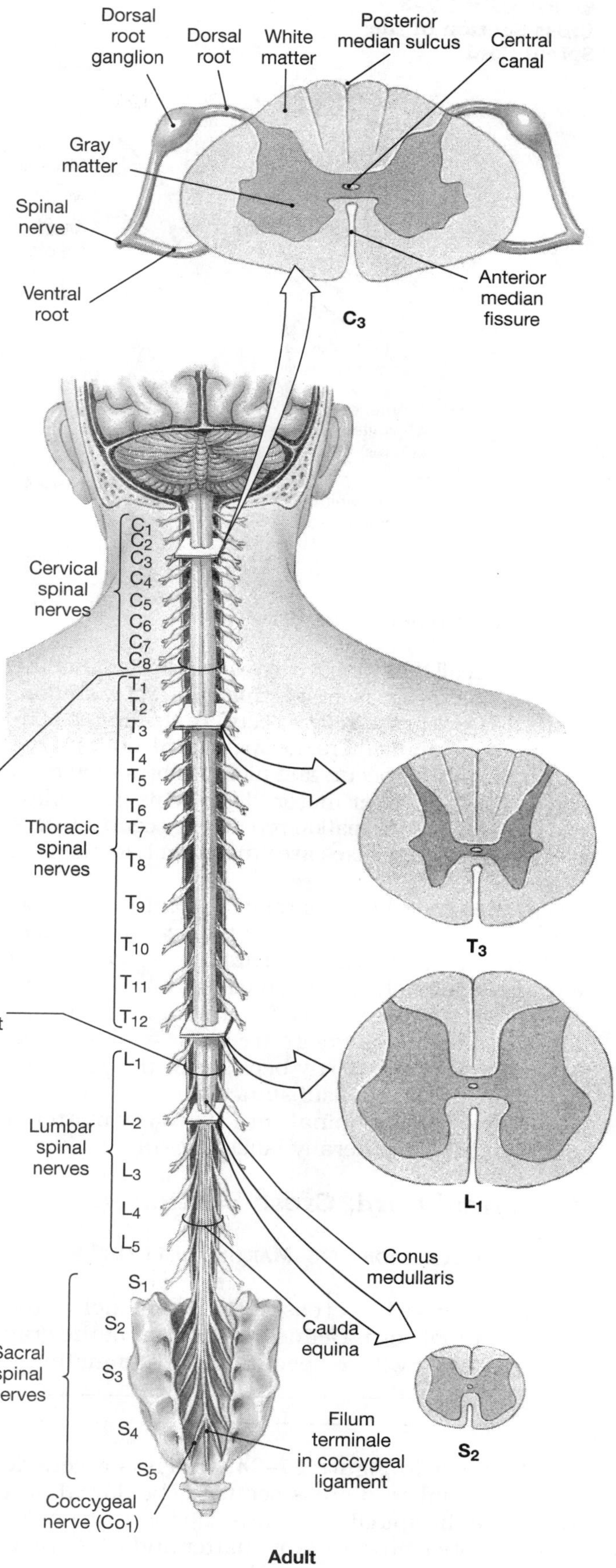

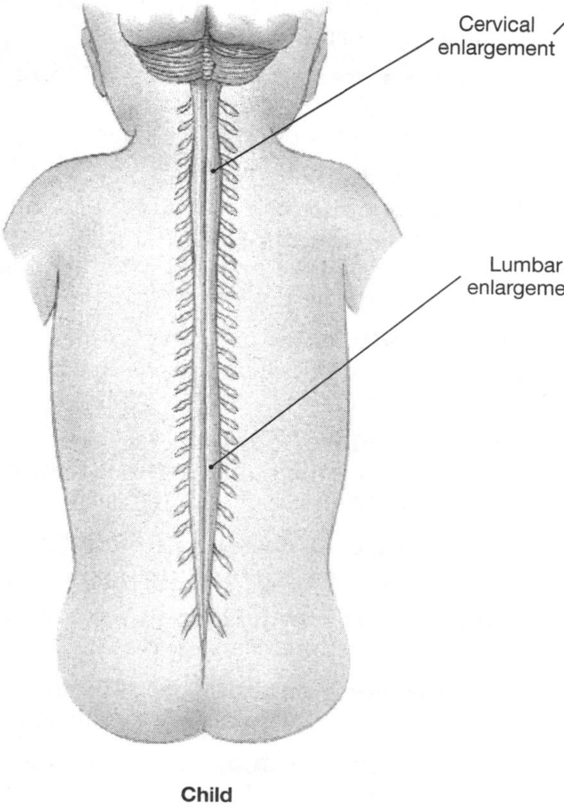

Dorsal root ganglion
Dorsal root
White matter
Posterior median sulcus
Central canal
Gray matter
Spinal nerve
Ventral root
Anterior median fissure
C_3
Cervical spinal nerves
C_1 C_2 C_3 C_4 C_5 C_6 C_7 C_8
Cervical enlargement
Thoracic spinal nerves
T_1 T_2 T_3 T_4 T_5 T_6 T_7 T_8 T_9 T_{10} T_{11} T_{12}
T_3
Lumbar enlargement
Lumbar spinal nerves
L_1 L_2 L_3 L_4 L_5
L_1
Conus medullaris
Cauda equina
Sacral spinal nerves
S_1 S_2 S_3 S_4 S_5
S_2
Filum terminale in coccygeal ligament
Coccygeal nerve (Co_1)
Child
Adult

● **FIGURE 27–2**
Gross Anatomy of the Spinal Cord.

● **FIGURE 27–3**
Cross Section of the Spinal Cord.

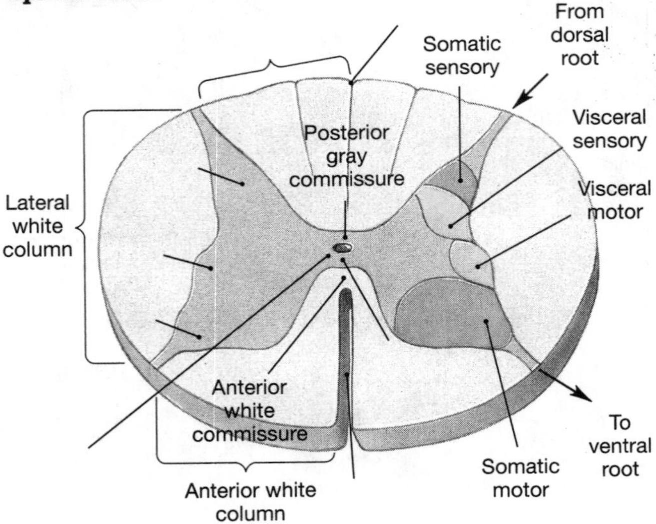

E. Functions of the Spinal Cord

1. Recall that we have already indicated that the spinal cord is an informational relay station. Two types of relays occur. In the first, a sensory input enters the spinal cord, one or more synapses occur, and motor information exits to the proper motor fibers. This is a reflex, and no information processing occurs. (In this manual reflexes are studied in Exercise 29.)

In the second type of relay, incoming information synapses to nerve tracts that carry the message to the brain. The information is acted upon either consciously or unconsciously.

The areas responsible for transmitting the information via nerve tracks are in the white matter. This area is white partly because of the presence of myelin, the lipoidal substance surrounding most of the axons of the spinal cord. The gray matter is gray because it generally lacks myelin.

II. Spinal Cord, Cross Section

A. Description [∞ Martini, p. 412]

1. Look at the cross-sectional model of the spinal cord. Define the location of the gray matter with respect to the white matter.

____

____

2. Refer to Figure 27–3●, which is a generic spinal cord cross section. The dorsal part of the spinal cord faces the top of the page. Notice how the gray matter and white matter seem to be neatly packaged into distinct areas. Notice, too, that the right and left halves are mirror images of each other; for the sake of clarity, we have drawn some lines on one side and some on the other.

3. Label the Figure 27–3● Spinal Cord, with the appropriate boldface terms. Read the descriptions carefully.

Central canal "Hole" in the center of the spinal cord, continuous with the 4th ventricle of the brain and filled with CSF.

Anterior median fissure Centrally located groove that begins on the anterior edge and continues toward the central canal.

Posterior median sulcus Centrally located groove that begins on the posterior edge and continues toward the central canal.

Posterior gray horn Area of gray material extending on a diagonal toward the lateroposterior aspect of the cord.

Anterior gray horn Area of gray material extending on a diagonal toward the lateroanterior aspect of the cord.

Lateral gray horn "Bulge" on the side of the gray matter.

Posterior white column (posterior funiculus) White area medial to and between the posterior gray horns.

Fasciculus gracilis Medial portion of posterior white column. (This name of an ascending spinal tract is often used when referring to this portion of the posterior white column.)

Fasciculus cuneatus Lateral portion of posterior white column. (This name of an ascending spinal tract is often used when referring to this portion of the posterior white column.)

Lateral white column (lateral funiculus) "Side" white area between the posterior and anterior gray horns.

Anterior white column (anterior funiculus) White area medial to and between the anterior gray horns.

Commissure Central bar, including both gray and white areas, surrounding the central canal.

Anterior gray commissure Gray strip immediately anterior to central canal.

Anterior white commissure White area immediately anterior to anterior gray commissure.

Posterior gray commissure Gray strip immediately posterior to central canal.

Posterior white commissure white area immediately posterior to posterior gray commissure. (This term is not commonly used.)

Origin		Destination[a]	Sensations/Actions	Comments
ASCENDING (SENSORY) TRACTS				
The Posterior Column Pathway:				
Fasciculus gracilis	Proprioceptors and fine touch, pressure, vibration receptors of lower body	Nucleus gracilis (medulla)	Position, fine touch, pressure, vibration	Ascends, on same side as stimulus
Fasciculus cuneatus	Proprioceptors and fine touch, pressure, vibration receptors of upper body	Nucleus cuneatus (medulla)	Position, fine touch, pressure, vibration	Ascends, on same side as stimulus
The Spinocerebellar Pathway:				
Posterior spinocerebellar	Interneurons relaying information from proprioceptors	Cerebellum	Proprioception	Ascends, on same side as stimulus
Anterior spinocerebellar	Interneurons relaying information from proprioceptors	Cerebellum	Proprioception	Crosses to ascend on side opposite stimulus
The Spinothalamic Pathway:				
Lateral spinothalamic	Interneurons relaying information from pain and temperature receptors	Thalamus (ventral nuclei)	Pain and temperature	Crosses to ascend on side opposite stimulus
Anterior spinothalamic	Interneurons relaying information from crude touch and pressure receptors	Thalamus (ventral nuclei)	Crude touch and pressure	Crosses to ascend on side opposite stimulus
Descending (Motor) Tracts				
Pyramidal Tracts:				
Lateral corticospinal	Primary motor cortex (cerebral hemispheres)	Motor neurons of anterior gray horns	Voluntary motor control of skeletal muscles	Crosses to opposite side before entering spinal cord
Anterior corticospinal	Same	Same	Same	Descends uncrossed but crosses to opposite side before synapsing
Extrapyramidal Tracts:				
Rubrospinal	Red nucleus (midbrain)	Motor neurons of anterior gray horns	Involuntary regulation of posture and muscle tone	Crosses to opposite side before entering spinal cord
Reticulospinal	Reticular formation (network of nuclei in brain stem)	Somatic and visceral motor neurons of anterior and lateral gray horns	Involuntary regulation of reflex activity and autonomic functions	Descends without crossing to opposite side
Vestibulospinal	Vestibular nucleus (near rostral border of medulla)	Motor neurons of anterior gray horns	Involuntary regulation of balance and muscle tone	Descends without crossing to opposite side
Tectospinal	Tectum (midbrain)	Motor neurons of anterior gray horns of cervical spinal cord	Involuntary regulation of eye, head, neck, and arm position in response to visual and auditory stimuli	Crosses to opposite side before entering spinal cord

[a]Location of first synapse.

● **FIGURE 27–4**
Principal Tracts of the Spinal Cord.

B. Slides

1➡ Obtain the slides of the serial sections of the spinal cord. Follow the major landmarks from the cervical region to the conus medullaris.

2➡ Observe the differences in the sizes and shapes of the various internal spinal cord structures from one section of the spinal cord to another.

3➡ Sketch several representative spinal cord sections. Use the extra space at the end of this chapter.

Spinal cord

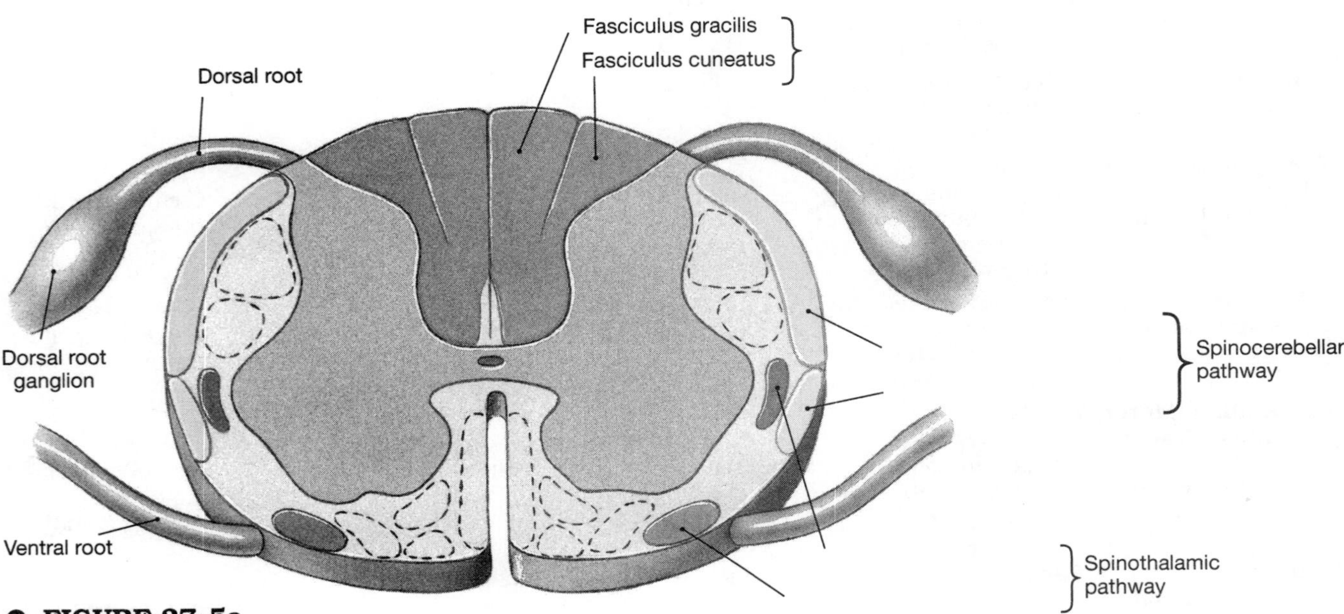

● **FIGURE 27–5a**
Ascending Tracts in the Spinal Cord.

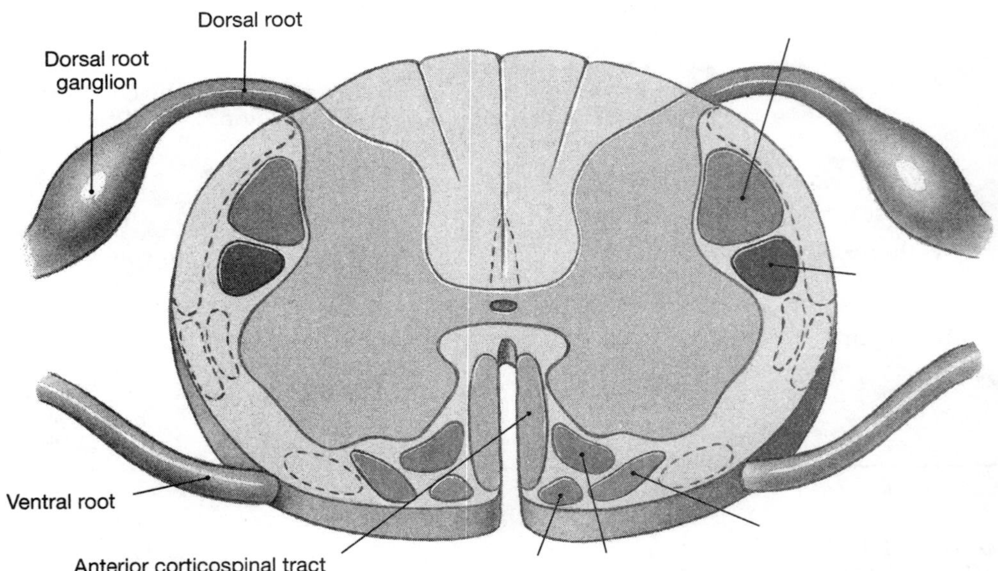

● **FIGURE 27–5b**
Descending Tracts in the Spinal Cord.

III. Spinal Tracts

A. General Description

1➡ The spinal tracts are the informational relay pathways between the spinal cord and the brain, composed of bundles of axonic and dendritic fibers, which transport electrical messages from one end of the system to the other.

2➡ The ascending tracts are the sensory tracts. That is logical if you think that the brain must receive a sensation message in order to integrate information, either voluntarily or involuntarily. The descending tracts are motor tracts. That, too, is logical; think of the brain as dispatching a message to a muscle.

B. Principal Tracts

1➡ Refer to Figure 27–4●, which delineates the principal tracts. Notice that the names of all the nerve tracts define both the origin and the destination of each bundle of fibers. For example, the spinothalamic tract runs from the spinal cord to the thalamus. Since "spino-" is the first part of the word, the tract originates in the spinal cord. Therefore, it must be an ascending (sensory) tract.

2➡ Read through Figure 27–4●.

3➡ Label the appropriate areas of Figure 27–5● with the names of the spinal tracts, referring as needed back to the labels you wrote on Figure 27–3●. In Figure 27–4● the right-side tracts are ascending tracts, and the left-side tracts are descending tracts. This is a matter of convenience; keep in mind that both sets of tracts are found on both sides of the spinal cord.

4➡ What difficulties did you have in labeling any of the tracts?

____

____

C. Crossing Over

The "Comments" column of Figure 27–4● mentions crossing over. This means that the fibers of several tracts cross from one side of the spinal cord to the opposite side.

List the tracts that cross over. ______________

____

____

❑ Additional Activities

1. Construct a model spinal cord demonstrating the spinal tracts that do and do not cross over.
2. Find out the similarities and differences between the spinal cord of humans and the spinal cords of other animals.
3. Research the embryological development of the spinal cord.

NOTES

❑ Lab Report

1. Referring to the numbered inquiries at the beginning of this exercise, complete the following box summary:

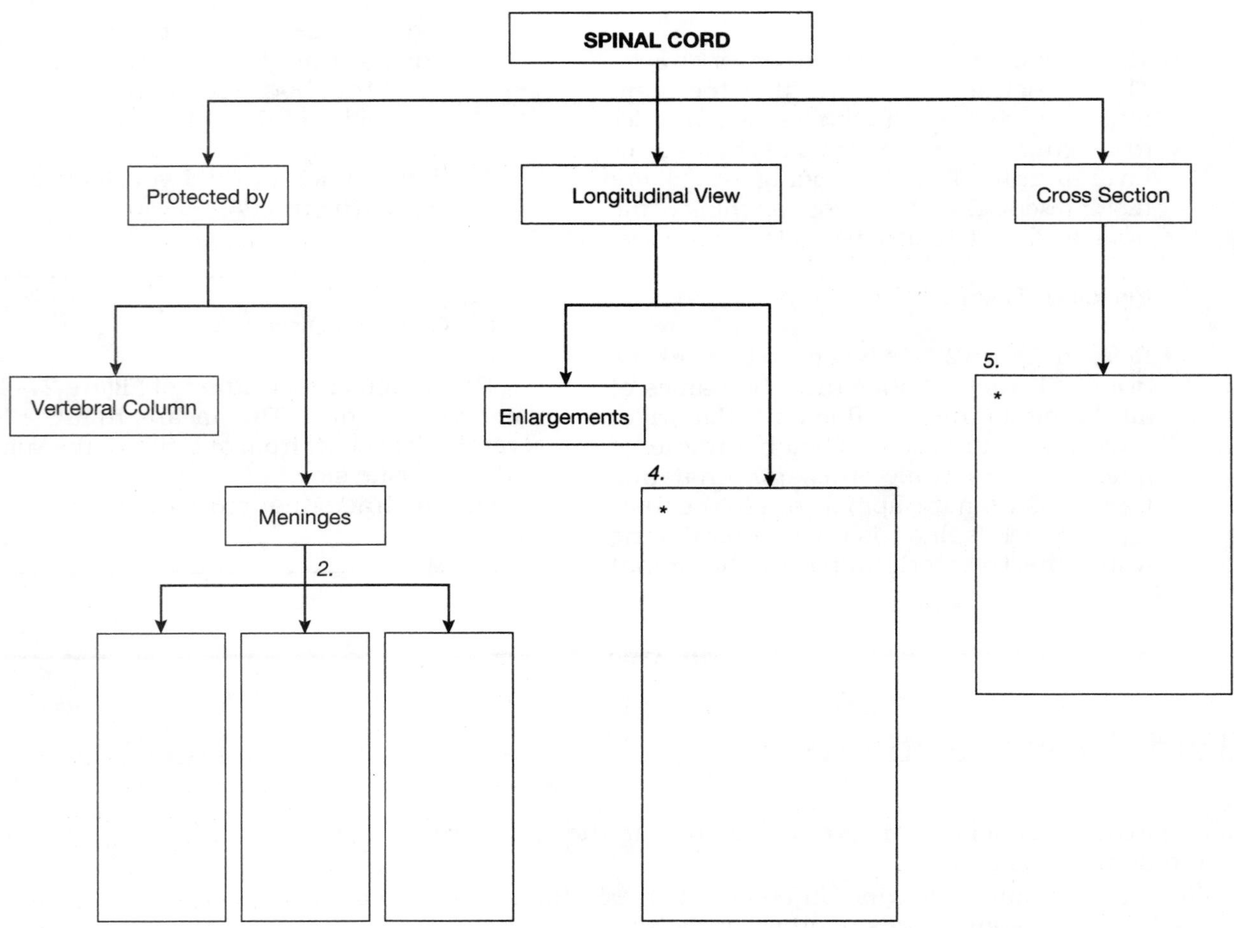

*List according to your instructor's directions.

General Questions

NOTES

1. What are the functions of the spinal cord?
2. List and describe the protective coverings of the spinal cord.
3. Compare the anatomy of the white matter with the anatomy of the gray matter.
4. What are the anatomical and physiological differences between the ascending and descending spinal tracts?
5. What are the similarities and differences between the spinal cord of the adult and the spinal cord of the child?

EXERCISE 28

Spinal Nerves

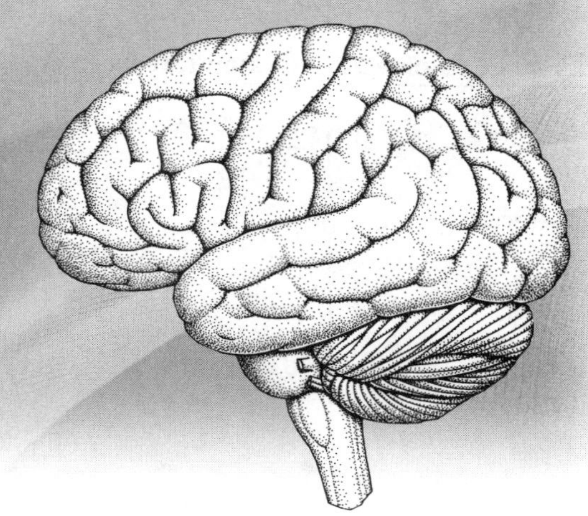

PROCEDURAL INQUIRIES

Preparation

1. How many spinal nerves are there?
2. How are the spinal nerves named?

Examination

3. What is the difference between the dorsal root and the ventral root of the spinal cord? How does this relate to the spinal nerves?
4. What are the dorsal and ventral rami?
5. What is a dermatome?
6. How are the dermatomes identified?
7. What are the three plexuses?
8. What are the divisions of the lumbosacral plexus?

Additional Inquiries

9. To which branch of the nervous system do the spinal nerves belong?
10. Embryologically, how are the spinal nerves formed?

Key Terms

Brachial Plexus
Cervical Plexus
Dermatome
Dorsal Ramus
Dorsal Root
Dorsal Root Ganglion
Gray Ramus
Intervertebral Foramina
Lumbar Plexus
Lumbosacral Plexus
Meningeal Branch
Plexus
Rami Communicantes
Ramus (pl., Rami)
Sacral Plexus
Ventral Ramus
Ventral Root
White Ramus

Materials Needed

Skeleton (or Vertebral Column)
Compound Microscope
Prepared Slides (some of these slides may not be available in your laboratory)
- Peripheral (Spinal) Nerve — cross section
- Dorsal Root Ganglion
- Sympathetic Ganglion
- Sensory Nerve Endings
- Motor Nerve Endings

In the previous exercise we examined the spinal cord, one of the components of the central nervous system. Intimately involved with the spinal cord are the spinal nerves, which are components of the peripheral nervous system.

Embryologically, the peripheral nervous system splintered from the central nervous system early in development. It is not yet completely understood what neuronal growth factors are involved as the embryonic motor and sensory fibers migrate toward their respective target tissues. Nevertheless, both the relative position of one nerve to another and the location of the particular embryonic tissue seem to be involved.

In this laboratory exercise we will examine the major components of the sensory and motor aspects of the spinal nerves and their peripheral projections.

❑ Preparation

I. Background

A. PHYSICAL POSITION

1➡ Thirty-one pairs of spinal nerves exit the spinal cord. The neurons of these spinal nerves synapse with various other neurons, forming the complex innervation scheme known as the peripheral nervous system.

2➡ Examine the classroom skeleton (or vertebral column). Note the **intervertebral**

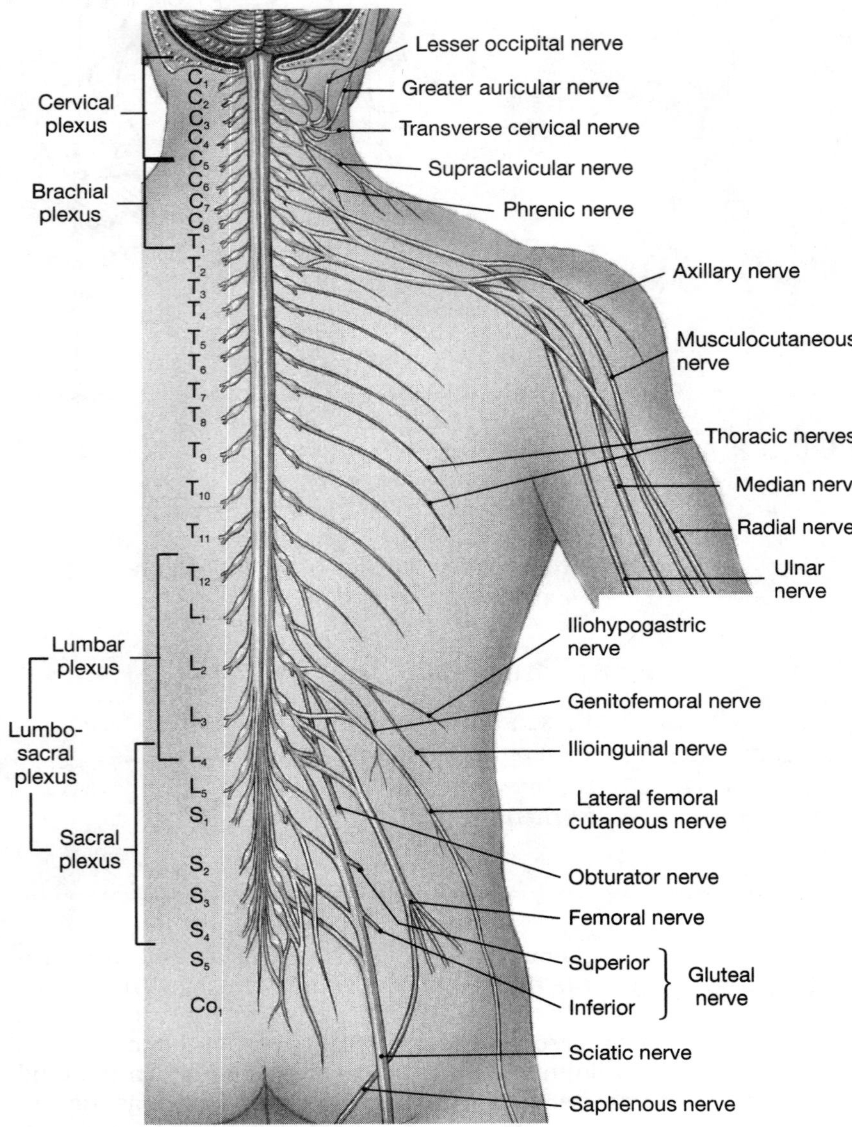

● **FIGURE 28–1**
Peripheral Nerves and Plexuses.

foramina, those lateral openings in the vertebral column from which the spinal nerves exit.

3➜ Compare your skeletal observations to Figure 28–1●.

____

____

B. Naming of the Spinal Nerves

1➜ The thoracic, lumbar, and sacral spinal nerves customarily take their names from the vertebrae immediately preceding them. Thus, lumbar nerve L_2 exits the vertebral column between vertebrae L_2 and L_3.

2➜ For the cervical vertebrae, the nerve numbers precede the vertebrae numbers because C_1 exits between the skull and the atlas. Nerve C_8 can be found between the last cervical vertebra and the first thoracic vertebra.

3➜ The pattern for the 31 human spinal nerves is cervical 8, thoracic 12, lumbar 5, sacral 5, coccygeal 1. (Some other animals have slightly different spinal nerve patterns, although the innervation principles seem to be constant throughout the animal world.)

❑ Examination

II. Branches

A. Dorsal and Ventral Roots

1➜ Look for the **dorsal** and **ventral roots** of the spinal cord on Figure 28–2●. The spinal nerves have two points of communication with the spinal cord, called the dorsal root and the ventral root due to their relative positions. The dorsal root is the sensory portion of the nerve, receiving only input signals from the periphery. The ventral root is the motor portion of the nerve, functioning in the transmission of signals from the cord to the body proper. The dorsal root is sometimes known as the posterior or sensory root; the ventral root is sometimes called the anterior or motor root.

2➜ Study Figure 28–2a●, the ventral root system of the spinal nerves. First observe how the motor signal leaves the gray matter of the spinal cord. The skeletal fibers innervate the appropriate muscles directly. Note that the visceral motor fiber enters the autonomic (sympathetic) ganglion. Also note the various synapses.

List the areas of innervation ______________

____

3➜ Identify the **dorsal root ganglion** on Figure 28–2b● or on your laboratory models. The dorsal root includes a swelling, the dorsal root ganglion. This ganglion contains the cell bodies of the peripheral sensory neu-

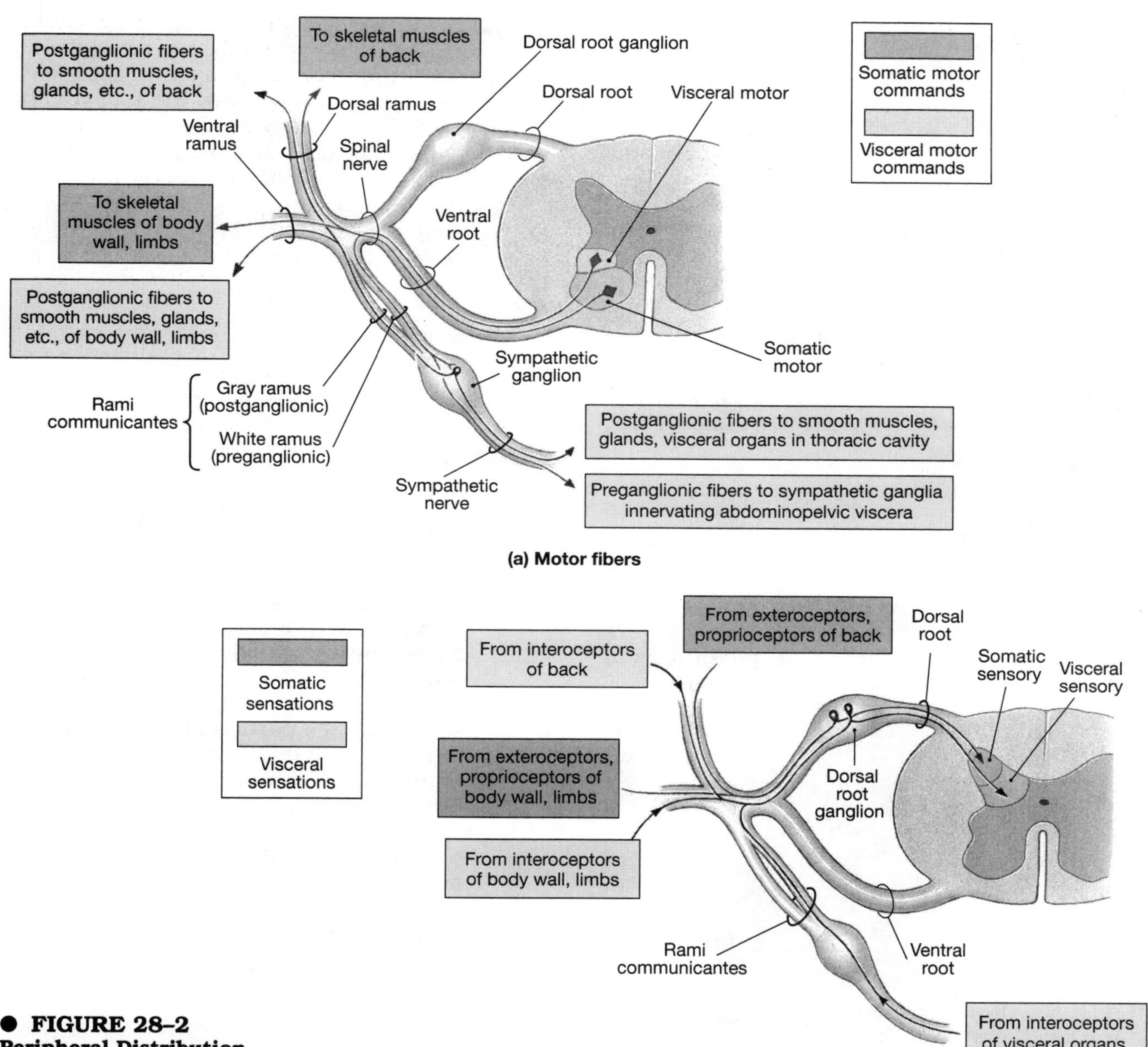

● **FIGURE 28–2 Peripheral Distribution of Spinal Nerves.**

rons. The swelling on the lower part of the Figure 28–2b● is the autonomic (sympathetic) ganglion. Note the five areas of the body transmitting signals into the spinal cord via this dorsal root.

List those areas ______________________
____

B. White and Gray Rami Communicantes

1➡ Locate the white and gray rami on Figure 28–2a●. We often use the term **ramus** (pl. rami) when referring to a particular branch of a spinal nerve. Ramus is defined as *branch.* The **white ramus** consists of the preganglionic motor fibers, while the **gray ramus** includes the postganglionic (autonomic) nerve fibers.

The white ramus and gray ramus together form the **rami communicantes**, a branch connecting the ventral rami with the sympathetic chain. (Not all spinal nerves have rami communicantes.)

2➡ Study Figure 28–3●. Note how the sympathetic (autonomic) ganglia form what looks like the beginning of a chain, and are thus often called the chain ganglia. Describe Figure 28–3● in your own words.
____
____

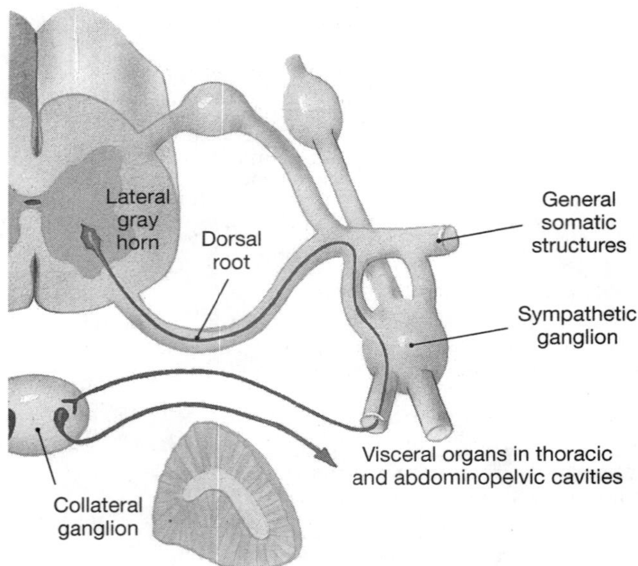

● **FIGURE 28–3**
Sympathetic Ganglion.

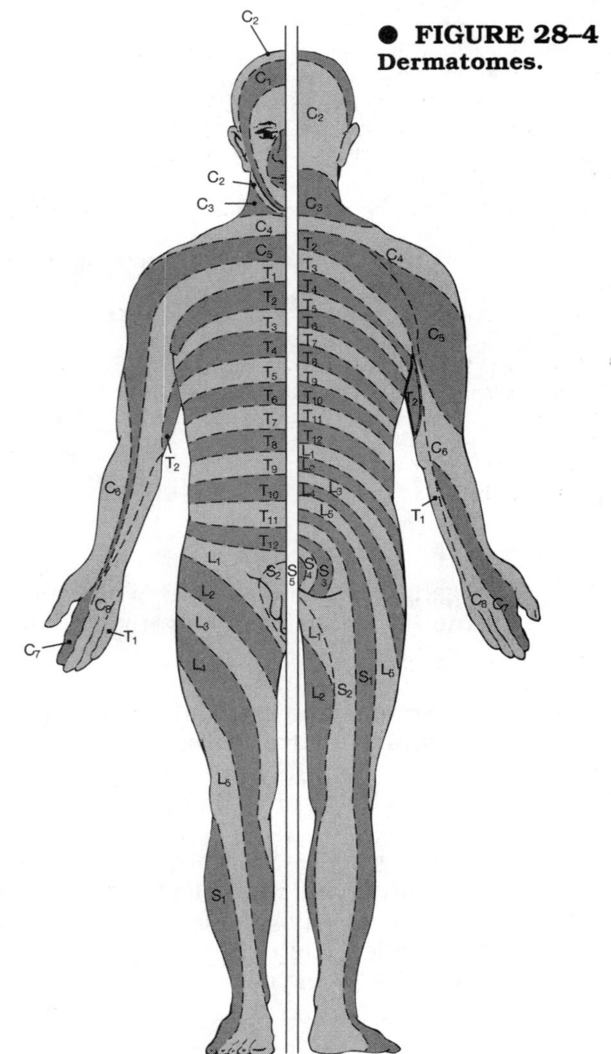

● **FIGURE 28–4**
Dermatomes.

3➔ Follow the **dorsal** and **ventral rami**. The dorsal ramus provides sensory and motor innervation to the skin and muscles of the back. The ventral ramus innervates the ventrolateral body surface, the structures in the body wall, and the limbs.

Another small ramus, the **meningeal branch** (not shown), circles back and innervates the vertebrae, vertebral ligaments, meninges, and spinal blood vessels.

4➔ Students sometimes confuse the meanings of root and ramus. Based on what you have learned, define each of these terms.

Concept Check 1
Root ______________________________
Ramus ______________________________

C. DERMATOMES [∞ MARTINI, P. 415]

1➔ Each pair of spinal nerves monitors a specific body region known as a **dermatome**. Study Figure 28–4●, which illustrates the various dermatomes. Dermatome identification can become important in the case of injury because loss of skin sensation in a particular dermatome will indicate probable injury to a particular spinal nerve or dorsal root ganglia. Although exact information cannot be obtained from a loss of sensation, certain potentially identifying parameters can be defined.

Concept Check 2
Based on Figure 28–4●, which nerve would be involved in a loss of sensation to each of these areas?

a. Thumb ______________________________
b. Cheek ______________________________
c. Little toe ______________________________
d. Medial side of knee ______________________________

III. Plexuses — Ventral Rami

A. PLEXUS

1➔ Refer back to Figure 28–2a● and note the routing of nerve impulses through the dorsal and ventral rami.

2➔ Note the ventral rami. Innervation would be relatively simple were it not for these meshwork-like configurations of the ventral rami of some of the spinal nerves. Each meshwork is known as a **plexus**, containing fibers from one or more spinal nerve. (Plexus means braid.) Since only the ventral rami are involved in the plexuses, the

Spinal Segment	Nerves	Distribution
Cervical Plexus		
C_1–C_4	Ansa cervicalis (superior and inferior branches)	Five of the extrinsic laryngeal muscles (geniohyoid, thyrohyoid, sternothyroid, sternohyoid, omohyoid)
C_2–C_3	Lesser occipital, transverse cervical, supraclavicular, and greater auricular nerves	Skin of upper chest, shoulder, neck, and ear
C_3–C_5	Phrenic nerve	Diaphragm
C_1–C_5	Cervical nerves	Levator scapulae, trapezius, scalenes, sternocleidomastoid
Brachial Plexus		
C_5,C_6	Axillary nerve	Deltoid and teres minor muscles Skin of shoulder
C_5–T_1	Radial nerve	Extensor muscles on the upper arm and forearm (triceps brachii, brachoradialis, extensor carpi radialis, and ulnaris) Digital extensors and abductor pollicis Skin over the posterolateral surface of the arm
C_5–C_7	Musculocutaneous nerve	Flexor muscles on upper arm (biceps brachii, brachialis, coracobrachialis)
C_6–T_1	Median nerve	Flexor muscles on forearm (flexor carpi radialis, palmaris longus) Pronators (p. quadratus and p. teres) Digital flexors Skin over lateral surface of hand
C_8, T_1	Ulnar nerve	Flexor muscles on forearm (flexor carpi ulnaris) Adductor pollicis and small digital muscles Skin over medial surface of hand
Lumbar Plexus		
T_{12},L_1	Iliohypogastric nerve	Abdominal muscles (external and internal obliques, transversus abdominis) Skin over lower abdomen and buttocks
L_1	Ilioinguinal nerve	Abdominal muscles (with iliohypogastric) Skin over medial upper thigh and portions of the external genitalia
L_1,L_2	Genitofemoral nerve	Skin over anteromedial surface of thigh and portions of external genitalia
L_2,L_3	Lateral femoral cutaneous nerve	Skin over anterior, lateral, and posterior surfaces of thigh
L_2–L_4	Femoral nerve	Anterior muscles of thigh (sartorius and quadriceps) Adductors of thigh (pectineus and iliopsoas) Skin over anteromedial surface of thigh, medial surface of leg and foot
L_2–L_4	Obturator nerve	Adductors of thigh (adductor magnus, brevis, longus) Gracilis muscle Skin over medial surface of thigh
L_2–L_4	Saphenous nerve	Skin over medial surface of leg
Sacral Plexus		
L_4–S_2	Gluteal nerves:	
	Superior	Adductors of thigh (gluteus minimus, gluteus medius, and tensor fasciae latae)
	Inferior	Extensor of thigh (gluteus maximus)
L_4–S_3	Sciatic nerve:	Two of the hamstrings (semimembranosus, semitendinosus) Adductor magnus (with obturator nerve)
	Tibial branch	Flexors of leg and plantar flexors of foot (popliteus, gastrocnemius, soleus, tibialis posterior) Flexors (plantar flexors) of toes Skin over posterior surface of leg, plantar surface of foot
	Peroneal branch	Biceps femoris of hamstrings Peroneus (brevis and longus) and tibialis anterior Extensors (dorsiflexors) of toes Skin over anterior surface of leg and dorsal surface of foot
S_2–S_4	Pudendal nerve	Muscles of perineum

● **FIGURE 28–5**
Nerve Plexus.

posterior portion of the body is relatively unaffected by the spinal nerve blending.

3➡ Reexamine Figure 28–1●. Note the locations of the **cervical**, **brachial**, **lumbar**, and **sacral** plexuses. Where specifically is each of these plexuses located?

____

____

____

4➡ Now study Figure 28–5●. Take each nerve on the chart and trace its path on Figure 28–1●. (Refer to Exercise 21 if necessary.)

IV. Microscopy

A. Slides

1➡ View whatever slides have been made available to you. These will probably include slides of different ganglia, different nerve endings, and a spinal nerve cross section.

2➡ Sketch and label your drawing according to your instructor's directions.

❑ Additional Activities

1. Compare the numbers and locations of the human spinal nerves with the numbers and locations of the spinal nerves in other animals.

2. Compare the sensory and motor innervation of the voluntary and involuntary muscles.

3. This information is based on some recent studies.

Geography (location), the relative position of one nerve to another, is important, but chemoaffinity, the chemical attraction of a nerve fiber for its target, will override relative position. In other words, fibers will grow toward their proper targets even when either the target or the nerve fibers are manually displaced.

Sensory fibers exhibit less specificity than do motor fibers. In development, motor fibers often grow toward their proper targets, but if the proper motor fibers are manually displaced, the sensory fibers will follow the nearest motor axon. And if motor axons are prevented from innervating a tissue, no sensory axons will innervate either.

What implication does this have for those studying abnormal development?

NOTES

Answers to Selected Concept Check Questions

NOTES

1. The root is the source or the beginning; the ramus is the branch.
2. These nerves would be involved:
   a. Thumb C_6
   b. Cheek C_2
   c. Little toe S_1
   d. Medial knee L_4

❑ Lab Report

1. Referring to the numbered inquiries at the beginning of this exercise, complete the following box summary:

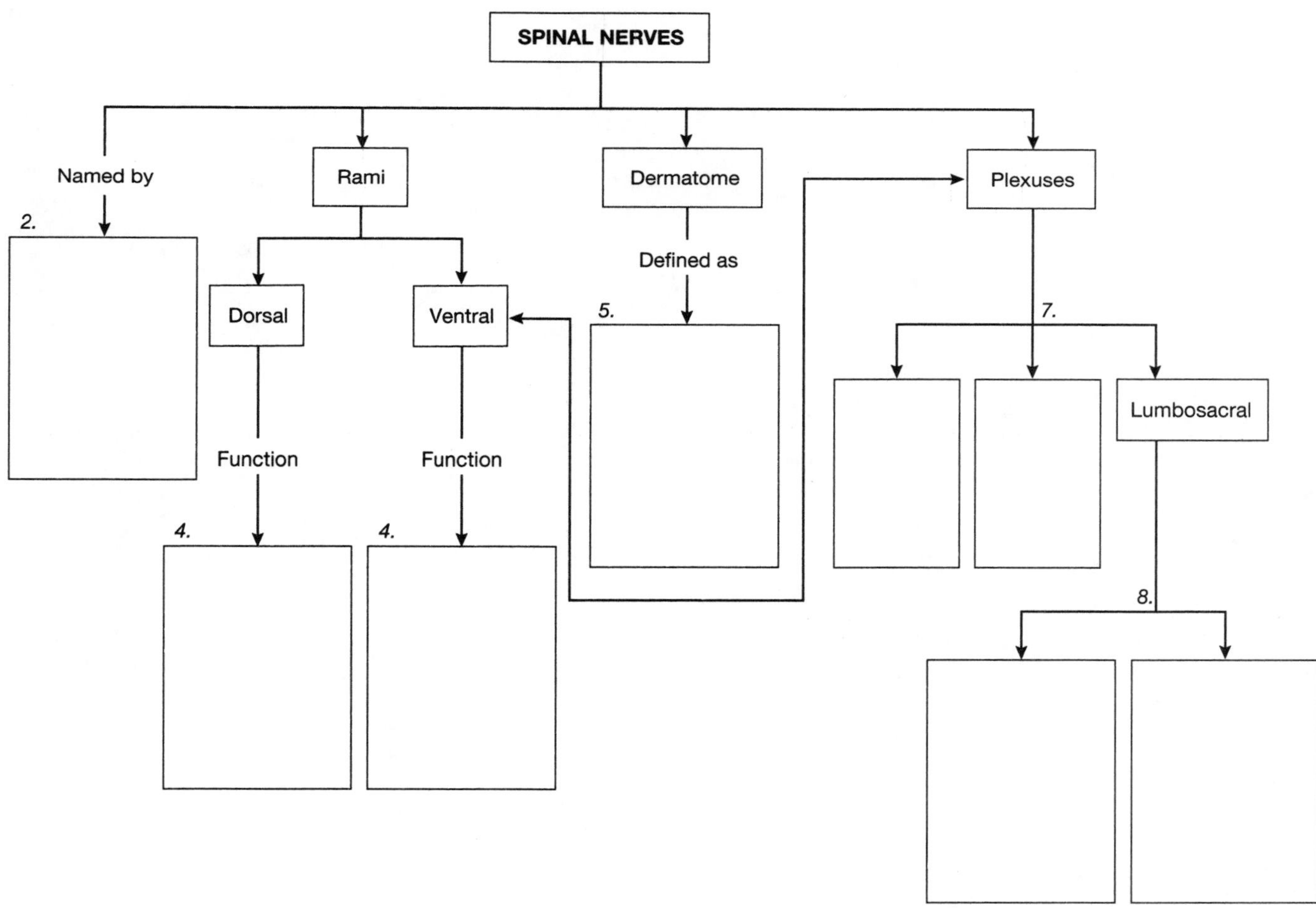

General Questions

NOTES

1. What is the numbering pattern for human spinal nerves?
2. Which spinal nerve(s) would be involved in each of these situations?
   a. Numbness in the middle toe.
   b. Itchy nose.
   c. Tingling from hitting the "funny bone."
   d. Lower back pain. (Several answers may be possible.)
   e. You don't feel the chigger bite just below your umbilicus.
3. What is the difference between
   a. Dorsal and ventral root of the spinal cord?
   b. Dorsal and ventral ramus?
4. Based on your knowledge of the sacral plexus, what is the problem with the old-timer in the movies, saying he is having a bout of "sciatica"?
5. Compare and contrast the plexuses. Why aren't the middle ten thoracic nerves involved in plexuses?

EXERCISE 29

Reflex Physiology

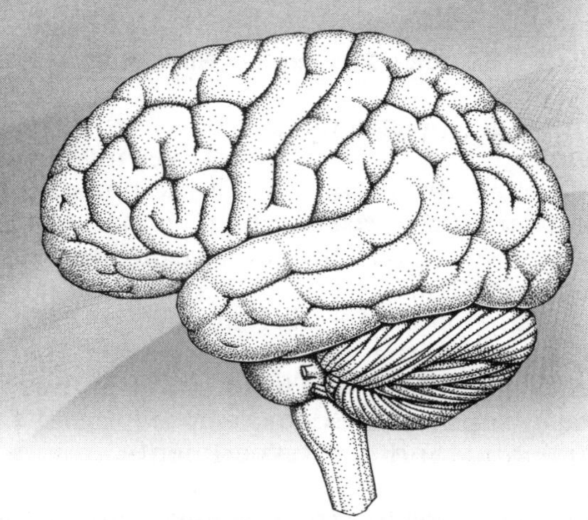

PROCEDURAL INQUIRIES

Preparation

1. What is the difference between an afferent and an efferent neuron?
2. What are the landmarks of the spinal cord?
3. What is the difference between a deep reflex and a superficial reflex?
4. What is a stretch reflex?
5. What is a withdrawal reflex?
6. What is a crossed extensor reflex?

Experimentation

For each reflex listed below, answer these questions:
How is the reflex test performed?
What does the reflex test demonstrate?

7. Biceps reflex
8. Triceps reflex
9. Brachioradialis reflex
10. Hoffmann's reflex
11. Patellar reflex
12. Achilles reflex
13. Babinski reflex
14. Crossed extensor reflexes
15. Glabellar reflex
16. Abdominal reflex

Additional Inquiries

17. Which neurons are involved in a reflex arc?
18. What is the difference between a visceral and a somatic reflex?
19. What is the difference between ipsilateral and contralateral?
20. What is a ganglion?

Key Terms

Afferent Neuron
Autonomic
Contralateral
Deep Reflex
Dorsal Root
Dorsal Root Ganglion
Efferent Neuron
Gray Matter
Hyperreflexia
Hyporeflexia
Interneuron
Ipsilateral
Jendrassic Maneuver
Polysynaptic
Reflex
- Abdominal Reflex
- Achilles Reflex
- Babinski Reflex
- Biceps reflex
- Brachioradialis reflex
- Crossed extensor reflex
- Glabellar reflex
- Hoffmann's reflex
- Patellar reflex
- Triceps reflex

Reflex arc
Skeletal
Somatic
Stretch Reflex
Superficial Reflex
Tendon Reflex
Ventral Root
Visceral
White Matter
Withdrawal Reflex

Materials Needed

Reflex Mallet
(You may substitute the medial side of your hand and a plastic pen cap.)
Ice Chips
Pin (or other sharp object)

A reflex is an involuntary response to a stimulus, which begins with a sensory input, continues through one or more neuronal synapses, and terminates in a motor reaction. The pathway is known as the **reflex arc.** [∞ *Martini, p. 422]*

Many types of reflexes can be found in the body, reflexes ranging from the automatic production of saliva upon seeing food to the automatic movement of a body part in the presence of pain.

Reflexes can be categorized in numerous ways but most commonly are divided into **visceral (autonomic)** reflexes and **somatic (skeletal)** reflexes.

Although the specific nervous pathways are different and the visceral reflexes involve two efferent neurons instead of the one efferent neuron characteristic of the somatic reflex, the arc concept is the same regardless of the type of reflex.

To demonstrate the reflex arc we will concentrate on the somatic reflexes. (Visceral reflexes involving the involuntary muscles of the eye and ear are demonstrated with the visual and auditory experiments found in Exercises 37 and 39, respectively. Visceral reflexes involving the digestive, circulatory, renal, and endocrine systems are not usually demonstrated in introductory courses.)

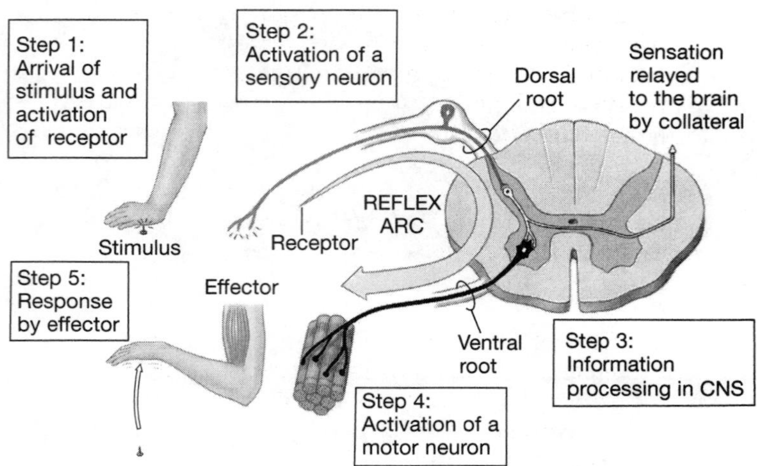

● **FIGURE 29–1**
Basic Reflex Arc.

❑ Preparation

I. Background

Before performing the reflex experiments, it is important that you understand the anatomy of the components of the **reflex arc**.

A. Overview

In the classic example of the somatic reflex, your hand touches a hot stove and, before your brain registers the heat involved, you pull your hand away. The sensory message (heat) has traveled via an **afferent (sensory) neuron** to the spinal cord. This neuron has synapsed with an **efferent (motor) neuron**, and the message for muscle contraction has traveled to the fibers of the biceps brachii, and you lifted your forearm.

In other words, the nervous impulse travels via the dendrite of an afferent (sensory) neuron from the stimulus toward the nerve cell body that is located in the dorsal root ganglion. (Recall that a ganglion is a conglomerate of nerve cell bodies located outside the central nervous system.) The dorsal root ganglia are located lateral to the spinal column. From the ganglion the message travels via the axon into the posterior horn of the spinal cord. In the spinal cord the axon may synapse directly with the efferent (motor) neuron or it may synapse first with an **interneuron** (association neuron), which in turn synapses with the motor neuron.

Recall the spinal tracts discussed in Exercise 27. The neurons of the reflex arc also synapse with the neurons of these tracts so the sensory message can be carried to the brain in order that information can be processed and additional motor message can be carried back from the brain. Note, however, that the synapse with the ascending and descending tracts is independent of the reflex arc, and thus reflexes happen without your conscious awareness of them.

B. Location of Structures

1➡ Use the diagram (Figure 29–1●) to orient yourself to the cross section of the spinal cord and its associated components. Locate the **dorsal root**, **ventral root**, **dorsal root ganglion**, **gray matter**, **white matter**, **afferent neuron**, **interneuron** (not shown), **efferent neuron**.

2➡ Use Figure 29–1● and trace a somatic reflex arc. Note the **sensory neuron** (coming from a stimulus), **interneuron** (located within the gray matter), and **motor neuron** (leading to a muscle).

3➡ Consider these questions in relation to each of the reflex experiments:
Where are the synapses?
What is the reflex arc pathway?
What is the relation of the brain to this particular reflex arc?

II. Reflex Test Preparation

A. Background

1➡ To perform the reflex tests, work with a partner. Be certain that both you and your partner have a good grasp of exactly what you are looking for before you begin any given test.

B. Terminology [∞ Martini, p. 425]

1➡ Be aware that in describing reflexes, the terms **deep (tendon)** and **superficial** are often used. Deep means that the reflex being tested is within the body and tendon means that a particular tendon is being stimulated and that the response will be the result of the action of the muscle associated with that particular tendon. In superficial reflexes, also called cutaneous reflexes, the stimulus is on the body surface.

2➡ Also note that deep somatic reflexes can be categorized as **stretch reflexes**, **withdrawal (flexor) reflexes**, and **crossed extensor reflexes**.

a. Stretch reflexes are monosynaptic reflexes. In other words, the sensory neuron synapses in the spinal cord with only one other neuron, the motor neuron. The action is **ipsilateral**, on the same side as the sensation.
b. The withdrawal reflex is also ipsilateral, although withdrawal reflexes are **polysynaptic**, meaning more than one synapse. Most withdrawal reflexes can also be called flexor reflexes because a limb is withdrawn from a stimulus because a muscle has been flexed.
c. Crossed extensor reflexes are polysynaptic **contralateral** reflexes. This means that more than one synapse is involved and that the sensation and the motor action are on opposite sides of the body.

C. Evaluating Responses

1➡ Unless your instructor directs you otherwise, use this somewhat subjective method to evaluate the deep tests:

++++ hyperreaction (you probably will not see a hyperreaction)
+++ strong reaction
++ weak reaction
+ very weak (barely noticeable) reaction
0 no reaction.

If you cannot classify the reaction, give a description of the reaction.

2➡ For the superficial tests, give a descriptive evaluation.

3➡ Be aware of the following:

a. Occasionally you may see "normal" and "abnormal" labels put on some of the reflex evaluations. We believe judgements like this cannot be made in the classroom laboratory setting. "Strange" or apparently abnormal reflex test reactions do not necessarily indicate that you have a problem. Remember, these tests are being performed in the classroom, which is hardly an ideal setting, by your classmates, who usually do not have a great deal of expertise in these matters. In "real life" situations, however, hypoactive or absent reflexes are usually indicative of peripheral nerve damage or ventral horn damage, while hyperactive responses tend to indicate corticospinal tract lesions (which upset normal equilibrium by decreasing inhibitory control on the reflexes). Hypoactive reactions are often termed **hyporeflexia**, and hyperactive responses are often called **hyperreflexia**.
b. Sometimes it is difficult to be certain whether a reaction is strong or weak strictly by observation. If you are having this trouble evaluating your partner's responses, try placing your hand lightly on the muscle involved while you do the test.

D. Overriding a Reflex

Some reflex tests can be influenced by a person's conscious or subconscious actions. In other words, you can sometimes "override" a reflex.

1➡ If you find yourself tensing up as your partner is about to use the reflex mallet on you, try a diversionary action. Sometimes all you need to do is close your eyes and consciously relax so that you are unaware of when your partner is going to test you.

2➡ On tense leg reflexes, try performing the **Jendrassic maneuver**, whereby you lock your fingers and pull your hands against each other.

3➡ If you are having trouble with the arm reflexes, try crossing your ankles and pulling your feet against each other. Close your eyes and concentrate on keeping your upper body as limp as possible.

4➡ If you are still having trouble with the reflex test, try sucking on some ice. Concentrate on the ice. Be careful not to relax too much!

❑ Experimentation

III. Reflex Tests

A. General Directions

1➡ Study Figure 29–2●, which includes sketches of some of the following reflex tests.

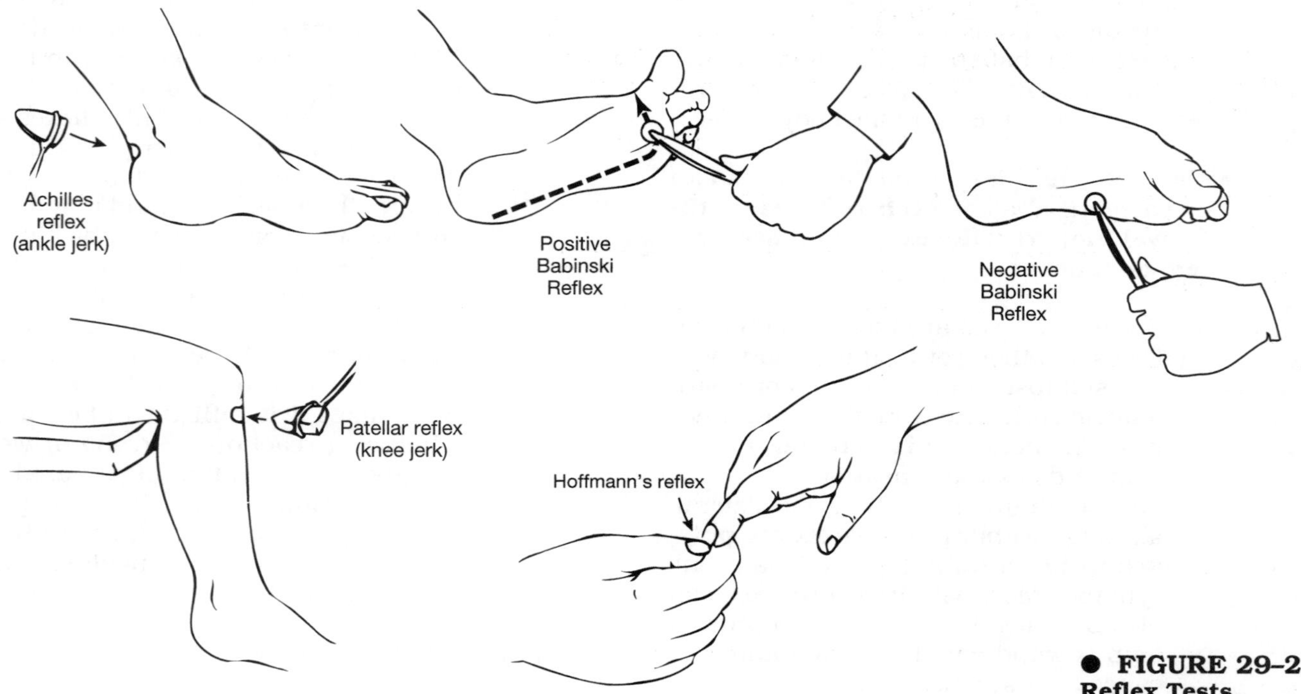

● **FIGURE 29–2**
Reflex Tests.

2➡ Record all your results on the summary chart, Figure 29–3●.

3➡ Test both the left and the right sides and note any differences in results in the space provided in Figure 29–3●.

B. Specific Tests

Arm and Hand

1➡ The **biceps reflex** tests the functions of cervical nerves 5 and 6 (C_5 and C_6). With your partner seated comfortably, take his/her arm and locate the tendon of the biceps brachii muscle. If you can't find this tendon easily, ask your partner to flex the biceps muscle. Follow the muscle down to the antecubital area (front of elbow). Place your thumb over the tendon and hold the elbow with the rest of your hand. Take your reflex mallet and, with the sharp end, strike your thumb. Note the response. The biceps brachii muscle flexes (via the musculocutaneous nerve), so you should notice the "upward" movement of the forearm.

2➡ The **triceps reflex** tests the functions of C_7 and C_8. Find the triceps brachii muscle on your partner's posterior arm. You should be able to locate the tendon just superior to the elbow. With the pointed end of the reflex mallet, strike a sharp blow to this tendon. Note the response. The triceps brachii muscle contracts (via the radial nerve), so you should notice the "downward" movement of the forearm.

3➡ The **brachioradialis reflex** also tests C_5 and C_6. However, this reflex is concerned with the efferent innervation of brachioradialis and supinator muscles. While seated comfortably, your partner should place his/her hand palm down on the thigh. Locate the approximate area of the brachioradialis tendon, about 2-1/2 cm above the radionavicular (wrist) joint. Strike this area with the broad side of the reflex mallet. The expected action is supination of the forearm and hand. You may also notice some flexion of the fingers.

4➡ **Hoffmann's reflex** tests for pyramidal tract lesions. Hold your partner's hand and flick the distal phalanx of the index finger. Except for the actual movement of the index finger, you should note little or no additional movement. In a medical setting, adduction and flexion of the thumb and twitching of the other fingers might indicate pyramidal tract lesion.

Leg

5➡ The **patellar reflex**, also known as the knee jerk reflex, the knee reflex, or the quadriceps reflex, tests the L_2, L_3, and L_4 tracts. Your partner should sit on the edge

Test	Results (Resting)		Results Varying Conditions	Comments
	Left	Right		
Biceps				
Triceps				
Brachioradialis				
Hoffmann's				
Patellar				
Achilles				
Babinski				
Crossed Extensor				
Glabellar				
Other				

● **FIGURE 29–3**
Reflex Test Results.

of a table with his/her leg dangling over the edge. Locate the patellar tendon, just below the knee cap. Strike this tendon with the sharp edge of the reflex mallet. Sometimes clonus (a series of jerk-like contractions) may be noticed after a normal response. Short-term clonus may be normal (due to test anxiety), but long-term clonus may be the result of central nervous system damage. *[∞ Martini, p. 426]*

6➡ The **Achilles reflex**, also known as the ankle jerk, tests the S_1 and S_2 tracts. For this test your partner may sit as for the patellar reflex test, or your partner may kneel on the edge of a chair. Grip your partner's foot and locate the calcaneal (Achilles) tendon. Strike this tendon forcefully with the broad end of the reflex mallet. In a normal action plantar flexion will occur because of contraction of the soleus muscle.

7➡ The **Babinski reflex,** known as the Babinski sign or plantar flexion, is a superficial reflex that like the Achilles reflex, tests the S1 and S2 tracts. To some extent the L4 and L5 tracts are also involved. To perform this test, have your partner remove his/her shoes. Take the handle of the reflex mallet and run it lightly from the heel along the lateral edge of the foot to the area of the tarsal-metatarsal joints and then across the foot toward the great toe. The negative (normal) response is plantar flexion. The positive (abnormal) response is dorsiflexion of the great toe and a spreading of the other toes. A positive Babinski is normal in infants in whom myelinization of the nerve fibers is incomplete. A positive Babinski in older children and adults is usually indicative of damage to the descending corticospinal tracts. [∞ Martini, p. 430]

Crossed Extensor Reflexes [∞ Martini, p. 428]

These crossed extensor reflexes involve polysynaptic neurons. Some of these neurons synapse across the central portion of the spinal cord so that the action occurs on the side of the body opposite the stimulus. These reflexes are often difficult to test because of the subject's sense of anticipation.

8➡ One test for cross reflexes involves the prick of a pin (or other sharp object). Ask your partner to sit quietly with eyes closed and arms resting on the desk or the lap. When your partner is least expecting it, prick the pad of his/her middle finger. (Do not draw blood!) Note the action in the other arm. You may have to ask your partner to describe any muscular feeling he/she experienced. In a normal response, the muscles in the opposite arm will tend to move the arm downward while the muscles in the arm involved will lift that arm upward.

9➡ Another way to check cross reflexes is to ask your partner to pick up a heavy object, such as a brief-case, with one arm. Notice how the opposite arm moves laterally to achieve balance.

Additional Reflex Tests

10➡ The **glabellar reflex** is also a superficial reflex. Ask your partner to sit quietly with eyes closed and glasses removed. With your middle finger, sharply tap the glabella (flattened portion of the forehead) and observe the results. You should notice such reactions as blinking (even though the eyes are closed).

11➡ The **abdominal reflex** is a superficial reflex that tests the lower thoracic spinal nerves — T_8, T_9, and T_{10} above the umbilicus and T_{10}, T_{11}, and T_{12} below the umbilicus. It is not normally practical to test the abdominal reflex in the classroom laboratory. In a positive reflex, after lightly stroking the abdominal skin, the abdominal muscles contract and the umbilicus moves toward the stimulus. A negative reflex (no movement) indicates some type of damage to one or more of the lower thoracic spinal nerves.

C. VARIATIONS

Try any or all of the preceeding tests under these nonresting conditions:

1➡ Run in place for a few minutes. Immediately repeat the tests.

2➡ Use ice to cool an area being tested. Hold the ice while the test is being repeated.

3➡ Record your results.

❑ Additional Activities

1. Use the library or available health care references to explore the field of reflexology.
2. Check with a neurologist or consult a neurological reference book to learn about additional reflex tests.
3. Explain why tumors, poisonings, stroke, or inflammation at specific points on the nerve tract might cause hyperreflexia or hyporeflexia.

❑ Lab Report

1. Referring to the numbered inquiries at the beginning of this exercise, complete the following box summary:

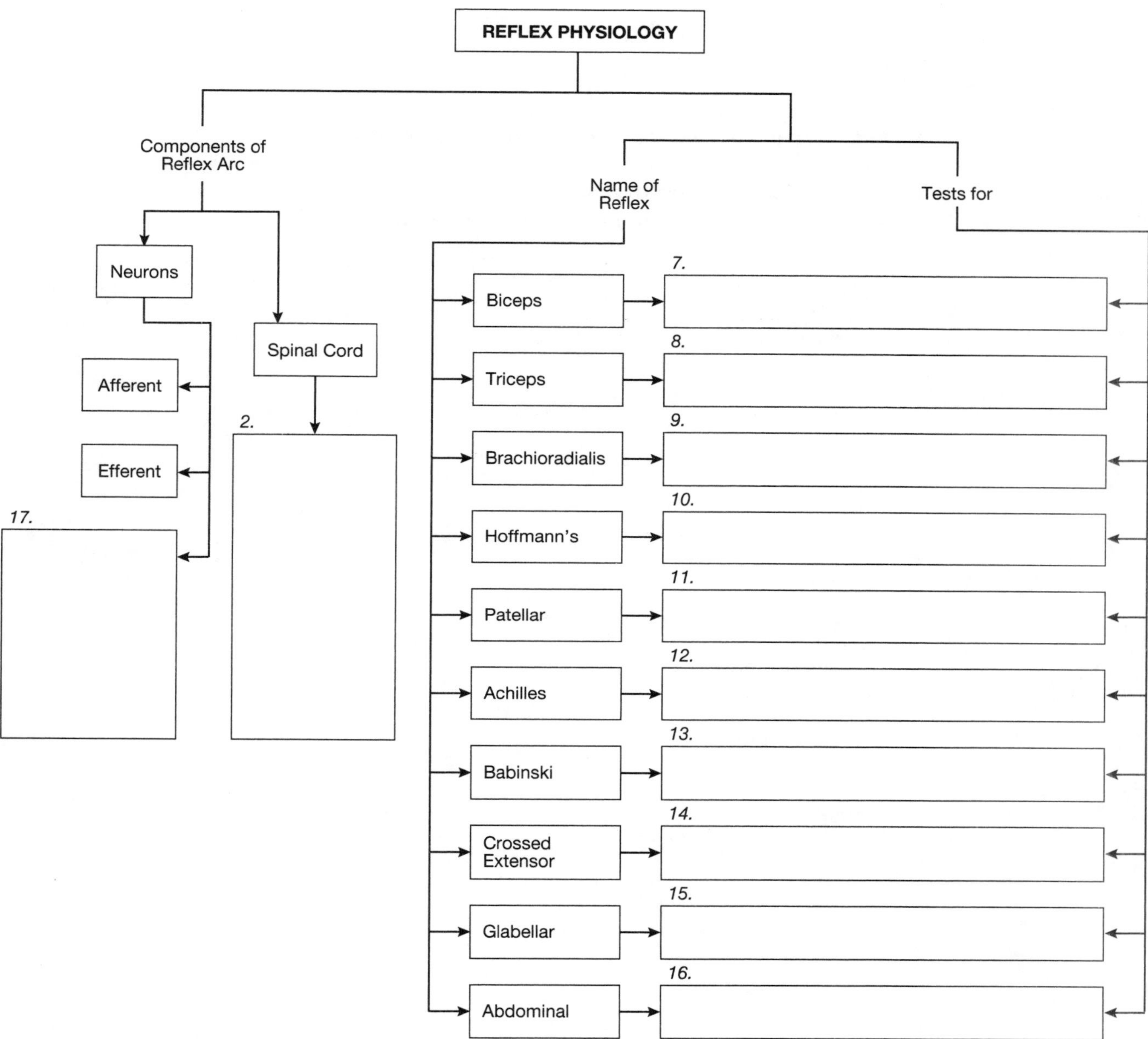

General Questions

NOTES

1. What is the difference between an afferent and an efferent neuron?
2. What is the difference between a deep reflex and a superficial reflex?

Give an example of each of the following:

3. Stretch reflex
4. Withdrawal reflex
5. Crossed extensor reflex
6. Contralateral reflex
7. Are the reflexes we tested for in this exercise considered to be visceral or somatic reflexes? Why?
8. What is a ganglion and how does it relate to a neuron?
9. Why is the Babinski sign normal in infants but pathological in adults?

EXERCISE 30

Anatomy of the Human Brain

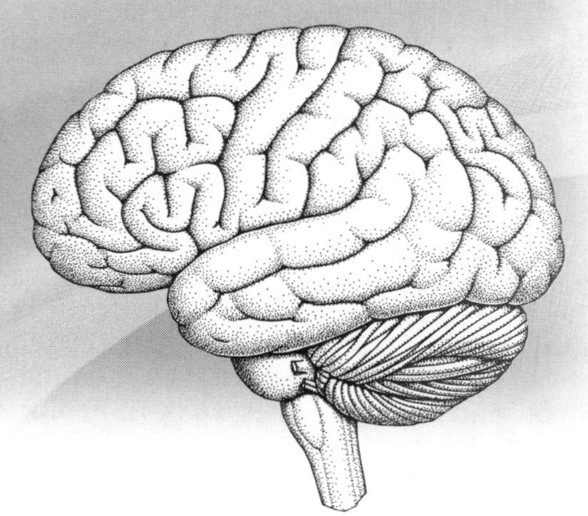

PROCEDURAL INQUIRIES

Preparation

1. What is the difference between a sulcus and a gyrus?
2. What is a fissure?

Examination

3. What are the three meningeal layers of the brain and spinal cord? Where is each located?
4. Where is the cerebrum located?
5. Where is the longitudinal fissure?
6. What are the lobes of the cerebrum?
7. Where is the cerebellum located?
8. Where is the vermis located?
9. What are the corpora quadrigemina and where are they located?
10. What is the infundibulum?
11. What is the mammillary body?
12. Where are the cerebral peduncles?
13. Where is the pons?
14. Where is the medulla oblongata?
15. Where is the corpus callosum?
16. What are the four ventricles and where are they located?
17. Where do each of the cranial nerves exit the brain?

Additional Inquiries

18. What are the cerebral nuclei?
19. In which part of the brain does most thinking occur?

Key Terms

Arachnoid (layer, or membrane, or mater)
Arbor Vitae
Central Canal
Central Sulcus
Cerebellum
Cerebral Cortex
Cerebral Nucleus
Cerebral Peduncles
Cerebrum
Colliculi
Corpora Quadrigemina
Corpus Callosum
Cranial Nerves (See Exercise 33)
Dura Mater
Falx
Fissure
Fornix
Gyrus
Infundibulum
Lateral Sulcus
Longitudinal Fissure
Lobes: Frontal, Parietal, Temporal, Occipital, Insula
Mammillary Body
Medulla Oblongata
Meninges
Pia Mater
Pons
Postcentral Gyrus
Precentral Gyrus
Septum Pellucidum
Sulcus
Transverse Fissure
Ventricles — 2 lateral, 3rd, 4th
Vermis

Materials Needed

Torso
Human Brain Models
Preserved Human Brain
  Preserved and Sectioned (if available)
Charts and Diagrams
Prepared Slides
  Cerebral Pyramidal Cells
  Cerebellar Purkinje Cells
  Other Brain Cell Slides (as available)

The brain has long been known as the seat of wisdom, the font of emotion, and the central processing unit of the body. In addition to support tissue, the brain contains approximately 35 billion neurons, which can handle perhaps 80,000 simultaneous synaptic spurts. The brain is the one organ required just to study the brain. And studying the brain is not limited to the biological or physical sciences. Psychologists, educators, and theologians are also interested in the intricacies of neuronal interaction and the impact the complexities of the brain has on their particular disciplines. It is said that the brain can store more information than is contained in all the libraries of the world and that we have only just begun to tap the powers of our sentient natures. The brain is probably the most remarkable organ in the entire body! In this laboratory exercise, however, we will limit ourselves to an examination of the gross anatomy and certain aspects of the histology of the human brain.

❑ Preparation

I. Background

A. Terminology

Before beginning this exercise, you should familiarize yourself with these terms:

Sulcus (pl., sulci) Shallow groove or furrow.

Gyrus (pl., gyri) Convolution or ridge.

Fissure Deep prominent groove or furrow.

Falx Curving extension of membranous tissue.

Cranial Nerve Any of 12 specific nerve tracts attached to the brain, normally abbreviated as N I, N II...N XII.

(Cerebral) Nucleus Mass of nerve cell bodies within the central nervous system (in this case, within the brain).

Cerebrum Largest, most prominent, anterior portion of the brain.

Cerebellum "Bulging" structure at the posterior base of the cerebrum.

Hemisphere Left or right half of the cerebrum or cerebellum.

Corpus Callosum Band of tissue (bundle of axons) connecting the centers of the right and left cerebral hemispheres.

Pons Posterior portion of the brain, anterior to the cerebellum.

Medulla Portion of the brain between the pons and the spinal cord.

B. Overview

1➡ Use any available brain models to locate the above structures.

❑ Examination

II. External Structures

A. Meninges: Coverings of the Brain and Spinal Cord [∞ Martini, p. 440]

1➡ Use the "take apart" models, charts, and diagrams that are provided for you to observe various aspects of brain anatomy. Begin your study by observing the brain within the skull. Note the space between the bone and the surface of the brain. Now examine the surface of the brain. The

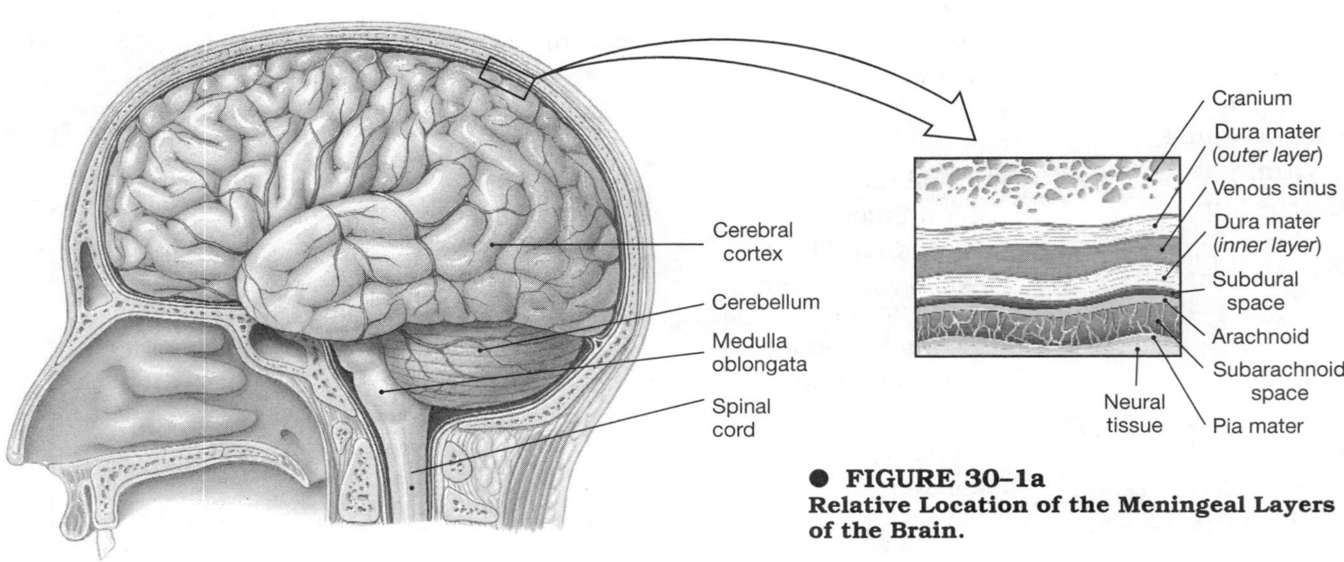

● **FIGURE 30–1a**
Relative Location of the Meningeal Layers of the Brain.

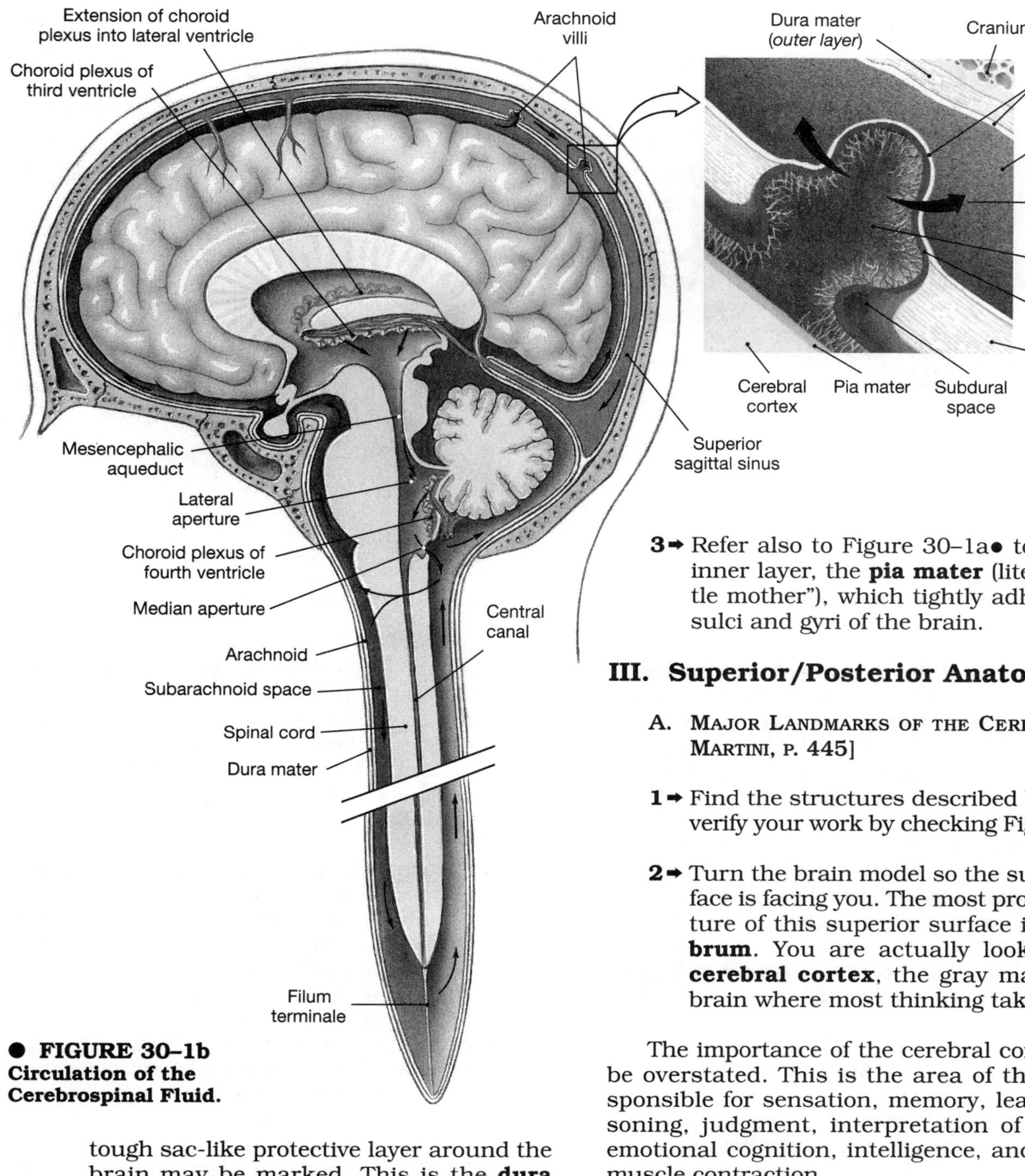

● **FIGURE 30–1b Circulation of the Cerebrospinal Fluid.**

tough sac-like protective layer around the brain may be marked. This is the **dura mater** (literally, "tough mother"), the outermost of the **meninges**, the three protective connective tissue membranes that surround the brain and the spinal cord.

2➡ Refer to Figure 30–1a● to find the second meningeal layer, the **arachnoid layer** (sometimes called arachnoid mater; literally, "spider mother"). The arachnoid layer, or **arachnoid membrane**, may not be identified on your models. Observe the delicate nature of this web-like meninx as shown in Figure 30–1a● and 30–1b●.

3➡ Refer also to Figure 30–1a● to locate the inner layer, the **pia mater** (literally, "gentle mother"), which tightly adheres to the sulci and gyri of the brain.

III. Superior/Posterior Anatomy

A. Major Landmarks of the Cerebrum [∞ Martini, p. 445]

1➡ Find the structures described below, then verify your work by checking Figure 30–2●.

2➡ Turn the brain model so the superior surface is facing you. The most prominent feature of this superior surface is the **cerebrum**. You are actually looking at the **cerebral cortex**, the gray matter of the brain where most thinking takes place.

The importance of the cerebral cortex cannot be overstated. This is the area of the brain responsible for sensation, memory, learning, reasoning, judgment, interpretation of sensation, emotional cognition, intelligence, and voluntary muscle contraction.

Meningitis is an inflammation of the meningeal layers. Why might meningitis cause damage to the cerebral cortex?

____

____

3➡ Observe the **longitudinal fissure**, which divides the cerebrum into the right and left cerebral hemispheres. Note the numerous sulci and gyri forming the surface of the brain. The **central sulcus (fissue of Rolando)** is the deep groove extending laterally from about the middle of the longitudinal

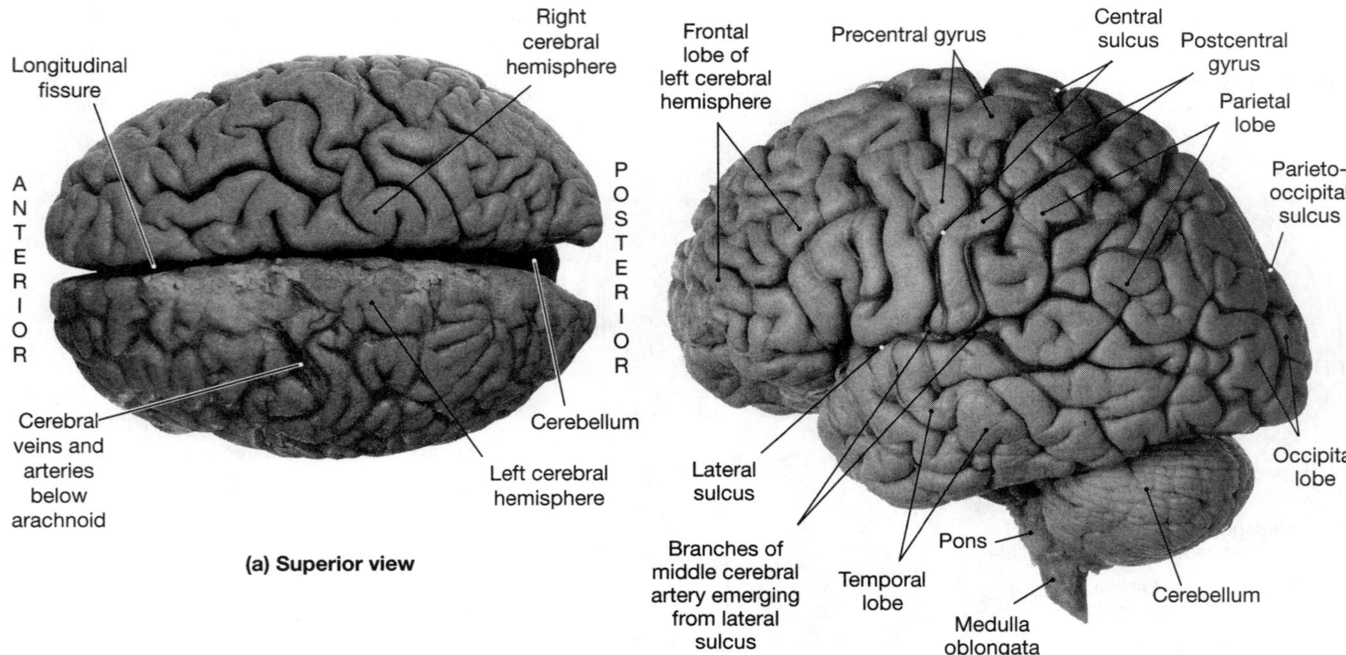

● **FIGURE 30–2**
Surface Anatomy of the Brain.

fissure, and the **lateral sulcus** (**fissure of Sylvius**) marks the inferior border of the frontal lobe. The **postcentral gyrus** is immediately posterior to the central sulcus and the **precentral gyrus** is immediately anterior to the central sulcus.

4➡ Identify the four lobes of each cerebral hemisphere: **frontal**, **parietal**, **temporal**, and **occipital**. These lobes are directly inferior to the corresponding cranial bones. Identify these lobes on your brain models. A fifth lobe, the **insula**, or **island of Reil**, is located deep within the lateral sulcus and cannot be seen from this superficial view.

B. CEREBELLUM [∞ MARTINI, P. 460]

1➡ Observe the **transverse fissure** separating the cerebrum from the **cerebellum**. Note that the sulci and gyri of the cerebellum seem to be much more tightly packed than the corresponding structures in the cerebrum.

2➡ Observe the **falx cerebelli**, which divides the cerebellum into hemispheres.

3➡ Find the **vermis**, the receding central area of the cerebellum, which can be found as a midline lobe between the cerebellar hemispheres.

4➡ If your model comes apart, detach the cerebellum from the brain stem before locating the three paired **cerebellar peduncles**. The superior cerebellar peduncles connect the cerebellum to the midbrain. The middle cerebellar peduncles connect the cerebellum to the pons. The inferior cerebellar peduncles connect the cerebellum to the medulla.

5➡ Return to the transverse fissure and identify the **corpora quadrigemina**, the four prominent swellings immediately anterior to the cerebellum. Although they are not part of the cerebellum, they may be easy to observe at this time. Depending on your model, you may have to use a sagittal section of the cerebrum for this identification. The paired swellings closest to the cerebellum are the **inferior colliculi**, which are involved with auditory reflexes. Anterior to the inferior colliculi are the larger **superior colliculi**, which are involved with visual reflexes. Centrally located at the anterior end of the superior colliculi is the **pineal body**, the endocrine gland that secretes melatonin.

C. SPINAL CORD

1➡ Find the **spinal cord**, which is inferior to the cerebellum.

2➡ Look for remnants of the cervical (spinal) nerves. Depending on the type of brain model you are using, one or more of these cervical (spinal) nerves may be observable.

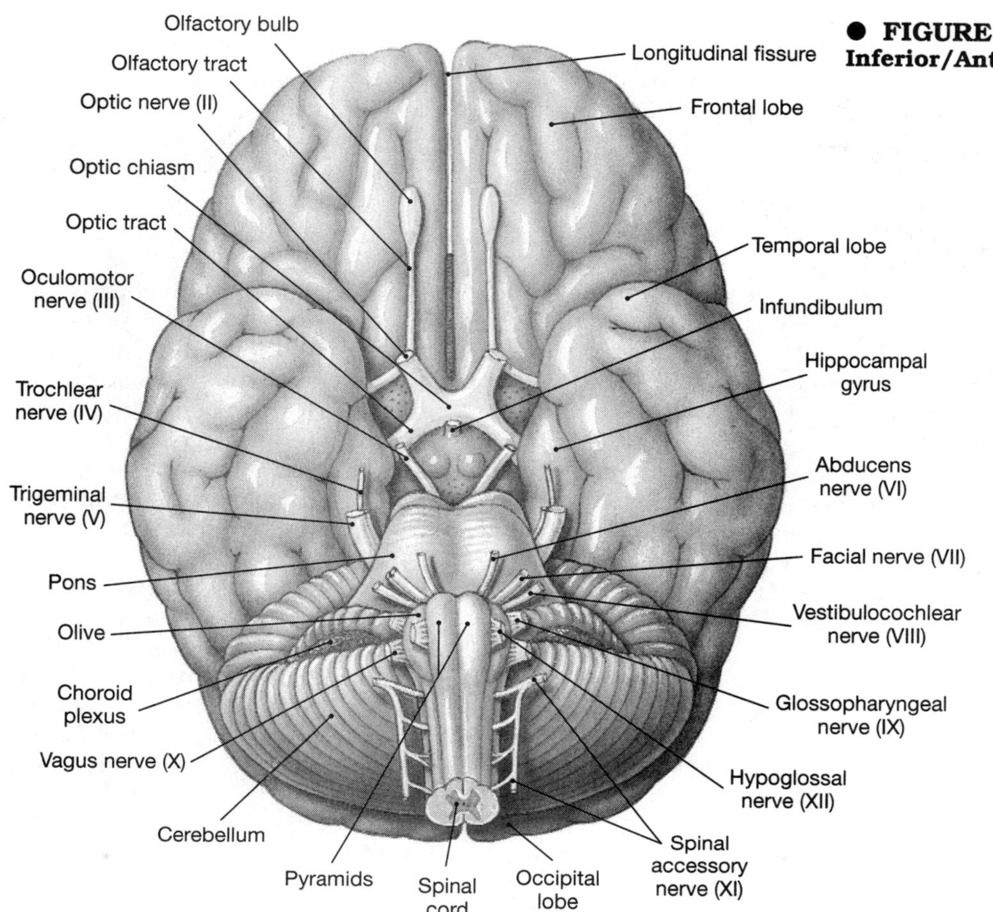

● **FIGURE 30–3**
Inferior/Anterior Anatomy of the Brain.

IV. Inferior/Anterior Anatomy

A. Cranial Nerves I and II

The cranial nerves are attached to the inferior/anterior aspect of the brain. All of the cranial nerves are paired.

1➡ Use Figure 30–3● to verify your identification of the structures listed in this section.

2➡ Begin your examination at the anterior end of the cerebrum. As you proceed posteriorly, the large paired extensions you see are the **olfactory bulbs**, which are components of the olfactory nerve tract, N I.

3➡ Observe an "X-like" structure posterior to the olfactory bulbs. The crossing point is the **optic chiasma**. The anterior portions of the X are the **optic nerves**, N II, while the portions receding from the chiasma into the brain are the **optic tracts**.

B. Infundibulum and Mammillary Body

1➡ Observe the structure that seems to originate under the optic chiasma, the **infundibulum**, or the stalk of the pituitary gland. In actuality, the infundibulum arises from the hypothalamus. (We will examine the hypothalamus further when we study the sagittal anatomy of the brain.) In preserved specimens the infundibulum is sometimes a bit "mushy" and thus is easily destroyed.

2➡ Posterior to the infundibulum find the double-lobed **mammillary body**. The mammillary body controls the feeding reflexes. The mammillary body is part of the hypothalamus.

C. Cerebral Peduncles and Cranial Nerve III

1➡ Find the **cerebral peduncles**, the fibrous tracts connecting the cerebrum and the medulla. These structures are located posterior to the mammillary body.

2➡ Locate the **oculomotor nerves**, N III, that exit from about the middle of the cerebral peduncles.

D. Pons and Associated Cranial Nerves

1➡ Locate the **pons,** which is posterior to the cerebral peduncles. The **trochlear nerves**,

N IV, can be found at the junction of the midbrain and the pons.

2➡ Follow Figure 30–3● to locate the **trigeminal nerves**, N V, the **abducens nerves**, N VI, and the **facial nerves**, N VII. All of these nerves arise from the pons. These nerves should all be indicated on your brain models.

E. Medulla Oblongata and Associated Cranial Nerves

1➡ Observe the **medulla oblongata** posterior to the pons. This structure is more commonly called the medulla.

2➡ Again, follow Figure 30–3● and locate the **vestibulocochlear nerves**, N VIII, the **glossopharyngeal nerves**, N IX, the **vagus nerves**, N X, the **accessory nerves**, N XI, and the **hypoglossal nerves**, N XII.

V. Sagittal Anatomy

A. Corpus Callosum

1➡ Manually separate the cerebral hemispheres to explore the internal structures of the brain. Note the thick fibrous tissue that would be holding the hemispheres together. This is the **corpus callosum**, the neural tissue that functions as a communicating band between the cerebral hemispheres. Refer to Figure 30–4● to verify the location of this structure.

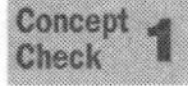

What will happen if the corpus callosum in a living human is cut?

____

____

2➡ Follow the fibrous band of the corpus callosum around posteriorly. The posterior end of the corpus callosum is known as the **splenium of the corpus callosum**. The fibrous band inferior to the corpus callosum is the **fornix**. Between the corpus callosum and the fornix is an opening. This opening is covered by a thin membrane known as the **septum pellucidum**.

B. Ventricles [∞ Martini, p. 439]

1➡ View the area lateral to the septum pellucidum. This is the **lateral ventricle**. If your model is large enough, put your finger or your probe into the ventricle and get a feel for the relative size of this cavity. The brain has a lateral ventricle in each cerebral hemisphere.

2➡ Inferior to the fornix find a large round area known as the **intermediate mass of the thalamus**. Along the underside of the thalamus a cavity, the **third ventricle**, may be

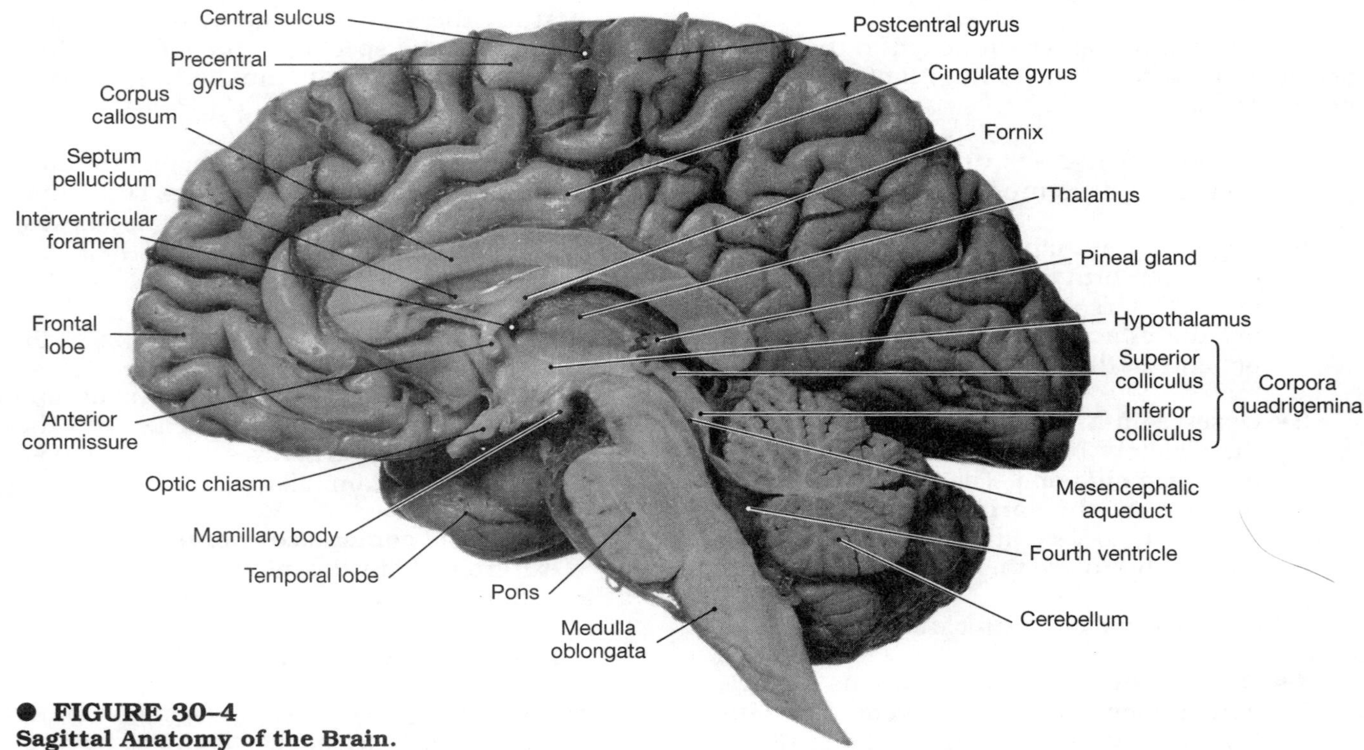

● **FIGURE 30–4**
Sagittal Anatomy of the Brain.

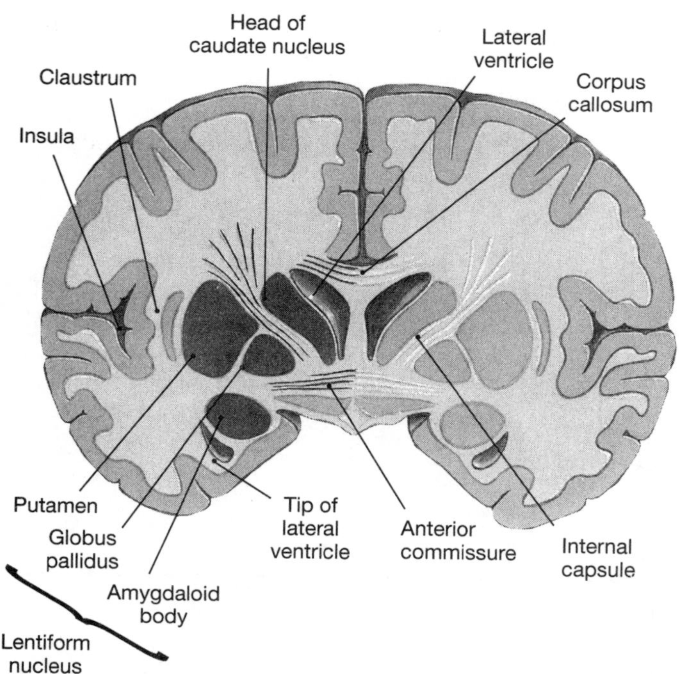

FIGURE 30–5
Coronal (Frontal) Anatomy of the Brain.

marked on the brain models. (The third ventricle is sometimes difficult to recognize as a cavity in preserved brains.) Between the lateral ventricles and also communicating with the third ventricle is the **interventricular foramen**, sometimes known as the **foramen of Munro**. This foramen may not be marked on your model. Nevertheless, you can identify its approximate location.

3➡ Trace a thin canal running posteriorly from the third ventricle toward the cerebellum. This canal, the **mesencephalic aqueduct**, which is also known as the **cerebral aqueduct** or the **aqueduct of Sylvius**, leads to the **fourth ventricle**, which is just posterior to the pons and medulla and just inferior to the cerebellum. Continuing on from the fourth ventricle is the **central canal** of the spinal cord.

4➡ Consider that the function of the ventricles and aqueducts is to maintain the supply and circulation of the cerebrospinal fluid. The cerebrospinal fluid is produced in the ependymal cells that line the ventricles. The ependymal cells producing the cerebrospinal fluid form the choroid plexus. The fluid circulates through the canal system and is eventually carried back into the venous system through the arachnoid villi. Study Figure 30–1b●. You will be asked in a Lab Report question to trace the path of the cerebrospinal fluid.

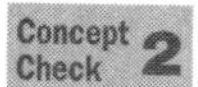
Concept Check 2 Why is the overproduction (or inadequate drainage) of cerebrospinal fluid a serious problem?

____

____

C. Previously Observed Structures

1➡ Relocate on the sagittal section those structures you found when you examined the superior and inferior views of the brain. Note particularly the hypothalamus, which is directly inferior to the thalamus and the cavity of the third ventricle.

2➡ Return to the cerebellum and observe the pattern formed by the white matter and gray matter on the longitudinal section. This pattern is known as the **arbor vitae**, the "tree of life."

VI. Coronal (Frontal) and Transverse Anatomy

A. Internal Anatomy of the Brain

1➡ Observe a coronal (frontal) section of the brain. Use whatever frontal sections are available to identify some or all of the structures in Figure 30–5●. Sections through the thalamus and hypothalamus are especially good for descriptive perspectives.

2➡ If your lab does not have access to coronal brain sections, use Figure 30–5● to give you this perspective. Note the relative size and position of the gray matter (the cerebral cortex) and the white matter.

3➡ Compare a transverse section of the brain (if available) with Figure 30–6●. Identify those structures you have previously observed. Make note of any structures you had not already identified. Make particular note of the ventricles and nuclei on the transverse section.

4➡ Study Figure 30–7● and locate the cerebral nuclei on your brain models. What are the functions of these nuclei? Use your lecture text if necessary.

____

____

VII. Microscopy

A. Histology of Brain Tissue

1➡ Examine prepared slides of the cerebrum, cerebellum, and any other structures your instructor may have set out for you.

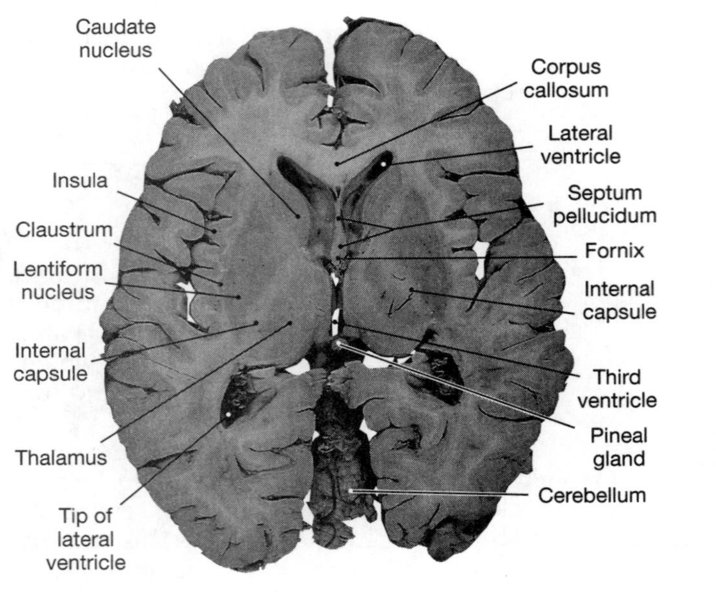

● **FIGURE 30–6**
Transverse Anatomy of the Brain.

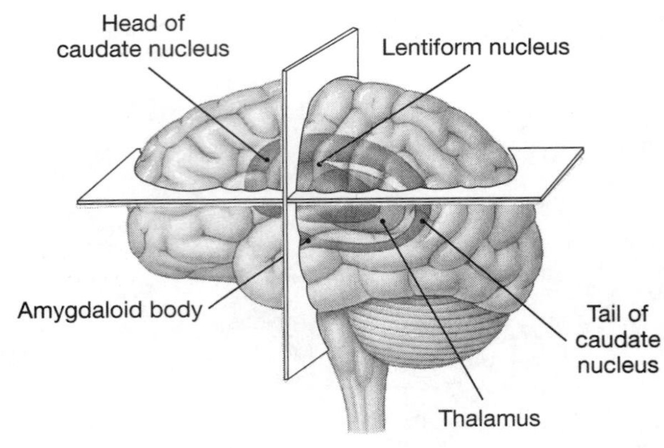

● **FIGURE 30–7**
Cerebral Nuclei.

2➡ Sketch and label what you observe. Additional space can be found at end of this exercise.

Brain tissue

❑ Additional Activities

1. Research the numbers and types of sulci and gyri in the brains of different animals. How does this information relate to any stereotype you might have about the particular animals?
2. Consult Martini, [∞ *pp. 438–439]*, and trace the development of the major brain structures. Relate that information to the anatomy you have just studied.
3. Research the embryology and morphology of the brains of different animals and compare that information with the embryology and morphology of the human brain.

Answers to Selected Concept Check Questions

1. Information is not passed between cerebral hemispheres. This procedure is occasionally performed medically to alleviate severe seizure disorders. Sometimes an injury to the corpus callosum results in the same inability to transfer information from one side of the brain to the other. This can result in a lack of mental and verbal coordination.
2. Brain damage can easily result from pressure caused by the fluid buildup within the brain.

❑ Lab Report

1. Referring to the numbered inquiries at the beginning of this exercise, complete the following box summary:

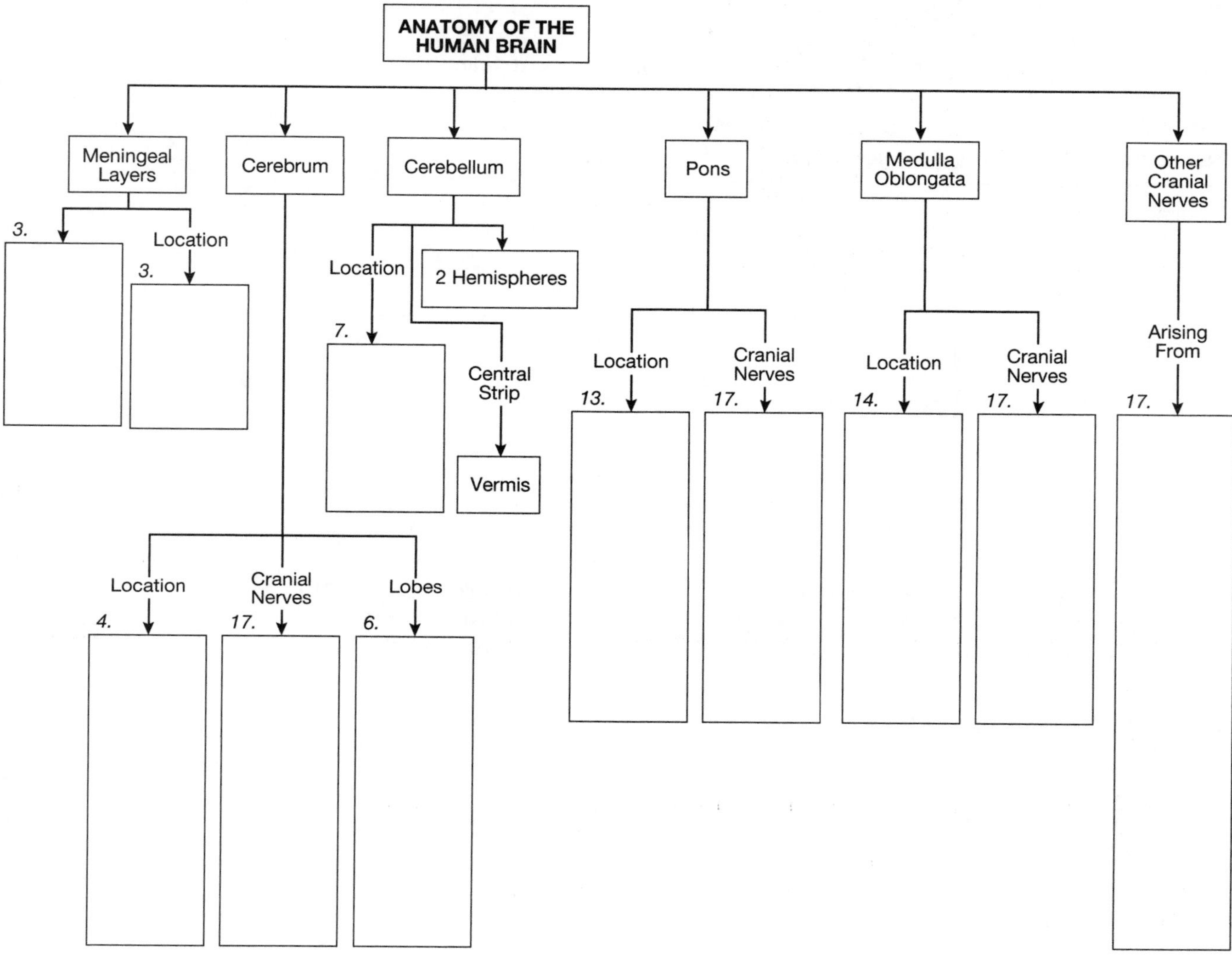

General Questions

NOTES

1. How would you define the term fissure?
2. How does the physical appearance of the cerebral sulci and gyri compare with the physical appearance of the cerebellar sulci and gyri?
3. What is the relationship between the longitudinal fissure and the corpus callosum?
4. Trace the path of the cerebrospinal fluid through the ventricles of the brain.
5. Fill in the following chart. If necessary, use your lecture text.

Structure	Location	Function
Corpora quadrigemina		
Infundibulum		
Mammillary body		
Cerebral peduncles		
Cerebral nuclei		

6. The olfactory bulbs of the sheep are proportionately much larger than the olfactory bulbs of the human. Why do you suppose this is? What other adaptations would you expect different animal brains to have?

EXERCISE 31

Electroencephalography

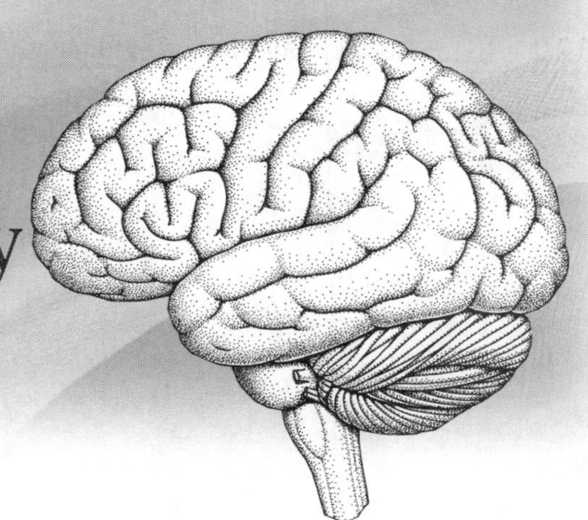

PROCEDURAL INQUIRIES

Preparation

1. What is the difference between an electroencephalogram and an electroencephalograph?
2. What does the horizontal axis of an EEG measure?
3. What does the vertical axis of the EEG measure?
4. From which portion of the brain does the EEG measure brain waves?
5. What are alpha waves and what do they signify?
6. What are beta waves and what do they signify?
7. What are delta waves and what do they signify?
8. What are theta waves and what do they signify?

Experimentation

9. How do you attach someone to the electroencephalograph in your particular lab?
10. What effect does opening or closing the eyes have on an EEG?
11. What effect does mental activity have on an EEG?
12. What effect does physical activity have on an EEG?

Additional Inquiries

13. What is the nature of a nerve impulse?
14. What are some practical uses of the EEG?

Key Terms

Alpha Wave
Amplitude
Beta Wave
Cycles Per Second
Delta Wave
Electroencephalogram
Electroencephalograph
Theta Wave

Materials Needed

Electroencephalograph

We established in our previous exercises that nerve impulses are electrical in nature. We know too that the brain is the central clearinghouse for nerve-based electrical activity. Therefore, by examining the electrical activity of the brain we can glean numerous clues as to what is happening in the nervous system. Electroencephalography is a method of studying the electrical activities of the brain. And in this laboratory exercise we will strive to gain an understanding of the basic principles of electroencephalography.

❑ Preparation

I. Background

A. Electroencephalography

1 ➡ Examine the **electroencephalograph**, the machine used to record the **electroencephalogram (EEG)**, which is the actual report of the electrical activity. Numerous types of electroencephalographs are in use around the country. These range from some very old physiographs and rather expensive oscilloscopes to some very modern computer-interfaced EEG modules. Your instructor will give you directions for the particular electroencephalograph you will be using.

Regardless of the type of encephalograph you are using, the principle will be the same. The electroencephalograph records the electrical activities in

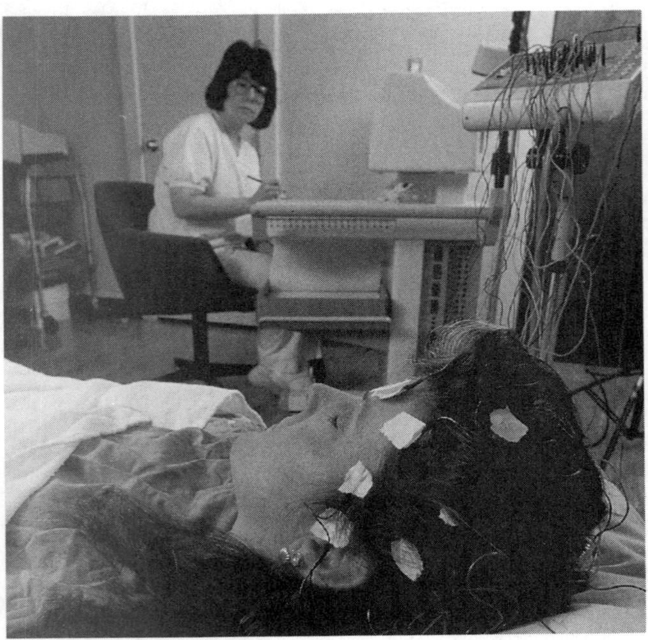

● **FIGURE 31–1**
Electroencephalography.

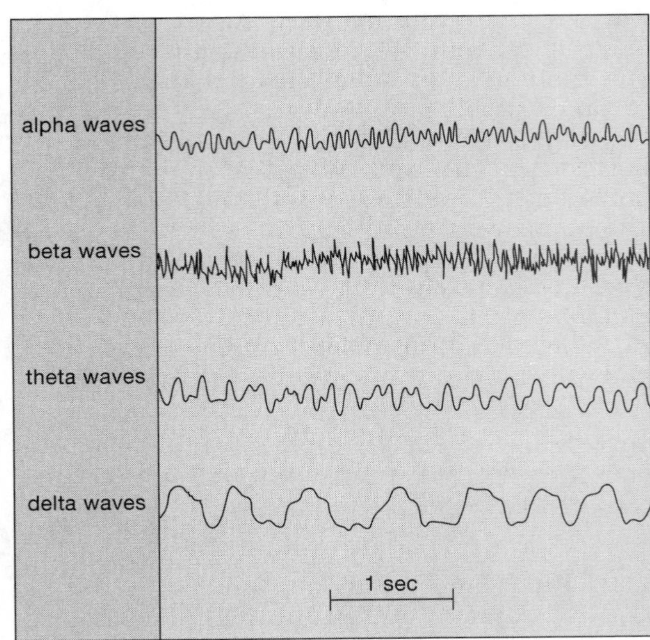

● **FIGURE 31–2**
Brain Waves

a particular part of the brain. In a professional or clinical setting, this may be done by placing between 16 and 23 electrodes at strategic points around the skull, either by needle electrodes directly into the scalp or by surface electrodes placed on the skin. See Figure 31–1●. Readings are then taken from each electrode, usually simultaneously, on some type of read-out device.

In the introductory classroom setting, readings are generally taken from one or two points, usually on the occipital or temporal bone. (Your instructor may direct you to take additional readings.)

2➡ Note that the EEG is traced on a computer screen or on quadrille lined paper. The computer screen may or may not be graphed, depending on the computer program. The darkened blocks on the paper strip are 5 mm square.

As the readout progresses, the horizontal axis of the strip measures time. The standard run time is 25 mm per second, although the tracing speed can be altered to examine some particular aspect of the EEG. Brain waves are measured in **cycles per second** (cps).

The vertical axis (**amplitude**) records electrical activity in microvolts. Recorded electrical discharge does not normally exceed 50 microvolts, so it is usually adequate to set the electroencephalograph to record 10 microvolts per millimeter.

3➡ Keep in mind that the EEG is a recording of spontaneous waves, primarily from the cortical region of the brain. Since the electricity generated by a single neuron would not register on standard equipment, these spontaneous waves come from a large number of neurons that partially discharge without generating an action potential. Electrical potentials are registered according to the changing degrees of excitability of the neurons. The registered waves are a mere fraction of the total waves generated by the electrical activity in the brain.

B. Brain Waves [∞ Martini, p. 493]

Each human being has a unique brain wave pattern. These electrical patterns observed in an EEG are identifiable from about 27 days gestation.

Wave patterns change with age, sensory stimuli, pathology, and the body's physiochemistry. Actually almost anything can change the character of the brain waves.

Much of the time, brain waves are irregular, nondescript, and carry no particular meaning.

1➡ Study Figure 31–2● and note that brain waves are generally classified into four major types.

Describe the differences you see in these four wave patterns.

____

____

____

____

Alpha waves are generally found in EEGs of awake, relaxed persons whose eyes are closed and who are thinking about nothing in particular. Alpha waves disappear during sleep. Alpha waves occur at a rate of about 8–13 cps and seem to be the result of stimulation of the general thalamic nuclei.

In all probability, the spontaneous activity of the thalamocortical region activates several million cortical neurons. Mental or physical stress (including mental or visual stimuli) decreases the amplitude (voltage) but increases the frequency of the alpha waves, eventually suppressing the alpha rhythm completely. This suppression is known as an alpha block.

What would this mean in terms of the height and width of the brain wave?

____

____

____

Beta waves emanate primarily from the parietal and frontal regions of the brain and seem to be activated by CNS tension or activity. Beta waves are faster than alpha waves, usually in the 14–25 cps range. These waves are also of lower amplitude. Beta waves occur in the attentive or alert state, both when a person is awake and when the person is in REM sleep. Some experts divide beta waves into different subtypes; we will not delve into the specifics of that classification scheme in this exercise.

Delta waves are large, slow (<4 cps) waves that become larger and slower as a person falls more and more deeply asleep. Infants also produce copious quantities of delta waves. Delta waves seem to be independent of the activities of the lower regions of the brain. Pathologically, delta waves can be found in serious organic brain disease.

Theta waves are sometimes the subject of much confusion. Theta waves, which have a standard frequency of about 4–7 cps, are characterized by their abnormal contour. Theta waves are normal in children. In adults, however, theta waves are nonpathological only during times of emotional stress, disappointment, and frustration. In the absence of these stresses, theta waves can be indicative of emotional problems (disorders) and certain neural imbalances.

2➡ Check Figure 31–3● for some common brain wave patterns. Can you discern any alpha, beta, delta, or theta waves?

____

____

C. Diagnostic Use of Electroencephalography

1➡ Consider that the EEG is especially useful in diagnosing tumors, abscesses, infections, and assorted brain lesions, particularly epileptic lesions. Since the EEG measures cortical activity almost exclusively, a subcortical lesion might not register on the EEG while still causing considerable havoc in the brain of the person involved.

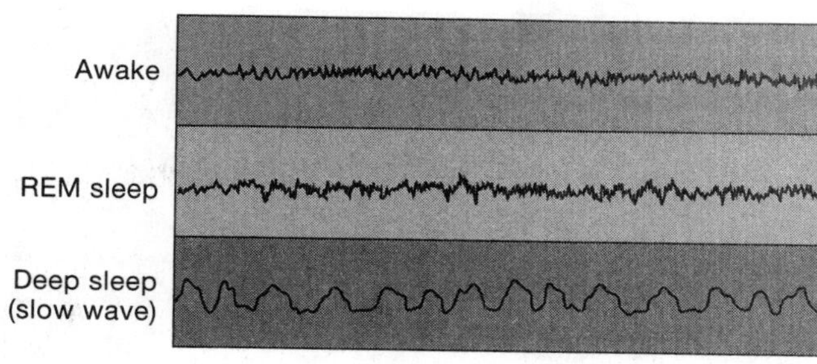

● **FIGURE 31–3**
Normal Electroencephalograms.

An "abnormal" EEG is not necessarily indicative of pathology. For instance, a percentage (between 10% and 25%, or up to 40%, depending on whose statistics are quoted) of people with "normal" brain waves experience the symptoms of epilepsy, while an equal number of people with "abnormal" brain waves have never had any epileptic symptoms.

Today it is more correct to say "seizure disorder" or "the epilepsies" rather than to say epilepsy. The reason is that the medical community recognizes over forty different types of seizure disorders, many of them unique in both etiology and pathology and yet many of them displaying seemingly identical symptoms.

2➡ Note that central nervous system depressants, such as alcohol and barbiturates, also give distinctive depressed brain waves; barbiturate intoxication produces a rapid, disoriented EEG. Numerous neurological problems produce distinctive abnormal EEGs.

Sometimes a multilead EEG will demonstrate the location, rather than the nature, of a brain lesion, because the EEG reading at the point of the lesion is not in context with the other readings.

Other wave patterns may be indicative of specific pathologies. See Figure 31–4● for some common abnormal EEG patterns.

As is often shown on the oscilloscopes on television dramas, the flat EEG is indicative of death, no brain activity.

❑ Experimentation

II. Obtaining the EEG

It is not the purpose of this laboratory exercise to give clinical interpretation to the EEGs that may come up in your class. Keep in mind that you are using classroom laboratory equipment that may not have the accuracy found in certified health care instruments. Also, classroom atmosphere can influence clinical readings. If you are worried about a reading obtained from these laboratory machines, please consult an appropriate health care professional.

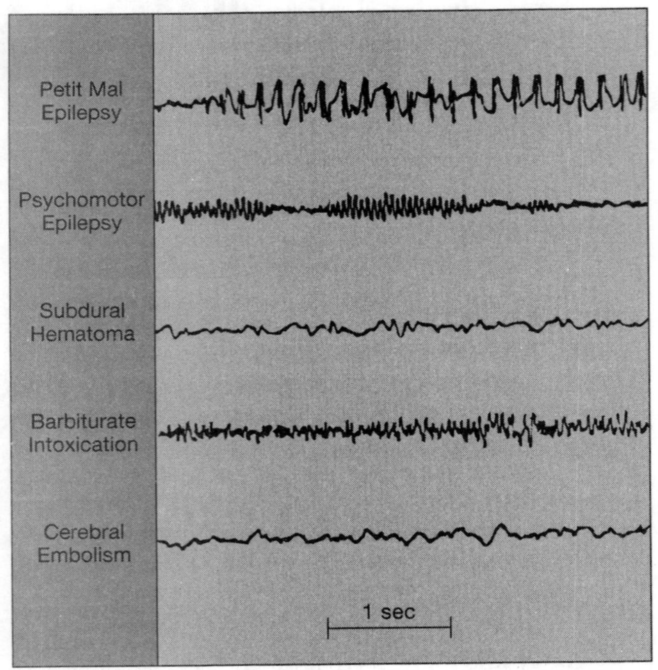

● FIGURE 31–4
Abnormal Brain Waves.

A. Set Up

1➡ Follow your instructor's directions for attaching yourself (or your lab partner) to the electroencephalograph.

2➡ Ideally, the person whose EEG is being taken will be reclining and relaxed in a room without bright lights, and classroom atmosphere will be subdued. Ideal situations rarely exist in the laboratory.

B. Demonstration

The person conducting the test is to run the electroencephalograph under each of the following situations. *Record on the readout* exactly what conditions the test subject is experiencing at the time. If you are using computerized equipment, follow your instructor's directions for recording test conditions.

1➡ Instruct subject to close eyes, relax, think of nothing.

2➡ Instruct subject to open eyes, gaze around lazily without concentrating on anything in particular.

3➡ Repeat #2 but change the room lighting (dramatically, if possible).

4➡ Instruct subject to close eyes and relax; then have someone read a series of simple mathematical problems for subject to think about, followed by a few more complicated problems. Instruct the subject to do the problems mentally while keeping eyes closed.

5➡ Repeat #4 with eyes open.

6➡ Startle subject, perhaps by slamming a door, clanging a bell, or dropping a heavy object.

7➡ Instruct the subject to hyperventilate while attached to the electroencephalograph.

8➡ Have subject blink slowly, then more quickly.

9➡ Tell subject to rotate eyeballs slowly, then more quickly. This is to be done first with eyes closed and then with eyes open.

10➡ Have subject run in place if possible while attached to the electroencephalograph. Do not do this one if your machine is too fragile.

C. Analysis

1➡ Briefly describe each of the patterns you found in IIB.

1. ______________________
2. ______________________
3. ______________________
4. ______________________
5. ______________________
6. ______________________
7. ______________________
8. ______________________
9. ______________________
10. ______________________

2➡ Analyze your results. What types of waves can you recognize? What changes occur with each of the different activities in IIB? What would account for these changes? Are the changes logical?

____
____
____
____

3➡ Compare classroom results. What general similarities and differences do you see with each situation?

____
____
____

❑ Additional Activities

NOTES

1. During activity registered EEG voltage (amplitude) falls despite increased brain activity. Why is that? (Hint: check synchronicity.)
2. Explore the history of brain studies. What factors have contributed to our understanding of the functioning of the brain?

❑ Lab Report

1. Referring to the numbered inquiries at the beginning of this exercise, complete the following box summary:

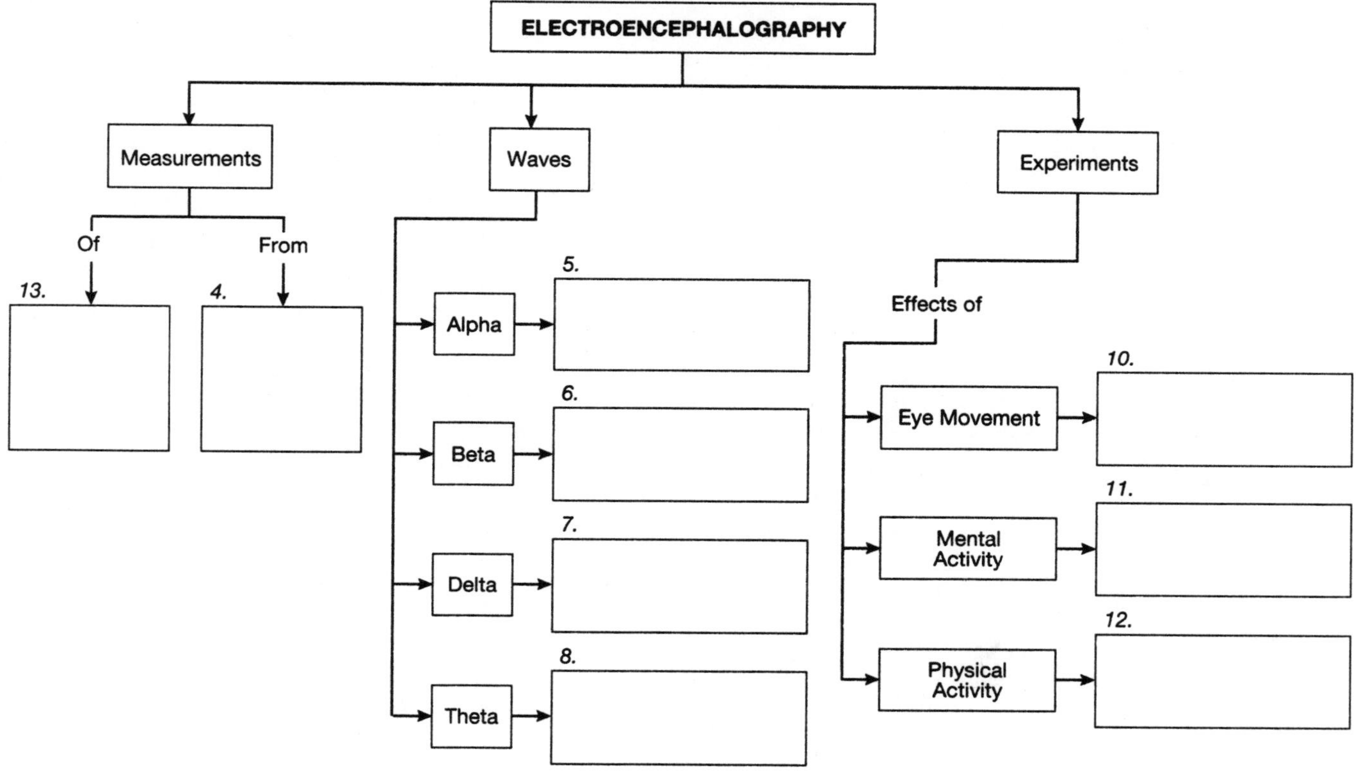

General Questions

NOTES

1. How is an electroencephalogram related to an electroencephalograph?
2. How do you set up your electroencephalograph?
3. Would you expect a tired person to have an EEG with greater or lesser amplitude? Why? What about the cps? Why?
4. How is the vertical axis on the electroencephalogram related to the horizontal axis?
5. Look at numbers 10, 11, and 12 in your Box Summary above. Why do you think you saw the effects you did?
6. What are some practical applications of electroencephalography?

EXERCISE 32

Sheep Brain Dissection

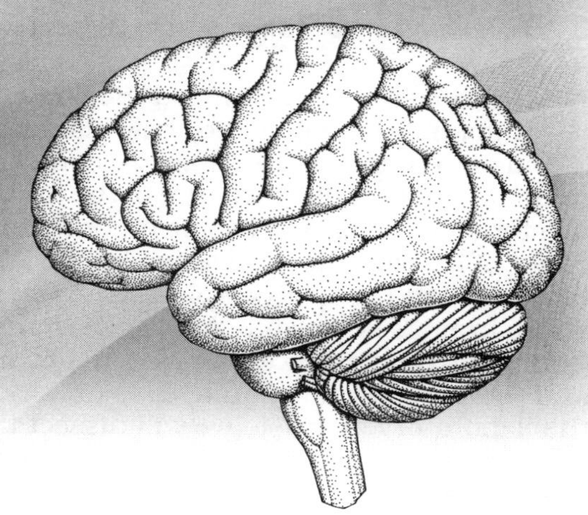

PROCEDURAL INQUIRIES

Preparation

1. What basic terminology should you be familiar with before beginning your brain dissection?
2. How should you prepare the sheep brain for dissection?

Dissection

3. What are the structures found when studying the dorsal anatomy of the sheep brain?
4. What are the structures found when studying the ventral anatomy of the sheep brain?
5. What are the structures found when studying the sagittal anatomy of the sheep brain?
6. What are the structures found when studying the coronal (frontal) anatomy of the sheep brain?

Additional Inquiry

7. What are the major similarities and differences between the sheep brain and the human brain?

Key Terms

Central Canal
Cerebellum
- Arbor Vitae
- Cerebellar Peduncles
- Vermis

Cerebrum
- Cerebral Cortex
- Frontal Lobe
- Longitudinal Fissure
- Occipital Lobe
- Parietal Lobe
- Temporal Lobe

Corpora Quadrigemina
- Inferior Colliculi
- Superior Colliculi

Fornix
Fourth Ventricle
Hypothalamus
Infundibulum
Interventricular Foramen
Lateral Ventricle
Mammillary Body
Medulla Oblongata
Meninges (sing., meninx)
- Arachnoid (Mater)
- Dura Mater
- Pia Mater

Mesencephalic Aqueduct
Nerve Tracts and Structures
- Abducens Nerve
- Accessory Nerve
- Cerebral Peduncles
- Corpus Callosum
- Facial Nerve
- Glossopharyngeal Nerve
- Hypoglossal Nerve
- Oculomotor Nerve
- Olfactory Bulbs
- Optic Chiasma
- Optic Nerve
- Optic Tracts
- Trigeminal Nerve
- Trochlear Nerve
- Vestibulocochlear Nerve

Pineal Body
Pituitary Gland
Pons
Septum Pellucidum
Spinal Cord
Thalamus
- Intermediate Mass of the Thalamus

Third Ventricle

Materials Needed

Preserved Sheep Brains
Dissecting Pan
Blunt Nose Probe
Forceps
Scalpel
Gloves
Human Brain Models or Charts and Diagrams (for comparison)

In Exercise 30 we examined the anatomy of the human brain. Since the basic plan of all mammalian brains is the same, we can expand our understanding of the structure of the human brain by studying the brain of some other mammal. Because of its size and relative availability, the sheep brain is often used for comparative dissection. The sheep brain is somewhat smaller than the human brain but the structures are essentially analogous and the sheep brain is quite easy to work with. In this laboratory exercise, we will dissect the sheep brain and examine those structures common to both the sheep and the human brain.

❑ Preparation

I. Background

A. Terminology

1➡ Before beginning this exercise, you should be familiar with basic brain terminology. If necessary, review this material in Chapter 30.

B. Preparing the Brain

1➡ Obtain a sheep brain and rinse it thoroughly but carefully under running water in order to remove as much of the preservative as possible.

2➡ Obtain your dissecting instruments.

C. Additional Preparation

1➡ Obtain a model of the human brain, or have ready charts and diagrams for your anatomical comparison.

❑ Dissection

II. Dorsal Anatomy

A. Meninges

1➡ Examine the surface of the brain (Figure 32–1●). You may be able to locate a tough sac-like protective layer around the brain. This is the **dura mater** (literally, "tough mother"), the outermost of the **meninges**, the three protective connective tissue membranes that surround the brain and the spinal cord. It is possible that the dura mater has been removed. If the dura mater is still present, gently remove it now. Be especially careful when separating the dura mater from the nerve tracts on the ventral surface of the brain.

2➡ Regardless of whether you have found a dura mater, you probably will not be able to locate the second meningeal layer, the **arachnoid** layer, or arachnoid membrane. The arachnoid layer is sometimes called the **arachnoid mater** (literally,"spider mother"). This web-like meninx does not withstand preservation well. (If you notice tufts of cottony tissue across certain brain fissures, however, you may have located remnants of the arachnoid mater.)

3➡ You should be able to locate the inner layer, the **pia mater** (literally, "gentle mother"), which tightly adheres to the sulci and gyri of the brain. Gently separate at least part of the pia mater from the surface of the brain by lifting it up with your forceps.

4➡ Describe the meningeal layers you were able to locate.

____

____

B. The Cerebrum

1➡ Read the descriptions of the structures of the dorsal anatomy first, then verify your work by checking Figures 32–1● and 32–2●.

2➡ Begin with your sheep brain ventral surface down in your dissecting pan. The most prominent feature of the dorsal surface of the brain is the **cerebrum**. The **longitudinal fissure** divides the cerebrum into the right and left cerebral hemispheres. Note the numerous sulci and gyri forming the surface of the brain. You are actually looking at the **cerebral cortex**, the gray matter of the brain where most voluntary thought processes take place.

3➡ Observe that each cerebral hemisphere is divided into four lobes: **frontal, parietal, temporal** and **occipital**. These lobes are directly inferior to the corresponding cranial bones. Identify these lobes on your sheep brain. Check them off as you find them.

4➡ How do these lobes compare with the human cerebral lobes?

____

____

C. The Cerebellum

1➡ Locate the **cerebellum** by finding a transverse fissure posterior to the cerebrum that

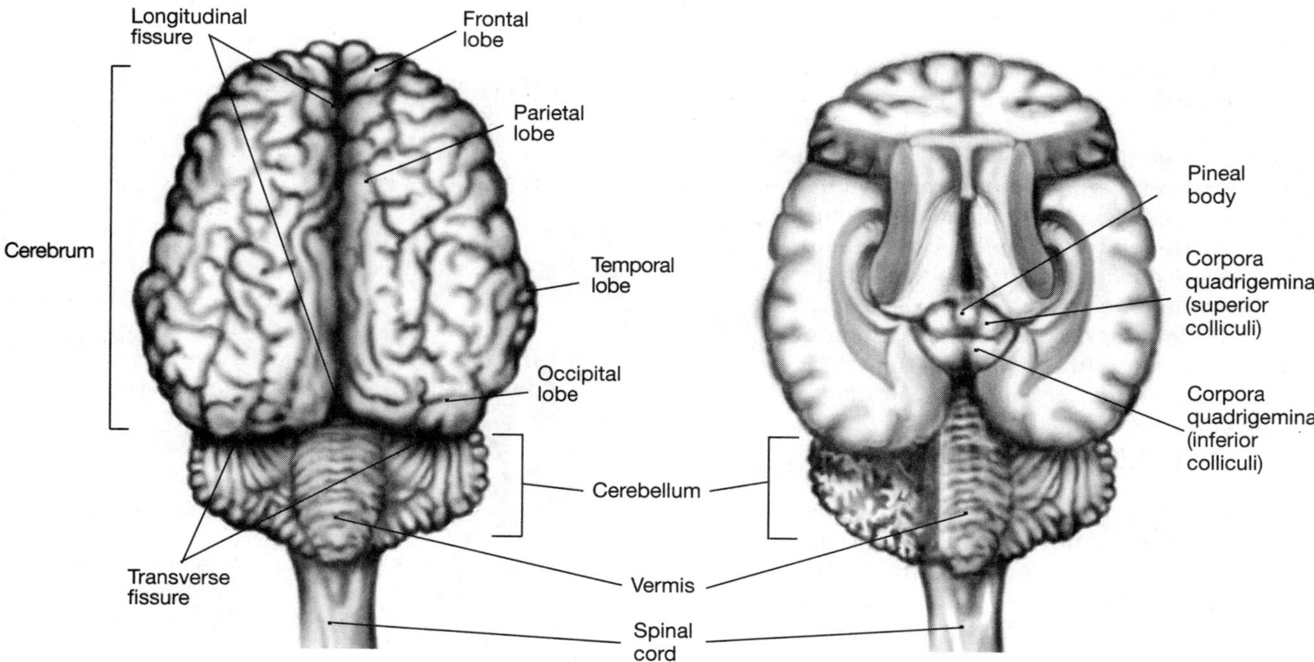

● **FIGURE 32–1**
Dorsal View of the Sheep Brain.

● **FIGURE 32–2**
Dorsal Midbrain — Exposed.

separates it from the cerebellum. Note that the sulci and gyri of the cerebellum seem to be much more tightly packed than the corresponding structures in the cerebrum.

2➡ Note the protruding central area of the cerebellum, the **vermis**. (This midline structure is also present in humans. In humans, however, the vermis is *reduced* in size and the cerebellum is divided into hemispheres separated by the **falx cerebelli.**)

3➡ Check back to Exercise 30 and record other observations about how the vermis of the sheep brain compares with the vermis of the human brain.

____

____

4➡ Pull the cerebellum dorsally away from the brain stem and see if you can locate the three paired **cerebellar peduncles**. The superior cerebellar peduncles connect the cerebellum to the midbrain. The middle cerebellar peduncles connect the cerebellum to the pons. The inferior cerebellar peduncles connect the cerebellum to the medulla.

D. The Spinal Cord

1➡ Locate the **spinal cord**, inferior to the cerebellum. Depending on the condition of the brain and how much of the spinal cord is present, you may be able to find remnants of one or more of the cervical (spinal) nerves.

2➡ What specific features can you identify on the spinal cord of the sheep?

____

____

E. The Corpora Quadrigemina

1➡ With your hands, bend the specimen so you separate the cerebrum from the cerebellum along the transverse fissure. Identify the **corpora quadrigemina**, the four prominent swellings immediately anterior to the cerebellum. The paired swellings closest to the cerebellum are the **inferior colliculi**, which are involved with auditory reflexes. Anterior to the inferior colliculi are the larger **superior colliculi**, which are involved with visual reflexes. Centrally located at the anterior end of the superior colliculi is the **pineal body**, the endocrine gland, which secretes the hormone melatonin.

2➡ Check off the colliculi as you locate them.

3➡ How do the colliculi and pineal body of the sheep compare with the colliculi and pineal body of the human?

____

____

____

III. Ventral Anatomy

A. Structures Related to Cranial Nerves I and II

1➜ Turn your sheep brain over and begin your ventral examination at the anterior end. The large paired extensions are the **olfactory bulbs**, components of the olfactory tract, or N I. All of the cranial nerves are paired. Use Figure 32–3● to verify your identification of these structures as well as the others listed in this section. *[∞ Martini, p. 465 for a human analog. Sheep brain anatomy is almost identical.]*

2➜ Posterior to the olfactory bulbs, you should notice an X-like structure. The crossing point is the **optic chiasma**. The anterior portions of the X are the **optic nerves**, N II, while the portions receding from the chiasma into the brain are the **optic tracts**.

3➜ Compare the olfactory and optic structures of the sheep and human brains.

____

____

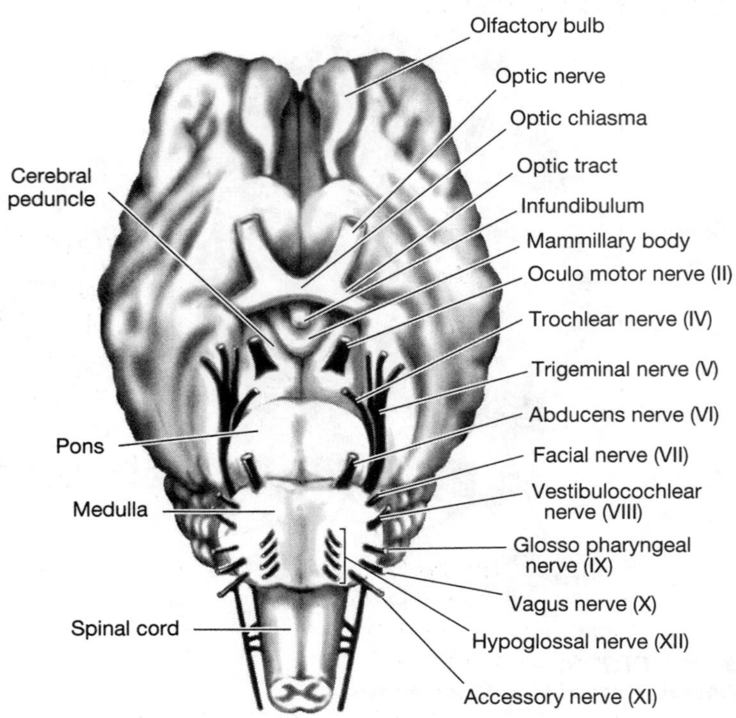

● **FIGURE 32–3**
Ventral View of Midbrain.

B. Structures of the Hypothalamus

1➜ Observe the structure that seems to originate under the optic chiasma, the **infundibulum**, or the stalk of the pituitary gland. In actuality, the infundibulum arises from the **hypothalamus**. Some specimens may still have the pituitary gland attached to the infundibulum. (We will examine the hypothalamus further when we study the sagittal anatomy of the brain.) If your sheep brain is overly preserved, you may find the infundibulum a bit mushy and thus easily destroyed. Posterior to the infundibulum is the single-lobed **mammillary body**. In humans the mammillary body is double-lobed. The mammillary body controls the feeding reflexes.

2➜ Compare these sheep structures with the corresponding human structures.

____

____

C. The Cerebral Peduncles and Cranial Nerve III

1➜ Find the **cerebral peduncles**, posterior to the mammillary body; these are fibrous tracts connecting the cerebrum and the medulla.

2➜ How do the sheep's cerebral peduncles compare with the human cerebral peduncles?

____

____

3➜ Depending on the state of your sheep brain, you may be able to locate the **oculomotor nerves**, N III, extending from about the middle of the cerebral peduncles.

D. The Pons and Cranial Nerves IV Through VII

1➜ Locate the **pons**, posterior to the cerebral peduncles. The **trochlear nerves**, N IV, can be found at the junction of the midbrain and the pons.

2➜ Compare the pons of the sheep with the human pons.

____

____

3➜ Follow Figure 32–3● to locate the **trigeminal nerves,** N V, the **abducens nerves**,

● **FIGURE 32–4**
Sagittal Section of Sheep Brain.

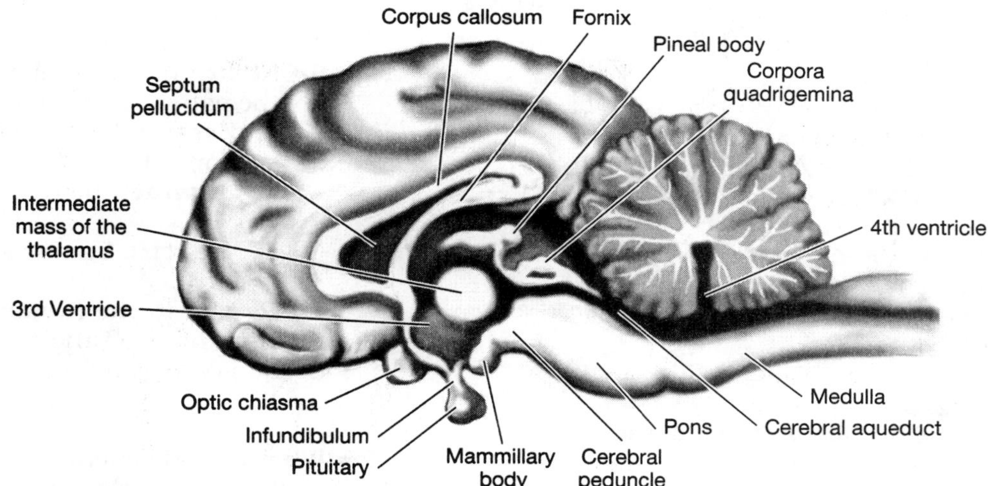

N VI, and the **facial nerves**, N VII. All of these nerves arise from the pons. (Finding these nerves will depend on the condition of your sheep brain.)

E. THE MEDULLA AND CRANIAL NERVES VIII THROUGH XII

1 ➡ Find the **medulla oblongata,** posterior to the pons. This structure is more commonly called the medulla.

2 ➡ Compare the sheep's medulla with the human medulla.

____

____

3 ➡ Refer again to Figure 32–3● to locate the **vestibulocochlear nerves**, N VIII; the **glossopharyngeal nerves**, N IX; the **vagus nerves**, N X; the **accessory nerves**, N XI; and the **hypoglossal nerves**, N XII. (Finding these nerves will depend on the condition of your sheep brain.)

IV. Sagittal Anatomy

A. PREPARATION OF THE SAGITTAL SECTION

1 ➡ To explore the internal structures of the brain, it will be necessary to separate the brain into left and right halves. Refer to Figure 32–4● as you work to identify the structures described.

2 ➡ Manually move the left and right cerebral hemispheres apart from each other. Note the thick fibrous tissue holding the hemispheres together. This is the **corpus callosum,** the nervous tissue that functions as a communicating band between the cerebral hemispheres.

3 ➡ With a sharp scalpel, carefully cut the corpus callosum. With equal care, continue bisecting the structures of the diencephalon. Bisect the cerebellum. Continue your incision through the pons, medulla, and other internal structures until you have completely separated the right and left sides of the brain.

B. INTERNAL STRUCTURES OF THE BRAIN

1 ➡ Examine either half of the brain. Locate the corpus callosum. Follow this fibrous band around posteriorly. The posterior end of the corpus callosum is known as the **splenium of the corpus callosum**. The fibrous band inferior to the corpus callosum is the **fornix**. Between the corpus callosum and the fornix is an opening. If your incision was good, this opening is covered by a thin membrane known as the **septum pellucidum**. If you destroyed the septum pellucidum, note where it should be.

2 ➡ The area behind the septum pellucidum is the **lateral ventricle**. Put your blunt nose probe into the ventricle and get a feel for the size of this cavity. The brain has a lateral ventricle in each cerebral hemisphere.

3 ➡ Inferior to the fornix you will find a large round area known as the **intermediate mass of the thalamus**. Run your probe along the underside of the thalamus and

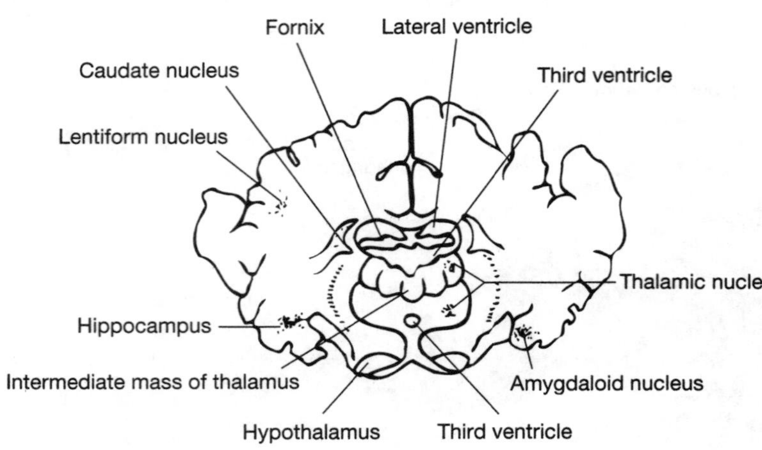

● **FIGURE 32–5**
Coronal (Frontal) Section of Sheep Brain.

see if you can locate a cavity. This is the **third ventricle**. Refer to Figure 32–5● to help you visualize this third ventricle. The third ventricle is sometimes difficult to recognize as a cavity in preserved brains. Between the lateral ventricles and also communicating with this third ventricle is the **interventricular foramen**, also known as the **foramen of Munro**. You will probably not be able to locate this foramen.

4➡ If you found the third ventricle, you may be able to trace a thin canal running posteriorly from the ventricle toward the cerebellum. This canal, the **mesencephalic aqueduct**, which is also known as the **cerebral aqueduct** or the **aqueduct of Sylvius**, leads to the **fourth ventricle**, which is just posterior to the pons and medulla and just inferior to the cerebellum. Continuing on from the fourth ventricle is the **central canal** of the spinal cord.

5➡ Relocate on the sagittal section those structures you found when you examined the dorsal and ventral views of the brain. Note particularly the hypothalamus, which is directly inferior to the thalamus and the cavity of the third ventricle. If your brain is well preserved, the hypothalamus will be a distinct V-shaped structure. The infundibulum extends from the hypothalamus. The **pituitary gland**, or hypophysis, is the knob-like structure at the end of the infundibulum.

6➡ Return to the cerebellum and observe the pattern formed by the white matter and gray matter on the longitudinal section. This pattern is known as the **arbor vitae**, the tree of life.

C. Sagittal Comparison

1➡ Return to the human models or diagrams. Check off the sheep structures in Section IV as you find them on the human model.

2➡ What type of generalization can you make about the sagittal anatomy of the sheep and the sagittal anatomy of the human?

____

____

V. Coronal (Frontal) Anatomy

A. Preparation of the Coronal (Frontal) Section

1➡ To gain a perspective of the internal anatomy of the brain, you will have to do a coronal (frontal) section. Do a practice section first by making a slice through one cerebral hemisphere about 2 cm from the anterior end. Examine this slice. Note the relative size and position of the gray matter (the cerebral cortex) and the white matter.

2➡ Make another frontal incision. Pass this incision through the thalamus and hypothalamus. Use just one half of your sheep brain. You may wish to explore further with the other half. Depending on your accuracy and on the condition of the brain, you will be able to identify some or all of the structures in Figure 32–5●.

3➡ What generalization can you make when comparing the coronal anatomy of the sheep and the human?

____

____

❑ Additional Activities

NOTES

1. Research the numbers and types of sulci and gyri in the brains of different animals. How does this information relate to any stereotype you might have about the particular animals?
2. Research the embryology and morphology of the brains of different animals and compare that information with the embryology and morphology of the human brain.

❑ Lab Report

1. Referring to the numbered inquiries at the beginning of this exercise, complete the following box summary:

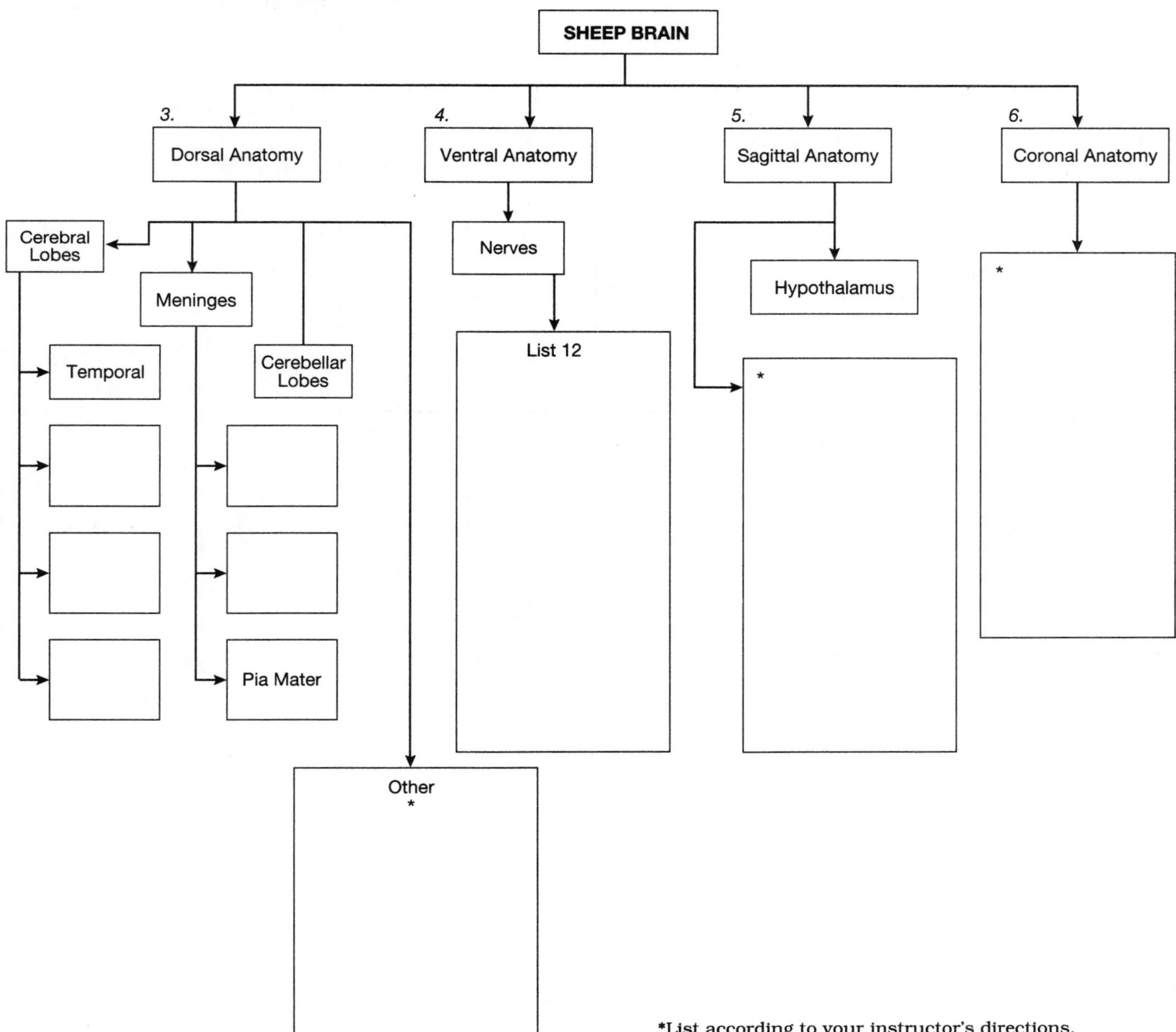

*List according to your instructor's directions.

General Questions

NOTES

1. Go back to Exercise 30. List the points of common terminology you find between these two exercises.
2. Based on your dissection, list at least three physical differences between the sheep brain and the human brain.
3. The olfactory bulbs of the sheep are proportionately much larger than the olfactory bulbs of the human. Why do you suppose this is?
4. Complete the following chart using the boldface terms (areas of the brain) from this lesson. Add to this chart by referring to the previous exercise or to your lecture text.

Brain Part	Location	Function

EXERCISE 33

Cranial Nerves

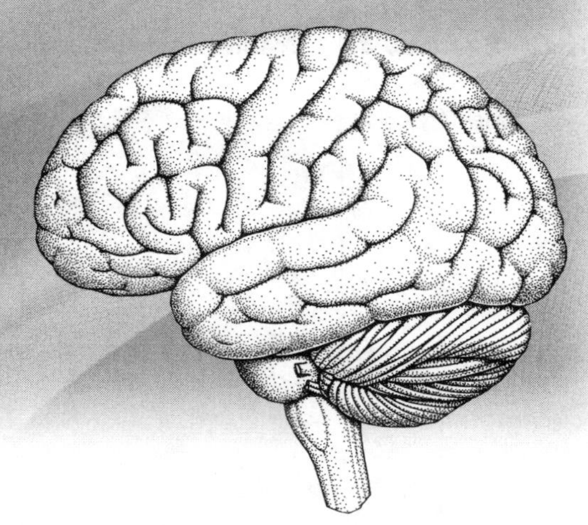

PROCEDURAL INQUIRIES

Examination

1. Where do we find the cranial nerves?
2. What are the names of the cranial nerves?
3. What is the physiological difference between the sensory and motor neurons?
4. What is a mixed nerve?

Experimentation

5. How can we test for the proper functioning of each of the cranial nerves?

Additional Inquiries

6. Why are the cranial nerves a part of the peripheral nervous system?
7. What problems might be detected by evaluating cranial nerve functions?

Key Terms

Cranial Nerves
Mixed Nerves
Motor Neurons
Peripheral Nervous System
Sensory Neurons

Materials Needed

Brain Models
Charts and Diagrams, as available
Skulls (for perspective)
Testing Instruments (if needed)

The **cranial nerves** have fascinated anatomists for centuries, not only because of their cerebral origin but also because of their impact on the body as a whole. The cranial nerves are components of the **peripheral nervous system** because they are connected to (rather than being a part of) the brain. These nerves arise from the brain and exit through foramina in the base of the skull. Occasionally you may hear a phrase like "the cranial nerves arise from the cerebrospinal center or cerebrospinal axis" used in reference to the way the cranial nerves are attached to the brain.

In this laboratory exercise we will identify the cranial nerves on a human brain model, trace the paths of these nerves on a standard diagram, and perform certain tests to observe the manifestations of cranial nerve function.

❑ Examination

I. Identification [∞ *Martini, pp. 464-474*]

A. Models and Charts

1➡ Obtain a classroom brain model or use the available charts and diagrams. By comparing your brain model with Figure 33–1•, identify the twelve pairs of cranial nerves as they proceed from the central nervous system. Use the skulls for orientation and perspective. (For a detailed description of the cranial nerves leaving the brain, see the laboratory exercise on the human brain, Exercise 30.)

2➡ Notice that the twelve cranial nerves are numbered in succession along the ventro-

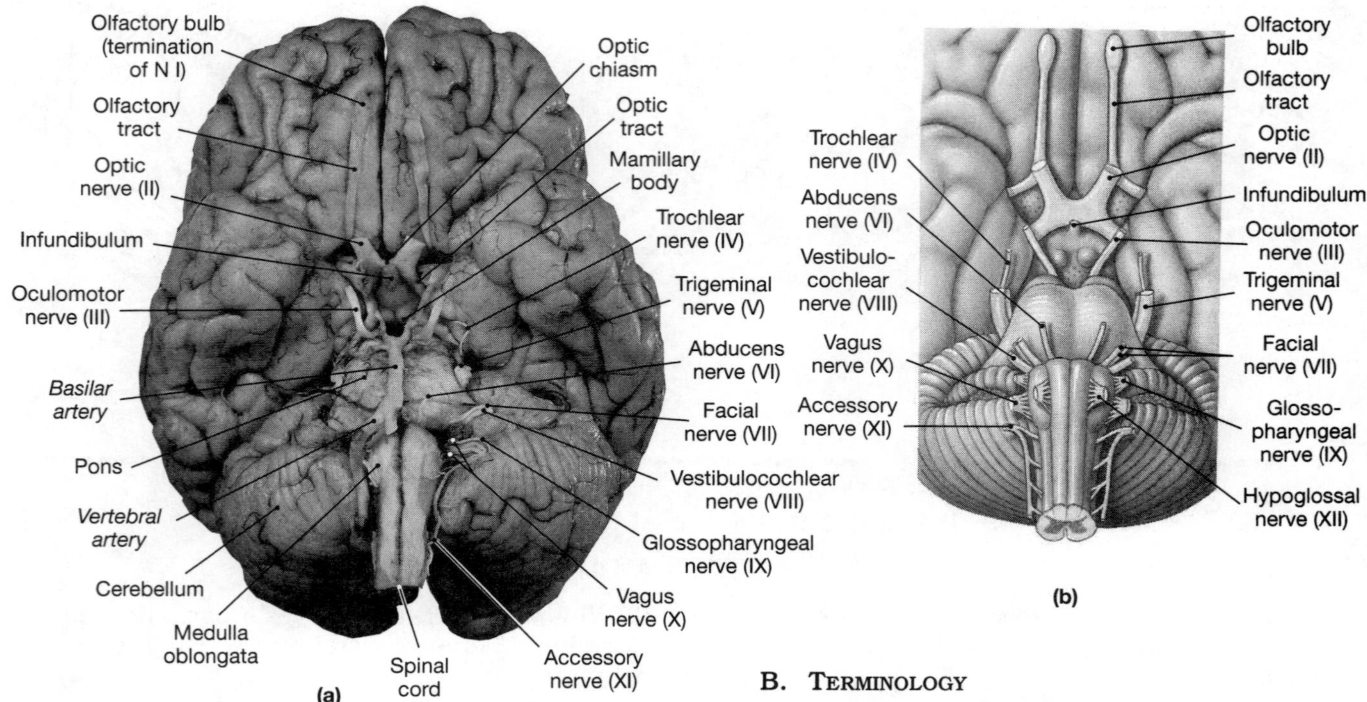

● FIGURE 33–1
Origins of the Cranial Nerves.

lateral surface of the brain. Keep in mind that these nerves are part of the peripheral nervous system. Only the brain and spinal cord form the central nervous system. Recall from Exercise 26 that a nerve is a bundle of nerve cell fibers and their connective tissue coverings outside the central nervous system. Each cranial nerve consists of thousands of individual neuronal fibers.

B. Terminology

1➡ Help yourself remember the names of the cranial nerves by learning a mnemonic. Figure 33–2● gives you what may well be the most famous one in anatomy. If you make up your own phrase, write it out here.

____

____

2➡ In working with the cranial nerves, remember that **sensory neurons** carry a message or sensation away from a stimulation point and toward a processing unit.

On	Old	Olympus'	Towering	Tops	A	Finn	And	German	Viewed	Some	Hops
l	p	c	r	r	b	a	c	l	a	p	y
f	t	u	o	i	d	c	o	o	g	i	p
a	i	l	c	g	u	i	u	s	u	n	o
c	c	o	h	e	c	a	s	s	s	a	g
t		m	l	m	e	l	t	o		l	l
o		o	e	i	n		i	p			o
r		t	a	n	s		c	h		A	s
y		o	r	a				a		c	s
		r		l				r		c	a
								y		e	l
								n		s	
								g		s	
								e		o	
								a		r	
								l		y	

● FIGURE 33–2
Cranial Nerve Mnemonic.

Cranial Nerve (Number)	*Sensory Ganglion*	*Branch*	*Primary Function*	*Foramen*	*Innervation*
Olfactory (I)			Special sensory	Cribriform plate of ethmoid	Olfactory epithelium
Optic (II)			Special sensory	Optic foramen	Retina of eye
Oculomotor (III)			Motor	Superior orbital fissure	Inferior, medial, superior rectus, inferior obliques and levator palpebrae muscles; intrinsic muscles of eye
Trochlear (IV)			Motor	Superior orbital fissure	Superior oblique muscle
Trigeminal (V)	Semilunar		Mixed		Areas associated with the jaws
		Ophthalmic	Sensory	Superior orbital fissure	Orbital structures, nasal cavity, skin of forehead, upper eyelid, eyebrows, nose (part)
		Maxillary	Sensory	Foramen rotundum	Lower eyelid; upper lip, gums, and teeth; cheek, nose (part), palate and pharynx (part)
		Mandibular	Mixed	Foramen ovale	*Sensory* to lower gums, teeth, lips; palate (part) and tongue (part) *Motor* to muscles of mastication
Abducens (VI)			Motor	Superior orbital fissure	Lateral rectus muscle
Facial (VII)	Geniculate		Mixed	Internal acoustic canal to facial canal; exits at stylomastoid foramen	*Sensory* to taste receptors on anterior 2/3 of tongue *Motor* to muscles of facial expression, lacrimal gland, submandibular gland, sublingual salivary glands
Vestibulocochlear (Acoustic) (VIII)			Special sensory	Internal acoustic canal	
		Cochlear			Cochlea (receptors for hearing)
		Vestibular			Vestibule (receptors for motion and balance)
Glossopharyngeal (IX)	Superior, inferior		Mixed	Jugular foramen	*Sensory* from posterior 1/3 of tongue; pharynx and palate (part); blood pressure and composition *Motor* to pharyngeal muscles, parotid salivary gland
Vagus (X)	Jugular, nodose		Mixed	Jugular foramen	*Sensory* from pharynx; pinna and external canal; diaphragm; visceral organs in thoracic and abdominopelvic cavities *Motor* to palatal and pharyngeal muscles, and visceral organs in thoracic and abdominopelvic cavities
Accessory (XI)		Medullary	Motor	Jugular foramen	Voluntary muscles of palate, pharynx, and larynx (via vagus nerve)
		Spinal	Motor	Jugular foramen	Sternocleidomastoid and trapezius muscles
Hypoglossal (XII)			Motor	Hypoglossal canal	Tongue musculature

● **FIGURE 33–3**
Cranial Nerves.

Motor neurons carry a message away from the processing unit and toward an action point (think: motor = action). **Mixed nerves** have both sensory and motor components. To summarize this information, see the notation at the bottom of Figure 33–3●.

3➡ As you learn the names of the cranial nerves, you may find other reference books using slightly different terminology for some of the nerves. For example:

Cranial nerve VIII, the acoustic nerve, is sometimes called the vestibulocochlear nerve or the auditory nerve. As you delve further into the nervous system, you will find that the vestibular nerve and the cochlear nerve are both branches of the acoustic nerve. Acoustic has to do with hearing, and each of these nerve branches is connected with a part of the inner ear apparatus. (See Exercise 37.)

Cranial nerve XI, the spinal accessory nerve, is sometimes simply called the accessory nerve (even though we need the S for our mnemonic).

(In older texts you may also find the trochlear nerve referred to as the pathetic; the trigeminal nerve called the trifacial; and the vagus nerve called the pneumogastric.)

II. Innervation

A. Models and Charts

1➡ Study Figure 33–3●. Review the listed functions and innervations with the brain model in front of you.

2➡ Figure 33–4● is a diagram of the distribution and functions of the cranial nerves. Use the information in Figure 33–3● to draw in the lines connecting the cranial nerve stems with their points of innervation. List any major branches of these nerves and note whether the innervation is sensory, motor, or mixed. Use other sources as necessary.

____

____

❑ Experimentation

III. Cranial Nerve Tests

Note: If you will be performing these and other tests later in the course as parts of the sensory and reflex exercises, your instructor may decide not to complete these modified versions of the cranial nerve tests at this time.

A. Observations

1➡ Work with a partner. Check each other's cranial nerve responses. Keep in mind that classroom tests should never be considered as diagnostic!

Nerve	*Test*
I	Sniff and identify unknown substances. Refer to Exercise 43.
II	Use standard eye charts. Check peripheral vision by bringing your finger around from behind your head until it is visible. Refer to Exercise 38.
III	Examine pupils for size, shape, and equality. Check ability to follow moving objects. Refer to Exercise 38.
IV	For screening purposes, nerve IV is tested with nerve III.
V	Sensory: Check the face for pain, touch, and temperature sensations using appropriate objects. Refer to Exercise 44.
	Motor: Check clenched teeth, opening mouth against resistance, and ability to move jaw equally from side to side.
VI	For screening purposes, nerve VI is tested with nerve III.
VII	Check standard tastes (sweet, salty, bitter, sour) and standard facial movements (smiling, closing eyes, etc.). Check tearing with ammonia or onion. Refer to Exercise 42.
VIII	Check hearing and bone conduction with standard hearing tests. Refer to Exercise 40.
IX	Check gag and swallowing reflexes. (Don't actually do this one—just recall what gagging is like!)
X	For screening purposes, nerve X is tested with nerve IX.
XI	Hold subject's head and shoulders. Check shrugging and rotating movement against resistance.
XII	Check protrusion and retraction of tongue.

B. Pathologies

Go back to Figure 33–3●. Make a list of problems you might find associated with damage to different cranial nerves.

____

____

____

____

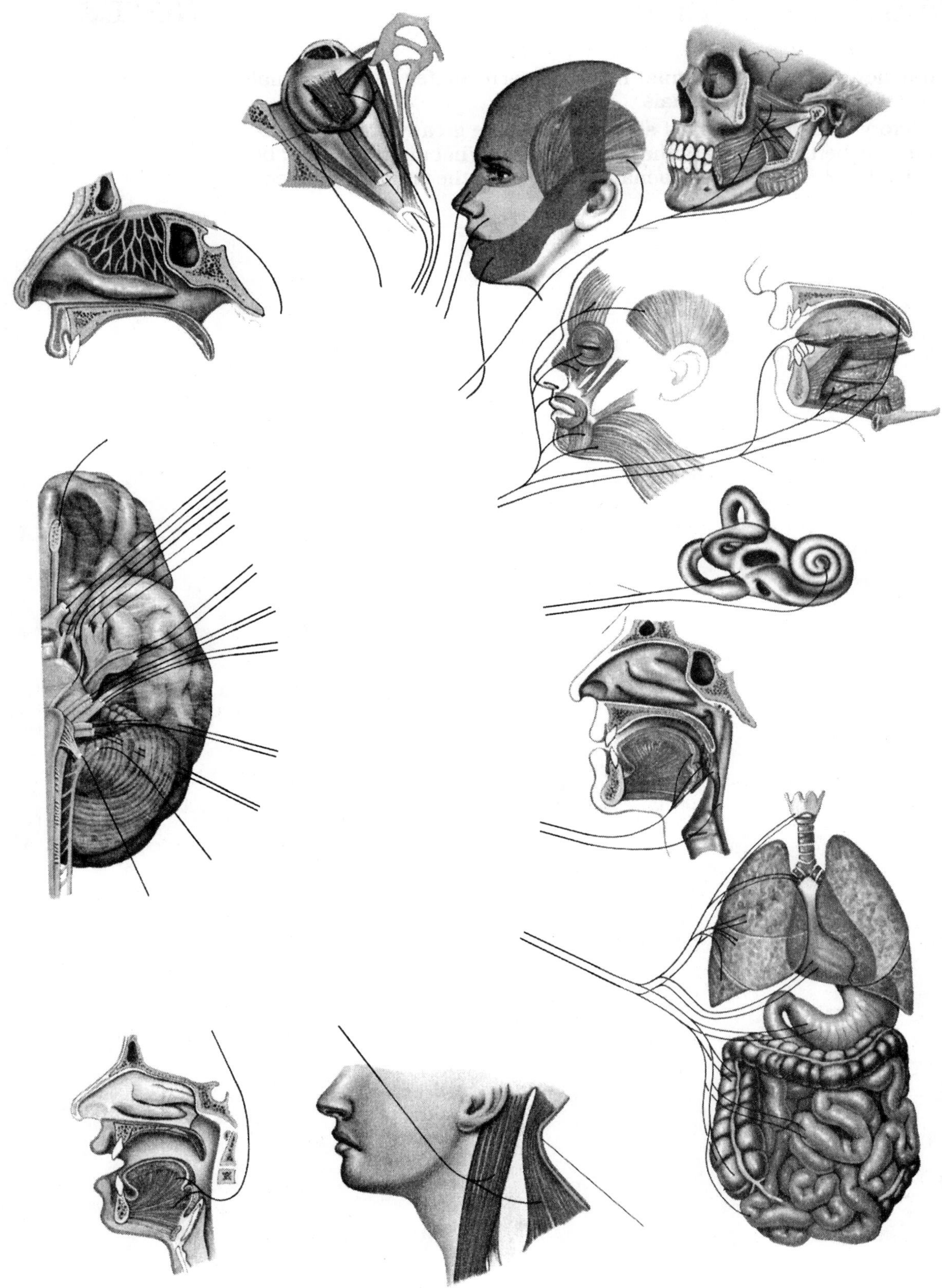

● **FIGURE 33–4**
Functional Distribution of the Cranial Nerves.

☐ Additional Activities

NOTES

1. Find out how the human cranial nerve pattern relates to the cranial nerve patterns of other mammals.
2. The olfactory and optic bulbs of some animals are greatly enlarged proportionately when compared to the human bulbs. What animals would be included in this? Is there a selectional advantage for the animals involved?

❑ Lab Report

1. Referring to the numbered inquiries at the beginning of this exercise, complete the following box summary:

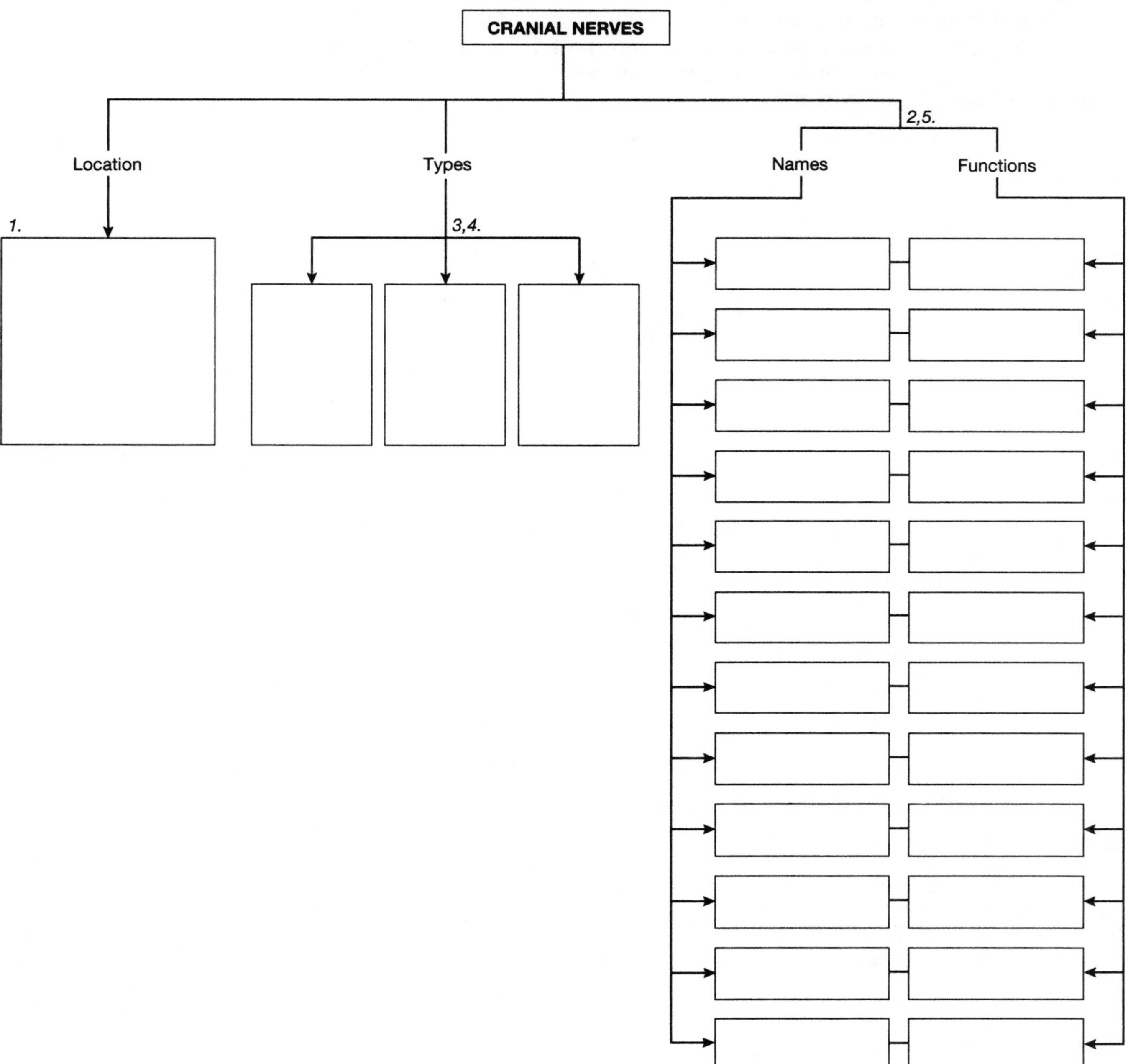

General Questions

NOTES

1. Why are nerves IV, VI, and X not tested independently on simple screening tests?
2. Explain how each of the sensory cranial nerves is related to specific motor nerves. What does this tell you about the concept of afferent and efferent informational integration?
3. List the functions of the vagus nerve. Do you think a person could survive if the vagus nerve were cut? If so, the motor or the sensory portions, or both? If not, why not?

EXERCISE 34

Autonomic Nervous System

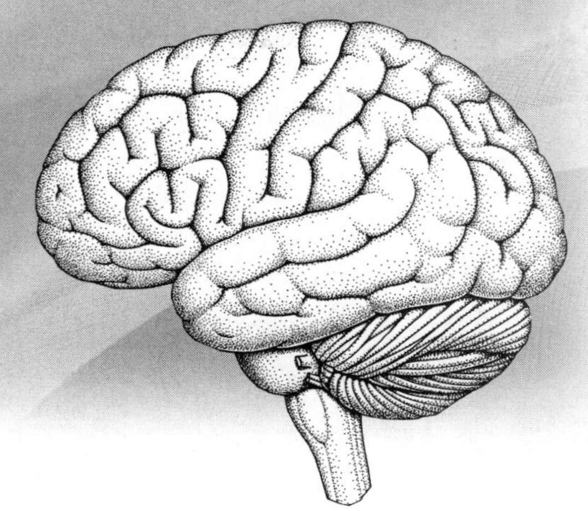

PROCEDURAL INQUIRIES

Preparation

1. What is the sympathetic division of the autonomic nervous system?
2. From which nerves does the sympathetic division arise?
3. What is the parasympathetic division of the autonomic nervous system?
4. From which nerves does the parasympathetic division arise?
5. What does preganglionic mean?
6. What does postganglionic mean?

Examination

7. Where does the information that causes the autonomic response come from?
8. How do the sympathetic and parasympathetic divisions function in unison to maintain homeostasis?

Additional Inquiries

9. What is an intramural ganglion?
10. What is a collateral ganglion?

Key Terms

Autonomic
Collateral Ganglion
Craniosacral Division
Intramural Ganglion
Parasympathetic Division
Postganglionic Fibers
Preganglionic Fibers
Sympathetic Division
Terminal Ganglion
Thoracolumbar Division

Materials Needed

Charts and Diagrams

The **autonomic** nervous system (ANS) is the "automatic" nervous system, the system responsible for the series of actions and reactions that take place primarily without the voluntary consent of the rest of the body. This system is the homeostatic system of checks and balances responsible for fine-tuning the functioning body, for maintaining normal fluid and electrolyte concentrations, for preserving the organism as a whole, and for serving as a counterbalance to adverse external conditions. You should be aware that the ANS is composed only of *efferent* or motor components.

These autonomic outputs are the end results of numerous inputs, integrated and processed at the subconscious level but profoundly influenced by the emotions and intellectual activities of the real and ethereal components of the nervous system as a whole.

In this laboratory exercise we will delineate the essentials of the autonomic nervous system by examining both the autonomic reflex and the autonomic response.

(It is not the purpose of this exercise to explain the fine detail of the autonomic nervous system. You will be reviewing those concepts in the lecture portion of your course.)

❑ Preparation

I. Background

A. Divisions of the ANS

1 ➡ Refer to Figure 34–1● for an overview of the autonomic nervous system. The autonomic nervous system is generally consid-

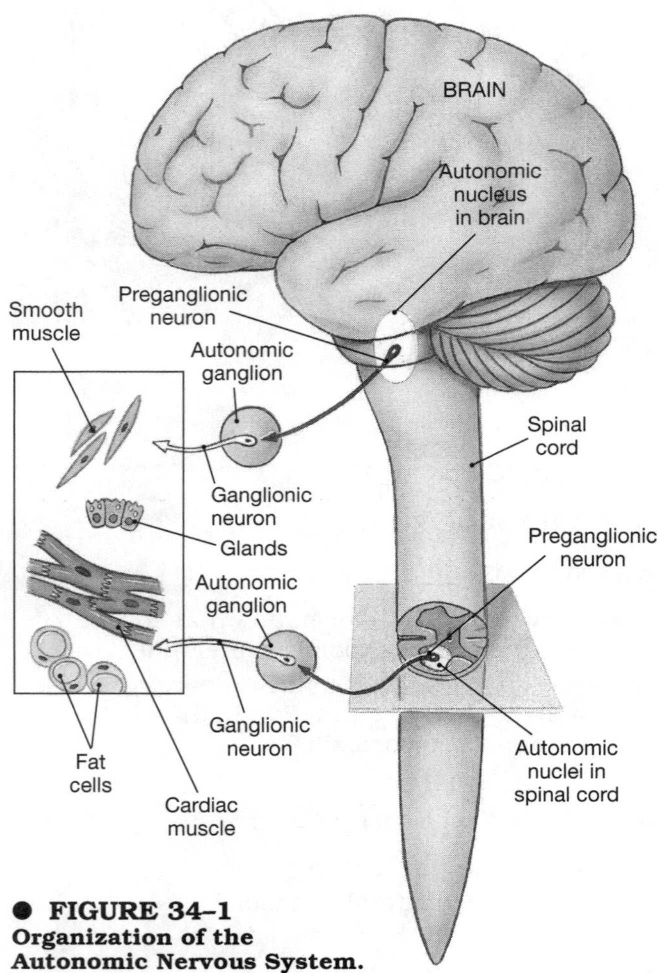

● **FIGURE 34–1**
Organization of the Autonomic Nervous System.

ered to be composed of two antagonistic divisions, the **sympathetic division** and the **parasympathetic division**. The sympathetic division arises from the thoracic and lumbar spinal nerves and is also known as the **thoracolumbar division**. The parasympathetic division, the **craniosacral division**, incorporates the cranial and sacral nerves.

2➡ Consider the term **intramural ganglion**, or **terminal ganglion**. These ganglia, associated with the parasympathetic division, are located at or near the target organ. The ganglia of the sympathetic division (the sympathetic chain ganglia and the collateral ganglia) are located near the spinal cord. *[∞ Martini, p. 513]*

B. Innervation

1➡ Compare Figures 34–2● and 34–3●. Also use similar figures as may be found on your laboratory charts and diagrams. Notice that many of the visceral organs are innervated by both the sympathetic and parasympathetic systems. This double innervation allows for continuous, finely tuned homeostatic adjustments.

Concept Check 1 Why is this logical? ______________________

____

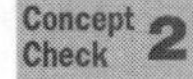
Based on the information in the various figures in this exercise, make a list of the organs that have both sympathetic and parasympathetic innervation.

____

____

C. Ganglionic Fibers

1➡ As you compare the figures, note the ganglia. Note also the use of the terms **preganglionic** and **postganglionic fibers.**

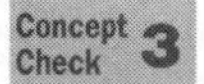
Based on logic, what do the terms preganglionic and postganglionic mean?

____

____

2➡ Note also Figure 34–2c●, the organization of the sympathetic division of the ANS. The **collateral ganglia** are conglomerates of nerve cell bodies whose postganglionic fibers extend throughout the abdominopelvic cavity. *[∞ Martini, p. 520]*

How does Figure 34–2c● relate to Figure 34–2b●?

____

____

❑ Examination

II. Visceral Reflex Arc *[∞ Martini, p. 519]*

A. Concept

1➡ Consider what you have already learned about the ganglia of the autonomic nervous system. It should seem fairly logical to you that the principal difference between the somatic and visceral reflex arcs is the two efferent neurons of the autonomic arc. Recall that efferent fibers take a message away from a central processing unit and toward an effector (muscle, gland, etc.). The somatic arcs do not have that extra ganglion. (Somatic arcs are covered in Exercise 29, the laboratory exercise on reflex physiology.)

2➡ Think about such visceral reflex arcs as found in the photopupillary reflex or the salivary reflex.

B. Components of the Arc

1➡ Go back to Figures 34–2● and 34–3● and locate the additional efferent neurons discussed in the previous sections.

2➡ Note that you have just identified efferent neurons and effector targets. But we have identified this section as a reflex arc. Where does the information come from, the information that causes the effector response in the autonomic nervous system? We have already said the autonomic system is strictly an efferent system.

Recall what we said at the beginning of this exercise about a variety of sources. The autonomic information comes from the spinal cord, the brain stem nuclei, the limbic system, the cerebral hemispheres, the emotions, and the intellect, to name a few. All of these sources represent the afferent part of the arc, which is not considered part of the ANS.

C. Examples

1➡ Describe the afferent and efferent components of the following:

a. You feel sleepy after eating a big meal.
Afferent ____________________
Efferent ____________________

b. You feel extremely hungry immediately after deciding to go on a diet.
Afferent ____________________
Efferent ____________________

c. You are thirsty on a hot summer day.
Afferent ____________________
Efferent ____________________

d. You breathe faster while exercising.
Afferent ____________________
Efferent ____________________

e. Your heart rate increases at a tense spot in the movie you are watching.
Afferent ____________________
Efferent ____________________

2➡ Based on these afferent and efferent components, do you see any physiological advantage to the second efferent ganglia?

____

____

III. The Autonomic Plexus

A. Definition

Nerve fibers branch, meet, and rebranch, forming complex nerve networks called plexuses. The autonomic plexus contains the sympathetic postganglionic and parasympathetic preganglionic fibers innervating a particular area, in addition to actual parasympathetic ganglia.

Restudy Figures 34–2● and 34–3●, checking this time for plexuses. Now examine Figure 34–4●. Compare the plexuses. What anatomical or physiological advantage do you see in having these plexuses?

____

____

IV. Practical Function

A. Opposing Effects of the Two Divisions

1➡ Study Figure 34–4● and note the contradictory effects of the two autonomic divisions. As a general rule, we can say that if an action is taking place, the sympathetic nervous system will support (or be in sympathy with) the action. This is the "fight or flight" mode, enhancing the action. Note that "fight" and "flight" are really synonymous terms, both implying that the body is swinging into the survival mode.

The parasympathetic nervous system will oppose the action; the parasympathetic division enhances the resting state, thus the "rest and repose" mode. "Rest" and "repose" are synonymous with the "non-survival-instinct" mode or state.

These two opposing divisions function in unison to maintain not only homeostasis but also survival.

2➡ Visualize the following situation. You are camped out and you have just finished a hearty meal cooked over your camp fire. You sit back and relax. What parasympathetic actions occur?

____

____

Use the list in Figure 34–4● to explain why each of the parasympathetic nonactions is taking place. Why is it not contradictory to the resting state to say that digestive and urinary activities are increased? (Hint: We are not continuous feeders, as are some animals. We take in food at intervals.)

Suppose, suddenly, a herd of wild elephants appears on the scene. The parasympathetic shuts down and the sympathetic revs up. What is the survival advantage for each of the sympathetic actions that is suddenly activated? Why do you need a faster heart rate? Why does your digestive system shut down? Don't you need food?

____

____

____

____

Sphenopalatine ganglion
N III
Lacrimal gland
Eye
Ciliary ganglion
N VII
Submandibular ganglion
Salivary glands
N IX
Otic ganglion
Vagus (N X)
Heart
Cardiac and pulmonary plexuses
Lungs
Celiac plexus
Liver and gallbladder
Stomach
Pancreas
Spleen
Hypogastric plexus
Large intestine
Small intestine
Pelvic nerves
Rectum
Spinal cord
S_2
S_3
S_4
Kidney
Urinary bladder
Ovary
Uterus
Penis
Scrotum

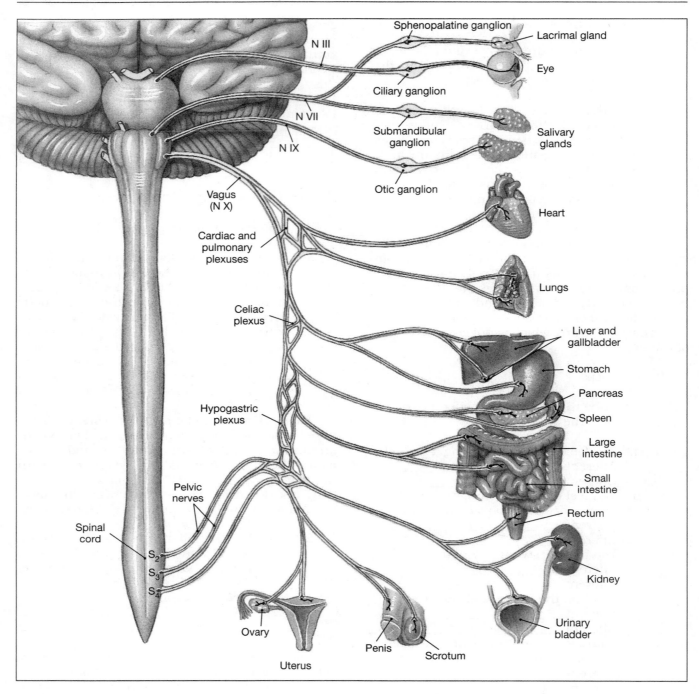

● **FIGURE 34–2a**
Parasympathetic Division of the Autonomic Nervous System.

3➡ Explain how the antagonistic sympathetic/parasympathetic systems constitute an intricate mesh of checks and balances.

____

____

____

4➡ Use your lecture text and explain the receptor types mentioned in Figure 34-4.

____

____

____

____

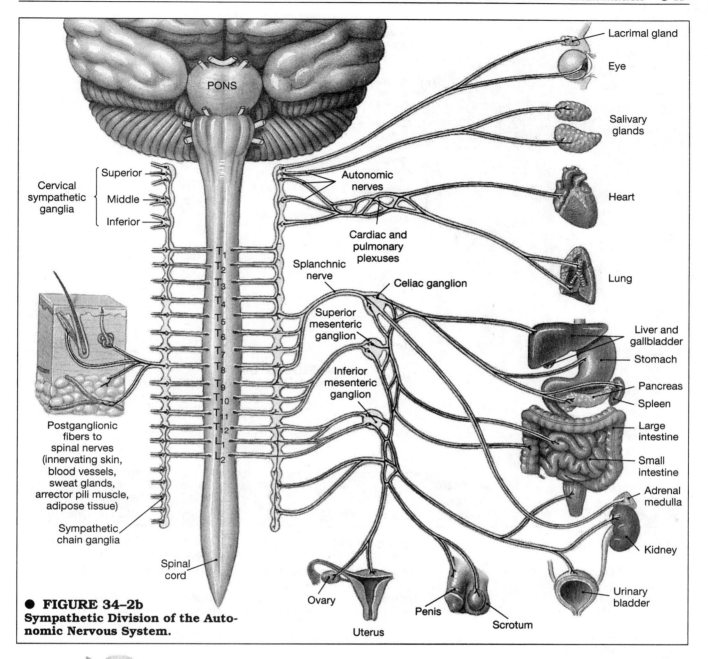

Lacrimal gland
Eye
Salivary glands
Heart
Lung
PONS
Cervical sympathetic ganglia
Superior
Middle
Inferior
Autonomic nerves
Cardiac and pulmonary plexuses
Splanchnic nerve
Celiac ganglion
Superior mesenteric ganglion
Inferior mesenteric ganglion
T_1 T_2 T_3 T_4 T_5 T_6 T_7 T_8 T_9 T_{10} T_{11} T_{12} L_1 L_2
Liver and gallbladder
Stomach
Pancreas
Spleen
Large intestine
Small intestine
Adrenal medulla
Kidney
Urinary bladder
Postganglionic fibers to spinal nerves (innervating skin, blood vessels, sweat glands, arrector pili muscle, adipose tissue)
Sympathetic chain ganglia
Spinal cord
Ovary
Uterus
Penis
Scrotum

● **FIGURE 34–2b**
Sympathetic Division of the Autonomic Nervous System.

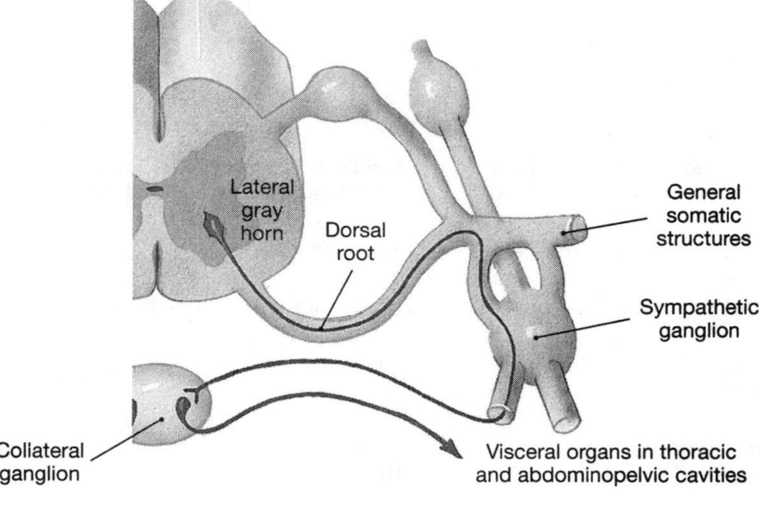

● **FIGURE 34–2c**
Organization of the Sympathetic Division.

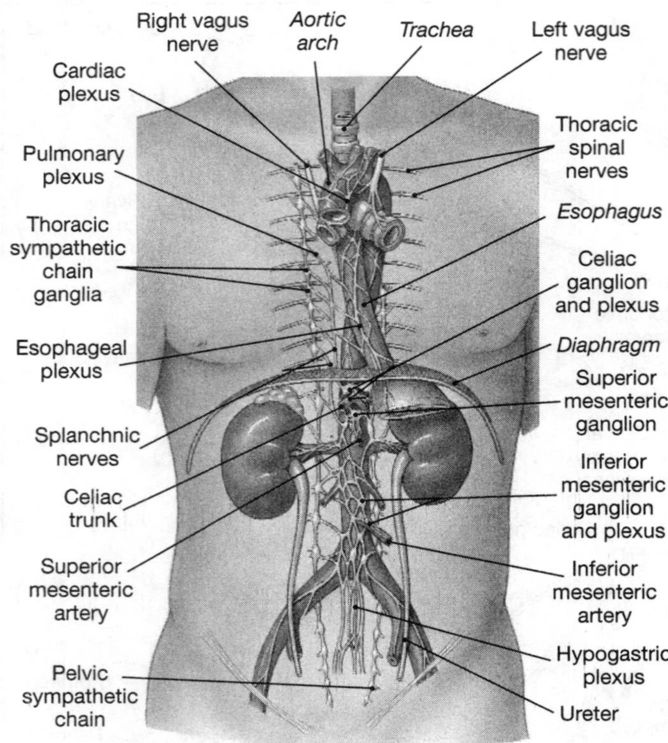

● **FIGURE 34–3**
The Peripheral Autonomic Plexuses.

Structure	*Sympathetic Receptor Type*	*Sympathetic Innervation Effect*	*Parasympathetic Innervation Effect (All muscarinic receptors)*
EYE	α_1	Dilation of pupil, accommodation for distance vision	Constriction of pupil, accommodation for close vision
SALIVARY GLANDS	α_1, β_1	Serous secretion stimulated	Watery secretion stimulated
SKIN			
Sweat glands	α_1	Increased secretion	None (not innervated)
Arrector pili	α_1	Contraction, erection of hairs	None (not innervated)
TEAR GLANDS		None (not innervated)	Secretion
CARDIOVASCULAR SYSTEM			
Blood vessels			None (not innervated)
To integument	α_1	Vasoconstriction	
To skeletal muscles	β_2	Vasodilation	
To heart	β_2	Vasodilation	
To digestive viscera	α_1	Vasoconstriction	
Veins	α_1, β_1	Constriction	
Heart	β_1	Increased heart rate, force of contraction, and blood pressure	Decreased heart rate, force of contraction, and blood pressure
ADRENAL GLAND		Secretion of epinephrine, norepinephrine by medulla	None (not innervated)
POSTERIOR PITUITARY	β_1	Secretion of ADH	None (not innervated)
RESPIRATORY SYSTEM			
Airways	β_2	Increased diameter	Decreased diameter
Respiratory rate		Increased	Decreased
DIGESTIVE SYSTEM			
Sphincters	α_1	Constriction	Dilation
General level of activity	α_2, β_2	Decreased	Increased

❑ Additional Activities

NOTES

1. Explore what factors might cause the human autonomic nervous system to be different from the autonomic nervous systems of some other mammals?
2. Research the role of the various transmitter substances in the functioning of the autonomic nervous system.

Answers to Selected Concept Check Questions

1. The antagonistic divisions act as a finely tuned system of checks and balances.
2. Double innervation can be found in the eye, the salivary glands, the heart, the respiratory system, the digestive system (most parts), the renal system, and the reproductive systems (both male and female).
3. *Pre*ganglionic means before the ganglion and *post*ganglionic means after the ganglion.

Structure	*Sympathetic Receptor Type*	*Sympathetic Innervation Effect*	*Parasympathetic Innervation Effect (All muscarinic receptors)*
Secretory glands	α_2	Inhibition	Stimulation
Liver	α_1, β_2	Glycogen breakdown, glucose synthesis and release	Glycogen synthesis
Pancreas	α_1	Decreased exocrine secretion	Increased exocrine secretion
	α_2	Decreased hormone (insulin) secretion	Increased hormone (insulin) secretion
SKELETAL MUSCLES	β_2	Increased force of contraction, glycogen breakdown	None (not innervated)
	α_2	Facilitation of ACh release at neuromuscular junction	None (not innervated)
ADIPOSE TISSUE	β_1, β_3	Lipolysis, fatty acid release	
URINARY SYSTEM			
Kidneys	β_2	Decreased urine production	Increased urine production
Bladder	α_1, β_2	Constriction of sphincter, relaxation of bladder	Tension of bladder, relaxation of sphincter to eliminate urine
MALE REPRODUCTIVE SYSTEM	α_1	Increased glandular secretion and ejaculation	Erection
FEMALE REPRODUCTIVE SYSTEM	α_1	Increased glandular secretion; contraction of pregnant uterus	Variable (depending on hormones present)
	β_2	Relaxation of nonpregnant uterus	Variable (depending on hormones present)

● **FIGURE 34–4**
Functional Comparison of the Sympathetic and Parasympathetic Divisions of the ANS.

❑ Lab Report

1. Referring to the numbered inquiries at the beginning of this exercise, complete the following box summary:

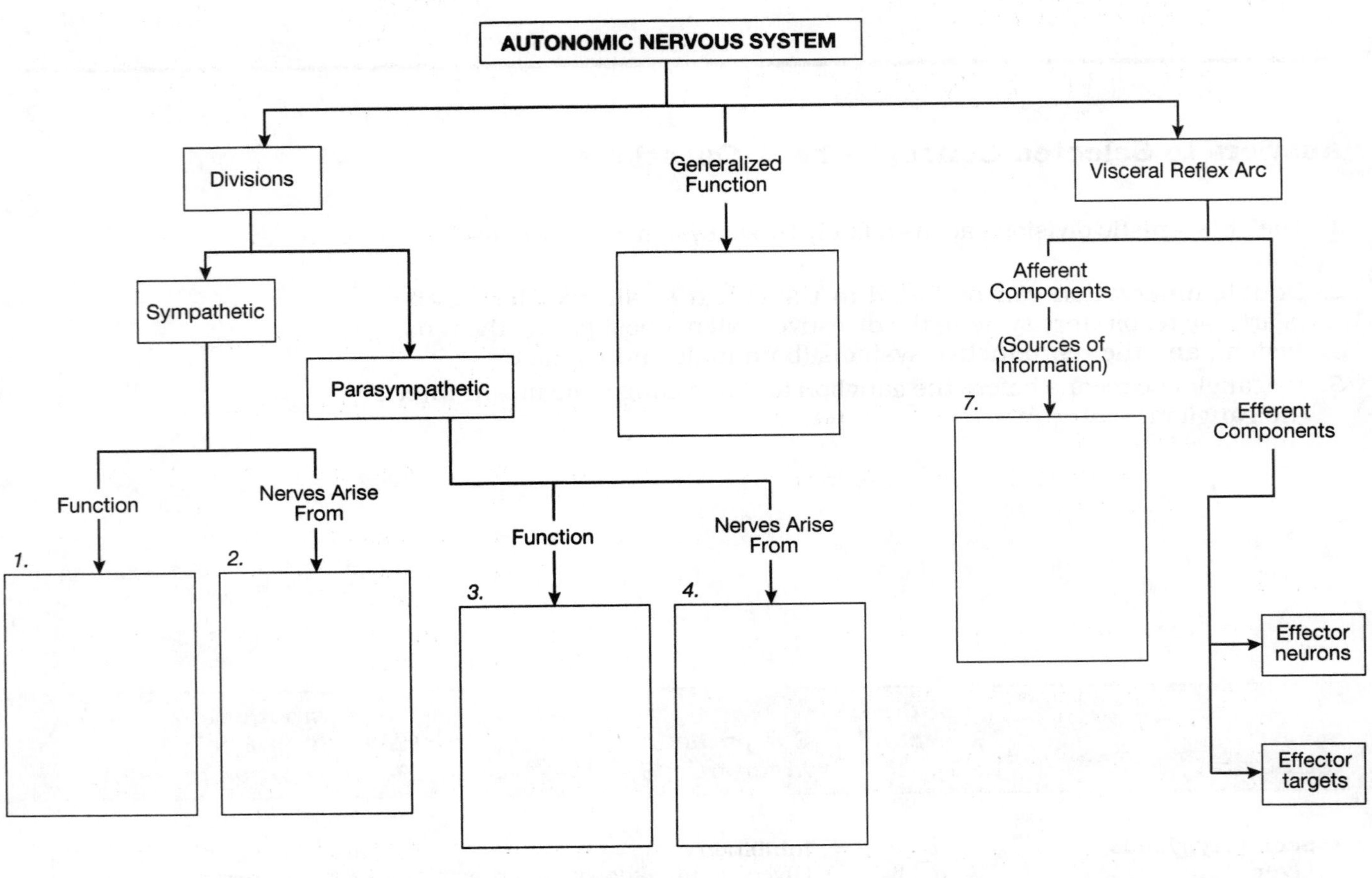

General Questions

1. What are the similarities and differences between the somatic system and the autonomic system?
2. Explain the autonomic reflex arc. Include both sympathetic and parasympathetic components.
3. List at least ten ways the sympathetic and parasympathetic divisions function to maintain homeostasis.
4. Where are the sympathetic ganglia located? The parasympathetic ganglia? What does this tell you about possible differences in preganglionic and postganglionic fibers?
5. What is the difference between the intramural ganglia and the collateral ganglia?

NOTES

EXERCISE 35

Dissection of the Mammalian Nervous System

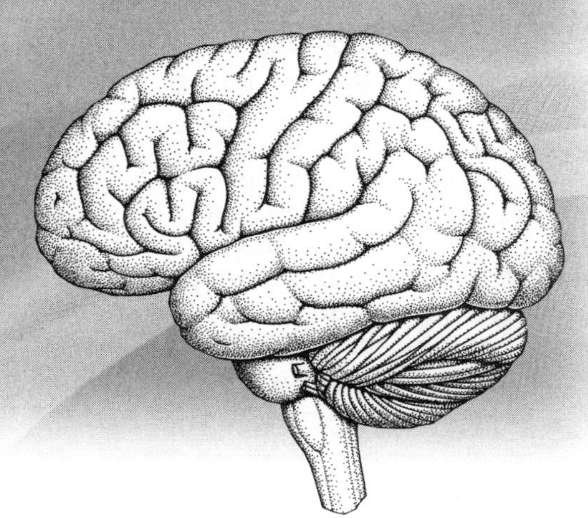

PROCEDURAL INQUIRIES

Preparation

1. Which dissecting instruments should be used in working with the mammalian nervous system?
2. What are the principal organs of the nervous system of your animal?

Dissection

3. What are the principal parts of the animal's brain?
4. What is the relationship between the brain and the floor of the cranium?
5. Where are the cranial nerves located?
6. Where does the brain/spinal cord exit the cranial cavity?
7. What are the meningeal layers? Where are they located?
8. What are the major landmarks of the spinal cord?
9. What are some common mammalian anomalies?
10. What are the principal parts of the spinal ganglia?
11. What are the major plexuses?
12. What are the major nerves of the brachial, lumbar, and sacral plexuses?
13. What are some of the organs innervated by the vagus nerve?

Additional Inquiries

14. What are some of the common nervous system factors found in all mammals?
15. What are the major types of vertebrae? How many of each type of vertebrae do the mammals have?

Key Terms

Arachnoid layer (Arachnoid mater)
Autonomic (sympathetic) ganglion
Cauda equina
Cervical enlargement
Conus medularis
Dorsal median sulcus
Dorsal root
Dorsal root ganglion
Dorsolateral sulcus
Dura mater
Femoral triangle
Filum terminale
Gray ramus
Lumbar enlargement
Meninges
Nerves (not listed separately)
Pia mater
Plexus
Tentorium cerebelli
Ventral median fissure
Ventral root
White ramus

Materials Needed

Preserved Animal
Dissecting Kit
Dissecting Microscope
Rongeurs (or Wire Cutters, or other Bone Cutting Instruments)
Dissecting Pan
Gloves
Disinfectant
Optional: Human Charts and Models for Comparative Purposes
Other Equipment as Indicated by Your Instructor
(Compound microscope)
(Prepared slides of animal nervous tissue; recommended if you are not completing Exercises 27 and 28 at this time.)

If you have not already completed Exercises 26–28, 30, and 33–34, we suggest you read through those lessons before beginning your dissection of the nervous system.

This exercise can be used for the dissection of any mammalian nervous system (and, though not included here, adaptations can be made for the dissection of many non-mammalian nervous systems, such as those of frogs, salamanders, and birds).

❑ Preparation

I. Background

A. Preparing the Animal

It is assumed that you have finished with any exploration of the muscles of the animal. It is also assumed that you have opened the thorax and abdomen of your animal. This will be necessary if you will be tracing the vagus nerve.

B. Organs of the Nervous System

To review the basics of the nervous system and of the nervous tissue, go back to Exercise 26 in this book or to the section on nervous system division in your lecture text *[∞ Martini, Chapters 12–16]*. These basics are common across the mammalian world.

C. Dissecting Hints

Your primary bone tool will depend on the animal whose brain and spinal cord you are dissecting. For cats, rats, and most adult animals you will need good bone clippers (rongeurs, wire cutters, or other bone cutting instruments). **DO NOT** use scissors for adult animals! You can ruin the scissors by trying to cut bone with them.

If you are dissecting a fetal animal, you can use scissors to cut the fibrocartilage sutures and the fontanelles between the bones of the skull. But, do not use scissors on ossified tissue.

The major problem you will encounter when dissecting the nervous system is the proximity of very hard tissue (bone) immediately next to very soft tissue (brain and nerves). The bones of the older animals in particular may be difficult to cut, even with the finest of bone clippers. Be careful that you do not tear the delicate nervous tissue as you break through the bones.

Use your blunt nose probe to separate the nerves from the surrounding tissue. Your sharp nose probe should be used sparingly, if at all. Sharp nose probes can easily damage nerves, particularly the more delicate ones. Sharp nose probes can also tear other tissues and make your identification more difficult.

❑ Dissection

II. Central Nervous System

The brain and the spinal cord of the animal can be dissected separately. If you are dissecting both, we recommend you start with the brain. Follow your instructor's directions concerning how much dissection you will be doing.

The dissection of the **meninges** is covered below in point D. The identification of these layers can be included as a part of your brain dissection and/or your spinal cord dissection.

A. Brain

1➡ Begin your dissection of the brain by removing the muscles from the dorsal and lateral sides of the head. For very young fetal animals, this may not be necessary. For older fetal and adult animals, removing the muscle tissue is necessary so that you can work with the bones of the skull.

2➡ Use whatever bone cutters you have at hand to make an opening in the skull. For larger and older animals, you may need a bone saw. If you can locate a suture line along the parietal bone, you may find that the easiest place to begin. Use extreme care in making this opening because the brain is directly beneath the bone.

3➡ Carefully chip away the bone from the lateral and dorsal areas of the skull. In strong mature animals, this bone is hard and tough. Be careful about becoming overly frustrated! In fetal and immature animals, you will find large sections of the skull bones still in an unossified state and you may be able to cut the strips of cartilage along the suture lines.

4➡ In some animals (such as the cat) you will find an ossified strip known as the **tentorium cerebelli** between the cerebrum and the cerebellum. Carefully remove this bone.

5➡ When the brain is completely exposed, note its relationship to the cranial floor. Now note how the brain stem becomes the spinal cord and exits the skull at the foramen magnum.

6➡ Sever the spinal cord transversely as close to the foramen magnum as possible.

7➡ Lift the brain from the floor of the cranium. Identify the cranial nerves as you do so. Refer to Point III below and to the Figures in Exercise 33 of this book. (The cranial nerves of most mammals are analogous.) Note particularly cranial nerves I, II, and X.

8➡ After you have identified the cranial nerves, sever them as far from the brain as possible. Note how and where each nerve exits the brain. Depending on the flexibility of your animal's tissues, you may have to sever as you identify.

9➡ To continue with your dissection, turn to Exercise 32 in this book and follow the directions for the dissection of the sheep brain. If you have already dissected the sheep brain, use the space below to record any outstanding similarities and differences between the sheep brain and the brain of the mammal you are dissecting.

____
____
____
____
____
____

B. Spinal Cord

If you have dissected the brain, you have already followed the brain stem past the point where it becomes the spinal cord. Continue your dissection from there.

If you have not dissected the brain and if you are dissecting the entire spinal cord, begin your dissection at the most cranial point possible.

If you are only dissecting a part of the spinal cord, follow your instructor's directions as to where you should begin.

1➡ Start your dissection of the spinal cord by removing the muscles surrounding the vertebral column.

2➡ Note how and where the spinal cord exits the foramen magnum. (Your instructor will indicate whether or not you should sever the spinal cord. If so, cut it as close to the brain as possible.)

3➡ Use whatever bone clippers (not scissors) you have to cut through the laminae of the vertebrae. Again, for larger or older animals you may need a bone saw. Before cutting the bone, note exactly where the spinal cord is located. Keep this in mind as you remove the bony tissue. The spinal cord is easily cut by overly anxious bone clippers!

4➡ When the spinal cord is entirely exposed, note the meninges. If you are examining the meninges here, refer to section D below. If you are not examining the meninges, you may need to tear back the tough outer coating of the spinal cord. This is the **dura mater**, the outer meningeal layer.

5➡ Count the vertebrae. The standard number of cervical vertebrae for virtually all mammals is seven. Cats usually have 7 cervical, 13 thoracic, 7 lumbar, 2–3 sacral, and a variable number of coccygeal vertebrae. (The longer the tail, the more coccygeal vertebrae the cat will have.)

The standard vertebral numbers for pigs are 7 cervical, 14–15 thoracic, 6–7 lumbar, 4 sacral, and 20–23 coccygeal. If you are dissecting some other mammal, check with your instructor or a veterinary science reference book for the standard vertebral numbers.

How many vertebrae does your animal have? It may not be practical to count the coccygeal vertebrae!

Cervical ____
Thoracic ____
Lumbar ____
Sacral ____
Coccygeal ____

6➡ As you expose the spinal cord, note which spinal nerves correspond with which vertebral region. As with humans, most mammals have 8 cervical spinal nerves.

7➡ Identify the **cervical enlargement** – found between about the fourth cervical nerve and the first thoracic vertebrae – and the **lumbar enlargement**, which begins about the middle of the lumbar region and extends toward the sacral region.

8➡ Find the point where the cord ceases. This tapering area is called the **conus medularis**. The more caudal nerve roots radiate inferiorly from the lower part of the cord. These fibers collectively are known as the **cauda equina** because they resemble a horse's tail.

9➡ If your animal's sacrum is still in place, note how the fibers of the cauda equina are given off through the sacral foramina. The last remaining piece of the spinal cord is the **filum terminale**, a fibrous strand composed mostly of the pia mater.

10➡ Now examine the cord itself. You should be able to locate the **dorsal median sulcus**, a shallow groove on the dorsal surface, and the **ventral median fissure**, a deep groove on the ventral surface. You may be able to find the **dorsolateral sulcus**, which is lateral and parallel to the dorsal median sulcus. If you have trouble locating the dorsolateral sulcus, follow a spinal nerve back to the spinal cord. The dorsolateral sulcus is the attachment point for the roots of the spinal nerves.

11➡ Take a transverse slice of the spinal cord to the dissecting microscope. Sketch what you see in the box below.

Compare your spinal cord section with Figure 27–2●. What similarities and differences do you see?

____

____

C. Anomalies

Most animals can suffer from the same basic vertebral and spinal anomalies that can be found in humans. Because the survival rate for animals with gross anomalies is relatively low, you are more likely to find severe but hidden defects in the fetal or young animals than in older or more mature animals. Nevertheless, domestic animals (such as cats) often survive quite well with major anomalies.

1➡ To find the most common vertebral and spinal anomalies, answer the following questions:

Does your animal have an extra cervical or thoracic vertebra? ______ Where? ________________ ______________________________ Extra vertebrae are not uncommon. You may also find a half vertebra (hemi-vertebra) at some point along the way.

Do you see any evidence of fused vertebrae (particularly in the cervical and lumbar regions)? _____ This condition may be more visible in older animals.

Look for gaps in the vertebral column. Spina bifida results from an incomplete or improper development and fusion of the vertebral arches. Do you find any evidence of spina bifida?

Does your animal have any kind of spinal curvature? Scoliosis is sometimes seen. You may also find lordosis or kyphosis, particularly in older animals, though these may be harder to identify in your preserved specimen.

Anomalies of the spinal cord itself will be harder to identify, particularly in a preserved animal. However, if you do find any questionable nervous tissue patterns, note them here.

____

____

D. Meninges

The meninges are the three layers of protective connective tissue membranes that surround the brain and spinal cord. You should have no trouble identifying these layers in older and larger animals. Discernment of the layers may be more difficult in fetal animals. In addition, in young fetal animals, the layers may be more difficult to find in examining the spinal cord than in examining the brain.

1➡ Examine the surface of the brain or spinal cord. You will notice both the brain and spinal cord are surrounded by a tough sac-like protective layer. This is the **dura mater** (literally, "tough mother"), the outermost of the meninges. Gently remove the dura mater. Be especially careful when separating the dura mater from the nerve tracts on the ventral surface of the brain.

2➡ You will probably not be able to locate the second meningeal layer, the **arachnoid** layer, or arachnoid membrane. The arachnoid layer is sometimes called the **arachnoid mater** (literally, "spider mother"). This web-like meninx does not withstand preservation well. If you notice tufts of cottony tissue across certain of the central nervous system fissures, however, you may have located remnants of the arachnoid mater.

3➡ You should be able to locate the inner layer, the **pia mater** (literally, "gentle mother"), which adheres tightly to the nervous tissue of the brain and spinal cord. Gently separate at least part of the pia mater from its surface by lifting it up with your forceps. It will not be necessary for you to remove the pia mater.

Describe the meningeal layers you were able to locate. If you located these layers in both the brain and the spinal cord, what differences did you notice?

____
____
____
____

III. Cranial Nerves

It is beyond the scope of this exercise to trace each of the cranial nerves from the brain to its specific destinations. However, we will examine certain selected points of interest. In section I above, you were instructed to note particularly cranial nerves I, II, and X.

1➡ Look at the floor of the cranium. Identify the points where cranial nerves I, II, and X exit the skull. You may also be able to locate the exit point for cranial nerve III. (Cranial nerve X exits at the jugular foramen.)

2➡ Identify any other cranial nerve landmarks that you may be able to find.

IV. Peripheral Nervous System

A complete dissection of the spinal nerves and the peripheral nervous system is beyond the scope of this book. We will examine the spinal ganglia, the spinal plexuses, and some representative nerves coming from the brachial, lumbar, and sacral plexuses.

A. Spinal Ganglia

Unless your instructor indicates otherwise, it will only be necessary for you to dissect a few of the spinal ganglia. (The information given here corresponds with the material in Exercise 28.)

1➡ Begin your dissection by carefully clearing away the tissue around the spinal nerves you are studying.

2➡ Refer back to Figure 28-2● and look for the **dorsal root** and the **ventral root** of the spinal cord. The dorsal root is the sensory portion of the nerve, receiving only input signals from the periphery. The ventral root is the motor portion of the nerve, functioning in the transmission of signals from the cord to the body proper. The dorsal root is sometimes known as the posterior or sensory root; the ventral root is sometimes called the anterior or motor root. Notice how the root fibers exit the spinal cord.

3➡ Identify the **dorsal root ganglion**. The dorsal root includes a swelling, the dorsal root ganglion, which contains the cell bodies of the peripheral sensory neurons. The other swelling associated with the spinal nerves, the swelling more distal from the spinal cord, is the **autonomic (sympathetic) ganglion**.

If your animal will fit under the dissecting microscope, examine the ganglia. Note any unusual findings.

4➡ Return to Figure 28-2● and locate the white and gray rami. We often use the term ramus (pl. rami) when referring to a particular branch of a spinal nerve. Ramus is defined as *branch*. The **white ramus** consists of the preganglionic motor fibers, while the **gray ramus** includes the postganglionic (autonomic) nerve fibers.

5➡ Locate the rami and follow one or more ramus as your instructor may indicate. Figure 28-4● shows the human dermatomes. The dermatomes for your particular animal will be more or less analogous to this diagram.

Record your observations here__________
____
____
____
____

B. Plexuses

Innervation would be relatively simple were it not for the meshwork-like configurations of the ventral rami of some of the spinal nerves. Each meshwork is known as a plexus and contains fibers from one or more spinal nerve. (Plexus means braid.) Only ventral rami are involved in the plexuses.

1➡ Reexamine Figure 28–1● and locate the cervical, brachial, lumbar, and sacral plexuses in the human figure. Your animal's plexuses will be analogous. Where specifically are these plexuses located?
cervical ______________________________
brachial ______________________________
lumbar ______________________________
sacral ______________________________

C. Brachial Plexus

1➡ Place the animal dorsal side down in your dissecting pan. Reflect the pectoralis muscles. You will find the brachial plexus in the axillary region.

2➡ Separate the nerves by clearing away the connective tissues. Note the proximity of the nerves to the corresponding blood vessels.

3➡ Follow the nerves of the brachial plexus as far back toward the spinal cord as possible.

4➡ Refer to Figure 35–1● and locate the following nerves.

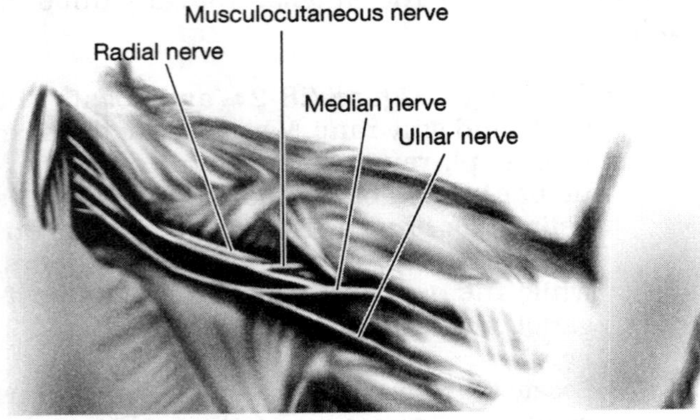

● **FIGURE 35–1**
Generalized Mammal. Major nerves of the brachial plexus (ventral).

Musculocutaneous nerve – This is the most superior nerve emanating from the brachial plexus. Note how this nerve splits and follows the coracobrachialis and biceps brachii muscles. Where do the fibers of this nerve terminate?______________________

____

Radial nerve – This large nerve is inferior to the musculocutaneous nerve. What are the radial nerve's points of innervation?______________________

____

Median nerve – This nerve runs approximately parallel to the brachial artery and vein. What are the median nerve's points of innervation?______________________

____

Ulnar nerve – This is the most posterior of the brachial plexus nerves. What are the ulnar nerve's points of innervation?______________________

____

To locate additional nerves, refer to your lecture text. The animal's nerves and innervation patterns are approximately the same as the human's.

D. Lumbar Plexus and Sacral Plexuses

1➡ Begin by placing the animal dorsal side down in your dissecting pan. Use Figures 35-2● and 35-3● as your guide.

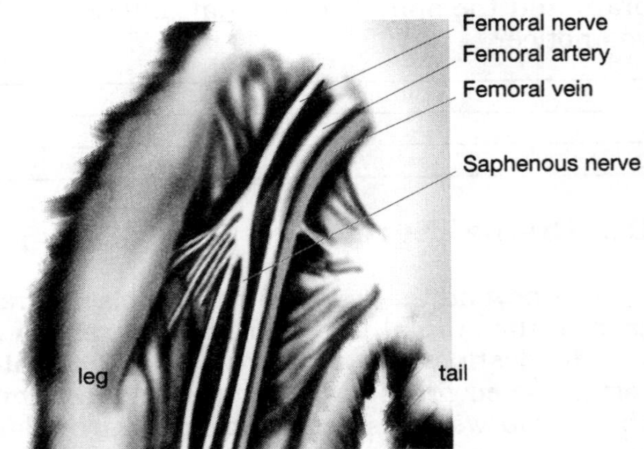

● **FIGURE 35–2**
Generalized Mammal. Femoral triangle (ventral).

As with the brachial plexus, separate the nerves by clearing away the connective tissues. Note the proximity of the nerves to the corresponding blood vessels.

2➡ Identify the **femoral triangle**, the region of the thigh bordered by the sartorius and adductor muscles. The large nerve you see there is the **femoral nerve**.

3➡ Trace the femoral nerve as it passes distally through the leg. The section of this nerve that is parallel to the great saphenous vein and artery is the **saphenous nerve**. What parts of the lower limb does this nerve service?______________________
What other branches of this nerve can you identify?______________________

4➡ Now place your animal ventral side down. Use Figure 35–3● as your guide.

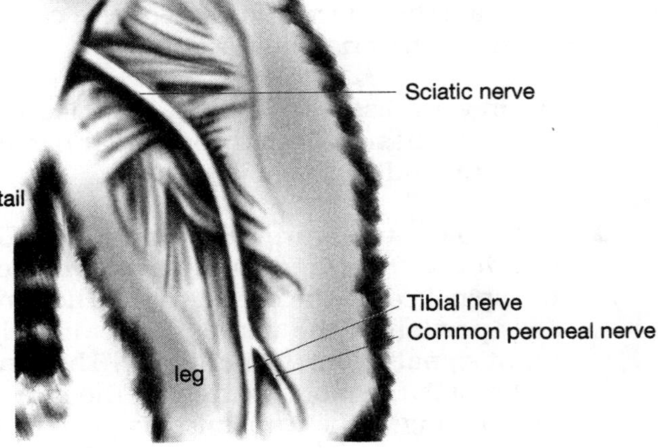

● **FIGURE 35–3**
Generalized Mammal. Sciatic nerve (dorsal).

Reflect the biceps femoris muscle and note the large **sciatic nerve**. This cord-like structure arises in the sacral plexus. Trace the sciatic nerve back toward the spinal cord as far as possible. How far were you able to go?____________________

5➡ Now follow the sciatic nerve as it travels distally. Just superior to the gastrocnemius muscle, the sciatic nerve divides into the **tibial nerve** and the **common peroneal nerve**. Which nerve is which?__________
What do you base your identification on?
____

V. Autonomic Nervous System

Review the discussion of the human autonomic nervous system as found in Exercise 34. Your animal's autonomic nervous system is comparable. Our autonomic nervous system dissection is limited to one representative nerve, cranial nerve X, the **vagus nerve**.

Vagus Nerve

It is beyond the scope of this exercise to examine the intricate details of the vagus nerve (cranial nerve X). However, by examining a few parts of this and related nerves, you should gain an appreciation for some of the complexity of the nervous system.

If you have opened the thorax and abdomen of your animal, you should be able to find the vagus nerve.

1➡ Locate the thin whitish fibrous strand of tissue in the area of the carotid artery and the jugular vein. Check Figures 53a-2a● and 53b-2a●. The strand shown there immediately lateral to the common carotid artery is the vagus nerve. Refer also to Figure 25a-5●.

2➡ Trace the vagus nerve cranially as far possible. The nerve exits the brain at the jugular foramen (though tracing it to the jugular and nodose ganglia and into the brain is not a part of this exercise).

3➡ Follow the vagus nerve caudally and notice how the various branches form. You may be able to find the pulmonary and cardiac plexuses. Notice too how part of the left and right vagus nerves unite near the root of the lungs while the other part of the vagus nerves unite near the diaphragm.

4➡ The vagus nerve passes through the diaphragm at the esophagus. You may be able to locate some of the plexuses and anastomoses that give rise to the nerves leading to the abdominal organs. (Note the diagrammatic representation in Figure 34–2●.)

VI. Prepared Slides

This section is recommended if Exercises 27 and 28 are not being completed at this time.

Examine slides of the animal nervous system according to your instructor's directions. Use the space at the end of this exercise to sketch what you see.

❑ Additional Activities

NOTES

1. Take a fresh or cooked chicken, turkey, or other bird and, using the information given in this exercise, dissect as much of the nervous system as possible. What similarities and differences do you notice in the vertebrae and in the spinal cord itself? Were you able to locate any of the ganglia or spinal nerves?
2. If you take part in the gutting of a freshly killed animal (whether farm chicken or hunted big game), see if you can find the vagus nerve, particularly the branch going directly to the heart. The cranial nerves, as shown in Figure 33–4● are analogous, regardless of the animal.
3. Design an experiment that could be used to test the different parts of the vagus nerve.

❑ Lab Report

1. Referring to the numbered inquiries at the beginning of this exercise, complete the following box summary.

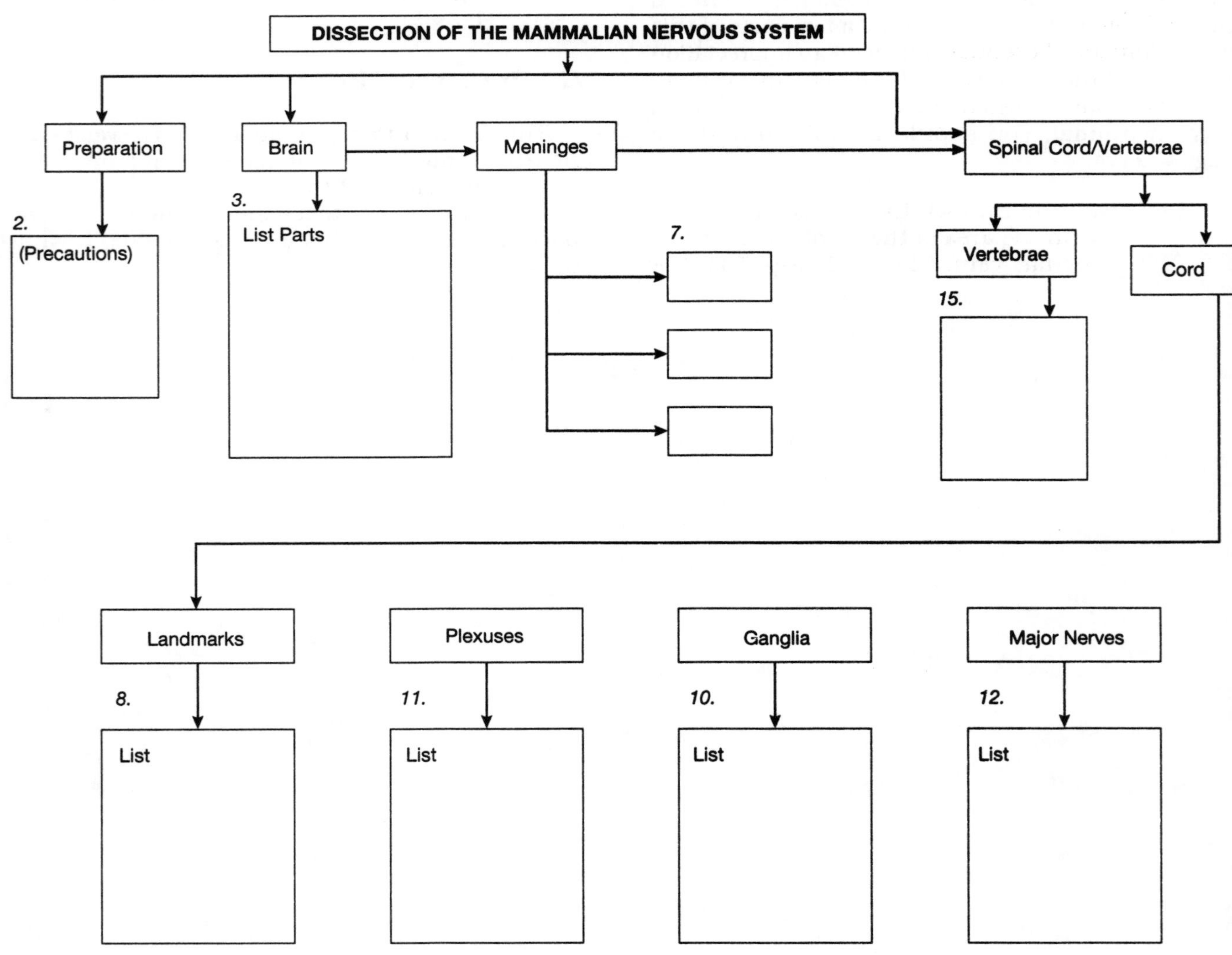

General Questions

NOTES

1. Why would a domestic animal be more likely than a feral animal to survive with gross vertebral anomalies?
2. What are the similarities and differences between the brain of your laboratory animal and the human brain?
3. What specific points in the skull and vertebral column are especially adapted for the protection of the nervous tissue?
4. Where does the spinal cord itself actually end? Do you think this point changes over the life of the animal (from fetal to adult)?
5. What is the function of each of the meningeal layers?
6. What would happen if the dorsal root ganglion of a given nerve were severed?
7. What relationship do you see between the four major plexuses?
8. Describe briefly the anatomical and physiological reaction if each of the following nerves were severed.

   Musculocutaneous
   Radial
   Median
   Ulnar
   Sciatic
   Saphenous
   Tibial
   Common Peroneal

EXERCISE 36

Anatomy of the Eye

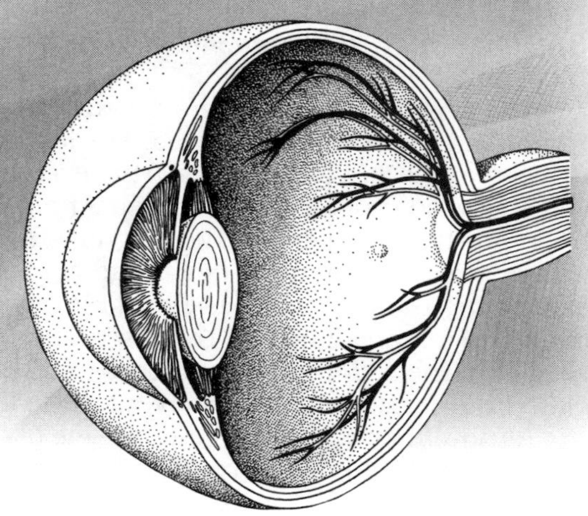

PROCEDURAL INQUIRIES

Examination

1. What are the accessory structures of the eye?
2. What are the extrinsic muscles of the eye?
3. What are the external structures of the eye?
4. What are the internal structures of the eye?
5. What are the layers of the retina?
6. What is the difference between rods and cones?

Dissection

7. What procedure should be used in dissecting the mammalian eye?
8. How does the laboratory mammalian eye relate to the human eye?

Additional Inquiries

9. What is the path the light follows on entering the eye?
10. What is the path of the nerve impulse from the retina to the brain?
11. What is the role of the fat around the eye?
12. What is the difference between the aqueous and vitreous humors?

Key Terms

Anterior Cavity
Anterior Chamber
Aqueous Humor
Choroid Coat
Ciliary Body
Cones
Conjunctiva
Cornea
Fovea
Iris
Lacrimal Apparatus
Lens
Optic Disc
Optic Nerve
Posterior Cavity
Posterior Chamber
Pupil
Retina
Rods
Sclera
Vitreous Chamber
Vitreous Humor

Materials Needed

Eyeball models
Eyeball charts and diagrams
Compound microscope
Prepared slides of the retina

For dissection

Cow (or Sheep) Eye
Dissecting Pan
Scalpel
Scissors
Forceps
Gloves
Stereoscopic Microscope

The visual process begins when the eye admits light from specific objects and the image of these objects is focused on the retina. Specialized nerve cells of the retina are stimulated by the light and respond by generating a nerve impulse. (In this way, light is transduced into electrochemical energy.) As the retinal cells are stimulated, the impulses are collated and transmitted via the optic nerve to the occipital lobe of the brain, where they are interpreted as vision.

The eyeball itself can be studied from both external and internal perspectives. The features readily visible when viewing the external eyeball include the extrinsic muscles, the cornea, the sclera, the iris, and the optic nerve. The internal structures of the eye are those features most readily identified by examining the interior of the eyeball model or the dissected mammalian eye. These structures include primarily the retina, the chambers and their associated humors, the ciliary body, and the lens.

Although eyes may be cited as one of our most uniquely individual features, the basic structure of the eye is the same across the spectrum of humanity. For that matter, the mammalian eyeball is so constant that we can examine and dissect some other mammal's eyeball and note an almost absolute correlation between that animal's eye parts and our own. Eye dissection in the anatomy and physiology laboratory is most often done with cow eyes or sheep eyes. These are usually inexpensive, readily available (either fresh from a slaughterhouse or preserved from a biological supply house), and large enough to work with easily.

Let us now examine the anatomy of this intricate organ. The physiology of vision will be considered in Exercise 37.

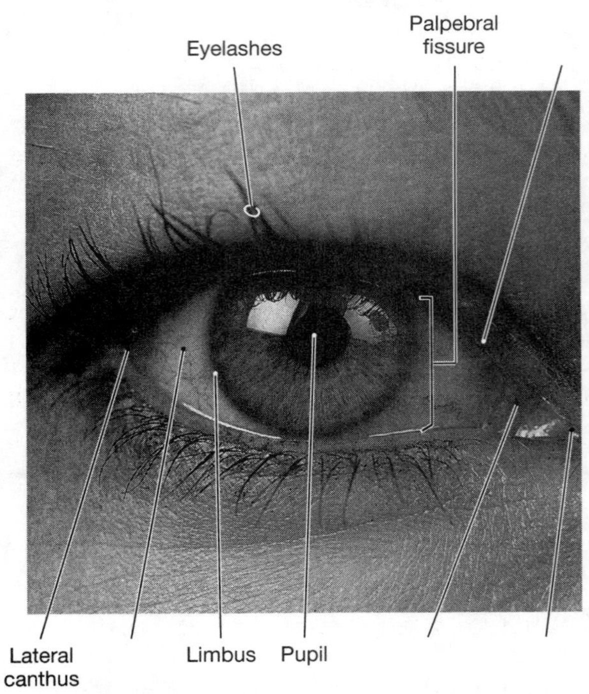

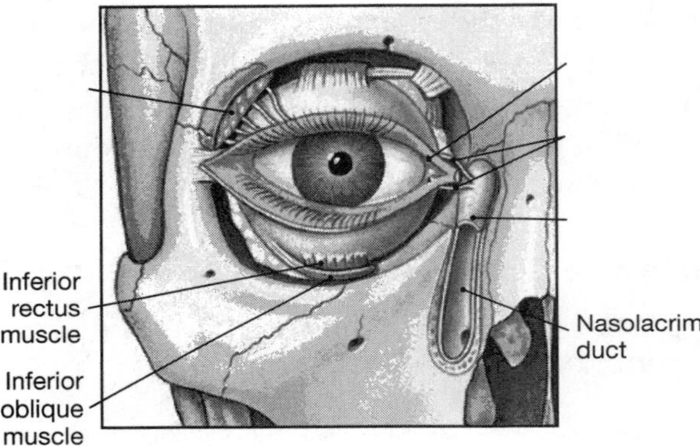

● **FIGURE 36–1**
Accessory Structures of the Eye.

❑ Examination

I. Structures Surrounding the Eye

[∞ *Martini, pp. 538–541*]

We will study the eye by looking at the charts and diagrams and by examining the model of the eyeball.

A. Bone and Fatty Tissue

1➡ Notice that the eye is located within a protective bony socket and is surrounded by other accessory structures that also serve to protect it. The eye socket is padded with fat, a cushion against the jolts and jars of daily living. (If you dissect a mammalian eye, Section V, you will notice these protective pockets of fat surrounding the eye.)

B. Eyelids

1➡ Use the following boldface terms to begin labeling Figure 36–1●. When you are finished, check your identification with your lecture text or the laboratory charts.

2➡ Identify the **palpebrae**, the eyelids that cover the eyeball. The palpebrae are connected at the corners by the **lateral canthus** and **medial canthus**. The space between the palpebrae is the **palpebral fissure** and the eyelashes are located along the palpebral margins. The **lacrimal caruncle**, the raised area containing secretory glands, is located at the medial canthus. Compare your answers with your lecture text or the laboratory charts.

3➡ Envision the **tarsal glands** (sebaceous ciliary glands associated with the eyelash follicles, also known as the glands of Zeis) and the **meibomian glands** (lipid glands embedded in the thick folds forming the border of the eyelid). They secrete protective substances specifically for the eyeball and the palpebrae. Although these glands are not specifically identified in Figure 36–1●, make note of their probable location.

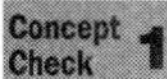

What do you suppose happens when palpebral glands become plugged or infected?

____
____
____

The internal membranous lining of the eyelids is the **conjunctiva**. An inflammation of the conjunctiva is known as conjunctivitis or "pink eye." Conjunctivitis is a symptom rather than a disease and may be caused by an external irritation (such as dust or pollen) or by some disease-causing agent (a bacterium, a virus, a protozoan, or even a parasitic worm).

C. Lacrimal Apparatus

1➜ Again, use the following boldface terms to complete labeling Figure 36–1●. When you are finished, check your identification with your lecture text or the laboratory charts.

2➜ Locate the **lacrimal gland**, which is just beneath the lateral border of the eyebrow. This gland, which can be divided into superior and inferior portions, secretes a slightly alkaline, antibacterial substance — commonly called tears — through about 12 ducts onto the anterior surface of the eyeball. Tears are antibacterial because they contain both antibodies and an antibacterial enzyme called lysozyme.

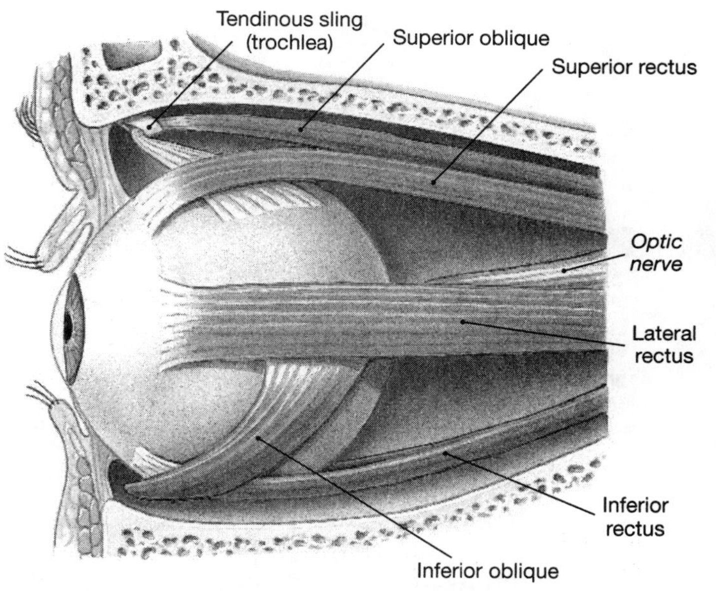

(a) Muscles on the lateral surface of the eye.

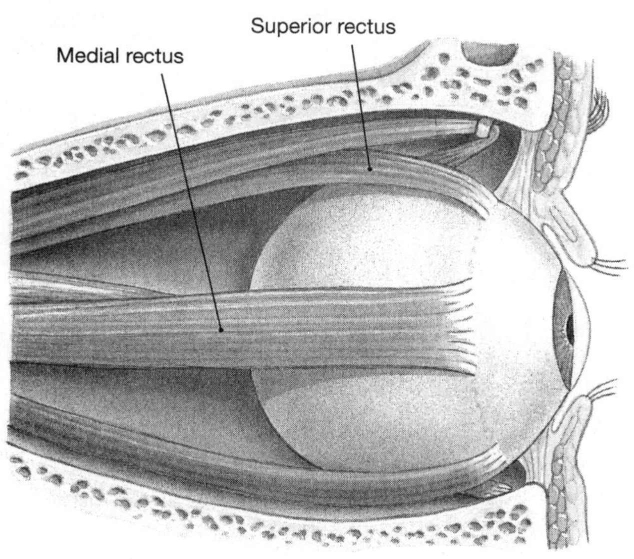

(b) Muscles on the medial surface of the eye.

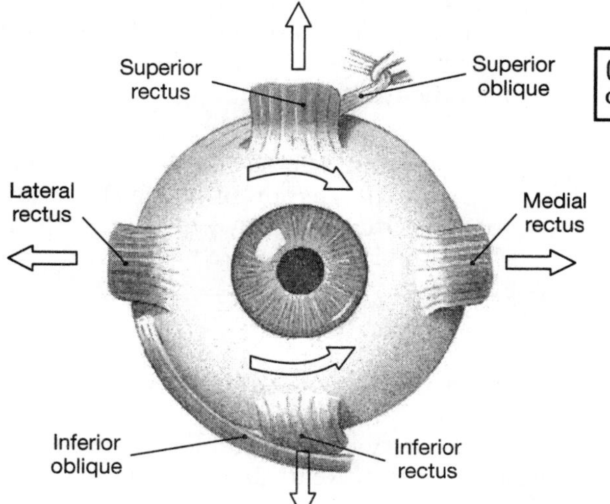

(c) An anterior view of the eye, showing the orientation of the oculomotor musces.

● **FIGURE 36–2**
External Muscles of the Eye.

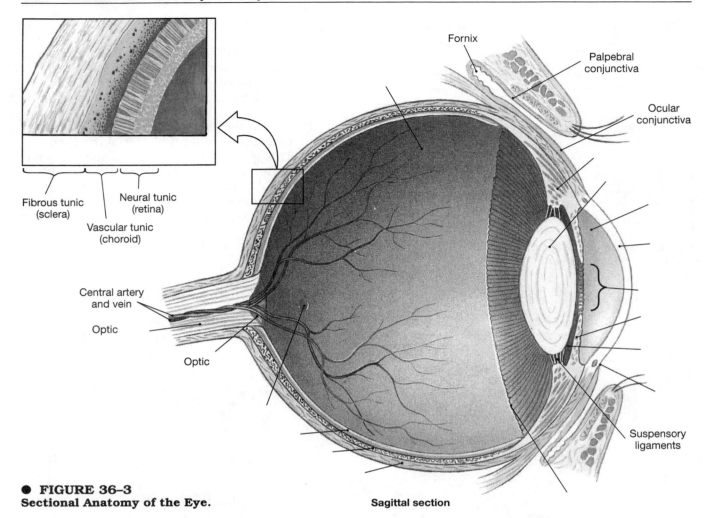

Fibrous tunic (sclera)
Vascular tunic (choroid)
Neural tunic (retina)
Central artery and vein
Optic
Optic
Fornix
Palpebral conjunctiva
Ocular conjunctiva
Suspensory ligaments
Sagittal section

● **FIGURE 36–3**
Sectional Anatomy of the Eye.

3 ➡ Identify the lacrimal structures by following the pathway of the tears. Tears accumulate at the medial canthus (in the **lacus lacrimalis**) and from there drain through two **puncta** (openings) into the **lacrimal canals**, which join and drain into the **lacrimal sac** and from there into the **nasolacrimal duct**. From there the tears are carried to the **inferior meatus** of the nose.

II. External Examination of the Eye

A. External (Extrinsic) Muscles

1 ➡ Use the eyeball model and Figure 36–2● to locate the muscles that surround the eye. The **superior rectus**, located above the eye, contracts to move the eyeball upward. The **inferior rectus**, located on the inferior surface of the eye, contracts to move the eyeball downward. The **medial rectus**, on the nasal side of the eye, contracts to move the eyeball medially, and the **lateral rectus**, on the temporal side of the eye, contracts to move the eyeball laterally. The **inferior oblique**, which hooks across the inferior portion of the eye, rotates the eyeball upward and laterally. The **superior oblique**, which hooks across the superior portion of the eye, rotates the eyeball downward and laterally. As you examine the superior oblique muscle, make note of the tendon of this muscle and observe how the **trochlea** of the superior oblique plays a role in its movement. Because of the synergism of these six muscles, your eyeball rolls smoothly.

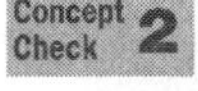

Look to your left. Which left eye muscle is primarily responsible for this movement? Which right eye muscle performs the same leftward gaze?

____

____

B. Externally Visible Structures of the Eye

You should now be able to identify the following exterior features. Read these descriptions and then label Figure 36–3●. Again, the terms in bold-

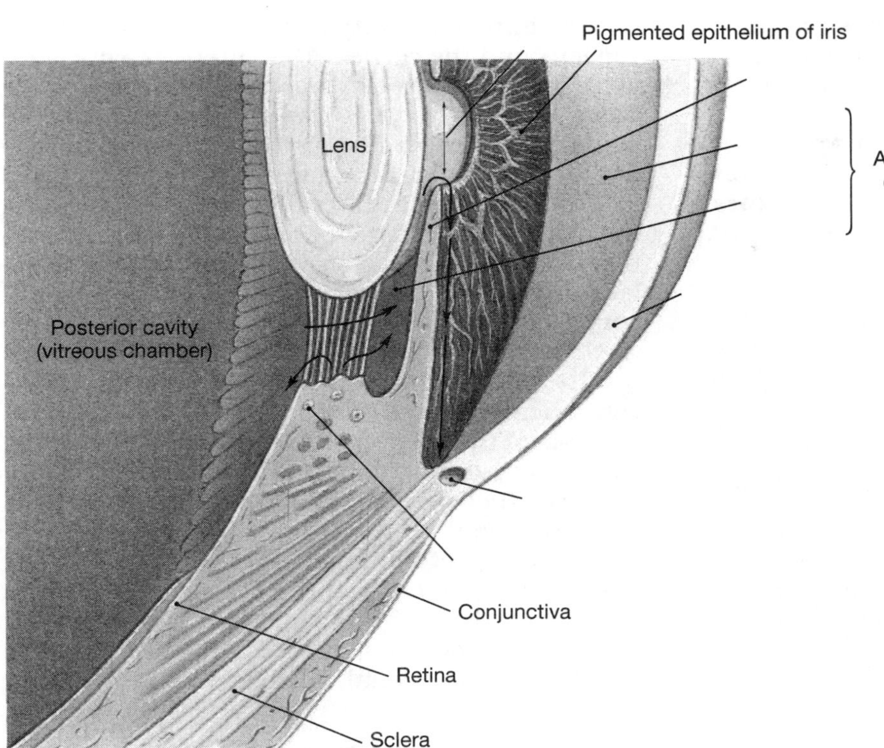

● **FIGURE 36–4**
Close-up of the Anterior Eye.

face should be used for your labels. You will be asked to label the remaining structures in Figure 36–3● when you work through Section III. When you are finished with both sections, check your labels with your lecture text or the laboratory charts.

1➡ Examine the **sclera**, the white outer coat covering the posterior three-fourths of the eye. The sclera is composed of dense fibrous collagen and elastin fibers. The **cornea** is the transparent structure continuous with the sclera, and the collagenous fibers of the cornea are organized into layers to allow the passage of light. The fibers of the sclera are not arranged in organized layers.

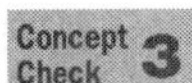
Concept Check 3 How does the structure of the sclera render it white rather than transparent?

____

____

2➡ Envision the mucous membrane covering of the outer surface of the eye — as well as the previously mentioned inner surface of the eyelids. This membrane is the **conjunctiva**. The conjunctiva will probably not be specifically identified on your models but make note of where it would be.

3➡ Note the **iris**, the pigmented anterior tissue that regulates the amount of light entering the eye through the pupil. The pupil is the opening in the center of the iris. The iris, which is continuous with the choroid coat, is a contractile membrane containing two muscles, the *circular pupillary sphincter* muscles and the *radial pupillary dilator* muscles.

4➡ Identify the **lens**, the rubbery, flexible structure posterior to the iris. The shape of the lens helps to determine the refraction of light passing through the eye. (Cataracts, which cause a cloudiness in the eye and thus a refractive visual distortion, are the result of the misalignment of protein in the lens or cornea. Cataracts are discussed further in Exercise 37.)

5➡ Find the **optic nerve**, the tough cord of nerve tissue extending from the posterior of the eye. This nerve is composed of millions of axons from neurons in the retina.

III. Internal Structures of the Eye

Open the model eyeball to examine the internal structures. Read through the following descriptions and, using the boldface terms, label Figure 36–4●, and the remainder of Figure 36–3●.

A. Posterior Section

1➡ Examine the posterior section first. In the living system, a gelatinous mass fills the **vitreous chamber** (or **posterior cavity**), which is the cuplike structure you are now studying.

This mass is called the **vitreous humor** or vitreous body. Your model may not include a structure representing the vitreous humor.

Look at the posterior portion of this chamber. On the surface is the **retina**, a clear thin layer containing the visual receptors, the **photoreceptors**. The layers of the retina are covered in Section IV.

In the living system, the **retina**, which may be connected to the rest of the eyeball only at the optic nerve, is easily detached by trauma or disease. The retina is kept in place partly as a result of pressure from the vitreous humor. A detached retina, often characterized by apparent spots of bright light in the visual field, is an abnormal condition requiring immediate medical attention. In the past a detached retina could lead to blindness; today, if detected soon enough, the retina can often be reattached easily with laser surgery.

How would a detached retina lead to blindness?

____

____

2➡ Note the **optic disc**, the area from which the retinal neurons leave the eyeball and form the optic nerve. Because the optic disc contains no photoreceptors, it cannot transmit images and is thus often called the blind spot. About 2 mm lateral to the optic disc is a yellowish area known as the **macula lutea**. In the center of the macula lutea is a shallow depression known as the **fovea**, or **fovea centralis**. The fovea is the point where the highest concentration of visual receptors is found. This fovea, or pit, is a tightly packed layer of cones. The area is depressed because the ganglia and bipolar cells of the retina have been pushed aside. This arrangement affords better light transmission and thus keener vision at the exact center of the eye. This fovea is at the terminal end of the visual axis of the eye.

Beneath the retina is the **vascular tunic** (also known as the **choroid coat** or the **uvea**). Beneath this tunic is the sclera (fibrous tunic), which you identified previously on your external examination.

In many animals you would find an iridescent structure called the **tapetum lucidum**, a layer of the choroid coat (not present in humans) that apparently reflects light back to the retina, thus increasing the animal's night vision. The tapetum lucidum is also the reflective structure that gives the animal's eyes that glowing iridescence.

B. Anterior Section

1➡ Continue labeling Figure 36–4● as you read through the following descriptions:

The **anterior cavity** is the entire area between the lens and the cornea. The anterior cavity is divided into the **anterior chamber**, the portion located between the iris and the cornea, and the **posterior chamber**, the portion located behind the iris (between the iris and the lens).

The **ora serrata** (literally, serrated mouth) is the black, pleated anterior edge of the sensory portion of the retina. This structure is posterior to the lens and the suspensory ligaments. The **suspensory ligaments** are the hairlike structures holding the lens in place and extending to the ciliary body.

The ciliary body is posterior to the iris and is continuous with the choroid coat. The ciliary body consists of the ciliary muscle (which regulates the suspensory ligaments) and the ciliary processes, whose epithelial cells produce the watery **aqueous humor**, which fills the entire anterior cavity.

The **canal of Schlemm**, sometimes called the *scleral venous sinus*, is the passageway at the iris-corneal junction, which allows the aqueous humor to return to the venous system. You may not be able to locate the canal of Schlemm on the model.

GLAUCOMA The ciliary process is continuously producing the aqueous humor, which is continuously drained into the venous system. A problem exists when production exceeds drainage. This imbalance could lead to a build-up of aqueous humor in the aqueous chamber of the eye. This build-up could lead to excessive pressure on the vitreous humor, which in turn would put excessive pressure on the retina. Excessive pressure on the retina can lead to a myriad of problems, culminating in atrophy of the optic nerve and eventual blindness. This condition of excessive pressure is known as glaucoma. With modern optometric procedures, glaucoma is usually easily detected and easily treated.

Acute (or narrow) angle glaucoma occurs when the angle between the iris and the lens (see Figure 36–4) becomes so narrow that the aqueous humor cannot reach the anterior chamber and thus builds up behind the iris. This rare form of glaucoma, affecting only about 2% of all glaucoma patients, is almost exclusively a problem of hyperopic people, and must be treated immediately or permanent damage or blindness will result.

IV. Retinal Layers [∞ *Martini, pp. 542–544*]

A. Macroscopic Examination

1➡ Use Figure 36–5● to identify the layers of the retina.

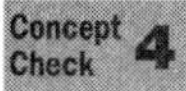

Name the layers of the retina.

____

____

2➡ Referring to the retinal layers you just identified, trace the path of light as it enters the

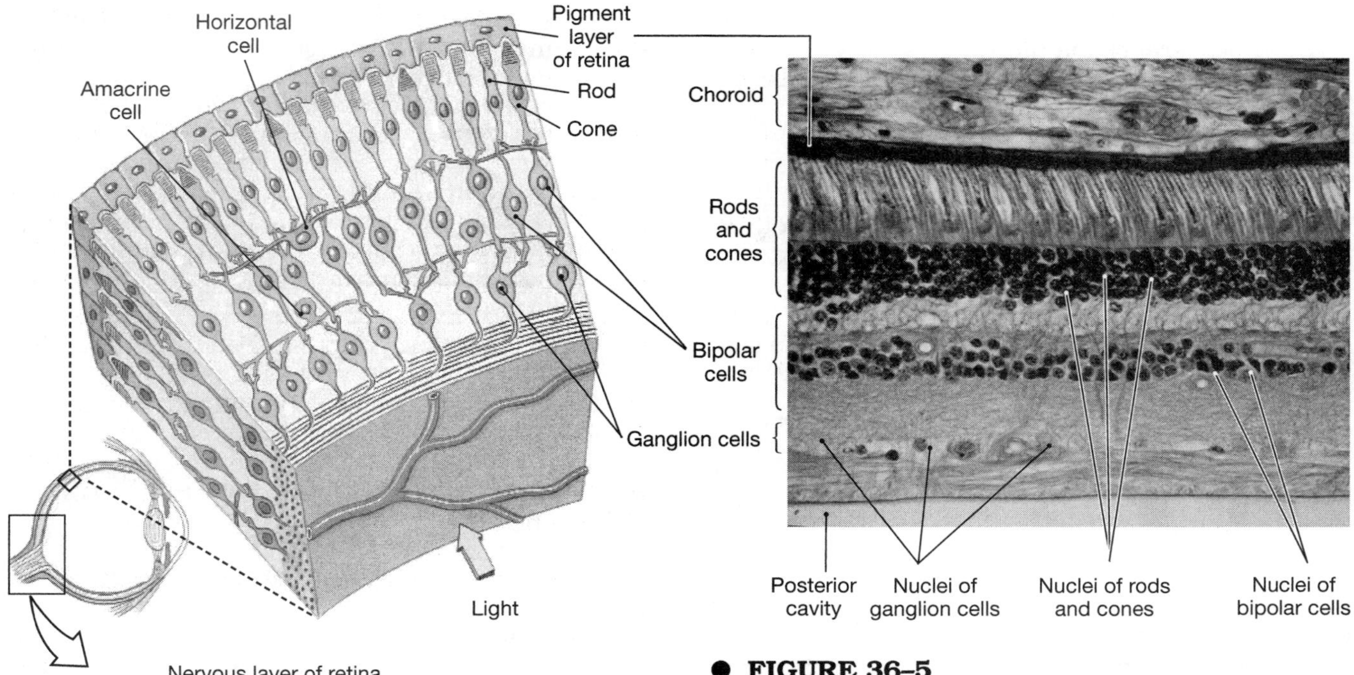

● **FIGURE 36–5**
Retinal Organization.

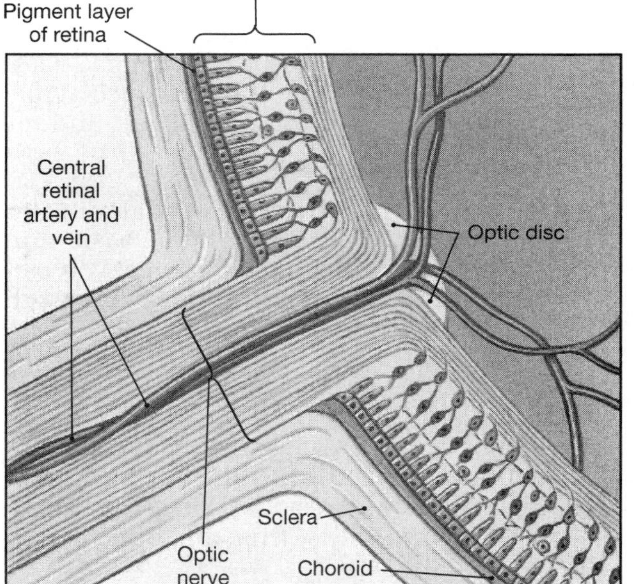

eyeball. Light must pass through the layers of the retina (except the pigment layer) before stimulating the visual receptors known as the **rods** and **cones**.

3 ➜ Locate the rods and cones in Figure 36–5●. There are three types of cones, red, blue, and green, each responding maximally to its respective color wavelength. There is only one type of rod.

Use your eyeball model to describe the location of the rods and cones in relation to the sclera, the pigment layer, and the optic nerve.

____

____

4 ➜ Locate the **bipolar cells** on Figure 36–5●. When light strikes the retinal receptors, the neurons generate an impulse that is transmitted to the intermediate neurons of the inner neural layer, known as bipolar cells. (The **horizontal cells** either inhibit or facilitate communication between the photoreceptors and the bipolar cells.). Locate these bipolar neurons on Figure 36–5●. From there the message is transmitted to the integrating neurons, also known as the **ganglion cells**. Locate the ganglion cells. (The amacrine cells modulate communication between bipolar and ganglion cells.) The axons from these ganglion cells converge and leave the eyeball as the optic nerve.

5 ➜ Recall the fovea centralis (Section III.A). The fovea centralis, the area of most acute vision, contains only cones. In this area there is almost a 1:1 ratio of photoreceptor cells to ganglion cells. Spreading out from the fovea, the cones become fewer and fewer while the rods become more and more numerous. Also in the outlying areas, the ratio of photoreceptor cells to ganglion cells gets much higher — 100:1 or higher. Therefore, in the peripheral retina, more light information from more receptors (literally up to hundreds of receptors) is collated in a single ganglion.

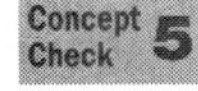

In terms of our ability to see bright and dim light, what does this high receptor-to-ganglion ratio in the peripheral retina mean?

____

____

B. Microscopic Examination

1➡ Examine the prepared slide of the retina. How many individual cell layers can you identify?

2➡ In the following space, sketch what you see. Compare your diagram with Figure 36–5●.

Cell layers of the retina

❑ Dissection

V. Dissection of the Mammalian Eye

A. Orientation

1➡ Obtain a preserved cow (or sheep) eye. Orient the eye so that the ventral surface is facing you. If the muscles are cropped close to the eyeball, at this point you probably will not be able to distinguish between the superior and inferior sides of the eye. Nevertheless, envision the eye in place in the animal's face and note where you would find the palpebrae, medial canthus, lacrimal sac, nasolacrimal duct, and lacrimal glands. Because all mammalian eyes are similar, you can use the human eye diagrams to guide you.

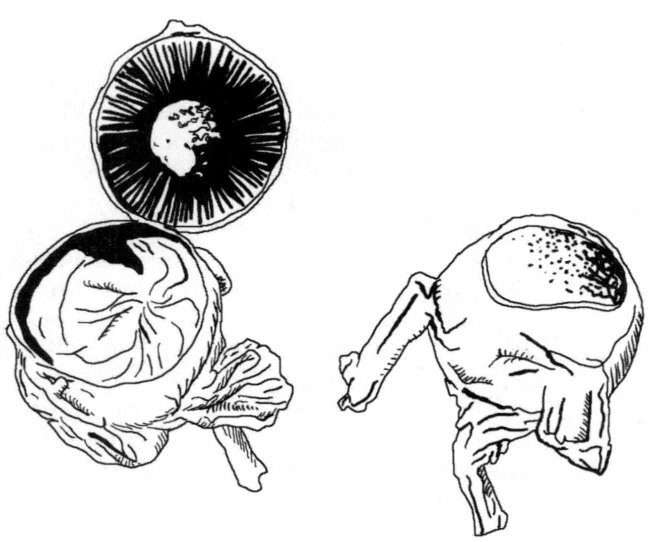

● **FIGURE 36–6**
Mammalian Eye.

2➡ Remove any excess fatty tissue from the exterior surface of the eye. You may have to use your forceps to dislodge some of this fat. What is the function of the fat?

____

____

3➡ You may also find some tough connective tissue around the eye. Remove this tissue as well but be careful not to cut the optic nerve or any of the muscles that surround the eye.

B. External (Extrinsic) Muscles

1➡ Note the muscles that surround the eye. Depending on the condition of your preserved eye, now that you have removed the fat you may be able to identify some or all of the six specific extrinsic eyeball muscles: superior rectus, inferior rectus, medial rectus, lateral rectus, inferior oblique, and superior oblique.

2➡ Identify as many of the external muscles as you are able. (Because of the basic similarity found in mammalian eyes, you can refer back to Section II.A to help you with this identification.)

C. External Examination of the Eye

1➡ Examine the external surface of the mammalian eye. You should now be able to identify these externally visible features: sclera, cornea, iris, lens, and optic nerve.

2➡ Use Section II.B as your guide and identify the same structures you identified in the human eye.

D. Internal Structures of the Eye

After you have identified the externally visible structures, you should be ready to dissect the eyeball to observe the internal structures.

1➡ Begin by noting an imaginary circular line about 1 cm posterior to the edge of the cornea (also known as the corneal margin). With your scissors carefully cut along this line. Use special caution with preserved eyeballs because the internal liquids contain preservatives.

2➡ Describe the consistency and opacity of the gelatinous vitreous humor (which should readily fall out into your dissecting pan).

____

____

3➡ Examine the posterior section first. The structures will be the same as the human eye structures described in Section III.A: retina, optic disc, macula lutea, fovea centralis, and vascular tunic (choroid coat). You will probably find the retina detached from the posterior surface, except at the optic nerve. Also, you may not be able to locate the macula lutea and the fovea centralis.

4➡ Recall the reference in Section III.A to the tapetum lucidum. See if you can identify this iridescent structure on your specimen.

5➡ Now examine the anterior portion of your eyeball. Begin on the exterior and locate the cornea, iris, lens, and pupil. The iris of your preserved mammalian eye will probably be dark.

6➡ Unless you have an exceptional specimen, you will not be able to observe the distinct areas of the anterior cavity. Further, you may not be able to discern the canal of Schlemm. However, restudy Figures 36–3● and 36–4● and identify where the aqueous chamber areas and the canal of Schlemm *should* be. Use your probe as an aid.

7➡ Examine a section of the retina from your specimen under the stereoscopic microscope. Take the section preferably from the lateral/posterior area. Do not use a section from the optic disc unless you wish to compare areas. Are you able to identify any specific retinal structures? If so, sketch what you see.

Retinal structures in dissected eye

8➡ Examine Figure 36–6●. Based on your study of the human eye and your own observations of the dissected mammalian eye, label the structures you are able to identify.

❑ Additional Activities

NOTES

1. Compare the primary and accessory structures of the human eye with those of various animals.
2. Research the chemical components of tears.
3. List the chemical and physical differences between the vitreous and aqueous humors.
4. Draw some conclusions about the nature of the "sleep" often found in the corners of your eyes when you awaken.
5. Research the changes that take place in the eye with aging.
6. Ophthalmoscopy is the examination of the retina and inside (fundus) of the eyeball using a hand-held, illuminated instrument known as an ophthalmoscope. Although all ophthalmoscopes operate on the same physical principles, several types of ophthalmoscopes do exist. If you have access to an ophthalmoscope, you should follow the specific directions for your instrument. You should be able to gain a rough determination of your far and near vision by adjusting the lenses of the ophthalmoscope. You should also be able to study the retinal blood vessels, the optic disc, the fovea centralis, and the macula lutea.

Answers to Selected Concept Check Questions

NOTES

1. They would cause a sty or some form of conjunctivitis. A sty is an inflammation or swelling of a sebaceous gland at the margin of the eyelid.
2. The principal muscles involved are:
   Left eye → lateral rectus
   Right eye → medial rectus
3. In the sclera the collagen fibers are not arranged in layers; thus light cannot pass through. Light will be reflected.
4. The layers from the external eyeball inward are: pigment layer, rod and cone (or visual photoreceptor) layer, horizontal cell layer, bipolar cell layer, amacrine cell layer, ganglion cell layer.
5. The higher the ratio, the dimmer the light that can be seen. Think in terms of spatial summation of nerve impulses. Smaller amounts of light from many receptor cells add up to a signal strong enough to register in the ganglion cell. In the fovea, although brighter light is required, the lower ratio allows for "fine tuning" or better visual resolution.

❑ Lab Report

1. Referring to the numbered inquiries at the beginning of this exercise, complete the following box summary:

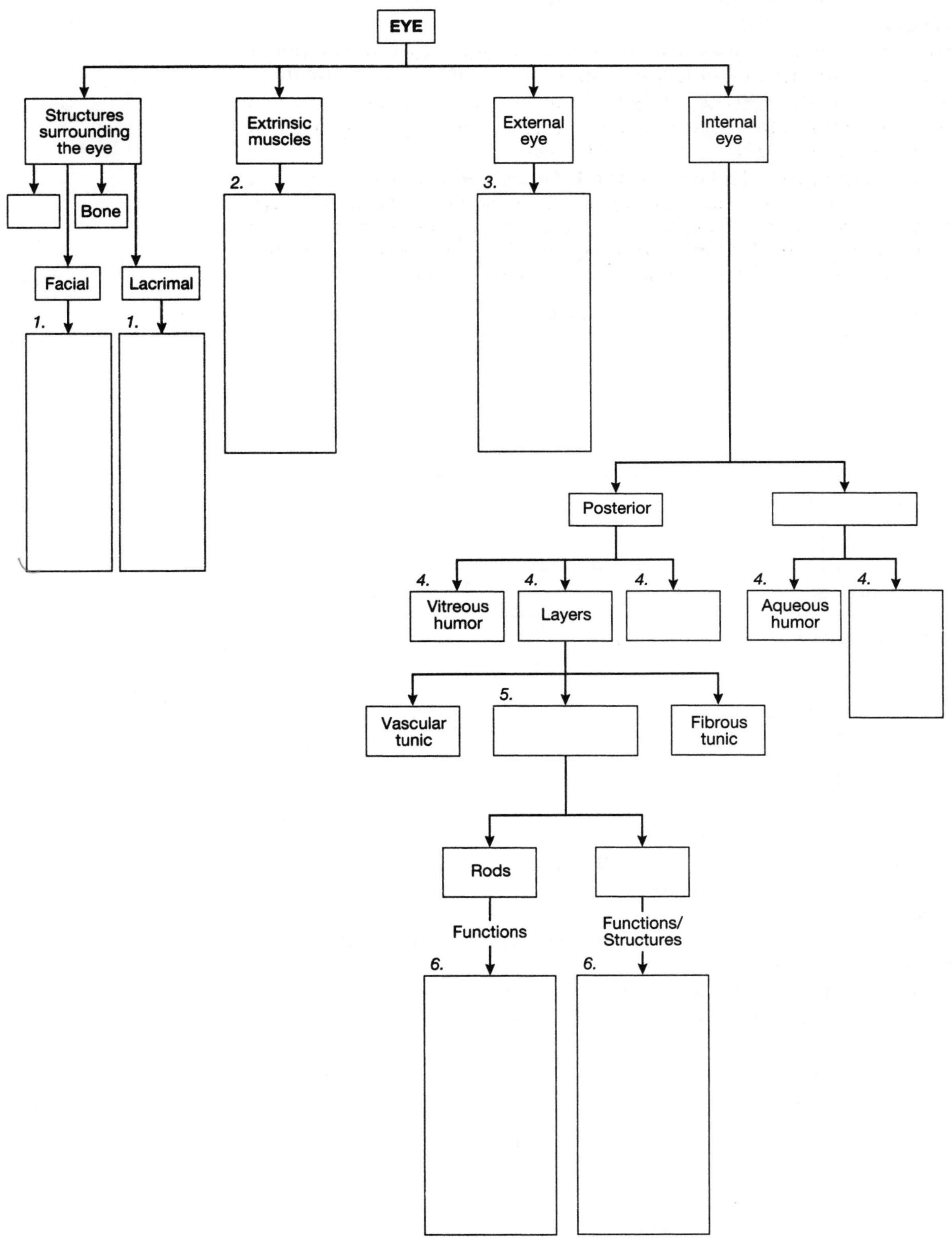

General Questions

NOTES

1. Why does your nose run so much when you are crying?
2. Glaucoma, which causes excessive pressure on the retina, can develop when the canal of Schlemm is blocked. How do you explain this phenomenon?
3. What evolutionary advantage is there in an animal having a tapetum lucidum?
4. Which extrinsic muscles are involved in each of the following actions? Specify left or right eye muscles as appropriate. (More than one muscle may be involved in a given action.)
   - Looking cross-eyed
   - Looking straight up
   - Looking at the ground over your right shoulder
   - Rolling your eyes
5. In dim light, why do you see objects more clearly when you do not look directly at them?
6. Fill in this chart:

Eye part	Location	Function
Sclera		
Iris		
Pupil		
Cornea		
Optic nerve		
Fovea centralis		
Medial rectus muscle		
Puncta		
Canal of Schlemm		
Suspensory ligaments		

7. The ora serrata is continuous with the ______________________.
8. The ciliary body is continuous with the ______________________.
9. Name and identify the function of two structures found in the ciliary body.
10. What is the difference in location between the aqueous and vitreous humors?
11. Describe the functional difference between the rods and the cones.
12. Trace the path of a tear from its production until it enters the nasal cavity.
13. Trace the path of a nerve impulse after a light ray strikes the retina until it exits the eyeball.

EXERCISE 37

Aspects of Vision

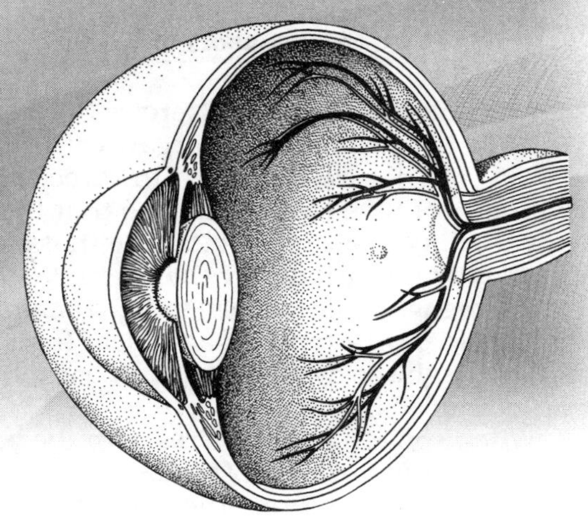

PROCEDURAL INQUIRIES

Preparation

1. What is the difference between myopia, hyperopia, and emmetropia?
2. What types of lenses should be used to correct myopia and hyperopia?
3. Through which structures does a light ray pass from the time it enters the eye until it strikes the retina?
4. What is the nervous pathway from the retina to the brain?

Experimentation

5. How is a Snellen eye chart test interpreted?
6. Why does a person with astigmatism see certain lines on the astigmatism chart as darker or clearer than other lines?
7. What does near point accommodation mean?
8. Which type of photoreceptors determine whether a person will be color blind?
9. Which structure is responsible for the presence of the blind spot?
10. What happens to the size of the pupil in the presence of bright light?
11. What happens to the size of the pupil when vision shifts from a near object to a distant object?
12. What is the approximate range of vision for the normal eye?
13. How does one test for eye dominance?

Additional Inquiries

14. What is a cataract?
15. How can the extrinsic eye muscles affect vision?

Key Terms

Amblyopia
Astigmatism
Blind Spot
Cataract
Color Blindness
Diopter
Diplopia
Flicker Fusion Frequency
Hyperopia
Lateral Genticulate Nucleus
Myopia
Near Point Determination
Optic Chiasma
Optic Nerve
Optic Tract
Peripheral Vision
Presbyopia
Projection Fibers
Pupillary Distance Reflex
Pupillary Light Reflex
Snellen Chart
Visual Acuity
Visual Cortex

Materials Needed

Assorted Concave and Convex Lenses
Assorted Glasses (from Members of the Class)
Snellen Eye Chart
Astigmatism Chart (or use the one in this manual)
Ishihara (or Stilling) Color Chart (or Assorted Strands of Colored Yarn)
Plain White Paper
Blind Spot "X" and "O" (or use the one in this manual)
Rulers (Metric)
Chalk or Pointer

Optional:

Flicker Fusion Frequency Determination (film and projector with accurate frames per second setting)

We established in Exercise 36 that the eye is a complicated organ with numerous structures involved in directing the incoming light rays to the retina so that they can be converted into electrical impulses that can be appropriately interpreted by the brain. Vision is the act of seeing; vision involves the transmission of the physical properties of an object from that object, through the eye, and (via electrical impulses) to the brain.

When something interferes with the proper reception or transmission of these visual stimuli, the electrical impulses are either not received or not recorded properly and a visual problem exists. If the eyeball itself is too long or too short, or if the lens is improperly shaped, or if the ciliary muscles do not contract as they should, a person will not see clearly.

What we see and how we see it are not always just a matter of physics. We know, for instance, that victims of war and other physical and mental terrors may develop a psychological blindness in the absence of any apparent physical abnormality. We also know that sometimes a refractive lens might be mechanically correct but that practically or psychologically it is not the right lens for the person being tested. For these reasons we can say that vision testing today is both an art and a science.

Many sophisticated instruments exist for measuring vision and many sophisticated techniques exist for correcting visual aberrations. For instance, a retinoscope (an instrument used for measuring the refractive powers of the eye) can be used to fit glasses to infants and other persons who cannot verbalize whether a lens is correct for them. Lens implants, often with a correction of +15 **diopters**, can be used following cataract removal.

The diopter is the unit used to measure the refractive power of a lens with a focal distance of one meter. In other words, with a one-diopter lens, the point of convergence or divergence for the light rays would be one meter from the object. Concave lenses (the diverging lenses used to correct **myopia**, or nearsightedness) are measured in negative diopters. Convex lenses (the converging lenses used to correct **hyperopia**, or farsightedness) are measured in positive diopters. Corrections are normally made in steps of quarter diopters. The correction on most people's lenses today is less than four diopters (positive or negative). For a more detailed explanation of refraction, consult a physics book.

Bifocals have been around since the time of Benjamin Franklin. Generally speaking, the bifocal is less concave (or more convex) than the regular lens. Thus, the numerical diopter measurement of the bifocal is less negative (or more positive) than the regular lens measurement. Trifocals offer three different lens strengths. The strength of the middle lens is usually half-way between the strengths of the other two lenses.

But visual problems are not limited to nearsightedness, farsightedness, or the need for bifocals. As we shall see as we work through this exercise, good vision also includes the ability to see color, the ability to regulate incoming light, the ability to react quickly to incoming stimuli, and the ability to focus evenly in all directions.

In this lesson we will examine some of the physiological aspects of vision. It should be mentioned that anomalies tested for here do not indicate the overall "health" of the eye. (Example: A person may be quite nearsighted while still having eyes that are physically strong, disease-free, and showing no signs of degeneration.)

❑ Preparation

I. Basic Vision *[∞ Martini, pp. 547–549]*

The focus of the following paragraphs is the lens. You should be aware, however, that one of the functions of the cornea is the refraction of light and that abnormalities in the shape or structure of the cornea can also cause visual disturbances.

A. Emmetropia

Emmetropia is the word used to describe the refraction in the "normal" eye.

1➡ Study Figure 37–1● and note that the visual image is focused on the retina.

B. Myopia

Myopia (nearsightedness) results when the parallel light rays of the image come to focus in front of the retina. (It should be remembered that when we speak here of "image" we are referring to a series of light points. Each focal point is projected individually onto the retina.)

1➡ Study the myopic eye (Figure 37–2a●). In the myopic eye, the eyeball is too long for

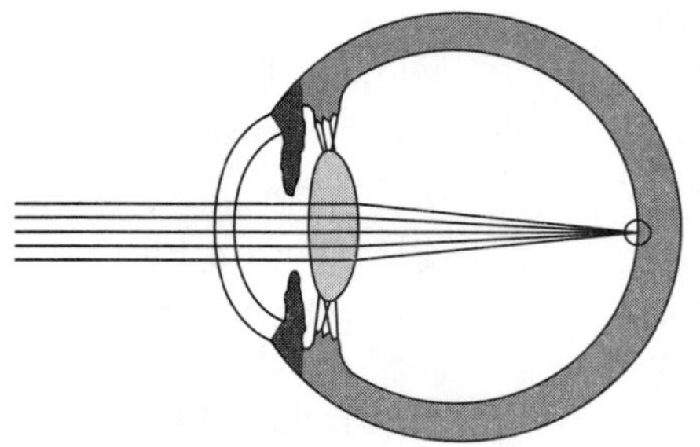

● **FIGURE 37–1**
Emmetropia.

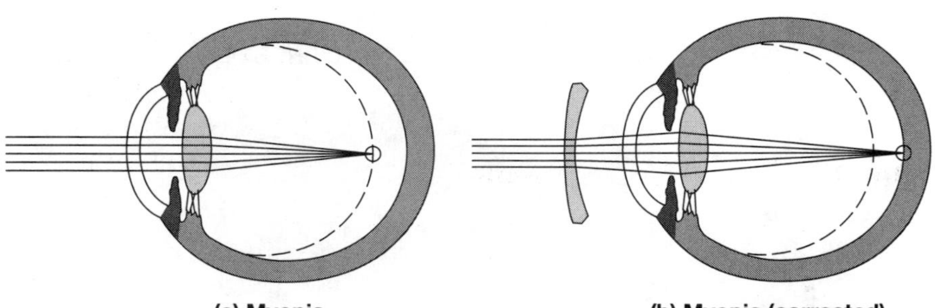

● **FIGURE 37–2**
(a) Myopia. **(b)** Myopia Corrected.

the adaptive power of the lens. The lens can become more spherical so that the person can focus clearly on close objects. The lens, however, cannot relax sufficiently to "flatten out" and allow focusing on distant objects.

2 ➡ Note that myopia is corrected with a concave lens (Figure 37–2b●). This type of lens will cause the entering light rays to diverge. Therefore, the light rays will be angled farther back on the eyeball, placing the point of maximum focus on the retina.

C. Hyperopia

Hyperopia (farsightedness) results when the parallel light rays of the image come to focus behind the retina.

1 ➡ Study the hyperopic eye (Figure 37–3a●). In the hyperopic eye, the eyeball is too short for the adaptive power of the lens. The lens can become more relaxed so that the person can focus clearly on distant objects. The lens, however, cannot become sufficiently more spherical to allow for focusing on near objects.

2 ➡ Note that hyperopia is corrected with a convex lens (Figure 37–3b●). This type of lens will cause the entering light rays to converge. Therefore, the light rays will be angled more anteriorly in the eyeball, placing the point of maximum focus on the retina.

D. Lenses

In the average anatomy/physiology laboratory, we cannot accurately test for myopia and hyperopia. We can, however, gain an insight into the lenses used to correct these conditions. (For a detailed examination of these phenomena, consult a physics book.)

Concept Check 1 Take some assorted concave and convex lenses, some plain white paper, and any desktop objects you choose, such as pens, books, or paper clips. Look at different near and distant objects with different lenses. Move the lenses closer and farther away from your eye. Note distances where images change focus, invert, disappear, etc. If you wear glasses, try this experiment with and without your glasses.

Record your comments and observations here.

____
____
____
____

2 ➡ Try focusing light rays through both the concave and convex lenses onto a piece of paper. Note your results. Examine what happens when you place concave and convex lenses together. If you wear glasses, compare your glasses with the lenses you are now working with. Are your lenses concave or convex? ______________

Numerous conclusions can be drawn from this experiment. Discuss these conclusions with your partner. Relate your conclusions to the discussion on myopia and hyperopia.

____
____
____

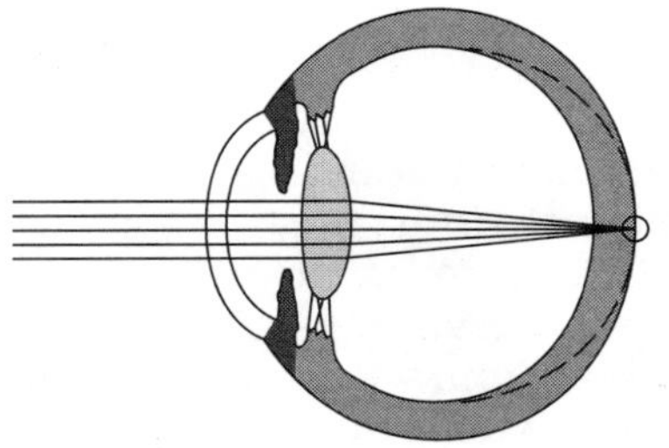

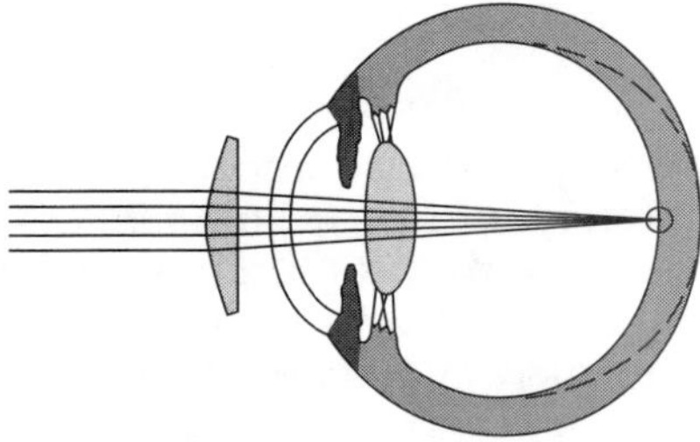

● **FIGURE 37–3**
(a) Hyperopia. **(b)** Hyperopia Corrected.

II. Visual Pathways

A. Within the Eye

1➡ Return to Figure 37–1●. Note that once past the lens, the light rays traverse the posterior cavity. Within the posterior cavity the vitreous humor also acts as a refractive medium helping to focus the light onto the retina.

2➡ Recall that in Exercise 36 we discussed how light must pass through the layers of the retina before stimulating the rods and cones. The rods and cones are often called the primary receptors. From the rods and cones the impulses are transmitted to the secondary, or bipolar, neurons, with the horizontal cells functioning as inhibitors or facilitators. The impulse then passes to the tertiary, or integrating, neurons, also known as the ganglion cells. In this case, the amacrine cells modulate the interneuronal communication. The axons of the ganglion cells converge and exit the eyeball at the optic disc. The nerve formed by this convergence is the **optic nerve** (N II). Write a sentence summarizing the path a light ray takes from an object to the optic nerve.

____

____

____

B. To the Brain

1➡ Study Figure 37–4●, which depicts the nerve pathway followed by N II. Notice how the medial axons from the eyeball cross at the **optic chiasma**. The lateral axons do not cross. Notice that the **optic tract** leads to the **lateral genticulate nucleus**, which is located in the thalamus. In this nucleus the nerve fibers synapse with **projection fibers**, and the impulse then travels to the **visual cortex** of the occipital lobe of the cerebrum.

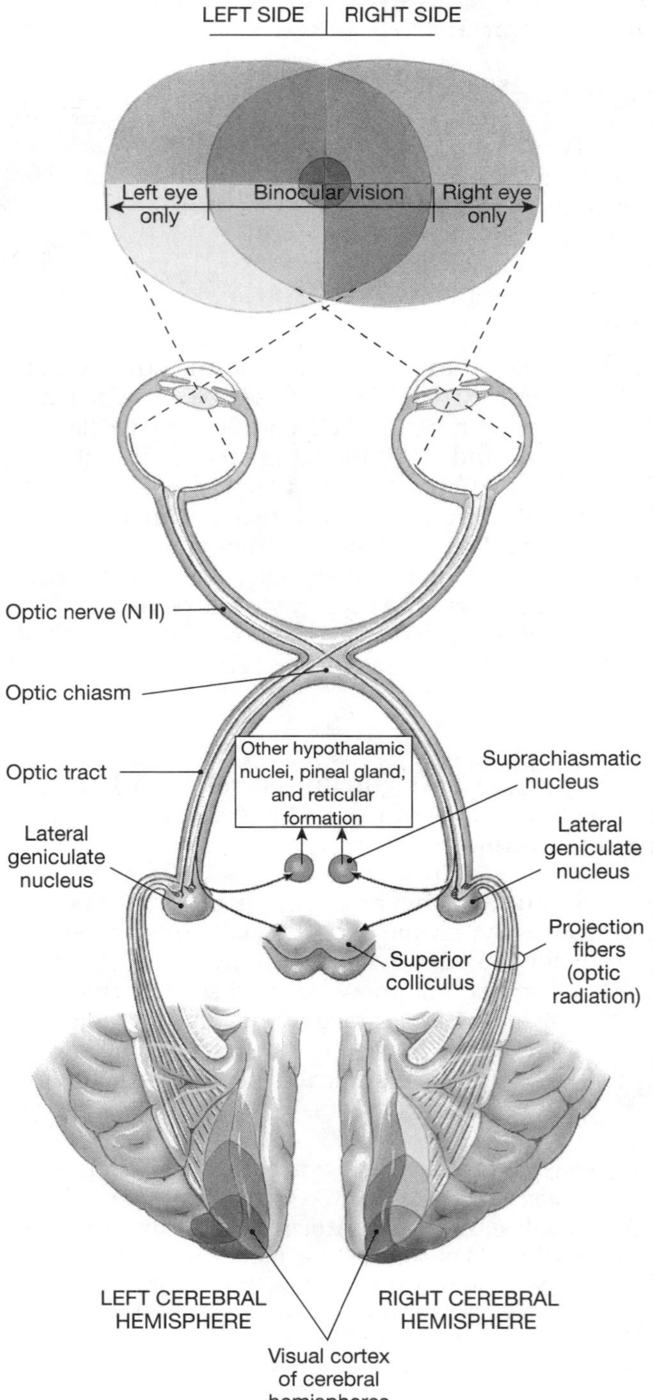

● **FIGURE 37–4**
Visual Pathways.

❑ Experimentation

III. Vision

Despite the complexity of visual examinations today, a number of tests can easily be run in the standard anatomy/physiology laboratory. It should be remembered, however, that tests performed in a school laboratory such as yours lack both the precision and the accuracy you would find in the professional laboratory of an optometrist or ophthalmologist. If after completing these tests you feel you have a vision problem, consult an appropriate health care professional.

Some of these exercises are more easily completed if you and your partner work together, although you should both do all the exercises. **As**

you go through these tests, record your results on the chart at the end of the exercise.

A. Visual Acuity

Visual acuity is a measure of the resolving power of the eye. The normal human eye can differentiate at a distance of 10 m two points that are 1 m apart. This is because of the retinal cones. If two photoexcited cones are separated by at least one nonexcited cone, the human brain can differentiate between the points of excitement. A person with less than "normal" visual acuity has some aberration that causes the excitation of all cones in a particular area. Pinpoints of light are not recognized as distinct. Thus, separate points are not recognized as separate and the images appear blurred.

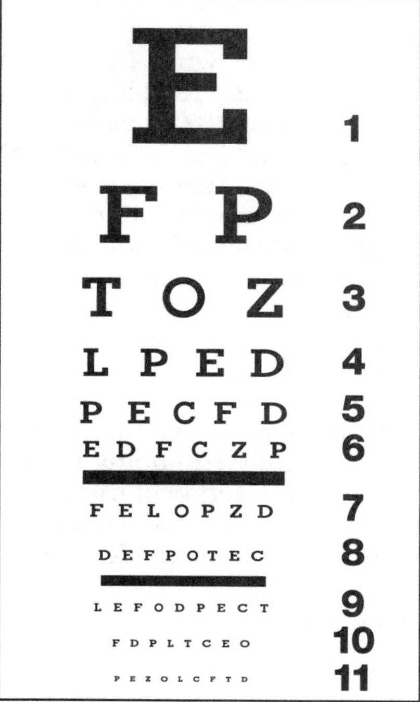

● **FIGURE 37–5**
Snellen Eye Chart.

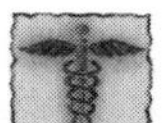

The **Snellen chart** (Figure 37–5●), named for Dutch ophthalmologist Herman Snellen (1834–1908), is a standard chart used for testing visual acuity. One problem with the Snellen test is that it tests ONLY for distance visual acuity. In other words, Snellen tests for myopia. The Snellen test does not test directly for hyperopia, and a person with simple moderate hyperopia might not be detected at all on the Snellen test. This person might well have a Snellen score of 20/20. Another person might be extremely farsighted and test out at 20/200 on the Snellen chart — not because the person is unable to focus "that far away" but rather because the person is unable to focus "that close"!

In addition to hyperopia, the Snellen chart also does not test for either astigmatism or presbyopia. (Astigmatism and presbyopia are discussed in the next sections.) Hyperopia, astigmatism, and presbyopia are essentially symptomless and present definite problems for vision-screening experts, particularly those screening the vision of young children.

Additionally, the Snellen test does not test for **amblyopia** (lazy eye) or any eye disease. *Amblyopia expanopsia* is the term commonly used for the inability to control the movements of one or both eyes. To avoid **diplopia** (double vision), vision in the amblyotic or "lazy" eye often becomes dim or suppressed. If not detected in the preschooler, optical muscles do not strengthen properly and vision may be permanently lost in the affected eye.

In school screenings about 80% of the referred vision problems are simple nearsightedness, even though in the general population about 60% of all people are hyperopic. (Very young children tend to be farsighted.) It is generally considered to be more difficult to function in a society with myopia than with hyperopia.

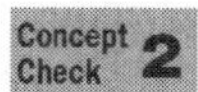

Concept Check 2 Why do you suppose that is?

____

____

In the past, educators have tended to dismiss vision problems that are not detected by the Snellen test. Many optometrists believe that even today too much absolute credence is given to the Snellen test, leaving undetected a number of easily correctable eye problems. (Example: A farsighted child may not be able to focus on his or her lessons but still tests out as having virtually perfect vision on the Snellen test.)

There are no standard tests in today's schools for detecting hyperopia, astigmatism, or presbyopia. Some schools today are incorporating the "Modified Clinical Technique" into the standard visual screening program. The Modified Clinical Technique includes tests for far and near acuity, depth perception, and color vision. The increased awareness and sensitivity of the public have also increased the detection of visual problems in children. Hyperopia, astigmatism, and presbyopia are all easily corrected by glasses even in the youngest children.

Somewhere in your laboratory you will find the Snellen eye chart. The Snellen chart tests for myopia. A point will be marked designating a distance of 20 feet (6.1 meters) from the chart. Stand at that point and test each eye separately.

1➡ Start at the top of the chart and read as many lines as you can. Your partner can stand close to the chart and tell you whether you are reading the letters correctly. Identify your last accurate line. Note the number on the eye chart that corresponds to this last line. Suppose that number is 40. That would mean that your vision in that eye is 20/40. You can read at 20 feet what the "normal" (emmetropic) eye can read at 40 feet. If your eye can read at 20 feet what the "normal" (emmetropic) eye can read at 20 feet, your eye is said to have 20/20 vision. If you wear glasses, try this test with and without your glasses. Again, test each eye separately. Record your results. Keep in mind that you are testing only for myopia. Visual aberrations other than myopia are sometimes detected by the Snellen test simply because other aberrations do cause a blurring of the vision.

[Radial keratotomy is a relatively new procedure that involves surgically reshaping the cornea and thus changing the refraction of the incoming light. See *[∞ Martini, p. 548]* for a discussion on radial keratotomy and photoreflective keratectomy.]

B. Astigmatism

Astigmatism is a condition caused by an uneven curvature of one of the rounded surfaces of the eye. Usually the lens is the part with the aberration; however, the cornea may also be afflicted with uneven curvature. (A misshaped retina will cause astigmatic vision, but this condition is not usually classified as true astigmatism.) Because of the uneven curvature, lines in one or more planes will be out of focus. In other words, the light rays striking one plane (or portion of a plane) will not be bent proportionately the same as light rays striking another plane (or portion of a plane).

Astigmatism can be simple (affecting only one meridian of the eye), compound (involving both horizontal and vertical curvatures of the eye), index (resulting from refractive indices inequalities in different parts of the lens), or mixed (causing one meridian to be hyperoptic and the other myopic).

1➡ Test for astigmatism one eye at a time. Look at the astigmatism test chart or use Figure 37–6●. Which lines seem darkest? Do all lines seem equally clear? Record your results.

Lines that are out of focus will determine the axis (or axes) where the astigmatism is to be found.

2➡ If your glasses correct for astigmatism, try the astigmatism test without your glasses, even though some other refractive abnormality may make the results less than accurate. (With your astigmatic correction, all lines should appear equally distinct.)

3➡ If someone in the class has glasses that correct for astigmatism, hold these glasses in front of the Snellen chart or the astigmatism chart or in front of some other

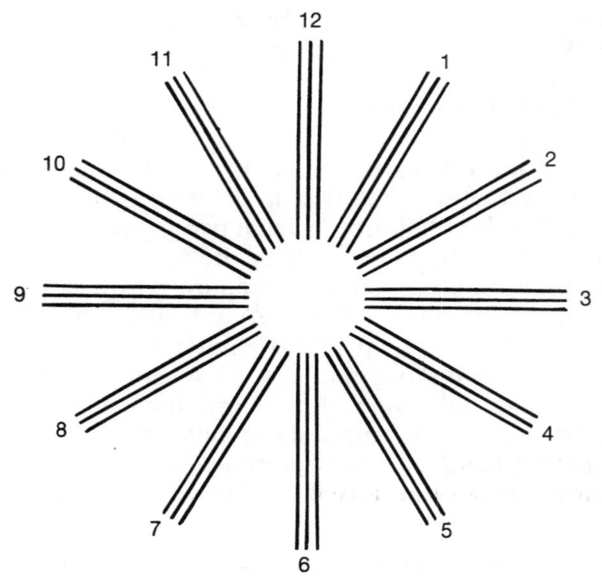

● **FIGURE 37–6**
Astigmatism Test Chart.

sharply defined object in the room and rotate the glasses in several directions. The apparent changes in the width and breadth of the items in front of you are caused by the glass or plastic cylinders used to correct the astigmatism. The cylindrical shape of the lens changes the focal plane of the incoming light rays to correspond with those light rays coming into the eye on different axes.

C. Near Point Determination

Near point determination is a test to measure the elasticity of the lens. Specifically, as you focus on closer objects, your lens must become more spherical than when you focus on more distant objects. The ability of the lens to accommodate this increased convexity decreases with age, and a condition known as **presbyopia** — meaning old eye — develops. (Presbyopia is not limited to older people. Optometrists today tell us that presbyopia is not the best term to use, especially since more and more children are being diagnosed with this condition. This may be due to an increased awareness of the problem rather than to an actual increase in presbyopia.) Near point is the measurement from the eye to the closest distance at which the subject can focus sharply.

1➡ Test each eye separately. Print a letter on a white paper (or use any page printed with sharp black letters on a white background). Start with the test paper at an arm's length from your eye. Slowly move the paper toward your eye. See Figure 37–7●. When the image

● **FIGURE 37–7**
Near Point Determination.

begins to blur, have your partner measure the distance from your eye to the paper. This is your near point. Record your results.

Average near points for different ages are given as follows:

Years	*cm*
10	7.0–7.5
20	9.0–10.0
30	11.5–13.0
40	17.2–20.0
50	45.0–52.5
60	83.3–90.0

D. Color Blindness

Color blindness, which can be any of several sex-linked hereditary conditions, is due to one or more defects in retinal cones. Recall that cones are responsible for color vision. Protanopia exists if red cones are lacking. Individuals with this condition tend to see blue-greens and purplish reds as grays. Deuteranopia exists when green cones are lacking. Such individuals see greens and purple-reds as gray. People with either of these conditions have red-green color blindness and have varying degrees of difficulty in differentiating between the reds and the greens. Red-green color blindness is technically called daltonism (after the English chemist, John Dalton, 1766–1844).

1➡ In addition, yellow weakness, blue weakness, and total color blindness exist. It also seems that some persons have a mild color loss. These people can distinguish the entire color spectrum, given sufficient time, light, and concentration. However, their color acuity does not seem to be as sharp as that of the general population. These individuals seem to have fewer, though equally efficient, cones than do other people.

2➡ Color blindness can best be tested with the Ishihara or Stilling color test plates. If your lab has these charts, follow the instructions that come with them. If your lab does not have the color charts, do the yarn test.

3➡ Gather a handful of assorted colored yarn samples and put them in piles according to like colors. See if your partner agrees with you. If you and your partner disagree, see what other teams have decided and discuss possible explanations for the discrepancies. Record your results. What might account for differences in yarn piles?

____

E. Blind Spot Determination

The **blind spot** determination test verifies that there are no photoreceptors in the optic disk. This is the point where the retinal nerve fibers leave the eye. (Check Figure 36–3● in the previous lesson.)

1➡ Keep your left eye closed, and use your right eye to focus on the dot in Figure 37–8●. The dot should be directly in front of your open eye. Slowly move this book directly toward your face until the cross disappears. Repeat this experiment with the right eye closed and the left eye open. Record your results. Discuss this phenomenon with your partner.

F. Pupillary Light Reflex

The **pupillary light reflex** (Figure 37–9●) demonstrates how quickly the pupil of the eye changes size in response to the amount of incoming light.

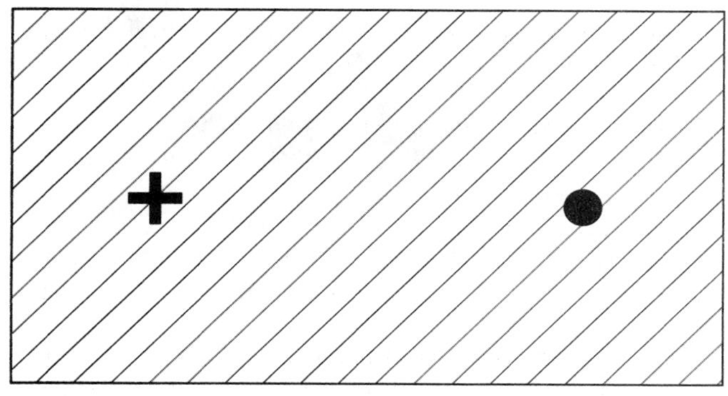

● **FIGURE 37–8**
Blind Spot Determination.

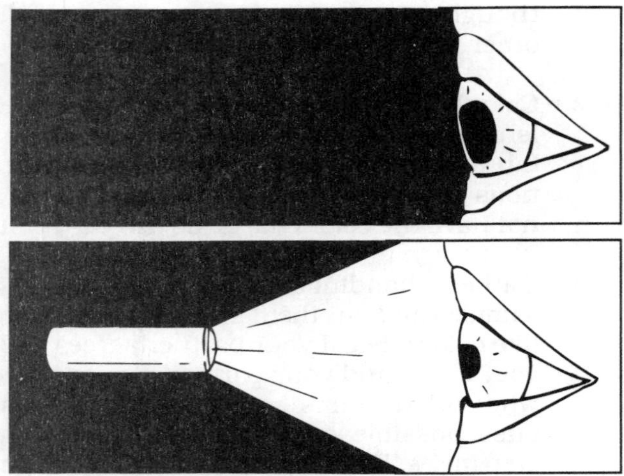

● **FIGURE 37–9**
Pupillary Light Reflex.

1➜ Hold one eye closed for at least one minute and then release the eye suddenly in the presence of a bright light. As you do this, have your partner observe the changes in your pupil size. Repeat this experiment with the other eye and have your partner note any differences in the reflex between one eye and the other eye. Record your results.

G. Pupillary Distance Reflex

The **pupillary distance reflex** is based on the need for more light to enter the eye to view more distant objects. Thus, to view these far objects, the pupil will be larger.

1➜ Test this reflex (Figure 37–10●) by concentrating on a close object for at least one minute. Suddenly switch your vision to a predetermined distant object at least 20 ft away and illuminated with the same light intensity as your nearby object. Have your partner observe changes in your pupils. This test can be done with each eye separately and with both eyes together. Be certain to rest your eyes between tests so that ocular fatigue does not influence your pupillary reflexes. Record your results.

H. Peripheral Vision

Peripheral vision is the name given to your "sideways" vision. Some people develop "tunnel vision," which is an inability to see lateral objects or events. If your peripheral vision is good, while looking straight ahead you should be able to see approximately 90 degrees to either side of your head. In other words, with you arms outstretched to the side and with both eyes open and focusing on a distant object, you should be able to discern the fingertips of both hands. Were your nose not in the way, each eye should actually have a range of vision of about 180 degrees. Although rather complicated machines for performing peripheral vision tests without human error do exist, you can use one of two laboratory methods to help you understand your own visual range.

1➜ For the more accurate method: Check the peripheral vision of each eye separately. You need a darkened room, a point straight in front of you to concentrate on, a slowly

● **FIGURE 37–10**
Pupillary Distance Reflex.

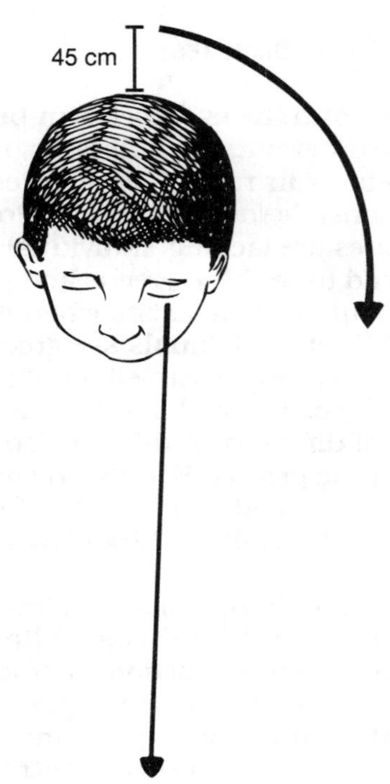

● **FIGURE 37–11**
Peripheral Vision Test.

moving white-tipped pointer or flashing light coming in a circular motion from the rear, and a method of accurately measuring the angle at which you first perceive the moving pointer (Figure 37–11●).

2➡ For the less accurate method: Test each eye separately. Sit in a somewhat darkened area and concentrate on a distant object directly in your line of vision. Have your partner slowly move his or her finger or a piece of white chalk around your head from the rear, maintaining a radius of about 45 cm. Stop your partner as soon as you see the finger or chalk. Measure the angle from the point of your forward vision. Record your results for each eye.

I. Eye Dominance

Eye dominance is the term used to describe whether you are "left-eyed" or "right-eyed." Although you may seem to see equally well with both eyes, your vision in one eye is probably dominant over your vision in the other eye. Being right- or left-eyed is totally independent from being right- or left-handed!

1➡ Test for eye dominance by focusing both eyes on an object somewhere across the room. Now, with your thumb or index finger quickly point to that object. Hold your hand still and look at the object with each eye separately. With one eye your finger should appear to be directly on the object. With the other eye your finger will be approximately 3 degrees to one side or the other of the object. Whichever eye is open when your finger seems to be on the object is your dominant eye. Record your results.

J. Cataract

A **cataract** is an opacity of the lens, the lens capsule, or both. Despite numerous varieties and etiologies for cataracts, the problem is that a misalignment of the protein strands prevents the passage of light through the eye. Cataracts are generally more common among older people, but anyone of any age can develop a cataract.

It is unlikely that someone in the class will either have cataracts or will volunteer to explain his or her cataracts. Nevertheless, you can often gain a perception of cataracts by examining the eyes of an older dog or cat. As these animals enter their early and mid teens, their eyes often begin to look milky or cloudy. These are cataracts. These animal cataracts are virtually the same as human cataracts. Record any comments you may have.

K. Extrinsic Conditions

The preceding tests have all been for intrinsic conditions. In addition, the extrinsic muscles of the eye may weaken. In a young child this can result in amblyopia (Section III.A). In older children and adults amblyopia or any of several other problems may occur. In severe cases the person may not have control of either or both eyes. In less severe cases the person may develop diplopia (particularly when tired) or may have persistent difficulty in focusing both eyes on near objects. Often this muscle weakness is more pronounced in one eye than in the other. Eye exercises can help strengthen the muscles. If the problem is nonpathological, the treatment may be the inclusion of prisms in the lenses of the glasses. These prisms redirect the light rays and compensate for the inability of the eyes to focus in unison.

❑ Additional Activities

1. **Flicker fusion frequency** (FFF) is the name given to the ability of the eye to see as a continuum individual frames of light. For example, if you are watching a movie, you know that that movie is actually a series of individual frames (or pictures). If the movie is being projected fast enough, you see a continuum of action rather than individual pictures. The flicks have fused. The FFF for humans is generally about 24 frames per second. If the movie is showing the individual pictures at less than 24 frames per second, you will probably be aware of a jerkiness in the film.

   To test your specific FFF, you will need a film projector that can be set to show a specific number of frames per second. In a darkened area have your partner show the film at specified numbers of frames per second.

NOTES

NOTES

Tell your partner exactly when you notice that the film action does not seem to be a continuum. Work quickly because visual fatigue can affect the accuracy of this test.

2. Research the types of astigmatism and find out the types of cylinders used to correct for these aberrations.
3. Research the types of amblyopia and find out about the different corrective lenses, different eye exercises, and different surgeries that can be used to correct the condition.
4. If you wear glasses or contact lenses, check your prescription and ask your optometrist or ophthalmologist to explain the numbers to you.

Answers to Selected Concept Check Questions

1. You should be able to tell if the lenses in your own glasses are concave or convex. You will probably notice that the "cave" or the "bulge" in your glasses is not nearly as dramatic as it is in the classroom lenses.
2. Survival often depends on our ability to see things at a distance. This was probably more true in prehistoric times than it is today.

Test Results: For each of the vision tests you did, write down your test results and draw any conclusion that seems logical to you.

1. Visual acuity (right eye) ______________________________
   ____
   Visual acuity (left eye) ______________________________
   ____
2. Astigmatism (right eye) ______________________________
   ____
   Astigmatism (left eye) ______________________________
   ____
3. Near point determination (right eye) ______________________________
   ____
   Near point determination (left eye) ______________________________
   ____
4. Color blindness ______________________________
   ____
5. Blind spot determination (right eye) ______________________________
   ____
   Blind spot determination (left eye) ______________________________
   ____
6. Pupillary light reflex (right eye) ______________________________
   ____
   Pupillary light reflex (left eye) ______________________________
   ____
7. Pupillary distance reflex ______________________________
   ____
8. Peripheral vision (right eye) ______________________________
   ____
   Peripheral vision (left eye) ______________________________
   ____

9. Eye dominance ____
____

10. Cataract ____
____

11. Other observations of interest ______________________________
____

If you did the FFF test, explain your results here.

NOTES

❑ Lab Report

1. Referring to the numbered inquiries at the beginning of this exercise, complete the following box summary:

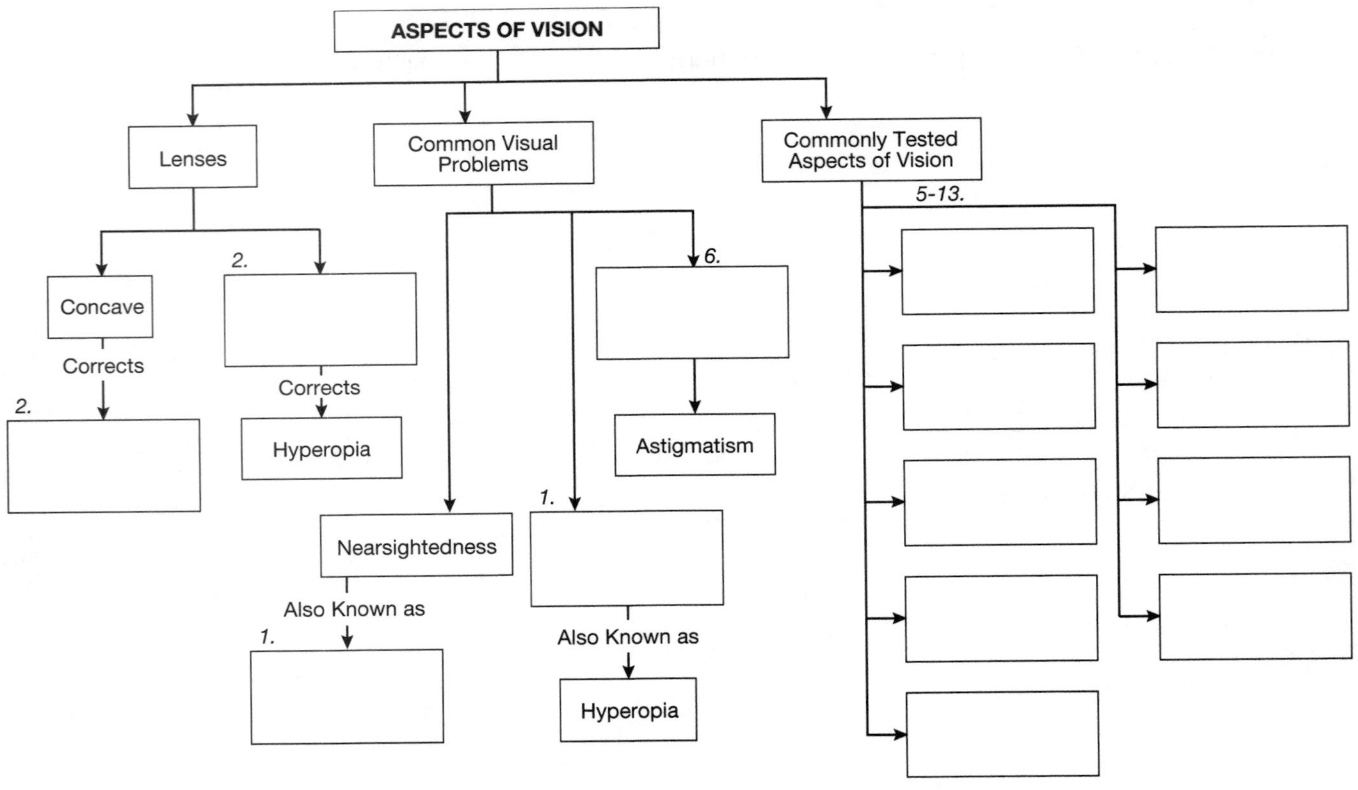

General Questions

NOTES

1. If the Snellen Eye Chart is used to screen potential vision problems in grade schools, why could Janey score 20/20 and still not be able to read because of her eyes? (Assume she does not have a muscle control problem.) What should be done about this?
2. What principle explains how concave and convex lenses correct myopic and hyperopic vision?
3. Where would the image be focused if a person with emmetropia wore glasses with concave lenses?
4. What can you conclude from handling the concave and convex lenses?
5. Identify the problem or the type of test used to determine the conditions listed.
   a. With his left eye George can only see about 45 degrees laterally.
   b. Ann has trouble adjusting to bright light.
   c. Fred must hold the book about 50 cm from his eyes in order to read.
   d. When she is tired, Linda is often afflicted with diplopia.
   e. Pat was told he had an uneven curvature of the cornea.
   f. Gina's lenses seem cloudy.

EXERCISE 38

Anatomy of the Ear

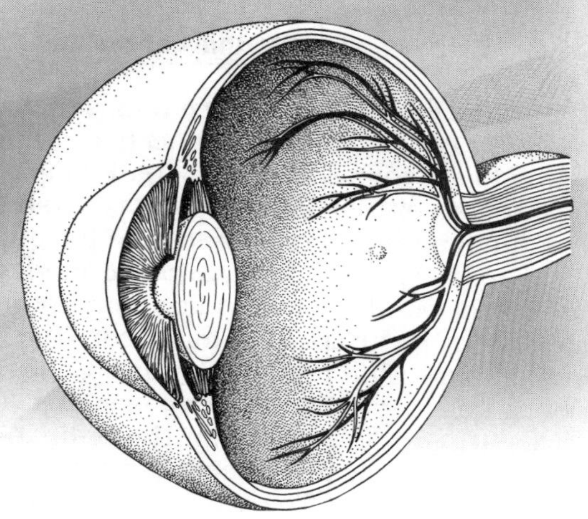

PROCEDURAL INQUIRIES

Examination

1. What are the major structures of the external ear?
2. What are the major structures of the middle ear?
3. What are the major structures of the inner ear?
4. What is the function of the external ear?
5. What is the function of the middle ear?
6. What are the functions of the inner ear?

Experimentation

7. How can straws be used to demonstrate the structures of the inner ear?

Additional Inquiries

8. What is the relationship between the mastoid region of the temporal bone and the middle ear?
9. What is the difference between endolymph and perilymph? Where is each found?
10. What are some of the accessory structures of the ear?

Key Terms

Ampulla
Auricle
Bony (Osseous) Labyrinth
Cochlea
Endolymph
External Auditory Meatus (and Canal)
External Ear
Incus
Inner Ear
Malleus
Membranous Labyrinth
Middle Ear
Organ of Corti
Oval Window
Perilymph
Pharyngotympanic Tube
Pinna
Round Window
Semicircular Canals
Stapes
Tympanic Membrane
Utricle
Vestibule

Materials Needed

Laboratory Diagrams and Charts of the Ear
Models (cutaway) of the Ear
Mirror
Straws — Narrow and Wide
(Thin plastic tubing (2 sizes) may be used)
Optional: Human Skull and Probe
Optional: Otoscope
Optional: Lab Animal, such as Cat

The ear, the functional unit of both hearing and equilibrium, is a complex organ readily divided into three major regions: the **external ear**, the **middle ear**, and the **inner ear**. The external ear channels sound waves from the environment in toward the nervous system receptors. The major external ear structures are the **pinna** (or **auricle**), the **external auditory meatus** (which leads into the **auditory**, or **ear**, **canal**), and the **tympanic membrane** (also called the **tympanum** or **ear drum**).

The accessory structures, such as the ceruminous and sebaceous glands and the ear canal hairs, are specialized parts of the integument, which covers the entire external ear apparatus. Elastic cartilage provides support and flexibility for the framework of the external ear, including the auditory canal. Fibrous connective tissue can be found in the thin, semitransparent tympanic membrane. The external ear functions in sound wave conduction and has no direct role in equilibrium.

The middle ear, or **tympanic cavity**, is primarily the conduit of sound transmission between the external ear and the inner ear. The principle structures within the middle ear are the three tiny bones known as the **auditory ossicles**. These ossicles are suspended within the tympanic cavity, which communicates directly with the pharynx by the **auditory tube**. The accessory structures of the middle ear include several suspensory muscles and ligaments. The middle ear functions only indirectly or to a minor extent in equilibrium.

The inner ear is usually divided into three parts, the **cochlea**, the **vestibule**, and the **semicircular canals**. The cochlea is responsible for the conversion of sound waves into identifiable electrical impulses. The vestibule and the semicircular canals function to maintain equilibrium. The inner ear is a continuum of fluid-filled membranous tubes housed within fluid-filled bony tubes.

Think about the sounds around you — sounds such as the traffic, or the wind in the trees, or the creak in your desk. The vibrations from these sounds were focused by and then traversed through the external ear. They were then conducted across the middle ear and become recognizable because of the cochlear portion of the inner ear. Meanwhile, your position in space — your equilibrium — was maintained because of the happenings in the vestibular and semicircular portions of your inner ear.

In this laboratory exercise we will examine the functional anatomy of the ear. The physiology of hearing and the physiology of equilibrium are covered in Exercises 39 and 40, respectively.

❑ Examination

I. External Ear *[∞ Martini, p. 557]*

A. Structural Orientation

All of the structures of the external ear aid in guiding the sound vibrations toward the ear drum.

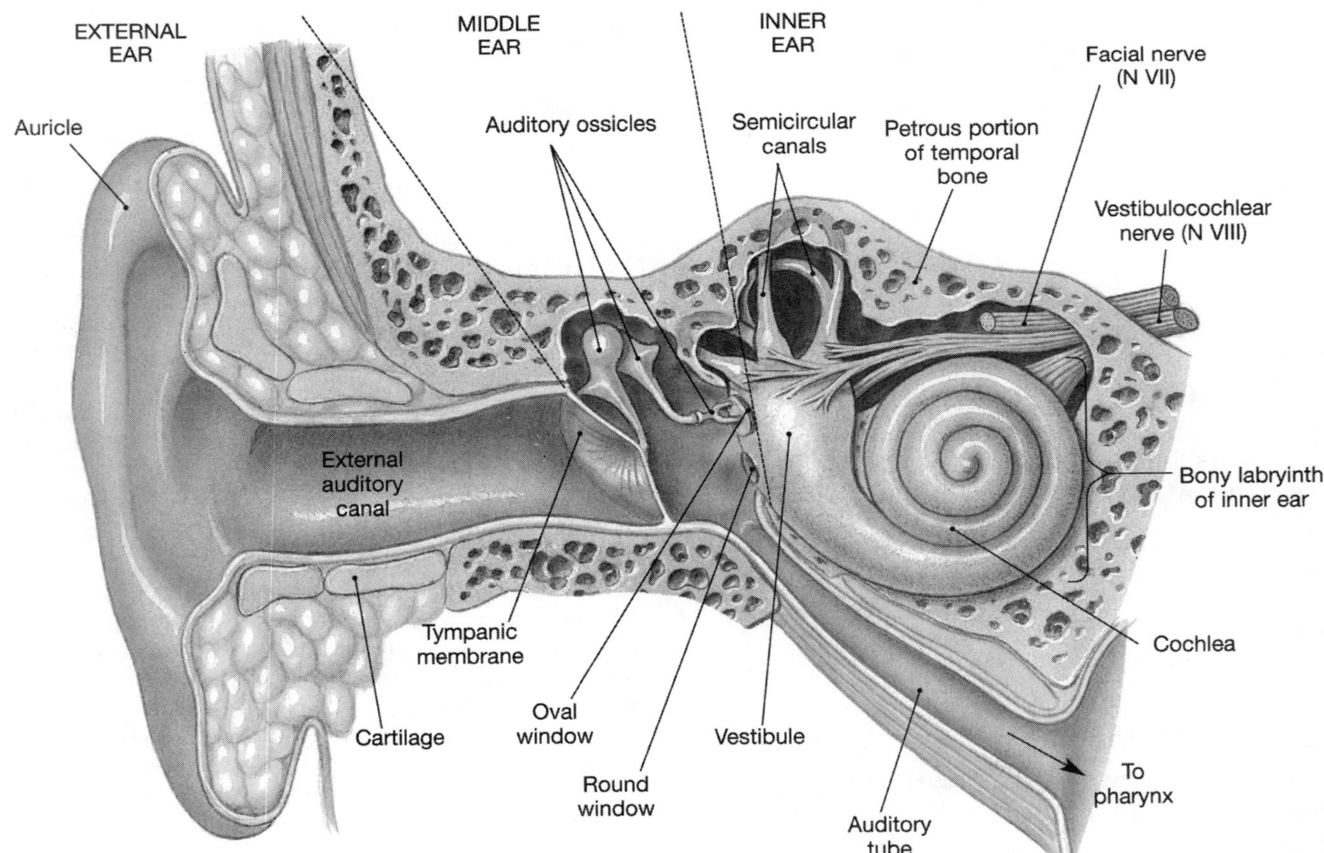

● **FIGURE 38–1**
Anatomy of the Ear. The orientation of the external, middle, and inner ear.

Use your own ear, Figures 38–1● and 38–2●, and the classroom models to work through the following descriptions of the external ear.

1➡ Begin by feeling the visible portion of your own outer ear, the pinna. Note the flexibility and texture and run your finger around the various ridges and depressions. Look in a mirror and see if you can find any hair on the auricle. You may find soft downy hair on the helix. Occasionally (rarely) men will have coarse hair on the pinna. This is a holandric condition (inherited on the Y chromosome). Record your findings.

____

____

2➡ Use your own ear or the available models and diagrams to locate the following auricular structures:

Helix Upper curved rim.

Lobule (Lobe) Inferior fleshy portion, not supported by cartilage.

Triangular Fossa Depression just inferior to the upper rim of the helix.

Antihelix Raised edge just inferior and posterior to the triangular fossa.

Concha Shell-like indentation inferior to the antihelix.

Tragus Anterior "flap" that can be "pushed in" to cover the ear canal.

Antitragus "Pseudoflap" opposite the tragus.

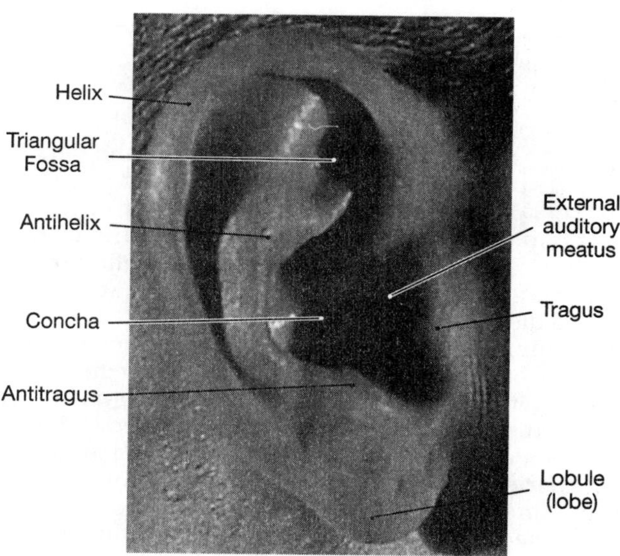

● **FIGURE 38–2**
External Ear.

Now examine Figure 38–2●. Were you able to identify all the listed structures? ______________

3➡ Identify the external auditory canal, which extends medially and at a slightly raised angle from the pinna for about 2.5 cm (1 inch) to the tympanic membrane. The canal is embedded within the temporal bone.

See if the model you are examining is detailed enough to observe such integumentary structures as the downy hair and the sebaceous (oil) glands that are located near the entrance to the canal. Deep within the canal you may find the ceruminous glands, which produce cerumen, the substance commonly known as earwax.

4➡ Locate the double-layered tympanic membrane, the structure marking the medial boundary of the external ear. The outer layer, the concave external surface of this epithelial structure, faces the canal, and the convex inner layer, the internal surface, faces the middle ear.

Study Figure 38–3● and identify the internal and the external layers of the tympanic membrane.

Examine the model of the ear. Note both the location of the tympanic membrane and its relation to the rest of the ear. How would you describe the tympanic membrane on the model you are studying?

____

____

II. Middle Ear [∞ *Martini, p. 558*]

A. Structural Orientation

The middle ear is a narrow cavity (often called the tympanic cavity) within the petrous portion of the temporal bone. The lateral boundary of the middle ear is the tympanic membrane, and the medial boundary is the inner ear.

Within the middle ear are three tiny bones, the auditory ossicles, which form an auditory lever system to transmit and amplify sound vibrations (as mechanical movements) from the tympanic membrane to the inner ear.

The middle ear has no direct function in equilibrium. Nevertheless, infections or other problems within the pharyngotympanic tube, or the middle ear cavity, may occasionally cause pressure or referred pain that in turn can affect equilibrium. (In referred pain, a pain sensation is perceived as arising from a source other than the actual source. See Exercise 43.)

1➡ Return to Figure 38–3● and use available models and diagrams to locate the three auditory ossicles. Note how the handle of

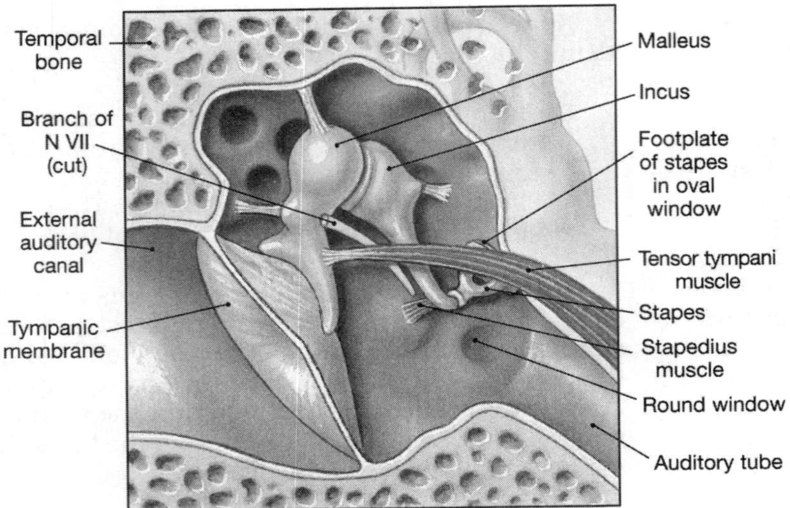

(a)

(b)

● **FIGURE 38–3**
The Middle Ear. (a) Detail of the structures associated with the middle ear. **(b)** Tympanic membrane and auditory ossicles. (Copyright Lennart Nilsson, BEHOLD MAN, Little Brown and Company.)

the **malleus** (hammer) is attached to the tympanic membrane, whereas the head of the malleus articulates with the **incus** (anvil), which then articulates with the **stapes** (stirrup). The stapes attaches to the oval window of the inner ear.

2➡ See if you can locate two ligaments extending between the malleus and the wall and one ligament between the stapes and the wall. The auditory ossicles are attached to the walls of the tympanic cavity by these tiny suspensory ligaments.

3➡ Identify the muscles of the middle ear, the **tensor tympani**, which originates on the petrous portion of the temporal bone and inserts on the handle of the malleus, and the **stapedius**, which originates on the posterior wall of the tympanic cavity and inserts on the stapes. These small skeletal muscles are responsible for the spatial orientation of the ossicles. If your ear models show these muscles, locate them and figure out the movement of the ossicles when these muscles contract.

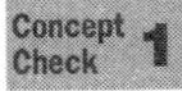
Concept Check 1 What will happen to sound conduction through the middle ear when these muscles contract?

____

____

____

4➡ Locate (if possible) the one or more small variable connections to the mastoid sinuses (the air sacs within the mastoid process of the temporal bone). On your models these may be indicated simply as small indentations on the posterior wall of the tympanic cavity.

A mastoid infection usually involves the air cells, those cells lining the air sacs of the mastoid process. If the air cells become inflamed, the passageways from the mastoid process to the middle ear could become blocked. Drainage would not occur and an infection could fester. Otitis mastoidea (mastoiditis) is an inflammation involving the mastoid spaces.

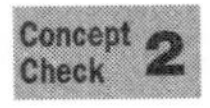
Concept Check 2 Why is a mastoid infection particularly serious?

____

____

Otitis is an inflammation of the ear. This inflammation can occur at any point along the auditory or equilibrium pathway. Otitis is usually designated as **externa**, **media**, or **interna**, according to the location of the inflammation. The inflammation can be the result of an irritation or a disease-causing agent. Otitis media is the term commonly used for the oft-occurring bacterial infections of the middle ear. Otitis externa can be especially interesting if it is caused by a fungus or a protistan. Aerotitis, an inflammation caused by blocked tubes and barometric pressure changes, is often experienced by pilots and deep-sea divers. Other common types of otitis include otitis parasitica (caused by a parasite), otitis labyrinthica (inflammation of the labyrinth of the inner ear), and otitis sclerotica (an inflammation accompanied by a hardening of the ear structures).

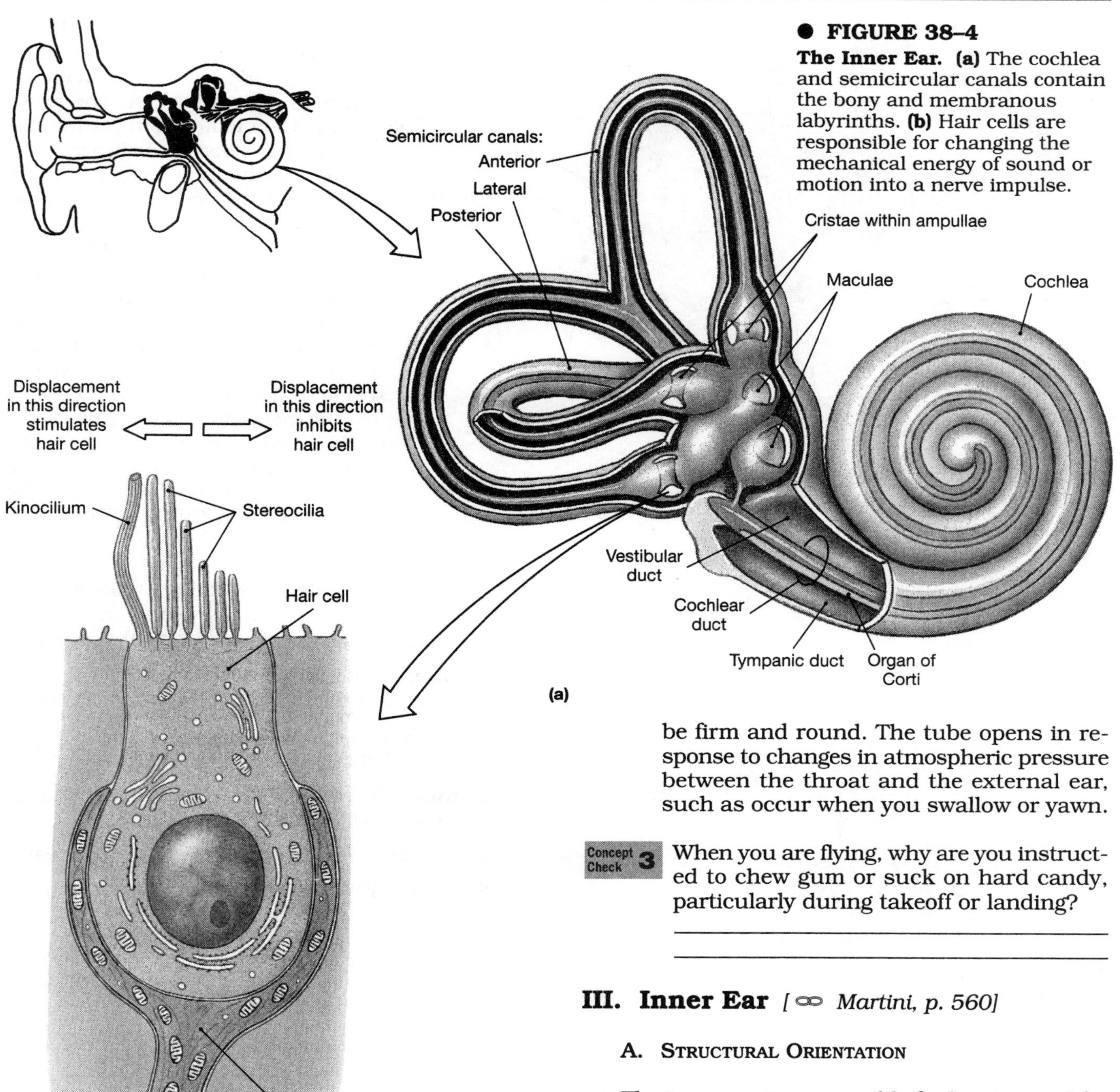

● **FIGURE 38–4**

The Inner Ear. (a) The cochlea and semicircular canals contain the bony and membranous labyrinths. **(b)** Hair cells are responsible for changing the mechanical energy of sound or motion into a nerve impulse.

5➡ Find the auditory tube (also called the pharyngotympanic tube or the Eustachian tube), which connects the middle ear with the pharynx. The mucosal lining of this tube is continuous with both the throat and the middle ear. Locate this tube on your models or diagrams. Although it does have cartilage supports, in the living person this tube is usually limp and closed — collapsed — but on your models it will be firm and round. The tube opens in response to changes in atmospheric pressure between the throat and the external ear, such as occur when you swallow or yawn.

Concept Check 3 When you are flying, why are you instructed to chew gum or suck on hard candy, particularly during takeoff or landing?

____

____

III. Inner Ear *[∞ Martini, p. 560]*

A. Structural Orientation

The inner ear is responsible for hearing and for some aspects of equilibrium. Although we will examine hearing and equilibrium in detail in the next two exercises, keep these two functions in mind as you work through this part of this exercise.

The inner ear, often called the internal ear or the labyrinth, is a collection of fluid-filled membranous tubes (the **membranous labyrinth**) contained within a collection of fluid-filled bony tubes (the **bony**, or **osseous**, **labyrinth**). The membranous labyrinth is filled with **endolymph**, a fluid chemically similar to the intracellular fluid. The bony labyrinth is filled with **perilymph**, a fluid similar to cerebrospinal fluid. See Figure 38–4●.

The cochlea, the anterior portion of the inner ear, is responsible for the conversion of the me-

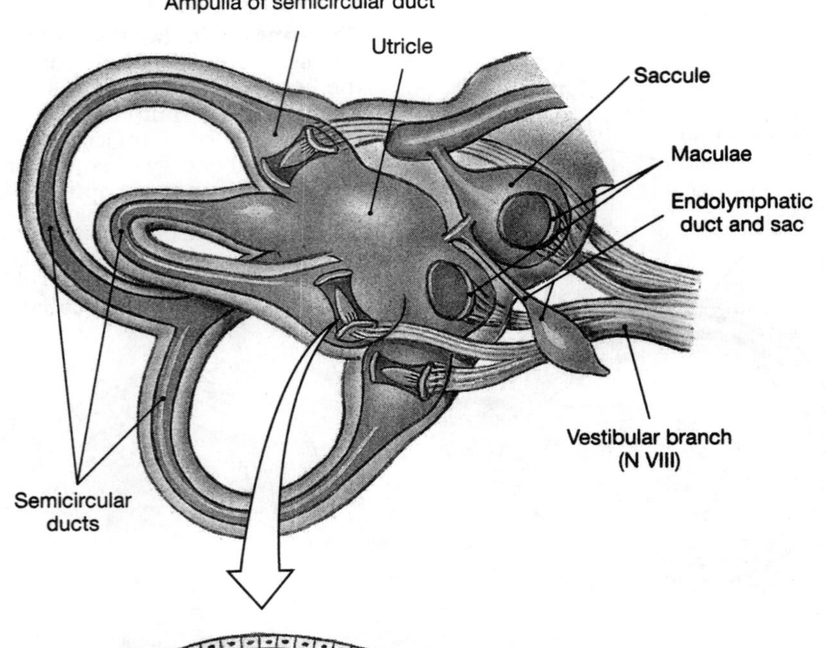

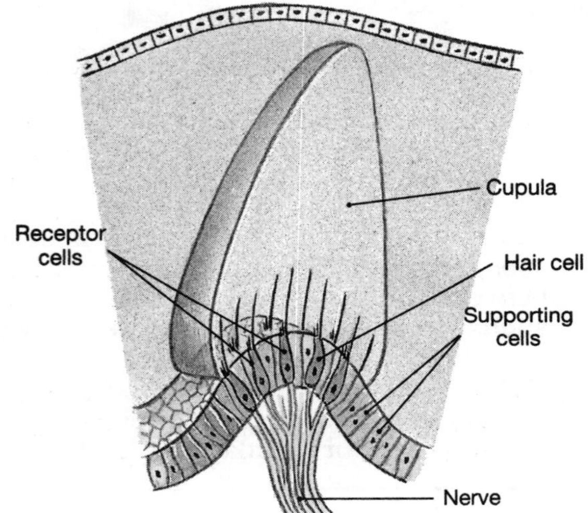

● **FIGURE 38–5**
The Vestibular Complex.

chanical sound waves (which arrived from the vibrating ossicles) into electrical impulses that leave the inner ear via the cochlear branch of the vestibulocochlear nerve (cranial nerve VIII).

The vestibule, which is the central portion of the inner ear, and the semicircular canals, which form the posterior portion of the inner ear, are responsible for static and dynamic equilibrium, respectively. Changes in equilibrium are transmitted to the brain via the vestibular branch of the vestibulocochlear nerve.

❑ Experimentation

1➡ Demonstrate the labyrinthic maze by taking a narrow straw and putting it inside a wide straw. If possible, fill the entire system with water and pinch off the ends. The larger straw is the bony labyrinth, and the water between the straws is the perilymph. The inner straw is the membranous labyrinth, and the water inside the inner straw is the endolymph. Note how this arrangement relates to Figure 38–4●.

You can use your straws to help you picture all the parts of the inner ear because all parts of the inner ear have the same tube within a tube structure. Jiggle your straws and notice how the membranous labyrinth seems to float in the perilymph.

The membranous labyrinth within the semicircular canals is often called the semicircular duct, and the membranous labyrinth within the cochlea is often called the cochlear duct.

2➡ Use the models and Figures 38–4● and 38–5● to locate the following structures. Keep the models in anatomical position.

Cochlea Snail-like bony structure. The bony core is sometimes called the **modiolus**.

Oval Window Membranous point of attachment between the stapes and the inner ear.

Round Window Membranous area inferior to the oval window.

Vestibule Central portion of the inner ear.

Saccule Enlarged area of the vestibule nearest to the cochlea.

Utricle Enlarged area of the vestibule at the base of the semicircular ampullae.

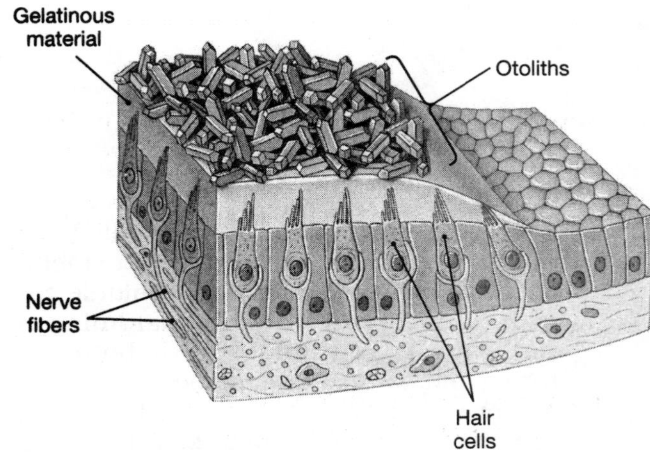

● **FIGURE 38–6**
Macular Structures.

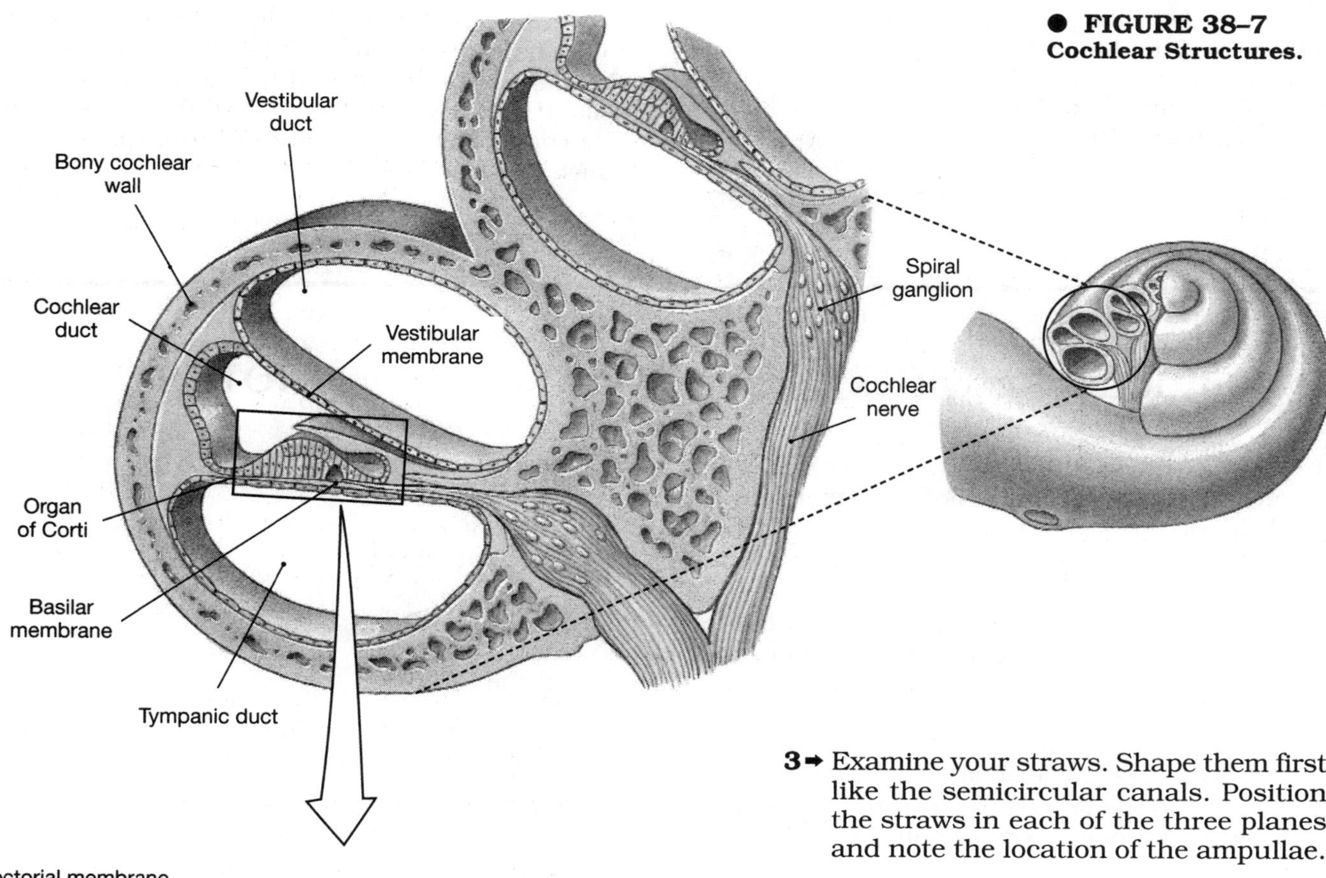

● **FIGURE 38–7**
Cochlear Structures.

Semicircular Canals Three planar bony protrusions forming the posterior of the inner ear.

Ampulla (pl., ampullae) Enlargement at the base of each semicircular canal containing the sensory receptors.

Anterior (sometimes called Superior or Frontal) **Canal** "Upper" Canal.

Posterior (sometimes called Inferior or Sagittal) **Canal** "Back" Canal.

Lateral (sometimes called Horizontal) **Canal** "Sideways" Canal.

3➡ Examine your straws. Shape them first like the semicircular canals. Position the straws in each of the three planes and note the location of the ampullae.

Now twist your straws into the shape of the cochlea. Note the location of the round and oval windows.

4➡ Use the classroom models and diagrams to locate the internal structures of the vestibule and cochlea.

Ampullar Structures — On the floor of the ampulla is the region of hair cells known as the crista (or crista ampullaris). This region, which consists of ciliated hair cells (hair tufts often called cilia and stereocilia) and supporting cells is topped by a gelatinous cap called a **cupula**. Gain a three-dimensional perspective of these structures, and then identify them in Figure 38–5●.

Vestibular Structures — The hair cells in both the saccule and utricle are clustered in **maculae**. Each macula consists of ciliated hair cells embedded in a gelatinous material and topped by densely packed mineral crystals known as **otoconia**, or **otoliths**. Identify these structures in Figure 38–6●. The vestibular structures are important in equilibrium, which is covered in Exercise 40

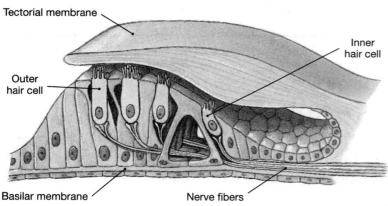

Cochlear Structures — Find the **tectorial membrane** and the **organ of Corti**. Note how these structures lie within the inner straw, the membranous labyrinth, which is now called the **cochlear duct** (or scala media). Note the area around the inner straw, the area filled with perilymph. The superior space (extending from the oval window) is known as the **vestibular duct**, or scala vestibuli, and the inferior space (extending from the round window) is known as the **tympanic duct**, or scala tympani. Twist your straws around again to gain a three-dimensional perspective on the parts of the cochlea. When you have this perspective, use Figure 38–7● to identify the tectorial membrane and the parts of the organ of Corti, such as the ciliated hair cells, the support cells, the basilar membrane, and the nerve fibers. The cochlear structures are important in hearing, which is covered in Exercise 39.

❑ Additional Activities

NOTES

1. Check a physics book and demonstrate why the design of the auricle maximizes the perception of incoming sound.
2. Examine the external auditory meatus and the auditory canal of a skull (either articulated or disarticulated) with your probe. Figure out where the tympanic membrane should be.
3. If an otoscope is available, follow your instructor's directions and examine a classmate's ear. As you view the tympanic membrane, you should be able to see the **umbo**, the concave depression that occurs at the tip of the handle of the malleus. (The malleus is attached directly to the internal surface of the tympanic membrane.) In the canal you may possibly be able to see some cerumen, which is usually dark amber in color.
4. Use any of the procedures mentioned thus far in this exercise to examine the ears of a laboratory animal.

Answers to Selected Concept Check Questions

1. Bones are shifted away from the membranes and do not vibrate as easily. This will protect the ear because loud noises will not be magnified to the same extent as they would with full contact with the membranes.
2. Infectious material has nowhere to go. The infection can spread throughout the bone.
3. Swallowing forces the pharyngotympanic tube to open, thus allowing for the equalization of pressure between the pharynx (or middle ear) and the outside environment. When flying, fluctuations in cabin pressure are most noticeable on takeoff and landing.

❑ Lab Report

1. Referring to the numbered inquiries at the beginning of this exercise, complete the following box summary:

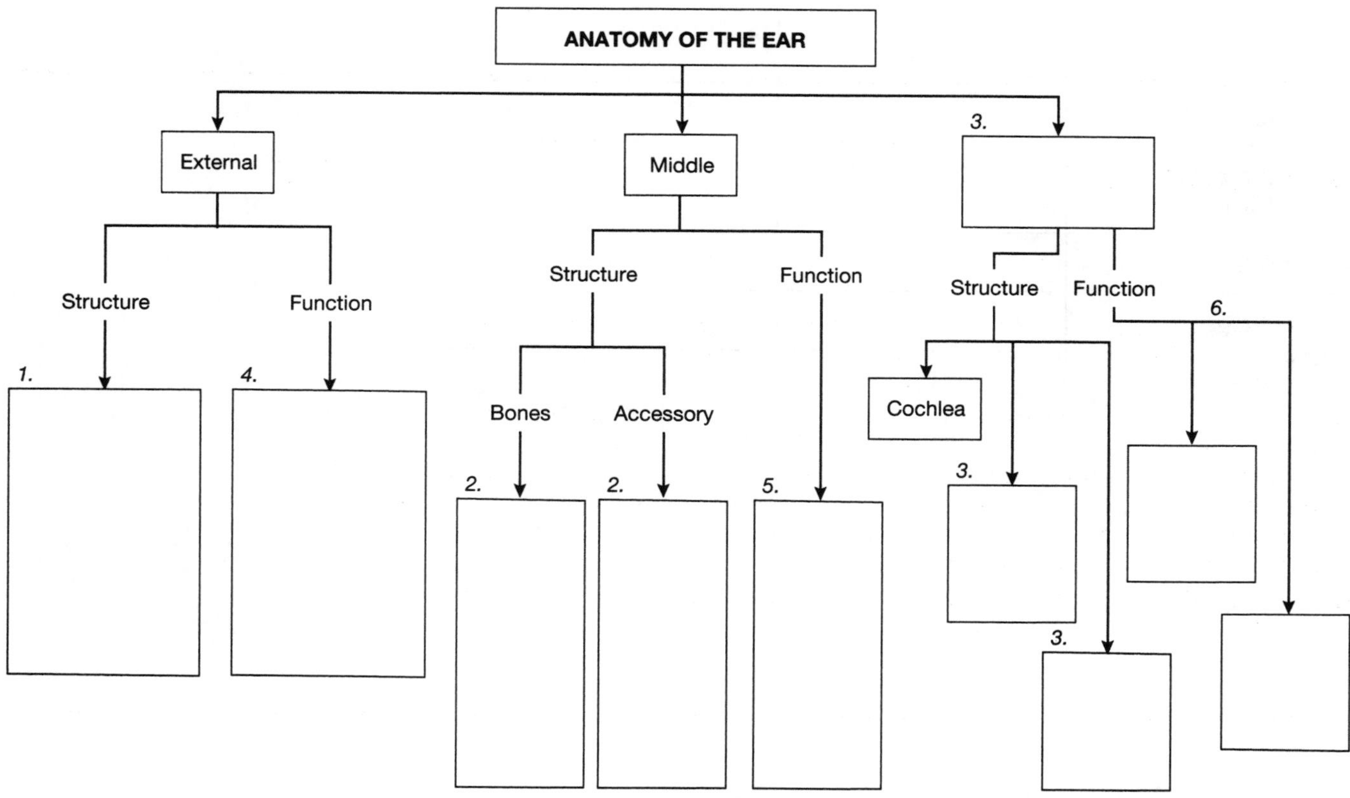

General Questions

NOTES

1. Explain how losing the pinna of the ear would affect hearing.
2. Name and locate the two types of glands that are accessory structures for the external ear.
3. What effect would a fusion of the ossicles have on hearing? Why?
4. What is the function of the ligaments that attach the ossicles to the walls of the tympanic cavity?
5. Where is the mastoid region and what is mastoiditis?
6. What is the role of the pharyngotympanic tube in equalizing air pressure on both sides of the tympanic membrane?
7. Endolymph is located in the ______________________ and is similar in composition to the ______________________. Perilymph is located in the ______________________ and is similar in composition to the ______________________.

NOTES

8. Fill in this chart:

Structure	**Location**	**Function**
Cochlea		
Ampulla		
Scala Vestibuli		
Scala Media		
Semicircular Canal		
Utricle		
Otoliths		

EXERCISE 39

Physiological Aspects of Hearing

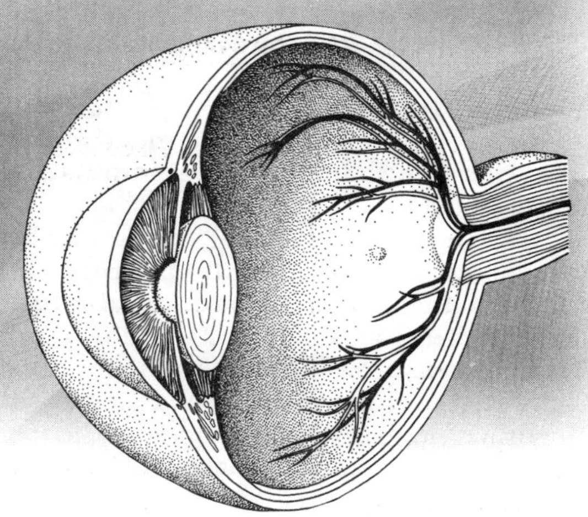

PROCEDURAL INQUIRIES

Examination

1. Which structures are responsible for channeling and amplifying sound from the exterior to the inner ear?
2. Which ear structures are responsible for hearing (changing mechanical sound waves into electrical nerve impulses)?
3. What is the micro-anatomy of the organ of Corti?
4. What properties of the basilar membrane enable it to differentiate between different sounds?
5. What are the fluids of the inner ear?

Experimentation

6. What is auditory acuity? How do we test for auditory acuity?
7. What is sound localization? How do we test for sound localization?
8. What is the purpose of the Rinne Test? How do we perform the Rinne Test?
9. What is the purpose of the Weber Test? How do we perform the Weber Test?

Additional Inquiries

10. What are some of the properties of sound?
11. What are the two major types of deafness? What are some of the causes of each?

Key Terms

Auditory Accommodation
Auditory Acuity
Basilar Membrane
Cochlea
Conduction Deafness
Endolymph
Hearing
Nerve Deafness
Organ of Corti
Ossicles
Oval Window
Perilymph
Round Window
Scala Media
Scala Tympani
Scala Vestibuli
Sound Localization
Sound Waves
Tectorial Membrane
Vestibulocochlear Nerve

Materials Needed

Ticking Watch
Tuning Forks — Several Frequencies
Rubber mallets for striking tuning forks (if available)
Cotton
Ruler (metric)
Model of Cochlea (if available)
Optional: Audiometer

Hearing is the process of perceiving sound. Sound waves are received by the auditory apparatus and, after a series of mechanical and electrical events, the particular auditory impressions are registered in the brain.

The auditory apparatus is the ear. Sound waves travel through the auditory canal of the external ear and strike the **tympanic membrane**. At the tympanic membrane the sound waves are amplified as the ossicles (*malleus, incus,* and *stapes*) are vibrated. Thus, mechanical energy of the sound is transferred across the middle ear from the tympanic membrane to the **oval window**, the connecting point between the stapes and the cochlea. The cochlea is that portion of the inner ear responsible for converting the mechanical movements into the electrical impulses that travel to the brain.

The bending of hair cells by fluid vibrations in the cochlea is central to the **hearing** process. We can trace the general path of **sound waves** through the inner ear as follows: As the oval window vibrates, pressure is applied to the fluids within the cochlea. Various pressure changes pass through the fluids and cause distortions of the membranes within the ducts of the cochlea. These distortions, which are directly related to the pitch and intensity of the original sounds, cause the sensory hair cells within the **organ of Corti** to bend, which triggers an electrical signal. The electrical impulses are then sent to the brain via the cochlear branch of the **vestibulocochlear nerve** (N VIII). Thus, the mechanical energy of sound waves — amplified by the **ossicles** in the middle ear and converted to electrical energy by the organ of Corti in the inner ear — is finally received and interpreted by the brain as sound.

In this laboratory exercise we will explore the principles of hearing physiology by observing and testing different aspects of hearing. (If it has been a while since you studied the functional anatomy of the ear as a whole, review Exercise 38 before continuing with this exercise.)

Overview of the Properties of Sound

Before examining the principles of hearing physiology, it is important that we recall some of the basic vocabulary of sound and sound waves. *[∞ Martini, p. 566]*

Pitch Subjective perception of different sound frequencies

Frequency Cycles per second (cps), measured in Hertz (Hz). The frequency range in which most people hear sound is between 20 Hz and 20,000 Hz with the most sensitive range being between 1,000 Hz and 4,000 Hz. The normal speech frequency range is 125 Hz to 8,000 Hz.

Loudness Subjective evaluation of the intensity of sound waves.

Intensity Objective measurement of nerve fiber stimulation, measured in decibels (dB). Sounds with greater intensity — i.e., where the sound waves have greater *amplitude* — stimulate more nerve fibers, so subjectively the sounds are louder. You should keep in mind that intensity as measured in dB, is logarithmic, while our subjective analysis of intensity (loudness) is arithmetic. This means a sound registering 20 dB is *10 times more intense* than a sound registering 10 dB, but it *sounds only twice as loud.*

❑ Examination

I. Hearing

A. Cochlear Structures: Ducts, Fluids, Membranes, and Hair Cells

1➜ Refer to Figure 39–1●, the cochlea. Notice the relative locations of the vestibular duct (**scala vestibuli**), the cochlear duct (**scala media**), and the tympanic duct (**scala tympani**). The organ of Corti rests on the **basilar membrane** within the scala media. The organ of Corti consists of several rows of delicate **hair cells**, plus the **tectorial membrane**, which forms a delicate canopy over the hair cells.

2➜ Study the hair cells in Figure 39–1●. Now refer back to the hair cell in Figure 38–4● (in the previous chapter). Note that each cell has a single **kinocilium** and a multitude of stereocilia. The stereocilia are arranged in densely packed rows and are interconnected by tip links and cross-fibers (lateral links). These are not shown in the figure.

During development the hair cells lose their kinocilia; however, the basal body (from which the kinocilium arises) remains.

3➜ If you have a classroom model showing the inner parts of the cochlea, locate the structures in the preceding paragraph on the model. Note that the vestibular duct is essentially continuous with the tympanic duct.

4➜ Note the different fluids within the cochlea and their names: the **perilymph**, which fills the vestibular and tympanic ducts and which surrounds the cochlear duct; and the **endolymph**, which fills the cochlear duct and surrounds the organ of Corti.

B. Function of the Organ of Corti: Detecting a Sound's Loudness and Pitch

1➜ Look at Figure 39–2●, the uncoiled cochlea. Trace the path of a vibration from the oval window to the **round window**. Movement of the perilymph in the vestibular duct exerts pressure on the endolymph. This sets up a traveling wave in the basilar membrane (not unlike the wave produced when you "flick" a rope from one end), which peaks at some point and causes the stereocilia of the hair cells to bend against the overlying tectorial membrane.

The stereocilia bend only at the base. If the stereocilia movement is toward the kinocilium (or its basal body), the ion channels

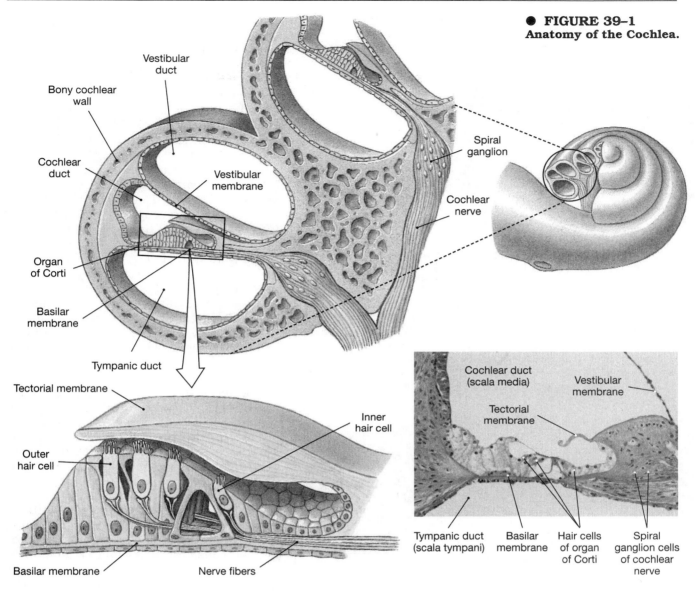

● FIGURE 39–1
Anatomy of the Cochlea.

open and the current through the cell membrane increases. If the stereocilia movement is away from the kinocilium, the current decreases. This bending in turn stimulates the nerve fibers of the cochlear branch of the vestibulocochlear nerve (N VIII), and the message is on its way to the brain. The stronger the sound wave, the higher the peak of the traveling wave and the more rows of hair cells bent at that point. The more hair cells stimulated, the louder the sound is perceived to be.

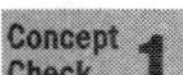

What would be the consequence of a sound wave so strong that it *breaks* the stereocilia of the hair cells?

____

____

2➡ Refer now to Figure 39–3●, the basilar membrane. The basilar membrane, which separates the tympanic duct and the cochlear duct, is only about 0.04 mm wide at the oval window. This membrane gradually widens to about 0.5 mm at the apex. This gradation in width seems to correlate with the relative sensitivity of the membrane to different sound waves. High frequency sound waves peak near the oval window. The lower the frequency of the sound, the farther down the basilar membrane the sound deflection seems to occur (Figure 39–3b●).

Thus, hair cells on the organ of Corti provide information on the frequency and intensity of a sound by recording *where* and *how much* distortion in the basilar membrane occurs.

II. Deafness

Problems with hearing loss can occur at any point along the auditory pathway. Generally, hearing

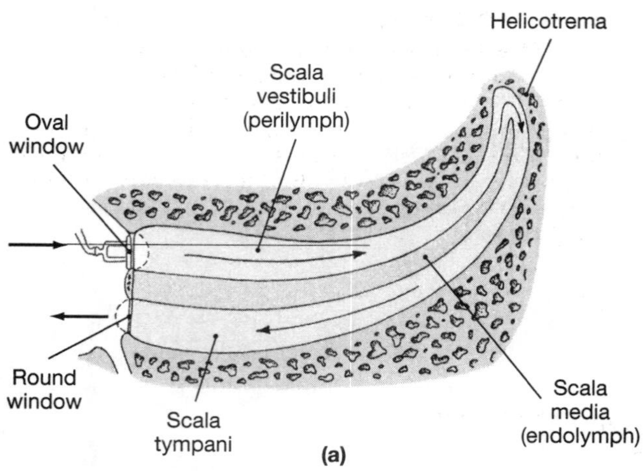

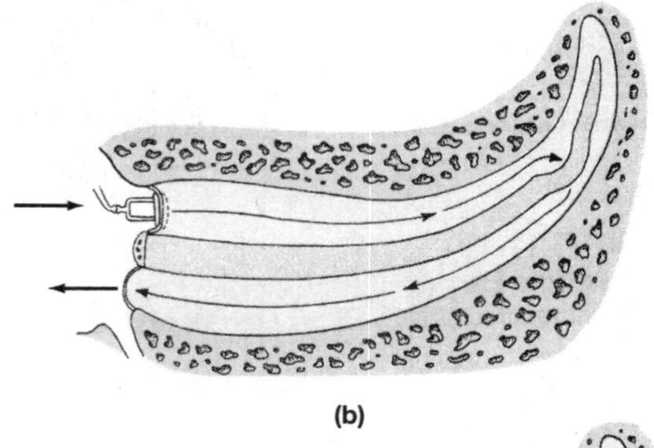

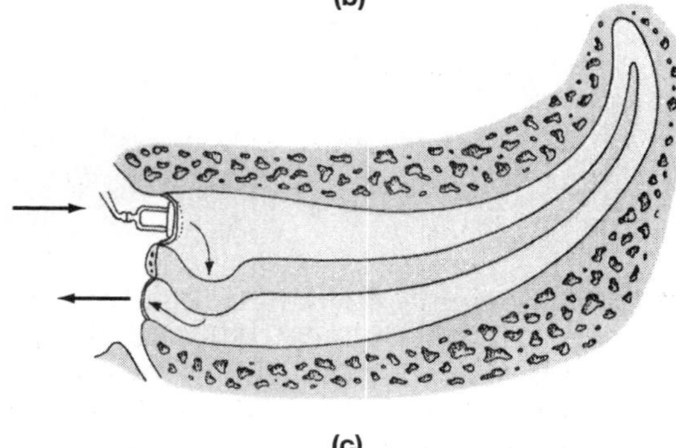

● **FIGURE 39–2**
Uncoiled Cochlea. **(a)** Pressure pathway; **(b)** Low frequency sound; **(c)** High frequency sound.

difficulties are classified as either *conduction deafness* or *nerve deafness*.

A. Conduction Deafness

1➡ Recall the physical structure of the ear (as found in Figure 38–3● in the previous exercise). In **conduction deafness**, the sound cannot be conducted (transmitted and/or amplified) properly as it traverses the auditory system. Conduction deafness usually involves the bone, though adverse conditions in the auditory canal (such as *otitis externa*, an inflammation of the auditory canal) or difficulties with the tympanic membrane can also be called conduction deafness.

2➡ Suppose the joint between the stapes and the incus became nonfunctional because the bones have fused. The sound waves (as mechanical vibrations) would not be amplified, the membranous oval window into the cochlea would not vibrate sufficiently, and sound perception would be inadequate. This is conduction deafness. This is also an example of *otosclerosis*, a condition often associated with hearing loss in older individuals. In addition to involvement of the ossicles, conduction deafness could also occur if damage were sustained by the bony cochlear labyrinth or, occasionally, by the petrous or mastoid portions of the temporal bone.

B. Nerve Deafness

In **nerve deafness**, damage occurs at some place along the nerve pathway from the internal cochlea on up to the acoustic areas of the brain located in the auditory cortex of the temporal lobe. Usually, nerve deafness occurs because of damage to the hair cells of the organ of Corti.

1➡ Think about the tectorial membrane (Figure 39–1●). If the tectorial membrane or some other portion of the inner cochlea is damaged, the hair cells are likewise not stimulated, and this too is nerve deafness. The afferent nerve fibers leading from the cochlea can also be affected.

2➡ Consider the problems of noise pollution. Some exposure to excessive noise can cause a temporary deadening or incapacitation of the hair cells. This will lead to a temporary insensitivity to certain sounds. You may have experienced this if you've ever come out of a loud concert and couldn't hear your friends talking. This type of auditory accommodation is a physical adaptation to sound waves.

Frequent exposure to loud music (of any type) can also cause permanent auditory nerve damage. In this case, the hair cells cannot recover. In fact, frequent exposure to any loud noise — jet engines, jackhammers, or gunshots, for example — can cause the same type of nerve deafness. *[∞ Martini, p. 571]*

[∞ Martini and AM for additional hearing-related pathologies.]

❑ Experimentation

III. Hearing Tests

Working with a partner, you can perform several simple experiments to test different aspects of your

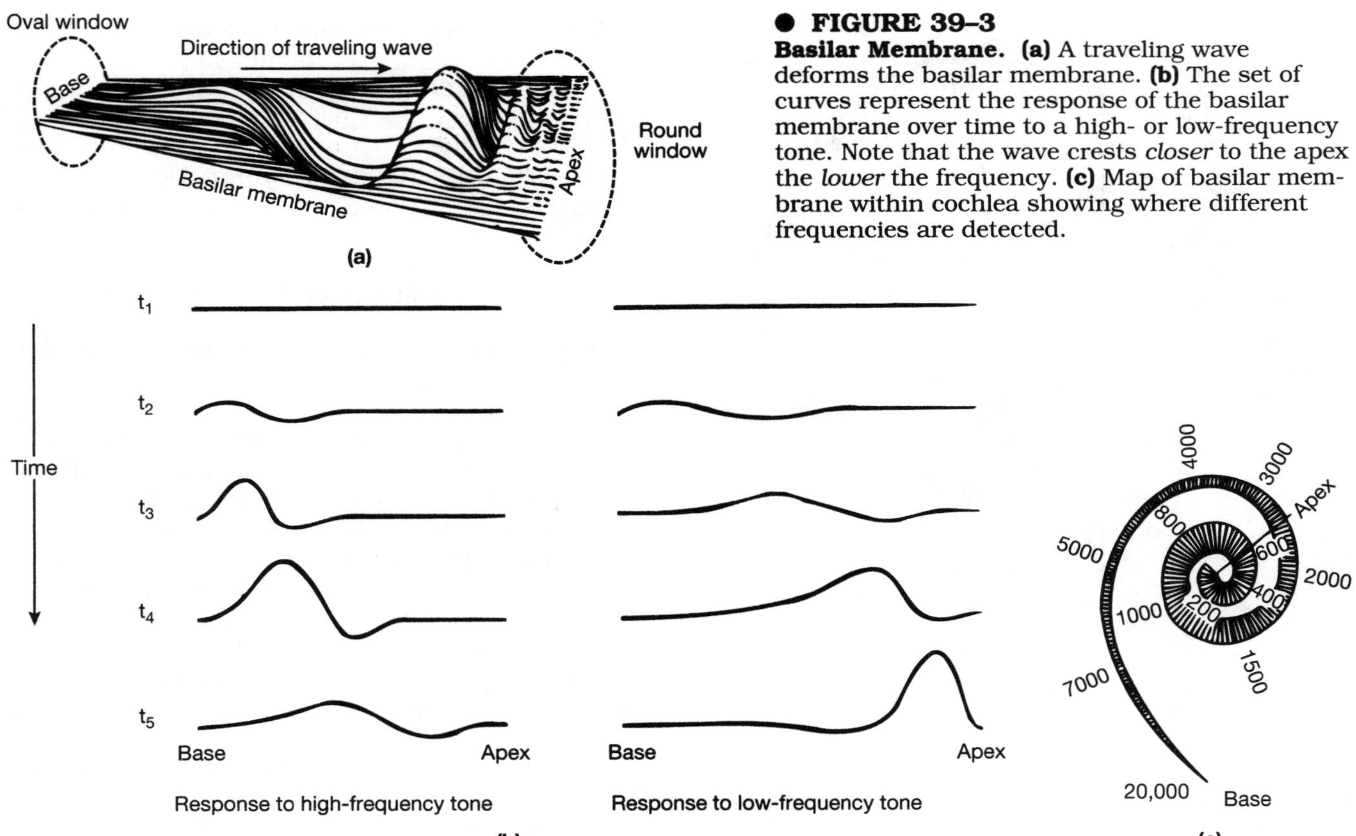

● **FIGURE 39–3**
Basilar Membrane. (a) A traveling wave deforms the basilar membrane. **(b)** The set of curves represent the response of the basilar membrane over time to a high- or low-frequency tone. Note that the wave crests *closer* to the apex the *lower* the frequency. **(c)** Map of basilar membrane within cochlea showing where different frequencies are detected.

From: *Perception* by R. Sekuler and R. Blake (©1994 McGraw-Hill, Inc.) Reproduced by permission.

hearing. Keep in mind that these tests are not conclusive. Your test results may not be particularly accurate because of both the test setting and your classroom instrumentation. If, after performing these hearing tests, you suspect you may have a hearing problem, consult an appropriate professional.

A. Preparation

1➜ Before you start the experiments, familiarize yourself with the equipment. Experiment informally with the tuning forks for a few minutes. Observe the numbers on the tuning forks. These forks produce apure sound with a single frequency. The numbers indicate the wave cycles per second given off when that particular fork is vibrated. (Note: Instead of cycles per second (cps), these numbers may be given as vibrations per second (vps).) The fork may also have a letter on the stem. This letter represents the note on the musical scale corresponding to the frequency of the fork.

2➜ Using either the base of your hand or a rubber mallet, try striking the forks separately and at the same time. (Figure out why you could damage the tuning forks by striking them on the side of the lab table.) Try touching the base of a vibrating tuning fork to your wrist, your forehead, and your work table.

3➜ Hold the watch near one ear and concentrate on the ticking while ignoring the sounds that may be coming into the other ear. You are probably quite familiar with the sound of a ticking watch, but it may take some practice to tune out the other sounds around you. Repeat this with the other ear.

4➜ When you are comfortable with the equipment, rest your ears for a few minutes while you and your partner get organized in a *quiet* place ready to begin the tests. You need to rest your hearing both now and between tests because you will have a tendency to develop **auditory accommodation**. This type of auditory accommodation, which is a mental (rather than physical) adaptation to certain sound waves, occurs when you are so accustomed to particular sounds that you no longer hear them. For instance, you are usually oblivious of the sounds emanating from your own refrigerator or furnace or the sound of your own heartbeat, though you can hear these sounds if you concentrate on them.

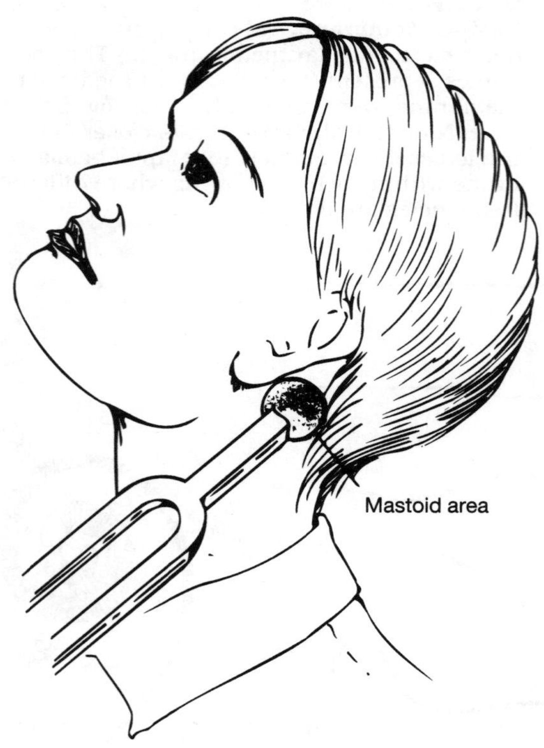

● **FIGURE 39–4**
Rinne Test.

B. Auditory Acuity

Auditory acuity is the threshold at which sounds are audible. To test auditory acuity we will perform two separate "ticking watch" tests. You and your partner will need a watch, some cotton, and a ruler. Plug one ear with cotton. (You may also need to cover that ear with your hand to help you concentrate on the sound coming into the unplugged ear.) Sit quietly with your eyes closed. Be prepared to signal with your hand rather than your voice.

Concept Check 2 Why should you signal with your hand instead of your voice?

1➡ For the first test your partner should hold the ticking watch a meter or so away from your unplugged ear and, at the rate of about 1 cm per second, should move the watch toward that ear. Give the hand signal the instant you hear the first tick. Your partner should then measure the distance from your ear. __________ cm

2➡ For the second test, start with the watch next to your ear. Signal the instant you hear the last tick. Measure the distance. __________ cm

3➡ Repeat the entire process (steps 1 and 2) with the other ear.

Watch moving toward ear.
__________ cm
Watch moving away from ear.
__________ cm

What differences did you notice?

____

____

C. Sound Localization

Sound localization tests determine your ability to locate the source of a sound.

1➡ Sit quietly with both ears unplugged and your eyes closed while your partner moves the ticking watch to various locations around your head (back, sides, front, above, below, etc.). Point to the locations. Use general descriptive terms to describe your accuracy.

Location #1 __________
Location #2 __________
Location #3 __________
Location #4 __________

Use descriptive terms to explain your results.

____

____

2➡ Repeat this experiment with one ear plugged. Now try it with the other ear plugged. What differences do you notice between the right and left ears? Compare your results with the results you obtained on the previous test.

____

____

D. Rinne Test

The **Rinne test** enables you to differentiate between nerve deafness and conduction deafness.

1➡ Plug one ear again. Your partner should vibrate the tuning fork and hold it about 10 cm from your unplugged ear. The instant you can no longer hear the vibrations, signal your partner to place the base of the fork against your mastoid process (Figure 39–4●). Conduction deafness may be present if you are able to hear the sound again.

What happened?

____

____

2➡ Repeat the procedure with the other ear.

What happened?

____

____

Concept Check 3 Why would the Rinne test be a logical test for conduction rather than nerve deafness?

____

____

3➡ Try the procedure with different tuning forks. What happened?

____

____

Concept Check 4 What might account for perceived sound differences between the tuning forks?

____

____

(Again, don't panic if your results are unexpected. Your laboratory is not an ideal test center.)

E. Weber Test

The Weber test can be used to determine relative conductive and neural deafness.

1➡ Sit quietly. Your partner should vibrate the tuning fork and place the base of the handle in the middle of your forehead. If the sound is equally loud in both ears, your hearing in both ears is either equally good or equally bad. What were your results?

____

____

If you have nerve deafness, you will not hear the sound in your affected ear. If you have conduction deafness, the sound will be louder in the weaker ear. Why would that be so?

____

____

2➡ Try the Weber test with different tuning forks. Why might your results be different?

____

____

❑ Additional Activities

1. Research various compensatory mechanisms that people with different types of hearing losses use in order to function in a hearing world.
2. Research different types of hearing aids. Discuss the advantages and disadvantages of each type for the various types of hearing loss.
3. An audiometer is an instrument designed to measure hearing acuity at different frequencies and at different intensities. Each ear can be tested independently and the results can be graphed. If your laboratory has either a manual audiometer or a program for computer-generated audiometry, follow the directions given by your instructor to check your hearing acuity.

Answers to Selected Concept Check Questions

1. Deafness could result.
2. You don't want the sound of your own voice influencing your perception of the sound of the ticking watch.
3. By going directly to the mastoid you are bypassing the external and middle ears — the places where conduction deafness occurs.
4. Different results could be noted for any number of reasons including: degree of conduction deafness, location of problem area, nerve deafness (with or without associated conduction deafness) concentrated at certain frequencies, noisy classroom or other distractions.

❑ Lab Report

1. Referring to the numbered inquiries at the beginning of this exercise, complete the following box summary:

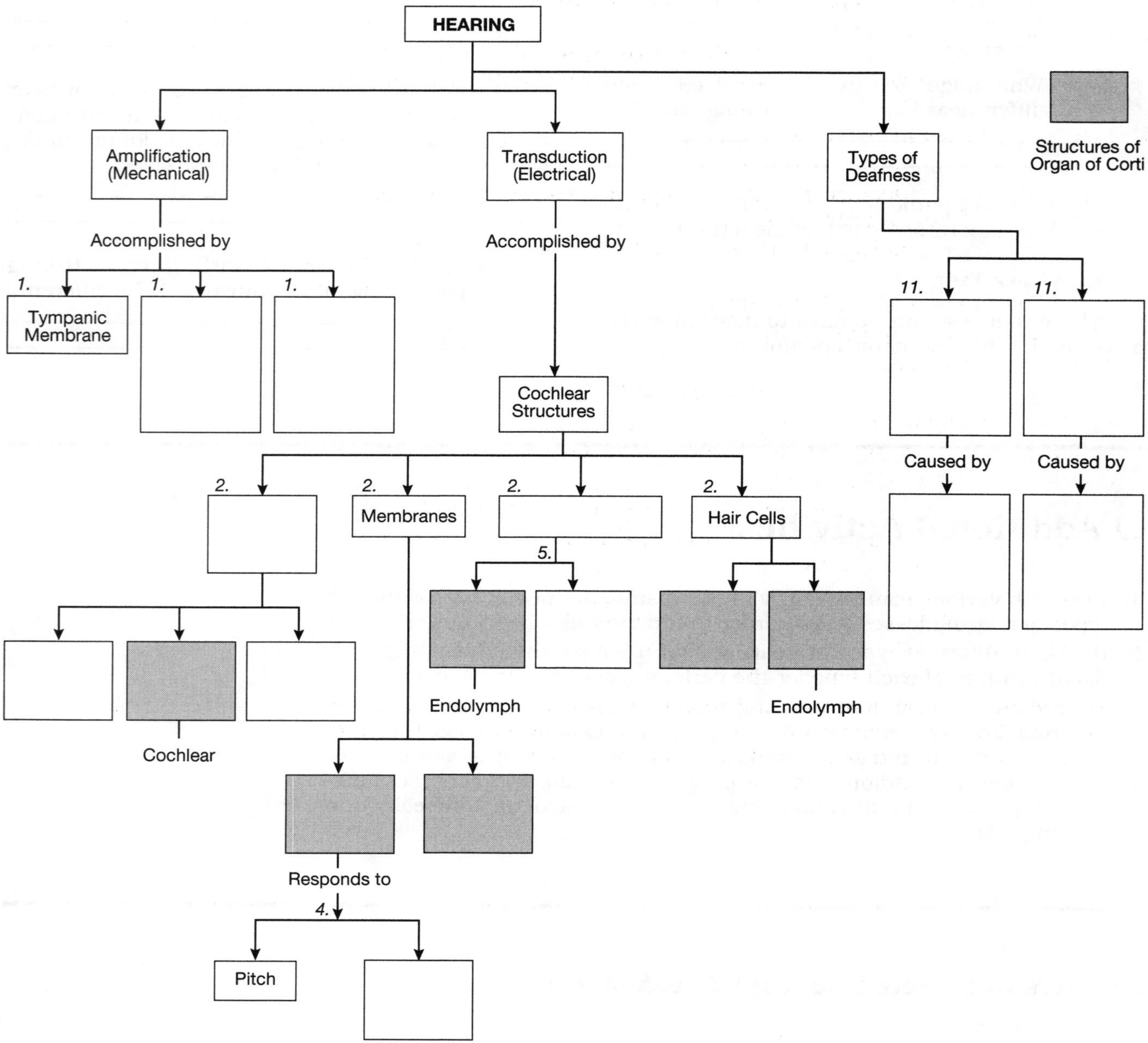

General Questions

NOTES

1. Regarding sound:
   a. How is frequency related to cycle?
   b. How is wavelength related to cycle?
   c. Differentiate between pitch, loudness, and intensity.
2. Regarding the acuity tests:
   Why might you have noticed a difference moving the watch toward your ear as opposed to moving it away from your ear? What factors might have been involved?
3. Regarding the sound localization test:
   a. Do you need both ears to localize sound?
   b. What compensatory mechanisms do people use when they have a problem with sound localization?
   c. How might you explain any problems you had in the sound localization test?
4. Explain the results of the Rinne test.
5. Explain the results of the Weber test.
6. Answer the following:
   a. The scala ____________ is continuous with the scala ____________.
   b. The scala tympani is filled with ______________________________.
   c. The pressure "safety valve" of the inner ear is the ______________.
   d. The organ of Corti is located ________________________________.
   e. The ______________ membrane forms the floor of the organ of Corti.
   f. High-frequency sounds are registered closer to the (proximal/distal) end of the tectorial membrane.
   g. The stereocilia are found on the ______________________________.
   h. The ossicles form the bony connection between the ________________ membrane and the ________________ window.
7. What is auditory accommodation?
8. In your own words, describe the process of hearing from the time the sound waves enter the ear until the registered sound is "heard" in the brain. Include the physiology of the cochlea.

EXERCISE 40

Physiological Aspects of Equilibrium

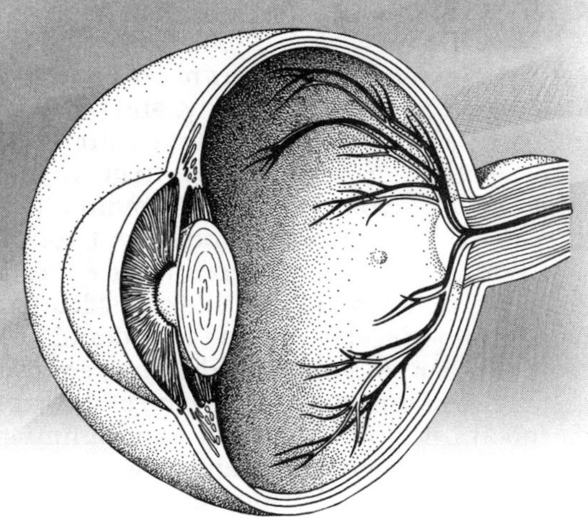

PROCEDURAL INQUIRIES

Examination

1. What are the anatomical structures associated with static equilibrium?
2. What are the anatomical structures associated with dynamic equilibrium?
3. Which cranial nerve is involved in equilibrium?

Experimentation

4. What is balance? What are three common balance tests?
5. What do the pointing tests check for? What are some common pointing tests?
6. What do the nystagmus tests check for? What are some common nystagmus tests?
7. What do the Romberg tests check for? How are the Romberg tests performed?
8. What are the common field tests used to check for drunk driving?

Additional Inquiries

9. What is equilibrium?
10. What is the difference between static and dynamic equilibrium?
11. What is the role of the proprioceptors in maintaining equilibrium?
12. What is the role of the tactile receptors in maintaining equilibrium?

Key Terms

Ampulla
Balance
Crista Ampullaris
Cupula
Dynamic Equilibrium
Equilibrium
Maculae
Nystagmus
Otoconia
Otoliths
Proprioceptors
Saccule
Semicircular Canals
Static Equilibrium
Tactile Receptors
Utricle
Vestibular Nuclei
Vestibule
Vestibulocochlear Nerve

Materials Needed

Models of Inner Ear
Revolving Stool
Cotton Swab
Ice Water
Blindfold
Ruler (metric)
Bright Light (such as a strobe lamp)
Penlight or Flashlight
Coins

Think of a time when you lost your balance. Perhaps you nodded off while sitting and suddenly caught yourself, or perhaps you turned quickly and almost fell. Do you remember the rapid series of reflex actions you underwent in order to right yourself? You were trying to regain your balance, but you were also trying to regain your sense of conventional equilibrium.

Equilibrium is the normal state of orientation between an organism and its environment. As the body moves, it must constantly adjust to changes in gravitational and spatial relationships. Equilibrium becomes the state of balance between opposing or divergent forces.

In your movements you must continuously adjust, align yourself with gravity, and position yourself in space so that you can continue with whatever task you are about. In relation both to gravity and to the world around you, you shift to maintain your balance. You unconsciously smooth out your actions.

We often classify equilibrium as either **dynamic** or **static**. (As we shall see, this classification scheme corresponds nicely with the functional anatomy of the inner ear.) Maintaining dynamic equilibrium involves a response to changes in rotational or angular motion. Maintaining static equilibrium involves a response to gravity or linear acceleration.

Notice that equilibrium always relates to change or potential change and that maintaining both static and dynamic equilibrium involves a compensatory action or a physical response to a change. If you are sitting or standing, you are vertical because your body is responding to a gravitational tendency to be horizontal. This is an example of linear acceleration. As you turn a corner, you are responding to a tendency to continue in a straight line. Turning a corner is rotational acceleration.

Maintaining equilibrium involves a series of complex, interrelated, physiological processes. Although we generally think of equilibrium as centered in the inner ear, most of the senses, along with **tactile receptors** and **proprioceptors**, play some role in our equilibrium. (Tactile receptors are the touch receptors. Proprioceptors monitor the position of the joints, the tensions in the tendons, and the contracting status of the skeletal muscle.)

Messages from these equilibrium centers are transmitted to the brain via the vestibular branch of the **vestibulocochlear nerve**. (Recall that this cranial nerve, N VIII, has two branches, the vestibular branch and the cochlear branch.) Synapses for integrating information are located at the boundary between the pons and the medulla. From the processing centers, the **vestibular nuclei**, efferent information travels to the oculomotor center, the head, the neck, and the appropriate spinal neurons. In other words, from the brain appropriate motor messages are sent to various parts of the body so that continuous compensation for opposition to equilibrium can be made. The nerve tract involved in affecting specific equilibrium is the vestibulospinal tract.

Go back to the introductory paragraph. If you start to doze and catch yourself or if you start to fall, the equilibrium centers register the change and appropriate reflex actions are initiated. Equilibrium is soon restored.

Because of its major role in equilibrium and because of its profound sensitivity to changes in the environment, the inner ear is a ready focal point for studying equilibrium. In this laboratory exercise we will examine the structure and function of those parts of the inner ear concerned with equilibrium.

(If necessary, review Exercise 38, before beginning this exercise.)

❑ Examination

I. Inner Ear [∞ *Martini, p. 560*]

The entire inner ear can be visualized as a tube within a tube. The fluid in the inner tube is endolymph and the fluid in the outer tube is perilymph. Shifts in these fluids (particularly the endolymph) stimulate nerve fibers and are registered as changes in equilibrium. Changes in equilibrium are transmitted to the brain via the vestibular branch of the vestibulocochlear nerve (N VIII).

A. Structures Involved in Dynamic Equilibrium

The **semicircular canals**, which form the posterior portion of the inner ear, are responsible for maintaining dynamic equilibrium, the rotational (or angular) movements of the head. The swollen region of each semicircular canal is known as the **ampulla**. Within the ampulla is the region of hair cells known as the **crista ampullaris** (crista within the ampulla). This region, which consists of ciliated hair cells (hair tufts often called cilia and stereocilia) and supporting cells resting on a basilar membrane, is topped by a gelatinous cap called a **cupula**.

Changes in the endolymph shift the pressure on the cupula. This stimulates the hair cells and, subsequently, the entire nervous pathway and compensatory motor changes are made.

1➡ Use a classroom inner ear model to help you see that the semicircular canals are oriented in three planes and are at right angles to each other. Locate the ampulla of each semicircular canal.

2➡ Identify the ampulla, the crista ampullaris, and the cupula in Figures 40–1● and 40–2●.

B. Structures Involved in Static Equilibrium

The **vestibule**, the central portion of the inner ear is composed of the **saccule** and **utricle**. Together these areas are responsible for static equilibrium, or compensations for gravity and linear acceleration. The hair cells in both the saccule and utricle are clustered in **maculae**. Each macula consists of ciliated hair cells (extensions of the nerve fibers) embedded in a gelatinous material and topped by densely packed mineral crystals known as **otoliths** or **otoconia**. Because of the inertia of the otoliths, changes in the movement of the head cause shifts in the gelatinous material, which in turn stimulate the nerve fibers.

1➡ Use Figure 40–1● and a classroom inner ear model to locate the parts of the vestibule.

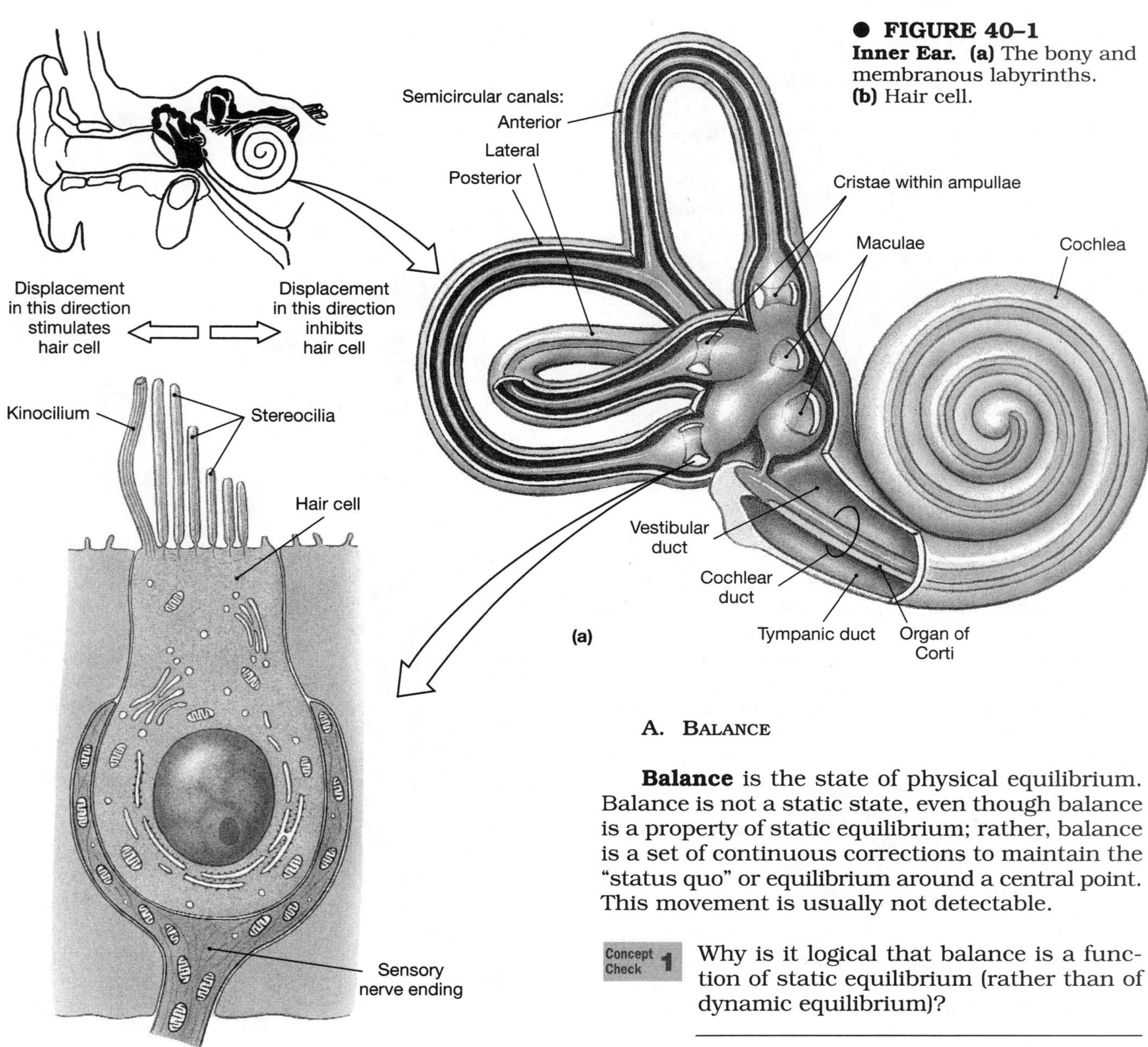

● **FIGURE 40–1**
Inner Ear. (a) The bony and membranous labyrinths. **(b)** Hair cell.

2➡ Identify the maculae and otoliths in Figure 40–2●.

3➡ Return to the model and note the relation of the saccule, utricle, semicircular canals, and vestibular branch of the vestibulocochlear nerve to the rest of the ear.

❑ Experimentation

II. Equilibrium Tests

Record your results from the following tests in the box, Figure 40–3●.

A. Balance

Balance is the state of physical equilibrium. Balance is not a static state, even though balance is a property of static equilibrium; rather, balance is a set of continuous corrections to maintain the "status quo" or equilibrium around a central point. This movement is usually not detectable.

Concept Check 1 Why is it logical that balance is a function of static equilibrium (rather than of dynamic equilibrium)?

____

____

1➡ Stand still with your eyes closed for the first balance test. Have your partner observe any movements you might make. To divert your attention you might dorsiflex your foot and splay your toes. If possible, your partner might measure your oscillations. Try this again with your eyes open. Record your results.

2➡ For another test stand on one foot, hold the other foot, and count to ten. Now try the same test with your eyes closed. Record your results. What differences did you notice?

____

____

3➡ The third balance test, the "walk the straight line" test, is described below under

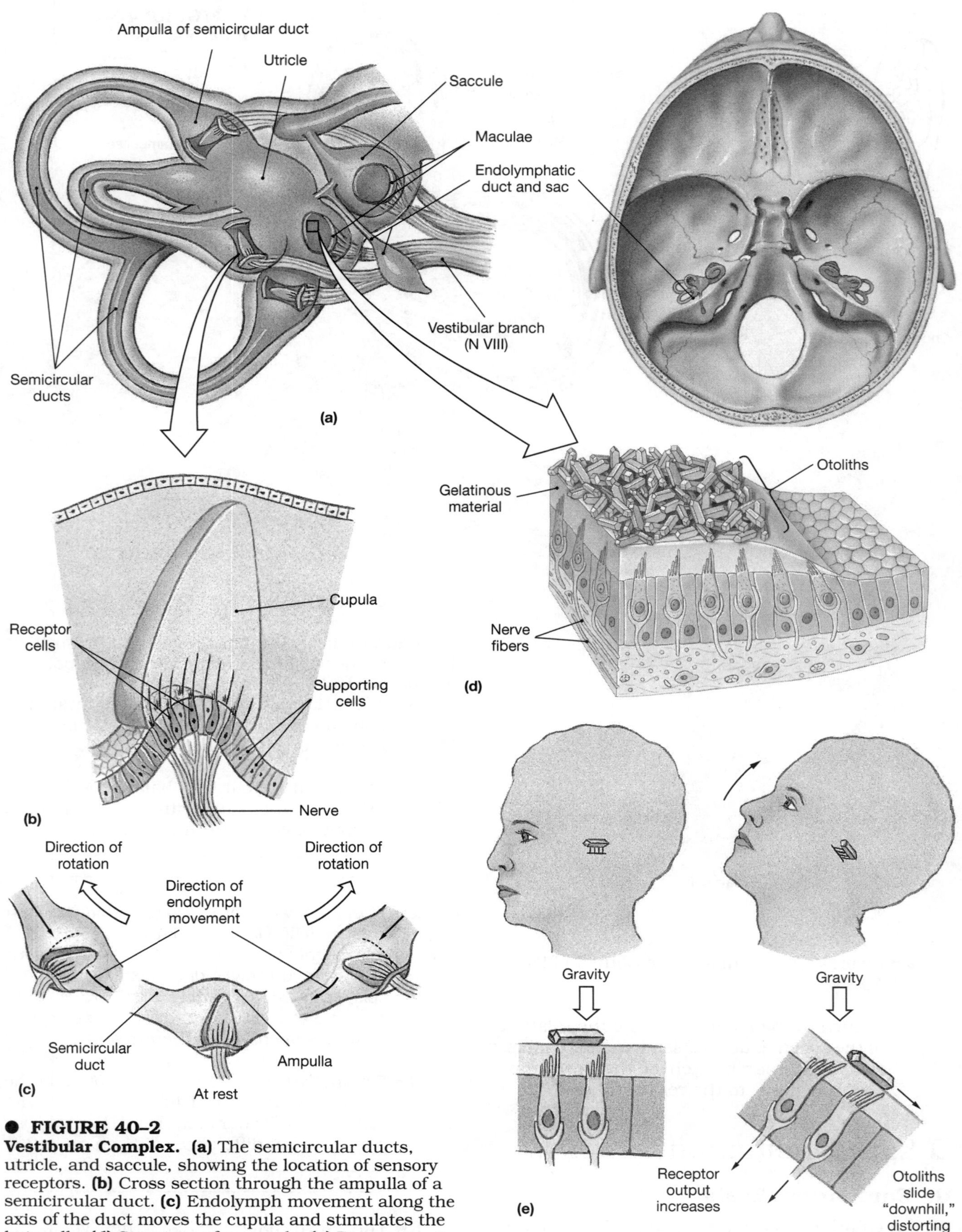

● **FIGURE 40–2**
Vestibular Complex. **(a)** The semicircular ducts, utricle, and saccule, showing the location of sensory receptors. **(b)** Cross section through the ampulla of a semicircular duct. **(c)** Endolymph movement along the axis of the duct moves the cupula and stimulates the hair cells. **(d)** Structure of a macula. **(e)** Diagrammatic view of macular function when the head is tilted back.

Test	Results
Balance	
1. Eyes open, both feet	
2. Eyes closed, both feet	
3. Eyes open, 1 foot	
4. Eyes closed, 1 foot	
5. Walk straight line	
Pointing	
1. Finger to nose	
2. Facial parts — Right arm	
3. Facial parts — Left arm	
Nystagmus	
1. Twirl stool, sitting straight	
2. Twirl stool, head on chest	
3. Twirl stool, head on lap	
4. Counter clockwise	
5. Eyes closed	
6. Eyes open, blindfold	
Ice Water	
1. Swab in right ear	
2. Swab in left ear	
Romberg	
1. Back to board	
2. Side to board	
3. Eyes open, one foot	
4. Eyes open, two feet	
5. Eyes closed, one foot	
6. Eyes closed, two feet	

● **FIGURE 40–3**
Equilibrium Test Results.

the drunk driving tests. Perform that test either now or when you get to Section III.

B. Pointing Tests

The pointing tests check your spatial awareness and spatial coordination. (In other words, how do you physically orient yourself in space?) Recall that proprioception is the internal monitoring of joints, tendons, and muscles. If your equilibrium is keen, your spatial awareness will be well coordinated with your proprioceptors. The proprioceptors monitor conditions but do not directly alter those conditions.

1➡ Try the simplest pointing test. Extend your arm straight out laterally, close your eyes, and bring your index finger to your nose. Record your results.

2➡ Again extend your arm straight out laterally and ask your partner to call out parts of your face. With your eyes still closed, quickly touch these parts. Try these tests with each arm separately. Record your results. What differences did you notice?

____

____

C. Nystagmus — The Banary Test

Nystagmus is a rapid, involuntary horizontal, vertical, or rotational movement or oscillation of the eye. Nystagmus occurs in response to a disturbance in the equilibrium of the semicircular canals. The eyes will move in a jerky fashion until normal equilibrium is reachieved.

Nystagmus can be induced by twirling on a revolving stool and stopping suddenly. (In addition to this "natural" condition, nystagmus can be caused by pathological conditions, which are beyond the scope of this laboratory exercise.)

(As you do the nystagmus tests, keep safety in mind.)

1 ➡ Sit straight, with your partner standing by; now swirl for a few seconds. Stop quickly. Your partner should check your eyes immediately. Record the information. Your partner will probably notice a horizontal movement of your eyes. This is because the lateral semicircular canal was stimulated. Go back to the ear model and identify the lateral semicircular canal. Why was this the only canal affected by your swirling?

____

____

2 ➡ Repeat the experiment with your head down so your chin is resting on your chest. This should stimulate the posterior canal and induce a rotary nystagmus. If you bend over from the waist so your head is on your lap while rotating, you will stimulate the anterior (sometimes called superior) canals, and your partner will notice a vertical nystagmus. Record the results.

3 ➡ Repeat these experiments using the following options: a) Rotate the chair clockwise as well as counterclockwise. b) Close your eyes. c) Use a blindfold but keep your eyes open. Record your results. What differences, if any, do you notice when using these variations?

____

____

D. Ice Water

Ice water increases the density of endolymph. Convection currents in the semicircular canals give the sensation of changes in equilibrium. This test is often done to check for vestibular nerve damage. If the nerve is damaged, no equilibrium reactions will occur.

1 ➡ Have your partner ready to observe your eyes and your upper body reflexes.

2 ➡ Dip a cotton swab in the ice water. When the swab is quite cold, carefully place the swab in your ear. Have your partner note all your immediate reactions. Record your results. Repeat this experiment with the other ear. Are there any differences between the left and right ears?

____

____

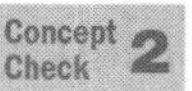

Why would ice water increase the density of endolymph?

____

____

E. Romberg Test

At first glance the Romberg Test seems much like the first balance test described in Section II.A. However, the Romberg Test checks the integrity of the dorsal white column of the spinal cord, which transmits impulses to the brain from the proprioceptors involved with posture. Recall the function of the proprioceptors.

1 ➡ Stand between a strong light source and a chalk board (so that your shadow can be measured). Do this first with your back to the board so your side-to-side movements can be measured.

2 ➡ Repeat everything while standing with your side to the board so your front-to-back movements can be measured.

3 ➡ Have your partner sketch the edges of your shadow and measure the front-to-back and side-to-side movements.

4 ➡ Do this under each of the following conditions: a) with your eyes open, standing on one foot, then on two feet. b) with your eyes closed, standing on one foot, then on two feet. Record your results. Explain the differences.

____

____

III. Drunk Driving Tests

Drunk driving tests are a practical application of the physiology of equilibrium. When a person has consumed certain amounts of alcohol, the central nervous system is depressed. The equilibrium apparatus is an extension of the central nervous system; decreased central nervous system response is readily discernible by observing changes in equilibrium.

A. Equilibrium Tests

Five equilibrium tests are commonly used by police officers in the field.

1 ➡ Review the standard balance test in Section II.A. You may have had some difficulty with this test, particularly the part where you stood on one foot. Many people do. For that reason, most police officers do not place much credence in that test alone. Usually the balance test is done in conjunction with other tests.

2 ➡ Recall the second common test, the pointing test in Section II.B. If a person's central nervous system is depressed, he/she may loose a sense of spatial orientation. The

pointing tests indirectly check a person's static equilibrium. If the static equilibrium component of the central nervous system is functioning normally, a person's proprioceptors give him/her an accurate sense of spatial orientation.

3➡ Try the field nystagmus test on your partner. The field nystagmus tests are usually called gaze nystagmus tests. The person performing the test asks the subject to follow a penlight. If the pupils are of unequal size or if any of the eye rotational movements discussed in Section II.C are present, the person is said to have gaze nystagmus. This indicates a decrease in the functioning of the semicircular ducts — often brought on by a central nervous system depressant.

4➡ The fourth test is the traditional "walk the straight line" test. In this test the person walks a straight line of about 15 feet by placing the heel of one foot directly in front of the toe of the other foot. The person then turns and walks back the same way. While walking, the sense of linear acceleration is observed. As the person turns, the officer may detect problems with rotational acceleration (the head is moving). If problems exist, the person may well lose balance. Try this test. How did you do?

____

____

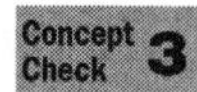

Concept Check 3: Does this test check for static or dynamic equilibrium or both?

____

5➡ Test your balance and spatial perception (as well as your manual dexterity). In the field, the police officer places three coins of varying size on the ground and asks the person to pick them up. The motion involved should be smooth, and the person should have no trouble either locating or manipulating each coin. Try this test now. While you are doing it, concentrate on the muscle coordination involved in picking up the coins. Did you have any trouble with this test?

____

____

B. Mental Acuity

The "count backwards from 10" test is usually more of a test of mental acuity than it is strictly an equilibrium test. Nevertheless, sometimes a person whose central nervous system is depressed may concentrate so hard on counting backwards correctly that he/she may actually lose balance, at least to some extent.

1➡ Try this test now. How fast were you able to count backwards from ten?

____

❑ Additional Activities

NOTES

1. Research the causes and effects of motion sickness. Drugs commonly used to overcome motion sickness apparently depress the action of vestibular nuclei. Find out the names and actions of specific drugs.
2. Find out what types of long-term equilibrium compensation can be made by persons having some type of inner ear damage.

Answers to Selected Concept Check Questions

1. Balance is a response to gravity.
2. Decreased temperature slows molecular motion, which increases density.
3. Both. The straight line is linear or static equilibrium. The turn is rotational or dynamic equilibrium.

☐ Lab Report

1. Referring to the numbered inquiries at the beginning of this exercise, complete the following box summary:

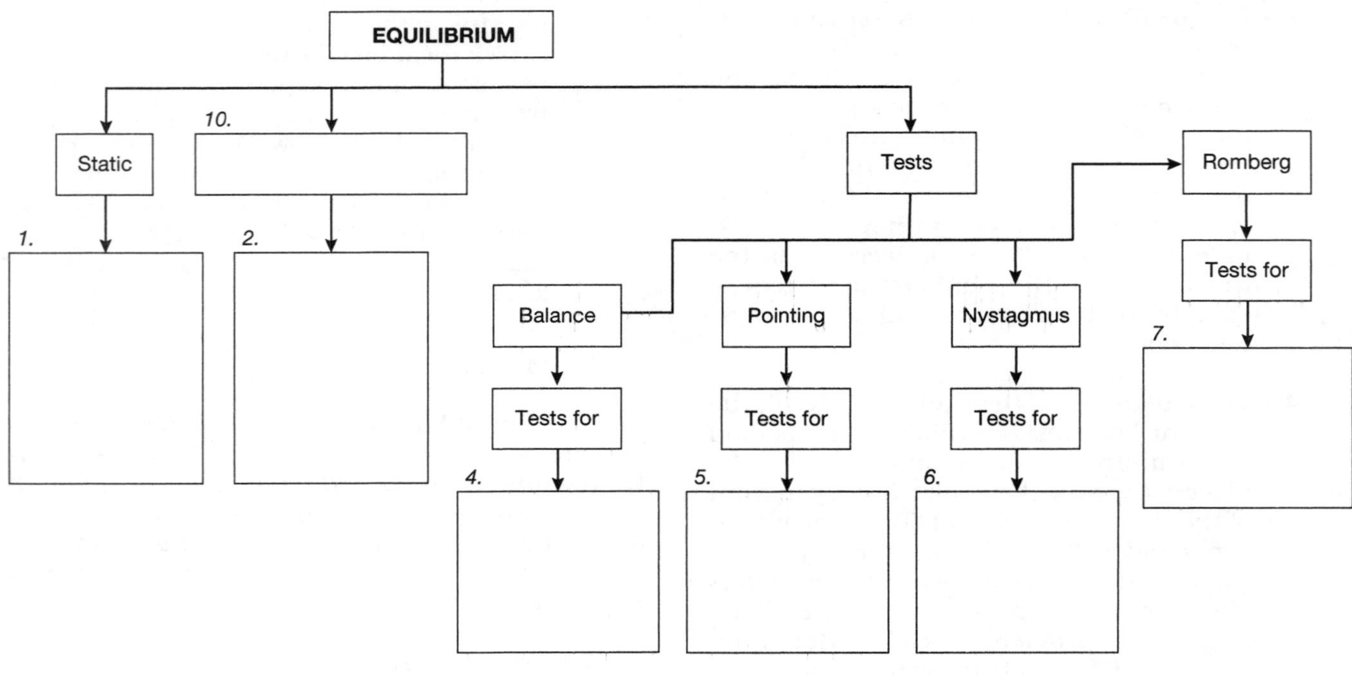

General Questions

NOTES

1. What is the role of vision in equilibrium? Why did you notice differences in your open-eyed and closed-eyed test reactions?
2. Why might you have had a problem with the balance test, even if you are perfectly healthy and perfectly sober?
3. Explain why coordination can be affected by central nervous system depressants.
4. Why are each of these tests used to check for drunk driving?
   a. Walk a straight line
   b. Field nystagmus
   c. Coins on the ground
   d. Count backwards from 10
5. Why might a person with an earache have trouble with the coin test?
6. Which test or tests would be used to check for each of these conditions?
   a. Vestibular nerve damage
   b. Damage to the dorsal white column of the spinal nerve
   c. Static equilibrium
   d. Decreased functioning of the proprioceptors
   e. Damage to the posterior semicircular canal
7. Refer to the tests listed in the bos summary. Briefly explain how each of these tests is performed:
   a. Balance
   b. Pointing
   c. Nystagmus
   d. Romberg

EXERCISE 41

Anatomy and Physiology of Taste

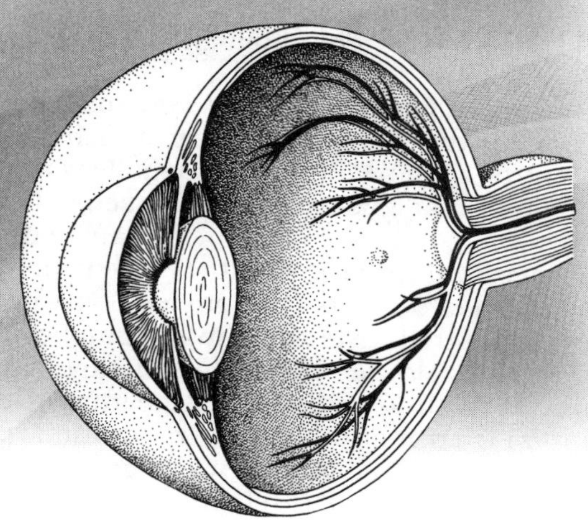

PROCEDURAL INQUIRIES

Examination

1. Where are the papillae located?
2. What are the major types of tastebuds?
3. Where is each type of tastebud found?
4. Which nerve pathways are involved in taste?
5. What are the principal taste sensations?

Experimentation

6. What is the purpose of the tongue map?
7. How do you make a tongue map?
8. How does temperature influence taste?
9. What is the relationship between taste itself and taste-related sensations?
10. What is the relationship between taste and smell?

Additional Inquiry

11. What is taste?

Key Terms

Circumvallate Papillae
Filiform Papillae
Fungiform Papillae
Gustation
Gustatory Cells
Papillae
Sensory Cells
Sustentacular Cells
Taste
Taste Buds
Taste Hairs
Taste Pores
Tongue Map

Materials Needed

Hand Lens
Concentrated Salt Solution (at least 5%)
Concentrated Sugar Solution (at least 5%)
Quinine Sulfate Solution (about 0.5%)
(regular [not diet] tonic water or epsom salts may be substituted)
Dilute Vinegar (about 1% Acetic Acid)
Sugar Crystals
Rock Salt Crystals
Cotton Swabs
Absorbent Cotton or Paper Towels
Fresh Ground Black Pepper
Jalapeño Peppers in Juice
Ice Chips
"Dry" Objects Such As:
Wood
Paper Clips
Plastic
Pieces of Apple, Potato (raw), Carrot, Onion
(a piece of pear may also be used)
Prepared Microscope Slides of Taste Buds
Optional: Laboratory Animal

Think about **taste** for a moment. Even as you read this and your mind turns to taste, you are probably beginning to salivate. This is the salivary reflex. Your mouth waters in preparation for incoming food because you can taste nothing that is not dissolved in water.

Now think about some of your favorite dishes. Perhaps you can "almost taste" some ice cream, or a steak, or a fresh tomato right from the garden. If you think about it long enough, you can literally conjure up the distinct tastes of hundreds of different foods. You can also modify those tastes right in your own mind. For example, you can imagine what the steak would be like if it were too salty, or what the tomato would be like if it were too bitter. You are recalling the specific chemistry of each of these substances. That recollection is usually very vivid.

Taste, or **gustation**, is the sense that allows us to respond to the chemical nature of the substances put into the mouth. Taste is probably the least understood of the senses. Some people also maintain that taste is the most complicated sense, and the relationship between taste and smell makes the understanding of the intricacies of gustation even more difficult. In addition, gustatory acuity tends to diminish with age.

Taste buds are the chemoreceptor sense organs located within the epithelial folds of **papillae** on the surface of the tongue, the soft palate, the epiglottis, and parts of the pharynx. Despite differences in the papillae, all taste buds are more or less identical.

Normally we think of taste in terms of four primary taste sensations: salty, sweet, bitter, sour. When the taste buds are stimulated, we perceive some combination (one or more) of these four fundamental tastes. Variations in the intensity of the stimuli and different combinations of signals allow us to be aware of an exceptionally wide range of taste sensations. (Some Japanese scholars now recognize a fifth taste — *umami* — the meaty or savory taste. The seasoning MSG is considered to have the umami taste.) Water is often thought to have no taste. However, the presence of water receptors in the pharynx has been demonstrated. Perhaps water should be considered as a separate taste.

In addition to the specific tastes, you are probably also very aware of the texture of the foods you are thinking about. Ask yourself if the steak would taste the same pureed as baby food or if the tomato would taste the same served as tomato juice. Even now your tongue should be feeling the differences between the solid and semisolid states of these foods. Also, you may know intellectually that there is no nutritional difference between fresh ice cream and refrozen ice cream, but the receptors in your mouth are perfectly aware of the textural differences, and you as a result have a very different concept of these two substances.

Physical awareness of, for example, sharpness or coldness, is not taste in the strictest sense of the word. We do not taste sharpness or coldness. Nevertheless, the taste-related perceptions, such as the size, shape, hardness, softness, and temperature, of a substance greatly influence our gustatory perceptions.

Taste, then, is primarily the chemical recognition of an ingested substance and secondarily a physical recognition of that same substance.

In this exercise we will gain an understanding of the anatomy and physiology of gustation by examining the macroscopic and microscopic organization of the taste bud and its accessory structures and then by performing certain experiments to study the manifestations of this gustatory sense.

❑ Examination

I. Tongue [∞ *Martini, pp. 537, 854*]

A. Papillae

The papillae (sing., papilla) are the epithelial projections that house the taste buds.

1➔ Use Figure 41–1● to identify the form and location of each of the three major types of human papillae.

Circumvallate Papillae These are the large, round, folded papillae, usually about 10 in number, which form a V near the posterior margin of the tongue. (If you examine your own tongue in the mirror, you may have trouble distinguishing these papillae because they are generally just beyond your range of vision.) Though few in number, the circumvallate papillae have the largest number of taste buds. These taste buds are embedded in the sides of the papillae.

Fungiform Papillae The most numerous of the papillae, the fungiforms are found scattered across the tongue, though concentrated on the sides and tip of the tongue. The taste buds of these knoblike elevations are located on the surface of the papillae. Some of the fungiform papillae do not have taste buds.

Filiform Papillae These threadlike projections are located on the anterior two-thirds of the tongue.

2➔ Use a hand lens to examine a classmate's tongue. See if you can identify the different types of papillae. Which types do you see? Where are they located?

____

____

B. Taste Bud

The taste bud is the concentrated cluster of chemoreceptor cells and their accessory structures that allow us to perceive the chemical nature of the substances we put in our mouth.

1➔ Use Figure 41–1c● to identify the constituent parts of each taste bud: the **sensory** (nerve) **cells**, the **gustatory** (taste receptor) **cells**, and the **supporting** (sustentacular) **cells**. The gustatory cells are actually neuroepithelial cells, and the free surface of each taste receptor cell includes dendritic **taste hairs**, extending through the inner **taste pore**.

2➔ Keep in mind that the taste bud is actually three-dimensional. You might try visualizing the taste bud as an upside-down

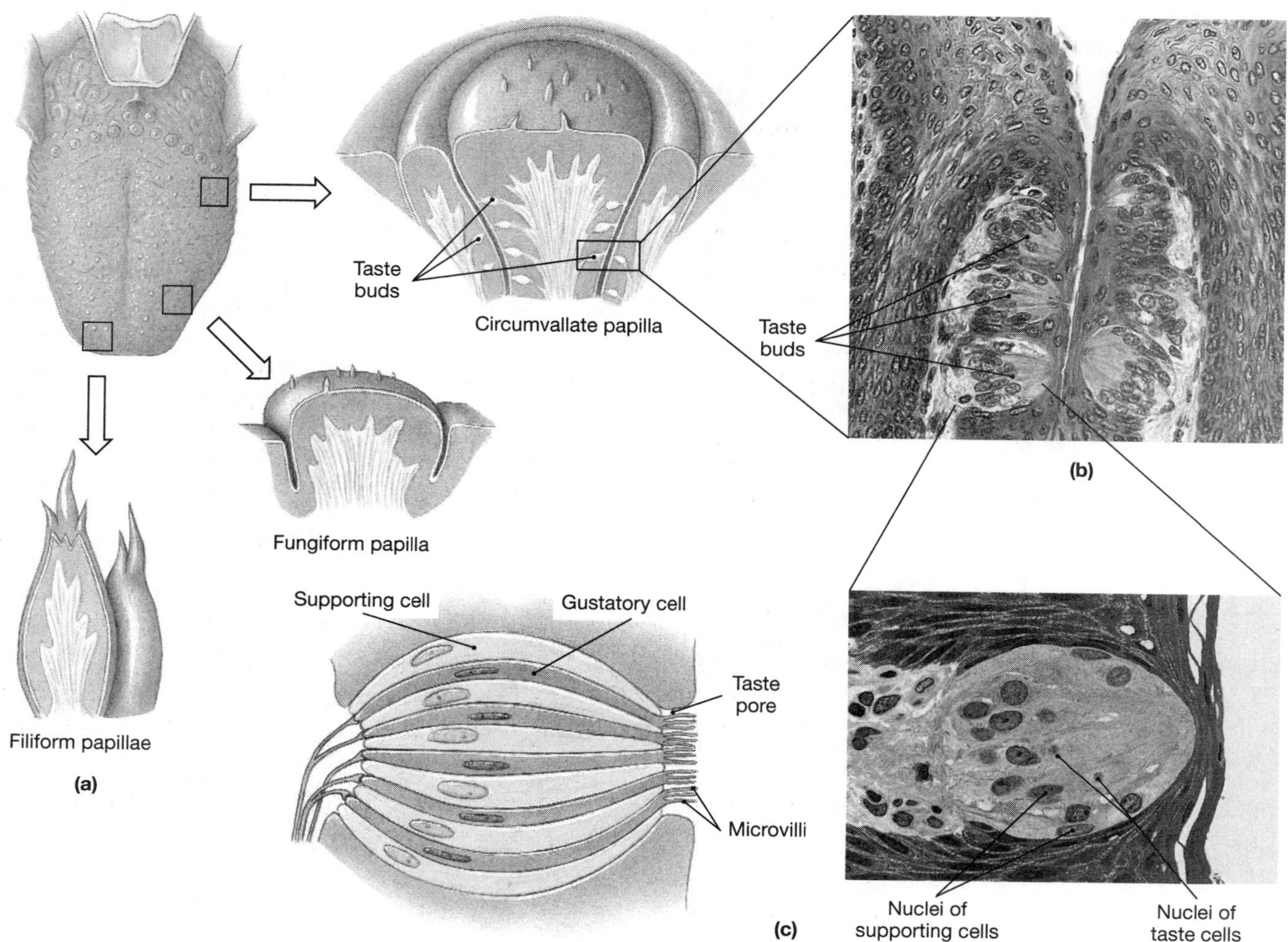

● **FIGURE 41–1**
Gustatory Reception. **(a)** Gustatory receptors are found in taste buds that form pockets in the epithelium of the tongue. **(b)** Photomicrograph of taste buds in circumvallate papilla. **(c)** A taste bud showing receptor cells and taste pore details.

onion, with the green stem representing the nerve fibers, the layers of the bulb analogous to the gustatory and sustentacular cells, and the root being the taste hairs.

3→ Examine the microscope slides of the taste buds. Identify the parts of the taste buds mentioned.

4→ Sketch and label what you see.

Taste buds

C. Nerve Pathways [∞ Martini, p. 536]

With each taste bud stimulation, a nerve impulse travels to the brain by one of three routes. The facial (or lingual) nerve (N VII) services taste buds on the anterior two-thirds of the tongue, whereas the glossopharyngeal nerve (N IX) carries taste bud nerve impulses from the posterior one-third of the tongue. The vagus nerve (N X) receives messages from the epiglottal and pharyngeal taste buds. Damage to N IX tends to decrease (but not eliminate) our perception of the bitter taste while damage to N VII tends to decrease (but not eliminate) our awareness of the sweet, sour, and salty tastes.

The cranial nerves synapse is the solitary nucleus of the medulla. From there the gustatory message is sent to the thalamic nucleus and then to the primary sensory cortex.

● **FIGURE 41–2**
Gustatory Pathways.

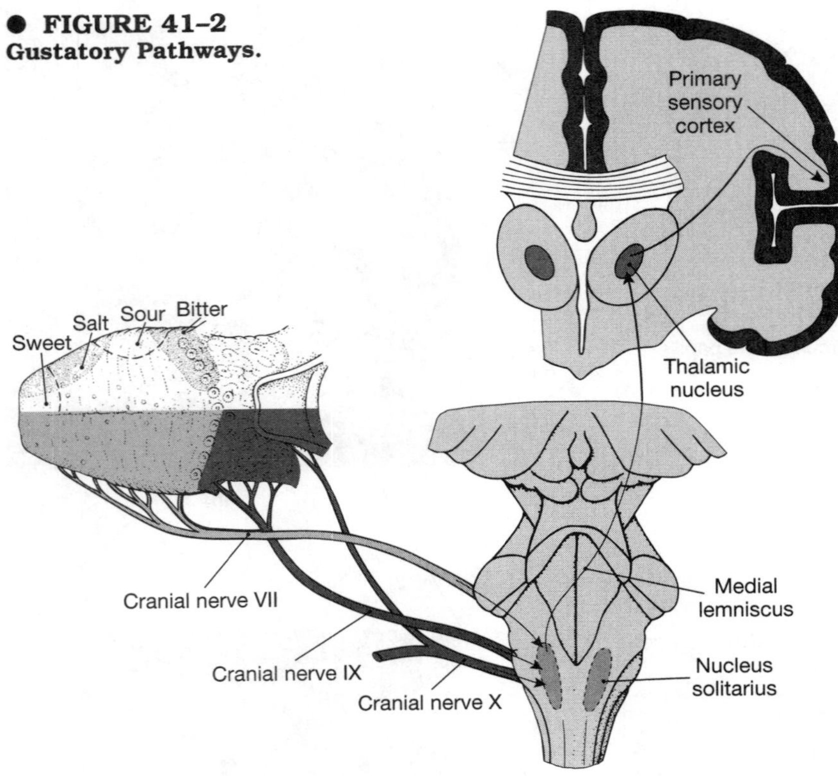

Concept Check 1 Would you expect a person with damage to the primary sensory cortex to want to use more or less salt on food? Why?

____

____

The trigeminal nerve (N V) innervates the tongue. This nerve, however, is not directly related to taste (gustation). Rather, the trigeminal nerve is responsible for texture and taste-related perceptions such as peppery and burning sensations. Because of the interrelatedness of taste and taste-related functions, damage to the trigeminal nerve can leave the impression of a decrease in gustatory function.

A dentist often deadens a branch of the trigeminal nerve. After you have had dental work, why might you think your dinner is extremely bland, even if it is highly seasoned?

____

____

Concept Check 2 Use Figure 41–2● to trace the paths of hypothetical gustatory nerve impulses. Summarize those pathways here:

N V ________________________________

N VII ______________________________

N IX _______________________________

N X ________________________________

❑ Experimentation

II. Gustatory Perception

A. Tongue Map

Today it is generally believed that all of the major taste sensations can be perceived on all parts of the tongue but that the intensity of that perception leads to the identification of certain tongue areas as salty, sweet, bitter, and sour. The **tongue map** shows general areas where the taste buds are most sensitive to a particular taste.

1 ➡ Work with a partner to map the modalities of taste. Rinse your tongue with plain water and dry it thoroughly with absorbent cotton or paper towels. Have your partner dip a clean cotton swab into one of the taste solutions (salty, sweet, bitter, or sour). Your partner should then carefully dab the swab onto the tip, back, sides, and center of your tongue. (Prevent contamination: do not re-dip the swab.)

Use only one of the "tongues" in Figure 41–3● to indicate where on your tongue you taste that solution. Rinse and dry your tongue and repeat this experiment with each of the other three solutions. Use the same "tongue" for all four solutions. Compare your map with the generalized tongue map (Figure 41–2●).

You may notice that you do not taste one particular substance at all. For example, some people may not be able to discern a 0.5% solution of quinine sulfate anywhere on the tongue. If you are such a person, ask your instructor if you might try a more concentrated quinine solution. (You may still not be able to taste it. However, you may also find that you are able to perceive the stronger solution.)

Try mixing a few drops of different solutions together and repeat the mapping. Use a clean swab and do not contaminate any of the original solutions. What changes do you notice in the taste sensations? Does this help you understand your own taste perception range?____________________

____

After you, your partner, and your classmates have completed this experiment, share your results. You should notice that certain trends exist as to which part of the tongue predominates in the tasting of which type of substance. Those trends may approximate Figure 41–3●. Nevertheless, you should also notice that no two maps are exactly

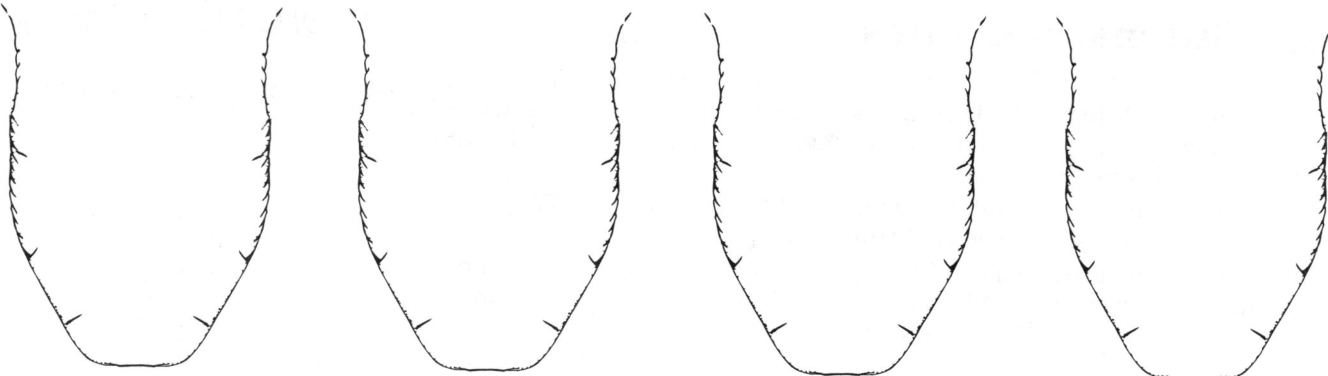

● **FIGURE 41–3**
Tongue Map.

the same; one person's map may actually contradict another person's map.

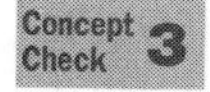

Based on what you learned in the first paragraph of Section II, why is it not surprising that two people's tongue maps may contradict each other.

2➡ Suck on some ice chips. Repeat the tongue mapping experiment, using another "tongue" in Figure 41–3●.

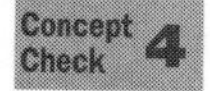

What differences do you notice? Were some tastes affected more than others? What role did time play in this experiment? What do you suppose the ice did? Can you relate the change in sensation with ice to the change in sensation you might experience with nerve damage?

3➡ Again start with a clean, dry tongue. Have your partner put a few sugar crystals onto a known "sweet" area of your tongue. (Refer to your original tongue map.) What happened to the taste sensation? If you noted a time lag, why was this? Repeat this experiment with a few salt crystals.

4➡ Again start with a clean, dry tongue. Plug your nose. Now test your awareness of the "dry" objects listed in the "Materials Needed" section of this exercise.

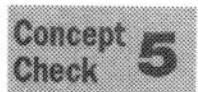

Are you actually tasting these objects or are you primarily aware of their texture?

5➡ For this experiment your tongue should be clean but not dry. (Salivation should supply enough moisture. If you are nervous, add a few drops of distilled water to your tongue.) Plug your nose and have your partner put some black pepper on various parts of your tongue.

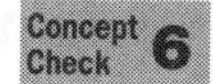

What do you taste? What do you feel? Would you classify the pepper taste as a true taste or a taste-related perception?

6➡ Try number 5 again. This time use the jalapeño pepper, or use the jalapeño pepper juice. What conclusions can you draw from this pepper?

B. Relation to Smell

Some experts believe that as much as 80% of our taste is actually smell. Regardless of the actual percentage, we do know that taste and smell are very closely related. Test your ability to taste by using only your taste buds.

1➡ Plug your nose, close your eyes, and have your partner put on your tongue, one at a time, equal-sized wedges of apple (or pear), raw potato, carrot, and onion. Can you identify which is which? If you were able to figure out which was which, did you do so on taste alone or were other factors — such as feel — also involved?

2➡ Try tasting the block of wood with your eyes closed and your nose plugged. What happened?

❑ Additional Activities

NOTES

1. Examine the tongue of a laboratory animal. Discuss the similarities and differences between the animal's tongue and papillae and the human tongue and papillae.
2. Research the use of artificial flavors in the foods we eat. Why do these flavors often taste like the real thing?
3. Go back to the tongue map. Use regular tonic water and diet tonic water to map the bitter areas of your tongue. Use the third tongue in Figure 41–3•. Explain any differences you might notice.

Answers to Selected Concept Check Questions

1. More salt. Sensory awareness would have decreased with the injury.
2. The gustatory pathways are as follows:
   N V — no direct relation to taste
   N VII — anterior two-thirds of tongue → solitary nucleus → thalamic nucleus → primary sensory cortex
   N IX — posterior one-third of tongue → solitary nucleus → thalamic nucleus → primary sensory cortex
   N X — epiglottis and pharynx → solitary nucleus → thalamic nucleus → primary sensory cortex
3. All major taste sensations can be perceived on all parts of the tongue, but the intensity varies. It is not surprising that one person may experience a greater intensity of one sensation in a particular place than another. These areas of intensity could vary from person to person.
4. Ice chips should lessen the taste sensation by temporarily deadening the nerve endings. This approximates nerve damage.
5. You are primarily aware of the texture.
6. This is a taste-related perception rather than taste.

❑ Lab Report

1. Referring to the numbered inquiries at the beginning of this exercise, complete the following box summary:

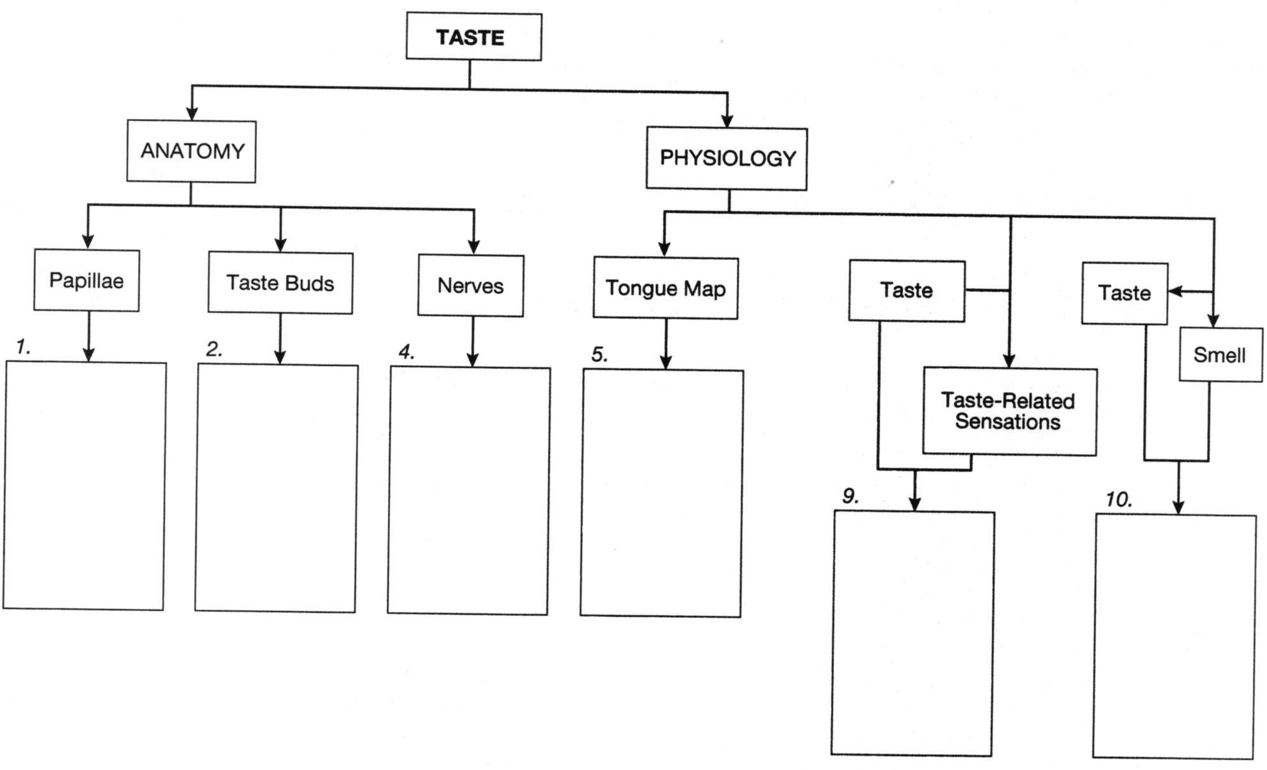

General Questions

1. Sometimes when you are ill you don't feel like eating because "nothing tastes good," even when your illness should not really prevent you from eating anything you want. What physiological explanation can you provide for this phenomenon?
2. Use your own tongue map and consider damage to each of the nerves of the tongue. Fill in the chart with what would happen to your own taste sensations.

Nerve	Taste Sensations
N V	
N VII	
N IX	
N X	

NOTES

3. Why might your tongue map differ from the generalized tongue map? Differ from your classmate's tongue map?
4. Consider what happened when you put sugar and salt crystals on a dry tongue. Speculate on what reflex action might have occurred.
5. Everything you actually tasted was in solution, either before being put in your mouth or when mixed with saliva. What conclusion can you draw about the importance of water in gustation?
6. What differences did you notice between the primary taste sensations and the taste-related sensations?
7. Can you really taste a block of wood?
8. List several reasons why you think the taste of steak is different from the taste of candy.
9. List five factors which might affect your gustatory perception.

NOTES

EXERCISE 42

Anatomy and Physiology of Smell

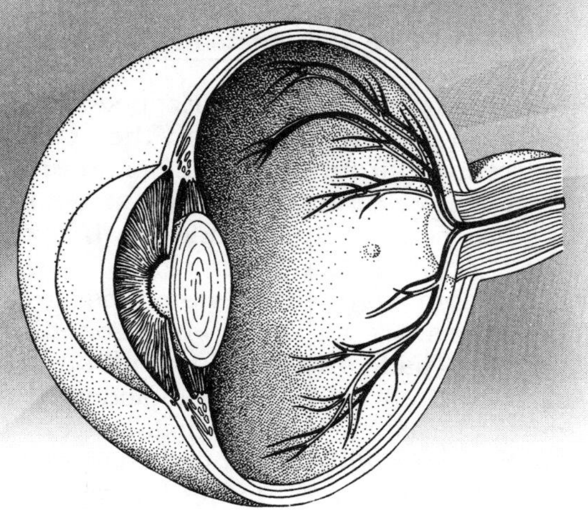

PROCEDURAL INQUIRIES

Examination

1. What are the principal anatomical components of the olfactory system?
2. What is the function of each of the olfactory components?
3. What is the relationship between the cribriform plate, the crista galli, and the olfactory foramina?
4. How do the olfactory bulbs relate to these structures?
5. What is the relationship between the olfactory epithelium, the basal cells, the olfactory receptor cells, and the supporting (sustentacular) cells?
6. What is the function of the olfactory glands?
7. What is the olfactory nerve pathway.

Experimentation

8. How did we perform the spice identification test?
9. How do you smell through your mouth?
10. What effect does cold have on our ability to smell?
11. How are smell and taste related?
12. Can we smell dry objects?

Additional Inquiries

13. What are the two principal theories of olfaction?
14. How can olfactory adaptation influence our smell perceptions?

Key Terms

Basal Cells
Chemical Theory
Cribriform Plate
Crista Galli
Lipofucsin
Nasal Epithelium
Nerve Pathways
Olfaction
Olfactory Adaptation
Olfactory Bulbs
Olfactory Cells
Olfactory Epithelium
Olfactory Fatigue
Olfactory Foramina
Olfactory Glands
Olfactory Nerves
Olfactory Receptor Cells
Physical Theory
Receptors
Supporting (Sustentacular) Cells

Materials Needed

Skull with open Cranium

Preserved Brain (if available) (animal brain may be substituted for human brain)

Prepared Microscope Slides of Nasal Epithelium

10 Numbered Bottles of Assorted Spices (key held by instructor; bottles should be dark and stoppered to disguise contents)

"Dry" Objects Such As:

- Wood
- Paper Clips
- Metal or Plastic

Ice Chips

Wedges of Apple (or Pear), Raw Potato, Carrot, and Onion (if this exercise will be used in conjunction with Exercise 41)

(Optional: Frozen Food and Microwave Oven)

Have you ever walked past a restaurant where something unbelievably good was being prepared? Your head turns, your nostrils constrict, and no matter how deep in thought you might be, your concentration is suddenly completely focused on the sensations you are receiving from your sense of **smell**.

Smell, or **olfaction**, is the sense that allows us to respond to the chemical nature of the substances coming into the nose.

In many ways smell is the most mysterious of the senses. We all know when something smells "good" or "bad," and we all know that many of our animals (such as dogs) use olfaction as a primary source of learning because their sense of smell is so much more highly developed than ours. We also know that smells can excite us, repulse us, bring out the "chemistry" between people, or stir up memories from the near or distant past. Despite what "we all know," we don't really know how smell works.

There is evidence that we do not all smell exactly the same things in exactly the same way. If we did, why would some of us be fond of one perfume or aftershave, while others of us do not care for the scent at all?

Over 50 primary smells have been identified, yet no apparent difference exists among the **olfactory cells** themselves. If we add our individual likes and dislikes, we find that smell discrimination is a very real mystery.

We do know, however, that we smell because molecules of a substance stimulate **receptors** in the nose. These receptors are a part of the **nerve pathways** (olfactory pathways) that allow us to register the nature of the incoming smells.

Think about your dog or cat for a moment. Your animals sniff openly and will often sniff out a worrisome odor long before you are even aware of it. We humans sniff too, though we usually sniff with a bit more social acumen than do our animals. When you walked past that restaurant in the first paragraph, your nostrils constricted because you were automatically sniffing for new information. You were concentrating the incoming smell and moving it with more force across the **nasal (olfactory) epithelium**. Sniffing increases the air flow across the nasal membranes, thus enhancing the stimulation of the olfactory receptors. *[∞ Martini, p. 534]*

Two general theories of **olfaction** exist. The first (the **chemical theory**) states that specific chemoreceptors exist on the membrane surface. Molecules of a substance with many of these receptors constitute an action potential and subsequently a specific smell perception. The **physical theory** maintains that the receptors respond to the physical shape rather than the chemical nature of the different molecules. Recall that in three-dimension each molecular structure has a unique shape. See Figure 42–1●.

The purpose of this laboratory exercise is to gain an understanding of the anatomy and physiology of olfaction by studying the physical characteristics of the olfactory structures and by performing certain experiments that demonstrate different aspects of our sense of smell.

❑ Examination

I. Anatomy

A. Internal Structures

We will use a skull and Figure 42–1● to locate the terms in boldface.

1➡ Take a skull and remove the calvarium. Look at the cranial floor and locate the **cribriform plate** of the ethmoid bone. The cribriform plate is the indented area with the small openings leading into the nose. The ethmoid bone also forms the roof of the nasal cavity.

The small openings in the cribriform plate are called the **olfactory foramina**. These foramina normally occur in three rows, though at least one row may be lost in the grooves of the septum. The **olfactory bulbs** lie on top of the olfactory foramina, one on each side of the **crista galli** (that upward protrusion in the center of the plate). The crista galli is so called because of its resemblance to the comb of a rooster. Without the olfactory foramina, the olfactory nerves would not be able to reach the nasal epithelium.

2➡ Recall that the olfactory bulb is the anterior extremity of the olfactory tract, the gross terminus of cranial nerve one (NI).

3➡ Examine any preserved brain you may have available. Notice the olfactory tracts and the olfactory bulbs. Do you see how these bulbs would fit onto the cribriform plate?

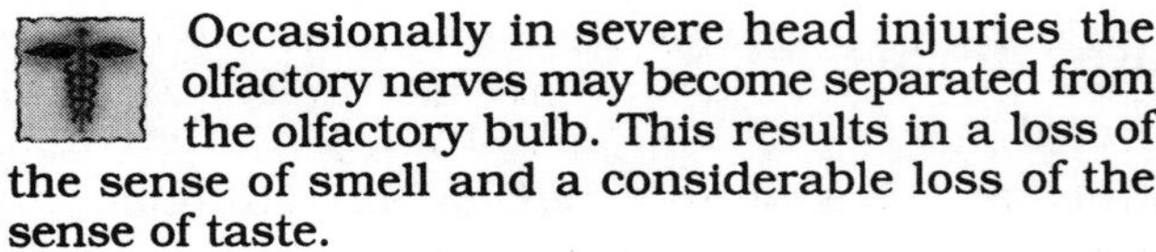

Occasionally in severe head injuries the olfactory nerves may become separated from the olfactory bulb. This results in a loss of the sense of smell and a considerable loss of the sense of taste.

4➡ As you locate the terms on Figure 42–1●, note the similarities between the olfactory structures and the gustatory structures examined in the previous exercise (Figure 41–1●).

B. Olfactory Pathway

The **olfactory nerves** extend from the olfactory bulbs through the olfactory foramina. These olfactory nerves divide into the **olfactory receptor cells**, which are part of the **olfactory epithelium** that lines the roof of the nasal cavity.

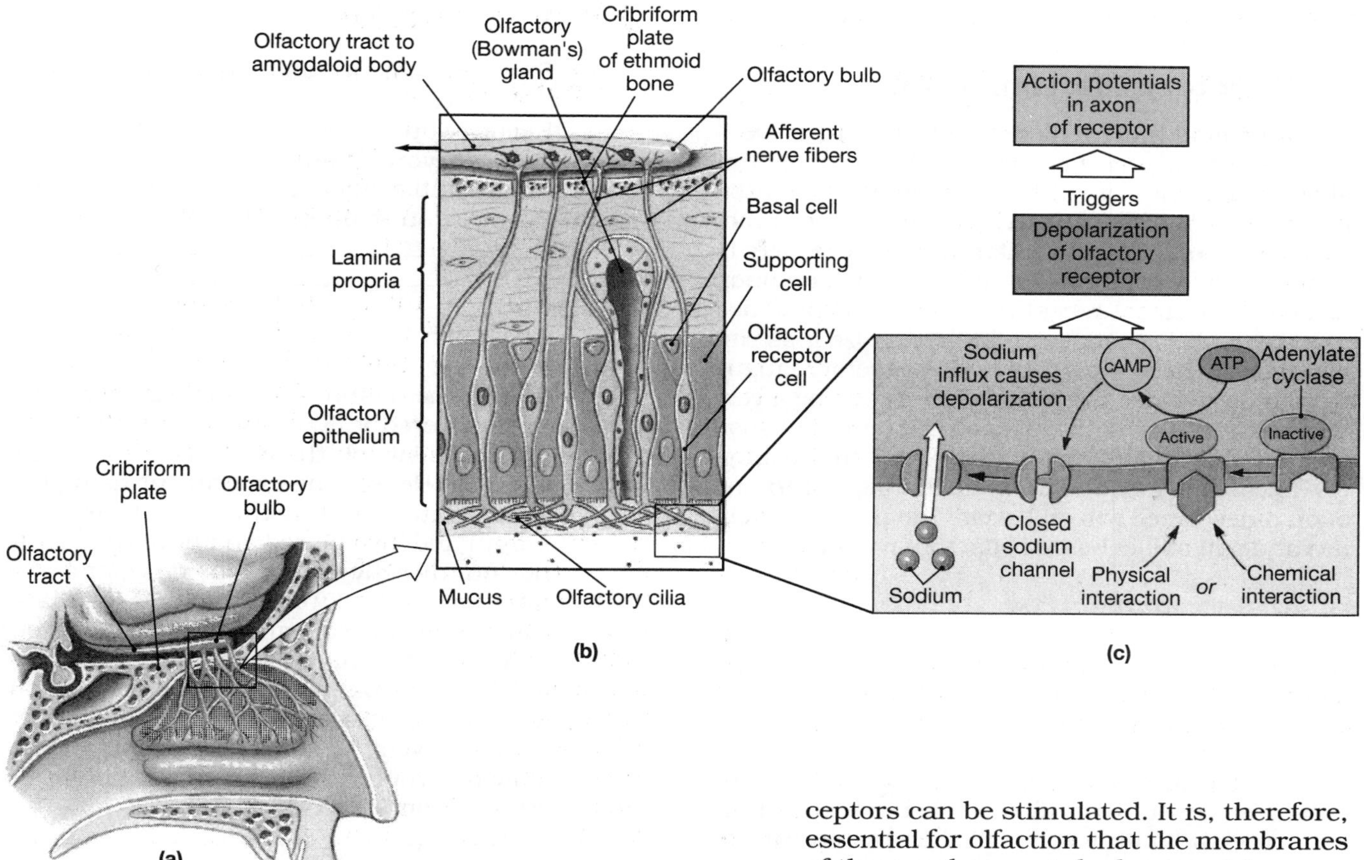

● **FIGURE 42–1**
Olfactory Structures. (a) The structure of the olfactory organ on the right side of the nasal septum. **(b)** An olfactory receptor is a modified neuron with multiple cilia extending from its free surface. **(c)** Steps in the transduction process.

Olfactory receptor cells are actually bipolar neurons, the only neuronal cells in the mature human that regularly reproduce. Along with the olfactory cells, the olfactory epithelium is also made up of **basal cells** (stem cells that divide by mitosis to produce additional olfactory receptors) and **supporting cells** (also known as **sustentacular** cells). The olfactory cells form knobs on the epithelial surface and from these knobs protrude olfactory cilia. *[∞ Martini, p. 535]*

Beneath the basal membrane of the epithelium are **olfactory glands**, which produce a yellow-brown mucus containing **lipofucsin**, a granule found in many of the neurons. Water-soluble and lipid-soluble compounds diffuse into this mucus before stimulating the olfactory receptors. The transduction of these stimuli generate an action potential.

1➡ Notice that incoming substances must be in a dissolved state before the olfactory receptors can be stimulated. It is, therefore, essential for olfaction that the membranes of the nasal passage be kept moist.

2➡ Obtain a microscope slide of the nasal epithelium. Identify the cells listed in the preceding paragraphs.

3➡ Sketch and label what you see.

Nasal epithelium cells

C. Summary

The olfactory nerve pathway can be summed up as follows. The olfactory receptor in the nasal epithelium is activated. The axons of these receptor cells collect in bundles and penetrate the cribriform plate and reach the olfactory bulbs, where they synapse with other neurons. Axons leaving the olfactory bulbs travel along the olfactory tract (N I) to the olfactory cortex, the hypothalamus, and portions of the limbic system. (Notice that, unlike other nerve tracts, no synapse occurs in the thalamus.) *[∞ Martini, p. 535]*

❑ Experimentation

II. Olfactory Discrimination

In performing these experiments, be sure to work quickly. Either pause between each experiment or alternate experiments with your partner. The reason for this is that **olfactory adaptation,** or accommodation, sometimes called **olfactory fatigue**, sets in very quickly. Olfactory adaptation is that condition whereby the nasal receptors become accustomed to a particular scent and fail to recognize it as unusual or different. You may notice olfactory fatigue in your own home. Suppose you are cooking a very distinct stew. You notice the aroma at first but soon you are oblivious of any distinct odor. Interestingly, however, while you may have adapted to one odor, a new smell will still excite the nasal epithelium and will still cause an olfactory response.

A. Spice Identification

1➡ Locate the selection of common household spices that have been placed in unmarked bottles. Without looking at the spice in the bottle and without conferring with your classmates, remove the stopper, smell the spice, and identify the substance by writing what you think it is in the space provided.

1. ____________
2. ____________
3. ____________
4. ____________
5. ____________
6. ____________
7. ____________
8. ____________
9. ____________
10. ____________

B. "Mouth Smelling"

1➡ Plug your nose and concentrate on breathing through your mouth. Now repeat the spice identification test. Can you identify any of the spices by picking up the smell through your mouth? Do any of the spices "smell" the same? What other impressions do you have?

____

____

C. Cold Identification

1➡ Suck on some ice chips until your mouth is quite cold. Repeat the above experiment. Record your impressions and any deviations in your results. You will probably have trouble smelling certain spices at all! Why do you suppose this is?

____

____

D. Smell and Taste

1➡ Review the relationship between smell and taste covered in Exercise 41, Section II.B. This time switch the emphasis from taste to smell. Repeat the experiment using the same materials, but this time keep your mouth and eyes closed and have your partner put directly under your nose, one at a time, equal-sized wedges of apple (or pear), raw potato, carrot, and onion. See if you can identify which is which. If you were able to figure out which was which, did you do so on smell alone or were other factors also involved? Record your observations. How did your results compare with your results from Exercise 41?

____

____

E. Dry Smelling

1➡ Take whichever "dry" objects (from the Materials Needed Section at the beginning of this exercise) you have chosen and repeat experiments A though D. Record your results.

____

____

____

____

F. Discussion

1➡ Discuss the results and the implications of these experiments, first with your partner and then with the entire class.

❑ Additional Activities

NOTES

1. Relate the results obtained in this exercise with the results from Exercise 41.
2. Take the frozen food. Close your eyes and try to smell the substance. Record your impressions. Also record how long it took you to reach your conclusions.
3. Now take the frozen food and place it in the microwave oven. Do not add water, butter, or anything else. Heat the food thoroughly. Repeat the smell test. What was the difference in time? Why did the difference exist?

❑ Lab Report

1. Referring to the numbered inquiries at the beginning of this exercise, complete the following box summary:

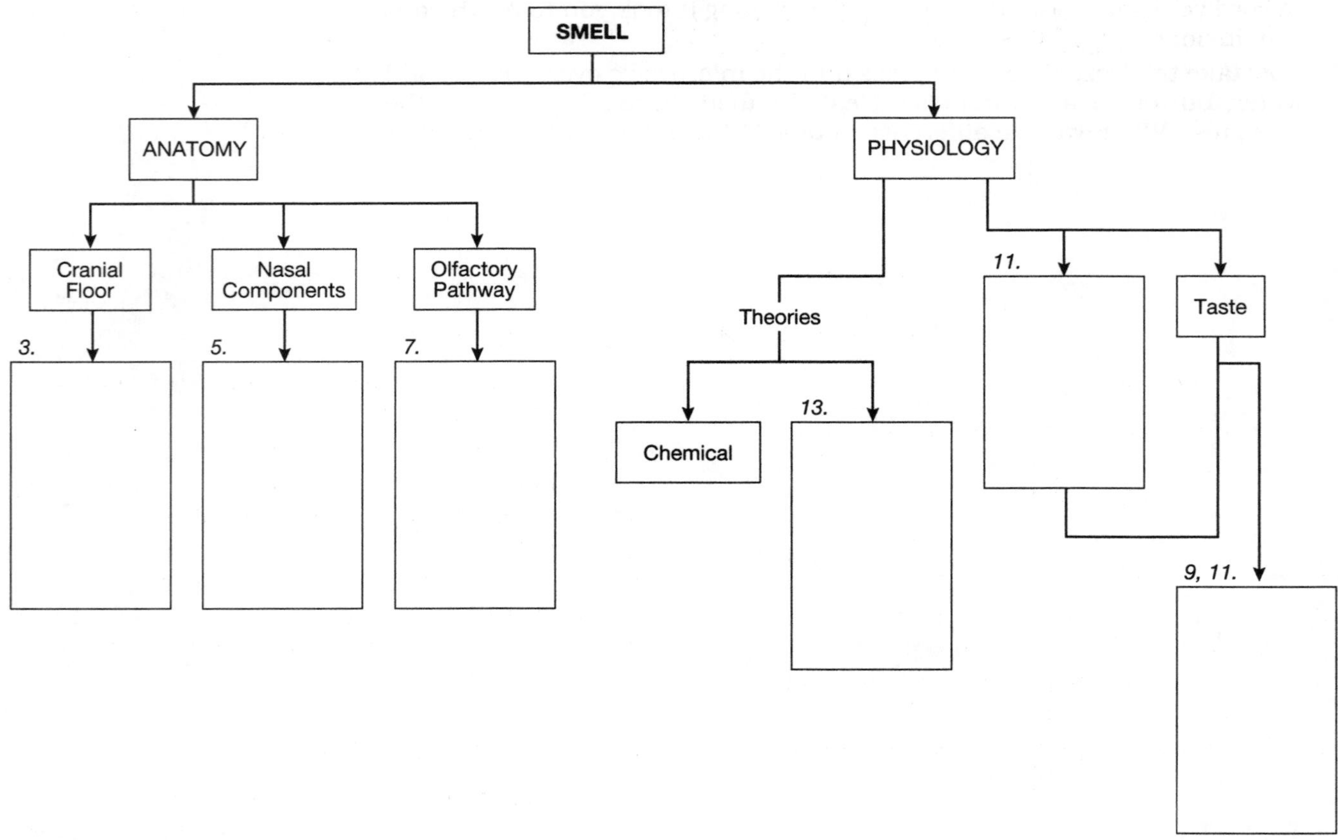

General Questions

NOTES

1. Some people believe that sense of smell is keener in the rain forest than in the desert. What would be the basis for such a belief?
2. What would happen to your sense of smell if the olfactory glands were inhibited from secreting the olfactory mucus?
3. What is the role of water in the sense of smell?
4. Identify each of the following:
   a. Cribriform plate
   b. Crista galli
   c. Olfactory bulbs
   d. Olfactory nerves
   e. Olfactory foramina
   f. Olfactory receptor cells
   g. Olfactory epithelium
   h. Olfactory glands
   i. Basal cells
   j. Supporting (sustentacular) cells
   k. Lipofucsin
5. When you did the spice test ("smell identification"), did you sniff? If so, what did the sniffing do?
6. Compare the results of the "cold identification" test with what you think would happen if you had olfactory nerve damage.
7. List five factors which might affect your olfactory perception.

EXERCISE 43

Anatomy and Physiology of Touch

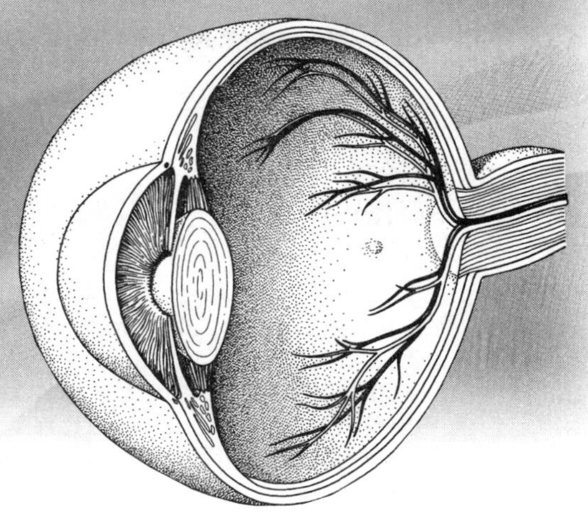

PROCEDURAL INQUIRIES

Preparation

1. What are the types of sensory receptors?
2. Which touch receptors can be found on a dermal cross section?

Examination

3. Which structures can be identified on a histological slide of the skin?
4. Which structures can be seen on a wet mount of a hair shaft?
5. Which nervous pathways are involved in our perception of touch?

Experimentation

6. What does the two-point discrimination test demonstrate?
7. How can we determine superficial tactile receptor density?
8. What does tactile localization demonstrate?
9. What does tactile adaptation demonstrate?
10. How do we map the density of cold receptors?
11. How do we map the density of heat receptors?
12. How do we demonstrate temperature adaptation?

Additional Inquiries

13. What do we mean by these terms?
    Adaptation, Referred Pain, Phantom Pain
14. How can we test for referred pain?

Key Terms

Adaptation
Cold
Corpuscles
Dermis
Encapsulation
Exteroceptors
Free Nerve Endings
Hair Follicle
Hair Shaft
Heat
Interoceptors
Meissner's Corpuscles
Merkel's Disks
Lamellated Corpuscles
Pain
Proprioceptors
Referred Pain
Root Hair Plexus
Ruffini Corpuscles
Sensory Receptors
Superficial Tactile Receptor Density
Tactile Adaptation
Tactile Localization
Temperature Adaptation
Thermoreceptors
Touch
Two-Point Discrimination

Materials Needed

Histological Slides of Sensory Receptors
- Lamellated Corpuscles
- Meissner's Corpuscles

Model of Skin (Cross Section)
Slides of Dermal Sections
Forceps
Slides and Cover Slips
Microscope
Hand Lens or Magnifying Glass
Caliper or Measuring Compass (or 2 pins)
Metric Ruler
Markers (Different Colors)
Watch with Second Hand
Metal Probe with Blunt Nose

Coins
Ice Water
Hot Water (about 45° C)
Room Temperature Water

Imagine yourself suspended in a vacuum without even a remnant of clothing touching your body. That absence of physical contact is difficult to imagine because no matter what we do we are in continuous communication with some aspect of the world around us. **Touch** is the sense that allows us to perceive our physical environment by direct physical contact. In most ways we are so well-adapted to our physical environment that unless we deliberately think about it, we ignore the normal and are only aware of a feeling that is out of the ordinary, such as a sharp pain or a sip of a hot drink we thought was cold.

In some ways studying touch is studying an extension of the integumentary system because so much of this sense is found in the **dermis**. (Refer back to Exercise 13.) We think of the **sensory receptors** as being a part of the integument. But we can also think of the sensory receptors in connection with the standard reflex arc. (See Exercise 29.) Upon stimulation these sensory receptors transmit electrical messages to the spinal cord. The sensory neurons synapse, either directly or indirectly, both with the motor neurons and with the ascending spinal tracts.

Touch is actually part of an entire class of cutaneous sense receptors. Sensory receptors are either **free** (bare) or **modified** dendritic endings. Recall that the dendrite is the "beginning point" of the sensory neuron—the beginning point of sensory transmission to the brain. Dendritic modifications include **encapsulation**. **Corpuscles** of the nervous system are, by definition, encapsulated.

We often lump the sense receptors together and call them all touch. In actuality, we should separate the cutaneous receptors into **touch**, **pain**, **cold**, and **heat**. The difference between touch and pain is often a matter of degree; nevertheless, we still have separate receptors sensitive to each sensation. Regarding heat and cold, although anatomically identical, different **thermoreceptors** are sensitive to different temperature ranges.

The generalization of lumping all sensations together as touch is not entirely incorrect, however, if we consider that we would not feel pain if we did not "touch" something. Furthermore, we usually perceive cold and heat because we "touch" something, whether that be the molecules from a blast of arctic air or the flame from a hot stove.

Touch is essential for us in knowing our world. We judge the texture of the skin and make decisions regarding the firmness of the foods we eat based on touch. Despite our immediate perceptions of discomfort, it is important to remember that our sensitivity to pain and temperature can mean the difference between life and death. For instance, if you step on a sharp object, the resulting pain is a signal of danger and you become aware of where you are walking. In the same way, our thermoreceptors can indicate potential dangers from temperature extremes.

In this laboratory exercise we will examine the anatomy and physiology of touch, first by studying the components of the sensory system and then by doing certain experiments to demonstrate the principles of physical sense perception.

❑ Preparation

I. Sensory Receptors

Sensory receptors can be divided into three functional categories. **Exteroceptors** provide information about the external environment. **Proprioceptors** provide information about the body's position and movement. **Interoceptors** (also known as enteroceptors or visceroceptors) provide information about the body's internal systems and internal operations. Proprioceptors are covered specifically in Exercise 40. Interoceptors, which are functions of the visceral sensory neurons, are discussed with the digestive, respiratory, cardiovascular, urinary, and reproductive exercises. *[∞ Martini, p. 530]*

In this exercise, we are concerned with the exteroceptors, particularly those exteroceptors responsible for our sensations of touch, pain, cold, and heat. (The exteroceptors for hearing and sight are covered in other laboratory exercises.)

A. Dermal Cross Section

1➡ Study Figure 43–1●. Note the location of the following structures:

Hair Shaft The hair shaft extends into a **hair follicle**. Use this apparatus as your point of reference.

Free Nerve Endings These superficial tactile receptors, found scattered throughout the lower epidermis, are primarily pain receptors. Free nerve endings also give us our awareness of articles in continuous contact with our bodies, such as our clothes. (Did you notice on reading that last sentence that you were suddenly aware of the feel of your clothing?) Note in the figure how many of these nerve endings are branched.

Root Hair Plexus These coiled free dendritic endings wrap around the bulb of the hair follicle and monitor distortions and movements across the body surface.

● **FIGURE 43–1**
Tactile Receptors in the Skin.

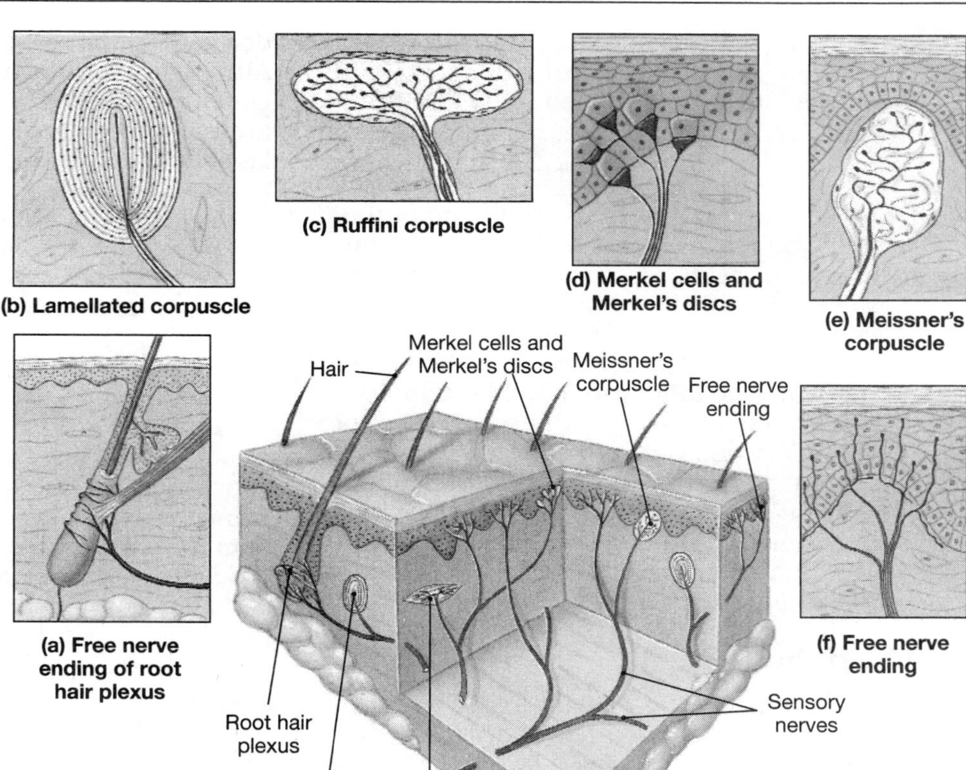

Merkel's Disks These flattened free dendritic endings, bound to the Merkel cells in the lower epidermis, are highly sensitive to light touch.

Meissner's Corpuscles These oval receptors, sometimes found in clusters, are surrounded by connective tissue and are most numerous in areas with little or no hair. These corpuscles are highly responsive to light (fine) touch and low-frequency vibration. They are also functional in determining the texture of a touched object.

Lamellated Corpuscles These corpuscles are layered (thus the name "lamellated") and are commonly found in joints, perimyseal tissue, certain visceral organs, the hands and feet, the wall of the urinary bladder, the breasts, and the external genitalia. Lamellated corpuscles are deep corpuscles and respond primarily to heavy pressure and internal vibrations.

Ruffini Corpuscles These dermal stretch and distortion receptors are encapsulated networks of dendrites intertwined with collagen fibers. Thermoreceptors are scattered beneath the surface of the skin, with warm receptors generally three or four times more numerous than cold receptors. No structural differences have been found between the free nerve endings of these thermoreceptors.

❑ Examination

B. Histological Observation

1 ➡ Study the laboratory skin model and then examine the histological slides of the skin. Note the different types of sensory receptors mentioned in the previous section. Identify these receptors.

2 ➡ Sketch and label what you see.

Skin receptors

C. Hair Follicle Observation

1 ➡ Take the hand lens or magnifying glass and scan your arm or leg. Note the density and textures of the hairs. Note too the external indentations marking the beginnings of the hair follicles.

2 ➡ Pluck a hair from either your head or your leg with the forceps, using a swift, straight movement. Be certain that you pluck the entire bulb. As you are plucking, make note of the point location of the pain. Check Figure 43–1● to ascertain the cause of your

pain. Which type of dendritic ending was probably involved? Did you notice any pain at a point other than the hair follicle?

____

____

3➡ Make a wet mount of the bulb end of this hair and examine this bulb under the microscope, using high power. Although you will not be able to see any actual free nerve endings, you should notice that the bulb is a slightly different color than the rest of the hair. The bulb also has a torn appearance. Quite literally you ripped the edges of the bulb when you yanked it away from its vascular and nervous plexes. Depending on your plucking technique, you may notice some other tissue just above the bulb. That may be part of a sebaceous gland, a sweat gland, or even a receptor corpuscle.

II. Physiology

A. Nervous Pathway

A sensory receptor is geared to transform an environmental stimulus into a nerve (electrical) impulse. The action potential generated by the environmental stimulus travels via the dendrite (more correctly, the peripheral branch) of a unipolar (pseudounipolar) neuron. The action potential bypasses the nerve cell body in the dorsal root ganglion and continues via the axon (more correctly, the central branch) on into the gray matter of the spinal cord. [∞ *Martini, p. 529*]

Within the spinal cord, the sensory neuron synapses with one or more additional neurons. One result of these synapses is a reflex response. (Refer back to Exercise 34.) For example, the free nerve endings of the thermoreceptors are stimulated as you touch a hot stove. As an immediate reflex, a motor neuron is stimulated and you move your hand. Another synaptic result is your conscious awareness of the hot stove. In this case, the electrical impulse travels to the brain via an ascending spinal tract. When the message reaches the cerebral cortex, you can make a conscious decision to activate other motor neurons and turn off the stove.

1➡ Use Figure 43–2● as your guide to write out the pathways involved in the major types of tactile perception.

____

____

____

____

B. Nervous Modification

Adaptation is the tendency for the rate of electrical discharge (generation of action potentials) to decrease in the presence of prolonged stimulation. Although you are in constant contact with your clothes, unless you consciously think about it, you are not aware of the supposed continuous stimulation of the touch receptors just beneath your skin.

1➡ Think about yourself sitting in your chair with your clothes on. Do you feel your clothes or your chair now that you are thinking about them?

____

____

Referred pain is the phenomenon that occurs when a person is aware of a sensation in a part of the body other than the part that was stimulated. For instance, have you ever experienced a cold associated pain in your eye after eating ice cream? __________

Some referred pain is highly predictable and can be used in medical diagnosis. For example, a person with gall bladder problems often experiences banded pain in the back and right shoulder. Pain perception down the medial side of the left arm may indicate a heart attack. Sometimes, however, referred pain can be medically misleading. Classic appendicitis produces a localized pain in the lower right quadrant (the exact area of the appendix) but some patients experience referred pain in the left quadrant, or even in one of the upper quadrants. The complete physiology of referred pain is not well understood.

(When you plucked the hair in the previous section of this exercise, you may have noticed a muscle twitch in a different part of your body. Although this is not exactly referred pain, the principle is the same. The motor response was projected, or referred, away from the sensory stimulus.)

Phantom pain occurs when a person experiences pain in a body part that has been amputated. After amputation the remaining parts of the sensory neurons seem to be intact but the reasons why pain is perceived is not known.

❑ Experimentation

III. Tests

The following tests have been designed to check your sensory perception. Work with a partner and **close your eyes for all tests**.

A. Two-Point Discrimination

Although touch receptors are found throughout the body, the density of these receptors varies considerably from one part of the body to another. **Two-point discrimination** is the ability to perceive the stimulation of two separate sensory receptors.

● FIGURE 43–2
Sensory Pathways.

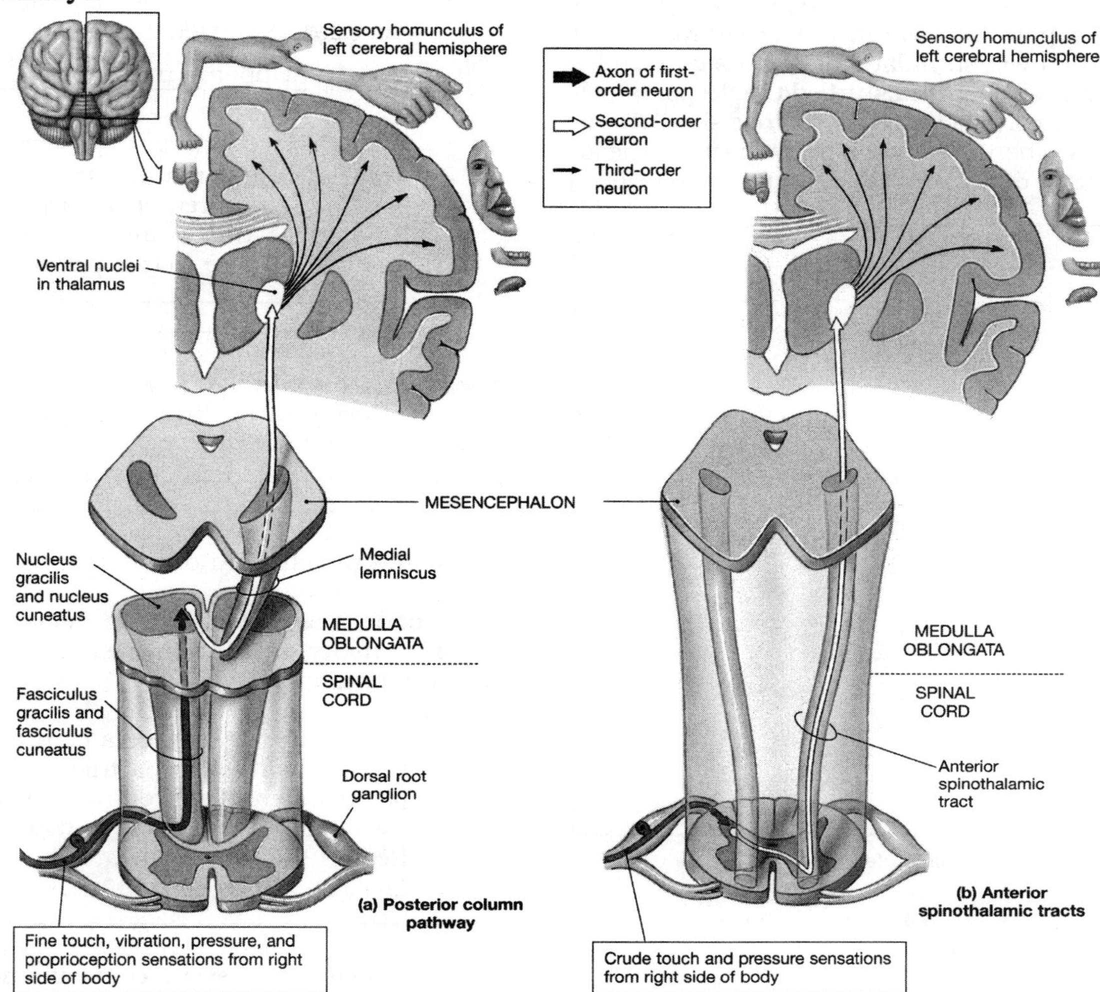

1➡ Use a caliper, measuring compass, or two pins to test the arm, neck (dorsal and ventral), hand (dorsal and ventral), fingertips, lips, dorsal calf, and any other surface point you may deem appropriate. Make sure your partner is comfortable. Start with the two points together and gradually move the stimuli away from each other. Your partner should tell you when (s)he feels the points as two separate sensory perceptions. Measure the distance between the points in millimeters. Record your data on the chart.

Location	**Distance between perceived points**
Arm	
Neck (Ventral)	
Neck (Dorsal)	
Hand (Ventral)	
Hand (Dorsal	
Fingertips	
Lips	
Dorsal Calf	
Other	

2➡ Compare class results. Which parts of the body seem to have the most sensory nerve endings? What advantage do you see in this?

____

____

B. Superficial Tactile Receptor Density

Superficial tactile receptor density can be determined by constructing a density map.

1➡ Draw a 2-cm square on your partner's anterior forearm. Use a sharp pin and very lightly tap around within the square. The light touch will stimulate Meissner's corpuscles but will leave undisturbed the deeper lamellated corpuscles. Your partner should indicate when a sensation is felt. Indicate the location of the corpuscles on the map below.

Corpuscle map

2➡ Compare class results. What conclusions can you draw?

____

____

3➡ If time permits, repeat this experiment on other parts of the body.

C. Tactile Localization

Tactile localization is the ability to perceive which part of the body has been touched.

1➡ Use a colored marker and be certain your partner has a marker of a different color. With your marker, touch a place on the palm of your partner's hand. Your partner should now try to touch that same place with his/her marker. Measure and record the error on the chart below. Repeat this test on the upper back, fingertip, ventral forearm, and ventral upper arm, plus any other place you deem appropriate.

Location	Error Distance
Palm of Hand	
Upper Back	
Fingertip	
Ventral Forearm	
Ventral Upper Arm	
Other	

2➡ Compare class results. In which parts of the body was greater accuracy achieved? What does this demonstrate? What advantage is there to this phenomenon?

____

____

Concept Check 1 If time permits, repeat the preceding test. Does your accuracy rating improve? Why or why not?

____

____

D. Tactile Adaptation

Tactile adaptation is the decrease in conscious sensory perception over a period of prolonged stimulation.

1➡ Check the time. Place a coin on the anterior surface of your partner's forearm. Your partner should tell you the moment (s)he is no longer aware of the coin sensation. Record the time below.

____

____

2➡ Try stacking several coins and determine if this increased weight alters the sensation duration. Also, try holding your partner's arm above, on, and below the coin. What changes do you notice?

____

____

3➡ If time permits, try this test on different parts of the body. Record your results.

Coin	Time
1 Coin — Forearm	
Several Coins — Forearm	
1 Coin — Hold arm above	
1 Coin — Hold arm on coin	
1 Coin — Hold arm below coin	
Other	

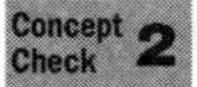

Without moving, think about your left foot. What do you feel? Verbalize the different sensations coming from different parts of your foot, particularly in relation to your shoe and sock. If you had not stopped to think about it, you probably would have been unaware of the sensations surrounding your foot. While you were not thinking about them, had those sensations ceased? Or had you simply lost awareness?

____

____

IV. Cold and Heat

From the standpoint of physics, cold and heat are terms relative to a specific number of calories in a given situation. From the standpoint of biological touch, cold and heat are phenomena perceived by two different sets of receptors, one on each side of a biological median. We can map the numbers and locations of these receptors.

A. Cold

Cold can be mapped in the same way we mapped the density of the tactile receptors.

1 ➡ Place a metal probe in the ice water while you check the map on your partner's forearm. You may use the same 2-cm square or you may wish to draw another map. When your probe is cold, quickly dry it and begin testing different points within the square. Do not apply deep pressure and do not hold the probe too long in one place. Remember, you are testing for an awareness of cold and not an awareness of touch. Map the points of cold perception as indicated by your partner. Keep the probe cold.

Cold receptor map

B. Heat

Heat mapping is done the same way as cold mapping.

1 ➡ Map these thermoreceptors using a probe heated to about 45 degrees and follow the directions given for the cold map. Keep the probe hot.

Heat receptor map

C. Temperature Adaptation

Temperature adaptation is the phenomenon we experience every summer and every winter. If the summer temperature dips to 40° (F), we feel cold. If the winter temperature rises to 40° (F), we feel warm. We can demonstrate this phenomenon more precisely in the laboratory.

1 ➡ Take a bowl of ice water, a bowl of hot water, and a bowl of room temperature water. Have your partner place one hand in the ice water and one hand in the hot water. Record his/her immediate sensations. Wait two minutes (while your partner leaves his/her hands in the water). Now what sensations does your partner perceive? Did one hand adapt more quickly than the other? Your partner should now simultaneously take both hands out of the hot and cold waters and place both hands in the same bowl of room temperature water and describe the immediate perceptions for each hand.

V. Pain

We will not test for direct pain, although any of the preceding tests could be used for pain perception. The difference between touch and pain is often a matter of degree. You should be aware that cold and heat receptors register pain rather than temperature if the stimulus is beyond certain limits.

A. Referred Pain

1➡ Recall what we said about the pain from plucking your hair. Although the principle was the same, that was not true referred pain. However, we can demonstrate a real referred pain.

2➡ Have your partner immerse one elbow in a bowl of ice water. Record his/her immediate sensations. Your partner should concentrate on any change in sensation. Over the next 2 minutes (s)he should notice pain. The ulnar nerve is involved and your partner should feel a definite discomfort in the medial two fingers and the medial side of the hand. Write down the sensations as your partner describes them.

❑ Additional Activities

NOTES

1. Research the life and work of
   a. Georg Meissner (1829–1905), German histologist.
   b. Filippo Pacini (1812–1883), Italian anatomist.
   c. Angelo Ruffini (1864–1929), Italian anatomist.
   d. Wilhelm J. F. Krause (1833–1910), German anatomist.
2. We have the same number of hair follicles as the ape. Compare the types of hair we have with the types of hair the ape has. What are the advantages and disadvantages of each?
3. Do a comparative study of sensory reception as found in different animals.
4. Trace the specific nervous pathways involved in different sensory reactions described in this lesson.

Answers to Selected Concept Check Questions

1. Your accuracy *might* improve as your conscious awareness increases. Some additional corpuscles *may be* activated. Increased accuracy will depend on your individual situation.
2. The sensation had not ceased. We selectively process important information and ignore the rest. This is tactile adaptation.

❑ Lab Report

1. Referring to the numbered inquiries at the beginning of this exercise, complete the following box summary:

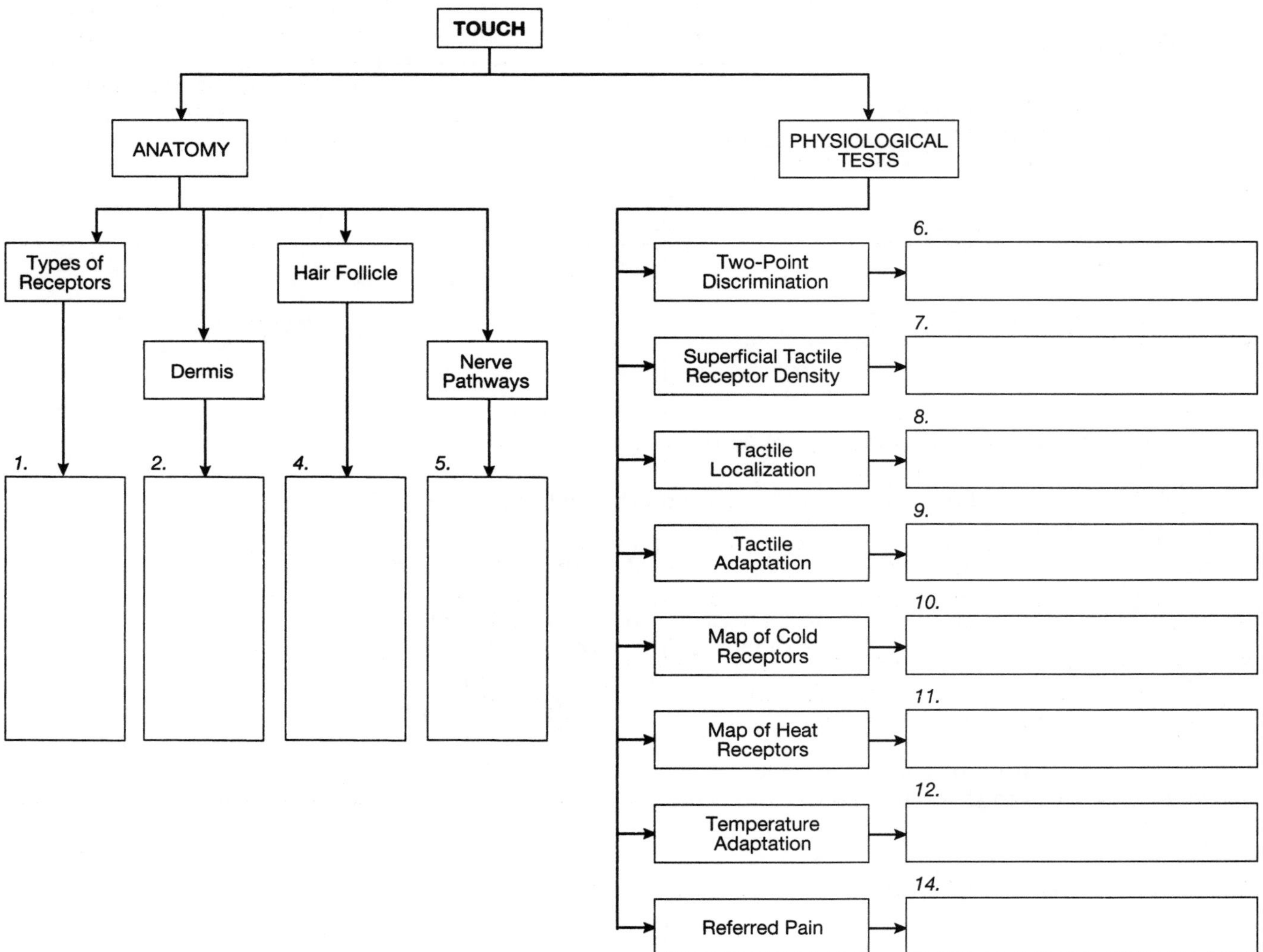

General Questions

NOTES

1. Fill in this chart

Nerve Tissue	Location	Function
Free Nerve Endings		
Merkel's Disks		
Meissner's Corpuscles		
Lamellated Corpuscles		
Ruffini Corpuscles		

2. How does the function of each type of sensory receptor relate to its location?
3. Which sensory receptors were you able to see on the histological slides? What prevented you from seeing all of them?
4. What is the psychological advantage of adaptation in the presence of extreme pain?
5. Consider the nerve impulse pathways involved in touching a hot stove and in feeling the texture of a piece of cloth. Compare these pathways.
6. What is the difference between referred pain and phantom pain?
7. Why were you instructed to close your eyes for the sensory perception tests?
8. Compare the results of your corpuscle, heat, and cold maps. Why are there differences in these three maps?

EXERCISE 44

Endocrine System: An Overview

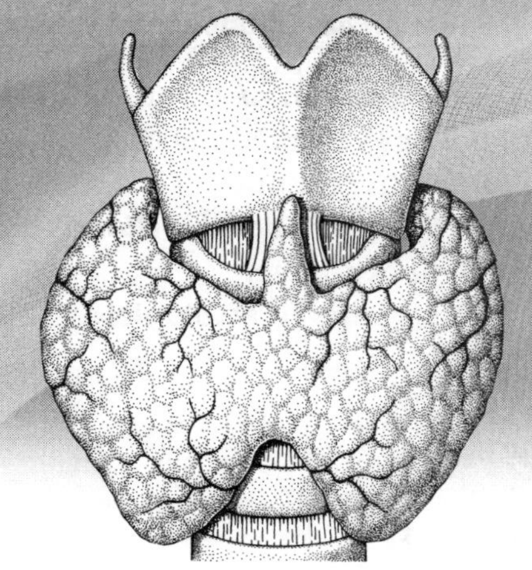

PROCEDURAL INQUIRIES

Preparation

1. What is the endocrine system?
2. What is a hormone?
3. What is an endocrine gland?

Examination

For each gland listed below, answer these questions:
Where is the gland located?
What are the principle hormones secreted by this gland?
What is the function of each of these hormones?

4. Pineal
5. Hypothalamus
6. Pituitary
7. Thyroid
8. Parathyroid
9. Thymus
10. Pancreas
11. Adrenal
12. Ovaries
13. Testes

Additional Inquiries

14. What are other organs with endocrine functions?
15. Which endocrine glands are actually two glands in one?

Key Terms

Adenohypophysis
Adrenal Gland
- Androgens
- Epinephrine
- Glucocorticoids
- Mineralocorticoids
- Norepinephrine

Anterior Pituitary
- Adrenocorticotrophic Hormone
- Follicle-Stimulating Hormone
- Growth Hormone
- Luteinizing Hormone or Interstitial Cell-Stimulating Hormone
- Melanocyte-Stimulating Hormone
- Prolactin
- Thyroid-Stimulating Hormone

Hypophysis
Hypothalamus
Neurohypophysis
Ovary
- Estrogens
- Progesterone

Pancreas
- Glucagon
- Insulin

Parathyroid
- Parathormone

Pars Distalis
Pars Intermedia
Pars Nervosa
Pineal
- Melatonin

Pituitary
Posterior Pituitary
- Antidiuretic Hormone
- Oxytocin

Testes (sing., testis)
- Inhibin
- Testosterone

Thymus
- Thymosins

Thyroid
- Calcitonin
- Thyroxine

Materials Needed

Models and Charts of Endocrine System
- Torso
- Trachea
- Kidney
- Stomach with Pancreas

Midsagittal Section of Mammalian Brain
(Model or Preserved Section)
Compound Microscope
Prepared Slides of Endocrine Glands and Structures
- Pineal (if available)
- Hypothalamus (if available)
- Pituitary
- Thyroid
- Parathyroid
- Thymus
- Pancreas
- Adrenal
  - Cortex
  - Medulla
- Ovary
- Testis

The endocrine system includes the cells and tissues of the body that secrete many of our most familiar hormones. The word "hormone," coined by Bayliss and Starling in 1902, is Greek for "I stir up" or "I stimulate." Classically, a hormone is a chemical messenger produced in one part of the body and carried, usually via the cardiovascular system, to a target organ (or functional site) in another part of the body. Today we recognize that certain other organs (such as the placenta, the stomach, the heart, and the organs producing the pheromones) have endocrine functions. In addition, numerous "local hormones,"or paracrine factors (such as prostaglandins, erythropoietin, and histamine) have also been identified.

Hormones up-regulate or down-regulate function but do not contribute energy to that function. Endocrine glands are ductless glands, glands with no tubular secretory system. (In contrast, exocrine glands empty their products directly into specialized tubes or ducts.) Some organs, such as the pancreas, have both endocrine and exocrine components.

In this exercise we will examine the anatomy and histology of the endocrine system by studying appropriate models and charts and by looking at prepared microscope slides of the endocrine glands.

❑ Preparation

I. Approaching the Endocrine System

A. Charts and Models

1➡ Refer to models delineating the human endocrine system. A model torso is excellent for gaining an overall perspective of where

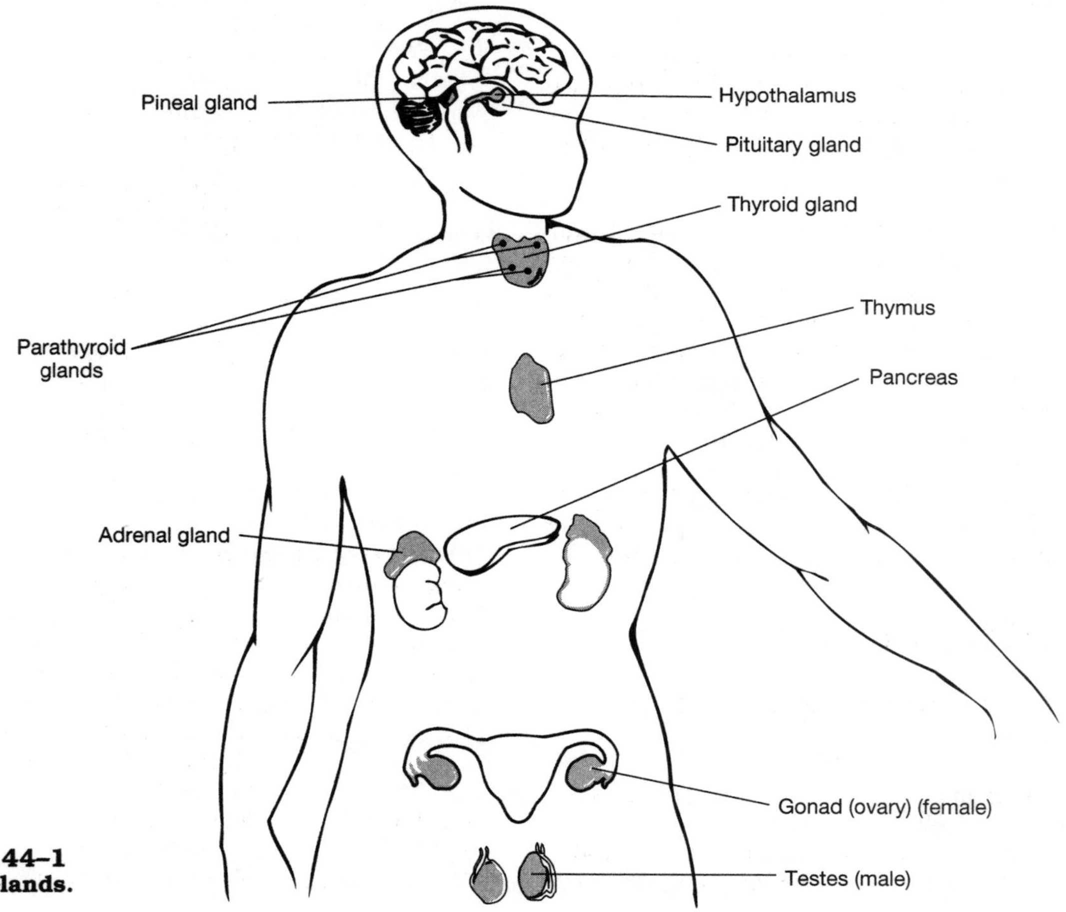

● **FIGURE 44–1**
Endocrine Glands.

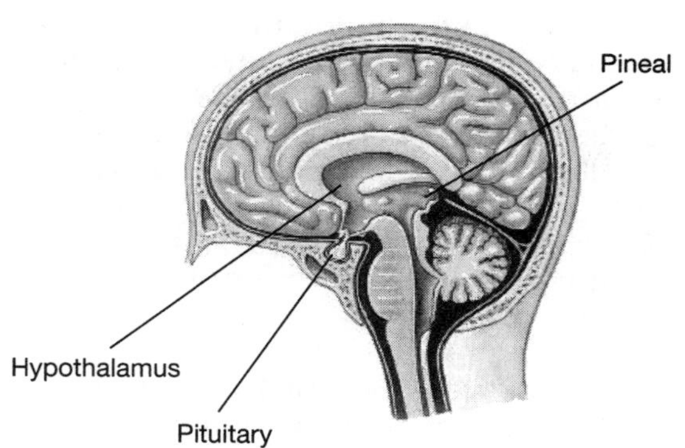

● **FIGURE 44–2**
Relative Locations of the Hypothalamus, Pineal Gland, and Pituitary Gland.

the endocrine glands are located. Models of the individual organs will help in gaining a perspective of the endocrine structures.

2➡ Study Figure 44–1●, which delineates the location of the principal endocrine structures in the human body. Refer also to other figures in this book or check other available resources for additional figures and charts of the endocrine system.

B. Slides

Since the endocrine structures are quite similar throughout the mammalian world, if you are viewing the endocrine system of some other mammal, those structures will be analogous to the human structures.

1➡ Use the prepared slides that your instructor has made available for you. View the slides first under low power and then under high power. It should not be necessary for you to use the oil lens for these slides, unless your instructor directs you to do so. Refer to the accompanying diagrams to identify the parts of the endocrine structures.

2➡ Sketch and label what you see. When you are finished, compare your sketches with the diagrams or histology plates in this book.

❑ Examination

II. Endocrine Structures *[∞ Martini, Chapter 18]*

A. The Pineal Gland [∞ Martini, p. 603]

The **pineal** gland, which is part of the epithalamus, is located in the roof of the third ventricle. This cone-shaped body lies between the corpora quadrigemina and the splenium (posterior portion) of the corpus callosum. In humans, as well as in other animals, the pineal gland begins as a relatively large structure, but as puberty approaches the pineal gland begins to atrophy and pineal sand (small deposits of calcium carbonate) begins to form. With age the gland becomes less glandular in appearance.

1➡ Locate the pineal gland on Figures 44–1● and 44–2● and then find it on the brain model. Although you are examining an adult brain, the human pineal gland does remain prominent. The pineal gland secretes **melatonin**, which inhibits the production of the pigment melanin. Melatonin also slows the maturation of eggs, sperm, and reproductive structures.

If you disected the sheep brain, Exercise 32, how do the size and location of the sheep pineal glands relate to the size and shape of the human pineal gland?

____

____

2➡ Observe a prepared slide of the pineal gland (Figure 44–3●) under low and high power on the microscope. Can you determine if the specimen is from a fetal, juvenile, or adult animal? ____________
Identify the pinealocytes and pineal sand (in adult specimens).

Sketch and label what you see.

Pineal gland

B. The Hypothalamus

The **hypothalamus** is not exactly a gland in the classical sense. The hypothalamus is a part of the brain that houses control and integrative centers for autonomic and emotional functions. Nevertheless, the hypothalamus also integrates with the endocrine system directly by producing two of the pituitary hormones — antidiuretic hormone (ADH) in the supraoptic nucleus and oxytocin in the paraventricular nucleus — and indirectly by producing numerous pituitary-regulating hormones. (The hypothalamus also has direct neural control over the adrenal medulla's secretion of epinephrine and norepinephrine.)

● **FIGURE 44–3**
Pineal Gland.

1➜ Note the location of the hypothalamus on Figure 44–2● then locate it on the brain model.

2➜ Observe the hypothalamus on a microscope slide. It may be on the same slide as the specimen of the pituitary gland.

3➜ Sketch and label what you see.

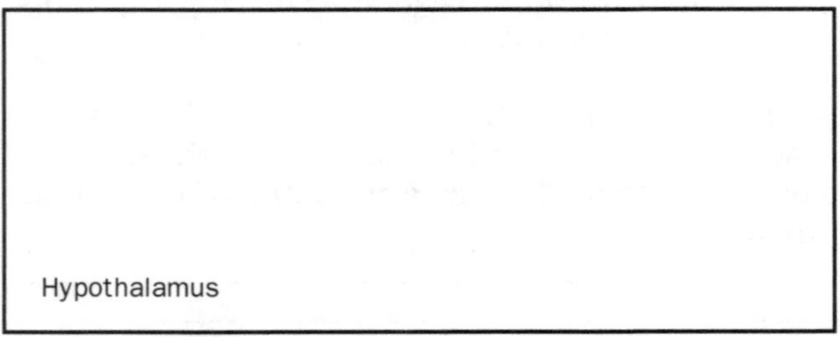

C. THE PITUITARY GLAND [∞ MARTINI, P. 586]

Based on embryology, the **pituitary (hypophysis)** gland is actually two glands. The **anterior** pituitary (also known as the **adenohypophysis** or the **pars distalis**) originates as Rathke's Pouch, an evagination of the oral epithelium that pinches off and migrates toward the floor of the brain. The **posterior** pituitary (also known as the **neurohypophysis** or the **pars nervosa**) results from an evagination of the hypothalamus. This region remains attached to the hypothalamus, and the stalk becomes known as the infundibulum. The **pars intermedia** is actually part of the anterior pituitary. The entire pituitary gland rests on the sella turcica, the saddle of the sphenoid bone. Refer back to Exercise 16 for a diagram of the sella turcica.

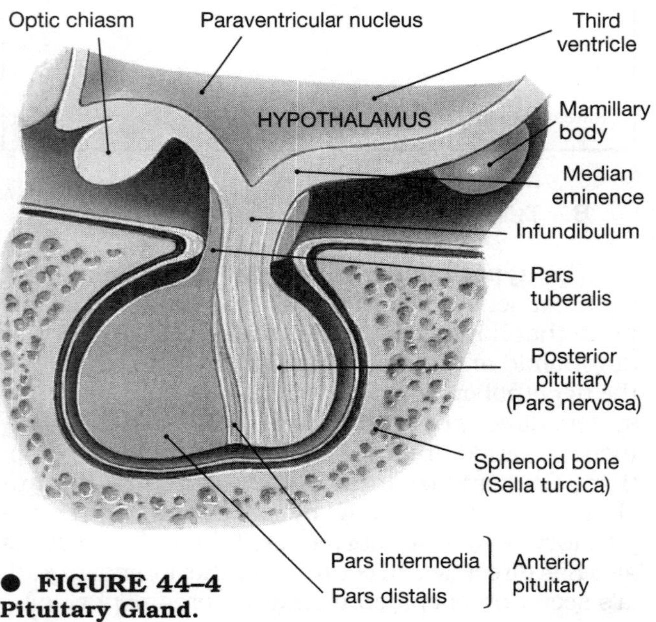

● **FIGURE 44–4**
Pituitary Gland.

1➜ Check Figure 44–2● for the location of the pituitary, and then locate it on the brain model.

2➜ Note Figure 44–4●, which delineates the parts of the pituitary gland in the human.

3➜ Viewing your slide of the pituitary gland, identify as many of the following structures as possible. Use Figure 44–5● as a reference.

a. Pars Distalis (Major Portion of the Anterior Pituitary)
   - Acidophil Cells*
   - Basophil Cells*
   - Chromophobe Cells*

b. Pars Intermedia (Part of Anterior Pituitary)

c. Pars Nervosa (Posterior Pituitary)
   - Nerve Fibers*
   - Pituicytes*

4➜ Sketch and label what you see.

Pituitary gland

5➜ Note that Figure 44–6● outlines the pituitary hormones. You will use this information in the Box Summary.

D. THE THYROID GLAND [∞ MARTINI, P. 593]

The **thyroid** gland is composed of two lobes connected by an **isthmus** with a central superior projection. In humans this gland is located immediately inferior to the thyroid cartilage of the larynx, although in some animals the gland begins at about the third or fourth tracheal ring. In older dogs and cats the thyroid isthmus may virtually

* These structures are not specifically identified in Figure 44–5●. Depending on the slides available in your lab, your instructor may be able to identify the different pituitary cells. Usually the acidophils will be a reddish-pink, the basophils a dark blue, and the chromophobes dusky or colorless. The nerve fibers are exactly that — fibers. The pituicytes will be the cells between the fibers. The pars intermedia is a fibrous band between the anterior pituitary and the posterior pituitary.

disappear. The thyroid is heavily encapsulated so it will appear to be a firm (almost tough) gland.

The thyroid follicles (shown in Figure 44–7●) manufacture, secrete, and store thyroglobulin, which is subsequently released into the circulation as **thyroxine**, also known as T_4, and **triiodothyronine**, T_3. These two hormones increase cellular metabolism.

The C cells of the thyroid secrete **calcitonin**, which decreases calcium ion concentration in the body fluids. (The antagonist of calcitonin is parathormone, secreted by the parathyroid gland.)

1➡ Check Figure 44–7● and locate the position of the thyroid gland on the model of the trachea, noting the lobes and isthmus.

2➡ On the thyroid slide you should be able to locate the following structures that are indicated in Figure 44–7●:

a. Thyroid Follicle
b. Lumen of Follicle
c. Follicular (Principal) Cells
d. C (Parafollicular) Cells
e. Capsule

3➡ Sketch and label what you see.

Thyroid gland

E. The Parathyroid Gland [∞ Martini, p. 597]

The **parathyroid** glands in humans (and most other animals) consist of two pairs of glands. In humans the parathyroids are embedded on the posterior surface of the thyroid gland. Pigs have only one pair of parathyroid glands. Two types of cells found in parathyroid are the **chief cells** and the **oxyphil cells.** The chief cells secrete **parathormone**, the hormone that increases calcium concentration in the body fluids. (The antagonist of parathormone is calcitonin.) The function of the oxyphil cells is not well understood.

1➡ Refer to Figure 44–8● and the trachea model to see the location of the parathyroid glands.

2➡ On your microscope slide find the chief (principal) cells and the oxyphil cells (not identified on Figure 44–8●).

3➡ Sketch and label what you see.

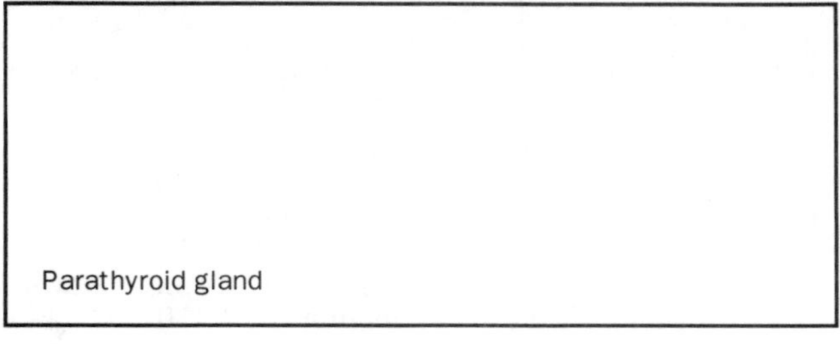

F. The Thymus Gland [∞ Martini, pp. 605, 758, 760]

The **thymus** gland is usually just posterior to the sternum. In fetal and young mammals, the thymus is relatively quite large. The thymus begins to atrophy after puberty and in the adult is small and quite inconspicuous.

The thymus is also part of the lymphatic system (and subsequently the immune system). The **thymosins**, the hormones of the thymus, are essential in the differentiation and development of the T cells. T cells are part of the immune system.

1➡ Locate the thymus on Figure 44–1● and on the trachea model. Your trachea model may not include a thymus. Nevertheless, you should be able to approximate its location.

2➡ If available, observe a microscope slide of the thymus.

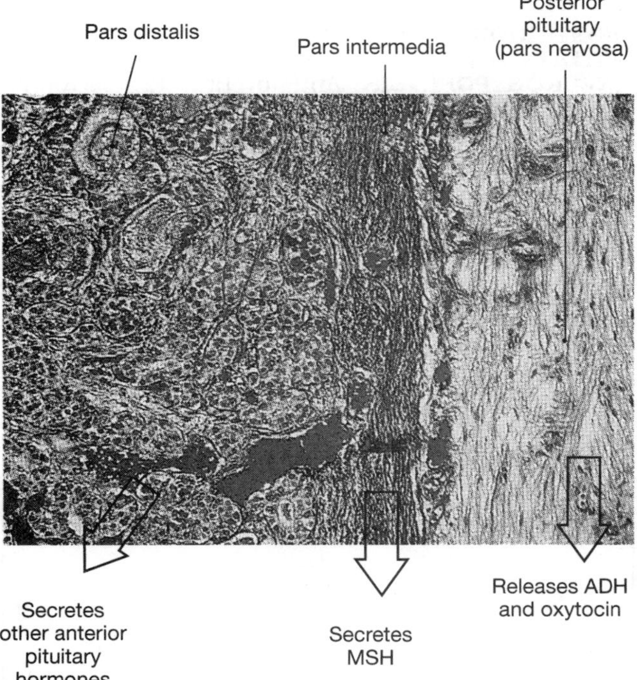

● FIGURE 44–5
Anterior Pituitary Gland — Microscopic View.

Region/Area	*Hormones*	*Targets*	*Hormonal Effects*	*Hypothalamic Regulatory Hormones*
ANTERIOR PITUITARY (ADENOHYPOPHYSIS)				
Pars distalis	Thyroid-stimulating hormone (TSH)	Thyroid gland	Secretion of thyroid hormones	Thyrotropin-releasing hormone (TRH)
	Adrenocorticotropic hormone (ACTH)	Adrenal cortex (fasciculata)	Glucocorticoid secretion	Corticotropin-releasing hormone (CRH)
	Gonadotropic hormones:			
	Follicle-stimulating hormone (FSH)	Follicle cells of ovaries	Estrogen secretion, follicle development	Gonadotropin-releasing hormone (GnRH)
		Sustentacular cells of testes	Sperm maturation	
	Luteinizing hormone (LH) or interstitial cell-stimulating hormone (ICSH)	Follicle cells of ovaries	Ovulation, formation of corpus luteum, progesterone secretion	As above
		Interstitial cells of testes	Testosterone secretion	As above
	Prolactin (PRL)	Mammary glands	Production of milk	Prolactin-inhibiting hormone (PIH), prolactin-releasing hormone
	Growth hormone (GH)	All cells	Growth, protein synthesis, lipid mobilization and catabolism	Growth hormone-releasing hormone (GH-RH), growth hormone-inhibiting hormone (GH-IH)
Pars intermedia (not active in normal adults)	Melanocyte-stimulating hormone (MSH)	Melanocytes	Increased melanin synthesis in epidermis	Melanocyte-stimulating hormone – inhibiting hormone (MSH-IH)
POSTERIOR PITUITARY (NEUROHYPOPHYSIS or PARS NERVOSA)	Antidiuretic hormone (ADH)	Kidneys	Reabsorption of water; elevation of blood volume and pressure	Transported over axons from supraoptic nucleus
	Oxytocin (OT)	Uterus, mammary glands (female); Prostate gland (male)	Labor contractions, milk ejection; Prostatic contractions	Transported over axons from paraventricular nucleus

● **FIGURE 44–6**
Pituitary Hormones.

3➜ If you observed a slide of the thymus, sketch and label what you saw.

Thymus gland

G. The Pancreas [∞ Martini, p. 605]

The **pancreas** is a long slender gland located near the border between the stomach and the small intestine. The clusters of endocrine cells, known as "islets of Langerhans,"are scattered among the exocrine cells. Alpha and beta cells are two of the four endocrine cells of the pancreas. The alpha cells release **glucagon**, the hormone that stimulates an increase in the blood glucose level. The beta cells release **insulin**, the hormone

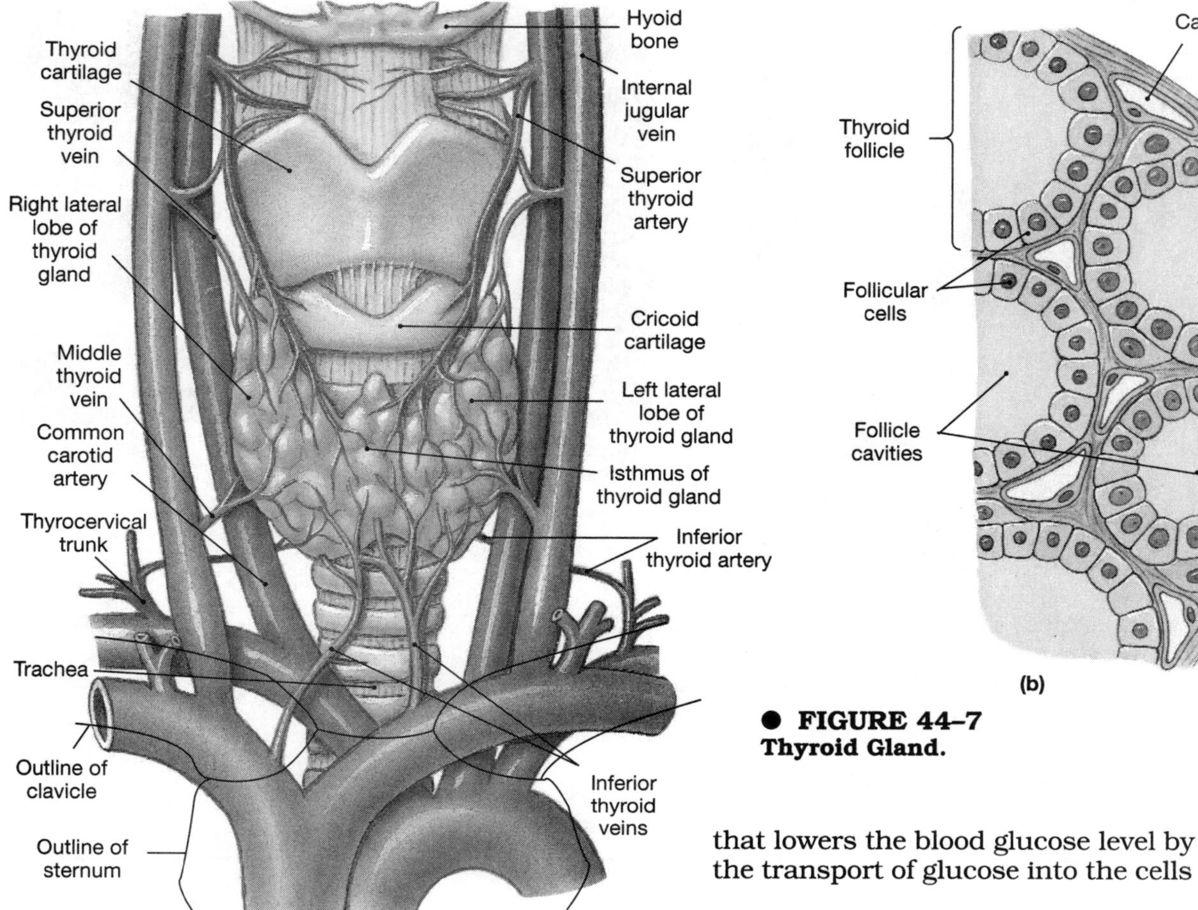

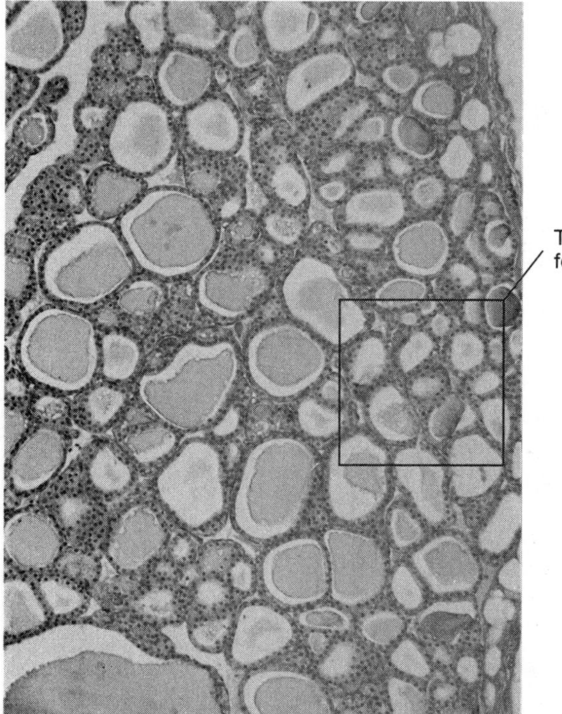

● **FIGURE 44–7**
Thyroid Gland.

that lowers the blood glucose level by assisting in the transport of glucose into the cells of the body.

1➡ Use Figure 44–9● in addition to the charts and diagrams to differentiate between the endocrine and exocrine cells.

2➡ On the slide of the pancreas, locate:
- a. Islets of Langerhans (including alpha and beta cells, if appropriately stained).
- b. Acinar cells (exocrine).

3➡ Sketch and label what you see.

Pancreas

H. The Adrenal Glands [∞ Martini, p. 600]

Each of the paired **adrenal** (or suprarenal) glands is actually two glands, the adrenal cortex and the adrenal medulla. Embryologically, the adrenal medulla develops as a functional part of the nervous system. In all mammals the medulla is embedded within the cortex. In humans the

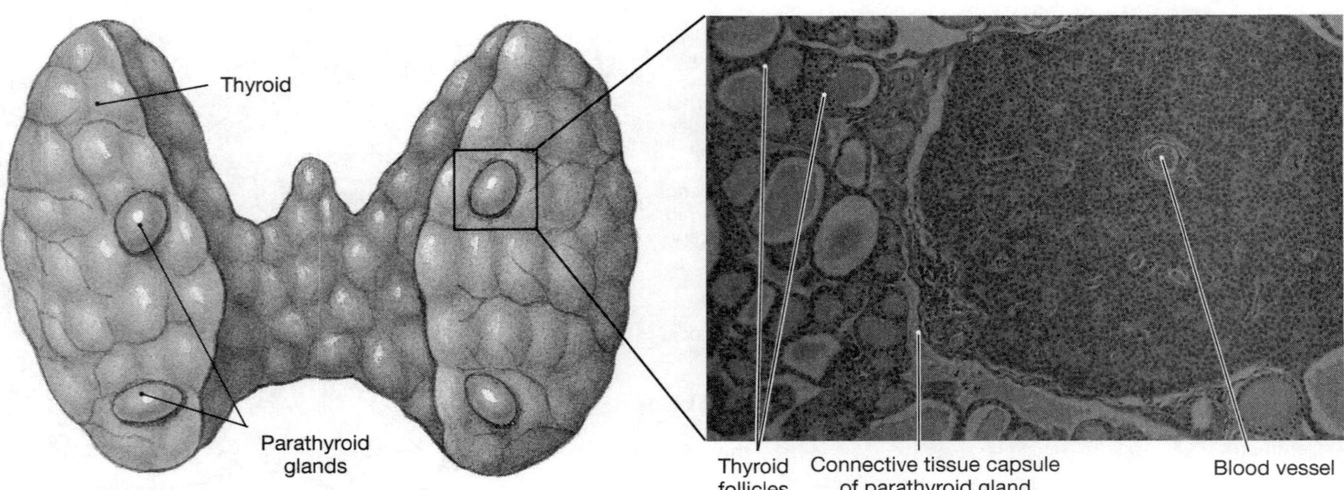

● **FIGURE 44–8**
Parathyroid Gland.

adrenal glands are located on the cranial border of the kidney — hence the name "ad + renal." (In certain other animals, such as the cat, the adrenal glands do not directly adhere to the kidneys.)

1➡ After checking Figure 44–10●, use the torso and the model kidney to locate the adrenal glands.

2➡ On your microscope slide, identify:
a. Capsule
b. Cortex
  Zona Glomerulosa
  Zona Fasciculata
  Zona Reticularis
c. Medulla

3➡ Sketch and label what you see.

Adrenal gland

4➡ Refer to Figure 44–11● as a summary of the functions of the adrenal hormones.

I. THE OVARIES [∞ MARTINI, P. 609]

The **ovaries** are the paired female gonads located in the lateral pelvic cavity in humans and

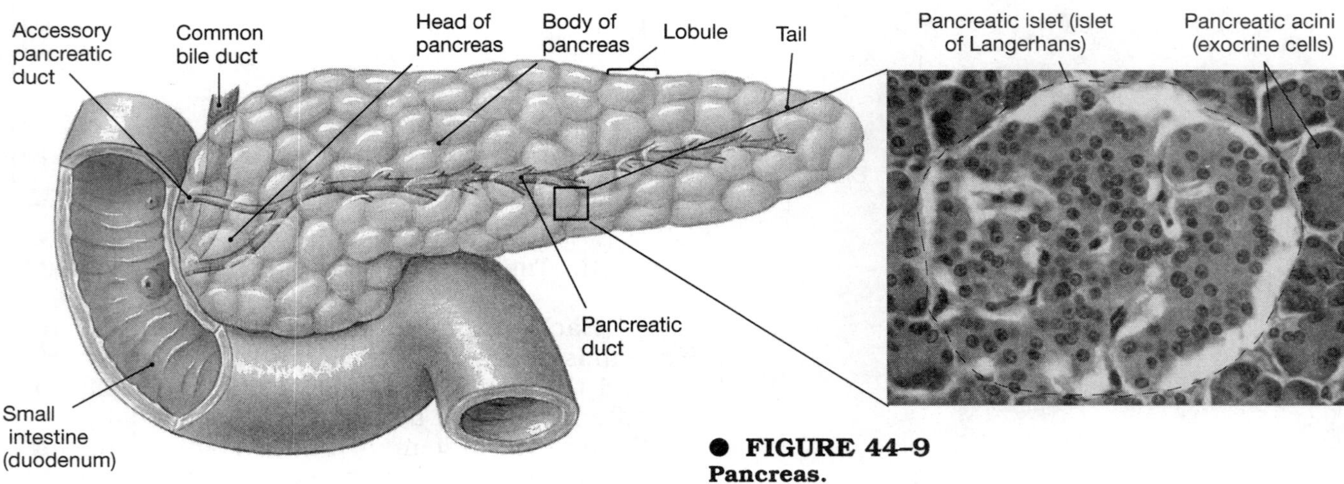

● **FIGURE 44–9**
Pancreas.

somewhat more cranial in most quadrupeds. If you have difficulty locating the ovaries on a model or diagram, find the body of the uterus, the pear-shaped organ located in the lower pelvic cavity. Follow the uterine tubes from the uterine body laterally to their termination point — where the fimbriae (finger-like projections) wrap around a small oval structure. The oval structure is the ovary. (The ovary is considered in greater detail in the reproductive and genetics exercises, Units XII and XIII.)

Your instructor may elect to wait until later in the course to assign this microscopic examination.

1 ➡ Locate the following on the slide of the ovary using Figure 44–12● as your guide:

a. Primary Follicle
b. Secondary Follicle
c. Graafian (Tertiary) Follicle

2 ➡ Sketch and label what you see.

Ovary

3 ➡ Refer to Figure 44–13●, which summarizes the hormones of the reproductive system.

J. THE TESTES [∞ MARTINI, P. 609]

The **testes** are the paired male gonads located in the scrotal sac external to the body proper in

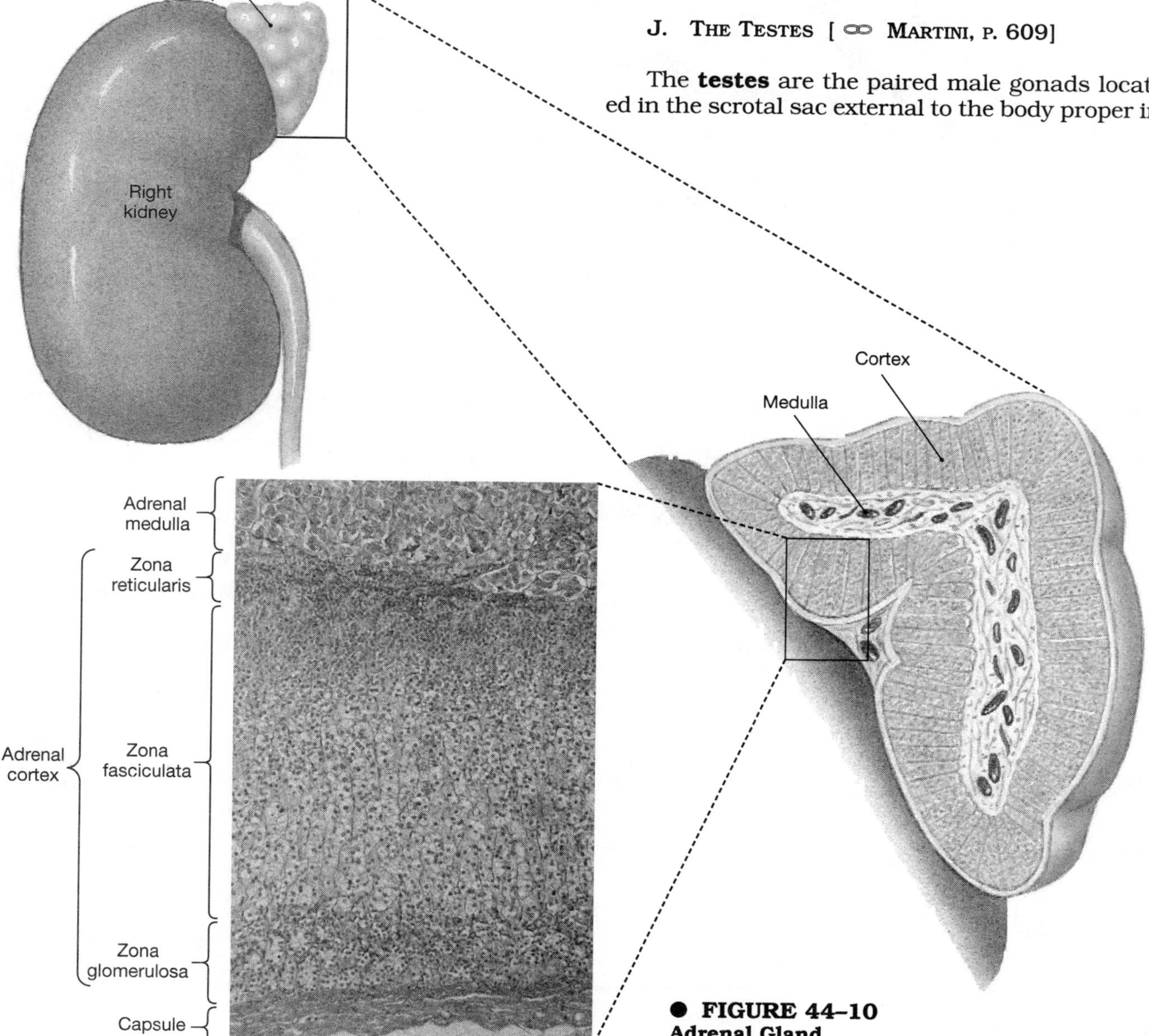

● **FIGURE 44–10**
Adrenal Gland.

● FIGURE 44–11
Ovary.

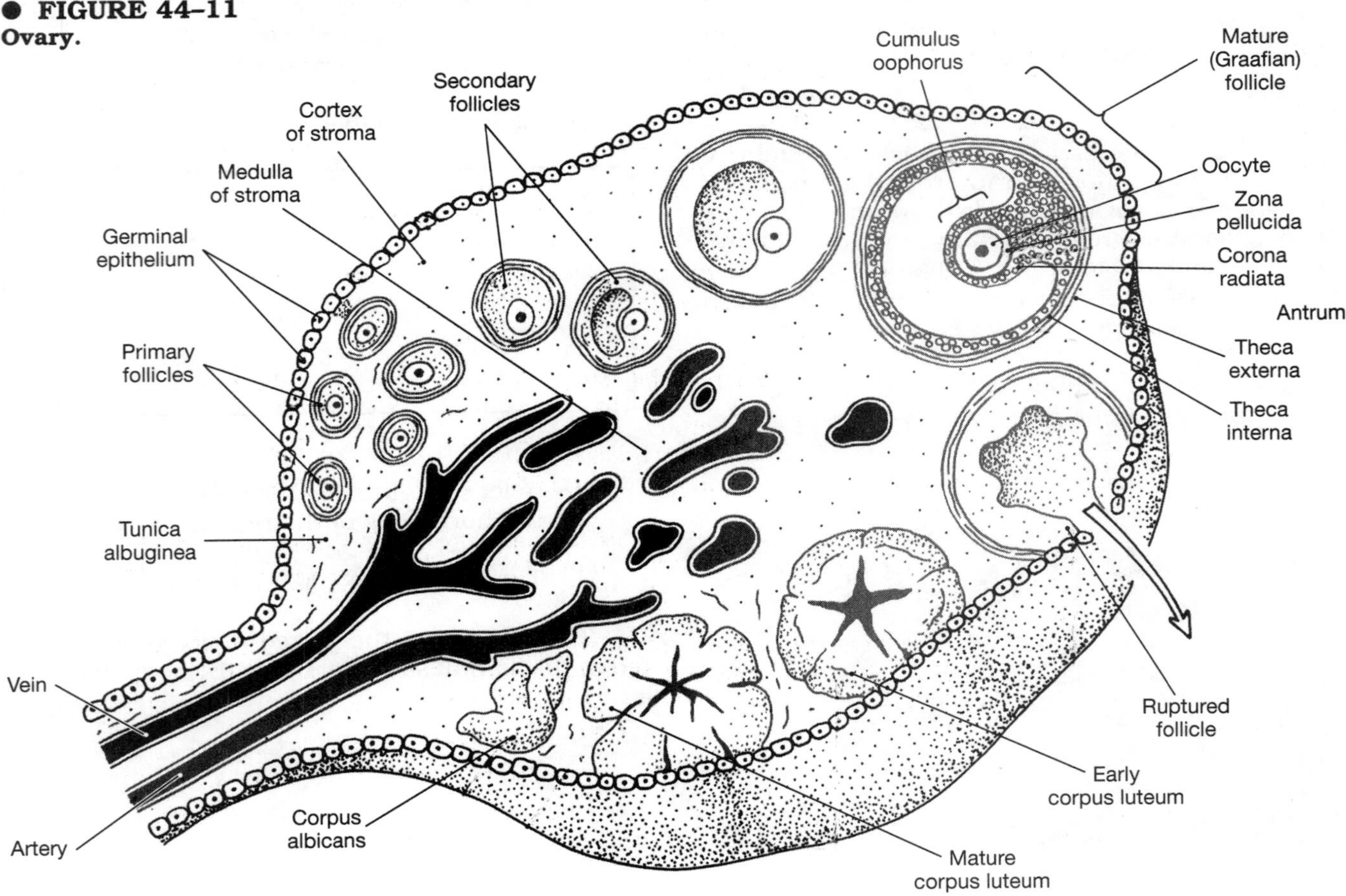

Region/Zone	*Hormone*	*Targets*	*Effects*	*Regulatory Control*
CORTEX				
Zona glomerulosa	Mineralocorticoids (MC), primarily aldosterone	Kidneys	Increases reabsorption of sodium ions and water from the urine	Stimulated by angiotensin II; inhibited by ANP
Zona fasciculata	Glucocorticoids (GC): cortisol (hydrocortisone), corticosterone, cortisone	Most cells	Releases amino acids from skeletal muscles, lipids from adipose tissues; promotes liver glycogen and glucose formation; promotes peripheral utilization of lipids anti-inflammatory effects	Stimulated by ACTH from anterior pituitary
Zona reticularis	Androgens		Uncertain significance under normal conditions	Stimulated by ACTH; significance uncertain
MEDULLA	Epinephrine (adrenaline, E), norepinephrine (noradrenaline, NE)	Most cells	Increased cardiac activity, blood pressure, glycogen breakdown, blood glucose; release of lipids by adipose tissue (see Chapter 16)	Stimulated during sympathetic activation by sympathetic preganglionic fibers

● FIGURE 44–12
Adrenal Hormones.

Structure/ Cells	Hormone	Primary Targets	Effects	Regulatory Control
TESTES				
Interstitial cells	Androgens (including (testos-terone)	Most cells	Support functional maturation of sperm, protein synthesis in skeletal muscles, male secondary sexual characteristics, and associated behaviors	Stimulated by LH from anterior pituitary
	Inhibin	Anterior pituitary	Inhibits secretion of FSH	Stimulated by FSH from anterior pituitary
OVARIES				
Follicular cells	Estrogens	Most cells	Support follicle maturation, female secondary sexual characteristics, and associated behaviors	As above
	Inhibin	Anterior pituitary	Inhibits secretion of FSH	As above
Corpus luteum	Progestins (including testos-terone)	Uterus, mammary glands	Prepare uterus for implantation; prepare mammary glands for secretory functions	Stimulated by LH from anterior pituitary
	Relaxin	Pubic symphysis, uterus, mammary glands	Loosens pubic symphysis, relaxes uterine muscles, stimulates mammary gland development	Stimulated by LH from the anterior pituitary and by placental hCG (human chorionic gonadotropin)

● **FIGURE 44–13**
Reproductive Hormones.

humans and in various stages of descent into the scrotal sac in fetal animals. (The testis is considered in greater detail in the reproductive and genetics exercises.)

Your instructor may elect to wait until later in the course to assign this microscopic examination.

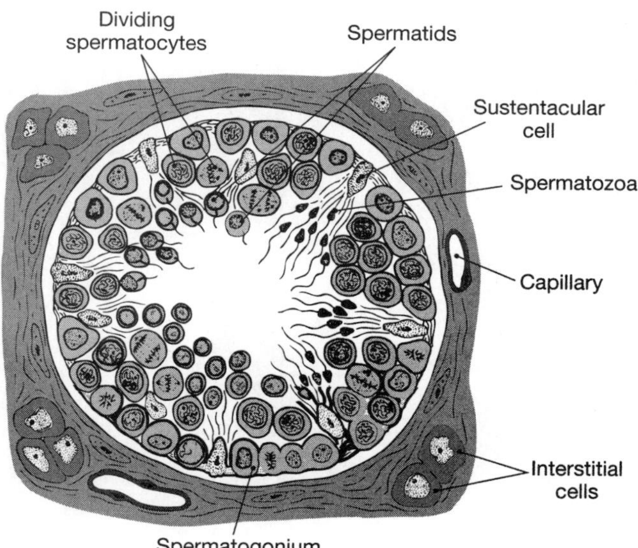

● **FIGURE 44–14**
Testis.

1 ➡ Locate the following structures on your slide of the testis, using Figure 44–14● as your guide:
   a. Seminiferous Tubule
   b. Interstitial Cells

2 ➡ Sketch and label what you see.

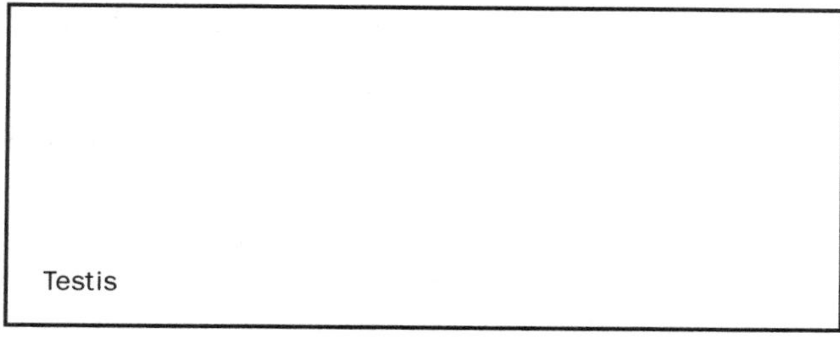

3 ➡ Refer to Figure 44–13● for a summary of the reproductive hormones.

III. Extending the Concept

A. Clinical Applications

1 ➡ Search through your lecture text. Make a chart listing the major hormones and describing the problems resulting from an overproduction and an underproduction of the hormone.

❑ Additional Activities

NOTES

1. Research the history of endocrinology.
2. Find out how the endocrine system of the mammal compares with the endocrine system of the bird, the amphibian, and the reptile.

❑ Lab Report

1. Referring to the numbered inquiries at the beginning of this exercise, complete the following box summary:

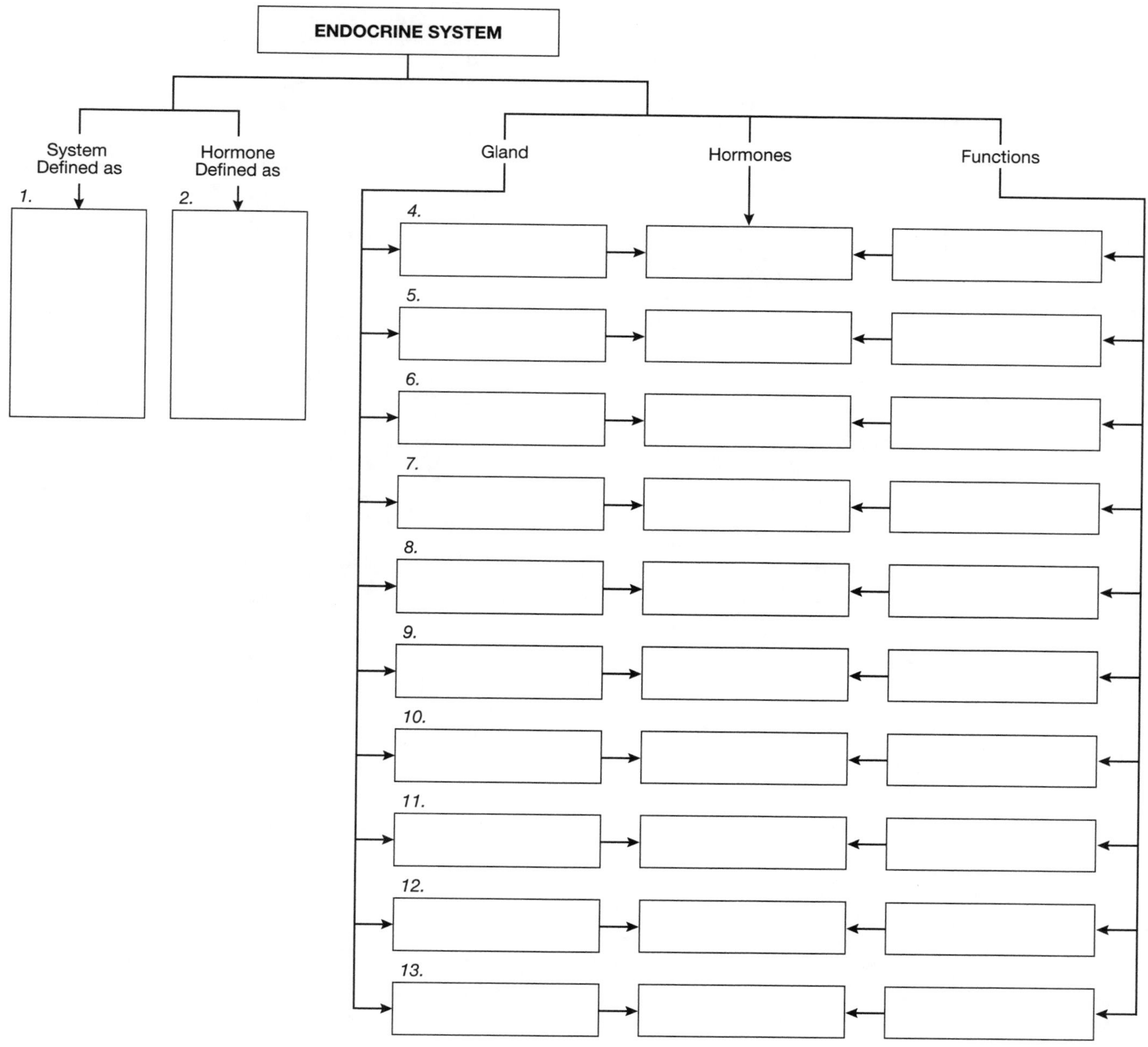

General Questions

NOTES

1. Which glands are actually two glands in one?
2. Which glands are directly related to the nervous system?
3. List several organs that are not primarily endocrine but that have endocrine functions.

EXERCISE **45a**

Dissection of the Endocrine System: Cat

PROCEDURAL INQUIRIES

Dissection

For each feline gland or structure listed below, answer these questions:
Where is the gland (or structure) located?
What is one interesting characteristic about the gland (or structure)?

1. Pineal
2. Hypothalamus
3. Pituitary
4. Thyroid
5. Parathyroid
6. Thymus
7. Pancreas
8. Adrenal
9. Ovary
10. Testes

Additional Inquiries

11. Why is it appropriate to study the endocrine glands of the cat as part of a human anatomy/physiology course?

HUMAN CORRELATION: AM Cadaver Atlas in Applications Manual.

Key Terms

Adenohypophysis
Adrenal
Anterior Pituitary
Hypophysis
Hypothalamus
Neurohypophysis
Ovaries
Pancreas
Parathyroid
Pars Distalis
Pars Intermedia
Pars Nervosa
Pineal
Pituitary
Posterior Pituitary
Testes
Thymus

Materials Needed

Models and Charts of Endocrine System
Midsagittal Section of Human (or Other Mammalian) Brain (Model or Preserved Section)
Cat
Gloves
Dissecting Kit
Dissecting Microscope
(Prepared slides of endocrine glands; recommended if you are not completing Exercise 44)

Exercise 45a is an ancillary chapter to Exercise 44. If you have already worked through Exercise 44, you can use that lesson as a base for your explorations in this lesson.

If you have not already completed Exercise 44, we suggest that you read through that lesson before beginning your endocrine dissection of the cat.

In this laboratory exercise we will examine the anatomy and histology of the principal endocrine structures by dissecting the endocrine system of the cat and by examining microscope slides of certain endocrine structures.

Because all mammals are anatomically similar, we can study the endocrine system of the cat and correlate that information with the structures we find in the human. In addition to analogous structures, we also find that the hormones are almost identical.

Mammalian hormones are so similar that we can use bovine (cow) or porcine (pig) insulin — a pancreatic hormone — to treat human diabetes. Because slight molecular variations in the hormones do exist, use of the native hormone is always recommended. This is particularly true with regard to the hormones regulating the cyclic functions of the reproductive systems. As a general rule, reproductive hormones do not cross species lines as well as certain other hormones.

❑ Preparation

I. Background

A. Endocrine Systems of Mammals

1➡ Study Figure 45a–1●, which delineates the location of the endocrine glands in the cat.

2➡ Compare Figure 45a–1● with Figure 44–1●. Based on these diagrams, list the similarities and differences you notice between the human and feline endocrine systems.
SIMILARITIES ______________________
____
DIFFERENCES ______________________
____

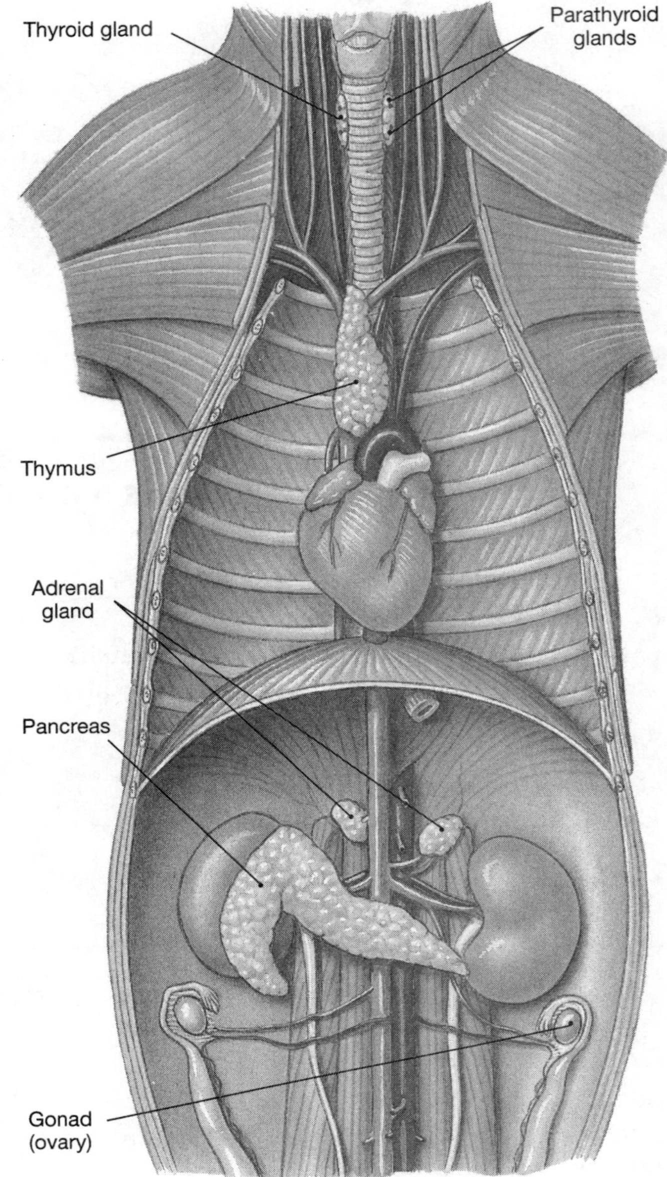

● **FIGURE 45a–1**
Endocrine System — Cat.

Dissection

II. Cat Dissection

It is assumed that you have already opened the thorax and abdomen of your cat. If you have not, please follow your instructor's directions.

A. Endocrine Structures

Use the human (or mammalian) brain models and Figure 45a–2● to identify the first three endocrine structures—the pineal gland, the hypothalamus, and the pituitary gland. Use your cat to locate the other endocrine structures. Refer to Exercise 44 for the functions of each of the hormones.

1➡ Find the **pineal** gland, which is part of the epithalamus, located in the roof of the third ventricle. This cone-shaped body lies between the corpora quadrigemina and the splenium (posterior portion) of the corpus callosum. If you were to find this gland in your cat, you would notice that the younger your cat, the larger and more glandular in appearance the pineal would be.

2➡ Locate the **hypothalamus**, a structure that is not exactly a gland in the classical sense. The hypothalamus is a part of the brain that houses control and integrative centers for autonomic and emotional functions. The hypothalamus also secretes ADH and oxytocin (which are released from the posterior pituitary) in addition to numerous hormones that regulate the actions and secretions of the anterior pituitary.

3➡ Locate the **pituitary** (**hypophysis**) gland. Compare the parts of this gland in the human (Figure 44–4● in the previous exercise) with those of the cat (Figures 45a–2● and 45a–3●).

The pituitary is actually two glands, the **anterior pituitary** and the **posterior pituitary.** The anterior pituitary is also known as the **adenohypophysis** or the **pars distalis**. The posterior pituitary is also known as the **neurohypophysis** or

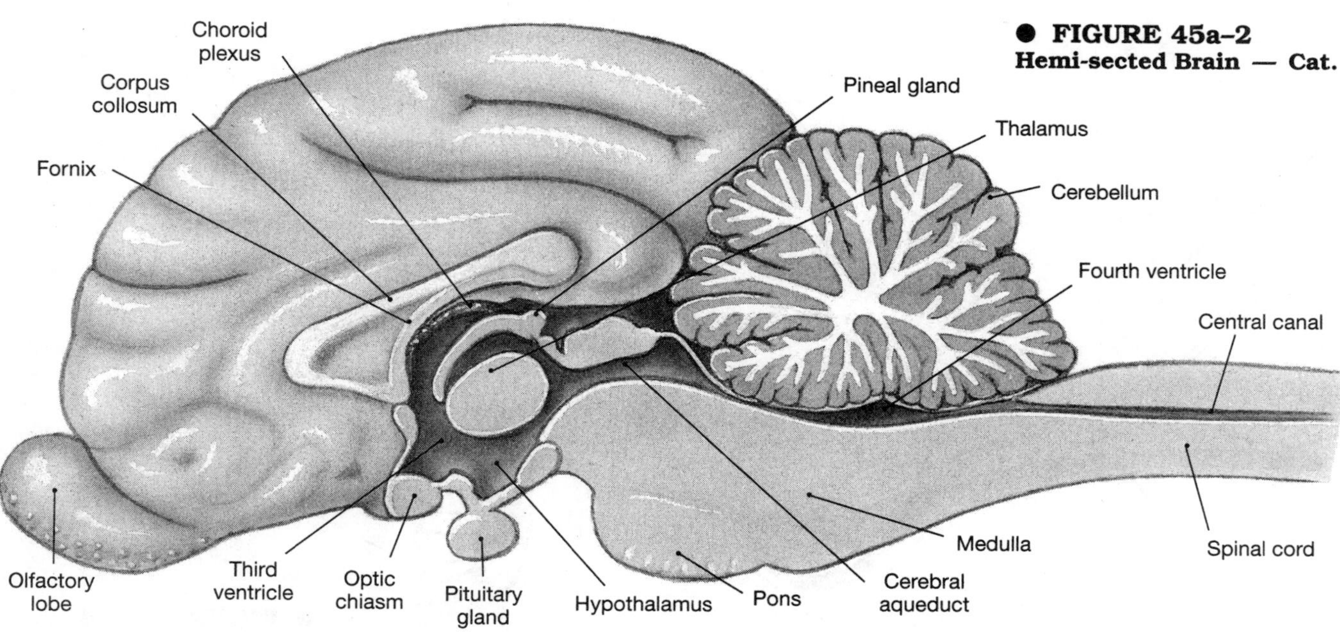

● **FIGURE 45a–2**
Hemi-sected Brain — Cat.

pars nervosa. The **pars intermedia** is part of the anterior pituitary.

4➡ Look for the **thyroid** gland on the human model. Also refer to Figure 45a–4●. This gland is similar in humans and felines. It is composed of two lobes connected by an isthmus. In humans this isthmus remains prominent whereas in cats the isthmus generally disappears quite early in life. You may not be able to find a true isthmus in your cat, in which case the lobes of the thyroid are found on the lateral surfaces of the trachea, beginning about a centimeter below the larynx. The thyroid is heavily encapsulated so it will appear to be a tough gland.

Remove and examine this gland. If you are careful, you should be able to identify the lobes, possibly the isthmus, and the major blood vessels of the thyroid. Use the dissecting microscope if necessary.

Sketch and label what you see.

Thyroid gland

5➡ Find the **parathyroid** glands (Figure 45a–3●). In humans (and in most other animals) the parathyroids consist of two pairs of glands. The human parathyroid glands are embedded on the posterior surface of the thyroid gland. In cats these glands are called the internal and external parathyroids, and at least one pair can usually be found on the back surface of the thyroid gland. The parathyroids are usually lighter in color than the surrounding thyroid tissue. Use the dissecting microscope and try to identify the parathyroid glands. Sketch and label what you see.

Parathyroid gland

6➡ Locate and examine the **thymus** gland. It is usually just posterior to the sternum. In fetal and young mammals, the thymus is relatively quite large. The thymus begins to atrophy after puberty and in the adult is small and inconspicuous. Usually the left portion of the thymus is larger than the right. Unless your cat is exceptionally old, you should have no difficulty in locating the spongy thymus of your cat. Examine the thymus under the dissecting microscope. Sketch and label what you see.

Thymus gland

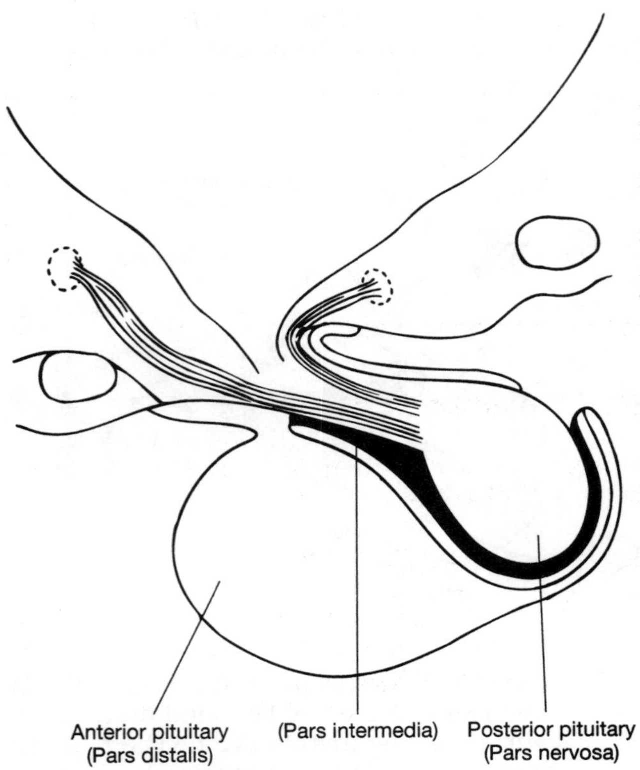

● **FIGURE 45a–3**
Pituitary Gland — Cat.

7➡ Find the **pancreas**, a long slender gland located near the border between the stomach and the small intestine. Put a piece of the pancreas under the dissecting microscope and try to differentiate between the endocrine and exocrine cells. The clusters of endocrine cells known as islets of Langerhans are scattered among the exocrine cells. The tubes you may see are the exocrine ducts. Refer to your charts to help you. Sketch and label what you see.

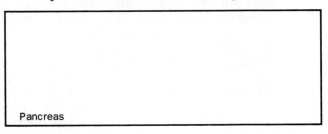

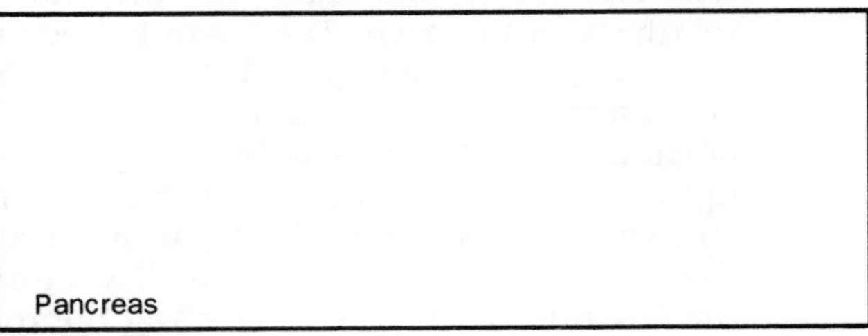

8➡ Locate the kidneys and the small **adrenal** glands embedded in fatty tissue at the cranial end of the kidneys. Unlike the human, in the cat the adrenal gland does not directly adhere to the kidney. Each of the paired adrenal glands is actually two glands, the adrenal cortex and the adrenal medulla. Remove one adrenal gland and examine it under the dissecting microscope. Bisect the gland and note the difference between the cortical and medullary cells. Sketch and label what you see.

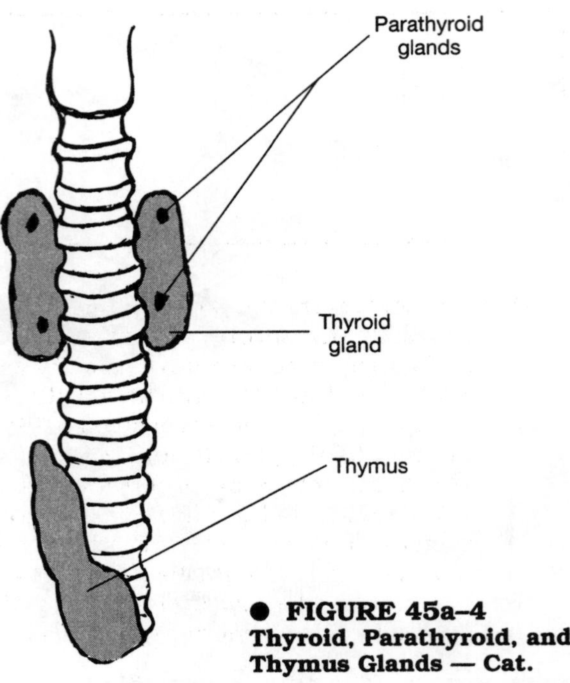

● **FIGURE 45a–4**
Thyroid, Parathyroid, and Thymus Glands — Cat.

Adrenal gland

9➡ The **ovaries** are the paired female gonads, located in the lateral pelvic cavity in humans and somewhat more cranial in most quadrupeds. If your cat is male, check the ovaries on a classmate's cat. If you have difficulty locating your cat's ovaries, find the body of the uterus, which is centrally located in the lower pelvic cavity. Follow the uterine horns to their termination point in the midlateral part of the animal's body. At the end of the uterine horn you will find a tiny, thinner tube and then a small oval structure, the ovary. If you have not yet studied the reproductive system, your instructor may direct you not to remove an ovary at this point. (The ovary is considered in greater detail in the reproductive and genetics exercises in Units XII and XIII.)

If you are studying this gland now, carefully remove one ovary and examine it under the dissecting microscope. Refer to Exercise 69a and bisect the gland. Sketch and label what you see.

Ovary

10➡ The **testes** are the paired male gonads, located in the scrotal sac external to the body proper in humans and in various stages of descent into the scrotal sac in fetal animals. The scrotum is quite well defined in the cat. If your cat is female, check the testes on a classmate's cat. If you have not yet studied the reproductive system, your instructor may direct you not to remove a testis at this point. (The testis is considered in greater detail in the reproductive and genetics exercises in Units XII and XIII.)

If you are studying this gland now, carefully remove one testis and examine it under the dissecting microscope. Refer to Exercise 69a and bisect the gland. Sketch and label what you see.

Testis

III. Prepared Slides

This section is recommended if Exercise 44 is not being completed at this time.

A. Endocrine Structues

1➡ Examine prepared slides of the endocrine structures described in this exercise. Use Exercise 44 as your guide as you sketch and label what you observe.

❑ Additional Activities

1. Find out how the endocrine system of the mammal compares with the endocrine system of the bird, the amphibian, and the reptile.
2. Research the history of endocrinology.

NOTES

❑ Lab Report

1. Referring to the numbered inquiries at the beginning of this exercise, complete the following Box Summary:

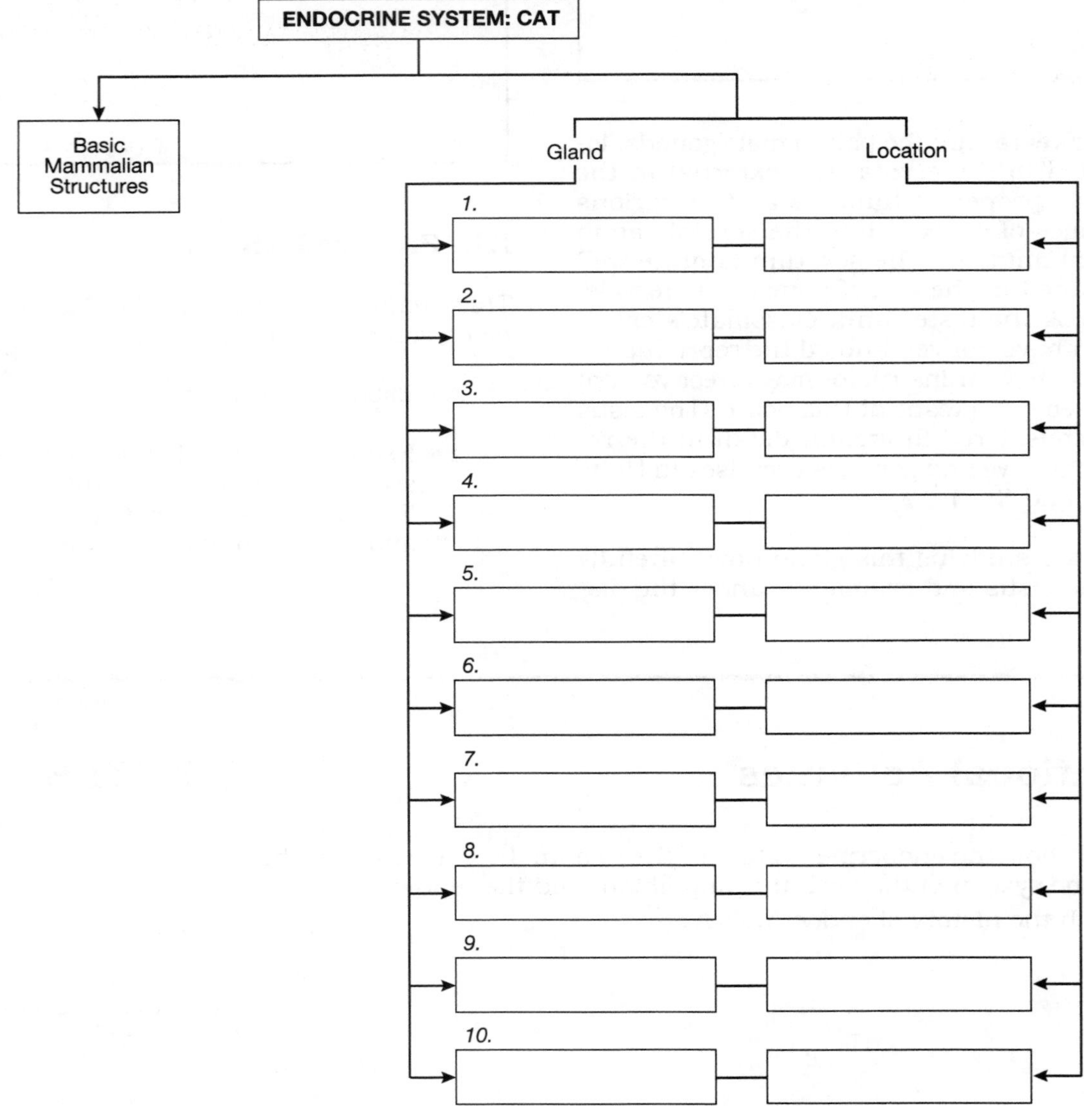

General Questions

1. Make a list of similarities and differences between the edocrine structures of the cat and the human. Refer back to Exercise 44 if necessary. What differences in particular do you notice between the thyroid glands, the parathyroid glands, and the adrenal glands?
2. What generalization can you make when comparing the endocrine function of the cat with the endocrine function of the human?
3. Which glands are actually two glands in one?
4. Why is it appropriate to study the feline endocrine system in a human anatomy/physiology class?

EXERCISE 45b

Dissection of the Endocrine System: Fetal Pig

PROCEDURAL INQUIRIES

Dissection

For each porcine gland or structure listed below, answer these questions:
Where is the gland (or structure) located?
What is one interesting characteristic about the gland (or structure)?

1. Pineal
2. Hypothalamus
3. Pituitary
4. Thyroid
5. Parathyroid
6. Thymus
7. Pancreas
8. Adrenal
9. Ovary
10. Testes

Additional Inquiries

11. Why is it appropriate to study the endocrine glands of the fetal pig as part of a human anatomy/physiology course?

HUMAN CORRELATION: AM Cadaver Atlas in Applications Manual.

Key Terms

Adenohypophysis
Adrenal
Anterior Pituitary
Hypophysis
Hypothalamus
Neurohypophysis
Ovaries
Pancreas
Parathyroid
Pars Distalis
Pars Intermedia
Pars Nervosa
Pineal
Pituitary
Posterior Pituitary
Testes
Thymus

Materials Needed

Models and Charts of Endocrine System
Midsagittal Section of Human (or Other Mammalian) Brain (Model or Preserved Section)
Fetal Pig
Gloves
Dissecting Kit
Dissecting Microscope
(Prepared slides of endocrine glands; recommended if you are not completing Exercise 44)

Exercise 45b is an ancillary chapter to Exercise 44. If you have already worked through Exercise 44, you can use that lesson as a base for your explorations in this lesson.

If you have not already completed Exercise 44, we suggest that you read through that lesson before beginning your endocrine dissection of the fetal pig.

In this laboratory exercise we will examine the anatomy and histology of the principal endocrine structures by dissecting the endocrine system of the fetal pig and by examining microscope slides of certain endocrine structures.

Because all mammals are anatomically similar, we can study the endocrine system of the fetal pig and correlate that information with the structures we find in the human. In addition to analogous structures, we also find that the hormones themselves are almost identical.

Mammalian hormones are so similar that we can use bovine (cow) or porcine (pig) insulin — a pancreatic hormone — to treat human diabetes. Because slight molecular variations in the hormones do exist, use of the native hormone is always recommended. This is particularly true with regard to the hormones regulating the cyclic functions of the reproductive systems. As a general rule, reproductive hormones do not cross species lines as well as certain other hormones.

❑ Preparation

I. Background

A. Endocrine Systems of Mammals

1➡ Study Figure 45b–1●, which delineates the location of the endocrine glands in the fetal pig.

2➡ Compare Figure 45b–1● with Figure 44–1●. Based on these diagrams, list the similarities and differences you notice between the human and porcine endocrine systems.
SIMILARITIES ______________________
____
DIFFERENCES ______________________
____

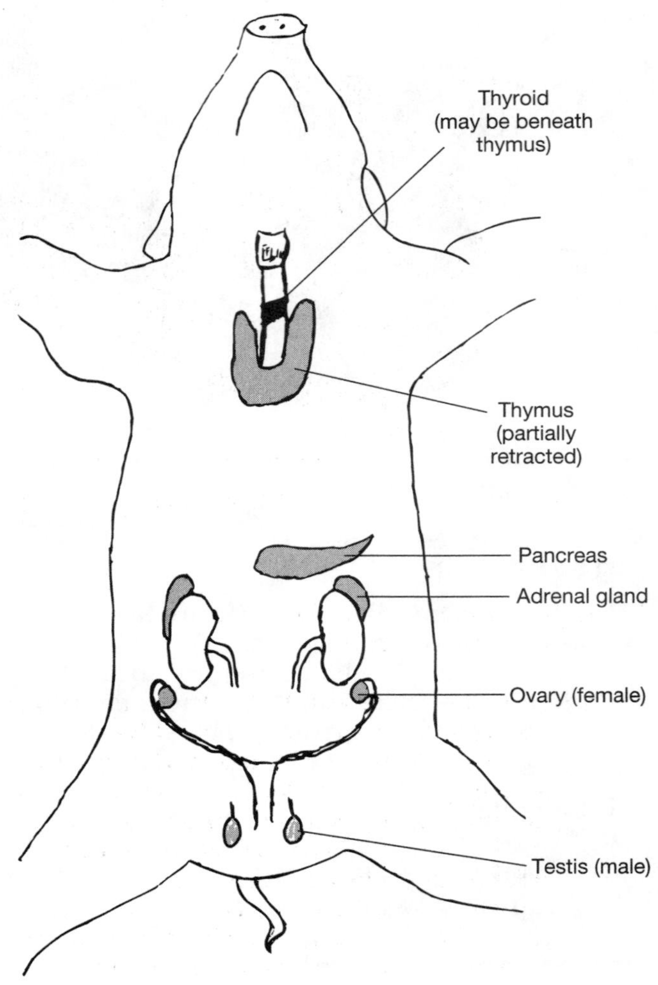

● **FIGURE 45b–1**
Endocrine System — Fetal Pig.

Dissection

II. Fetal Pig Dissection

It is assumed that you have already opened the thorax and abdomen of your fetal pig. If you have not, please follow your instructor's directions.

A. Endocrine Structures

Use the human (or mammalian) brain models to identify the first three endocrine structures — the pineal gland, the hypothalamus, and the pituitary gland. Use your fetal pig to locate the other endocrine structures. Refer to Exercise 44 for the functions of each of the hormones.

1➡ Find the **pineal** gland, which is part of the epithalamus, located in the roof of the third ventricle. This cone-shaped body lies between the corpora quadrigemina and the splenium (posterior portion) of the corpus callosum. If you were to find this gland in your fetal pig, you would notice that it is relatively large and quite glandular in appearance. Note the location on Figure 45b–2●.

2➡ Locate the **hypothalamus,** a structure that is not exactly a gland in the classical sense. The hypothalamus is a part of the brain that houses control and integrative centers for autonomic and emotional functions. The hypothalamus also secretes ADH and oxytocin (which are released from the posterior pituitary) in addition to numerous hormones that regulate the actions and secretions of the anterior pituitary.

3➡ Locate the **pituitary (hypophysis)** gland. Compare the parts of this gland in the human (Figure 44–4● in the previous exercise) with those of the fetal pig (Figure 45b–3●).

The pituitary is actually two glands, the **anterior pituitary** and the **posterior pituitary.** The anterior pituitary is also known as the **adenohypophysis** or the **pars distalis.** The posterior pituitary is also known as the **neurohypophysis** or **pars nervosa.** The **pars intermedia** is part of the anterior pituitary. As the pig matures, the area la-

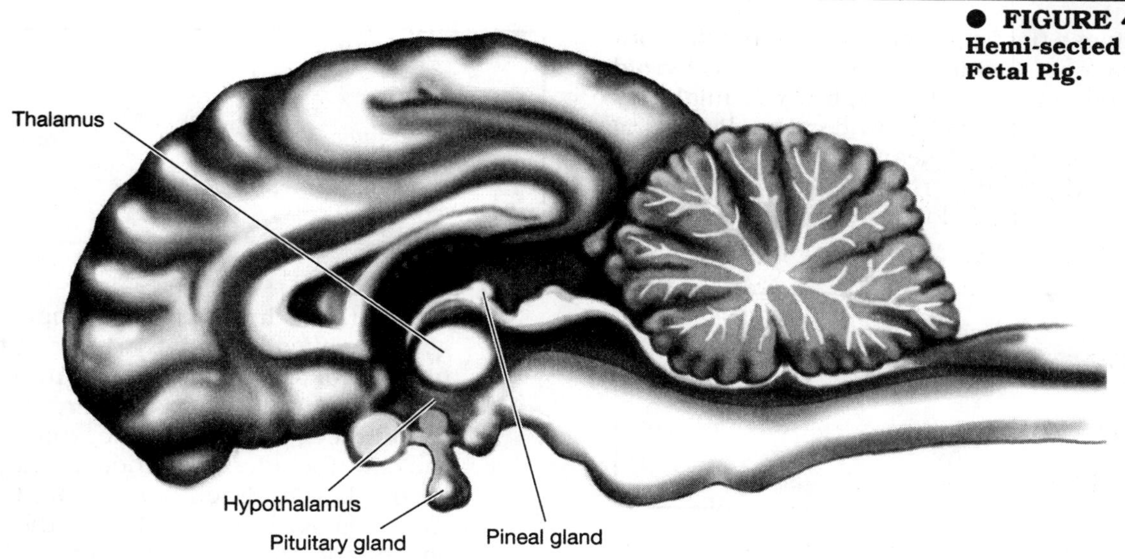

● **FIGURE 45b–2**
Hemi-sected Brain — Fetal Pig.

beled pars intermedia will concentrate on the anterior-posterior interface.

4➡ Look for the **thyroid** gland on the human model. Also refer to Figure 45b–4●. This gland is similar in humans and porcines. It is composed of two lobes connected by an isthmus with a central superior projection. In humans this gland is located immediately inferior to the larynx. In fetal pigs the gland begins at about the third or fourth tracheal ring. The thyroid is heavily encapsulated, so it will appear to be a tough gland.

Remove and examine this gland. If you are careful, you should be able to identify the lobes, the isthmus, and the major blood vessels of the thyroid. Use the dissecting microscope if necessary.

Sketch and label what you see.

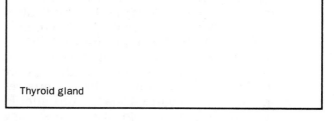

5➡ Find the **parathyroid** glands (Figure 45b–4●). In humans (and in most other animals) the parathyroids consist of two pairs of glands. The human parathyroid glands are embedded on the posterior surface of the thyroid gland. Pigs have only one pair of parathyroid glands. These glands are cranial to the thyroid gland and are embedded in the thymus in fetal animals and in

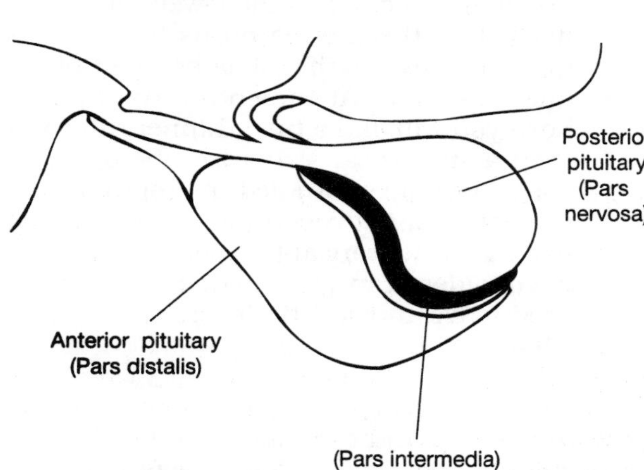

● **FIGURE 45b–3**
Pituitary — Fetal Pig.

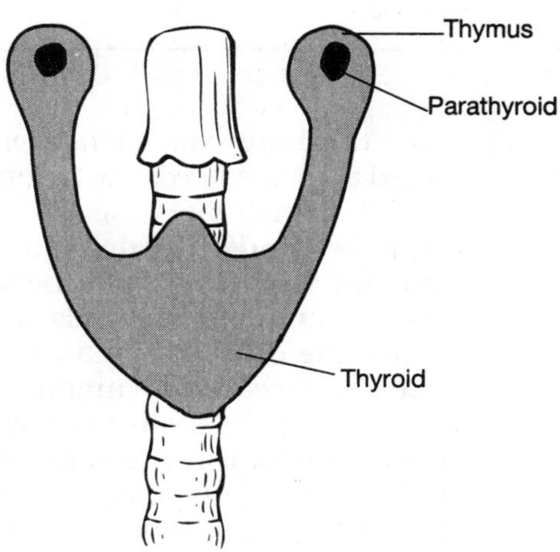

● **FIGURE 45b–4**
Thyroid, Parathyroid, and Thymus Glands — Older Pig.

adipose tissue in older pigs. You will probably not be able to locate the parathyroid glands in your fetal pig, but you might try putting the thymus under the dissecting microscope and searching for them anyway. If you find the parathyroid glands, sketch and label what you see.

Parathyroid gland

6➡ Locate and examine the **thymus** gland. It is usually just posterior to the sternum. In fetal and young mammals, the thymus is relatively quite large. The thymus begins to atrophy after puberty and in the adult is small and inconspicuous. The thymus is quite large in the fetal pig and will continue to grow until the animal reaches about nine months of age. In the adult pig only microscopic remnants of the thymus remain. Check Figure 45b–4●. You should have no difficulty in locating the spongy thymus of your fetal pig. Examine the thymus under the dissecting microscope. Sketch and label what you see.

Thymus gland

7➡ Find the **pancreas,** a long slender gland located near the border between the stomach and the small intestine. Put a piece of the pancreas under the dissecting microscope and try to differentiate between the endocrine and exocrine cells. The clusters of endocrine cells known as islets of Langerhans are scattered among the exocrine cells. The tubes you may see are the exocrine ducts. If you scrape the gland carefully with your scalpel, you may be able to find the ducts without the microscope. Sketch and label what you see. Refer to your charts to help you.

Pancreas

8➡ Locate the kidneys and the small **adrenal** glands embedded in fatty tissue at the cranial end of the kidneys. As in humans, the porcine adrenal glands adhere directly to the kidney. In certain other animals (like the cat), the adrenal glands are not on the kidney border. Each of the paired adrenal glands is actually two glands, the adrenal cortex and the adrenal medulla. Remove one adrenal gland and examine it under the dissecting microscope. Bisect the gland and note the difference between the cortical and medullary cells. Sketch and label what you see.

Adrenal gland

9➡ The **ovaries** are the paired female gonads, located in the lateral pelvic cavity in humans and somewhat more cranial in most quadrupeds. If your fetal pig is male, check the ovaries on a classmate's fetal pig. If you have difficulty locating your fetal pig's ovaries, find the body of the uterus, which is centrally located in the lower pelvic cavity. Follow the uterine horns to their termination point in the midlateral part of the animal's body. At the end of the uterine horn you will find a tiny, thinner tube and then a small oval structure, the ovary. If you have not yet studied the reproductive system, your instructor may direct you not to remove an ovary at this point. (The ovary is considered in greater detail in the reproductive and genetics exercises in Units XII and XIII.)

If you are studying this gland now, carefully remove one ovary and examine it under the dissecting microscope. Refer to Exercise 69b and bisect the gland. Sketch and label what you see.

Ovary

Testis

10➡ The **testes** are the paired male gonads, located in the scrotal sac external to the body proper in humans and in various stages of descent into the scrotal sac in fetal animals. The scrotum, though present, is not well defined in the fetal pig. If your fetal pig is female, check the testes on a classmate's fetal pig. If you have not yet studied the reproductive system, your instructor may direct you not to remove a testis at this point. (The testis is considered in greater detail in the reproductive and genetics exercises in Units XII and XIII.)

If you are studying this gland now, carefully remove one testis and examine it under the dissecting microscope. Refer to Exercise 69b and bisect the gland. Sketch and label what you see.

III. Prepared Slides

This section is recommended if Exercise 44 is not being completed at this time.

A. Endocrine Structures

1➡ Examine prepared slides of the endocrine structures described in this exercise. Use Exercise 44 as your guide as you sketch and label what you observe.

❑ Additional Activities

NOTES

1. Find out how the endocrine system of the mammal compares with the endocrine system of the bird, the amphibian, and the reptile.
2. Research the history of endocrinology.

❑ Lab Report

1. Referring to the numbered inquiries at the beginning of this exercise, complete the following box summary:

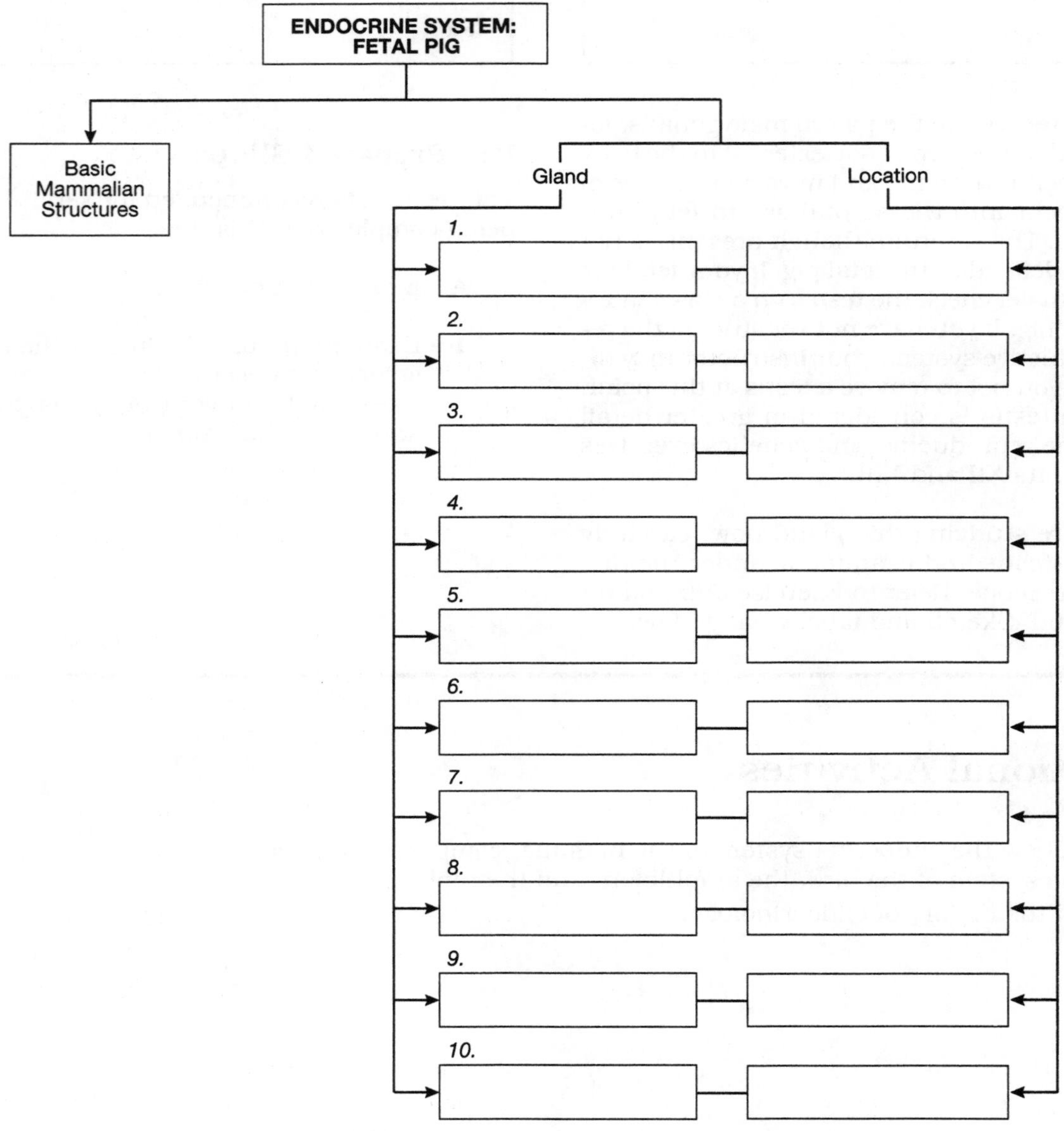

General Questions

NOTES

1. Make a list of similarities and differences between the endocrine structures of the fetal pig and the human. Refer back to Exercise 44 if necessary. What differences in particular do you notice between the thyroid glands, the parathyroid glands, and the adrenal glands?
2. What generalization can you make when comparing the endocrine function of the fetal pig with the endocrine function of the human?
3. Which glands are actually two glands in one?
4. Why is it appropriate to study the porcine endocrine system in a human anatomy/physiology class?

EXERCISE 46

Anatomy of the Blood

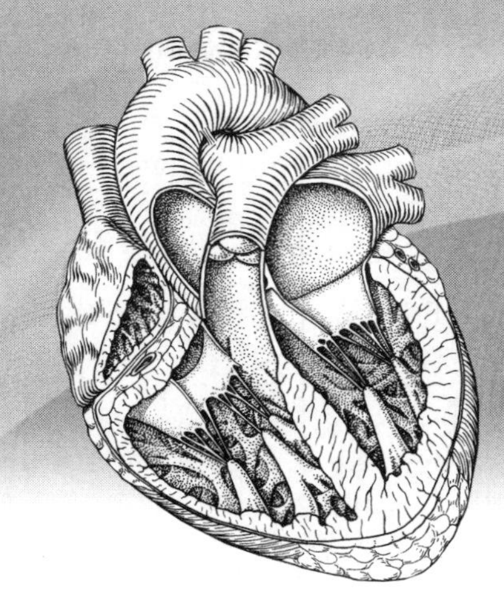

PROCEDURAL INQUIRIES

Preparation

1. What safety precautions must be followed in handling human blood?

Examination

2. What is blood?
3. What is plasma?
4. What are the components of plasma?
5. What are the formed elements?
6. What is the primary function of the red blood cells?
7. What are the major categories of white blood cells?
8. What is the function of each of the white blood cell types?
9. What are platelets?
10. What is the primary function of the platelets?
11. Developmentally, how are the formed elements related?
12. What is Wright's staining procedure?
13. What is a differential white blood cell count?
14. In sickle cell anemia, why do the red blood cells sickle?
15. How are the leukemias classified?
16. What is mononucleosis?

Additional Inquiries

17. Why can animal blood be substituted for human blood in this exercise?
18. What are the basic functions of blood?

Key Terms

Agranulocytes
Basophils
Blood
Differential White Blood Cell Count
Eosinophils
Erythrocytes
Formed Elements
Granulocytes
Leukocytes
Lymphocytes
Lymphogenous Leukemia
Megakaryocytes
Monocytes
Mononucleosis
Myelogenous Leukemia
Neutrophils
Plasma
Platelets
Red Blood Cells
Sickle Cell Anemia
White Blood Cells
Wright's Stain

Materials Needed

Compound Microscope

Prepared Blood Slides

- Standard Smear (Human, or other mammalian, blood)
- Sickle Cell Anemia
- Myelogenous Leukemia
- Lymphogenous Leukemia
- (Other slides as available)

Clean Slides

Alcohol Swabs or 70% Alcohol

Lancets

Wright's Stain

Distilled Water

Optional:

- Hemacytometer
- WBC Diluting Pipette
- WBC Diluting Fluid (1% Acetic acid)

RBC Diluting Pipette
RBC Diluting Fluid (Hayem's fluid)
Tubing

Probably no single human tissue throughout history has been the subject of more praise, degradation, analogy, allegory, metaphor, insult, injury, and acclamation than blood.

You might be a red-blooded American, but if your family is of royal descent, you are definitely blue-blooded. If your temper flares, you are hot-blooded, and if you become really angry, your blood might even boil. If you are the criminal type, you are cold-blooded. Of course, fear might make your blood run cold (which has nothing to do with being a criminal). And if you simply become an unfeeling person, your blood has turned to ice.

If you are from proper society, you've got good blood in your veins, while that lowlife your cousin hangs around with obviously inherited bad blood. The list is endless.

Although certain human attributes throughout history have been incorrectly attributed to blood, even the ancients recognized the vital role blood plays in the life of every human being.

Blood is the river of life — the communication and transportation network uniting and integrating all the systems of the human body.

In this laboratory exercise we will limit our exploration to the anatomy and histology of this fascinating substance.

❑ Preparation

I. Safety

The importance of proper handling of blood and blood-testing equipment cannot be stressed too often. You should understand and always keep the following in mind when completing this and any other exercise involving blood. Proper blood-handling procedures protect not only against HIV but also against hepatitis, Epstein-Barr, and many other pathogenic blood-borne diseases.

A. Nonhuman Blood (Mammalian) Blood (If your lab uses animal blood)

1➡ For any experiment in this exercise, animal blood can be substituted.

2➡ If your lab uses animal blood purchased from a supply house, you do not have to worry about the same types of contamination that might occur with human blood. Nevertheless, nonhuman blood can become contaminated. Always handle any blood as if it were contaminated.

3➡ Animal (mammalian) blood has all the same cells and same physical properties as human blood. Some minor differences do exist (such as sheep erythrocytes being about half the size of human erythrocytes), but the basic principles are identical.

B. Human Blood (If your lab uses human blood)

1➡ DO *ALL* OF YOUR OWN BLOOD HANDLING. This means do *all* of your own lancet stabbing, swabbing, plate counting, and clean-up. Do not help your friends; do not share anything with anyone.

2➡ Do not touch anyone's blood or blood-testing equipment.

3➡ Use *everything* only ONCE. If you make a mistake, start over with fresh equipment.

4➡ Do not set on the lab table anything that has touched blood.

5➡ Be aware of the location of every lancet, every slide, and every piece of equipment so you do not risk leaning into or bumping some dirty equipment.

6➡ Dispose of *everything immediately* in the proper tubs or autoclave bags as directed by your instructor.

7➡ If you are not completely certain about what you are doing, do not do it until you ask questions and get all your questions answered completely.

8➡ Clean up and disinfect *exactly* according to your instructor's directions.

It is assumed that if you have a known blood disease, serious clotting disorder, or some blood-letting dysfunction, you will not participate in any lab exercise requiring the drawing of your blood.

❑ Examination

II. Composition of Blood

A. Fluid Organization [∞ Martini, p. 624]

Blood, the fluid portion of the vascular system, is a specialized connective tissue with functions ranging from transport and regulation to restriction of fluid loss, defense against toxins and pathogens, and stabilization of body temperature.

The average person has approximately 5.5 liters of blood (men 5–6 liters, women 4–5 liters).

Plasma is the liquid portion, the matrix, of the blood and the **formed elements** are the blood cells and blood cell fragments.

B. Plasma [∞ Martini, p. 626]

Plasma, which constitutes about 55% of the blood volume, is acellular and, when isolated, is straw-colored and slightly more viscous than plain water. ("Slightly" is a relative term because the viscosity of whole blood is actually about 5 times that of water; however, antifreeze is 20 times more viscous than water, and molasses can be more than 300 times as viscous.)

Plasma plays a role in the three major functions of blood: transport, regulation, and protection.

Plasma transports well over 100 substances, including gases, nutrients, electrolytes, wastes, hormones, and enzymes. We sometimes forget that gases can be dissolved in another substance. The gases of interest here include oxygen and carbon dioxide. About 1.5% of the oxygen and about 7% of the carbon dioxide in the system are circulating as dissolved gases. (The remainder of the oxygen is transported by the protein hemoglobin, a component of the red blood cell. The remainder of the carbon dioxide is transported by either the hemoglobin molecule or as part of the carbonate ion.)

Nitrogen is also transported dissolved in the plasma. However, since we cannot utilize free nitrogen, any nitrogen that is dissolved in our plasma becomes irrelevant under normal conditions. Dissolved nitrogen is a problem only during times of drastic pressure changes, such as surfacing after deep-sea diving (causing a condition known as the bends) and flying too high in an unpressurized plane.

Because of substances carried in the plasma, cellular homeostasis is maintained. For example, plasma carries the buffers; plasma absorbs and redistributes heat, allowing us to maintain a con-

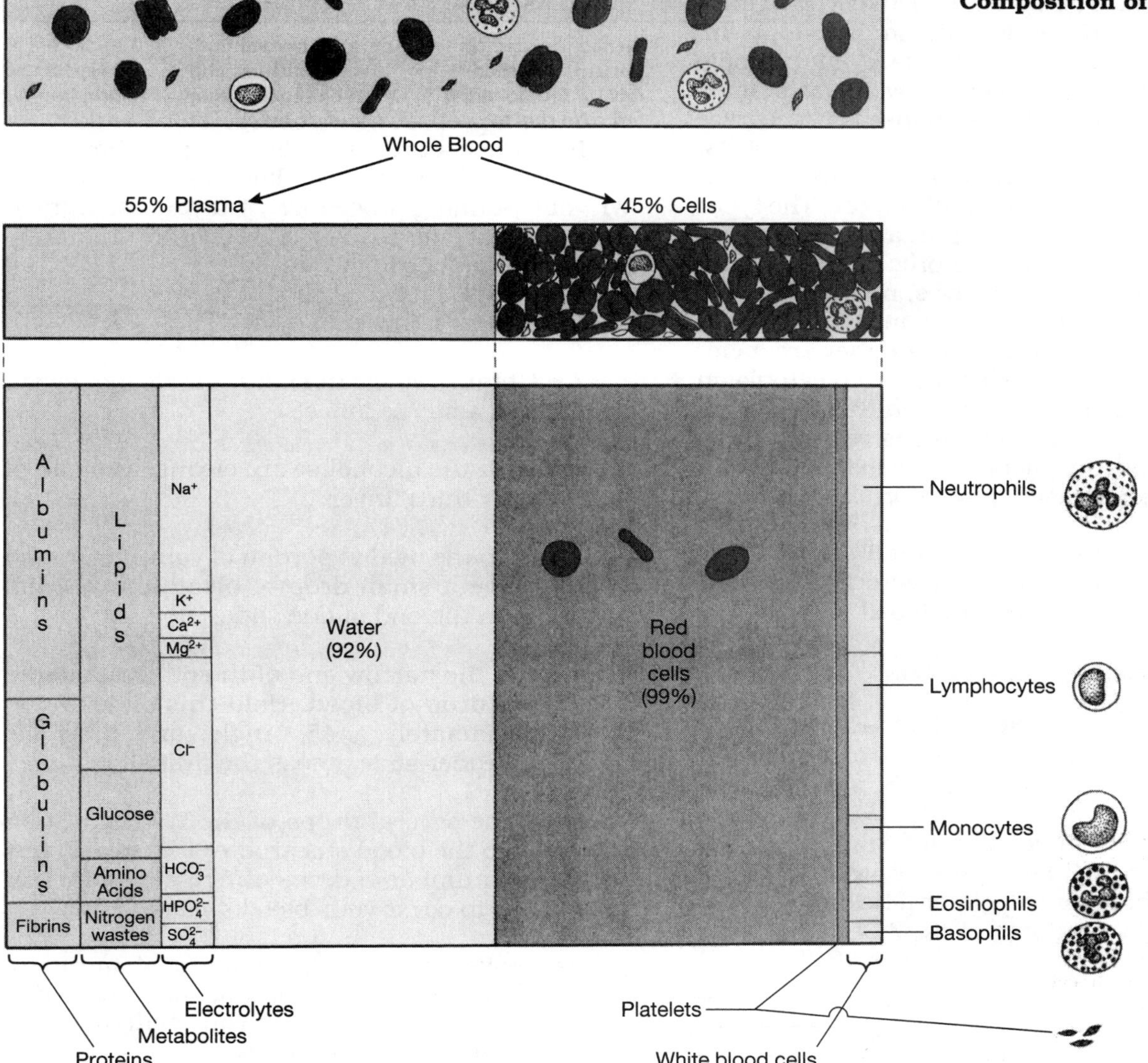

● **FIGURE 46–1**
Composition of Whole Blood.

stant body temperature; and plasma provides the forum for the transport of the formed elements.

1➡ Study Figure 46–1● to gain a perspective on the nature of the plasma components.

2➡ Write a brief summary of the plasma section of Figure 46–1●.

____

____

C. Formed Elements

The formed elements, which make up the other 45% of the blood, are the cell and cell fragment components of the blood. The two major types of blood cells are the **red blood cells (erythrocytes)** and the **white blood cells (leukocytes)**.

The red blood cells function primarily in gas transport, and the white blood cells function in defense. *[∞ Martini, p. 628]*

We generally identify two subgroups of white blood cells, **granulocytes** (or granular leukocytes) and **agranulocytes** (or agranular leukocytes). Granulocytes have distinctive granules in the cytoplasm, as you can observe with standard microscopic equipment. Agranulocytes have few, if any, stained granules under normal conditions. We recognize three types of granulocytes, **neutrophils**, **eosinophils**, and **basophils**, and two types of agranulocytes, **monocytes** and **lymphocytes**. The types of granulocytes are so named because of the dyes they retain in certain staining procedures: neutral dyes, acidic dyes, and basic dyes, respectively.

Platelets are also formed elements. However, platelets are not truly cells. Instead, they are membrane-enclosed packets of enzyme-rich cytoplasm shed from large marrow cells called **megakaryocytes**. Platelets are somethimes referred to by their older name, thromobcytes. Platelets function primarily in hemostasis. *[∞ Martini, p. 644]*

1➡ Study Figure 46–2●, which summarizes the functions and characteristics of the formed elements of the blood.

D. Exercises

1➡ Study Figure 46–3●. Verbally explain to your lab partner the common origin of the different formed elements, blood cells, and platelets.

What is the value of this information in the health care setting? For instance, what might a health care professional look for if he or she found an abundance of megakaryocytes? Of monocytes? Of neutrophils? Refer also to Figure 46–2●.

____

____

____

2➡ Go back to Figure 46–1●. Construct a pie diagram showing the composition of the blood.

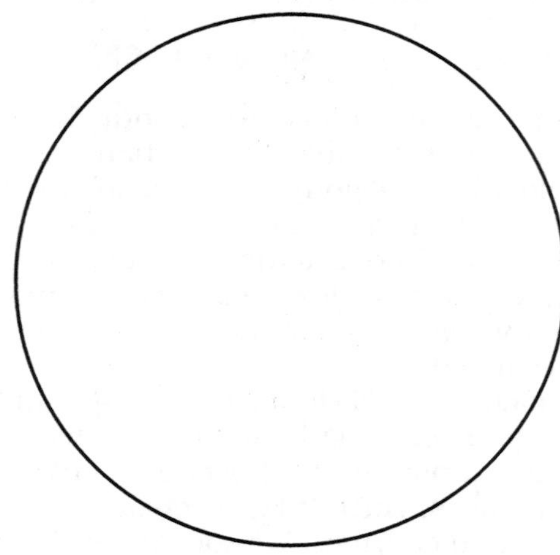

III. Microscopy

A. Slides

Your instructor will direct you either to make your own blood slide (with human or animal blood) or to use a prepared microscope blood slide.

If you use a prepared slide, skip section B.

If you make your own slide, use the following Wright's Staining procedure. ** If you use human blood, reread the CAUTION section of this exercise before proceeding.

B. Wright's Stain

1➡ Obtain two clean slides, an alcohol swab, and a sterile lancet.

2➡ With the alcohol swab, cleanse the pad of your third finger.

3➡ Stab the medial portion of your finger, and place a small drop of blood about 2 cm from the end of one slide.

4➡ Put the narrow end of the other slide into the drop of blood. Hold this slide at approximately a 45° angle, and drag the spreader slide across the first slide.

5➡ Place several drops of the Wright's Stain onto the blood smear after it is dry. Count the number of drops of Wright's Stain you use to cover your blood smear.

6➡ Leave the stain in place for 4 minutes. Then add the same number of drops of distilled water and allow your slide to stand for another 10 minutes.

Cell	Abundance (Average per μl)	Appearance in Blood Smear	Functions	Remarks
RED BLOOD CELLS	5.2 million (range: 4.4–6.0 million)	Flattened, circular cell; no nucleus, mitochondria, or ribosomes; red in color	Transport oxygen from lungs to tissues and carbon dioxide from tissues to lungs	Remain in circulation; 120-day life expectancy; amino acids and iron recycled; produced in bone marrow
WHITE BLOOD CELLS	7000 (range: 6000–9000)			
Neutrophils	4150 (range: 1800–7300) Differential count: 50–70%	Round cell; nucleus lobed and may resemble a string of beads; cytoplasm contains large, pale inclusions	Phagocytic: Engulf pathogens or debris in tissues, release cytotoxic enzymes and chemicals	Move into tissues after several hours; may survive minutes to days, depending on tissue activity; produced in bone marrow
Eosinophils	165 (range: 0–700) Differential count: 2–4%	Round cell; nucleus usually in two lobes; cytoplasm contains large granules that are usually stained bright red	Phagocytic: Engulf antibody-labeled materials, release cytotoxic enzymes, promote inflammation	Move into tissues after several hours; survive minutes to days, depending on tissue activity; produced in bone marrow
Basophils	44 (range: 0–150) Differential count: <1%	Round cell; nucleus usually cannot be seen because of dense, blue granules in cytoplasm	Enter damaged tissues and release histamine and other chemicals that promote inflammation	Survival time unknown; assist mast cells of tissues in producing inflammation; produced in bone marrow
Monocytes	456 (range: 200–950) Differential count: 2–8%	Very large cell; kidney bean–shaped nucleus; abundant pale cytoplasm	Enter tissues to become free macrophages; engulf pathogens or debris	Move into tissues after 1–2 days; survive months or longer; primarily produced in bone marrow
Lymphocytes	2185 (range: 1500–4000) Differential count: 20–30%	Usually round cell, slightly larger than RBC; round nucleus; very little cytoplasm	Cells of lymphatic system, providing defense against specific pathogens or toxins	Survive for months to decades; circulate from blood to tissues and back; produced in bone marrow and lymphatic tissues
PLATELETS	350,000 (range: 150,000–500,000)	Spindle-shaped cytoplasmic fragment; contains enzymes and proenzymes; no nucleus	Hemostasis: Clump together and stick to vessel wall (platelet phase); activate intrinsic pathway of coagulation phase	Remain in circulation or in vascular organs; remain intact 7–12 days; produced by megakaryocytes in bone marrow

● **FIGURE 46–2**
Formed Elements of the Blood.

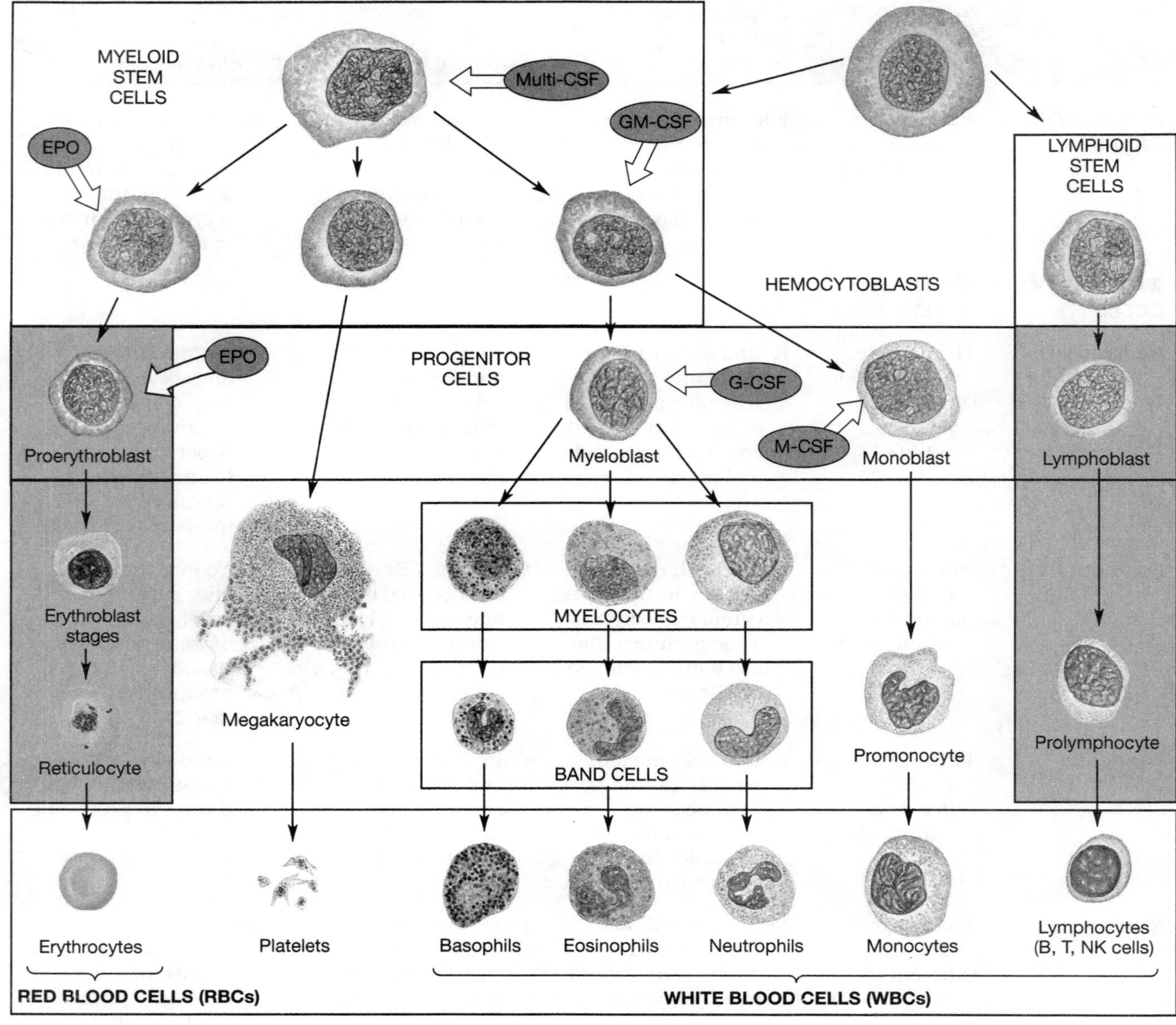

● **FIGURE 46–3**
Origin and Differentiation of Blood Cells.

7 ➡ Carefully rinse the slide under running water for 30 seconds; shake and blot the slide with bibulous paper (blotting paper). Make certain your slide is dry before proceeding.

C. PRELIMINARY EXAMINATION

1 ➡ Use your blood smear (or obtain a standard smear blood slide) and a compound microscope.

2 ➡ Focus with the oil immersion lens, and use the information in Figure 46–2● and the pictures in Figures 46–3● and 46–4● to identify the erythrocytes, neutrophils, eosinophils, basophils, monocytes, lymphocytes, and platelets.

3 ➡ Sketch and label what you see.

Blood

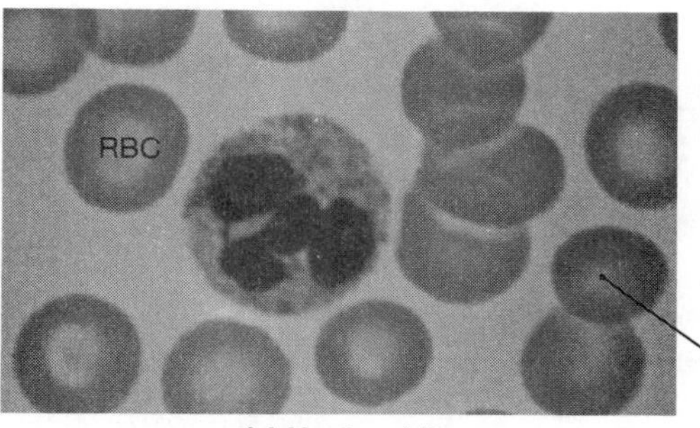

(a) Neutrophil

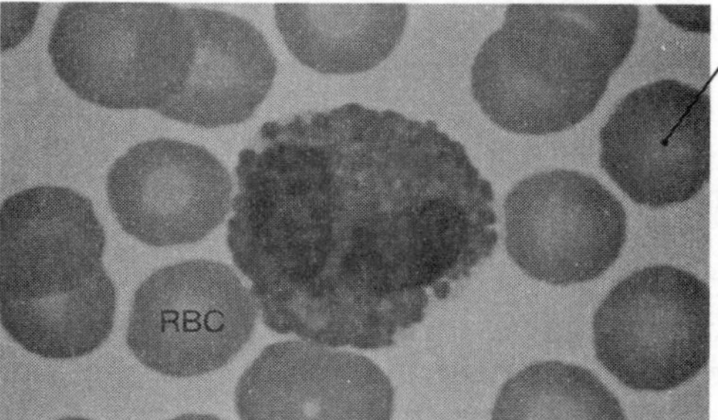

(b) Eosinophil

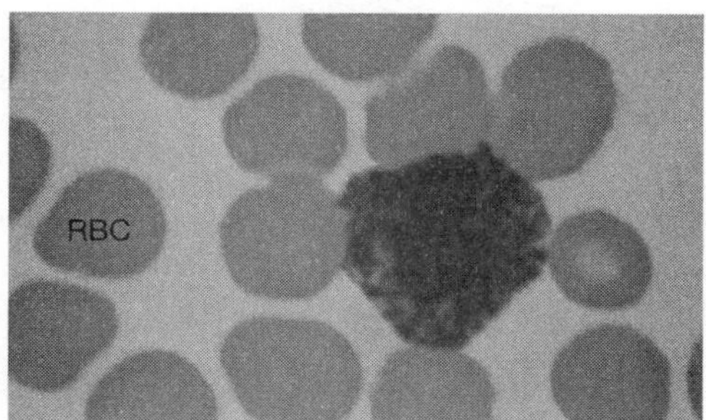

(c) Basophil

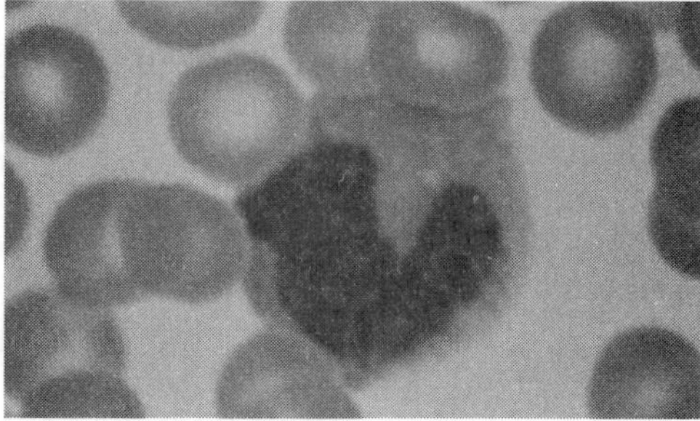
(d) Monocyte

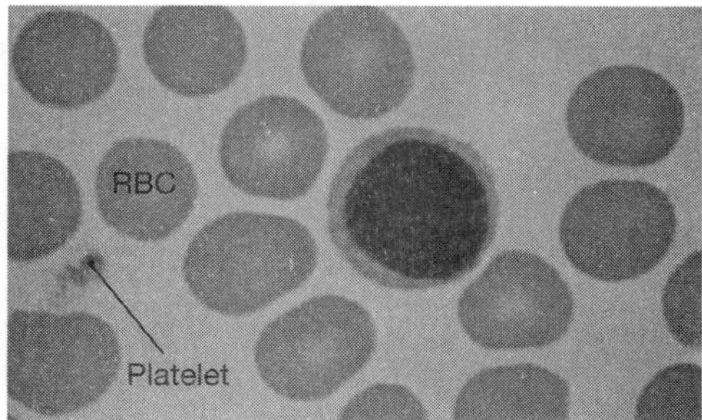

(e) Lymphocyte

● **FIGURE 46–4**
Blood Cells.

D. Differential White Blood Cell Count
[∞ Martini, p. 641]

A **differential white blood cell count** can be performed on any blood smear.

1➡ In this procedure use the oil immersion lens, and move your slide in the pattern indicated in Figure 46–5●.

2➡ Count the first 100 white blood cells you see and classify each according to type. Use caution so that you do not recount the same area.

3➡ Record your results on the following chart.

Neutrophil	*Eosinophil*	*Basophil*	*Monocyte*	*Lymphocyte*

4➡ Convert your results to percentages, and compare those percentages with the information in Figure 46–2●. Keep in mind that the percentages in the table are average values.

IV. Blood Abnormalities

A. Sickle Cell Anemia [∞ AM]

1➡ Obtain a slide of a sickle cell anemia blood smear. Examine this slide under oil.

2➡ Notice the unusual shape of many of the red blood cells. These cells look like old-fashioned sickles. An error in the basic genetic code for the hemoglobin molecule (the molecule that carries the oxygen in red blood cells) causes the amino acid valine to be substituted for the amino acid glutamic acid. This switch causes a geometric change as the hemoglobin protein folds over itself when oxygen concentration is low. Thus, the cell sickles. This sickled cell

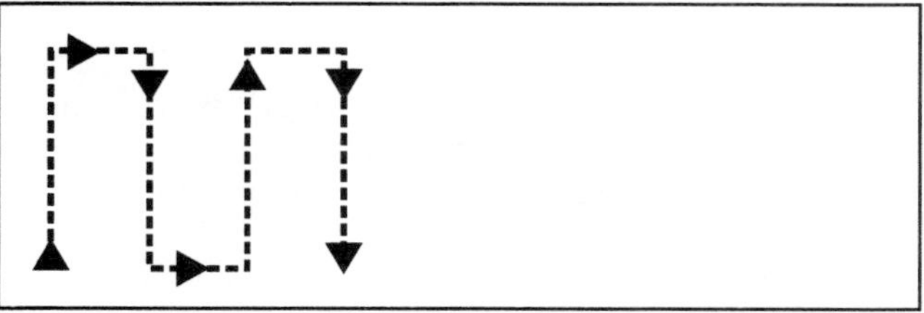

● **FIGURE 46–5**
Differential White Blood Cell Count Procedure.

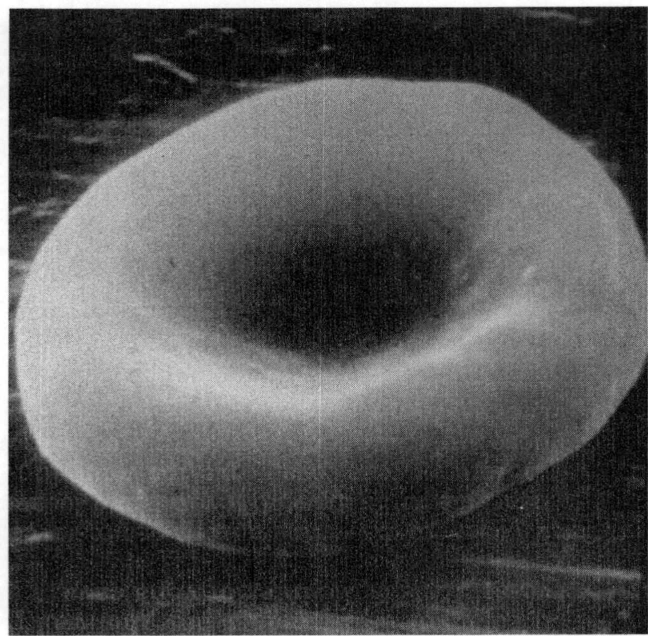

(a) Normal RBC

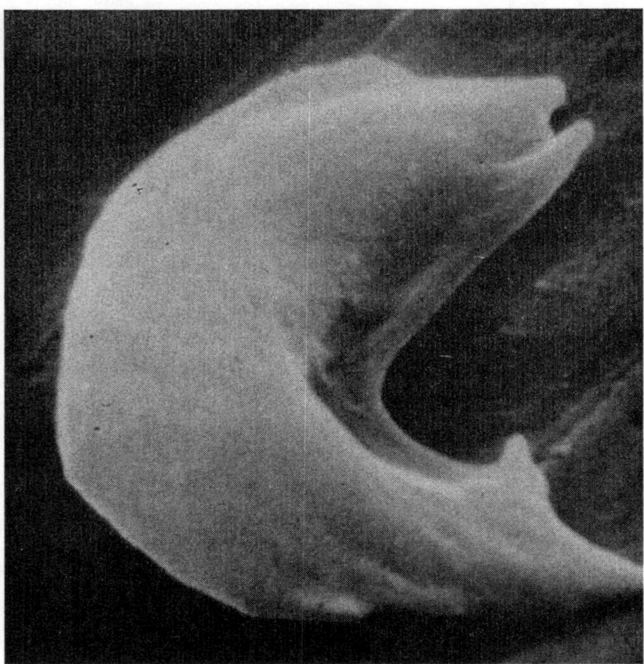

(b) Sickled RBC

● **FIGURE 46–6**
Normal and Sickling Red Blood Cells.

is fragile and easily damaged. Also, it does not fold well when going through capillaries. This damages the capillaries and subsequently causes oxygen deprivation to the tissue.

3➡ Sketch and label what you see. Use Figure 46–6● as your guide.

Sickle cell anemia

About 90% of the people worldwide who are afflicted with sickle cell anemia are black. The remaining 10% are of other racial origins, most of which originate in the Mediterranean regions.

B. Leukemia

Leukemia is the term used for any of several malignant conditions resulting in abnormal white blood cell production. As a generality we can say that a leukemic blood smear shows an increase in the number of white blood cells and that many of those circulating cells are immature and/or abnormal.

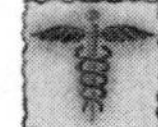

Persons with infections or inflammatory responses may often have elevated white blood cell counts. Leukemic white blood cells are often immature and/or abnormal.

Leukemia is classified according to the dominant cell type and the severity of the disease. **Myelogenous (granulocytic)** leukemia shows an increase in the number of granular leukocytes, **lymphogenous** leukemia is characterized by increased numbers of lymphocytes. Both types of leukemia show immature white blood cells.

1➡ Obtain myelogenous and lymphogenous leukemia slides. View these slides under oil.

2➡ Sketch and label what you see.

Leukemia

C. Mononucleosis

Mononucleosis is a viral disease primarily affecting the lymphoidal tissues of the body, but it is also able to affect most other systems, too. A mononucleotic blood smear will show an increased number of lymphocytes, actually B lymphocytes. (B lymphocytes and T lymphocytes, the two major lymphocytic components of the immune system, cannot be distinguished with Wright's Stain.) This is because the Epstein-Barr virus (EBV), causative agent of infectious mononucleosis, affects only this particular leukocyte, the B lymphocyte. The speedy replication of the virus within the white blood cell accounts for the rapid proliferation of this type of cell. (Incidentally, this condition is sometimes difficult to diagnose because the blood smear often resembles a blood smear of acute leukemia.)

It is not true that people over thirty do not get mononucleosis. Anyone can get this disease. Some experts estimate that by adulthood 80% or more of the population has actually had mononucleosis. Most of these cases would be subclinical.

1➡ Obtain a mononucleosis blood smear. Compare the numbers and types of white blood cells with the numbers and types of white blood cells on other blood smears you have viewed.

2➡ Write out the similarities and differences you notice between the white blood cells on these different slides.

____

____

____

❑ Additional Activities

NOTES

1. Use the hemacytometer and the diluting pipettes listed in the Materials Needed section to do a blood count.

   If you use human blood for this procedure, reread the CAUTION section of this exercise before beginning.

   Because several types of equipment exist, your instructor will have specific directions for you. Record your count below.

   RBC/0.1 ml of blood ______________________________

   WBC/ 0.1 ml of blood ______________________________
2. Bone marrow (red marrow) found in the long bones of the body is the site of formed element production in the body. Without bone marrow, we would not live. Examine the principles involved in bone marrow transplants. Do you think that we need to match marrow donors and recipients (as is done with blood donors and recipients)? Explain your answer. *[∞ Martini, pp. 627, 633, 643]*
3. AIDS (Acquired Immune Deficiency Syndrome) is a deadly disease caused by the HIV virus. What changes, if any, would you expect to see in the blood of an AIDS patient?
4. Research the current methods used for studying blood gases (including the oxymeter), cholesterol counts, blood profiles, chemical analyses, and other up-to-date methods of studying the state of the human system by analyzing the river of life.

❑ Lab Report

1. Referring to the numbered inquiries at the beginning of this exercise, complete the following box summary:

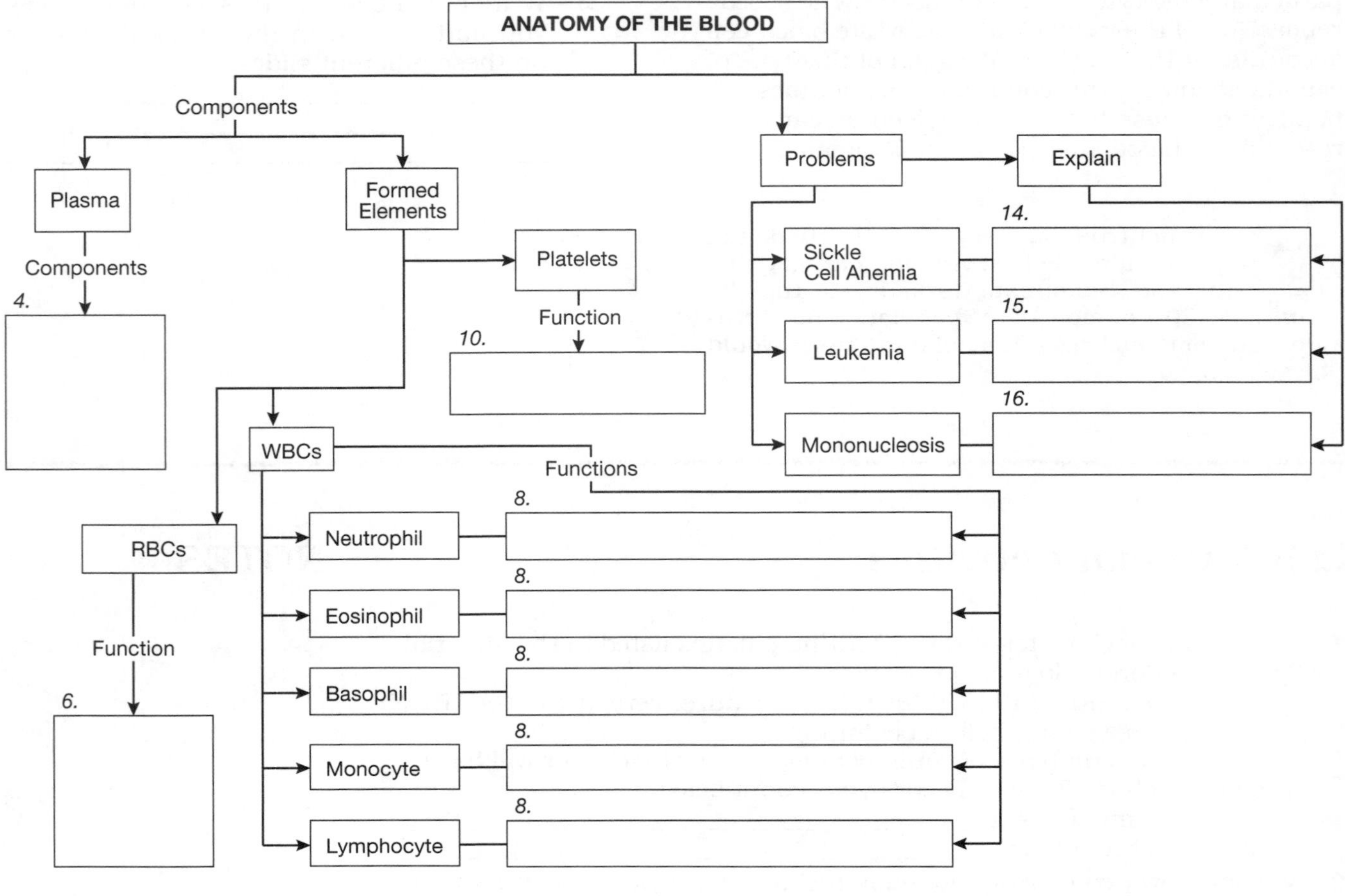

General Questions

NOTES

1. What safety rules must be followed when handling human blood?
2. What are the basic functions of blood?
3. Explain the developmental relationship among the formed elements.
4. Why can animal blood be substituted for human blood in these investigations?
5. What is Wright's Staining procedure and why is it used?
6. How do you do a differential white blood cell count? What is the purpose of doing this count?
7. Your blood volume in liters is approximately 7% of your body weight in kilograms. Calculate your own blood volume. (1 kg = 2.2 lb). Since this value is an estimate, what factors might cause the actual volume to be higher or lower than expected?
8. Using your answer to the preceding question, how much circulating plasma do you have? How much circulating water? (Recall that plasma is about 55% of the blood volume, and water constitutes about 92% of the plasma volume.)

EXERCISE 47

Blood Testing Procedures

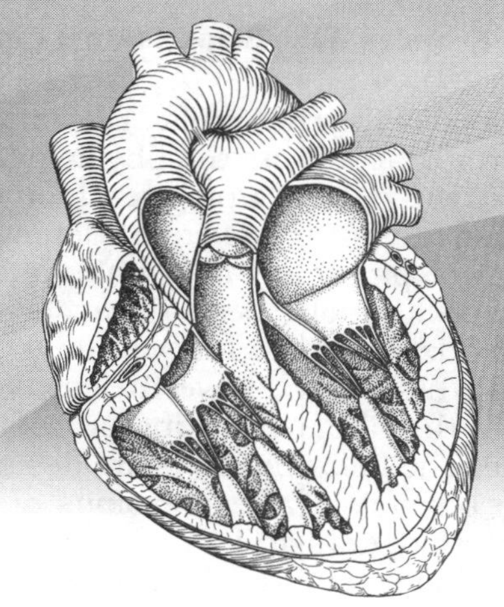

PROCEDURAL INQUIRIES

Preparation

1. What safety precautions must be followed when handling human blood?
2. How do agglutinogens and agglutinins relate to antibodies and antigens?

Experimentation

For each of these tests answer these questions:
What does the test check for?
How do you perform the test?
What is the principle behind the test?
Why is the test important?

3. Sedimentation rate
4. Bleeding time
5. Clotting time
6. Hematocrit
7. Hemoglobin
8. Blood typing

Additional Inquiries

9. What types of substances found in the blood can be tested for?
10. What types of tests cannot be done successfully with prepared animal blood? Why not?

Key Terms

Agglutination
Agglutinin
Agglutinogen
Anemia
Antibody
Anticoagulant
Antigen
Antigenicity
Bleeding Time
Blood Typing
Clot
Clotting Time
Coagulate
Hematocrit
Hemoglobin
Hemoglobinometer
Heparin
PCV
Platelets
Polycythemia
Rouleaux Formation
Sedimentation Rate
Tallquist Method

Materials Needed

Watch with Second Hand
Lancets
Alcohol Swabs or 70% Alcohol
Capillary Tubes
Heparinized
Unheparinized
Capillary Tube Sealer or Clay
Hematocrit Reader (if available) or Ruler
Tallquist Paper
Tallquist Scale
Hemoglobinometer (if available)
Landau SED-rate pipette
Landau Rack
5% Sodium Citrate
Mechanical Suction Device
Filter Paper
Cotton Swabs or Facial Tissue
Centrifuge
Blood Typing Kits
Clean Slides (2 per student)
Wax Pencils
Toothpicks (3 per student)
(Optional: Warming Tray)

Blood is the river of life. Blood is also the mirror that reflects the physiological condition of the body at any given moment in time.

Blood tests today can detect even trace amounts of almost any imaginable substance. Testing for specific molecules — from cholesterol to carbon dioxide to potassium cyanide — is beyond the scope of this lesson. Nevertheless, you should be aware of the range of testing procedures that can be done. The following tests are designed to examine the physiological expressions of certain basic blood components and give you a hands-on explanation of some standard blood tests.

❑ Preparation

I. Safety

In today's world we find a great deal of concern about the potential dangers of dealing with blood and blood products. The present AIDS epidemic certainly justifies this concern. Another concern is the Hepatitis B virus, which many professionals believe is actually easier to contract than the AIDS virus.

The importance of the proper handling of blood testing equipment cannot be stressed too often.

You should understand this concern and always keep the following in mind when completing this and any exercise involving blood.

A. Human Blood

If you use certain common sense safety precautions, you will virtually eliminate, to the greatest possible extent, the possiblity of spreading blood-transmitted diseases, including AIDS. (There are no absolute guarantees.)

1➡ DO ALL OF YOUR OWN BLOOD HANDLING. That means do your own lancet stabbing, swabbing, blood testing, and clean-up. DO NOT help your friends; DO NOT share anything with anyone.

2➡ Do not touch anyone's blood or blood-testing equipment.

3➡ Use *everything* only once. If you make a mistake, start over with fresh equipment.

4➡ DO NOT set anything that has touched blood on your lab table or lab bench.

5➡ Be aware of the location of every lancet, slide, and piece of equipment so you do not risk leaning into or bumping some dirty equipment.

6➡ Dispose of *everything immediately* in the proper tubs or autoclave bags as directed by your instructor.

7➡ If you are not completely certain about what you are doing, do not do it until you ask questions and get all your questions answered completely.

8➡ Clean up and disinfect *exactly* according to your instructor's directions.

It is assumed that if you have a known blood disease, serious clotting disorder, or some blood letting dysfunction, you will not participate in any lab exercise requiring the drawing of your blood.

B. Animal Blood

1➡ For most of the experiments in this exercise, animal (nonhuman) blood from a biological supply house can be substituted. The *clotting time* experiment will require fresh blood because blood from the supply houses has been treated with an anticlotting substance, such as heparin. We recommend you read through this part of the experiment to gain an understanding of what clotting is all about.

2➡ Blood types do exist in the other members of the animal kingdom. These types differ from human blood types. Your school laboratory is probably not equipped to do animal blood typing, so we recommend you simulate (or simply discuss) the human blood typing experiment.

3➡ If your laboratory uses animal blood purchased from a supply house, you do not have to worry about the same types of contamination, including AIDS that might occur with human blood. Nevertheless, nonhuman blood can become contaminated. Always handle any blood as if it were contaminated.

4➡ Animal (mammalian) blood has all the same cells and same physical properties as human blood. Some minor differences do exist (for instance, sheep erythrocytes are much smaller than human erythrocytes), but the basic physiological principles are the same.

II. Background Information

A. Vocabulary

In order to understand the blood testing procedures, you should familiarize yourself with the following terms. As you learn this material, be on guard because many of the words are so similar!

Agglutination Clumping caused by the complexing of agglutinogens and agglutinins (antigens and antibodies).

Agglutinin Antibody, or similar substance, which produces agglutination when complexing with an agglutinogen (antigen). These agglutinins are examples of a class of proteins called immunoglobulins.

Agglutinogen Cell membrane molecules (or antigens) that interact with specific agglutinins (antibodies).

Anemia Decrease in the oxygen carrying capacity of the blood. This condition, which is a symptom rather than a disease, is associated either with a decrease in the number of red blood cells or with an abnormality in the ability of the red blood cells to transport oxygen.

Antibody In the case of blood, antibody is the same as agglutinin, a substance that will react with a specific cell membrane molecule (agglutinogen).

Anticoagulant Substance that prevents or slows clot formation by interfering with the clotting mechanism.

Antigen A substance capable of inducing the production of antibodies. In this case the antigen, also known as an agglutinogen, is a molecule located on the plasma membrane of the red blood cells.

Antigenicity The relative ability of a substance to induce the production of antibodies. Certain antigens are far more antigenic than others. That is to say, they cause a greater production of antibodies.

Clot Network of fibrin fibers and trapped blood cells.

Coagulate Form a clot.

Heparin Anticoagulant released by activated basophils, mast cells and certain body tissues.

Platelets Membranous packets of cytoplasm containing clotting-response enzymes.

Polycythemia Elevated numbers of red blood cells. Polycythemia may occur normally (erythrocytosis) when a person goes from an area of lower elevation to an area of higher elevation. Polycythemia vera is a disease state.

❑ Experimentation

III. Physiological Characteristics of Blood

A. Sedimentation Rate

(We recommend you set up this experiment first because once set up it must remain undisturbed for exactly one hour.)

Sedimentation rate (SED rate) is the rate at which red blood cells fall out of solution when allowed to stand in a verticle tube. **Rouleaux formation** is the term used for the alignment of red blood cells in vertical stacks.

The normal adult SED rate is 0–6 mm/hr. For children the normal rate is 0–8 mm/hr. Pregnant and menstruating women generally have rates above 8 mm/hr. This nonspecific diagnostic blood test is of immense value in determining whether additional tests should be run to pinpoint the cause of specific problems. Acute infections and malignancies accelerate the SED rate increases, presumably because the negative charge on the surface of the red blood cells is reduced. This reduces intercellular repulsion; the cells settle more readily; the SED rate increases. Sickle cell anemia decreases the SED rate because the red blood cells are abnormally shaped. Polycythemia also decreases the SED rate, and the anemias increase the SED rate.

Note: You will need over one hour to complete this test.

1➡ Obtain a sterile lancet, alcohol swab, Landau SED-rate pipette, Landau rack, 5% sodium citrate, and mechanical suction device.

2➡ Your instructor will demonstrate how to use the mechanical suction device and how to mix the blood and citrate solution.

3➡ Draw the sodium citrate solution up into the Landau pipette to the first mark. (Sodium citrate is an anticoagulant.)

4➡ Puncture your finger, wipe off the first drop of blood, then draw blood into the pipette until the *sodium citrate* solution reaches the second line.

5➡ Mix the solution thoroughly. Draw the mixture up into the bulb and force it back down into the lumen six times.

6➡ Remove the suction device as directed. Keep the lower end tightly sealed. Stand the pipette in the Landau rack. The tube must be *exactly vertical.*

Record the time: ____________

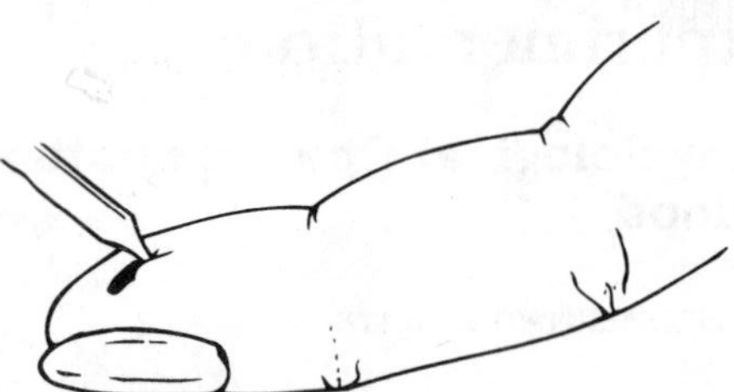

● **FIGURE 47–1**
Stabbing the Side of the Finger.

7➡ Allow the pipette to stand undisturbed for exactly one hour. At the end of the hour, measure in millimeters the amount of clear plasma at the top of the tube.
What is your sedimentation rate? ______________

B. BLEEDING TIME

Bleeding time is the time necessary for a small incision to stop bleeding. Bleeding time within the normal range of 1–3 minutes is indicitive of normal platelet function.

1➡ Obtain a lancet, alcohol swab, and filter paper.

2➡ Cleanse the ball and sides of your middle finger. You should keep your hand low so that blood flow is enhanced by gravity. Do not squeeze or massage the finger. Squeezing and massaging the finger increase the free interstitial fluid in the area. This will dilute the blood.

3➡ See Figure 47–1● for a demonstration schematic of the stabbing procedure.

4➡ Note the time as you puncture the medial side of the finger deeply enough so the blood flows freely. (Your blood draw will be more productive if you stab the side, rather than the bulb, of the finger.)
Time of puncture: ______________

5➡ Lightly blot the blood from the wound at 30-second intervals until the bleeding stops. Approximately how many seconds elapsed from the time of the puncture until the bleeding stopped?

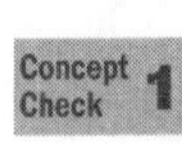
Concept Check 1 Explain why the bleeding time is indicative of adequate platelet function.

____
____
____
____

C. CLOTTING TIME

Clotting time is the time required for a blood sample outside the body to form a clot. Primarily, if clotting time values fall within the normal range of 2–4 minutes, sufficient clotting factors are present in the plasma. Persons with clotting disorders, such as hemophilia, will have much longer clotting times. Indirectly, clotting time is also a determination of platelet formation.

1➡ Obtain a lancet, alcohol swab, and an *unheparinized* capillary tube.

2➡ Cleanse you finger. Again, you should keep your hand low to enhance blood flow. Do not massage the finger.

3➡ Puncture the finger and dispose of the first drop of blood.

4➡ Place one end of the unheparinized capillary tube into the next drop of blood. Hold the finger at a downward angle and hold the capillary tube parallel to the finger. See Figure 47–2● for a schematic of filling a capillary tube.

5➡ Note the time the tube was filled.
Time tube was filled: ______________

6➡ After 1 minute, break off a *small* piece of the tube and slowly pull the parts of the tube away from each other. If a thread of coagulated blood can be seen between the pieces of tube, coagulation has occurred.

7➡ Repeat step 6 every 30 seconds until the blood thread (clot) is noted.
Approximately how long did it take to form the clot? ______________

Concept Check 2 What is the difference between bleeding time and clotting time?

____
____
____

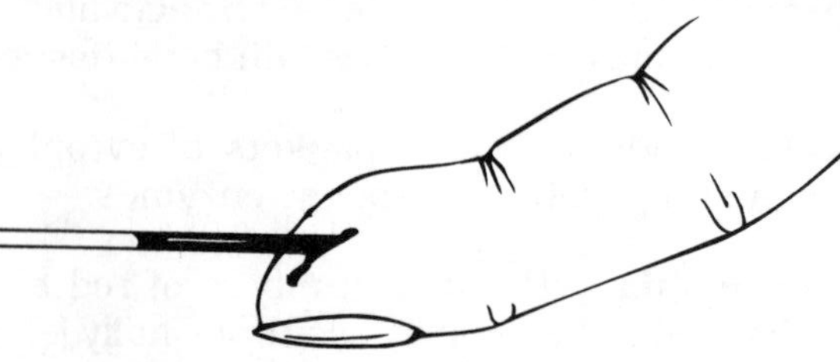

● **FIGURE 47–2**
Filling a Capillary Tube.

How are bleeding time and clotting time related?

____

____

____

D. Hematocrit [∞ Martini, p. 634]

Hematocrit, also called packed cell volume (PCV), is the measure of the percentage of red blood cells in a given volume of blood. A person's hematocrit can be used to determine abnormalities in red blood cell volume, abnormalities such as certain anemias or polycythemias.

1➔ Obtain a lancet, alcohol swab, heparinized capillary tube, capillary tube sealer, and hematocrit reader.

2➔ Cleanse your finger. Again, you should keep your hand low to enhance blood flow. It is especially important that you do not massage the finger. Massaging can change the red blood cell concentration in the given sample. You want your sample to be representative of the hematocrit throughout your body.

3➔ Fill the heparinized capillary tube about two-thirds full and seal the clean end. (Any volume is valid but the fuller the tube, the more accurate your reading will be.) Refer again to Figure 47–2● for a schematic of filling a capillary tube.

4➔ Place your capillary tube in the centrifuge. Follow your instructor's directions for balancing the centrifuge. Spin the tubes for about 4 minutes at a medium speed.

5➔ Remove the capillary tube and determine the hematocrit by using the hematocrit reader. If a hematocrit reader is unavailable, use the ruler. Measure both the total volume in the tube and the volume of red blood cells. Compute the percentage.

What is your hematocrit? __________

Normal values for men are between 40% and 55%. Normal values for women are between 37% and 47%.

How do you compare with the norms?

____

E. Hemoglobin [∞ Martini, p. 634]

Hemoglobin is the oxygen-carrying molecule within the red blood cell. Thus, the amount of hemoglobin present is indicative of the oxygen-carrying capacity of the blood. Blood hemoglobin usually ranges between 12 and 16 gm/100 ml blood. The norm for men is usually slightly above the norm for women.

1➔ Determine your hemoglobin count using this ballpark method. Take the hematocrit value you obtained in the previous test and divide by 3. This method is not completely accurate because it does not accurately consider variations in blood volume. Nevertheless, if your hematocrit was approximately normal, this method of determining hemoglobin will be reasonably accurate. What is your hemoglobin count using this method? __________

2➔ Now use the old but simple method of determining hemoglobin, the **Tallquist Method**. The Tallquist Method is based on the supposition that the higher the hemoglobin concentration, the darker the blood sample. Again, this method is reasonably accurate if your basic blood properties are within a normal range. Problems with blood volume and blood concentration can affect your results. Also, the Tallquist Method does not indicate the true oxygen-carrying capacity of the blood. For instance, if you have sickle cell anemia, you might have an adequate quantity of hemoglobin, but the oxygen-carrying capacity of your hemoglobin might be severly reduced.

To perform the Tallquist Method, follow these directions.

a. Obtain a sterile lancet, alcohol swab, piece of Tallquist Paper, and Tallquist Scale.
b. Puncture your finger.
c. Place the second drop of blood on the Tallquist Paper and wait about 15 seconds.
d. Compare the Tallquist Paper with the Tallquist Scale.

What is your hemoglobin using the Tallquist Method? __________

How do your two hemoglobin counts compare?

____

3➔ The **hemoglobinometer** is a hand-held instrument designed to give a fast and easy hemoglobin determination. If you have access to a hemoglobinometer, follow the specific directions for your instrument. To operate most hemoglobinometers, you simply focus a beam of light through a drop of blood and read the value either on a set scale or by comparing colors with a standard scale.

F. Blood Typing [∞ Martini, p. 634]

Blood typing is the process of determing the specific agglutinogens (antigens) found on the surface of red blood cells. There are numerous blood groupings, but we are primarily interested in two blood groupings—the ABO blood group and the Rh blood group.

The agglutinogens (antigens) A and B are molecules located on the membrane of the red blood cell. If the red blood cells in your system have the A agglutinogen (antigen), you have type A blood. If you have B agglutinogen (antigens), you have type B blood. If you have both A and B agglutinogens (antigens), your blood type is AB. If you have neither the A agglutinogen (antigen) nor the B agglutinogen (antigen), your blood type is O.

Genetically, agglutinogens (antigens) A and B are generated by codominant genes. Type O is recessive. The genetics of this process is described in Exercise 71.

If you have the A agglutinogen (antigen) on your red blood cells, you have anti-B agglutinins (antibodies) in the plasma portion of your blood. If you have the B agglutinogen (antigen), your plasma contains anti-A agglutinins (antibodies). Type AB people have neither anti-A nor anti-B agglutinins (antibodies) in the plasma. Type O people have both anti-A and anti-B agglutinins (antibodies).

If red blood cells with A agglutinogens (antigens) come in contact with anti-A agglutinins (antibodies), agglutination will occur. Even in a few drops of blood, this complexing will be visible as small clumps.

The Rh factor (named for the rhesus monkey in which it was discovered) is also a molecule located on the plasma membrane of the red blood cell. The Rh antigen is genetically independent of the ABO antigens. The gene for the Rh factor is even located on a different chromosome. If you are Rh+, you have this Rh agglutinogen (antigen). You do not have the agglutinogen (antigen) if you are Rh–.

(It should be noted that there are actually three Rh antigens — C, D, and E. All three antigens are inherited as dominants. D is the antigen with the greatest antigenicity and thus is the one we generally test for.)

Your blood testing kit probably contains three test substances. One is labeled anti-A, the second labeled anti-B, and the third labeled anti-D (for the D antigen of the Rh group). Blood typing is based on the antigen-antibody theory.

1➡ Obtain a blood typing kit, two clean slides, sterile lancet, alcohol swab, wax pencil, three toothpicks.

2➡ With the wax pencil draw two circles on one slide and one circle on the other slide. Label the two circles A and B, and label the one circle D.

3➡ Clean your middle finger. Stab the side of the finger with the lancet and put a drop of blood in each of the three circles.

4➡ Place a drop of anti-A on the blood in the A circle, a drop of anti-B on the blood in the B circle, and a drop of anti-D on the blood in the D circle.

5➡ Using a *different toothpick for each circle,* gently stir the blood and antibody mixture.

6➡ If you have access to a warming tray, place the D slide on the warmer for a few minutes. If you do not have a warmer and if you do not see an immediate reaction, wait a few extra minutes before reading the D test.

7➡ To read your blood tests, examine the circles. If you see any agglutination (clumping), you have a positive test. If you are in doubt, use a stereoscopic or compound microscope to check for agglutination.

Explanation Keep in mind that blood typing is based on the antigen/antibody concept.

Type A: If the anti-A antibody agglutinates the blood, you have the A antigen. Therefore, your blood is type A.

Type B: If the anti-B antibody agglutinates the blood, you have the B antigen. Therefore, your blood is type B.

Type AB: If both the anti-A and the anti-B agglutinate your blood, you have type AB blood because you have both antigens.

Type O: If neither the anti-A nor the anti-B causes agglutination, you have neither antigen so your blood is type O.

Rh+/–: If agglutination occurs in the D circle, you have the Rh antigen and you are Rh+.

If no agglutination occurs, you are Rh–.

What is your ABO blood type?: __________

What is your Rh type?: __________

Population	*Percentage with Each Blood Type* O	A	B	AB	Rh+
U.S. (average)	46	40	10	4	85
Caucasian	45	40	11	4	85
African-American	49	27	20	4	95
Chinese	42	27	25	6	100
Japanese	31	39	21	10	100
Korean	32	28	30	10	100
Filipino	44	22	29	6	100
Hawaiian	46	46	5	3	100
Native North American	79	16	4	<1	100
Native South American	100	0	0	0	100
Australian Aborigines	44	56	0	0	100
Class	___	___	___	___	___

● **FIGURE 47–3**
Blood Type — Population Distribution Percentages.

Record your class figures on the world figures chart (Figure 47–3●).

It is possible the blood type you obtained from this classroom test is different from the results of a previous blood typing test. Keep in mind that this is a classroom setting and accuracy is expectedly less than would be found in a medical setting.

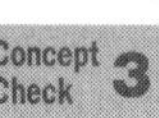

Logically, what types of problems would occur as a result of a mismatch on a blood transfusion?

____

____

____

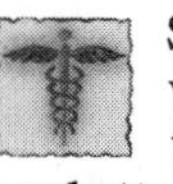

Sometimes a mistyping occurs because of a very low titer of the particular blood agglutinogen. This is particularly true of the A agglutinogen. Some people who test type O on simple tests such as this one are genetically type A. This would become evident on more controlled and more sophisticated blood tests. In addition, some blood types, particularly type A, have several recognized subtypes. This can become confusing to those who do not understand what blood type really means.

❑ Additional Activities

NOTES

1. Research the H antigen and explain how H relates to the ABO group.
2. Find out how the Bombay type relates to the ABO blood grouping. What did the discovery of the Bombay type do to ABO theories?
3. Many of the other blood groupings are not highly antigenic. Make a list of these blood groupings and find out what tests are used to determine these groupings. Many of these antigens have low antigenicity. What does this mean in terms of blood transfusions? How does the "type and cross match" theory fit into the discussion of the other blood groupings?
4. Find out what is meant by the "major side" and the "minor side" in terms of blood transfusions.
5. Fibromyalgia is a condition diagnosed by exclusion. Research the symptoms of fibromyalgia and explain which diagnostic blood tests are used to eliminate other conditions—conditions whose symptoms might be the same as the symptoms of fibromyalgia.

Answers to Selected Concept Check Questions

1. Platelets become ensnared on the edges of the cut and form an initial platelet plug.
2. Bleeding time is indicative of platelet formation. Platelets are formed elements in the blood. Clotting time is an indication of the clotting factors in the plasma.
3. The agglutinins (antibodies) in the plasma will react with the agglutinogens (antigens) on the red blood cells. This clumping or agglutination within the blood vessels could lead to vascular destruction or even death.

❑ Lab Report

1. Referring to the numbered inquiries at the beginning of this exercise, complete the following box summary:

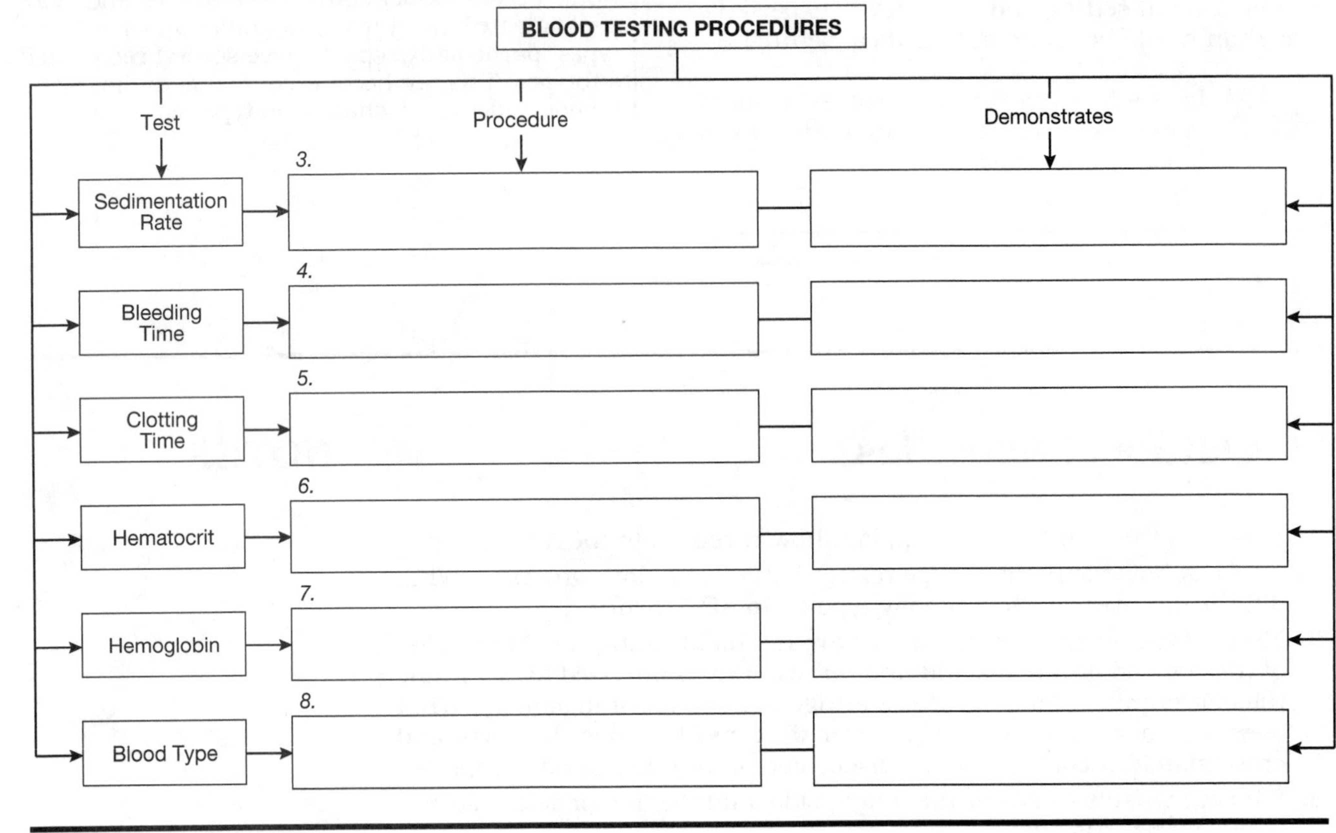

General Questions

NOTES

1. What safety precautions must always be followed when handling blood?
2. Why are blood tests so valuable today?
3. What is the difference between hematocrit and hemoglobin concentration?
   a. How are the two similar?
   b. How are the two different?
4. Fill in this chart:

Blood Type	Substances on RBC	Substances in Plasma
A		
B		
AB		
O		

5. How are ABO and Rh similar? How are they different?
6. Sketch the major events in antigen-antibody reactions in the blood.
7. Menstrual blood contains tissue heparin. Why is this important?
8. What similarities and what differences can be found between animal blood and human blood when doing blood tests?

EXERCISE 48

Anatomy of the Human Heart

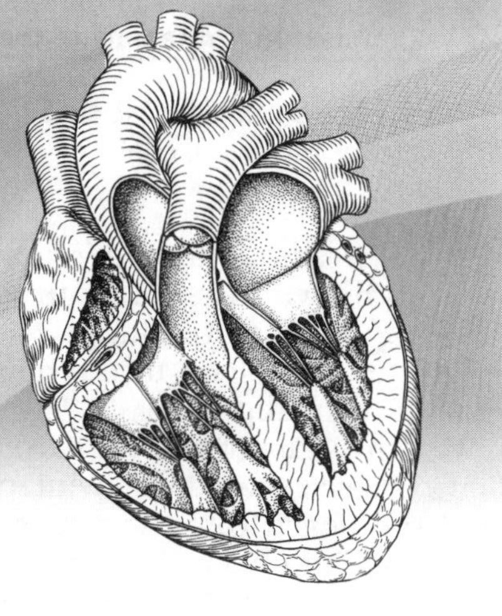

PROCEDURAL INQUIRIES

Preparation

1. How can you describe the human heart?
2. What is the path of blood through the heart?

Examination

3. How can you distinguish the anterior surface of the heart from the posterior surface?
4. What are the layers of the heart?
5. What external structures are located between the atria and the ventricles?
6. Which blood vessels are associated with each chamber of the heart?
7. What is the ligamentum arteriosum?
8. What are the internal walls of the heart? Where is each located?
9. What are the valves of the heart? Where is each located?
10. What are the internal muscles and tendons of the heart?
11. What is the moderator band?
12. What is the coronary sinus?
13. Where would you find the coronary arteries?
14. What are the components of the coronary circulatory system?
15. What are the nodes of the heart?

Additional Inquiries

16. What is the physical relationship of the atria to the ventricles?
17. How do you define sulcus, anastomosis, and intercalated disk?

Key Terms

Anastomoses
Aorta
Aortic Arch
Aortic Semilunar Valve
Apex
Atria
Atrioventricular Node
Atrioventricular Septa
Auricle
Bicuspid Valve
Chordae Tendinae
Coronary Artery
Coronary Sinus
Ductus Arteriosus
Endocardium
Epicardium
Foramen Ovale
Fossa Ovalis
Infarct
Interatrial Septum
Intercalated Disks
Interventricular Septum
Ligamentum Arteriosum
Moderator Band
Myocardium
Papillary Muscles
Pectinate Muscle
Pericardial Sac
Pulmonary Arteries
Pulmonary Semilunar Valve
Pulmonary Trunk
Pulmonary Veins
Sinoatrial Node
Trabeculae Carnae
Tricuspid Valve
Vena Cava
Ventricles

Materials Needed

Torso
Heart Models
Charts and Diagrams
Preserved Human Heart (if available)
Dissecting Microscope
Compound Microscope
Prepared Slides of Heart
Cross Section, Cardiac Wall

Cardiocytes (may be labeled Cardiac Cells, or Heart Muscle Cells)

The heart is a four-chambered double pump with the dual purpose of circulating blood to the lungs for gas exchange and circulating blood throughout the body for metabolic exchange.

We can summarize these cardiac functions as follows. Deoxygenated blood enters the right atrium of the heart from the superior and inferior vena cavae and from the coronary sinus. From the right atrium the blood travels through the tricuspid valve into the right ventricle. From the right ventricle the blood is pumped through the pulmonary semilunar valve into the pulmonary trunk, which subsequently branches into the right and left pulmonary arteries that then carry the blood to the lungs.

Following gas exchange in the lungs, the oxygenated blood is returned to the heart via the four pulmonary veins, which empty into the left atrium. From the left atrium the blood passes through the bicuspid (mitral) valve into the left ventricle. From the left ventricle, the blood passes through the aortic semilunar valve into the aorta, and from the aorta the oxygenated blood is carried throughout the body. In the body proper the metabolic exchange occurs — oxygen is exchanged for carbon dioxide. Deoxygenated blood is then returned to the heart. And the cycle begins again.

The heart is located in the mediastinum, the central area between the pleural cavities, and is usually situated at an angle with the flattened **base** approximately medial to the right breast and the angular **apex** pointing toward the left hip. The upper chambers, the right and left **atria** (sing., **atrium**), are superior to the right and left **ventricles**. The term **auricle** should be used only to identify the flappy appendages of the atria.

This exercise, which is an overview of the basic anatomy of the human heart, can be completed in conjunction with Exercise 49.

❑ Preparation

I. Background

A. Models and Charts

1➜ It is assumed you have access to well-defined "take-apart" models of the heart. Use these models liberally. Augment your examination with careful analyses of the charts and diagrams.

2➜ If your laboratory does not have heart models, use the charts and diagrams, plus your imagination, to gain a perspective of the anatomy of the human heart.

❑ Examination

II. Gross External Examination

A. Orientation

1➜ Begin your study with the anterior surface of the heart facing you. The anterior surface is readily discernible because of the large diagonal line of fat that seems to extend from your upper right to your lower left. Beneath the layer of fat is the **anterior interventricular sulcus**. Your model may be constructed so that you can remove some of the (plastic) fat and expose the **anterior interventricular artery** and the **great cardiac vein**. A similar line on the posterior aspect of the heart, known as the **posterior interventricular sulcus**, seems to be vertical.

2➜ Examine the position of the heart in the torso. Also, check the charts and diagrams. The heart is surrounded by a thick fibrous membrane known as the **pericardial sac** or the **parietal pericardium**. This pericardium is actually composed of an outer **fibrous layer** and an inner **serous layer** that secretes the pericardial fluid.

3➜ Examine Figure 48–1● for the layers of the heart. Your model may have labeled the **visceral pericardium (epicardium)**, which forms the outer surface of the heart. Beneath this single squamous cell layer is the hefty **myocardium**, the muscular layer we generally think of as heart muscle. The inner layer of the heart is the **endocardium**, which is continuous with the endothelium of the blood vessels. The endocardium may not be identified on your models, but you should have no difficulty locating it on your charts and diagrams.

B. Anterior Examination [∞ Martini, p. 657]

1➜ Read through the following descriptions. Label Figure 48–2● with the boldface terms.

2➜ With the anterior surface of the heart facing you, the thicker left ventricle will be on your right. In the living heart the left ventricular wall will be firmer than the right ventricular wall because the thicker left ventricle is required to pump the blood throughout the body. Externally the left and right ventricles

● **FIGURE 48–1**
Layers of the Heart.

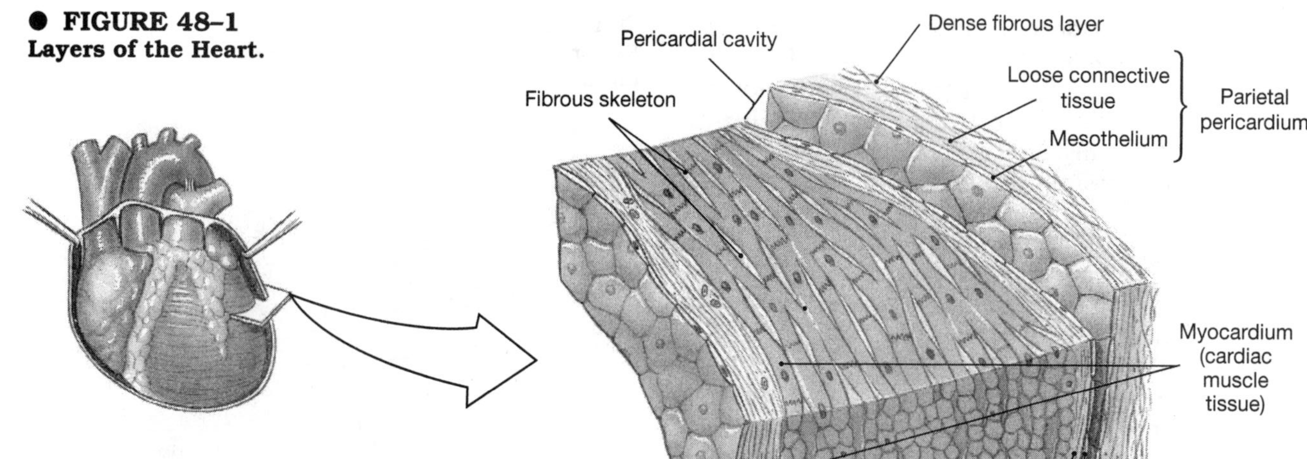

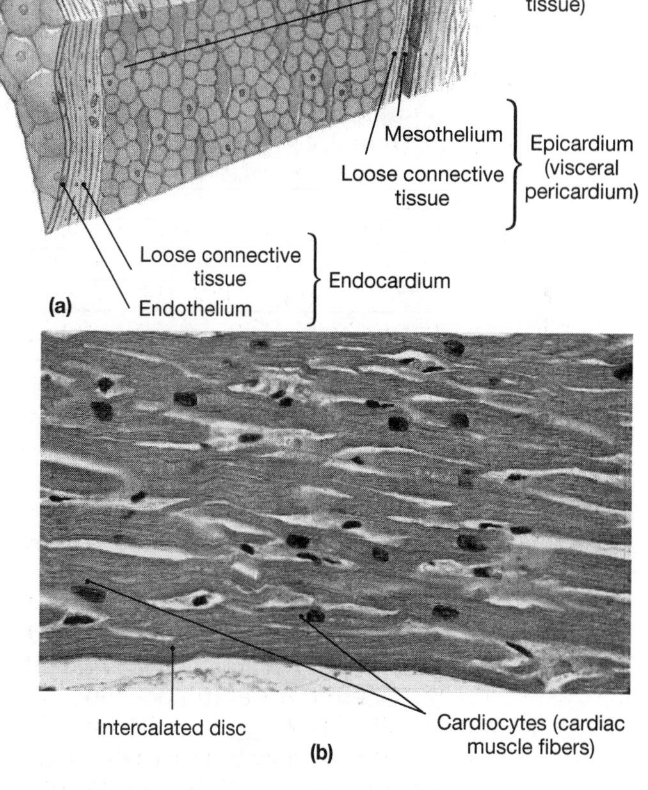

are separated by the anterior and posterior **interventricular sulci**. (A sulcus is a deep groove.) Find these sulci.

3➡ The right and left atria will be superior to the ventricles. With the posterior surface facing you, you should be able to locate a fat-filled horizontal line. Beneath this fatty tissue is the **atrioventricular sulcus**, which separates the atria from the ventricles. Find this sulcus. (On most heart models the atria are represented as being full and extended. In the "real" heart, the atria are often collapsed.)

4➡ Return to the anterior base of the heart. Note the large thick vessel arising from the center of the base (actually arising from the left ventricle) and hooking sharply toward the anatomical left side. The hooked area is the **aortic arch**, and the entire vessel is the **aorta**.

Three major blood vessels exit the aortic arch: the brachiocephalic, left common carotid, and left subclavian arteries. These arteries will be examined in Exercise 50.

5➡ Anterior to the aorta is the **pulmonary trunk**. The pulmonary trunk divides into the **right** and **left pulmonary arteries**. If the vessel on your heart model is of sufficient length, you can trace this division.

6➡ Locate the tough band of tissue connecting the aortic arch with the pulmonary trunk. This is the **ligamentum arteriosum**. In the fetus the ligamentum arteriosum is known as the **ductus arteriosus** and functions as a shunt between the pulmonary trunk and the aorta to bypass pulmonary circulation. This shunt closes shortly after birth and the vessel atrophies, forming the ligamentum arteriosum. This information can be corollated with Exercise 54.

C. Posterior Examination

1➡ Read through the following descriptions. Label Figure 48–2● with the boldface terms.

2➡ Now examine the posterior aspect of the heart. You should see a large thin-walled vessel that seems to be opening straight up from the right side of the base of the heart. This is the **superior vena cava**. Also on the right, just superior to the atrioventricular sulcus, is another large thin-walled vessel. This is the **inferior vena cava**.

3➡ If you were dissecting a fresh or preserved heart (as in Exercise 49), you could locate the inferior vena cava by placing your blunt nose probe into the superior vena cava and gently following the posterior wall of the vessel until it passed though the right atrium and exited through the inferior vena cava.

4➡ Just superior to the inferior vena cava you should see two thin-walled vessels. Directly across the heart (on the left) you should see two more thin-walled vessels. These four laterally directed vessels are the **pulmonary veins**, and they enter the left atrium from the lungs. In an animal specimen you could run your blunt nose probe through these veins also, although this might be more difficult because these veins are much smaller than the other vessels mentioned thus far.

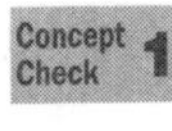

Concept Check 1 Compare the structures and thickness of the aorta and the vena cavae. Do you notice a difference? What is that difference? What do you suppose is the reason for this difference? Explain your answer.

____

____

____

____

____

____

III. Interior Examination

A. Orientation

1➡ Read these descriptions and label Figure 48–3● with the boldface terms.

2➡ Your model heart probably opens on a midcoronal plane, bisecting both atria and both ventricles. Does your heart model open on a midcoronal plane? __________ If not, use anatomical terminology to describe the way it does open.

B. Ventricular Structures [∞ Martini, p. 662]

1➡ Open the heart. You should immediately notice the **interventricular septum**, the wall between the two ventricles.

2➡ Look at Figure 48–4●. Identify the right and left ventricles. Keep these landmarks in mind as you continue with your examination of the heart.

3➡ In the healthy heart, the left and right ventricles are volumentrically the same. The left ventricle may appear larger than the right ventricle because the myocardium of the left ventricle is considerably thicker than the right ventricle. Also, the left ventricle is rounder than the right. See Figure 48–4●.

● **FIGURE 48–2**
Superficial Anatomy of the Heart.
(a) Anterior view. **(b)** Posterior view.

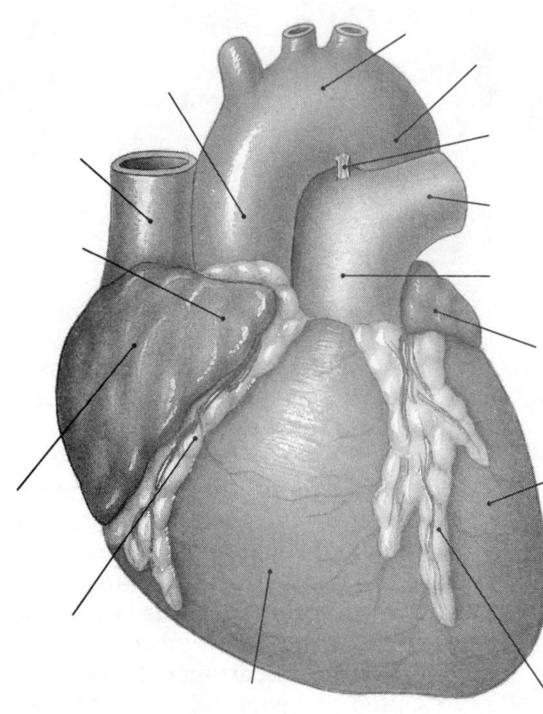

(a)

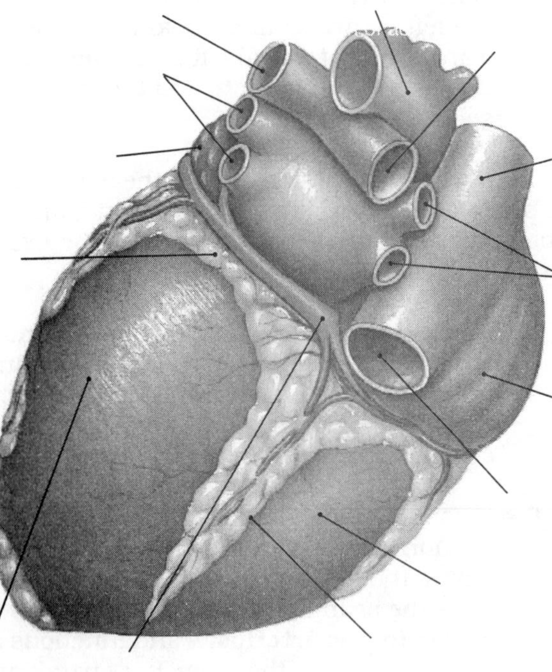

(b)

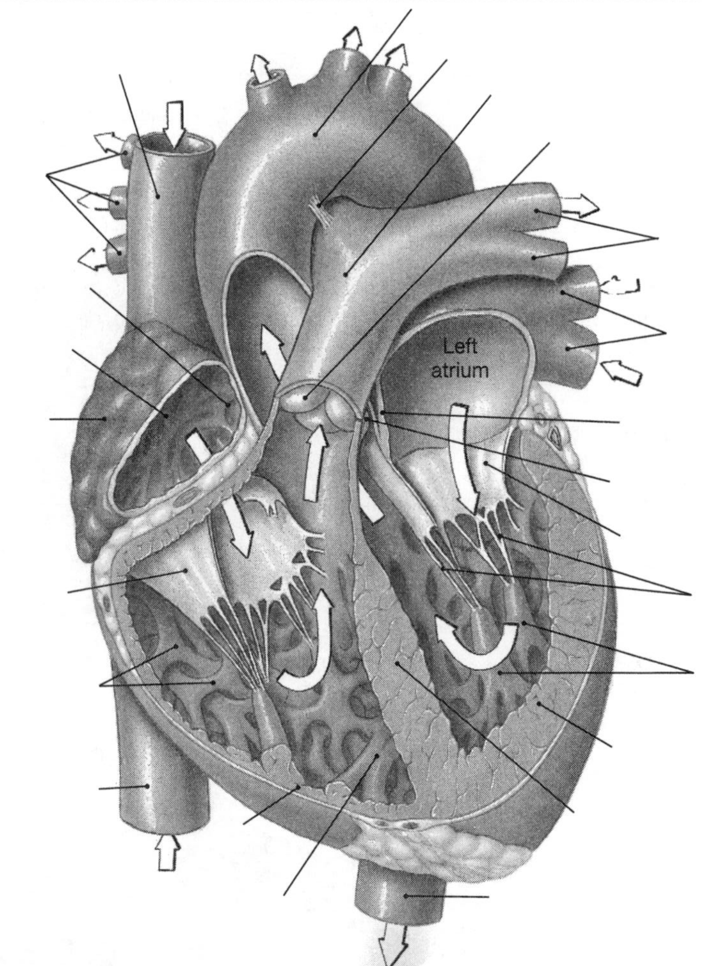

● **FIGURE 48–3**
Sectional Anatomy of the Heart. These structures are not shown on the diagram:
- Atrioventricular Septum
- Coronary Sinus

Concept Check 2 Based on your knowledge of blood flow through the heart, explain why the left ventricle is thicker and more muscular than the right ventricle.

____

____

____

____

4➡ Identify the muscles of the ventricular walls, the **papillary muscles**. The chordae tendinae (Section IV.A) attach to the heart wall via these papillary muscles. The papillary muscles in both ventricles can be subdivided into anterior and posterior papillary muscles. The folds and grooves on the internal surface of the ventricle are the **trabeculae carnae**.

5➡ At about the middle of the right ventricle, locate a tough diagonal cord running between the interventricular septum and the ventricular wall. This is the **moderator band**, which functions both as myocardial reinforcement and as a part of the cardiac conduction system.

C. Atrial Structures
[∞ Martini, pp. 660, 662]

1➡ Examine the right and left atria by first noting the **atrioventricular septa**, the walls between the atria and the ventricles These walls are primarily taken up with cusps. Observe the **interatrial septum**, the wall between the atria. Depending on your model, you may be able to reach up and bring your thumb and index finger together on either side of the interatrial septum. If you could feel this area, you would note an oval depression. This is the **fossa ovalis**. In the fetus this fossa ovalis was the **foramen ovale**, an opening between the right and left atria, a shunt to bypass pulmonary circulation. This foramen closes at birth because of a shift in blood pressure. See Exercise 54.

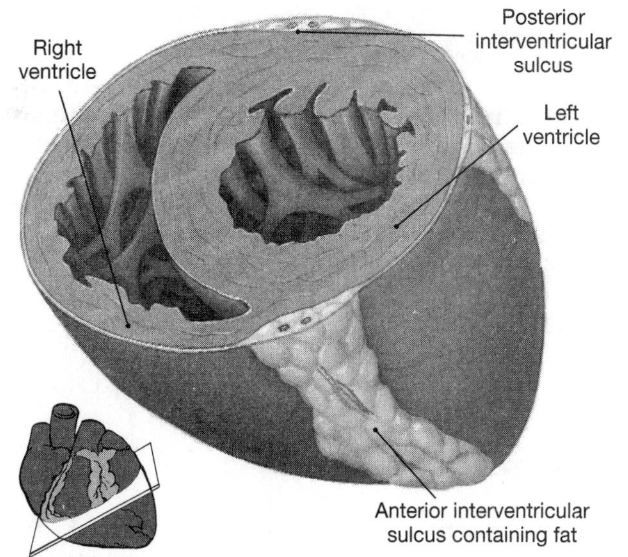

● **FIGURE 48–4**
Structural Differences between the Left and Right Ventricles.

2➡ Note the interior walls of the atrium. The ridge-like surface is called the **pectinate muscle** (from *pecten*, "comb"). Pectinate muscle tissue in the left atrium is generally smaller and less abundant than pectinate muscle tissue in the right atrium.

3➡ On the posterior wall of the right atrium you should be able to locate both the superior and inferior vena cavae. Slightly medial from the entrance of the inferior vena cava into the right atrium you should notice a "hole" on the posterior wall. The opening leads to the **coronary sinus**, the vein that drains the cardiac muscle. If you were examining an actual heart instead of a model, you could put your probe into the opening and feel the exterior surface of the heart because you would actually be following the posterior atrioventricular sulcus.

4➡ In the left atrium you should be able to locate the points where the pulmonary veins enter the heart. On some heart models these points will be indicated by small painted circles.

IV. Heart Valves

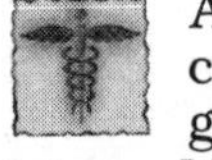

Any of the heart valves discussed below can develop defects. These defects may be genetic or may be the result of disease or injury. Minor defects in the length or shape of the cusps are often not serious. These defects can lead to the heart murmurs doctors often tell us not to worry about. Stenosis (scarring) of the valve, or any of the toxic or autoimmune valvular conditions, can be serious or even life-threatening.

A. Atrioventricular Valves

1➡ Three **cusps** (look like thin flaps) — the **anterior**, **medial**, and **posterior** — form the **tricuspid valve** between the right atrium and right ventricle, and two cusps — **anterior** and **posterior** — form the **bicuspid valve** between the left atrium and left ventricle.

2➡ Note the **chordae tendinae**, the string-like attachments running from the cusps of the valves to the muscles on the ventricle walls.

B. Semilunar Valves

1➡ Return to the left ventricle and locate the area between the bicuspid valve and the interventricular septum where the aorta leaves the heart. Identify the **aortic semilunar valve** at the entrance to this great vessel.

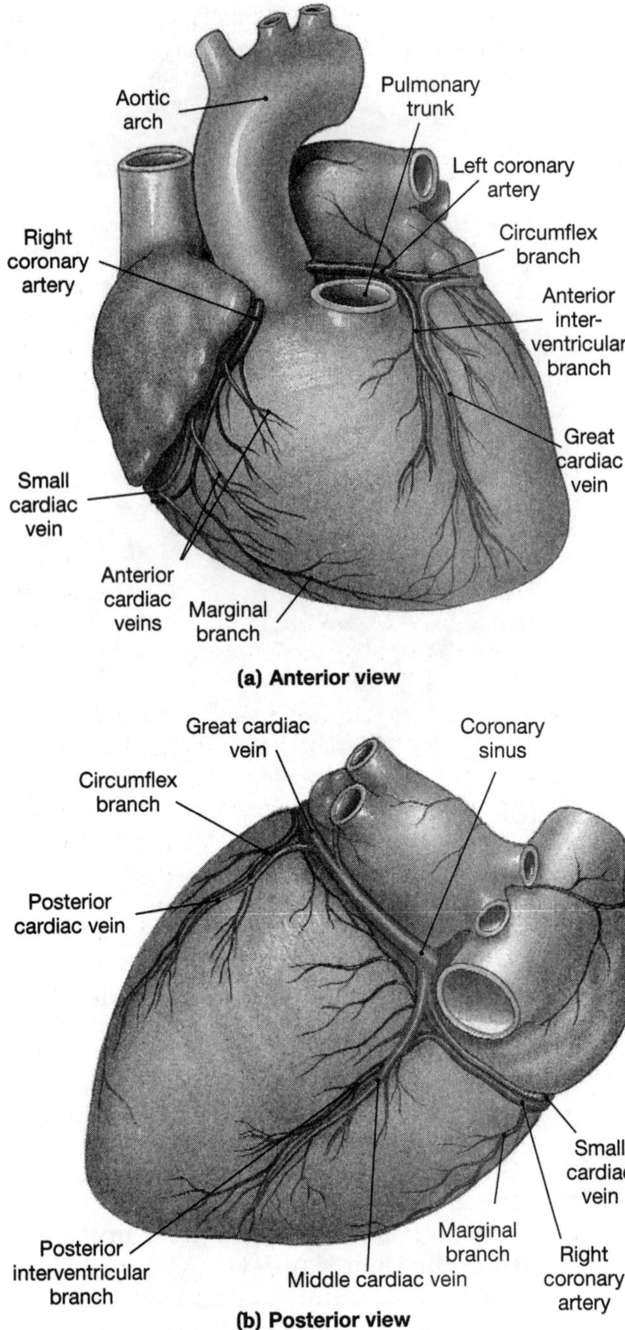

● **FIGURE 48–5**
Coronary Circulation. **(a)** Anterior view. **(b)** Posterior view.

2➡ In the walls of the aorta just superior to the aortic semilunar valve, find the two small openings to the **coronary arteries**. These arteries supply blood to the heart itself.

3➡ Return to the right ventricle and locate the area between the tricuspid valve and the interventricular septum where the pulmonary trunk leaves the heart. Identify the **pulmonary** (or **pulmonic**) **semilunar valve** near the base of the pulmonary trunk. Your model should delineate the three cusps of this valve.

V. Coronary Circulation [∞ *Martini, p. 664*]

A. Definition

The coronary circulatory system consists of those blood vessels servicing the heart itself. The cells of the heart are not nourished by the blood passing through it. Rather, the heart has its own series of arteries and veins.

B. Coronary Arteries

1➡ Return to the coronary arteries as they exit the aorta. If you are unable to follow the coronary circulatory system on your model, use Figure 48–5● to locate the arteries and veins listed in this section.

2➡ Find the **right coronary artery**, which gives rise to both the **marginal branch**, which extends along the right border, and the **posterior interventricular branch**, which runs toward the apex within the posterior interventricular sulcus.

3➡ Now locate the **left coronary artery**, which forms both the **circumflex branch**, which curves left around the left coronary sulcus, and the **anterior interventricular branch** (also known as the **left anterior descending artery**), which curves around the pulmonary trunk and follows the anterior surface within the anterior interventricular sulcus.

4➡ Examine the circumflex branch that fuses with branches of the right coronary artery, as do branches of the anterior interventricular branch. Interconnections between the arteries are called **anastomoses**. The function of these cardiac anastomoses is to maintain a constant blood supply at a constant pressure at the cellular level. These anastomoses become particularly important should a coronary vessel become blocked.

Deviations in this blood supply or blood pressure can lead to cell death. A nonfunctional area is an **infarct**. A myocardial infarction is also known as a heart attack. [∞ *Martini, p. 678*]

C. Coronary Veins

1➡ Locate the **great** and **middle cardiac veins**, which carry blood away from the coronary capillaries that drain into the coronary sinus. You located the coronary sinus when you examined the right atrium. As you examine Figure 48–5●, make note of the venous branches draining into the cardiac veins.

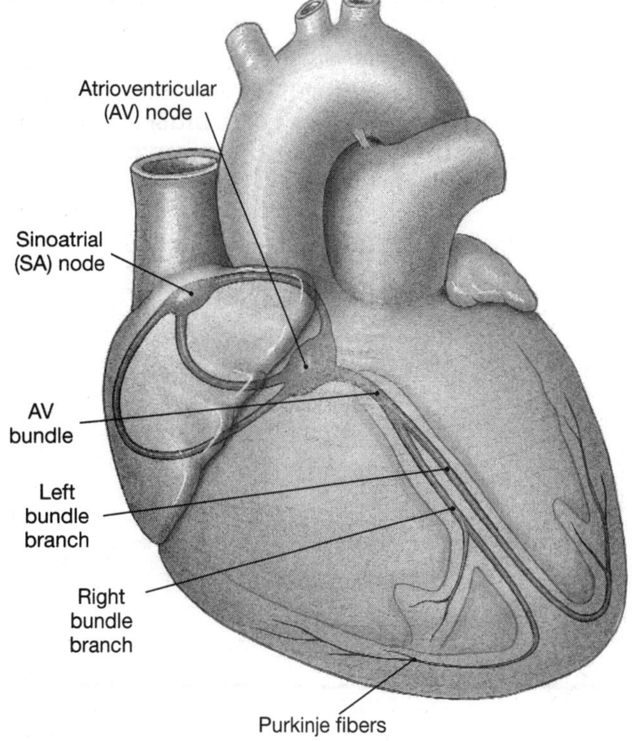

● **FIGURE 48–6**
Cardiac Nodes.

VI. Cardiac Nodes

A. SA Node and AV Node

1➡ Check your heart model for markings identifying the **sinoatrial (SA) node**, the pacemaker of the heart, which is located in the right atrium near the opening of the superior vena cava. The **atrioventricular (AV) node** is located on the inferior portion of the interatrial septum. If you are able to find either of these nodes, note them on Figure 48–6●.

2➡ Be aware that these nodes are part of the electrical conduction system of the heart. This system, which generates and circulates electrical impulses throughout the heart, will be examined in Exercise 51.

VII. Histology

A. Directions

1➡ As you work through this section, use Figure 48–1● as your guide.

B. MICROSCOPY USING THE DISSECTING MICROSCOPE

1➡ Use the dissecting microscope to view the cross section of the heart muscle. Be able to identify the epicardium, myocardium, and endocardium.

2➡ Sketch and label what you see.

Cross section of heart muscle

C. MICROSCOPY USING THE COMPOUND MICROSCOPE.

1➡ Obtain a compound microscope and a prepared slide of cardiac muscle. Note that the heart cells are uninucleate and striated. The cells are separated by the interdigitation of the cell membranes, known as **intercalated disks**. Identify the cardiocyte, nucleus, branching pattern, and intercalated disk. Refer back to Exercise 11 if necessary

2➡ Sketch and label what you see.

Cardiac muscle

❑ Additional Activities

1. If a preserved human heart is available, view that heart according to your instructor's directions.
2. The anatomy of the heart is relatively constant throughout the mammalian world. Nevertheless, certain subtle differences do exist. Research some of these differences and figure out what types of adaptational advantages these differences have allowed.
3. Regarding coronary bypass surgery, answer the following questions:
   What is coronary bypass surgery?
   When is coronary bypass surgery used?
   Which blood vessels are used in this surgery?
   What are some of the alternative procedures to coronary bypass surgery (are there alternatives)?
4. By constructing or sketching your own model of the heart, you can gain a much greater degree of understanding of the structures that make a heart. Using clay or a sketch pad, create or sketch a model of the interior heart. Identify the major structures.

NOTES

Answers to Selected Concept Check Questions

1. The aorta should be much thicker and much firmer than the vena cavae. This is directly related to the pressure on the vascular system. The thick walls sustain the higher pressure of the arterial system and are able to rebound to increase the flowing capacity. This will become clearer as you study cardiovascular physiology.
2. The left ventricle must be more forceful to accomplish systemic circulation. The right ventricle "only" has to push the blood to the lungs.

❑ Lab Report

1. Referring to the numbered inquiries at the beginning of this exercise, complete the following box summary:

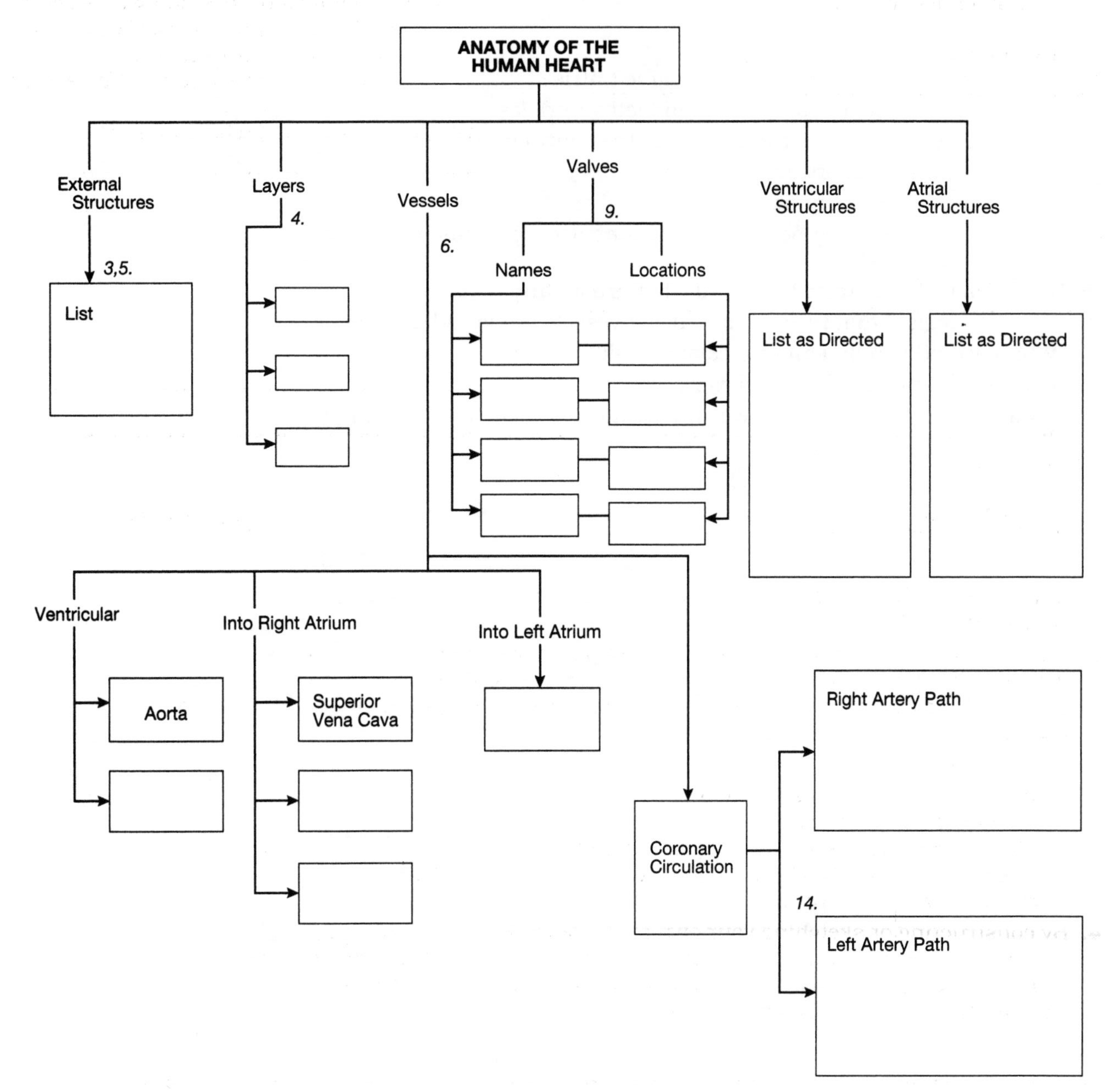

General Questions

NOTES

1. Give an overall description of the human heart.
2. Trace the path of blood through the heart.
3. What is the relationship between the different external sulci and the internal divisions of the heart?
4. Regarding the walls of the heart:
   a. Which wall separates the right and left atria?
   b. Which wall separates the right and left ventricles?
   c. Which wall separates the atria from the ventricles?
5. Where are the coronary arteries?
6. Where are the coronary nodes?
7. What is the difference between the ligamentum arteriosum and the ductus arteriosus?
8. Where is the moderator band and what are its functions?
9. Why is it logical that the coronary blood vessels are found in sulci?
10. Where would you find an intercalated disk?
11. Give two examples of anastomoses.
12. What is the difference between a pectinate muscle and a papillary muscle?

EXERCISE **49**

Dissection of the Sheep Heart

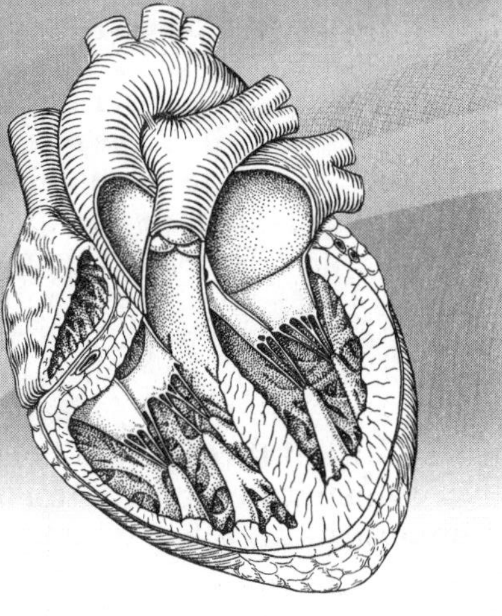

PROCEDURAL INQUIRIES

Preparation

1. How can you distinguish the anterior surface of the sheep heart from the posterior surface?
2. What anatomical landmarks will you find as you prepare for bisection?

Dissection

3. What are the layers of the heart?
4. Which blood vessels are associated with each chamber of the heart?
5. What is the ligamentum arteriosum?
6. What are the internal walls of the heart? Where is each located?
7. What are the valves of the heart? Where is each located?
8. What are the internal muscles and tendons of the heart?
9. What is the moderator band?
10. What is the coronary sinus?
11. Where would you find the coronary arteries?
12. What are the nodes of the heart?

Additional Inquiries

13. What is the path of blood through the heart?
14. What is the physical relationship of the atria to the ventricles?

Key Terms

Anterior Interventricular Artery
Anterior Interventricular Sulcus
Aorta
Aortic Arch
Apex
Atrioventricular Node (AV Node)
Atrioventricular Septum
Atrioventricular Sulcus
Base
Bicuspid Valve
Chordae Tendinae
Coronary Artery
Coronary Sinus
Ductus Arteriosus
Foramen Ovale
Fossa Ovalis
Great Cardiac Vein
Interatrial Septum
Interventricular Septum
Ligamentum Arteriosum
Moderator Band
Myocardium
Papillary Muscle
Parietal Pericardium
Pectinate Muscle
Posterior Interventricular Sulcus
Pulmonary Artery
Pulmonary Trunk
Pulmonary Vein
Semilunar Valve
Sinoatrial Node (SA Node)
Trabeculae Carnae
Tricuspid Valve
Vena Cava
Visceral Pericardium

Materials Needed

Sheep Heart (Fresh or Preserved)
(Other mammalian hearts may be substituted)
Dissecting Pan
Blunt Nose Probe
Scalpel
Gloves
Pluck, if available

This exercise is a corollary to Exercise 48. It is assumed that you have already worked through Exercise 48 or that you are working through Exercises 48 and 49 simultaneously. Necessarily, a great deal of overlap exists between Exercises 48 and 49. In dissecting the sheep heart, you are encouraged to refer back to the information on the human heart as often as necessary.

Because of its size, availability, and similarity to the human heart, the sheep heart is an ideal organ to dissect. The sheep heart, like the human heart, is a four-chambered double pump (Figure 49–1•) with the right and left **atria** (sing., **atrium**) superior to the right and left **ventricles**.

Before beginning your dissection, review the course of blood through the heart. Deoxygenated blood enters the right atrium of the heart from the superior and inferior vena cavae and from the coronary sinus. From the right atrium the blood travels through the tricuspid valve into the right ventricle. From the right ventricle the blood is pumped through the pulmonary semilunar valve into the pulmonary trunk, which subsequently branches into the right and left pulmonary arteries, which then carry the blood to the lungs.

Following gas exchange in the lungs, the oxygenated blood is returned to the heart via the four pulmonary veins, which empty into the left atrium. From the left atrium the blood passes through the bicuspid (mitral) valve into the left ventricle. From the left ventricle the blood passes through the aortic semilunar valve into the aorta, and from the aorta the oxygenated blood is carried throughout the body. In the body proper the metabolic exchange occurs: oxygen is exchanged for carbon dioxide. Deoxygenated blood is then returned to the heart, and the cycle begins again.

In this exercise we will dissect the sheep heart, although it should be noted that the same instructions would apply to the dissection of any mammalian heart.

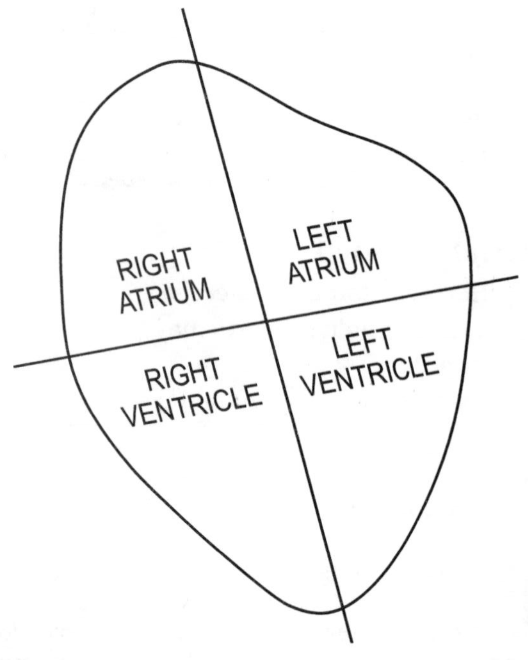

● **FIGURE 49–1**
Chambers of the Heart.

❑ Preparation

I. Preliminary Instructions

A. Identification

1➡ As you work through this exercise, identify the structures (in boldface) based on the following descriptions.

2➡ Label Figures 49–2● and 49–3● as you work through your dissection. Refer to Exercise 48 as necessary.

B. Preparation of the Heart

1➡ Obtain a sheep heart and rinse thoroughly under running water in order to remove as much of the preservative as possible. Run water through the blood vessels if possible to remove fluids that may remain in the cavities.

2➡ Place the heart in a dissecting pan, with the anterior (ventral) surface facing you. The anterior surface is readily discernible because of the large diagonal line of fat that seems to extend from your upper right to your lower left. Beneath the layer of fat is the **anterior interventricular sulcus**.

3➡ If you are able to trim away the fat with your blunt nose probe, you will expose the **anterior interventricular artery** and the **great cardiac vein**. A similar line on the posterior aspect of the heart, known as the **posterior interventricular sulcus**, seems to be vertical. Keep in mind that with the anterior surface facing you, the animal's left side is on your right side.

❑ Dissection

II. External Examination

A. Anterior

1➡ Place the heart so that the anterior position is facing you. The pointed area at the bottom of the heart is the **apex**, while the broad, flattened area at the top of the heart is the **base**.

2➡ Note that extending from the base of the heart are several vessels. These vessels may be surrounded by excessive amounts of loose, globular fat. Depending on how carefully the heart was removed from the animal, you may find superior to the vessels and fat a piece of cartilagenous material. This is a part of the trachea. (If a pluck is available for demonstration, observe the relation of the trachea to the heart.)

3➡ Around the origins of the large blood vessels, look for traces of the relatively thick fibrous membrane known as the **parietal pericardium.** Much of this membrane is probably missing.

4➡ Depending on the condition of your specimen, you may be able to locate and isolate the **visceral pericardium (epicardium)** from the outer surface of the heart. Beneath this single squamous cell layer is the **myocardium**, the hefty section we generally think of as heart muscle.

5➡ With the anterior surface facing you, gently squeeze various parts of the heart. The animal's left ventricular wall should feel firmer than the right ventricular wall. The thicker left ventricle is required to pump the blood throughout the body. Describe the differences between different parts of the heart.

____

____

The left and right ventricles will be separated by the interventricular sulci, mentioned in Section I.B.

6➡ With the posterior (dorsal) surface facing you, locate a fat-filled horizontal line. Beneath this fatty tissue is the **atrioventricular sulcus**, which separates the atria from the ventricles.

7➡ Return to the anterior base of the heart and, with your blunt nose probe, remove as much of the fatty tissue as possible. You should now be able to distinguish the various blood vessels.

8➡ Examining the heart from the anterior, note the large thick vessel arising from the center of the base (actually arising from the left ventricle) and hooking sharply to-

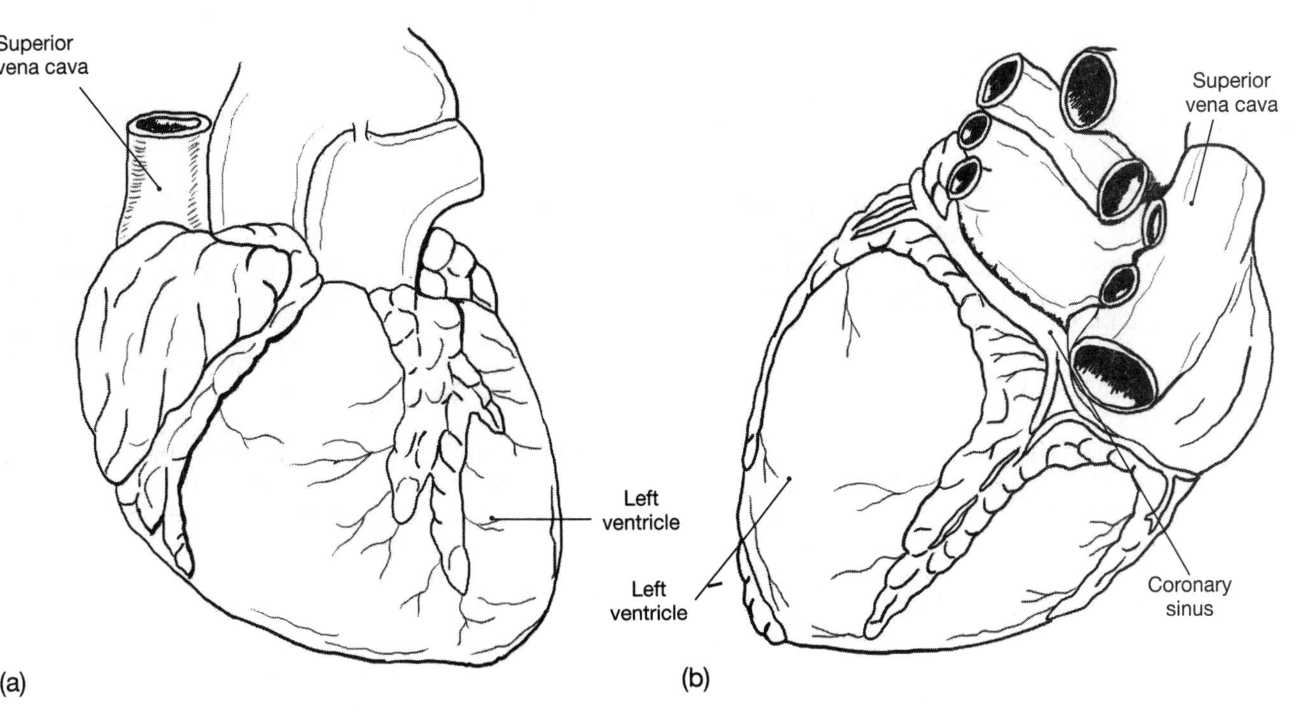

● **FIGURE 49–2**
Sheep Heart —
Exterior View.
(a) Anterior **(b)** Posterior.

● **FIGURE 49–3**
Sheep Heart — Anterior View.

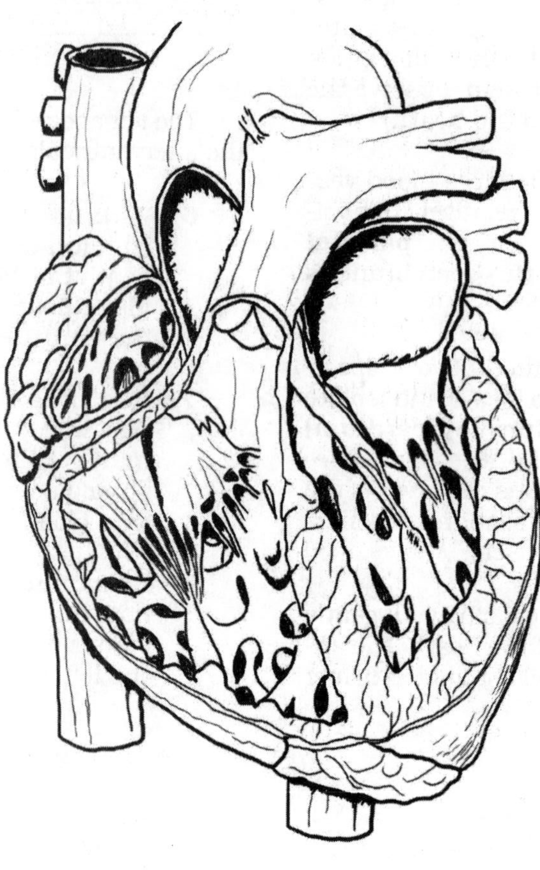

ward the animal's left side. The hooked area is the **aortic arch**, and the entire vessel is the **aorta**. Put your blunt nose probe into the aorta and gently guide it into the ventricle. You may wish to repeat this action later when the heart is open to help you identify various internal structures.

9➡ Refer back to Figures 48–2● and 48-3● and compare with your sheep heart. One vascular difference you should notice is that the human heart has three vessels leaving the aortic arch (the brachiocephallic, left common carotid, and left subclavian arteries) while most other mammalian hearts have only two exiting arteries, the brachiocephallic and the left subclavian arteries. Identify these vessels on your sheep heart.

10➡ Anterior to the aorta is the **pulmonary trunk**. The pulmonary trunk divides into the **right and left pulmonary arteries**. If the vessel on your heart is of sufficient length, you can trace this division. You should be able to put your blunt nose probe into the left pulmonary artery and push it across the pulmonary trunk and out the right pulmonary artery. You should also be able to put your blunt nose probe into either of the pulmonary arteries and angle it gently down the pulmonary trunk into the right ventricle.

11➡ Note the tough band of tissue connecting the aortic arch with the pulmonary trunk. This is the **ligamentum arteriosum**. In the fetus the ligamentum arteriosum is known as the **ductus arteriosus** and functions as a shunt between the pulmonary trunk and the aorta to bypass pulmonary circulation. This shunt closes shortly after birth and the vessel atrophies, forming the ligamentum arteriosum.

B. Posterior

1➡ Now examine the posterior aspect of the heart. On the animal's right you should see a large thin-walled vessel that seems to be opening straight up from the base of the heart. This is the **superior vena cava.** Also on the right, just superior to the atrioventricular sulcus, is another large thin-walled vessel. This is the **inferior vena cava.**

2➡ If you are having difficulty locating the inferior vena cava, place your blunt nose probe into the superior vena cava and gently follow the posterior wall of the vessel. Your probe will pass though the right atrium and exit via the inferior vena cava.

3➡ Just superior to the inferior vena cava you should see two thin-walled vessels. Directly across the heart (on the animal's left) you should see two more thin-walled vessels. These four vessels are the **pulmonary veins**, and they enter the left atrium. You can run your blunt nose probe through these veins also, although this is sometimes more difficult because these veins are much smaller than the other vessels mentioned thus far. If the cut of the heart makes finding these vessels difficult, come back to them when you examine the interior aspects of the left atrium. You may then be able to find these vessels by running your probe around the left atrial wall.

III. Internal Examination

Refer to Exercise 48 as often as necessary.

A. Directions for Dissection

1➡ Although several methods of dissecting the heart exist, we will do an anterior/posterior bisection, beginning at the apex.

2➡ Before beginning your bisection, carefully plan the location of your incision. On each side of the heart project an imaginary lateral line from the apex to the approximate area of the atrioventricular sulcus. With your scalpel begin at the apex and carefully cut the myocardium bisecting each ventricle. (If you use scissors instead of the scalpel, keep the blunt end of the scissors on the inside of the heart.) Now cut the **interventricular septum**, the wall between the two ventricles.

B. Ventricular Structures

1➡ Lift the anterior section of the heart and observe the interior of the ventricles. The myocardium of the left ventricle is thicker than the myocardium of the right ventricle. Refer to Figure 48–4●. The left ventricle will appear to be considerably larger than the right ventricle, though the lumina of the ventricles are actually the same size.

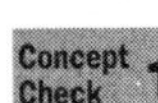

Why would you expect these differences in the myocardial walls? Why would you expect the actual lumen size to be the same?

____

____

2➡ Note the **chordae tendinae**, the string-like attachments running from the **cusps** of the valves to the muscles on the ventricle walls. Three cusps — the **anterior, medial, and posterior** — form the **tricuspid valve** between the right atrium and right ventricle, and two cusps — **anterior and posterior** — form the **bicuspid valve** between the left atrium and left ventricle. (The bicuspid valve was formerly known as the mitral valve.) The muscles of the ventricular walls are the **papillary muscles**. The chordae tendinae anchor into these papillary muscles. The papillary muscles in both ventricles can be subdivided into anterior and posterior papillary muscles. The folds and grooves on the internal surface of the ventricle are the **trabeculae carnae.**

3➡ At about the middle of the right ventricle, locate a tough diagonal cord running between the interventricular septum and the ventricular wall. This **moderator band** functions both as myocardial reinforcement and as a part of the cardiac conduction system.

C. Atrial Structures

1➡ Continue your lateral incisions up into the atria. Carefully cut the **atrioventricular septa,** the walls between the atria and the ventricles, but leave the **interatrial septum** intact.

2➡ Using your thumb and index finger reach up and bring your fingers together on either side of the interatrial septum. As you feel this area, you will note an oval depression. This is the **fossa ovalis**. In the fetus this fossa ovalis was the **foramen ovale,** an opening between the right and left atria, a shunt to bypass pulmonary circulation. This foramen closes at birth. See Exercise 54 for a discussion on the analogous human structures.

3➡ Note the interior walls of the atrium. The ridge-like surface is called the **pectinate muscle** (from *pecten*, "comb"). On the posterior wall of the right atrium you should be able to locate both the superior and inferior vena cavae. You may have to run your blunt nose probe through the vena cavae from the outside in order to be certain.

4➡ Slightly medial from the entrance of the inferior vena cava into the right atrium you should notice a "hole" on the posterior wall. Put your blunt nose probe into this opening. Your probe, which you should be able to feel on the exterior surface of the heart, will follow the posterior atrioventricular sulcus. The opening is the opening to the **coronary sinus,** the vein that drains the cardiac muscle.

5➡ In the left atrium you should be able to locate the points where the pulmonary veins enter the heart. You may need your probe to verify that you have found these small openings.

D. Structures of the Aorta

1➡ Locate the **aortic semilunar valve** by tracing the aorta back into the left ventricle.

2➡ In the walls of the aorta just superior to the aortic semilunar valve, you should be able to find two openings to the **coronary arteries.** Examine these openings with your blunt nose probe. Were you able to find the coronary arteries? _______ What is the physical relationship between the semilunar valve and the coronary arteries?

____

____

What is the functional relationship between the semilunar valve and the coronary arteries?

____

____

E. Pulmonary Semilunar Valves

1➡ Locate the **pulmonary** (or pulmonic) **semilunar valve** by tracing the pulmonary trunk back into the right ventricle. You should be able to see the three cusps of this valve.

F. Nodal Tissue

1➡ Depending on the condition of your heart, you may possibly be able to locate the **sinoatrial (SA) node**, the pacemaker of the heart, which is located in the right atrium near the opening of the superior vena cava. The **atrioventricular (AV) node** is located on the inferior portion of the interatrial septum. If you are able to find either of these nodes, the tissue will feel like a hard nodule.

❑ Additional Activities

NOTES

1. Compare the sheep heart with other mammalian hearts. What similarities and differences do you see?
2. Use the dissecting microscope to examine parts and sections of the sheep heart. Try to get a close-up of the valves and vessels. What similarities and differences do you notice between the atrioventricular valves and the semilunar valves?

Answers to Selected Concept Check Questions

1. The ventricles pump the same amount of blood so the lumen volumes should be the same. The walls of the left ventricle are thicker because more contractile force is necessary to pump the blood throughout the body.
2. This relationship prevents backflow and excessive pressure in the coronary arteries while assuring continuous blood flow to the heart muscle.

❑ Lab Report

1. Referring to the numbered inquiries at the beginning of this exercise, complete the following box summary:

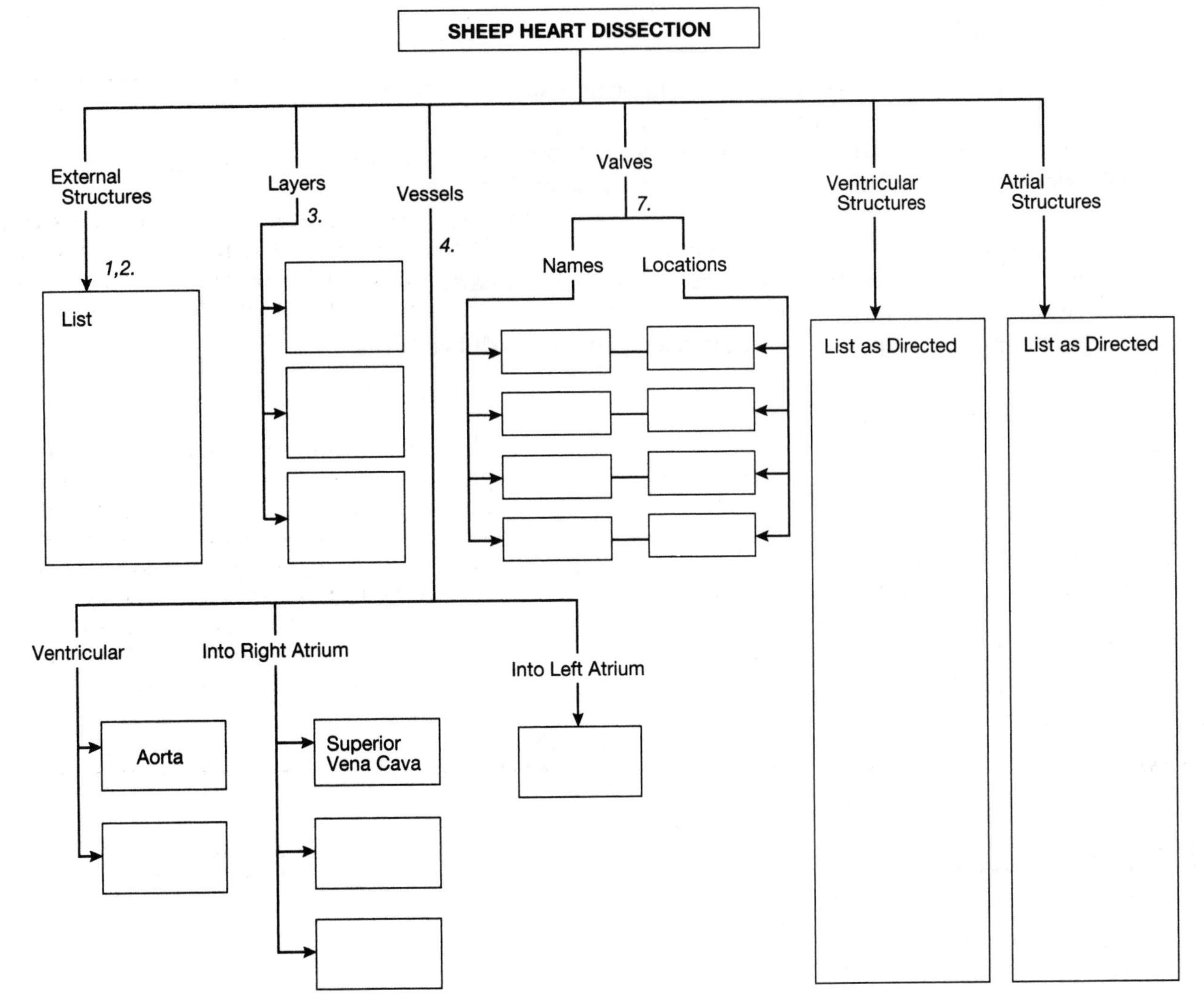

General Questions

NOTES

1. Regarding the walls of the heart:
   a. Which wall separates the right and left atria?
   b. Which wall separates the right and left ventricles?
   c. Which wall separates the atria from the ventricles?
2. Where are the coronary arteries?
3. Where are the coronary nodes?
4. What is the difference between the ligamentum arteriosum and the ductus arteriosus?
5. Where is the moderator band and what are its functions?
6. What is the difference between the papillary muscle and the pectinate muscle?
7. What is the path of blood through the heart?
8. Make a chart comparing the human heart and the sheep heart. Use descriptive terms to describe similarities and differences.
9. Make a list of anatomical points that surprised you as you did your dissection.

EXERCISE 50

Human Vascular System

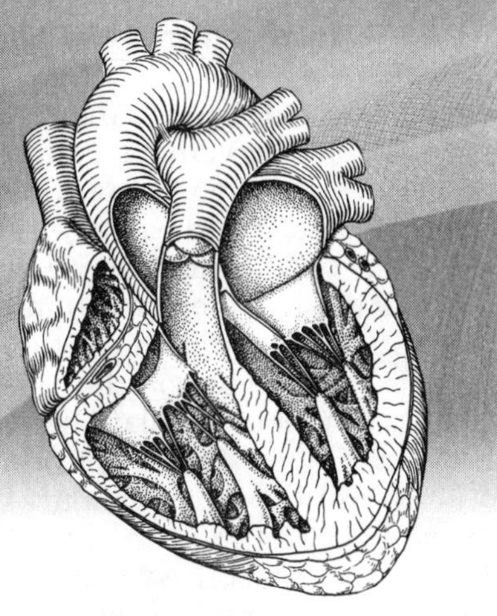

PROCEDURAL INQUIRIES

Preparation

1. What strategies should you follow for learning the names of the major blood vessels?
2. What path does cardiovascular dissection normally follow?

Examination

3. What are the layers of the vascular wall?
4. How can you distinguish arteries from veins from capillaries?
5. What are the arteries of the heart?
6. What are the arteries of the head and neck?
7. What are the arteries of the upper extremity?
8. What are the arteries of the thorax?
9. What are the arteries of the abdomen?
10. What are the arteries of the lower extremity?
11. What are the veins of the heart?
12. What are the veins of the head and neck?
13. What are the veins of the upper extremity?
14. What are the veins of the thorax?
15. What are the veins of the abdomen?
16. What are the veins of the lower extremity?

Additional Inquiries

17. How are most blood vessels named?
18. What is the generalized vascular route blood follows from the heart back to the heart?

Key Terms

Specific blood vessels are not listed here. These can be found in boldface within the body of the exercise.

Arteries and Veins of the Abdomen
Arteries and Veins of the Head and Neck
Arteries and Veins of the Heart
Arteries and Veins of the Lower Extremity
Arteries and Veins of the Thorax
Arteries and Veins of the Upper Extremity

Materials Needed

Compound Microscope

Prepared Slides

Cross Sections of Different Blood Vessels

Longitudinal Sections of Different Blood Vessels

Torso

Heart Models

Other Models (as available)

Charts and Diagrams

The cardiovascular system includes the pumps and vessels directly responsible for the organized movement of fluid throughout the body. The principle transport components include the heart and blood vessels, the arteries, arterioles, capillaries, venules, and veins.

Oxygenated blood leaves the left ventricle of the heart via the aorta and travels through the arterial system to each of the organs and tissues of the body. This is systemic circulation. The arteries are well-defined structures carrying blood away from the heart. Arteries are generally named. At the tissue level, the arteries undergo extensive subdivision to form the arterioles (which are less well defined and are usually not named). The arterioles in turn branch to form the capillary beds. Gas and nutrient exchange takes place at the capillary level. The capillaries converge to form the venules (analogs of the arterioles). The venules converge to form the veins. The entire venous system carries the deoxygenated blood back to the right atrium of the heart.

From the right atrium the deoxygenated blood passes into the right ventricle and from there travels via the pulmonary circuit to the lungs. The pulmonary arteries undergo extensive subdivision to form the pulmonary capillary beds, where gas exchange takes place. The capillaries then converge, eventually forming the pulmonary veins returning the blood to the left atrium of the heart.

Notice that in both systemic and pulmonary circulation the arterial system subdivides to form a capillary network that converges to form a venous system.

In this laboratory exercise we will examine the microanatomy of the arteries, veins, and capillaries and then identify the major arteries and veins of the human cardiovascular system.

❑ Examination — Microscopic

I. Microscopy [∞ *Martini, p. 692*]

Before beginning the identification of the specific blood vessels of the human cardiovascular system, let us first examine the microanatomy of the arteries, veins, and capillaries.

A. Comparison of Arteries, Veins, and Capillaries in Cross Section

1➡ Obtain a compound microscope and slides of cross sections of different arteries, veins, and capillaries. Study Figure 50–1●. Notice that the arteries and veins have the same layers but that the thickness of these layers is different. If your slides are labeled so you know which particular arteries and veins you are examining, corollate this information with the information in your Martini text, pages 692–697.

2➡ Identify on your slides the blood vessel layers shown in Figure 50–1●. Sketch and label what you see.

Artery Vein Capillary

In your own words describe the structures of each of the major blood vessel types:

Artery ______________________

Vein ______________________

Capillary ______________________

What is the functional advantage of the makeup of each layer in these vessels? Refer to your lecture text if necessary.

Artery ______________________

Vein ______________________

Capillary ______________________

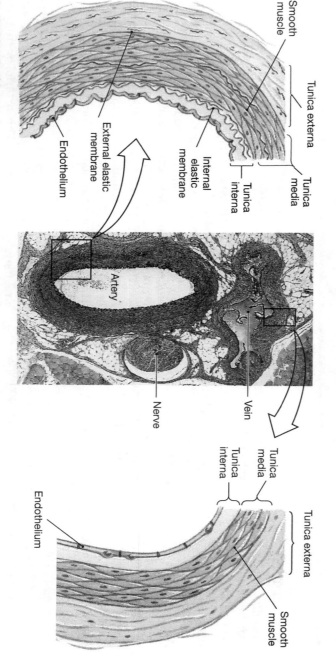

● **FIGURE 50–1a**
A Comparison of a Typical Artery and a Typical Vein. (LM × 74)

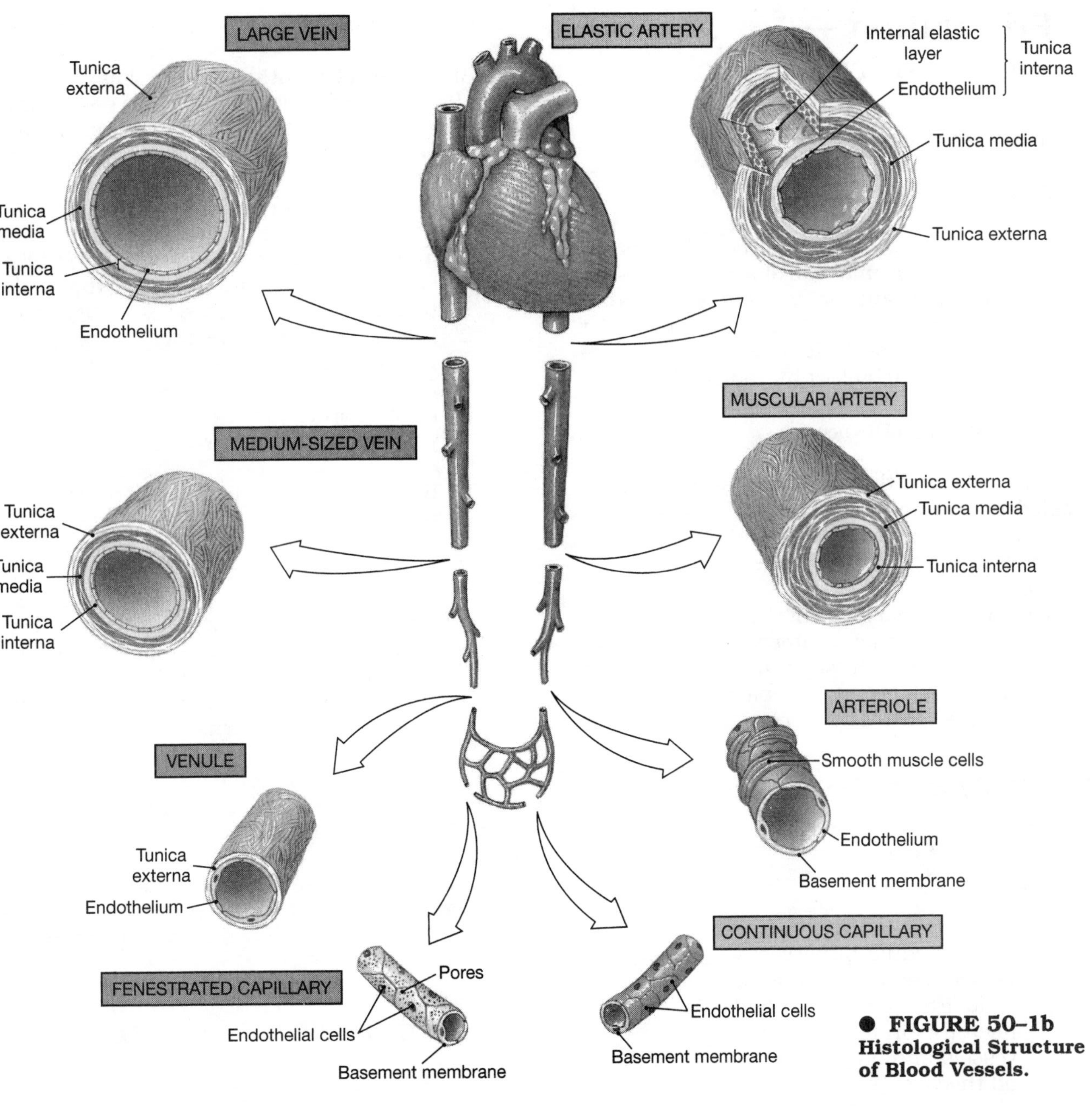

● **FIGURE 50–1b Histological Structure of Blood Vessels.**

Concept Check 1 What is the distinguishing characteristic of the capillary layer?

____

____

B. Comparison of Arteries and Veins in Longitudinal Section

1 ➡ Obtain a slide of a longitudinal section of various arteries and veins. Compare the layers in these vessels. Notice the valves in the larger veins.

Concept Check 2 What is the function of these valves?

____

____

C. Blood Supply for the Great Blood Vessels

Many of the great blood vessels have their own blood supply — small blood vessels actually penetrating the vessel walls. These vessels are called the **vasa vasorum**.

Concept Check 3 Based on what you have just learned about the layers of tissue found in the different vessels, why do you think these vessels have their own vascular system?

____

____

____

❑ Preparation — for Identification

II. Background

A. Identification Procedures

1➡ Work with a partner (if possible) as you study the human cardiovascular system. It is important that you and your partner share your ideas as you identify the different blood vessels. It is also important that you quiz each other as you go along.

B. Charts and Diagrams:

It is assumed that you have access to well-labeled charts and diagrams to help you work through this exercise. Concentrate on gaining a total perception of the cardiovascular system.

1➡ Use the torso to orient yourself to the body. Locate the major internal organs: lungs, heart, stomach, small intestine, large intestine, kidney, spleen, pancreas, liver, gall bladder, testes or uterus.

III. Study Hints for Naming the Blood Vessels

A. Location

1➡ Pay particular attention to the location of the individual blood vesssels. Most blood vessels are named either for their location or for the part of the body they service. Keep this in mind as you trace the major vessels and as you identify the branches off the vessels.

2➡ Please note*: A number of the smaller blood vessels that you may not be able to find on your charts are included in this write-up. Each of these vessels, **marked with an asterisk**, is here for your information and not necessarily for your positive identification. Your instructor will advise as to which, if any, of these vessels you will be required to know.

Also, this exercise does not purport to include all vessels you might find on a given chart or diagram. (Additional note of caution: certain vessels do have more than one common name. If you encounter this, examine where the vessel originates and make note of its apparent function. The multiple names should then make sense.)

B. Orientation

Blood vessel identification normally begins with the heart and follows the major paths of the great vessels to and from the head, arms, and lower body. We will follow that pattern here.

1➡ Become acquainted with the diagrams you will be studying. Follow these diagrams and illustrations in identifying the blood vessels. Do not limit yourself to the pictures, however. Read the descriptions carefully and coordinate these descriptions with the illustrations. You should also feel free to use any additional sources you may have in order to verify or expand on what you are presently learning.

2➡ Be aware of anomalies. Even the classroom models may not be in total agreement on the exact location of certain blood vessels. In real life the blood vessels, particularly those of the venous system, may exhibit some interesting branching patterns. If you have any doubt about this, compare the back of your hand with the back of your lab partner's hand. Notice the differences in the venous patterns.

If your class will be doing one of the cardiovascular dissection exercises, someone will most certainly find some interesting blood vessel anomalies in the laboratory animals.

You should be aware that variations do exist. Your goal, however, should be to understand the pathways of the major blood vessels, regardless of the anomalies.

❑ Examination — Macroscopic

IV. Cardiopulmonary System

A. Heart

1➡ Note the heart. The principles of the cardiopulmonary system were covered in detail in Exercises 48 and 49. If necessary, go back and review the major structures.

2➡ Be able to identify the major heart structures and blood vessels on the models provided.

B. Coronary Arteries and Veins

1➡ Check the red vessels (indicating oxygenated blood) on the surface of the heart. These are the **coronary arteries**, which supply blood to the heart muscle itself. These arteries arise from the base of the aorta.

2➡ Find the blue surface vessels (indicating deoxygenated blood). These are the **coronary veins**, which drain the heart muscle. The coronary veins culminate in the **coronary sinus**, which empties directly into the right atrium on the dorsal aspect of the heart.

V. Arterial System [∞ *Martini, p. 724*]

A. Nomenclature

Since arteries carry blood away from the heart and toward an organ or body region, we often use terms such as "supplies" or "services" or "goes to" when discussing the arterial system.

1➡ Use Figures 50–2● to 50–6● as your guide for arterial examination. Review these figures briefly before beginning your investigation.

B. Aortic Arch

1➡ Begin your examination of the arterial system by returning to the heart and locating the **aorta**. The portion of the aorta leaving the heart is the **ascending aorta**. The hooked portion is the **aortic arch**, and the portion continuing caudally along the dorsal wall of the thorax is the **descending** or **thoracic aorta**. When the aorta passes into the abdominal cavity, it becomes known as the **abdominal aorta**.

2➡ Observe the three vessels exiting the aortic arch. The right vessel (the model's right) is the **brachiocephalic** or **innominate artery**, which gives rise to the **right common carotid** and **right subclavian arteries**. The second branch off the aortic arch is the **left common carotid**. The third (left) branch exiting the aortic arch is the **left subclavian artery**. Note that as each subclavian artery

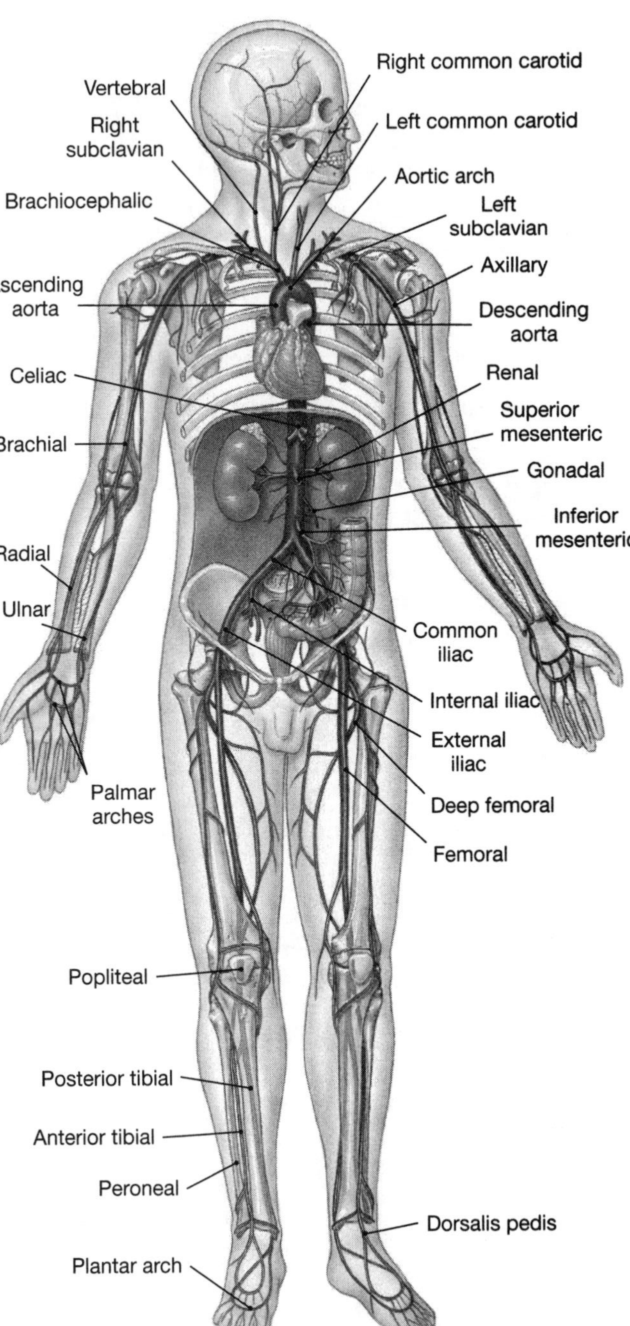

● **FIGURE 50–2**
The Arterial System.

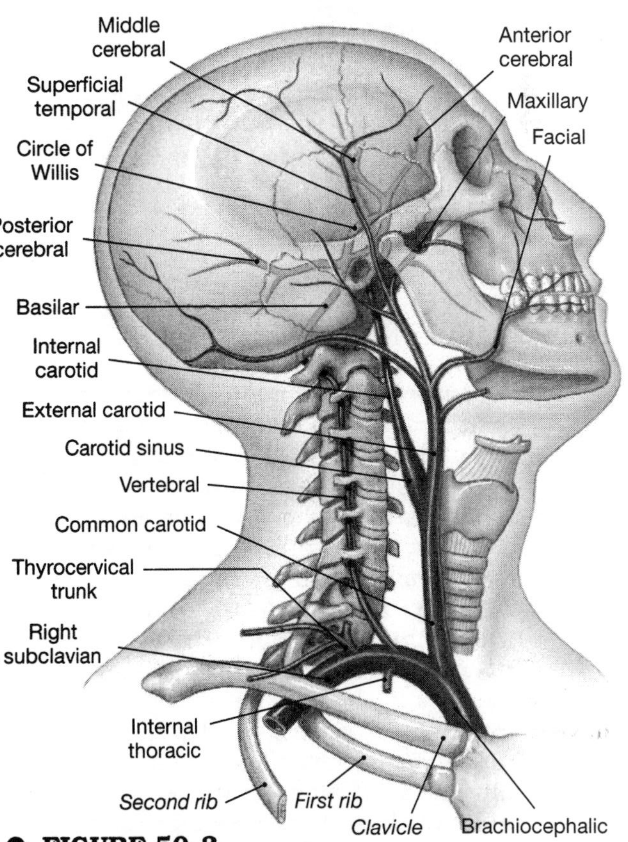

● **FIGURE 50–3**
Arteries of the Head and Neck.

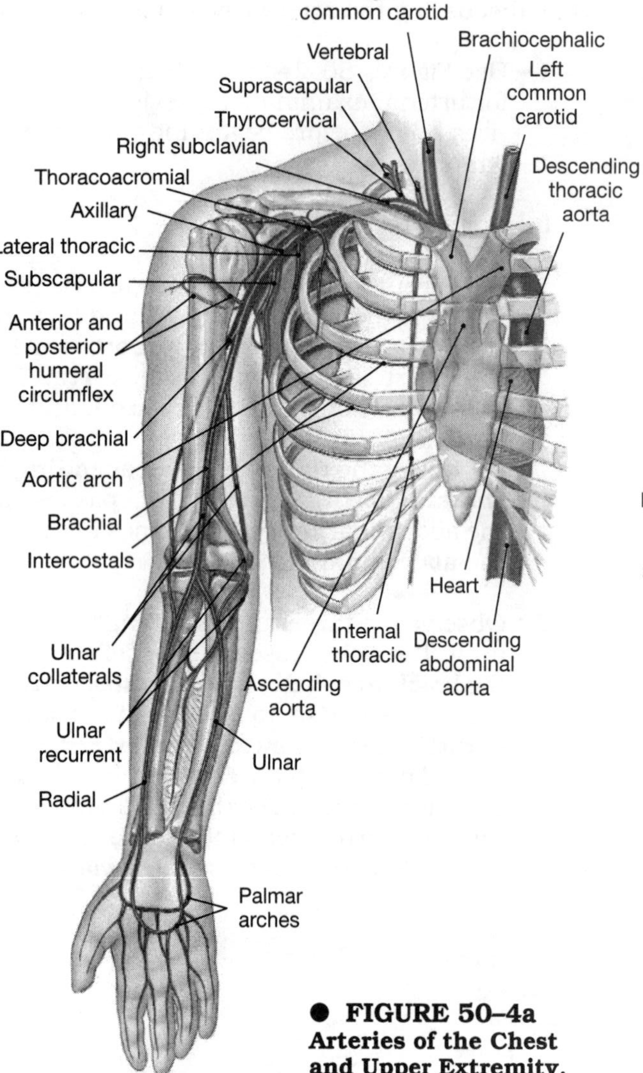

● **FIGURE 50–4a Arteries of the Chest and Upper Extremity.**

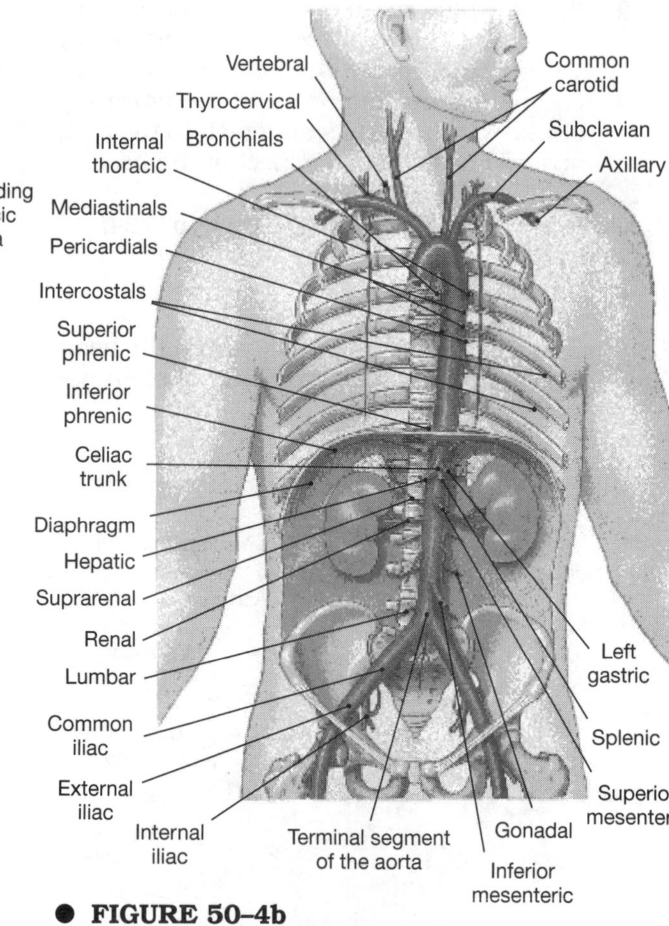

● **FIGURE 50–4b Major Arteries of the Trunk.**

turns laterally toward the arm, the branching patterns become nearly identical.

C. Head and Neck

1➡ Return to the common carotid artery. Unlike the common carotid arteries of many other mammals, the human common carotid usually gives off no branches prior to bifurcation (although occasionally a thyroid, laryngeal, pharyngeal, or vertebral branch may be found). At the upper border of the thyroid cartilage, the common carotid artery bifurcates to form the **external** and **internal carotid arteries**.

2➡ Locate the external carotid artery on Figure 50–2●. The external carotid artery generally gives off eight branches. Locate as many of these branches as possible: **superior thyroid**,* **lingual**,* **facial**,* **occipital**,* **posterior auricular**,* **ascending pharyngeal**,* **superficial temporal**,* **internal maxillary**.* (It is beyond the scope of this exercise to explore the more intricate branchings of these arteries.)

The internal carotid artery in the adult is about the same size as the external carotid. In the child, however, the internal carotid is the larger of the two vessels. The internal carotid supplies the eye, forehead, nose, and brain via the cerebral arterial circle, or the circle of Willis (Figure 50–5●). The circle of Willis is an interesting anastamosis encircling the infundibulum and pituitary gland and enabling the neural tissue to receive blood from either the carotid arteries or the vertebral arteries. This is a backup system — a survival mechanism — whereby the brain is guaranteed an alternate route for receiving blood should a major vessel become blocked. [∞ *Martini, p. 729*]

3➡ From the figure, identify the vessels making up the Circle of Willis.

D. Upper Extremity

1➡ Return to the subclavian artery and locate the first branch off the subclavian artery.

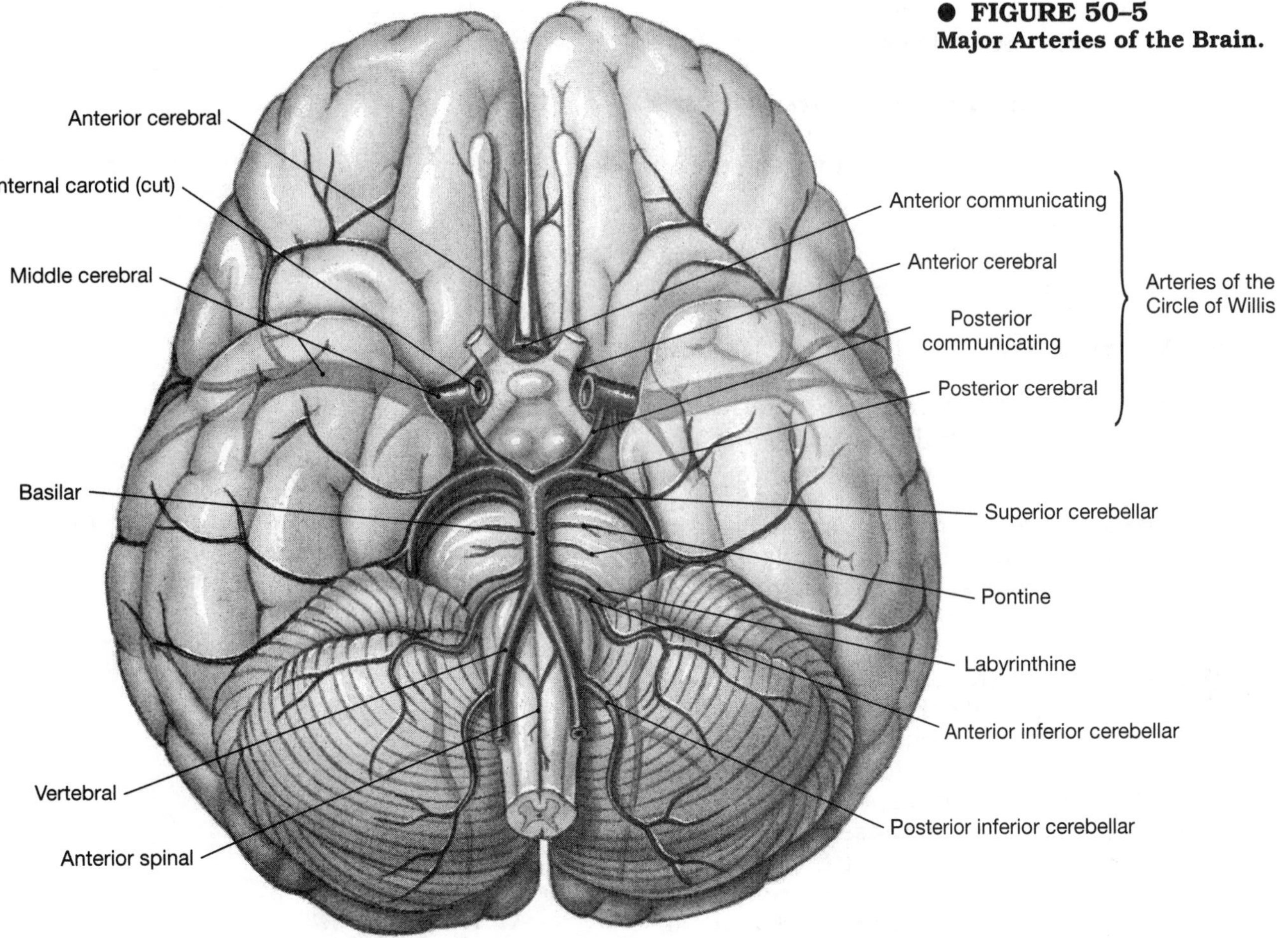

● **FIGURE 50–5**
Major Arteries of the Brain.

This is the **vertebral artery**, which passes through the transverse foramen of the cervical vertebrae and services the brain.

The **thyrocervical artery*** supplies the deep muscles of the neck, back, and intercostal areas. The **internal mammary artery*** supplies the ventral body wall and the pericardium.

The **costocervical trunk*** arises from the cranial surface of the subclavian and supplies blood to the neck and shoulders. At the scapula the thyrocervical artery becomes known as the **transverse scapular artery**.*

At the level of the first rib, the subclavian artery becomes the **axillary artery**.

The first branch off the axillary artery is the **lateral thoracic artery**,* which supplies the pectoral muscles. The largest branch off the axillary artery is the **subscapular artery**, which gives rise to the **thoracodorsal artery*** and the **posterior humeral circumflex artery**.*

The **subscapular artery** supplies the scapula, shoulder, and pectoral regions, as does the **thoracoacromial artery**.* The **thoracodorsal artery*** services the latissimus dorsi. Distal to the subscapular artery, at the level of the teres major, the axillary artery becomes the **brachial artery**. It is the brachial artery you work with when taking a blood pressure.

The first branch of the brachial artery is the **anterior humeral circumflex artery**,* which you probably will not be able to locate. The **deep brachial artery** passes under the belly of the biceps muscle. Distal to the elbow the brachial artery divides into the **radial artery** and the **ulnar artery**. The radial and ulnar arteries continue down into the palm of the hand, where they form the **deep palmar arch** and the **superficial palmar arch**. Various **digital arteries** branch off these arches to supply the thumb and fingers.

E. Thoracic Aorta

1 ➡ Return to the aorta and follow the aorta caudally through the thorax. Locate as many of the following arteries as possible.

The **posterior intercostal arteries** supply the intercostal muscles as well as the spinal cord, back muscles, body wall, and skin. (These arteries anastomose with the anterior intercostal arteries — coming from the mammary artery. This is an example of collateral circulation.)

The **bronchial arteries*** supply the bronchi as they enter the lung. Several **esophageal arteries*** of varying origins supply the esophagus. Numerous **mediastinal*** and **pericardial arteries*** supply the mediastinum and pericardium, respectively, while the **superior phrenic arteries** service the diaphragm.

At approximately the second lumbar vertebra, the thoracic aorta passes through the diaphragm at the aortic hiatus and becomes the abdominal aorta.

F. Abdominal Aorta

1➡ Continue to follow the aorta into the abdomen. Locate the branches of the abdominal aorta.

The **inferior phrenic arteries*** are small arteries servicing the inferior surface of the diaphragm. These arteries may not be identified on your models. The phrenic arteries are not uniform in origin, sometimes even arising from the renal artery.

The first major branch inferior to the diaphragm is the unpaired **celiac artery**. The celiac artery has three major branches, the **hepatic**, **left gastric**, and **splenic** arteries. Identify these arteries by tracing them to their respective organs.

The unpaired **superior mesenteric artery** supplies the small intestine and the first two-thirds of the large intestine. The **suprarenal (adrenal)** arteries supply the suprarenal (adrenal) glands and anastamose with branches of the phrenic and renal arteries.

The paired **renal arteries** supply the kidneys. The renal arteries also branch to the suprarenal capsule. Often two arteries to the same kidney are found. The paired **spermatic (testicular) arteries** in the male, or paired **ovarian arteries** in the female, arise near the mid or lower region of the kidneys. The spermatic arteries service the scrotum as well as the testes, and the ovarian arteries supply the ovaries as well as the uterus.

The single **inferior mesenteric artery** branches to form the **left colic artery**,* which supplies the descending colon, and the **sigmoidal*** and **superior rectal arteries**,* which supply the sigmoid colon and rectum, respectively. The four pairs of small **lumbar arteries** supply the abdominal wall.

At this point the abdominal aorta bifurcates, forming the **common iliac arteries**. The **middle sacral artery**,* arising as a tail from the iliac bifurcation, supplies blood to the sacrum, coccyx, and rectum. In many animals this terminal vessel of the abdominal aorta is known as the **caudal artery**.*

Each common iliac artery bifurcates to form the **internal** and **external iliac arteries**. The large paired external iliac arteries pass through the body wall and become the **femoral arteries** in the legs. The internal iliac arteries, the caudalmost paired arteries of the pelvic region, supply the gluteal muscles, the skin, and the urinary and reproductive organs.

G. Lower Extremity

1➡ Return to the external iliac artery and locate the major arteries of the lower extremity. Use Figure 50–6● as your guide.

As stated, as the external iliac artery leaves the body wall, it becomes the femoral artery. About 5 cm distal to this emergence is a lateral branch, the **deep femoral artery**. Find this artery, which continues to branch to supply the bladder, ventral abdominal wall, and some of the femoral muscles.

Several branches exit the femoral artery to serve the thigh. When the femoral artery reaches the knee, it becomes known as the **popliteal artery**. The popliteal artery bifurcates to form the **anterior** and **posterior tibial arteries**, which supply the lower leg and foot.

At the ankle the anterior tibial artery becomes the **dorsalis pedis artery**, which continues into the foot

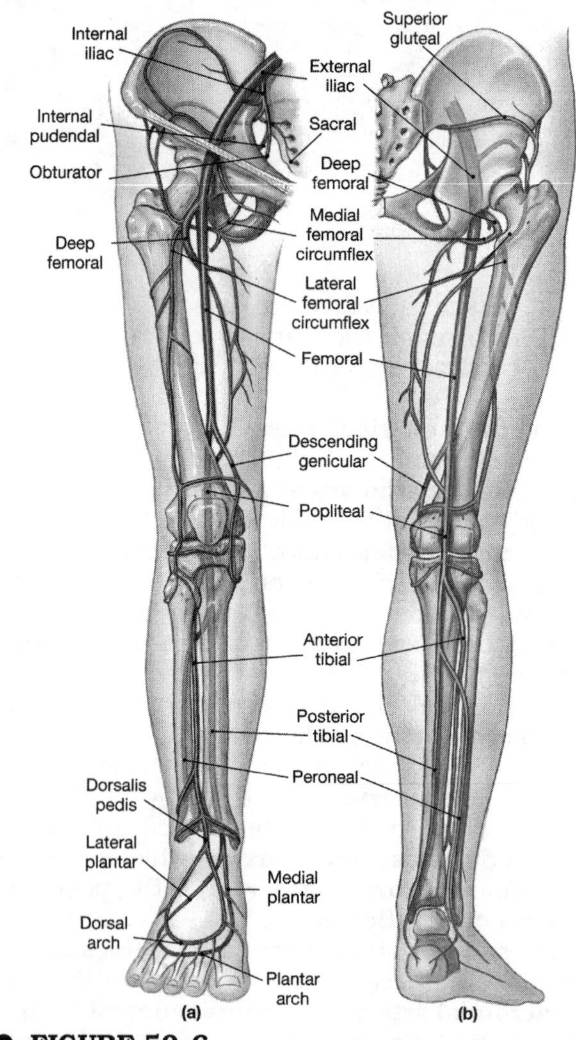

● **FIGURE 50–6 Arteries of the Lower Extremity.**

to form part of the **plantar arch**.* The posterior tibial artery gives off the **peroneal artery**. At the ankle the posterior tibial artery bifurcates, forming the **lateral** and **medial plantar arteries**. These service the foot and give rise to parts of the plantar arch. The **digital arteries*** arise from the plantar arch to serve the toes.

VI. Venous System [∞ *Martini, p. 734]*

A. Nomenclature

Because veins carry blood toward the heart from an organ or region of the body, we often use terms like "entering" or "drains" or "empties into" when describing the venous system.

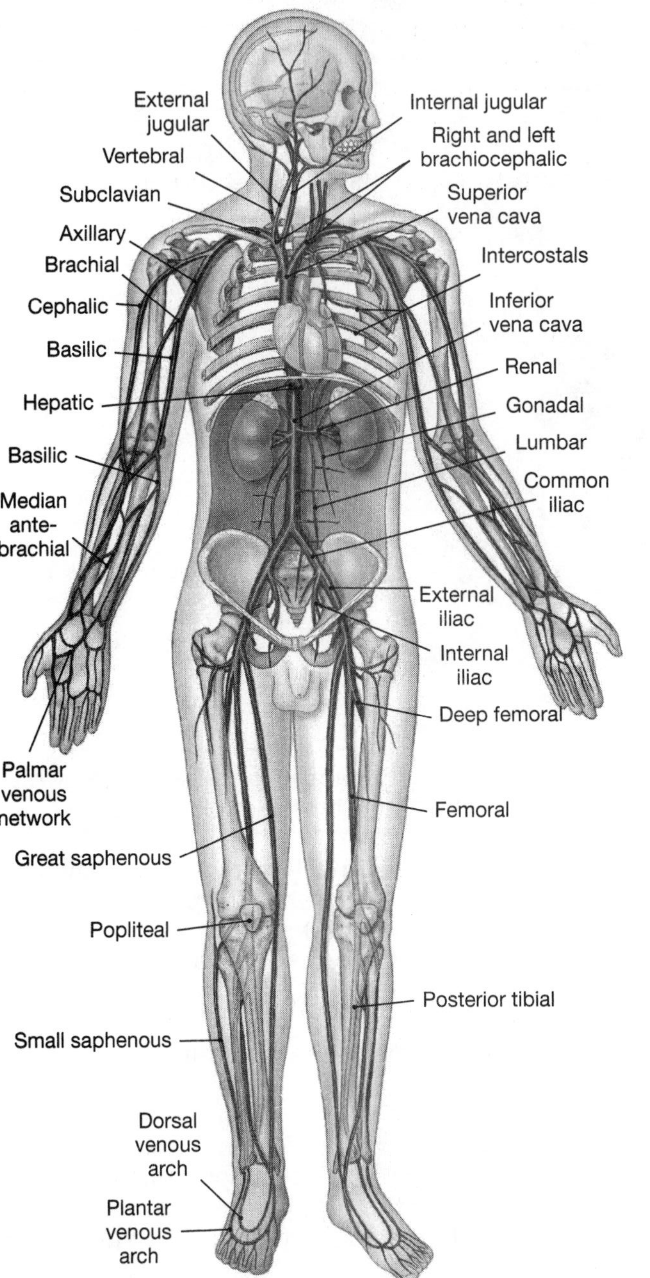

● **FIGURE 50–7**
Venous System.

1 ➡ Use Figures 50–7● to 50–11● as your guide for examining the venous system. Review these figures briefly before beginning your investigation.

B. Superior Vena Cava

1 ➡ Begin your examination of the venous system by locating the large vein draining the head region and entering the right side of the heart (right atrium). This is the **superior vena cava**.

The first vessel entering the superior vena cava is the **azygous vein**, a multibranched vessel draining the **intercostal**, **mediastinal**,* **esophageal**,* and **bronchial veins**.* While you are examining the azygous vein, note the **hemiazygous** vein, which enters the azygous about half-way between the superior vena cava and the diaphragm. (In some animals, the hemiazygous empties directly into the vena cava.)

The **internal thoracic (sternal) vein*** enters the superior vena cava from the ventral wall. The superior vena cava is formed from the junction of two short vessels, the **brachiocephalic (innominate) veins**.

C. Head and Neck

1 ➡ Study the brachiocephalic vein, which is formed by the junction of the **internal jugular** and **subclavian veins**. Look carefully at Figure 50–6● and note the following pattern: the **superior sagittal sinus** merges with the **straight sinus** to form the **transverse sinus**. As the transverse sinus passes through the jugular foramen, it becomes the internal jugular vein and is joined by the **facial vein** shortly thereafter.

The first major vein entering the subclavian vein is the **external jugular**. The **temporal** and **maxillary veins** are the major vessels entering the external jugular. Immediately lateral to the external jugular vein is the **vertebral vein**, which drains the cranium, spinal cord, and vertebrae.

D. Upper Extremity

1 ➡ Return to the subclavian vein and follow it laterally. At the bifurcation the superior branch is the superficial **cephalic vein**. The inferior bifurcation branch is the **axillary vein**, which becomes the **brachial vein** in the upper arm.

Branching off the medially located brachial vein is the **basilic vein**. The basilic vein anastomoses

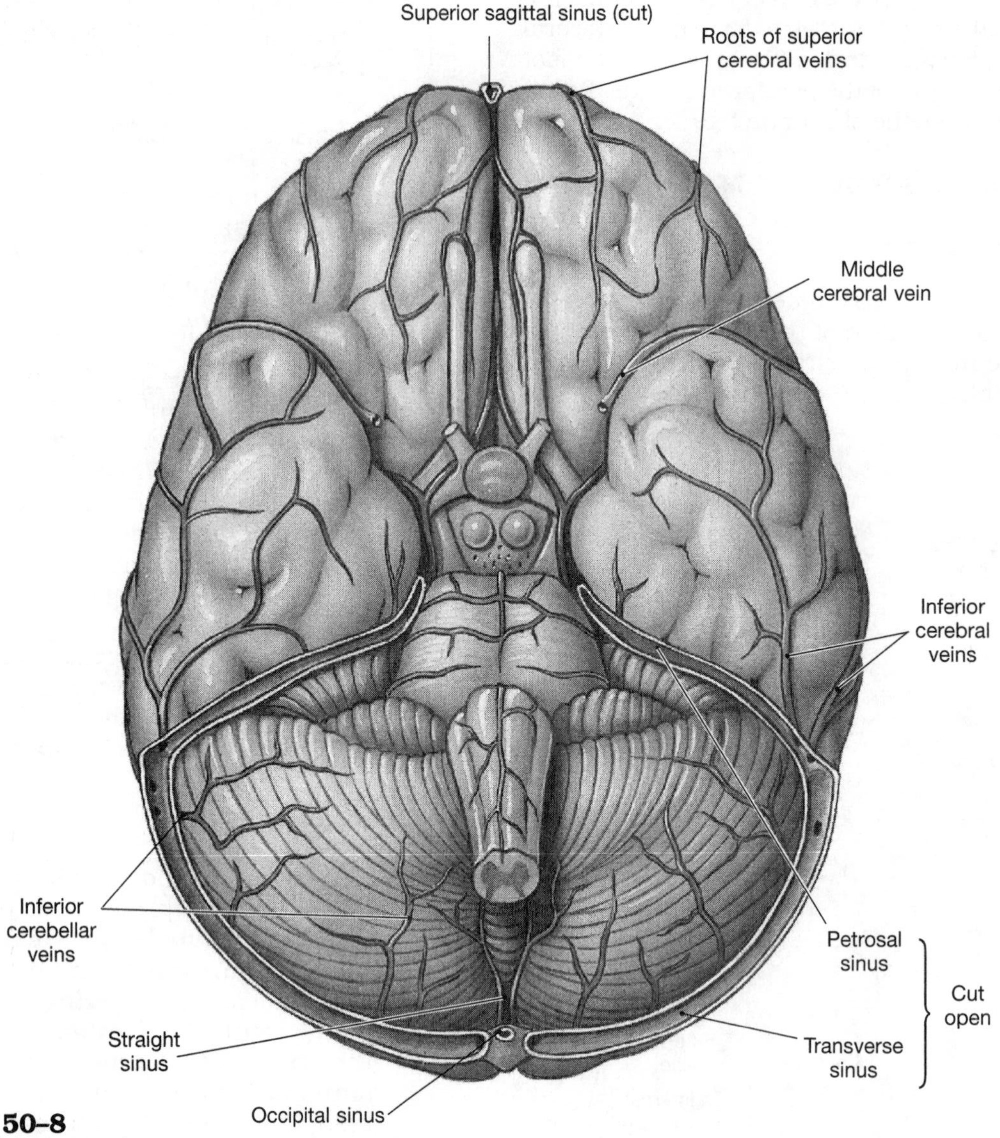

● **FIGURE 50–8**
Major Veins of the Brain.

with the cephalic vein via the **median cubital vein** at the anterior elbow. A more distal branch off the basilic vein is the **median antebrachial vein**, one of the veins that drain the forearm.

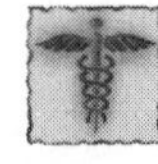

In some fair-skinned people and body builders, the basilic and the cephalic veins are often quite visible. The median cubital vein is the most common blood sampling site.

The brachial vein proper is formed by the union of the **radial vein** and the **ulnar vein**. These veins are drained by the common **superficial palmar arch**.

E. THORACIC VEINS

1 ➡ Return to the heart and relocate the superior vena cava. You have already noted the veins that drain into the azygous, which empties into the superior vena cava.

2 ➡ Identify the **inferior vena cava**, which enters the heart (right atrium) from the inferior portion of the body. No branches of note enter the thoracic portion of the inferior vena cava.

F. INFERIOR VENA CAVA (ABDOMINAL)

1 ➡ Follow the inferior vena cava into the abdominal cavity. The inferior vena cava passes through the diaphragm and usually lies to the right of the abdominal aorta. Locate the major abdominal veins.

Immediately inferior to the diaphragm are the paired **hepatic veins**, which begin deep within the

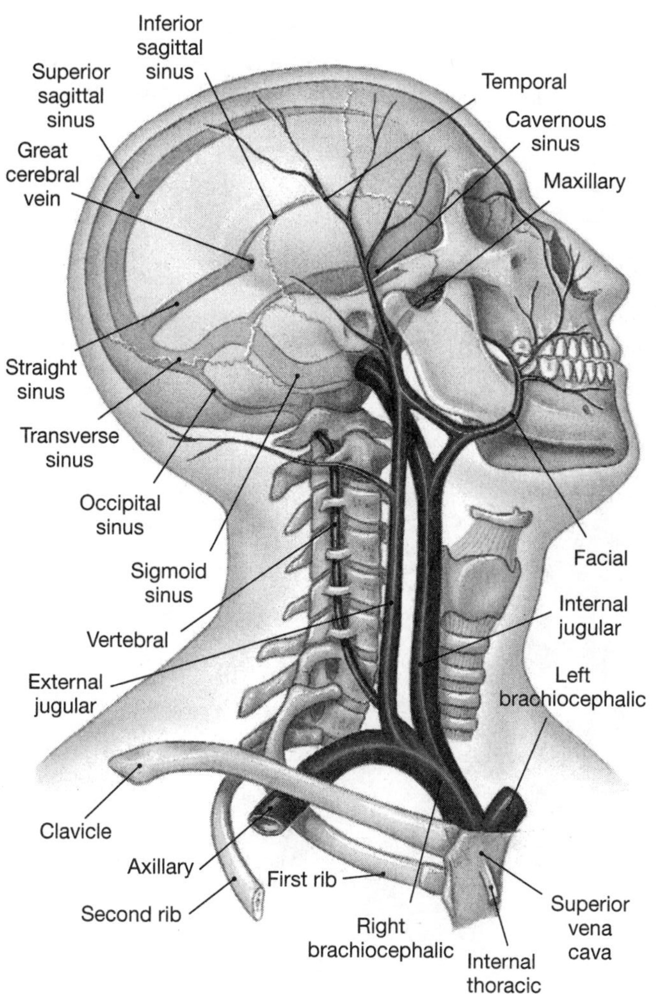

● **FIGURE 50–9**
Veins of the Head and Neck.

liver. You may be able to locate the paired **inferior phrenic veins**,* which drain the inferior diaphragm. The paired **suprarenal veins (adrenal veins)**, which drain the adrenal glands, are not actually parallel as the right vein empties directly into the vena cava while the left drains into either the renal vein or the phrenic vein.

The paired **renal veins** drain the kidneys. The right renal vein enters the vena cava at a point caudal to the left renal vein. The paired **ovarian** or **testicular veins** are also not parallel. The left vein empties into the renal vein, and the right vein empties directly into the vena cava. The four **lumbar veins** drain the posterior abdominal wall.

Caudal to the lumbar veins you will notice that the inferior vena cava is formed by the convergence of two large veins, the **right** and **left common iliac veins**.

The **middle sacral vein (caudal vein*** in some animals) branches off the left common iliac. This vein drains the sacral regions of the body. Follow the right common iliac vein toward the right leg to the point of bifurcation.

The caudal branch is the **internal iliac vein**, which drains the gluteal region and the pelvic organs. The larger branch of the common iliac vein is the **external iliac vein**, which continues on into the thigh.

G. Lower Extremity

1 ➡ Return to the external iliac vein and locate the major veins of the lower extremity. Use Figure 50–11● as your guide.

In the thigh the external iliac vein becomes the **femoral vein**, which is proximally joined by the **great saphenous vein** on the medial aspect of the thigh. The **popliteal vein** becomes the femoral vein in the deep region behind the knee.

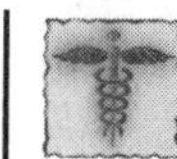
As a point of interest, the great saphenous vein is the vein of choice for heart bypass surgery.

The **peroneal vein** and **small saphenous vein** enter the popliteal vein proximal to the knee. The **anterior** and **posterior tibial veins** converge to form the popliteal vein just medial to the knee.

H. Hepatic Portal System

1 ➡ Approximate the location of the hepatic portal system, that part of the venous system that allows for the transport of nutrients from the digestive system directly to the hepatic system for immediate processing.

The **hepatic portal vein** is formed by the convergence of the splenic vein and the superior mesenteric vein. The **splenic vein** drains the spleen, parts of the stomach, and pancreas.

The **superior mesenteric vein** drains the small intestine, ascending and transverse regions of the large intestine, and the stomach. The **inferior mesenteric vein** drains the distal portion of the large intestine and the rectum before emptying into the splenic vein.

The hepatic portal blood flows through the sinusoids of the liver and enters the hepatic veins. As previously seen, the hepatic veins empty into the inferior vena cava.

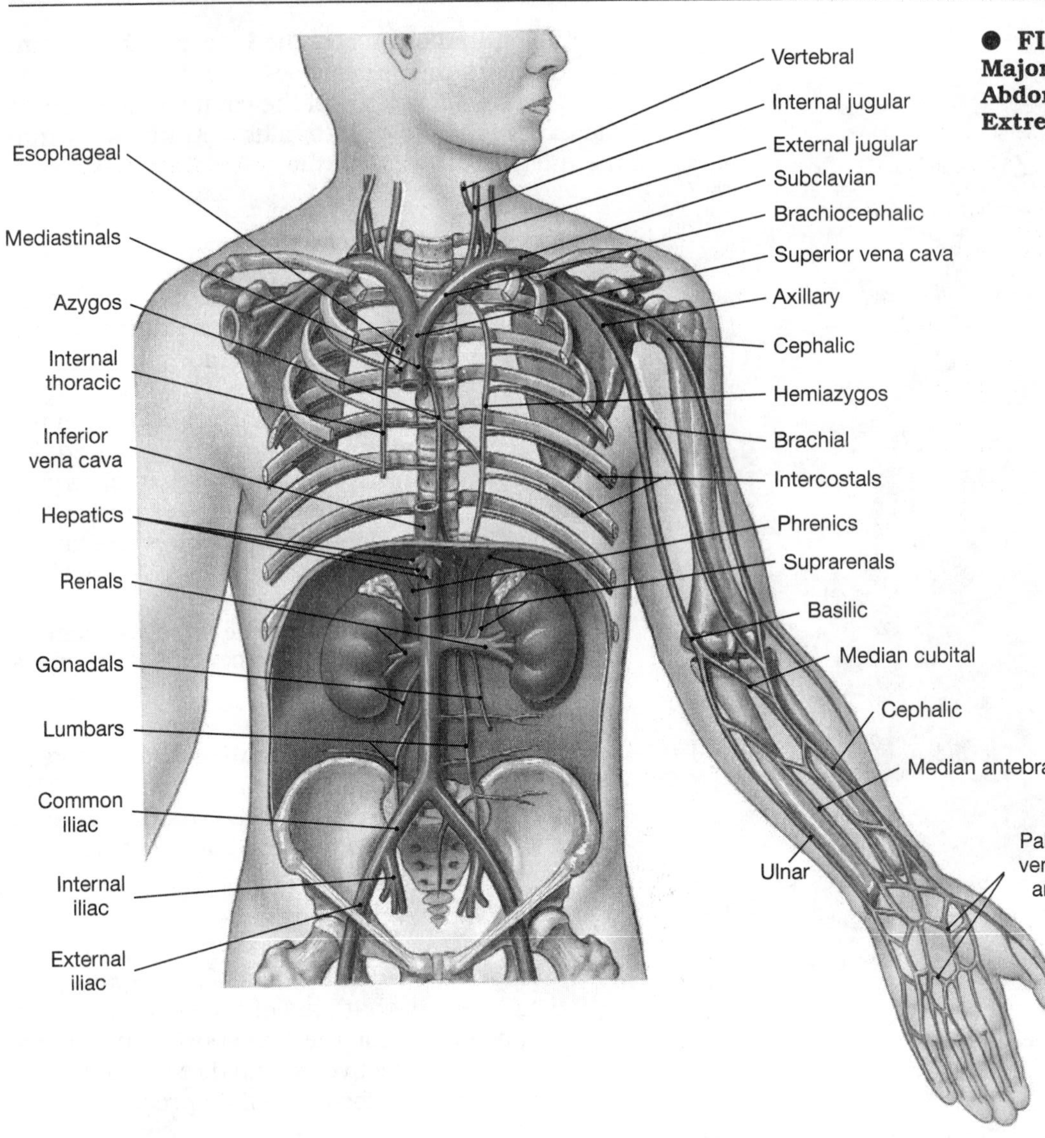

● **FIGURE 50–10 Major Veins of the Chest, Abdomen, and Upper Extremity.**

❑ Additional Activities

1. Research some of the branches of the major blood vessels, especially those branches in the brain.
2. Although the cardiovascular system is rather consistent throughout the mammalian world, certain subtle differences do exist. Find out about some of these differences and decide if those differences have any evolutionary advantage for the particular animal you are studying.

NOTES

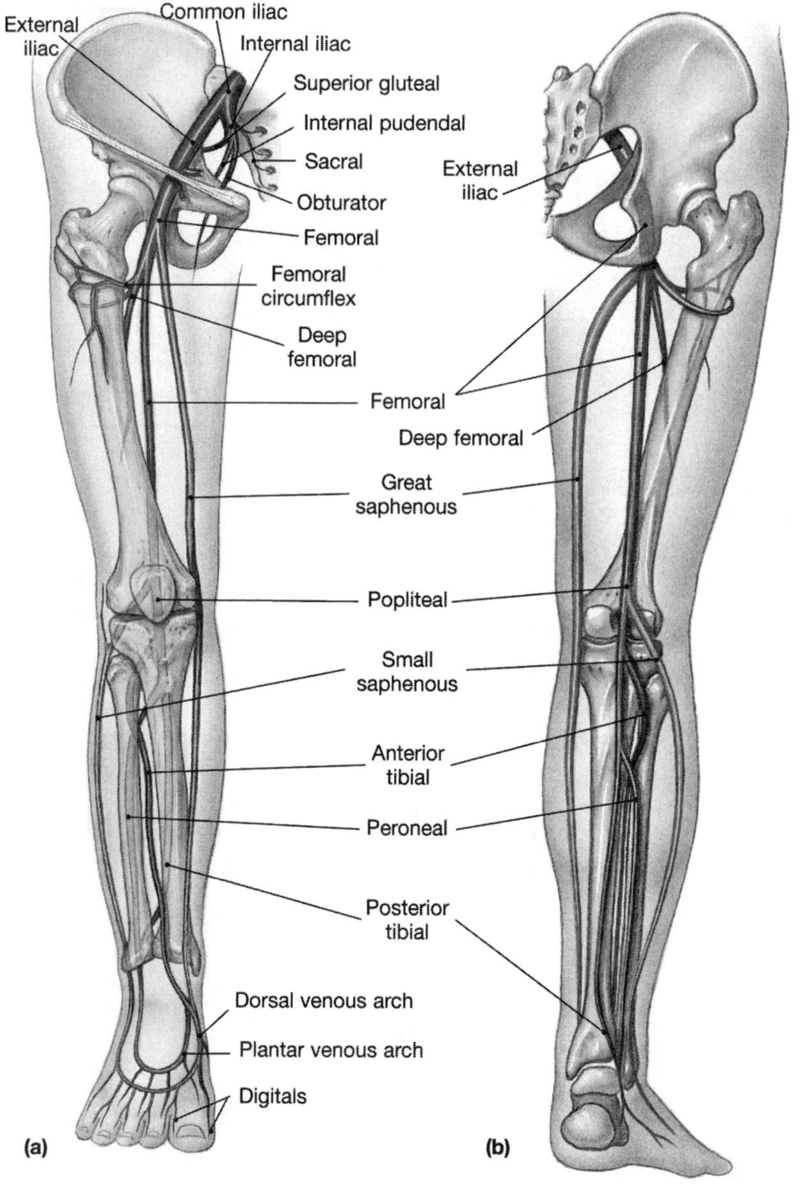

● **FIGURE 50–11**
Veins of the Lower Extremity.

Answers to Selected Concept Check Questions

1. The capillary wall is only one cell thick. The capillary wall is also a continuation of the endothelium throughout the rest of the cardiovascular system.
2. The function of the venous valves is to prevent the backflow of blood.
3. These vessels are too thick to be nourished by the blood passing through them.

❑Lab Report

1. Referring to the numbered inquiries at the beginning of this exercise, complete the following box summary:

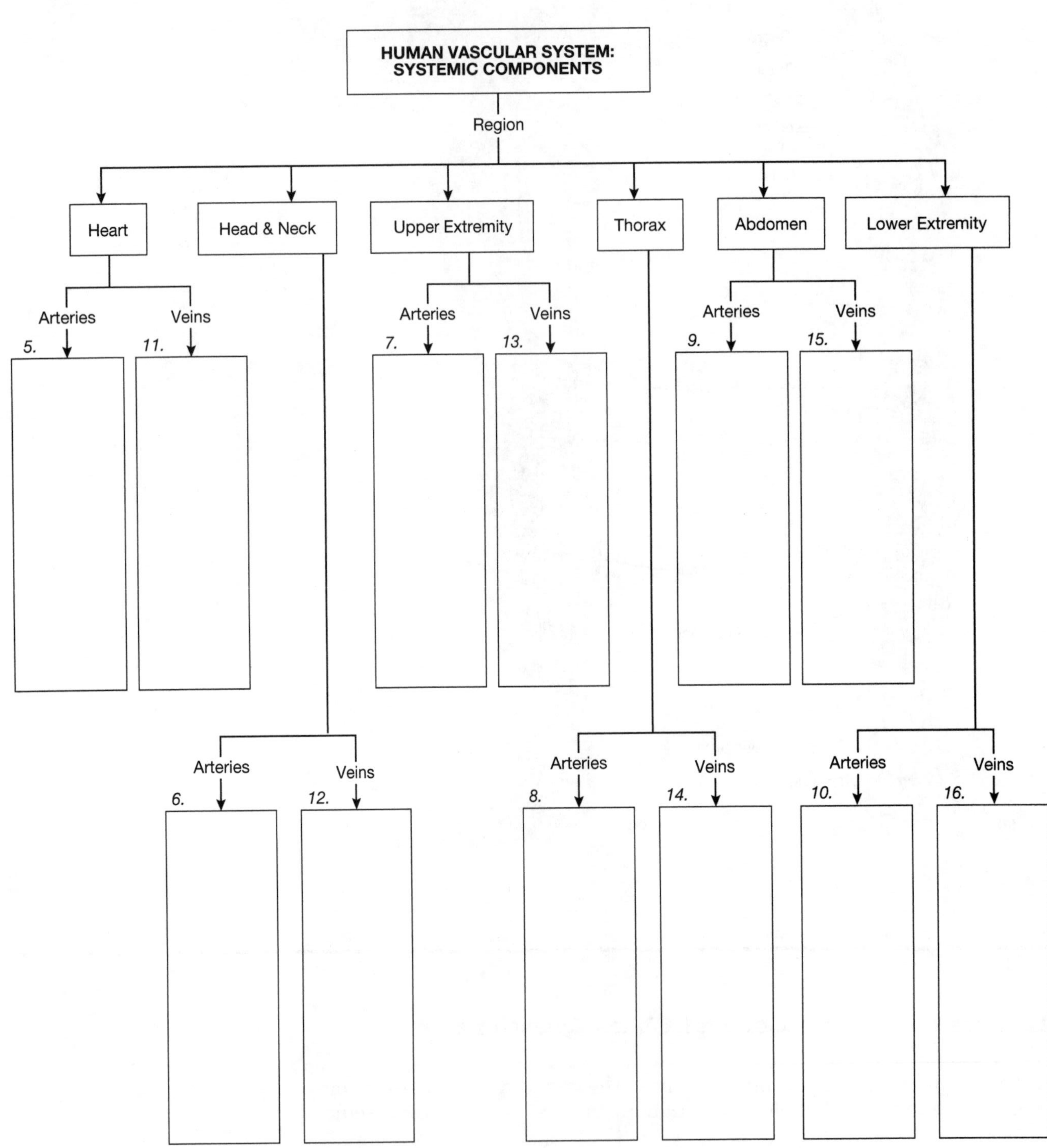

General Questions

NOTES

1. Fill in the following chart comparing the arteries, veins, and capillaries

Vessel	Name and Description of Vessel Layer	Name and Description of Vessel Layer	Name and Description of Vessel Layer
Artery			
Vein			
Capillary			

2. In terms of capillaries, arteries, and veins, how does the functional anatomy systemic circulatory system compare with that of the pulmonary circulatory system?
3. How can we correlate the route blood takes through the cardiovascular system with our basic blood vessel dissection plan?
4. Trace a drop of blood: (for some of these situations, more than one pathway may be possible)
   a. From the femoral artery to the femoral vein
   b. Aortic arch to the ulnar artery
   c. Brachiocephalic vein to the external carotid artery
   d. Hemiazygous vein to the right subclavian artery
5. In general, how can you distinguish an artery from a vein?
6. In general, how are blood vessels named?
7. Which strategies did you find most effective in helping you learn the names of the blood vessels?
8. In the past certain anatomy professors claimed that students should not be given diagrams to aid them in their identification of internal structures. Rather, students should describe and identify all structures strictly from the written word. What advantages and disadvantages do you see to the “nondiagram” method of learning the names of the blood vessels?

EXERCISE 51

Cardiovascular Physiology

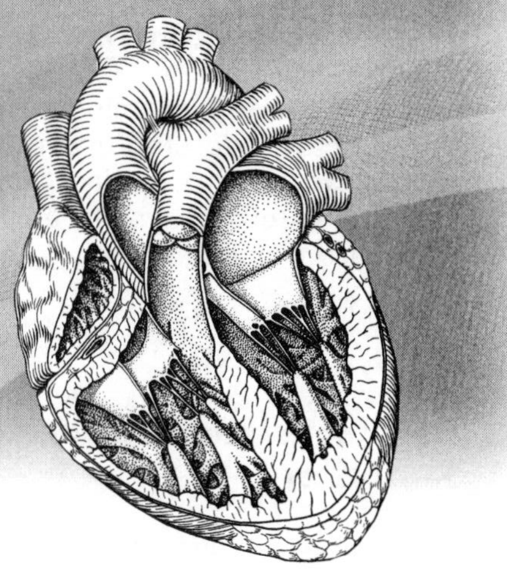

PROCEDURAL INQUIRIES

Preparation

1. What is the path blood takes through the heart?
2. What do we mean when we say the heart is two functional syncytia?
3. What are the parts of the electrical conduction system through the heart?
4. What are the parts of the cardiac cycle and how is each part identified?

Experimentation

5. What is auscultation?
6. What are the four heart sounds?
7. What is the proper way to use a stethoscope?
8. What is the proper way to take a pulse?
9. What are the most easily studied pulse points?
10. What are some important pulse characteristics?
11. What is the proper way to take blood pressure?
12. What is the difference between systolic and diastolic blood pressure?
13. How do you determine the pulse pressure?
14. What does the cold pressor test demonstrate?
15. What is an easy way to take a venous blood pressure?
16. What is cardiovascular efficiency?
17. How do we test for cardiovascular efficiency?

Additional Inquiries

18. How would you describe a typical heart cell?
19. What is a general fatigue curve?
20. What is hypertension?

Key Terms

Amplitude
Arterial Blood Pressure
Atrioventricular (AV) Node
Auscultation
AV Bundle
Blood Pressure
Bundle Branches
Bundle of His
Cardiovascular Efficiency
Cold Pressor
Diastole
Diastolic Pressure
General Fatigue Curve
Hypertension
Korotkoff's Sound
Lubb, dup
Pulse
Pulse Pressure
Purkinje Fibers
Rate
Rhythm
Sinoatrial (SA) Node
Sphygmomanometer
Syncytium
Systole
Systolic Pressure
Tension

Materials Needed

Sphygmomanometer
Stethoscope
Watch (or Clock) with Second Hand
Pan of Ice Water
Compound Microscope
Prepared Slide of Cardiac Muscle

The heart is a four-chambered double pump composed of uninucleate striated cells. The cells of the upper chambers, the atria, contract in unison, as do the cells of the lower chambers, the ventricles. It is this synchronous contraction that allows the cardiac muscle to exert enough pressure to pump blood as needed throughout the body.

In this laboratory exercise we will try to develop a practical understanding of cardiovascular physiology by examining certain physiological characteristics of the system and by performing selected cardiovascular tests.

❑ Preparation

I. Cardiovascular Structure and Function

A. Cardiac Review — Gross Structure and Function

1➡ Review the basic anatomical landmarks of the heart (Exercise 48) before you begin this exercise. Note the particularly relative positions of the atria, the ventricles, the atrioventricular valves, the semilunar valves, the **sinoatrial (SA) node**, the **atrioventricular (AV) node**, and the interventricular septum. Refer to Figure 51–1●.

2➡ Deoxygenated blood enters the right atrium of the heart from all parts of the body and begins flowing into the right ventricle. Upon atrial contraction, additional blood passes through the tricuspid valve into the right ventricle. With ventricular contraction, the blood leaves the right ventricle through the pulmonary semilunar valve and travels via the pulmonary trunk and pulmonary arteries to the lungs, where gas exchange takes place.

3➡ From the lungs the oxygenated blood returns to the heart via the pulmonary veins and enters the left atrium. This blood begins flowing into the left ventricle and, upon atrial contraction, additional blood passes from the left atrium through the bicuspid valve into the left ventricle. Upon ventricular contraction, the blood exits the left ventricle through the aortic semilunar valve and out via the aorta to all parts of the body.

4➡ Note that each part of the double pump has a separate function. The atrial **syncytium** contracts just ahead of the ventricular syncytium, and blood flow throughout the system is continuous. (In a syncytium, all cells function as one unit).

B. Cardiac Review — Tissue

1➡ Briefly review the anatomy of cardiac tissue (Exercise 48), recalling in particular the uninucleate, branched cardiac cells and the intercalated disks. *[∞ Martini, p. 659]*

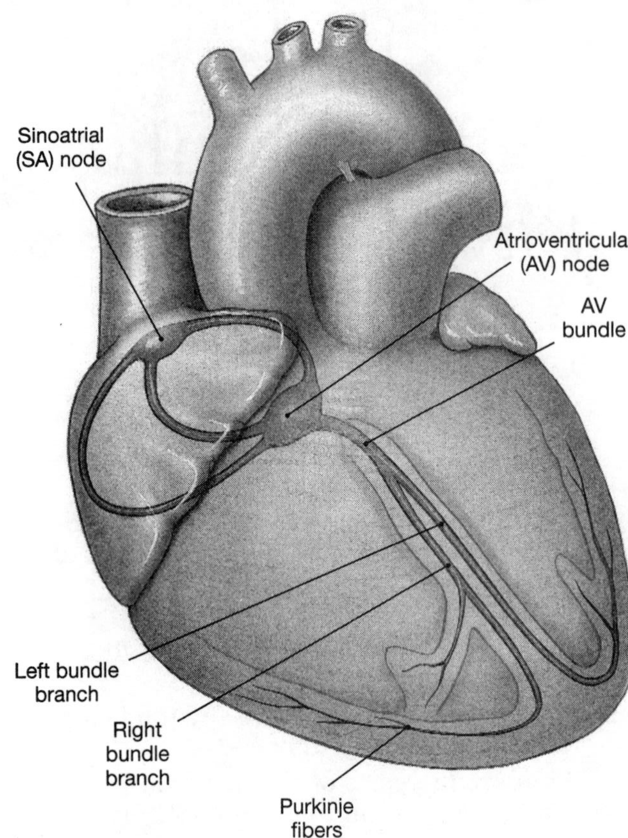

● **FIGURE 51–1**
Conduction System of the Heart.

2➡ Obtain a compound microscope and a prepared slide of cardiac muscle. Return to Exercise 48 and compare your slide with Figure 48–1●. Identify the branched cardiac cells, the single nucleus, and the intercalated disks. The intercalated disks are the membranes of adjoining cells that fuse to form interdigitating junctions highly permeable to freely flowing ions.

C. Electrical Conduction [∞ Martini, p. 670]

The SA node is situated in the posterior (dorsal) wall of the right atrium near the entrance to the superior vena cava. Check Figure 51–1●. The SA node is composed of specialized muscle fibers. These nodal fibers are continuous with the atrial fibers, thus ensuring any action potential will spread immediately throughout the atria.

Electrical conduction in the heart begins with the self-excitation of the fibers of the SA node. Although most cardiac cells are self-excitatory, the SA nodal cells display the greatest degree of autorhymicity and therefore are responsible for setting the pace for the cardiac cycle. The SA node is called the pacemaker of the heart.

After the SA node self-excites, the action potential spreads across the atria. The atria, sometimes

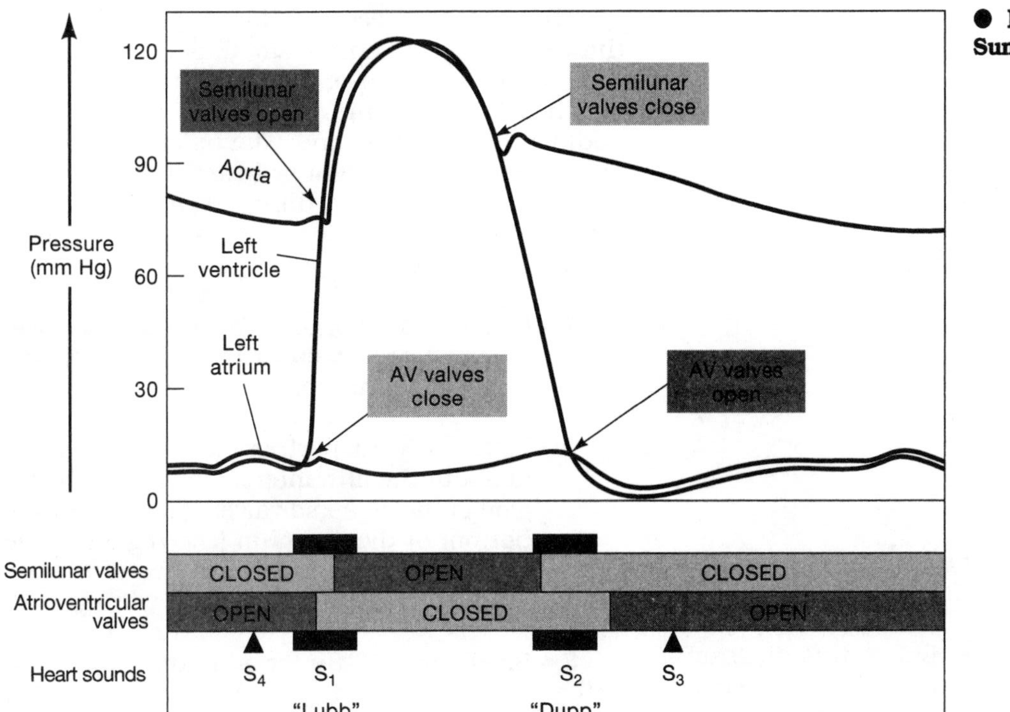

● **FIGURE 51–2**
Summary of Cardiac Events.

called the primer pumps, form a functional syncytium and thus all atrial cells contract together.

Meanwhile the action potential spreads via the internodal pathways to the AV node, which is situated on the floor of the right atrium near the opening of the coronary sinus. The atrial fibers form a junctional complex with certain transitional fibers that in turn join with the specialized fibers of the AV node. The junctional complex of the atrial fibers going into the AV node slows the action potential down sufficiently so that the atrial contraction is complete before the ventricular syncytium is stimulated.

As the electrical stimulation traverses the AV node, the action potential proceeds down the interventricular septum via the **AV bundle**, also known as the **bundle of His**.

Partway down the interventricular septum the AV bundle divides, forming the **bundle branches**, which spread across the inner surfaces of the ventricles. The bundle branches subdivide into specialized **Purkinje fibers**, which convey the impulses to the contractile cells of the ventricular syncytium. The intercalated disks enhance the spread of the stimulus and the ventricles contract in unison. The ventricles are sometimes called the power pumps.

1➡ Locate the preceeding boldface terms in Figure 51–1●.

2➡ Describe in your own words the route of the electrical stimulation through the cardiac conduction system.

____

____

3➡ Check your lecture text (or Exercise 34 in this book). What is the role of the vagus nerve (NX) in cardiac conduction?

____

____

D. The Cardiac Cycle [∞ Martini, p. 674]

The cardiac cycle is the summation of events between the beginning of one heartbeat and the beginning of the next heartbeat. The cardiac cycle includes alternating periods of contraction and relaxation.

The cycle begins with the stimulation of the SA node and subsequent spread of the atrial impulse. Although blood has been passively draining through the flaccid valves from the atria into the ventricles, the atrial contraction (atrial **systole**) forces the blood remaining in the atria into the ventricles. Atrial relaxation (atrial **diastole**) follows atrial systole.

When the AV node is stimulated, the impulse spreads across the ventricles, and the cells contract in unison. This contraction, **ventricular systole**, forces the blood against the valves, closing the bicuspid and tricuspid valves, thus preventing regurgitation of blood into the atria. The semilunar valves are forced open and the blood is ejected into the pulmonary trunk and ascending aorta. **Ventricular diastole** follows. With relaxation, back pressure from the blood in the aorta and the pulmonary artery causes the semilunar valves to close, and pressure from blood flowing into the atria causes the atrioventricular valves to reopen.

Repolarization takes place during relaxation — diastole — as blood pours into the right atrium from the body and into the left atrium from the lungs

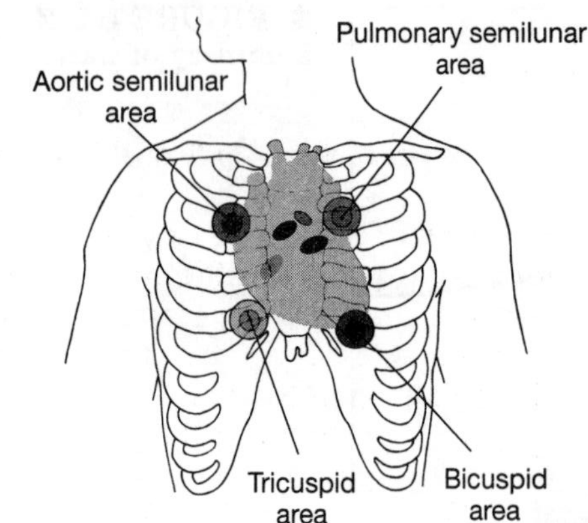

● **FIGURE 51–3**
Auscultation Areas.

and begins pouring into the ventricles from the atria. The cycle is about to begin again.

1➡ Study Figure 51–2●. Relate this diagram to the information you have just learned and to the information in the next section.

Describe in your own words what is happening to the pressure in the left atrium, the left ventricle, and the aorta.

____

What is happening to each valve? Why?

____

❑ Experimentation

Work with a partner and, unless directed otherwise, both you and your partner should do each of the experiments.

II. Heart Sounds [∞ *Martini, p. 677*]

A. Auscultation

As the blood traverses the heart and the valves open and close, certain heart sounds are produced and can be identified through a stethoscope. Characteristically, the cardiac cycle sounds like "LUBB, DUP, PAUSE." The first heart sound is the **lubb**. This sound, marking the beginning of ventricular systole, occurs as the atrioventricular valves close. At this point the ventricles are contracting and the semilunar valves are closed for a period of isovolumetric contraction. As the pressure rises, the semilunar valves open and the blood is ejected.

The **dup**, the second heart sound, occurs at the beginning of ventricular diastole when the semilunar valves close. Just after the "dup" you may hear a faint sound. This is the sometimes inaudible third heart sound. The usually inaudible fourth heart sound might be heard just before the "lubb" sound. Both the third and fourth sounds are the result of systolic and diastolic blood flow rather than valve closings.

1➡ Obtain a stethoscope. Your instructor will demonstrate **auscultation** (the proper way to use a stethoscope).

2➡ Listen to your partner's chest, preferably at about the fifth intercostal space (the region of the bicuspid valve) and at about the bottom of the sternum (the region of the tricuspid valve). These are two of the asculation areas.

3➡ See Figure 51–3● for additional auscultation sites. List the major auscultation points. _______________

III. Pulse

A. Taking the Pulse

Pulse, the pressure surges resulting from the rhythmic flow of blood through a vessel, is a manifestation of the pumping heart. Blood moves through all blood vessels, and the pressure surge can technically be quantified and qualified in any vessel. However, for most purposes, we are interested in the pulse as measured in the major arteries.

1➡ The pulse can be easily studied in five major arteries. Use your three middle fingers and very light pressure to locate these pulse points, preferably on your lab partner:

Temporal Side of the head, temporal region.

Carotid Neck, about 1 cm caudal and slightly medial to the angle of the jaw. (Do not try both carotid pulses at the same time!)

Radial Lateral wrist.

Popliteal Back of knee. (Occasionally this pulse is hard to locate as the popliteal artery is deeper in some people than in others.)

Dorsalis Pedis Dorsum of foot, usually best felt slightly medial to the proximal end of the first metatarsal.

2➡ In the next section of this exercise you will find the **brachial** pulse in order to take your partner's blood pressure. You can also sometimes find the **facial** pulse just anterior to the masseter muscle, or the **femoral** pulse in the groin area.

B. Pulse Characteristics

Finding the pulse is usually not sufficient. You will want to know certain characteristics of the pulse. These include:

Rate How many beats per minute? The most accurate measurement seems to be to count for one full minute. For normally healthy people who have not been performing any vigorous exercises, it is usually adequate to count for 15 seconds and multiply by 4. (In compromised patients, this may not be adequate.)

In the cardiovascular efficiency test later in this exercise, you will be asked to count for 10 seconds and multiply by 6. The reason for this is that you will have been exercising and you want an immediate number. Over the course of a full minute, the pulse may slow down considerably, so a full minute count will not be as accurate as a 10-second count.

A resting pulse usually averages about 70–80 beats per minute. Due to individual differences, average pulse rates between 40 and 110 are usually considered normal.

An accelerated pulse rate is known as **tachycardia**, and an unusually slow rate is **bradycardia**. These conditions may or may not be pathological. If either condition is noted for an extended period of time, the cause should be determined by an appropriate health care professional.

Rhythm Does the pulse maintain a regular rhythm or does it speed up or slow down or skip beats? For checking rhythm, the longer pulse count will be more accurate because you will have less chance of missing an irregularity. (Be assured that everyone has an occasional rhythmic irregularity!)

Tension Does the pulse feel firm and strong, moderate, or weak and wimpy? Are there any changes in tension over the time span?

Amplitude Can you see the pulse throbbing? If so, is it high (visible), moderate, or low (invisible)? All of the pulses can be visible on occasion. Often the easiest one to see is the radial pulse. Relax your arm and hyperextend your wrist slightly. Sometimes if you can see the radial pulse this way, you can also see the ulnar pulse.

For best visual results, should the arm be above or below the level of the heart? Try it and see.

It is not unusual for other pulses to be visible on occasion.

C. Factors Influencing the Pulse Characteristics

1 ➡ For each of the above characteristics, various factors might influence your pulse. To test the influence of certain factors, have your partner take your pulse after each of the following activities:

- Lying perfectly still for 3–5 minutes
- Sitting perfectly still for 3–5 minutes
- Standing perfectly still for 3–5 minutes
- Running in place for 3–5 minutes
- Fill in the following chart with the data from at least two of the above activities. For additional space, use the last page of this exercise.

Pulse	*Rate*	*Rhythm*	*Tension*	*Amplitude*
Temporal				
Carotid				
Radial				
Popliteal				
Dorsalis Pedis				

IV. Blood Pressure

A. Blood Pressure Characteristics

The force exerted by the blood against the blood vessel wall is the **blood pressure**. Technically, every vessel in the body has a blood pressure. See Figure 51–4●.

The blood pressure we normally think of is the pressure exerted on the walls of the aorta by the blood as it exits the heart. This pressure is measured as close to the heart as possible (at the same level as the heart so that hydrostatic pressure is not a factor), normally on the brachial artery. This pressure is known as **arterial blood pressure**.

The highest, or peak, blood pressure occurs as the blood is ejected into the arteries. This is the **systolic pressure**. Systolic pressure corresponds with ventricular systole, the contraction that ejects blood from the ventricle into the aorta.

The lowest pressure recorded during the cardiac cycle is the **diastolic pressure**, the amount of "non-push" pressure that is always present in the vessel. Diastolic pressure is a measure of the condition of the blood vessels. Diastolic pressure corresponds with the last part of ventricular diastole, the relaxation phase of the cardiac cycle.

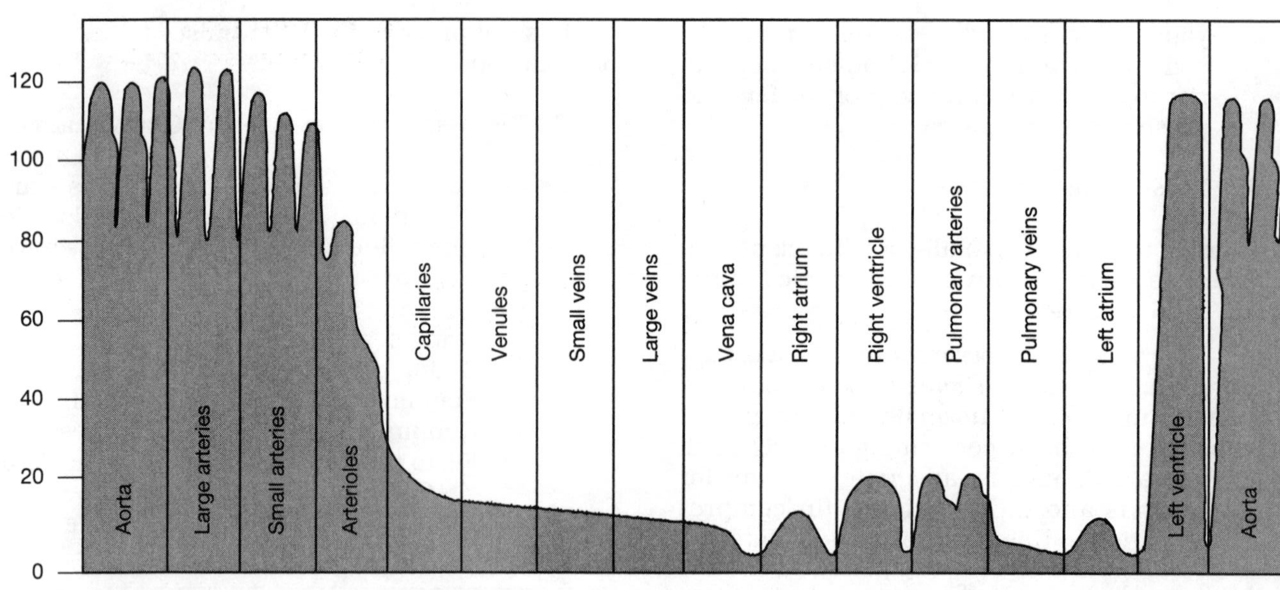

● **FIGURE 51–4**
Pressures within the Circulatory System.

Blood pressure is measured in millimeters mercury (mm Hg). If a pressure number is 75, this means that the pressure exerted by the blood is sufficient to push up a column of mercury 75 mm (at STP, standard temperature and atmospheric pressure).

If your blood pressure is 120/80, the 120 is the systolic pressure, 120 mm Hg, the force on the blood vessel as the blood is being ejected. The 80 is the diastolic pressure, 80 mm Hg, the amount of pressure exerted by the resting blood.

B. Taking Blood Pressure

The **sphygmomanometer** is the instrument used for taking the blood pressure. Several different types of sphygmomanometers exist, so you should follow your instructor's directions for the specifics on how to operate yours. Some models have a "built-in stethoscope." For others you will need to operate the stethoscope.

Regardless of the type of instrument, the principle behind taking the blood pressure is constant. The cuff is inflated beyond the point that cuts off circulation in a given artery. The cuff pressure is measured on a gauge or a digital terminal.

1➜ Obtain a sphygmomanometer (and a stethoscope, if required) and examine the instruments carefully.

2➜ Have your partner lie down for the blood pressure test. If this is not possible, your partner can sit quietly.

3➜ Put the sphygmomanometer cuff on your partner and position the stethoscope over the brachial artery. See Figure 51–5•, a diagram depicting the proper position of the sphygmomanometer on the arm.

4➜ Inflate the cuff, usually to about a reading of about 160. If your partner has a history of high blood pressure, you may have to inflate the cuff higher than 160. Remember, you want to cut off the circulation in that brachial artery.

5➜ The cuff is then gradually deflated (at the rate of about 2 mm Hg per second) while you listen to the artery through the stethoscope. When the pressure in the cuff drops below the pressure in the artery, blood will be ejected into the distal part of the vessel. This is the systolic pressure. You will hear a distinct thumping sound known as **Korotkoff's sound**. If you do not have a digital display, make mental note of the pressure value as soon as you hear it.

6➜ Continue to deflate the cuff slowly. The point where the thumping stops is the diastolic pressure. Make note of the diastolic reading. Be aware that after the diastolic pressure point has been reached, a clicking sound corresponding to the heartbeat may sometimes be heard. The standard average blood pressure is 120/80 mm Hg. Keep in mind that this is an average.

7➜ Record the blood pressure. ____________

Concept Check 1 What factors might influence blood pressure readings?

____

____

____

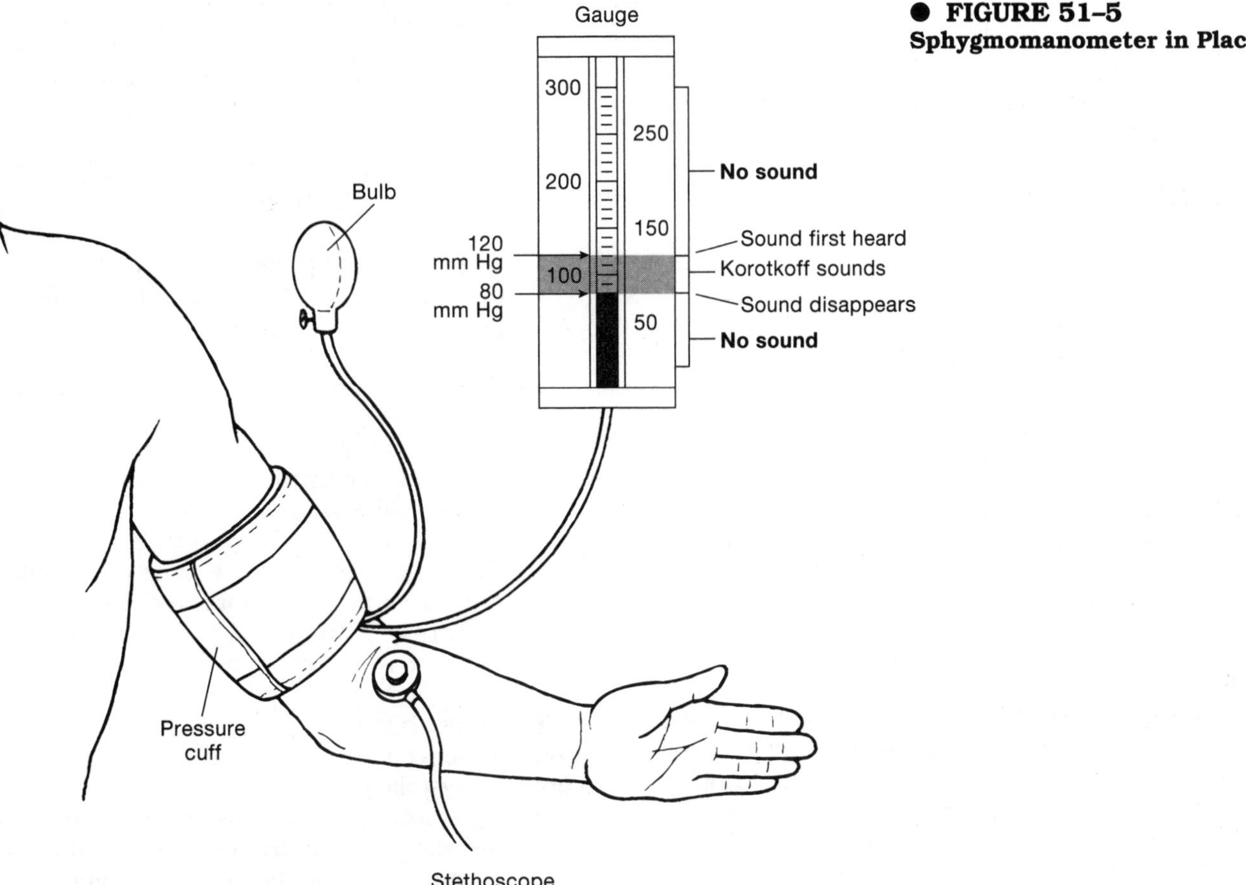

● **FIGURE 51–5**
Sphygmomanometer in Place.

C. Pulse Pressure

Pulse pressure is the mathematical difference between the systolic and diastolic readings. If your blood pressure is 120/80, your pulse pressure is 40.

1➡ Record your own pulse pressure.________

What value might there be in knowing the pulse pressure? ______________________________

____

____

D. Hypertension

Hypertension is the term used for significantly elevated blood pressure. Although factors such as age and physical condition may influence "normal" blood pressure values, hypertension is usually considered if the resting systolic pressure is above 140 and the resting diastolic pressure is above 90. (In clinical settings you may hear different systolic and/or diastolic readings classified as the lower limits of hypertension.) Of course, no diagnosis can be made on one reading alone! Nor can a diagnosis be made without considering the influences of stress, diet, personal habits, and environment.

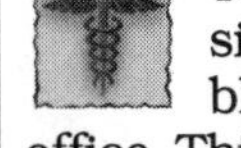

You may have heard of "white coat hypertension," the phenomenon whereby a person's blood pressure only goes up in the doctor's office. This very real phenomenon demonstrates the role of the psyche in physiological events. Sometimes people with severe "white coat hypertension" must have their blood pressures taken outside the medical setting.

E. Blood Pressure in Regions of the Body

Certain pathologies can be identified by checking blood pressures in different parts of the body. For instance, if all factors are equal, the blood pressure in the right arm should be approximately equal to the blood pressure in the left arm. Large discrepancies — as may occur after a stroke — should always be investigated.

If you are lying down, the ratio of your popliteal blood pressure to your brachial blood pressure should be approximately one. Gross discrepancies could indicate an arterial obstruction. Also, the pressure in your right leg should be approximately equal to the pressure in your left leg. Your instructor will indicate whether you should take the blood pressure on both arms and/or the legs.

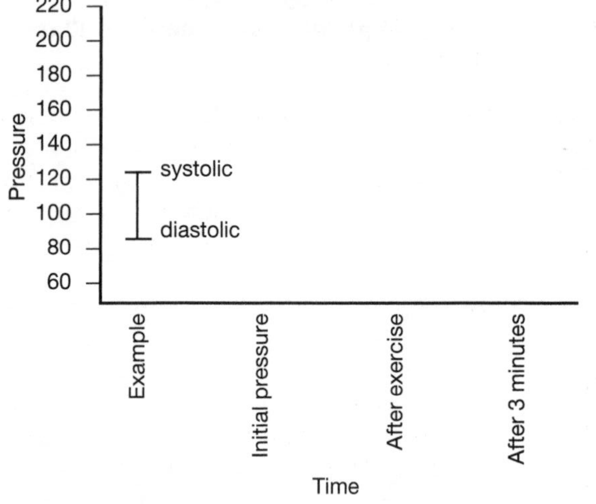

● **FIGURE 51–6**
Effects of Exercise on Blood Pressure.

F. Exercise and Blood Pressure

1➜ Examine the effects of exercise on blood pressure by wrapping the cuff around your partner's arm so that it is snugly in place but not so tight that circulation is cut off. (You want to be ready to take the blood pressure immediately without donning the cuff.)

2➜ Take your partner's resting blood pressure.

3➜ Now have your partner run vigorously in place for 2 minutes.

4➜ As soon as the run is complete, take your partner's blood pressure.

5➜ Take the blood pressure again in 3 minutes.

6➜ Graph the results. Use a bar to show both systolic and diastolic pressure. (See the example in Figure 51–6●.)

Was your partner's blood pressure back to normal after 3 minutes? If it was not, what factors do you think could account for the difference?

___

___

___

G. Cold Pressor

(Students with known hypertension should not participate in this experiment.)

The **cold pressor** test is an examination of the stability and sensitivity of the vasomotor center, which is located in the medulla of the brain.

1➜ Have your partner lie down for 5 minutes.

2➜ After your partner has been lying down for 5 minutes, determine his/her blood pressure and pulse rate. This is a baseline value. Leave the cuff in place.

3➜ Your partner should now put his/her free hand in a pan of ice water.

4➜ Take the blood pressure again after 30 seconds and again after 60 seconds. Take the pulse again.

5➜ Your partner can now remove his/her hand.

6➜ Take blood pressure and pulse readings every minute until the pressure returns to the baseline value.

7➜ Graph your results. Use a bar for the blood pressure and points for the pulse. Connect the pulse points to show continuity. Use Figure 51–7● for the graph.

If this test has been performed correctly and if the vasomotor center is functioning normally, in most people the systolic pressure will climb under the influence of the ice. This is because the cold shock activated the sympathetic division of the autonomic nervous system, causing the release of epinephrine.

The diastolic pressure should increase about 10% because of increased vasoconstrictor activity (because of the activity of the autonomic nervous system). The heart rate will also increase due both to the epinephrine and to the inhibition of the car-

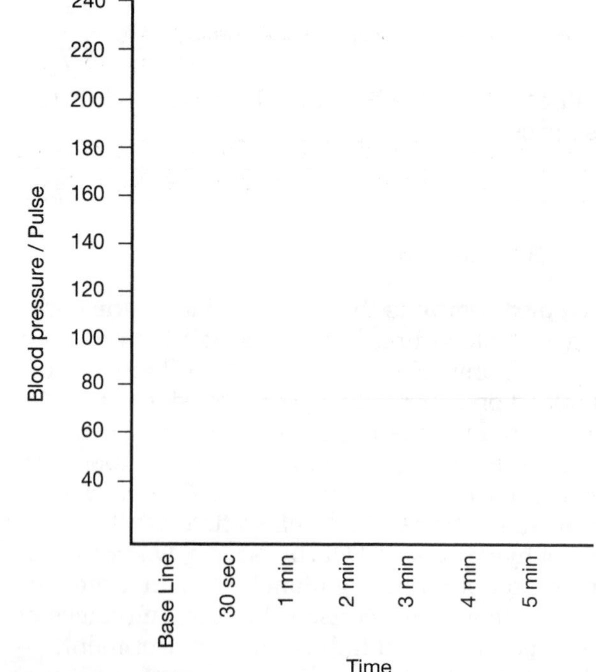

● **FIGURE 51–7**
Effects of Cold on Blood Pressure.

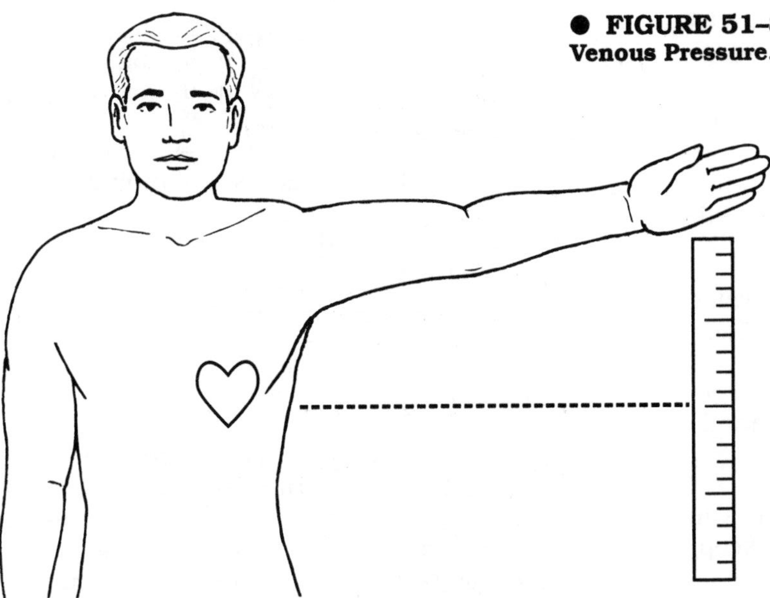

● **FIGURE 51–8**
Venous Pressure.

dioinhibitory center. Almost everyone will experience the diastolic increase, although some women experience little or no systolic, or heart rate change.

H. Venous Blood Pressure

Venous blood pressure cannot be measured directly with a standard sphygmomanometer. We do, however, have a rather accurate indirect method of measuring venous blood pressure. (Because of vasoconstriction in colder temperatures, this test works best in a warmer room.)

1➡ Hold your hand down and observe the veins on the dorsal surface.

2➡ Practice raising your hand slowly to a position above your head. Notice how the veins collapse.

3➡ Perform the actual test by extending your hand straight out on a level with your heart. Your partner can note the position. Slowly raise your hand until the veins disappear.

4➡ Your partner should now measure in centimeters the distance your hand traveled. See Figure 51–8●. This is the point where the venous pressure is overcome by gravity.

5➡ To convert this figure to blood pressure in mm Hg, use the fact that each 13.6 mm your hand rises above your heart will be represented by a 1 mm rise in the mercury. Therefore: mm Hg = cm x 10/13.6

6➡ Record your venous pressure. ____

Your venous pressure should be between 0 and 10 mm Hg. How does your venous pressure correlate with the information in Figure 51–4●?

____

____

V. Cardiovascular Function

A. Cardiovascular Efficiency

One function of the heart is the transport of oxygen to energy-producing cells. When energy requirements increase, the heart may have to beat faster to circulate the blood so that those cells are supplied with the oxygen necessary for energy production. Pulse rate should increase to supply the oxygen demanded by the cell. When the demand goes down, the pulse rate should decrease accordingly. In the healthy person, the pulse should return to a baseline value within minutes after exertion.

A **cardiovascular efficiency** test is a test that indicates how well the heart is doing its job. For this particular test you will need a watch (or clock)

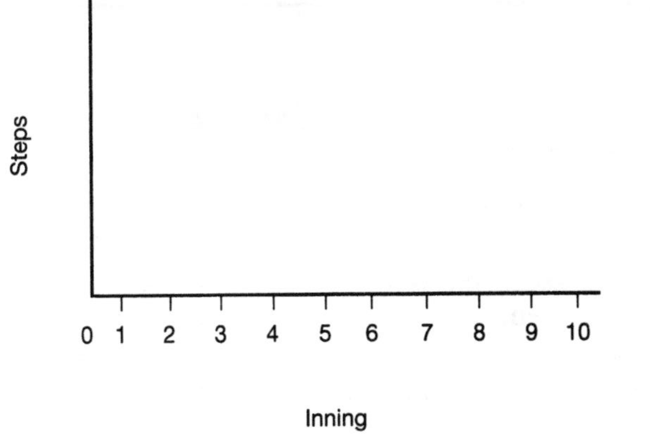

● **FIGURE 51–9a**
Cardiovascular Efficiency.

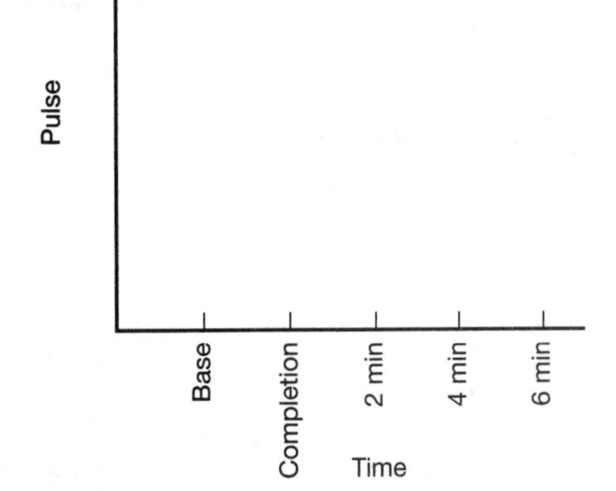

● **FIGURE 51–9b**
General Fatigue Curve.

with a second hand. Although this test may be done with you and your partner working by yourselves, as an alternative the entire class can do the test at the same time, with your instructor calling the time.

1➡ Begin by sitting in a relaxed position while your partner takes a 10-second carotid pulse.

2➡ Multiply this number by 6. Record this value on the following chart as your base pulse.

3➡ You will be running in place for 10 ten-second innings, with 10-second rest periods between the innings.

4➡ Your partner will record the data and will keep time (unless your instructor is keeping time for the entire class).

5➡ As you run in place, count the number of times your right foot strikes the floor. Run as fast as you can. As you finish each inning, tell your partner how many right-foot hits you had. Your partner can record this number in the appropriate place.

6➡ Rest for 10 seconds and repeat.

7➡ Ten seconds after finishing the 10th inning, your partner should take your carotid pulse again. Your partner should again take the carotid pulse at 2 minutes, 4 minutes, and 6 minutes after the exercise is complete. Record the data.

Base Pulse _____ Steps Inning 1 _____
After Pulse _____ Steps Inning 2 _____
2-Min. Pulse _____ Steps Inning 3 _____
4-Min. Pulse _____ Steps Inning 4 _____
6-Min. Pulse _____ Steps Inning 5 _____
Steps Inning 6 _____
Steps Inning 7 _____
Steps Inning 8 _____
Steps Inning 9 _____
Steps Inning 10 _____

8➡ In the first graph (Figure 51–9a●), plot your number of steps against the innings.

What fluctuations do you observe? How can you explain any gross fluctuations?_____

B. General Fatigue Curve

1➡ Using the data you just collected, create a second graph (Figure 51–9b●) by plotting the pulse rate against time. This is a **general fatigue curve**. Your "before exercise" pulse should be considered your baseline value.

Does your curve surprise you? Why or why not?

2➡ At about minute 4 you may notice a dip below the baseline level.

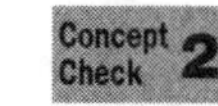
What do you suppose accounts for this dip?

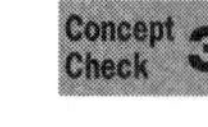
What does it mean if your pulse rate is not back to normal after 6 minutes?

How does your cardiovascular efficiency stack up against the cardiovascular efficiencies of your classmates? Did anything surprise you about this test?

❑ Additional Activities

1. Examine slides of the SA and AV nodes. Compare these fibers with the fibers of the cardiac muscle fibers.
2. Explore the roles of fast sodium channels, slow calcium-sodium channels, and potassium channels in the cardiac electrical system.
3. Research venous pressure. Find out the different standard values of venous pressure based on hydrostatic pressure. Where might you find negative hydrostatic pressure in the body?
4. Find out how the "blood pressure machines" found at many pharmacies and department stores operate. How accurate are these machines?

NOTES

5. Find out the physiological reason why a standard sphygmomanometer cuff should not be used for a small child or a very large person.

NOTES

Answers to Selected Concept Check Questions

1. Blood pressure can be influenced by almost anything — physical condition, diet, emotional stress, illness, environmental factors.
2. Your body is "over-compensating" for the shift in activity. This homeostatic mechanism is usually normal — an example of the body striving toward maintaining equilibrium.
3. It may mean nothing. Or it may mean you are out of condition. Other factors, some as simple as the common cold, can influence your return to "normal."

❑ Lab Report

1. Referring to the numbered inquiries at the beginning of this exercise, complete the following box summary:

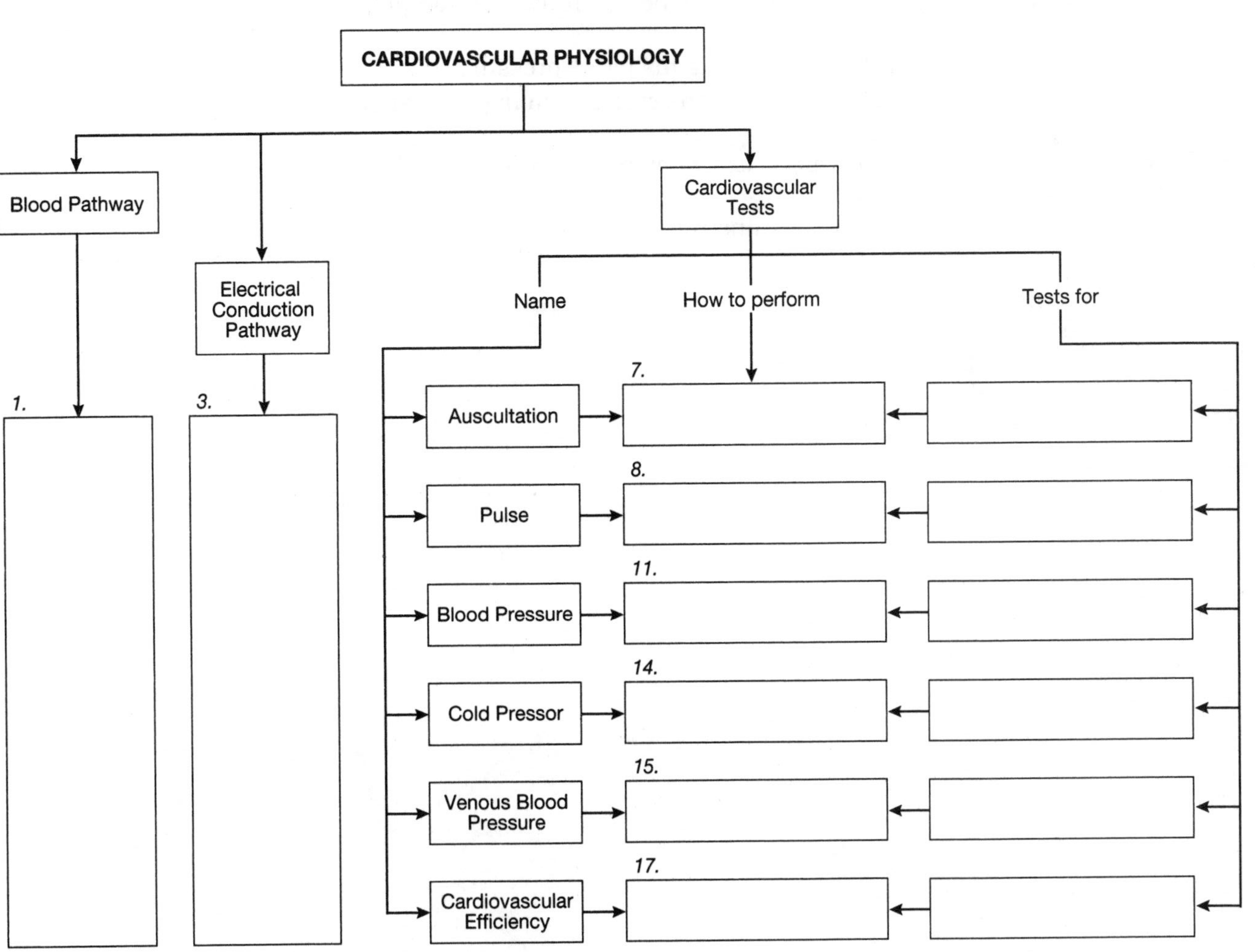

General Questions

NOTES

1. How many times per year does your heart beat?
2. Fill in this chart:

	Parts of Cycle	Events of Cycle
1		
2		
3		
4		

3. What are the names of the most easily studied pulse points?
4. In addition to the number of beats per minute, what does a diagnostician check for when taking a pulse?
5. Why can the diastolic pressure never be higher than the systolic pressure?
6. Dave's blood pressure is 160/105. What is his pulse pressure?
7. Annie's blood pressure is 170/110. Does she suffer from hypertension? Explain.
8. How does the anatomy of the typical heart cell help us understand the syncytial nature of the heart?
9. How does a general fatigue curve demonstrate cardiovascular efficiency?

EXERCISE 52

Electrocardiography

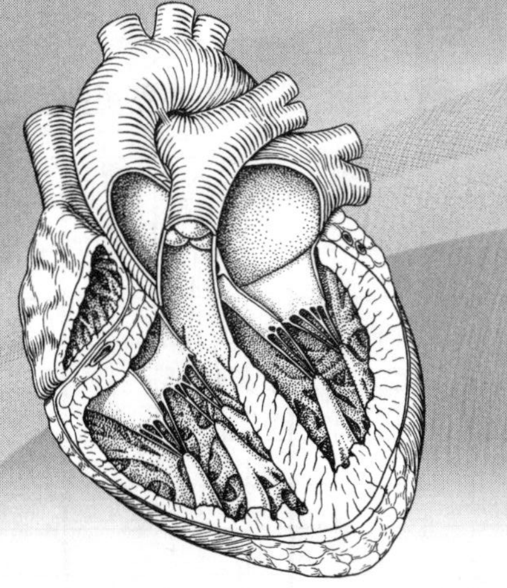

PROCEDURAL INQUIRIES

Preparation

1. What is the difference between an electrocardiogram and an electrocardiograph?
2. What does the horizontal axis of an ECG measure?
3. What does the vertical axis of an ECG measure?
4. What do P, Q, R, S, and T stand for?
5. What are the major landmarks of the ECG? What does each signify?

Experimentation

6. What is a bipolar limb lead?
7. What is Einthoven's Triangle?
8. What are the three standard leads and how is each set up?
9. What is Einthoven's Law?
10. What is the electrical axis of the heart?
11. How can you determine the electrical axis of the heart?

Additional Inquiries

12. What type of stimulation is responsible for cardiac contraction?
13. What are the precordial leads?

Key Terms

Bipolar Limb Leads
Einthoven's Law
Einthoven's Triangle
Electrocardiogram
Electrocardiograph
PQRST
Precordial
Standard Lead

Materials Needed

Electrocardiograph

In Exercise 51, we established that the heart cells contract because of stimulation from the electrical conduction system. Although modulated by the autonomic nervous system, this stimulation originates at the intrinsic pacemaker, the sinoatrial (SA) node.

The SA node initiates the depolarization, which spreads across the atria. The atrioventricular (AV) node is stimulated and depolarization spreads throughout the ventricles. Meanwhile the atria repolarize. The ventricles repolarize after ventricular polarization is complete.

By examining the electrical activity of the heart we can gain an insight into the particular events of cardiovascular physiology. Electrocardiography is a method used for studying the electrical nature of the heart. In this laboratory exercise we will become familiar with the essentials of electrocardiography.

❑ Preparation

I. Background

A. Electrocardiography

The **electrocardiograph** is the machine used to record the **electrocardiogram** (**ECG** or **EKG**). An ECG is a statement of the electrical activity of the heart. The ECG does not record the specific mechanical events perpetrating that activity. Therefore, any diagnosis made from an ECG is necessarily made by inference.

Numerous types of electrocardiographs are in use around the country. These range from some very old physiographs and rather expensive oscilloscopes to some very modern computer-interfaced ECG modules. Your instructor will explain the type of electrocardiograph you will be using. Regardless of how new or how old your machine is, the basic principle of measuring electrical conduction is the same.

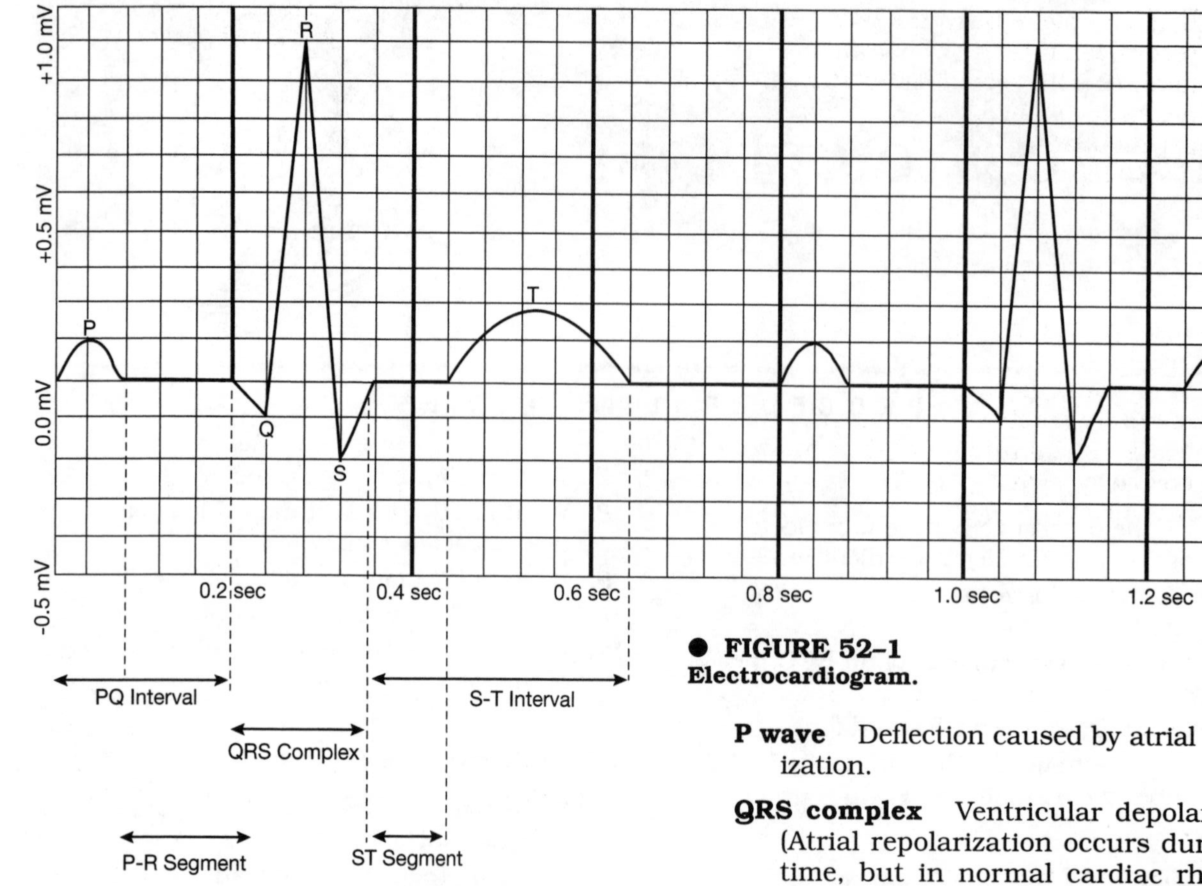

● **FIGURE 52–1**
Electrocardiogram.

P wave Deflection caused by atrial depolarization.

QRS complex Ventricular depolarization. (Atrial repolarization occurs during this time, but in normal cardiac rhythm is masked by the ventricular event.)

T wave Deflection caused by ventricular repolarization.

P–Q interval (also known as P-R interval) Time between beginning of atrial depolarization and beginning of ventricular depolarization.

Q–T interval Time for singular ventricular depolarization/repolarization cycle.

S–T segment Time of depolarized state.

The ECG is traced on quadrille lined paper or on a computer screen. The computer screen may or may not be graphed, depending on the computer program. The darkened blocks on the paper strip are 5 mm square.

As the readout progresses, the horizontal axis of the strip records time. The standard run time is 25 mm per second, although the tracing speed can be altered to examine some particular aspect of the ECG.

The vertical axis records electrical activity in millivolts. The baseline value — as the pen records no activity — measures the relative electronegativity of the resting heart muscle. As the ECG progresses, an upward deflection records a decrease in electronegativity (the membrane becomes more positive); a downward deflection records an increase in electronegativity (the membrane becomes more negative). Electrocardiographs are regularly set so that each 10 mm vertical deflection represents 1 millivolt. Electrical sensitivity can also be altered for specific purposes.

B. Reading the ECG [∞ Martini, p. 673]

1➔ You are undoubtedly familiar with the basic ECG recording. Study the generic tracing in Figure 52–1● and acquaint yourself with the major landmarks of the ECG. Make note of what each section means in terms of time and electricity.

C. Calculations from an ECG Recording

1➔ If the run speed on this ECG tracing is 25 mm per second (which is 25 1mm squares per second), calculate the duration of each following wave and interval.

P wave ______________
QRS complex ______________
T wave ______________
P-Q interval ______________
Q-T interval ______________
S-T segment ______________

2➔ The generic heartbeat lasts 0.8 seconds. How many blocks on the graph would be used per beat? Per minute?

Blocks/beat: ______________
Blocks/minute: ______________

3➡ If a complete beat takes 30 blocks to record, what is the heart rate in beats per minute? If the recording registers 20 blocks per beat, what is the heart rate in beats per minute?

30 blocks to record: ________ beats per minute

20 blocks to record: ________ beats per minute

4➡ What problems might you foresee in reading the ECG from someone experiencing bradycardia (slow rate, below 60 beats per minute) or tachycardia (fast rate, over 100 beats per minute)? (Neither bradycardia nor tachycardia is necessarily indicative of pathology.)

____

____

____

5➡ Based on the information in the preceding sections, how many millivolts of electricity are represented by each deflection (wave) on the ECG?

P wave ________

QRS complex ________

T wave ________

❑ Experimentation

II. Set Up

It is not the purpose of this laboratory exercise to give clinical interpretations to the ECGs that may come up in your class. Keep in mind that you are using classroom laboratory equipment that may not have the accuracy found in certified health care instruments. Also, classroom atmosphere can influence clinical readings. If you are worried about a reading obtained on these laboratory machines, please consult an appropriate health care professional.

A. Arrangement of the Limb Leads

Follow your instructor's directions for attaching yourself (or your lab partner) to the electrocardiograph.

Three standard limb leads can be used with most people. (Metallic prostheses and artificial joints invalidate the limb leads. Alternatives for obtaining ECGs do exist. For instance, the precordial leads mentioned below can be used.)

1➡ For this exercise we will be using **bipolar limb leads**. We will be attaching electrodes to two different body limbs so that our ECG tracing will record the relative difference in electronegativity between these two points.

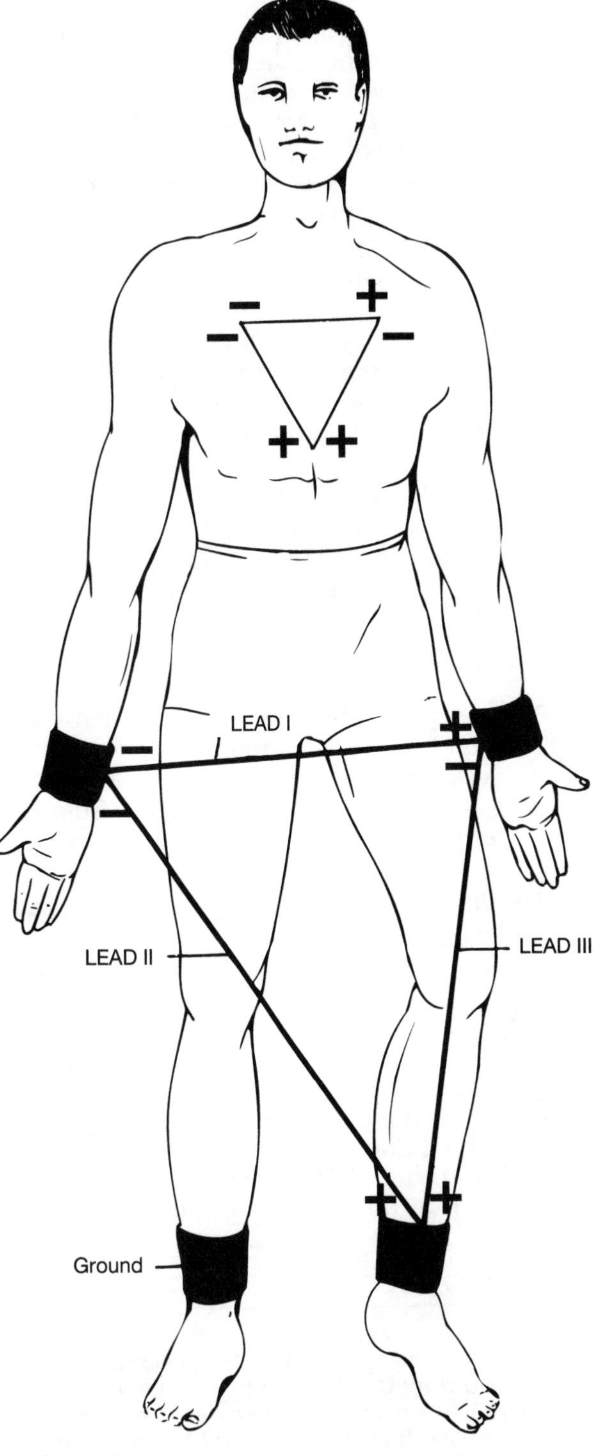

● **FIGURE 52–2**
Einthoven's Triangle and Standard Lead Triangle.

2➡ Please see Figure 52–2●. This figure summarizes the leads. Notice the triangle around the figure's heart. This is **Einthoven's Triangle**, a diagrammatic illustration of the arms and left leg connecting electrically with the fluids surrounding the heart.

3➡ Pay attention to the "+" and "–" signs on the lower triangle. These signs show how you should connect the electrodes to achieve the **standard lead** in question.

Lead I Right arm negative; left arm positive.

Lead II Right arm negative; left leg positive.

Lead III Left arm negative; left leg positive.

4➡ Remember, you are only connecting two of the three points of the triangle. With each of these leads the right leg is always connected to the ground wire.

5➡ Try taking a reading with only one lead connected at a time (make sure the ground is connected, or you won't get a proper reading). Make note of which reading is which. SAVE these readings for a later exercise in Section IIIB.

6➡ Set up the three leads (lead I, then lead II, then lead III) and ground (always right leg) according to Figure 52–2● and your instructor's directions. Take readings from each lead.

7➡ What would happen if you reversed the polarity on the leads? Try it and see.

a. How do you explain this phenomenon?

______

______

______

8➡ What would happen if you grounded your left leg and used your right leg as part of the triangle? Try it and see.

a. How do you explain this phenomenon?

______

______

______

9➡ If your electrocardiograph is computerized, you may be able to superimpose your own ECGs on one another. If your electrocardiograph is not equipped for superimposition, compare the readings by hand. Superimpose your ECG on a classmate's ECG from the same lead.

a. What similarities and differences do you notice?

______

______

______

B. Einthoven's Law

Einthoven's Law states that if the electrical potential of any two of the three bipolar leads are known, their algebraic sum will equal the electrical potential of the third lead. (Stress algebraic sum; you may have to switch a sign.) Study your tracings; count the squares; make note of positive and negative electrodes.

1➡ Can you validate Einthoven's Law? In the space below, show your calculations.

______

______

______

C. Additional Leads

Sometimes in addition to limb leads, chest wall (**precordial**) leads are used. For this type of reading, the chest electrode is connected to the positive terminal of the electrocardiograph while the negative electrode is simultaneously connected to the right arm, left arm, and left leg. Readings from six major chest points are normally taken. The readings are not identical to our standard lead readings, but the information can be extrapolated. We will not be studying this precordial setup.

III. Electrical Axis of the Heart

A. "Normal" Electrical Axis

By using the standard leads, you can determine the electrical axis of your heart. If your heart is angled like the textbook diagrams, your mean electrical axis will be about 59 degrees off the horizontal, although any axis from 20 degrees to 100 degrees may be normal. Check Figure 52–3●. (*Note:* If the numbers seem "backwards" to you, recall Einthoven's Triangle. The angle of the axis is in relation to lead I.)

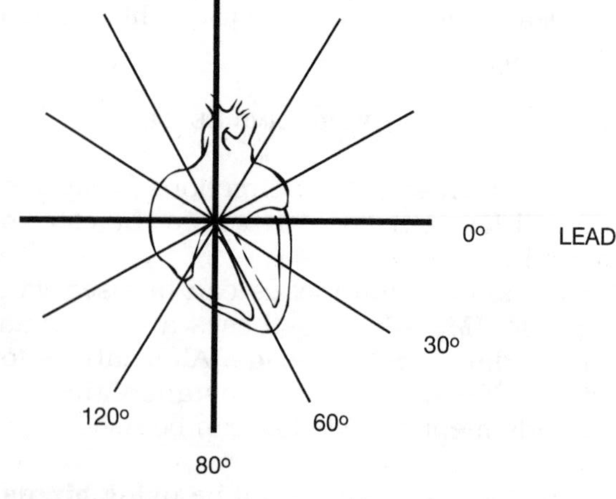

● **FIGURE 52–3**
Electrical Axis of the Heart.

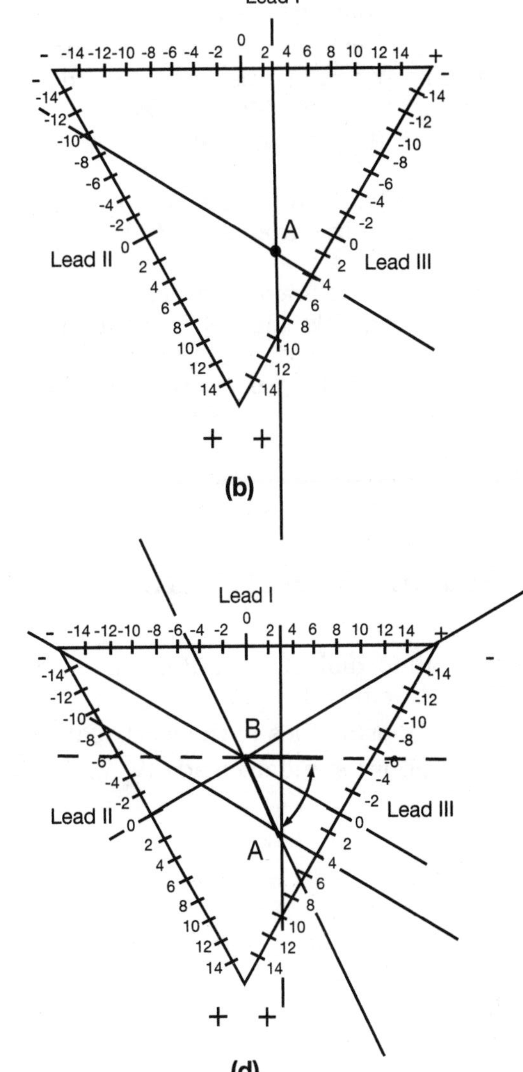

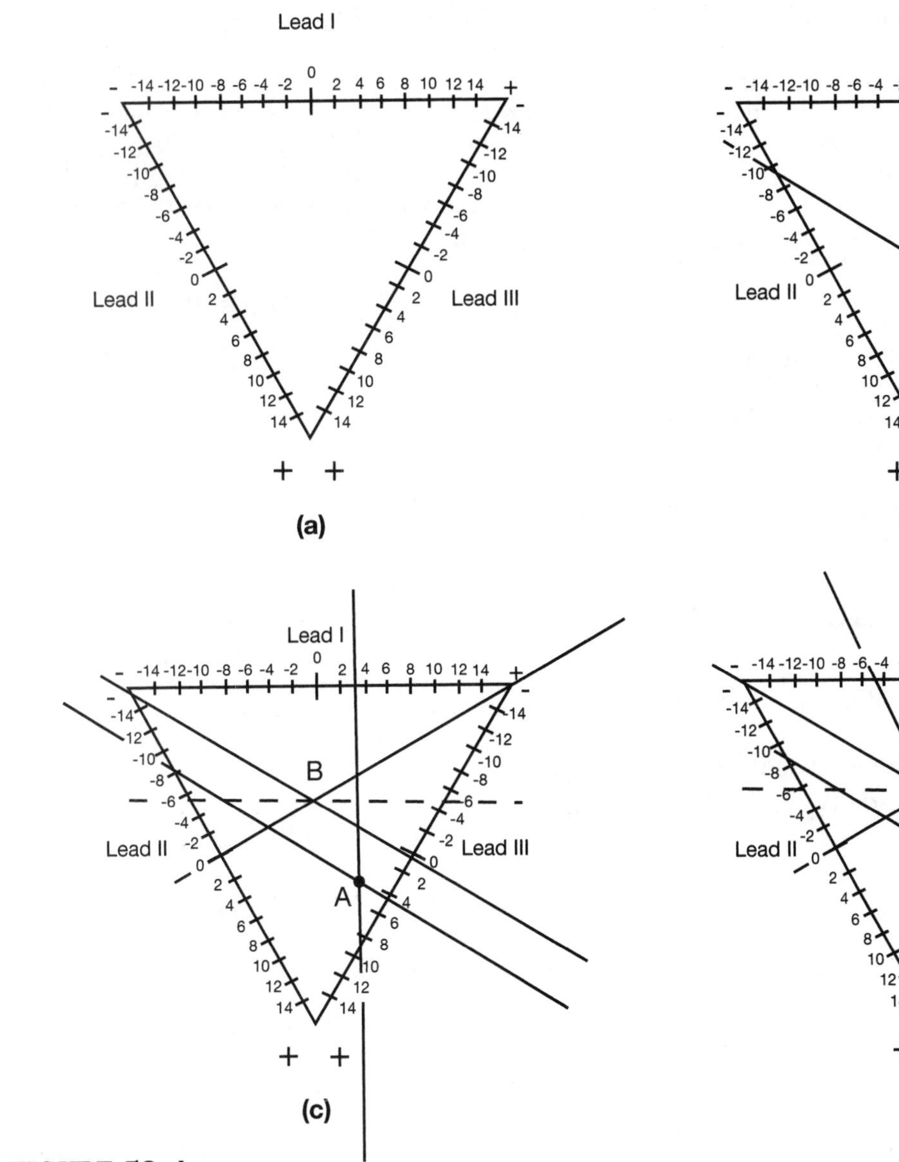

● **FIGURE 52–4**
Computing the Electrical Axis of the Heart.

Normal deviations from the standard can be based on anatomical differences, anomalies in the distribution of the Purkinje fibers, unusual angulation of the heart and lungs, or even on unusual physique. *Abnormal* deviations can be caused by such pathological conditions as ventricular hypertrophy or bundle branch blockage.

B. CALCULATING THE ELECTRICAL AXIS: AN EXAMPLE

1➡ To determine the electrical axis of your heart, you only need the values from two leads. For this example we will arbitrarily use lead I and lead III.

2➡ Draw an equilateral triangle. Bisect each side and label as indicated in Figure 52–4a●. (Your unit values should all be equidistant on the triangle, but any arbitrary measurement unit will do.)

3➡ Take the QRS complex from lead I. Let us assume the negative deflection (the lowest point below the baseline) is 0.2 mV. Let us further assume the greatest positive deflection (the highest point above the baseline) is 0.5 mV. Our net deflection then is +0.3 mV.

4➡ Since 1 mV = 10 mm, 0.3 mV = 3 mm. Mark the point +0.3 mm on the lead I line. Drop a perpendicular line.

5➡ Repeat this procedure with your net value from lead III. For this example, our net value is +0.4 mV or 4 mm.

6➡ Label the point where the two perpendiculars cross as point A. Draw a line perpendicular to lead II through point A (Figure 52–4b●).

7➡ Now determine the geometric center of your triangle. This can be done by bisecting the vertices and extending the line to the midpoint on the opposite side of the triangle. Label this geometric center as point B (Figure 52–4c●).

8➡ Draw a line parallel to lead I through point B. Draw a line from A to B.

9➡ Measure the angle between the lines you just drew. This angle is the mean electrical axis of your heart. In our example the angle is about 64 degrees (Figure 52–4d●).

C. Determine the Electrical Axis of Your Own Heart

1➡ Determine the electrical axis of your own heart by following the preceding directions. If your heart is functioning normally, this axis will correspond with the anatomical angle of your heart.

2➡ Do a class comparison of the electrical heart axes.

❑ Additional Activities

1. Explore abnormal ECG readings. Learn why certain abnormalities give the ECG tracings they do.
2. Explore abnormalities in the electrical axis of the heart.
3. Find out about standard leads IV, V, and VI.

❑ Lab Report

1. Referring to the numbered inquiries at the beginning of this exercise, complete the following box summary:

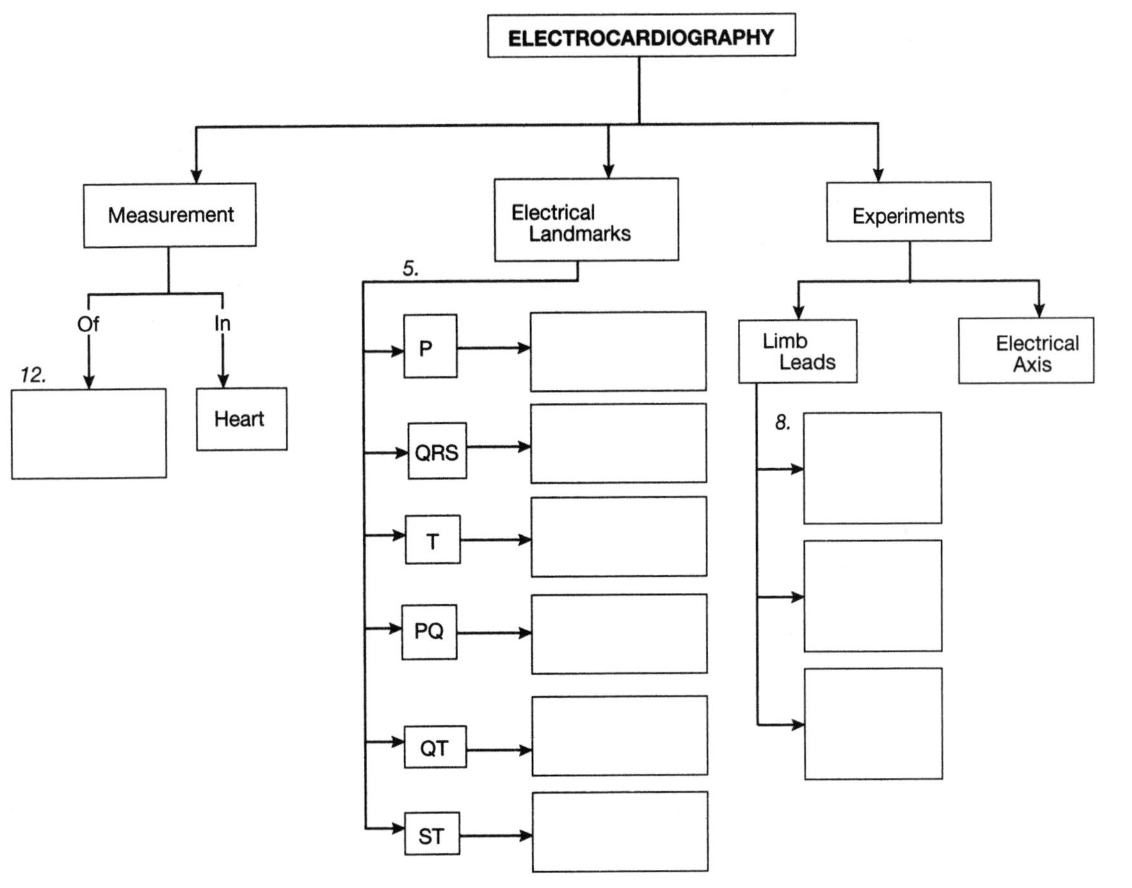

General Questions

1. How do you set up the electrocardiograph in your laboratory?
2. Fibrillation occurs when the heart is undergoing rapid uncontrolled contractions. Fibrillation can be atrial or ventricular. How could you identify fibrillation on an ECG? How does a defibrillator work?
3. What is Einthoven's Triangle? Einthoven's Law?
4. What is a bipolar limb lead?
5. What is a precordial lead?
6. Explain the principles involved in taking standard lead measurements.
7. Explain the principle behind identifying the electrical axis of the heart. Of what value is this information?
8. What are some practical applications of our knowledge of electrocardiography?

EXERCISE **53a**

Dissection of the Cardiovascular System: Cat

PROCEDURAL INQUIRIES

Preparation

1. How should you prepare your cat for cardiovascular dissection?
2. Which instruments should be used for dissection?

Dissection

3. What are the major vessels of the heart?
4. What are the major arteries associated with the aortic arch?
5. What are the major arteries of the head and neck?
6. What are the major arteries of the forelimb?
7. What are the major arteries of the thorax?
8. What are the major arteries of the abdomen?
9. What are the major arteries of the hind limb?
10. What are the major veins associated with the superior vena cava?
11. What are the major veins of the head and neck?
12. What are the major veins of the forelimb?
13. What are the major veins of the thorax?
14. What are the major veins of the abdomen?
15. What are the major veins of the hind limb?

Additional Inquiries

16. How are blood vessels named?
17. How do the blood vessels of the cat compare with the blood vessels of the human?

HUMAN CORRELATION: AM Cadaver Atlas in Applications Manual.

Key Terms

No specific key terms are given for this exercise. The key terms are the major blood vessels. Your instructor will indicate which vessels you are responsible for.

Materials Needed

Gloves
Preserved Cat (Double-Injected)
Dissecting Tray
Dissecting Kit

The cardiovascular system, which includes the heart, the blood vessels, and certain accessory structures, is essentially the same in all mammals.

Oxygenated blood leaves the heart by the aorta and travels through specific parts of the arterial system to each of the organs and tissues of the body. At the tissue level, the arteries break down to form a capillary bed. Gas and nutrient exchange takes place at the capillary level. The capillaries converge to form the venous system, which carries the blood back to the heart.

In this laboratory exercise we will further our understanding of cardiovascular anatomy by dissecting the arterial and venous systems of the cat.

You should have no particular difficulty in identifying the major arteries and veins and in locating the principal branches off these vessels. Refer back to Exercise 50 for comparative human vessels.

❑ Preparation

I. Background

You will probably be working with a partner in your dissection of the circulatory system of the cat. It is important that you and your partner share your ideas as you identify the different blood vessels.

It is also important that you quiz each other as you go along. Work on gaining a total perception of cardiovascular anatomy.

A. Preparing the Cat

1➡ Obtain your cat and prepare it according to your instructor's directions. Place the cat ventral side up on your dissecting tray.

2➡ It is assumed that you have already made your initial incision and reflected back at least part of the ventral skin. If you have not previously done so, you should follow these procedures to cut the muscle layers and open the thoracic and abdominal cavities.

3➡ Begin your incision in the groin area. Snip the abdominal muscle layers and make a transverse cut across the groin, keeping the blunt end of your scissors on the inside of the animal's body. Make a midline ventral incision from the pubic bone to the diaphragm. Note Figure 53a–1●.

4➡ It may be necessary to rinse the abdominal cavity because of an accumulation of preservatives or other injected fluids. **Do not** dispose of solid products in the sink.

5➡ When you reach the diaphragm, note its domed appearance. Continue your incision slightly to the right of an imaginary midventral line. (Directly in the center is the sternum, which may be difficult to cut with your regular scissors.)

6➡ Make a transverse incision across the shoulders.

7➡ To open the thoracic cavity, it will be necessary to cut the diaphragm. Note both surfaces of the diaphragm. Now, make a transverse incision along the ventral wall on each side of the cat. Cut the diaphragm as necessary.

8➡ Spread the rib cage apart to reveal the pleural and cardiac cavities. When you begin working in the thoracic cavity, you will probably notice that the cavity does not stay open very well. You can solve this problem by breaking the ribs. Stand above your specimen and, with the heels of your hands on the spread-apart ribs, press downward forcefully several times.

B. Major Organs

1➡ Orient yourself to your cat. Locate the major internal organs: lungs, heart, stomach, small intestine, large intestine, kidney, spleen, pancreas, liver, gallbladder, testes or uterus.

2➡ Note the glandular organ covering part of the trachea above the heart. This is the **thymus**. Unless your instructor directs otherwise, you can remove the thymus.

C. Dissecting Hints

The major impediment to good cardiovascular dissection is separating the vessels from the connective tissue. Be certain you understand your instructor's directions about other tissues and systems you may be working with in later exercises. (For instance: Should you be concerned about innervation? Or will you be required to locate any lymphatic vessels?)

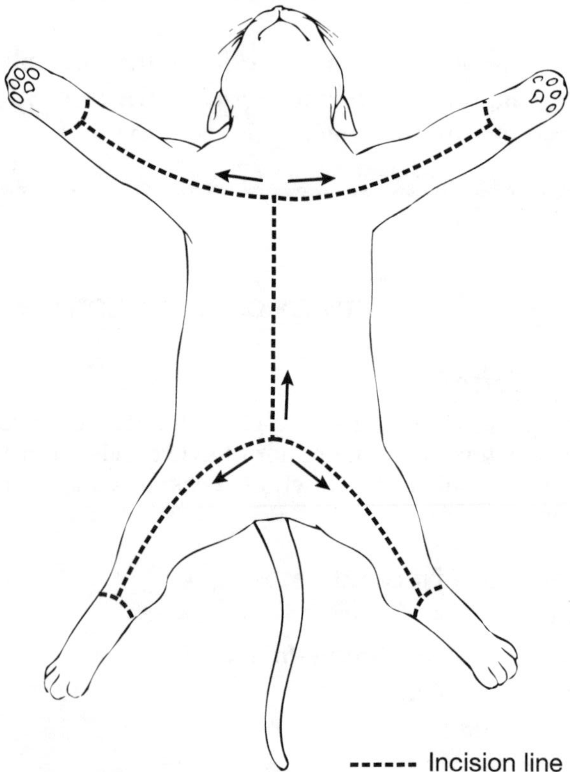

● **FIGURE 53a–1**
Opening the Cavities of the Cat.

Your cat has probably been injected with latex — red latex for the arterial system and blue latex for the venous system. The red and blue latex do not mix because the vessels of the capillary bed are too narrow to permit the passage of the latex.

The usual points of injection are the neck and the groin. You may find the blood vessels in these regions to be slightly disturbed. You should be able to work around this with no particular difficulty. Occasionally a blood vessel will explode from the latex injection. If your cat has this problem, simply clean out the latex and continue with your dissection. If the problem is serious (which is highly unlikely), show your instructor. Certain blood vessels do not inject as well as others, though usually this will cause you no particular problem. If you do have difficulty, inform your instructor and look at a classmate's cat.

In this laboratory exercise a number of blood vessels that you may not be able to find are included in the write-up. These are generally the smaller vessels that do not always inject well. These vessels are included here for your information — because you might find them and because they are necessary vessels — and not for your positive identification. **These vessels are marked with asterisks.**

Unless you are directed otherwise, it will only be necessary for you to dissect the vessels on one side of the body.

It is important for you to follow the diagrams and illustrations in identifying the blood vessels. Do not limit yourself to the pictures, however. Read the descriptions carefully and coordinate the written word and the illustrations with what you observe on your cat.

D. Isolation of the Vessels

Once you are certain what you are required to do, work at separating the blood vessels from the surrounding tissue. The blood vessels are quite elastic. If you use your blunt nose probe to clear out the connective tissue, you can actually be quite forceful without harming the vessel. The smaller vessels are more delicate.

Your blunt nose probe is your instrument of choice. You will also be able to pull the vessels as well as slide the probe under and lift the vessels without any particular danger of causing irrevocable damage.

Your sharp nose probe should be used sparingly, if at all. Sharp nose probes can easily damage vessels. Sharp nose probes can also tear other tissues and make your identification more difficult.

E. Nomenclature

Most blood vessels are named either for their location or for the part of the body they service. Keep this in mind as you trace the major vessels and as you identify the branches off the vessels.

F. Anomalies

The blood vessels of the cat, particularly those of the venous system, may exhibit some interesting anomalies. For instance, the common iliac veins may bifurcate (split or divide) in the renal area instead of the pelvic area. Or, the internal jugular vein or the brachiocephalic artery may be severely reduced (or even absent). You should be aware that anomalies can exist. Your goal, however, should be to understand the pathways of the major blood vessels, regardless of anomalies.

Dissection

II. Cardiac Dissection

A. Heart Review

Cardiovascular dissection normally begins with the heart and follows the major paths of the great vessels to and from the head, forelimbs, and lower body. That pattern is followed here.

1➡ Note the heart. The principles of cardiopulmonary anatomy are covered in Exercises 48 and 49. If necessary, go back and review the major structures. You should be able to identify these structures in your cat.

B. Heart Observations

1➡ Observe the heart from the ventral aspect. Remove the thin, tough **pericardial sac** that surrounds the heart. Note any particular points of attachment.

2➡ Locate the cardiac chambers and the great vessels as found in Exercises 48 and 49. Identify the **ligamentum arteriosum**.

3➡ Note the red vessels on the surface of the heart. These are the **coronary arteries**, which supply blood to the heart muscle itself. These arteries arise from tiny vessels that exit from the base of the aorta.

The blue surface vessels are the **coronary veins**, which drain the heart muscle. The coronary veins culminate in the **coronary sinus**, which empties directly into the right atrium on the dorsal aspect of the heart.

4➡ If your instructor so directs, make a frontal (coronal) slice through the heart starting at the apex (bottom) and ending just before the great vessels.

5➡ Identify the internal structures of the heart.

III. Arterial System

***You may not be able to find the vessels marked with an asterisk.**

A. Terminology

Since arteries carry blood away from the heart and toward an organ or a body region, we often use terms such as "supplies" or "services" or "goes to" when discussing the arterial system.

1➔ Use Figures 53a–2a● and 53a–2b● to help you identify the major arteries.

B. Aortic Arch

1➔ To begin your study of the arterial system, return to the heart and locate the **aorta**. The portion of the aorta leaving the heart is the **ascending aorta**. The hooked portion is the **aortic arch**, and the portion continuing caudally along the dorsal wall is the **descending** or **thoracic aorta**. When the aorta passes into the abdominal cavity, it becomes known as the **abdominal aorta**.

2➔ Observe the two vessels exiting the aortic arch. The right vessel (the cat's right) is the **brachiocephalic** or **innominate artery**. Follow this vessel cranially. Note the bifurcation where the vessel splits to form the **left** and **right common carotid arteries**, which travel along either side of the trachea.

The third branch of the brachiocephalic artery is the **right subclavian artery**, which in many animals appears to branch off the right common carotid artery.

3➔ Please note that in humans, three arteries branch off the aortic arch: the left subclavian; the left common carotid; and the brachiocephalic, which gives rise to the right common carotid and right subclavian arteries.

4➔ The left branch exiting the aortic arch is the **left subclavian artery**. Note that as the subclavian arteries turn toward the arms, the branches off these vessels become approximately identical.

C. Head and Neck

1➔ Return to the common carotid artery. You may notice the **inferior (posterior) thyroid artery**, which arises just cranial to the carotid origin.

The **superior (anterior) thyroid artery** arises near the thyroid cartilage and supplies the thyroid gland and certain ventral muscles.

Immediately anterior to the superior thyroid artery is the **occipital artery**, which supplies certain neck muscles.

2➔ You will notice that at the laryngeal (larynx) border the common carotid artery bifurcates to form the **external** and **internal carotid arteries**. The larger external carotid artery supplies the external cranial structures via the **lingual**,* **auricular**,* and **temporal arteries**,* while the internal carotid artery branches to supply the eye, forehead, nose, and brain via the Circle of Willis. You will not be able to find most of the latter arteries.

D. Forelimb

1➔ Return to the subclavian artery. The first branch off the subclavian artery is the **vertebral artery**. Students are sometimes amazed at the relatively small size of this artery, especially considering that it passes through the cervical vertebrae and services the brain.

The **costocervical artery** supplies the deep muscles of the neck and back. Branches of this artery supply the first two intercostal spaces.

The **internal mammary artery** supplies the ventral body wall.

The **thyrocervical artery** arises from the cranial surface of the subclavian and supplies blood to the neck and shoulders. At the scapula the thyrocervical artery becomes known as the **transverse scapular artery**.

2➔ Look at the level of the first rib, where the subclavian artery becomes the **axillary artery**. The first branch off the axillary artery is the **ventral external thoracic artery**, which supplies the pectoral muscles.

The **long thoracic artery** passes caudally to the pectoral and latissimus dorsi muscles.

The large **subscapular artery** gives rise to the **thoracodorsal artery** and the **posterior humeral circumflex artery**.

3➔ Distal to the subscapular artery, at the level of the teres major muscle, the axillary artery becomes the **brachial artery**.

The first branch of the brachial artery is the **anterior humeral circumflex artery**,* which you probably will not be able to locate. The **deep brachial artery*** passes under the belly of the biceps muscle.

● **FIGURE 53a–2a**
Arteries of the Chest and Neck.

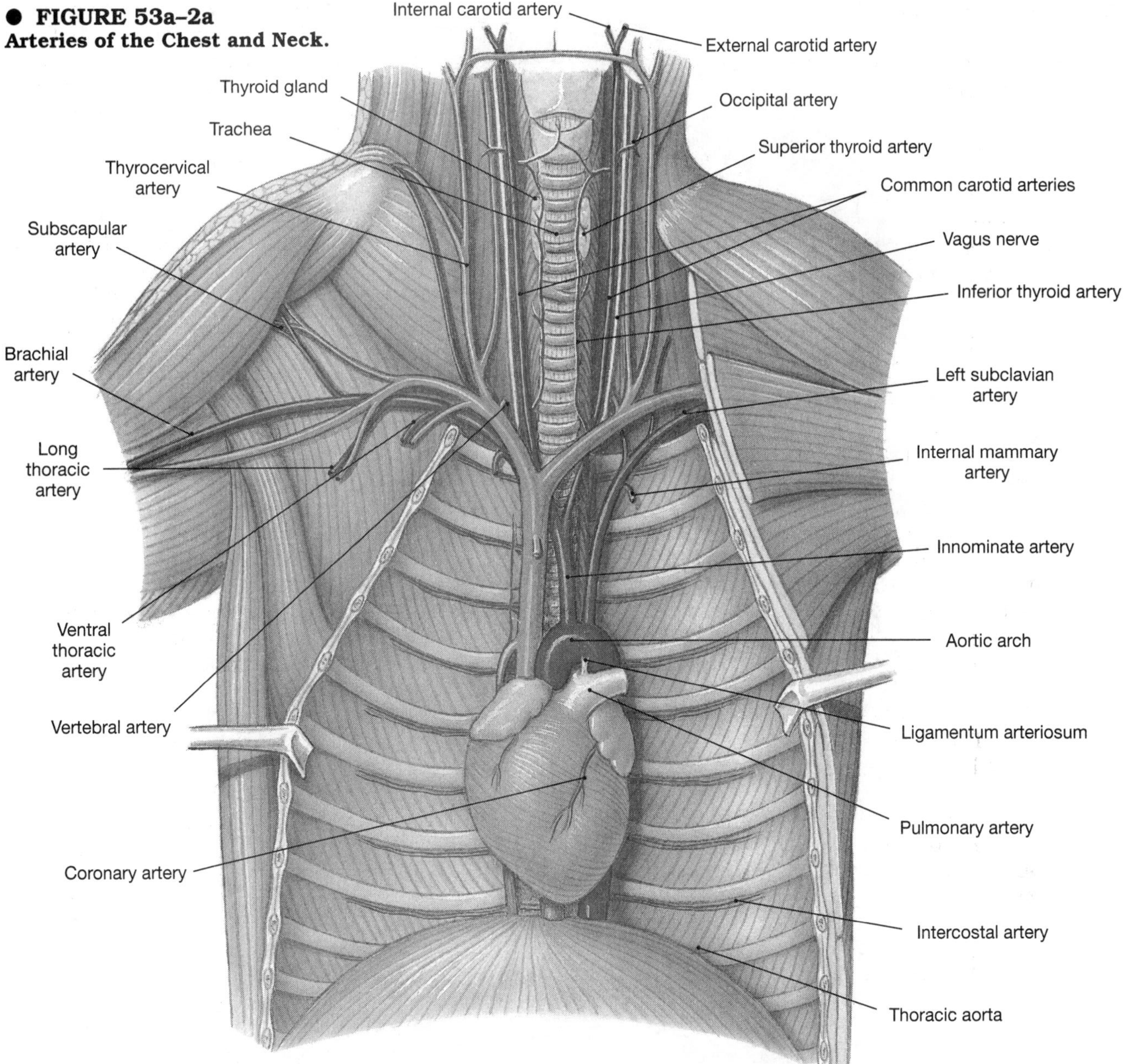

Proximal to the elbow you may or may not be able to locate the **superior radial collateral artery*** and the **ulnar collateral artery**.* Distal to the elbow the brachial artery becomes the **radial artery**,* which subsequently gives rise to the **ulnar artery**.*

How does this compare with the human arterial system?

____

____

E. Thoracic Aorta

1 ➡ Return to the aorta. Pull the thoracic visceral organs to the right (the cat's right) and follow the aorta caudally through the thorax.

2 ➡ Notice the ten pairs of **intercostal arteries** that supply the intercostal muscles between the last eleven ribs. The first and second intercostal spaces are supplied by the costocervical artery. The **bronchial arteries** supply the bronchi as they enter the lung. Several **esophageal arteries** of varying origins supply the esophagus.

3 ➡ Continue to follow the aorta caudally. At approximately the second lumbar vertebra the thoracic aorta passes through the diaphragm at the aortic hiatus and becomes the abdominal aorta.

F. Abdominal Aorta

1 ➡ As you follow the aorta into the abdominal cavity, the first major branch inferior to the diaphragm is the unpaired **celiac artery**. The

● **FIGURE 53a–2b**
Arteries of the Abdominal Cavity.

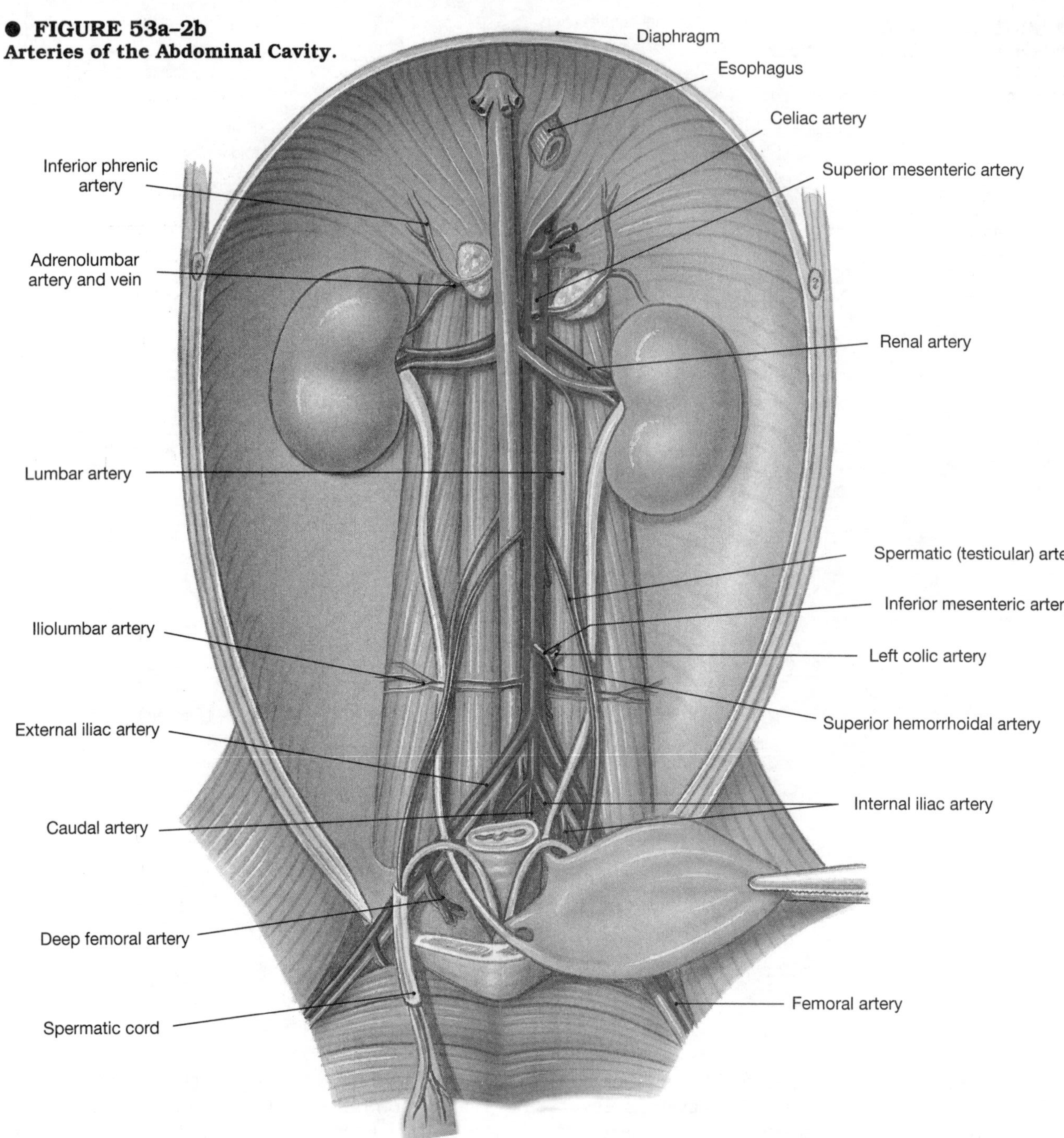

celiac artery has three major branches, the **hepatic**, **left gastric** and **splenic** arteries.

2➡ Identify the above arteries by tracing them to their respective organs. Green vessels in this area are bile ducts; they tear easily.

3➡ Continue caudally. The unpaired **superior mesenteric artery**, the next larger branch off the aorta after the celiac, supplies the small intestine and portions of the large intestine.

4➡ The paired **adrenolumbar arteries** supply the adrenal glands (**adrenal artery**), the diaphragm (**inferior phrenic artery**), and certain muscles of the dorsal body wall. Variations in the branching patterns of the adrenolumbar arteries are common.

The paired **renal arteries** supply the kidneys and sometimes the adrenal glands.

The paired **spermatic (testicular) arteries** in males or paired **ovarian arteries** in females arise near the posterior end of the kidneys. The spermatic arteries service the scrotum as well as the testes, and the ovarian arteries supply the ovaries as well as the uterus (by anastomosing with the **uterine artery**, via the **middle hemorrhoidal artery**,* both deep branches off the internal iliac artery). Depending on your cat, you may or may not be able to identify these branches.

The single **inferior mesenteric artery** branches to form the **left colic artery**, which supplies the descending colon, and the **superior hemorrhoidal artery**,* which supplies the rectum.

The seven pairs of small **lumbar arteries** supply the abdominal wall. The large paired **iliolumbar arteries** supply dorsal abdominal wall muscles.

The large paired **external iliac arteries** pass through the body wall and become the **femoral arteries** in the legs.

The caudal-most paired arteries are the **internal iliac arteries,** which supply the gluteal muscles, the rectum, and the uterus in the female.

The abdominal aorta continues on into the tail as the **caudal** (or **median sacral**) **artery**.

How do the human iliac arteries compare with the feline iliac arteries? ______________________

____

____

G. Hind Limb

1 ➡ Return to the external iliac artery. Just before the artery leaves the body cavity, find the branch, the **deep femoral artery**, which continues to branch to supply the bladder, ventral abdominal wall, and some of the femoral muscles.

Several branches, the **lateral femoral circumflex artery**,* the **superior articular artery**, and the **saphenous artery**, exit the femoral artery to serve the upper leg. When the femoral artery reaches the posterior knee, it becomes known as the **popliteal artery**. The popliteal artery bifurcates to form the **anterior** and **posterior tibial arteries**, which supply the lower leg and foot.

IV. Venous System

***You may not be able to find the vessels marked with an asterisk**.

A. Terminology

Because veins carry blood toward the heart from an organ or region of the body, we often use terms like "entering" or "drains" or "empties into" when describing the venous system.

1 ➡ Use Figures 53a–3a● and 53a–3b● as your guide to venous dissection.

B. Superior Vena Cava (Precava)

1 ➡ Begin your dissection of the venous system of the cat by locating the large vein coming from the head region and entering the right side of the cat's heart (right atrium). This is the **superior vena cava**, also known in the cat as the **precava**. (A general term for the superior vena cava in the carnivore is **cranial vena cava**.)

The first vessel entering the superior vena cava from the animal's right is the **azygos vein**, a multibranched vessel draining the **intercostal veins**, the abdominal wall muscles, the **esophageal veins**, and **bronchial veins**.

The **internal mammary (sternal) vein** enters from the ventral wall. The superior vena cava is formed from the junction of two short vessels, the **brachiocephalic** or **innominate veins**.

The **right vertebral vein** junctions with the **costocervical vein** before entering the superior vena cava. The **left vertebral** and **costocervical veins** junction and enter the left brachiocephalic vein.

2 ➡ Follow the brachiocephalic vein cranially and note that it is formed by the junction of the **external jugular vein**, the largest vein of the upper body, and the **subclavian**, a rather short vessel extending toward the cat's arm.

C. Head and Neck

1 ➡ Return to the external jugular vein. Notice a small vein entering the external jugular vein on the medial side. This is the **internal jugular vein**, which runs parallel to the carotid artery and drains the venous sinuses of the brain, the vertebral column, and the back of the head. (In humans the internal jugular vein, which is generally larger than the external jugular, empties directly into the brachiocephalic vein.)

Return to Exercise 50. What differences do you notice between the feline and human jugular systems? ______________________

____

____

The large vein emptying into the lateral side of the external jugular just below the shoulder is the **transverse scapular vein**. You may also notice one or more small unnamed vessels that drain nearby muscles.

2 ➡ As you proceed up toward the face, note that the external jugular vein is formed by the union of three veins, the **posterior facial vein**,* which drains the side of the head; the **anterior facial vein**,* which drains the

● **FIGURE 53a–3a**
Veins of the Chest and Neck.

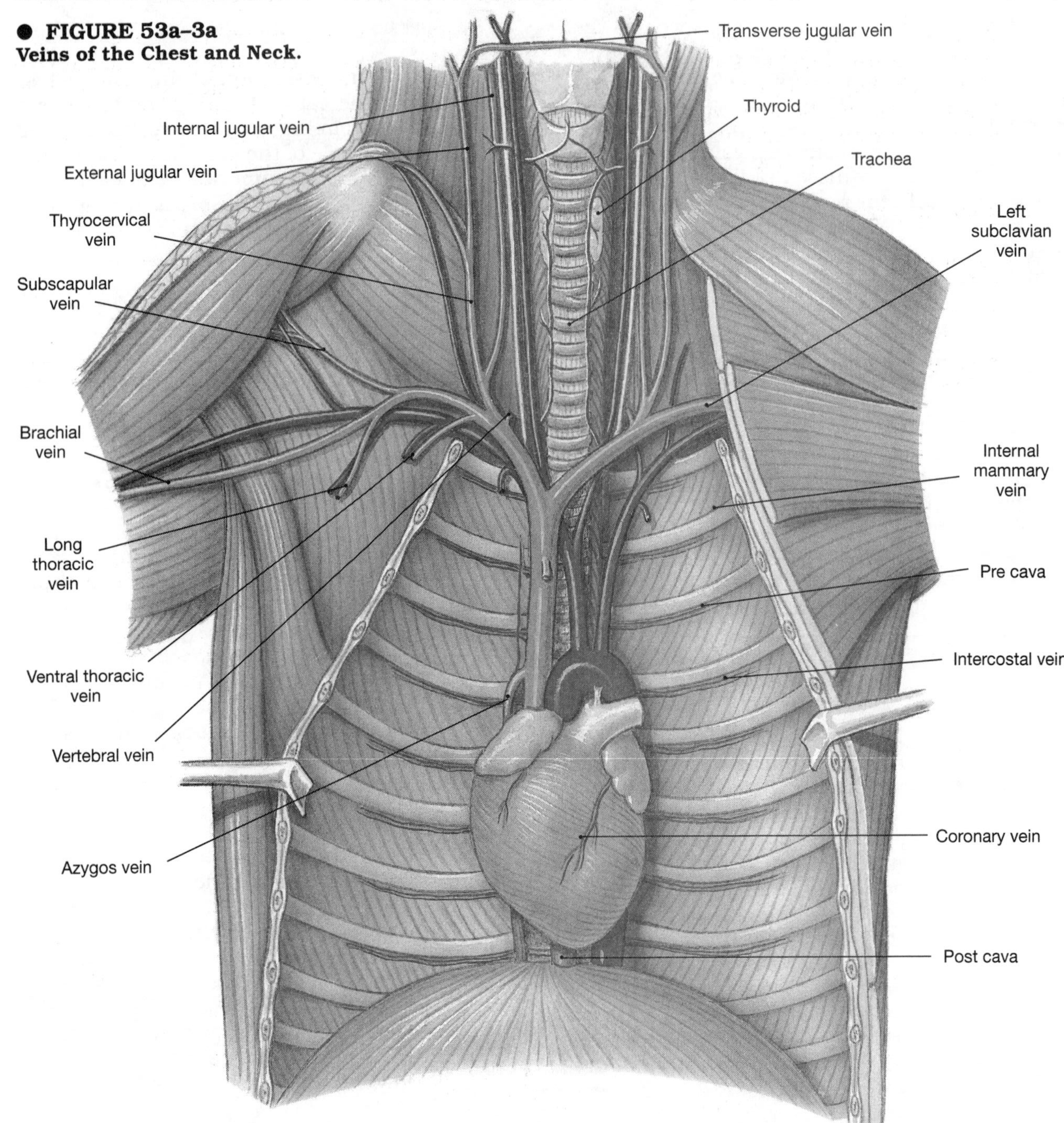

mouth, lips, and teeth; and the **transverse jugular vein,*** which connects the external jugular veins at the base of the chin.

D. Forelimb

1➡ Return to the subclavian vein and follow it laterally. At the bifurcation the superior branch is the **subscapular vein**.

2➡ Follow the subscapular vein out toward the shoulder. The first branch off the subscapular vein is the **humeral circumflex vein,*** which extends cranially to drain the upper brachium, and the second branch is the **thoracodorsal vein,*** which drains the teres major and the subscapularis muscles.

3➡ Return to the inferior bifurcation of the subclavian branch, the **axillary vein**. Emptying into the axillary vein are the **ventral thoracic vein** and the **long thoracic vein**.

The axillary vein becomes the **brachial vein** as it begins to follow the brachium. (Recall that many vessels are named for their location.) You may or may not be able to find the **deep brachial vein,*** which parallels the deep brachial artery.

4➡ If you look toward the elbow, you will find three veins draining into the brachial vein. The **cephalic vein*** connects with the brachial vein via the short **median cubital vein**.* The **radial*** and **ulnar veins*** converge to form the actual brachial vein.

E. Thoracic Veins

1➡ Return to the heart. Move the heart and internal viscera to the left (the cat's left) and relocate the azygos vein. Follow the vein down the posterior thorax and note the **intercostal**, **esophageal**, and **bronchial veins** that empty into the azygos.

2➡ Examine the **inferior vena cava**, also known as the **postcava** or the **caudal vena cava**. It enters the heart (right atrium) from the posterior portion of the body. No branches of note enter the thoracic portion of the inferior vena cava.

F. Inferior Vena Cava (Abdominal)

1➡ Follow the inferior vena cava as it passes through the diaphragm and lies to the right of the abdominal aorta.

2➡ As you descend further down the inferior vena cava, be able to identify the following

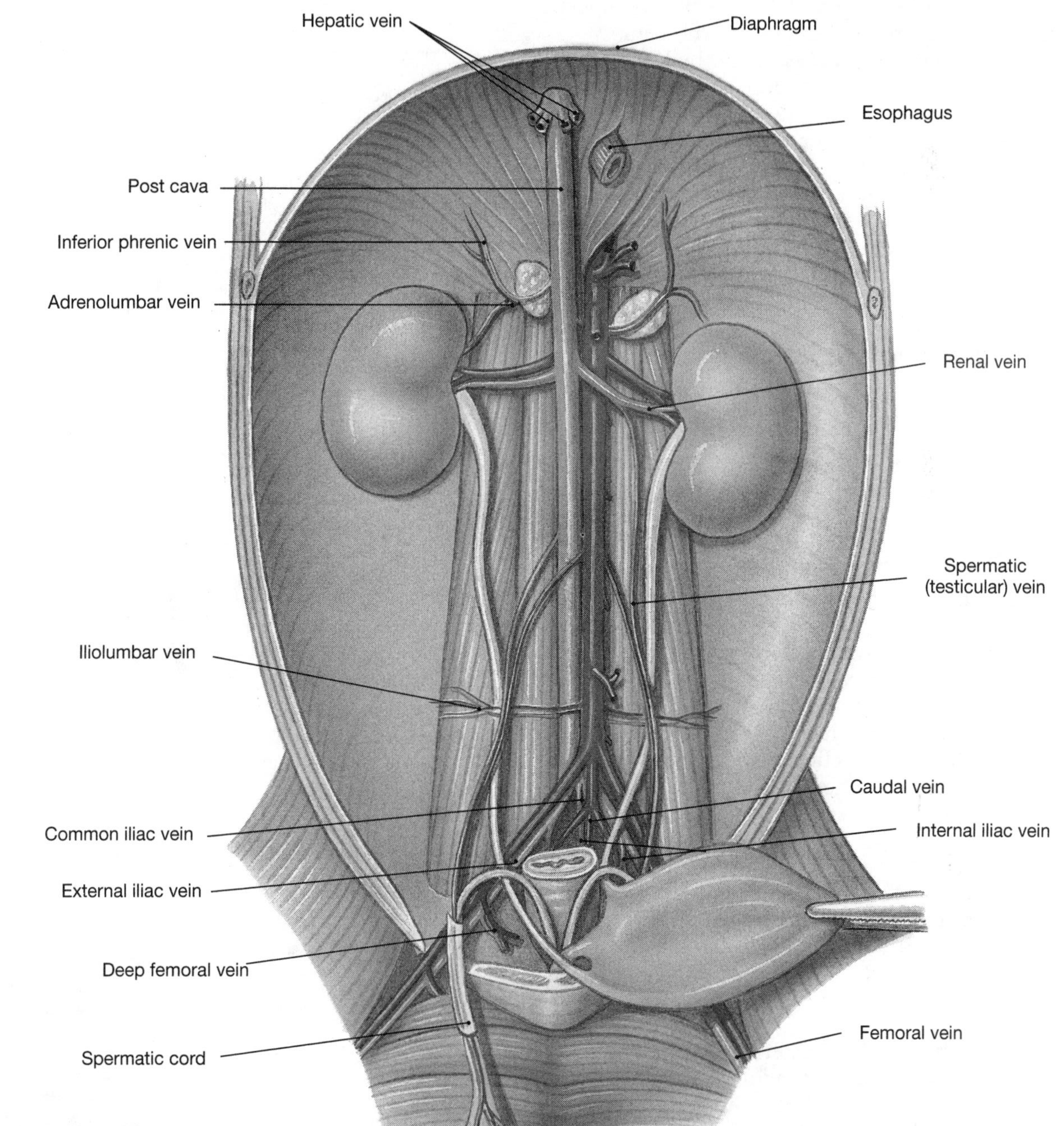

● **FIGURE 53a–3b**
Veins of the Abdominal Cavity.

veins. (Remember, the veins are all draining organs or tissues, and eventually they all converge either directly or indirectly into the inferior vena cava, which goes to the heart.)

Immediately inferior to the diaphragm are the paired **hepatic veins**, which begin deep within the liver. To get a true picture of these veins you will have to scrape away some of the hepatic tissue.

You may be able to locate the paired **inferior phrenic veins**, which drain the inferior diaphragm.

The paired **adrenolumbar veins**, which drain the adrenal glands and the dorsal lumbar region, are not actually parallel as the right vein empties directly into the vena cava while the left drains into the renal vein. Depending on the condition of your cat, you may be able to identify the specific adrenal and lumbar branches of the veins.

The paired **renal veins** drain the kidneys. The right renal vein, which is often double, enters the vena cava at a point cranial to the left renal vein.

The paired **ovarian** or **testicular veins** are also not parallel. The left vein empties into the renal vein while the right vein empties directly into the vena cava. The large paired **iliolumbar veins** drain the posterior abdominal wall. You may also find several small, unpaired **lumbar veins*** in this region.

3➡ Caudal to the iliolumbar veins you will notice that the inferior vena cava is formed by the convergence of two large veins, the **right** and **left common iliac veins**. Follow the right common iliac vein toward the right leg to the point of bifurcation. The caudal branch is the **internal iliac vein**, which drains the gluteal region and the pelvic organs. You probably will not be able to locate the **medial hemorrhoidal vein**,* which branches off the internal iliac.

The larger branch of the common iliac vein is the **external iliac vein**. Just before exiting the body cavity you will find the **deep femoral vein**, which receives blood from the **posterior epigastric vein**,* which receives blood from the posterior abdominal wall, the external genitalia, and the urinary bladder.

4➡ Return to the left common iliac vein and notice the unpaired **caudal vein (medial sacral vein)**, which normally forms the first branch off the left common iliac.

G. Hind Limb

1➡ In the thigh the external iliac vein becomes the **femoral vein**, which is formed by the convergence of the **saphenous vein** on the medial aspect of the thigh and the **popliteal vein** from the deep region behind the knee. About midthigh you may find a small medial vein, the **proximal caudal femoral vein**.* The **caudal femoral (sural) vein** enters the popliteal vein proximal to the knee. The **anterior*** and **posterior tibial veins*** converge to form the popliteal vein just medial to the knee.

H. Hepatic Portal System

The hepatic portal system is that part of the venous system that allows for the transport of nutrients from the digestive system directly to the hepatic system for immediate processing.

What advantage is there to immediate processing? Why shouldn't the nutrients remain in the circulatory system?

____

____

1➡ As you dissect the hepatic portal system, make note of any special directions your instructor may have regarding the displacement of specific abdominal tissues. Use Figure 53a–4●.

The **hepatic portal vein** in the cat is formed by the union of the **gastrosplenic vein**, which drains the spleen and stomach, and the **superior mesenteric vein**, which drains the large and small intestines and the pancreas. (In the human the hepatic portal vein is formed by the conver-

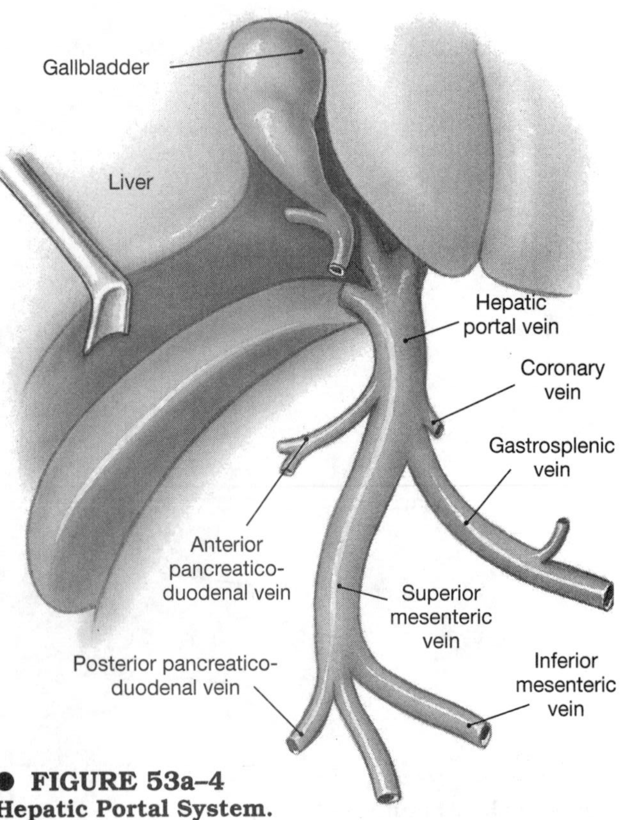

● FIGURE 53a–4
Hepatic Portal System.

gence of the splenic vein and the superior mesenteric vein.)

The **inferior mesenteric vein** parallels the inferior mesenteric artery and drains the left colic and superior hemorrhoidal veins. The **coronary vein** drains the lesser curvature of the stomach.

The **anterior pancreaticoduodenal vein** empties directly into the hepatic portal vein, and the **posterior pancreaticoduodenal vein** empties into the superior mesenteric vein. (In humans, both pancreaticoduodenal veins empty into the superior mesenteric vein.)

The hepatic portal blood flows through the sinusoids of the liver and enters the hepatic veins. The hepatic veins empty into the inferior vena cava.

❑ Additional Activity

1. Research how the cardiovascular system of the cat compares with the cardiovascular systems of other mammals. You have already seen a few differences between the cat and the human.

NOTES

❑ Lab Report

1. Referring to the numbered inquiries at the beginning of this exercise, complete the following box summary:

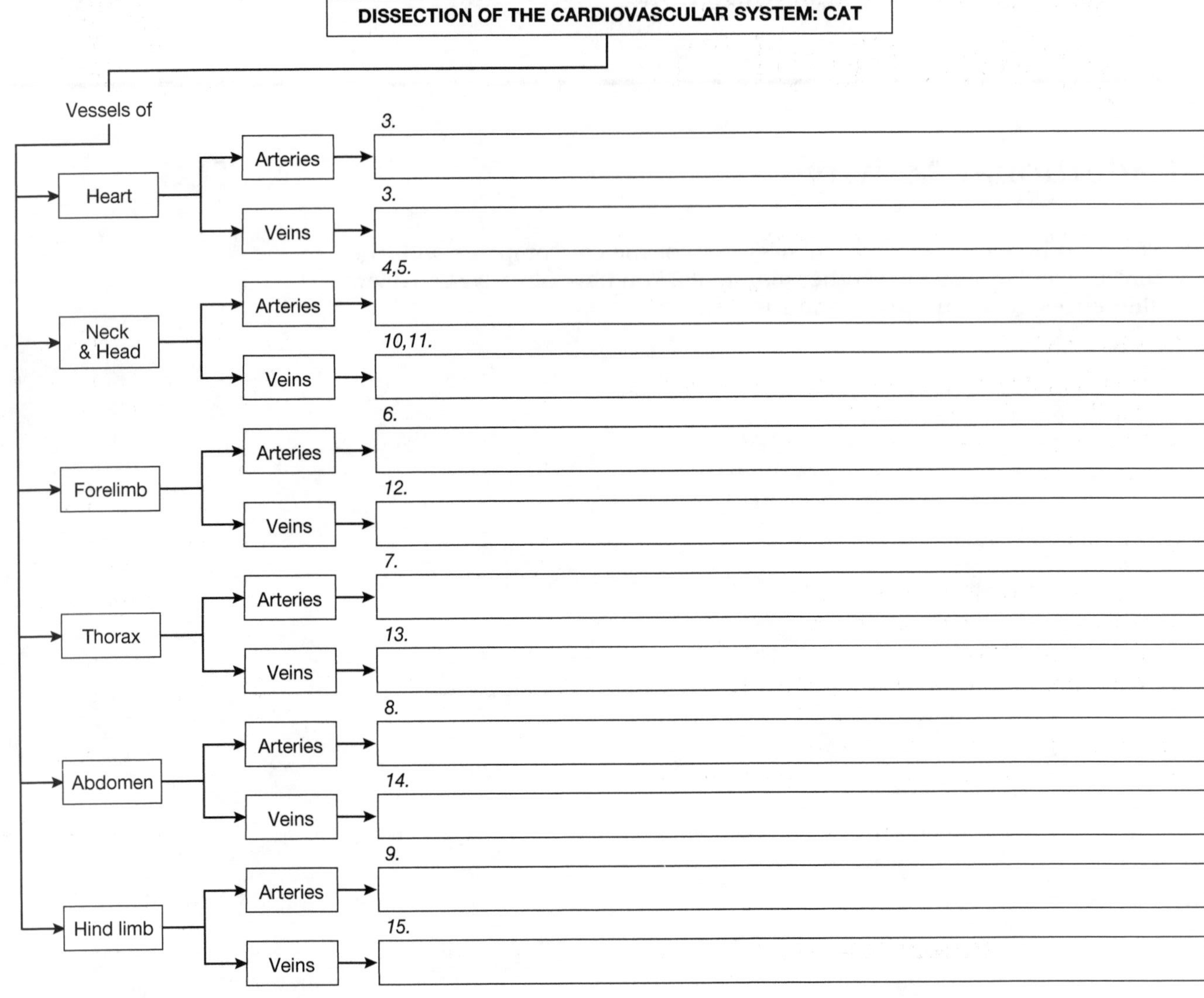

You may need additional paper.

General Questions

1. Compare the cardiovascular system of the cat with the cardiovascular system of the human (Exercise 50). What similarities and differences do you see between:
   a. Great vessels of the heart (carotids, subclavians, pulmonary vessels, vena cavae)
   b. Internal and external jugular veins
   c. Internal and external iliac arteries
2. How are the blood vessels named?
3. How should you prepare your cat for cardiovascular dissection?

NOTES

EXERCISE **53b**

Dissection of the Cardiovascular System: Fetal Pig

PROCEDURAL INQUIRIES

Preparation

1. How should you prepare your fetal pig for cardiovascular dissection?
2. Which instruments should be used for dissection?

Dissection

3. What are the major vessels of the heart?
4. What are the major arteries associated with the aortic arch?
5. What are the major arteries of the head and neck?
6. What are the major arteries of the forelimb?
7. What are the major arteries of the thorax?
8. What are the major arteries of the abdomen?
9. What are the major arteries of the hind limb?
10. What are the major veins associated with the superior vena cava?
11. What are the major veins of the head and neck?
12. What are the major veins of the forelimb?
13. What are the major veins of the thorax?
14. What are the major veins of the abdomen?
15. What are the major veins of the hind limb?

Additional Inquiries

16. How are blood vessels named?
17. How do the blood vessels of the fetal pig compare with the blood vessels of the human?

HUMAN CORRELATION: AM Cadaver Atlas in Applications Manual.

Key Terms

No specific key terms are given for this exercise. The key terms are the major blood vessels. Your instructor will indicate which vessels you are responsible for.

Materials Needed

Gloves

Preserved Fetal Pig (Double-Injected)

Dissecting Tray

Dissecting Kit

The cardiovascular system, which includes the heart, the blood vessels, and certain accessory structures, is essentially the same in all mammals.

Oxygenated blood leaves the heart by the aorta and travels through specific parts of the arterial system to each of the organs and tissues of the body. At the tissue level, the arteries break down to form a capillary bed. Gas and nutrient exchange takes place at the capillary level. The capillaries converge to form the venous system, which carries the blood back to the heart.

In this laboratory exercise we will further our understanding of cardiovascular anatomy by dissecting the arterial and venous systems of the fetal pig. You should have no particular difficulty in identifying the major arteries and veins and in locating the principal branches off these vessels. Refer back to Exercise 50 for comparative human vessels.

❑ Preparation

I. Background

You will probably be working with a partner in your dissection of the circulatory system of the fetal pig. It is important that you and your partner share your ideas as you identify the different blood vessels.

It is also important that you quiz each other as you go along. Work on gaining a total perception of cardiovascular anatomy.

A. Preparing the Fetal Pig

1➡ Obtain your fetal pig and prepare it according to your instructor's directions. Place the fetal pig ventral side up on your dissecting tray and fasten the legs.

2➡ It is assumed that you have already made your initial incision and reflected back at least part of the ventral skin. If you have not previously done so, you should follow these procedures to cut the muscle layers and open the thoracic and abdominal cavities.

3➡ Begin your incision in the groin area. Snip the abdominal muscle layers and make a transverse cut across the groin, keeping the blunt end of your scissors on the inside of the animal's body. Make a midline ventral incision from the pubic bone to the diaphragm. Note Figure 53b–1●.

4➡ It may be necessary to rinse the abdominal cavity because of an accumulation of preservatives or other injected fluids. **Do not** dispose of solid products in the sink.

5➡ When you reach the diaphragm, note its domed appearance. Continue your incision slightly to the right of an imaginary midventral line. (Directly in the center is the sternum, which may be difficult to cut with your regular scissors.)

6➡ Make a transverse incision across the shoulders. With small pigs your transverse incision may be more of an arc.

7➡ To open the thoracic cavity, it will be necessary to cut the diaphragm. Note both surfaces of the diaphragm. Now, make a transverse incision along the ventral wall on each side of the fetal pig. Cut the diaphragm as necessary.

8➡ Spread the rib cage apart to reveal the pleural and cardiac cavities. When you begin working in the thoracic cavity, you will probably notice that the cavity does not stay open very well. You can solve this problem by breaking the ribs. Stand above your specimen and, with the heels of your hands on the spread-apart ribs, press downward forcefully several times. Your instructor may suggest that you remove the ventral rib cage.

B. Major Organs

1➡ Orient yourself to your fetal pig. Locate the major internal organs: lungs, heart, stomach, small intestine, large intestine, kidney, spleen, pancreas, liver, gallbladder, testes or uterus.

2➡ Note the glandular organ covering part of the trachea above the heart. This is the **thymus**. Unless your instructor directs otherwise, you can remove the thymus.

C. Dissecting Hints

The major impediment to good cardiovascular dissection is separating the vessels from the connective tissue. Be certain you understand your instructor's directions about other tissues and systems you may be working with in later exercises. (For instance: Should you be concerned about innervation? Or will you be required to locate any lymphatic vessels?)

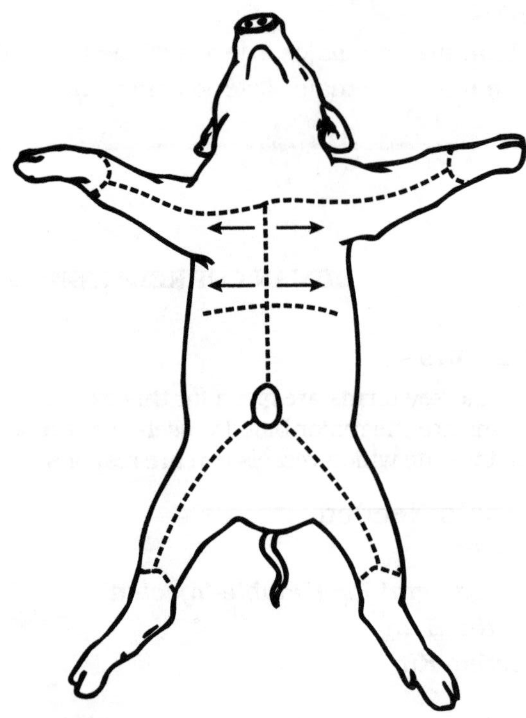

● **FIGURE 53b–1**
Opening the Cavities of the Fetal Pig.

Your fetal pig has probably been injected with latex — red latex for the arterial system and blue latex for the venous system. The red and blue latex do not mix because the vessels of the capillary bed are too narrow to permit the passage of the latex.

The usual points of injection are the neck and the groin. You may find the blood vessels in these regions to be slightly disturbed. You should be able to work around this with no particular difficulty. Occasionally a blood vessel will explode from the latex injection. If your fetal pig has this problem, simply clean out the latex and continue with your dissection. If the problem is serious (which is highly unlikely), show your instructor. Certain blood vessels do not inject as well as others, though usually this will cause you no particular problem. If you do have difficulty, inform your instructor and look at a classmate's fetal pig.

In this laboratory exercise a number of blood vessels that you may not be able to find are included in the write-up. These are generally the smaller vessels that do not always inject well. These vessels are included here for your information — because you might find them and because they are necessary vessels — and not for your positive identification. **These vessels are marked with asterisks.**

Unless you are directed otherwise, it will only be necessary for you to dissect the vessels on one side of the body.

It is important for you to follow the diagrams and illustrations in identifying the blood vessels. Do not limit yourself to the pictures, however. Read the descriptions carefully and coordinate the written word and the illustrations with what you observe on your fetal pig.

D. Isolation of the Vessels

Once you are certain what you are required to do, work at separating the blood vessels from the surrounding tissue. The blood vessels are quite elastic. If you use your blunt nose probe to clear out the connective tissue, you can actually be quite forceful without harming the vessel. The smaller vessels are more delicate.

Your blunt nose probe is your instrument of choice. You will also be able to pull the vessels as well as slide the probe under and lift the vessels without any particular danger of causing irrevocable damage.

Your sharp nose probe should be used sparingly, if at all. Sharp nose probes can easily damage vessels. Sharp nose probes can also tear other tissues and make your identification more difficult.

E. Nomenclature

Most blood vessels are named either for their location or for the part of the body they service. Keep this in mind as you trace the major vessels and as you identify the branches off the vessels.

F. Anomalies

The blood vessels of the fetal pig, particularly those of the venous system, may exhibit some interesting anomalies. For instance, some atrial anomalies are actually quite ordinary. In addition, the common iliac veins may bifurcate (split or divide) in the renal area instead of the pelvic area. Also, the brachial vein may be double. Some arteries or veins may be severely reduced (or even absent). You should be aware that anomalies can exist. Your goal, however, should be to understand the pathways of the major blood vessels, regardless of anomalies.

Dissection

II. Cardiac Dissection

A. Heart Review

Cardiovascular dissection normally begins with the heart and follows the major paths of the great vessels to and from the head, forelimbs, and lower body. That pattern is followed here.

1➜ Note the heart. The principles of cardiopulmonary anatomy are covered in Exercises 48 and 49. If necessary, go back and review the major structures. You should be able to identify these structures in your fetal pig.

2➜ Exercise 54 covers human fetal circulation. The circulatory adaptations of the fetal pig are approximately the same as those of the human. Your instructor may wish to combine parts of Exercise 54 with this exercise on the fetal pig. As you examine the great vessels of the heart, notice the bypass between the pulmonary trunk and the aortic arch. This shunt is the **ductus arteriosus**, which is quite prominent in the fetal animal. The ductus arteriosus carries most of the blood from the pulmonary trunk to the aorta. Because of postnatal changes in pressure, after birth the ductus arteriosus becomes nonfunctional and gradually atrophies to form the **ligamentum arteriosum**.

B. Heart Observations

1➜ Observe the heart from the ventral aspect. Remove the thin, tough **pericardial sac** that surrounds the heart. Note any particular points of attachment.

2➜ Locate the cardiac chambers and the great vessels as found in Exercises 48 and 49. Identify the **ductus arteriosus**.

3➜ Note the red vessels on the surface of the heart. These are the **coronary arteries**, which supply blood to the heart muscle itself. These arteries arise from tiny vessels that exit from the base of the aorta.

The blue surface vessels are the **coronary veins**, which drain the heart muscle. The coronary veins culminate in the **coronary sinus**, which empties directly into the right atrium on the dorsal aspect of the heart.

4➜ If your instructor so directs, make a frontal (coronal) slice through the heart starting at the apex (bottom) and ending just before the great vessels.

5➜ Identify the internal structures of the heart.

III. Arterial System

***You may not be able to find the vessels marked with an asterisk.**

A. Terminology

Since arteries carry blood away from the heart and toward an organ or a body region, we often use terms such as "supplies" or "services" or "goes to" when discussing the arterial system.

1➜ Use Figures 53b–2a● and 53b–2b● to help you identify the major arteries.

B. Aortic Arch

1➜ To begin your study of the arterial system, return to the heart and locate the **aorta**. The portion of the aorta leaving the heart is the **ascending aorta**. The hooked portion is the **aortic arch**, and the portion continuing caudally along the dorsal wall is the **descending** or **thoracic aorta**. When the aorta passes into the abdominal cavity, it becomes known as the **abdominal aorta**.

2➜ Observe the two vessels exiting the aortic arch. The right vessel (the fetal pig's right) is the **brachiocephalic** or **innominate artery**. Follow this vessel cranially. Note the bifurcation where the vessel splits to form the **bicarotid trunk (artery)*** and **right subclavian artery**. The bicarotid trunk further bifurcates to form the **right** and **left common carotid arteries,** which rise on either side of the trachea.

3➜ Please note that in humans, three arteries branch off the aortic arch, the left subclavian; the left common carotid; and the brachiocephalic, which gives rise to the right common carotid and right subclavian arteries.

4➜ The left branch exiting the aortic arch is the **left subclavian artery**. Note that as the subclavian arteries turn toward the arms, the branches off these vessels become approximately identical.

C. Head and Neck

1➜ Return to the common carotid artery. The **superior (anterior) thyroid artery** arises at variable levels and supplies the thyroid gland and certain ventral muscles in addition to giving off laryngeal, esophageal, tracheal, and pharyngeal branches.

(Sometimes the superior thyroid artery is found only as a branch of the left common carotid artery. In many animals, such as the cat, another branch off the common carotid, the **inferior (posterior** or **caudal) thyroid artery,** can be found. In pigs, the right inferior thyroid artery usually branches off the **thyrocervical trunk**. The left inferior thyroid artery is not predictable.)

You should be able to locate the **occipital artery**, which often arises from a common trunk with the internal carotid artery and supplies certain neck muscles.

2➜ You will notice that at the laryngeal (larynx) border the common carotid artery bifurcates to form the **external** and **internal carotid arteries**. The larger external carotid artery supplies the external cranial structures via the **lingual**,* **auricular**,* and **temporal arteries**,* while the internal carotid artery branches to supply the eye, forehead, nose, and brain via the circle of Willis. You will not be able to find most of the latter arteries.

Go back to Exercise 50. How do the pig's head and neck vessels compare with the human head and neck vessels? ______________________

____

____

D. Forelimb

1➜ Return to the subclavian artery. From the left subclavian artery, you will notice the **costocervical trunk**, which is usually absent on the right. It will not be necessary for you to trace all the tiny branchings that you will notice on both subclavian arter-

● **FIGURE 53b–2a**
Arteries of the Chest and Neck.

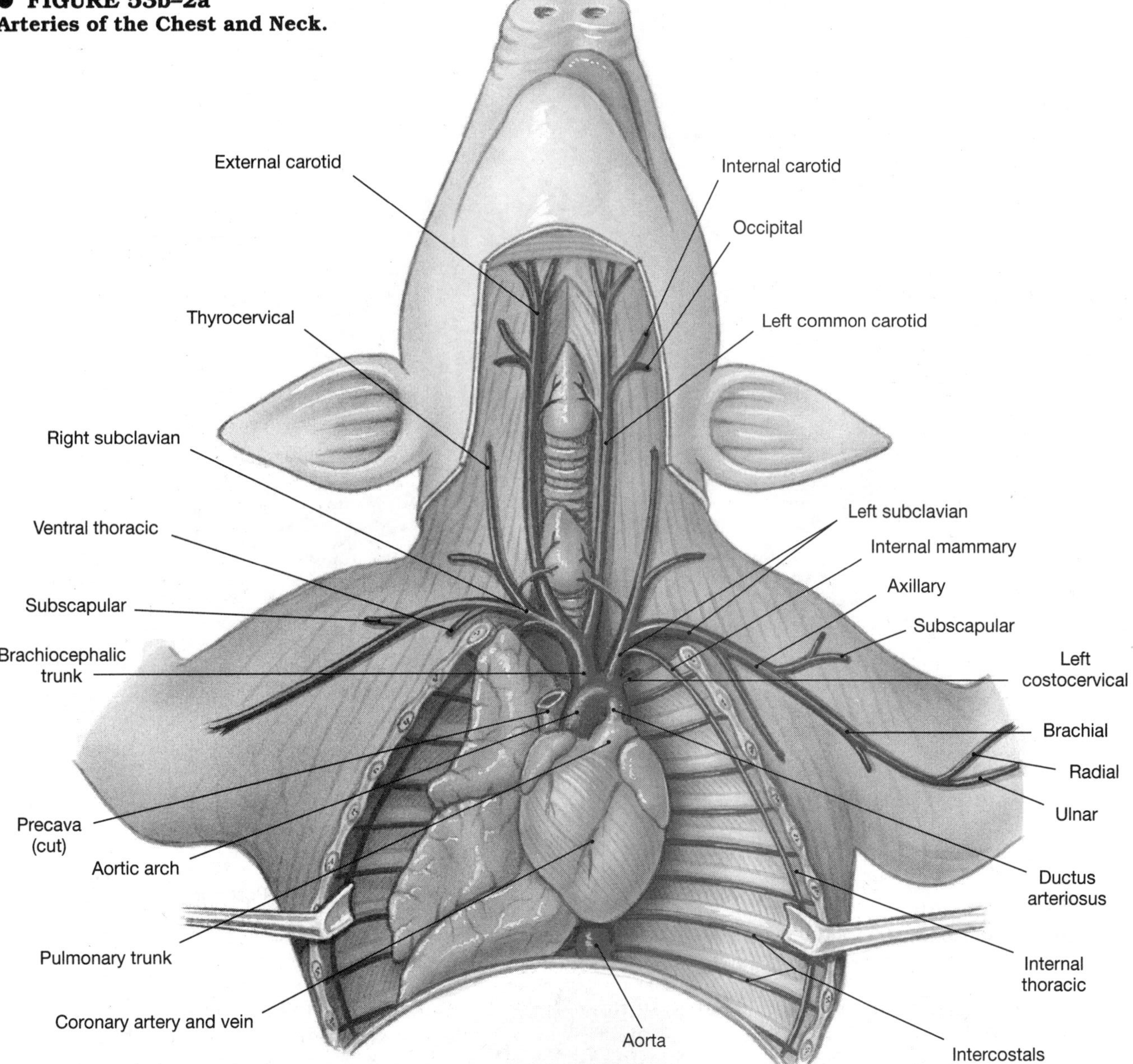

ies. Nevertheless, you should be aware of the **vertebral artery**, an artery of relatively small size that passes through the cervical vertebrae and services the brain.

The **costocervical artery** supplies the deep muscles of the neck and back.

The **internal mammary artery** supplies the ventral body wall.

The **thyrocervical artery (trunk)** arises from the cranial surface of the subclavian and supplies blood to the neck and shoulders. At the scapula the thyrocervical artery becomes known as the **transverse scapular artery**.

2➡ Look at the level of the first rib, where the subclavian artery becomes the **axillary artery**. At approximately that point you may notice the large **subscapular artery.** The first branch off the axillary artery is the **ventral (external) thoracic artery**, which supplies the pectoral muscles.

The large **subscapular artery** gives rise to the **thoracodorsal artery*** and the **posterior humeral circumflex artery.***

3➡ Distal to the subscapular artery, as the vessel parallels the brachium, the axillary artery becomes the **brachial artery.**

The first branch of the brachial artery is the **anterior humeral circumflex artery**,* which you probably will not be able to locate. The **deep brachial artery**, which passes under the belly of the biceps muscle, is usually fairly large.

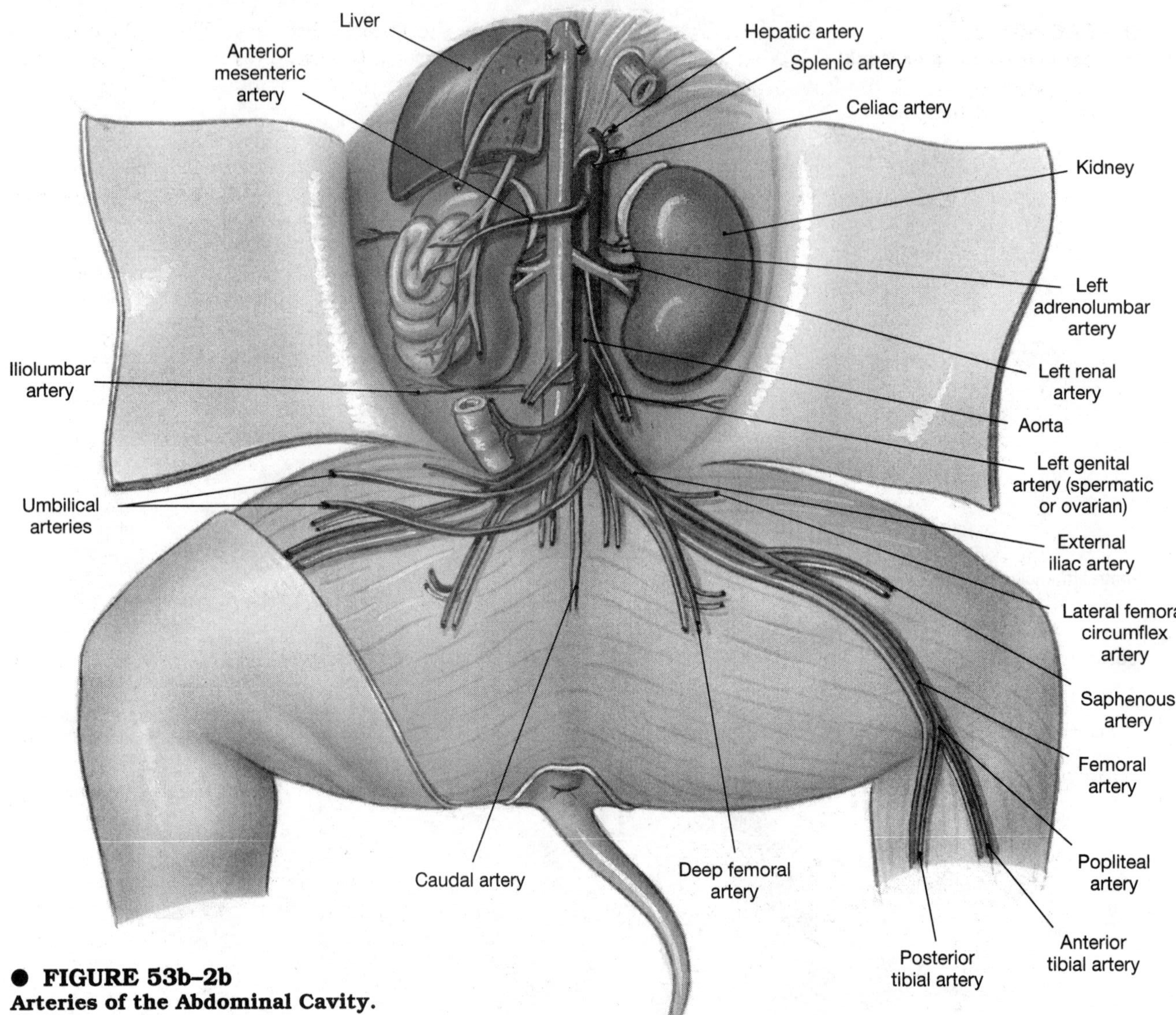

● **FIGURE 53b–2b**
Arteries of the Abdominal Cavity.

Proximal to the elbow you may or may not notice several arterial branches. You should, however, be able to identify the **radial** and **ulnar arteries.**

How does this compare with the human arterial system?______________________________

____

____

E. Thoracic Aorta

1➡ Return to the aorta. Pull the thoracic visceral organs to the right (the fetal pig's right) and follow the aorta caudally through the thorax.

2➡ Notice the eight or nine pairs of **intercostal arteries** that supply the intercostal muscles. The **bronchial arteries*** supply the bronchi as they enter the lung. Several **esophageal arteries*** of varying origins supply the esophagus.

3➡ Continue to follow the aorta caudally. At approximately the third lumbar vertebra the thoracic aorta passes through the diaphragm at the aortic hiatus and becomes the abdominal aorta.

F. Abdominal Aorta

1➡ As you follow the aorta into the abdominal cavity, the first major branch inferior to the diaphragm is the unpaired **celiac artery (trunk)**. The specific branching of the celiac artery is highly variable. The two major branches are the **hepatic artery** and the **splenic artery**.

2➡ Other arteries branch either from the celiac artery itself or from the major celiac branches. Locate as many of these arteries as you can by tracing them to their respective organs: stomach, liver, gallbladder, pancreas, spleen.

3➡ Continue caudally. The unpaired **superior (anterior** or **cranial) mesenteric artery** supplies the small intestine and portions of the large intestine.

4➡ The paired **adrenolumbar arteries** supply the adrenal glands (**adrenal artery**), the diaphragm (**inferior phrenic artery**), and certain muscles of the dorsal body wall. Variations in the branching patterns of the adrenolumbar arteries are common.

The paired **renal arteries** supply the kidneys and sometimes the adrenal glands.

The paired **spermatic (testicular) arteries** in males or paired **ovarian arteries** in females arise near the posterior end of the kidneys. The spermatic arteries service the scrotum as well as the testes, and the ovarian arteries supply the ovaries as well as the uterus (by anastomosing with the **uterine artery**,* via the **middle hemorrhoidal artery**,* both deep branches off the internal iliac artery). Depending on your fetal pig, you may or may not be able to identify these branches.

The single **inferior (posterior** or **caudal) mesenteric artery** branches to form the **left colic artery**,* which supplies the descending colon, and the **superior hemorrhoidal artery**,* which supplies the rectum.

The large paired **iliolumbar arteries** supply dorsal abdominal wall muscles.

The large paired **external iliac arteries** pass through the body wall and become the **femoral arteries** in the legs.

The caudal-most paired arteries are the **internal iliac arteries**, which supply the gluteal muscles, the rectum, and the uterus in the female.

Arising from the internal iliac arteries are the **umbilical arteries**, which you can trace to the umbilicus proper.

The abdominal aorta continues on into the tail as the **caudal** (or **median sacral**) **artery**.

How do the human iliac arteries compare with the porcine iliac arteries? ______________________

____

____

G. Hind Limb

1➡ Return to the external iliac artery. Just before the artery leaves the body cavity, find the branch, the **deep femoral artery**, which continues to branch to supply the bladder, ventral abdominal wall, and some of the femoral muscles.

Several branches, the **lateral femoral circumflex artery*** and the **saphenous artery**, exit the femoral artery to serve the upper leg. When the femoral artery reaches the posterior knee, it becomes known as the **popliteal artery**. The popliteal artery bifurcates to form the **anterior** and **posterior tibial arteries**, which supply the lower leg and foot.

IV. Venous System

***You may not be able to find the vessels marked with an asterisk**.

A. Terminology

Because veins carry blood toward the heart from an organ or region of the body, we often use terms like "entering" or "drains" or "empties into" when describing the venous system.

1➡ Use Figures 53b–3a• and 53b–3b• as your guide to venous dissection.

B. Superior Vena Cava (Precava)

1➡ Begin your dissection of the venous system of the fetal pig by locating the large vein coming from the head region and entering the right side of the fetal pig's heart (right atrium). This is the **superior (anterior) vena cava**, also known in the fetal pig as the **precava**. (A general term for the superior vena cava in the carnivore is **cranial vena cava**.)

Another vessel that usually enters the right atrium of the pig is the **hemiazygos vein**, a multi-branched vessel draining the **intercostal veins**, the **esophageal veins**, and the **bronchial veins**. Sometimes the hemiazygos enters the inferior (posterior) vena cava.

The **costocervical trunk** is usually the first vessel entering the anterior vena cava cranial to the heart. Of the three major veins emptying into the costocervical trunk, you should be able to identify the **vertebral vein**, which drains the cervical vertebral area.

The **internal mammary (sternal) vein** enters from the ventral wall, usually superior to but occasionally inferior to the costocervical vein.

The anterior vena cava itself is formed from the junction of the two **brachiocephalic** or **innominate veins**.

2➡ Follow the brachiocephalic vein cranially and note that it is formed by the junction of the **internal jugular vein**, the **external jugular vein**, and the **subclavian**, a rather short vessel extending toward the fetal pig's forelimb.

C. Head and Neck

1➡ Return to the external jugular vein. Notice a small vein entering the external jugular

● **FIGURE 53b–3a**
Veins of the Chest and Neck.

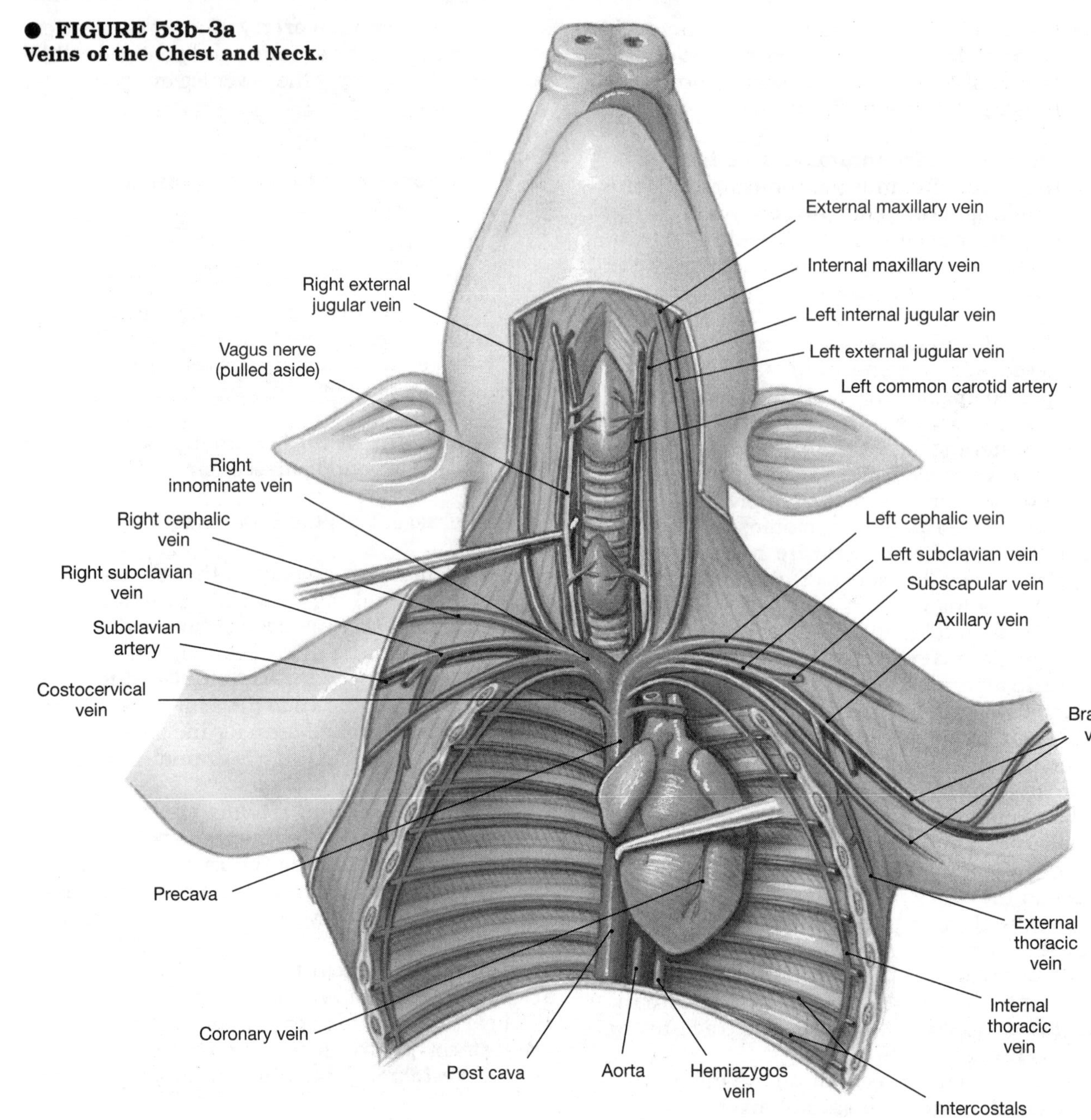

vein on the medial side. This is the **internal jugular vein**, which runs parallel to the carotid artery and drains the venous sinuses of the brain, the vertebral column, and the back of the head. (In humans the internal jugular vein, which is generally larger than the external jugular, empties directly into the brachiocephalic vein.)

2➡ Return to the internal jugular veins, which seem to straddle the trachea. A number of veins, such as the **inferior cervical**,* **lingual**,* and **occipital veins**,* which drain the brain and adjacent areas, empty into the internal jugular veins.

Lateral to the internal jugular veins, find the external jugular veins, which are formed by the union of the **external*** and **internal maxillary veins**.* The external jugulars drain the upper and lower jaws, the face, and the tongue.

3➡ Now locate the **cephalic vein**, which drains the superficial forelimb and empties into the base of the external jugular.

4➡ Return to Exercise 50. What differences do you notice between the porcine and human jugular systems?________________

____

____

D. Forelimb

1➡ Return to the subclavian vein and follow it laterally. At the bifurcation the superior branch is the **subscapular vein**.

2➡ Return to the inferior bifurcation of the subclavian branch, the **axillary vein**. The axillary vein becomes the **brachial vein** as it begins to follow the brachium. (Recall that many vessels are named for their location.) You may or may not be able to find the **deep brachial vein**,* which parallels the deep brachial artery.

3➡ If you look toward the elbow, you will find a number of anastomosing veins draining into the brachial vein. Often this anastomosing system is actually two parallel brachial systems. At the elbow itself note the convergence of the **radial*** and **ulnar veins*** to form the actual brachial vein.

E. Thoracic Veins

1➡ Return to the heart. Move the heart and internal viscera to the left (the fetal pig's left) and relocate the azygos vein. Follow the vein down the posterior thorax and note the **intercostal**, **esophageal**, and **bronchial veins** that empty into the azygos.

2➡ Examine the **posterior vena cava**, also known as the **postcava** or the **inferior vena cava** as it enters the heart (right atrium) from the posterior portion of the body.

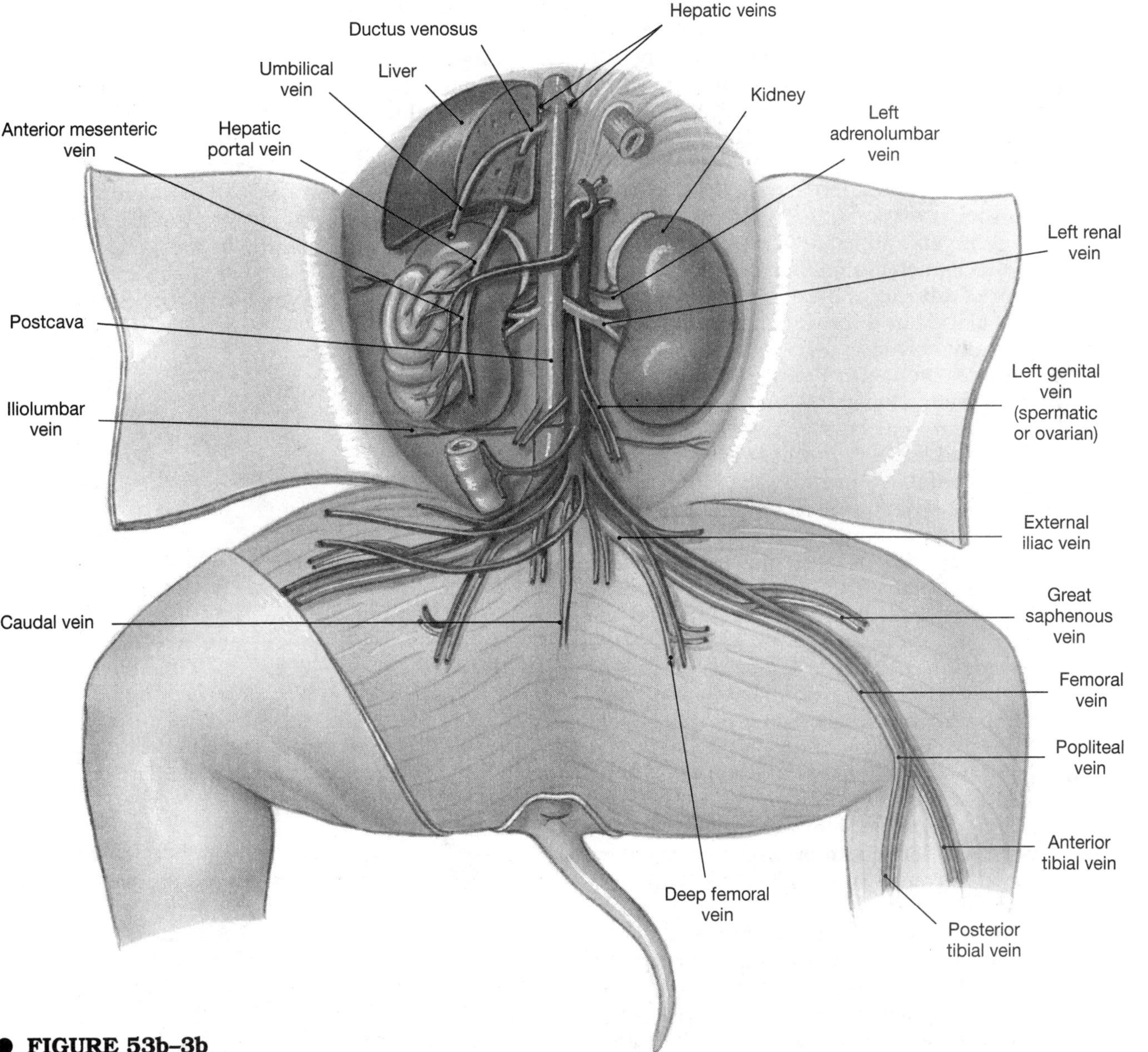

● **FIGURE 53b–3b**
Veins of the Abdominal Cavity.

No branches of note enter the thoracic portion of the inferior vena cava.

F. Inferior Vena Cava (Abdominal)

1➡ Follow the posterior vena cava as it passes through the diaphragm at about the third lumbar vertebra.

2➡ As you descend further down the inferior vena cava, be able to identify the following veins. (Remember, the veins are all draining organs or tissues, and eventually they all converge either directly or indirectly into the posterior vena cava, which goes to the heart.)

Immediately inferior to the diaphragm are three or four **hepatic veins**, which begin deep within the liver. To get a true picture of these veins you will have to scrape away some of the hepatic tissue.

Identify the **umbilical vein**, which carries nutrients and oxygen from the placenta directly into the liver. In the liver the umbilical vein is the **ductus venosus**. (The umbilical vein and the ductus venosus are found only in the fetal animal. After birth both atrophy to form the round ligament and the ligamentum venosum, respectively. See Exercise 54 for the corresponding human structures.)

You may be able to locate the paired **inferior phrenic veins**, which drain the inferior diaphragm.

The paired **adrenolumbar veins**, which drain the adrenal glands and the dorsal lumbar region, are not actually parallel as the right vein empties directly into the vena cava while the left drains into the renal vein. Depending on the condition of your fetal pig, you may be able to identify the specific adrenal and lumbar branches of the veins.

The paired **renal veins** drain the kidneys. The right renal vein, which is often double, enters the vena cava at a point cranial to the left renal vein.

The paired **ovarian** or **testicular veins** are also not parallel. The left vein empties into the renal vein while the right vein empties directly into the vena cava.

The large paired **iliolumbar veins** drain the posterior abdominal wall. You may also find several small, unpaired **lumbar veins*** in this region.

3➡ Caudal to the iliolumbar veins you will notice that the posterior vena cava is formed by the convergence of two large veins, the **right** and **left common iliac veins**. Follow the right common iliac vein toward the right leg to the point of bifurcation. The medial branch is the **internal iliac vein**, which drains the gluteal region and the pelvic organs. The lateral branch of the common iliac vein is the **external iliac vein**. Just before exiting the body cavity you will find the **deep femoral vein**, which receives blood from the posterior abdominal wall, the external genitalia, and the urinary bladder.

4➡ Return to the left common iliac vein and notice the unpaired **caudal vein (medial sacral vein)**, which enters the tail as a continuation of the posterior vena cava.

G. Hind Limb

1➡ In the thigh the external iliac vein becomes the **femoral vein**, which is formed by the convergence of the **great saphenous vein** on the medial aspect of the thigh and the **popliteal vein** from the deep region behind the knee. Several additional veins, such as the small medial **proximal caudal femoral vein*** and the **caudal femoral (sural) vein*** may be found, but these veins are often too small to locate on the fetal pig.

2➡ The **anterior*** and **posterior tibial veins*** converge to form the popliteal vein just medial to the knee.

H. Hepatic Portal System

The hepatic portal system is that part of the venous system that allows for the transport of nutrients from the digestive system directly to the hepatic system for immediate processing.

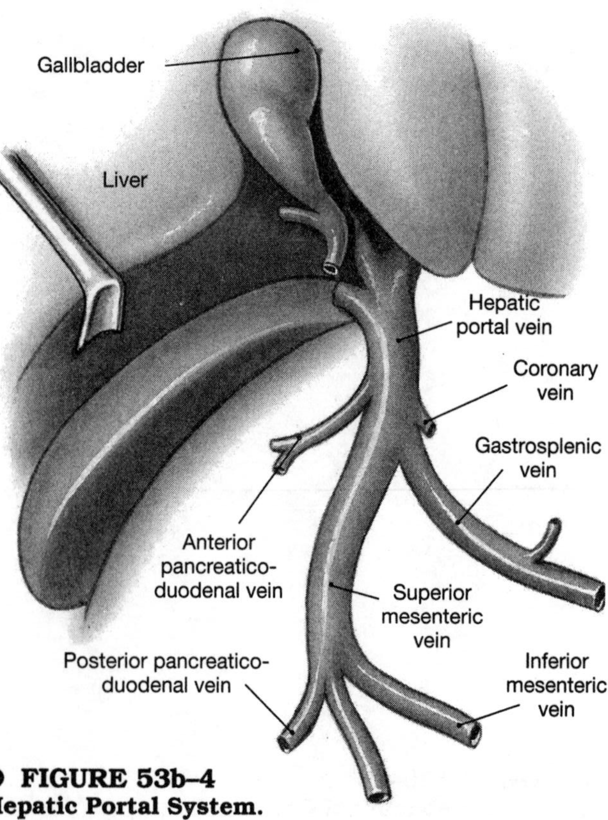

● **FIGURE 53b–4**
Hepatic Portal System.

What advantage is there to immediate processing? Why shouldn't the nutrients remain in the circulatory system?____________________

____

1➡ As you dissect the hepatic portal system, make note of any special directions your instructor may have regarding the displacement of specific abdominal tissues. Use Figure 53b–4●.

Relocate the ductus venosus and trace it posteriorly to the **hepatic portal vein**. Follow the hepatic portal vein toward the intestines. The hepatic portal vein in the fetal pig is formed by the union of the **gastrosplenic vein**, which drains the spleen and stomach, and the **superior (anterior) mesenteric vein**, which drains the large and small intestines and the pancreas. (In the human the hepatic portal vein is formed by the convergence of the splenic vein and the superior mesenteric vein.)

The **inferior (posterior) mesenteric vein** parallels the inferior mesenteric artery and drains the large intestine.

Since the fetal hepatic portal system does not inject well, you probably will not be able to locate additional branches of the hepatic portal system.

❑ Additional Activity

1. Research how the cardiovascular system of the fetal pig compares with the cardiovascular systems of other mammals. You have already seen a few differences between the fetal pig and the human.

NOTES

❑ Lab Report

1. Referring to the numbered inquiries at the beginning of this exercise, complete the following box summary:

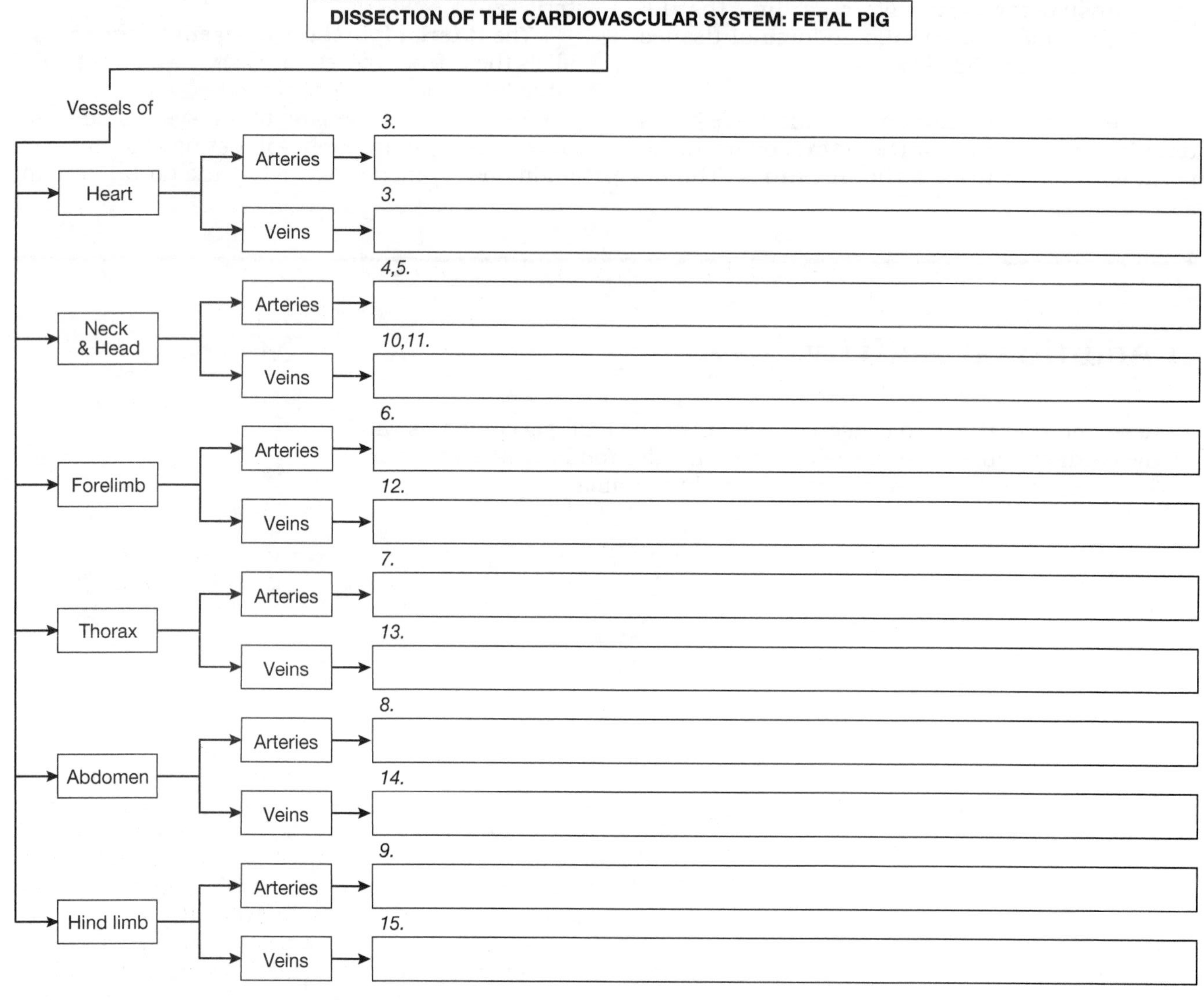

You may need additional paper.

General Questions

1. Compare the cardiovascular system of the fetal pig with the cardiovascular system of the human (Exercise 50). What similarities and differences do you see between:
   a. Great vessels of the heart (carotids, subclavians, pulmonary vessels, vena cavae)
   b. Internal and external jugular veins
   c. Internal and external iliac arteries
2. How are the blood vessels named?
3. How should you prepare your fetal pig for cardiovascular dissection?

NOTES

EXERCISE **54**

Fetal Circulation

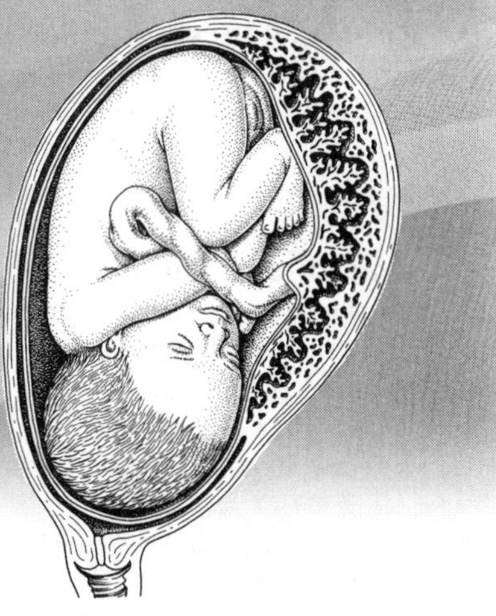

PROCEDURAL INQUIRIES

Preparation

1. What is the basic path of blood from the left ventricle back to the left ventricle?

Examination

2. Where are the umbilical arteries and what is their function?
3. Where is the umbilical vein and what is its function?
4. What are the fetal blood vessels of the liver?
5. What are the structures of the atrial shunt?
6. What are the structures of the pulmonary bypass?
7. What is the basic fetal circulatory route, including the bypasses?

Additional Inquiries

8. What is the function of the placenta?
9. What is the role of pressure in closing the shunts?

Key Terms

Arterio-Capillary-Venous Network
Ductus Arteriosus
Ductus Venosus
Endometrial Spiral Arteries
Foramen Ovale
Fossa Ovalis
Internal Iliac Arteries
Intervillous Space
Lateral Umbilical Ligaments
Ligamentum Arteriosum
Ligamentum Teres
Ligamentum Venosum
Placenta
Portal Sinus
Portal Vein
Septum Primum
Septum Secundum
Umbilical Arteries
Umbilical Vein

Materials Needed

Charts and Diagrams

Models (if available)

Since the fetus does not have direct contact with the external environment, fetal respiratory and digestive structures are generally nonfunctional. Nevertheless, the fetal circulatory system must transport the same gases, nutrients, and waste products to and from the individual cells that the postnatal system must transport. To accomplish this end, fetal circulation is replete with distinct modifications allowing it to exchange nutrients and wastes with the maternal system via the **placenta**.

In this laboratory exercise we will conceptualize the basics of fetal circulation by examining the anatomical and physiological structures of the fetus and by comparing prenatal structures with their postnatal analogues.

❑ Preparation

I. Background

A. Postnatal Heart

It is assumed that you have already studied the adult (postnatal) heart and the adult (postnatal) circulatory system. The fetal circuit is the same as the postnatal circuit, with a few functional additions and modifications to be discussed in the succeeding sections of this exercise.

B. Charts and Diagrams

1 ➡ Work through this section, using whatever charts, diagrams, and models you have available, in addition to the figures included here, so that you are able to conceptualize the differences between prenatal and postnatal circulation.

C. Review of the Basic Circuit of Blood Flow

1 ➡ Keep in mind the basic circuit of blood flow: left ventricle → aorta → arteries → arterioles → capillaries → venules → veins → vena cavae → right atrium → right ventricle → pulmonary arteries → lungs → pulmonary veins → left atrium → left ventricle. (Refer to Exercise 50 if necessary.)

❑ Examination

II. Prenatal Circulation [∞ *Martini, p. 742]*

A. Systemic Circuit

In the fetus, blood leaves the left ventricle via the aorta and follows the standard systemic circuit. Because the lungs of the fetus are not functional, this blood is not specifically oxygenated.

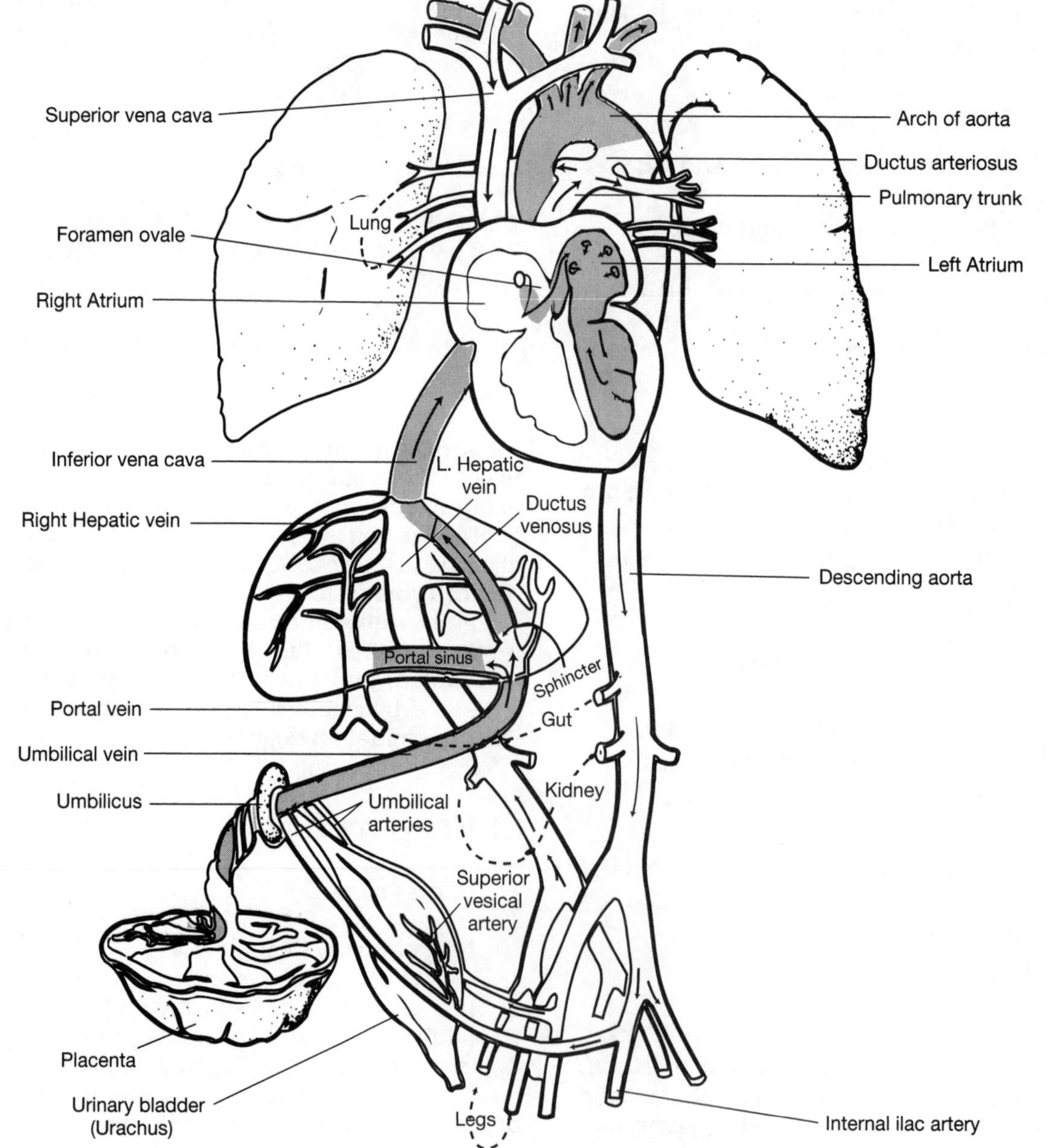

● **FIGURE 54–1**
Fetal Circulation.

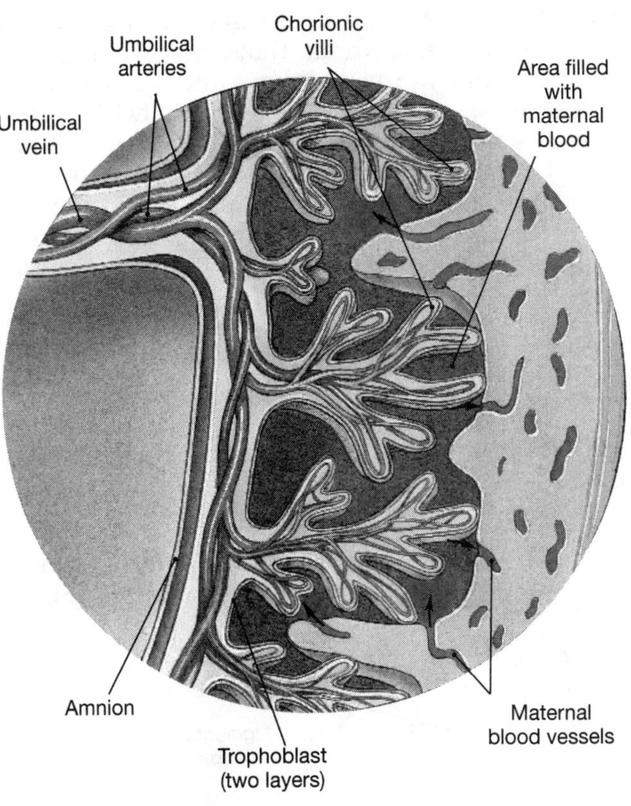

● **FIGURE 54–2**
Placental Organization.

1➡ Examine Figure 54–1●. Note how the fetal systemic blood circuit reaches all necessary body areas.

B. Blood Flow and the Placenta

Unoxygenated, nutrient-poor blood is carried in the systemic circuit from the two **internal iliac arteries** via the paired **umbilical arteries** to the placenta. (The umbilical arteries are arteries because they carry blood away from the heart.) A single **umbilical vein** returns the oxygen and nutrient-rich blood from the placenta to the body of the fetus. (The umbilical vein is a vein because it carries blood toward the heart.)

1➡ Follow the path of these structures on Figure 54–1●.

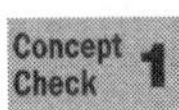

In your own words, describe the anatomical relationship between the umbilical arteries and the umbilical vein both in the umbilical cord and in the body proper.

____

____

____

2➡ Now study Figure 54–2●, the placenta. Within the placenta the umbilical arteries break down to form an **arterio-capillary-venous** network. Notice the **intervillous space** between the umbilical blood vessels and the **endometrial spiral arteries** of the maternal tissues.

The function of the placenta is nutrient and waste exchange. Gas exchange occurs by simple diffusion across the placental concentration gradient. Nutrients, electrolytes, hormones, wastes, drugs, and infectious agents are also exchanged across the placental barrier. Molecular weight, concentration, polarity, and intermolecular interactions all affect the rate of diffusion. Also, the oxygen/carbon dioxide exchange is facilitated by fetal hemoglobin, which has a greater oxygen-carrying capacity than does maternal (adult) hemoglobin. *[∞ Martini, p. 630]*

What can we say about the physical relationship between the umbilical vessels and the placenta?

____

____

C. Blood Flow from the Placenta

From the placenta, blood travels through the umbilicus and toward the liver via the umbilical vein. About half the blood enters the liver through the **portal sinus**, which subsequently empties into the **portal vein**. The rest of the blood bypasses the liver and travels via the **ductus venosus** directly to the **inferior vena cava**.

1➡ Locate and follow the blood flow from the placenta and through these structures on Figure 54–1●.

D. Superior Vena Cava

Oxygenated, nutrient-rich blood travels back to the heart and enters the right atrium. Also entering the right atrium is the unoxygenated blood from the **superior vena cava**.

1➡ Follow this course on Figure 54–1●.

E. Two Fetal Shunts

1➡ As you study Figure 54–1●, notice the hole between the atria. This is the **foramen ovale**. The flap on the left atrial side is the **septum primum**, or the valve of the foramen ovale. The smaller flap on the right atrial side is the **septum secundum**.

Blood entering the right atrium of the heart is under more pressure than blood entering the left atrium. This pressure forces open the septum primum. As a result, much of the blood entering the right atrium bypasses the pulmonary circuit by entering the left atrium directly from the right atrium through the foramen ovale.

2➡ Follow the blood flow through the fetal foramen ovale on Figure 54–1●.

Some blood does enter the right ventricle and is subsequently pumped out through the pulmonary trunk toward the lungs. Most of this blood never reaches the lungs, however, because the lungs are nonfunctional and require very little blood. Instead, most of the blood in the pulmonary trunk is shunted to the aorta through the **ductus arteriosus**.

3➡ Identify this structure on Figure 54–1●.

4➡ Be able to trace the blood flow through this shunt.

To summarize fetal circulation: Blood enters the left atrium from either the right atrium or the pulmonary veins and empties into the left ventricle. From the left ventricle the blood is pumped to the body by way of the ascending aorta. This is the standard systemic circuit. Blood entering the right atrium (from the vena cavae and the coronary sinus) passes either to the left atrium (where it becomes a part of the systemic circulation) or it enters the right ventricle in preparation for the pulmonary circuit. Blood is pumped from the right ventricle via the pulmonary trunk toward the lungs. Most of this blood bypasses the lungs and enters the aorta via the ductus arteriosus.

5➡ Review the entire circuit of both shunts with a classmate.

III. Pressure

A. Pressure and Fetal Circulation

As already seen, pressure is the primary factor in maintaining the fetal circulatory system. Pressure is

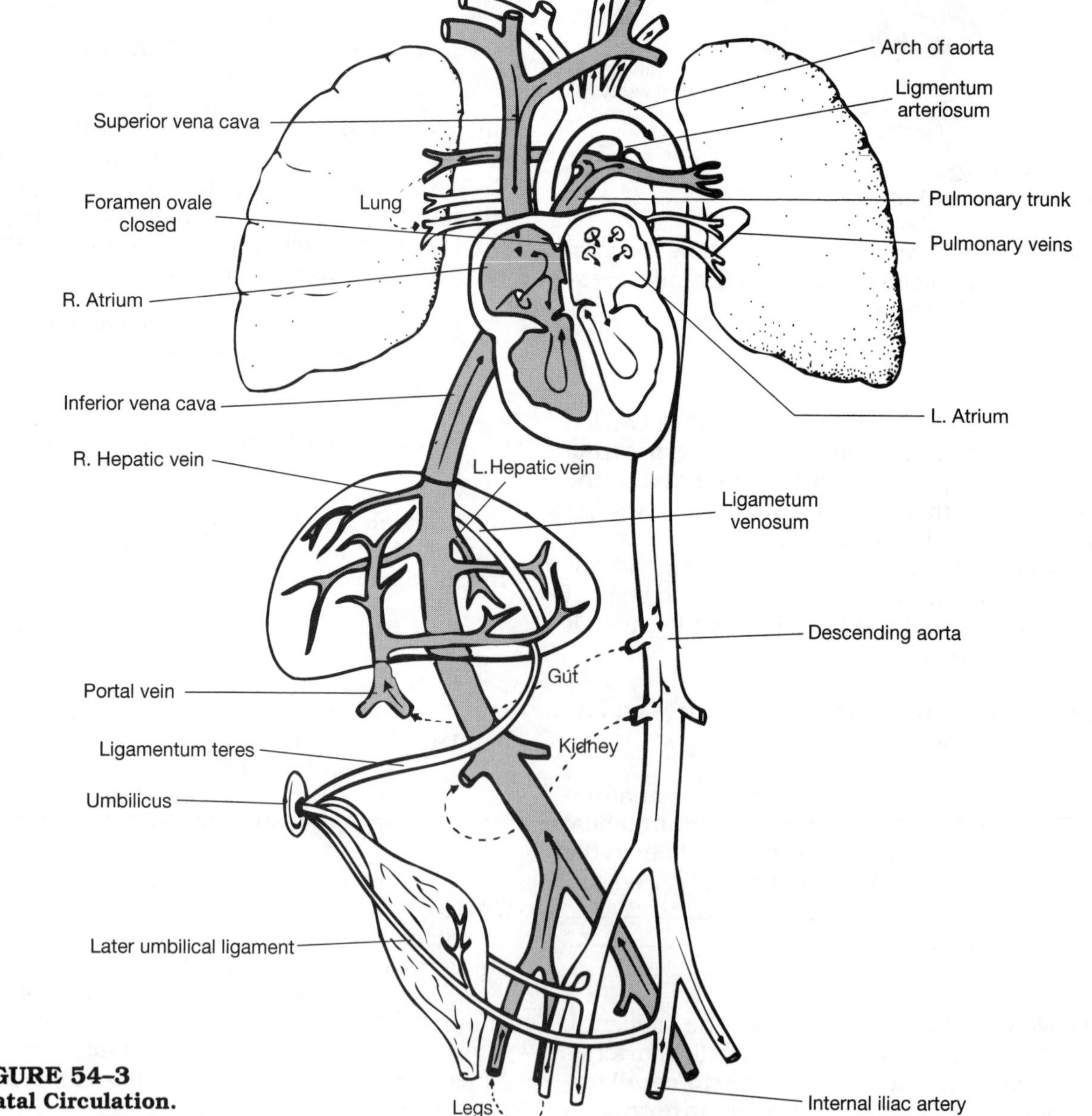

● **FIGURE 54–3**
Postnatal Circulation.

also the major factor in transforming the fetal circulatory system into the postnatal circulatory system.

B. Closing of the Shunts

At birth the lungs become functional. Because the lungs are functioning, the atrial pressure gradient is shifted so that pressure is higher in the left atrium than in the right atrium. The shift in pressure closes the septum primum. During the first few months of life this functional closure will become permanent as tissue proliferation and adhesion seal the two septa together. The closed area will form a permanent depression known as the **fossa ovalis**.

Sometimes the two septa do not seal properly, resulting in a common type of "hole in the heart." This hole might be as tiny as a pinprick or as large as the original foramen ovale. The pinhole-size hole might not be cause for concern. The large hole might be life-threatening.

The shift in pulmonary pressure is also responsible for the closure of the ductus arteriosus, a process normally taking about three months to complete. The ductus arteriosus is then known as the **ligamentum arteriosum**.

1 ➡ Identify these structures on Figure 54–3●.

C. Atrophy of the Umbilical Structures

At birth the umbilical cord becomes nonfunctional. Because blood is no longer pulsating through the umbilical arteries and umbilical vein, these vessels atrophy, becoming the **lateral umbilical ligaments** and the **ligamentum teres**, respectively.

Because the portal sinus receives blood from other vessels, it continues to function. The ductus venosus, however, because it receives blood only from the umbilical vein, ceases to function, atrophies, and becomes known as the **ligamentum venosum**.

1 ➡ Study Figures 54–1● and 54–3●. Compare these prenatal structures with their postnatal structure counterparts.

IV. Postnatal Circulation

A. Mechanism of Switch

Postnatal circulation is the result of the changes in blood pressure mentioned in the previous section. Prenatal circulation is a functional adaptation that is no longer essential after birth.

The change from fetal to postnatal circulation is not a sudden occurrence. Although the foramen ovale becomes nonfunctional at birth, as the endothelial and fibrous connective tissues build up, it is still possible for the foramen to open and function to an extent on certain occasions. This is normally not a problem. Gradual atrophy is also the rule with the formation of the ligaments. Loss of function is gradual with the ductus arteriosus.

1 ➡ Return to Figure 54–3● and trace the postnatal circulatory circuit.

❑ Additional Activities

NOTES

1. Research the timeframe during which each of the prenatal vessels ceases functioning.
2. Research the embryological development of the major fetal vascular anomalies and abnormalities. What types of problems result for the postnatal individual? What types of treatments are possible?

Answers to Selected Concept Check Questions

1. In the body the umbilical arteries come from the lower abdomen to the umbilical cord, while the umbilical vein goes from the umbilical cord to the liver. Within the umbilical cord the umbilical arteries wrap around the umbilical vein.
2. The maternal and placental vessels are very close, but they do not touch.

❑ Lab Report

1. Referring to the numbered inquiries at the beginning of this exercise, complete the following box summary:

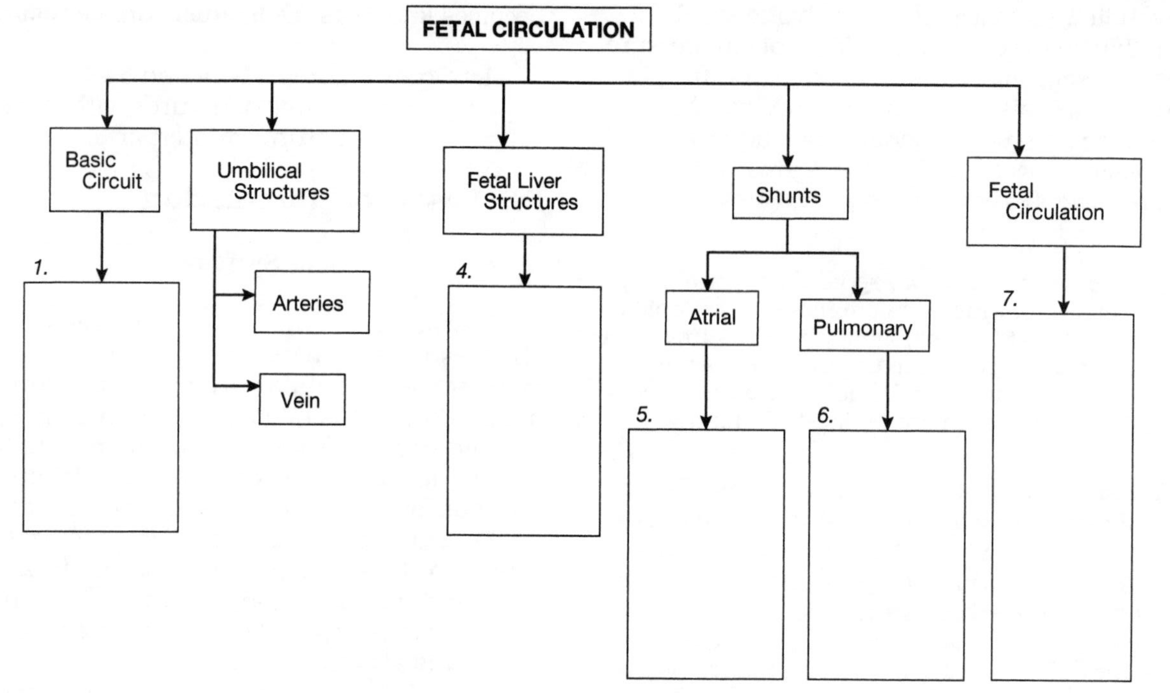

General Questions

1. List and explain the structural differences between the prenatal and postnatal circulatory systems.
2. List and explain the functional differences between the prenatal and postnatal circulatory systems.
3. What functional advantages and disadvantages do you see in the structure of the placenta? In the relationship between the placenta and the maternal tissues?
4. What are the functions of the following:
   a. Umbilical arteries
   b. Umbilical vein
   c. Prenatal pressure
   d. Postnatal pressure
5. What is the fate of each of the following:
   a. Umbilical cord
   b. Ductus venosus
   c. Septum primum
   d. Foramen ovale
   e. Ductus arteriosus

EXERCISE 55

Anatomy of the Respiratory System

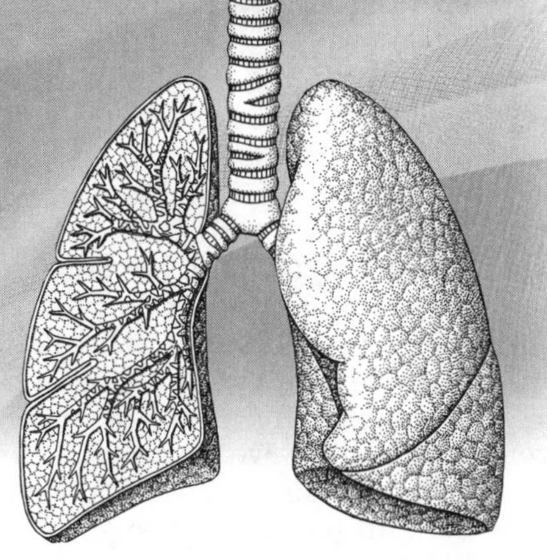

PROCEDURAL INQUIRIES

Preparation

1. What is the difference between pulmonary ventilation and external respiration?
2. What is the difference between inhalation and exhalation?
3. What are the major organs of the respiratory system?

Examination

4. What are the inner structures of the nose?
5. What are the sinuses and where are the major sinuses located?
6. What are the areas of the pharynx?
7. What are the cartilages of the trachea and larynx?
8. Where are the vocal cords?
9. How many sets of vocal cords do we have?
10. What are the branches of the bronchial tree?
11. What are the names of the lobes of the lungs?

Additional Inquiries

12. How is the cardiovascular system related to the respiratory system?
13. What is the function of surfactant?

Key Terms

Alar Cartilage
Alveolar Duct
Alveolus (pl., Alveoli)
Arytenoid Cartilage
Bronchus (pl., Bronchi)
Carina
Corniculate Cartilage
Cricoid Cartilage
Cuneiform Cartilage
Epiglottis
Esophagus
Exhalation
Glottis
Inhalation
Laryngopharynx
Larynx
Lateral Nasal Cartilage
Lobe
Lobule
Lung
Meatus
Nares
Nasal Cavity
Nasal Concha (pl., Conchae)
Nasal Septum
Nasopharynx
Oropharynx
Pulmonary Ventilation
Respiration
Respiratory Bronchiole
Sinus
Surfactant
Terminal Bronchiole
Thyroid Cartilage
Trachea
Tracheal Cartilage
Vocal Cords

Materials Needed

Torso

Models (as available)
- Skull (Whole and Sagittal)
- Trachea
- Lungs
- Alveolar Apparatus

Charts and Diagrams

Dissecting Microscope

Gross Sections of Respiratory Organs (if available)

Compound Microscope

Prepared Slides (as available)
- Respiratory Epithelium
- Tracheal Cartilage
- Bronchiole Cross Section
- Alveolar Tissue
  - Slice (if available)
  - Cross Section

Every cell in the human body needs oxygen to complete the breakdown of foodstuffs and to produce the usable energy needed for daily living. Every cell in the human body also needs to get rid of carbon dioxide, the gaseous waste product of foodstuff breakdown. Since cells do not communicate directly with the environment, a means must exist to facilitate this oxygen/carbon dioxide gas exchange.

The gas exchange network involves both the circulatory system, which functions in gas transport, and the respiratory system, which functions as the conduit for gas exchange.

Respiration is technically the exchange of gases between the living cell and the environment. Respiration includes pulmonary ventilation, external respiration, internal respiration, and cellular respiration. Thus respiration means a great deal more than just breathing. When we speak of the **respiratory system**, which consists of those organs directly involved in gas exchange between the cardiovascular system (and subsequently the body as a whole) and the external environment, we are generally referring only to those organs and tissues involved in **pulmonary ventilation (breathing)** and **external respiration** (gas exchange between the alveolar sacs and the circulatory system).

Pulmonary ventilation begins with **inhalation** (breathing in), as air from the environment passes from outside the body through the nose, pharynx, larynx, trachea, and bronchial system. This process continues as the air moves into the alveolar sacs (the functional units of the lungs), where oxygen is passively exchanged for carbon dioxide. This gas exchange is external respiration. Pulmonary ventilation concludes with **exhalation** (breathing out), as the carbon dioxide-rich air passes back to the environment through the same series of respiratory structures.

In this exercise we will examine the anatomy and histology of the structures involved in pulmonary ventilation and external respiration, first by examining various models, charts, and diagrams and then by studying histological sections and slides of selected respiratory structures.

❑ Preparation

I. Identification

A. Gross Methodology

1➡ Begin by reviewing the definitions included in the introduction to this exercise.

2➡ Identify the terms in boldface as you work through the following descriptions. If you are working with a partner, you and your partner should quiz each other as you go along.

3➡ Handle the models in your laboratory and identify the structures out loud so that you gain an additional perspective of the different parts of the respiratory system.

B. Microscopic Methodology

1➡ After you have studied the slides, relate your microscopic findings to the structures you identified on your gross examination.

❑ Examination

II. Gross Examination

A. Respiratory Structures of the Nose and Pharynx

1➡ Study Figure 55–1● to orient yourself to the relative positions of the nasal passages, the tracheal system and bronchial tree, and the pulmonary structures.

2➡ Refer to Figure 55–2●. Locate the **nostrils** (or **external nares**) and the **nasal septum**, the internal plate between the sides of the nose. If available, locate the bony portion of the nasal septum on a skull.

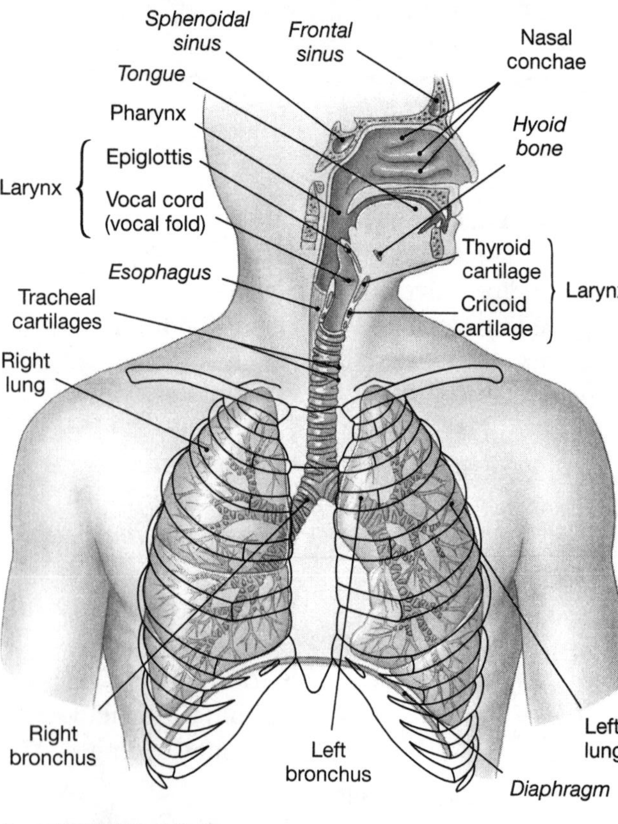

● **FIGURE 55–1**
Components of the Respiratory System.

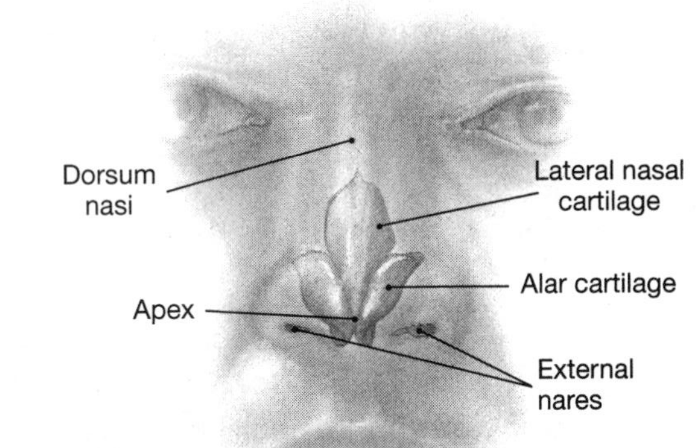

● **FIGURE 55–2**
Nasal Cartilages.

3➡ Note the location of the **lateral nasal cartilage** and the **alar cartilage** in Figure 55–2●. Feel your own nose and see if you can distinguish these cartilages. (Often when people say they have a broken nose, what they really have is a dislocation of one or both of these cartilages.)

Were you able to distinguish each of these cartilages on your own nose? ________________

4➡ Study Figure 55–3a●. The **vestibule** is the area immediately inside the nares. Note the relative size of the **nasal cavity**. Three lateral bony projections extend into the nasal cavity. These are the **superior**, **middle**, and **inferior nasal conchae**. The superior and middle conchae are parts of the ethmoid bone while the inferior concha is a separate facial bone. The conchae are sometimes called the **turbinate** bones. Locate the conchae on the skull model.

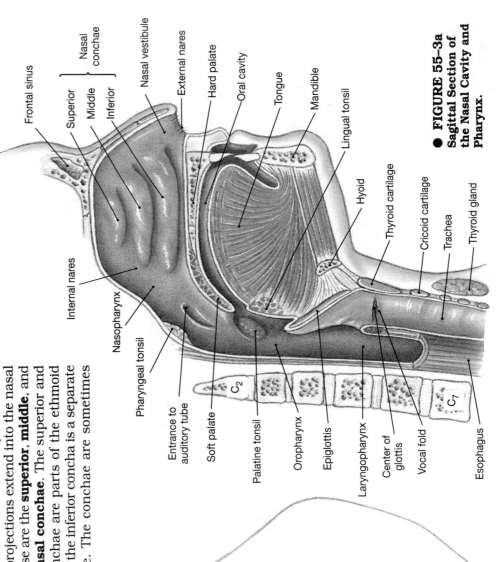

Concept Check 1 What is the function of these nasal conchae?

____

____

5➡ On Figure 55–3b●, locate the passages between the conchae and the lateral nasal wall. These are the **superior**, **middle**, and **inferior meatuses**.

6➡ Use both Figures 55–3a● and 55–3b● to locate the **frontal**, **sphenoid**, **maxillary**, and **ethmoid sinuses**. A sinus is a pocket or cavity within a bone. Note how these sinuses drain into the nasal cavity, which is exactly what happens when you encounter some sort of sinus irritant or allergen! Identify these areas on the skull. Does anything surprise you about the size or location of these allergy-prone sinuses?

____

____

● **FIGURE 55–3a**
Sagittal Section of the Nasal Cavity and Pharynx.

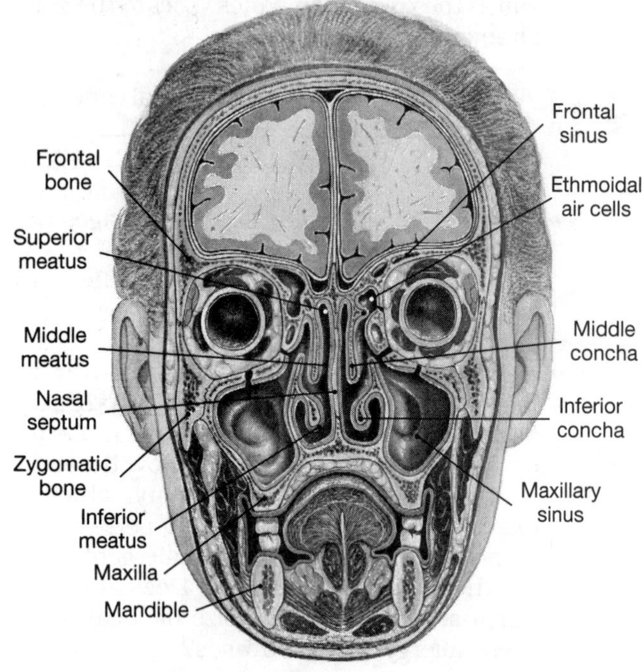

● **FIGURE 55–3b**
Frontal Section Showing Nasal Conchae, Meatuses, and Sinuses.

7➡ Again refer to Figure 55–3a●, and observe the narrowing of the posterior portion of the nasal cavity forming the "opening" between the nose and the throat. The paired openings (one on each side) are the **internal nares**. If possible, locate these on the skull model. You may wish to look in your own mouth and visualize the relationship between the internal nares and the soft palate.

8➡ Refer back to Figure 55–1●. The area immediately posterior to the nose is the **nasopharynx**, which is the cranialmost of the three pharyngeal (throat) areas. The **oropharynx** is immediately inferior to the nasopharynx, and the **laryngopharynx** is the caudalmost of the pharynges. Additional information about the pharyngeal areas can be found in Exercise 58.

B. The Trachea and Larynx

1➡ Refer to Figure 55–4● to see the structure of the **trachea**, the ringed tube extending from behind the mouth caudally to the pulmonary bifurcation. If a tracheal model is available, use this also for your study. At the cranial end of the trachea is the **larynx**, the sturdy cartilaginous structure sometimes known as the voice box (or Adam's apple). In the human body, just below the larynx, is the **thyroid gland** (which may not be identified on your laboratory models). The thyroid gland consists of two lobes, one on either side of the trachea, connected by a glandular isthmus with a cranial projection. (Refer to Exercise 44.)

2➡ Immediately posterior to the trachea is the **esophagus**, the flexible tube extending

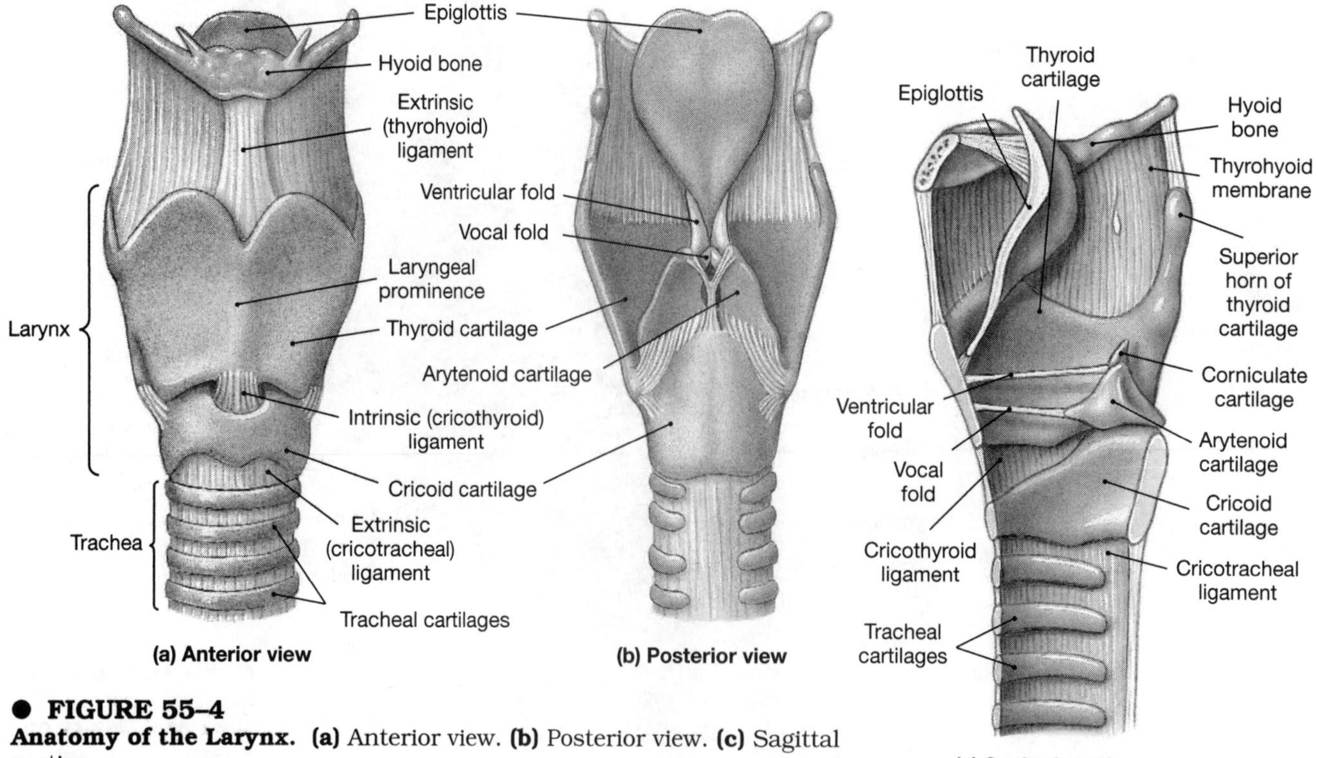

● **FIGURE 55–4**
Anatomy of the Larynx. **(a)** Anterior view. **(b)** Posterior view. **(c)** Sagittal section.

from the pharynx to the stomach. Note the relationship between the trachea and the esophagus.

3➡ Use Figure 55–4● for a more detailed view of the larynx. Identify the **thyroid cartilage** (the large, ventral cartilage that makes up most of the external larynx), the **cricoid cartilage** (the only complete cartilage ring located just inferior to the thyroid cartilage), and the **epiglottis** (the leaf-like flap superior to the thyroid cartilage). Locate the three paired laryngeal cartilages: the **arytenoid cartilages**, the **corniculate cartilages**, and the **cuneiform cartilages**.

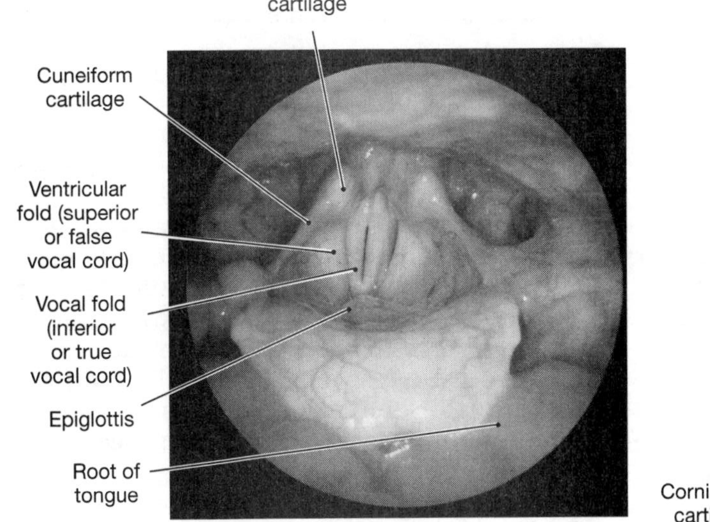

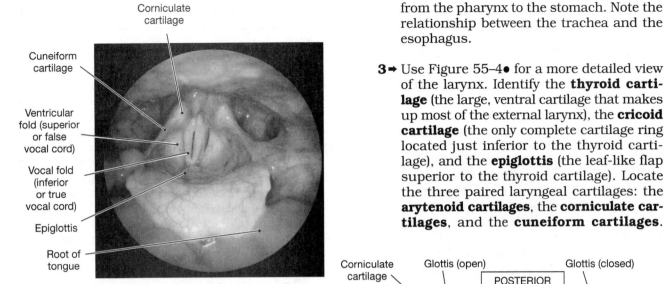

● **FIGURE 55–5a**
Vocal Cords and Surrounding Tissues.

● **FIGURE 55–5b**

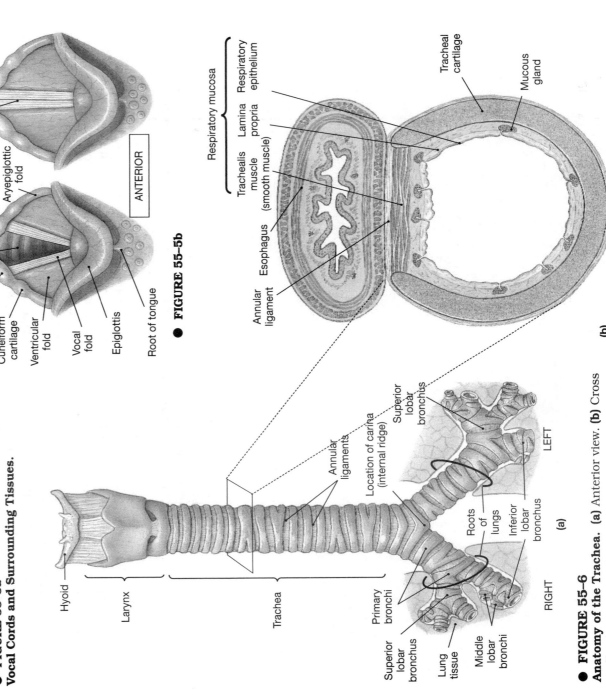

● **FIGURE 55–6**
Anatomy of the Trachea. (a) Anterior view. **(b)** Cross section.

(The cuneiforms are found in the human and dog, but not in the pig or cat.)

4➡ Refer to Figures 55–5a● and 55–5b● and review the labeled structures. Use the tracheal model to look down into the larynx. The hole you see is the **glottis**. (Keep in mind that the epiglottis is the cartilagenous flap that covers the glottis during swallowing.) On either side of the glottis you should be able to find two strips of membranous tissue, often described as membranous folds. These are the **vocal cords**. The **superior** or **false vocal cords** are above and lateral to the **inferior** or **true vocal cords**.

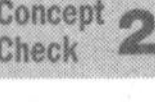

Concept Check 2 Which set of vocal cords is directly responsible for sound production?

____

5➡ Follow the trachea down toward the lungs. Notice the approximately 20 incomplete **tracheal cartilages**, which help prevent the trachea from collapsing. The incomplete portion — the ligamentous/muscular (posterior) portion — of the tracheal rings (C-rings), which lies anterior to the esophagus, makes swallowing food a bit easier. Why would that be true?

____

____

6➡ Study Figures 55–6a● and 55–6b●. What is the function of each tracheal layer listed in Figure 55–5b●?

____

____

C. The Bronchi and Their Subdivisions
[∞ Martini, p. 808]

1➡ Return to the trachea proper. Examine the bifurcation and identify the **right** and **left primary bronchi** (sing., **bronchus**). Note the location of the **carina,** the line of separation between the two bronchi. The **secondary bronchi** branch off the primary bronchi. Each secondary bronchus extends into a specific **lobe** of the **lung.** Occasionally in real life some variation in the branching patterns of the secondary bronchi can be found.

2➡ Use the models and Figure 55–7● to identify the lobes of the human lungs. The three right lobes are the **superior, middle**, and **inferior** lobes, respectively. The two left lobes are the **superior** and **inferior** lobes.

3➡ Refer to Figure 55–8● and continue on down the bronchial tree. You should be able to identify **tertiary** and perhaps even some **quaternary** bronchial branches.

4➡ As you explore farther and farther down the bronchial tree, you will notice that the cartilaginous rings become less and less obvious, eventually becoming blocks of cartilage rather than actual cartilage rings.

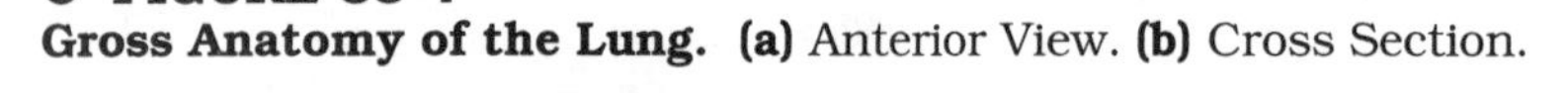

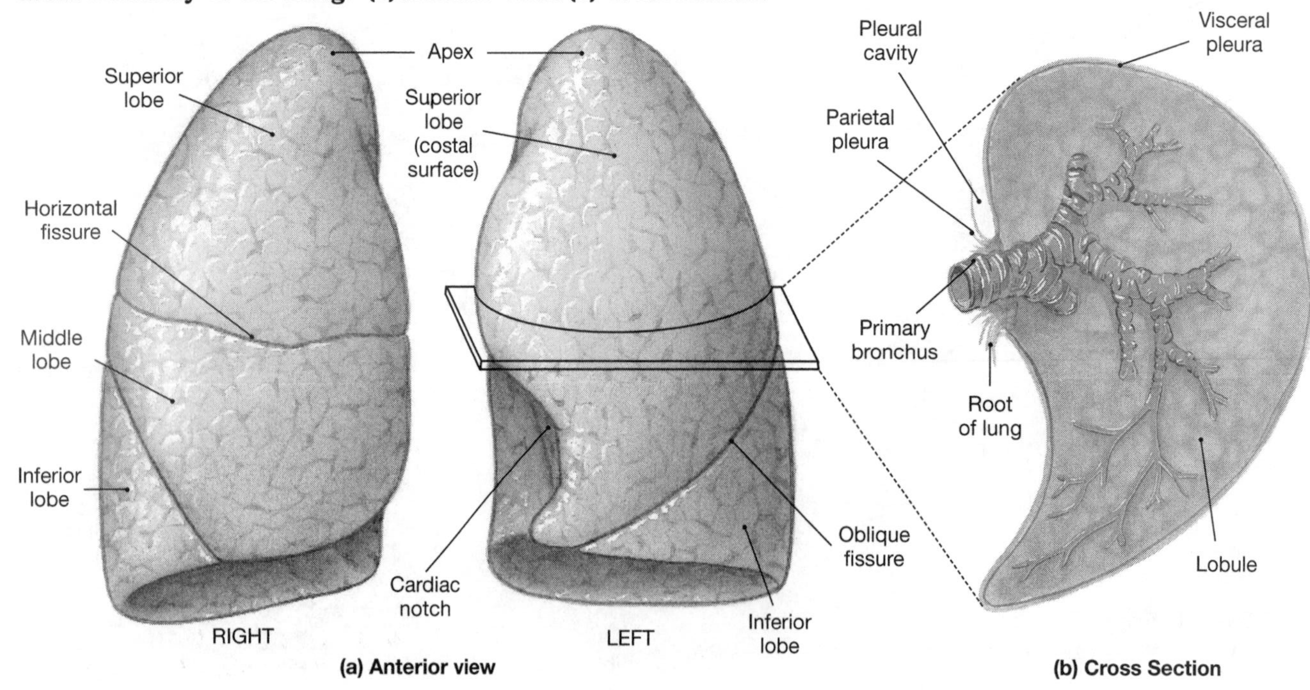

● **FIGURE 55–7**
Gross Anatomy of the Lung. **(a)** Anterior View. **(b)** Cross Section.

These smaller branches also have proportionately more and more smooth muscle.

5➡ When the bronchial lumen has narrowed to about 1 mm, cartilage is no longer present and the tube is now known as a **bronchiole**. Each bronchiole enters a **lobule**, where it further branches into **terminal bronchioles**. The terminal bronchioles give rise to **respiratory bronchioles**, which open into chambers called **alveolar ducts**. The blind pockets of these chambers are the **alveoli** (sing., **alveolus**).

D. The Alveolus [∞ Martini, p. 811]

1➡ Refer to the drawing of an alveolus in Figure 55–9●. Note that that drawing is a continuation from Figure 55–8●. The alveolus is the functional unit of the lung. In the alveolus, gas exchange between the environment and the circulatory system takes place.

Concept Check 3 Check your lecture text. Approximately how many alveoli can be found in each lung?

____

2➡ Within each alveolar sac are **septal** or **surfactant** cells scattered among the squamous epithelial cells. The septal cells secrete a phospholipid mixture known as surfactant, which functions in maintaining the integrity of the alveolus. Also within the alveoli are wandering alveolar macrophages, the finely tuned clean-up crew of the lower respiratory tract.

● **FIGURE 55–8**
Organizational Pattern of the Lung.

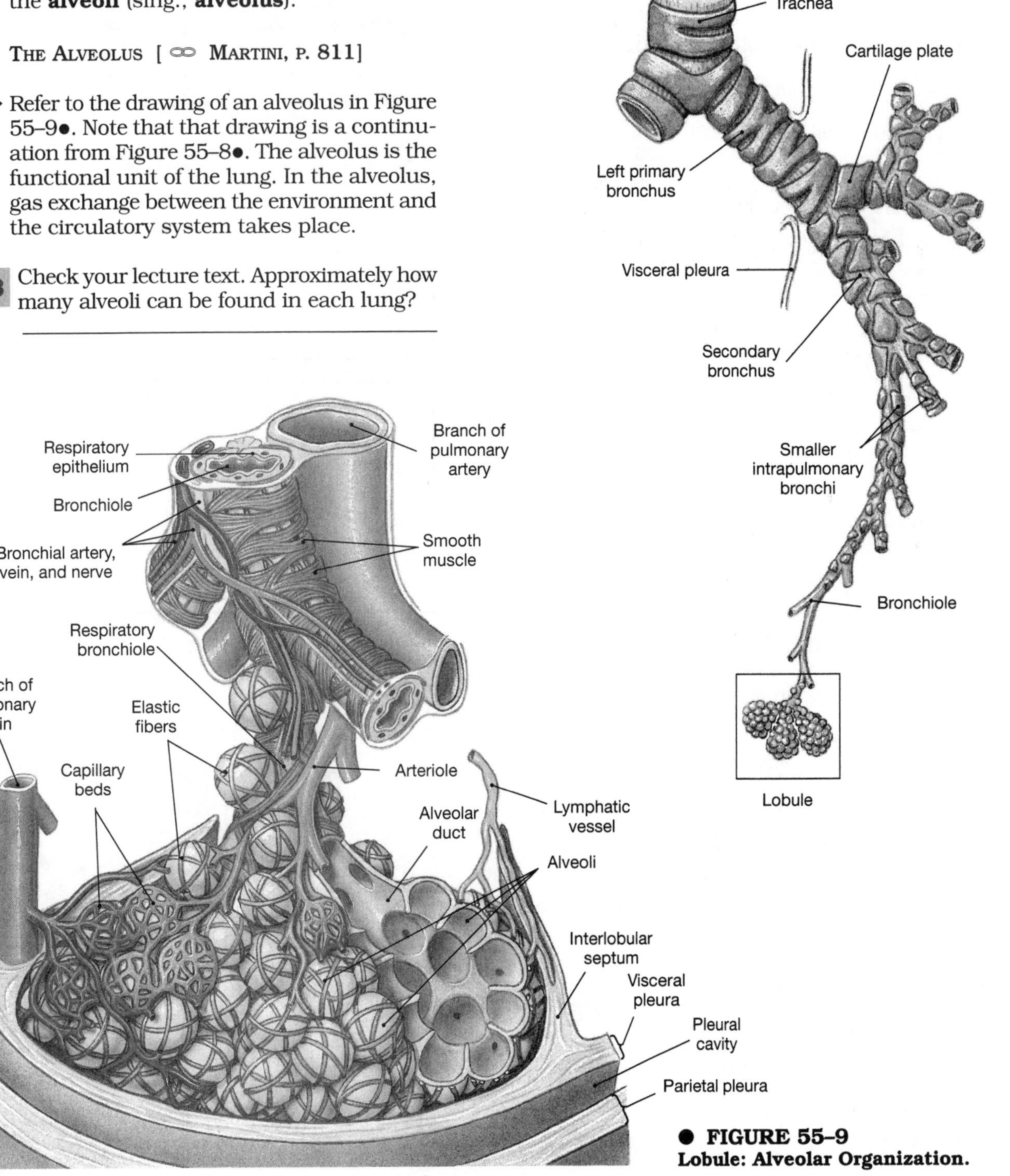

● **FIGURE 55–9**
Lobule: Alveolar Organization.

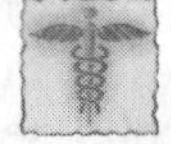

Premature babies often suffer from inadequately developed respiratory systems. This can result in any of several potentially lethal conditions. One common problem is neonatal respiratory distress syndrome (NRDS), sometimes known as hyaline membrane disease (HMD). With this syndrome, surfactant levels fail to reach normal levels. Sometimes premature babies will produce adequate amounts of surfactant but the biochemical structure of the surfactant will be abnormal. With inadequate or abnormal surfactant, the alveolar sacs lose their integrity and are subject to collapse. When the alveolar sacs collapse, gas exchange is not possible.

III. Microscopic Examination

A. Dissecting Microscopy

1 ➡ If any gross sections of the respiratory system are available, examine these under the dissecting microscope. Your lab may possibly have laryngeal, tracheal, or bronchial cross sections for you to study. If such structures are available, sketch and label what you see.

B. Prepared Slides

1 ➡ Obtain a compound microscope and prepared slides of the columnar (respiratory) epithelium, hyaline (tracheal) cartilage, bronchiole cross section, and alveolar section. (You may have examined some of these slides in previous exercises. If so, reexamine the structures with an eye toward anatomical location and physiological function.)

2 ➡ Identify the following structures. Use Figure 55–10● as your guide.

Epithelium Goblet cells, cilia.

Hyaline Cartilage Matrix, chondrocytes (not identified below).

Bronchiole Cross Section Respiratory epithelium, smooth muscle.

Alveolus Bronchiole, arteriole, alveolus, alveolar duct.

Sketch and label what you see.

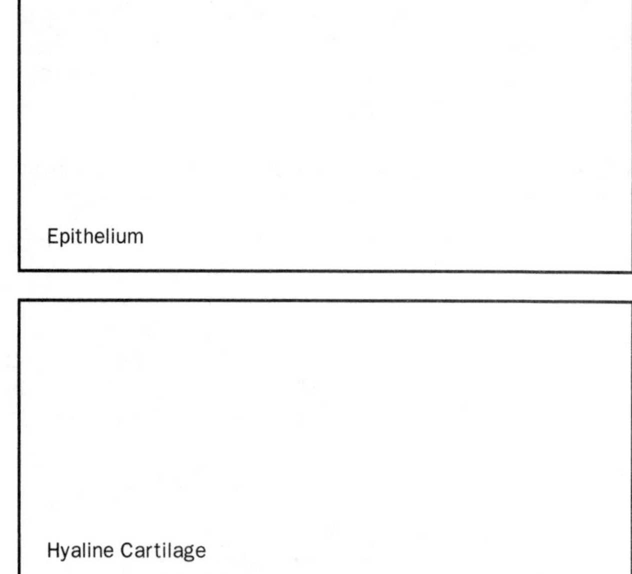

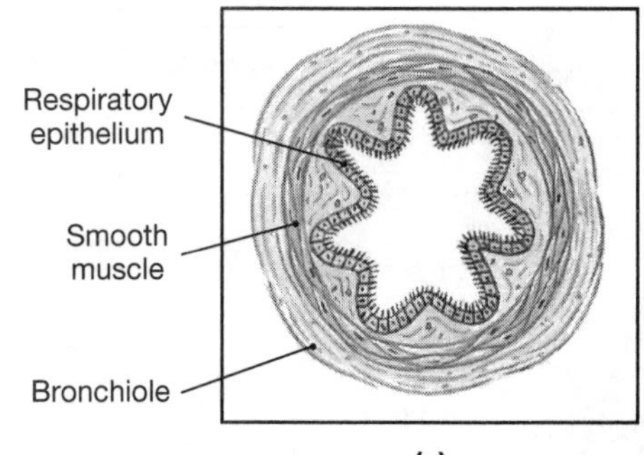

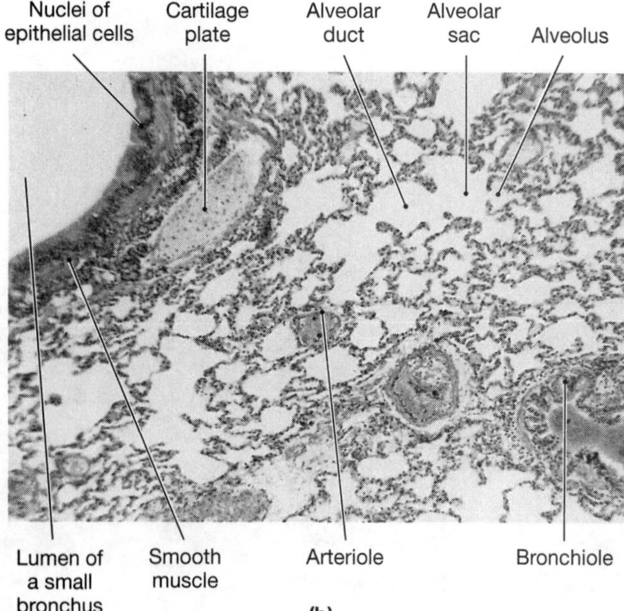

● **FIGURE 55–10a,b**
Microscopic Organization. (a) Bronchiole. **(b)** Alveolar structures (continued on next page).

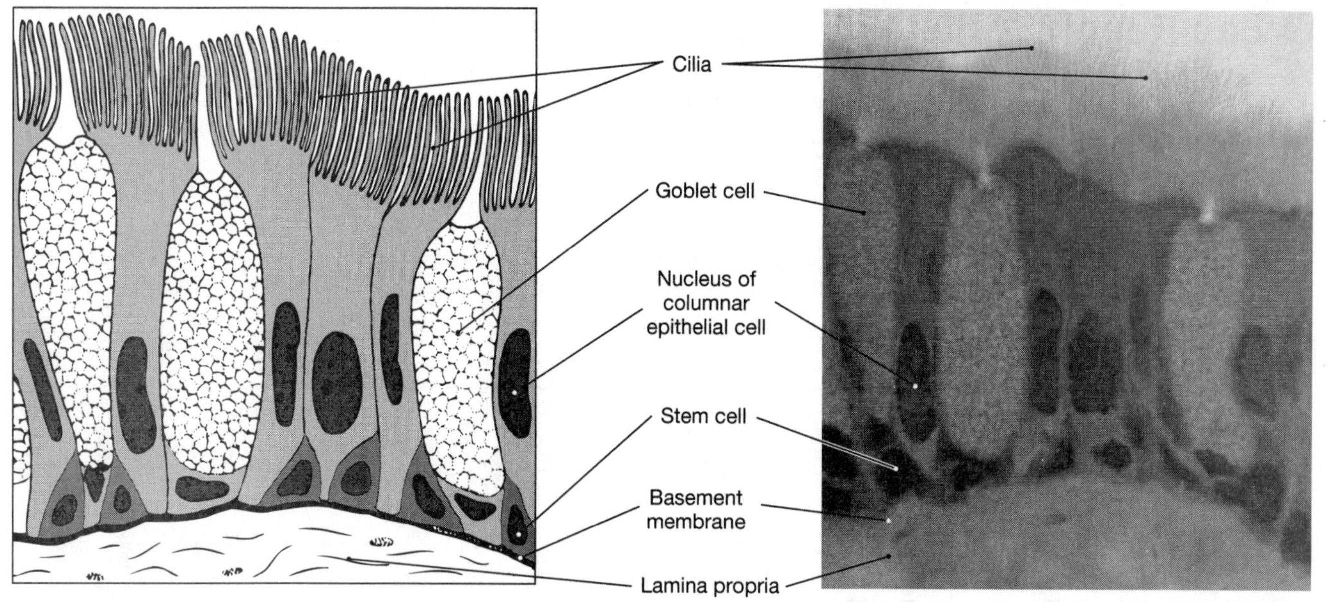

● **FIGURE 55–10c**
Microscopic Organization. **(c)** Respiratory epithelium.

Bronchiole

Alveolus

❑ Additional Activities

NOTES

1. Research and sketch the ligaments and skeletal muscles found in the region of the larynx and trachea.
2. Practice making different sounds. What are you doing to the vocal cords?

Answers to Selected Concept Check Questions

1. The nasal conchae increase the surface area within the nasal cavity. This aids in the humidification and warming of the air as it passes to the lungs.
2. The inferior or true vocal cords are responsible for sound production.
3. Each lung is composed of about 150 million alveoli.

❑ Lab Report

1. Referring to the numbered inquiries at the beginning of this exercise, complete the following box summary:

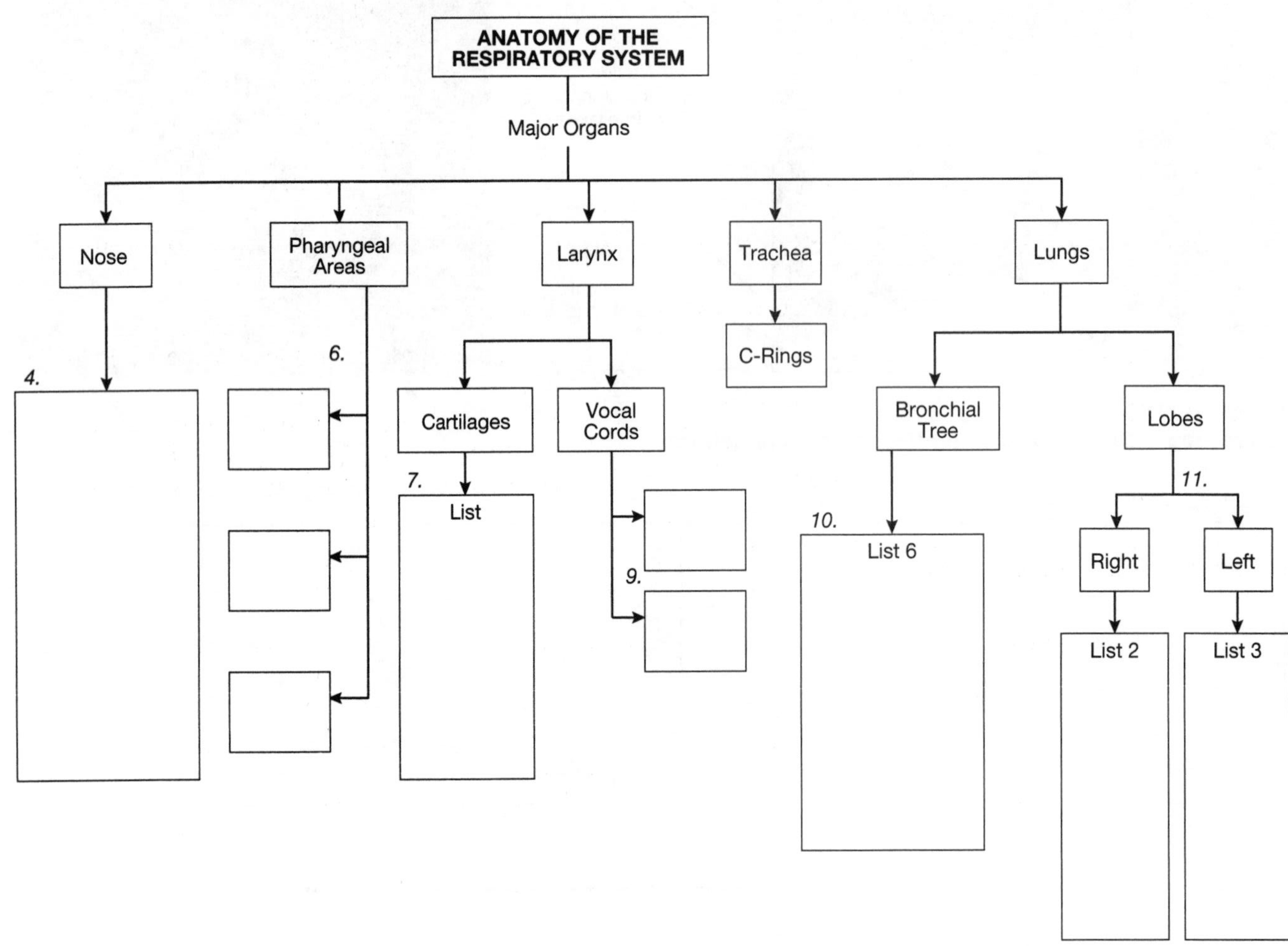

General Questions

1. What is the function of the epiglottis?
2. Do you see any anatomical reason why the right lung has more lobes than the left lung — and is subsequently larger and heavier?
3. What physiological advantage do you see in the changing proportions of muscle and cartilage as you descend the bronchial tree?
4. What are the sinuses and where are the major sinuses located?
5. What is the difference between pulmonary ventilation and external respiration?
6. What is the difference between inhalation and exhalation?
7. How is the cardiovascular system related to the respiratory system?
8. What is the function of surfactant?

EXERCISE **56**

Respiratory Physiology

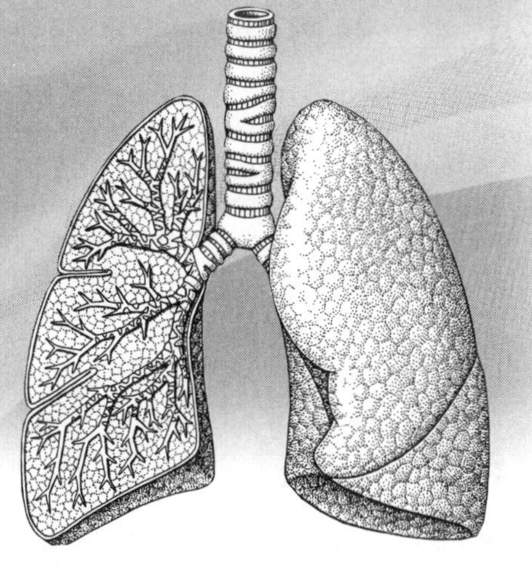

PROCEDURAL INQUIRIES

Preparation

1. What are the different types of respiration?
2. What is ventilation?
3. What is the role of the diaphragm in inspiration and expiration?
4. What is the role of the intercostal muscles in inspiration and expiration?

Experimentation

5. How does Boyle's Law relate to breathing?
6. How is Dalton's Law related to the events taking place in the alveolus?
7. What is spirometry?
8. What are some common volumetric terms related to breathing?
9. What can be determined by the forced expiratory volume demonstration?
10. How can we demonstrate carbon dioxide as a respiratory waste product?

Additional Inquiries

11. What is the function of each of the respiratory centers?
12. What is Boyle's Law?
13. What is Dalton's Law?
14. How do emphysema and lung cancer affect respiration?

Key Terms

Apneustic Center
Cellular Respiration
Diaphragm
Expiration
Expiratory Reserve Volume
External Respiration
Forced Expiratory Volume (FEV_T)
Hypercapnia
Hyperventilation
Inspiration
Inspiratory Reserve Volume
Internal Respiration
Minimal Volume
Pneumograph
Pneumotaxic Center
Pulmonic Expiration
Pulmonic Inspiration
Residual Volume
Respiration
Respiratory Center
Rhythmicity Center
Spirometry
Tidal Volume
Ventilation
Vital Capacity

Materials Needed

Plastic Lung Apparatus
Spirometer
Beaker (or Cup)
Water
Phenolphthalein
(0.1 N NaOH, if needed)
Straw
Watch (Clock) with Second Hand
Cloth Tape Measure
Compound Microscope
Prepared Slides
Normal Lung Tissue
Emphysema
Lung Cancer
Optional: Pneumograph

In the previous exercise, we examined the anatomy of the respiratory system, the system responsible for gas exchange between the environment and the inner parts of the body. In this exercise, we will examine some of the physiological mechanisms involved in the passage of oxygen from the atmosphere into the body proper. We will do this first by looking at the physiological principles of pressure gradients and equilibrium and then by examining some common respiratory measurements.

❑ Preparation

I. Background

A. Vocabulary

Respiration Gas exchange — at any level, including within the cell in the production of energy. In the colloquial sense, respiration is the act of breathing. In the strictest sense, however, breathing has very little to do with respiration. (What we colloquially call breathing is scientifically called ventilation.) If respiration is gas exchange, we must follow a gas (oxygen) from the environment, through the lungs, through the blood stream, into a given cell where it is utilized in the molecular breakdown of food in order to give off energy, and where it is exchanged for another gas (carbon dioxide) that then passes back through the blood vessels, through the lungs, and out into the environment.

Ventilation Breathing — the mechanics of expanding and contracting the lungs, allowing for the physical movement of air into and out of the lungs.

External Respiration Gas exchange between the lungs (the external environment) and the interstitial fluids of the body circulatory system.

Internal Respiration Gas exchange at the level of the individual cell.

Cellular Respiration The consumption and utilization of oxygen and the concurrent generation of carbon dioxide by the living cell. This is possible because of gas exchange between the interstitial fluid and the individual cell.

Each of these respiratory concepts involves gas exchange. Gases are moving in both directions; one gas is utilized, another gas is generated.

B. Anatomy

We covered the anatomy of the respiratory system in Exercise 55. If it has been some time since you studied this topic, go back and review basic respiratory anatomy before completing the following physiological exercises.

II. Mechanics of Ventilation

A. Diagrams [∞ Martini, p. 814]

1➜ Study Figure 56–1●. Identify the dome-shaped **diaphragm**. The diaphragm is composed of skeletal muscle. When the diaphragm contracts, it flattens out and the volume of the pleural cavity increases. Air rushes in. This is **inhalation** or **inspiration**, the process of taking air into the lungs.

The diaphragm is not the only muscle of inhalation. The external intercostal muscles, plus numerous accessory muscles of the neck and thorax, also contract to increase the volume of the pleural cavity.

2➜ When the diaphragm relaxes, it resumes its dome-shaped appearance, and the volume of the pleural cavity decreases. Air rushes out. This is **exhalation** or **expiration**, the process of expelling air from the lungs. The internal intercostals plus several of the muscles of the thorax and abdomen contract to aid in exhalation. Both inspiration and expiration are the passive results of movement down a pressure gradient. (Keep in mind nature's tendency toward equilibrium.)

3➜ Consider the muscle involvement again. Primary contraction is accomplished by the diaphragm. Upon contraction, the diaphragm causes an increase in the vertical measurement, and consequently the volume, of the thoracic cavity.The external intercostal and accessory muscles also contract, allowing for lateral and anteroposterior expansion of the thoracic cavity through their lifting and expanding action. Thus, on inspiration, we accomplish a three-dimensional volumetric expansion of the thoracic cavity.

On expiration, the diaphragm relaxes, and the internal intercostals and accessory muscles all contract, causing a three-dimensional volumetric decrease in the thoracic cavity.

❑ Experimentation

B. Pressure Change Demonstration

1➜ Examine the plastic lung apparatus (see Figure 56–2●). Note that if you pull down on the rubber "diaphragm" at the base of the structure, you increase the volume of the "chest cavity" without increasing the number of molecules of air present. This

● **FIGURE 56–1**
Respiratory Anatomy.

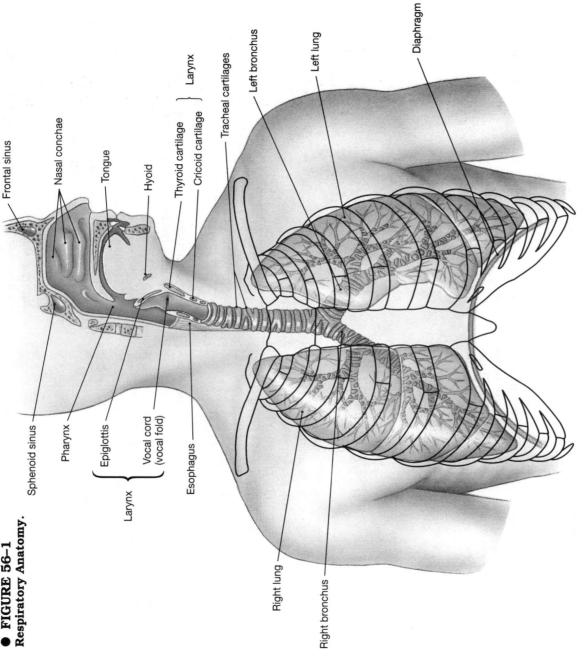

decreases the internal pressure. Air rushes into the lungs down the pressure gradient, equalizing the pressure between the chest cavity and the environment. The lungs inflate. You have simulated **pulmonic inspiration.**

2➡ When you release the “diaphragm,” the internal “chest cavity” pressure builds up. This pressure exerts a force on the lungs, and the air rushes out of the lungs down a pressure gradient, equalizing the pressure between chest cavity and the environment. The lungs deflate. You have simulated **pulmonic expiration**.

C. Application of Boyle's Law

1➡ You have also just demonstrated Boyle's Law, named after the British physicist Robert Boyle (1627–1691), who showed that

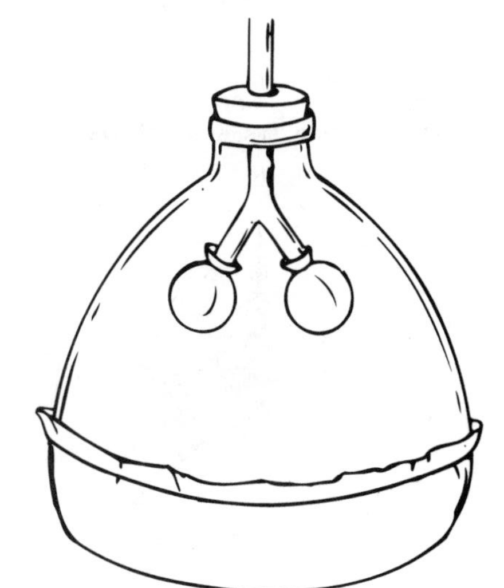

● **FIGURE 56–2**
Simulated Respiratory Apparatus.

the volume of a given mass of gas (at a constant temperature) varies inversely with the pressure put upon it. If you have had a course in chemistry, you may recall doing problems involving Boyle's Law. Although it is beyond the scope of this laboratory exercise to do exact computations involving Boyle's Law, you can easily measure the differences in the circumference of your chest (at about the level of the 6th rib) during forced inspiration and forced expiration.

What is your 6th rib chest circumference at maximum forced inspiration? ______

What is your 6th rib chest circumference at maximum forced expiration? ______

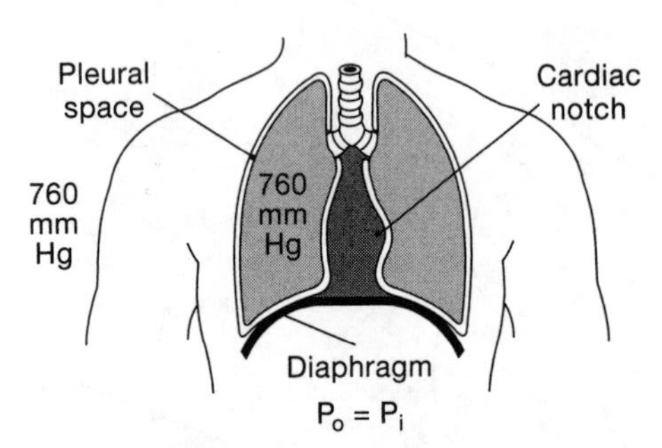

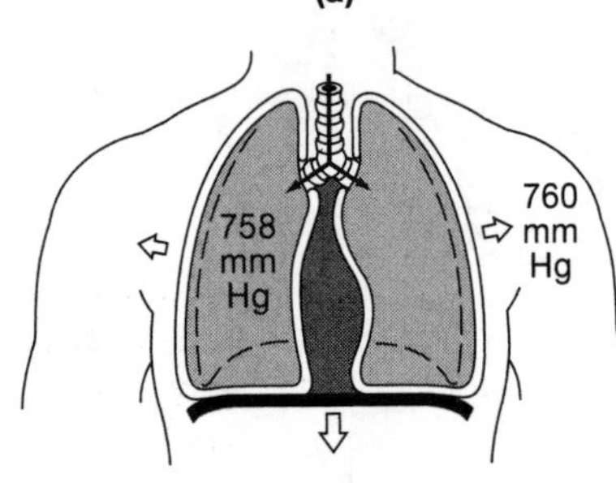

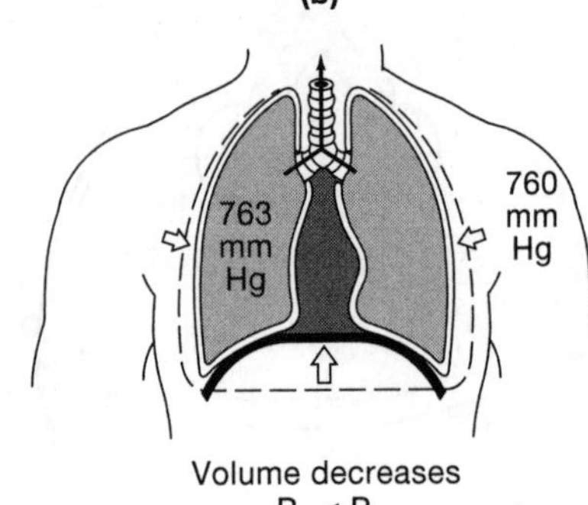

● **FIGURE 56–3**
Pulmonary Pressures.

What is the difference in these two measurements? ______

Compare your difference with the differences obtained by some of your classmates.

2➡ Follow the changes in pulmonary pressure shown in Figure 56–3● and the resulting movements of air in and out of the lungs. The pressures shown are some commonly used intrapulmonic and extrapulmonic pressure values. Look up the definitions of intrapulmonic and extrapulmonic.

Intrapulmonic ______

Extrapulmonic ______

For biological purposes, pressure is generally measured in mm Hg (millimeters mercury). Standard atmospheric pressure is 760 mm Hg. Note that all that is needed to achieve inspiration is a difference of a few mm Hg.

III. Passive Diffusion

A. Dalton's Law [∞ Martini, p. 822]

We have demonstrated that air moves in and out of the lung by moving down a pressure gradient toward equilibrium. Your lab is probably not equipped to demonstrate a respiratory concentration gradient concept. Nevertheless, it is important that you conceptualize this idea.

Dalton's Law (named after the English chemist John Dalton, 1766–1844) states that the total pressure exerted by a volume of gas is equal to the sum of all the partial pressures of the gases in the mixture. Figure 56–4● gives an example of how the partial pressures of carbon dioxide and of oxygen vary in different locations and the effect this has on the direction of movement of these gases. The lowercase "p" identifies the partial pressure of the gas in question. Note that carbon dioxide and oxygen always move toward equilibrium, from an area of higher concentration to an area of lower concentration. This is true for both gases. Note that to follow this concentration gradient, carbon dioxide and oxygen are moving in opposite directions!

1➡ Study Figure 56–4● and explain in your own words what happens to oxygen at the alveolar and at the cellular levels. Do the same for carbon dioxide. Note the role of partial pressure as it applies to external, internal, and cellular respiration.

______

______

______

______

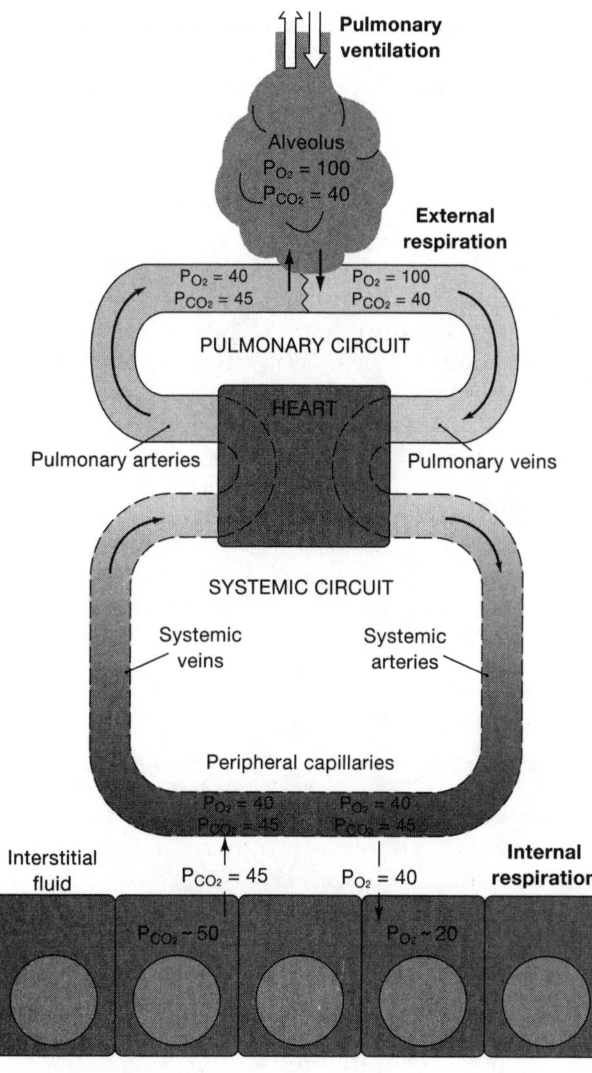

● **FIGURE 56–4**
Oxygen/Carbon Dioxide Partial Pressures.

IV. Spirometry

A. The Spirometer

1➡ **Spirometry** is used to demonstrate volumes involved in the passive movement of gases (air) down a pressure gradient. Spirometry can be used to measure whether adequate inspirational and expirational volumes are achieved in breathing. Adequate volumes are, of course, based on a person's age, sex, and height.

B. Performance/Volume Relationships
[∞ Martini, p. 821]

In order to understand the following volumes involved in spirometry, study Figure 56–5●, which demonstrates respiratory performance/volume relationships. Make note of each of the following definitions.

Tidal Volume The amount of air inhaled or exhaled in a normal breathing cycle.

Inspiratory Reserve Volume The amount of air beyond the tidal volume that can be forcefully inhaled.

Expiratory Reserve Volume The amount of air beyond the tidal volume that can be forcefully exhaled.

Vital Capacity The tidal volume plus the inspiratory reserve plus the expiratory reserve.

Residual Volume The amount of air remaining in the lungs after the expiratory reserve is depleted.

Minimal Volume The amount of air remaining in the lungs if they are allowed to collapse because of the alveolar surfactant coating. This volume remains regardless of other volumes or actions taken.

C. Demonstration of the Spirometer

Several types of spirometers exist, some with digital, some with mechanical, and some with computerized volume recorders. Most, but not all, can be attached to a chart recorder. Your instructor will give you specific directions on how to read and record the values obtained from your spirometer. Regardless of the type of spirometer in your lab, however, your function in operating the device will be approximately the same. Figure 56–6● is a diagram of a generic spirometer.

Perform these volume tests *three times* and then find the average value for each.

1➡ Obtain a tight-fitting mouthpiece. Stand (so that your lungs can expand completely) and breath normally. When you are relaxed, take a normal inspiration, plug your nose tightly, place the mouthpiece in your mouth, and forcefully perform a maximum exhalation. Remove the mouthpiece. The value on the spirometer will be the sum of your tidal volume plus your expiratory reserve. Record your value on Figure 56–7●.

2➡ Breathe normally and, after a normal exhalation, put the mouthpiece back in your mouth and perform a maximum exhalation. This value represents your expiratory reserve. Subtract this number from the value obtained in the first step and you have your tidal volume. Record your value on Figure 56–7●.

3➡ Breathe normally for a few moments, then perform a maximum inspiration. Insert the mouthpiece and perform a maximum exhalation. The value obtained will be your vital

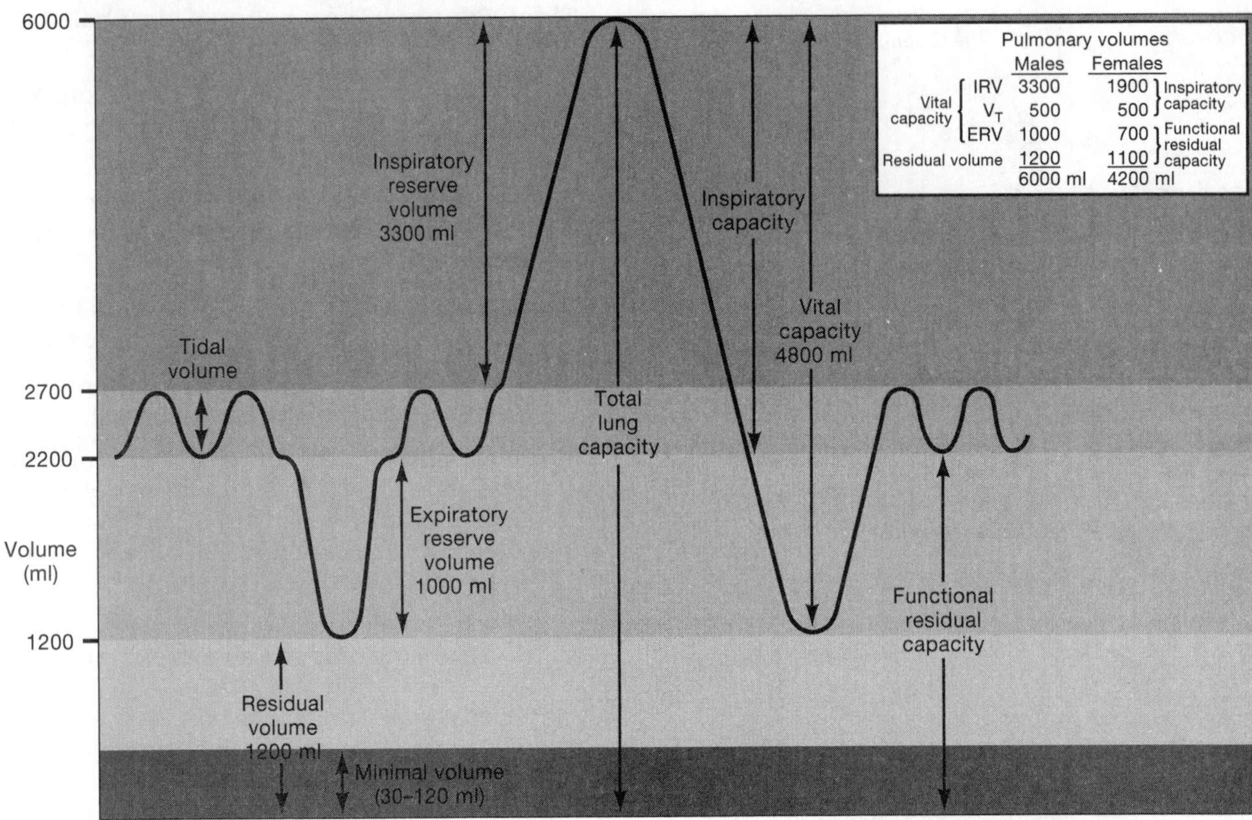

● **FIGURE 56–5**
Respiratory Performance and Volume Relationships.

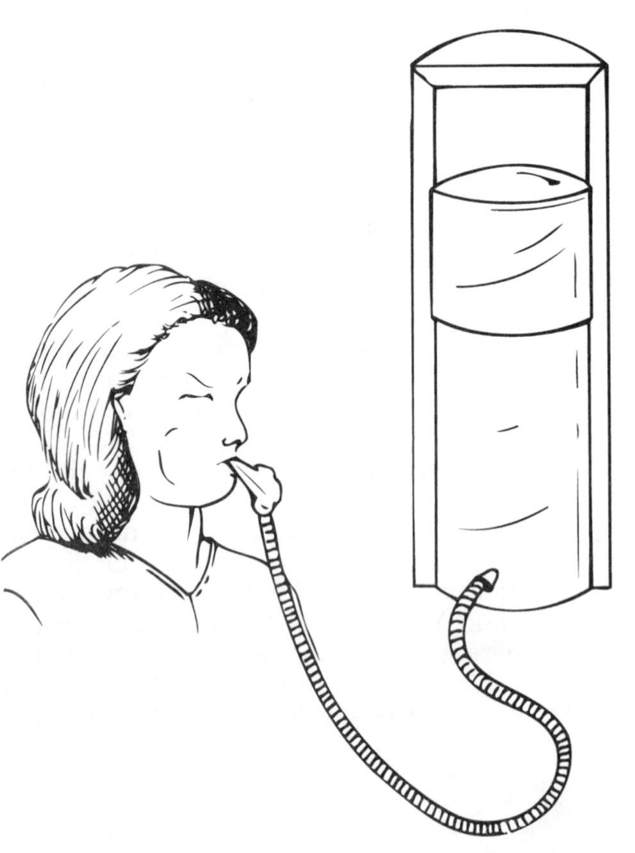

● **FIGURE 56–6**
Diagrammatic Spirometer.

capacity. By subtracting the value obtained in step one from this value, you will have your inspiratory reserve volume. Record your value on Figure 56–7●.

D. Normative Comparisons

1➡ Find the vital capacity for a person of your age, sex, and height on the vital capacity norm chart, Figure 56–8●.

2➡ Compute your own vital capacity as a percentage of the norm by dividing your average by the number you located on the chart. ____________________

3➡ Note the results recorded for each of the three trials in Figure 56–7●. List some reasons why your three trials might produce very diverse results.

____
____
____

4➡ Use the space at the end of this exercise to construct a diagram like Figure 56–5●. Insert your own (average) results in place of the normative results.

	Tidal Volume & Expiratory Reserve	*Expiratory Reserve*	*Tidal Volume*	*Vital Capacity*	*Inspiratory Reserve*
Trial #1					
Trial #2					
Trial #3					
Average					

● **FIGURE 56–7**
Spirometry Calculations.

E. Forced Expiratory Volume Demonstration

1➡ If your spirometer has a timer or a timed recorder, you may be able to perform the **forced expiratory volume (FEV_T)** test. Forced expiration is equivalent to vital capacity, the amount you are able to forcibly expel after a maximum inspiration. The T represents a given time. The lung associations often do this test to check the amount of air exhaled during the first second of forced expiration. To figure your FEV_1, FEV_2, and FEV_3 use the timer on your spirometer and note the amount of air expelled in the first second, the second second, and the third second, respectively. Record those amounts here.

1st ______ 2nd ______ 3rd ______

Depending on the type of spirometer you have, you may have to take temperature into account when computing your FEV. Follow your instructor's directions.

2➡ Take your FEVs as a percentage of your vital capacity. Normal individuals should be able to expel about 83% during the first second, 94% after two seconds, and 97% after three seconds. If your spirometer does not have a timer, you might be able to obtain a rough estimate of the value with a good stopwatch and an observant partner. Record your results here.

1. ______ 2. ______ 3. ______

V. Physiological Mechanisms

A. Demonstration of Carbon Dioxide as a Waste Product

1➡ Gas exchange is strictly passive. However, the mechanics for triggering ventilation are not passive. To begin our examination of the physiological triggers, let us first demonstrate that, despite passive movement in and out of the lungs, gas exchange is taking place.

2➡ In nature, if carbon dioxide is added to water, the following reaction takes place:

$$CO_2 + H_2O \longrightarrow H_2CO_3$$

The carbonic acid then dissociates, forming the hydrogen ion and the bicarbonate ion:

$$H_2CO_3 \longrightarrow H^+ + HCO_3^-$$

This reaction increases the hydrogen ion in solution, thus lowering the pH.

If you start with a basic solution and add enough hydrogen ions, your solution will become acidic.

3➡ Take a beaker of plain water and add a few drops of phenolphthalein. Phenolphthalein has a magenta color in a basic solution. (If your water is not slightly alkaline, add a few drops of NaOH.) Is your solution basic?

____

4➡ Check the time. Use the straw and blow into the water until the color disappears. How long did it take? ______________

Concept Check 1 Since we have already stated that carbon dioxide is a waste product, what must be happening for the water to become clear when you blew into it?

____

____

5➡ If we want to demonstrate that this carbon dioxide was the result of some specific body function, we could do something to raise the carbon dioxide level. Run in place or perform some other vigorous exercise for about two minutes. Now repeat the experiment. How long did it take for the water to become clear? ______________

Concept Check 2 How can you account for the difference in times? ______________

____

	Height in centimeters and inches																		
cm	152	154	156	158	160	162	164	166	168	170	172	174	176	178	180	182	184	186	188
Age in.	59.8	60.6	61.4	62.2	63.0	63.7	64.6	65.4	66.1	66.9	67.7	68.5	69.3	70.1	70.9	71.7	72.4	73.2	74.0
16	3,070	3,110	3,150	3,190	3,230	3,270	3,310	3,350	3,390	3,430	3,470	3,510	3,550	3,590	3,630	3,670	3,715	3,755	3,800
17	3,055	3,095	3,135	3,175	3,215	3,255	3,295	3,335	3,375	3,415	3,455	3,495	3,535	3,575	3,615	3,655	3,695	3,740	3,780
18	3,040	3,080	3,120	3,160	3,200	3,240	3,280	3,320	3,360	3,400	3,440	3,480	3,520	3,560	3,600	3,640	3,680	3,720	3,760
20	3,010	3,050	3,090	3,130	3,170	3,210	3,250	3,290	3,330	3,370	3,410	3,450	3,490	3,525	3,565	3,605	3,645	3,695	3,720
22	2,980	3,020	3,060	3,095	3,135	3,175	3,215	3,255	3,290	3,330	3,370	3,410	3,450	3,490	3,530	3,570	3,610	3,650	3,685
24	2,950	2,985	3,025	3,065	3,100	3,140	3,180	3,220	3,260	3,300	3,335	3,375	3,415	3,455	3,490	3,530	3,570	3,610	3,650
26	2,920	2,960	3,000	3,035	3,070	3,110	3,150	3,190	3,230	3,265	3,300	3,340	3,380	3,420	3,455	3,495	3,530	3,570	3,610
28	2,890	2,930	2,965	3,000	3,040	3,070	3,115	3,155	3,190	3,230	3,270	3,305	3,345	3,380	3,420	3,460	3,495	3,535	3,570
30	2,860	2,895	2,935	2,970	3,010	3,045	3,085	3,120	3,160	3,195	3,235	3,270	3,310	3,345	3,385	3,420	3,460	3,495	3,535
32	2,825	2,865	2,900	2,940	2,975	3,015	3,050	3,090	3,125	3,160	3,200	3,235	3,275	3,310	3,350	3,385	3,425	3,460	3,495
34	2,795	2,835	2,870	2,910	2,945	2,980	3,020	3,055	3,090	3,130	3,165	3,200	3,240	3,275	3,310	3,350	3,385	3,425	3,460
36	2,765	2,805	2,840	2,875	2,910	2,950	2,985	3,020	3,060	3,095	3,130	3,165	3,205	3,240	3,275	3,310	3,350	3,385	3,420
38	2,735	2,770	2,810	2,845	2,880	2,915	2,950	2,990	3,025	3,060	3,095	3,130	3,170	3,205	3,240	3,275	3,310	3,350	3,385
40	2,705	2,740	2,775	2,810	2,850	2,885	2,920	2,955	2,990	3,025	3,060	3,095	3,135	3,170	3,205	3,240	3,275	3,310	3,345
42	2,675	2,710	2,745	2,780	2,815	2,850	2,885	2,920	2,955	2,990	3,025	3,060	3,100	3,135	3,170	3,205	3,240	3,275	3,310
44	2,645	2,680	2,715	2,750	2,785	2,820	2,855	2,890	2,925	2,960	2,995	3,030	3,060	3,095	3,130	3,165	3,200	3,235	3,270
46	2,615	2,650	2,685	2,715	2,750	2,785	2,820	2,855	2,890	2,925	2,960	2,995	3,030	3,060	3,095	3,130	3,165	3,200	3,235
48	2,585	2,620	2,650	2,685	2,715	2,750	2,785	2,820	2,855	2,890	2,925	2,960	2,995	3,030	3,060	3,095	3,130	3,160	3,195
50	2,555	2,590	2,625	2,655	2,690	2,720	2,755	2,785	2,820	2,855	2,890	2,925	2,955	2,990	3,025	3,060	3,090	3,125	3,155
52	2,525	2,555	2,590	2,625	2,655	2,690	2,720	2,755	2,790	2,820	2,855	2,890	2,925	2,955	2,990	3,020	3,055	3,090	3,125
54	2,495	2,530	2,560	2,590	2,625	2,655	2,690	2,720	2,755	2,790	2,820	2,855	2,885	2,920	2,950	2,985	3,020	3,050	3,085
56	2,460	2,495	2,525	2,560	2,590	2,625	2,655	2,690	2,720	2,755	2,790	2,820	2,855	2,885	2,920	2,950	2,980	3,015	3,045
58	2,430	2,460	2,495	2,525	2,560	2,590	2,625	2,655	2,690	2,720	2,750	2,785	2,815	2,850	2,880	2,920	2,945	2,975	3,010
60	2,400	2,430	2,460	2,495	2,525	2,560	2,590	2,625	2,655	2,685	2,720	2,750	2,780	2,810	2,845	2,875	2,915	2,940	2,970
62	2,370	2,405	2,435	2,465	2,495	2,525	2,560	2,590	2,620	2,655	2,685	2,715	2,745	2,775	2,810	2,840	2,870	2,900	2,935
64	2,340	2,370	2,400	2,430	2,465	2,495	2,525	2,555	2,585	2,620	2,650	2,680	2,710	2,740	2,770	2,805	2,835	2,865	2,895
66	2,310	2,340	2,370	2,400	2,430	2,460	2,495	2,525	2,555	2,585	2,615	2,645	2,675	2,705	2,735	2,765	2,800	2,825	2,860
68	2,280	2,310	2,340	2,370	2,400	2,430	2,460	2,490	2,520	2,550	2,580	2,610	2,640	2,670	2,700	2,730	2,760	2,795	2,820
70	2,250	2,280	2,310	2,340	2,370	2,400	2,425	2,455	2,485	2,515	2,545	2,575	2,605	2,635	2,665	2,695	2,725	2,755	2,780
72	2,220	2,250	2,280	2,310	2,335	2,365	2,395	2,425	2,455	2,480	2,510	2,540	2,570	2,600	2,630	2,660	2,685	2,715	2,745
74	2,190	2,220	2,245	2,275	2,305	2,335	2,360	2,390	2,420	2,450	2,475	2,505	2,535	2,565	2,590	2,620	2,650	2,680	2,710

● **FIGURE 56–8a**
Predicted Vital Capacities — Females.

B. THE RESPIRATORY CENTERS

1 ➡ The **respiratory centers** of the brain control the rate and rhythm, pace and depth of respiration. The **medullary rhythmicity center** establishes the pace, the **pneumotaxic** and **apneustic centers** modify the pace.

The pace of respiration is influenced by the metabolic or chemical activities of the neural tissues. These influences can range from changes in body temperature to the introduction of stimulants or depressants. Mechanical reflexes such as those activated by the stretch receptors play a major role in respiratory pace. If the blood pressure falls (decreasing the "stretch" on the stretch receptors), the respiratory rate will climb.

The carbon dioxide level in the blood is the trigger for the chemoreceptor reflex. **Hypercapnia**, which is an increase in the partial pressure of carbon dioxide, stimulates the peripheral chemoreceptors of the carotid and aortic bodies. The carbon dioxide will also cross the blood-brain barrier, causing an increase in the carbon dioxide concentration in the cerebrospinal fluid. This will stimulate the central chemoreceptive neurons in the medulla. This will cause **hyperventilation**, an increase in the rate and depth of respiration. [∞ *Martini, p. 835]*

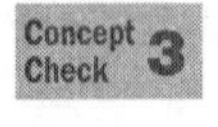

From the standpoint of carbon dioxide levels, why do you breathe more heavily after strenuous exercise?

____

____

If someone starts to "hyperventilate" in the colloquial sense, explain why that person is advised

	Height in centimeters and inches																		
cm	152	154	156	158	160	162	164	166	168	170	172	174	176	178	180	182	184	186	188
Age in.	59.8	60.6	61.4	62.2	63.0	63.7	64.6	65.4	66.1	66.9	67.7	68.5	69.3	70.1	70.9	71.7	72.4	73.2	74.0
16	3,920	3,975	4,025	4,075	4,130	4,180	4,230	4,285	4,335	4,385	4,440	4,490	4,540	4,590	4,645	4,695	4,745	4,800	4,850
18	3,890	3,940	3,995	4,045	4,095	4,145	4,200	4,250	4,300	4,350	4,405	4,455	4,505	4,555	4,610	4,660	4,710	4,760	4,815
20	3,860	3,910	3,960	4,015	4,065	4,115	4,165	4,215	4,265	4,320	4,370	4,420	4,470	4,520	4,570	4,625	4,675	4,725	4,775
22	3,830	3,880	3,930	3,980	4,030	4,080	4,135	4,185	4,235	4,285	4,335	4,385	4,435	4,485	4,535	4,585	4,635	4,685	4,735
24	3,785	3,835	3,885	3,935	3,985	4,035	4,085	4,135	4,185	4,235	4,285	4,330	4,380	4,430	4,480	4,530	4,580	4,630	4,680
26	3,755	3,805	3,855	3,905	3,955	4,000	4,050	4,100	4,150	4,200	4,250	4,300	4,350	4,395	4,445	4,495	4,545	4,595	4,645
28	3,725	3,775	3,820	3,870	3,920	3,970	4,020	4,070	4,115	4,165	4,215	4,265	4,310	4,360	4,410	4,460	4,510	4,555	4,605
30	3,695	3,740	3,790	3,840	3,890	3,935	3,985	4,035	4,080	4,130	4,180	4,230	4,275	4,325	4,375	4,425	4,470	4,520	4,570
32	3,665	3,710	3,760	3,810	3,855	3,905	3,950	4,000	4,050	4,095	4,145	4,195	4,240	4,290	4,340	4,385	4,435	4,485	4,530
34	3,620	3,665	3,715	3,760	3,810	3,855	3,905	3,950	4,000	4,045	4,095	4,140	4,190	4,225	4,285	4,330	4,380	4,425	4,475
36	3,585	3,635	3,680	3,730	3,775	3,825	3,870	3,920	3,965	4,010	4,060	4,105	4,155	4,200	4,250	4,295	4,340	4,390	4,435
38	3,555	3,605	3,650	3,695	3,745	3,790	3,840	3,885	3,930	3,980	4,025	4,070	4,120	4,165	4,210	4,260	4,305	4,350	4,400
40	3,525	3,575	3,620	3,665	3,710	3,760	3,805	3,850	3,900	3,945	3,990	4,035	4,085	4,130	4,175	4,220	4,270	4,315	4,360
42	3,495	3,540	3,590	3,635	3,680	3,725	3,770	3,820	3,865	3,910	3,955	4,000	4,050	4,095	4,140	4,185	4,230	4,280	4,325
44	3,450	3,495	3,540	3,585	3,630	3,675	3,725	3,770	3,815	3,860	3,905	3,950	3,995	4,040	4,085	4,130	4,175	4,220	4,270
46	3,420	3,465	3,510	3,555	3,600	3,645	3,690	3,735	3,780	3,825	3,870	3,915	3,960	4,005	4,050	4,095	4,140	4,185	4,230
48	3,390	3,435	3,480	3,525	3,570	3,615	3,655	3,700	3,745	3,790	3,835	3,880	3,925	3,970	4,015	4,060	4,105	4,150	4,190
50	3,345	3,390	3,430	3,475	3,520	3,565	3,610	3,650	3,695	3,740	3,785	3,830	3,870	3,915	3,960	4,005	4,050	4,090	4,135
52	3,315	3,353	3,400	3,445	3,490	3,530	3,575	3,620	3,660	3,705	3,750	3,795	3,835	3,880	3,925	3,970	4,010	4,055	4,100
54	3,285	3,325	3,370	3,415	3,455	3,500	3,540	3,585	3,630	3,670	3,715	3,760	3,800	3,845	3,890	3,930	3,975	4,020	4,060
56	3,255	3,295	3,340	3,380	3,425	3,465	3,510	3,550	3,595	3,640	3,680	3,725	3,765	3,810	3,850	3,895	3,940	3,980	4,025
58	3,210	3,250	3,290	3,335	3,375	3,420	3,460	3,500	3,545	3,585	3,630	3,670	3,715	3,755	3,800	3,840	3,880	3,925	3,965
60	3,175	3,220	3,260	3,300	3,345	3,385	3,430	3,470	3,500	3,555	3,595	3,635	3,680	3,720	3,760	3,805	3,845	3,885	3,930
62	3,150	3,190	3,230	3,270	3,310	3,350	3,390	3,440	3,480	3,520	3,560	3,600	3,640	3,680	3,730	3,770	3,810	3,850	3,890
64	3,120	3,160	3,200	3,240	3,280	3,320	3,360	3,400	3,440	3,490	3,530	3,570	3,610	3,650	3,690	3,730	3,770	3,810	3,850
66	3,070	3,110	3,150	3,190	3,230	3,270	3,310	3,350	3,390	3,430	3,470	3,510	3,550	3,600	3,640	3,680	3,720	3,760	3,800
68	3,040	3,080	3,120	3,160	3,200	3,240	3,280	3,320	3,360	3,400	3,440	3,480	3,520	3,560	3,600	3,640	3,680	3,720	3,760
70	3,010	3,050	3,090	3,130	3,170	3,210	3,250	3,290	3,330	3,370	3,410	3,450	3,480	3,520	3,560	3,600	3,640	3,680	3,720
72	2,980	3,020	3,060	3,100	3,140	3,180	3,210	3,250	3,290	3,330	3,370	3,410	3,450	3,490	3,530	3,570	3,610	3,650	3,680
74	2,930	2,970	3,010	3,050	3,090	3,130	3,170	3,200	3,240	3,280	3,320	3,360	3,400	3,440	3,470	3,510	3,550	3,590	3,630

Predicted Vital Capacities **(a)** Female **(b)** Male. From Archives of Environmental Health, Volume 12, pp 146–189, February 1966. Reprinted with permission of the Helen Dwight Reid Educational Foundation. Published by Heldref Publications, 4000 Albemarle St., N.W., Washington, D.C. 20016. Copyright © 1966.

● **FIGURE 56–8b**
Predicted Vital Capacities — Males.

to breathe into a paper bag. ________________

____

VI. Microscopy

A. Normal Lung Tissue

1 ➡ Obtain a compound microscope and a slide of normal lung tissue. Identify the alveoli, paying special attention to the walls of the alveolar sacs. Refer back to Exercise 55 if necessary.

B. Pathological Lung Tissue

1 ➡ Study the slide of the emphysema lung. What differences do you notice in the walls of the alveolar sacs? How would this affect respiration?

____

____

____

2 ➡ Study a lung cancer slide. What differences do you notice in the walls of these alveolar sacs? If your slide includes a tumor section, what types of cells do you see? From a respiratory point of view, what do you suppose are the major problems connected with lung cancer?

____

____

____

❑ Additional Activities

NOTES

1. Research these questions: Why does a dog pant? What mechanisms are involved? When a dog is panting, does it inhale and exhale through its nose, mouth, or both? Why? Are there any differences between human panting and canine panting?
2. Read about Robert Boyle and John Dalton. Discuss their contributions to science.
3. A **pneumograph** records variations in breathing patterns due to chemical or physical activities. Obtain a pneumograph and use it, according to your instructor's directions, to demonstrate respiratory differences in reading, laughing, yawning, talking, swallowing, and whatever other activities you may be inclined to check.
4. Refer to Figures 56–5•, 56–7•, and 56–8•. What types of statistical values can you obtain from the information given in these figures? What importance might some of these statistical values have to a respiratory physiologist?

Answers to Selected Concept Check Questions

1. Carbon dioxide must be combining with water, and the carbonic acid molecules must be dissociating, resulting in an increase in the hydrogen ion concentration and a lowering of the pH.
2. More carbon dioxide is present.
3. There is more carbon dioxide to expel from the system.

❑ Lab Report

1. Referring to the numbered inquiries at the beginning of this exercise, complete the following box summary:

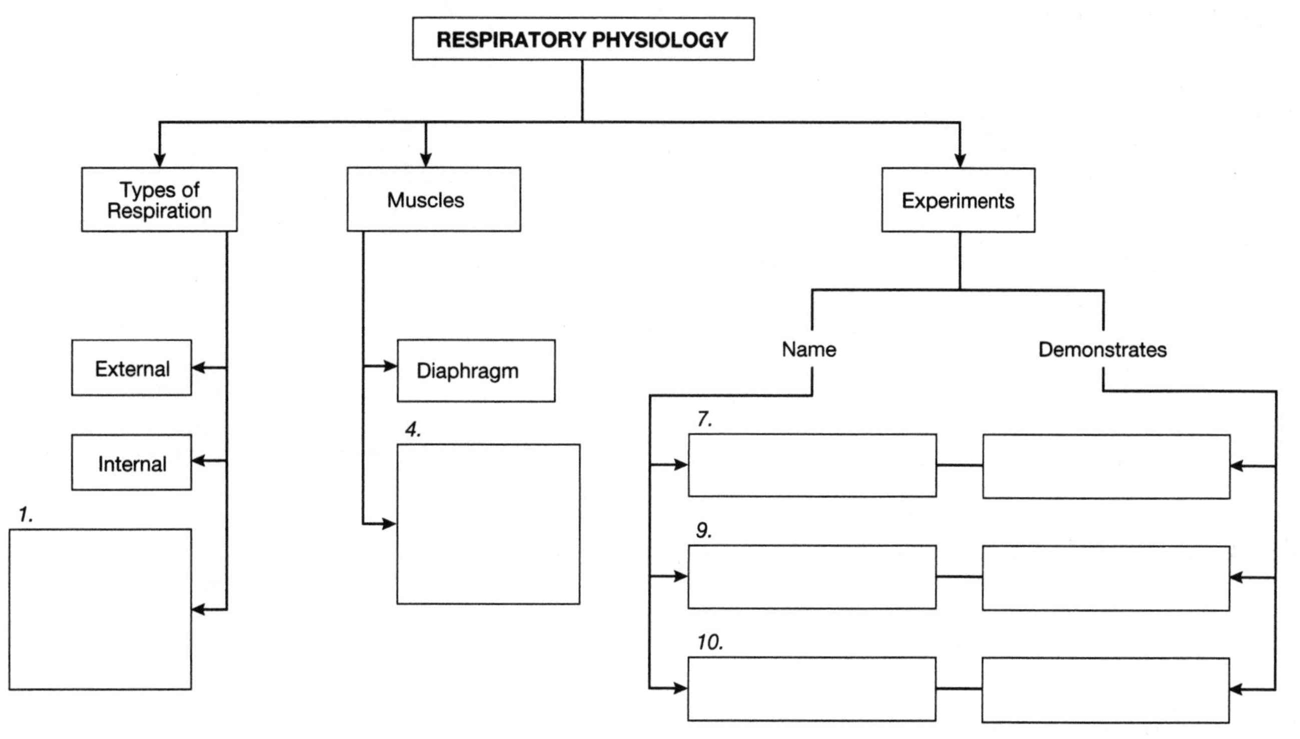

General Questions

1. What happens to the respiratory apparatus during the cough reflex?

2. What FEV values would you expect from a person with emphysema or asthma? A person with lung cancer?

3. During pulmonic inspiration, why does the size of your abdomen increase?

4. Why is the average child in no particular danger if (s)he decides to hold his/her breath as long as possible?

5. What is the difference between Boyle's Law and Dalton's Law?

6. How do the diaphragm and the intercostal muscles function together in breathing? What is the difference in the roles of the internal and external intercostal muscles?

7. What is the difference between:

   a. Inspiratory reserve volume and expiratory reserve volume?
   b. Residual volume and minimal volume?
   c. Tidal volume and vital capacity?

EXERCISE **57a**

Dissection of the Respiratory System: Cat

PROCEDURAL INQUIRIES

Dissection

1. What are the structures found in the cat's nasal area?
2. What are the divisions of the cat's pharynx?
3. What are the cartilages of the respiratory system?
4. What are the divisions of the bronchial tree?
5. What are the names of the lobes of the cat's lungs?

Experimentation

6. Why does lung tissue from a preserved adult cat float?
7. What does blowing into a straw inserted in the cat's trachea demonstrate?

Additional Inquiries

8. Where is the esophagus in relation to the trachea?
9. What is the physiological advantage to having incomplete cartilage rings around the trachea?
10. What is the relationship between the glottis and the epiglottis?
11. What is the difference between the true vocal cords and the false vocal cords?

HUMAN CORRELATION: AM Cadaver Atlas in Applications Manual.

Key Terms

Alveolar Duct
Alveolus (pl., Alveoli)
Arytenoid Cartilage
Bronchiole
Bronchus (pl., Bronchi)
Carina
Corniculate Cartilage
Cricoid Cartilage
Epiglottis
Esophagus
External Nares
Glottis
Laryngopharynx
Larynx
Lobe
Lobule
Lung
Nasal Septum
Nasopharynx
Nose
Nostrils
Oropharynx
Pulmonary Arteries
Pulmonary Veins
Thyroid Cartilage
Thyroid Gland
Trachea
Vocal Cords

Materials Needed

Cat
Dissecting Pan
Dissecting Kit
Gloves
Straws or Glass Pipettes
Beaker of Water
Dissecting Microscope or Hand Lens
Human Models (if available)
Sheep Pluck (if available)

The respiratory system consists of those organs involved in gas exchange between the external environment and the cardiovascular system. Upon inhalation (inspiration), environmental air passes from outside the body, through the nose, pharynx, larynx, trachea, bronchi, and bronchiole system, and into the alveolar sacs (the functional units of the lungs), where gas exchange takes place. Oxygen moves passively from the alveolar sacs into the blood, and carbon dioxide moves passively from the blood into the alveolar sacs. Upon exhalation (expiration) the carbon dioxide-rich air is passed back to the environment through the same set of respiratory structures.

In this laboratory exercise we will study the anatomy of the respiratory system by dissecting the respiratory system of the cat and by briefly comparing the feline respiratory system to the human respiratory system.

Dissection

I. Dissection

A. Nasal and Pharyngeal Areas

1➡ Place the cat ventral side up in your dissecting pan. Carefully examine the **nose**. You should be able to locate the **nostrils** (or **external nares**) and the **nasal septum**, the internal plate between the sides of the nose. We will not cut the nose to examine the internal nasal structures.

2➡ The area immediately behind the nose is the **nasopharynx**, which is the cranialmost of the three pharyngeal (throat) areas. The **oropharynx** is immediately inferior to the nasopharynx, and the **laryngopharynx** is the caudalmost of the pharynges. Directions for opening the pharyngeal area can be found as a part of Exercise 61a. Your instructor will decide whether the pharynx should be examined now or when you are studying the digestive system.

B. Upper Chest and Neck

1➡ If you have not already opened the upper chest and neck of your cat, you should do so now by extending your midventral incision toward the jaw. Note the **trachea**, the ringed tube extending from behind the mouth caudally to the pulmonary bifurcation. At the cranial end of the trachea is the **larynx**, the sturdy cartilaginous structure sometimes known as the voice box (or Adam's apple). Just below the larynx, beginning at about the second tracheal ring, you should find the **thyroid gland**. The thyroid gland consists of two lobes, one on either side of the trachea, connected by a glandular isthmus. It is easy to destroy the thyroid isthmus, particularly if your cat is older or has been preserved for a long time. (Refer back to the thyroid identification section of Exercise 45a.)

What in particular did you notice about your cat's thyroid gland? ____________________

____

2➡ Immediately posterior to the trachea is the **esophagus**, the flexible tube extending from the pharynx to the stomach. Gently separate the larynx and trachea from the esophagus and surrounding tissue. Cut the larynx from the pharynx so that the upper portion of the tube is free in your hand. Check with your instructor so that you know whether you should separate the esophagus from the trachea.

Record any additional instructions here. ____

____

____

3➡ Examine the larynx. Compare your cat's larynx with the anterior and posterior laryngeal diagrams. (Figure 57a–1●). Identify the **thyroid cartilage** (the large, ventral cartilage that makes up most of the external larynx), the **cricoid cartilage** (the only complete cartilage ring just below the thyroid cartilage), and the **epiglottis** (the leaf-like flap superior to the thyroid cartilage). If your instructor so indicates, locate the two paired laryngeal cartilages, the **arytenoids**, and the **corniculates**. (Another paired cartilage set, the **cuneiforms**, is found in the human and dog but not in the pig or cat.)

Which laryngeal cartilages were you able to find? ____________________

____

4➡ Look down into the larynx. The hole you see is the **glottis**. (Recall that the epiglottis is the flap of cartilage that covers the glottis.) On either side of the glottis you should be able to find two strips of mem-

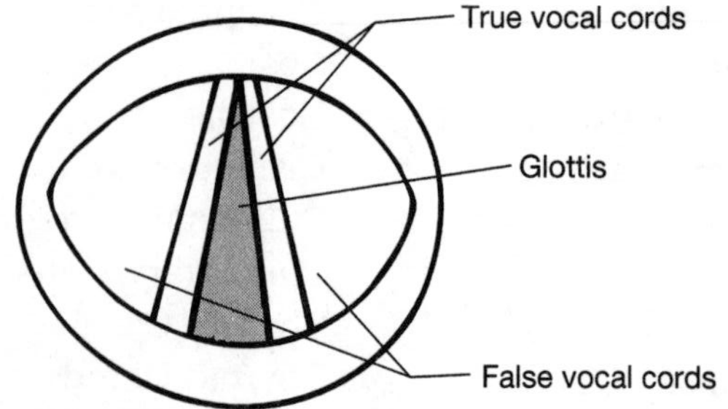

● **FIGURE 57a–1**
Larynx of the Cat.

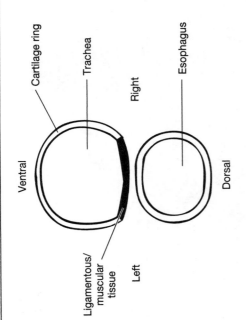

● **FIGURE 57a–2**
Cartilage Rings and Esophagus.

branous tissue, often described as membranous folds. These are the **vocal cords**. The **superior** or **false vocal cords** are above and lateral to the **inferior** or **true vocal cords**. See these structures labeled in Figure 57a–1●.

5➡ Follow the trachea down toward the lungs. Notice that the 38–43 tracheal rings (C-rings) are incomplete. The incomplete portion — the ligamentous/muscular (posterior) portion — of the trachea lies on top of the esophagus. Observe this relationship in Figure 57a–2●. With your gloved finger or blunt nosed probe, identify the parts of these tracheal rings. (Humans generally have 16–20 tracheal rings.)

C. Related Circulatory Structures

1➡ If you have not previously done so, try to locate the **pulmonary arteries** (which are probably blue) and the **pulmonary veins** (which are probably pink).

Concept Check 1 What are the functions of these arteries and veins? ______________________

____

____

D. Lungs

1➡ If you are so instructed, remove the heart and the proximal portions of the great vessels from the mediastinum. Examine the bifurcation and identify the **right** and **left primary bronchi** (sing., **bronchus**). Note the **carina**, the line of separation between the two bronchi. To actually see the carina, you would have to cut into the bronchial bifurcation. The **secondary bronchi** branch off the primary bronchi. Each secondary bronchus extends into a specific **lobe** of the **lung**.

2➡ Identify the lobes of the cat's lungs. The four right lobes are the **right cranial**, **right middle**, **right caudal**, and **right accessory** lobes. (These lobes are sometimes referred to as the **anterior**, **medial**, **posterior**, and **cardiac** lobes respectively.) The three left lobes are the **left cranial** (anterior), **left middle** (medial), and **left caudal** (posterior) lobes. These are indicated on Figure 57a–3●.

Which lobes were you able to identify? ______

____

____

Concept Check 2 Turn back to Exercise 55. Compare the names and locations of the human lung lobes to those of the cat. ____________

____

____

3➡ Separate the pulmonary tissue from the bronchi and see how much branching you can identify. The easiest way to separate this tissue is with your gloved fingers. However, you may feel more comfortable scratching away the tissue with a blunt nose probe or perhaps gently scraping the tissue aside with a scalpel. Whichever way you choose to gently pull apart the pulmonary tissue, start with the secondary bronchi and work on into the lungs. If possible, do this under a dissecting microscope or use a hand lens. You should be able to detect the bronchial paths. Specifically, you should also be able to identify **tertiary** and perhaps even some **quaternary** bronchial branches. All of these branches are encircled with cartilaginous rings. You will probably not be able to identify down to the seventh degree bronchial branches.

How many branches were you able to discern?

____

____

4➡ As you explore farther and farther down the bronchial tree, notice that the cartilaginous rings become less and less obvious, eventually becoming blocks of cartilage rather than actual cartilage rings. These smaller branches also have proportionately more and more smooth muscle.

When the bronchial lumen has narrowed to about 1 mm, cartilage is no longer present, and

● **FIGURE 57a–3**
Respiratory System of the Cat. Right accessory lobe posterior to heart not shown.

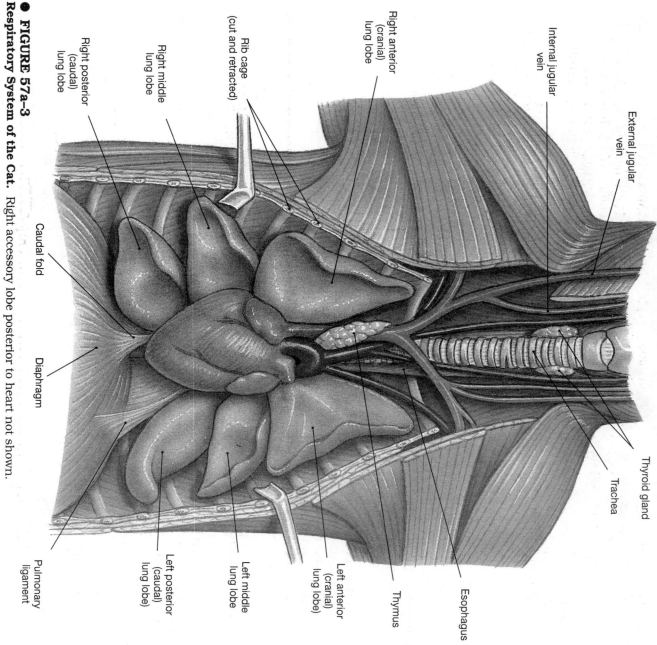

the tube is now known as a **bronchiole**. Each bronchiole enters a **lobule**, where it will further branch into **terminal bronchioles**. The terminal bronchioles give rise to **respiratory bronchioles**, which open into chambers called **alveolar ducts**. The blind pockets of these chambers are the **alveoli** (sing., **alveolus**). In all probability you will not be able to find the bronchioles in your dissection unless you use a dissecting microscope.

Concept Check 3 Return again to Exercise 55. How do the cat's bronchioles compare with the human's bronchioles? ______________
____
____

❑ Experimentation

II. Experiments

A. Demonstration of the Effect of Air in Lung Tissue

1➡ Cut a small piece of lung tissue from one lobe and float this tissue in a beaker of water. This should demonstrate that air may still be present in the lungs even in the presence of preservatives.

Did your lung piece float? ______________

If the pulmonary tissue of a fetal animal is available, try floating a piece of this lung. Did this piece of lung float? ________________ Keep in mind that the fetal lung tissue has never been inflated.

B. DEMONSTRATION OF THE INFLATION OF THE LUNGS

1➡ Take a drinking straw or glass pipette and insert it into the trachea. Do not push the straw all the way to the carina. Seal the trachea tightly around the straw by grasping the C-rings between your fingers. Blow into the straw and see if you can force the lungs to inflate. (Use caution so you do not accidentally inhale any preservative.) If the diameter of the straw is too large for the larynx, you can perform a tracheotomy by cutting the trachea about 0.5 cm below the larynx. Insert your straw into the incision and continue with your experiment.

Describe what happened to the lungs. ______

____

III. Comparison

If a sheep pluck or a model of the human system is available, examine it carefully and fill in the following box comparing your findings with what you learned about the cat's respiratory system.

	Human	*Sheep Pluck*	*Cat*
Nose			
Larynx			
Trachea			
Bronchi			
Lung Lobes			
Other Points			

❑ Additional Activities

NOTES

1. Examine and identify additional nasal structures found in the cat. Compare these structures to the human.
2. Research and sketch the ligaments found in the larynx and trachea of the cat. Compare these structures to the human.
3. This activity is similar to Additional Activity #1 in Exercise 56. Find out why and how dogs pant. What parts of the respiratory system are involved in panting? Do cats or humans pant? If so, is there a difference between human or feline panting and canine panting? What is the physiological trigger for panting?

Answers to Selected Concept Check Questions

1. The arteries carry the blood from the heart to the lungs, the veins carry the blood from the lungs to the heart.
2. The human left lung lobes are the superior lobe and the inferior lobe; the human right lung lobes are the superior lobe, the middle lobe, and the inferior lobe.
3. The cat's bronchioles are essentially the same as the human bronchioles.

❑ Lab Report

1. Referring to the numbered inquiries at the beginning of this exercise, complete the following box summary:

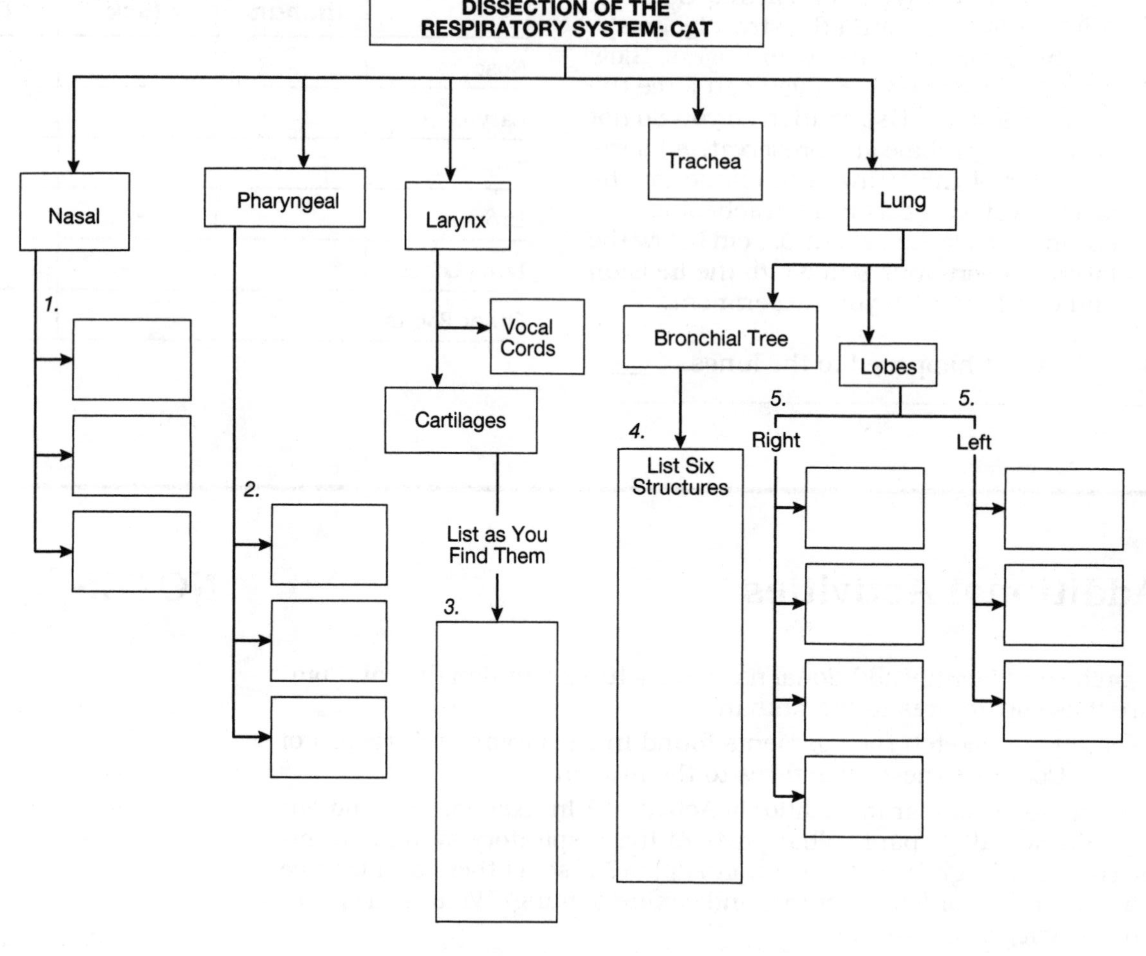

General Questions

NOTES

1. What is the function of the epiglottis?
2. Which set of vocal cords is responsible for sound?
3. Do you see any reason why the right lung has more lobes than the left lung?
4. What physiological advantage do you see in the incomplete tracheal rings?
5. Why did the piece of lung tissue float? Why would a piece of fetal lung tissue have sunk?
6. What specifically happened when you blew into the straw inserted in the cat's trachea?

EXERCISE **57b**

Dissection of the Respiratory System: Fetal Pig

PROCEDURAL INQUIRIES

Dissection

1. What are the structures found in the fetal pig's nasal area?
2. What are the divisions of the fetal pig's pharynx?
3. What are the cartilages of the respiratory system?
4. What are the divisions of the bronchial tree?
5. What are the names of the lobes of the fetal pig's lungs?

Experimentation

6. Why does lung tissue from a preserved fetal pig sink but lung tissue from an adult animal float?
7. What does blowing into a straw inserted in the fetal pig's trachea demonstrate?

Additional Inquiries

8. Where is the esophagus in relation to the trachea?
9. What is the physiological advantage to having incomplete cartilage rings around the trachea?
10. What is the relationship between the glottis and the epiglottis?
11. What is the difference between the true vocal cords and the false vocal cords?

HUMAN CORRELATION: AM Cadaver Atlas in Applications Manual.

Key Terms

Alveolar Duct
Alveolus (pl., Alveoli)
Arytenoid Cartilage
Bronchiole
Bronchus (pl., Bronchi)
Carina
Corniculate Cartilage
Cricoid Cartilage
Epiglottis
Esophagus
External Nares
Glottis
Laryngopharynx
Larynx
Lobe
Lobule
Lung
Nasal Septum
Nasopharynx
Nose
Nostrils
Oropharynx
Pulmonary Arteries
Pulmonary Veins
Snout
Thyroid Cartilage
Thyroid Gland
Trachea
Vocal Cords

Materials Needed

Fetal pig
Dissecting Pan
Dissecting Kit
Gloves
Straws or Glass Pipettes
Beaker of Water
Dissecting Microscope or Hand Lens
Human Models (if available)
Sheep Pluck (if available)

The respiratory system consists of those organs involved in gas exchange between the external environment and the cardiovascular system. Upon inhalation (inspiration), environmental air passes from outside the body, through the nose, pharynx, larynx, trachea, bronchi, and bronchiole system, and into the alveolar sacs (the functional units of the lungs), where gas exchange takes place. Oxygen moves passively from the alveolar sacs into the blood, and carbon dioxide moves passively from the blood into the alveolar sacs. Upon exhalation (expiration) the carbon dioxide-rich air is passed back to the environment through the same set of respiratory structures.

In this laboratory exercise we will study the anatomy of the respiratory system by dissecting the respiratory system of the fetal pig and by briefly comparing this porcine respiratory system to the human respiratory system.

Dissection

I. Dissection

A. Nasal and Pharyngeal Areas

1➡ Place the fetal pig ventral side up in your dissecting pan. Carefully examine the **nose**, usually called a **snout** or **rostrum** in the pig. You should be able to locate the **nostrils** (or **external nares**) and the **nasal septum**, the internal plate between the sides of the nose. To examine the internal nasal structures, make an incision along the lateral wall of the snout. Note particularly the **conchae** (turbinate bones), which project from the lateral wall and extend almost to the nasal septum.

2➡ The area immediately behind the nose is the **nasopharynx**, which is the cranialmost of the three pharyngeal (throat) areas. The **oropharynx** is immediately inferior to the nasopharynx and superior to the **laryngopharynx**. Directions for opening the pharyngeal area can be found in Exercise 61b. Your instructor will decide whether the pharynx should be examined now or when you are studying the digestive system.

B. Upper Chest and Neck

1➡ If you have not already opened the upper chest and neck of your fetal pig, you should do so now by extending your midventral incision toward the jaw. Note the **trachea**, the ringed tube extending from behind the mouth caudally to the pulmonary bifurcation. At the cranial end of the trachea is the **larynx**, the sturdy cartilaginous structure sometimes known as the voice box (or Adam's apple). Just below the larynx, beginning at about the third or fourth tracheal ring, you should find the **thyroid gland**. The thyroid gland consists of two lobes, one on either side of the trachea, connected by a glandular isthmus that includes a central superior projection. It is easy to destroy the thyroid isthmus, particularly if your fetal pig has been preserved for a long time. (Refer back to Exercise 45b.)

What in particular did you notice about your fetal pig's thyroid gland? ________________

____

2➡ Immediately posterior to the trachea is the **esophagus**, the flexible tube extending from the pharynx to the stomach. Gently separate the larynx and trachea from the esophagus and surrounding tissue. Cut the larynx from the pharynx so that the upper portion of the tube is free in your hand. Check with your instructor so that you know whether you should separate the esophagus from the trachea.

Record any additional instructions here. ____

____

____

3➡ Examine the larynx. Compare your fetal pig's larynx with the anterior and posterior laryngeal diagrams. (Figure 57b–1●). Identify the **thyroid cartilage** (the large, ventral cartilage that makes up most of the external larynx), the **cricoid cartilage** (the only complete cartilage ring just below the thyroid cartilage), and the **epiglottis** (the leaflike flap superior to the thyroid cartilage). If your instructor so indicates, locate the two paired laryngeal cartilages, the **arytenoids**; the **corniculates**; and the unpaired interarytenoid cartilage. (Another paired cartilage set, the **cuneiforms**, is found in the human and dog but not in the pig or cat.)

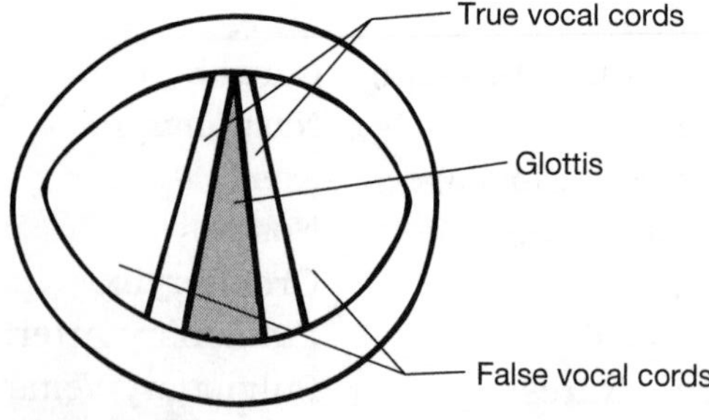

● **FIGURE 57b–1**
Larynx of the Fetal Pig.

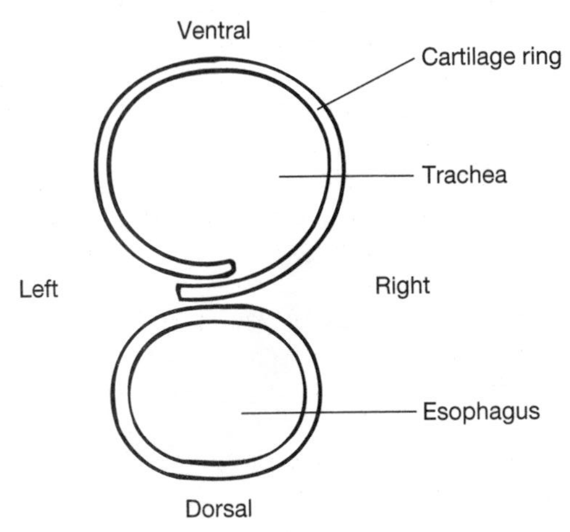

● **FIGURE 57b–2**
Overlapping of Cartilage Rings.

Which laryngeal cartilages were you able to identify? ______

4➡ Look down into the larynx. The hole you see is the **glottis**. (Recall that the epiglottis is the flap of cartilage that covers the glottis.) On either side of the glottis you should be able to find two strips of membranous tissue, often described as membranous folds. These are the **vocal cords**. The **superior** or **false vocal cords** are above and lateral to the **inferior** or **true vocal cords**. See these structures labeled in Figure 57b–1●.

5➡ Follow the trachea down toward the lungs. Notice the 32–36 tracheal rings (C-rings), cartilagenous plates of the trachea. Run your probe or your gloved finger along the posterior portion of the trachea. The rings appear complete but you can feel that the right portion of the ring overlaps the left portion (Figure 57b–2●). This overlapping portion of the trachea lies ventral to the esophagus (Figure 57b–3●). (The 16-20 human tracheal rings are C-shaped and do not overlap.)

C. Related Circulatory Structures

1➡ If you have not previously done so, try to locate the **pulmonary arteries** (which are probably blue) and the **pulmonary veins** (which are probably pink).

Concept Check 1 What are the functions of these arteries and veins? ______

D. Lungs

1➡ If you are so instructed, remove the heart and the proximal portions of the great vessels from the mediastinum. Examine the bifurcation and identify the **right** and **left primary bronchi** (sing., **bronchus**). Note the **carina**, the line of separation between the two bronchi. To actually see the carina, you would have to cut into the bronchial bifurcation. The **secondary bronchi** branch off the primary bronchi. Each secondary bronchus extends into a specific **lobe** of the **lung**.

2➡ Identify the lobes of the fetal pig's lungs. The four right lobes are the **right cranial** (apical), **right middle** (diaphragmatic), **right caudal** (diaphragmatic), and **right accessory** (intermediate) lobes. (These lobes are sometimes referred to as the **superior**, **medial**, **posterior**, and **cardiac** lobes, respectively.) The three left lobes are the **left cranial** (apical or superior), **left middle** (medial), and **left caudal** (diaphragmatic or posterior) lobes. These are indicated on Figure 57b–3●.

Which lobes were you able to identify? ______

Concept Check 2 Turn back to Exercise 55. Compare the names and locations of the human lung lobes with those of the fetal pig. ______

3➡ Separate the pulmonary tissue from the bronchi and see how much branching you can identify. The easiest way to separate this tissue is with your gloved fingers. However, you may feel more comfortable scratching away the tissue with a blunt nose probe or perhaps gently scraping the tissue aside with a scalpel. Whichever way you choose to gently pull apart the pulmonary tissue, start with the secondary bronchi and work on into the lungs. If possible, do this under a dissecting microscope or use a hand lens. You should be able to detect the bronchial paths. Specifically, you should also be able to identify **tertiary** and perhaps even some **quaternary** bronchial branches. All of these branches are encircled with cartilagenous rings. You will probably not be able to identify down to the seventh degree bronchial branches.

How many branches were you able to discern? ______

● **FIGURE 57b–3**
Respiratory System of the Fetal Pig.

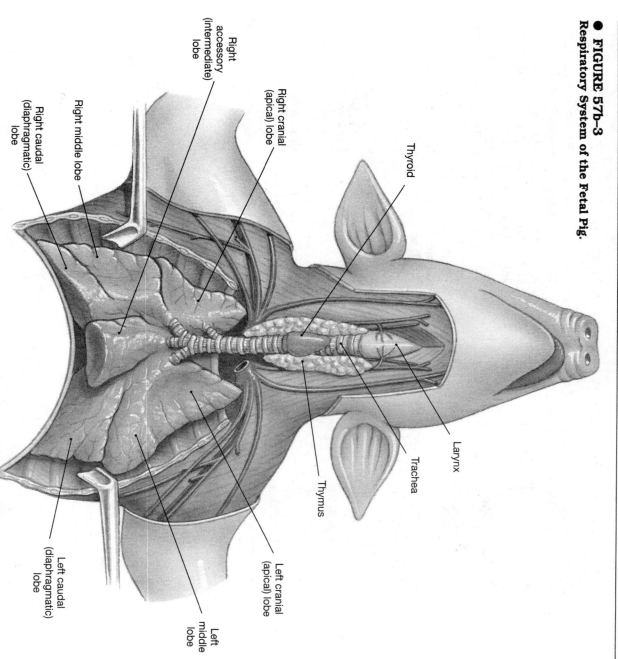

Larynx
Thyroid
Trachea
Thymus
Right cranial (apical) lobe
Left cranial (apical) lobe
Right accessory (intermediate) lobe
Left middle lobe
Right middle lobe
Left caudal (diaphragmatic) lobe
Right caudal (diaphragmatic) lobe

4➡ As you explore farther and farther down the bronchial tree, notice that the cartilaginous rings become less and less obvious, eventually becoming blocks of cartilage rather than actual cartilage rings. These smaller branches also have proportionately more and more smooth muscle.

In the adult animal, when the bronchial lumen has narrowed to about 1 mm, cartilage is no longer present, and the tube is now known as a **bronchiole**. Each bronchiole enters a **lobule**, where it will further branch into **terminal bronchioles**. The terminal bronchioles give rise to **respiratory bronchioles**, which open into chambers called **alveolar ducts**. The blind pockets of these chambers are the **alveoli** (sing., **alveolus**). In all probability you will not be able to find the bronchioles in your dissection unless you use a dissecting microscope.

Return to Exercise 55. How do the fetal pig's bronchioles compare with the human bronchioles? ______________
____
____

❑ Experimentation

II. Experiments

A. Demonstration of the Effect of Air in Lung Tissue

1➡ Cut a small piece of lung tissue from one lobe and try floating this tissue in a beaker of water. Did it float? __________ Keep in mind that this lung tissue has never been inflated; it should sink. If a preserved post-birth animal is present, try floating some of its pulmonary tissue in the beaker. This latter lung tissue should float because air may still be present in the lungs even in the presence of preservatives.

Record your observations. ______________

B. DEMONSTRATION OF THE INFLATION OF THE LUNGS

1 ➡ Take a drinking straw or glass pipette and insert it into the trachea. Do not push the straw all the way to the carina. Seal the trachea tightly around the straw by grasping the C-rings between your fingers. Blow into the straw and see if you can force the lungs to inflate. (Use caution so you do not accidentally inhale any preservative.) If the diameter of the straw is too large for the larynx, you can perform a tracheotomy by cutting the trachea about 0.5 cm below the larynx. Insert your straw into the incision and continue with your experiment. (If your fetal pig is very young, you may have some difficulty with this.)

Describe what happened to the lungs. ______

III. Comparison

If a sheep pluck or a model of the human system is available, examine it carefully and fill in the following box comparing your findings with what you learned about the fetal pig's respiratory system.

	Human	*Sheep Pluck*	*Fetal Pig*
Nose			
Larynx			
Trachea			
Bronchi			
Lung Lobes			
Other Points			

❑ Additional Activities

1. Examine and identify additional nasal structures found in the fetal pig. Compare these structures to the human.
2. Research and sketch the ligaments found in the larynx and trachea of the fetal pig. Compare these structures to the human.
3. This activity is similar to Additional Activity # 1 in Exercise 56. Find out why and how dogs pant. What parts of the respiratory system are involved in panting? Do adult pigs or humans pant? If so, is there a difference between pig or human panting and canine panting? What is the physiological trigger for panting?

NOTES

Answers to Selected Concept Check Questions

1. The arteries carry the blood from the heart to the lungs, the veins carry the blood from the lungs to the heart.
2. The human left lung lobes are the superior lobe and the inferior lobe; the human right lung lobes are the superior lobe, the middle lobe, and the inferior lobe.
3. The fetal pig's bronchioles, although somewhat less developed, are essentially the same as the human bronchioles.

☐ Lab Report

1. Referring to the numbered inquiries at the beginning of this exercise, complete the following box summary:

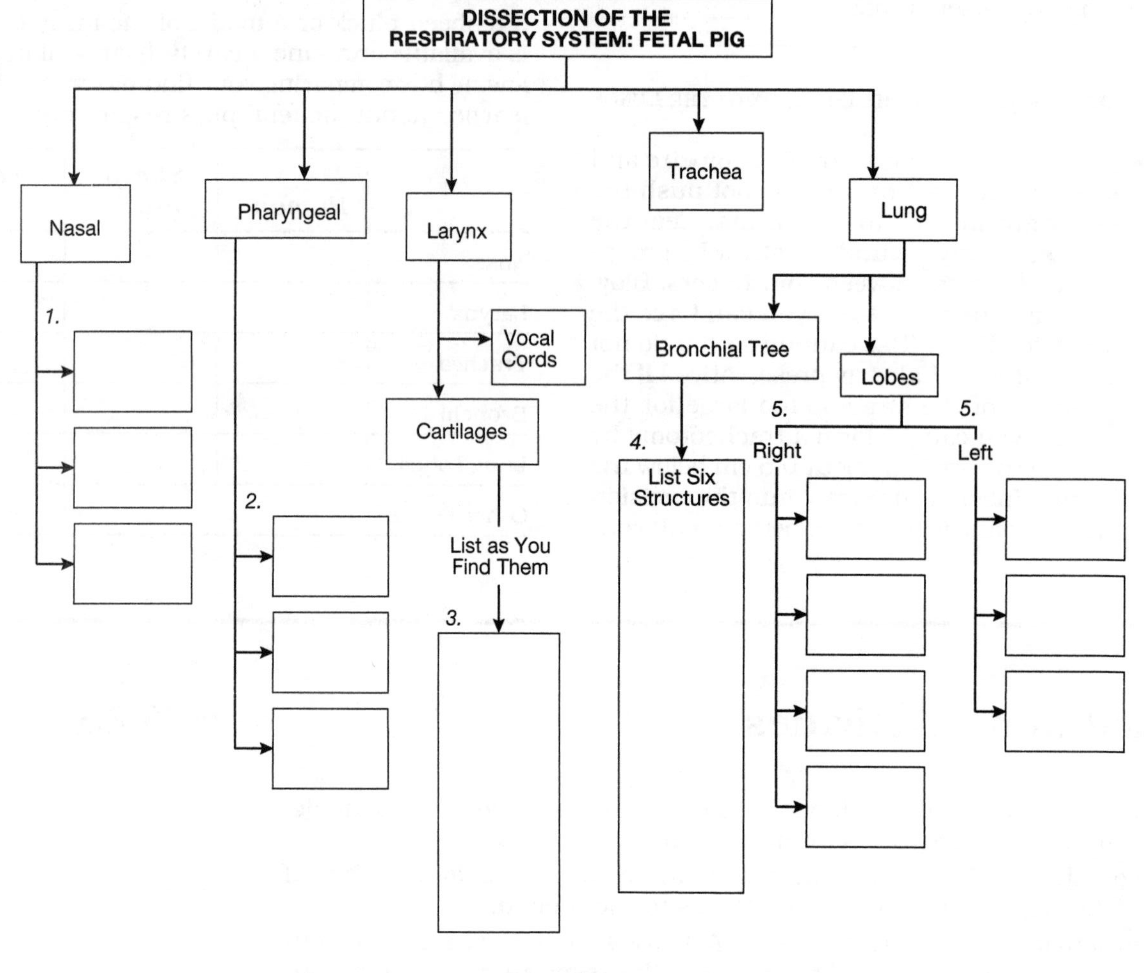

General Questions

1. What is the function of the epiglottis?
2. Which set of vocal cords is responsible for sound?
3. Do you see any reason why the right lung has more lobes than the left lung?
4. What physiological advantage do you see in the incomplete but overlapping tracheal rings?
5. Why did the piece of fetal pig lung tissue sink?
6. What specifically happened when you blew into the straw inserted in the fetal pig's trachea?

NOTES

EXERCISE 58

Anatomy of the Digestive System

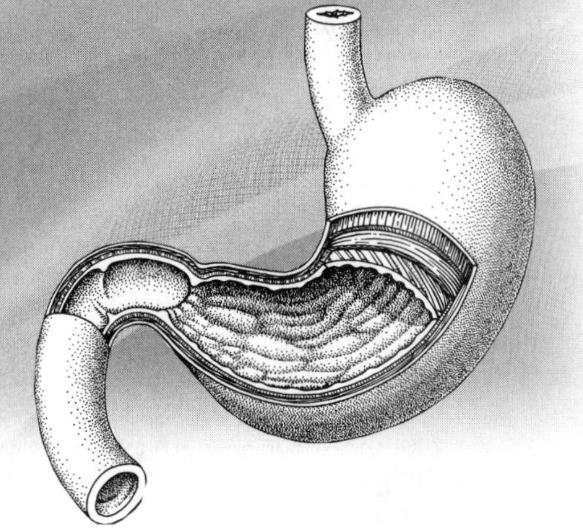

PROCEDURAL INQUIRIES

Examination

1. What are the principal parts of the mouth?
2. What are the types of teeth and what is the function of each type?
3. What are the major glands of the mouth?
4. What are the parts of the pharynx?
5. What is the difference between the visceral peritoneum and the parietal peritoneum?
6. Where are the greater and lesser omenta found?
7. What are the lobes of the liver?
8. Where is the gallbladder?
9. What is the relationship between the cystic duct, the hepatic duct, and the common bile duct?
10. What are the major landmarks of the stomach?
11. What are the major divisions of the small intestine?
12. What are the major divisions of the large intestine?
13. What are the rectal sphincters?

Additional Inquiries

14. What is the alimentary canal?
15. Which digestive system organs are accessory organs?

Key Terms

Accessory Organs
Alimentary Canal
Anus
Appendix
Cheeks
Colon
Common Bile Duct
Cystic Duct
Dentition Pattern
Duodenum
Esophagus
Hard Palate
Hepatic Duct
Jejunum
Large Intestine
Larynx
Ileum
Lingual Frenulum
Lips
Liver
Mesentery
Mouth
Omentum
Oral Cavity
Pancreas
Papillae
Parotid Gland
Peritoneum
Pharynx
Rectum
Rugae
Small Intestine
Soft Palate
Stensen's Duct
Stomach
Submandibular Gland
Teeth
Tongue
Tonsils
Uvula
Vestibule

Materials Needed

Torso
Models (as available)
Charts and Diagrams
Tooth Cutaway
Compound Microscope
Dissecting Microscope
Prepared microscope slides of
- Glands — Parotid, Submaxillary
- Stomach — Pyloric and Cardiac Regions
- Small Intestine — Duodenum, Jejunum, Ileum

Large Intestine — Colon, Appendix
Liver
Pancreas

The digestive system consists of all the organs and tissues involved either directly or indirectly in the ingestion of foods, the digestion or chemical breakdown of these foods, the absorption of the breakdown products into the system, and the egestion of the digestive waste products.

The principal digestive pathway is the **alimentary canal**, the continuous passage that begins at the mouth, includes the esophagus, stomach, small intestine, and large intestine, and ends at the anus. Exocrine and endocrine glands, whose products affect the functioning of the principal organs, form the **accessory organs**, as do the teeth and tongue.

In this laboratory exercise we will study the macro- and microanatomy of the human digestive system by examining appropriate models and by viewing relevant digestive system slides.

❑ Examination

I. Gross Anatomy

A. Orientation

1➡ Begin your study of the digestive system by orienting yourself to the relative position of the major organs along the alimentary tract. Use the torso to help you. Refer back to Exercise 7 if necessary.

B. Mouth [∞ Martini, p. 853]

1➡ You may wish to use a mirror and study the **mouth** by examining your own oral structures. As you look at the mouth, pay particular attention to the **cheeks** and **lips**. Notice the space between the cheeks and lips and the teeth. This is the **vestibule**.

2➡ To appreciate the relative size of the vestibule, run your washed finger or your tongue around the teeth and lips.

3➡ Examine the **teeth**. You are a **heterodont** (as are most mammals), meaning that your teeth differ in structure to perform different functions. Make note of the **dentition pattern**, or dental formula, the respective numbers of **incisors** (for cutting food), **canines** (or cuspids) for tearing and shredding food), and **premolars** (bicuspids) and **molars** (for crushing and grinding food). Your adult dental formula is the same for both sides of your upper and lower jaws, 2-1-2-3. This means in each quadrant of your mouth you have 2 incisors, 1 canine, 2 premolars, and 3 molars. You may wonder if your dental formula is different if you have had your wisdom tooth removed. Generally we think of dentition patterns in terms of species, so we can say the human formula is 2-1-2-3, regardless of the action of the dentist.

4➡ Use the tooth cutaway as your guide to the location of the tooth structures identified on Figure 58–1●. Quiz your partner. *[∞ Martini, p. 856]*

5➡ The **oral cavity proper** is the area behind the teeth and in front of the throat. The floor of the oral cavity is occupied primarily by the **tongue**. Locate the **lingual frenulum**, which attaches the tongue to the floor of the mouth.

6➡ Run your finger in several directions over the superior surface of the tongue and feel the **papillae**, the elevations, many of which contain the taste buds. (We discussed the papillae in Exercise 41.) Note the orientation or direction of the papillae. (This orientation is more pronounced in many mammals, allowing them to drink more efficiently.)

7➡ Examine the roof of the mouth. The roof of the mouth is composed of the **hard** and **soft palates**. Notice the **ridges** (also called rugae), the horizontal ridges on the bony surface of the hard palate. The hard palate and the muscular soft palate form the separation between the nasal cavity and the oral cavity.

C. Glands

Draining into the mouth are the secretions of numerous serous and mucousal glands. Serous glands produce thick secretions; mucousal glands produce watery saliva. Some of these glands may not be well delineated on your models.

1➡ Use Figure 58–2● as your guide to locate the following glands. Just below the ear you should be able to find the large glandular mass, the **parotid gland**. The **parotid duct** or Stensen's duct lies across the masseter muscle (the large transverse "chewing" muscle of the cheek). Inferior to the parotid gland and lying just beneath the jugular vein is the **submandibular gland**, which is identified on Figure 58–2●. The **submandibular duct** can be found at the lower edge of the masseter muscle. The **sublingual glands** are located beneath the mucous membrane of the floor of the mouth. Several ducts lead from the sublingual gland to the mouth proper.

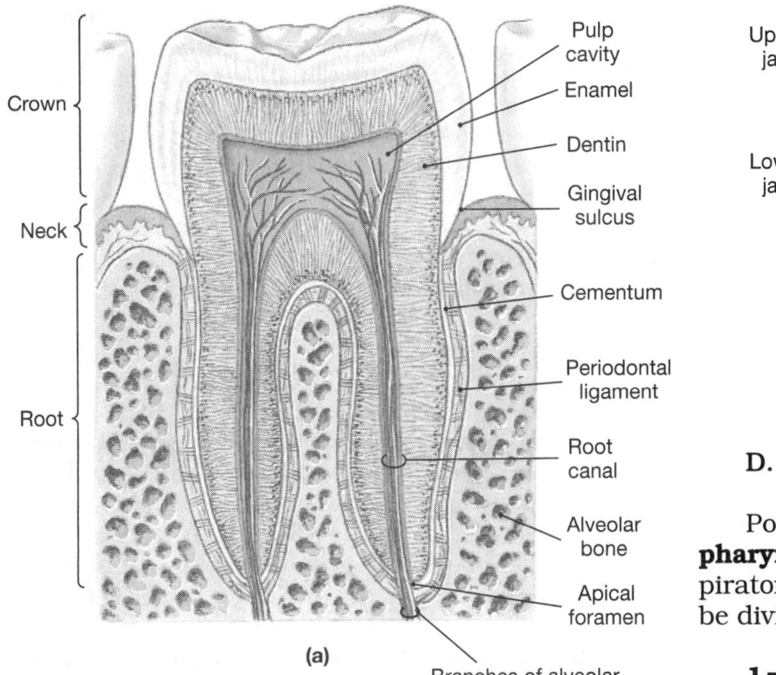

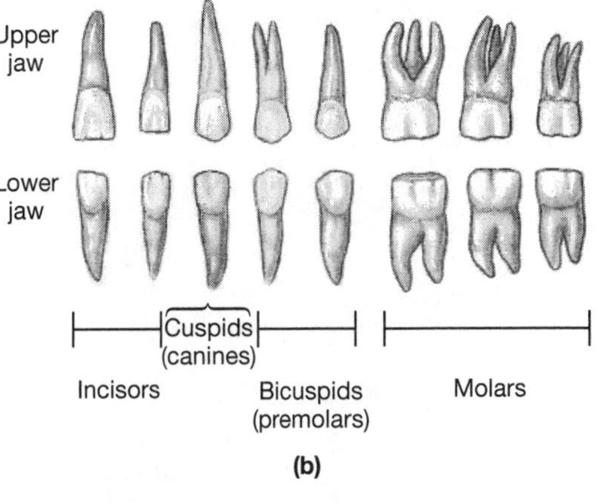

● **FIGURE 58–1**
Teeth. (a) Typical Tooth. **(b)** Adult Dentition Pattern (left side only).

2➡ Be aware of several other oral glands not shown on Figure 58–2●. The pea-sized **labial glands** are situated around the orifice of the mouth. The **buccal glands** are between the buccinator muscle and the mucous membrane. The ducts of the larger buccal glands open into the mouth opposite the last molars. These glands are the **molar glands**.

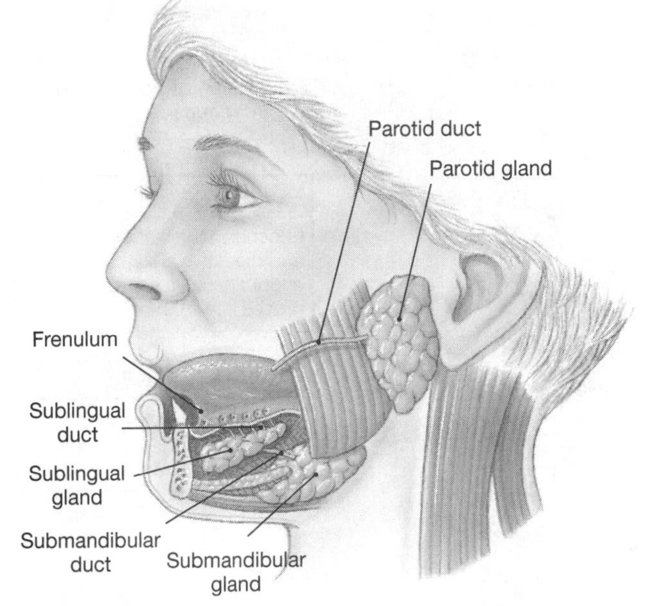

● **FIGURE 58–2**
Salivary Glands and Ducts.

D. Pharynx

Posterior to the soft palate you will find the **pharynx**, the passageway common to both the respiratory and digestive systems. The pharynx can be divided into three regions.

1➡ Examine Figure 58–3● and locate the following structures and regions. The **nasopharynx**, the upper portion of the pharynx, can be studied on a diagram or a cutaway model. If you have such an aid, you may be able to locate the **internal nares (choanae)**, which form the internal exit from the nose proper into the pharynx, and the **auditory (eustachian) tubes**, the slitlike openings extending laterally toward the middle ear.

2➡ Identify the **oropharynx**, the area of the throat immediately inferior to the soft palate (the part you can see in the mirror) and extending to the level of the hyoid bone. Anterior to the oropharynx you should see the **palatoglossal arches**. In the center of the oral cavity, continuous with the palatoglossal arches, is the **uvula**,the punching bag-like structure that prevents food from going up into the nose when you swallow. On the lateral walls behind the arches are the **pharyngeal tonsils**. This lymphoidal tissue is what we commonly refer to simply as the tonsils. Posterior to the **pharyngeal tonsils** are the **palatopharyngeal arches**. The tonsils are thus between the two sets of arches.

3➡ Locate the **laryngopharynx** beneath the level of the hyoid bone, below the epiglottis, and extending down to the **larynx** and **esophagus**.

E. Esophagus [∞ Martini, p. 858]

The esophagus is the muscular tube running posterior to the trachea from the pharynx down

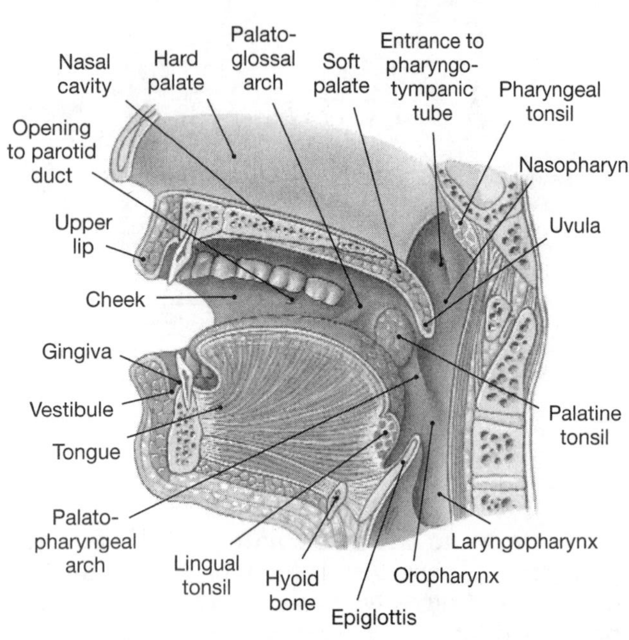

● **FIGURE 58–3**
Oral Cavity—Sagittal Section.

through the diaphragm (at an opening called the **esophageal** or **diaphragmatic hiatus**) and into the **stomach**. The superior third of the esophagus is composed of skeletal muscle, the middle third is part skeletal muscle and part smooth muscle, and the lower third is entirely smooth muscle.

1➡ Note the relation of the esophagus to the other organs of the thorax. Locate the preceding structures on Figure 58–4●.

F. ABDOMINAL CAVITY

1➡ As points of reference use Figures 58–4● and 58–5● to locate the **liver**, **spleen**, **stomach**, **pancreas**, **small intestine**, and **large intestine**. We will examine these organs in greater detail as we continue this study. (The spleen is deep to the stomach and is therefore not shown on Figure 58–4●.)

2➡ Examine the smooth serous epithelial membranes of the abdominal cavity. The

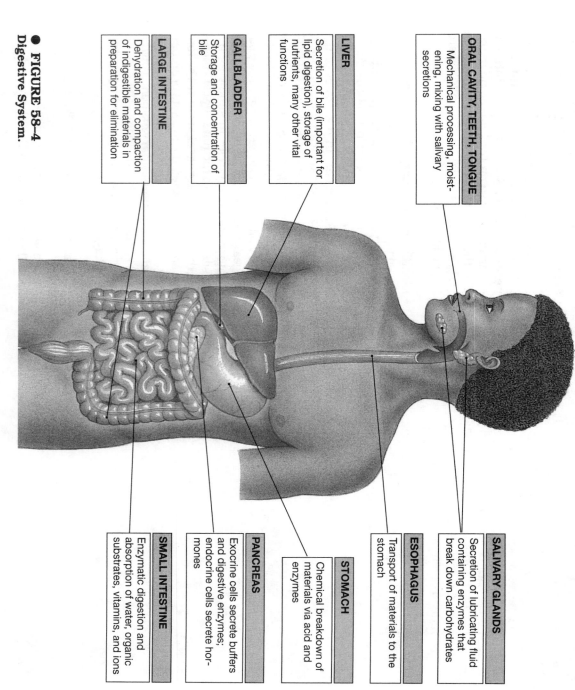

● **FIGURE 58–4**
Digestive System.

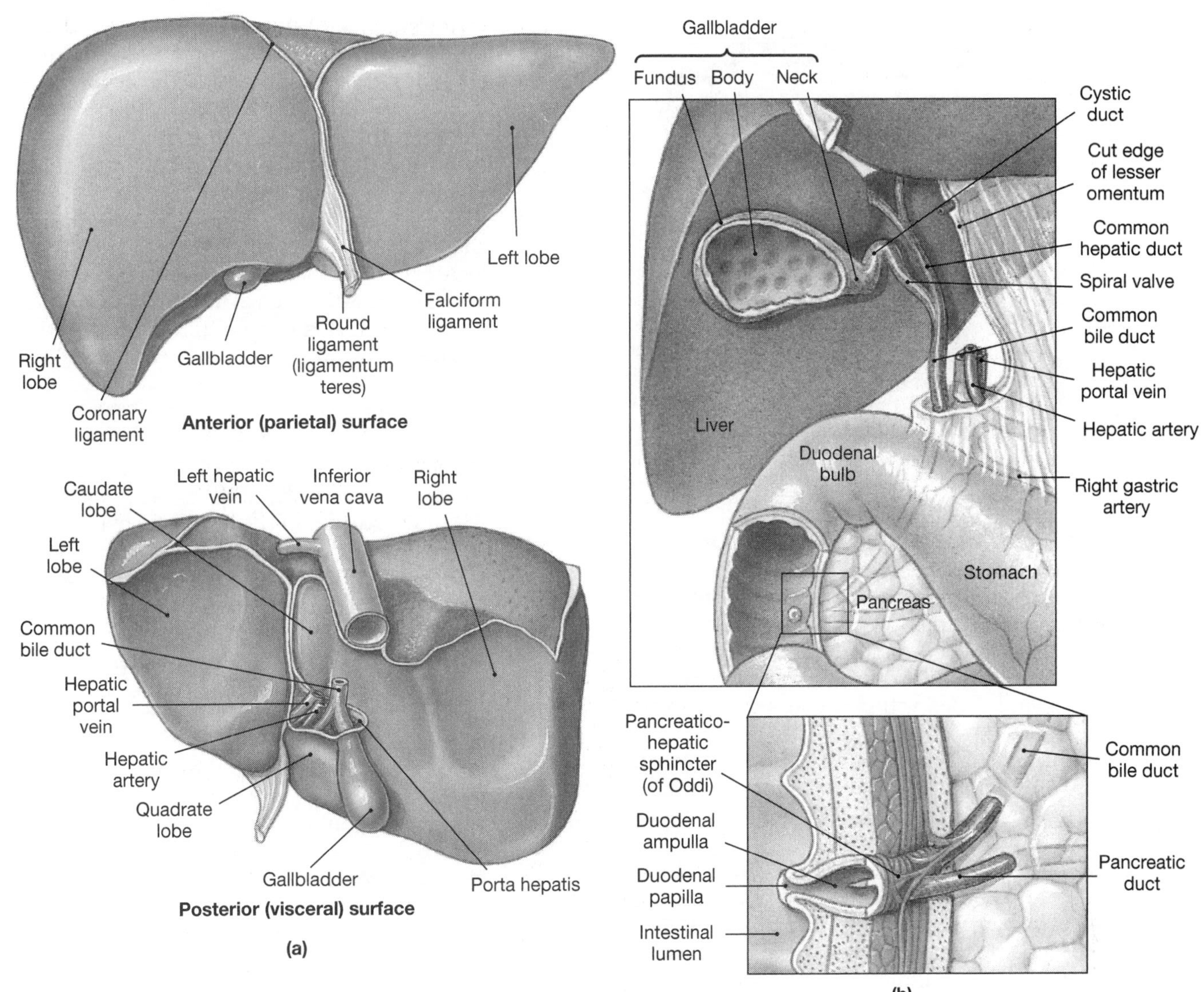

● **FIGURE 58–5**
Accessory Organs of the Digestive System.
(a) Liver; **(b)** Gallbladder; **(c)** Pancreas.

parietal peritoneum lines the abdominal wall, and the **visceral peritoneum** covers the abdominal viscera.

3➡ In Figure 58–6● locate the **greater omentum**, the double sheet apron-like structure anterior to the viscera that contains vast amounts of adipose tissue. The greater omentum is attached to the greater curvature of the stomach and encloses the spleen, part of the pancreas, and most of the intestinal tract and terminates on the dorsal abdominal wall. The omental connection between the stomach and the spleen is the **gastrosplenic ligament**.

4➡ Also in Figure 58–6● find the **lesser omentum**, which extends from the lesser curvature of the stomach and the duodenum to the

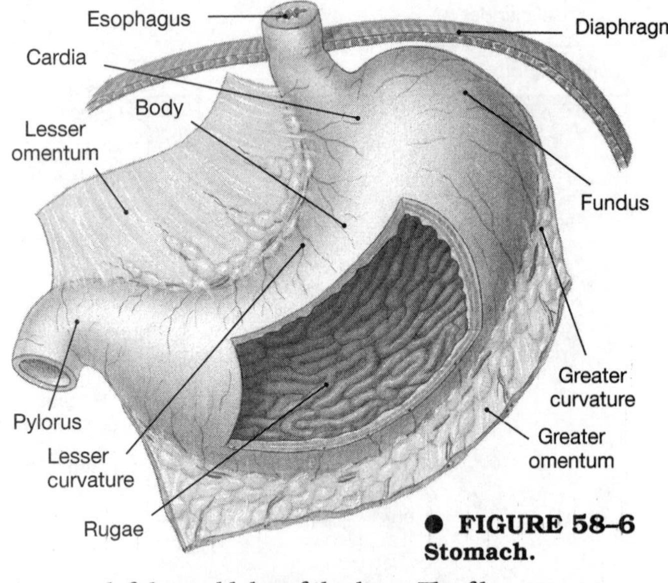

● **FIGURE 58–6 Stomach.**

left lateral lobe of the liver. The fibrous omental connection between the stomach and the liver is the **hepatogastric ligament**, and the connection between the duodenum and the liver is the **hepatoduodenal ligament**. The **common bile duct**, described in Section H, is embedded within the lesser omentum.

In addition to the adipose tissue, within the omenta are blood vessels, nerves, and lymphatic tissues.

5➡ Locate the structures of the mesenterial sheet. Note the extensive network of veins and arteries in the mesenteries. Some distinct mesenteric lymph nodes may be identified on your models. The mesentery is actually two separate structures: the **mesentery proper** (or true mesentery), which can be found supporting the coils of the small intestine, and the **mesocolon**, which connects the large intestine and the posterior body wall.

G. Liver [∞ Martini, p. 874]

1➡ Now return to the liver, the largest abdominal organ, which is located just inferior to the diaphragm and seems to drape over the right side of the remaining abdominal viscera. Recall from Exercise 11 that the liver is composed of reticular connective tissue.

2➡ You should be able to find the four lobes of the human liver. Use Figure 58–5a● as your guide. On the superior aspect are the larger right lobe and smaller left lobe. On the inferior aspect are the caudate lobe (near the vena cava) and the quadrate lobe (below the caudate lobe and near the gall bladder). Older texts often identify a fifth lobe, the **spigeli lobe**, situated on the posterior surface of the liver and "looking" directly backward. This spigeli lobe is now considered to be part of the right lobe.

3➡ Between the right and left lobes you should find the **falciform ligament**, which holds the liver in place by attaching to the ventral wall. If you examine the free edge of the falciform ligament, you may be able to locate a fibrous strand of tissue. This is the vestigial **umbilical vein**, which atrophies after birth and is known in the adult as the **round ligament** (or ligamentum teres).

H. Gallbladder

1➡ Examine the gallbladder, the greenish sac located in a depression (the fossa vesicalis) on the undersurface of the right lobe of the liver.

2➡ Use Figure 58–5b● to locate the **cystic duct**, which leads out of the gallbladder. The cystic duct joins with the **common hepatic duct** to form the **common bile duct**. Trace the common hepatic duct back up toward the liver and note its branches, the right and left hepatic ducts. Trace the common bile duct down to the **duodenal ampulla** (ampulla of Vater), the enlarged area where the duct enters the small intestine. Guarding this passageway is the muscular **pancreaticohepatic sphincter** (or sphincter of Oddi).

I. Pancreas [∞ Martini, p. 872]

1➡ Return to the pancreas and notice the portion of the organ lying medial to the duodenum. This is the **head**. The rest of the pancreas is known as the **body**, and the tapered end is called the **tail**. Refer back to Figure 58–5c● and locate the **pancreatic duct**. This duct often unites with the common bile duct just before entering the small intestine at the duodenal ampulla. Sometimes the merger of the ducts occurs after they penetrate the wall of the small intestine.

J. Stomach [∞ Martini, p. 861]

1➡ Return to the stomach and use Figure 58–6● to help you locate the following landmarks:

Cardiac Region (Cardia) Adjacent to the esophagus.

Fundus Domed portion to the left of the cardiac area.

Greater Curvature Longer, left (lateral or inferior) margin of the stomach.

Lesser Curvature Shorter, right (medial or superior) margin of the stomach.

Body Large, dominant portion of the stomach.

Rugae Longitudinal folds of mucosa lining the inside of the body of the stomach.

Pyloric Region (Pylorus) Narrow portion where the stomach empties into the duodenum.

Pyloric Sphincter Muscular ring between the pylorus and the duodenum. The pyloric sphincter is not identified on Figure 58–6●.

K. Spleen

1➡ Locate the spleen, the elongate reddish brown organ on the left side of the stomach. The spleen is actually part of the lymphatic system and not the digestive system. Nevertheless, it is important to gain a perspective on the location of the spleen as you examine the gross anatomy of the digestive organs.

L. Small Intestine

The small intestine, 6 m long, is divided into three sections: the **duodenum**, the **jejunum**, and the **ileum**.

1➡ Observe the first section, the duodenum, which is about 25 cm long. When the duodenum turns left, usually about the end of the pancreas, the duodenum becomes the jejunum. The duodenum is normally thinner and firmer than the jejunum.

2➡ Follow the jejenum. The jejunum runs for about 2.5 m before becoming the ileum. On gross examination you probably would not be able to discern the difference between the jejunum and the ileum, although you would see structural differences on microscopic examination. (See Section II.B of this exercise). The ileum averages about 3.5 meters in length.

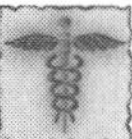

Occasionally extending from the ileum about one meter before its terminus is a blind diverticulum or tube. This is **Meckel's diverticulum**, a remnant of the vitelline duct connecting the umbilical duct with the alimentary canal in early fetal life.

M. Large Intestine

1➡ Use Figure 58–7● as your guide as you identify the parts of the large intestine. The sphincter located where the small intestine empties into the large intestine is known as the **ileocecal valve**. In the living or preserved human body, this sphincter can be felt.

2➡ Note the large intestinal pouch caudal to the sphincter. This is the **cecum**. In the human, a vestigial organ, the **appendix**, extends from the end of the cecum. Most other mammals do not have an appendix, although occasionally, in some mammals, a thickened tissue or tiny knot may be found at the end of the cecum. (As a point of interest, the rabbit has a large appendix.)

3➡ Follow the large intestine as it extends cranially on the right side, becoming the **ascending colon**. The structure bends left at the **hepatic flexure**, and as it crosses the abdomen it is known as the **transverse colon**. This colon bends downward at the **splenic flexure**, and as it continues down the left abdomen it is known as the **descending colon**. At its base the descending colon in the human bends medially and becomes known as the **sigmoid** (or **S**) **colon**. (Except for the horse, domestic animals do not have a true, distinct sigmoid colon, although sometimes the medial bend in the animal's colon is referred to as sigmoid.)

4➡ Notice that the colon is pouched. These pouches, called **haustrae**, keep their "pouchiness" because of three longitudinal strips of muscle known as the **taenia coli**. These bands are a continuation of the outer layer of the muscularis externa.

5➡ The colon empties into the **rectum**, which opens to the exterior at the **anus**. Identify the **anorectal canal**, the last portion of the rectum. The **internal anal sphincter**, an involuntary smooth muscle layer, is located here. The **external anal sphincter**, which is caudal to the internal anal sphincter, is under voluntary control as it is skeletal muscle.

II. Micro-Anatomy

A. Layers of the Digestive Tract

1➡ Begin your study of the microanatomy of the digestive system by examining Figure 58–8●. Note the layers, beginning on the inside.

- Mucosa
  - Mucosa epithelium
  - Lamina propria
- Muscularis mucosae
  - Smooth muscle and elastic fibers
- Submucosa
  - Loose connective tissue
- Muscularis externa
  - Circular layer
  - Longitudinal layer

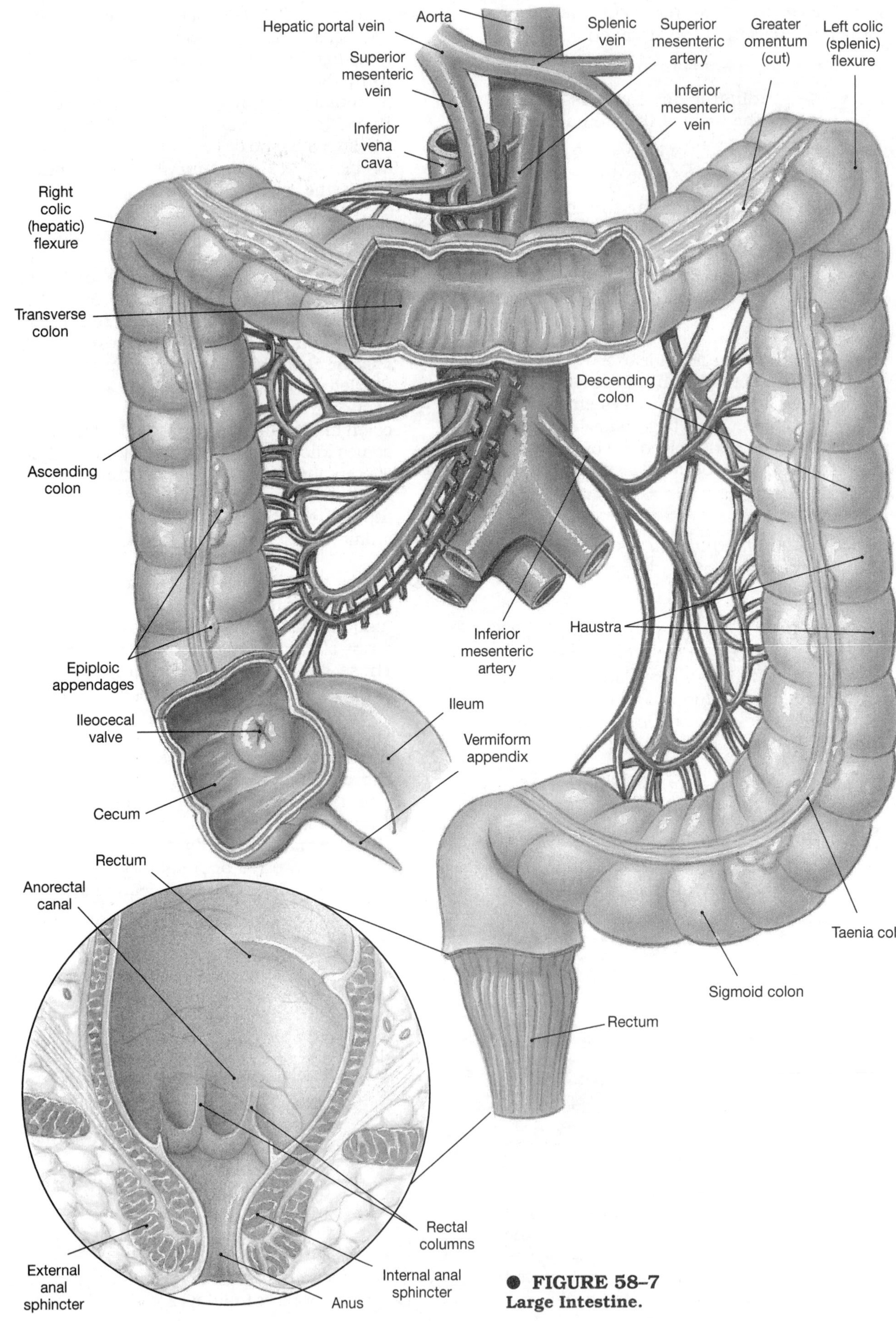
Hepatic portal vein
Aorta
Splenic vein
Superior mesenteric artery
Greater omentum (cut)
Left colic (splenic) flexure
Superior mesenteric vein
Inferior mesenteric vein
Inferior vena cava
Right colic (hepatic) flexure
Transverse colon
Descending colon
Ascending colon
Haustra
Inferior mesenteric artery
Epiploic appendages
Ileum
Ileocecal valve
Vermiform appendix
Cecum
Rectum
Anorectal canal
Taenia coli
Sigmoid colon
Rectum
Rectal columns
External anal sphincter
Internal anal sphincter
Anus

● **FIGURE 58–7**
Large Intestine.

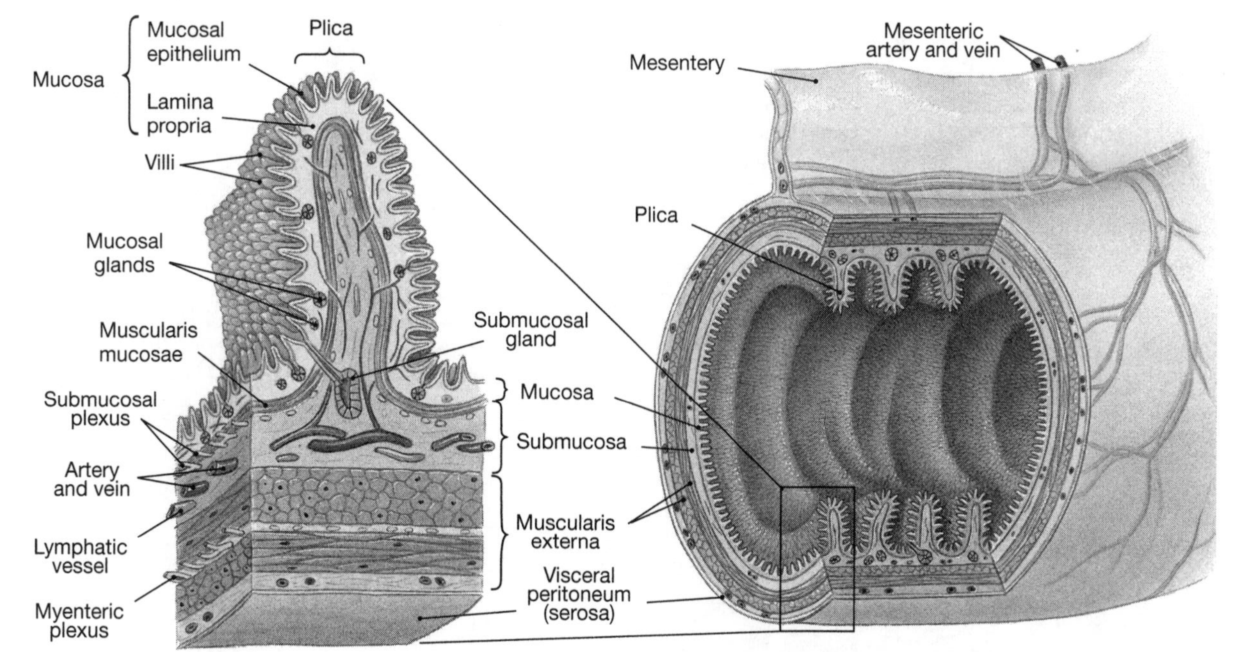

● **FIGURE 58–8**
Structural Organization of the Digestive Tract. Diagrammatic view of the small intestine.

Adventitia (surrounding structures outside the peritoneal cavity)
Serosa (surrounding structures within the peritoneal cavity)

2➡ Note the submucosal plexus (plexus of Meissner) and the myenteric plexus (plexus of Auerbach). These plexuses are intrinsic nerve networks within the digestive tract.

B. MICROSCOPY

To examine the following slides, follow your instructor's directions as to whether you should use the standard microscope or the dissecting microscope.

1➡ Study the slides of the stomach and small intestine. Keep the preceding outline pattern in mind as you sketch the *layers* of both the stomach and small intestine.

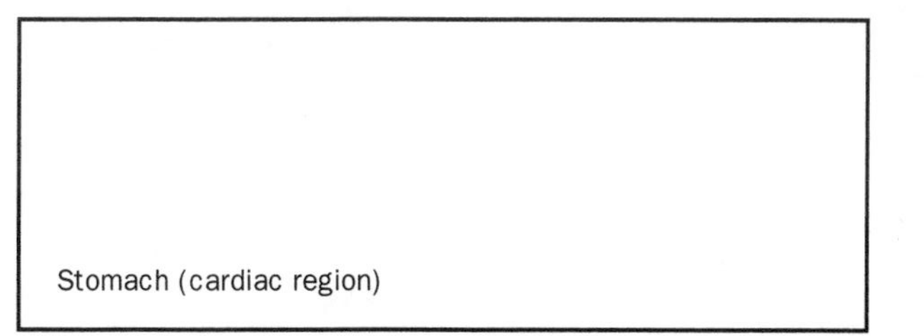

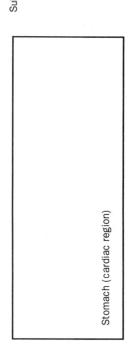

Stomach (cardiac region)

Stomach (pyloric region)

Small intestine layers

What differences do you see between the two regions of the stomach? ______________________

____

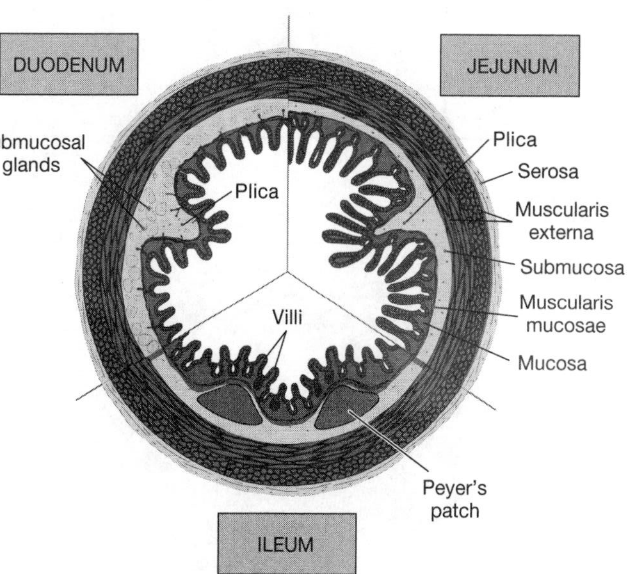

● **FIGURE 58–9**
Regional Histology of the Small Intestine.

2➡ Use Figure 58–9● to differentiate between the parts of the small intestine. Notice that this figure is divided into three sections, each delineating the unique structural characteristics of a given part of the small intestine.

List these differences.

______

______

3➡ Sketch the parts of the small intestine as you see them on your slides.

Small intestine

4➡ Examine the sides of other digestive organs. Compare these slides to Figures 58–10a● through 58–10f●. Identify the

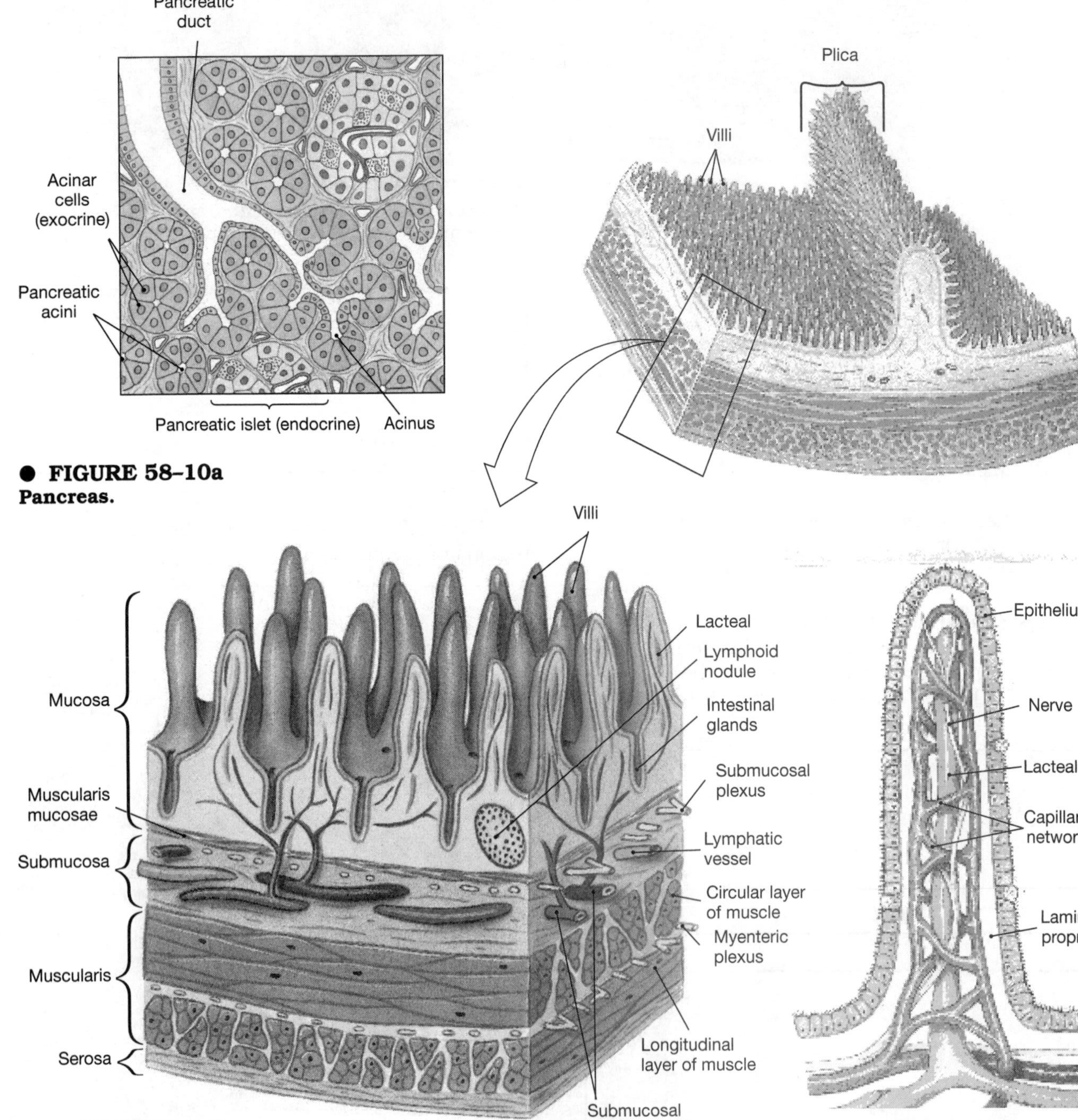

● **FIGURE 58–10a**
Pancreas.

● **FIGURE 58–10b**
Intestinal Wall.

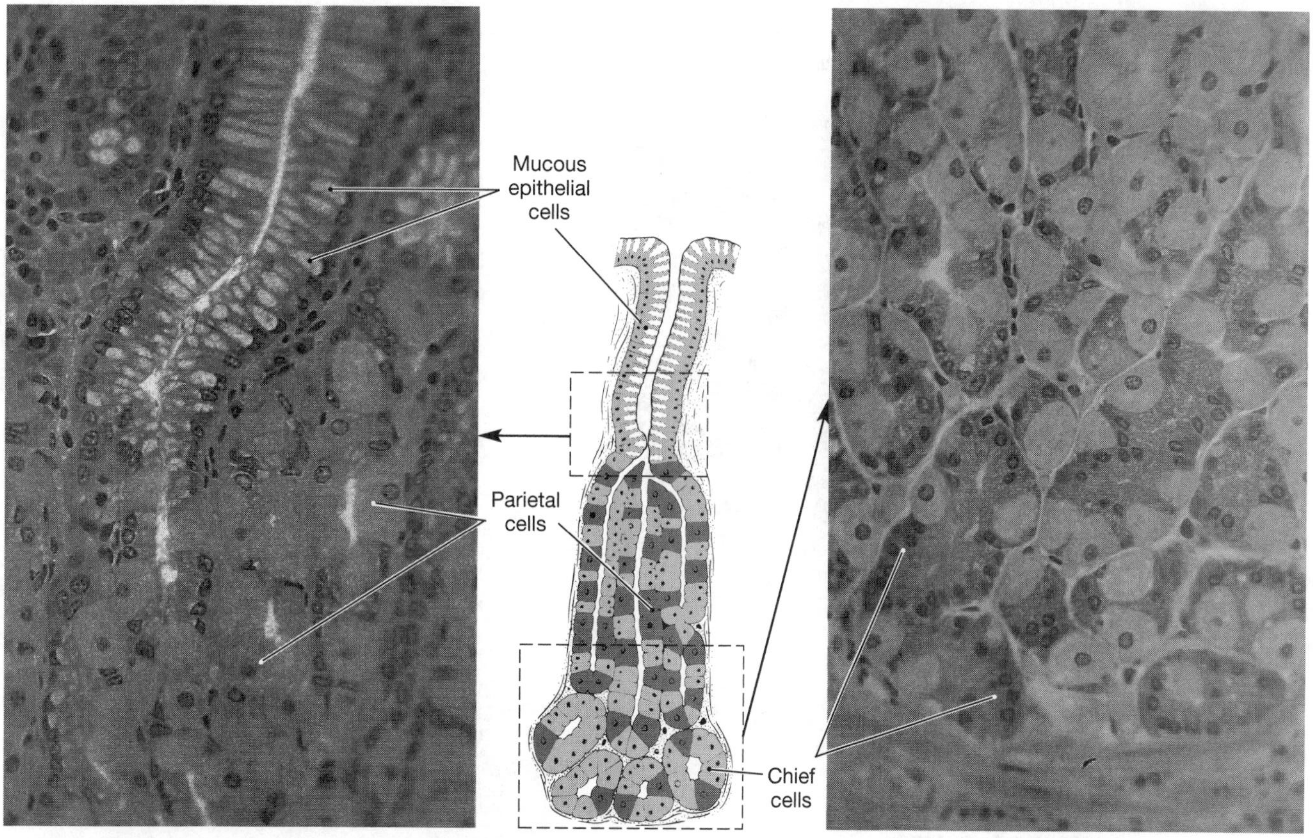

Mucous epithelial cells
Parietal cells
Chief cells

● **FIGURE 58–10c**
Gastric (Stomach) Glands.

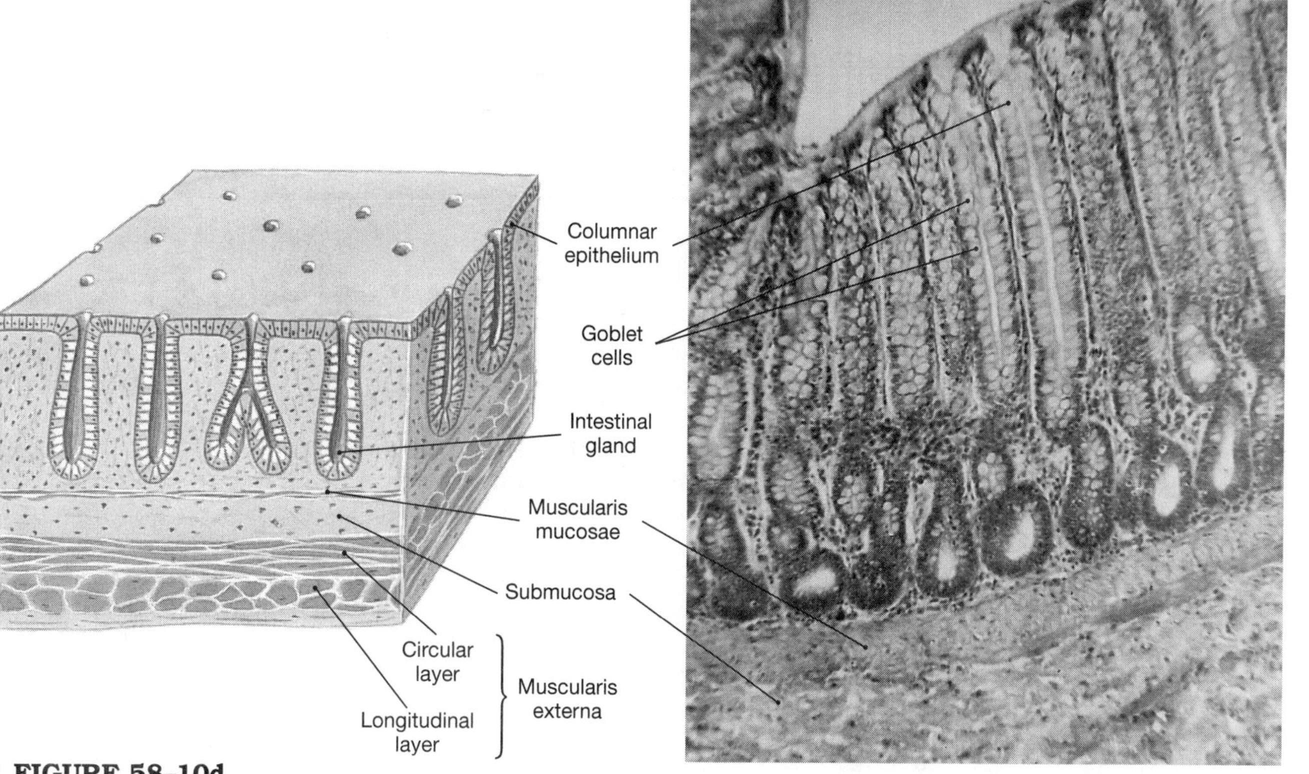

● **FIGURE 58–10d**
Large Intestine.

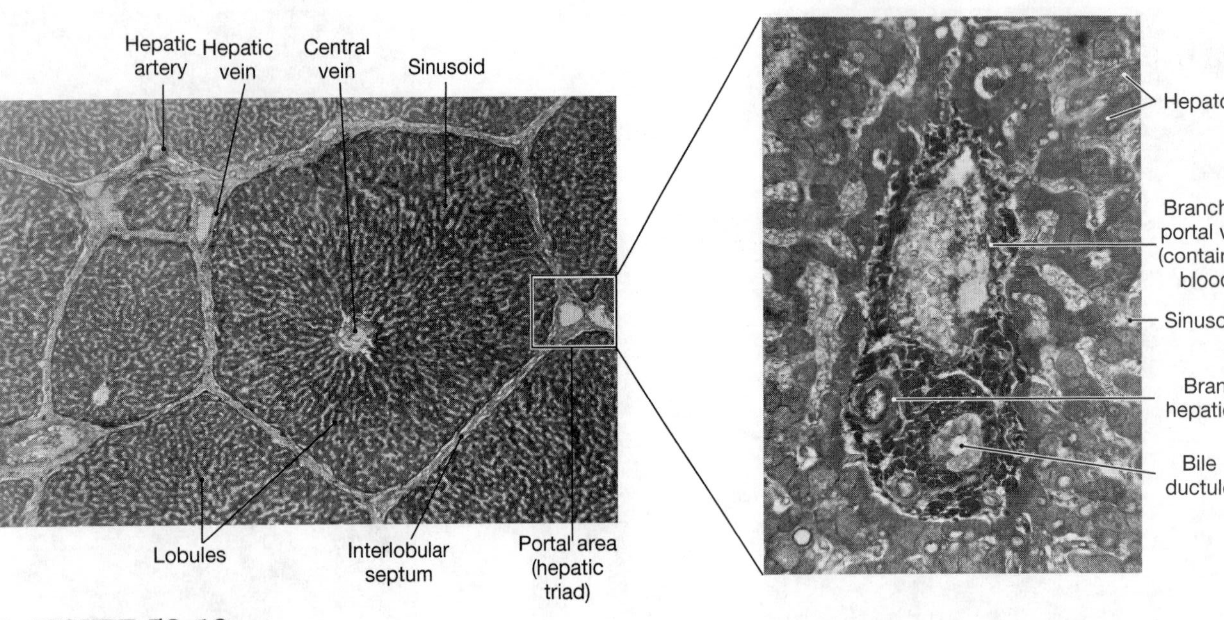

● **FIGURE 58–10e**
Liver.

structures indicated on these figures. Check off the organs as you view them.

Salivary glands _______
Pancreas _______
Intestinal mucosa _______
Gastric glands _______
Large intestine _______
Liver _______

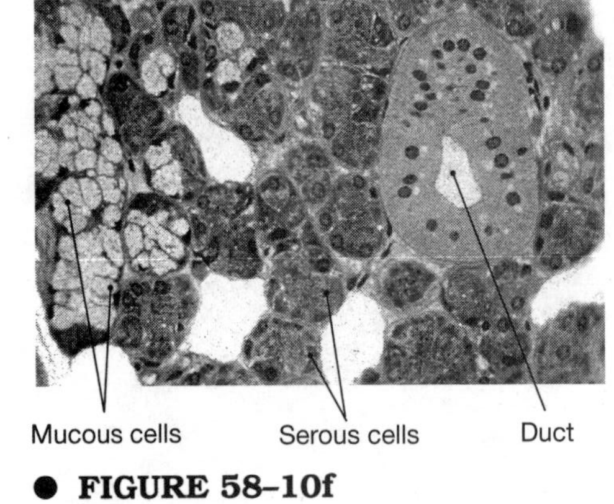

● **FIGURE 58–10f**
Salivary Gland.

❑ Additional Activities

1. Compare and contrast the anatomy and histology of the human digestive system with that of other mammals.
2. Find out what happens to the digestive system when the human (or animal) is raised gnotobiotically (in a completely sterile environment). David, the boy in the bubble, was an example of someone raised gnotobiotically.

NOTES

❑ Lab Report

1. Referring to the numbered inquiries at the beginning of this exercise, complete the following box summary:

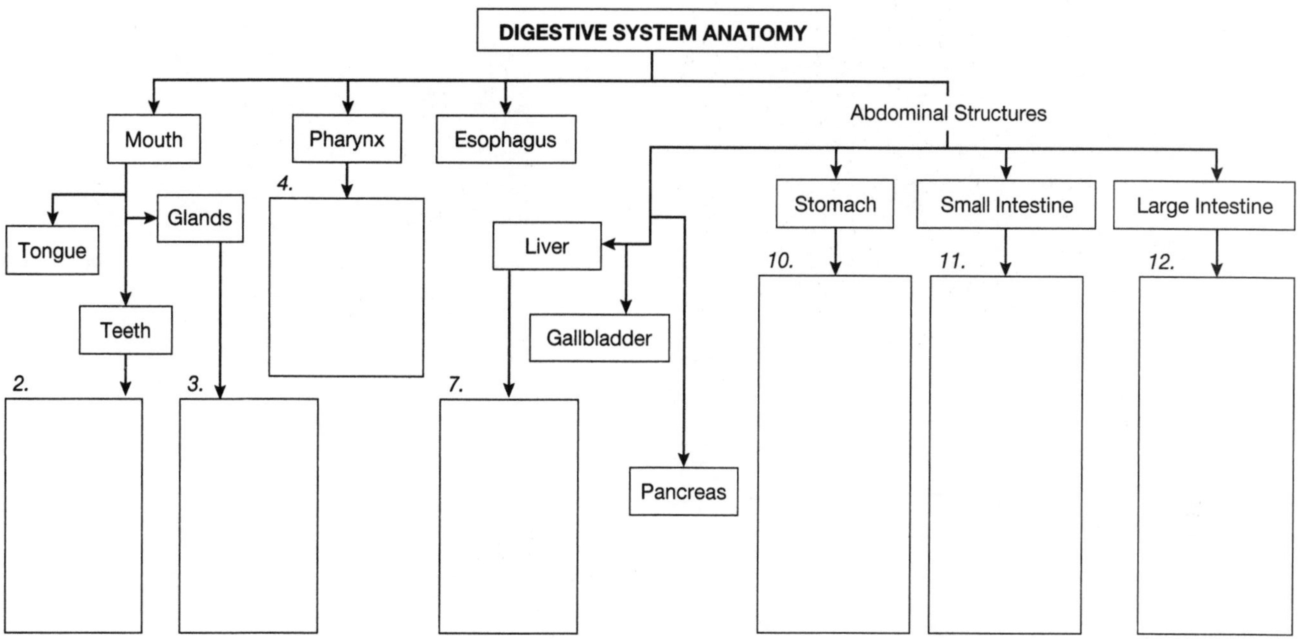

General Questions

NOTES

1. What are the principal organs of the alimentary canal? What are the accessory organs?
2. What is the difference between the visceral peritoneum and the parietal peritoneum?
3. Where are the omenta, and what is the function of each?
4. Where is the gallbladder in relation to the liver?
5. Anatomically, how are the cystic duct, the hepatic duct, and the common bile duct related?
6. How do the sizes and shapes of the human heterodont teeth relate to the omnivorous diet of most humans?
7. Name the principal valves and sphincters of the digestive system. What common function do they all have? What differences do you note between them?
8. List the similarities and differences you noted in the microscopic structure of the different parts of the digestive system.
9. How do the salivary glands compare with the intestinal glands? With the pancreas?

EXERCISE **59**

Digestive Physiology

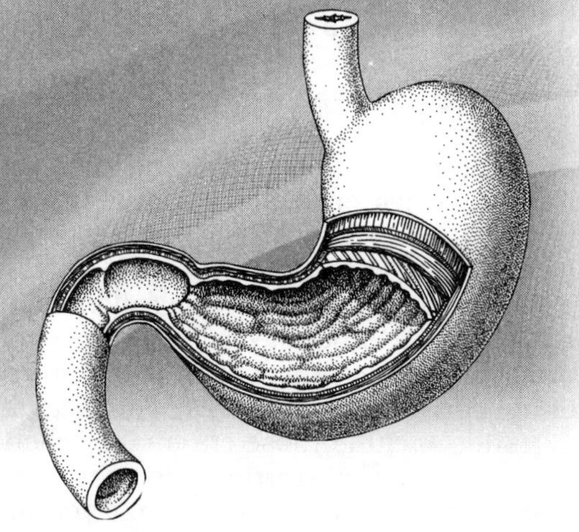

PROCEDURAL INQUIRIES

Preparation

1. What are the major structures of the alimentary canal?
2. What do we mean by the physical and chemical components of digestion?

Examination

3. What are the functions of the mouth?
4. What is a bolus?
5. What is peristalsis and what is the role of peristalsis in digestion?
6. What digestive events occur in the stomach?
7. What is chyme?
8. What digestive events occur in the duodenum?
9. What digestive events occur in the ileum and jejunum?

For each of the following, answer these questions:
Where does digestion take place?
What are the enzymes involved in digestion?
Where does absorption take place?
How does absorption take place?

10. Carbohydrates
11. Proteins
12. Lipids
13. What digestive events occur in the large intestine?
14. What factors are involved in the final movement?

Additional Inquiries

15. What is digestion?
16. What is mastication?
17. What is deglutition?
18. What is the intestinal feedback mechanism?
19. What are normal flora?

Key Terms

Alimentary Canal
Bile Salts
Bolus
Carbohydrates
Cecum
Chief Cells
Cholecystokinin
Chylomicrons
Chyme
Defecation Reflex
Deglutition
Enteroendocrine Cells
Fatty Acids
Final Movement
Gastric Contractions
Gastrin
Haustration
Ileocecal Valve
Lower Esophageal Sphincter
Mass Movements
Mastication
Micelles
Mixing Waves
Monoglycerides
Myenteric Reflexes
Normal Flora
Parietal Cells
Pepsin
Peristalsis
Proteins
Secretin
Segmentation
Taenia Coli
Triglycerides

Materials Needed

Torso
Charts and Diagrams

Digestion, the process of converting foodstuffs into products usable by the body, requires both physical and chemical action. Food particles must be broken down and transformed into these usable components while being moved along the digestive tract.

The physical aspects of digestion include those activities that either mechanically break down food products or move the food products from one end of the canal to the other. The chemical aspects of digestion include those enzymatic reactions that prepare potential nutrients for passage into the system and those reactions that actually transport substances across the alimentary barrier.

In this laboratory exercise we will examine selected physical and chemical aspects of digestion. This exercise is designed to solidify your background and reinforce your physiological understanding of digestion. Your instructor may assign this exercise for completion outside of laboratory time.

❑ Preparation

I. Introduction

A. Background

1➡ Review Exercise 58 to conceptualize the gross anatomy of the digestive system.

2➡ Study the torso to gain a perspective of the major structures of physiological interest along the **alimentary canal**: the mouth, esophagus, stomach, small intestine, large intestine, and rectum.

3➡ Think about how when food is ingested, some of that food enters the cardiovascular system and is utilized by the cells, while the rest is passed out of your body as waste.

4➡ For a functional overview of digestive physiology, refer back to Exercise 58, particularly Figure 58–4●.

B. Procedure

1➡ Acquaint yourself with the generalities of the digestive process by working through the explanations and simulated experiments given here. Since digestion has both physical and chemical components, it is important that you understand how these aspects work both individually and as a unit.

2➡ Based on your present understanding, begin this exercise by writing out an example of a physical aspect of digestion and an example of a chemical aspect of digestion.

____

____

❑ Examination

II. Action

A. In the Oral Cavity

1➡ To study the digestive process, start at the mouth with a hypothetical bite of food. Your bite should be replete with **carbohydrates**, **proteins**, and **fats**. Think of a food fitting that description, perhaps a piece of hamburger on a bun with lettuce and tomato. Or a bite of taco pizza. In this exercise we will explore the physical and chemical aspects of this bite's journey through the alimentary canal.

Concept Check 1 Think about the bite of food in your mouth. Compose a list of the physical and chemical functions the mouth performs when dealing with incoming food.

____

____

____

____

____

____

Now check the suggested answer at the end of this exercise. You probably have the most obvious action of the mouth is chewing (**mastication**). Were you surprised at how many additional functions of the mouth you were able to think of?

____

Can you think of any oral cavity functions not on the list at the end of this exercise? __________

____

____

____

According to your hypothetical bite, which foodstuffs does carbohydrate digestion include?*

____

____

____

Suffice it here that initial carbohydrate hydrolysis is taking place.

Liquids, of course, bypass the mouth, or do they? What do you think? __________

____

____

____

*The specifics of this carbohydrate digestion will be discussed in Exercise 60.

B. Deglutition (Swallowing) [∞ Martini, p. 860]

The tongue guides the food toward the throat. The food is now known as a **bolus**.

1➡ Think about your hypothetical bite, now a bolus. Hypothetically swallow your bolus and list the actions involved in **deglutition** as you perceive them.

____

____

____

2➡ Now study Figure 59–1●. Describe in writing and out loud, either to yourself or to your lab partner, exactly what is happening with the mouth, throat, epiglottis, and esophagus as your bolus heads toward the stomach.

____

____

____

____

Did you realize swallowing was so complicated?

____

3➡ Did you know that you could not swallow that bolus if it were not moist? Friction between a dry bolus and the walls of the esophagus would make peristalsis ineffective. In case the bolus is not moist enough, you do have a few glands along the esophagus. No chemical action to speak of takes place on the way down to the stomach.

C. Toward the Stomach

Your digestive system is a one-way track. The basic design is such that substances go in one end of the chute and are mechanically passed to the other end. Mechanical action along the chute prevents regurgitation or backflow within the system.

1➡ The major propelling motion of the digestive tract is **peristalsis**. Study Figure 59–2● and note how peristaltic action dictates the one-way trip the bolus is taking.

2➡ Describe in writing and out loud what happens in peristalsis. Peristalsis is found along the entire length of the canal.

____

____

____

____

In addition to peristalsis, the sphincters of the digestive tract also keep the bolus moving in one direction. The trigger for the **lower esophageal**

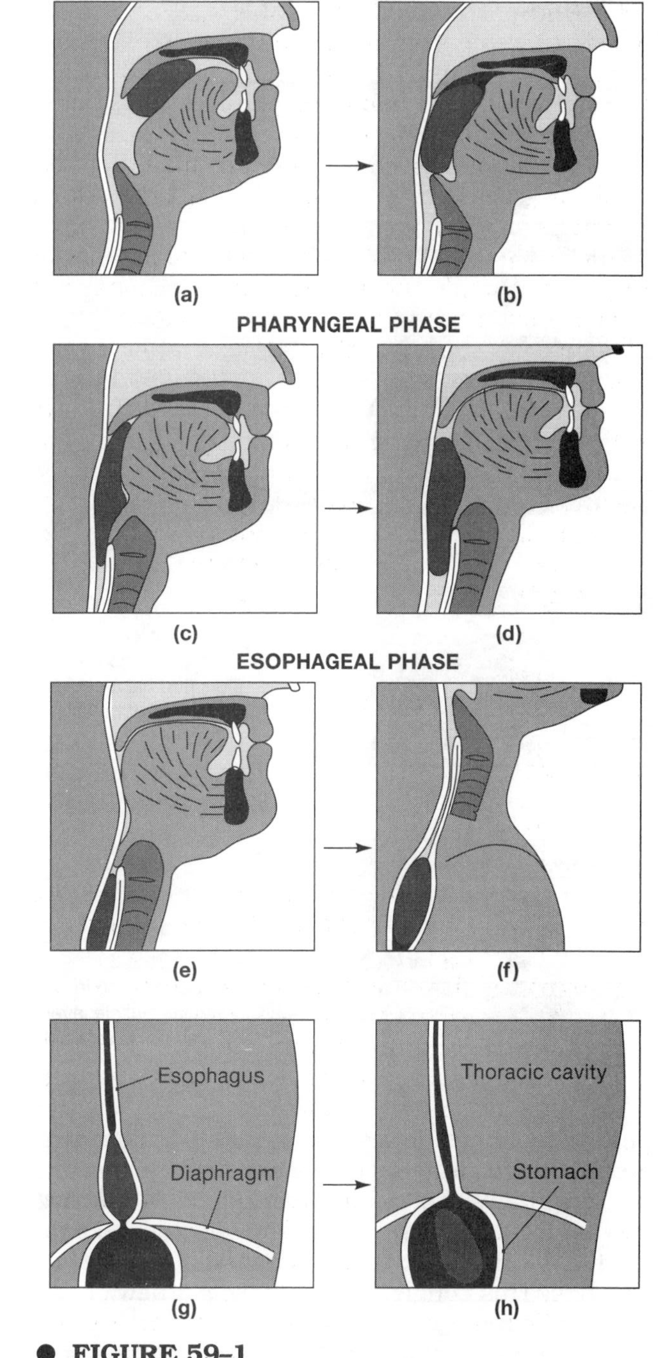

● **FIGURE 59–1**
Swallowing Process.

sphincter is peristaltic movement in the esophagus (not the stomach). That means that vomiting is against the norm.

Myenteric reflexes, stimuli originating from the myenteric plexus (the nerve net within the muscularis externa), continue to produce peristaltic contractions along the rest of the digestive tract.

D. In the Stomach

Once your bolus reaches the stomach, the real action begins. Of course, peristalsis is present but

● FIGURE 59–2
Peristalsis.

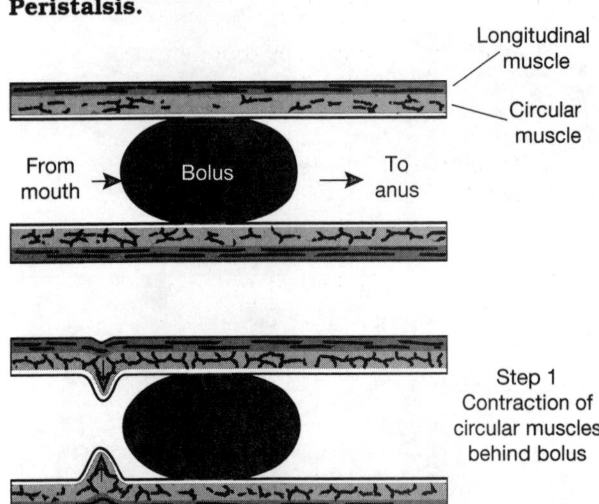

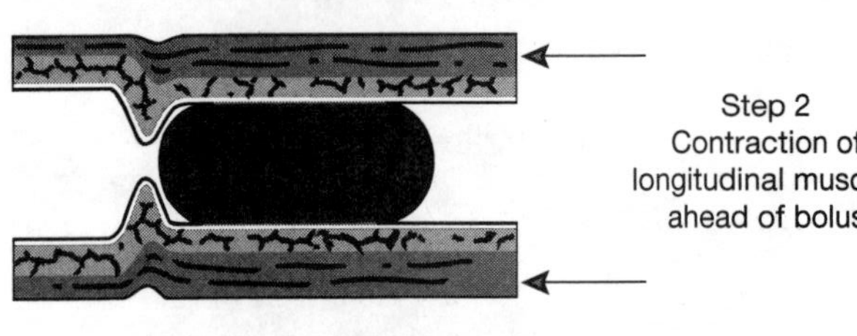

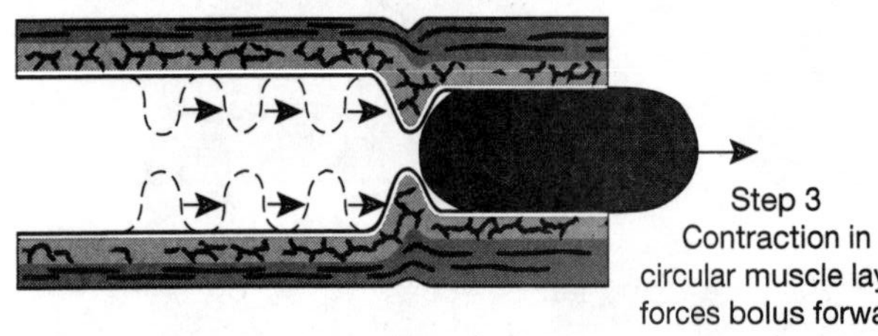

Step 3
Contraction in circular muscle layer forces bolus forward

peristalsis almost takes a back seat to the **gastric contractions**. Initially the contractions are weak, but gradually the pulsations turn into **mixing waves** strong enough to swirl and churn the bolus, mixing it surreptitiously with assorted gastric secretions. This conglomeration forms a new entity called **chyme**.

Gastric secretions are of varying origins. Gastric glands are responsible for much of what we think of as the "stomach juices."

1➡ Examine Figure 59–3●. Each gastric gland includes numerous **parietal cells**, which secrete both **intrinsic factor** (to facilitate the intestinal absorption of vitamin B_{12}) and **hydrochloric acid** (to lower the pH and activate **pepsinogen**). Pepsinogen is produced by the **chief cells**, the other cells of the gastric glands. Activated pepsinogen is **pepsin**, an enzyme for breaking down proteins. Pepsin is functional at very low pHs. *[∞ Martini, p. 864]*

● FIGURE 59–3
Gastric Gland.

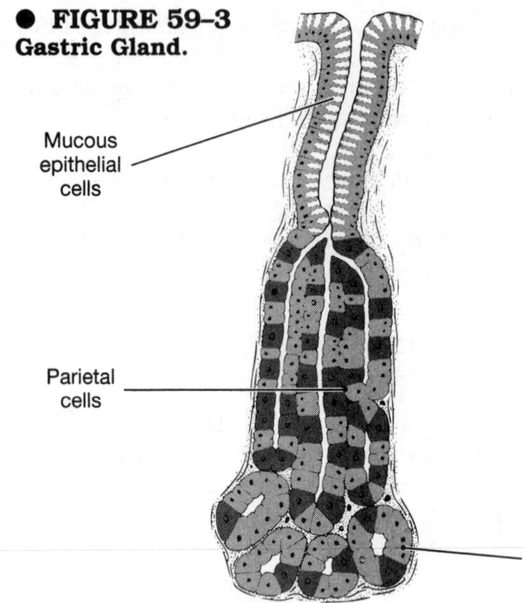

Enteroendocrine cells are scattered throughout the gastric region. These cells secrete at least six different local hormones, only one of which, **gastrin**, has a known function. Gastrin stimulates both the parietal cells and the chief cells.

2➡ Study Figure 59–4● for a summary of gastrin and the other enterogastric hormones.

Except for alcohol and a few other similar compounds, virtually nothing is absorbed from the stomach into the circulatory system.

E. In the Small Intestine

Your highly liquified chyme is now ready to be released into the small intestine. As you follow this chyme through the small intestine, keep in mind that the majority of the digestion in the small intestine takes place in the duodenum. Absorption into the body proper occurs mostly from the jejunum and the ileum. Refer back to Exercise 58 as necessary.

1➡ Study Figure 59–5●. Chyme enters in squirts, a few mls at a time, which helps to maintain maximum efficiency in the small intestine.

The feedback mechanism involved here includes the release of the duodenal hormones, **secretin** and **cholecystokinin (CCK)**, both of which inhibit gastrin and gastric activity. If the stomach is on temporary shutdown, the chyme cannot enter the intestine. As the intestinal chyme moves on, the inhibition is lifted and gastric activity continues. Another squirt of chyme enters the intestine. *[∞ Martini, p. 867]*

Hormone	Stimulus	Origin	Target	Effects
Gastrin	Vagal stimulation or arrival of food in the stomach	Stomach	Stomach	Stimulates production of acids and enzymes, increases motility
Enterocrinin	Arrival of acid chyme in the duodenum	Duodenum	Duodenal (Brunner's) glands	Stimulates production of alkaline mucus
Secretin	Arrival of acid chyme in the duodenum	Duodenum	Pancreas	Stimulates alkaline buffer production
			Stomach	Inhibits gastric secretion and motility
Cholecystokinin (CCK)	Arrival of acid chyme containing lipids and partially digested proteins	Duodenum	Pancreas	Stimulates production of pancreatic enzymes
			Gallbladder	Stimulates contraction of gallbladder
			Duodenum	Causes relaxation of sphincter at base of bile duct
			Stomach	Inhibits gastric secretion and motion
Gastric-inhibitory peptide (GIP)	Arrival of chyme containing large quantities of glucose	Duodenum	Pancreas	Stimulates release of insulin by pancreatic islets
Vasoactive intestinal peptide (VIP)	Arrival of chyme in the duodenum	Duodenum	Duodenal glands, stomach	Stimulates buffer secretion, inhibits acid production, dilates intestinal capillaries
Gastrin	Arrival of chyme containing large quantities of undigested proteins	Duodenum	Stomach	Stimulates gastric secretion and motion (as above)

● **FIGURE 59–4**
Important Gastrointestinal Hormones and Their Primary Effects.

Meanwhile, the pancreatic fluids raise the pH of the environment to approximately 8. The pancreatic and intestinal enzymes are functional in this alkaline environment. Additional alkaline secretions are produced by the duodenal (Brunner's) glands. Refer back to Figure 58–10●. *[∞ Martini, p. 871]*

2➡ Study Figure 59–6●. This table summarizes the digestive enzymes and their functions. Notice the different pancreatic and intestinal enzymes. (These concepts will be demonstrated in Exercise 60.)

3➡ Correlate Figure 59–6● with 59–7●. Protein digestion, which began in the stomach with the activation of the pepsinogen to pepsin, continues in the small intestine with the further breakdown of the polypeptides.

Concept Check **2** What is the compositional unit of a peptide? ______________________

____

4➡ Study Figures 59–6● and 59–7●. Where do the different components of carbohydrate digestion take place and what are the carbohydrate breakdown products?

____

____

____

Why is it necessary to break down both the peptides and the carbohydrates into their fundamental units? ______________________

____

____

5➡ To summarize the digestive processes covered thus far, explain out loud to your lab partner how the particular carbohydrates and proteins in your original bite of food get from the digestive tract into the circulatory system.

How do carbohydrates enter the cardiovascular system? ______________________

____

____

6➡ Now concentrate on the lipids. The lipids travel a slightly different route. Lipase con-

THE PHASES OF GASTRIC SECRETION

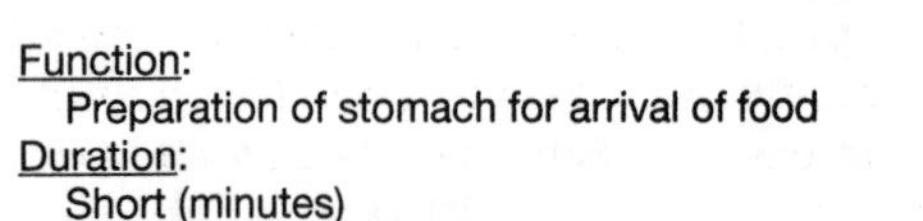

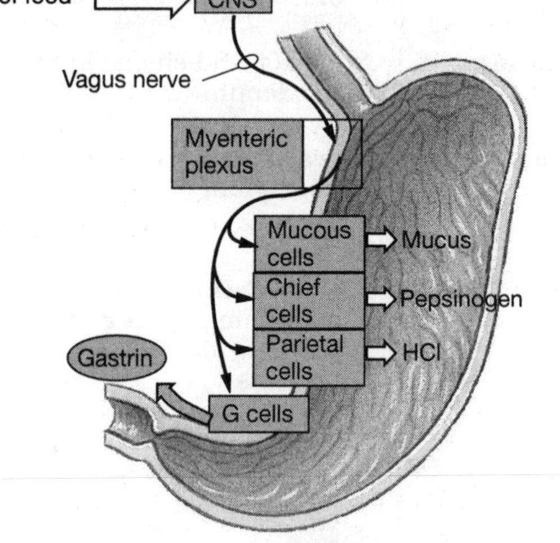

Function:
Preparation of stomach for arrival of food
Duration:
Short (minutes)
Mechanism:
Neural, via preganglionic fibers in vagus nerve and synapses in myenteric plexus
Actions:
Primary: increased volume of gastric juice by stimulating mucus, acid, and enzyme production
Secondary: stimulates release of gastrin by G cells

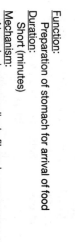

(a) The Cephalic Phase

Functions:
Enhance secretion started in cephalic stage; homogenization and acidification of chyme; initiate digestion of proteins by pepsin
Duration:
Long (3-4 hours)
Mechanisms:
Neural: short reflexes triggered by
(1) stimulation of stretch receptors as stomach fills
(2) stimulation of chemoreceptors as pH increases
Hormonal: stimulation of gastrin release by G cells by parasympathetic activity and presence of peptides and amino acids in chyme
Local: release of histamine by mast cells as stomach fills (not shown)
Actions:
Increased acid and pepsinogen production, increased motility and initiation of mixing waves

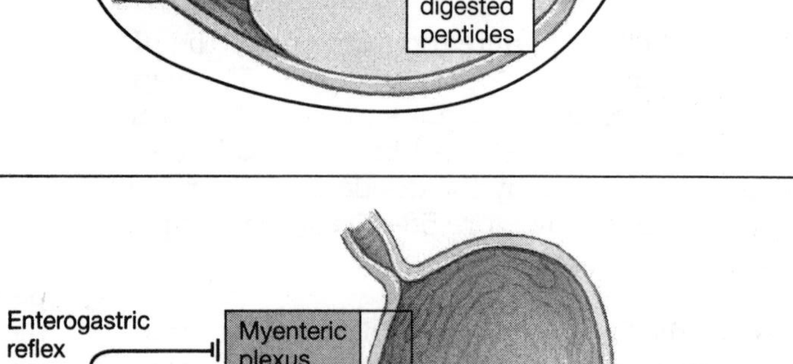

(b) The Gastric Phase

Functions:
Control rate of chyme entry into duodenum
Duration:
Long (hours)
Mechanisms:
Neural: short reflexes (enterogastric reflex) triggered by extension of duodenum
Hormonal:
Primary: stimulation of GIP, secretin, CCK release by presence of acid, carbohydrates, and lipids
Secondary: release of gastrin stimulated by presence of undigested proteins and peptides (not shown)
Actions:
Feedback inhibition of gastric acid and pepsinogen production, reduction of gastric motility

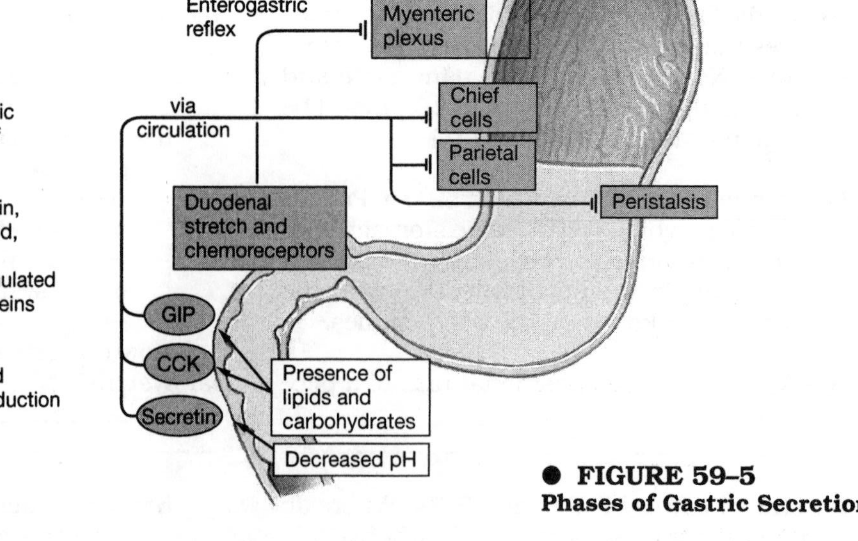

(c) The Intestinal Phase

● **FIGURE 59–5**
Phases of Gastric Secretion.

verts the **triglycerides** (the most abundant of the dietary lipids) into **fatty acids** and **monoglycerides**. These interact with **bile salts**, forming complexes called **micelles**. Because of the micelle complex, fats mix more readily with water, and the lipids can diffuse across the intestinal cell membrane. In the intestinal cell, the new triglycerides, coated with proteins to make them more water-soluble, are formed and as such are excreted into the interstitial fluid. *[∞ Martini, p. 890]*

Because they are generally too large to be picked up by the capillaries, the coated triglycerides, now called **chylomicrons**, enter the circulatory system by way of the lacteals of the lymphatic system. *[∞ Martini, Chapter 22]*

7➡ As you examine Figure 59–7●, verbalize to your partner which foods from your hypothetical bite contained lipids. Describe how your particular lipid entered your circulatory system.

Enzyme (Proenzyme)	*Source*	*Optimal pH*	*Target and Action*	*Products*	*Remarks*
Alpha-amylase	Salivary glands, pancreas	6.7–7.5	Breaks bonds between carbohydrate molecules	Disaccharides and trisaccharides	
Lingual lipase	Glands of tongue	6.7–7.5	Triglycerides	Fatty acids and monoglycerides	
Pepsin (Pepsinogen)	Chief cells of stomach	1.5–2.0	Breaks bonds between amino acids in proteins	Short-chain polypeptides	Secreted as proenzyme, pepsinogen; activated by H^+ in stomach acid
Renin			Coagulates milk proteins		Secreted only by stomachs of infants
Trypsin (Trypsinogen)	Pancreas	7–8	Proteins, polypeptides	Short-chain peptides	Proenzyme activated by enterokinase; activates other pancreatic proteases
Chymotrypsin (Chymotrypsinogen)	Pancreas	7–8	Proteins, polypeptides	Short-chain peptides	Activated by trypsin
Carboxypeptidase (Procarboxypeptidase)	Pancreas	7–8	Proteins, polypeptides	Short-chain peptides and amino acids	Activated by trypsin
Elastase (Proelastase)	Pancreas	7–8	Elastin	Short-chain peptides	Activated by trypsin
Pancreatic lipase	Pancreas	7–8	Triglycerides	Fatty acids and monoglycerides	Bile salts must be present for efficient action
Nuclease	Pancreas	7–8	Nucleic acids	Nitrogenous bases and simple sugars	Includes ribonuclease for RNA and deoxyribonuclease for DNA
Enterokinase	Intestinal glands	7–8	Trypsinogen	Trypsin	Reaches lumen through disintegration of shed epithelial cells
Maltase, sucrase, lactase	Brush border of small intestine	7–8	Maltose, sucrose, lactose	Monosaccharides	Found in membrane surface of microvilli
Exopeptidases	Brush border of small intestine	7–8	Dipeptides, tripeptides	Amino acids	Found in membrane surface of microvilli

● **FIGURE 59–6**
Digestive Enzymes.

● FIGURE 59–7
Events of Chemical Digestion.

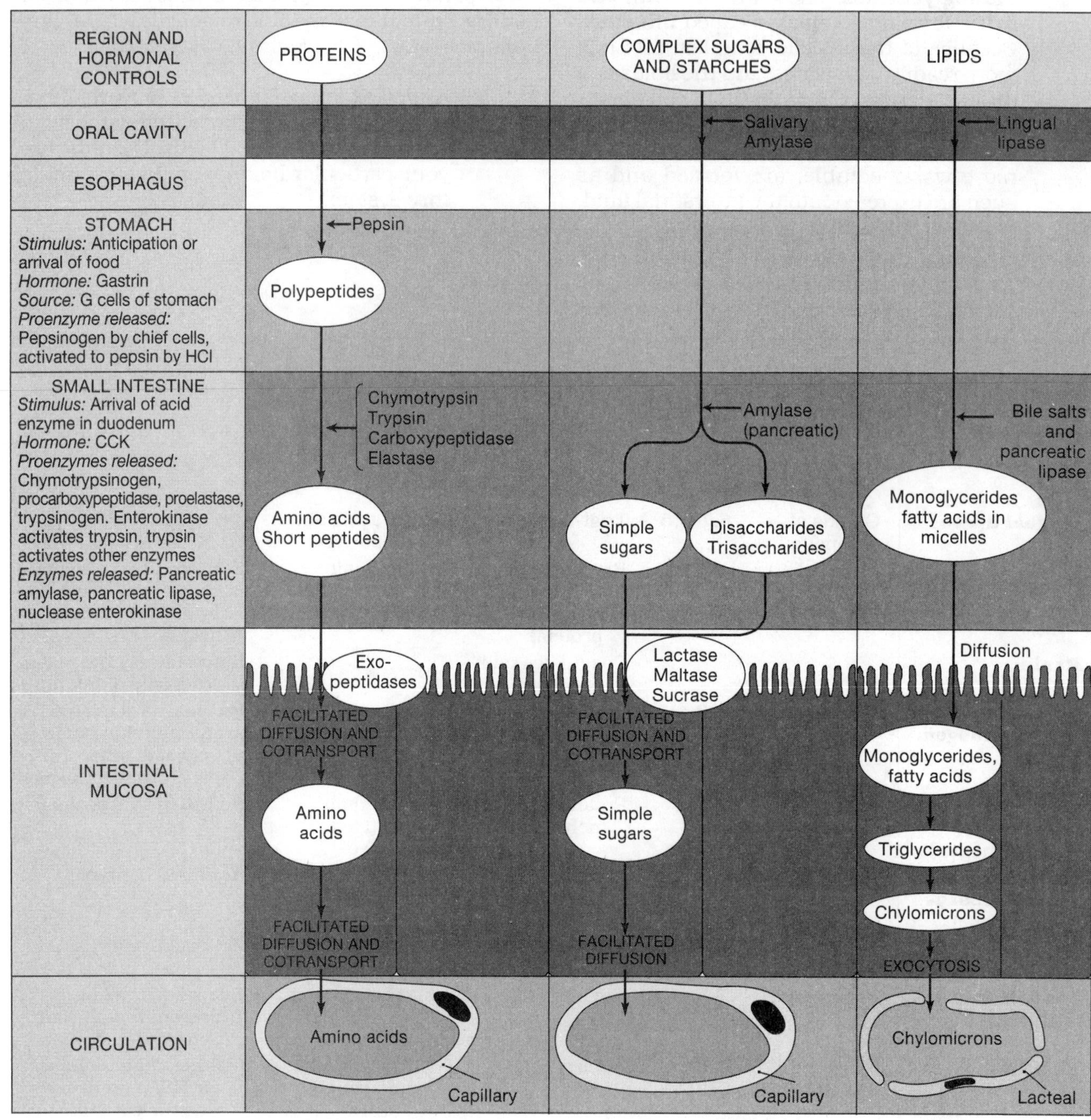

In addition to the enzymatic activity, a great deal of physical activity is also taking place. Continuous physical movement is essential for keeping your chyme well mixed and well exposed to the enzymes.

8➡ Study Figure 59–8● and notice that in addition to peristalsis, the small intestine undergoes **segmentation**. Basically, segmentation churns and fragments your chyme.

Describe segmentation as you see it in Figure 59–8●. ______________________

____

F. In the Large Intestine

Finally, your chyme reaches the end of its 6-meter small intestine journey and is ready to enter the large intestine. By this time all the basic foodstuffs have been absorbed, and the entire chute has just about churned its last. It is peristalsis that pushes the chyme past the relaxed **ileocecal valve**, which is really a sphincter, and into the rounded pouch called the **cecum**.

● **FIGURE 59–8**
Segmentation Movements.

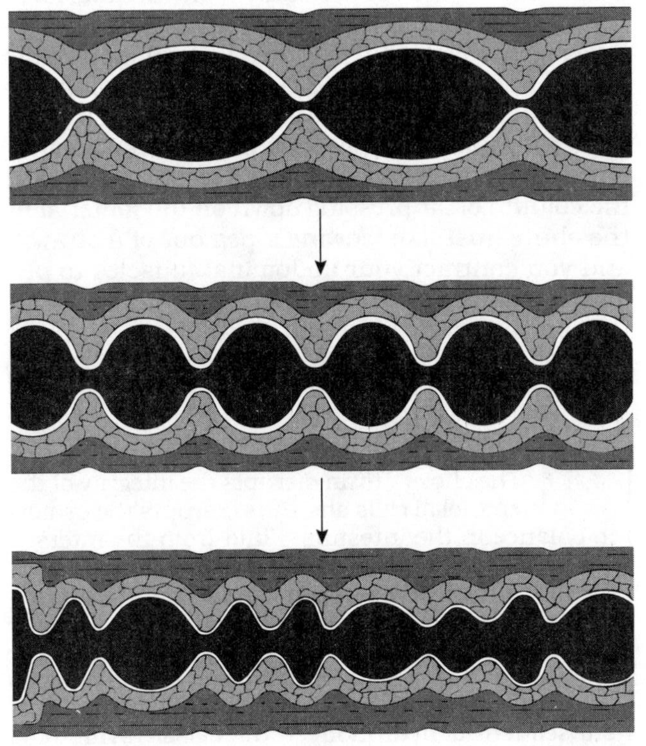

The primary functions of the large intestine are the resorption of water and the compaction of fecal material. Some vitamins (such as vitamin K) from bacterial action are also absorbed and, of course, fecal material is stored temporarily in the large intestine, while awaiting a time appropriate for defecation.

Movements are roughly the same in the large intestine as in the small intestine, although they are usually more sluggish. However, several times each day massive peristaltic contractions occur. These are called **mass movements**.

Another large intestinal movement is a **haustration**. Recall that the large intestine is a series of pouches, haustrae, striped with three longitudinal strips of muscle, the **taenia coli**. These muscles act like elastic. When they contract (upon stimulation from the distention of the filled haustrae), additional pocketed movement takes place.

The water resorption from the large intestine is basic osmosis. Remember that water always moves passively. *[∞ Martini, p. 891]*

1 ➡ Examine Figure 59–9●. Are you amazed at the tremendous conservative capacity of the large intestine? ______ Almost 10,000 ml enter the alimentary canal; less than 2% becomes feces!

The osmotic gradient is a concentration gradient. Nutrient and electrolyte movements can be either active or passive. Water follows either way. For osmotic purposes, the nature of the solute is irrelevant. For metabolic purposes the nature of the solute is critical. Despite years of research, scientists still do not understand it all.

● **FIGURE 59–9**
Secretion and Water Absorption Along the Digestive Tract.

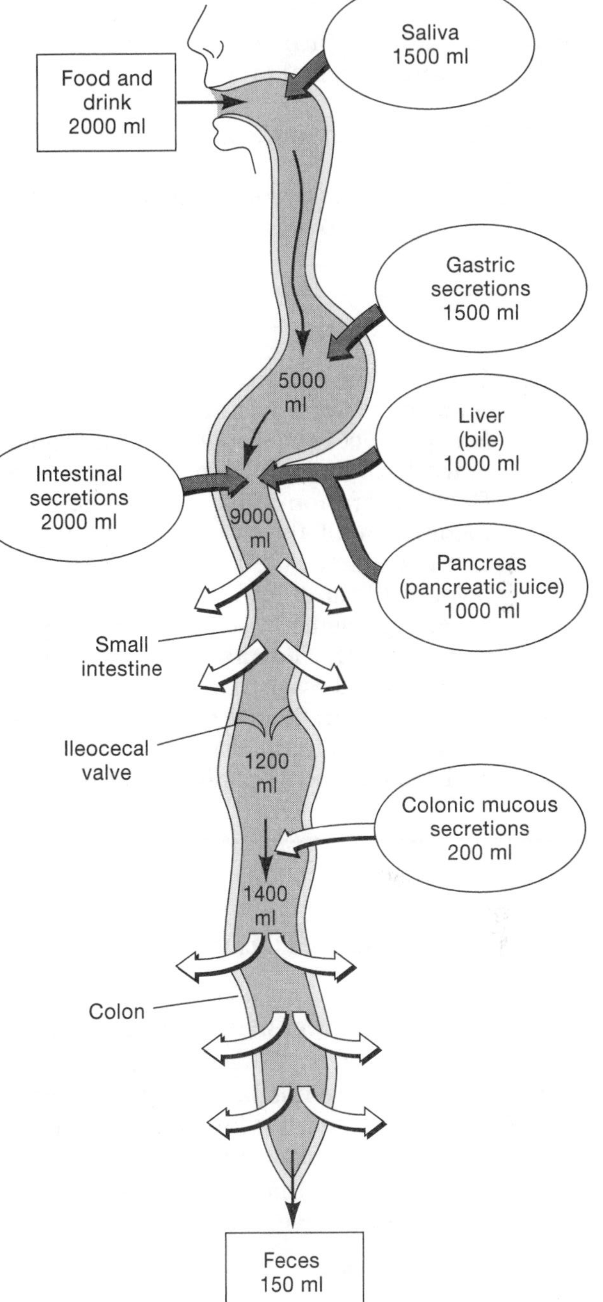

G. Final Processes

So, now you have gotten just about everything out of your bite of food, and the waste products are about to leave your body. Aside from the water, about half the waste products includes substances that cannot be digested (such as cellulose), substances that passed through too quickly to be digested, substances that were not completely digested, sloughed epithelial cells, and assorted metabolic and epithelial products.

1➡ Think about your food.

Which parts of your hypothetical bite are probably about to be shed?

____

____

____

2➡ Consider the bacterial components of your final movement. In addition to remaining food products and sloughed cells, about half of the fecal material is made up of bacteria. Bacteria are found in all areas of the alimentary canal, the fewest being in the stomach because of the low pH, and the most being in the large intestine. The bacteria and other organisms living in and on the body constitute the **normal flora**. The normal flora are a necessary part of our symbiotic relationship with the rest of the world. We supply our intestinal flora with a food source and some of them supply us with vitamin K.

Peristaltic action forces fecal materials into the **rectum**, or rectal chamber. This **final movement** triggers the **defecation reflex**. Stretch receptors perpetrate a series of peristaltic contractions forcing the feces toward the anus. This movement relaxes the involuntary internal anal sphincter which, in turn, forces the closing of the voluntary external anal sphincter. Under normal circumstances, you voluntarily open the external sphincter to defecate. You also close the glottis to increase the column of air pressing down on the material in the chute (just like blowing a pea out of a straw!), and you contract your abdominal muscles to put additional force on the fecal material. The rectum, the end of the alimentary canal, is emptied through its end point, the anus.

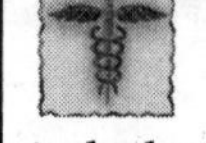

Cholera is a disease of the small intestine The cholera toxin disrupts the integrity of the epithelial cells and thus disrupts the osmotic balance in the intestine. Fluid from the interstitium rushes toward the epithelium and subsequently into the intestinal lumen. The large intestine continues to function normally. However, because of the tremendous overload of fluid coming from the small intestine, the large intestine cannot resorb the water quickly enough and massive diarrhea (and subsequent dehydration) — the classic symptoms of cholera — soon follow.

❑ Additional Activities

1. Research and answer in detail some of the thought questions in this exercise.
2. We as humans are not particularly conservative when it comes to water. Some animals resorb a far greater proportion of their intestinal water than we do. How do they manage that?
3. Explore the normal flora. What functional advantage do different intestinal bacteria have?

NOTES

Answers to Selected Concept Check Questions

NOTES

1. Functions of the mouth — possible list:
   - Enclosed chamber for initial food preparation.
   - Temporary storage facility.
   - Expandable area for churning.
   - Teeth for shredding or tearing (mastication).
   - Teeth for grinding (mastication).
   - Tongue for moving food from one side to another and wadding the food up for swallowing (deglutition).
   - Tongue for guiding food down throat.
   - Sensory receptors for touch, taste, and gastric response.
   - Thermoreceptors to warn of temperature problems.
   - Glands for moistening food.
   - Glands for lubricating food.
   - Glands for secreting salivary amylase to begin carbohydrate (starch) digestion.
2. The compositional unit of a peptide is an amino acid.
3. Answers will vary according to the components of the original bite. However, the answers should include any particles containing cellulose and foods traversing the system too quickly to be digested.

❑ Lab Report

1. Referring to the numbered inquiries at the beginning of this exercise, complete the following box summary:

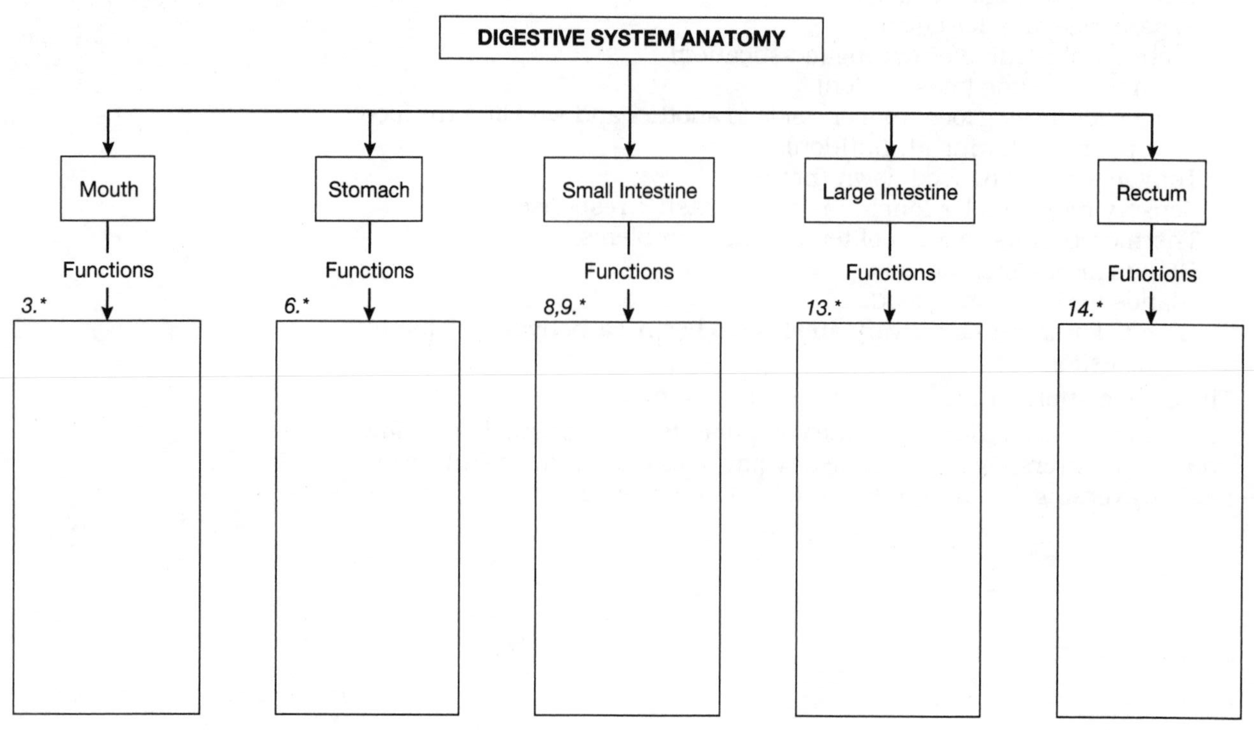

*List according to your instructor's directions.

General Questions

NOTES

1. What is the difference between the chemical and physical components of digestion?
2. Where in the digestive system would you find peristalsis?
3. For what types of substances would deglutition without mastication be possible?
4. What is the intestinal feedback mechanism and why is this important to digestion?
5. What is the role of the normal flora in the digestive processes?
6. What is the difference between a bolus and the chyme?
7. Can you think of any situations where a person would produce no fecal material?
8. Why can't you breathe while you are swallowing?
9. Can you digest food while standing on your head?
10. Complete this chart: (Use additional paper if necessary.)

	Digestion Begins	Enzymes Involved	Where Absorption Takes Place	How Absorption Takes Place
Carbohydrates				
Proteins				
Lipids				

EXERCISE 60

Enzymatic Action in Digestion

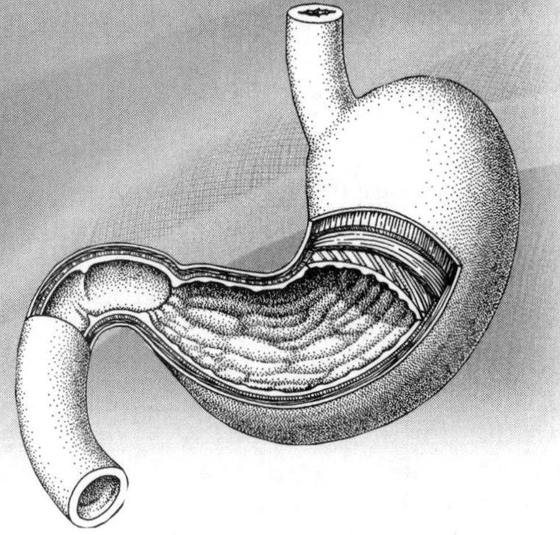

PROCEDURAL INQUIRIES

Preparation

1. What is hydrolysis?
2. Which major molecules must be hydrolyzed in digestion?
3. How do we simulate body conditions in setting up digestive experiments?

Experimentation

4. How do we test for starch?
5. What is the principle behind the starch test?
6. How do we test for reducing sugars?
7. What is the principle behind the reducing-sugar test?
8. How do we test for proteins?
9. What is the principle behind the protein test?
10. How do we test for lipids?
11. What is the principle behind the lipid test?

Additional Inquiries

12. What is digestion?
13. What is the role of enzymes in the digestive process?
14. What are bile salts?

Key Terms

Amylase
Benedict's Test
Bile
Biuret's Test
Carbohydrate
Hydrolysis
Lipid
Lugol's Iodine
Micelles
Pancreatin
Protein
Reducing Sugar
Starch

Materials Needed

Water Bath — 37°C
Water Bath — 100°C
Test Tubes
Test Tube Racks
Micropipettes or Droppers
Tongs
Spot Plates (microscope slides or watch glasses may be substituted)
Beaker of 1% Starch Solution
1–2% Maltose Solution
1–2% Glucose Solution
Water Buffered to pH 8
Iodine
Benedict's Solution
Protein Source — Albumin or Nutrient Broth
5% Pancreatin Solution (pH 8)
Biuret's Solution
Cream (or Half and Half)
Bile Salts
Litmus Solution (litmus paper may be substituted)

The foods we eat are composed primarily of carbohydrates, proteins, and lipids. In foods, however, these building blocks are often not in their most usable form and therefore must be digested before they can be transformed into whatever substance the body may need. Digestion is the process of breaking foods down into that usable form.

The digestive process is a series of enzyme-catalyzed reactions. Recall that an enzyme is an organic catalyst, a molecule at least part protein whose function is to change the rate of a chemical reaction.

Enzymes do not do anything that does not happen naturally. For instance, if you had had a steak for dinner last night, you really would not have needed those digestive enzymes for the digestive processes to have taken place. The only problem is that you would have starved to death waiting for digestion! Without digestive enzymes, what normally occurs in a few hours (digesting the steak) would take about 50 years to happen.

Time is the justifying factor for enzyme catalysis. Enzymes, by today's definition, increase the rate of biochemical reactions without being changed themselves. In this laboratory exercise we will demonstrate the principles of enzymatic action.

❑ Preparation

I. Introduction

A. Hydrolysis

Hydrolysis, the basic digestive breakdown mechanism, involves the splitting of a biological molecule by the addition of part of a water molecule.

1➡ Study Figure 60–1● so that you understand exactly what is happening to the molecules being hydrolyzed. If necessary, go back to your lecture text [∞ *Martini, Chapter 2*] or to your chemistry book to review the molecular intricacies of hydrolysis.

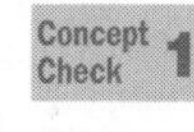

Concept Check 1: What are the major molecules that must undergo hydrolysis in the digestive processes?

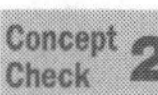

Concept Check 2: In digestive hydrolysis, what happens to the water molecule?

B. Overview of the Experiments

Specific enzymes act on each of the major biochemical molecules so that those molecules can be broken down into usable components. We will be demonstrating this biochemical action. Because each enzyme has its own optimum temperature and pH, we will be performing these experiments at 37°C to simulate the conditions of the body and at pH 8 to simulate the specific conditions in the small intestine where most digestion takes place.

For each of the following tests we will:

1. Demonstrate or identify the substance.
2. Simulate digestion under specific conditions.
3. Demonstrate the substance is no longer present after digestion.

❑ Experimentation

Be certain you have all your equipment ready for a particular experiment before beginning that part of the exercise.

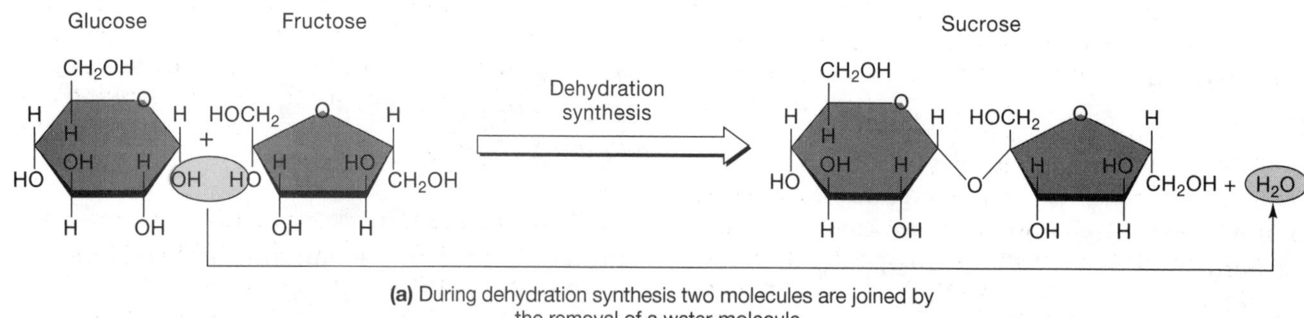

(a) During dehydration synthesis two molecules are joined by the removal of a water molecule.

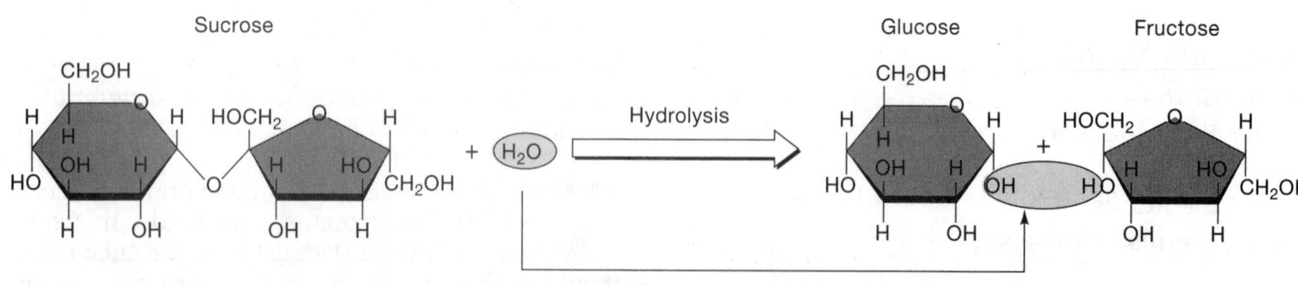

(b) Hydrolysis reverses the steps of dehydration synthesis; a complex molecule is broken down by the addition of a water molecule.

● **FIGURE 60–1a,b**
Starch Molecule. **(a)** Dehydration Synthesis. **(b)** Hydrolysis.

● FIGURE 60–1c,d,e
(c) Carbohydrate Hydrolysis;
(d) Lipid Hydrolysis;
(e) Dipeptide Hydrolysis.

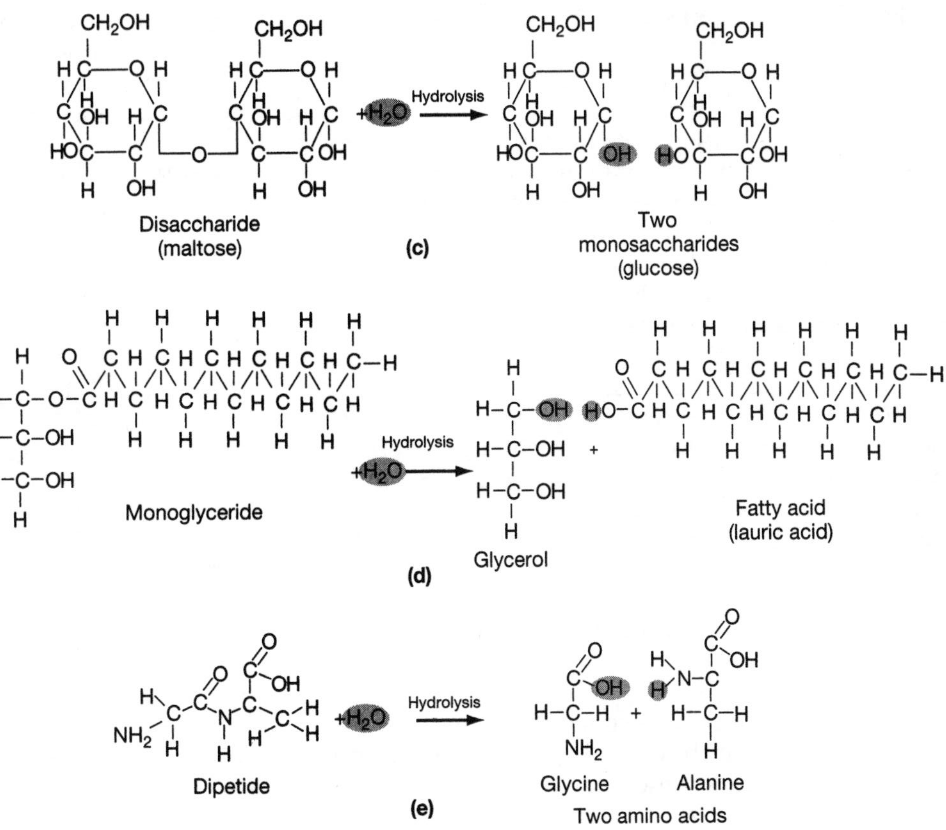

II. Carbohydrates

A. Background

Starch, a basic storage molecule of plants, is a long chain of glucose molecules that occurs in two forms, amylose and amylopectin, the latter being more highly branched and actually more common in the everyday diet. (As a comparison, glycogen is the animal storage molecule. Glycogen is a highly branched glucose chain.)

The same enzyme, **amylase**, is required to break down either of these starches. Amylase is found in the saliva as **salivary amylase** and in the small intestine as **pancreatic amylase**. Pancreatic amylase, which is produced in the pancreas, is virtually identical to salivary amylase.

1➡ Refer back to Figure 60–1●. Note that the principle breakdown products of amylase action are disaccharide maltose and certain other short-chain glucose molecules.

B. Iodine Test for Starch Breakdown

If you wish to test a substance for starch, add Lugol's iodine (IKI). If starch is present, the geometric pattern of the starch molecules will interact with the iodine and give a purplish-black color. If no starch is present, no geometric interaction can occur and the substance in question will not become dark.

1➡ Obtain a beaker of starch solution. (Technically you could use any starch or starch-like substance, including a hamburger bun or the paper in this book. However, for scientific validity, you will want to use the same starch supply for all your starch experiments. Be careful of your clothes because this experiment even works accidentally on cotton shirts!)

2➡ Also obtain: maltose solution, glucose solution, water (pH 8), Lugol's iodine, and a spot plate. (A clean microscope slide or chemistry watch glass can be substituted for a spot plate.)

3➡ On the spot plate put separate drops of starch, maltose, glucose, and water. To each spot add a drop of Lugol's iodine. What happened?

Starch ______________________
Maltose ______________________
Glucose ______________________
Water ______________________

Neither water, maltose, nor glucose has the geometric configuration to affect a color change.

Only the starch drop should have changed color.

C. Benedict's Test

Benedict's solution tests for the presence of reducing sugars. Glucose, maltose, and fructose are reducing sugars. Sucrose is not a reducing sugar. Neither is starch. Go back to the molecular structures of these carbohydrates to see if you can figure out why certain molecules are reducing sugars while other molecules are not.

____

____

Reducing sugars react with Benedict's solution to form the insoluble red precipitate, cuprous (Copper I) oxide. If only small amounts of a reducing sugar are present, the solution will be green. With more and more reducing sugar, the solution will be yellow, orange, or red.

1➡ Obtain four test tubes, labeled #1, #2, #3, #4.

2➡ Also obtain: glucose solution, maltose solution, starch solution, water (pH 8), and Benedict's solution.

3➡ Into test tube#1: 10 drops glucose, 10 drops Benedict's solution.
test tube #2: 10 drops maltose, 10 drops Benedict's solution.
test tube #3: 10 drops starch, 10 drops Benedict's solution.
test tube #4: 10 drops water, 10 drops Benedict's solution.

4➡ Place the tubes in the boiling water bath for 30 seconds. Remove and read your results.
#1 __________ #3 __________
#2 __________ #4 __________

Glucose and maltose should have given you a positive Benedict's test. Starch and water should have yielded a negative test. Why? __________

____

D. Experiment

In this experiment we are demonstrating the hydrolysis of starch. As starch is hydrolyzed, more and more maltose and other short-chain glucose molecules should be present.

1➡ Obtain two test tubes, labeled #1 and #2.

2➡ Also obtain: starch solution, water pH 8, iodine solution, Benedict's solution, and saliva (salivary amylase).

3➡ Into test tube #1: 10 drops starch, 10 drops water
test tube#2: 10 drops starch, 10 drops saliva (salivary amylase!)

4➡ Place these tubes in the 37° water bath for 30 minutes. Remove from the water bath.

5➡ Place 1 drop from each tube onto your spot plate. Add iodine. Record your results.
#1 __________ #2 __________

6➡ Place 9 drops of Benedict's solution in each tube and place the tubes in the boiling water for 30 seconds. Remove and read your results.
#1 __________ #2 __________

E. Explanation

If starch was hydrolyzed, you should get no color change upon adding iodine because no starch should be present.

If starch was hydrolyzed, you should get a positive Benedict's test because maltose is a breakdown product of starch and maltose is a reducing sugar. Glucose and other short-chain glucose molecules that may be present in your hydrolyzed starch solution are also reducing sugars.

Do your results match the predictions? If they do not, can you figure out what happened?

____

____

(Some people do not produce salivary amylase. If you are such a person, you will still have starch present and thus will have a positive iodine test and a negative Benedict's test. The inability to produce salivary amylase is not usually a problem because most amylase action comes from the pancreatic amylase. Before assuming you did not produce salivary amylase, check to make certain you did not mix up your test tubes.)

III. Proteins

A. Background

Proteins are long chains of amino acids held together by peptide bonds. Pepsin, secreted by the stomach, usually only begins the process of protein digestion by splitting the long amino acid chains into shorter amino acid chains.

1➡ Refer back to Figure 60–1●. Most protein digestion takes place in the small intestine under the influence of the pancreatic enzymes — trypsin, chymotrypsin, and carboxypolypeptidase. Trypsin and chymotrypsin split the amino acid chains into short polypeptides. Carboxypolypeptidase cleaves individual amino acids from the polypeptide. (Several additional enzymes can also be found along the brush border of the epithelium.)

B. Biuret's Test

Biuret's solution is a mixture of sodium hydroxide (NaOH) and copper sulfate ($CuSO_4$). In the presence of protein, the copper sulfate reacts with the peptide bonds of the protein, causing the mixture to turn a deep violet color. If protein was never present or if protein has been completely digested, no color change will occur.

1➡ Obtain two test tubes, labeled #1 and #2.

2➡ Also obtain: protein solution, water (pH 8), pancreatin (a mixture of pancreatic enzymes), and Biuret's solution.

3➡ Into test tube #1: 10 drops protein solution
test tube #2: 10 drops water (pH 8)

4➡ Add Biuret's solution to both tubes, drop by drop until you observe a definite, deep color change. Did the protein solution in tube #1 turn violet? ________ Tube #2 should have remained a light blue.

C. Experiment

1➡ Obtain two test tubes, labeled #1 and #2.

2➡ Also obtain: protein solution, water (pH 8), protein solution, and pancreatin.

3➡ Into test tube #1: 10 drops protein solution, 10 drops water (pH 8)
test tube #2: 10 drops protein solution, 10 drops pancreatin

4➡ Place these tubes in the 37° water bath for 30 minutes. Remove from the water bath.

5➡ Add about 5 drops of Biuret's solution and gently shake the tubes. Record the color.
#1 ____________ #2 ____________

D. Explanation

Test tube #1 should have changed color because no protein was digested; the color changed because protein was still present.

Test tube #2 may have changed color. If the color change was pronounced, your proteins were not digested. Head back to the drawing board and try to figure out what happened! If you had no color change, your proteins were completely digested. If you had some color change (say to a light pink or a slight purple), you had some protein digestion but you still have some protein present.

IV. Lipids

A. Background

Triglycerides are the primary lipids in the average diet. If necessary, review the chemistry of the triglycerides so that you have a picture of what we are doing in this experiment.

1➡ Refer back to Figure 60–1● for lipid breakdown.

Although an insignificant amount of lipid is digested in the mouth by lingual lipase and in the stomach by gastric lipase, virtually all fat digestion takes place in the small intestine.

The digestive enzymes involved in fat breakdown are water-soluble. Unfortunately, fats themselves are not water-soluble. Therefore, if the fat traversed the intestine in "glob" form, very little of it would be digested before the "glob" reached the ileocecal valve.

Enter **bile**, a secretory product of the liver. Bile does *not* contain any enzymes so it cannot function as an organic catalyst. Bile is an emulsifier. In other words, bile physically separates parts of the lipid into smaller and smaller pieces. Bile increases the surface area of the lipid, thus enhancing the possibility of lipase action.

The polar ionic and carboxyl parts of the bile salt molecule are highly water-soluble, while the sterol portion of the molecule is highly lipid-soluble. The lipid-soluble portion of the bile dissolves the fat, while the polar portion extends into the water-based fluid. This lowers the interfacial tension of the lipid, making the molecules far less cohesive. In other words, the lipid becomes more soluble in water. The agitation of the small intestine fragments the lipid, thus facilitating digestion.

As stated, the lipases are water-soluble and can only catalyze reactions on the surface of the fat globule. Since bile greatly increases the surface area of the fat, fat digestion is accelerated.

Bile salts also form aggregates of 20 to 40 molecules called **micelles**. In a micelle, all hydrophobic sterol units adhere at a central point while the hydrophilic tails extend out into the fluid.

Triglycerides are split into free fatty acids and monoglycerides, which are quickly dissolved into the micelle aggregates. The micelles ferry these products to the brush border of the intestinal epithelium.

B. Experiment

1➡ Obtain three test tubes, labeled #1, #2, #3.

2➡ Also obtain: cream, water (pH 8), litmus solution, and bile salts.

3➡ Into test tube #1: 2 ml cream, 2 ml water (pH 8), pinch of bile salts, 15 drops litmus solution
test tube #2: 2 ml cream, 2 ml pancreatin solution (pH 8), pinch of bile salts, 15 drops litmus solution
test tube #3: 2 ml cream, 2 ml pancreatin solution (pH 8), 15 drops litmus solution

If you are using litmus paper instead of a litmus solution, ignore the directions for adding litmus. When the experiment is complete, dip your litmus paper in the solution and read the color.

4➡ If the above solutions are not a bluish purple, add more litmus.

5➡ Place these tubes in the 37° water bath for 30 minutes. Remove from the water bath. Record the color changes.
#1 ____________ #3 ____________
#2 ____________

C. Explanation

Test tube #1 should have shown no change. Bile salts do not contain digestive enzymes. Although your bile salts emulsified the fats, the fats were not digested.

Test tube #2 should have turned a nice shade of pink. The bile salts emulsified the cream, and the enzymes in the pancreatin hydrolyzed the fats. You created an acidic environment in the process.

Test tube #3 should have turned a color somewhere between #1 and #2. You put the enzymes in but you did not add the bile salts. Thus, the digestive process was proceeding but at a much slower rate.

❑ Additional Activities

NOTES

1. Use the above tests on selected foods from your home or school.
2. Do these experiments, using temperature as a variable. Suggested temperatures: 1°C, 10°C, 40°C, 60°C, 100°C. Keep track of the time for the reactions to occur.
   (If time does not permit you to carry out this experiment, mentally simulate the procedure and explain the predicted results.)
3. Do these experiments, using pH as a variable. Suggested pH values: 4, 6, 8, 10. Keep track of the time for the reactions to occur.
   (If time does not permit you to carry out this experiment, mentally simulate the procedure and explain the predicted results.)

Answers to Selected Concept Check Questions

1. The major molecules are carbohydrates, proteins, and lipids. All of these must undergo hydrolysis.
2. In hydrolysis, both the compound molecule and the water molecule are split. The hydrogen ion combines with one part of the compound molecule while the hydroxyl ion combines with the other part.

❑ Lab Report

1. Referring to the numbered inquiries at the beginning of this exercise, complete the following box summary:

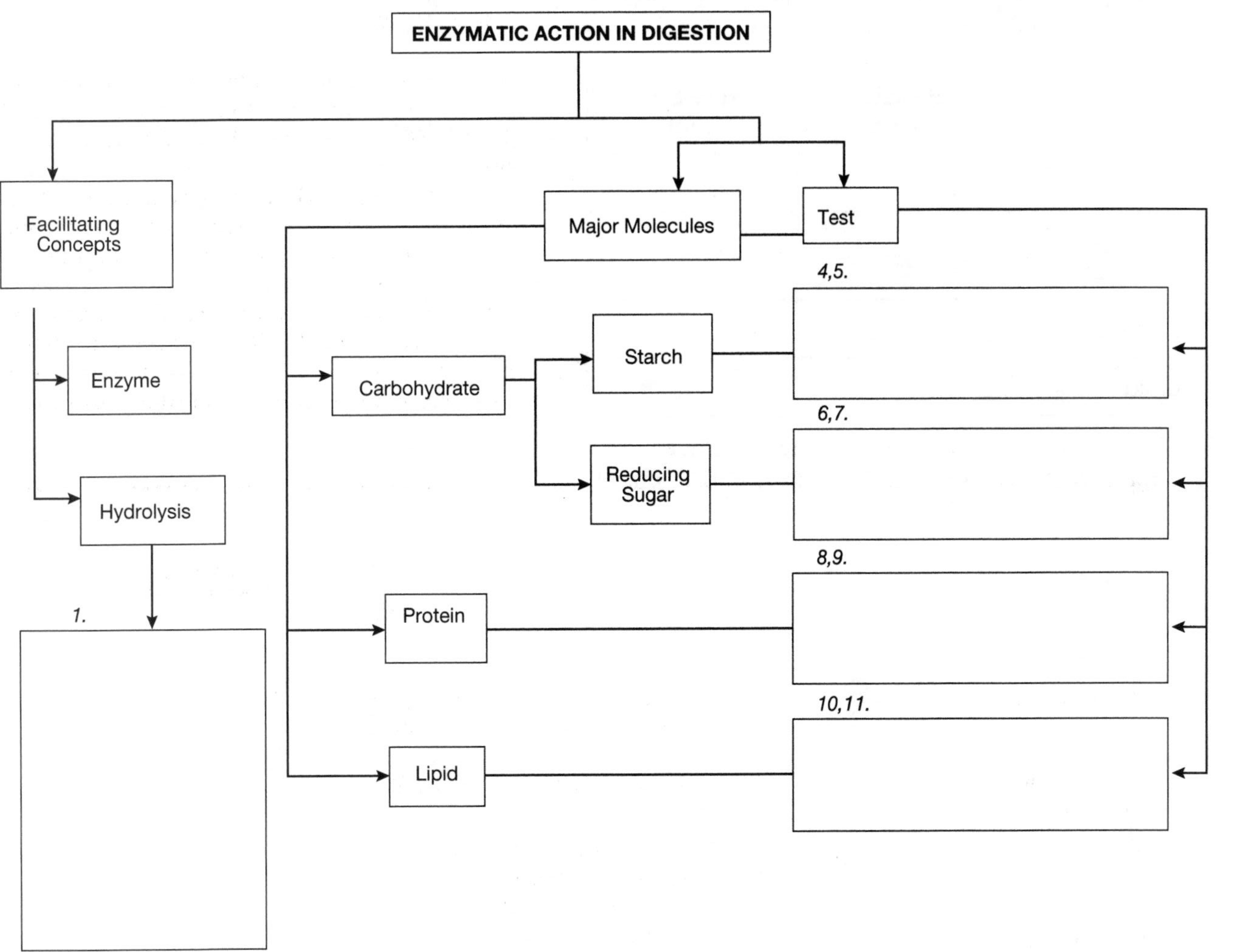

General Questions

NOTES

1. What is digestion?
2. Specifically, how is hydrolysis related to digestion?
3. We did these experiments under specific temperature and pH conditions. What were those conditions and why did we choose those parameters?
4. Complete the following chart:

	Positive Test	Negative Test	Explanation
Starch			
Reducing Sugar			
Protein			
Lipid			

5. What is the difference between the Benedict's test and the Biuret's test?
6. What types of enzymes will you find in bile salts?

EXERCISE 61a

Dissection of the Digestive System: Cat

PROCEDURAL INQUIRIES

Dissection

1. What are the principal parts of the mouth?
2. What are the types of teeth and what is the function of each type?
3. What are the major glands of the mouth?
4. What are the parts of the pharynx?
5. What is the difference between the visceral peritoneum and the parietal peritoneum?
6. Where are the greater and lesser omenta?
7. What are the lobes of the liver?
8. Where is the gallbladder?
9. What is the relationship between the cystic duct, the hepatic duct, and the common bile duct?
10. What are the major landmarks of the stomach?
11. What are the major divisions of the small intestine?
12. What are the major divisions of the large intestine?
13. What are the rectal sphincters?

Additional Inquiries

14. What is the alimentary canal?
15. Which digestive system organs are accessory organs?

HUMAN CORRELATION: AM Cadaver Atlas in Applications Manual.

Key Terms

Accessory Organs
Alimentary Canal
Anus
Cheeks
Colon
Common Bile Duct
Cystic Duct
Dentition Pattern
Duodenum
Esophagus
Hard Palate
Hepatic Duct
Ileum
Jejunum
Large Intestine
Larynx
Lingual Frenulum
Lips
Liver
Mesentery
Mouth
Omentum
Oral Cavity
Pancreas
Papillae
Parotid Gland
Peritoneum
Pharynx
Rectum
Rugae
Small Intestine
Soft Palate
Stenson's Duct
Stomach
Submaxillary Gland
Teeth
Tongue
Tonsils
Vestibule

Materials Needed

Cat
Gloves
Dissecting Pan
Dissecting Kit
Bone Cutter

*Prepared Microscope Slides**

Glands — Parotid, Submaxillary
Stomach — Pyloric and Cardiac Regions
Small Intestine — Duodenum, Jejunum, Ileum
Large Intestine — Colon, Appendix
Liver
Pancreas

Compound Microscope
Dissecting Microscope

*Your instructor may direct that these slides be set up as a demonstration so that you can see the microscopic parts of the digestive system as you work through your dissection.

The digestive system consists of all the organs and tissues involved either directly or indirectly in the ingestion of foods, the digestion or chemical breakdown of these foods, the absorption of these breakdown products into the system, and the egestion of the digestive waste products.

The principal digestive pathway is the **alimentary canal**, the continuous passage that begins at the mouth, includes the esophagus, stomach, small intestine, and large intestine, and ends at the anus. Along with the teeth and tongue, the exocrine and endocrine glands whose products affect the functioning of the principal organs are the **accessory organs**.

In this laboratory exercise we will study the macro- and microanatomy of the digestive system by dissecting the digestive system of the cat, examining relevant digestive system slides, and relating the knowledge gained to the digestive system of the human.

Dissection

I. Gross Anatomy

A. Salivary Glands

1➜ Begin your digestive system dissection by examining the glands of the mouth of the cat. Gently skin the cheek and neck on one side of the cat.

2➜ Study Figure 61a–1●. You should expose an area from above the ear medially to about 1 cm above the lateral angle of the eye and straight down to about 2 cm medial of the angle of the mouth. Carefully pull away the connective tissue so that you expose the ducts, glands, and superficial blood vessels of the side of the face.

3➜ Find the large glandular mass just below the ear. This is the **parotid gland**. The parotid duct, or **Stenson's duct**, lies across the masseter muscle, the large transverse "chewing" muscle of the cheek. Inferior to the parotid gland and lying just beneath the jugular vein is the **submaxillary gland**. The **submaxillary duct** (which is easily destroyed in dissection) can be found at the lower edge of the masseter muscle. The **sublingual gland** is located at the base of the tongue. You may not be able to find the very small **molar gland**, which is located between the masseter muscle and the mandible, near the angle of the mouth. (Refer again to Figure 61a–1●).

B. Mouth

In order to study the upper portions of the digestive system, it will be necessary for you to cut the jaw.

1➜ First find the angle of the jaw and cut away as much muscle and connective tissue as possible. You will now need a bone cutter. Carefully but firmly cut through the mandible. Extend your incision back to the pharynx and expose the internal oral structures.

2➜ Examine the **mouth** of the cat, paying particular attention to the **cheeks** and **lips** of the animal. Notice the space between the cheeks and lips and teeth. This is the **vestibule**. To appreciate the relative size of the vestibule, run your probe around the cat's teeth and lips.

3➜ Examine the **teeth** and make note of the **dentition pattern** (dental formula), the respective numbers of **incisors** (for cutting food), **canines** (for tearing and shredding food), and **premolars** and **molars** (for crushing and grinding food). This dental formula is 3-1-3-1 for the upper jaw and 3-1-2-1 for the lower jaw. In other words, the cat has 3 incisors, 1 canine, 3 premolars, and 1 molar for each half of the upper jaw, and 3 incisors, 1 canine, 2 premolars, and 1 molar for each half of the lower jaw. The human deciduous and adult formulas are 2-1-2-0 and 2-1-2-3, respectively, for both jaws. (Deciduous teeth are the juvenile teeth.) The adult cat has 30 teeth, compared to our 32.

4➜ Locate the **oral cavity proper**, the area behind the teeth and in front of the throat. The floor of the oral cavity is occupied primarily by the **tongue**. Locate the **lingual frenulum**, which attaches the tongue to the floor of the mouth. Run your finger over

● FIGURE 61a–1
Facial Glands — Cat.

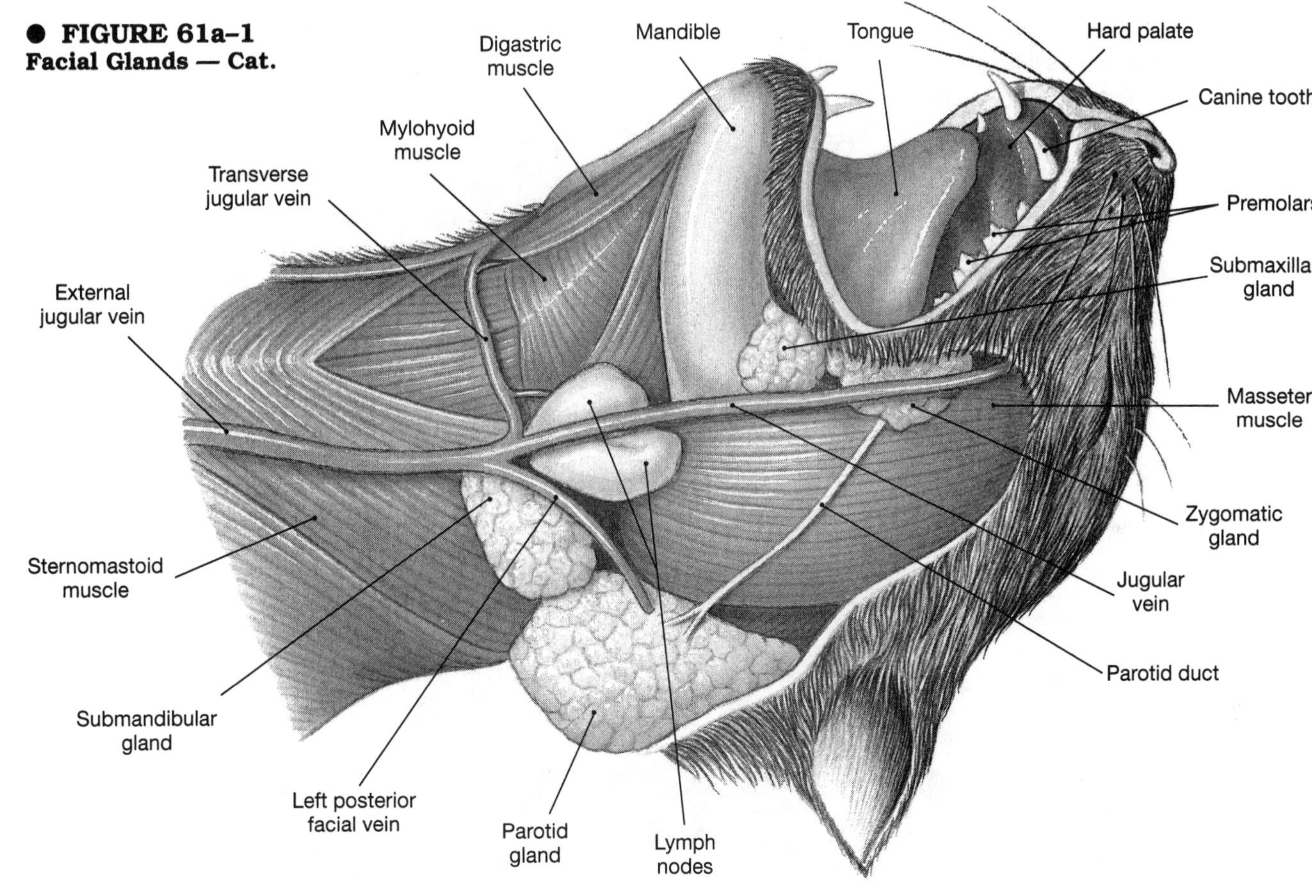

the dorsal surface of the tongue and feel the **papillae**, the elevations, many of which contain the taste buds. Note the orientation, or direction, of the papillae.

How does this papillar orientation allow the animal to drink efficiently? ________________

____

5➡ Observe that the roof of the mouth is composed of the **hard** and **soft palates**. Notice the **rugae**, the horizontal ridges on the bony surface of the hard palate. The hard palate along with the muscular soft palate form the separation between the nasal cavity and the oral cavity.

C. Pharynx

Posterior to the soft palate you will find the **pharynx**, the passageway common to both the respiratory and digestive systems. The pharynx is divided into three sections.

1➡ Locate the **nasopharynx**. This upper portion of the pharynx can be studied by placing your probe up behind the cat's nose. You may be able to locate the **internal nares (choanae)**, which form the internal exit from the nose proper into the pharynx, and the **auditory (Eustachian) tubes**, the slitlike openings extending laterally toward the middle ear.

The area immediately caudal to the soft palate and extending to the level of the hyoid bone is the **oropharynx**. On the lateral walls of the oropharynx you can find the small glandular **palatine tonsils**. You may notice that, unlike humans, the cat does not have a uvula.

Beneath the level of the hyoid bone and extending down to the **larynx** and **esophagus** is the **laryngopharynx**.

D. Esophagus

The esophagus is the muscular tube running posterior to the trachea from the pharynx down through the diaphragm (at an opening called the **diaphragmatic hiatus**) and into the **stomach**. As with the human, the cranial third of the cat's esophagus is composed of skeletal muscle, the middle third is both skeletal and smooth muscle, and the caudal third is composed entirely of smooth muscle.

1➡ Note the relation of the esophagus to the other organs of the thorax. Follow the esophagus caudally to the stomach.

● FIGURE 61a–2 Structures of the Digestive System — Cat. (Caudate lobe of liver not shown.)

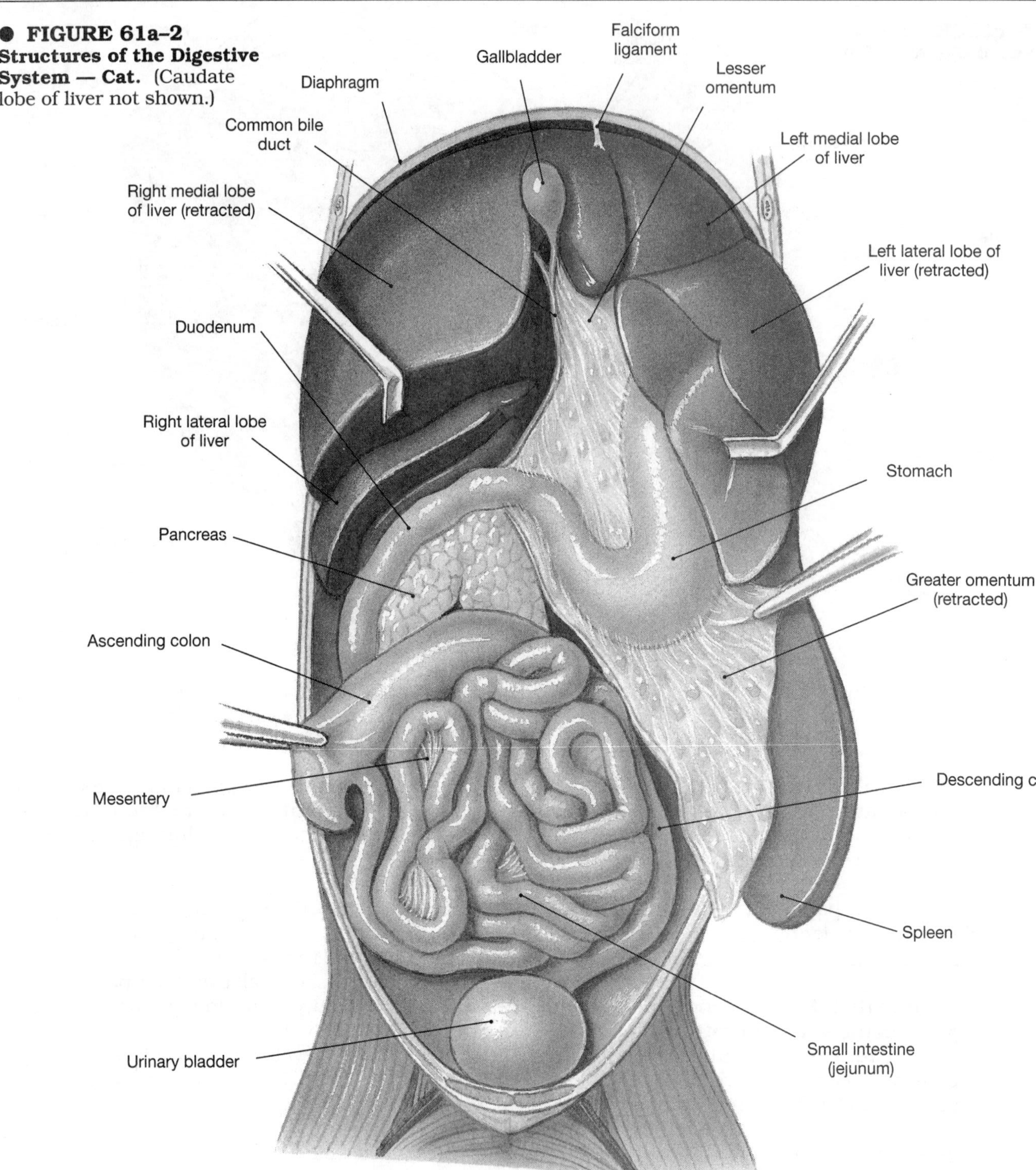

E. Abdominal Cavity

1➡ Examine the serous linings of the abdominal cavity. The **parietal peritoneum** is the smooth epithelial membrane lining the abdominal wall, and the **visceral peritoneum** is the smooth epithelium covering of the abdominal viscera. Extending from and between the peritonea are several connective structures: the **greater omentum**, the **lesser omentum**, the **mesentery**, and several connective ligaments. Within these structures are blood vessels, nerves, and lymphatic tissues.

2➡ As points of reference, locate the **liver**, **spleen**, **stomach**, **pancreas**, **small intestine**, and **large intestine**. We will examine these organs in greater detail as we continue this study. (See Figure 61a–2●.)

F. Greater Omentum

1➡ Locate the greater omentum, the double-sheet apronlike structure anterior to the viscera that contains vast amounts of adipose tissue. The greater omentum is attached to the greater curvature of the stomach; encloses the spleen, part of the pancreas, and most of the intestinal tract; and terminates on the dorsal abdominal wall. The omental connection between the stomach and the spleen is the **gastrosplenic ligament**. Follow your instructor's directions as to whether you should cut out the greater omentum.

G. Lesser Omentum

1➡ Find the lesser omentum, which extends from the lesser curvature of the stomach and the duodenum to the left lateral lobe of the liver. The connection between the stomach and the liver is the **hepatogastric ligament**, and the connection between the duodenum and the liver is the **hepatoduodenal ligament**. Later, when you are trying to locate the **common bile duct**, you will be pulling apart sections of the lesser omentum.

H. Mesentery

1➡ Examine the mesentery. The mesentery is actually two separate structures — the **mesentery proper** (or true mesentery), which can be found supporting the coils of the small intestine, and the **mesocolon**, which connects the large intestine and the dorsal body wall. Note the extensive network of veins and arteries in the mesenteries. You may also be able to find some distinct mesenteric lymph nodes.

I. Liver

1➡ Now return to the liver, the largest abdominal organ, which is located just inferior to the diaphragm and seems to drape over the right side of the remaining abdominal viscera. The cat's liver consists of five lobes: the **right lateral lobe**, the large lobe that seems to be divided into two lobes (although close examination will reveal these "pseudolobes" have a common origin); the **right medial lobe**, the lobe that may appear to be split by the **gallbladder**; the **left medial lobe**; the **left lateral lobe**; and the **caudate lobe**. You will have to lift the other lobes to find the caudate lobe, the small tail-like extension of hepatic tissue that seems to be directly behind the left lateral lobe.

(The human liver has four lobes: anteriorly the larger right lobe and smaller left lobe and posteriorly the caudate lobe — near the vena cava — and the quadrate lobe — below the caudate lobe and near the gallbladder.)

2➡ Look between the right and left medial lobes of the cat's liver and you should find the **falciform ligament**, which holds the liver in place by attaching to the ventral wall. If you examine the free edge of the falciform ligament, you may be able to locate a fibrous strand of tissue. This is the vestigial **umbilical vein**, which atrophies after birth and is known in the adult as the **round ligament** (or **ligamentum teres**).

J. Gallbladder

1➡ Examine the gallbladder, the greenish sac located in a depression on the ventral surface of the right medial lobe of the liver. Separate the connective mesentery away from the gallbladder and locate the **cystic duct**, which leads out of the gallbladder. The cystic duct joins with the **hepatic duct** to form the common bile duct. Trace the hepatic duct back up into the liver. Trace the common bile duct down to the **ampulla of Vater**, the point where the duct enters the small intestine. Guarding this passageway is the muscular **sphincter of Oddi**. Although you probably will not be able to identify this sphincter, make note of its probable location.

K. Pancreas

1➡ Return to the pancreas. Relate the size and location of the feline pancreas to that of the human pancreas. Notice the portion of the organ lying medial to the duodenum. This is the **head**. The rest of the pancreas is known as the **body**, and the tapered end is usually called the **tail**. By carefully teasing away the pancreatic tissue at the head, you may be able to locate the **pancreatic duct**. This duct often unites with the common bile duct just before entering the small intestine at the ampulla of Vater. Sometimes the merger of the ducts occurs after they penetrate the wall of the small intestine.

L. Stomach

1➡ Return to the stomach and use Figure 61a–3● to help you locate the following landmarks:

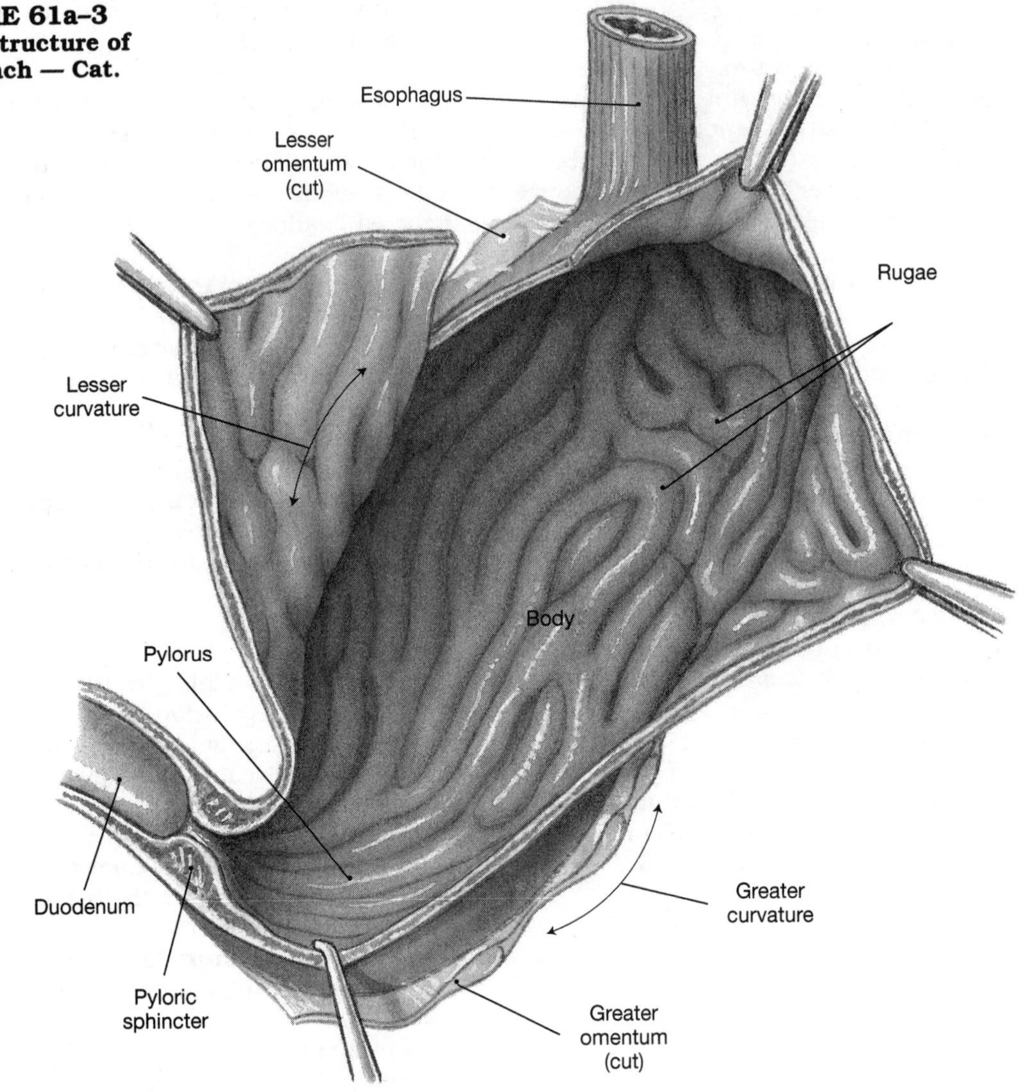

● **FIGURE 61a–3**
Internal Structure of the Stomach — Cat.

Cardiac Region (cardia) Adjacent to the esophagus (not identified on Figure 61a–3●).

Fundus Domed portion to the left of the cardiac area (not identified on Figure 61a–3●).

Greater Curvature Longer, left (lateral) margin of the stomach.

Lesser Curvature Shorter, right (medial) margin of the stomach.

Body Large, dominant portion of the stomach.

Rugae Longitudinal folds lining the inside of the body of the stomach. You will have to cut open the stomach to see the rugae.

Pyloric Region (Pylorus) Narrow portion where the stomach empties into the duodenum.

Pyloric Sphincter Muscular ring, which feels like a hard mass, between the pylorus and the duodenum. You can see this sphincter if you make an incision through the pyloric region.

M. Spleen

1➡ Locate the spleen, the elongate reddish brown organ on the left side of the stomach. You may have to move the stomach to the right to see the spleen. The spleen is actually part of the lymphatic system and not the digestive system.

N. Small Intestine

1➡ Locate the three sections of the small intestine. The **duodenum** is the first section, and in the cat this region is about 15 cm long. When the duodenum turns left, usually about the end of the pancreas, the duodenum becomes the **jejunum**. In a young, well-preserved animal you should be able to discern the difference in texture between these two regions. The duodenum is normally thinner and firmer than the jejunum. About half the distance from the duodenum to the large intestine is the jejunum. The remaining part of the small intestine is the **ileum**.

2➡ You will probably not be able to feel a difference between the jejunum and the ileum. If you were to examine the three sections histologically, you would see distinct differences. Refer back to Exercise 58, Figure 58–9●, to see the corresponding human differences.

O. Large Intestine

1➡ The sphincter located where the small intestine empties into the large intestine is known as the **ileocecal** (or ileocaecal) **valve**. You can feel this sphincter. Note the large intestinal pouch caudal to the sphincter. This is the **cecum** (caecum). (In the human, a vestigial organ, the **appendix**, extends from the end of the cecum. The cat does not have an appendix, although occasionally a tiny knot may be felt at the end of the cecum.)

2➡ Follow the large intestine as it extends cranially on the right side and is known as the **ascending colon**. The structure bends left at the **hepatic flexure**, and as it crosses the abdomen it is known as the **transverse colon**. This colon bends downward at the **splenic flexure**, and as it continues down the left abdomen it is known as the **descending colon**. (In the human the descending colon bends medially and becomes known as the **sigmoid** (or **S**) **colon**. The cat does not have a true, distinct sigmoid colon, although sometimes the medial bend in the cat's colon is referred to as sigmoid.) The colon empties into the **rectum**, which opens to the exterior at the **anus**.

II. Microanatomy

A. Prepared Slides

1➡ Examine any prepared slides your instructor may have set out for you. Use the information in Exercise 58, particularly along with Figures 58–10a● to 58–10f●, as your guide.

B. Anatomical Sections

1➡ Take representative samples from each of the different organs of the digestive system and examine those specimens under the dissecting microscope.

❑ Additional Activities

1. Write a comparison explaining the similarities and differences between the digestive system of the cat and the digestive system of the human.
2. Find out how the human cecum compares with the feline cecum in both structure and function.

NOTES

❑ Lab Report

1. Referring to the numbered inquiries at the beginning of this exercise, complete the following box summary:

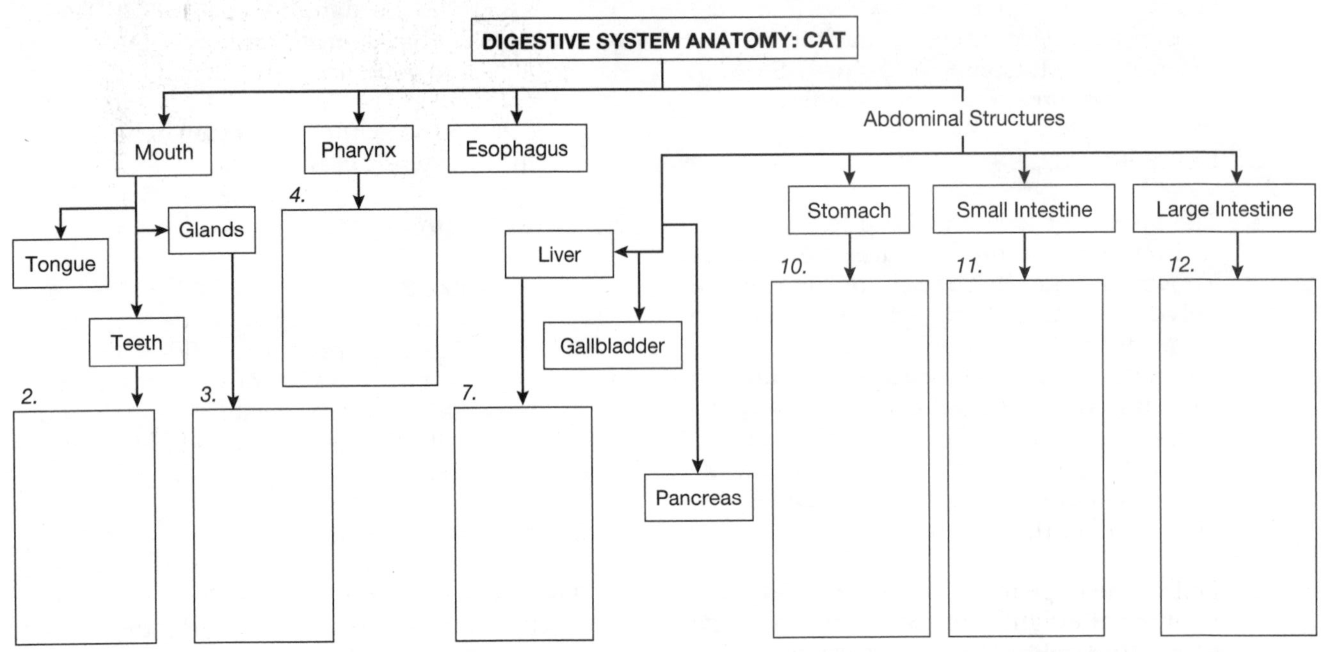

General Questions

NOTES

1. What are the principal organs of the alimentary canal? What are the accessory organs?
2. What is the difference between the visceral peritoneum and the parietal peritoneum?
3. Where are the omenta, and what is the function of each?
4. Where is the gallbladder in relation to the liver?
5. Anatomically, how are the cystic duct, the hepatic duct, and the common bile duct related?
6. How do the sizes and shapes of the feline heterodont teeth relate to its primarily carnivorous diet?
7. Name the principal valves and sphincters of the digestive system. What common function do they all have? What differences do you note between them?
8. List the similarities and differences you noted in the microscopic structure of the different parts of the digestive system.
9. How do the salivary glands compare with the intestinal glands? With the pancreas?
10. Why are the cat's incisors relatively small while its canine teeth tend to be quite long and sharp?
11. Compare the location of the oral glands of the cat with the location of the oral glands of the human.
12. Consider the cat's dentition pattern. How does the dentition pattern of the cat compare with the dentition pattern of the human?
13. What is the orientation of the papillae on your tongue? What physiological reason might cause the cat's tongue to be different from yours?

EXERCISE **61b**

Dissection of the Digestive System: Fetal Pig

PROCEDURAL INQUIRIES

Dissection

1. What are the principal parts of the mouth?
2. What are the types of teeth and what is the function of each type?
3. What are the major glands of the mouth?
4. What are the parts of the pharynx?
5. What is the difference between the visceral peritoneum and the parietal peritoneum?
6. Where are the greater and lesser omenta?
7. What are the lobes of the liver?
8. Where is the gallbladder?
9. What is the relationship between the cystic duct, the hepatic duct, and the common bile duct?
10. What are the major landmarks of the stomach?
11. What are the major divisions of the small intestine?
12. What are the major divisions of the large intestine?
13. What are the rectal sphincters?

Additional Inquiries

14. What is the alimentary canal?
15. Which digestive system organs are accessory organs?

HUMAN CORRELATION: AM Cadaver Atlas in Applications Manual.

Key Terms

Accessory Organs
Alimentary Canal
Anus
Cheeks
Colon
Common Bile Duct
Cystic Duct
Dentition Pattern
Duodenum
Esophagus
Hard Palate
Hepatic Duct
Ileum
Jejunum
Large Intestine
Larynx
Lingual Frenulum
Lips
Liver
Mesentery
Mouth
Omentum
Oral Cavity
Pancreas
Papillae
Parotid Gland
Peritoneum
Pharynx
Rectum
Rugae
Small Intestine
Soft Palate
Stenson's Duct
Stomach
Submandibular Gland
Teeth
Tongue
Tonsils
Uvula
Vestibule

Materials Needed

Fetal Pig
Gloves
Dissecting Pan
Dissecting Kit
Bone Cutters

Prepared Microscope Slides*

Glands — Parotid, Submaxillary
Stomach — Pyloric and Cardiac Regions
Small Intestine — Duodenum, Jejunum, Ileum
Large Intestine — Colon, Appendix
Liver
Pancreas

Compound Microscope

Dissecting Microscope

*Your instructor may direct that these slides be set up as a demonstration so that you can see the microscopic parts of the digestive system as you work through your dissection.

The digestive system consists of all the organs and tissues involved either directly or indirectly in the ingestion of foods, the digestion or chemical breakdown of these foods, the absorption of these breakdown products into the system, and the egestion of the digestive waste products.

The principal digestive pathway is the **alimentary canal**, the continuous passage that begins at the mouth, includes the esophagus, stomach, small intestine, and large intestine, and ends at the anus. Along with the teeth and tongue, the exocrine and endocrine glands whose products affect the functioning of the principal organs are the **accessory organs**.

In this laboratory exercise we will study the macro- and microanatomy of the digestive system by dissecting the digestive system of the fetal pig, examining relevant digestive system slides, and relating the knowledge gained to the digestive system of the human.

Dissection

I. Gross Anatomy

As you dissect your fetal pig, you should remember that, although all structures are present in the postembryonic animal, the physical maturity of your specimen may determine the ease with which you locate specific body parts.

A. Salivary Glands

1➡ Begin your digestive system dissection by examining the glands of the mouth of the fetal pig. Gently skin the cheek and neck on one side of the fetal pig.

2➡ Study Figure 61b–1●. Reflect the skin back over the neck. Carefully pull away the connective tissue so that you expose the ducts, glands, and superficial blood vessels of the side of the face.

3➡ Locate the masseter muscle, the large "chewing" muscle extending from below the ear to the angle of the mouth. Just caudal to the masseter muscle you should be able to find a large glandular mass. The large triangular portion of this mass is the **parotid gland**, and its duct, the parotid duct or **Stenson's duct**, lies across the middle of the gland and angles up along the oral border of the masseter muscle. The **submandibular gland** is a small oval gland, posterior and partially inferior to the parotid gland.

The **sublingual gland** is located at the base of the tongue, anterior to the submandibular gland. You probably will not be able to locate the ducts to these glands. Several other small glands also exist, but it is not necessary for you to identify them at this time. (Refer again to Figure 61b–1●).

B. Mouth

In order to study the upper portions of the digestive system, it will be necessary for you to cut the jaw.

1➡ First find the angle of the jaw and cut away as much muscle and connective tissue as possible. You will now need a bone cutter. Carefully but firmly cut through the mandible. Extend your incision back to the pharynx and expose the internal oral structures.

2➡ Examine the mouth of the fetal pig, paying particular attention to the **cheeks** and **lips** of the animal. Notice the space between the cheeks and lips and teeth. This is the **vestibule**. To appreciate the relative size of the vestibule, run your probe around the fetal pig's teeth and lips. Two rows of buccal glands are located on the mucous membrane of the cheek. You may not be able to find these glands in the young fetal animal.

Observe that the pig's upper lip is short and thick and seems to blend into the snout. The lower lip tends to be small and pointed.

3➡ Examine the **teeth** and make note of the **dentition pattern** (dental formula), the respective numbers of **incisors** (for cutting food), **canines** (for tearing and shredding food), and **premolars** and **molars** (for crushing and grinding food). For the fetal pig the deciduous (juvenile) dental formula is 3-1-4-0, meaning 3 incisors, 1 canine, 4 premolars, and no molars for each side of both the upper and lower jaws. (According to some authorities, the deciduous formula should be 3-1-3-0.) In the adult pig the for-

● FIGURE 61b–1
Facial Glands — Fetal Pig.

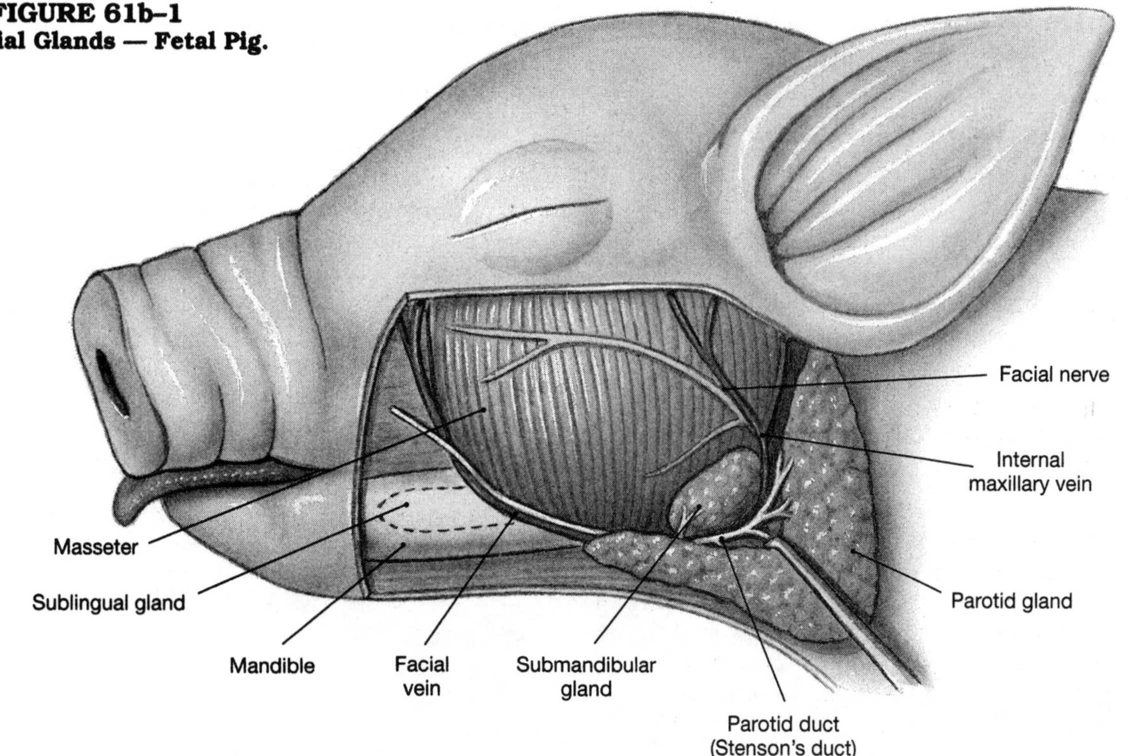

mula is 3-1-4-3. The human deciduous and adult formulas are 2-1-2-0 and 2-1-2-3 for each side of each jaw. The adult pig, therefore, has 44 teeth compared to our 32.

In the pig, the deciduous corner incisor and canine teeth are present at birth. If your fetal pig was close to birth, you should have no trouble identifying these teeth. To locate all deciduous teeth, carefully cut into the gum tissue. If yours is a very young fetal pig and you are not able to identify all teeth, at least make note of where each tooth should be. Check your classmates' specimens, as the teeth may be better developed in older animals.

4➡ Locate the **oral cavity proper**, the area behind the teeth and in front of the throat. The floor of the oral cavity is occupied primarily by the **tongue**. Locate the **lingual frenulum**, which attaches the tongue to the floor of the mouth. The frenulum is double in the pig, although you may not be able to notice this. Run your finger over the dorsal surface of the tongue and feel the **papillae**, the elevations, many of which contain the taste buds. Note the orientation, or direction, of the papillae.

How does this papillar orientation allow the animal to drink efficiently? ____________________

____

5➡ Observe that the roof of the mouth is composed of the **hard** and **soft palates**. Notice the **rugae**, the horizontal ridges on the bony surface of the hard palate. The hard palate along with the muscular soft palate form the separation between the nasal cavity and the oral cavity. The soft palate is relatively quite thick. The **uvula**, the median prolongation of the soft palate, may or may not be present in the fetal pig.

C. Pharynx

Posterior to the soft palate you will find the **pharynx**, the passageway common to both the respiratory and digestive systems. The pharynx is divided into three sections.

1➡ Locate the **nasopharynx**. This upper portion of the pharynx can be studied by placing your probe up behind the fetal pig's nose. You may be able to locate the **internal nares (choanae)**, which form the internal exit from the nose proper into the pharynx, and the **auditory (Eustachian) tubes**, the slitlike openings extending laterally toward the middle ear.

The area immediately caudal to the soft palate and extending to the level of the hyoid bone is the **oropharynx**. On the lateral walls of the oropharynx you can find the small glandular **palatine tonsils**. Tonsillar tissue can also be found at other points along the oropharynx.

Beneath the level of the hyoid bone and extending down to the **larynx** and **esophagus** is the **laryngopharynx**.

● FIGURE 61b–2
Structures of the Digestive System — Fetal Pig.

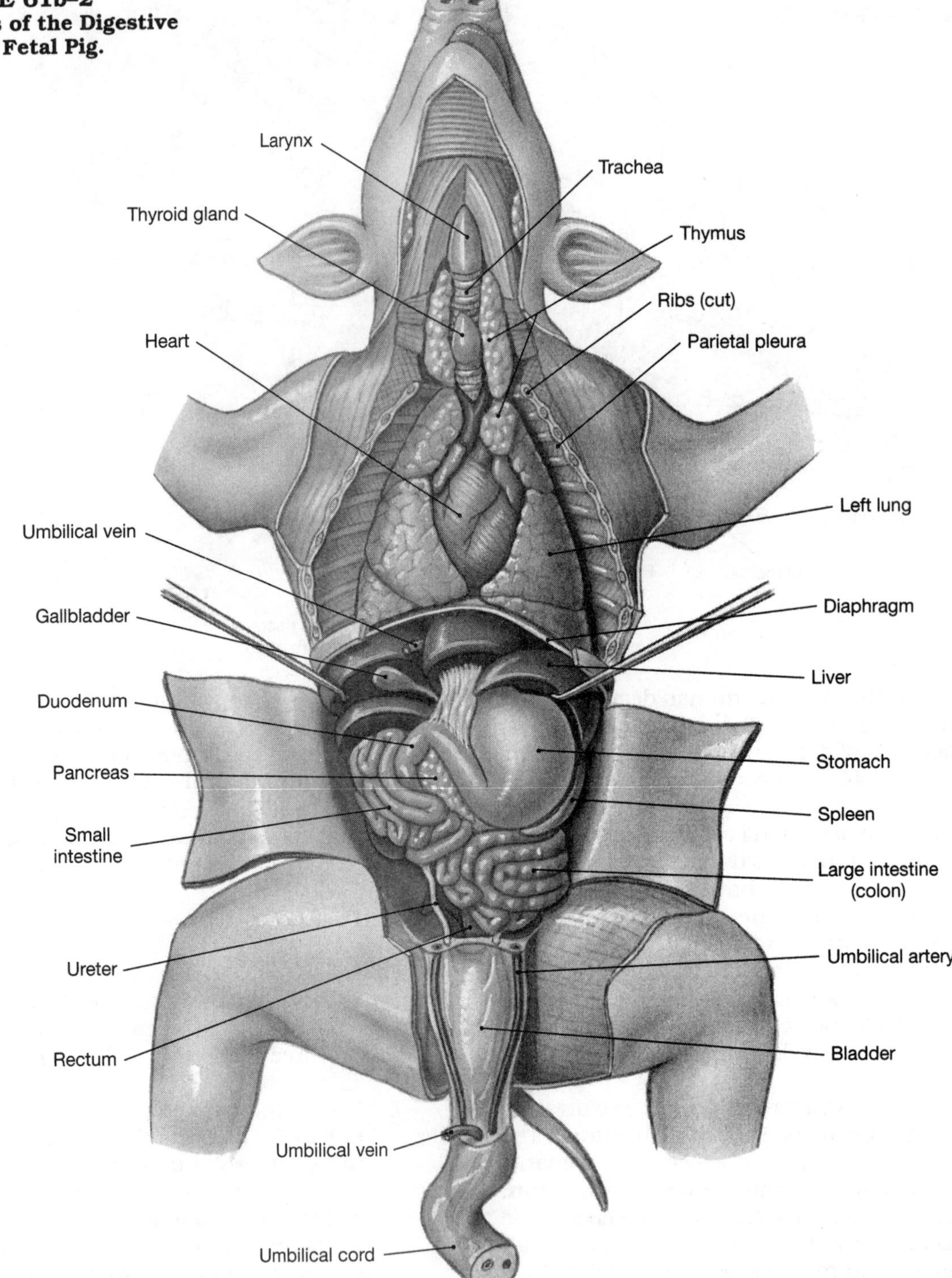

D. Esophagus

The esophagus is the muscular tube running posterior to the trachea from the pharynx down through the diaphragm (at an opening called the **diaphragmatic hiatus**) and into the **stomach**. As with the human, the cranial third of the fetal pig's esophagus is composed of skeletal muscle, the middle third is both skeletal and smooth muscle, and the caudal third is composed entirely of smooth muscle.

1➡ Note the relation of the esophagus to the other organs of the thorax. Follow the esophagus caudally to the stomach.

E. Abdominal Cavity

1➡ Examine the serous linings of the abdominal cavity. The **parietal peritoneum** is the smooth epithelial membrane lining the abdominal wall, and the **visceral peritoneum** is the smooth epithelium covering of the abdominal viscera. Extending from and between the peritonea are several connective structures: the **greater omentum**, the **lesser omentum**, the **mesentery**, and several connective ligaments. Within these structures are blood vessels, nerves, and lymphatic tissues.

2➡ As points of reference, locate the **liver**, **spleen**, **stomach**, **pancreas**, **small intestine**, and **large intestine**. We will examine these organs in greater detail as we continue this study. (See Figure 61b–2●.)

F. Greater Omentum

1➡ Locate the greater omentum, the double-sheet apronlike structure anterior to the viscera that contains vast amounts of adipose tissue. The greater omentum is attached to the greater curvature of the stomach; encloses the spleen, part of the pancreas, and most of the intestinal tract; and terminates on the dorsal abdominal wall. The omental connection between the stomach and the spleen is the **gastrosplenic ligament**. Follow your instructor's directions as to whether you should cut out the greater omentum.

G. Lesser Omentum

1➡ Find the lesser omentum, which extends from the lesser curvature of the stomach and the duodenum to the left lateral lobe of the liver. The connection between the stomach and the liver is the **hepatogastric ligament**, and the connection between the duodenum and the liver is the **hepatoduodenal ligament**. Later, when you are trying to locate the **common bile duct**, you will be pulling apart sections of the lesser omentum.

H. Mesentery

1➡ Examine the mesentery. The mesentery is actually two separate structures — the **mesentery proper** (or true mesentery), which can be found supporting the coils of the small intestine, and the **mesocolon**, which connects the large intestine and the dorsal body wall. Note the extensive network of veins and arteries in the mesenteries. Depending on the maturity of your fetal pig, you may also be able to find some distinct mesenteric lymph nodes.

I. Liver

1➡ Now return to the liver, the largest abdominal organ, which is located just inferior to the diaphragm and seems to drape over the right side of the remaining abdominal viscera. The fetal pig's liver consists of five lobes: the **right lateral lobe**, the large lobe that seems to be divided into two lobes (although close examination will reveal these "pseudolobes" have a common origin); the **right medial lobe**, the lobe that may appear to be split by the **gallbladder**; the **left medial lobe**; the **left lateral lobe**; and the **caudate lobe**. You will have to lift the other lobes to find the caudate lobe, the small taillike extension of hepatic tissue that seems to be directly behind the left lateral lobe. (You may or may not be able to identify another lobe, the **quadrate lobe**, which lies medial to the gallbladder.)

(The human liver has four lobes: anteriorly the larger right lobe and smaller left lobe and posteriorly the caudate lobe — near the vena cava — and the quadrate lobe — below the caudate lobe and near the gallbladder.)

2➡ Look between the right and left medial lobes of the fetal pig's liver and you should find the **falciform ligament**, which holds the liver in place by attaching to the ventral wall. (Interestingly, this falciform ligament is often absent in the adult pig.) If you examine the free edge of the falciform ligament, you may be able to locate a fibrous strand of tissue that extends from the liver to the umbilicus. This is the **umbilical vein**, which atrophies after birth and is known in the adult as the **round ligament** (or **ligamentum teres**). You may have already cut the umbilical vein when you were opening the abdominal cavity.

J. Gallbladder

1➡ Examine the gallbladder, the greenish sac located in a depression on the ventral surface of the right medial lobe of the liver. Separate the connective mesentery away from the gallbladder and locate the **cystic duct**, which leads out of the gallbladder. The cystic duct joins with the **hepatic duct** to form the common bile duct. Trace the hepatic duct back up into the liver. Trace the common bile duct down to the **ampulla of Vater**, the slightly enlarged point where the duct enters the small intestine. Guarding this passageway is the muscular **sphincter of Oddi**. Although you probably will not be able to identify this sphincter, make note of its probable location.

● **FIGURE 61b–3**
Internal Structure of the Stomach.

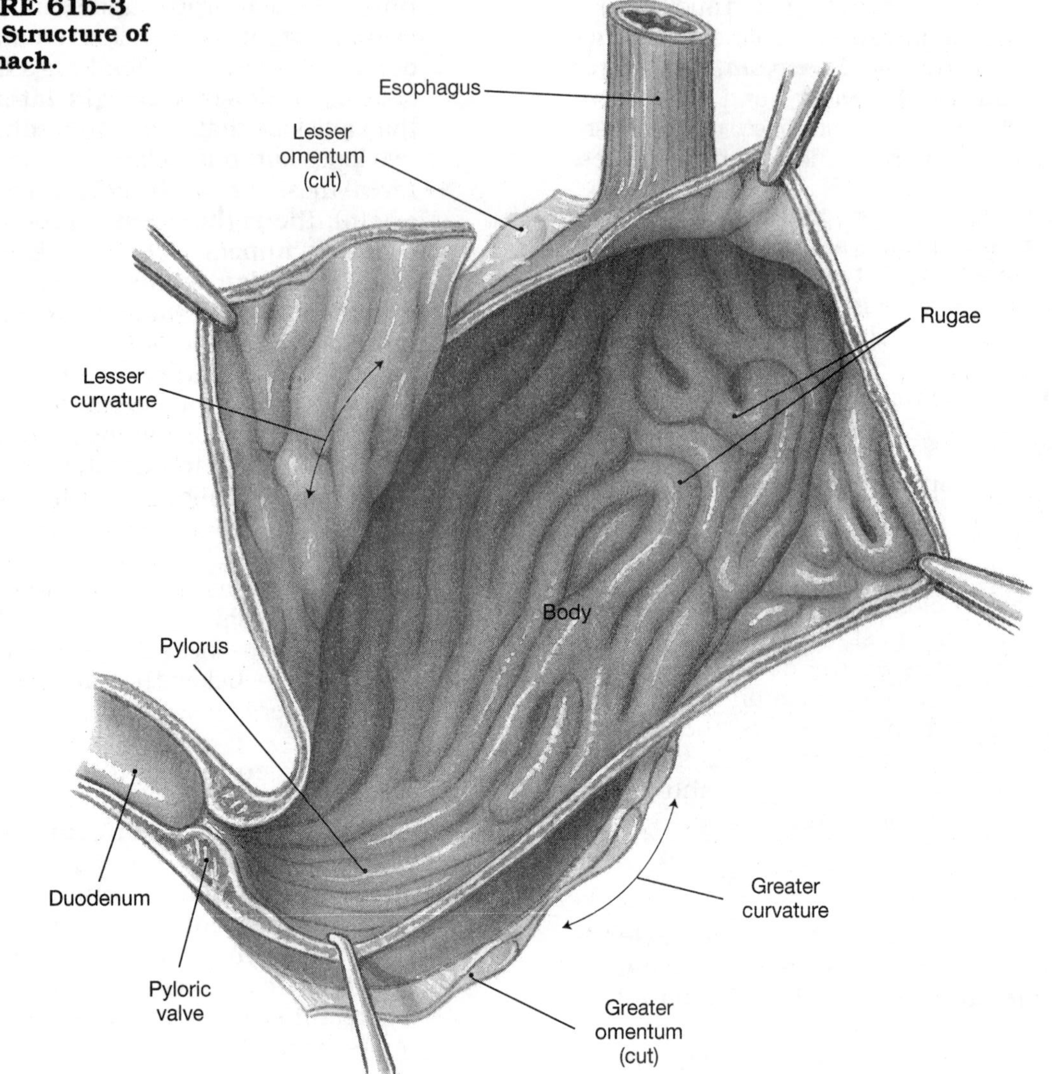

K. PANCREAS

1➡ Return to the pancreas. Relate the size and location of the porcine pancreas to that of the human pancreas. Notice the portion of the organ lying medial to the duodenum. This is the **head**. The rest of the pancreas is known as the **body**, and the tapered end is usually called the **tail**. By carefully teasing away the pancreatic tissue at the head, you may be able to locate the **pancreatic duct**. This duct often unites with the common bile duct just before entering the small intestine at the ampulla of Vater. Sometimes the merger of the ducts occurs after they penetrate the wall of the small intestine.

L. STOMACH

1➡ Return to the stomach and use Figure 61b–3● to help you locate the following landmarks:

Cardiac Region (cardia) Adjacent to the esophagus.

Fundus Domed portion to the left of the cardiac area.

Greater Curvature Longer, left (lateral) margin of the stomach.

Lesser Curvature Shorter, right (medial) margin of the stomach.

Body Large, dominant portion of the stomach.

Rugae Longitudinal folds lining the inside of the body of the stomach. You will have to cut open the stomach to see the rugae.

Pyloric Region (Pylorus) Narrow portion where the stomach empties into the duodenum.

Pyloric Sphincter Muscular ring, which feels like a hard mass, between the pylorus and the duodenum. You can see this sphincter if you make an incision through the pyloric region.

If you open the stomach, you will notice a green substance, the **meconium**, a fetal substance composed of assorted secretions, amniotic fluid, and sloughed epithelial cells.

M. Spleen

1➡ Locate the spleen, the elongate reddish brown organ on the left side of the stomach. You may have to move the stomach to the right to see the spleen. The spleen is actually part of the lymphatic system and not the digestive system.

N. Small Intestine

1➡ Locate the three sections of the small intestine. The **duodenum** arises on the animal's right side and extends medially for several centimeters. When the small intestine turns left, usually about the end of the pancreas, the duodenum becomes the **jejunum**. In a well-preserved animal you should be able to discern the difference in texture between these two regions. The duodenum is normally thinner and firmer than the jejunum. About half the distance from the duodenum to the large intestine is the jejunum. The remaining part of the small intestine is the **ileum**. Refer back to Exercise 58, Figure 58–9●, to see the histological distinction.

O. Large Intestine

1➡ The sphincter located where the small intestine empties into the large intestine is known as the **ileocecal** (or ileocaecal) **valve**. You can feel this sphincter. Your instructor may direct you to cut this valve and examine it with a hand lens or under the dissecting microscope.

2➡ Note the large intestinal pouch caudal to the sphincter. This is the **cecum** (caecum).

(In the human, a vestigial organ, the **appendix**, extends from the end of the cecum. The fetal pig does not have an appendix, although occasionally a tiny knot may be felt at the end of the cecum.)

3➡ Follow the large intestine. As the large intestine extends cranially on the right side of most animals, it is known as the **ascending colon**. In the pig, however, the ascending colon is actually arranged in three double-spiral coils. At the end of the third coil the structure bends left and crosses the abdomen as the **transverse colon**. This colon bends caudally, and as it continues down the left abdomen it is known as the **descending colon**. The colon empties into the **rectum**, distinguished in older animals by a quantity of fat. The rectum opens to the exterior at the **anus**.

(In the human the descending colon bends medially and becomes known as the **sigmoid** (or **S**) **colon**. The fetal pig does not have a true, distinct sigmoid colon, although sometimes the medial bend in the fetal pig's colon is referred to as sigmoid.)

II. Microanatomy

A. Prepared Slides

1➡ Examine any prepared slides your instructor may have set out for you. Use the information in Exercise 58, particularly along with Figures 58–10a● to 58–10f●, as your guide.

B. Anatomical Sections

1➡ Take representative samples from each of the different organs of the digestive system and examine those specimens under the dissecting microscope.

❑ Additional Activities

1. Write a comparison explaining the similarities and differences between the digestive system of the fetal pig and the digestive system of the human.
2. Find out how the human cecum compares with the porcine cecum in both structure and function.
3. Make a wet mount slide of the meconium and see if you can identify any cells or pigments.
4. Since the intestine of the pig is usually about 15 times the length of the body, measure your specimen and see if this rule of thumb holds true.

NOTES

❑ Lab Report

1. Referring to the numbered inquiries at the beginning of this exercise, complete the following box summary:

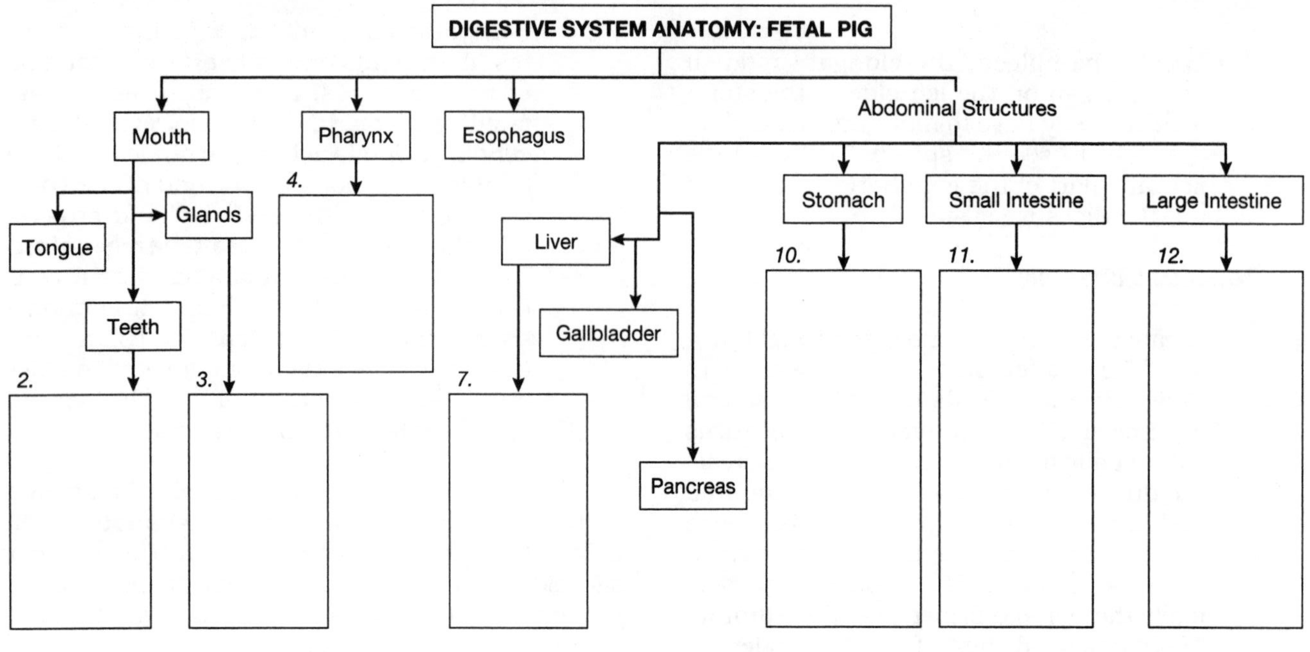

General Questions

NOTES

1. What are the principal organs of the alimentary canal? What are the accessory organs?
2. What is the difference between the visceral peritoneum and the parietal peritoneum?
3. Where are the omenta, and what is the function of each?
4. Where is the gallbladder in relation to the liver?
5. Anatomically, how are the cystic duct, the hepatic duct, and the common bile duct related?
6. From what you can discern, how do the sizes and shapes of the porcine heterodont teeth relate to the omnivorous diet of most pigs?
7. Name the principal valves and sphincters of the digestive system. What common function do they all have? What differences do you note between them?
8. List the similarities and differences you noted in the microscopic structure of the different parts of the digestive system.
9. How do the salivary glands compare with the intestinal glands? With the pancreas?
10. Compare the location of the oral glands of the fetal pig with the location of the oral glands of the human.
11. Consider the fetal pig's dentition pattern. How does the dentition pattern of the fetal pig compare with the dentition pattern of the human?
12. What is the orientation of the papillae on your tongue? What physiological reason might cause the fetal pig's tongue to be different from yours?
13. How does the porcine large intestine differ from the human large intestine?

EXERCISE 62

Anatomy of the Urinary System

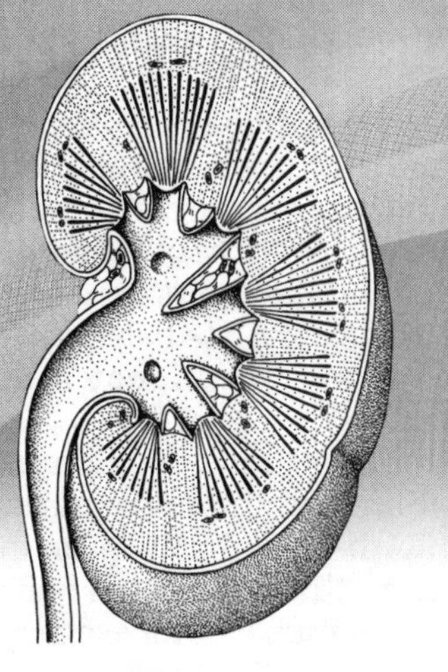

PROCEDURAL INQUIRIES

Preparation

1. Where are the kidneys?
2. What are the names of the renal borders?

Examination

3. What are the names of the capsules of the kidney?
4. What are the structures found in the renal cortex?
5. What are the structures found in the renal medulla?
6. What are the structures found in the renal pelvis?
7. What are the blood vessels of the kidney?
8. What is the role of the ureter?
9. What is the role of the urinary bladder?
10. What is the role of the urethra?

Additional Inquiries

11. What are the physiological functions of the kidney?
12. What are the major differences between the male and female urinary systems?

Key Terms

Kidney (External Structures)
- Adipose Capsule
- Hilum
- Lateral Border
- Medial Border
- Renal Capsule
- Renal Fascia

Kidney (Internal Structures)
- Bowman's Capsule
- Calyx
- Collecting Duct
- Cortex
- Distal Convoluted Tubule
- Glomerulus
- Loop of Henle
- Medulla
- Medullary Rays
- Nephron
- Proximal Convoluted Tubule
- Renal Column
- Renal Papilla
- Renal Pelvis
- Renal Pyramid

Ureter

Urethra

Urinary Bladder

Urogenital System

Vascular system
- Afferent Arteriole
- Arcuate Artery and Vein
- Efferent Arteriole
- Glomerular Capillary Vessels
- Interlobular Artery and Vein
- Renal Artery and Vein
- Segmental Artery and Vein

Materials Needed

Torso

Models of Kidney

Charts and Diagrams

Compound Microscope

Prepared Histological Slides:
- Kidney
- Ureter
- Bladder
- Urethra

The urinary system is the epicenter of homeostatic balance within the organism. The first function of the urinary system is the maintenance of fluid and electrolyte balance within the organism. The urinary system is also responsible for the removal of waste products and foreign substances from the circulatory system. In addition, the urinary system includes several diverse endocrine components for the maintenance of fluid and electrolyte balance as well as red blood cell production. The **kidney** is the principal organ of the urinary system.

With more than a million microscopic filters, each kidney exhibits a remarkable efficiency in cleansing the blood and in maintaining chemical equilibrium throughout the body. The anatomy of this organ, which we will examine in this and subsequent exercises, is even more amazing.

The paired retroperitoneal (behind the peritoneum) kidneys are located in the dorsal part of the abdominopelvic area, generally in the superior lumbar region. Exiting from each kidney is a **ureter**, a tube composed of a fibrous outer coat, two smooth muscle layers, and an inner mucous membrane lining. The ureters enter a common holding area called the **urinary bladder**. Caudal to the urinary bladder is a single tube, the **urethra**, which exits the body in one of two ways, depending on the sex of the individual. The ureter, the urinary bladder, and the urethra are said to be accessories to the kidney.

Because of a common embryological origin, the urinary and reproductive systems of many mammals are often considered together under the combined heading of the **urogenital system**. It is important for us to realize, though, that in the postembryonic animal, the urinary and reproductive functions are separate, although some overlap in structure does exist. In humans this overlap in structure is found in the male but not in the female.

❑ Preparation

I. Orientation

A. Torso Model and Drawings

This laboratory exercise assumes that you are working with a multifunctional torso and with well-marked kidney models. You are encouraged to touch and handle these models in order to familiarize yourself with the intricacies of renal anatomy. If your laboratory is not equipped with models, work on visualizing the three-dimensional aspects of the drawings and photographs included in this exercise.

1➡ Begin your examination of the urinary system by opening the abdominal cavity of the torso and locating the kidneys. Note that these organs are retroperitoneal (behind the peritoneum). The peritoneum is a thin, filmy sheet of transparent tissue. The ureters, urinary bladder, and urethra are also retroperitoneal.

2➡ Before dislodging the kidneys, observe that these paired structures are not parallel. The right kidney is usually anterior to the twelfth rib, making it slightly caudal to the left kidney, which is usually anterior to both the eleventh and twelfth ribs.

Notice that the kidney is embedded in adipose tissue, nature's cushion against the jars and jolts of everyday life.

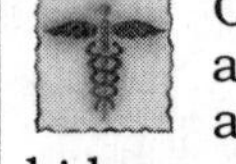

Occasionally a kidney will become dislodged and drop into a more inferior position in the abdominal cavity. This is called a floating kidney, and if the floating is severe, kinking of the ureter (which can block urine flow) can occur. The term used for a dropped or floating organ is ptosis.

Renal anomalies are not uncommon. A person may have kidneys fused at the base. Third kidneys, which may or may not be functional, are sometimes found in people who never imagined they had more than two. And as witnessed by kidney removals, a person actually functions quite normally with only one good kidney.

3➡ Identify the **adrenal glands**, which are located directly on the superior surface of the kidney.

4➡ Note that the **lateral border** of the kidney is convex, while the **medial border** is concave. The notch or depressed area of the medial border is the renal **hilus** (or **hilum**). You should notice three vessels exiting from the hilum. The ureter will be the largest and most posterior of these vessels. It will also have the thickest walls. The thin-walled **renal vein** will be the most anterior. The third vessel is the **renal artery**.

Trace these blood vessels back to the abdominal aorta and inferior vena cava, respectively. Junctioning with the aorta, the right renal artery is somewhat caudal to the left; junctioning with the inferior vena cava, the right renal vein is slightly caudal to the left.

5➡ Return to the renal hilus. Also exiting the kidney at the hilus are nerves and lymphatic vessels. These will probably not be identifiable on your models.

❑ Examination

II. Kidney [∞ *Martini, p 943*]

A. Kidney Model and Diagrams

1➡ Refer to Figure 62–1● and available models to help you in your identification of the kidney structures in this section. Visualize those structures not specifically identified on your models.

2➡ Find the three capsules surrounding the kidney. The outermost capsule, the **renal fascia**, serves to anchor the kidney both to the abdominal wall and to the peritoneum. The middle capsule, the **adipose capsule**, is the mass of fatty tissue surrounding the kidney. The innermost capsule, the **renal capsule**, is a tough, fibrous layer that adheres to the surface of the kidney.

3➡ Open the kidney and notice three distinct internal regions: first, a light reddish-brown outer area (just beneath the capsule); then, a ring of dark triangles; and finally, a fibrous area extending toward the hilus. The outer area is the **cortex**. It is in the cortex that there are up to one million **renal corpuscles**. Each renal corpuscle (sometimes loosely called a glomerular apparatus) consists of **Bowman's capsule** and the **glomerulus**. The glomerular capsule collects the filtrate from the glomerulus. The glomerulus is a minute capillary bed — the filter — through which pass in dilute form the materials that will eventually comprise the urine.

4➡ Refer to the labeled structures in Figure 62–2●. The functional unit of the kidney is the **nephron**. The nephron consists of the glomerular capsule, the **proximal convoluted tubule**, the descending limb of the nephron loop, the ascending limb of the nephron loop, and the **distal convoluted tubule**. The descending and ascending limbs of the nephron loop are often called the **loop of Henle**. Fluids equivalent to the plasma minus the plasma proteins are filtered into Bowman's capsule. As these fluids travel through the tubules, substances such as glucose, water, and salts are removed and ions are exchanged. Additional water is removed from the collecting duct. The end product is **urine**.

Except for a small portion of the nephron loop, the **cortical nephrons** are found entirely in the outer two-thirds of the cortex. The inner third of the cortex houses the **juxtamedullary** renal corpuscles. The loops of the juxtamedullary nephrons dip deep into the medulla.

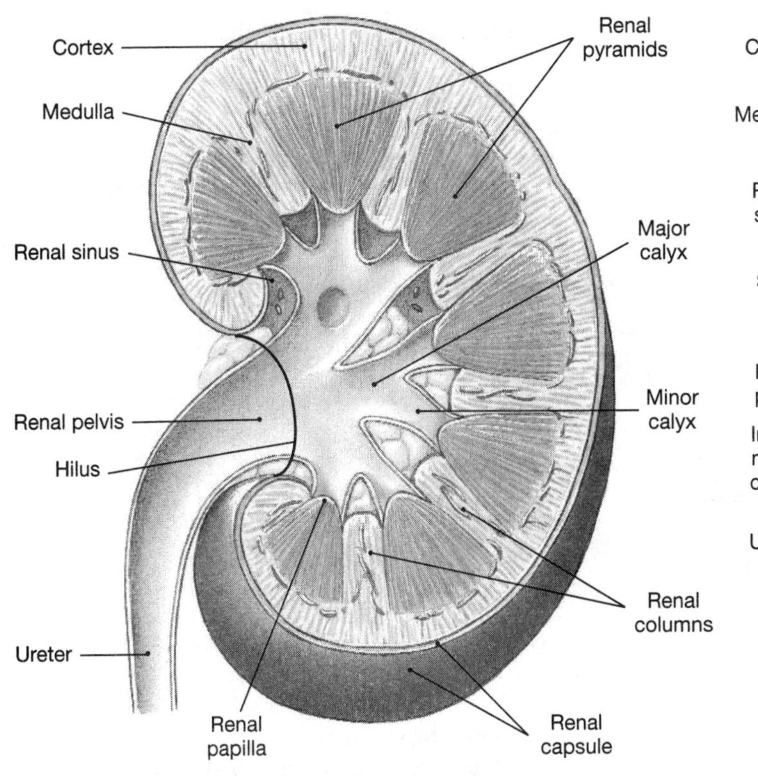

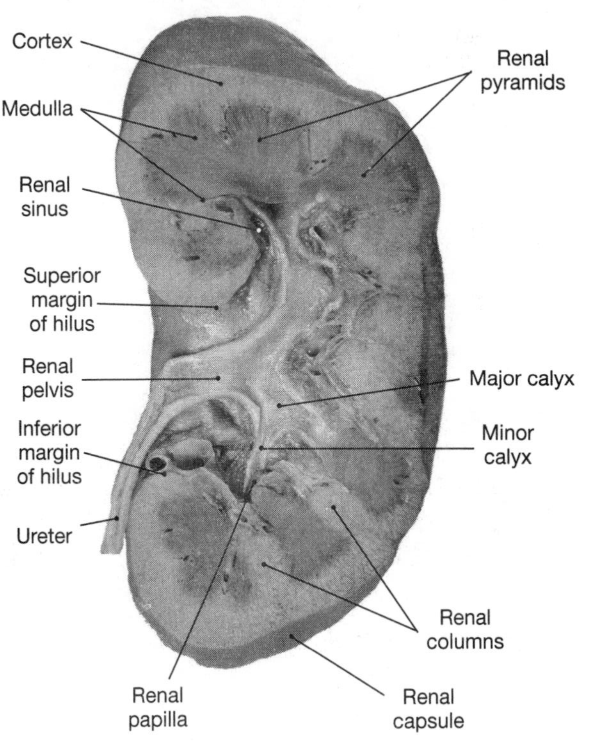

● **FIGURE 62–1**
Anatomy of the Kidney.

● FIGURE 62–2
Nephron Arrangement.

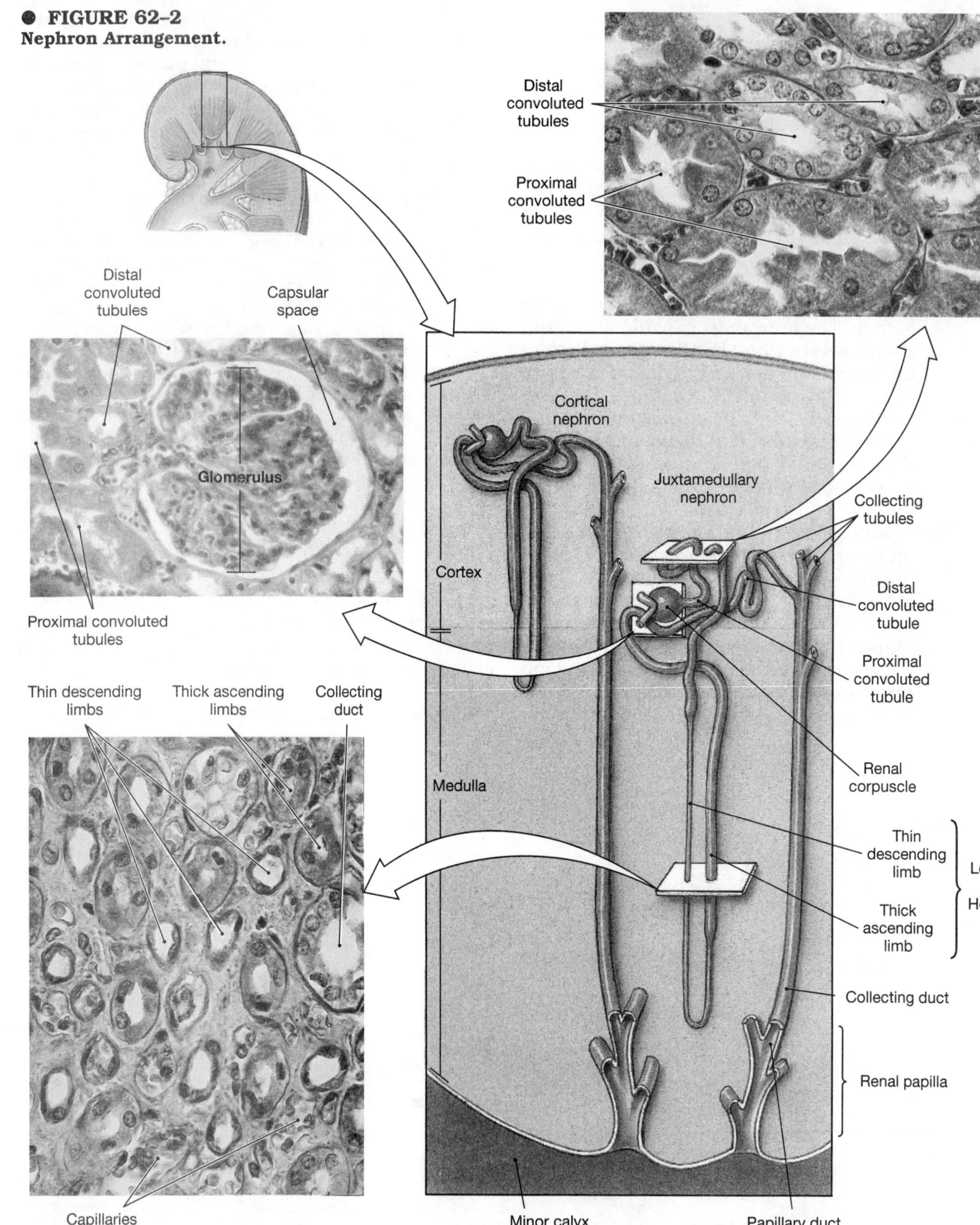

5➡ Find the **medulla** in Figure 62–1●. The medulla is deep to the cortex. The human kidney medulla consists of 6 to 18 conical (pyramidal) structures known as **renal pyramids**. The striations on the renal pyramids are known as **medullary rays**. The renal pyramids are separated by **renal columns**. The renal pyramid is in actuali-

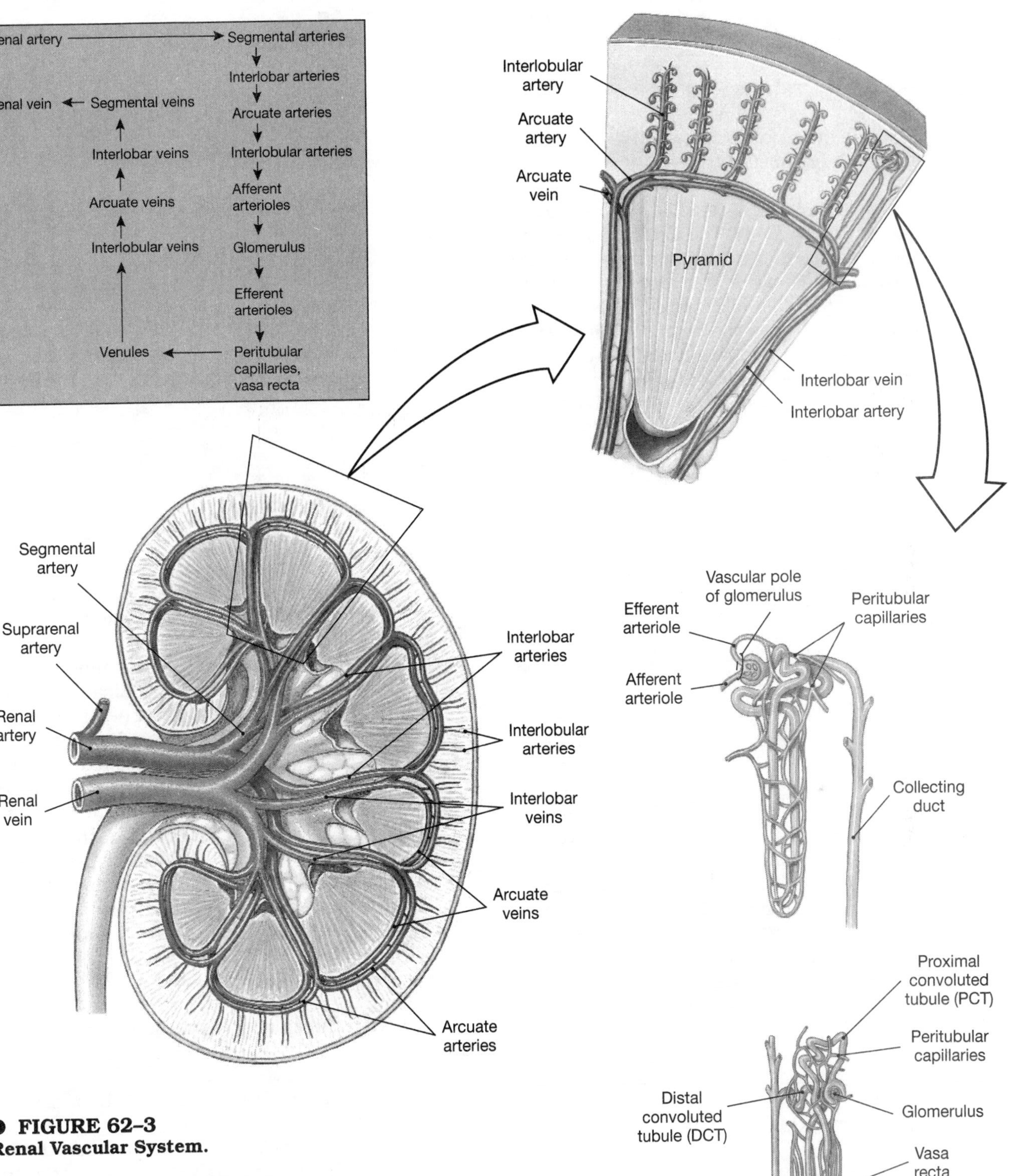

● **FIGURE 62–3**
Renal Vascular System.

ty a conglomerate of nephron loops and **collecting ducts**. The collecting ducts merge, forming a few terminal ducts at the apex (the pointed area), known as the **renal papilla**. It is here that the urine will leave the renal pyramid. The urine from the papilla empties into a small, cup-shaped depression known as a **minor calyx**. Several minor calyces merge, forming a **major calyx**. Three to six major calyces unite to form the **renal pelvis**. From the renal pelvis the urine will exit the kidney via the ureter.

III. Vascular System [∞ *Martini, p. 946*]

A. Pathway of Blood from the Renal Artery to the Glomerulus

1➡ Trace the following vessels on your model and on Figure 62–3●. The **renal artery** enters the kidney and divides into the **segmental arteries.** The segmental arteries break up into several **interlobar arteries**, which are in the renal columns. The interlobar arteries branch into the **arcuate arteries**, which are on the corticomedullary boundary. Off the arcuate arteries branch **interlobular arteries,** which ascend into the cortex. From the interlobular arteries arise the **afferent arterioles**. The afferent arteriole enters the glomerular capsule and divides into the capillary bed known as the glomerulus. (Remember, all glomeruli are in the cortex.)

Which arteries did you find? ______________

B. Pathway of Blood from the Glomerulus to the Renal Vein

1➡ Note that the **glomerular capillary vessels** merge to exit the capsule as one vessel, the **efferent arteriole**. The efferent arteriole leaves the glomerulus and divides into another capillary network, the **peritubular capillary bed**, which surrounds the nephron loop. The peritubular capillaries merge to form the **interlobular veins**, which are parallel to the interlobular arteries. The **arcuate veins**, **interlobar veins**, **segmental veins**, and the single **renal vein** are also parallel to their arterial counterparts.

Which veins did you find? ______________

IV. Excretory Structures [∞ *Martini, p. 972*]

A. The Ureters and the Urinary Bladder

1➡ Relocate the ureters on the torso. Follow the ureters caudally and note that in the pelvic region the ureters turn anteriorly and enter the posterior aspect of a single, firm, centrally located pouch. This structure is the urinary bladder. In the human male the urinary bladder lies between the rectum and the symphysis pubis. In the human female the urinary bladder is inferior to the uterus and anterior to the vagina.

● FIGURE 62–4
Anterior View of Male Bladder.

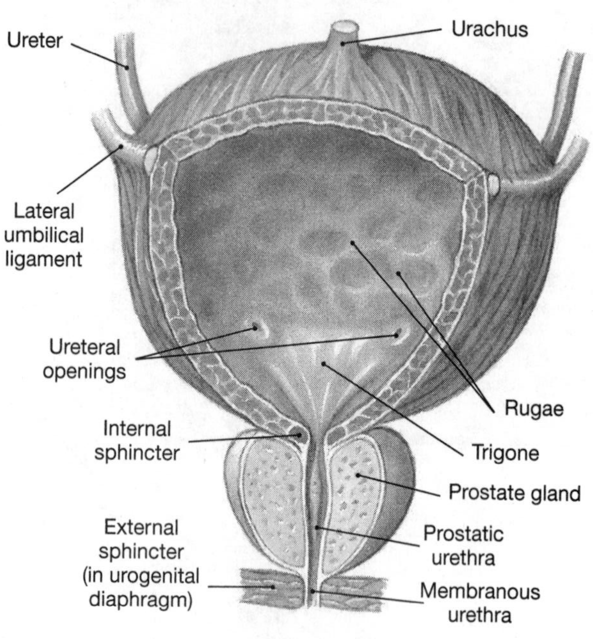

2➡ Turn your attention to the support structures. Your models may indicate the ligament connecting the bladder to the abdominal wall. This is the **median umbilical ligament**. Prior to birth this structure was the **urachus**, a part of the fetal **allantoic bladder**, which extended up into the umbilical cord. After birth the posterior portion of the allantoic bladder became the urinary bladder. In addition to support by this ligament, there are two **lateral ligaments** that hold the sides of the bladder to the posterior body wall. Locate the urachus and the lateral ligaments on Figure 62–4●.

Describe the location of these structures. ______________

B. The Urinary Bladder and the Urethra

1➡ Find the caudal exit of the bladder, the **internal urethral orifice**, which marks the beginning of the urethra, the structure carrying urine to the outside of the body.

In the area of the internal urethral orifice is the smooth muscle stricture known as the **internal sphincter**, which provides involuntary control over the discharge of urine from the bladder.

2➡ Study Figure 62–4● again and identify the **trigone**, the imaginary triangle formed by the two **ureteral orifices**, the entry points

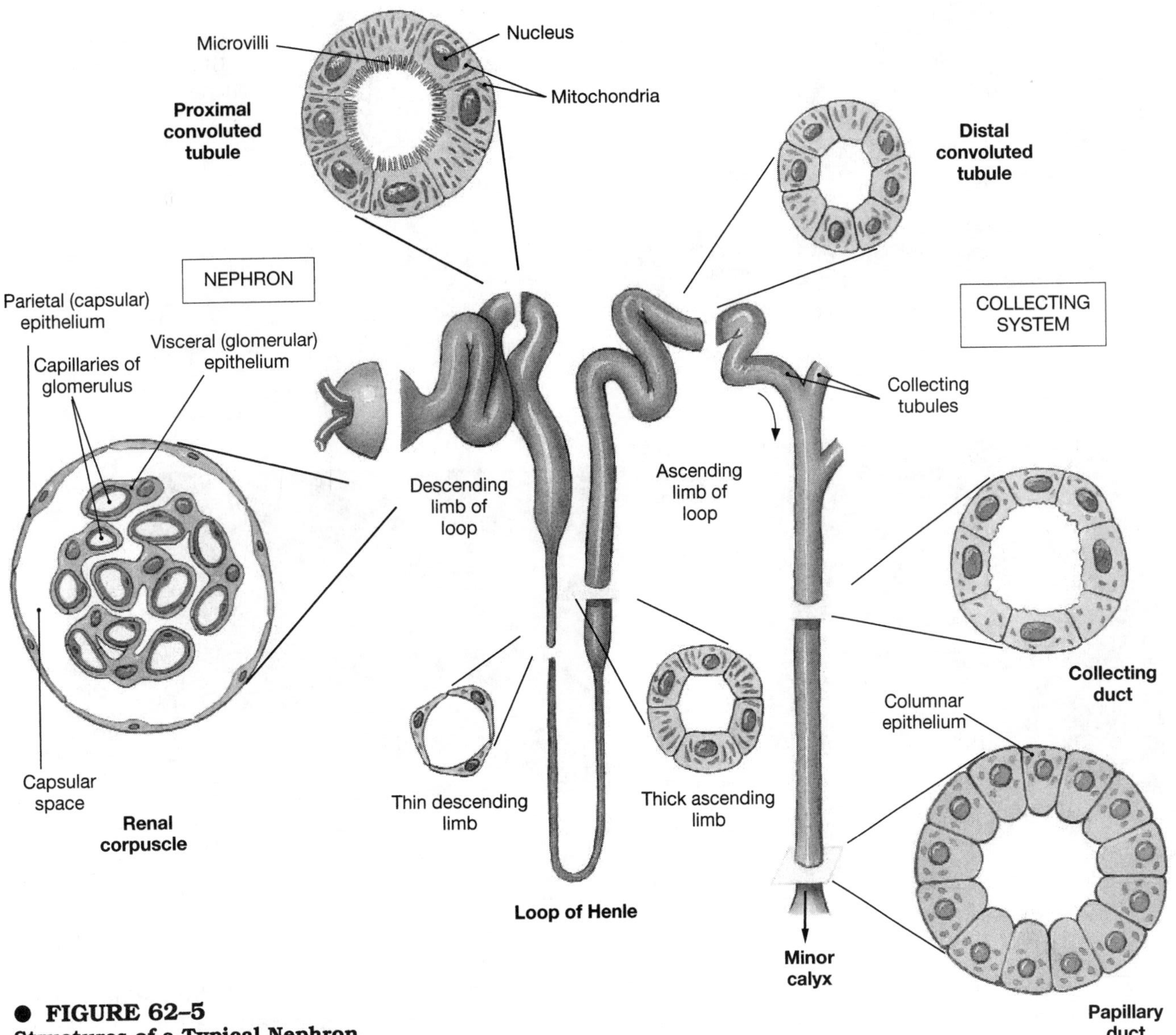

● **FIGURE 62–5**
Structures of a Typical Nephron.

from the ureters into the bladder, and the internal urethral orifice.

3➡ Note that in the female, the urethra follows a very short (25–30 mm) straight line to the **vestibule**. In the male, the urethra merges with the two spermatic cords to form the **urogenital canal**. The urogenital canal usually keeps the name urethra but is often divided into the **prostatic, membranous,** and **penile** urethras. This latter portion continues on into the **penis**. The **external urogenital opening** (also called the **preputial orifice**) is located at the distal end of the penis. The total length of the male urethra is usually about 18–20 cm.

4➡ Identify the muscular floor of the pelvic cavity known as the **urogenital diaphragm** in both the male and female. Surrounding the urethra in this area is the **external sphincter**, which is composed of voluntary skeletal muscle. The external sphincter permits voluntary control over the flow of urine.

V. Histological Slides

A. The Nephron and Associated Blood Vessels

1➡ Examine a microscope slide of the renal cortex. Use Figures 62–2● and 62–5● as your guide. The renal corpuscle should be identifiable under low power. Under high power you should be able to identify the glomerulus, the glomerular capsule, the afferent and efferent arterioles, and various tubules. Sketch and label the structures you are able to identify.

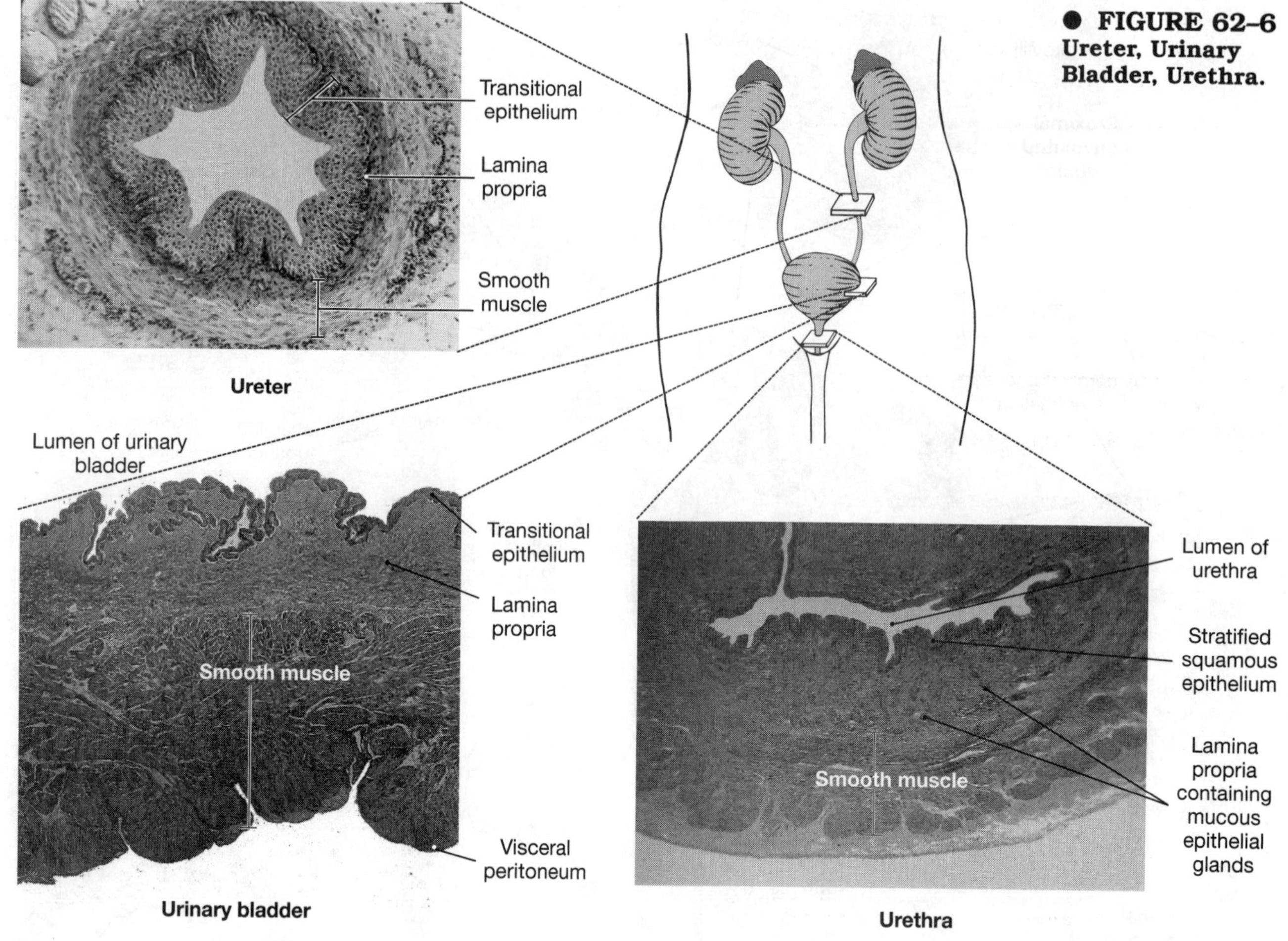

● **FIGURE 62–6**
Ureter, Urinary Bladder, Urethra.

Renal cortex

2➡ If you have a microscope slide of the renal medulla, try to identify the nephron loops, the collecting ducts, and the blood vessels. Sketch and label what you see.

Renal medulla

3➡ While examining your kidney slides, make note of these key features:

a. Relative diameter
b. Presence or absence of microvilli
c. Density of stain
d. Shape of epithelium

Can you use these characteristics to help you distinguish the renal corpuscle, proximal convoluted tubule, distal convoluted tubule, loop of Henle, and the collecting duct? Compare your slide with Figures 62–2● and 62–5●. Depending on your slide, you may or may not be able to find the collecting ducts and papillary ducts.

B. The Kidney Tubule

1➡ Locate the kidney tubule.

2➡ Examine the cross-sectional slide of the kidney tubule. Note the single layer of cuboidal epithelial cells surrounding the circular lumen. Sketch and label what you see. Again compare your slide with Figures 62–2● and 62–5●.

Kidney tubule

Urinary bladder

C. Cross-Sectional Slides

1➡ Examine cross-sectional slides of the ureter, urinary bladder, and urethra. Compare your slide with Figure 62–6● and identify the structures listed. Sketch and label what you see.

Urethra

Ureter

❑ Additional Activities

1. Research the life of Sir William Bowman (1816–1892), the English anatomist for whom Bowman's capsule was named.
2. Research the life of Friedrich G. J. Henle (1809–1885), the German anatomist for whom the loop of Henle was named.
3. Research the embryological development of the mammalian kidney. When does the kidney start functioning? Why is this important?

NOTES

❑ Lab Report

1. Referring to the numbered questions at the beginning of this exercise, complete the following box summary:

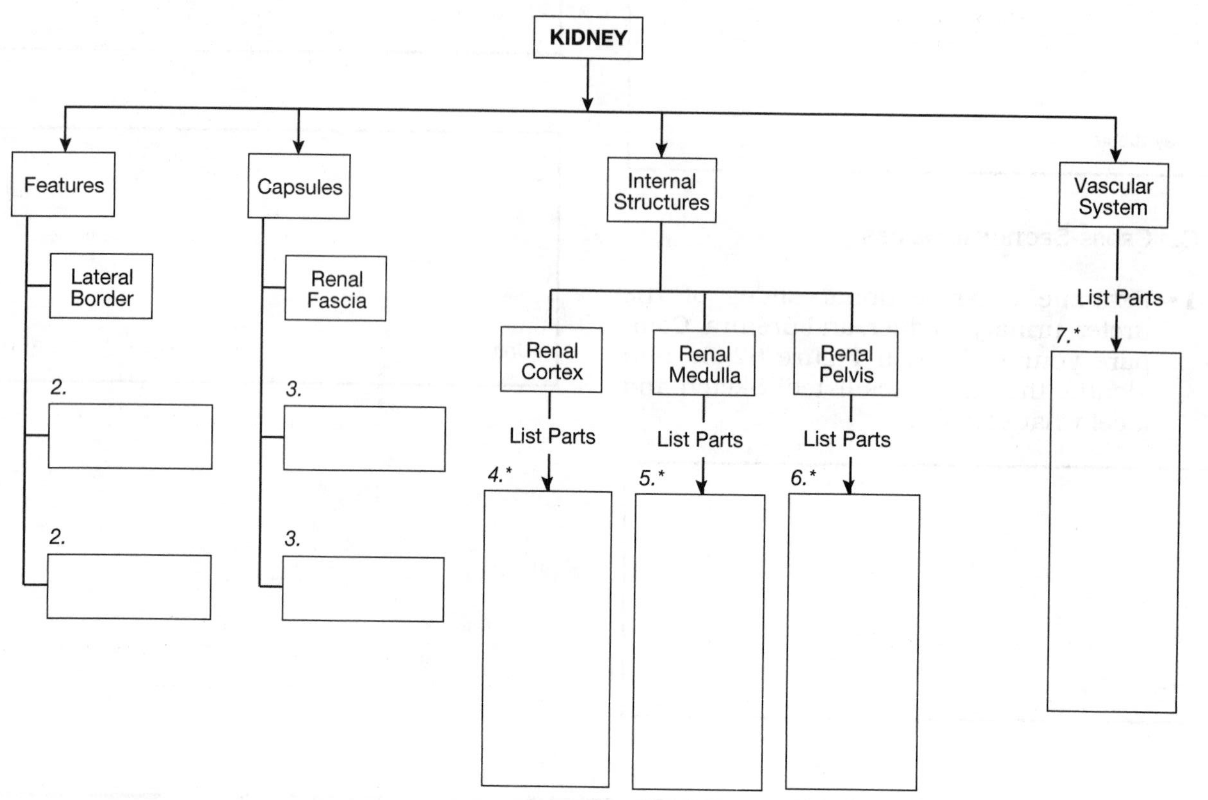

*List according to your instructor's directions.

General Questions

1. Name the three renal capsules and give the function of each.
2. Explain the microscopic makeup of the cortex and the medulla.
3. Can you think of a physiological advantage to the kidney being retroperitoneal?
4. Sketch and label the parts of the nephron.
5. Sketch and label the parts of the renal vascular system.
6. How are the ureter, the urinary bladder, and the urethra related?
7. Explain the role of the sphincters.
8. Explain the similarities and differences between the male and the female urinary systems.

NOTES

EXERCISE 63

Dissection of the Mammalian Kidney

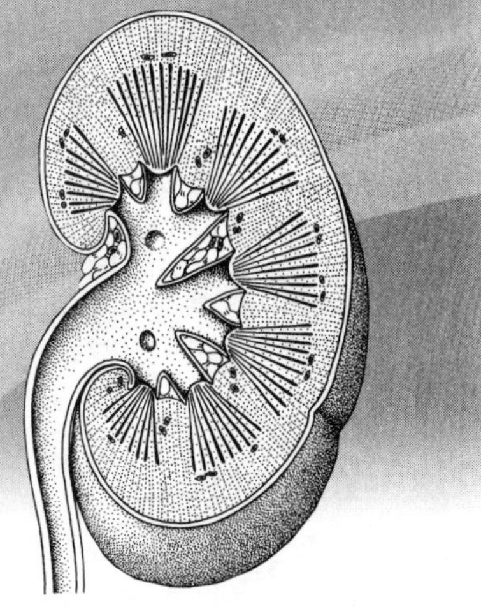

PROCEDURAL INQUIRIES

Examination

1. What are the names of the renal borders?
2. Which vessels can be found at the renal hilum?

Dissection

3. What are the names of the capsules of the kidney?
4. What are the structures found in the renal cortex?
5. What are the structures found in the renal medulla?
6. What are the structures found in the renal pelvis?
7. What are the blood vessels of the kidney?

Additional Inquiries

8. Where are the kidneys?
9. What are the two types of nephrons?
10. Why is it valid to study the kidneys of other mammals in order to gain a perspective on the human renal system?

Key Terms

Kidney
(External Structures)
- Adipose Capsule
- Hilum
- Lateral Border
- Medial Border
- Renal Capsule
- Renal Fascia
- Renal Sinus

Kidney
(Internal Structures)
- Calyx
- Collecting Duct
- Cortex
- Glomerular (Bowman's) Capsule
- Glomerulus
  - Loop of Henle
- Medulla
- Medullary Rays
- Nephron
- Renal Collumn
- Renal Papilla
- Renal Pelvis
- Renal Pyramid

Ureter

Vascular System
- Afferent Arteriole
- Arcuate Artery and Vein
- Efferent Arteriole
- Interlobular Artery and Vein
- Renal Artery and Vein

Materials Needed

Pig or Sheep Kidney — Double-Injected (if possible)
Dissecting Tray
Dissecting Kit
Gloves
Dissecting Microscope
Histological Slides
Compound Microscope

The typical mammal has two kidneys located in the superior lumbar region of the abdominal cavity. In humans this is generally between the T_{12} and L_3 vertebrae. Fatty deposits protect the kidney from physical agitation and also secure the organ in its retroperitoneal location. ("Retroperitoneal" means behind the peritoneal lining of the abdominal cavity.) In humans the right kidney is often slightly lower than the left kidney. In the cat the left kidney is usually somewhat lower, and in the fetal pig the kidneys are generally at about the same level.

Although all mammalian kidneys are quite similar, both the pig kidney and the sheep kidney are easy to work with and extremely close to the human kidney in the relative location of the internal structures. While pig kidneys tend to be longer and flatter than sheep kidneys, working with either kidney will give you a good overall renal perspective. (Certain other mammalian kidneys are either somewhat small for careful dissection or have environmental adaptations making them not totally analogous to the human kidney.)

❑ Examination

I. Orientation

A. External Structures

1➡ Obtain a kidney and, unless instructed otherwise, rinse it thoroughly under running water. Place the kidney in your dissecting tray and examine the external features. Use your blunt nose probe to explore the different regions of the kidney. See Figure 63–1a●.

Summarize your observation. ____________

____

2➡ Note that the **lateral border** is C-shaped, and the **medial border** is concave. The notch or depressed area of the medial border is the **hilum** (or **renal hilus**). You should notice three vessels exiting from the hilum. The **ureter** will be the largest of these vessels. It will also have the thickest walls and, depending on the preservative, may also have a blanched appearance. The thin-walled **renal vein** will probably be collapsed, though you may find traces of blue latex holding it firm. The third vessel is the **renal artery**, in which you may or may not find traces of pink latex. Anatomically, the artery is superior to the vein that is superior to the ureter. However, in your specimen, these vessels may have been cut so that the external location cannot be adequately determined. Which vessels were you able to identify? ____________

____

3➡ Also exiting the kidney at the hilum are nerves and lymphatic vessels. You will probably not be able to identify these structures easily.

4➡ Place your probe in the ureter. Note the feel of space within the kidney. This area is the **renal sinus**. You may wish to repeat this after you have opened the kidney. Were you able to identify the renal sinus? ________

____

5➡ Three capsules surround the kidney. The outermost capsule, the **renal fascia**, serves to anchor the kidney to the abdominal wall and to the peritoneum. You may only be able to find tufts of this capsule in the vicinity of the hilum. If your class will be dissecting a laboratory animal such as the cat or the fetal pig, you will be able to find the renal fascia quite easily. In the preserved kidney, such as the one you are working with now, the renal fascia may look or feel like a dried sausage skin. Did you locate any of the renal fascia? ______

____

The middle capsule, the **adipose capsule**, is the mass of fatty tissue surrounding the kidney. You should be able to see the remains of this capsule as the yellowish globular material found at the hilum. How much of the adipose material did you locate? ____________

If you will be dissecting the urinary system of the cat, keep this adipose capsule in mind. Some of the cats have extensive adipose capsules, while others have practically none. If you will be dissecting the urinary system of the fetal pig, you will find very little renal adipose capsule.

The innermost layer, the **renal capsule**, is a tough, fibrous layer that you should be able to separate from the kidney proper with your forceps and probe. In the preserved kidney, the renal capsule may not remain in one piece as you pull at it, but you should definitely be able to notice how it adheres to the surface of the kidney (Figure 63–1b●). Describe the renal capsule on your particular kidney.

____

____

❑ Dissection

B. Internal Structures

After you have identified the external structures, prepare to dissect the kidney. Plan the incision so that you will be dividing the kidney into anterior and posterior halves. This is a longitudinal or coronal incision. Use a sharp scalpel and begin at the superior border. Continue your incision around the lateral border. Your instructor may suggest that you leave your kidney halves attached at the hilum.

1➡ Use Figures 63–1c● and 63–1e● to help you identify the following internal structures of the kidney.

2➡ Internally you should immediately notice three distinct regions: first, a light reddish-brown outer area (just beneath the cap-

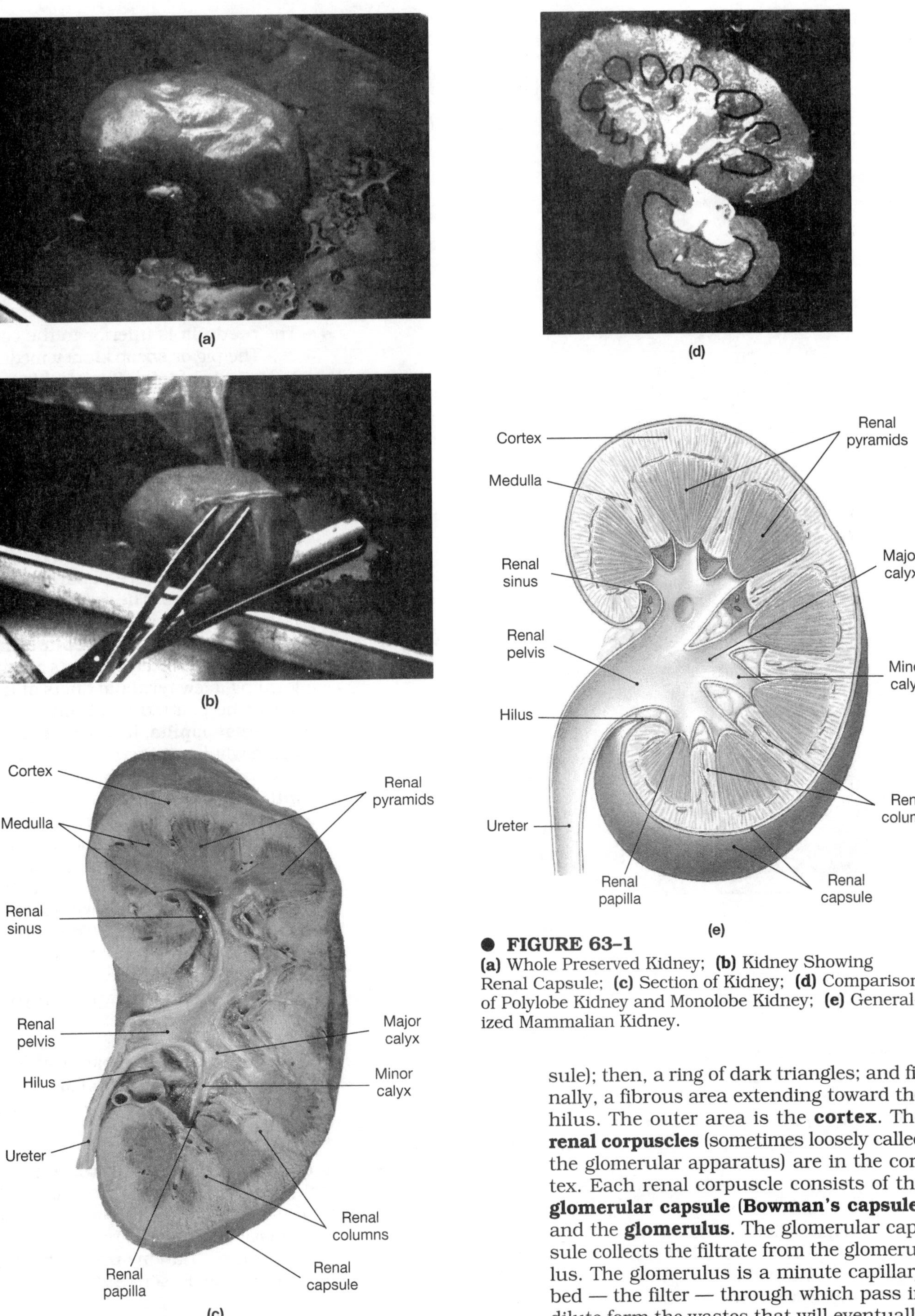

● **FIGURE 63–1**
(a) Whole Preserved Kidney; **(b)** Kidney Showing Renal Capsule; **(c)** Section of Kidney; **(d)** Comparison of Polylobe Kidney and Monolobe Kidney; **(e)** Generalized Mammalian Kidney.

sule); then, a ring of dark triangles; and finally, a fibrous area extending toward the hilus. The outer area is the **cortex**. The **renal corpuscles** (sometimes loosely called the glomerular apparatus) are in the cortex. Each renal corpuscle consists of the **glomerular capsule (Bowman's capsule)** and the **glomerulus**. The glomerular capsule collects the filtrate from the glomerulus. The glomerulus is a minute capillary bed — the filter — through which pass in dilute form the wastes that will eventually

● **FIGURE 63–2**
Nephron.

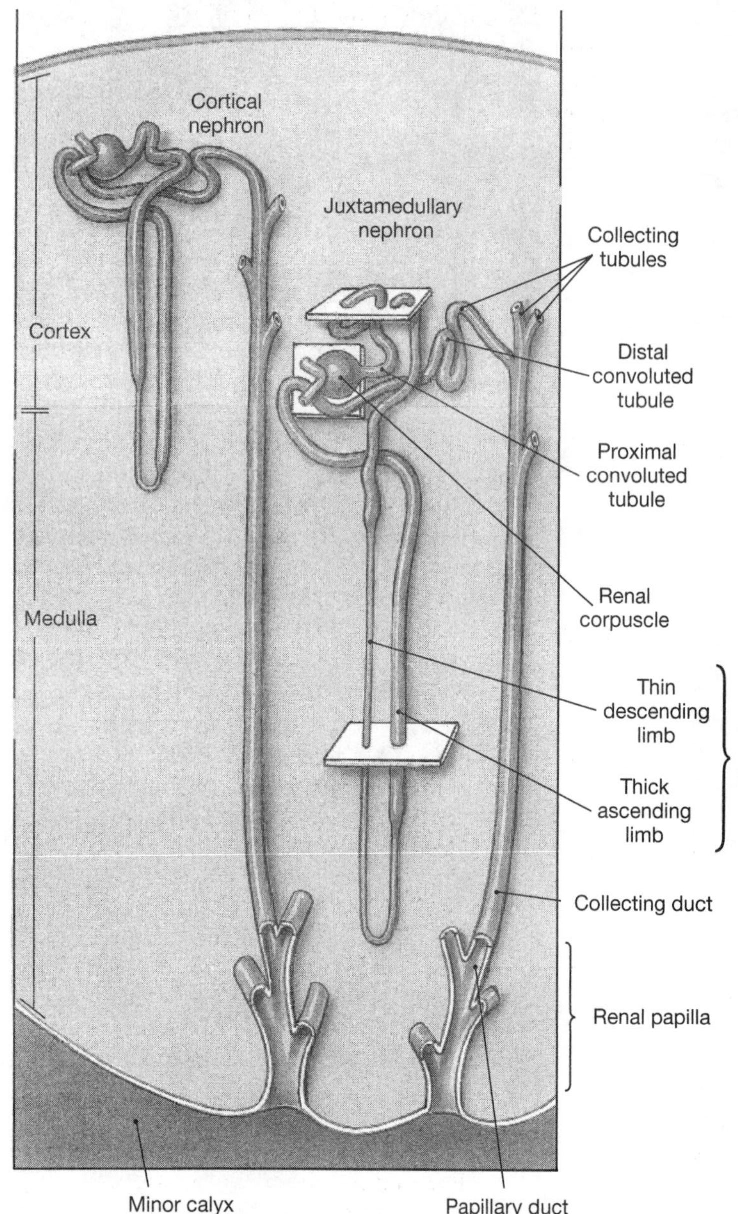

comprise the urine. Over one million such filters are found in each kidney.

3➜ Envision the functional unit of the kidney, the **nephron**. Refer to the labeled structures in Figure 63–2●. The nephron consists of the glomerular capsule, the proximal convoluted tubule, the descending limb of the nephron loop, the ascending limb of the nephron loop, and the distal convoluted tubule. The descending and ascending limbs of the nephron loop are often called the **loop of Henle**. Waste products are collected in the glomerular capsule. As these wastes travel through the tubules, glucose, water, and salts are removed. Additional water is removed from the collecting duct. The end product is urine.

Except for a small portion of the nephron loop, the **cortical nephrons** are located entirely within the cortex. In general, the renal corpuscles of the cortical nephrons are located in the outer two-thirds of the cortex. The inner third of the cortex houses the **juxtamedullary** renal corpuscles. The loops of the juxtamedullary nephrons dip deep into the medulla.

4➜ The **medulla** is inferior to the cortex. The pig or sheep kidney medulla consists of 6 to 18 conical (pyramidal) structures known as **renal pyramids**. (The human kidney also has 6 to 18 pyramids, although the cat kidney usually has only one pyramid culminating in a single papilla — see Figure 63–1d●.) The striations on the renal pyramids are known as **medullary rays**. The renal pyramids are separated by **renal columns**. The renal pyramid is in actuality a conglomerate of nephron loops and **collecting ducts**. The collecting ducts merge forming a few terminal ducts at the apex (the pointed area), known as the **renal papilla**. It is here that the urine will leave the renal pyramid. The urine from the papilla empties into a small depression known as a **minor calyx**. Several minor calyces merge forming a **major calyx**. Three to six major calyces unite to form the **renal pelvis**. From the renal pelvis the urine will exit the kidney via the ureter. The fibrous tissue forms the renal pelvis proper; the lumen is the renal sinus that you identified earlier on your external examination.

5➜ Use your probe to check the pathway from the renal pelvis to the ureter. This is the reverse of what you did when you probed the ureter from the outside.

6➜ Again using your probe, trace the edges of the major and minor calyces. Now trace the edges of the renal papilla. Pop a pyramid out of its place between the renal columns and examine the striations under the dissecting microscope. Describe what you see.

____

____

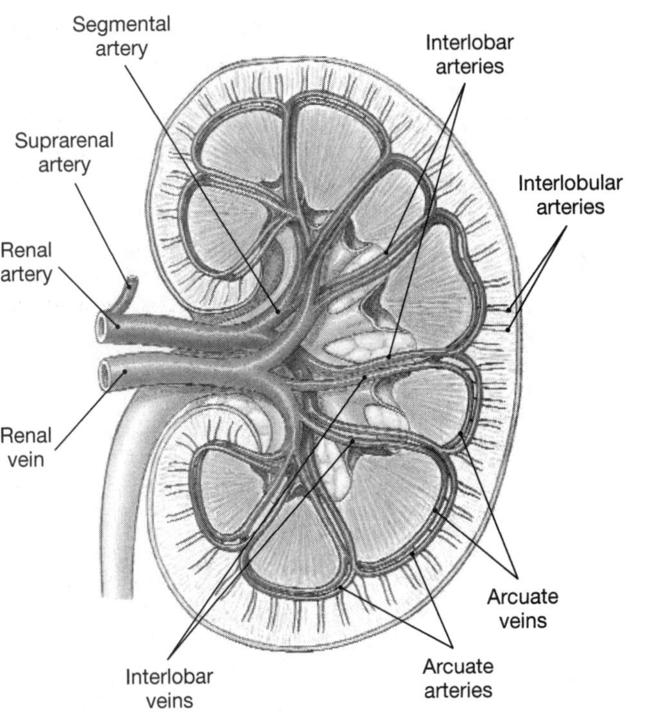

● **FIGURE 63–3**
Renal Vascular Arrangement.

C. Vascular System

1➡ If your kidney is double-injected, you should be able to trace most of the vascular pathway. The arterial portions should be pink and the venous portions blue.

Use Figure 63–3● to help you trace the flow of blood through the kidney.The **renal artery** enters the kidney and breaks up into several **interlobar arteries**, which can be found in the renal columns. The interlobar arteries branch into the **arcuate arteries**, which are found on the corticomedullary boundary. The arcuate arteries are probably the smallest arteries you will see on the "real" kidneys. Off the arcuate arteries branch **interlobular arteries**, which ascend into the cortex. From the interlobular arteries arise the **afferent arterioles**. The afferent arterioles enter the glomerular capsule and break down into the capillary bed, known as the glomerulus. (Remember, all glomeruli are in the cortex.)

Which arteries were you able to find? ______

______

2➡ The vessels of the glomerulus merge, forming the **efferent arteriole**. The efferent arteriole leaves the glomerulus and breaks down into another capillary network, the **peritubular capillary bed**, which surrounds the nephron loop. The peritubular capillaries merge to form the **interlobular veins**, which are parallel to the interlobular arteries. The **arcuate veins**, the **interlobar veins**, and the single **renal vein** are also parallel to their arterial counterparts. Which veins were you able to find? ______

______

II. Microscopic Inspection

A. The Nephron and Associated Blood Vessels

1➡ If so instructed, go back and review the microscope slides listed in Exercise 62. The same principles and directions will apply to this lab exercise.

B. Identification Using the Dissecting Microscope

1➡ Use the dissecting microscope to examine the external and internal structures of the kidney. What differences do you note in the texture of the different structures? ______

______

2➡ As you are examining the structures with the dissecting microscope, identify any blood vessels you see. Which vessels were you able to identify? ______

______

❑ Additional Activities

1. Compare the kidneys of different laboratory animals.
2. Find out about some renal adaptations found in different domestic and wild animals.

NOTES

❑ Lab Report

1. Referring to the numbered inquiries at the beginning of this exercise, complete the following box summary:

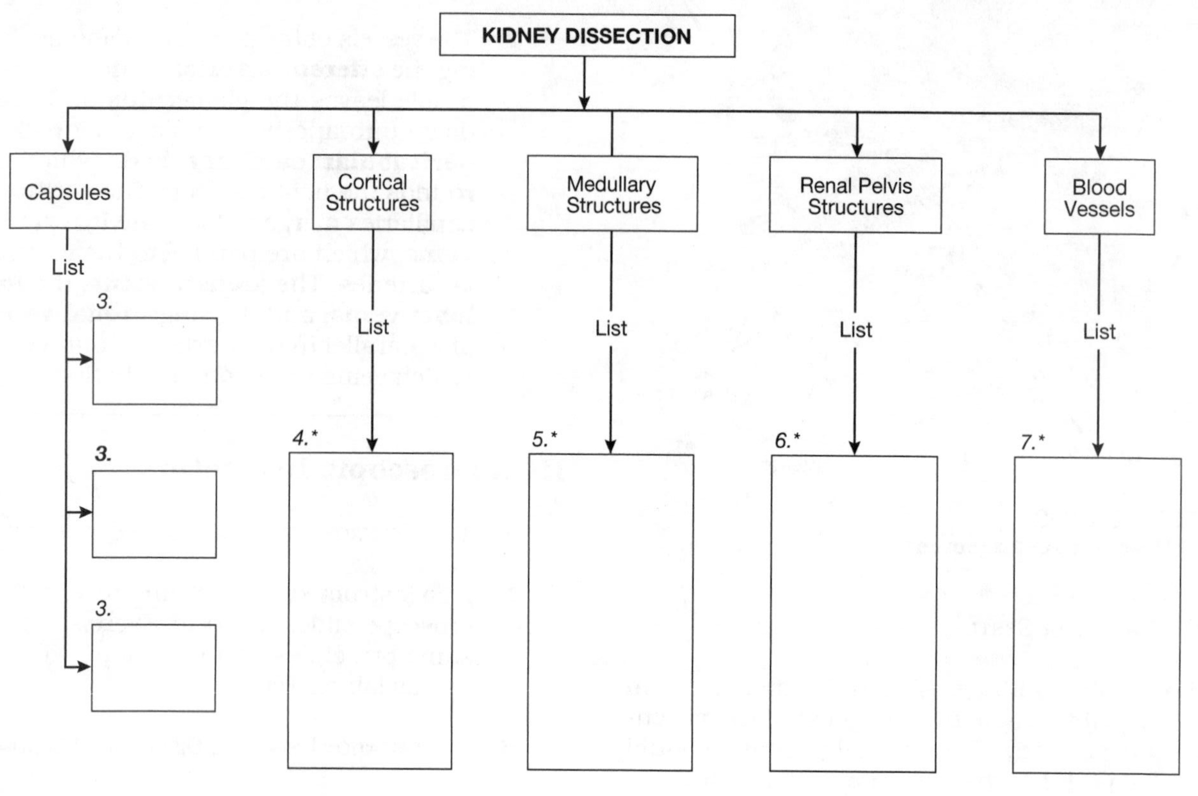

*List according to your instructor's directions.

General Questions

NOTES

1. Go back to Exercise 62. What similarities and differences do you find between the human kidney and the mammalian kidney you just dissected?
2. Where are the kidneys located?
3. What are the renal borders and where is each located?
4. Which vessels can be found at the renal hilum?

EXERCISE 64

Urinalysis

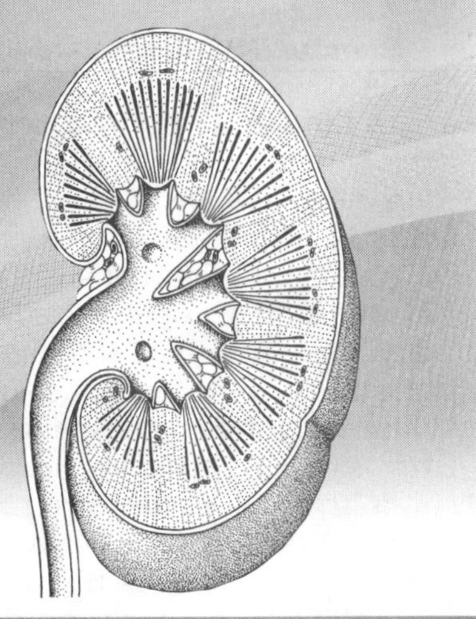

PROCEDURAL INQUIRIES

Preparation

1. What are some common terms used to describe urine?
2. How do you prepare a urine sample?
3. What are some urinalysis safety precautions?

Experimentation

4. What are normal colors of urine?
5. What might different urine odors mean?
6. What is the normal pH range of urine?
7. What is the normal specific-gravity range of urine?
8. What are some common inorganic urine tests?
9. How do you test for protein in the urine?
10. How do you test for glucose in the urine?
11. How do you test for ketones in the urine?
12. How do you test for blood in the urine?
13. What types of sediments might you find doing a microscopic examination of the urine?

Additional Inquiries

14. What is the general composition of urine?
15. Why does the urine image the events of the living system?

Key Terms

Anuria
Cast
Crystal
Dysuria
Enuresis
Inorganic Tests
Ischuria
Odor
Oliguria
pH
Polyuria
Renal Threshold
Sediments
Specific Gravity
Strangury
Test Strip
Urinometer
Urochrome
Void

Materials Needed

Urine Cup (Paper Cup or Beaker)
Urinometer
Urine Cylinder (or Deep Narrow Tube)
Thermometer (Celsius)
Test Tubes
pH Hydrion Paper
Dilute Hydrochloric Acid
10% Barium Chloride
3% Silver Nitrate
Lab Stix (several brands available)
Centrifuge
Appropriate Test Tubes
Glucose Tes-Tape
Stain
Sedi-Stain or
Steinheimer-Malbin or
Methylene Blue
Pipet or Dropper (Calibrated)
Slides and Cover Slips
Gloves
Compound Microscope
Optional:
Dilute Nitric Acid
Dilute Ammonium Molybdate
Hot Water Bath

The kidney filters about 180 l of plasma per day as the blood traverses the glomeruli of the renal system. Urine is the end product, the 1% of the filtrate not reabsorbed and readied for reuse by the body as a whole. *[∞ Martini, p. 947 and AM, Urinalysis]*

Urine is about 95% water and 5% dissolved solids. About 60 grams of solid products are filtered by the kidney and excreted from the bladder each day. Just over 40% of those 60 grams are in inorganic form. These inorganic substances include the sulfates, chlorides, phosphates, and ammonia.

Urea is one of the components of urine. By itself, urea, the final breakdown product of protein metabolism, is a colorless, odorless, crystalline solid. In solution urea is the major nitrogenous constituent of urine. Urea is produced in the liver, either indirectly from carbon dioxide and ammonia (formed from the deamination of amino acids) or directly from the amino acid arginine.

Uric acid, another component of urine, is a resultant of the oxidation of purines in the body. One uric acid product, sodium urate, forms an insoluble crystal that can invade the joints and result in gouty arthritis. This is why patients with gout are advised to limit their intake of foods rich in purines.

Creatinine, the hydrated form of creatine, is excreted at the rate of about 0.02 gm per kg per day. Creatine combines with phosphate to form creatine phosphate, a high-energy storage product found in many organs and various fluids of the body. Muscle is the most significant source of urine creatinine.

Detectable amounts of glucose and/or amino acids should not be found in the urine, although occasionally strange diets can lead to an abundance of these substances even in the absence of any pathology. Most other substances found in the urine will be the direct result of diet, pathology, or both.

Because the kidney filters the blood, the urine becomes a mirror imaging the events of the living system. By examining the urine we can identify numerous normal and pathological happenings within the body.

In this laboratory exercise we will examine various aspects of urinalysis by performing certain common diagnostic urine tests.

❑ Preparation

I. Terminology

A. Basic Vocabulary

Be familiar with these terms before beginning your urine tests.

Crystal Particles of insoluble ionic compounds. Limited numbers of crystals are normal; excessive quantities are abnormal.

Cast Mass of plastic or fibrous matter encased in a limiting membrane and formed in various body cavities. Casts usually have square ends and a constant diameter. Casts normally do not twist or bend. An occasional cast may be normal but usually significant numbers of casts indicate a disease state.

Renal Threshold Plasma concentration of a substance, beyond which that substance will appear in the urine. The renal threshold is the transport maximum.

Void Urinate.

Anuria Cessation of urine production.

Dysuria Difficult (painful) urination.

Enuresis Inability to retain urine (incontinence).

Ischuria Urine retention.

Oliguria Diminished urine output.

Polyuria Increased urine output.

Strangury Painful and spasmodic urination.

B. Derivation of Related Terminology

In addition to the common words listed above, almost any "substance" word can be used with the suffix "-uria" to designate that particular substance as being found in the urine. For example: "galactosuria" would signify galactose in the urine, and "bilirubinuria" would signify bilirubin in the urine. Also, adjectives can precede "-uria" to indicate a particular condition of the urine. For example: "alkalinuria" indicates that the urine is alkaline.

II. Procedure

A. Safety

You are working with body fluids in this lab exercise. It is imperative that you follow all safety regulations. Handle only your own specimens, clean up your own materials, and disinfect and dispose of wastes according to your instructor's directions.

B. Preliminaries

1➡ Please keep in mind that certain pathological (or potentially pathological) conditions will be mentioned but the results you obtain in the classroom laboratory are not necessarily either accurate or precise. If you have any concerns about results you obtain here, consult an appropriate health care professional.

2➡ Record the results of your urinalysis on the chart Figure 64–3 at the end of this laboratory exercise. In the space provided, make note of any irregularities or unusual occurrences in the testing procedures.

C. COLLECTION

1 ➡ Take your urine cup to the rest room and carefully void in the cup until the cup is reasonably full. It is not necessary to collect your specimen under sterile (or aseptic) conditions unless your instructor wishes you to culture your urine for possible microbial invasion.

2 ➡ After you return to the lab, fill your urine cylinder about two-thirds full. If any appreciable amount of foam is present on the top of your sample, remove this foam by skimming the surface with a piece of filter paper or paper towel. The presence of foam may make certain tests more difficult to read.

(If you are not running your urine tests immediately, cover the cup and store the sample in a cool place.)

❑ Experimentation

III. Analysis

A. APPEARANCE

1 ➡ Examine the appearance of your urine. What color is it? (Record the color on the chart at the end of this exercise.) The color of normal urine ranges from light straw to amber, depending on the concentration of the pigment **urochrome**. Urochrome is the end product of hemoglobin breakdown:

Hemoglobin ⟶ Hematin ⟶ Bilirubin ⟶ Urochromogen ⟶ Urochrome

Pale yellow indicates a very dilute urine. This could indicate diabetes insipidus, granular kidney, or the ingestion of copious amounts of water.

Milky white colors might signify fat globules or pus corpuscles. Pus could indicate an infection in the genitourinary tract.

Reddish colors are often confusing. Many a parent has rushed a young child to the emergency room because of red urine, only to discover that the child had eaten beets a few hours earlier. In addition to certain food pigments, red can also indicate certain drugs or blood in the urine. Blood can be misleading in women, particularly because menstrual blood may appear in the urine even when the woman is certain none is in her collection cup.

Greenish colors indicate either bile pigment (jaundice) or certain bacterial infections (such as those caused by *Pseudomonas sp.*).

Brown-black urine can indicate phenol or metallic poisonings or hemorrhages from such conditions as malaria or renal injury.

Although the urine is usually clear immediately after collection, cloudiness in older urine samples is not pathological but rather is caused by mucin from the urinary tract. Milky or turbid urine, however, can be pathological.

B. ODORS

Normal urine has a "peculiar" aromatic **odor** that can vary greatly according to both diet and pathology. An ammonia smell may result from certain foodstuffs, while a fishy smell may indicate cystitis. A fecal smell could indicate an intestinal-urinary tract fistula, and the smell of overripe apples usually means acetonuria. A new-mown hay smell could be from diabetes, which would also give a sweet-tasting urine.

1 ➡ Use the space provided at the end of the exercise to describe the odor of your urine.

C. pH

The **pH** of normal urine ranges from 4.5 to 8.0. The urine of vegetarians tends to be more alkaline, and the urine of athletes and those who eat highly acidic food tends to be more acidic.

Pathological conditions can also affect the pH of urine. Fevers and acidosis lower the pH, and anemia, vomiting, and ischuria raise the pH.

1 ➡ Use the pH Hydrion paper to check the pH of your urine. If your urine has been standing for a while, the pH will be higher than when you first voided because bacterial decomposition of urea can lead to an accumulation of ammonia. Record the pH of your urine on the chart at the end of this exercise.

D. SPECIFIC GRAVITY

1 ➡ See Figure 64–1• for a diagram of a standard **urinometer**. A urinometer is actually a hydrometer specifically gauged, or calibrated, to register the **specific gravity** of urine. Specific gravity is a measure of the density of a substance in gm/ml as compared to the density of water, which has a specific gravity of 1.00 gm/ml. The specific gravity of urine usually ranges between 1.003 and 1.030, although numbers slightly higher or slightly lower may be normal for persons whose diets are either very high or very low in fluid content. Generally, specific gravity is inversely proportional to urinary volume. Would you

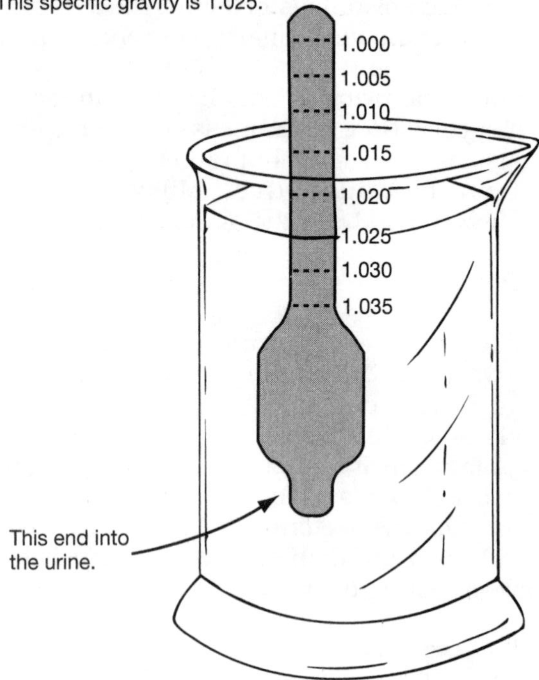

● **FIGURE 64–1**
Reading a Urinometer.

expect the specific gravity of a water binger to be high or low? ____________
____

A pathological low specific gravity indicates nephritis, whereas a pathological high specific gravity indicates either nephritis or diabetes mellitus.

2➡ Take the urinometer and spin or twist it as you place it in the urine cylinder. (Spinning helps prevent adhesion of the urinometer to the cylinder wall.) Be certain the urinometer is floating freely. Read the value where the urine level crosses the calibrated scale to three decimal places. Record this value here ____________

3➡ Specific gravity is sensitive to temperature. Unless you are instructed otherwise, your urinometer was calibrated to read correctly at 15°C. Therefore, you must make an adjustment for temperature.

Use the thermometer and determine the temperature (in celsius) of your urine specimen. Record that temperature here. ________. For every 3 degrees above 15°, add 1 to the last decimal place of your specific gravity. For every 3 degrees below 15°, subtract 1 from the last decimal place. For instance, if the specific gravity is 1.019 and the temperature of the urine is 21°C, you would add 2 to the last decimal place, making the corrected specific gravity 1.021.

Now calculate and record your adjusted specific gravity at the end of the exercise.

E. Inorganic Tests

The **inorganic tests** listed here are but two of many that could be performed on your urine sample. Most inorganic tests involve precipitation or color change.

1➡ Sulfates: Into a clean test tube put 5 ml of urine, a few drops of hydrochloric acid, and 2 ml of barium chloride. The presence of the hydrochloric acid assures an abundance of chloride ions. Barium chloride dissociates. Barium sulfate is insoluble in aqueous solutions, so a white precipitate (barium sulfate) indicates the presence of sulfate ions in the urine. Record your results.

2➡ Chlorides: Into a clean test tube put 5 ml of urine and a few drops of silver nitrate. Since silver chloride is insoluble, the presence of an off-white precipitate (silver chloride) will indicate that the chloride ion is present in the urine. Record your results.

F. Test Strip (Lab Stix)

Test Strips or Lab Stix are strips of treated plastic or acetate that will register particular chemical reactions when exposed to specific aqueous solutions. These strips or stix are used for analyzing several urinary components at the same time. Different types of Test Strips and Lab Stix are available. It is therefore important that you follow the specific directions for the type found in your lab. The following information will help you understand what normal and abnormal readings on standard Test Strips or Lab Stix mean.

Use this as a guide and make note of your particular readings for each of the tests included on your Test Strip or Lab Stix.

1➡ Protein: Although minute amounts of protein (10–100 mg/day) can be found in normal urine because of certain diets, exercise regimes, and/or mucus secretions, abnormally high levels of protein (particularly albumins) usually indicate altered renal functions, either a renal pathology or a systemic disease such as diabetes. Globulinuria often indicates multiple myeloma or pyelonephritis.

2➡ Glucose: Glucose reabsorption is carrier-mediated. That is, carrier molecules transport glucose from the glomerular ultrafiltrate (urine) back into the blood. The renal threshold for glucose, based on the number of carrier molecules available, is about 180 mg/100 ml. Normal blood glucose levels are usually 80–120 mg/100 ml. Therefore, under normal circumstances, glucose is virtually completely reabsorbed. (A minute amount, usually less than 1 gm/day, is excreted. This amount is not detectable by methods used in this laboratory.)

If the plasma levels of glucose exceed the renal threshold, glucose will be detected in the urine. Although you can artificially exceed the renal threshold for glucose by "pigging out" on doughnuts and "real" soft drinks shortly before the urine test, usually glycosuria is indicative of diabetes mellitus, hyperpituitarism, hyperthyroidism, or liver disease.

3➡ Ketones: Ketones will appear in the urine when carbohydrate metabolism is inadequate or when fatty acid utilization is incomplete. Ketones are intermediary breakdown products. Three ketone bodies causing ketonuria are acetoacetic acid (diacetic acid), acetone, and beta hydroxybutyric acid.

Ketonuria can result when a person suddenly begins a very low carbohydrate diet. Pathologically, however, ketonuria indicates the progressive diabetic acidosis of diabetes mellitus.

4➡ Blood: Blood is not normally present in the urine, although, as we have seen, occasionally blood of nonrenal origin (i.e., menstrual blood) may be found in the urine. Since blood normally does not enter the nephron, blood in the urine is usually indicative of either disease or trauma. Direct injury to the kidney can result in glomerular breakdown. Usually blood cells are hemolyzed in the urine, so tests are performed to detect the presence of hemoglobin rather than the red blood cells themselves.

G. Tes-Tape

Tes-Tapes are often used for immediate detection of glucose in the urine. To use most brands of Tes-Tapes, dip the tape in the urine, wait a short period of time (according to package directions), and read the results by comparing the Tes-Tape with the scale on the side of the box.

1➡ Perform this test and record your results.

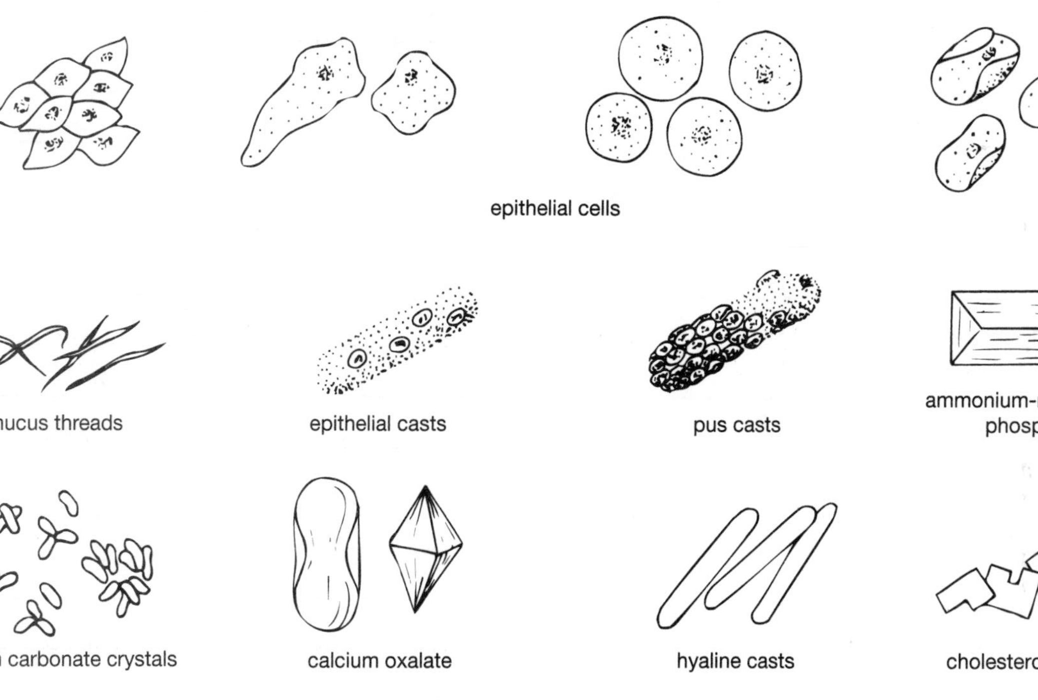

FIGURE 64–2
Common Sediments Found in the Urine.

H. Microscopy

A microscopic examination of the urine is vitally important. It is beyond the scope of this laboratory exercise to examine all aspects of urinary microscopy. Nevertheless, you can get a feel for this diagnostic tool by completing the following procedure and then examining the **sediments** in your urine sample.

1➡ Fill a centrifuge tube about two-thirds full and centrifuge at about 1500 rpm for about 5 minutes. Be sure to balance the centrifuge according to your instructor's directions.

2➡ Carefully pour off the supernatant (the liquid on top of the sediment) and add a drop of Sedi-Stain or Sternheimer-Malbin Stain to the sediment. Mix by swirling gently.

3➡ Using a transfer pipet or a dropper, place a drop of the mixture on a microscope slide and cover with a cover slip.

(If neither stain is available, methylene blue may be substituted. Place a drop of sediment on a microscope slide. If necessary, add a tiny drop of water. Cover the sediment with a cover slip. Place a drop of methylene blue on the outside edge of the cover slip and allow it to seep under the cover slip. Wait briefly while the methylene blue spreads evenly around the sediment.)

4➡ Examine the slide under low power. Manipulate the light until you achieve the best visual effects. Use Figure 64–2● to identify the various cells, crystals, and casts.

Test	*Range of Normal*	*Result*	*Comments or Explanations*
Color			
Odor			
pH			
Specific Gravity			
Inorganic			
Sulfate			
Chloride			
Test Strip (Lab Stix)			
Protein			
Glucose			
Ketone			
Blood			
(Other)			
(Other)			
Tes-Tape			
Other Tests			
—			
—			
—			
—			
Microscopy			

● **FIGURE 64–3**
Urine Test Results.

❑ Additional Activities

NOTES

1. Additional Tests
   A. Phosphates: Into a clean test tube put 5 ml of urine, a few drops of nitric acid, and 3 ml of ammonium molybdate. Mix well and heat gently in a hot water bath. A yellow precipitate will indicate the presence of phosphates.

      What chemical reaction took place? Why should you or why should you not expect phosphates to be in the urine?
   B. PKU (phenylketonuria) is a condition indicated by the presence of phenyl pyruvic acid in the urine. PKU is a failure of the body to produce the enzyme necessary to oxidize phenylalanine to tyrosine. PKU is a recessive genetic trait occurring at the rate of 1:40 000 in the United States and 1:25 000 in England. PKU causes physical and mental retardation. By reducing or eliminating phenylalanine from the diet, retardation does not occur. Because the phenylketones appear in the urine, infants with PKU often have diapers with distinctive odors. (This was noted by Swedish mothers long before the scientific community accepted the possibility of testing for PKU by checking the urine.) Today PKU is routinely tested for in many states. Explore the problem of PKU.

      What test is performed to check for PKU? What chemical principles are involved in the PKU test? What are the prospects for a child diagnosed with PKU today? What about the children of those afflicted with PKU?
   C. Do a 24-hour urinalysis test. Record what you eat. Start by voiding. Discard this original sample, then collect urine specimens and perform the basic urine tests on the samples at periodic intervals (such as every 3 hours) over the next 24 hours. Chart (graph) the collection data over time. (If you wish to run the tests simultaneously, cover each sample and store in a cool place.)

      What trends do you see? Why were you instructed to discard the original sample?
2. Research
   A. Find out about some of the common drug tests, such as those done for anabolic steroids, illegal substances, or alcohol. Explain the principles behind these drug tests. How accurate are these tests? Why?
   B. Explore the breakdown products of such foods as asparagus, and explain why this vegetable gives urine a distinctive odor.
   C. Explain why a urinometer and an anti-freeze tester are said to work on the same principle.

❑ Lab Report

1. Referring to the numbered inquiries at the beginning of this exercise, complete the following box summary:

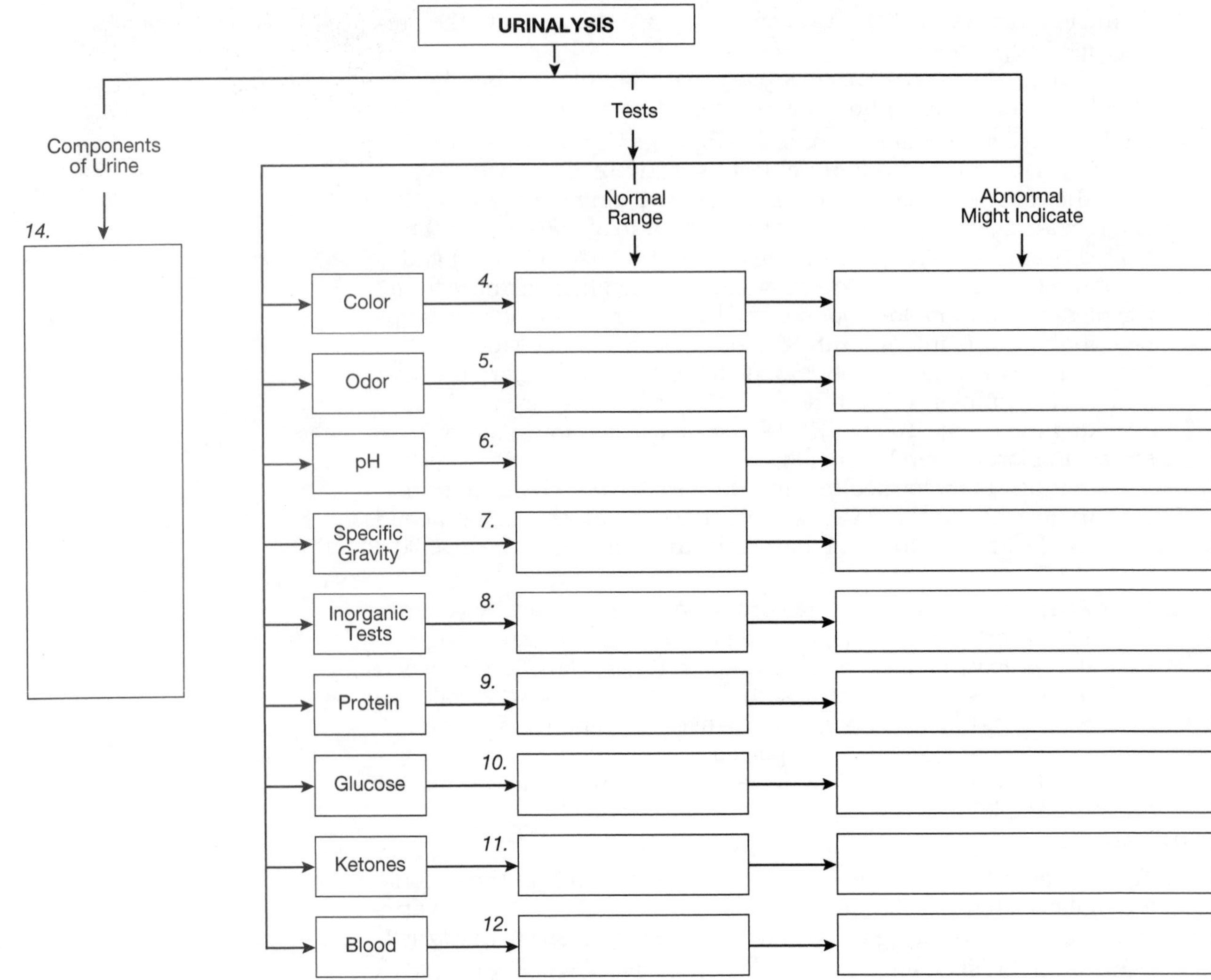

General Questions

1. If 200 l of blood pass through a certain person's kidney's per day, about how much urine does that person produce?
2. How would you interpret these test results?
   a. Specific gravity 1.018, ketonuria, glucosuria
   b. Specific gravity 1.003, negative ketones, negative glucose
   c. Specific gravity 1.020, negative ketones, negative glucose
3. Would a person suffering from dehydration have a high or low specific gravity?
4. Blood in the urine can indicate a serious kidney problem. Why are kidney problems so serious?
5. What types of sediments can be found in a normal urine sample?

NOTES

EXERCISE **65a**

Dissection of the Urinary System: Cat

PROCEDURAL INQUIRIES

Preparation

1. What are the major organs of the urinary system?
2. How should the cat be prepared for dissection of the urinary system?

Dissection

3. Why do we say the kidneys are retroperitoneal?
4. Where are the cat's adrenal glands located in relation to the kidneys?
5. Where are the ureters?
6. What are the major structural landmarks of the urinary bladder?
7. What are the anatomical differences between the urinary systems of the male and female cat?
8. What are the principal structures of the cat's kidney?
9. What are the major blood vessels of the kidney?

Additional Inquiries

10. How does the urinary system of the cat relate to the urinary system of the human?

HUMAN CORRELATION: AM Cadaver Atlas in Applications Manual.

Key Terms

Adrenal Glands
Allantoic Bladder
Cortex
Hilus
Internal Urethral Orifice
Kidney
Lateral Border
Lateral Ligaments
Medial Border
Median Umbilical Ligament
Medulla
Papilla
Renal Pelvis
Renal Pyramid
Spermatic Cord
Subcortex
Trigone
Urachus
Ureter
Ureteral Orifice
Urethra
Urinary Bladder
Urogenital System

Materials Needed

Cat (double-injected)
Dissecting Tray
Dissecting Kit
Gloves
Dissecting Microscope
Prepared Histological Slides
  Kidney
  Bladder
Compound Microscope

All functions of the urinary system center around homeostatic balance within the organism. The first major function of the urinary system is the removal and excretion of waste products from the circulatory system. The second function, highly correlated to the first, is the maintenance of fluid and electrolyte balance within the organism. The organ responsible for the accomplishment of these purposes is the **kidney**. The other structures of the urinary system are accessories to the kidney.

In the mammal, paired kidneys are located in the dorsal region of the abdominopelvic cavity, generally in the superior lumbar region. Exiting from each kidney is a tube known as a **ureter**, a tube composed of a fibrous outer coat, two smooth muscle layers, and an inner mucous membrane lining. The ureters enter a common holding area called the **urinary bladder**. Caudal to the urinary bladder is a single tube, the **urethra**, which exits the body in one of several ways, depending on the sex and species of the mammal.

Because of a common embryological origin, the urinary and reproductive systems of many mammals are often considered together under the combined heading of **urogenital system**. It is important for us to realize, though, that in the postembryonic animal, the urinary and reproductive functions are separate, although some overlap in structure does exist. In the cat, the overlap in structure is more pronounced in the male than in the female.

In this laboratory exercise we will study the anatomy of the urinary system by dissecting the urinary system of the adult cat.

❑ Preparation

I. Gross Dissection

A. Orientation

1➡ To begin your dissection of the urinary system of the cat, prepare your specimen by placing the animal dorsal side down in the dissecting tray. Open the abdominal cavity and reflect the visceral organs away from the midline of the body.

2➡ Follow the diagrams carefully and be aware of the similarities and differences between the male and female cat as well as between the cat and the human (see Figures 65a–1● and 65a–2●). Refer to Exercise 62 for the human system.

Dissection

B. Examination of the Kidney

1➡ Locate the kidneys. Note that these organs are retroperitoneal (behind the peritoneum). The peritoneum will be a thin, filmy sheet of transparent tissue. The ureters are also retroperitoneal, although you may have difficulty seeing this if you have already dissected the digestive system of your cat.

2➡ Carefully remove the peritoneum. Before dislodging the kidneys, observe that these paired structures are not parallel. The right kidney is usually ventral to the transverse processes of the first four lumbar vertebrae. The left kidney is usually ventral to the transverse processes of the second through fifth lumbar vertebrae. In addition to being more caudal, the left kidney is also somewhat ventral to the right kidney. In the human it is the right kidney that is usually caudal to the left kidney.

In the live animal, both kidneys are easily palpable, and the left kidney is also quite free-floating. (This occasionally leads cat owners to think their pet has a tumor or is in some odd state of pregnancy.) The mobility of this kidney does not cause a problem for the cat. In the human, however, a free-floating kidney can lead to serious problems, including a possible kinking of the ureter.

3➡ Notice that the kidney is embedded in adipose tissue, nature's cushion against the jars and jolts of everyday life. Carefully clean away this fatty material. (Some well-fed cats may have more than an ideal amount of renal fat.) Locate the **adrenal glands**, which can be found between the superior medial border of the kidney and the great dorsal blood vessels. Normally the feline adrenal glands are not in contact with the kidney. (In the human the adrenal glands are located directly on the superior surface of the kidney.)

4➡ Do not confuse the adrenal glands with the adipose tissue. The adipose tissue is probably a yellowish white, and the adrenal glands probably have a red-orange tinge. Also, the adrenal glands will feel firmer than the "mushy" adipose tissue, and they have a distinct circulatory supply, which you may or may not be able to discern.

5➡ Identify the **lateral** and **medial borders** of the kidney. Also, identify the **hilus**, the indented region on the medial border where you will find the blood vessels and the ureters.

C. Major Blood Vessels Supplying the Kidneys

1➡ Identify the renal arteries (pink latex injection) and the renal veins (blue latex injection). A single renal artery may enter each kidney or you may notice that the renal artery divides into several branches just medial to the hilus. Occasionally you may find two distinct left renal arteries. You may also find two distinct right renal veins exiting the right kidney. Normally only one renal vein is found on the left.

2➡ Trace the renal arteries and renal vein back to the abdominal aorta and inferior (posterior) vena cava, respectively. At its point of juncture with the aorta, the right

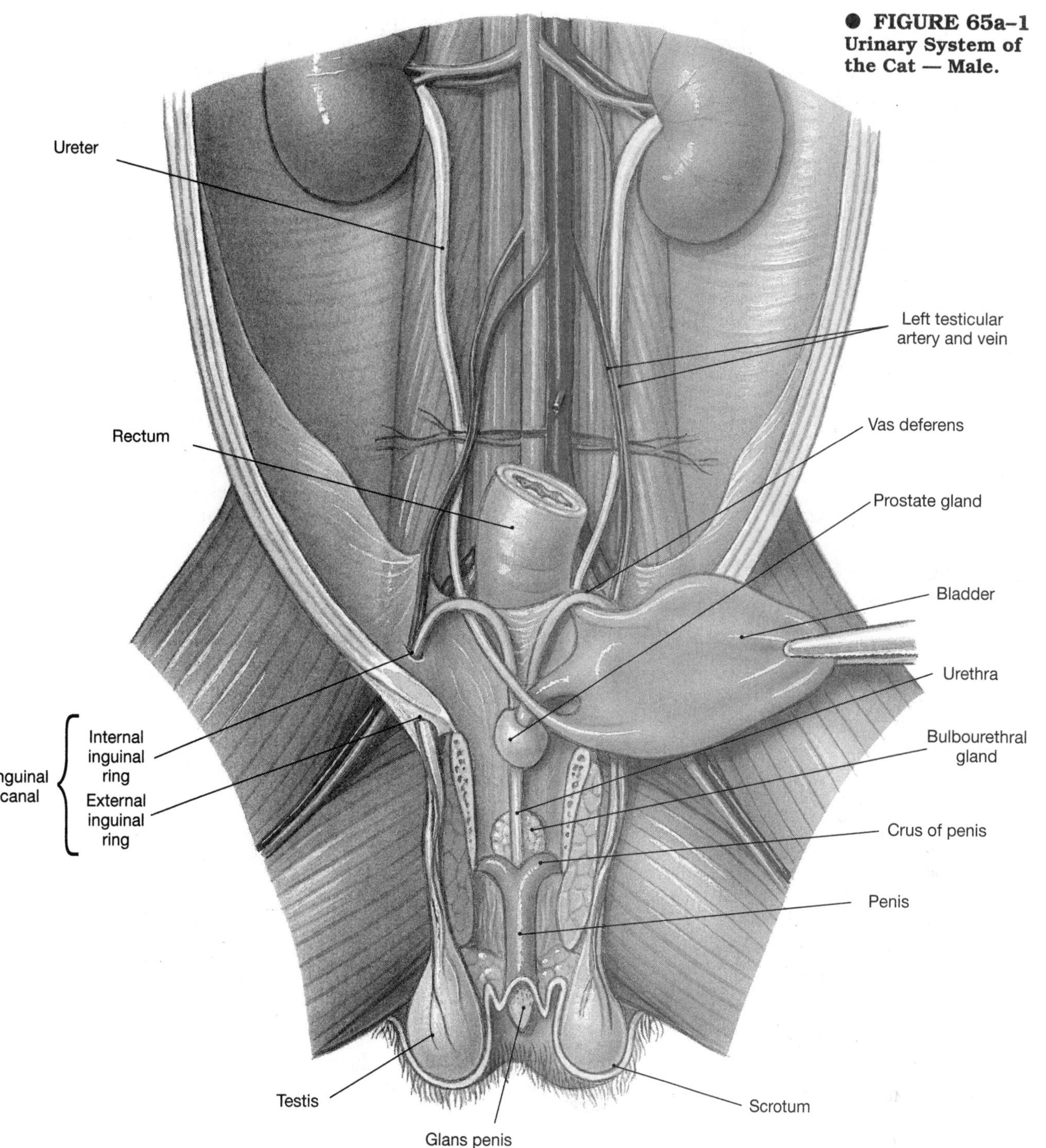

● **FIGURE 65a–1 Urinary System of the Cat — Male.**

renal artery is somewhat cranial to the left. At its point of juncture with the posterior vena cava, the right renal vein is slightly caudal to the left.

D. The Adrenal Blood Supply

1 ➡ Depending on the quality of the latex injections, you may be able to locate the adrenal arteries and adrenal veins. The exact branching of these blood vessels is highly variable. The arteries may enter the adrenal glands directly from the aorta, from a branch off the adrenolumbar artery, or from a branch off the renal artery, the phrenic artery, the celiac artery, or even the superior mesenteric artery. Occasionally more than one arterial branch can be found.

2 ➡ Often, too, you will find more than one adrenal vein. The adrenal veins usually drain directly into the adrenolumbar vein, which then drains into the posterior vena cava. Variations are common.

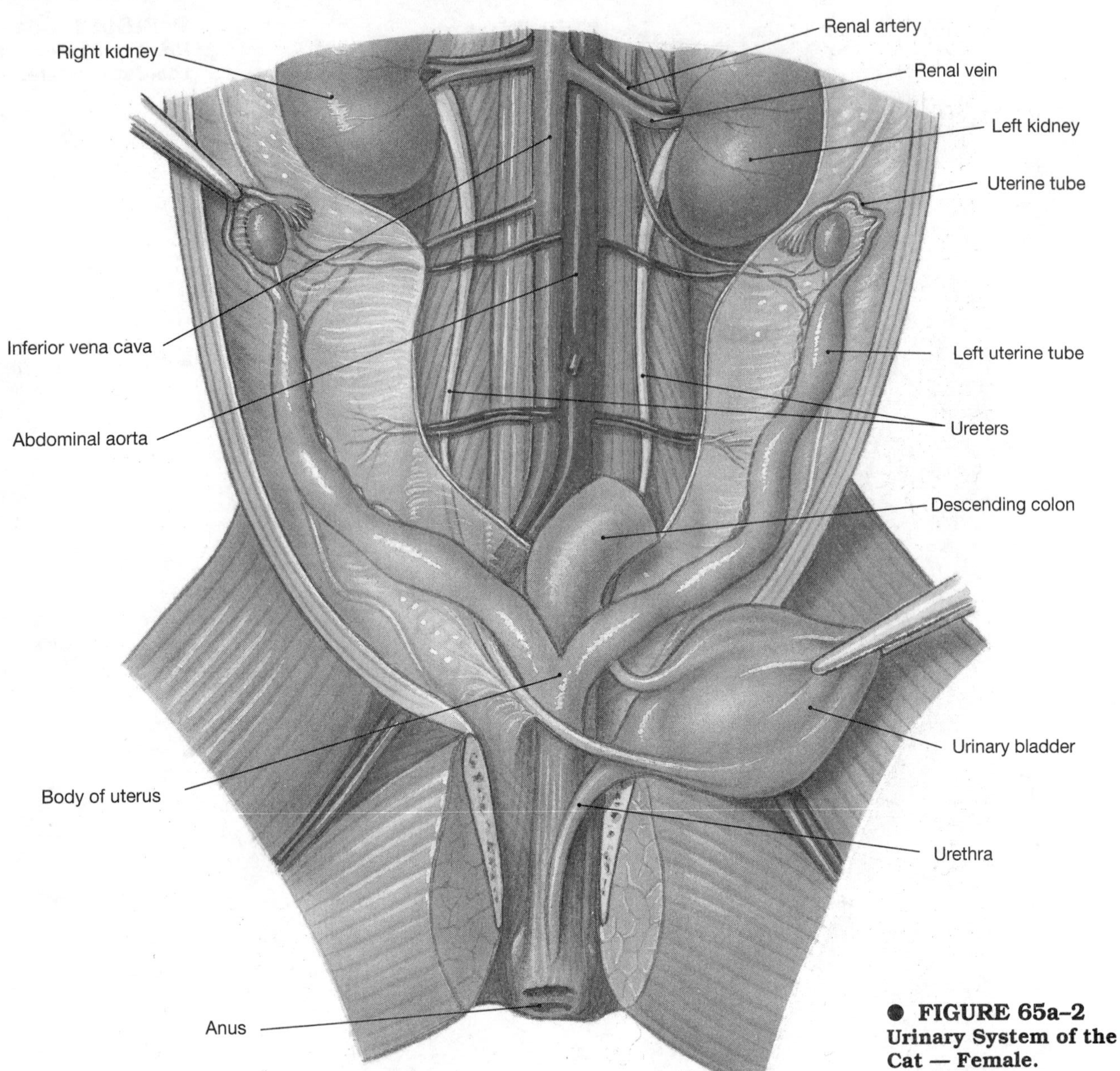

● **FIGURE 65a–2 Urinary System of the Cat — Female.**

3➡ Carefully sever the adrenal blood vessels. Remove one adrenal gland and set it aside for later inspection.

E. THE URETERS

1➡ Locate the ureters, the thin, lightly colored muscular fibrous tubes exiting from the hilus of the kidney. Follow the ureters caudally and note that in the pelvic region the ureters turn ventrally and enter the dorsal aspect of a single, firm, centrally located pouch. This structure is the urinary bladder.

1a➡ If your cat is a female, be sure to trace the ureter directly from the kidney because it is occasionally easy to confuse the ureter with the uterine horn, another thin tube located in the same general area.

1b➡ If your cat is a male, be particularly careful in tracing the ureter because just dorsal to the bladder you will notice a "stringy" structure looping over the ureter. This is the **spermatic cord**, which is sometimes embedded in a great deal of fascia or adipose tissue.

F. THE URINARY BLADDER

1➡ Depending on where you made your initial incision into the abdominal cavity, you may be able to find remnants of the ligament connecting the bladder to the abdominal wall. This is the **median umbilical liga-**

● FIGURE 65a–3
Kidney of the Cat.

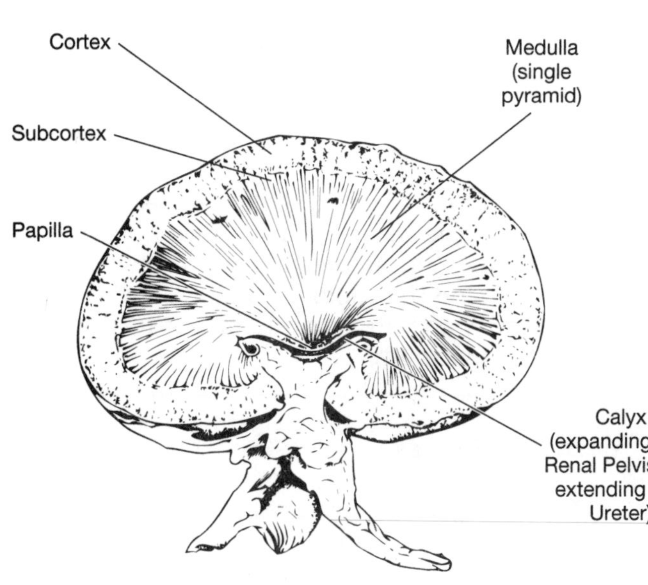

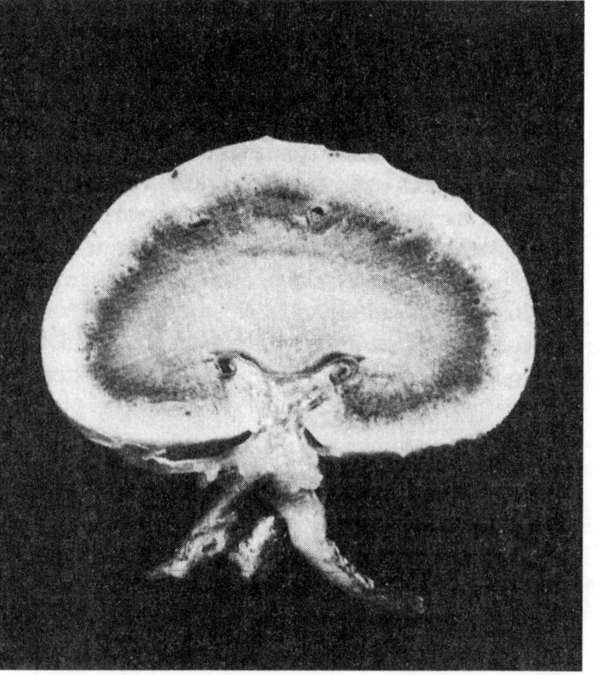

ment. Prior to birth this structure was the **urachus**, a part of the fetal **allantoic bladder**, which extended up into the umbilical cord. After birth the remainder of the allantoic bladder became the urinary bladder.

2➡ Two **lateral ligaments**, which you probably will not be able to identify, hold the sides of the bladder to the dorsal body wall. These are the remnants of the umbilical arteries that delivered blood to the placenta during development.

3➡ Find the caudal exit of the bladder, the **internal urethral orifice**, which marks the beginning of the urethra, the structure for carrying urine to the outside of the body.

4➡ Cut the ventral bladder wall in the area of the internal urethral orifice and see if you can locate the internal sphincter, the smooth-muscled stricture located at the beginning of the urethra. Also, cut a piece of the bladder wall and set it aside for later examination.

5➡ See if you can identify the **trigone**, the imaginary triangle formed by the two **ureteral orifices**, the entry points from the ureters into the bladder, and the internal urethral orifice. Refer back to Exercise 62 and identify the corresponding structures in the human. Did you locate all structures? ____________________

G. UROGENITAL STRUCTURES

1➡ Take your blunt nose probe and *internally* follow the course of the urethra to its terminus or as far as possible without tearing a structure.

2a➡ In the female cat, the urethra follows a short straight line to the **urogenital sinus**. The urogenital sinus is that area in the female animal common to both the urinary tract and the vaginal tract. To differentiate these two areas, insert your probe from the outside, through the **urogenital aperture**, into the urogenital sinus. Slide the probe along the ventral wall. You should soon see the probe in the opening you made in the bladder. Unless your instructor directs you otherwise, do not dissect the urethra at this time.

2b➡ In the male cat, the urethra merges with the two spermatic ducts to form the **urogenital canal**. The urogenital canal often keeps the name urethra. The urethra continues on into the **penis**. The point of exit is the **external urogenital opening**. Extend your probe into the male urethra as far as is possible without ripping any tissue. Do not attempt to insert your probe into the penis. Unless your instructor directs you otherwise, do not dissect the urethra or penis at this time.

3➡ Examine the urinary system of a cat of the opposite sex from the one you worked on.

H. Examination of the Kidney

1➡ Go back to the kidney. Choose one kidney and carefully sever the blood vessels and ureter.

2➡ Section a piece of the ureter and set it and the kidney aside for further examination.

II. Microscopic Dissection

A. The Kidney

1➡ For this part of the exercise use the dissecting microscope or a hand lens.

2➡ Take the cat kidney and carefully bisect it, starting at the lateral border. Do not cut completely through. Follow Figure 65a–3● to identify these structures:

Cortex Outer region of the kidney.

Subcortex Area just beneath the cortex. You may not be able to discern this region.

Medulla Region beneath the subcortex (or cortex, if you cannot distinguish a subcortical region).

Renal Pyramid Dark area in the medulla where the nephronic loops and collecting ducts are found. The cat generally has only one renal pyramid.

Papilla Internal tip of the pyramid.

Renal Pelvis Large smooth area into which the papilla empties and from which the ureter exits. In the cat the renal pelvis actually extends down into the ureter.

Sketch and label what you see.

Kidney of the cat

3➡ Note the branching pattern of the blood vessels entering the kidney. Compare your internal observation of the blood vessels with the blood vessels as you observed them on your external examination. Return to Exercise 63, Section I.C. How many of the blood vessels can you identify on your feline kidney?

____

____

4➡ After you have gained a perspective, complete the bisection of the kidney and examine the structures more closely. Try popping the renal pyramid out of the kidney. Can you separate any of the parts of the pyramid?

____

B. The Ureter

1➡ Examine the ureter tissue. Try to determine any outstanding characteristics of the tissue.

2➡ Put your probe into the ureteral lumen and, with your scalpel, slice the wall of the ureter. Continue your examination. Sketch and label what you see.

Ureter

C. The Bladder

1➡ Examine the bladder tissue. Can you detect the heavy muscular wall of the bladder? Try examining the dorsal, medial, and ventral aspects of the bladder slice. What similarities and differences do you notice?

____

Sketch and label what you see.

Urinary bladder

D. The Adrenal Gland

1 ➡ Place the adrenal gland under the microscope. Hold the gland with your forceps in one hand and your scalpel in the other. Try to bisect the adrenal gland. Once you have bisected the gland, observe the inner portions. Can you differentiate between the cortex (outer area) and the medulla (inner area)? ______________________________

Sketch and label what you see.

Adrenal gland

III. Prepared Slides

If you did not examine the histological slides described in Exercise 62, Section V, your instructor may direct you to do so now. Use the information in that exercise as your guide.

❑ Lab Report

1. Referring to the numbered questions at the beginning of this exercise, complete the following box summary:

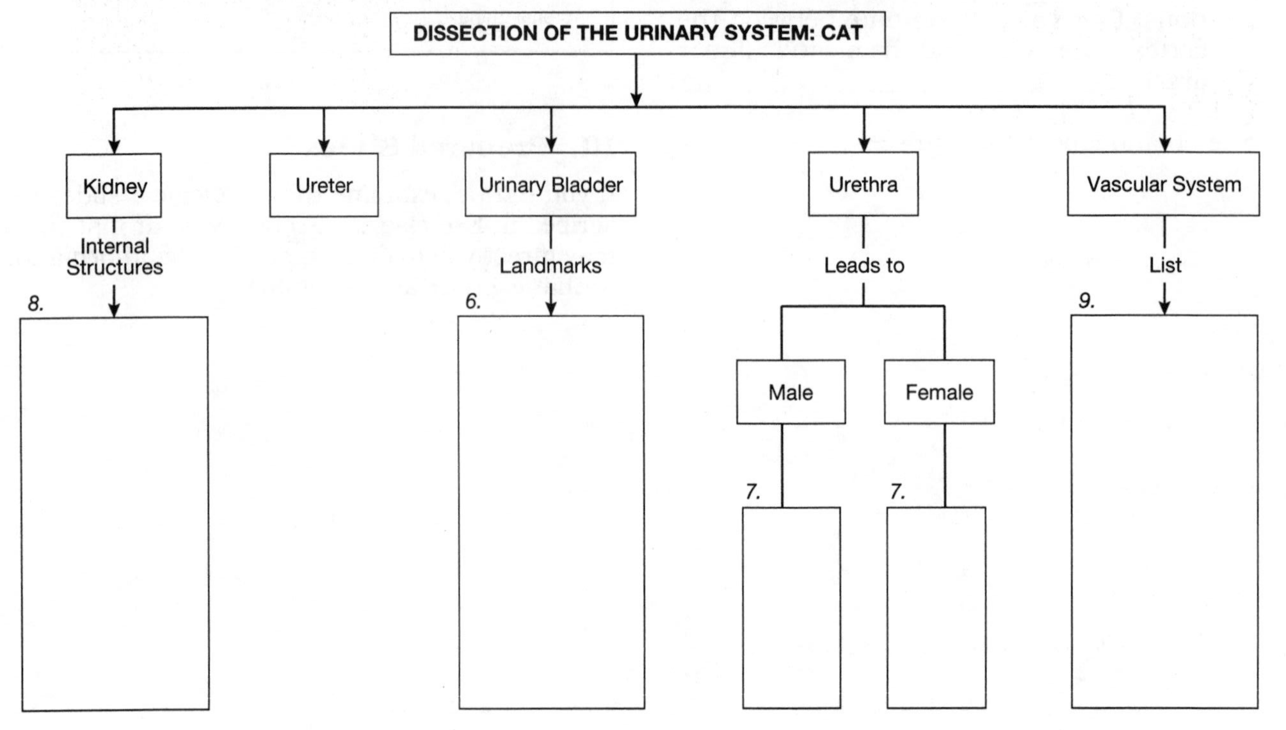

General Questions

NOTES

1. How should you begin your urinary tract dissection?
2. What similarities and differences did you find between the human and feline urinary systems?
3. Trace the path of a drop of urine from the time the original filtrate is produced until the urine leaves the body.

EXERCISE **65b**

Dissection of the Urinary System: Fetal Pig

PROCEDURAL INQUIRIES

Preparation

1. What are the major organs of the urinary system?
2. How should the fetal pig be prepared for dissection of the urinary system?

Dissection

3. Why do we say the kidneys are retroperitoneal?
4. Where are the fetal pig's adrenal glands located in relation to the kidneys?
5. Where are the ureters?
6. What are the major structural landmarks of the urinary bladder?
7. What are the anatomical differences between the urinary systems of the male and female fetal pig?
8. What are the principal structures of the fetal pig's kidney?
9. What are the major blood vessels of the kidney?

Additional Inquiries

10. How does the urinary system of the fetal pig relate to the urinary system of the human?

HUMAN CORRELATION: AM Cadaver Atlas in Applications Manual.

Key Terms

Adrenal Glands
Allantoic Bladder
Cortex
Genital Papilla
Hilus
Internal Urethral Orifice
Kidney
Lateral Border
Lateral Ligaments
Medial Border
Median Umbilical Ligament
Medulla
Papilla
Renal Pelvis
Renal Pyramid
Spermatic Cord
Trigone
Urachus
Ureter
Ureteral Orifice
Urethra
Urinary Bladder
Urogenital System

Materials Needed

Fetal pig (double-injected)
Dissecting Tray
Dissecting Kit
Gloves
Dissecting Microscope
Prepared Histological Slides
  Kidney
  Bladder
Compound Microscope

All functions of the urinary system center around homeostatic balance within the organism. The first major function of the urinary system is the removal and excretion of waste products from the circulatory system. The second function, highly correlated to the first, is the maintenance of fluid and electrolyte balance within the organism. The organ responsible for the accomplishment of these purposes is the **kidney**. The other structures of the urinary system are accessories to the kidney.

In the mammal, paired kidneys are located in the dorsal region of the abdominopelvic cavity, generally in the superior lumbar region. In the adult pig (porcine animal), the kidneys are usually ventral to the transverse processes of the first four lumbar vertebrae. In the fetal pig, the kidneys may not yet have achieved their adult location.

Exiting from each kidney is a tube known as a **ureter**, a tube composed of a fibrous outer coat, two smooth muscle layers, and an inner mucous membrane lining. The ureters enter a common holding area called the **urinary bladder**. Caudal to the urinary bladder is a single tube, the **urethra**, which exits the body in one of several ways, depending on the sex and species of the mammal.

Because of a common embryological origin, the urinary and reproductive systems of many mammals are often considered together under the combined heading of **urogenital system**. It is important for us to realize, though, that in the postembryonic animal, the urinary and reproductive functions are separate, although some overlap in structure does exist. In the fetal pig, the overlap in structure is more pronounced in the male than in the female.

In this laboratory exercise we will study the anatomy of the urinary system by dissecting the urinary system of the fetal pig. Dissecting this system will afford us a unique opportunity to examine the basic urinary structures as well as the prebirth mammalian urinary system.

❑ Preparation

I. Gross Dissection

A. Orientation

1➡ To begin your dissection of the urinary system of the fetal pig, prepare your specimen by placing the animal dorsal side down and securing the limbs to the dissecting tray. Open the abdominal cavity and reflect the visceral organs away from the midline of the body.

2➡ Follow the diagrams carefully and be aware of the similarities and differences between the male and female fetal pig as well as between the fetal pig and the human. (Follow Figures 65b–1● and 65b–2●.) Refer to Exercise 62 for the human system.

Dissection

B. Examination of the Kidney

1➡ Locate the kidneys. Note that these organs are retroperitoneal (behind the peritoneum). The peritoneum will be a thin, filmy sheet of transparent tissue. The ureters are also retroperitoneal, although you may have difficulty seeing this if you have already dissected the digestive system of your fetal pig.

2➡ Carefully remove the peritoneum. Before dislodging the kidneys, observe that these paired structures are approximately even with each other. In the adult porcine, however, the left kidney is often somewhat cranial to the right.

3➡ Notice that the kidney is embedded in adipose tissue, nature's cushion against the jars and jolts of everyday life. Carefully clean away this fatty material.

4➡ Do not mistake for fat the small glandular pieces of tissue located on the superior-medial (anteromedial) border of the kidney. These are the **adrenal glands**. (In the human the adrenal glands are located directly on the superior surface of the kidney. In cats the adrenal gland is not usually in direct contact with the kidney.)

The adipose tissue is probably a yellowish white, and the adrenal glands probably have a red-orange tinge. Also, the adrenal glands will feel firmer than the "mushy" adipose tissue. The adrenal glands will also have a distinct circulatory supply, which you may or may not be able to discern.

5➡ Identify the **lateral** and **medial borders** of the kidney. Also, identify the **hilus**, the indented region on the medial border where you will find the blood vessels and the ureters.

C. Major Blood Vessels Supplying the Kidneys

1➡ Identify the renal arteries (pink latex injection) and the renal veins (blue latex injection). A single renal artery may enter each kidney or you may notice that the renal artery divides into several branches just medial to the hilus. You may find two distinct renal veins exiting the left kidney. Normally only one renal vein is found on the right. (This is the opposite of what you might find in the cat.)

2➡ Trace the renal blood vessels back to the abdominal aorta and inferior (posterior) vena cava. At its point of juncture with the aorta, the right renal artery is somewhat cranial to the left. At its point of juncture with the posterior vena cava, the right renal vein is slightly caudal to the left.

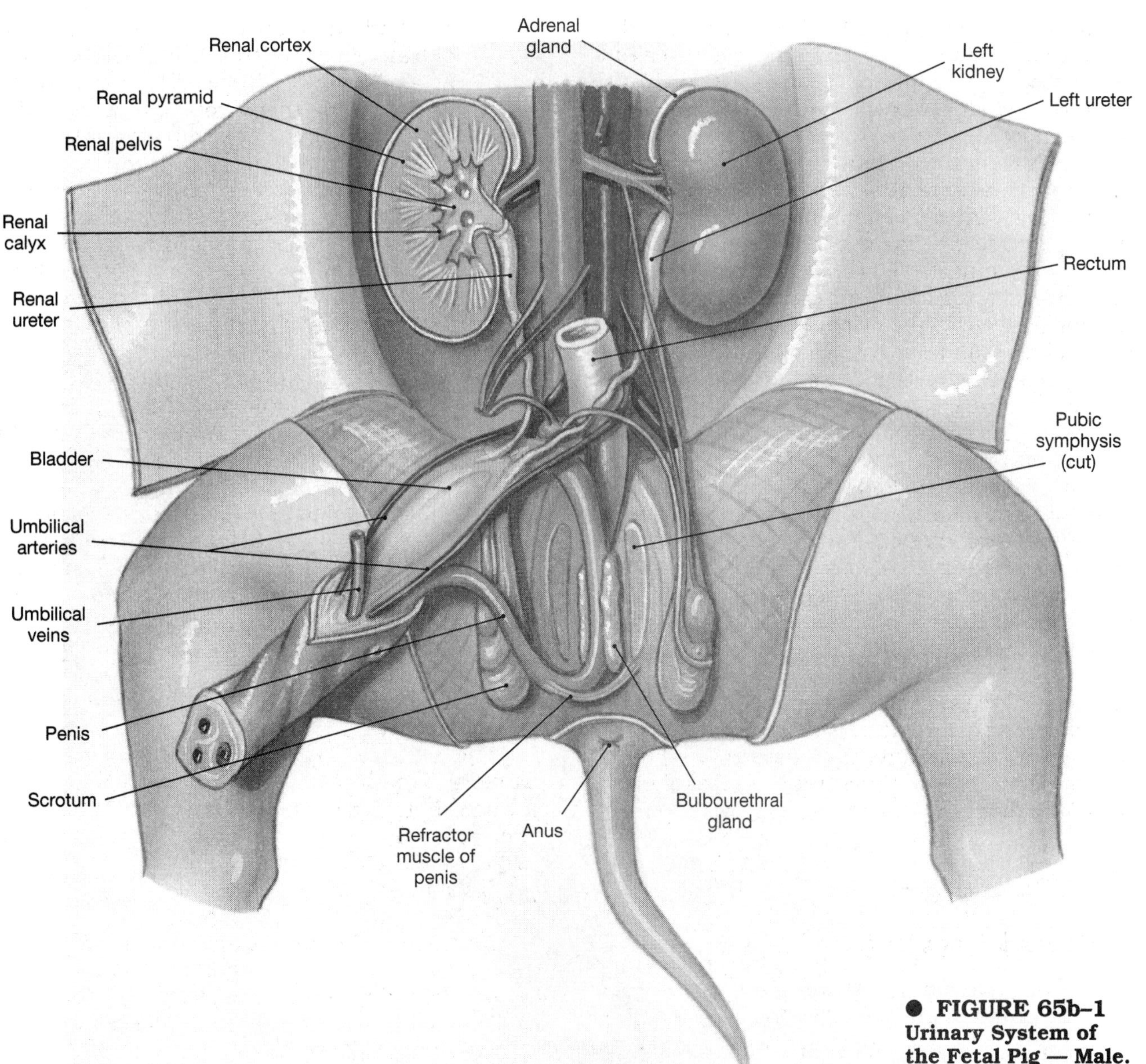

● **FIGURE 65b–1**
Urinary System of the Fetal Pig — Male.

D. The Adrenal Blood Supply

1➡ Depending on the age of your fetal pig and the quality of the latex injections, you may be able to locate the adrenal arteries and adrenal veins. The exact branching of these blood vessels is highly variable. The arteries may enter the adrenal glands directly from the aorta, from a branch off the adrenolumbar artery, or from a branch off the renal artery. Occasionally more than one arterial branch can be found.

2➡ Often, too, you will find more than one adrenal vein. The right adrenal veins usually drain directly into the posterior vena cava, and the left adrenal veins usually drain into the left renal vein. Variations, however, are common.

3➡ Carefully sever the adrenal blood vessels. Remove one adrenal gland and set it aside for later inspection.

E. The Ureters

1➡ Locate the ureters, the thin, lightly colored fibrous tubes exiting from the hilus of the kidney. Follow the ureters caudally and note that in the pelvic region the ureters turn ventrally and enter the dorsal aspect of a single, firm, centrally located pouch. This seemingly disproportionate structure is the **allantoic bladder**, or the fetal urinary bladder.

1a➡ If your fetal pig is a female, be sure to trace the ureter directly from the kidney because it is occasionally easy to confuse the ureter

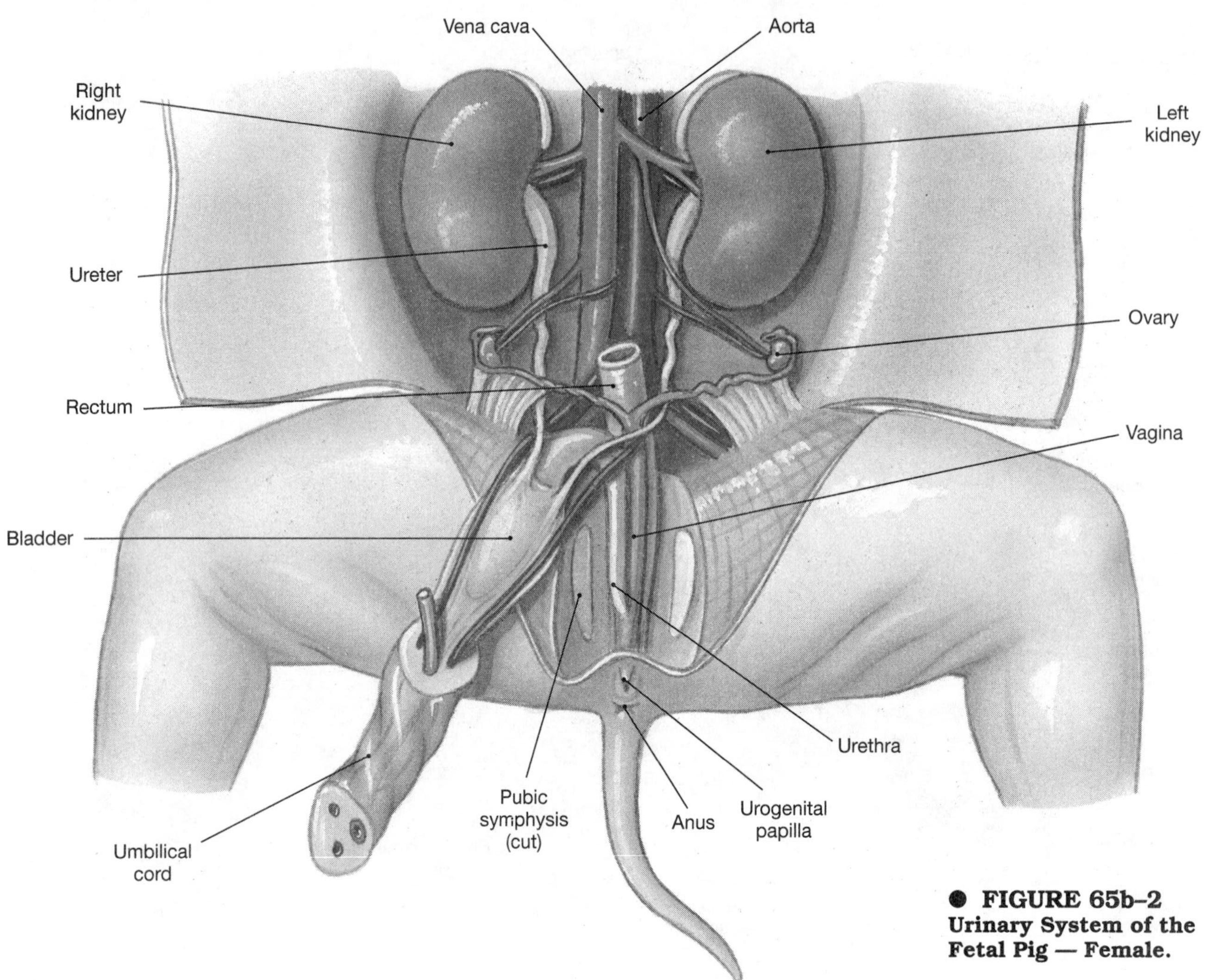

● **FIGURE 65b–2 Urinary System of the Fetal Pig — Female.**

with the uterine horn, another thin tube located in the same general area.

1b➡ If your fetal pig is a male, be particularly careful in tracing the ureter because just dorsal to the bladder you will notice a "stringy" structure looping over the ureter. This is the **spermatic cord**, which is sometimes embedded in fascia or adipose tissue.

F. The Allantoic Bladder

1➡ Follow the alantoic bladder ventrally to the point where it seems to enter the abdominal wall and continue on into the umbilical cord. This ventral portion of the allantoic bladder is known as the **urachus**. In the umbilical cord the allantoic bladder becomes the allantoic stalk. (In mammals the allantois, one of the extra-embryonic membranes, is generally small and nonfunctional as a membrane. Nevertheless, the allantois does produce early blood cells and does give rise to the umbilical cord, thus supplying blood vessels to the placenta.)

After birth the allantoic bladder becomes the urinary bladder. The urachus ceases functioning and becomes the **median umbilical ligament**. In the postbirth mammal (including the human), this ligament attaches the urinary bladder to the ventral abdominal wall. Return to Figure 62–4● and note the position of the urachus in the human.

2➡ Trace the allantoic bladder caudally to the point where it narrows. From this point until it exits the body, this tube is known as the urethra.

3➡ The caudal exit of the bladder, the **internal urethral orifice**, marks the beginning of the urethra, the structure for carrying urine to the outside of the body.

4➡ Cut the ventral bladder wall in the area of the internal urethral orifice and see if you

● **FIGURE 65b–3**
Mammalian Kidney.

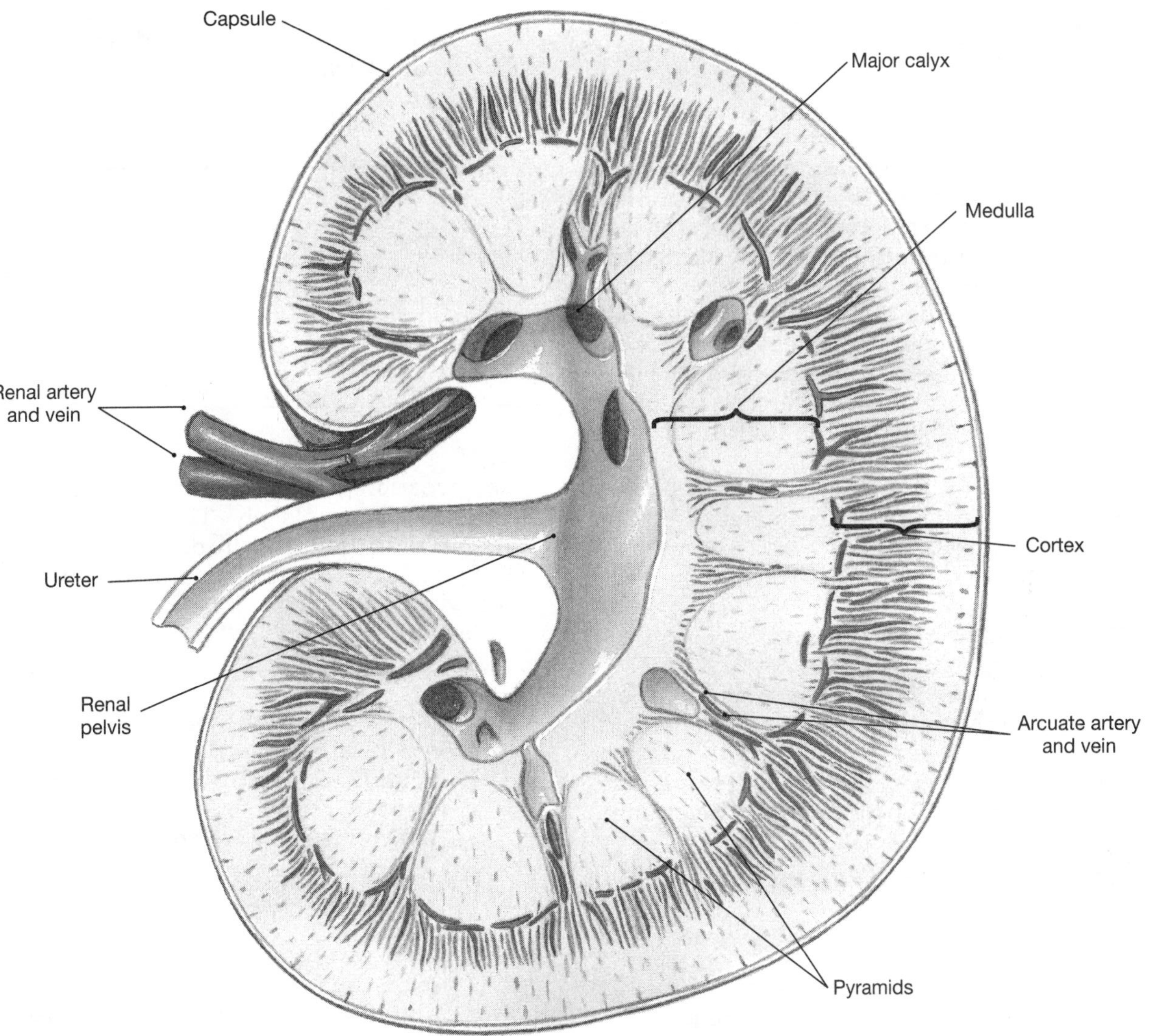

can locate the internal sphincter, the smooth-muscled stricture located at the beginning of the urethra. Also, cut a piece of the bladder wall and set it aside for later examination.

5. See if you can identify the **trigone**, the imaginary triangle formed by the two **ureteral orifices**, the entry points from the ureters into the bladder, and the internal urethral orifice. Refer back to Exercise 62 to identify the corresponding structures in the human. From the human description, can you locate the external urethral sphincter in the fetal pig? ____________

6. Identify the umbilical arteries, which are located on either side of the bladder. After birth these artieries, which deliver blood to the placenta during development, will atrophy and form the **lateral ligaments**.

G. Urogenital Structures

1. Take your blunt nose probe and *internally* follow the course of the urethra to its terminus or as far as possible without tearing a structure.

2a. In the female fetal pig, the urethra follows a short straight line to the **urogenital sinus** (about 7 or 8 cm in the adult animal). The urogenital sinus is that area in the female animal common to both the urinary tract and the vaginal tract. To differentiate these two areas, insert your probe from the outside, through the **genital papilla**, into the urogenital sinus. Slide the probe along the ventral wall. You should soon see the probe in the opening you made in the bladder. Unless your instructor directs you otherwise, do not dissect the urethra at this time.

2b➡ In the male fetal pig, the urethra loops down low in the pelvic cavity and then turns upward and toward the ventral wall. Here it becomes the penis. The point of exit, the **external urogenital opening** (also called the **preputial orifice**), is located just below the umbilical cord. Extend your probe into the male urethra as far as is possible without ripping any tissue. Do not attempt to insert your probe into the penis. Unless your instructor directs you otherwise, do not dissect the urethra or penis at this time.

3➡ Examine the urinary system of a fetal pig of the opposite sex from the one you worked on.

H. Examination of the Kidney

1➡ Go back to the kidney. Choose one kidney and carefully sever the blood vessels and ureter.

2➡ Section a piece of the ureter and set it and the kidney aside for further examination.

II. Microscopic Dissection

A. The Kidney

1➡ For this part of the exercise use the dissecting microscope or a hand lens.

2➡ Take the fetal pig kidney and carefully bisect it, starting at the lateral border. Do not cut completely through. Follow Figure 65b–3● to identify these structures:

Cortex Outer region of the kidney.

Medulla Region beneath the subcortex (or cortex, if you cannot distinguish a subcortical region).

Renal Pyramid Dark triangular areas in the medulla where the nephronic loops and collecting ducts are found. (Usually about 6 to 18 pyramids can be found in the fetal pig.)

Papilla Internal tip of the pyramid.

Medullary Ray (Renal Column) Area between the pyramids.

Minor Calyx Indented area surrounding the papilla.

Major Calyx Conglomerate of several minor calyces.

Renal Pelvis Large smooth area into which the papilla empties and from which the ureter exits.

Sketch and label what you see.

Kidney of the fetal pig

After you have gained a perspective, complete the bisection of the kidney and examine the structures more closely. The extent of your examination will depend in part on the age of the fetal pig. Try popping a renal pyramid out of the kidney. Can you separate any of the parts of the kidney?

____

3➡ Depending on the age of your pig and the quality of the injection, you may or may not be able to identify renal blood vessels. If you can, return to Exercise 63, Section I.C. Which vessels were you able to identify?

____

____

4➡ After you have gained a perspective, complete the bisection of the kidney and examine the structures more closely. Try popping the renal pyramid out of the kidney. Can you separate any of the parts of the pyramid? ______________

____

____

B. The Ureter

1➡ Examine the ureter tissue. Try to determine any outstanding characteristics of the tissue.

2➡ Put your probe into the ureteral lumen and, with your scalpel, slit open the ureter. Look at this opened tissue microscopically. How much you will be able to see will again depend on the age of your fetal pig. Nevertheless, try to determine any outstanding characteristics of the tissue.

Sketch and label what you see.

Ureter

C. The Bladder

1 ➡ Examine the bladder tissue. Can you detect the heavy muscular wall of the bladder? Although the prenatal bladder holds only very dilute urine, and that for only short periods of time, the postnatal bladder is much more specialized. What early indications of this more complicated function do you see in the fetal pig? ________

____

Try examining the dorsal, medial, and ventral aspects of the bladder slice. What similarities and differences do you notice? ____________________

____

Sketch and label what you see.

D. The Adrenal Gland

1 ➡ Place the adrenal gland under the microscope. Hold the gland with your forceps in one hand and your scalpel in the other. Try to bisect the adrenal gland. (This may not be possible with the adrenal gland of the very young fetal pig.) Once you have bisected the gland, observe the inner portions. Can you differentiate between the cortex (outer area) and the medulla (inner area)? ____________________

Sketch and label what you see.

Adrenal gland

III. Prepared Slides

If you did not examine the histological slides described in Exercise 62, Section V, your instructor may direct you to do so now. Use the information in that exercise as your guide.

❑ Lab Report

1. Referring to the numbered questions at the beginning of this exercise, complete the following box summary:

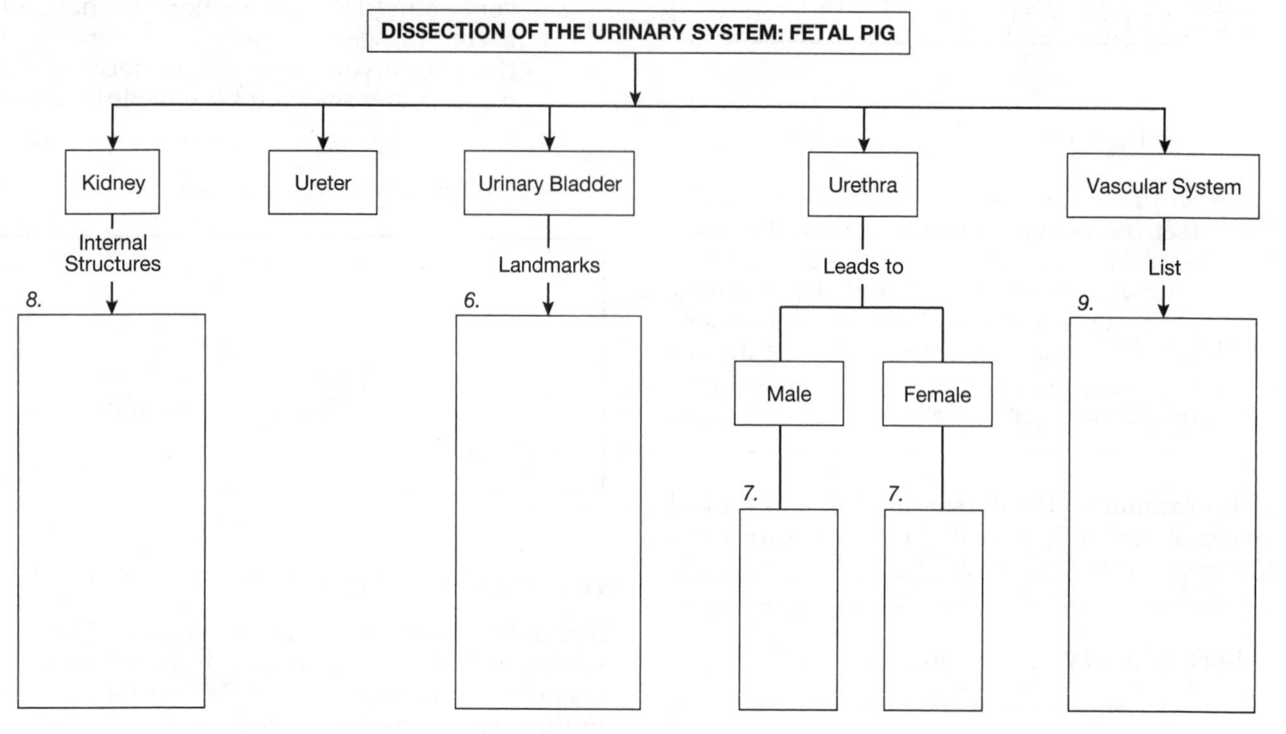

General Questions

NOTES

1. How should you begin your urinary tract dissection?
2. What similarities and differences did you find between the human and porcine urinary systems?
3. Trace the path of a drop of urine from the time the original filtrate is produced until the urine leaves the body.

EXERCISE 66

Anatomy of the Male Reproductive System

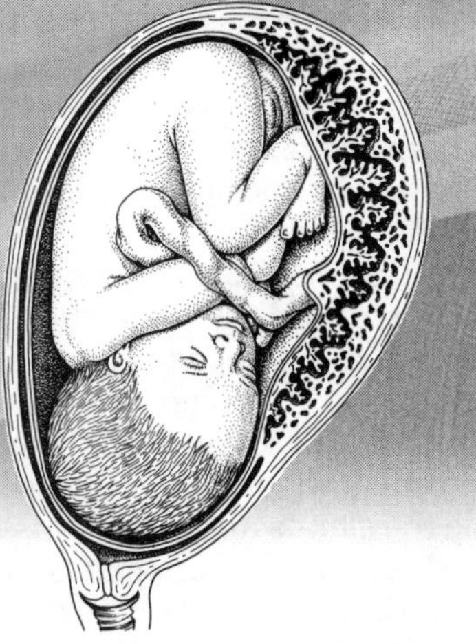

PROCEDURAL INQUIRIES

Preparation

1. What are the overall functions of the male reproductive system?

Examination

2. What are the major organs of the male reproductive system?

3. What are the structures located within the scrotum?

4. What are the structural components of the testis?

5. Where is the epididymis?

6. What is the path of the sperm from the testis to the penis?

7. What are the divisions of the urethra?

8. What are the glands of the male reproductive system?

9. What are the internal structures of the penis?

10. What are the external structures of the penis?

11. What physiological processes are necessary for ejaculation to occur?

12. How is erection of the penis maintained?

Additional Inquiries

13. What is the function of the reproductive accessory structures?

14. What is the function of the sustentacular cells?

15. What is the function of the interstitial cells?

Key Terms

Ampulla
Bulbocavernosus Muscle
Bulbourethral (Cowper's) Glands
Corpora Cavernosa
Corpus Spongiosum
Ductus Deferens
Efferent Duct
Epididymis
External Urethral Meatus
Glans Penis
Inguinal Canal
Interstitial Cells
Ischiocavernosus Muscles
Os Penis
Prepuce
Prostate Gland
Rete Testes
Semen
Seminal Vesicle
Seminiferous Tubules
Sperm
Spermatazoa
Spermatic Cord
Spermatid
Spermatocyte
Spermatogonia
Straight Tubules
Sustentacular Cells
Testis
Tunica Albuginea
Tunica Vaginalis
Urethra
Vas Deferens

Materials Needed

Torso
Organ Models
Charts and Diagrams
Dissecting Microscope
Compound Microscope
Prepared Slides (as available)
- Penis
- Testes
- Epididymis
- Ductus Deferens
- Male Accessory Glands
- Spermatozoa

The reproductive system is a complex of organs and accessory structures whose hormonally induced function is the propagation of the species.

In the human male, the reproductive structures are geared toward the production and maintenance of sperm and the deposition of that sperm into the female vagina. The major organs of the male reproductive system are the testes, the ductus deferens, the urethra, and the penis. The accessory structures include the glands and support tissues that facilitate the production and transport of the sperm.

In addition, certain reproductive structures have endocrine components which operate not only within the reproductive system but also within the body as a whole.

In this laboratory exercise, we will examine the major reproductive organs and their corollating accessory structures.

❑ Preparation

I. Introduction

A. Systematic Procedure

1➡ Use the available models and charts to gain a perspective of the structures printed in boldface. If you have models that can be taken apart, take them apart and put them back together again.Work toward conceptualizing the three-dimensional picture and the general organization of the reproductive system.

If you do not have classroom models, pay particular attention to any available diagrams and charts to help you gain a perspective on the system as a whole.

2➡ Refer to the textbook in places marked with Concept Links.

B. Labeling

1➡ As you use the descriptions to label the figures associated with each section of this exercise, cross-reference as often as necessary.

❑ Examination

II. External Reproductive Structures

These descriptions should be used to label Figures 66–1• and 66–2•.

A. Scrotal Structures

1➡ Begin your examination of the reproductive system by studying the external genitalia. Locate the skin-covered sac, the **scrotum**. Beneath the tough skin and connective tissue of the scrotum you will find the paired **testes**. The left testis usually hangs somewhat lower than the right testis because the left spermatic cord is usually somewhat longer than the right.

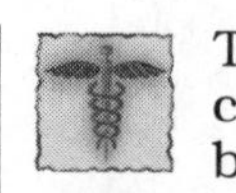

The testes are held outside the main body cavity because the temperature within the body is too high for the survival of healthy sperm. Men who take excessive numbers of very hot baths or very hot saunas and men who wear underwear that holds the testes too close to the body for excessive periods of time tend to have decreased viable sperm counts. (This condition is reversible over time upon lowering the testicular temperature.)

2➡ If you were to remove the scrotal sac, you would see that the entire testicular structure is covered with a layer of serous membrane known as the **tunica vaginalis**. The tunica vaginalis is composed of two layers, the external **parietal layer** and the internal **visceral layer**.

3➡ Removing the tunica vaginalis will expose a convoluted duct on the posterior surface of the testis. This is the **epididymis**. Notice how the epididymis arises from inside the testis.

B. Testes [∞ Martini, p. 1017]

1➡ Examine the individual testis. You will see the testis itself is surrounded by another coat, the **tunica albuginea**. Fibrous extensions of the tunica albuginea dip into the testis separating it into incomplete **lobules**. Each lobule contains **seminiferous tubules**. Notice the coiled nature of these tubules.

C. Sperm Production [∞ Martini, p. 1021]

1➡ Look closely at the diagrams. You can find sperm stem cells, called **spermatogonia**, toward the exterior of each seminiferous tubule. Also in this area of the tubule are the **sustentacular** (or **Sertoli**) **cells**, which supply nutrients to the developing sperm and form the blood-testis barrier. The sustentacular cells also produce the hormones inhibin, ABP (androgen-binding hormone), and MIF (Müllerian-inhibiting factor).

2➡ Notice that the developing sperm cells do not all look the same. This is because the sperm are produced in successive stages beginning in the outer regions of the seminiferous tubule.

The spermatogonia undergo continuous mitosis. One daughter cell remains as an undifferen-

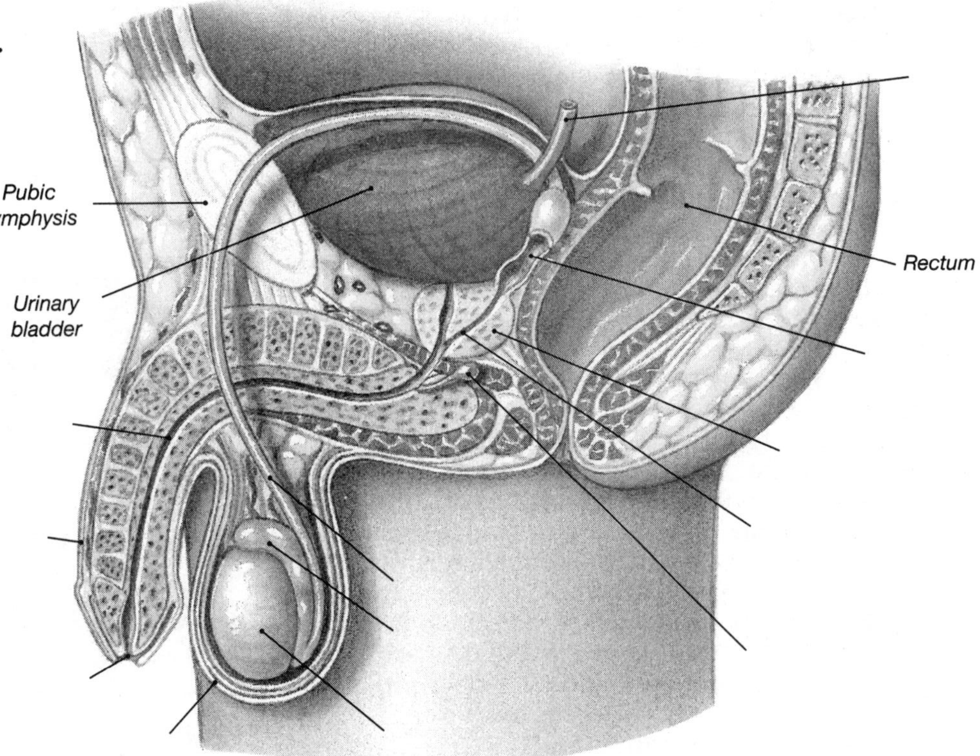

● **FIGURE 66–1**
Male Reproductive System.

tiated stem cell while the other differentiates to become a **primary spermatocyte**. This primary spermatocyte goes through **meiosis I**, producing a pair of **secondary spermatocytes. Meiosis II** then produces four **spermatids.** These spermatids develop tails and become functional **spermatazoa**. (Refer also to Exercises 68 and 70.)

When fully formed (although still immature) the sperm enter the lumen of the tubule.

3➡ Between the seminiferous tubules, notice the **interstitial** (or **Leydig**) **cells**. These specialized endocrine cells produce the male sex hormone **testosterone**.

Testosterone promotes the functional maturation of the spermatozoa, maintains the male reproductive accessory glands, determines the male secondary sexual characteristics, stimulates metabolism and muscle protein synthesis, and stimulates brain development to program certain sexual behaviors.

E. Path of the Sperm

1➡ Use the following descriptions to trace the route sperm must follow upon leaving the seminiferous tubules. Sperm leave the seminiferous tubules and enter first the **straight tubules** and then a mesh-like network known as the **rete testis**. The rete testis empties into a series of small ducts known as the **efferent ducts**, or the *ductus efferens*. These ducts unite to form the head of the epididymis, a structure you have already located.

2➡ Continue to follow the epididymis from its head at the cranial end of the testis to its tail located at the base of the testis. At the base of the testis, the epididymis turns cranially and becomes the **ductus deferens,** which is part of the **spermatic cord**. Within the spermatic cord are the ductus deferens, also known as the **vas deferens**, assorted blood vessels, and branches of the pelvic nerve. The outer covering of the spermatic cord is continuous with the tunica vaginalis.

3➡ Trace the spermatic cord up into the inguinal area, through the **inguinal canal**, and into the abdominopelvic cavity. We will continue this examination from the internal perspective.

III. Internal Reproductive Structures

These descriptions should be used to label Figure 66–3●.

A. Internal Orientation

1➡ Orient yourself to the internal structures of the male reproductive system by first locating the urinary bladder. Identify the ureter, which extends from the kidney to the bladder. Notice that the ureters enter the posterior caudal portion of the bladder.

2➡ Identify the tubular structure that loops over the ureter. This is the ductus deferens (vas deferens), which you have already identified. Trace the ductus deferens back into the spermatic cord and back through the inguinal canal. Notice that only the ductus deferens continues cranially from the spermatic cord. (The fates of the blood vessels and nerves of the spermatic cord can be traced in the exercises on the human cardiovascular system and the spinal nerves respectively.)

B. Glands of the Reproductive System
[∞ Martini, p. 1028]

1➡ Follow the ductus deferens after it loops over the ureter and continues medially toward the urethra. In humans (and in many other mammals) you will find an enlarged area, the **ampulla**, at the end of the ductus deferens immediately before it enters the urethra. In men also you will find a **seminal vesicle (vesicular gland)**, which empties into the ductus deferens (vas deferens) just before the urethral junction.

(Neither the cat nor the fetal pig has an ampulla and the cat does not have a seminal vesicle. In the pig, as well as in certain other mammals, the vesicular glands are located after the urethral junction.)

Semen is the entirety of the male reproductive fluid; semen includes the sperm and all the glandular secretions. The seminal vesicle secretions contribute about 60% of the semen volume. These secretions are high in fructose, which is easily metabolized by the sperm. Metabolizing this fructose gives added motility and agility to the flagellated sperm.

2➡ Notice the round glandular structure surrounded by a thin fibrous capsule located at the base of the urinary bladder somewhat inferior to the level of the symphysis pubis at the junction of the ductus deferens and the urethra. This is the **prostate gland**, which is basically divided into two lateral lobes and a small middle lobe. These lobes form a ring around the urethra. The prostate gland produces alkaline secretions contributing about 30% to the volume of the

● **FIGURE 66–2a**
Testicular Structure.

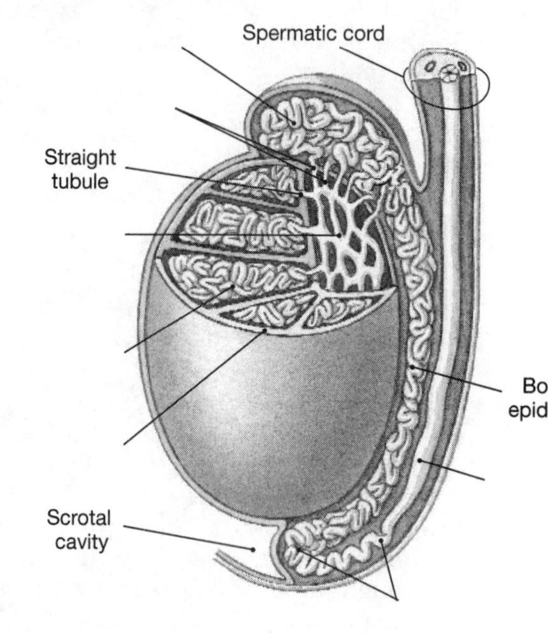

● **FIGURE 66–2b**
Seminiferous Tubule Cross Section.

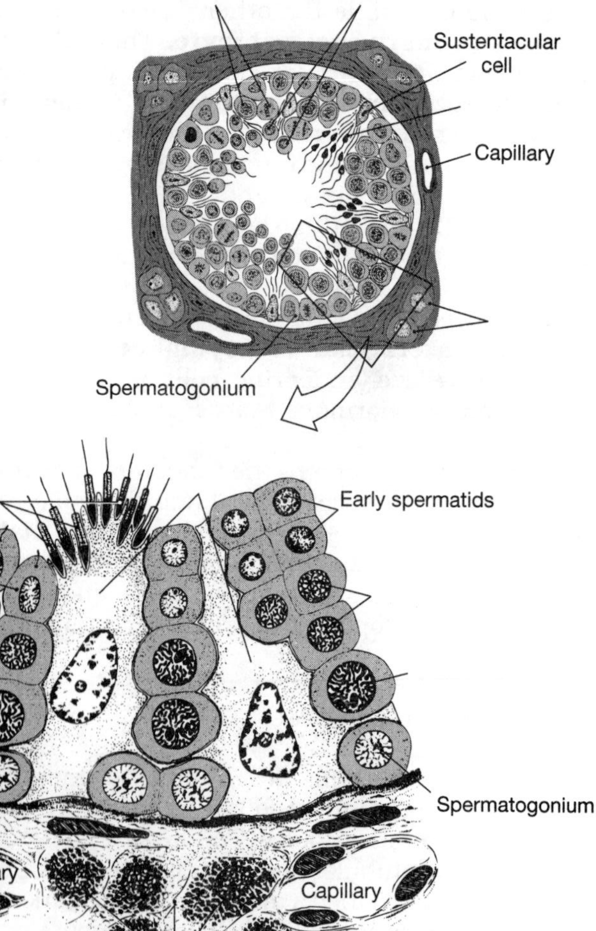

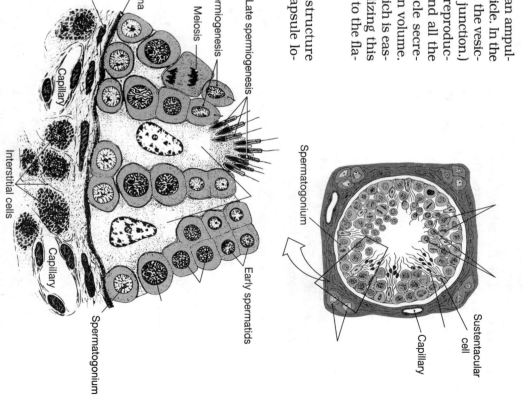

semen. Prostate gland size is highly variable and in many animals, including men, an enlarged prostate gland is not uncommon. In this area the urethra is known as the **prostatic urethra**.

Concept Check 1 What types of problems might an enlarged prostate gland cause?

____

____

____

____

3➡ Follow the urethra caudally. Here the urethra is known as the **membranous urethra**. Near the center of the membranous urethra you should be able to find the small paired **bulbourethral glands**, also known as **Cowper's glands**. These glands secrete a thick lubricating mucus with distinct alkaline properties. As the urethra passes through the **urogenital diaphragm**, it enters the **penis** and becomes known as the **penile urethra** (or **cavernous urethra**).

C. INTERNAL STRUCTURES OF THE PENIS
[∞ MARTINI, P. 1030]

1➡ Examine the internal structures of the penis. The shaft of the penis is composed of three masses of highly vascular erectile tissue. The paired cylindrical **corpora cavernosa** extend from the ischium and pubis of the pelvic girdle to the narrow **neck** of the penis. Each corpus cavernosum surrounds a central artery. (Note: corpora is the plural of corpus.)

2➡ Notice that the urethra is surrounded by additional erectile tissue, the **corpus spongiosum**, which extends from the urogenital diaphragm to the tip of the penis, where it expands to form the **glans penis**.

3➡ Find the dense network of elastin fibers internal to the skin of the penis and the internal layer of loose connective tissue. This dense collagenous sheath surrounds each of the corpora.

D. EXTERNAL STRUCTURES OF THE PENIS
[∞ MARTINI, P. 1030]

1➡ Examine the external structures of the penis. The **prepuce**, or foreskin, is the fold of skin that attaches to the penile neck and continues over the expanded end of the penis at the glans (or glans penis).

The urethral opening is also known as the **external urethral meatus** or the **urogenital opening**.

IV. Sexual Physiology

The physiological process of ejecting the semen from the male reproductive tract is usually considered in three stages: erection, emission, and ejaculation. We will consider each stage separately.

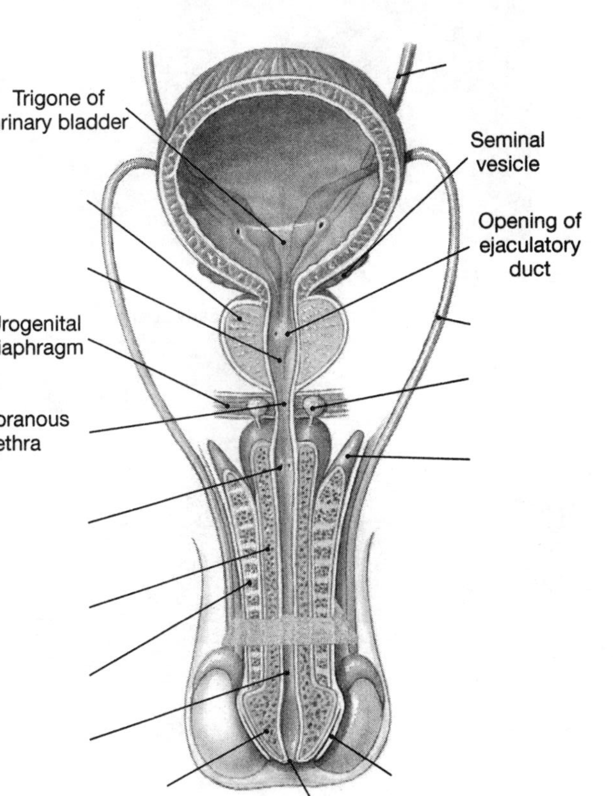

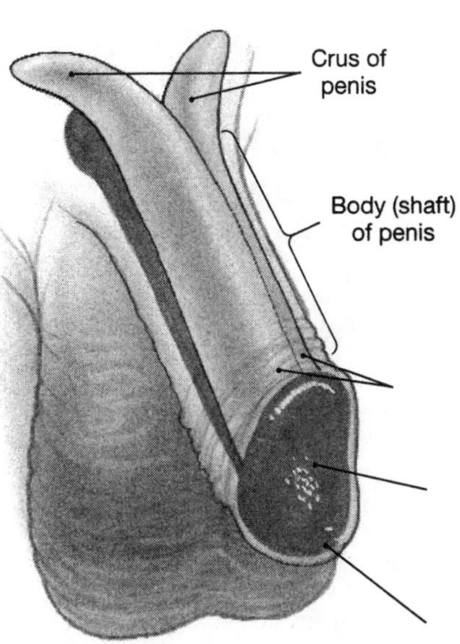

● **FIGURE 66–3**
Penis and Associated Structures.

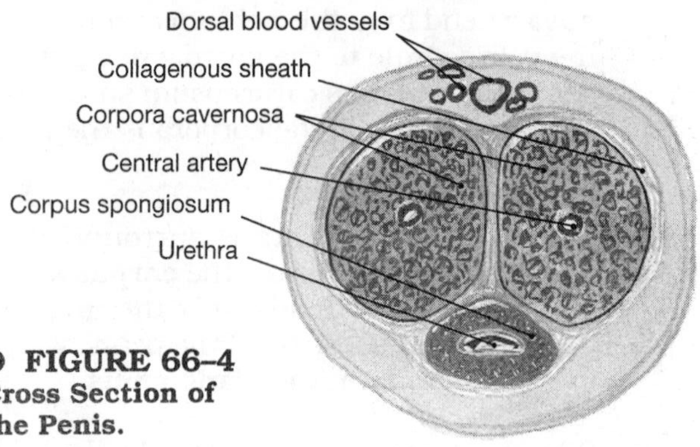

Dorsal blood vessels
Collagenous sheath
Corpora cavernosa
Central artery
Corpus spongiosum
Urethra

● **FIGURE 66–4**
Cross Section of the Penis.

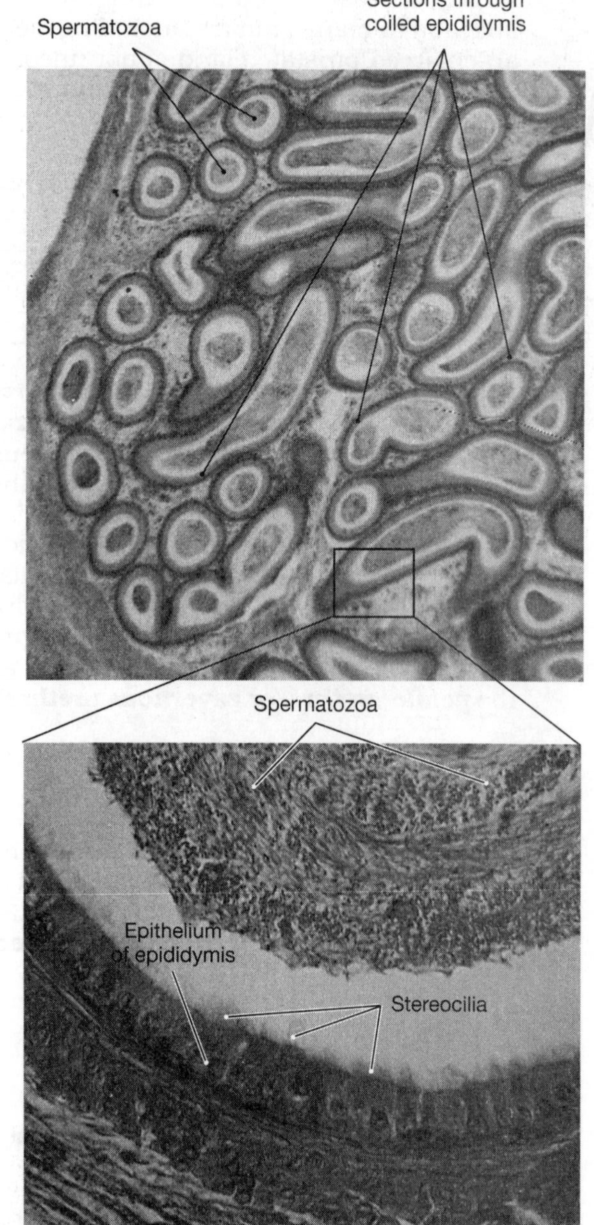

● **FIGURE 66–6**
Epididymis Section.

A. Erection

The erectile tissue of the penis is normally flaccid because the smooth muscle cells of the arterial branches are constricted, thus reducing blood flowing into these areas. During sexual excitement, however, these smooth muscle cells relax and the arterial branches become engorged with blood. This occurs primarily in the corpora cavernosum because engorgement of the corpus spongiosum would constrict the urethra. The increased blood supply in the penis along with the contraction of the superficial **ischiocavernosus muscles** cause the extended rigidity of the penis.

Concept Check 2 How do the opposing functions of two different groups of smooth muscle cells aid in the transport of sperm to its destination in the vagina?

____

____

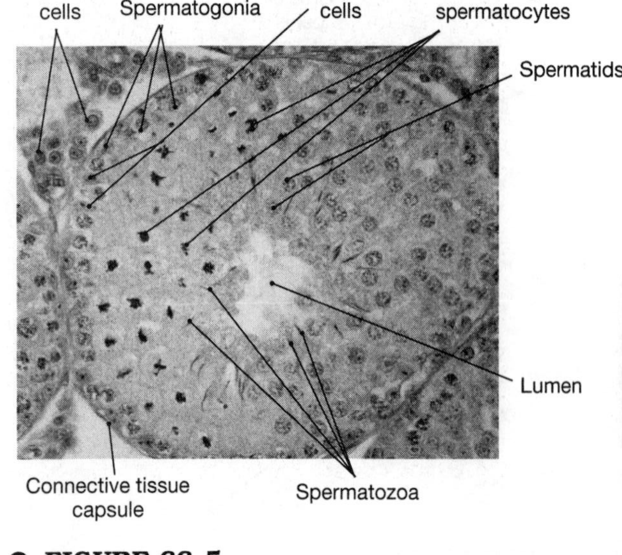

● **FIGURE 66–5**
Seminiferous Tubule.

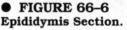

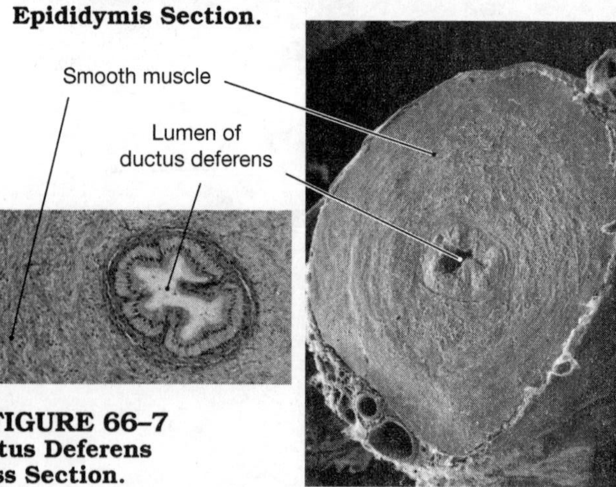

● **FIGURE 66–7**
Ductus Deferens Cross Section.

Source: Dr. Richard Kessel and Dr. Randy Kardon. Reproduced from R. G. Kessel and R. H. Kardon, *Tissues and Organs: A Text-Atlas of Scanning Electron Microscopy*, W.H. Freeman & Co., 1979

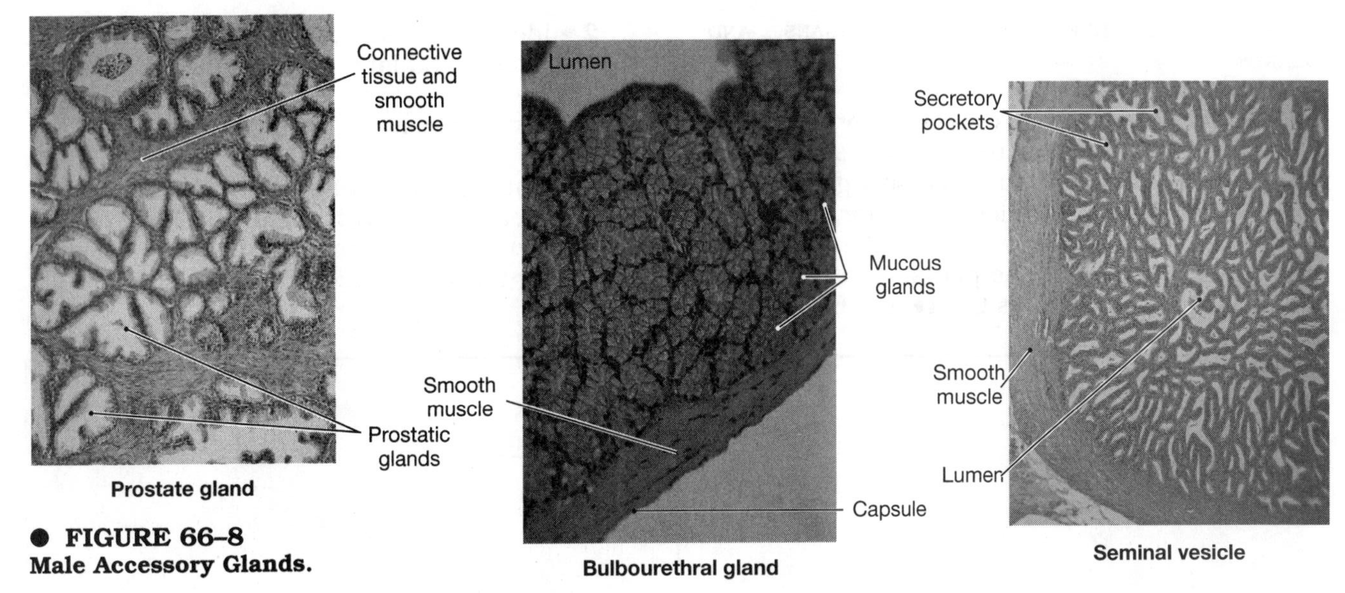

● **FIGURE 66–8**
Male Accessory Glands.

B. Emission

During the emission process, several distinct physiological events occur. First, peristaltic contractions of the ampulla of the ductus deferens push fluid and spermatozoa into the prostatic urethra. Fluid from the seminal vesicles and the prostate gland are forced into the urethra. Meanwhile the sphincter muscles at the beginning of the urethra contract, preventing the passage of seminal fluid into the bladder.

C. Ejaculation

Ejaculation occurs as the **bulbocavernosus muscle,** which wraps around the base of the penis, and the ischiocavernosus muscles contract and push the semen toward the external urethral orifice.

V. Microscopy

Use the space provided at the end of this exercise for your sketches.

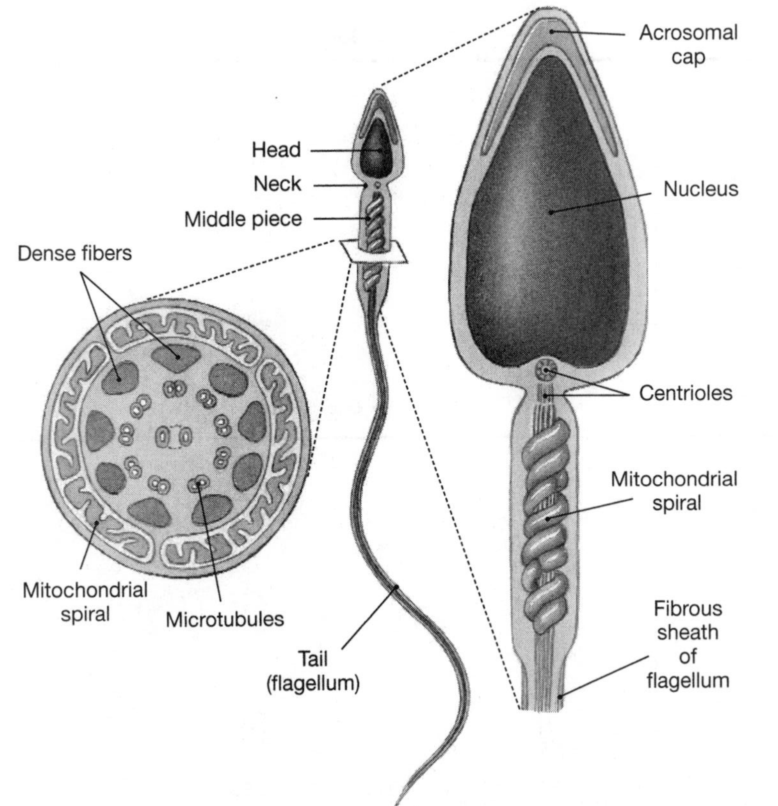

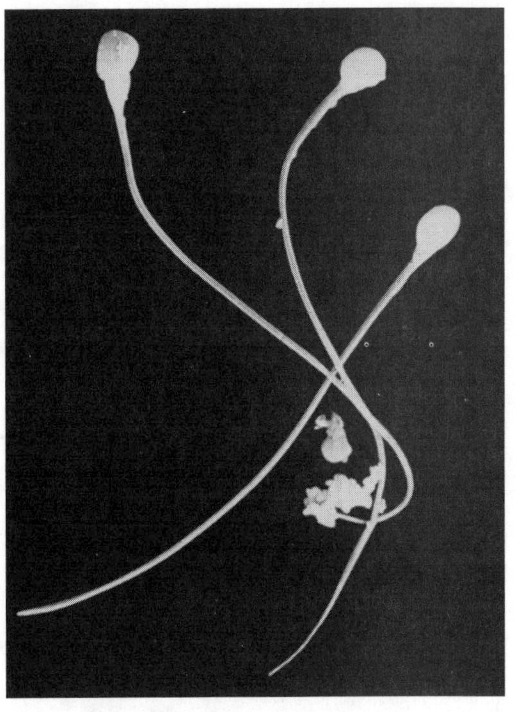

● **FIGURE 66–9**
Human Spermatozoa.

A. Reproductive Structures, Organs, and Glands

1➡ Obtain prepared cross-sectional microscope slides of the penis, testis, epididymis, ductus deferens and accessory glands. Use the compound microscope for all structures except the penis. Use the dissecting microscope for the penis. Compare your slides with Figures 66–4● and 66–8●.

2➡ Identify the structures listed in this text. Sketch and label what you see.

B. Spermatozoa

1➡ Obtain a slide of spermatozoa. Compare these sperm with Figure 66–9●. Your instructor may direct you to use the oil lens for this observation.

TISSUE SAMPLES	TISSUE SAMPLES
TISSUE SAMPLES	TISSUE SAMPLES
TISSUE SAMPLES	TISSUE SAMPLES
TISSUE SAMPLES	TISSUE SAMPLES

❑ Additional Activities

NOTES

1. Although the major reproductive structures of all mammals are basically the same, certain species differences do exist. For instance, the cat has cornified spines on the glans penis and many mammals have a penile bone known as the **os penis** or **baculum**. Research these differences and find out the advantage of each of these "unusual" structures.
2. Obtain spermatozoa slides of different animals. Observe the similarities and differences between these cells.

Answers to Selected Concept Check Questions

1. An enlarged prostate gland might cause a urethral constriction. This could prevent the proper flow of urine or semen. Depending on the reason for the enlargement, an enlarged prostate gland might also cause either an overproduction or an underproduction of prostatic secretions. This could cause problems with sperm viability. An enlarged prostate gland of itself is not necessarily pathological. Nevertheless, the etiology of the enlarged prostate should always be investigated.
2. The sphincter muscles at the beginning of the urethra contract to prevent seminal backflow into the bladder while the cavernosal smooth muscles relax, allowing engorgement and penile erection.

❑ Lab Report

1. Referring to the numbered inquires at the beginning of this exercise, complete the following box summary:

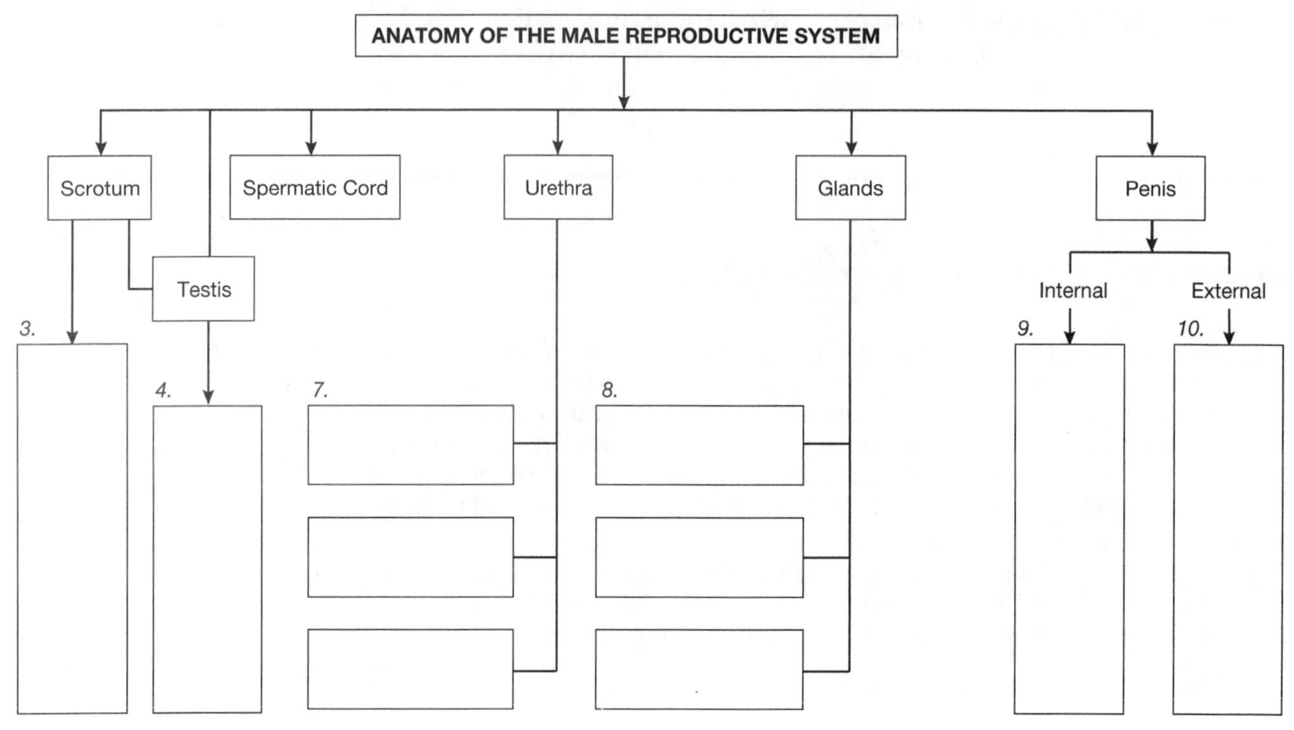

General Questions

NOTES

1. What are the overall functions of the male reproductive system?
2. How would you describe the relationship between the major organs of the reproductive system and the accessory organs of the reproductive system?
3. Trace the course of a sperm cell from the stem cell in the seminiferous tubule through ejaculation. What anatomical and physiological changes take place in the cell? Why are these changes important?
4. Where are the head and tail of the epididymis in relation to the testis?
5. How is the function of the sustentacular cells related to the function of the interstitial cells?
6. Refer to the glands you included in your box summary. Explain why each of these glands is important to the reproductive process.
7. For ejaculation to occur, events must take place in the urethra and in the penis.
   a. What happens in the urethra?
   b. What happens in the vascular portions of the penis?
   c. What happens for the final expulsion of semen?

EXERCISE **67**

Anatomy of the Female Reproductive System

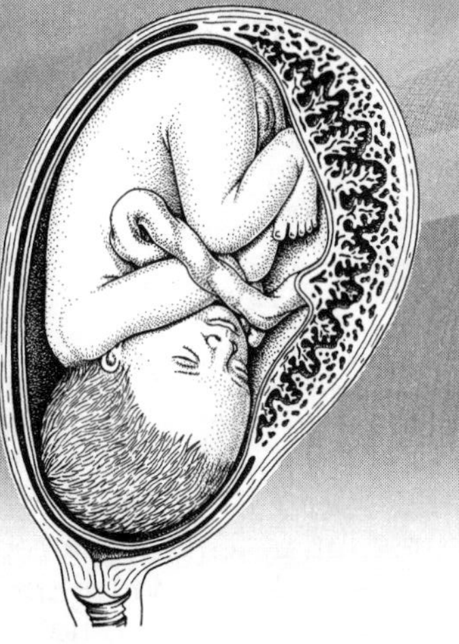

PROCEDURAL INQUIRIES

Preparation

1. What are the overall functions of the female reproductive system?

Examination

2. What are the major organs of the female reproductive system?
3. Where is the uterus?
4. What are the layers of the uterus?
5. Where are the uterine tubes?
6. What are the parts of the uterine tubes?
7. Where are the ovaries?
8. What are the principal parts of the ovary?
9. What are the principal ligaments of the female reproductive system, and where is each located?
10. What is the anatomical relationship among the uterus, the cervix, and the vagina?
11. What are the principal structures of the external female genitalia?
12. What are the principal glands of the external female genitalia?
13. What are the principal structures of the breast?

Additional Inquiries

14. What is the function of the reproductive secretory structures?
15. Why would it be equally appropriate to study the breast with the integument instead of with the reproductive system?

Key Terms

Ampulla
Areola
Basilar Zone
Bladder
Breasts
Broad Ligament
Cervical Canal
Cervix
Clitoris
Endometrium
Epithelial Lining
Fimbriae
Infundibulum
Labia Majora
Labia Minora
Lactiferous Ducts
Lobes
Lobules
Mons Pubis
Myometrium
Nipple
Ostium
Ovarian Follicles
Ovarian Ligament
Ovary
Perimetrium
Round Ligament
Suspensory Ligament
Urethral Orifice (External Opening)
Uterine Tube
Uterosacral Ligament
Uterus
Vagina
Vaginal Entrance
Vestibule
Vestibular Glands
Vulva

Materials Needed

Torso
Organ Models
Charts and Diagrams
Dissecting Microscope
Compound Microscope
Prepared Slides (as available)
- Ovary
- Corpus Luteum
- Corpus Albicans
- Uterus — Different Stages

Mammary Gland

Optional:

Uterine Tube
Vagina
Accessory Glands

The reproductive system is a complex of organs and accessory structures whose hormonally induced function is the propagation of the species.

In the human female, the reproductive structures are geared to produce the egg, receive the sperm, foster fertilization, house the developing fetus, and produce postnatal nutrition for the young child. The accessory structures include those glands and support tissues that facilitate the primary reproductive functions.

In addition, certain reproductive structures (the ovaries in particular) have an endocrine component, which operates not only within the reproductive system but also within the body as a whole.

In this laboratory exercise we will examine the major female reproductive organs and their corollary accessory structures.

❑ Preparation

I. Introduction

A. Systematic Procedure

1➡ Use the available models and charts to gain a perspective of the structures printed in boldface. If you have models that can be taken apart, take them apart and put them back together again. Work toward conceptualizing the three-dimensional picture and the general organization of the reproductive system.

If you do not have classroom models, pay particular attention to any available diagrams and charts to help you gain a perspective of the system as a whole.

2➡ Refer to the textbook in places marked with Concept Links.

B. Labeling

1➡ As you use the descriptions to label the figures associated with each section of this exercise, cross-reference as often as necessary.

❑ Examination

II. Internal Reproductive Structures

These descriptions should be used to label Figures 67–1● and 67–2●. A given structure may be found on both diagrams. Starred items are not on the diagrams.

A. Orientation

Begin your study of the female reproductive structures by identifying the **urinary bladder**. The bladder is centrally located in the anterior part of the abdominopelvic cavity. Immediately posterior

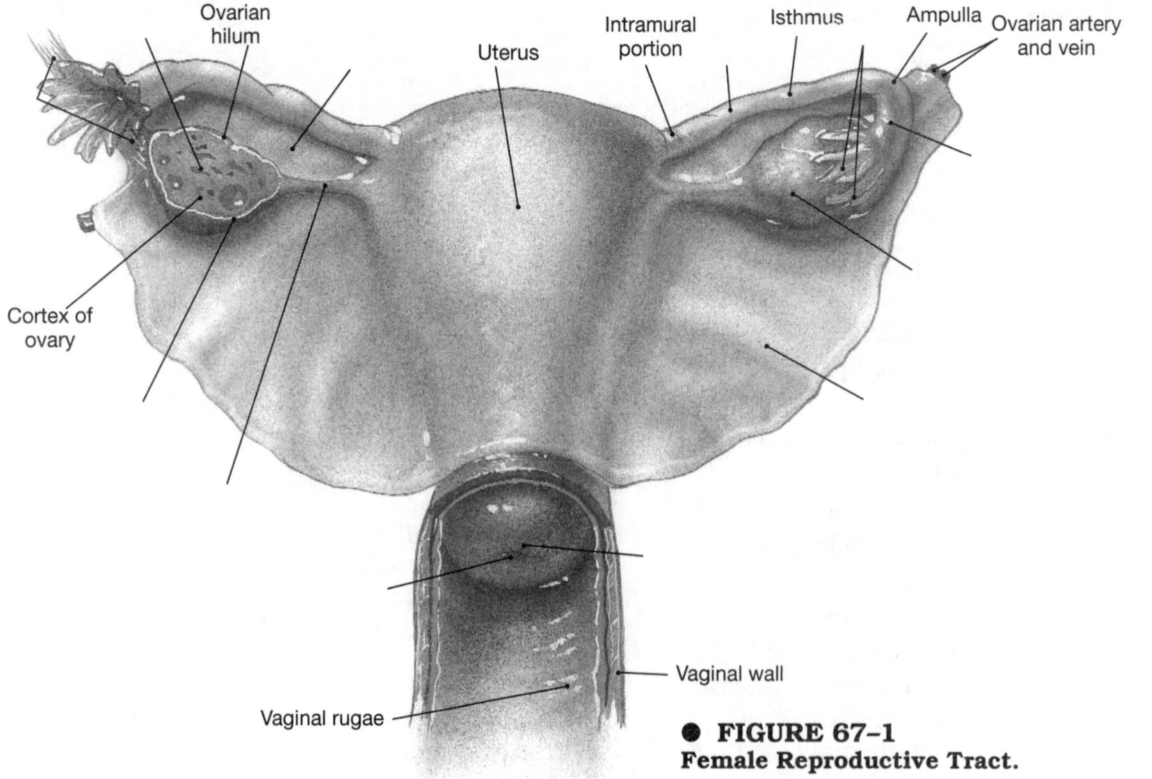

● **FIGURE 67–1**
Female Reproductive Tract.

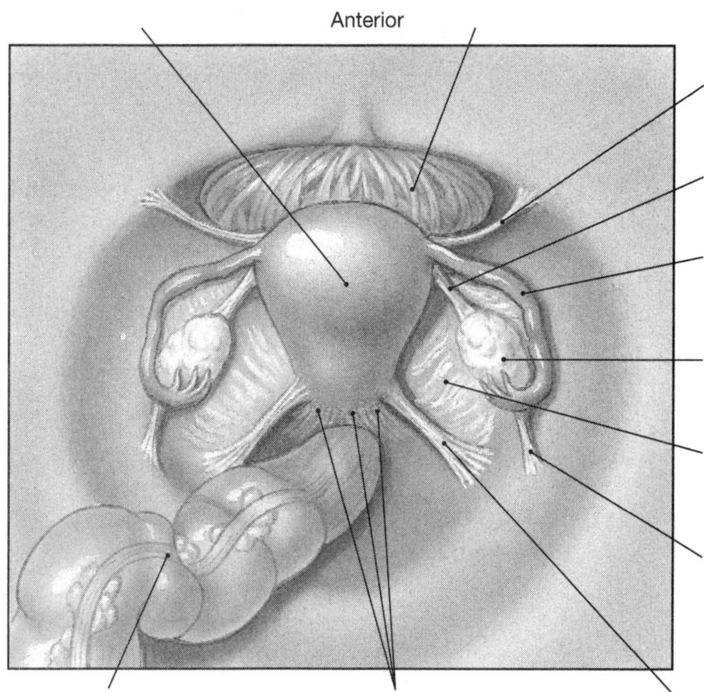

● **FIGURE 67–2a**
Uterus and Stabilizing Ligaments.

and somewhat superior to the bladder is the muscular, pear-shaped **uterus**.

B. Uterine Tube [∞ Martini, p. 1039]

Observe the narrow tubes that extend laterally from the superior portion of the uterine body. Each of these narrowed strucures is a **uterine tube**, also known as a **Fallopian tube**, or an **oviduct**. The portion of the uterine tube passing through the uterine wall is the **intramural** section. Proceeding laterally you will find a narrow **isthmus** constituting about one-third of the tube, followed by an elongated **ampulla**. The lateral end of the ampulla forms an expanded funnel, or **infundibulum**, replete with fingerlike projections, **fimbriae**, that seem to reach out and wrap around an oval **ovary**. Between the fimbriae is an opening extending down into the ampulla. This opening is the **ostium**. The ostium is not identified on the diagrams.

C. Ovarian Structures [∞ Martini, p. 1035]

1➡ Note the position of the ovary in the abdominal cavity. The exact location of the ovary varies somewhat with the specific tilt of the uterus. Nevertheless, the ovary is attached to the pelvic wall by the **suspensory ligament**, which also houses the major ovarian blood vessels, and to the uterine

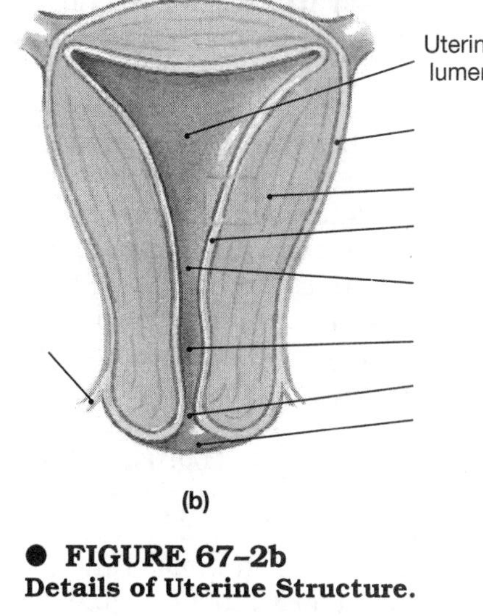

● **FIGURE 67–2b**
Details of Uterine Structure.

body by the thick, cordlike **ovarian ligament**. The ovarian **hilum** is the point where the blood vessels enter the ovary.

2➡ Examine the ovary more closely. The entire ovary is covered by a single layer of columnar cells, the **visceral peritoneum**, sometimes called the germinal epithelium or the germinal epithelium of Waldeyer. (The visceral peritoneum is not identified on the diagrams.)

The ovary proper has two major parts: (1) the external **tunica albuginea**, the fibrous capsule of the ovary, and (2) the internal **stroma**, which can be divided indistinctly into the **cortex** and the **medulla**. The **ovarian follicles**, sacs housing developing ova, are within the cortex. The lumpiness of the ovary is caused by projecting follicles. (The stroma is not identified on the diagrams.)

3➡ Look more closely at the ovarian follicles. These are the ova (eggs) that are released cyclically at ovulation and are swept into the ovarian tube by the fimbriae. The ova pass through the ovarian tube and into the uterus. This process is explained more fully in Exercise 68.

D. Suspensory Structures [∞ Martini, pp. 1035, 1041]

1➡ Notice a filmy tissue attaching the uterine tube and uterine body to the abdominal wall and to the cranial part of the vagina. This tissue is the **broad ligament**, also known as the **mesometrium**. (Technically the mesometrium is only part of the group of broad ligaments that also include the mesovarium and the mesosalpinx. The

mesovarium is a posterior extension of the broad ligament that attaches to the ovary.)

2➡ Observe the thickened ridge of heavier tissue beginning just below the uterine tube and extending laterally and anteriorly toward the abdominal wall. This cordy ridge is the **round ligament**. The **lateral ligaments** extend from the uterus and vagina to the lateral pelvic walls, and the **uterosacral ligaments** attach the uterus to the anterior surface of the sacrum.

E. Uterus [∞ Martini, p. 1040]

1➡ Return to the body of the uterus. If you were to cut into the wall of the uterine body, you would notice that it is quite muscular. The innermost layer is the **endometrium**, the layer that lines the uterus. The majority of the tissue is the **myometrium**, which is actually several interwoven smooth muscular layers. An incomplete serous layer, the **perimetrium**, covers the structure and is continuous with the mesothelium of the broad ligament.

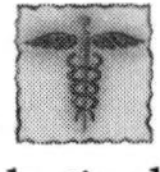

Adenomyosis is a condition caused by the benign invasive growth of endometrial tissue into the myometrium of the uterus. Histologically the endometrium often appears to form swirls or whirlpools of tissue deep into the uterine wall. Although adenomyosis is a noncancerous condition, its presence can cause excessive and sometimes continuous menstrual flow and can lead to anemia.

2➡ Consider the uterine body. The upper part of the uterine body includes a relatively large lumen. This lumen narrows caudally, forming a rather distinct canal, the **cervical canal**. The constriction at the beginning of this canal is the **isthmus**. The opening from the interior into this canal is the **internal os**. The external opening is the **external os**. The entire canal area is the **cervix**.

3➡ Identify the cervix. The caudal part of the cervix marks the end of the uterus and the beginning of the **vagina**, the thick, muscular tube extending from the uterus to the **external genitalia**. Because it is highly distensible to accommodate the penis during intercourse and to accommodate the infant during childbirth, the resting length and width of the vagina are highly variable. Nevertheless, the length usually averages about 7.5–9 cm.

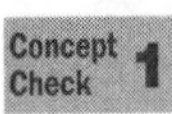

During childbirth, the vagina has to expand to allow passage of the child. What other structures must stretch or widen to "give way"?

____

____

4➡ Keep in mind that, in addition to the child, the unfertilized ovum and the menstrual products also pass from the uterus through the cervix, through the vagina, and out of the body through the external genitalia. (As a point of interest, the unfertilized ovum is passed from the body at midcycle. Normally this cell disintegrates before leaving the body, but occasionally it can be found intact in normal mid-cycle secretions.)

III. External Reproductive Structures

[∞ Martini, p. 1046]

These descriptions should be used to label Figure 67–3●.

A. Vulva

1➡ Examine the external genitalia, commonly called the **vulva**, in the female. Locate the **vaginal entrance**. The central space immediately surrounding the entrance is the **vestibule**. Immediately anterior to the vestibule is a small raised area that houses the **urethral opening** (or **urethral orifice**).

2➡ Observe the smooth hairless skin flaps on either side of the vestibule, called the **labia minora**. Just posterior to the anterior junction of the labia minora and projecting into the vestibule is the **clitoris**, which is analogous to the erectile tissue of the penis. The clitoris is composed of a small glans which contains the erectile tissue proper and fibrous extensions of the labia minora, that encircle the clitoral body forming a **prepuce**.

3➡ Note the lateral folds marking the external limits of the vulva. These folds are known as the **labia majora** and the prominent anterior bulge is the **mons pubis** (sometimes known as the mound of Venus, the *mons veneris*).

B. Glands in the Vulvar Region

1➡ Consider the numerous glands located in the vulvar region. The **lesser vestibular glands** keep the surface of the vestibule moist while the **greater vestibular glands** (also known as **Bartholin's glands**) discharge their mucousal secretions during

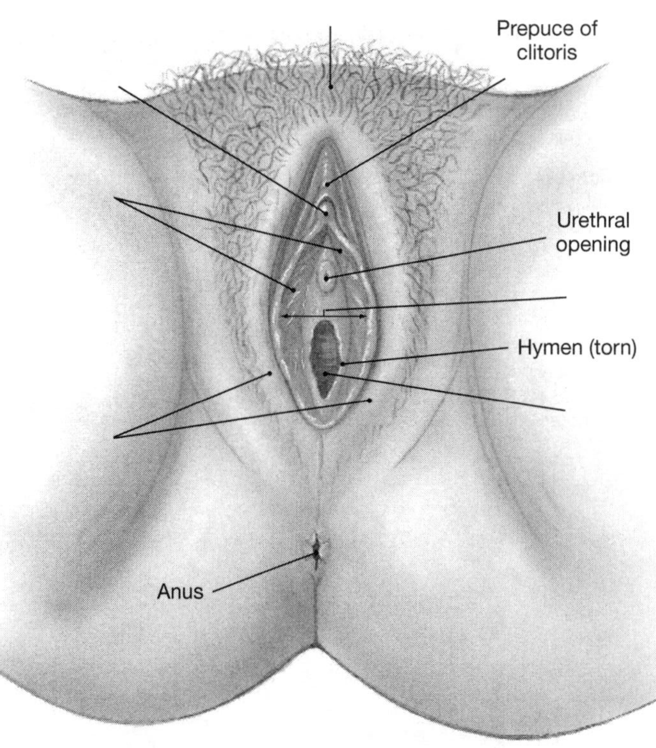

● **FIGURE 67–3**
External Female Genitalia.

sexual arousal. The greater vestibular glands are located near the posterolateral margins of the vaginal entrance. These glands are not identified on Figure 67–3●.

Concept Check 2 What types of problems might ensue if the vestibular glands were not producing enough lubricant?

____

____

____

IV. Mammary Structures

[∞ *Martini, p. 1047*]

These descriptions should be used to label Figure 67–4●.

The breast could well be studied with the epithelium rather than with the reproductive system because the breast is a modified apocrine sweat gland. Because of our sensual nature and because of the nutrient values of the breast milk, we tend to discuss the mammary structures with the reproductive system instead of with the integument.

A. Breast

1➡ Observe the paired eminences, known as the **breasts**, located on the anterior surface of the body between the 2nd or 3rd and 6th or 7th rib. (The human female generally has only one pair of breasts, although occasionally a vestigal remnant of one or more mammary structures can be found along a para-sagittal line running through an existing breast.) The left breast is usually slightly larger than the right breast.

2➡ Examine the surface of the breast. This surface consists of a conical projection, the **nipple**, surrounded by a brownish area, the **areola**. The **mammary glands**, the subcutaneous glandular tissue of the breast, consist of about 15 to 20 **lobes**, each containing several secretory **lobules**. The lobules secrete milk into small ducts, which converge to form **lactiferous ducts**. The lactiferous duct expands to form a **lactiferous sinus** (or **ampulla**), which opens onto the surface of the nipple.

3➡ Keep in mind that the terms "breast" and "mammary gland" are not synonymous. The mammary glands are structures within the breast. Nevertheless, the term "mammary" is often used in connection with any or all of the breast structures.

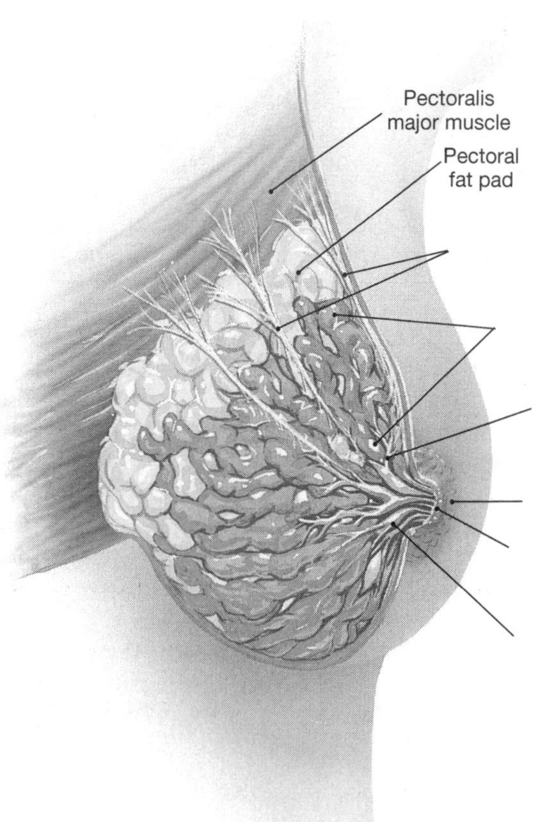

● **FIGURE 67–4**
Anatomy of the Breast.

4➡ Be aware that the size and shape of the breast is determined by the **adipose tissue**, which underlies the mammary lobes. No fat can be found immediately beneath the nipple or areola.

During pregnancy and lactation, it is the mammary glands, not the underlying adipose tissue, that hypertrophy. After menopause some mammary atrophy is common.

B. Breast Ligaments

1➡ Notice that no skeletal muscle tissue can be found within the breast. Instead, the breasts are held in place by bands of connective tissue, the **suspensory ligaments of Cooper,** which extend from the skin to the deep fascia overlying the pectoralis muscle.

V. Microscopy

Your instructor may choose to postpone this microscopy section until you work through Exercise 68.

Do not be surprised if you are viewing nonhuman structures in this section. The basic female reproductive structures of most mammals are the same, so you can take the information learned from some other mammal and apply it to the human.

A. Ovary

We will be examining the ovary in greater detail in Exercise 68. At this point, however, you should examine the more prominent ovarian features.

1➡ Obtain a slide of the ovary and, using the dissecting microscope, identify the ovarian structures in Section II.C. Use the box at the end of this exercise to sketch and label what you see.

B. Uterus

1➡ Obtain slides from different stages of the uterine cycle. Use the dissecting microscope to identify the structures in Section IIE. Use the box at the end of this exercise to sketch and label what you see.

The endometrium is the mucosal layer, which includes the **epithelial lining** of the uterine cavity and the underlying connective tissues of the

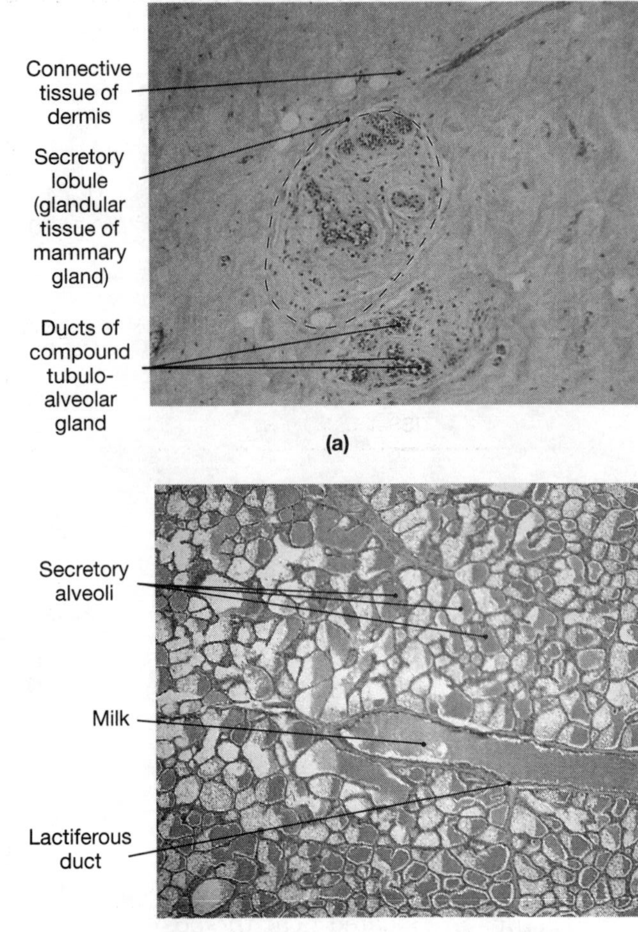

● **FIGURE 67–5**
Histological Organization of the (a) Resting Mammary Gland. **(b)** Active Mammary Gland.

lamina propria. Deep **uterine glands** are part of the epithelial lining. The endometrium can also be divided into a superficial **functional zone** and a relatively constant **basilar zone.** We will identify these structures in Exercise 68. The functional layer undergoes hormone-induced cyclic changes and is periodically shed as a characteristic of the **menstrual cycle**. If pregnancy occurs, both the functional and basilar zones will be involved.

C. Breast

1➡ Obtain and observe slides of breast and mammary gland tissue. Use the box at the end of this exercise to sketch and label what you see. Use Figure 67–5● as your guide.

TISSUE SAMPLES	TISSUE SAMPLES
TISSUE SAMPLES	TISSUE SAMPLES
TISSUE SAMPLES	TISSUE SAMPLES
TISSUE SAMPLES	TISSUE SAMPLES

❑ Additional Activities

NOTES

1. Examine cross-sectional slides of different parts of the uterine tube, different parts of the vaginal wall, and different female reproductive accessory glands. For the uterine tube, pay close attention to the structural differences in the different parts of the tube. Sketch and label what you see.
2. Compare the analogous structures between the male and female reproductive systems. This would be: clitoris vs. penis, female glands vs. male glands, and ovary vs. testes, etc.
3. Observe slides made of cancerous reproductive tissue. What differences do you see when comparing these structures with the norm?

___

___

___

Answers to Selected Concept Check Questions

1. Most lower abdominal structures must stretch or give way for the birth of a child. This includes particularly the uterine os, the cervix, the symphysis pubis, the vaginal opening, and the vagina proper.
2. Lack of vulvar lubricant can cause irritation, pain, itching, and general soreness. These conditions can exist in the presence or absence of sexual intercourse.

❑ Lab Report

1. Referring to the numbered inquires at the beginning of this exercise, complete the following box summary.

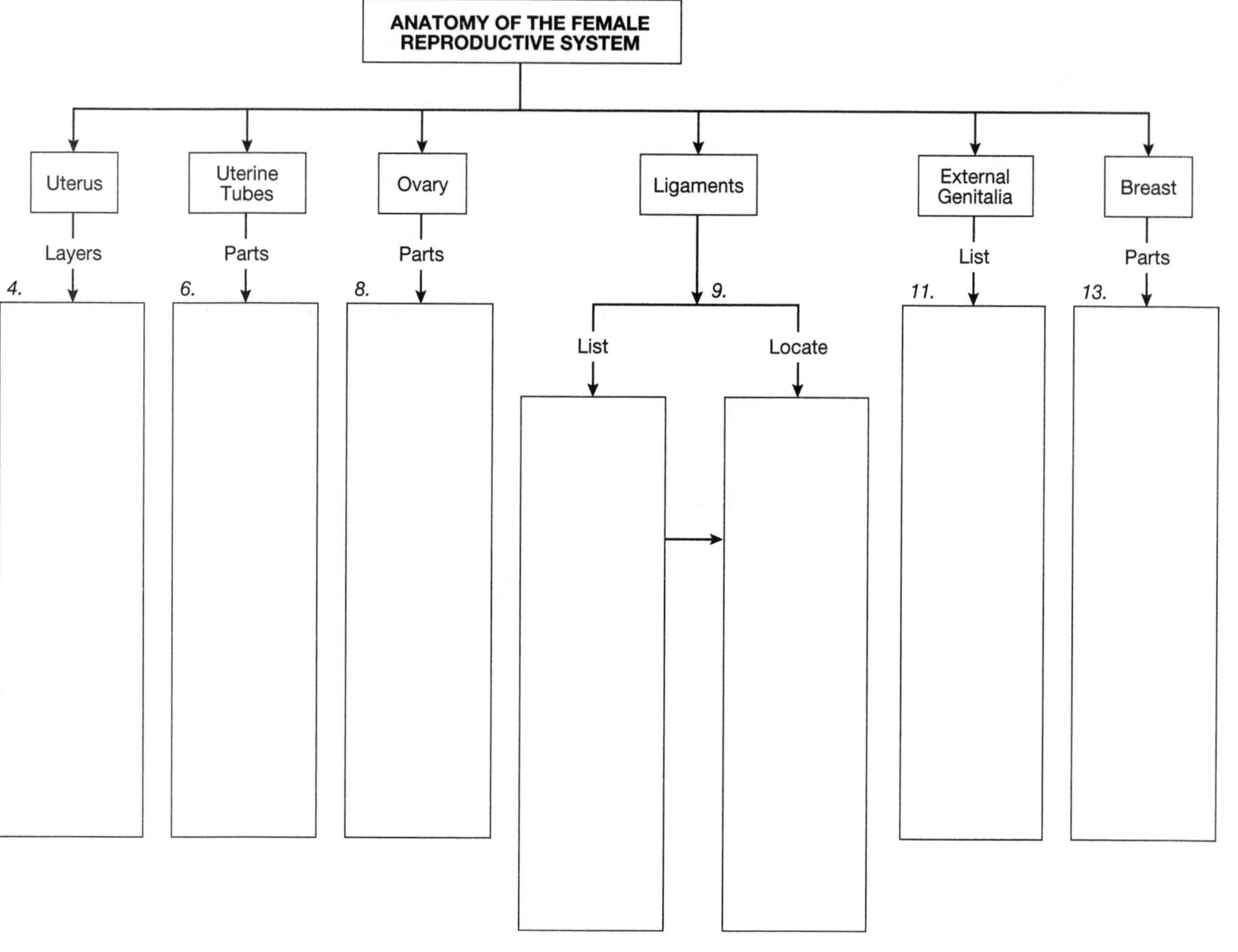

General Questions

NOTES

1. What are the major functions of the female reproductive system?
2. Schematically sketch the location of each of the reproductive structures.
3. How many ligaments support the uterus? Why might so many ligaments be needed?
4. What is the function of each of the glands of the external genitalia?
5. Compare the muscular nature of the uterus and the vagina. Refer to your text if necessary.
6. What is the relationship between the breast and the integumentary system?
7. Trace the path of an ovum from its point of production until it exits the body.

EXERCISE 68

Reproductive Physiology

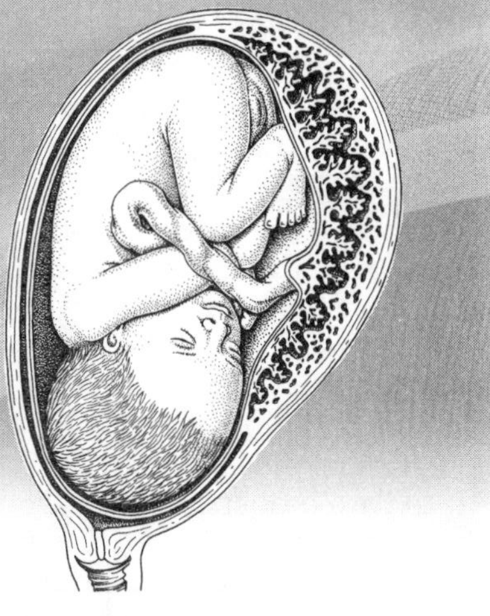

PROCEDURAL INQUIRIES

Preparation

1. What is the ovarian cycle?
2. What is the uterine cycle?

Examination

3. Where are the ovarian follicles?
4. What is the developmental sequence from oogonium to released ovum?
5. How does one distinguish between a primary, secondary, and tertiary follicle?
6. What are the parts of the tertiary follicle?
7. What is the difference between the corpus luteum and the corpus albicans?
8. What is menses?
9. What is the proliferation phase?
10. What is the secretory phase?

Additional Inquiries

11. What does male reproductive physiology include?
12. What does female reproductive physiology include?
13. How should we define reproductive physiology?
14. What is the function of the follicle?
15. Where is FSH produced and what is its function?
16. Where is LH produced and what is its function?
17. Where is estrogen produced and what is its function?
18. Where is inhibin produced and what is its function?
19. Where is progesterone produced and what is its function?

Key Terms

Antrum
Corona Radiata
Corpus Albicans
Corpus Luteum
Cumulus Oophorus
Estrogen
First Polar Body
Follicle Stimulating Hormone (FSH)
Follicular Cells
Follicular Fluid
Inhibin
Luteinizing Hormone (LH)
Menses
Oogonia
Ovarian Cycle
Ovarian Follicles
Ovulation
Primordial Follicle
Primary Oocytes
Progesterone
Proliferative Phase
Secondary Follicle
Secondary Oocyte
Secretory Phase
Tertiary (Graafian) Follicle
Theca Externa
Theca Interna
Uterine Cycle
Zona Pellucida

Materials Needed

Dissecting Microscope (or Hand Lens)
Compound Microscope
Prepared slides, (as available)
- Ovary
  - Corpus Luteum
  - Corpus Albicans
- Uterus — Different Stages

In the past, reproductive physiology often centered almost exclusively on the more obvious cyclic aspects of the female system, meaning our studies of reproductive physiology dealt with the menstrual cycle. Today, however, we consider reproductive physiology from both the male and female perspectives and we view this physiology as a very complex series of interrelated events. For the female, not only must the ovum (egg) be produced, it must also be nourished, maintained, and released into a hospitable environment.

If the ovum is to be fertilized, male physiological events must also be considered. The events of male reproductive physiology — the production and release of the sperm — were covered in Exercise 66. If necessary, review that exercise before proceeding.

If the ovum is fertilized, it must be transported to a place where it can implant. Following implantation, the site must be maintained for the nourishment and development of the child.

If pregnancy does not occur, the ovum and the structures prepared for its implantation must be shed so the cycle can begin again.

In this laboratory exercise we will concentrate on female reproductive physiology by examining the ovarian and uterine cycles both individually and as interrelated functions.

❑ Preparation

I. Procedure

A. Systematic Approach

1➡ If necessary, review the anatomy of the female reproductive structures as described in Exercise 67.

2➡ Keep in mind that the ovaries and the uterus are separate structures with overlapping functions. The **ovarian cycle** is the periodic chain of events occurring in the ovary and including the monthly release of the ovum and the cyclic progression of ovarian hormones. These events influence the **uterine cycle**, which is the periodic building up and shedding of the uterine lining (including whether or not the cycle is interrupted by pregnancy).

3➡ Use the interaction between the ovarian and uterine cycles as your key to understanding the structural and functional events of female reproductive physiology.

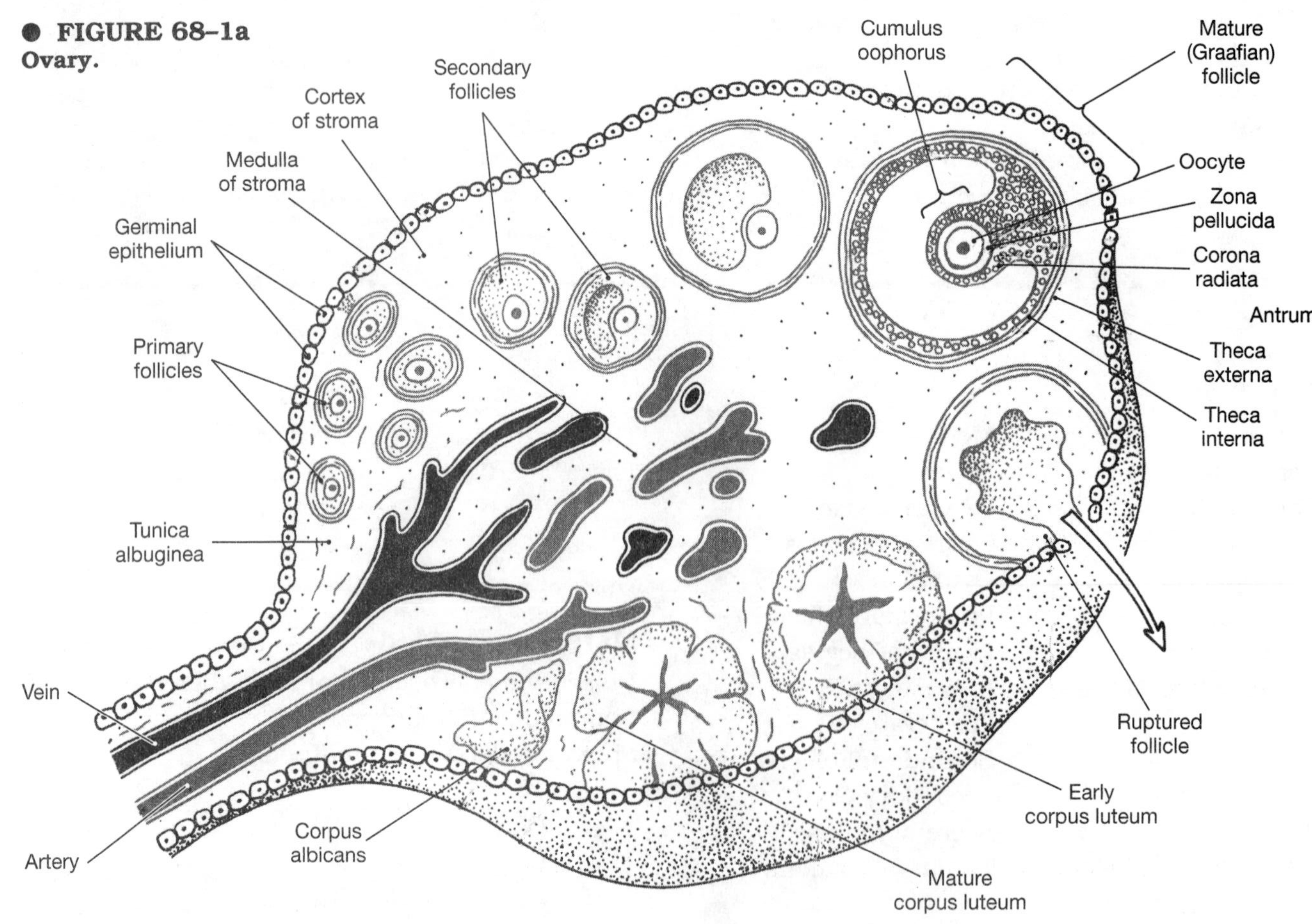

● **FIGURE 68–1a**
Ovary.

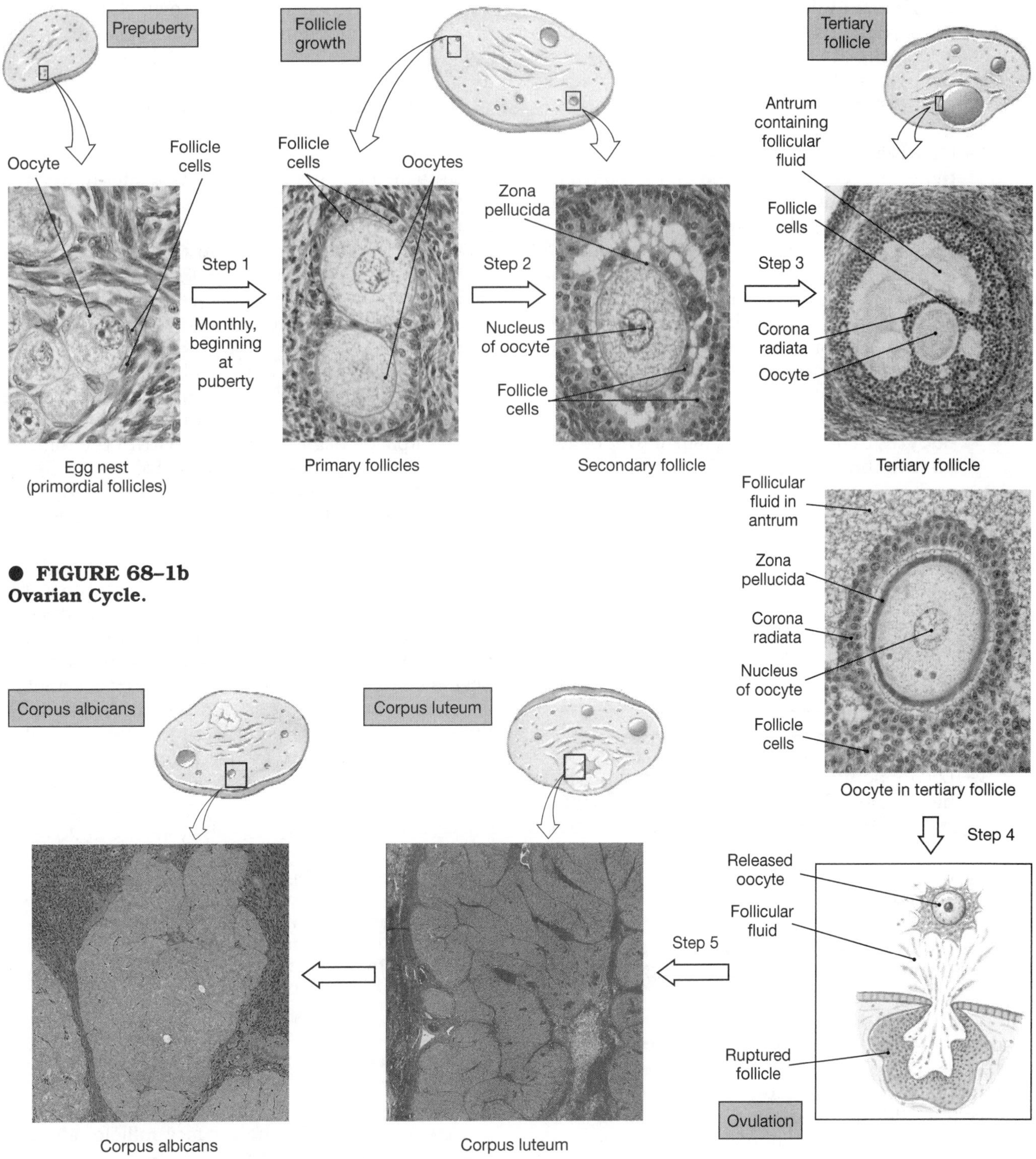

● **FIGURE 68–1b**
Ovarian Cycle.

❑ Examination

II. Ovarian Cycle *[∞ Martini, p. 1037]*

A. Ovary

1➡ Obtain a prepared slide of the ovary.

2➡ Use the low power objective to locate the tunica albuginea, the stroma, the cortex, and the medulla. (These structures are described in Exercise 67.) If necessary, study your slide with the dissecting microscope in addition to the compound microscope. Use Figure 68–1a, b● as your guide.

B. Ovarian Follicles

1➡ Concentrate on the cortex — the outer portion of the ovary — where you will find the

ovarian follicles. The ovarian follicles are the structures in which the eggs (ova) develop prior to being released from the ovary. On your slide you will see ovarian follicles with ova in various stages of development.

2➡ Observe these ovarian follicles. The primitive stem cells, the **oogonia** (sing., oogonium) are located in the ovarian follicles in the germinal epithelium. In the female fetus these oogonia undergo massive mitosis so that by birth the average female has about 400,000 such cells. These oogonia become encapsulated by **follicular cells**, which nourish and protect the developing eggs. At this stage the entire structure (oogonium and follicular cells) is a **primordial follicle**, and the oogonia are known as **primary oocytes**. The primordial follicle will develop into the primary follicle.

You will probably not be able to distinguish any primitive oogonia because any remaining stem cells cease to exist before or shortly after birth.

Contrast the timing of egg production in a woman with sperm production in a man. How do the male and female stem cells differ?

____

____

3➡ Locate some primary follicles. You should have no trouble distinguishing the primary follicles. The primary oocyte found within the follicle is a diploid cell, which has not yet proceeded beyond the prophase I stage of meiosis.

C. Follicular Development and Estrogen Production [∞ Martini, p. 1037]

As the primary follicle develops under the influence of **follicle stimulating hormone (FSH)**, a hormone of the anterior pituitary, the follicular cells begin producing **estrogen**. Estrogen is necessary to build up the uterine lining. (Estrogen is also critical in body development, calcium regulation and sexual response. The functions of both FSH and estrogen are covered in Exercise 44.)

The other ovarian hormone is **inhibin**, which is secreted by the granulosa cells of the ovary. Inhibin (along with estrogen) inhibits the secretion of gonadotropin releasing hormone (GnRH) and FSH. Note this feedback system.

1➡ Study Figure 68–2● and note the correlation between the increasing size of the primary follicle (in the ovarian cycle row) and the increasing levels of estrogen (as shown in the ovarian hormone level row). Note also the fluctuations in the inhibin levels during the different parts of the ovarian cycle.

D. Secondary and Tertiary Follicle Formation

As estrogen is produced, the oocyte completes its first meiotic division. This is a highly disproportionate division, leaving a large **secondary oocyte** enclosed in a large **secondary follicle** and a rather tiny **first polar body**, which gradually disintegrates. As you study your slide you will notice that the secondary follicles are considerably larger than the primary follicle.

1➡ Locate a more advanced (larger) follicle. As more and more estrogen is produced, the secondary follicle continues to enlarge and spaces begin to develop within the follicle. These spaces, filled with **follicular fluid**, or *liquor folliculi*, gradually merge, forming a pocket between the outer and inner layers of follicular cells.

The structure is now known as a **tertiary follicle** or **Graafian follicle**. (Follicular development is a continuum, so it is difficult to define the precise moment a follicle ceases being secondary and begins being tertiary. The oocyte, however, remains secondary until it undergoes the second meiotic division.) Meiosis I is completed just before ovulation, after the tertiary follicle is formed.

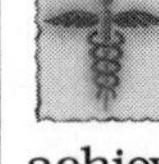

For any given ovarian cycle, about 20 or so oocytes begin the journey from the primordium, but normally only one per month achieves tertiary status. If two ova are released, both could be fertilized and twins would result. Many of the fertility drugs increase the possibility of multiple births by increasing the number of ova maturing to reach ovulation.

E. Structures of the Tertiary Follicle [∞ Martini, p. 1038]

1➡ As you study the tertiary follicle, identify the following structures:

Cumulus Oophorus Pedestal of cells supporting the oocyte.

Corona Radiata Layer of follicular cells immediately surrounding the oocyte.

Zona Pellucida Gel-like layer between the corona radiata and the oocyte

Oocyte Largest cell in the follicle. This will be the ovum, the egg.

Antrum Cavity within the follicle.

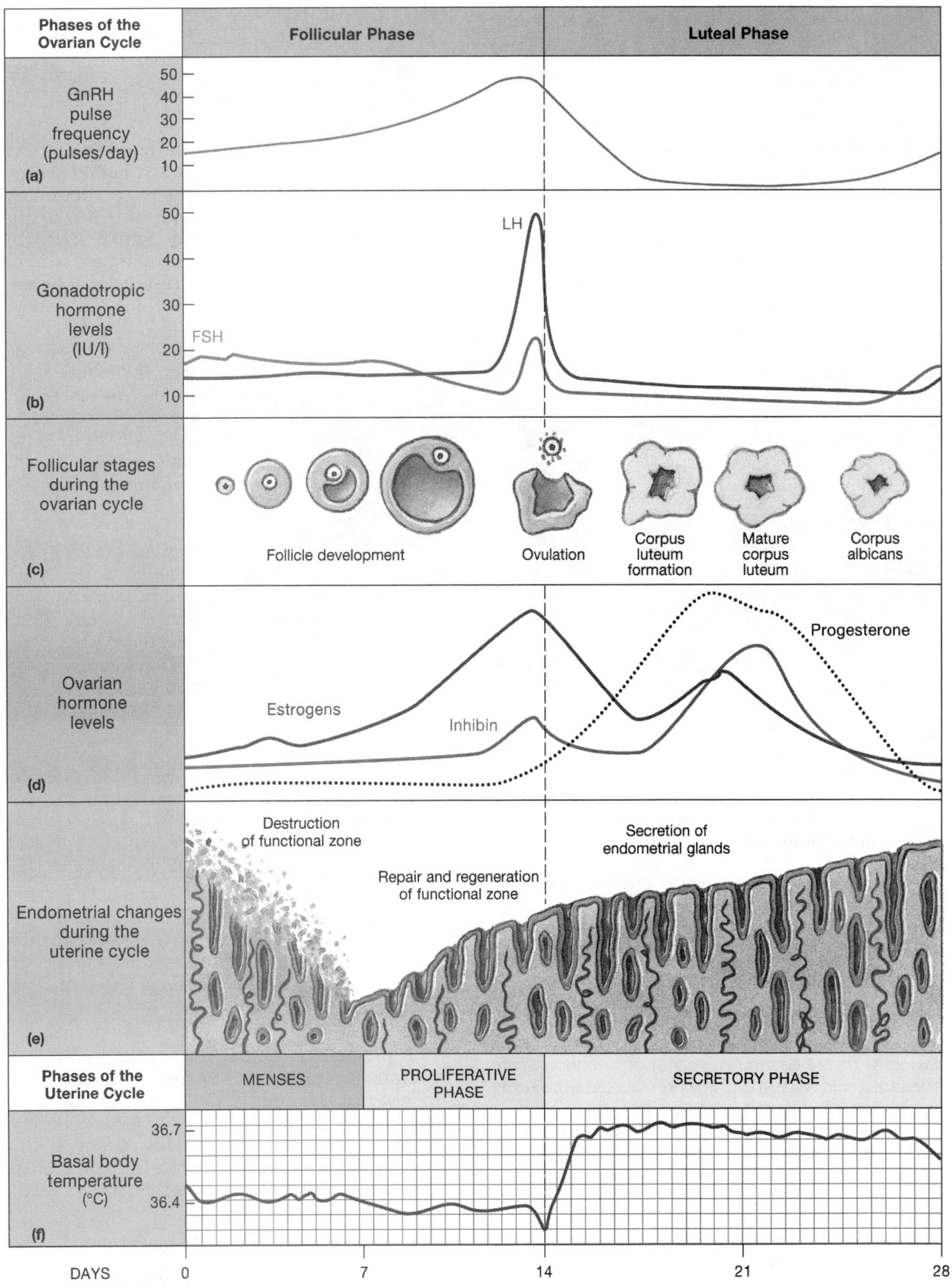

● **FIGURE 68–2**
Ovarian and Uterine Cycles.

(a) Menses

(b) Proliferative Phase

(c) Secretory Phase

(d)

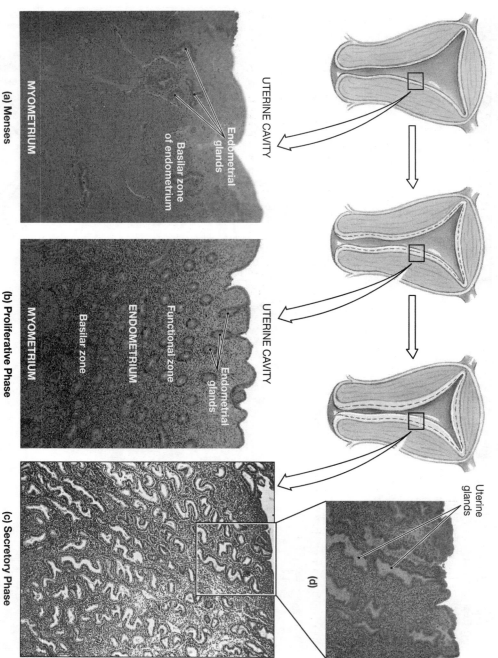

● **FIGURE 68–3**
Uterine Cycle.

Theca interna Layer of cells in the ovarian stroma immediately surrounding the follicle. "Theca" means "sheath" or "coating."

Theca Externa Layer of cells surrounding the theca interna.

F. Ovulation and Formation of the Corpus Luteum

A sharp rise in the level of luteinizing hormone about day 14 of the cycle causes **ovulation**. Ovulation is the release of the secondary oocyte from the ovary. This oocyte will then travel via the uterine (fallopian) tube to the uterus.

The empty follicle, under the influence of luteinizing hormone, begins to produce **progesterone** as well as estrogen, and it is now known as a **corpus luteum**.

1➜ Again study Figure 68–2● and note the correlation between the event of ovulation (in the ovarian cycle row) and the sharp rise in **luteinizing hormone (LH)** (in the gonadotropic hormone level row). Luteinizing hormone is produced by the anterior pituitary (see Exercise 44).

2➜ Now return to your ovary slide and locate a corpus luteum. You should have no trouble distinguishing this large glandular structure, particularly if you use the dissecting microscope to gain a perspective on the relative locations of different ovarian structures.

If pregnancy occurs, the corpus luteum will continue producing progesterone to maintain the integrity of the uterus for about three or four months until the placenta can completely take over progesterone production. This ensures that menses will not occur and that the lining of the uterus, along with the embryo, will not be shed.

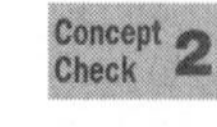

Based on the information graphed in Figure 68–2●, do you think a pregnancy has occurred? Why or why not?

____

____

(Although we stated that ovulation occurs about the 14th day of the cycle, it is more accurate to say that ovulation occurs 14 days before the onset of the next cycle, which often coincides with the 14th day of the present cycle. This is assuming a 28-day cycle. The trigger for the start of any given cycle — the trigger for menses to begin — is the fact that ovulation occurred 14 days before and that progesterone levels are not being maintained. This 14-day time period is quite constant. The first 14-day span is not constant. Thus, if a woman has a 35-day cycle, ovulation occurs on day 21.)

G. Corpus Albicans

If pregnancy does not occur, the corpus luteum gradually disintegrates to form the **corpus albicans**. The corpus albicans is essentially a nonfunctional pale scar.

1➡ Begin by returning to Figure 68–2●. The corpus albicans develops as the corpus luteum continues to disintegrate (on the far right of the ovarian cycle row).

2➡ As you look at your slide, you should notice that the corpus albicans is somewhat smaller and far less glandular than the corpus luteum. Use the dissecting microscope for this comparison. (Depending on the slides available in your laboratory, your instructor may have set out separate slides for observing the corpora lutea and the corpus albicans.)

What similarities and differences do you notice between the corpora?

____

____

III. Uterine Cycle *[∞ Martini, p. 1043]*

A. Overview of the Phases

The endometrium of the uterus can be divided into a superficial functional zone (sometimes called the functional layer) and a deep basilar zone. This basilar zone, which maintains a constant composition, is the stem cell layer for the development of the functional zone. It is the functional zone that undergoes the cyclic changes we commonly think of with menstruation. It is also the functional zone that we consider when discussing the phases of the uterine cycle.

1➡ To study the uterine cycle, first examine Figure 68–2●, which delineates the **menses**, and the **proliferative** and **secretory phases** of the menstrual (uterine) cycle. The first part of the cycle, the menses, begins with the shedding of the functional zone of the uterus. This is day 1 of the menstrual cycle.

The proliferative phase deals with the endometrial events. As estrogen is released from the developing follicle (ovarian cycle), the functional zone is again built up. Uterine glands develop and vascularization occurs. (In other words, there is a proliferation of tissue. The uterus is being prepared for pregnancy.)

The third part of the cycle, the secretory phase, begins at ovulation (and thus is sometimes called the **postovulatory** or **luteal phase**). The estrogen levels drop and the progesterone influence takes over. Progesterone maintains the status quo in the uterus. If pregnancy does not occur and the corpus luteum stops producing progesterone, the progesterone levels fall sharply and the functional zone of the uterus disintegrates. Menstruation ensues.

2➡ Obtain slides of the uterus during different parts of the uterine cycle. Sketch and label what you see, using Figure 68–3● as your guide.

What types of differences do you see?

____

____

____

____

____

Humans are monotocous (generally producing only one offspring at a time), non-seasonal, and noninduced ovulators. In humans ovulation — including the entire ovulatory-menstrual cycle — occurs at regular intervals regardless of the season of the year or the presence or absence of external ovulatory inductors. With some animals, such as the cat and the rabbit, ovulation does not take place unless sexual or mechanical stimulation has occurred.

NOTES

❑ Additional Activities

NOTES

1. Research the concept of the reproductive cycle in the male and compare that with the reproductive cycle in the female.
2. Research the concept of ovarian and uterine cycles in different animals (rats and sheep are particularly interesting), and compare that information with the human cycles.
3. Research the use of hormones in the action of birth control pills.

Answers to Selected Concept Check Questions

1. The female has all the eggs she will ever have seven months before she is born. With the male, sperm are continuously being produced. The female stem cell disintegrates around the time of birth. The male stem cell may be functional for 80 or more years.
2. A pregnancy has not occurred here because the progesterone level has fallen.

❑ Lab Report

1. Referring to the numbered inquiries at the beginning of this exercise, complete the following box summary.

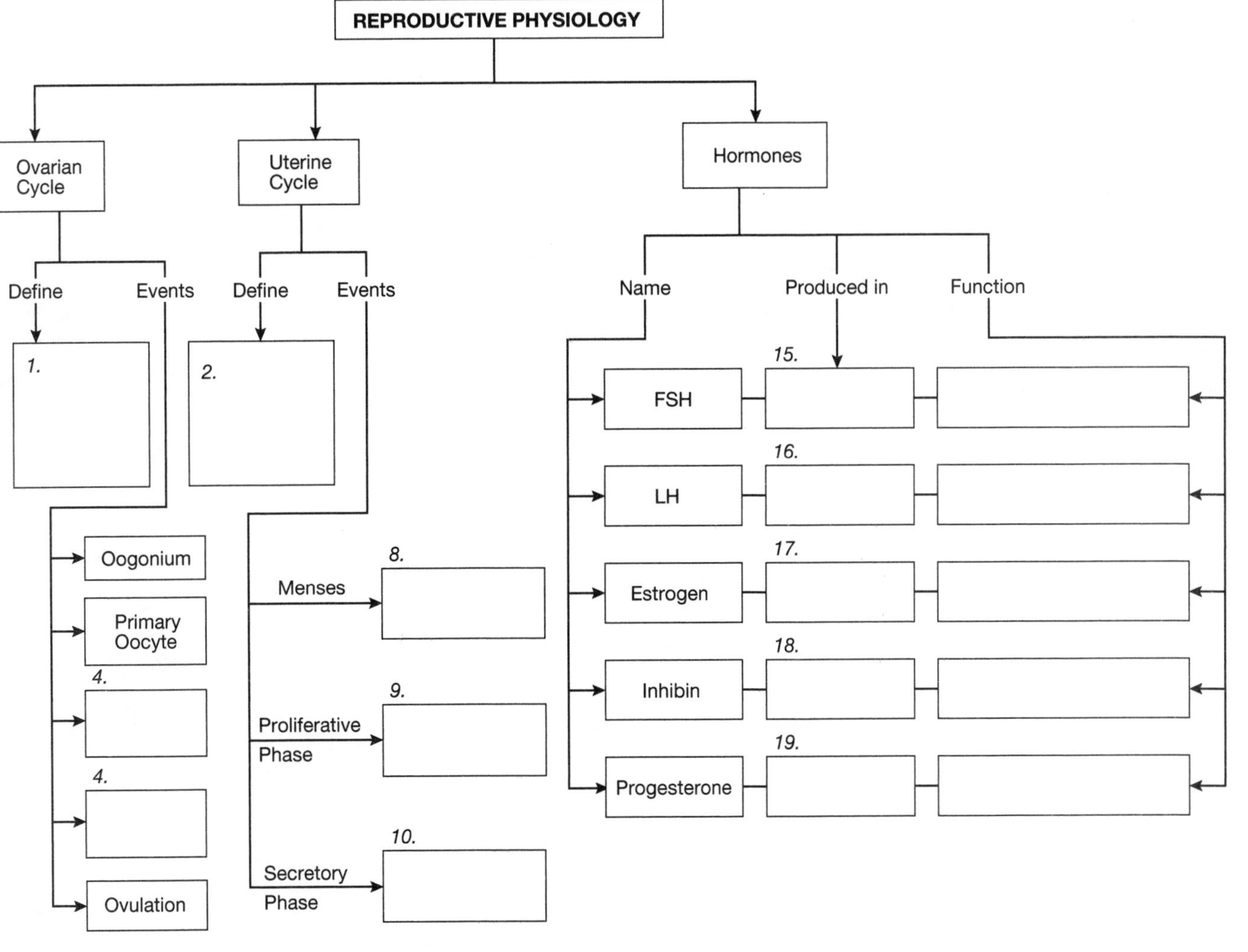

General Questions

NOTES

1. How should we define reproductive physiology in terms of the male? In terms of the female?
2. What is the function of the ovarian follicle?
3. What is the difference between the follicle and the oocyte?
4. Anatomically, where are the ovarian follicles located?
5. Sketch a tertiary follicle and label the principal parts.
6. Why are both the corpus luteum and the corpus albicans important in the beginning and maintainance of pregnancy?
7. Trace the course of an ovum from its primordial precursor to fertilization. What changes take place in both the cell and its surrounding tissue? What is the significance of these changes?
8. In your own words, explain the relationship between the ovarian cycle and the uterine cycle.

EXERCISE **69a**

Dissection of the Reproductive System: Cat

PROCEDURAL INQUIRIES

Preparation

1. How do you prepare your cat for the dissection of the reproductive system?
2. Where are the cat's reproductive organs located?

Dissection

3. What are the parts of the uterus of the female cat?
4. What are the parts of the ovary of the female cat?
5. What are the common urogenital structures in the female cat?
6. What are the parts of the external genitalia of the female cat?
7. What are the parts of the testis of the male cat?
8. What are the parts of the duct system of the male cat?
9. What are the reproductive glands of the male cat?
10. What are the parts of the penis of the male cat?
11. What are the parts of the external genitalia of the male cat?

Additional Inquiries

12. What are the major blood vessels found in both the male and female cat?
13. How should you dissect a pregnant cat uterus?
14. How would you describe feline placentation?
15. What are the feline mammary structures?
16. What comparisons can be made between the human and feline reproductive systems?

HUMAN CORRELATION: AM Cadaver Atlas in Applications Manual.

Key Terms

Ampulla
Bicornuate
Bladder
Broad Ligament
Bulbourethral (Cowper's) Glands
Cervix
Clitoris
Corpora Cavernosa
Corpus Spongiosum
Efferent Ductules
Epididymis
External Genitalia
Fimbriae
Glans Penis
Infundibulum
Interstitial Cells
Labia Majora
Mammae
Os Penis
Ostium
Ovarian Ligament
Ovary
Prostate Gland
Rete Testis
Round Ligament
Seminal Vesicle
Seminiferous Tubules
Sertoli Cells
Spermatic Cord
Spermatogonia
Suspensory Ligament
Testes
Tunica Albuginea (Male and Female)
Tunica Vaginalis
Urethra
Urethral Orifice
Urogenital Sinus
Uterine Tube

Uterus
Vagina
Vas Deferens
Vestibule
Zonary Placenta

Materials Needed

Preserved Cat
Gloves
Dissecting Pan
Dissecting Kit
Bone Cutters
Dissecting Microscope
Histological Slides
- Ovary
- Uterus
- Testes
- Penis
- Compound Microscope

Dissecting the reproductive system of the cat affords us a unique opportunity to study major reproductive structures common to all mammals. We can then use this information to better understand the anatomy of the human reproductive system.

As you work through this basic overview of the feline reproductive system, you will find that although much of the system is analogous to the human system, some significant differences do exist. These differences will be pointed out in the text. Your emphasis, however, should be on the similarities between the systems.

❑ Preparation

I. Introduction

A. Procedure

1➜ Regardless of whether you are working with a male cat or a female cat, you should begin by familiarizing yourself with the reproductive structures in the appropriate diagrams (Figures 69a–1● and 69a–2●).

2➜ When you have finished studying the reproductive anatomy of your own cat, you should examine the reproductive structures of a cat of the opposite sex. Throughout this exercise you should locate and/or identify each boldfaced item listed.

B. Anatomical Preparation

1➜ Prepare your cat for the dissection of the reproductive system by placing the animal ventral side up and opening the abdominal cavity.

2➜ Note the location of the digestive and urinary structures. Use this information for an anatomical perspective as you work through your dissection.

Dissection

II. Female Cat

As you work through this part of the exercise on the female cat, you are encouraged to refer often to Exercise 67 on the human female reproductive system.

A. Uterus

1➜ Begin your study of the female cat by first locating the urinary **bladder**. In the female cat the urinary bladder is in the ventral part of the lower central abdominal cavity. Immediately dorsal to the bladder is a muscular Y-shaped tube. The caudal arm of the Y is the **uterine body**. The two cranial arms of the Y are the **uterine horns**. The uterine horns are also known as **cornua** (sing., cornu). Hence, the cat uterus with two horns is **bicornuate**. The feline embryos implant in the cornua and not in the uterine body. (The human uterus has no discernible horns, so the human embryo normally implants in the body of the uterus.)

B. Uterine Tubes

1➜ Follow one uterine horn as it ascends the posterior abdominal wall. Toward the end of the horn, you will notice that the tube narrows appreciably. This narrowed portion is the **uterine tube**, known more familiarly in the human as the **fallopian tube** or the **oviduct**. Observe the flared end of the uterine tube, the **infundibulum**, as it wraps around an oval glandular structure. The finger-like projections of the infundibulum end are the **fimbriae**, and the glandular structure is the **ovary**.

2➜ Gently dislodge an ovarian tube (a term used sometimes for the most distal part of the uterine tube) and see if you can locate the **ostium**, the hole going down between the fimbriae. This hole leads into the lumen of the ovarian tube and subsequently into the uterine tube.

C. Uterine Ligaments

1➜ Carefully clear away unnecessary fascia, particularly from the posterior and medial

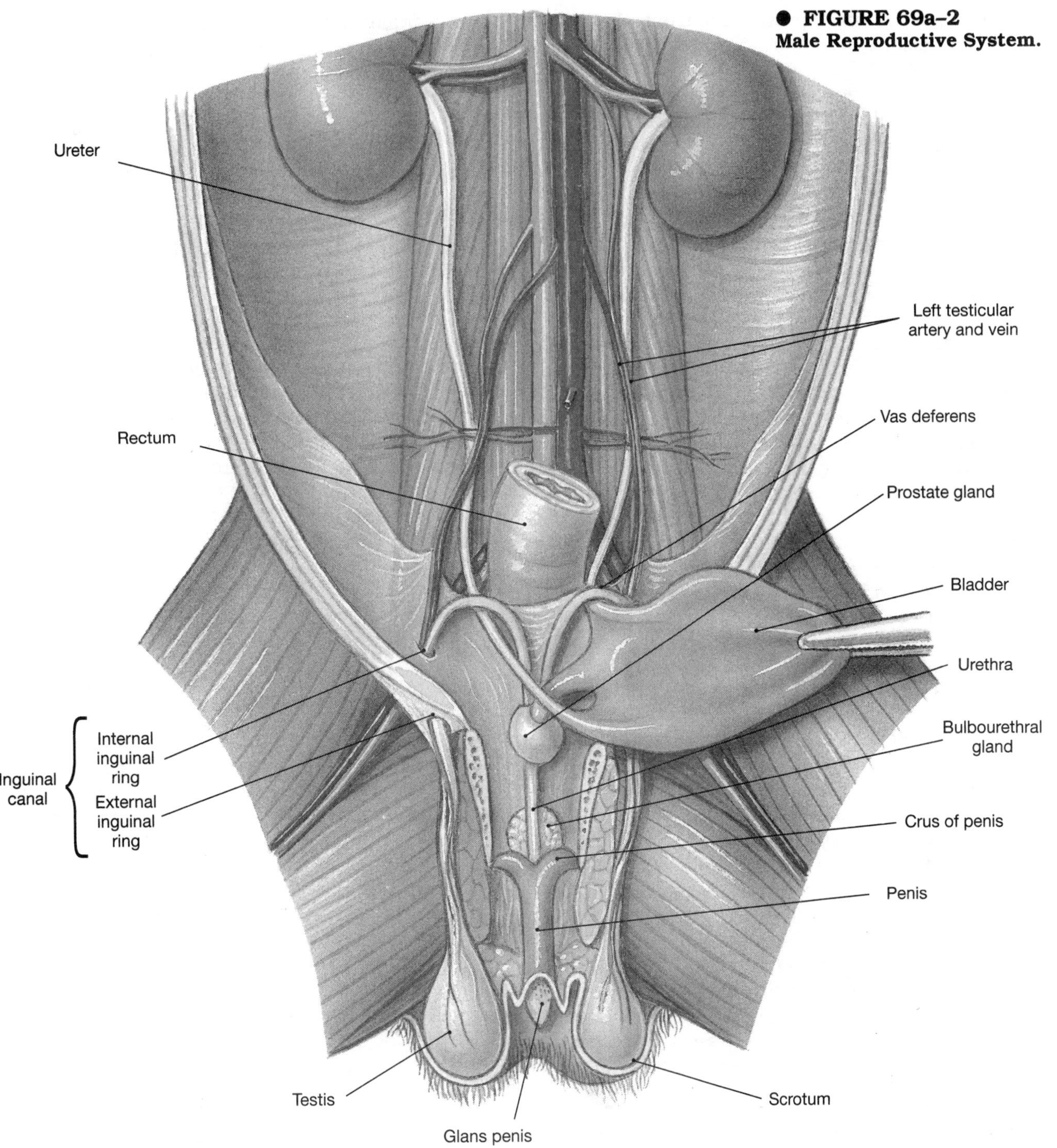

● **FIGURE 69a–2**
Male Reproductive System.

C. Testes

The testis itself is surrounded by another coat, the **tunica albuginea**. Fibrous extensions of the tunica albuginea dip into the testis, separating it into incomplete **lobules**. Each lobule contains the **seminiferous tubules**. Sperm stem cells, **spermatogonia**, are found toward the exterior of the seminiferous tubules. Also found in this area are the **Sertoli cells**, which supply nutrients to the developing sperm.

Between the seminiferous tubules are the **interstitial (Leydig) cells**, specialized interstitial endocrine cells that produce male sex hormones.

1 ➡ Although you will not be able to identify the preceding testicular structures on gross examination, after you have completed your dissection of the male cat, remove a testis and bisect it under the dissecting microscope. Most gross structures should be readily visible. Sketch what you

see in the space provided at the end of this exercise.

In Exercise 66 you examined a prepared slide of the human male seminiferous tubules. The male cat's seminiferous tubules look virtually the same.

D. Duct System

Sperm are produced in successive stages, beginning in the outer regions of the seminiferous tubules. When fully formed (although still immature) the sperm enter the lumens of the tubules. Sperm then move through the seminiferous tubules to a mesh-like network known as the **rete testis**. The rete testis empties into a series of small ducts known as the **efferent ductules**, or the **ductus efferens**. These ductules unite to form the head of the epididymis, the structure you have already located.

1➡ Follow the tail of the epididymis. At the base of the testis, the epididymis turns cranially and becomes part of the **spermatic cord** as the vas deferens. Within the spermatic cord are the vas deferens, assorted blood vessels, and branches of the 10th thoracic nerve. The outer covering of the spermatic cord is continuous with the tunica vaginalis.

E. Into the Body Cavity

1➡ When you began studying the male cat, you located the vas deferens as it looped over the ureter. You are now approaching the vas deferens from the other end. Your goal at this point is to trace the spermatic cord up into the inguinal area, through the inguinal canal, and into the abdominopelvic cavity. Use caution in cutting away the skin and connective tissue. Notice that after the inguinal canal only the vas deferens continues cranially.

(Human anatomy to this point is approximately the same as that of the cat.)

F. Separation of the Os Coxae

1➡ To continue your study of the male reproductive system, it will be necessary for you to separate the os coxae. Use the bone cutters and make your cut through the symphysis pubis, the fibrous pad between the sections of the pelvic bone. If yours is an older cat, this will take some energy. Use caution so that you cut only the bone and muscle. Once you have cut through the bone, your instructor may direct you to break the pelvis so you can better see the internal genitalia.

G. Glands

1➡ Clear away any connective tissue so that the entire tract of the urethra is exposed. Go back to the bladder and identify the urethra as it extends caudally toward the floor of the pelvic cavity. Follow the vas deferens as it continues medially toward the urethra.

(In humans and in many other mammals you will find an enlarged area, the **ampulla**, at the end of the vas deferens immediately before it enters the urethra. The cat does not have an ampulla. In humans also you will find a **seminal vesicle** — **vesicular gland** — which empties into the vas deferens just before the urethral junction. Cats do not have seminal vesicles either. In certain other mammals the vesicular glands are located after the urethral junction.)

2➡ Notice the round glandular structure at the junction of the vas deferens and the urethra at the approximate level of the symphysis pubis. This is the **prostate gland**, which is basically divided into two lateral lobes. Prostate gland size is highly variable and, as in humans, an enlarged prostate gland is not uncommon. In this area the urethra is known as the **prostatic urethra**.

3➡ As you continue caudally, the urethra is technically known as the **membranous urethra**. At the end of the membranous urethra you should be able to find the small paired **bulbourethral glands**, also known as **Cowper's glands**. (Human Cowper's glands are more cranial than the cat's.)

H. Penis

At this point the urethra enters the **penis** and becomes known as the **penile urethra**, or **cavernous urethra**. The penis of the cat is retracted into a sac of skin known as the **prepuce**; however, the penis is not suspended by the prepuce from the ventral abdomen as it is in other domestic animals. The area formed interior to the prepuce (where the penis is located) is the **preputial cavity**. At the distal end of the penis, the opening of the urethra to the external environment is the **urogenital opening**.

1➡ If you have not already done so, cut the tough skin of the prepuce and expose the **glans penis**, the enlarged end of the penis. Notice the cornified spines on the glans penis of the cat. These spines serve to stimulate ovulatory response in the female. (The female cat is an induced ovulator, so if pregnancy is to occur, ovulation must be induced.)

(Humans and other domestic animals do not have these cornified spines on the glans penis.)

I. Penis Cross Section

On cross section the penis can be divided into a spongy area, the **corpus spongiosum**, in the center of which can be found the urethra and paired cavernous areas, the **corpora cavernosa**, which house the deep arteries of the penis. The corpora cavernosa are separated by a **median septum**. (These structures are analogous to human structures.)

The cat also has a penile bone known as the **os penis** or **baculum**, a structure that facilitates copulation. (Humans do not have this bone.) Most authorities regard the os penis as a calcified part of the corpus cavernosum.

1 ➜ On a microscope slide of the penis, identify the preceding structures. (You may not specifically have a slide of a cat penis, so there may not be an os penis present. Other structures will be virtually the same.) Sketch what you see in the space at the end of this exercise.

TISSUE SAMPLES	TISSUE SAMPLES
TISSUE SAMPLES	TISSUE SAMPLES
TISSUE SAMPLES	TISSUE SAMPLES
TISSUE SAMPLES	TISSUE SAMPLES

J. Blood Supply

As with the female reproductive anatomy, you should take a few minutes to familiarize yourself with the blood vessels described in the following brief review. Knowing the blood vessels in this area provides both landmarks for identification and knowledge of how the reproductive structures receive their blood supply.

1➡ Locate the **aorta**. Just caudal to the renal arteries are the paired **testicular arteries**. The testicular arteries supply blood to the testes. The penile blood supply comes from the pudendal branch of the femoral artery.

2➡ Locate the **inferior vena cava**. Notice that the **right internal testicular vein** enters the vena cava directly caudal to the renal vein. The **left internal testicular vein**, however, drains into the left renal vein. The venous return from the penis occurs through drainage into the great saphenous vein and through the prostatic plexus.

IV. Reproductive Histology

A. Slides

Reproductive histology was examined in Exercises 66, 67, and 68. Now that you have completed your reproductive dissection, your instructor may direct you to view some additional reproductive slides. Also, this is a good time to examine any of the additional tissue samples you removed from your cat.

❑ Additional Activities

1. Research the embryological development of the cat. Compare the similarities and differences between the male and female.
2. If your instructor so directs, take additional tissue samples from your cat and examine them under the dissecting microscope.

NOTES

❑ Lab Report

1. Referring to the numbered inquiries at the beginning of this exercise, complete the following box summary:

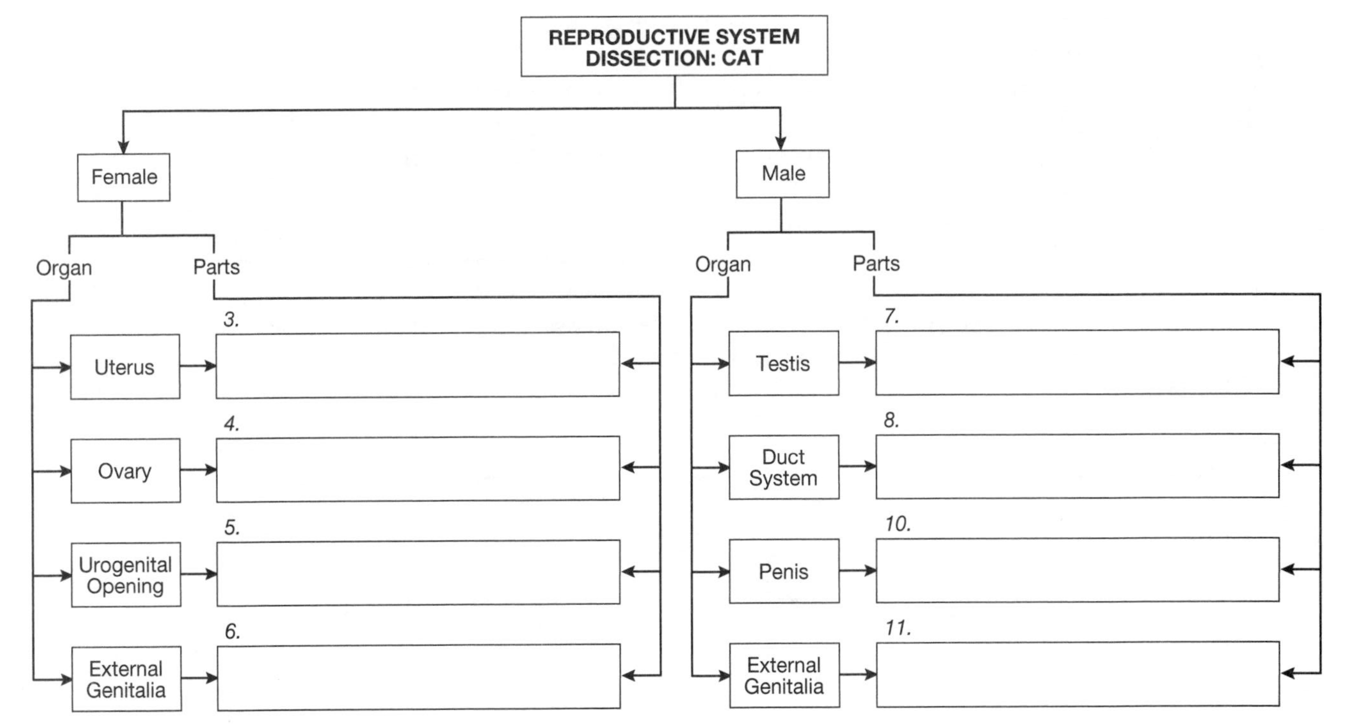

General Questions

NOTES

1. Trace the path of a sperm and an ovum from the testes and ovary through fertilization, implantation, development, and birth. Refer to Exercise 68 if necessary. Corollate that information with what you learned in this exercise.
2. What advantage is there to having the extensive arterial system to supply blood to the uterus?
3. How are both the male and female cat adapted to the female being an induced ovulator?
4. List the differences in reproductive tract glands between the human male and the feline male.
5. How is the human penis different from the feline penis?
6. Use the following chart to list the similarities and differences between the human and feline:

	Similarities	**Differences**
a. Uterus		
b. Ovary		
c. Urogenital sinus/ opening(s)		
d. Ligaments and other support tissue		
e. Placental components		

7. What procedure should you use in dissecting a pregnant cat uterus?
8. What are the feline mammary structures?

EXERCISE **69b**

Dissection of the Reproductive System: Fetal Pig

PROCEDURAL INQUIRIES

Preparation

1. How do you prepare your fetal pig for the dissection of the reproductive system?
2. Where are the fetal pig's reproductive organs located?

Dissection

3. What are the parts of the uterus of the female fetal pig?
4. What are the parts of the ovary of the female fetal pig?
5. What are the common urogenital structures in the female fetal pig?
6. What are the parts of the external genitalia of the female fetal pig?
7. What are the parts of the testis of the male fetal pig?
8. What are the parts of the duct system of the male fetal pig?
9. What are the reproductive glands of the male fetal pig?
10. What are the parts of the penis of the male fetal pig?
11. What are the parts of the external genitalia of the male fetal pig?

Additional Inquiries

12. What are the major blood vessels found in both the male and female fetal pig?
13. What is the urachus and what are the fates of the parts of the urachus?
14. What comparisons can be made between the human and porcine reproductive systems?

HUMAN CORRELATION: AM Cadaver Atlas in Applications Manual.

Key Terms

Allantoic Bladder
Ampulla
Bicornuate
Broad Ligament
Bulbourethral (Cowper's) Glands
Cervix
Clitoris
Corpora Cavernosa
Corpus Spongiosum
Efferent Ductules
Epididymis
External Genitalia
Fimbriae
Glans Penis
Infundibulum
Interstitial Cells
Labia Majora
Ostium
Ovarian Ligament
Ovary
Prostate Gland
Rete Testes
Round Ligament
Seminal Vesicle
Seminiferous Tubules
Sertoli Cells
Spermatic Cord
Spermatogonia
Suspensory Ligament
Testes
Tunica Albuginea (male and female)
Tunica Vaginalis
Urachus
Urethra
Urethral Orifice
Urogenital Sinus
Uterine Tube
Uterus
Vagina
Vas Deferens
Vestibule

Materials Needed

Preserved Fetal Pig
Gloves
Dissecting Pan
Dissecting Kit
Bone Cutters
Dissecting Microscope
Histological Slides
- Ovary
- Uterus
- Testes
- Penis

Compound Microscope

Dissecting the reproductive system of the fetal pig affords us a unique opportunity to study major reproductive structures common to all mammals. We can then use this information to better understand the anatomy of the human reproductive system.

As you work through this basic overview of the porcine reproductive system, you will find that although much of the system is analogous to the human system, some significant differences do exist. These differences will be pointed out in the text. Your emphasis, however, should be on the similarities between the systems.

❑ Preparation

I. Introduction

A. Procedure

1➡ Regardless of whether you are working with a male fetal pig or a female fetal pig, you should begin by familiarizing yourself with the reproductive structures in the appropriate diagrams (Figures 69b–1● and 69b–2●).

2➡ When you have finished studying the reproductive anatomy of your own fetal pig, you should examine the reproductive structures of a fetal pig of the opposite sex. Throughout this exercise you should locate and/or identify each boldfaced item listed.

B. Anatomical Preparation

1➡ Prepare your fetal pig for the dissection of the reproductive system by placing the animal ventral side up and opening the abdominal cavity.

2➡ Note the location of the digestive and urinary structures. Use this information for an anatomical perspective as you work through your dissection.

Dissection

II. Female Fetal Pig

As you work through this part of the exercise on the female fetal pig, you are encouraged to refer often to Exercise 67 on the human female reproductive system.

A. Uterus

1➡ Begin your study of the female fetal pig by first locating the urinary **bladder**. In the female fetal pig the urinary bladder is a cumbersome structure centrally located in the ventral part of the lower central abdominal cavity. In the fetal animal, the urinary bladder is usually referred to as the **allantoic bladder.** The ventral part of this structure is the **urachus.** The urachus extends to the umbilicus and seems to leave the body as part of the umbilical cord. After birth the urachus atrophies to become the **median umbilical ligament.** The remainder of the allantoic bladder becomes the urinary bladder.

Immediately dorsal to the bladder is a muscular Y-shaped tube. The caudal arm of the Y is the **uterine body**. The two cranial arms of the Y are the **uterine horns**. The uterine horns are also known as **cornua** (sing., cornu). Hence, the fetal pig uterus with two horns is **bicornuate**. The porcine embryos implant in the cornua and not in the uterine body. (The human uterus has no discernible horns, so the human embryo normally implants in the body of the uterus.)

B. Uterine Tubes

1➡ Follow one uterine horn as it ascends the posterior abdominal wall. Toward the end of the horn, you will notice that the tube narrows appreciably. This narrowed portion is the **uterine tube**, known more familiarly in the human as the **fallopian tube** or the **oviduct**. Observe the flared end of the uterine tube, the **infundibulum**, as it wraps around an oval glandular structure. The finger-like projections of the infundibulum end are the **fimbriae,** and the glandular structure is the **ovary**.

2➡ Gently dislodge an ovarian tube (a term used sometimes for the most distal part of the uterine tube) and see if you can locate the **ostium**, the hole going down between the fimbriae. This hole leads into the lumen of the ovarian tube and subsequently into the uterine tube.

● **FIGURE 69b–1**
Female Reproductive System.

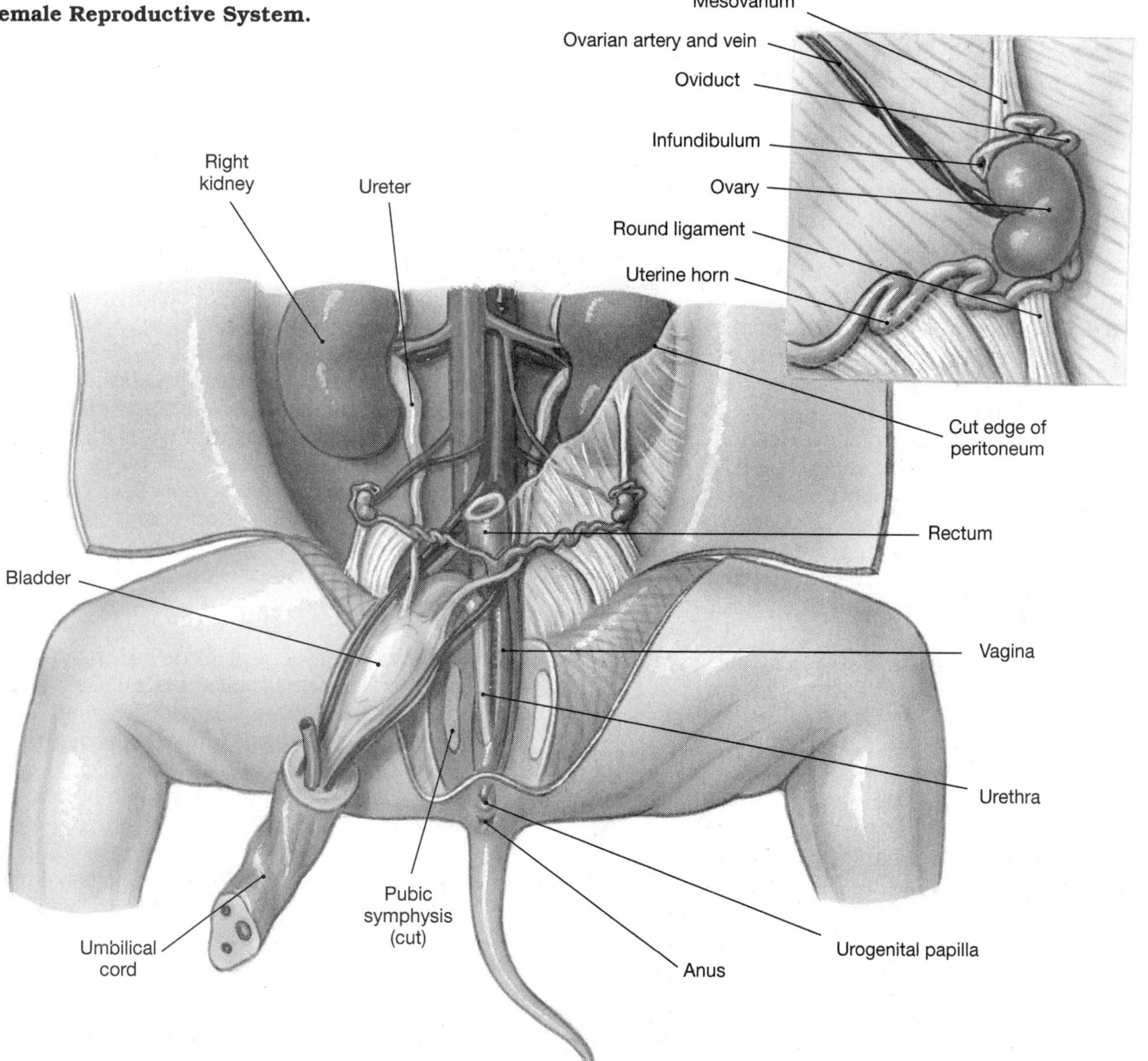

C. Uterine Ligaments

1➔ Carefully clear away unnecessary fascia, particularly from the posterior and medial aspects of the uterine horn. Then gently lift the uterine horn and observe the filmy tissue attaching the uterine tube and uterine body both to the abdominal wall and to the cranial part of the vagina. This tissue is the **broad ligament**, also known as the **mesometrium**. (Technically the mesometrium is only part of the group of broad ligaments, which also include the mesovarium and the mesosalpinx. The **mesovarium** is a posterior extension of the broad ligament that attaches to the ovary.)

2➔ Notice the thickened ridge of heavier tissue beginning just below the uterine tube and extending laterally and caudally toward the abdominal wall. You can hold up this ridge and see that it anchors additional filmy tissue to the abdominal wall. This cordy ridge is the **round ligament**. (Although the human does not have uterine horns, the uterine ligament support structures are approximately identical to those of the fetal pig.)

D. Ovary

1➔ As a rule of thumb, the porcine ovary is situated at or near the lateral margin of the pelvic inlet. In practice, however, particularly with adult animals that have borne young, the position of the ovary is quite variable. Identify the **ovarian ligament**, the thick cord attaching the ovary to the uterine horn. You may be able to locate the suspensory ligament holding the ovary to the posterior abdominal wall.

2➡ The entire ovary is covered by a single layer of cells, the **germinal epithelium**. The ovary proper has two major parts: (1) the external **tunica albuginea**, the fibrous capsule of the ovary, and (2) the internal **stroma**, which itself can be divided indistinctly into the **cortex** and the **medulla**. The **ovarian follicles**, sacs housing developing ova, are in the cortex. In the fetal pig, the ovary is quite nondescript. In the sow, however, the ovary resembles a large cluster of grapes.

3➡ You will not be able to distinguish the finer detail and subsections of the ovary by an external examination. Nevertheless, when you have completed your dissection, go back and remove one ovary and examine it under a dissecting microscope. Try bisecting the ovary. Depending on the age of the animal and the state of preservation, you may be able to distinguish the ovarian parts. Sketch what you see in the space provided at the end of this exercise.

(Although less grapelike, the human ovary is similar to the adult porcine ovary.)

E. Blood Supply

If it has been some time since you last studied the blood vessels in this area of the fetal pig, take a few minutes to refamiliarize yourself with the following blood vessels.

1➡ Locate the **aorta**. Just caudal to the renal arteries are the paired **ovarian arteries**. The ovarian arteries supply blood to the ovaries and to the uterus. The uterus is further supplied by the **internal iliac arteries**.

2➡ Locate the **inferior vena cava**. Notice that the **right internal ovarian vein** enters the vena cava directly caudal to the renal vein. The **left internal ovarian vein**, however, drains into the left renal vein. The uterus itself is drained by the **internal iliac vein**.

(The human vascular system for the reproductive organs is approximately the same as that of the fetal pig.)

F. Lower Reproductive Structures

1➡ Return to the uterus. To study the lower reproductive structures, you will need to take a pair of bone cutters and cut through the **symphysis pubis**, the ventral fibrous pad between the os coxae. In some fetal pigs cutting the bone takes a good deal of leverage. Be careful not to cut anything except the muscle and bone. Your instructor may direct you to break the pelvic bone.

2➡ Follow the uterus caudally. Depending on the age and condition of your fetal pig, you may be able to feel a depression in the ventral wall. This is the **cervix**, which marks the end of the uterus and the beginning of the **vagina**. The long fibrous vagina is thick and muscular. (The muscular nature of the vagina is more easily recognized in the sow than in the piglet.) The urethra, coming from the urinary bladder, opens into the ventral vaginal wall at the **urethral orifice**. The **vestibule**, or **urogenital sinus**, is the area common to the openings of both the vagina and the urethra.

(In the human no common urogenital sinus exists. Rather, the urethra and vagina each open separately to the outside.)

G. Vulva

1➡ Follow the urogenital sinus out through the opening, the **urogenital orifice**. The structure that seems to hold the urogenital orifice is the **genital papilla**. Examine the **external genitalia**, the **vulva**. The longitudinal folds are the **labia majora**. The **clitoris**, which is analogous to erectile tissue of the same name in the human, is, relatively speaking, long and flexous. Because of the small size of the fetal pig, however, you may have difficulty recognizing the clitoris, and you may not notice that the clitoral glans covers a fossa. If you are able to locate the clitoris on your fetal pig, be careful not to mistake the fold of mucous membrane that extends over the clitoris for the glans itself.

The labia minora are present, but you will probably not be able to distinguish these inner vulvar ridges.

2➡ Refer back to Exercise 67 and note the similarities between the fetal pig and human. On the material covered thus far, what point of comparison has interested you the most?

____

____

H. Layers of the Uterus

1➡ If your instructor so directs, cut the lateral wall of the urogenital sinus and vagina from the orifice to the uterine body. Although you can make this incision with your scissors, it may be easier for you to place your blunt nose probe into the orifice and, while holding it firmly against the *lateral wall*, use your scalpel to cut the tissue along the edge of the probe. (By placing your probe against the lateral wall, you ensure that you will not cut the urethral orifice.) Fold this tissue back and examine the structures you located in the previous sections.

On the floor of the vestibule you may notice a longitudinal fold and longitudinal ducts called the **canals of Gartner**.

2➡ If you cut into the wall of the uterine body, you should notice that it is quite muscular. In the fetal pig, it is often difficult to distinguish the layers of the uterus. Nevertheless, be aware that the inner layer is the **endometrium**. The majority of the tissue is the **myometrium**, which is actually three muscular layers. A serous layer, the **perimetrium**, covers the entire structure. Cut a small tissue sample from the uterine wall for examination under the dissecting microscope.

Sketch what you see in the space provided at the end of this exercise.

III. Male Fetal Pig

As you work through this part of the exercise on the male fetal pig, you are encouraged to refer often to Exercise 66 on the human male reproductive system.

A. Orientation

1➡ Orient yourself to the male reproductive system by first locating the bladder. In the male fetal pig the urinary bladder is a cumbersome structure centrally located in the ventral part of the lower central abdominal cavity. In the fetal animal, the urinary bladder is usually referred to as the **allantoic bladder**. The ventral part of this structure is the **urachus**. The urachus extends to the umbilicus and seems to leave the body as part of the umbilical cord.

2➡ Identify the ureter, which extends from the bladder to the kidney. Carefully clear away any loose connective tissue. Notice where the ureter enters the posterior caudal portion of the bladder.

3➡ Identify the tubular structure that loops over the ureter. This is the **vas deferens**, also known as the **ductus deferens**. Note that the vas deferens extends down into the pelvic cavity.

B. External Genitalia

1➡ Now shift your examination to the external genitalia. Locate the skin-covered sac, the **scrotum**, just ventral to the anus. (Even in the adult pig, the scrotum is not as well defined as it in in the human or in many domestic animals.)

Just caudal to the umbilical cord on the ventral surface of the animal, locate the well-defined opening, the **preputial orifice**. This orifice leads into an area known as the **preputial cavity**. The tissue surrounding the cavity is the **prepuce**. In the porcines, a **preputial gland** is located on the interior surface of the preputial cavity.

In humans as well as in most adult mammals, the **penis** is free within the preputial cavity. (In humans this means the prepuce is retractable; in some animals this means the penis can extend beyond the prepuce.) In the fetal pig, the epithelial surfaces of the prepuce and the penis are adhered. This shared structure is called the **balanopreputial fold**. (As the animal matures, cytolytic processes under the influence of the androgens will cause the epithelial surfaces to separate.)

2➡ If you have not done so previously, make a rectangular incision around the umbilical cord and the preputial orifice. Continue your incision toward the scrotum.

3➡ As you pull away the scrotal skin, you will expose the paired **testes**. The entire testicular structure is covered with a layer of white connective tissue known as the **tunica vaginalis**. The tunica vaginalis is composed of two layers, the external **parietal layer** and the internal **visceral layer**. You probably will not be able to distinguish these layers.

Cut a small sample of the tunica vaginalis for examination under the dissecting microscope. Sketch what you see in the space provided at the end of this exercise.

4➡ Remove the tunica vaginalis and expose the convoluted duct on the dorsal surface of the testis. This is the epididymis. Notice how the **epididymis** arises from inside the testis.

C. Testes

The testis itself is surrounded by another coat, the **tunica albuginea**. Fibrous extensions of the tunica albuginea dip into the testis, separating it into incomplete **lobules**. Each lobule contains the **seminiferous tubules**. Sperm stem cells, **spermatogonia**, are found toward the exterior of the seminiferous tubules. Also found in this area are the **Sertoli cells**, which supply nutrients to the developing sperm.

Between the seminiferous tubules are the **interstitial (Leydig) cells**, specialized interstitial endocrine cells that produce male sex hormones.

1➡ Although you will not be able to identify the preceding testicular structures on gross examination, after you have completed your dissection of the male fetal pig, remove a testis and bisect it under the dissecting microscope. Most gross structures should be readily visible. Sketch what you see in the space provided at the end of this exercise.

D. Duct System

Sperm are produced in successive stages, beginning in the outer regions of the seminiferous tubules. When fully formed (although still immature), the sperm enter the lumens of the tubules. Sperm then move through the seminiferous tubules to a mesh-like network known as the **rete testis**. The rete testis empties into a series of small ducts known as the **efferent ductules**, or the **ductus efferens**. These ductules unite to form the head of the epididymis, the structure you have already located.

1➡ Follow the tail of the epididymis. At the base of the testis, the epididymis turns cranially and becomes part of the **spermatic cord** as the vas deferens. Within the spermatic cord are the vas deferens, assorted blood vessels, and branches of the 10th thoracic nerve. The outer covering of the spermatic cord is continuous with the tunica vaginalis.

E. Into the Body Cavity

1➡ When you began studying the male fetal pig, you located the vas deferens as it looped over the ureter. You are now approaching the vas deferens from the other end. Your goal at this point is to trace the spermatic cord up into the inguinal area, through the inguinal canal, and into the abdomino-pelvic cavity. Use caution in cutting away the skin and connective tissue. Notice that after the inguinal canal only the vas deferens continues cranially.

(Human anatomy to this point is approximately the same as that of the fetal pig.)

F. Separation of the Os Coxae

1➡ To continue your study of the male reproductive system, it will be necessary for you to separate the os coxae. Use the bone cutters and make your cut through the symphysis pubis, the fibrous pad between the sections of the pelvic bone. In some fetal pigs, this will take some energy. Use caution so that you cut only the bone and muscle. Once you have cut through the bone, your instructor may direct you to break the pelvis so you can better see the internal genitalia.

G. Glands

1➡ Clear away any connective tissue so that the entire tract of the urethra is exposed. Go back to the bladder and identify the urethra as it extends caudally toward the floor of the pelvic cavity. Follow the vas deferens as it continues medially toward the urethra.

(In humans and in many other mammals you will find an enlarged area, the **ampulla**, at the end of the vas deferens immediately before it enters the urethra. The fetal pig does not have an ampulla. In humans also you will find a **seminal vesicle** — **vesicular gland** — which empties into the vas deferens just before the urethral junction. In fetal pigs and in certain other mammals the vesicular glands are located after the urethral junction.)

2➡ Notice the round glandular structure at the junction of the vas deferens and the urethra at the approximate level of the symphysis pubis. This is the **prostate gland**, which is basically a two-part gland — a body and a disseminate prostate. Prostate gland size is highly variable and, as in humans, an enlarged prostate gland (particularly in the adult) is not uncommon. Sometimes the gland is so small in the fetal pig that it is difficult to find. In the area of the prostate the urethra is known as the **prostatic urethra** or the **pelvic urethra**.

3➡ Below the prostate gland are the paired **bulbourethral glands**, also known as **Cowper's glands**. These glands are relatively large (elongated) in the fetal pig and remain large and cylindrical throughout life. (Cowper's glands in the human are in approximately the same relative position as they are in the fetal pig.) At this point the urethra is technically known as the **membranous urethra**.

4➡ Note that the urethra turns cranially and becomes part of the **penis**. In the adult pig the penis has a sigmoid flexure. It is the straightening of this flexure that allows protrusion of the penis. You will not see this flexure in the fetal pig. (The human penis does not have a flexure.)

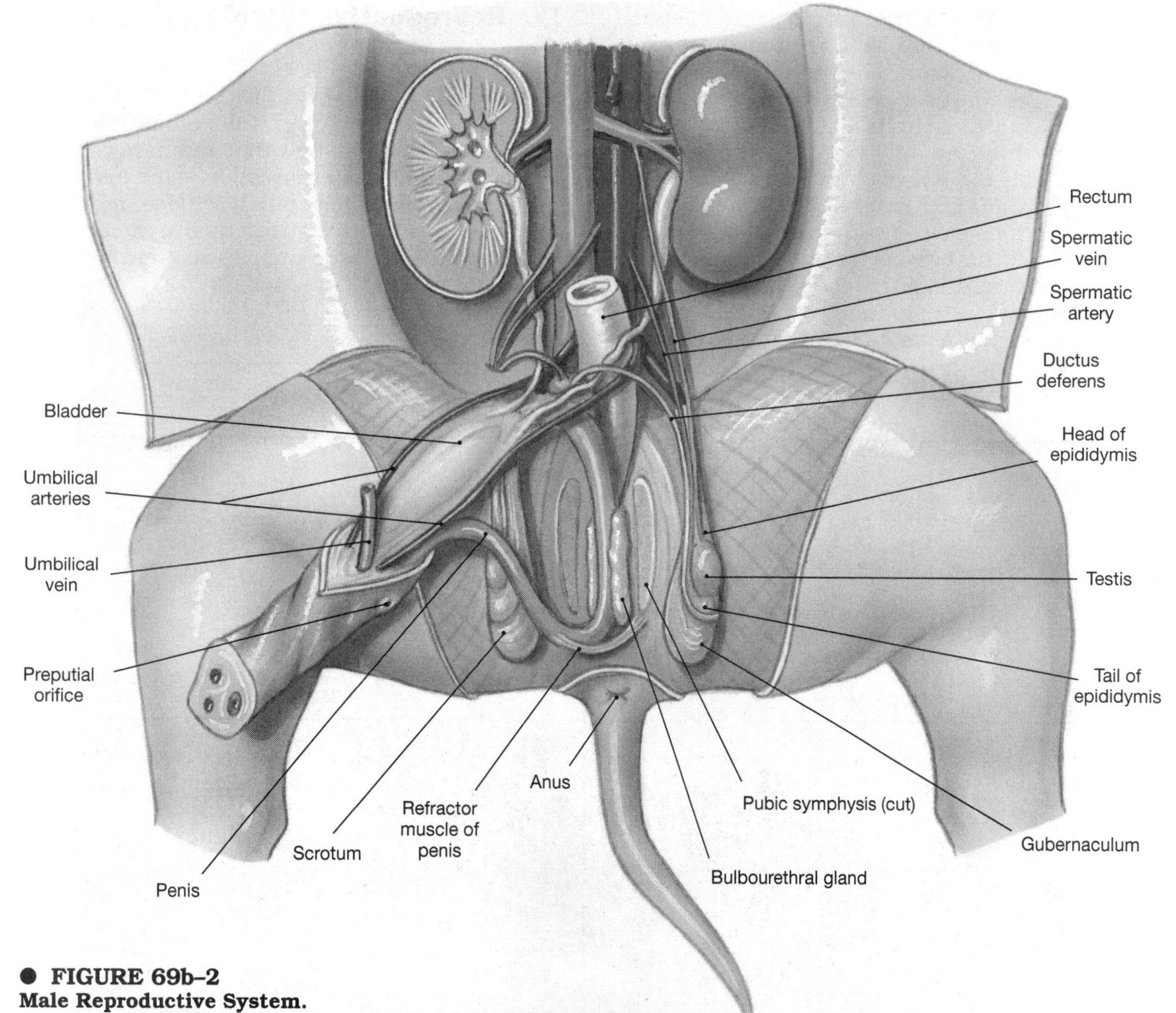

● **FIGURE 69b–2**
Male Reproductive System.

H. Penis

At this point the urethra enters the penis and becomes known as the **penile urethra**, or **cavernous urethra**.

1➡ If you have not already done so, cut the tough skin of the prepuce and expose the **glans penis**, the enlarged end of the penis.

I. Penis Cross Section

On cross section the penis can be divided into a spongy area, the **corpus spongiosum**, in the center of which can be found the urethra and paired cavernous areas, the **corpora cavernosa**, which house the deep arteries of the penis. The corpora cavernosa are separated by a **median septum**. (These structures are analogous to human structures.)

1➡ Locate the opening of the urethra to the external environment, the **urogenital opening**.

2➡ On a microscope slide of the penis, identify the preceding structures. (You may not specifically have a slide of a fetal pig penis, but in most mammals the penile structures are analogous.) Sketch what you see in the space at the end of this exercise.

J. Blood Supply

As with the female reproductive anatomy, you should take a few minutes to familiarize yourself with the blood vessels described in the following brief review. Knowing the blood vessels in this area provides both landmarks for identification and knowledge of how the reproductive structures receive their blood supply.

1→ Locate the **aorta**. Just caudal to the renal arteries are the paired **testicular arteries**. The testicular arteries supply blood to the testes. The penile blood supply comes from the pudendal branch of the femoral artery.

2→ Locate the **inferior vena cava**. Notice that the **right internal testicular vein** enters the vena cava directly caudal to the renal vein. The **left internal testicular vein**, however, drains into the left renal vein. The venous return from the penis occurs through drainage into the great saphenous vein and through the prostatic plexus.

IV. Reproductive Histology

A. Slides

Reproductive histology was examined in Exercises 66, 67, and 68. Now that you have completed your reproductive dissection, your instructor may direct you to view some additional reproductive slides. Also, this is a good time to examine any of the additional tissue samples you removed from your fetal pig.

TISSUE SAMPLES	TISSUE SAMPLES
TISSUE SAMPLES	TISSUE SAMPLES
TISSUE SAMPLES	TISSUE SAMPLES
TISSUE SAMPLES	TISSUE SAMPLES

❑ Additional Activities

NOTES

1. Research the embryological development of the fetal pig. Compare the similarities and differences between the male and female.
2. If your instructor so directs, take additional tissue samples from your fetal pig and examine them under the dissecting microscope.

❑ Lab Report

1. Referring to the numbered inquiries at the beginning of this exercise, complete the following box summary:

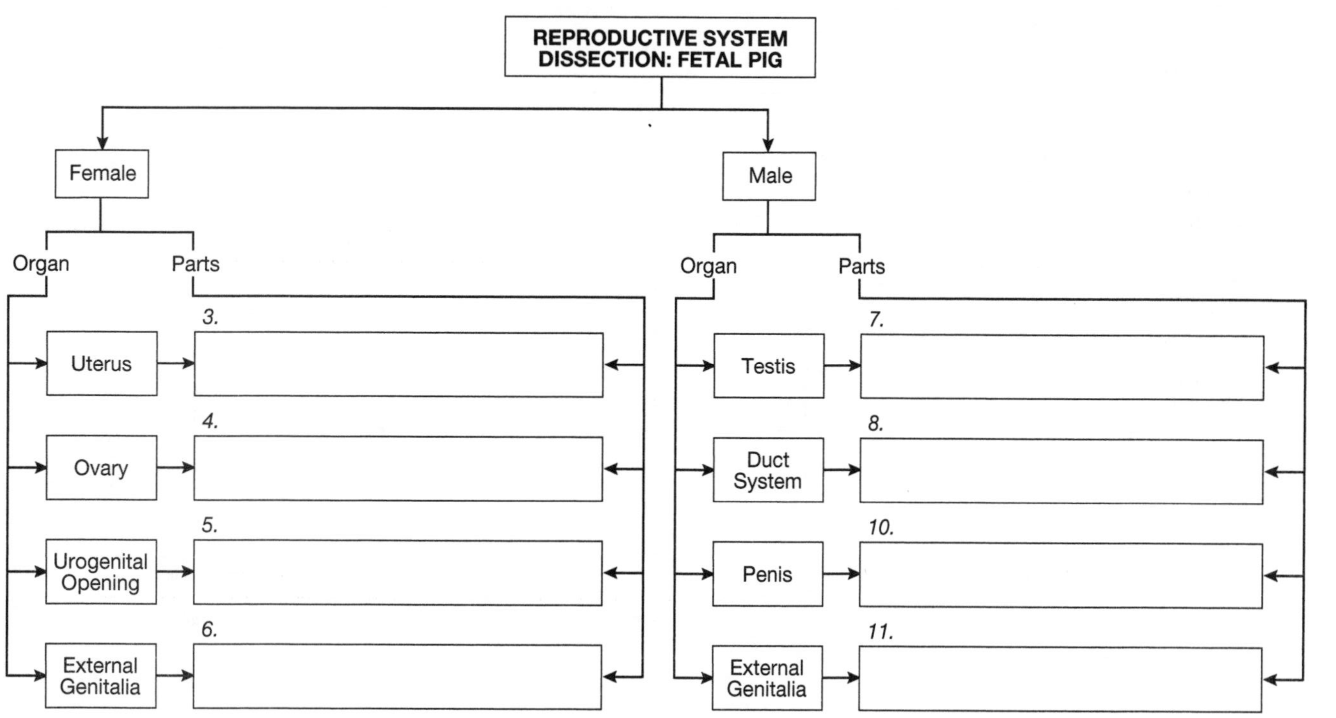

General Questions

NOTES

1. Trace the path of a sperm and an ovum from the testes and ovary through fertilization, implantation, development, and birth. Refer to Exercise 68 if necessary. Correlate that information with what you learned in this exercise.
2. What advantage is there to having the extensive arterial system to supply blood to the uterus?
3. What is the difference between the fetal bladder and the adult bladder? What are the fates of the parts of the fetal bladder? How is this information relevant to our study of the reproductive system?
4. List the differences in reproductive tract glands between the human male and the porcine male.
5. How is the human penis different from the porcine penis?
6. Use the following chart to list the similarities and differences between the human and porcine:

	Similarities	Differences
a. Uterus		
b. Ovary		
c. Urogenital sinus/ opening(s)		
d. Ligaments and other support tissue		

EXERCISE 70

Fertilization and Early Development

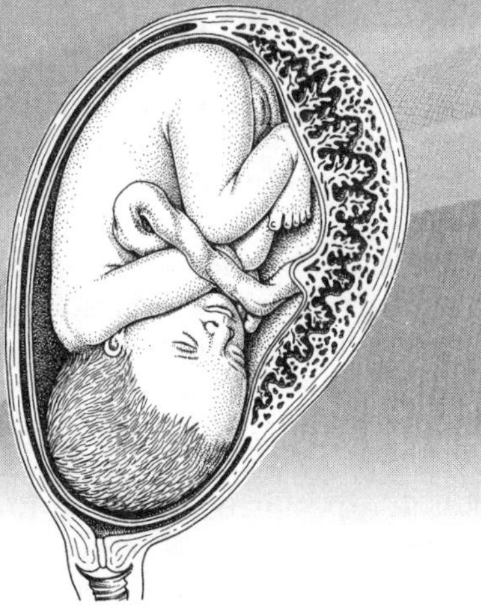

PROCEDURAL INQUIRIES

Preparation

1. What is meiosis?
2. Why is meiosis necessary?
3. What is gametogenesis? (What are gametes?)

Examination

4. What are the parts of the sperm cell?
5. What is fertilization?
6. What is capacitation?
7. What is amphimixis?
8. How does amphimixis relate to zygote formation?
9. What is the difference between an embryo and a fetus?
10. What is cleavage?
11. What are the parts of the blastocyst?
12. What are the parts and fates of the trophoblast?
13. What are the parts and fates of the inner cell mass?
14. What are the other human embryonic and extra-embryonic tissues?
15. What are the fates or functions of the embryonic and extra-embryonic tissues?

Additional Inquiries

16. What is epigenesis?
17. What is a discoidal placenta?

Key Terms

Allantois
Amnion
Amphimixis
Blastocoele
Blastocyst
Blastomere
Capacitation
Chorion
Chorionic Villi
Cleavage
Corona Radiata
Decidua Basalis
Decidua Capsularis
Decidua Parietalis
Discoidal Placenta
Embryo
Epigenesis
Fertilization
Fetus
Gamete
Gastrulation
Hyaluronidase
Inner Cell Mass
Morula
Ovulation
Pronucleus
Trophoblast
Yolk Sac
Zygote

Materials Needed

Charts and Diagrams
Models
Preserved Specimens, as Available
Compound Microscope
Prepared Slides
- Sperm (Mammalian)
- Unfertilized Eggs
- Fertilized Eggs
- Early Cleavage Stages

Every human being begins life as a single cell formed by the fusion of an ovum and a sperm. The process of uniting the two gametes has been the subject of controversy for millennia.

Aristotle believed the female contributed the physical form to the new human while the male contributed the vital force or soul. At different times during the succeeding centuries, various preformation theories were in the forefront. Preformation theories, which state the new human being is present (preformed) in miniature in either the egg or the sperm, were strongly advocated throughout the 17th and 18th centuries. In 1694 Dutch histologist Niklass Hartsoeker went so far as to sketch a preformed human infant in the head of a sperm cell.

Not everyone accepted preformation. In 1759 German embryologist Kaspar Wolff demonstrated that in the early developmental stages of the chick undifferentiated material gradually differentiated into specific layers of tissue and that from these layers of tissue an **embryo** gradually formed. Wolff called this progression **epigenesis**. Our present understanding of fertilization and physical development stems from epigenetic ideas.

In this laboratory exercise we will examine some of the essentials of fertilization and embryogenesis.

❑ Preparation

I. Background

A. Meiosis

Meiosis is the process whereby one diploid cell undergoes reduction/division to form four haploid cells. In humans, the functional haploid cells are the **gametes**, the egg (produced by the female) and the sperm (produced by the male). These gametes unite to form the new individual.

The human diploid number is 46. Every nongametic human cell has 46 chromosomes. We can easily see that if the egg and sperm were each to contain 46 chromosomes, the new individual would have 92. And the following generation's number would be 184. Then 368, 736, and so on.

The haploid cell has half the diploid number of chromosomes. We know that meiosis is necessary to reduce the number of chromosomes so that when the gametes (the egg and the sperm) unite, the resultant cell and all the cells of the new individual will have the correct diploid number of chromosomes.

The integrity of the chromosome number is not the onlyl reason for meiosis. Meiosis also allows for genetic recombination and thus for a healthier and more genetically diverse population.

B. Gametogenisis

From a meiotic standpoint, gametogenesis, the production of gamete cells, is essentially the same for both the egg and the sperm. This pattern is summarized in Figure 70–1●. Compare this figure with Figure 70–2●, which shows the differences between spermatogenesis (sperm production) and oogenesis (ovum production).

Certain differences do exist between egg production and sperm production. Notice that from each male primary spermatocyte, four functional spermatozoa are produced, while from each female primary oocyte only one functional ovum is obtained. (Four female haploid cells are actually or potentially produced, but only one is functional). The functional haploid cell, the ovum, receives most of the cytoplasm, organelles, and nutrients. The other haploid cells receive minimal cellular material.

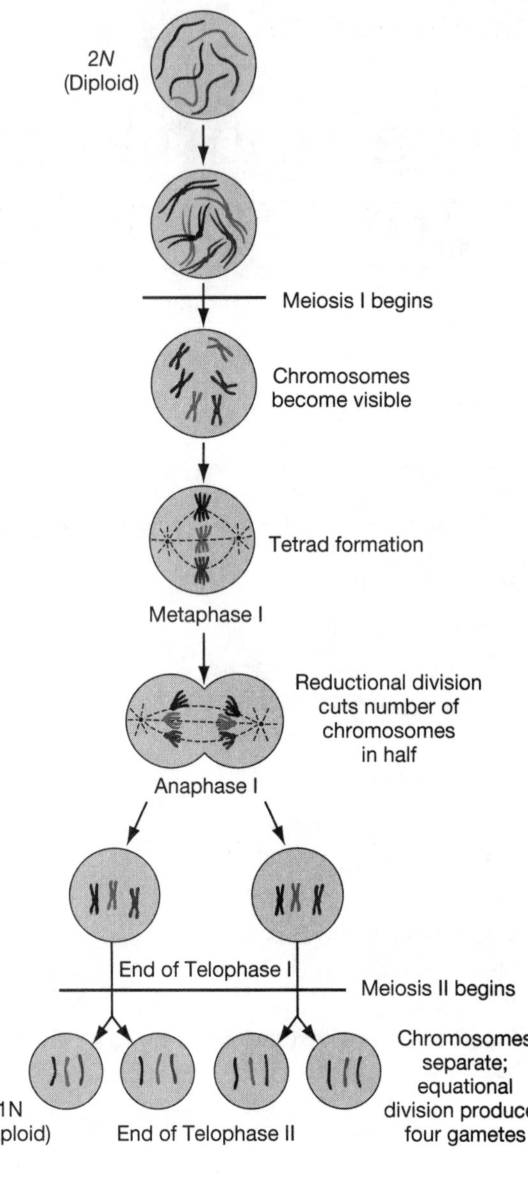

● **FIGURE 70–1**
General Pattern of Meiosis.

❑ Examination

C. Sperm

1➔ Obtain a compound microscope and slides of sperm. Your instructor may direct you to use the oil immersion lens to view the slide.

Note the head, neck, and tail of the sperm. Use Figure 70–3● as your guide. Sketch and label what you see.

Sperm cells

The sperm cells you are viewing may or may not be human. The sperm cells of each mammal are unique, varying in such characteristics as the shape

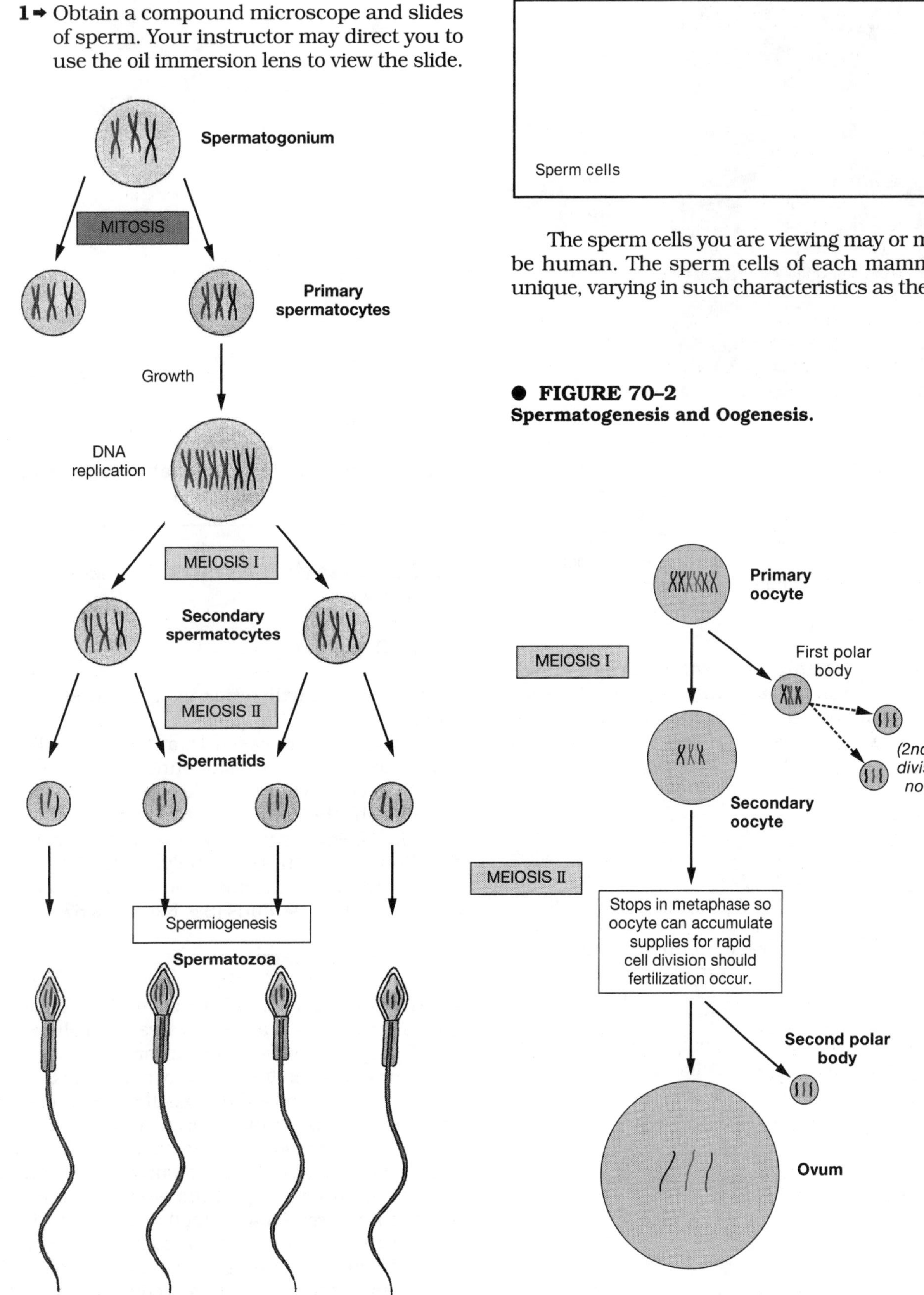

● **FIGURE 70–2**
Spermatogenesis and Oogenesis.

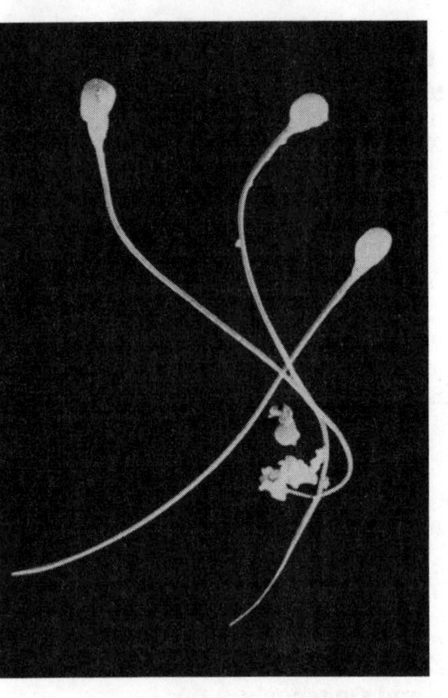

(a)

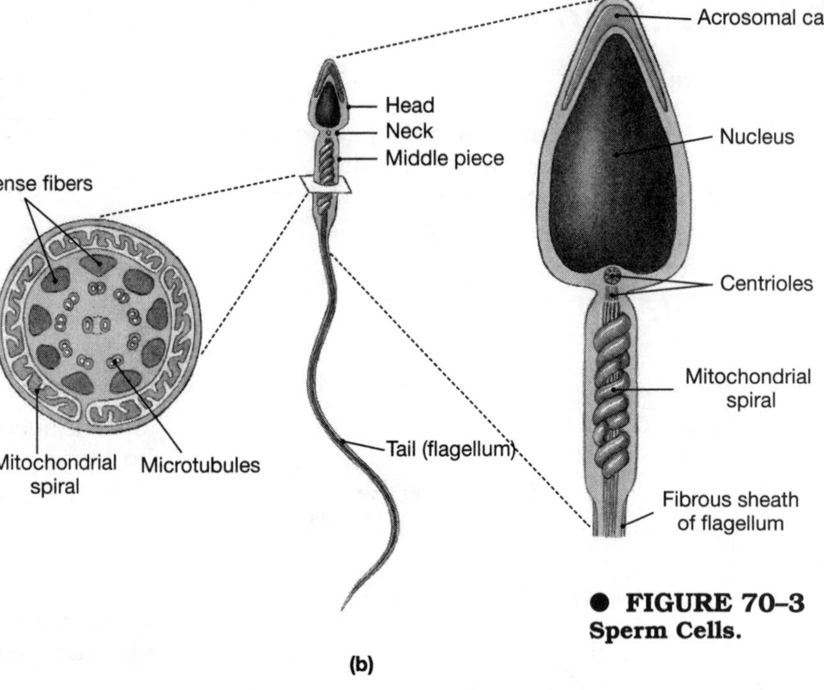

(b)

● **FIGURE 70–3 Sperm Cells.**

of the head, the length of the tail, or the construct of the acrosome. Sperm are also species-specific in their capacity to fertilize an ovum. However, if human sperm cells are not available, you will be able to gain a good perspective of the structure of the sperm by viewing the sperm of most any mammal.

D. Egg (Ovum)

You will probably be viewing starfish or sea urchin eggs, although any other ova may be substituted. (Mammalian ova are usually not available, primarily because mammals do not undergo massive ovulation. Again, the principles are the same, regardless of the type of ova.)

1 ➡ Obtain a slide of unfertilized eggs and one of fertilized eggs. You may be given a slide of mixed fertilized and unfertilized eggs. The nucleus will be distinct in the unfertilized eggs. In the fertilized eggs you should notice an extra membrane around the outside of the cell. This is the fertilization membrane, which we will discuss momentarily. You may also notice a small "blip" on the side of one or more fertilized egg. This "blip" is the second polar body. Refer to Figure 70–4●. Sketch and label what you see.

Ova

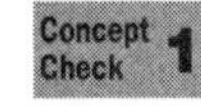

Concept Check 1 What is the functional advantage of the ovum being so large?

____

II. Fertilization Events [∞ *Martini, p. 1065*]

A. Ovulation

For normal **fertilization** to occur, several events must take place. First, an ovum (still suspended in metaphase II) is released from the ovary. This is **ovulation**. This ovum, surrounded by a crown of follicular cells known as the **corona radiata**, is picked up by the fimbriae and swept into the oviduct to begin its perilous journey toward the uterus. Fertilization, if it occurs, normally takes place in the distal third of the uterine tube, often between the ampulla and the isthmus (See Exercises 67 and 68).

B. Sperm Capacitation

When the sperm are deposited in the vagina, they are motile but are incapable of fertilizing the egg. By mechanisms still unclear, the sperm undergo **capacitation** (the capacity to fertilize the egg). The sperm move toward the egg. If no egg is present, the movement is far more random than if an egg is present. This would indicate a chemotactic attraction, a theory that has recently been demonstrated.

The acrosomal cap of the sperm contains the enzyme **hyaluronidase**, which creates gaps in the cement between the cells of the corona radiata, thus allowing the sperm to come in contact with the ovum. Although it takes a considerable number of sperm cells to break down the corona, only

● **FIGURE 70–4**
Ova.

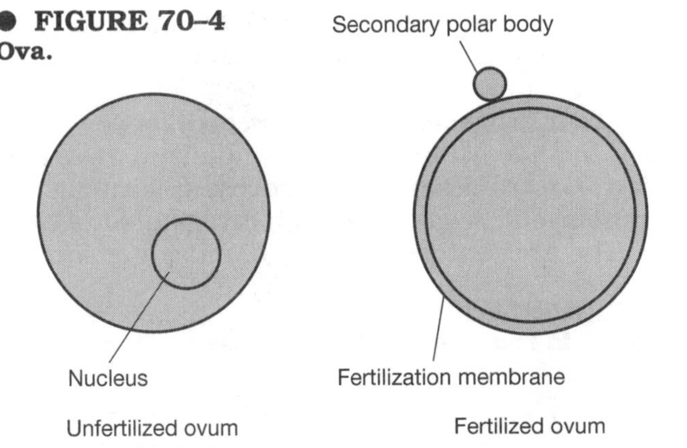

one spermatozoan accomplishes fertilization and activates the oocyte. The membranes of the egg and sperm fuse and the one sperm enters the ooplasm. A fertilization membrane forms around the ovum, preventing the admission of any other sperm.

C. FORMATION OF THE ZYGOTE [∞ MARTINI, P. 1067]

Fusion of the egg and sperm triggers completion of meiosis II, and the second polar body forms. The remaining nuclear material reorganizes as the **female pronucleus**. Meanwhile, the nuclear material of the spermatozoan becomes the **male pronucleus**. Each pronucleus forms an aster and spindle fibers. The pronuclei fuse in a process called **amphimixis**. Fertilization is complete. The new cell, now with the diploid chromosome compliment of 46, is known as a **zygote**.

III. Early Development

A. DEFINITION

Embryology deals with the development of the organism from the zygote stage until all major structures are formed. This is the initial stage of the development of the embryo. In humans this period lasts about two months. Beginning with the ninth week of development, the embryo is properly called a **fetus**, meaning "young one".

B. EARLY ZYGOTE DIVISIONS [∞ MARTINI, P. 1068]

Cleavage is the process whereby the zygote undergoes a series of mitotic divisions, which increase the cell number without increasing the cell mass. (The mass cannot increase because there is no source of nutrition or incoming raw materials.) Cleavage begins as fertilization is complete and continues as the ball of cells, known as a **morula**, traverses the uterine tube and reaches an implantation point, usually on the upper posterior wall of the uterine body. Each individual cell is a **blastomere**.

● **FIGURE 70–5**
Cleavage and Blastocyst Formation.

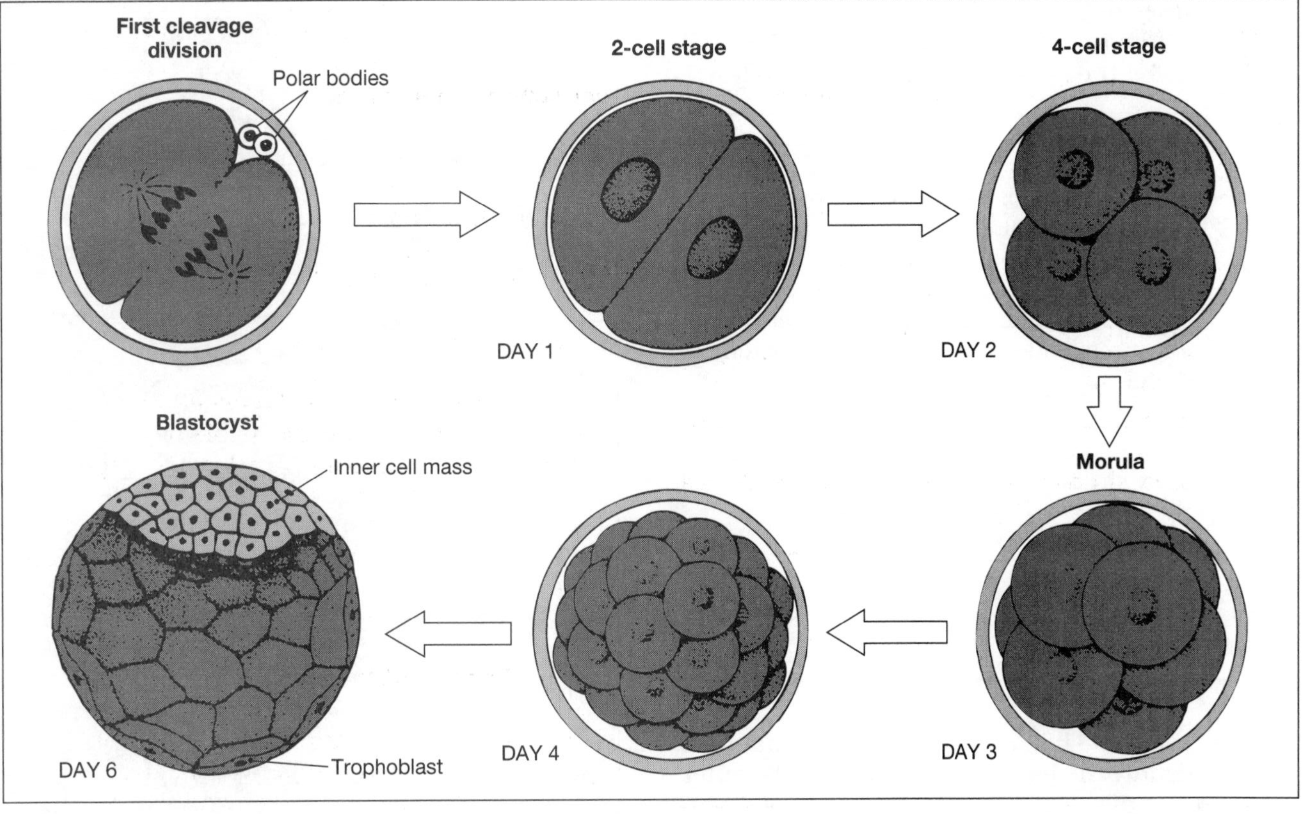

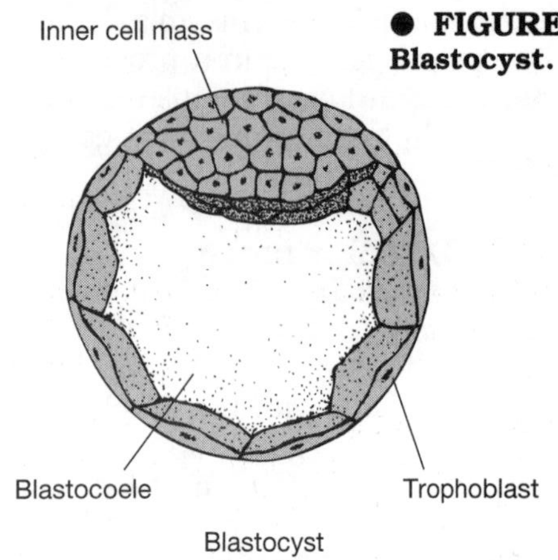

● **FIGURE 70–6**
Blastocyst.

1➡ Follow the process of cleavage and morula development in Figure 70–5●.

The journey to the uterus takes from five to seven days. About the sixth or seventh day, the morula begins to hollow out forming a **blastocyst** (blastula in some other organisms). Several distinct blastocyst structures can be identified.

2➡ Examine Figure 70–6● for blastocyst development and identify the following structures. The cavity is the **blastocoele**. The outer part of the circle is the **trophoblast**. The trophoblast will be involved with implantation, hormone production, and placenta formation. The thickened area is the **inner cell mass**. This structure will develop into the embryo. In your own words, describe the appearance of this blastocyst.

____

____

3➡ Obtain a slide of the early cleavage stages. Count the numbers of cells in each ball of cells. You probably will not be viewing mammalian cell masses. At these early stages, however, cleavage and blastulation are virtually identical for all deuterostomes. Sketch and label what you see.

Early cleavage stages

4➡ Obtain a slide of blastocysts or blastulas. Identify the hollowed-out ball and the outer layer of cells. Following this stage of development, certain major differences do exist among the species.

(Your instructor may have set out for you slides that include echinoderm or amphibian **gastrulation**. Gastrulation is the process of forming the germ layers. A gastrula is an invaginated blastula. The **archenteron** is the primitive gut formed

● **FIGURE 70–7**
Stages in the Implantation Process.

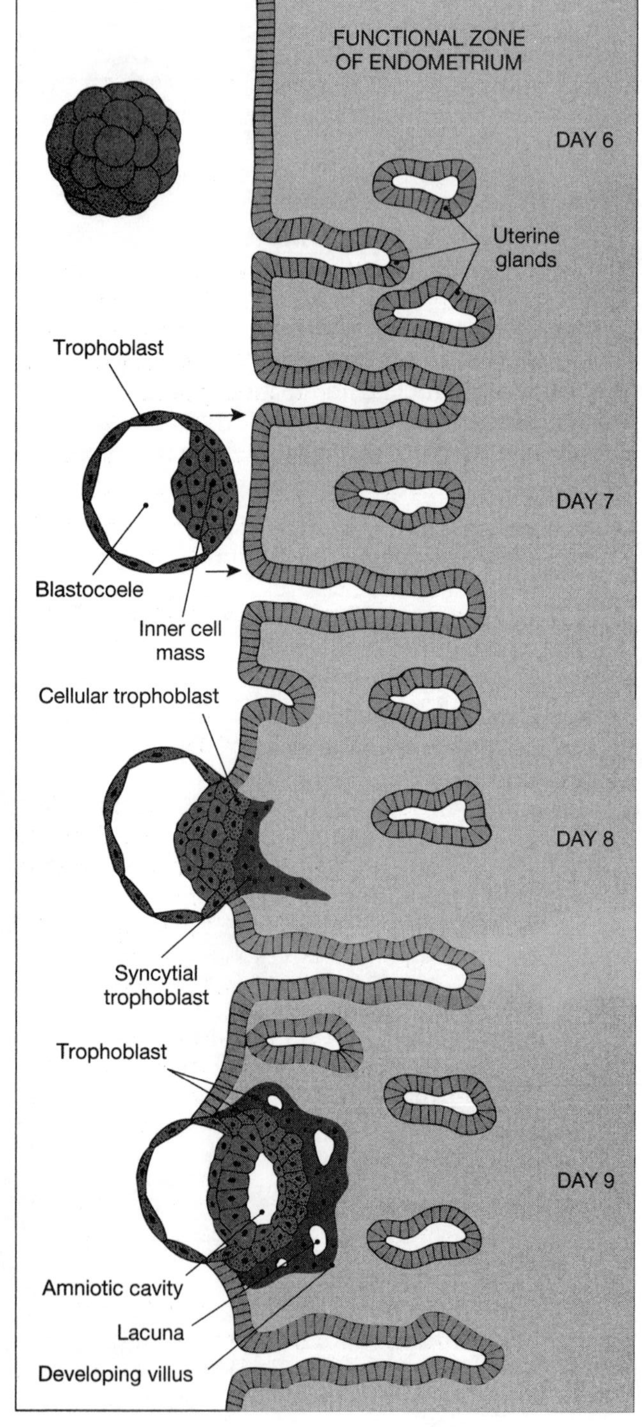

in lower deuterostomes by this invagination. Humans do not form an archenteron. Nevertheless, it is often valuable to study this structure to gain a perspective on development and structural differentiation.)

Sketch and label what you see when viewing the blastocysts or blastulas.

Blastocyst/blastula

IV. Implantation

A. Implantation Process

1➡ Once the blastocyst reaches the uterus, implantation must occur. The process begins when the blastula contacts the uterine lining. Study Figure 70–7●. In your own words, explain what is happening in this process.

____

____

____

____

V. Placentation (Development of the Placenta)

A. Extraembryonic Membranes (Placenta Tissues) [∞ Martini, p. 1072]

1➡ Refer to the charts, diagrams, and models in your laboratory. Trace the development of the extraembryonic membranes (or placental tissues): yolk sac, amnion, allantois, chorion, chorionic villi, decidua capsularis, decidua basalis, decidua parietalis. Label these structures on Figure 70–9●.

Yolk sac First extraembryonic membrane, responsible for early red blood cell production (probably beginning about gestational eighth day). The human yolk sac is not a source of nutrition.

Amnion Transparent sac covering the fetus. The amnion and the chorion fuse in the human.

Allantois Extraembryonic structure developing into the urachus and umbilical cord. In some animals, the allantois fuses with the chorion.

Chorion Extraembryonic structure developing into the fetal placenta. The amnion and the chorion fuse in the human.

Chorionic villi Mesodermal extensions forming an intricate network with the maternal endometrium.

Decidua capsularis Thin, external portion of the endometrium, not embedded in the uterine wall, surrounding the developing fetus.

Decidua basalis Disc-shaped portion of the endometrium where the placental functions are concentrated.

Decidua parietalis Uterine endometrium having no contact with the chorion.

B. Embryonic Sac

The human embryonic sac is a double-layered structure with an amniotic portion and a chorionic portion. The human placenta is called **discoidal** because the placenta proper is limited to a disc-shaped placental area in the maternal endometrium. (See the section on dissecting the pregnant feline uterus in Exercise 69a.)

In some animals, up to six layers of tissue separate the mother and fetus. In the human, only three layers — the fetal chorion, connective tissue, and endothelium — separate the mother and fetus.

Concept Check 2 When a pregnant woman's "water" breaks, what fluid forms that water?

____

VI. Development

Although it is beyond the scope of this exercise to explore the later stages of human development, it is important for you to gain a sense of the developmental continuum. Figure 70–8● will give you a chronology of major embryological and fetal events. Study this chart and then answer these questions:

1➡ Which systems develop first? Why is this logical?

____

____

2➡ Which systems develop last? Why is this logical?

____

____

3➡ Which systems are complete at birth?

____

____

4➡ Which systems are incomplete at birth?

____

____

Gestational Age (Months)	*Length and Weight*	*Integumentary System*	*Skeletal System*	*Muscular System*	*Nervous System*	*Special Sense Organs*
1	5 mm 0.02 g		(b) Somite formation	(b) Somite formation	(b) Neural tube	(b) Eye and ear formation
2	28 mm 2.7 g	(b) Nail beds, hair follicles, sweat glands	(b) Axial and appendicular cartilage formation	(c) Rudiments of axial musculature	(b) CNS, PNS organization, growth of cerebrum	(b) Taste buds, olfactory epithelium
3	78 mm 26 g	(b) Epidermal layers appear	(b) Ossification centers spreading	(c) Rudiments of appendicular musculature	(c) Basic spinal cord and brain structure	
4	133 mm 150 g	(b) Hair, sebaceous glands (c) Sweat glands	(b) Articulations (c) Facial and palatal organization	Fetus starts moving	(b) Rapid expansion of cerebrum	(c) Basic eye and ear structure (b) Peripheral receptor formation
5	185 mm 460 g	(b) Keratin production, nail production			(b) Myelination of spinal cord	
6	230 mm 823 g			(c) Perineal muscles	(b) CNS tract formation (c) Layering of cortex	
7	270 mm 1492 g	(b) Keratinization, nail formation, hair formation				(c) Eyelids open, retina sensitive to light
8	310 mm 2274 g		(b) Epiphyseal plate formation			Taste receptors functional
9	346 mm 2912 g					
Postnatal development		Hair changes in consistency and distribution	Formation and growth of epiphyseal plates continue	Muscle mass and control increase	Myelination, layering, CNS tract formation continue	

● **FIGURE 70–8**
Overview of Early Development.

Endocrine System	*Cardiovascular and Lymphatic Systems*	*Respiratory System*	*Digestive System*	*Urinary System*	*Reproductive System*
	(b) Heartbeat	(b) Trachea and lung formation	(b) Intestinal tract, liver, pancreas (c) Yolk sac	(c) Allantois	
(b) Thymus, thyroid, pituitary, adrenal glands	(c) Basic heart structure, major blood vessels, lymph nodes and ducts (b) Blood formation in liver	(b) Extensive bronchial branching into mediastinum (c) Diaphragm	(b) Intestinal subdivisions, villi, salivary glands	(b) Kidney formation (adult form)	(b) Mammary glands
(c) Thymus, thyroid gland	(b) Tonsils, blood formation in bone marrow		(c) Gallbladder, pancreas		(b) Definitive gonads, ducts, genitalia
	(b) Migration of lymphocytes to lymphatic organs, blood formation in spleen			(b) Degeneration of embryonic kidneys	
	(c) Tonsils	(c) Nostrils open	(c) Intestinal subdivisions		
(c) Adrenal glands	(c) Spleen, liver, bone marrow	(b) Alveolar formation	(c) Epithelial organization, glands		
(c) Pituitary gland			(c) Intestinal plicae		(b) Testes descend
		Complete pulmonary branching and alveolar formation		Complete nephron formation at birth	Descent complete at or near time of delivery
	Cardiovascular changes at birth; immune system becomes operative thereafter				

Note: (b) = begin formation; (c) = complete formation.

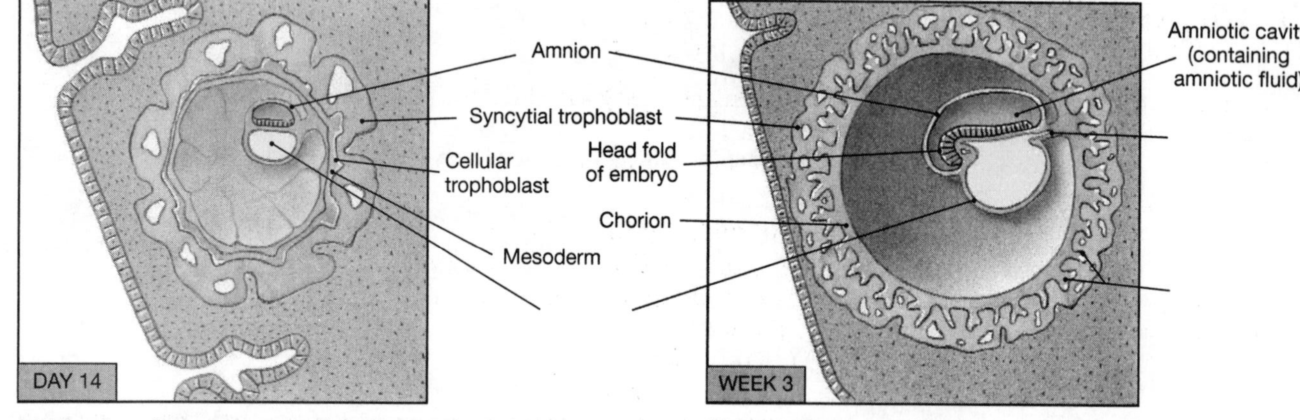

(a) Migration of mesoderm around the inner surface of the trophoblast creates the chorion. Mesodermal migration around the outside of the amniotic cavity, between the ectodermal cells and the trophoblast, creates the amnion. Mesodermal migration around the endodermal pouch below the blastodisc creates the definitive yolk sac.

(b) The embryonic disc bulges into the amniotic cavity at the head fold. The allantois, an endodermal extension surrounded by mesoderm, extends toward the trophoblast.

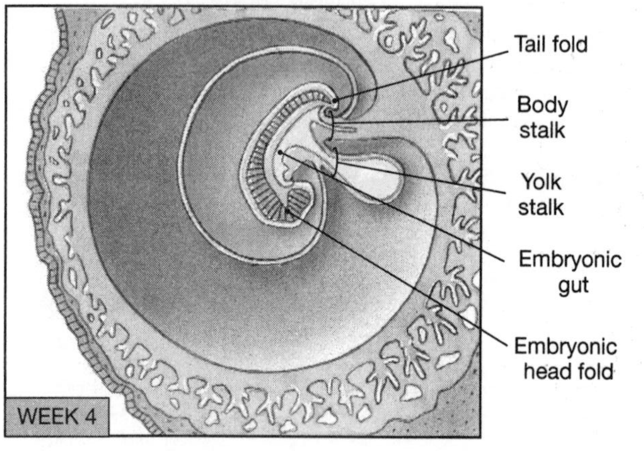

(c) The embryo now has a head fold and a tail fold. Constriction of the connection between the embryo and the surrounding trophoblast constricts the yolk stalk and body stalk.

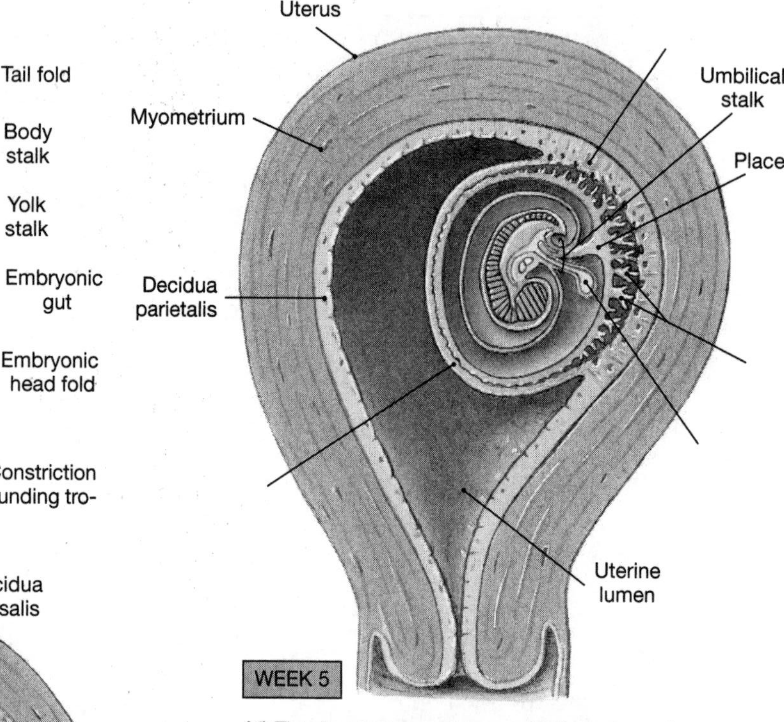

(d) The developing embryo and extraembryonic membranes bulge into the uterine cavity. The trophoblast pushing out into the uterine lumen remains covered by endometrium, but no longer participates in nutrient absorption and embryo support. The embryo moves away from the placenta, and the body stalk and yolk stalk fuse to form an umbilical stalk.

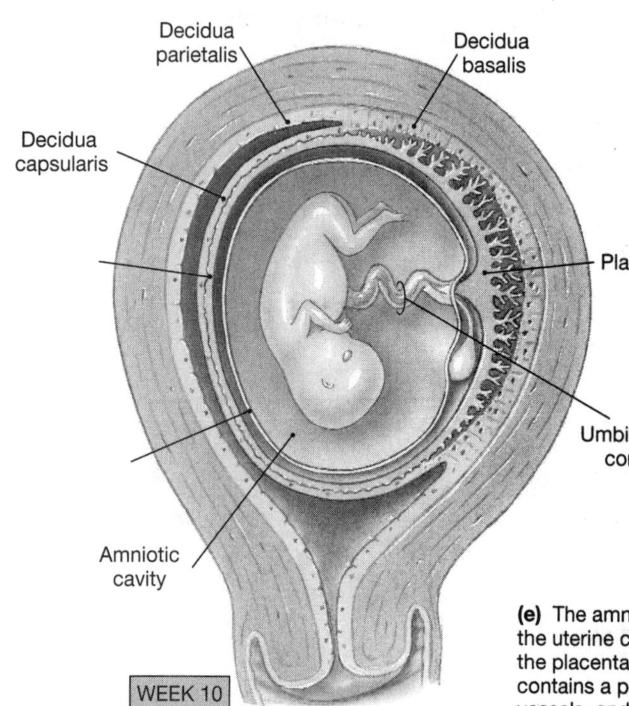

(e) The amnion has expanded greatly, filling the uterine cavity. The fetus is connected to the placenta by an elongate umbilical cord that contains a portion of the allantois, blood vessels, and the remnants of the yolk stalk.

● FIGURE 70–9 Embryonic Membranes and Placenta Formation.

❑ Additional Activities

NOTES

1. Compare the fertilization and embryological processes in different animals with those processes in humans.
2. Research cellular competence and induction. What factors determine the fate of a particular cell?
3. Research the types of placentas found in different animals.
4. Find out why some twins (identical as well as fraternal) are born with one placenta while others are born with two. (Ususally fraternal twins do have two placentas; occasionally they do not.)

Answers to Selected Concept Check Questions

1. The ovum has stockpiled the ribosomes, RNA's, mitochondria, nutrients, and other cellular organelles and products. These will be necessary after fertilization when cell division is occuring too rapidly for transcription and translation to keep up.
2. The "water" is the amniotic fluid.

❑ Lab Report

1. Referring to the numbered inquires at the beginning of this exercise, complete the following box summary:

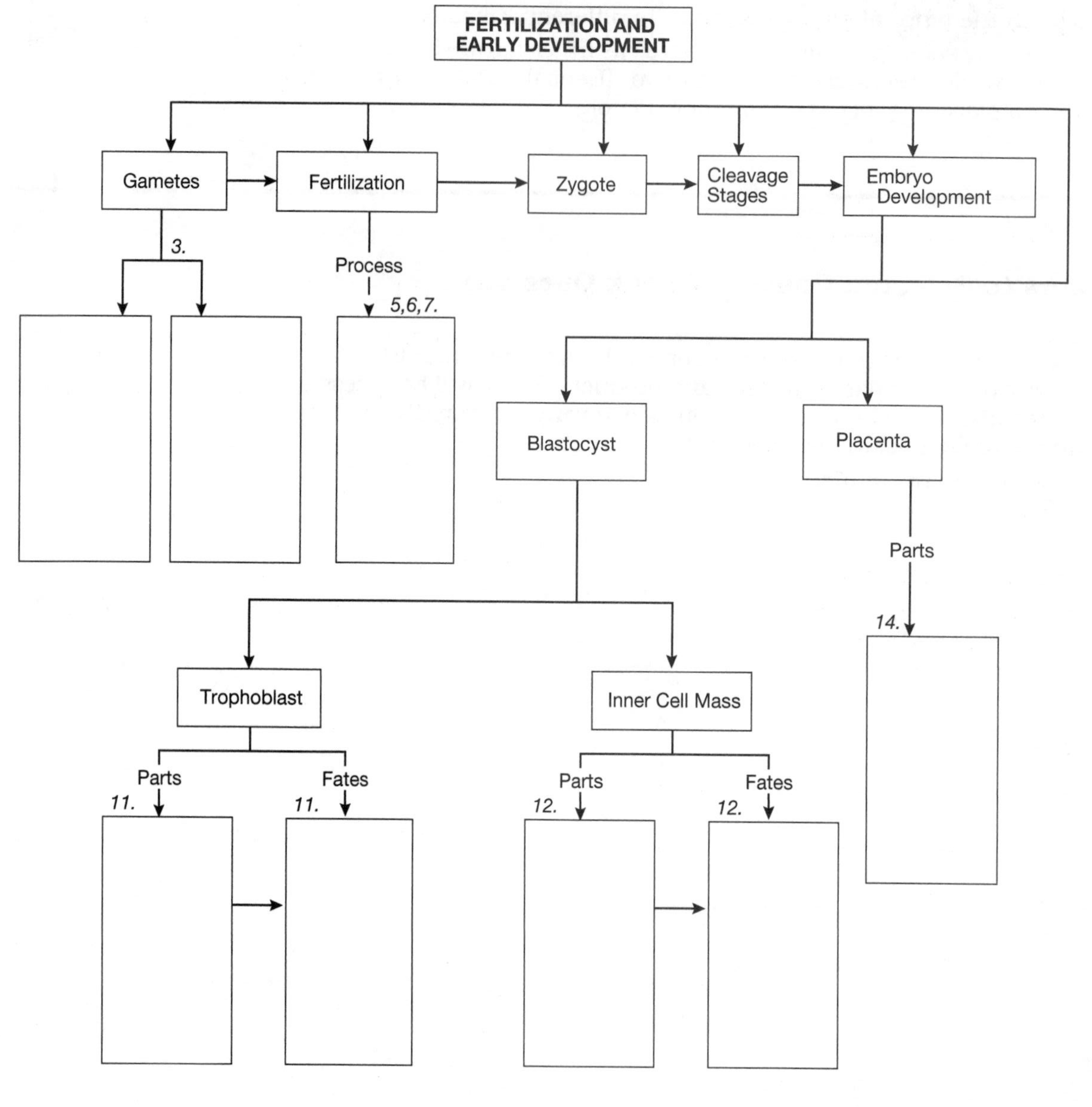

General Questions

NOTES

1. Explain the functional advantage of meiosis.
2. List the events of fertilization and implantation.
3. Explain the function of each of the extra-embryonic membranes.
4. Compare the physical appearance of the egg and the sperm.
5. How does epigenesis relate to our modern knowledge of embryology?
6. What is the difference between a zygote, an embryo, and a fetus?
7. Why is the human placenta called discoidal?

EXERCISE **71**

Heredity: Principles of Inheritance

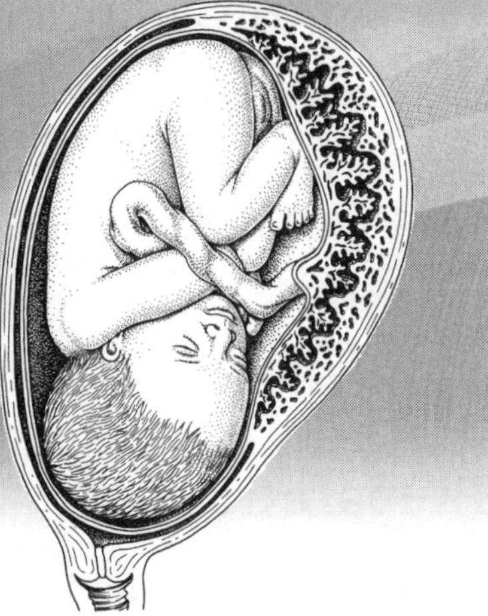

PROCEDURAL INQUIRIES

Preparation

1. What is the difference between genotype and phenotype?
2. What is the difference between a dominant trait and a recessive trait?
3. What is codominance?
4. What is partial dominance?
5. What is lack of dominance?
6. What is epistasis?
7. What is the difference between simple inheritance and polygenic inheritance?

Practice

8. How do you set up a Punnett Square?
9. What is the difference between a monohybrid cross and a dihybrid cross?
10. What do we mean by sex-influenced characteristics?
11. What do we mean by sex-linked characteristics?

Additional Inquiries

12. What is the difference between haploid and diploid?
13. What is the relationship between a gene, a locus, and an allele?
14. What are holandric genes?
15. What is a "carrier"?

Key Terms

Allele
Autosome
Carrier
Codominance
Dihybrid Cross
Diploid
Dominant
Epistasis
Gene
Genotype
Haploid
Heterosome
Heterozygous
Holandric
Homozygous
Incomplete Dominance
Monohybrid Cross
Multifactorial Inheritance
Phenotype
Polygenic Inheritance
Punnett Square
Recessive
Sex-Influenced
Sex-Linked
Simple Inheritance
X-Linked

Materials Needed

None

Fold your hands together. Which thumb is on top, left or right? __________ That is genetics in action; the thumb placement pattern is an inherited characteristic. Genetics is the means for passing biological traits from one generation to the next.

In eukaryotic cells, chromosomes occur in pairs. Humans have 23 pairs of chromosomes. Twenty-three is the **haploid** number, the number representing one member of each chromosome pair. (Recall meiosis.) Forty-six is the **diploid** number, the number representing two members of each pair, the complete chromosome complement in humans.

Of the 23 chromosome pairs, 22 are homologous pairs, that is, the two members of the pair are identical on gross examination. The homologous chromosomes are often called **autosomes**. The remaining pair is the sex-determining pair. The human female sex pair consists of two "X" chromosomes. The human male sex pair is composed of one "X" chromosome and one "Y" chromosome. The X and Y chromosomes are not homologous because they do not have corresponding components and are thus sometimes called **heterosomes.**

In this laboratory exercise we will demonstrate some of the basic principles of heredity.

❑ Preparation

I. Genetic Background

A. Genes

As you know, each chromosome is a molecule of DNA. A functional unit of the DNA molecule is a **gene**. The genes code for proteins and are ultimately responsible for the manifestation of the characteristics of the organism.

The location a gene occupies on a chromosome is the **locus**, and the genes on homologous chromosomes have corresponding loci.

Genes often come in different forms called **alleles**. Alleles differ in their nitrogen base sequence, and thus each allele will produce a functionally different protein.

(We often speak colloquially of a **blue eye gene** or a **brown eye gene** or some other color gene. Scientifically, we should be speaking of "the allele that codes for" whatever characteristic we are discussing.)

B. Genetic Inheritance

1➡ Be familiar with the terms in this section.

Since homologous chromosomes are paired, each body cell that is not a gamete will contain two alleles for any given trait. The pair of inherited alleles may be the same or they may be different. If a cell has two copies of the same allele, that cell is **homozygous**. If the alleles are different, the cell is **heterozygous**. Heterozygous individuals are often termed **carriers**, that is, carriers of the recessive trait even though the trait is not expressed.

Genotype is a term referring to which specific genes for a trait are present; a genotype can be symbolized using letters to represent the alleles. **Phenotype** refers to the physical manifestation of the particular genotype.

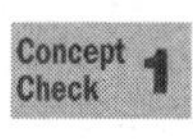

When you folded your hands — as per the opening instructions — and you noted that either your left or right thumb was on top, was this left or right situation an example of your genotype or your phenotype?

____

C. Dominant/Recessive Theory

1➡ Consider these points:

a. Traits are usually discussed in terms of their expression in the individual. Often, this discussion is limited to **dominant** and **recessive**. In this context, dominant means the trait is expressed if that allele is present, and recessive means the trait is expressed only if the dominant allele is not present.

b. According to classical (post-Mendelian) dominant/recessive theory, genetic traits are expressed as: (1) complete dominance, (2) codominance, (3) partial dominance, (4) lack of dominance.

In the case of complete dominance, when two different alleles are present, only one allele is expressed. Codominance means that both allelic traits are expressed equally (as in the AB blood type). Partial (or incomplete) dominance is a blending of the traits, and lack of dominance states that neither allele demonstrates dominance. (Some people do not make a distinction between partial dominance and lack of dominance.)

c. The dominant/recessive terminology used today often invites students to think of dominant traits as strong or aggressive and of recessive traits as weak or passive. Modern research demonstrates that this view is simply not true. For instance, type O blood is recessive, yet this is the most common blood type. (Refer back to the blood type population distribution chart in Exercise 47 of this book.)

D. Gene Expression

Modern research also indicates that dominant genes do not "turn off" or "shut down" recessive genes. Rather, in most cases, dominant and recessive alleles are expressed or not expressed; polypeptides are coded for or not coded for, according to the quantities expected from two separate alleles. The difference can be seen in the manifestation of this expression.

1➡ Let us look at two examples:

a. Ehlers-Danlos syndrome is an inherited disorder of the elastic connective tissue. The afflicted person has fragile, soft, hyperelastic skin, hyperextensible joints, and assorted other visceral malformations, scars and pseudotumors.

Ehlers-Danlos syndrome type IV is often considered to be a dominant trait. In the normal individual, the normal allele codes for a polypeptide in the collagenous connective tissue. The abnormal allele codes for an abnormal polypeptide that prevents proper connective tissue formation. In analyzing the heterozygous person, both normal and abnormal collagen can be found. This indicates again that both allelic traits are expressed.

b. Tay-Sachs disease, which primarily affects those of Eastern European Ashkenazi Jewish extraction, causes progressive neurological deterioration and results in mental and physical retardation and eventually in death, usually before the age of 3.

Tay-Sachs disease is often considered to be a recessive trait. The afflicted person would have to have two copies of the Tay-Sachs allele.

But Tay-Sachs disease is caused by the *absence* of the enzyme hexoseaminidase, an enzyme responsible for the metabolism of certain lipids (gangliosides). In the Tay-Sachs homozygous individual, hexoseaminidase is not coded for, and the myelin sheaths in the nervous system are destroyed because of the buildup of the gangliosides.

The person who is heterozygous produces hexoseaminidase. On gross examination, the heterozygous person cannot be distinguished from the nonafflicted, noncarrier person. However, upon serum analysis, we find that the heterozygous individual produces only about 50% of the normal amount of the enzyme. (This decreased quantity is quite sufficient for normal needs.) Obviously, in the homozygous "normal" individual, both alleles are coding. In the heterozygous individual, one allele is coding while the other is not coding. In what ways does this explanation alter our opinion of dominant/recessive characteristics? ______

______

How are Ehlers-Danlos syndrome and Tay-Sachs disease similar? ______

______

E. Additional Types of Gene Expression

1 ➡ Consider the following variations on simple genetic expression.

a. **Epistasis** (sometimes referred to as masking) occurs when the action or expression of one gene influences the action or expression of another gene. For instance, a particular gene may code for a protein that binds to a place on a chromosome and prevents another gene from being coded. Or an epistatic gene might code for a protein that inactivates another protein. The list of possibilities is really quite long.

(An epistatic condition worth mentioning here concerns eye color. Brown eyes are dominant; blue eyes are recessive. From a straight genetics standpoint, two blue-eyed individuals cannot have a brown-eyed child. However, the difference between brown and blue is the location of the pigment melanin. In brown-eyed individuals, the melanin is found toward the front of the iris; in blue-eyed individuals, the melanin is toward the back. According to some experts, in about 8 of 1,000 cases the gene dictating the location of the melanin expression is masked. Thus, a person may be phenotypically blue-eyed but genotypically brown-eyed. This epistatic condition might be passed on. The "blue-eyed" person, who should really be brown-eyed might well pass along the brown gene to his or her offspring. Following this logic, two phenotypically blue-eyed people could indeed have a brown-eyed child if at least one of the parents was genotypically brown-eyed. (Do you see why — despite pleas of innocence — murders, divorces, and lawsuits have occurred over this blue-eyed/brown-eyed problem?)

b. **Simple inheritance** is the situation existing when only one gene is completely responsible for the expression of a trait. In reality, probably very few traits are the result of pure, simple inheritance. Some traits, however, are probably only indirectly influenced by other gene actions.

c. **Polygenic inheritance** (or **multifactorial** or **multiple allelic inheritance**) occurs when more than one gene is responsible for the expression of a single trait. Most traits are probably at least indirectly the result of multifactorial inheritance.

❑ Practice

II. Classical Dominant/Recessive Situations

A. Working with Dominant/Recessive Traits

This present thinking does not invalidate our classical definitions of dominant and recessive. Rather, this new thinking should strengthen our understanding of why dominant and recessive exist. When working with dominant and recessive alleles, we should keep in mind the more scientific explanation of dominant and recessive coding. Redefine dominant and recessive. ______

______

______

Trait — Dominant	Trait — Recessive	Notations
Achondroplasia	Normal long bone cartilage	Defective cartilage formation
Adult Polycystic Kidney Disease	Normal kidney	
Anaridia	Normal eye iris	No iris
Astigmatism	No astigmatism	
Bent little finger	Straight little finger	
Brachydactyly	Normal digits and limbs	Short digits and limbs
Broad lips	Narrow lips	
Brown eyes	All other eye colors	Influencing factors
Cataract	No cataract	Not all cataracts inherited
Cleft chin	No chin dimple	
Color vision	Color blindness	Sex linked
Curly hair	Straight hair	Technically lack of dominance
Dark brown hair	All other hair colors	Melanin production; 3 gene pairs
Dark skin color	Light skin color	Several gene pairs involved
Diabetes mellitus	Normal insulin	Not all diabetes mellitus inherited
Dimples (cheek)	No dimples	
Double-jointed thumb	Normal thumb joint	Hitchhiker's thumb, > 45 degrees
Dwarfism	Normal growth	Not all dwarfism inherited
Ehlers-Danlos syndrome	Normal connective tissue	
Fragile bones	Normal bone strength	Not all fragile bones inherited
Freckles	No freckles	
Free earlobes	Attached earlobes	
Glaucoma	No glaucoma	Not all glaucoma inherited
Hematuria	No RBCs in urine	Sex linked; not all hematuria inherited
High, narrow nose bridge	Low, broad nose bridge	
Huntington's Chorea	No Huntington's Chorea	
Hyperopia	Normal near/distance vision	Farsightedness
Lack of musical ability	Musicial ability	
Large eyes	Small eyes	
Left/right interlocking thumbs	Right/left	Opposite "feels" wrong
Long eyelashes	Short eyelashes	
Marfan's Syndrome	Normal connective tissues	
Mid-digital hair	No mid-digital hair	Probably several dominant alleles
Migraine	No migraine	Not all migraine inherited
Mongolian eye fold	No extra eye fold	Extra fold of skin
Myopia	Normal near/distance vision	Nearsightedness
No anemia	Beta-thalassemia	
No palmaris longus muscle	Palmaris longus muscle	Third wrist tendon
No phenylketonuria (PKU)	PKU	Inability to oxidize phenylalanine
No tooth enamel	Tooth enamel	
Normal blood clotting	Hemophilia	Sex linked, some forms
Normal cranium	Microcephaly	
Normal exocrine secretion	Cystic fibrosis	
Normal foot	Club foot	Some forms not inherited
Normal foot arches	Flat foot	
Normal ganglioside metabolism	Tay-Sachs disease	Cell membrane lipid
Normal genitalia	Testicular feminization	Sex linked
Normal hearing	Deafness	Not all deafness inherited
Normal mental abilities	Retardedness	Not all retardedness inherited
Normal mentation	Schizophrenia	Perhaps not all forms inherited
Normal muscle development	Duchenne Muscular Dystrophy	Sex linked
Normal muscle development	Gowers' Muscular Atrophy	Sex linked
Normal pigmentation	Albinism	At least 2 recessive types known
Normal RBC morphology	Sickle cell anemia	
Normal retinal pigmentation	Ocular albinism	Sex linked
Pattern baldness	No baldness	Sex-influenced trait
Polydactyly	Normal digitation	Extra digits
PTC taster	PTC nontaster	Phenylthiocarbamide
Rh factor (Rh +)	No Rh factor (Rh–)	
Right handed	Left handed	Other factors often involved
Short second finger	2nd and 4th finger same length	Sex influenced
Shortness	Tallness	At least 10 gene pairs involved
Sweat glands	Absence of sweat glands	
Symphalangy	Normal digital joints	Fused digital joints
Syndactyly	Normal digitation	Webbed digits
Tongue roller	Tongue does not roll	
Type A blood: type B blood	Type O blood	Codominant
Widow's peak	Straight hair line	

● **FIGURE 71–1**
Common Genetic Traits.

B. KNOWN DOMINANT/RECESSIVE TRAITS

1→ Study Figure 71–1●. How many of these traits are you familiar with? ___________

C. THE PUNNETT SQUARE

Dominant/recessive genetics problems are usually done with a **Punnett square**. A Punnett square is a graphic representation of allelic combinations possible when given gametes (parental egg and sperm) unite. Work through the following example.

1→ Use the trait of freckles as an example of how the Punnett square works. (We will assume that everything is simple inheritance.)

2→ To set up a Punnett square, first recall that meiosis is the process whereby one diploid cell undergoes reduction/division to produce four haploid cells. Each haploid cell (gamete) contains the haploid chromosome complement. In humans this haploid cell has 23 chromosomes, one member of each pair.

3→ Choose the letter F to represent the freckle trait.

Genetic traits are commonly designated by one (or more) letter, with uppercase letters used for the dominant trait and lowercase letters used for the recessive trait. Some traits by convention have standard letter symbols. For other traits you can choose almost any letter you like.

4→ Note from the chart (Figure 71–1●) that having freckles is dominant to not having freckles. A person who is homozygous dominant for freckles has the genotype FF. The heterozygous individual is Ff and the homozygous recessive combination is ff. According to present dominant/recessive theory, what would this mean?

____

5→ Now recall that since alleles come in pairs, any given egg or sperm will contain either one or the other member of the allelic pair. The person who is heterozygous for freckles can pass on either the dominant allele (F) or the recessive allele (f).

The Punnett square in Figure 71–2● demonstrates the freckle possibilities among possible offspring of two heterozygous individuals. The phenotype of each of these parents would be freckled but the genotype would be Ff. We cross-multiply to find the progeny. This is a **monohybrid cross** because only one trait is considered.

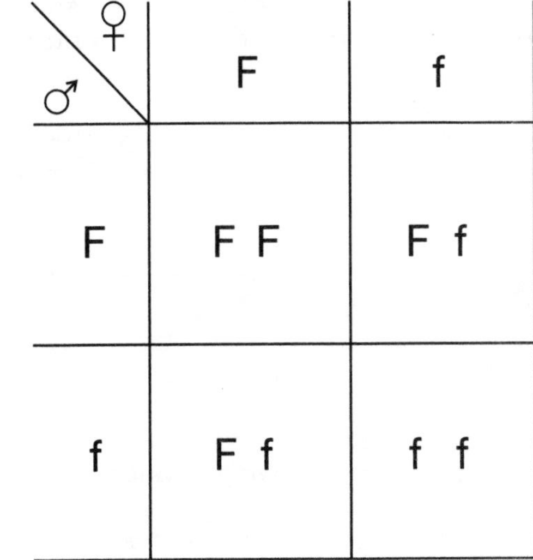

♂ \ ♀	F	f
F	F F	F f
f	F f	f f

● **FIGURE 71–2**
Monohybrid Cross — Two Individuals Each Heterozygous for Freckles.

6→ Note that genotypically each child has one chance in four of being homozygous dominant, one chance in four of being homozygous recessive, and two chances in four (1 in 2) of being heterozygous. Phenotypically, any given child has three chances out of four of having freckles.

Construct Punnett squares for each of the following situations. Give the genotypic and phenotypic ratios for the first three. Use Figure 71–1●. (If you are unfamiliar with ratios, simply state how many of each genotype or phenotype you would find represented in a given Punnett square.)

a. Man homozygous for long eyelashes; woman with short lashes.
b. Two people without middigital hair.
c. Woman of normal skin pigmentation whose mother was albino; man of normal pigmentation whose known relatives all were also normally pigmented.
d. Possible genotypes of parents of a non–tongue roller.
e. Possible genotypes of parents of a person heterozygous for sickle cell anemia.

D. DIHYBRID CROSS

In a **dihybrid cross,** two traits are considered. Figure 71–3● is a pattern that shows how the alleles for two independent traits located on separate pairs of chromosomes might combine to be passed on in gametes. Consider that the person here is heterozygous for fragile bones (Ff) and widow's peak (Ww).

1 ➡ Work through Figure 71–4●, the dihybrid cross of two people who are heterozygous for both of these traits.

Concept Check 3 What are the resultant phenotypic and genotypic ratios? ____________________

2 ➡ Set up Punnett squares to demonstrate each of the following situations:

a. Two nearsighted people without a palmaris longus muscle, both of whom had a nearsighted parent who had the muscle.
b. A schizophrenic who is heterozygous for hyperopia and a homozygous nonschizophrenic who is also heterozygous for hyperopia.
c. Two Rh negative persons who are both also PTC nontasters.
d. Possible parental genotypic combinations for a small-eyed person with inherited migraine.
e. Possible parental genotypic combinations for an astigmatic person with a bent little finger.

E. More Complex Crosses

Trihybrid and other polyhybrid crosses can certainly be done by hand. We usually use the computer for a complete printout or an expanded algebraic formula for a specific trait combination.

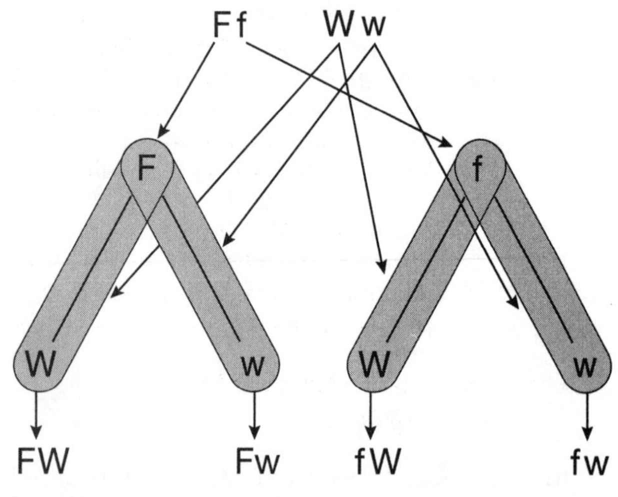

● **FIGURE 71–3**
Tree for Deteermining Allelic Combinations for Dihybrid Cross.

III. Additional Classical Genetics Problems

A. Solving Incomplete Dominance Problems

Incomplete dominance problems are done exactly like any dominant/recessive problem, using a Punnett square. We must simply remember that the heterozygous combination is a blend of the two traits when interpreting the resulting genotype. For instance, if a red flower is crossed with a white flower the progeny might be pink (a blend). Curly hair crossed with straight hair gives wavy hair (a blend). In many animal species, long tails crossed with short tails give medium-length tails.

Based on the discussion at the beginning of this exercise, we can easily see that many traits are blends because rarely does the phenotype of the progeny exactly copy the phenotype of the parent.

B. Solving Codominance Problems

Codominance problems are also done exactly like any other dominant/recessive problem. In this case, however, we must remember that when both alleles are listed, both are expressed.

The classic example of codominance is the ABO blood grouping. Types A and B are both dominant. (Type A is often represented as I^A and type B is often represented as I^B. Type O is "i." Sometimes ABO are simply written as A, B, O. Sometimes Type O is written i^O.)

The A and B are specific membrane molecules. The presence of either molecule does not interfere with the presence of the other molecule.

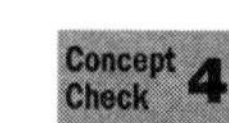

Concept Check 4 Study the Punnett squares in Figure 71–5● and then answer the following questions. Construct additional squares as necessary.

a. What are the chances a man heterozygous for type A blood and a woman heterozygous for type B blood could have a child with type O blood? With type AB blood?
b. If a child has type B blood, could a man with type O blood be the father?
c. A woman has type AB blood, and claims that a man with type B blood is the father of her blood-type-A child. Is this possible?
d. A woman with type AB blood claims a man with type O blood is the father of her blood-type-O child. Is this possible? If so, what is the problem?
e. The Rh factor is inherited independently of the ABO blood group. Set up a Punnett square to demonstrate

Two individuals heterozygous for both fragile bones and widow's peaks

♂ \ ♀	FW	Fw	fW	fw
FW	FFWW	FFWw	FfWW	FfWw
Fw	FFWw	FFww	FfWw	Ffww
fW	FfWW	FfWw	ffWW	ffWw
fw	FfWw	Ffww	ffWw	ffww

● **FIGURE 71–4 Punnett Square: Dihybrid Cross.**

the possible genotypes among offspring of a woman heterozygous for both type A blood and the Rh factor and a man heterozygous for both B blood and the Rh factor.

As with incomplete dominance, most traits exhibit some type of codominance. In cases of regular dominance the coding of the dominant allele does not negate the coding of the recessive allele; the dominant allelic expression is simply more obvious on a phenotypic level. With codominance both coded products are phenotypically obvious.

IV. Sex-Influenced Characteristics

A. Definition

1➡ **Sex-influenced** means that the expression of a characteristic will vary with the sex of the individual. This should not be surprising when we remember the role of epistasis (masking) in phenotypic expression. In other words, a trait located on an autosome is influenced by the specific heterosome combination: XX or XY. Sex-influenced traits are "influenced by the sex of the individual."

2➡ The most famous example of a sex-influenced trait in humans is (male) pattern baldness. This type of baldness is dominant in men but recessive in women. This means that a bald man might be heterozygous or homozygous for the trait but a bald woman must be homozygous in order to be bald. For a man not to be bald, he must be homozygous for the recessive trait.

B. Sex-Influenced Problems

Concept Check 5 Construct a Punnett square to show the possible baldness pattern for the offspring of a bald woman and a hairy man. What would the genotypic and phenotypic ratios be? ____________________
____
____

1➡ If a woman has a short second finger, what do you know about the genotypes of her parents? ____________________
____

V. Sex-Linked Characteristics

[∞ *Martini, p. 1095*]

A. Definition

1➡ **X-linked** characteristics are carried on the X chromosome and were formerly called

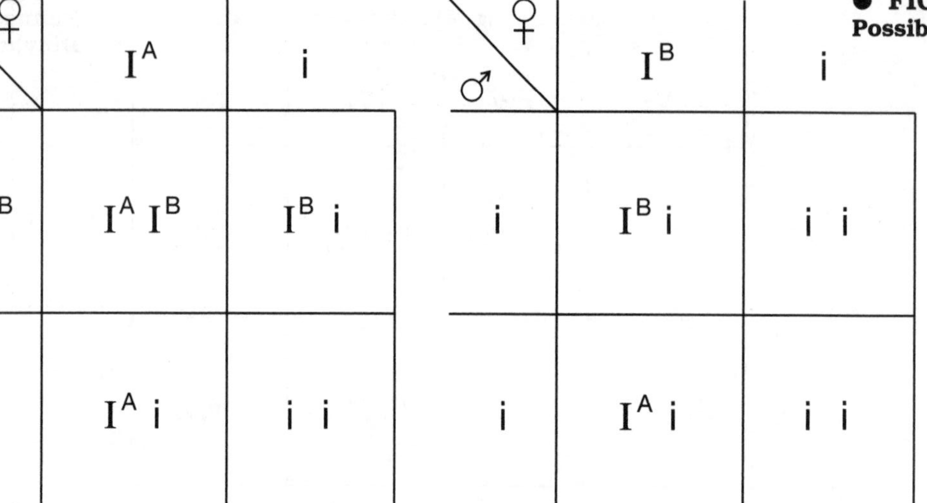

♀
♂
I^A
i
I^B
i
$I^A I^B$
I^B i
I^A i
i i
Individual heterozygous for Type A with Individual heterozygous for Type B
♀
♂
I^B
i
i
i
I^B i
i i
I^A i
i i
Individual heterozygous for Type B with Individual heterozygous for Type O

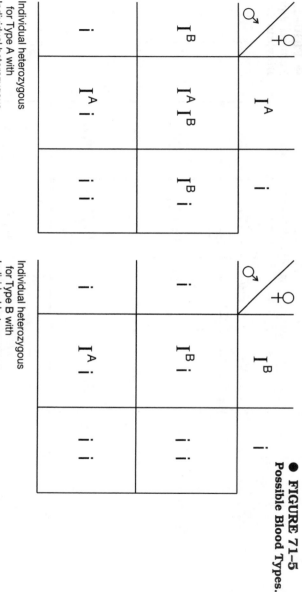

● **FIGURE 71–5**
Possible Blood Types.

sex-linked characteristics. It was previously believed that no genes were carried on the Y chromosome and that therefore all sex-linked characteristics were necessarily carried on the X chromosome. Scientists have now identified a number of **holandric**, Y chromosome, traits. As a result, what were previously called sex-linked traits are now called X-linked traits.

B. Solving Sex-Linked Problems

1➡ When discussing sex-linked (or X-linked) characteristics, we demonstrate the inheritance of the entire chromosome. The male can pass on an X chromosome or a Y chromosome. The female can pass on either one of her X chromosomes or her other X chromosome. We then indicate the trait with an uppercase letter (for a dominant trait) or a lowercase letter (for a recessive trait) as superscripts on the X. We do not include a superscript on the Y chromosome because we are not dealing with homologous or analogous alleles. Nevertheless, we must still include the Y chromosome in our Punnett square.

2➡ Hemophilia is an example of a trait that is located on the X chromosome. Study the Punnett squares for hemophilia in Figure 71–6●.

3➡ Construct similar squares to show the following situations. Explain the genotypic and phenotypic ratios.

a. Possible offspring of a hemophiliac man and a woman carrying the trait.
b. Possible offspring of a normal man and a woman carrying the trait for hemophilia.

4➡ Solve these problems.

a. A man without hemophilia claims a hemophiliac woman could not possibly be his daughter. Could he be right?
b. The presence of coarse body hair on the pinna of the ear is a holandric trait. Could a man exhibiting this trait have inherited it from his mother's father?
c. What are the possible genotypes for the parents of a woman with ocular albinism?
d. What are the possible genotypes for the parents of a color-blind man?

VI. Additional Practice

A. Mendelian Inheritance

1➡ Work with a partner and go through the chart in Figure 71–1●. How many of these traits can you identify either from your own phenotype or from members of your family? Your instructor may choose to do some class tallies for some of these traits.

2➡ Choose five or six traits from Figure 71–1● and identify whether you and/or your parents must be homozygous or heterozygous for these traits. For example: Suppose you

♂ \ ♀	X^H	X^h
X^H	X^HX^H	X^HX^h
Y	X^HY	X^hY

♂ \ ♀	X^H	X^h
X^h	X^HX^h	X^hX^h
Y	X^HY	X^hY

H = normal
h = hemophilia

● **FIGURE 71–6**
Hemophilia: Possible Combinations.

are right-handed but one of your parents is left-handed. You know that your left-handed parent must be homozygous recessive and that your right-handed parent must be at least heterozygous for handedness. You must be heterozygous because you are right-handed but you could only inherit the left-handed allele from your left-handed parent.

3➡ Note: Left- or right-handedness can be influenced by injury, necessity, adaptation, and social custom. Most people are at least somewhat ambidextrous. How does this fit into our discussion?

____

____

____

❑ Additional Activities

1. Research the lives and careers of:
   E. Ehlers, 1863–1937, Danish dermatologist
   H. A. Danlos, 1844–1912, French dermatologist
   Antoine Bernard-Jean Marfan, 1858–1942, French physician
   George Huntington, 1850–1916, American physician
   Sir William R. Gowers, 1845–1915, British neurologist
   Reginald C. Punnett, 1875–1967, British geneticist
2. Select a few of the traits listed in Figure 71–1● and explain how the present thinking on allelic action could be responsible for a trait being either dominant or recessive.

NOTES

Answers to Selected Concept Check Questions

1. This is an example of phenotype because the thumb placement is a physical manifestation of a trait. Incidently, about 60% of the class should be left over right.

2.

(a)

♂ \ ♀	ℓ	ℓ
L	L ℓ	L ℓ
L	L ℓ	L ℓ

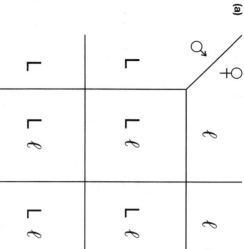

Genotype: All Lℓ
Phenotype: All long lashes

(b)

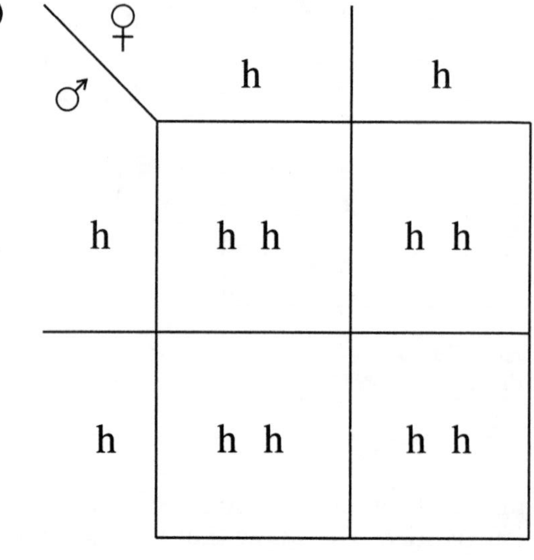

This is a recessive trait.
Genotype: All hh
Phenotype: All without mid-digital hairs

(c)

♂ \ ♀	A	a
A	A A	A a
A	A A	A a

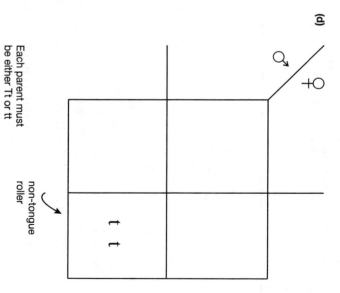

Genotype: 1:1 AA to Aa
Phenotype: All normal pigment
(half will be carriers)

(d)

♂ \ ♀		
		t t

non-tongue roller

Each parent must be either Tt or tt

(e)

Three possible combinations of parents who could produce Ss:

Parents must be SS × ss
Ss × Ss
Ss × ss

	S	S
s		Ss
s	Ss	

	S	s
S		Ss
s	Ss	

	S	s
s	Ss	
s		

3. The phenotypic ratios are as follows: 9 exhibiting fragile bones and widow's peak, 3 exhibiting fragile bones and straight hairline, 3 exhibiting normal-strength bones and widow's peak, 1 exhibiting normal-strength bones and straight hairline. This phenotypic ratio is often designated: 9:3:3:1.

   The genotypic ratio is: 1 FFWW, 2 FFWw, 2 FfWW, 1 FFww, 4 FfWw, 2 Ffww, 2 ffWw, 1 ffWW, 1 ffww

4. a. The chances for a type O child would be one in four. The chances for a type AB child would be one in four.
   b. Yes, a man with type O blood could father a child with type B blood, provided the mother was either type B or type AB.
   c. Yes, it is possible for the man with type B blood to be the father of the type A child, provided he is heterozygous for type B blood.
   d. A man with type O blood can father a child with type O blood, but a woman with type AB blood cannot be the mother of a type O child.
   e.

♂ \ ♀	AR	Ar	OR	Or
BR	ABRR	ABRr	BORR	BORr
Br	ABRr	ABrr	BORr	BOrr
OR	AORR	AORr	OORR	OORr
Or	AORr	AOrr	OORr	OOrr

NOTES

*Technically the following substitutions would make this chart more correct.

$A = I^A$

$B = I^B$

$O = i$

We have used the informal abbreviations for convenience in working with a dihybrid cross.

NOTES

5. Answer:
   B1 = Hairy
   B2 = Bald
   B1B1 = Hairy either sex
   B1B2 = Hairy female, bald male
   B2B2 = Bald either sex

♂ \ ♀	B2	B2
B1	B1B2	B1B2
B1	B1B2	B1B2

All offspring will be heterozygous. The boys will be bald, the girls hairy.

❑ Lab Report

1. Referring to the numbered inquiries at the beginning of this exercise, complete the following box summary:

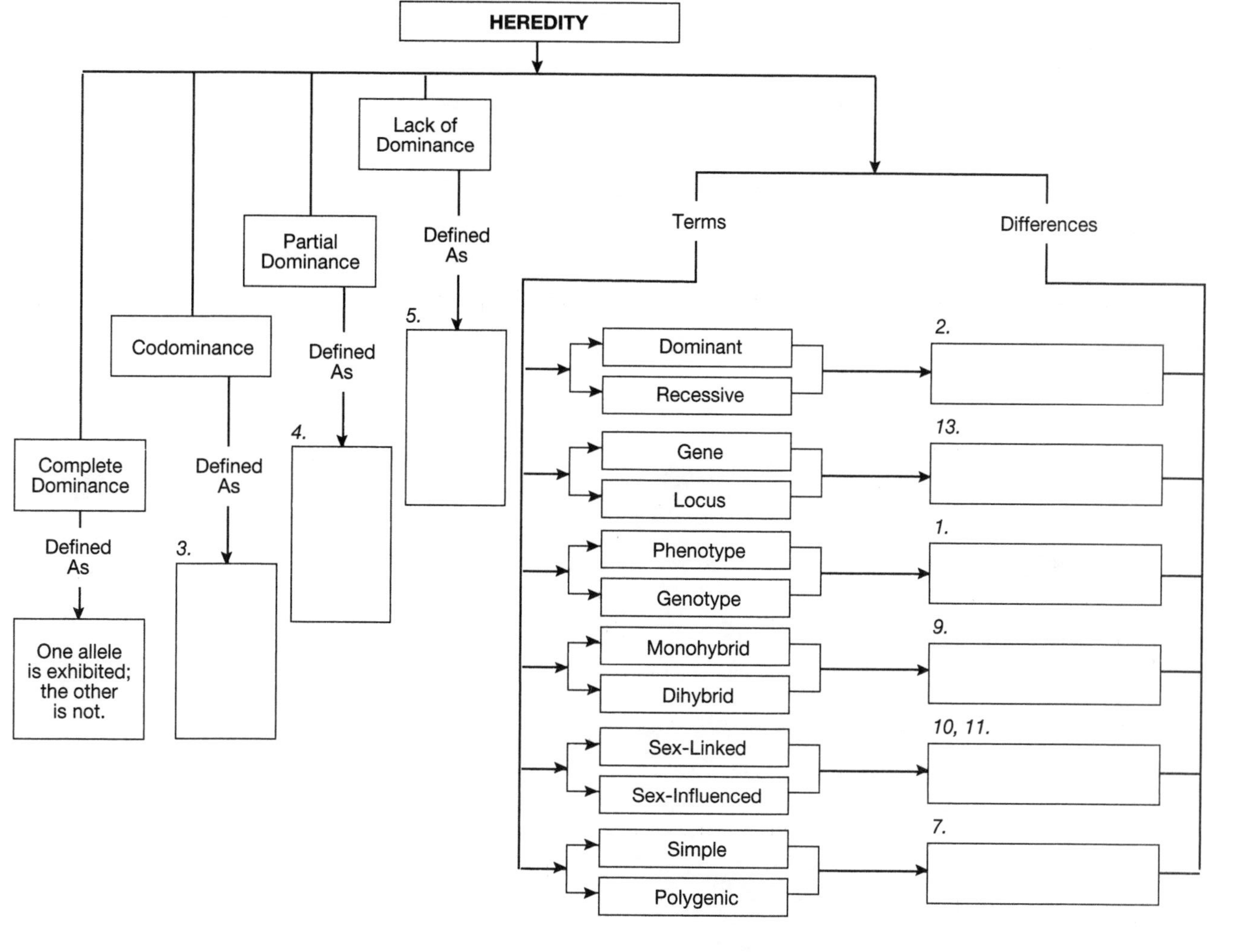

General Questions

1. Why are the terms haploid and diploid important to our discussion of inheritance?
2. How would you explain the superb musical talent of a child both of whose parents are professional musicians? (Check the chart before answering!)
3. Why is epistasis important?
4. If a woman exhibits a particular trait:
   a. Could this be a holandric trait?
   b. Could she be a carrier?

NOTES

Photo Credits

Exercise 3

3-1 Science VU/Visuals Unlimited
3-2 Science VU/Visuals Unlimited
3-3 Science VU/Visuals Unlimited

Exercise 9

9-7a David M. Phillips/Visuals Unlimited
9-7b David M. Phillips/Visuals Unlimited
9-7c David M. Phillips/Visuals Unlimited

Exercise 10

10-3a—e Ed Reschke/Peter Arnold, Inc.

Exercise 12

12-2a Frederic H. Martini
12-2b Frederic H. Martini

Exercise 13

13-2b Frederic H. Martini
13-2c Frederic H. Martini
13-3 Frederic H. Martini
13-4a Frederic H. Martini
13-4b Frederic H. Martini

Exercise 14

14-2a Dr. Richard Kessel and Dr. Randy H. Kardon
14-2b Dr. Richard Kessel and Dr. Randy H. Kardon
14-2c Ralph T. Hutchings
14-6 Arnold & Brown

Exercise 16

16-8 Ralph T. Hutchings

Exercise 17

17-1a Bates/Custom Medical Stock Photo, Inc.
17-1b Ralph T. Hutchings
17-2a Ralph T. Hutchings
17-2b Ralph T. Hutchings
17-2c Ralph T. Hutchings

Exercise 22

22-1a Ward's Natural Science Establishment, Inc.
22-1b Dr. Don Fawcett/Photo Researchers, Inc.
22-1c-B J. J. Head/Carolina Biological Supply/ Phototake/
22-1c-T J. J. Head/Carolina Biological Supply/ Phototake/

Exercise 26

26-3-R Dr. Richard Kessel/Dr. Randy H. Kardon
26-5 Frederic H. Martini
26-6 John D. Cunningham/Visuals Unlimited

Exercise 30

30-2a Ralph T. Hutchings
30-2b Ralph T. Hutchings
30-4 Ralph T. Hutchings
30-6 Pat Lynch/Photo Researchers, Inc.

Exercise 31

31-1 Larry Mulvehill/Photo Researchers, Inc.

Exercise 33

33-1 Ralph T. Hutchings

Exercise 36

36-1 Ralph T. Hutchings
36-5 Ed Reschke/Peter Arnold, Inc.

Exercise 38

38-3b Lennart Nilsson/Bonnier Alba, AB, BEHOLD MAN, pp. 206–207.

Exercise 39

39-01 Ward's Natural Science Establishment, Inc.

Exercise 41

41-1b Frederic H. Martini
41-1c G. W. Willis, MD/Terraphotographics/ Biological Photo Service

Exercise 44

44-5 Manfred Kage/Peter Arnold, Inc.
44-7c Frederic H. Martini
44-8 Frederic H. Martini
44-9b Ward's Natural Science Establishment, Inc.
44-10 Ward's Natural Science Establishment, Inc.

Exercise 46

46-4a Ed Reschke/Peter Arnold, Inc.
46-4b Ed Reschke/Peter Arnold, Inc.
46-4c Ed Reschke/Peter Arnold, Inc.
46-4d Ed Reschke/Peter Arnold, Inc.
46-4e Ed Reschke/Peter Arnold, Inc
46-6a Stanley Flegler/Visuals Unlimited
46-6b Stanley Flegler/Visuals Unlimited

Exercise 48

48-1b Phototake NYC

Exercise 50

50-1a Michael J. Timmons

Exercise 55

55-5 Phototake NYC
55-10b Ward's Natural Science Establishment, Inc.
55-10c Frederic H. Martini

Exercise 58

58-10c-L Frederic H. Martini
58-10c-R Frederic H. Martini
58-10d Ward's Natural Science Establishment, Inc.
58-10e-L Michael J. Timmons
58-10e-R Michael J. Timmons
58-10f Frederic H. Martini

Exercise 62

62-1b Ralph T. Hutchings
62-6-BL Frederic H. Martini
62-6-BR Frederic H. Martini
62-6-T Ward's Natural Science Establishment, Inc.

Exercise 63

63-1a Roberta M. Meehan
63-1b Roberta M. Meehan
63-1c Ralph T. Hutchings
63-1d Roberta M. Meehan

Exercise 66

66-5 Ward's Natural Science Establishment, Inc.

66-6-B Frederic H. Martini
66-6-T Ward's Natural Science Establishment, Inc.
66-7-L Ward's Natural Science Establishment, Inc.
66-7-R Dr. Richard Kessel/Dr. Randy H. Kardon
66-8-L Frederic H. Martini
66-8-M Frederic H. Martini
66-8-R Frederic H. Martini
66-9 David M. Phillips/Visuals Unlimited

Exercise 67

67-5a Ward's Natural Science Establishment, Inc.
67-5b Frederic H. Martini

Exercise 68

68-1ba Frederic H. Martini
68-1bb Frederic H. Martini
68-1bc Frederic H. Martini
68-1bd Frederic H. Martini
68-1bf G. W. Willis, MD/Biological Photo Service
68-1bg G. W. Willis, MD/Terraphotographics/Biological Photo Service
68-3a Frederic H. Martini
68-3b Frederic H. Martini
68-3c Michael J. Timmons
68-3d Frederic H. Martini

Exercise 70

70-3a David M. Phillips/Visuals Unlimited

Histology Atlas

Plate A, bottom Frederic H. Martini
Plate A, top Ward's Natural Science Establishment, Inc.
Plate B, bottom Frederic H. Martini
Plate C, bottom Frederic H. Martini
Plate C, top Frederic H. Martini
Plate D Frederic H. Martini
Plate E, left Frederic H. Martini
Plate E, right Frederic H. Martini
Plate F Ward's Natural Science Establishment, Inc.
Plate G, bottom Frederic H. Martini
Plate G, top John D.Cunningham/Visuals Unlimited
Plate H Frederic H. Martini
Plate I bottom Ed Reschke/Peter Arnold, Inc.
Plate I middle Science Source/Photo Researchers, Inc.
Plate I top Robert Brons/ Biological Photo Service
Plate J Frederic H. Martini
Plate K, left Ed Reschke/Peter Arnold, Inc.
Plate K, right Ed Reschke/Peter Arnold, Inc.
Plate L, bottom Phototake NYC
Plate L, top Eric Grave/Phototake NYC